W0227711

Functional and Stereotactic Neurosurgery

Stereotactic neurosurgery requires accuracy, a steady hand, and sound judgment.

Functional and Stereotactic Neurosurgery

Edward I. Kandel

Professor, Doctor of Medical Sciences
Director, Neurosurgical Clinic
Institute of Neurology
Academy of Medical Sciences of the USSR
Moscow, USSR

Translated from Russian by
George Watts

Translation edited by
A. Earl Walker

Professor Emeritus of Neurosurgery
The Johns Hopkins University School of Medicine
Baltimore, Maryland

Adjunct Professor, Neurosurgery/Surgery
University of New Mexico School of Medicine
Albuquerque, New Mexico

Plenum Medical Book Company
New York and London

Library of Congress Cataloging in Publication Data

Kandel, E. I. (Eduard Izrailevich)
[Funktsional'naia i stereotaksicheskaia neirokhirurgiia. English]
Functional and stereotactic neurosurgery / Edward I. Kandel; translated from Russian by George Watts; translation edited by A. Earl Walker.
p. cm.
Translation of rev. and updated ed. of: Funktsional'naíà i stereotaksicheskaíà neirokhirurgiíà. 1981.
Includes bibliographies and index.

1. Nervous system—Surgery. 2. Stereoencephalotomy. I. Walker, A. Earl (Arthur Earl), 1907- . II. Title.
[DNLM: 1. Nervous System—surgery. WL 368 K16f]
RD593.K2613 1989
617'.48
DNLM/DLC 88-39617
for Library of Congress CIP

ISBN-13: 978-1-4612-8040-8 e-ISBN-13: 978-1-4613-0703-7
DOI: 10.1007/978-1-4613-0703-7

Softcover reprint of the hardcover 1st edition 1989

This volume is based on *Funktsional'naya i stereotaksicheskaya neirokhirurgiya* by E. I. Kandel, published in 1981 by Meditsina, Moscow. The original Russian work has been completely revised and updated by the author for this edition.

233 Spring Street, New York, N.Y. 10013

Plenum Medical Book Company is an imprint of Plenum Publishing Corporation

In memory of my parents

To my wife, Helen

Foreword

Soon after neurosurgery had advanced past the stage of removing lesions on the surface of the brain, it became apparent that subcortical diseased tissue could not be excised safely by the usual surgical techniques because of the risk of damaging overlying normal structures. Various means of reaching deep-seated lesions were devised, most of which attempted to approach the pathological tissue through "silent areas" of the brain. However, these operations often resulted in serious neurological deficits. Spiegel and Wycis's modification of the Horsley–Clarke apparatus to reach targets deep in the human brain introduced a new approach to subcortical surgery. True, as Professor Edward Kandel relates, Russian surgeons had pioneered in the field, but it was the persistent endeavors of Spiegel and Wycis that popularized the technique in the various countries of the world. In the 1950s and 1960s, stereotactic surgery began to be used by many neurosurgeons for the treatment of movement disorders, uncontrollable aggression, inaccessible or difficult to reach tumors, and vascular lesions. The subject was sufficiently popular that a journal was founded to publish the advances.

It was at this time that Professor Kandel began his work with stereotactic neurosurgery. He realized that the stereotactic instrumentation then in use was either too complicated or too simple to adjust to brains of various sizes and shapes. To correct these discrepancies, he devised a stereotactic instrument based on novel principles. With this new technique he has treated many conditions with exceptionally good results.

In this book Professor Kandel gives a detailed account of his experiences with stereotactic operations and reviews the work of brain surgeons in different countries of the world. He has produced a unique record of the advances made by neurosurgeons of international stature. Although stereotactic surgery may not replace open operative procedures, Professor Kandel has shown that it has an important place in the neurosurgical armamentarium for the treatment of many conditions previously considered inoperable. Granted that older neurosurgeons will consider their craniotomies quite adequate for the relief of many neurological disorders that Professor Kandel shows so clearly to be amenable to stereotactic intervention, there are many lesions that undoubtedly can be reached more easily and with less risk to life and limb by stereotactic than by open procedures.

This book is not just a description of operative procedures, although it does give clear accounts of surgical techniques. It presents the postoperative histories of patients who have been cured or markedly relieved of longstanding afflictions; these persons have been followed for 10 to 15 or more years, so that the results may be considered more or less permanent. Thus, Professor Kandel has presented a chronicle of the benefits or, alas, in some cases, failures that may be expected as the result of modern refined stereotactic surgery for the relief of afflictions, particularly of the deep-seated nuclei of the brain.

Certainly he has broadened the scope of stereotactic surgery and has improved the traditional techniques for operations on the basal ganglia and thalami. Utilizing stereotactic principles and his new methods for eradicating deep-seated tumors, aneurysms, and angiomas with a permissible mortality, he has given long-term benefit to patients formerly considered to be suffering from inoperable conditions. Such novel techniques have promise of further application in neurological practice.

All who have to deal with patients suffering from diseases of the nervous system will find the insight that this book gives into the treatment of some neurological diseases by stereotactic surgery a very enlightening experience.

A. Earl Walker, M.D.
Division of Neurosurgery
University of New Mexico
School of Medicine
Albuquerque, New Mexico

Preface

Albert Einstein once said that the most amazing thing about this world is that it is comprehensible. It is no less amazing that we comprehend the human brain, which not only perceives this world but also remakes it.

It is necessary to understand the structures and functions of the brain in order to treat its disorders. The modern methods of diagnosis and treatment represent a most powerful and effective stimulus for the development of medicine in general and neurosurgery in particular. Over the course of several decades, hardly anyone expected that the experimental stereotactic method would become an effective instrument in the hands of surgeons in their fight against many grave diseases of the brain. Approximately a half century was needed for stereotaxis to emerge from the neurophysiological laboratory and embark on a fruitful path in the neurosurgical clinic. The indisputable credit for introducing this method in modern neurosurgery belongs to two American scientists, neurologist Spiegel and neurosurgeon Wycis, who in the 1950s designed a stereotactic apparatus for operating on the human brain and were the first to perform stereotactic operations on subcortical structures in man.

More than 35 years have gone by since the first stereotactic operation on deep-seated cerebral structures was performed. Just like angiography, radioactive isotopes, or microsurgery, the stereotactic method instantly raised neurosurgery to a qualitatively new level of development.

This new method, which initially, as so often is the case, was met with a certain distrust, has since then not only confirmed its right to existence but has become a generally acknowledged breakthrough in modern neurosurgery. This method brought neurosurgery substantially closer to its coveted goal—the most effective operation with the least damage to the brain. To a great extent, stereotaxis meets the main requirements for neurosurgical operations as formulated by N. N. Burdenko—"anatomically accessible, technically feasible, and physiologically permissible."

The stereotactic trend in neurosurgery was the result of a synthesis of several scientific and research currents flowing from various branches of science. The merging of this trend became possible through the integration of the achi vements of present-day engineering (stereotactic apparatuses, x-ray machines, computed tomography, nuclear magnetic resonance equipment, and instruments for various methods of destruction), neurophysiology (stereotactic method and the study of the functions of subcortical structures), neuromorphology (stereotactic atlases, the study of interrelationships, spatial localization, and variability of subcortical structures), neurology (the study of the etiology, pathogenesis, and clinical aspects of hyperkinesias, pain syndromes, epilepsy, vascular lesions of the brain, and other disorders), and neurosurgery (the development of techniques and methods of surgical operations on deep-seated cerebral structures).

Tens of thousands of stereotactic operations have been performed in many countries of the world during these 35 years. This trend proved to be a principally new and highly effective method in the management of many neurological diseases for which there is either no conservative treatment or the current treatment is of little effect. By its success, stereotactic surgery cannot be compared to any of the previously employed methods. The list of diseases in which stereotactic operations are very effective today includes more than 30 nosological forms, which is reflected in contents of this book. The "sphere of action" of the stereotactic method is continuing to expand steadily and gradually to embrace more and more nosological forms. One cannot but note that the "objective" of stereotaxis includes such pathological processes so different and seemingly far from each other as parkinsonism and pain, cerebral palsy and cerebral aneurysms, and brain tumors and temporal lobe epilepsy.

The stereotactic method in neurosurgery is rapidly and steadfastly advancing. Stereotactic operations on the cerebellum and the spinal cord have appeared quite recently and are being successfully performed. New and improved stereotactic apparatuses are being tested. Cryosurgery—a turbulently developing "offspring" of stereotaxis—is being employed more extensively in many branches of clinical medicine. Stereotactic biopsy of tumors and other deep-seated pathological processes in the brain is coming in for broader application. Microelectrode equipment, electronic amplifiers, computers, and, first and foremost, computed tomography, positron emission tomography, and nuclear magnetic resonance have become important elements of the stereotactic technique.

The International Society for Research in Stereoencephalotomy was founded in 1961, and in 1975 it became the World Society for Stereotactic and Functional Neurosurgery. Altogether, the Society held nine World Congresses, the last in Toronto in 1985.

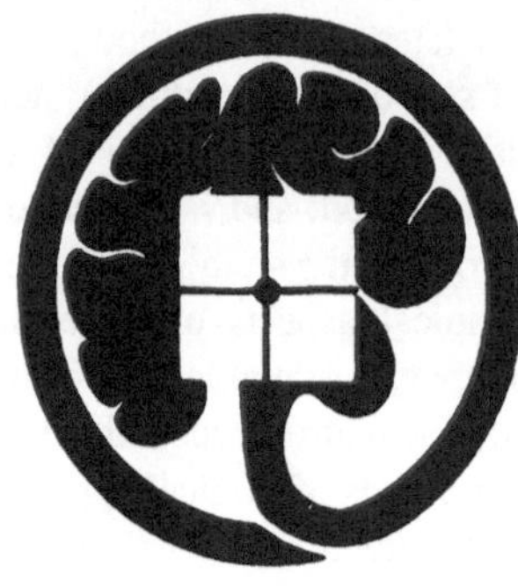

The emblem of the World Society for Stereotactic and Functional Neurosurgery

It should be stressed that all the great therapeutic achievements of the stereotactic method have not exhausted its significance and unique possibilities. This method also plays an exceptionally important part in the advancement of human neurophysiology. The development of stereotactic neurosurgery has primarily opened up new prospects for neurophysiology. One of the most important tasks of functional and stereotactic neurosurgery is to study the structures and functions of the human CNS. Undoubtedly, this task cannot be resolved only by extrapolating the data obtained in neurophysiological experiments on animals. The stereotactic method, which has dovetailed with many electrophysiological methods, has today become one of the main and most fruitful ways of understanding the functions of the human brain. This method has yielded much new information about the functional organization of subcortical structures, the interaction and functions of which have been studied far too insufficiently.

The clinical application of stereotactic and functional operations has made it possible to obtain a wealth of new and very useful data on the pathophysiological mechanisms of many cerebral disorders and on their etiology and pathogenesis. Many operations are performed under local anesthesia, making it possible to observe various effects of stimulation and destruction of discrete structures of the CNS as well as to obtain a verbal account from the patient about his sensations. Therefore, according to the "feedback" principle, these operations contribute greatly to the study of the pathogenesis of many neurological diseases, and this in turn offers greater possibilities and enhances the efficacy of new surgical techniques.

During the past two decades, one other term—"functional neurosurgery"—has acquired, as it were, "citizenship rights." As far back as almost a century ago, John Hughlings Jackson emphasized that the term "functional" was synonymous with "physiological." Consequently, "functional neurosurgery" is a method in the management of physiological disorders caused by various pathological processes. The term "functional surgery" was first proposed by René Leriche in reference to operations on the sympathetic nervous system in pain and vascular disorders. Subsequently, Wertheimer introduced the term "functional neurosurgery." Today, this generally acknowledged term is used for designating a trend that, over the course of several decades, had been gradually developing within the framework of general neurosurgery but only in recent years has evolved as a separate branch from that specialty. Functional neurosurgery may be defined as an aggregate of methods for surgical action on roots, pathways, and neuronal structures of the CNS based on the anatomophysiological properties of pathological processes in the CNS and having as its purpose to change these processes to obtain a therapeutic effect.

At this moment, it is still too early to speak about the total merging of the two trends—stereotactic and functional neurosurgery; this, in all likelihood, is a matter for the future. Nevertheless, it is quite apparent that the concept "functional neurosurgery" is much broader, for it includes, as is shown in this book, a great number of surgical methods not directly related to stereotaxis. At the same time, there can be no doubt that stereotactic neurosurgery represents the main component of the concept of functional neurosurgery.

A tremendous number of publications has accumulated in literature over the past quarter of a century. That is why the main goal of this book is to systematize and critically analyze this huge wealth of world literature and to review the current state of the art, i.e., to present the state of knowledge and development in the field of stereotactic and functional neurosurgery. The second objective of this book is to summarize the 30 years of experience of the author and his associates in the area of functional and stereotactic neurosurgery. Besides covering the experience gained from performing more than 2000 operations, the book sums up the results of the scientific and research work of the author in collaboration with representatives of many theoretical and clinical subjects.

This volume consists of two parts. The first part (Chapters 1–5) offers contemporary data on the stereotactic anatomy of the central nervous system, on the neurophysiological mechanisms of the main pathological phenomena and processes, on the fundamentals of the stereotactic technique, on the main stages of stereotactic operations, on the methods of destruction of deep cerebral structures, and so on.

The second part (Chapters 6–20) presents descriptions of various diseases of the brain and spinal cord in which functional and stereotactic operations are indicated and have been shown to be effective. On each of these nosological forms, there is brief but, in the main, updated information on their pathogenesis, pathological anatomy, clinical aspects, and diagnosis. In each of these chapters, the main emphasis is placed on questions of surgical treatment—indications for functional operations, their methods and techniques, and long-term results.

The monograph is addressed, first of all, to neurosurgeons and neuropathologists; however, it may prove interesting to representatives of other professions—neurophysiologists, electrophysiologists, x-ray specialists, psychiatrists, endocrinologists, and neuro-oncologists.

The author expresses cordial appreciation to his associates and many representatives of other medical and technical branches with whom joint work was done over the course of many years and thanks to whom the appearance of this book became possible.

The author also expresses his profound appreciation to the esteemed Professor A. Earl Walker for his vital and valuable advice in the preparation of this book for print.

E. Kandel

Moscow, USSR

Contents

Abbreviations

AL	Ansa lenticularis
Am	Nucleus amygdalae, amygdala
Br Con	Brachium conjunctivum
CA	Commissura anterior
CC	Corpus callosum
Cer	Cerebellum
CF	Campus Foreli
CI	Capsula interna
CL	Claustrum
CM	Centrum medianum
CP	Commissura posterior
DREZ	Dorsal root entry zone
F	Fornix
FL (H_2)	Fasciculus lenticularis
FM	Foramen Monroi
Fm	Fastigium
FR	Formatio reticularis
FT (H_1)	Fasciculus thalamicus
GC	Gyrus cinguli
GCC	Genu corporis callosi
GlP	Glandula pinealis
GP	Globus pallidus, pallidum
Hipp	Hippocampus
Hypoth	Hypothalamus
LI	Linea intercommissuralis
LMI	Lamina medullaris interna
MI	Massa intermedia
NC	Nucleus caudatus
ND	Nucleus dentatus
NL	Nucleus lenticularis
NR	Nucleus ruber
NS	Nucleus subthalamicus, corpus Luysii
PAG	Periaqueductal gray matter
Pul	Pulvinar
Put	Putamen
SN	Substantia nigra
Str	Striatum
Subth	Regio subthalamica
Th	Thalamus
V	Ventriculus lateralis
V_3	Ventriculus tertius
V_4	Ventriculus quartus
Vc	Nucleus ventrocaudalis
Vc pc	Nucleus ventrocaudalis parvocellularis
Vim	Nucleus ventralis intermedius
VL	Nucleus ventralis lateralis
VOa	Nucleus ventrooralis anterior
VOp	Nucleus ventrooralis posterior
VPL	Nucleus ventralis posterior lateralis
VPM	Nucleus ventralis posterior medialis
ZI	Zona incerta

Functional and Stereotactic Neurosurgery

1

Anatomy of Subcortical Structures Related to Stereotaxy

1. Extrapyramidal System

The concept of an extrapyramidal system was developed by O. and C. Vogt (Vogt and Oppenheim, 1911; Vogt and Vogt, 1919–1920, 1948), who emphasized that this system had no direct ties with the cortical motor centers but that its morphological substrate was the basal ganglia.

The ultimate rationale for the majority of functional and stereotactic operations is the destruction of discrete structures of the basal ganglia for the treatment of disorders of the extrapyramidal system. Accordingly, it is appropriate to describe briefly the morphology of the basal ganglia, their pathways, and their complex interrelationships. Because a detailed description would go far beyond the framework of this book, we shall present the morphology, interrelationships, and variability of only those subcortical structures that are most frequently the objects of stereotactic operations (Figs. 1 and 2).

Until recently, the basal ganglia represented an area of the CNS that had not been studied sufficiently, the *terra incognita* in neurology. In spite of the fact that a great deal of data has accumulated about their functional organization, there is still no agreement on which subcortical structures comprise the basal ganglia. There can certainly be no doubt that NC, Put, GP, and CL belong, but there is still no uniform opinion regarding the NS, SN, NR, Am, and Hipp. Bucy's and Case's (1939) opinion that these nuclei should be included because of their close anatomic and functional interrelationships with Str and GP seems to be quite convincing. Moreover, these latter structures play a very important role in human locomotion. In principle, one may also include such "minute" nuclei as Brock's diagonal nucleus, the nuclei of septum pellucidum, substantia innominata, and a few others. There can be no doubt that Th and its numerous nuclei represent an important afferent substrate for the functional activity of the extrapyramidal system. Consequently, the term "basal ganglia" usually incorporates structures localized not only in the telencephalon but also in the diencephalon and mesencephalon.

Phylogenetically, the extrapyramidal system is quite ancient. This system is found in lower vertebrates in which the pyramidal system is very poorly developed.

The first indications that the basal ganglia are connected with voluntary movements appeared as far back as the end of the 18th century. Experiments by Magendie (Denny-Brown, 1946) revealed that a lesion of NC in animals triggered an uncontrollable desire to run forward.

After the motor zones of the cerebral cortex were discovered, several authors in the second half of the nineteenth century demonstrated that whereas undercutting of these zones in dogs causes only transitory paralysis, section of the brain at the level of the basal ganglia leads to total and irreversible loss of voluntary movement.

The role of the basal ganglia in regulating motility was established by the investigations of C. Vogt (1911), Bechterev (1912), C. Vogt and O. Vogt (1919–1920), and others. For the first time, attention was focused on the tremendous difference in the behavior of animals after the extirpation of the cerebral

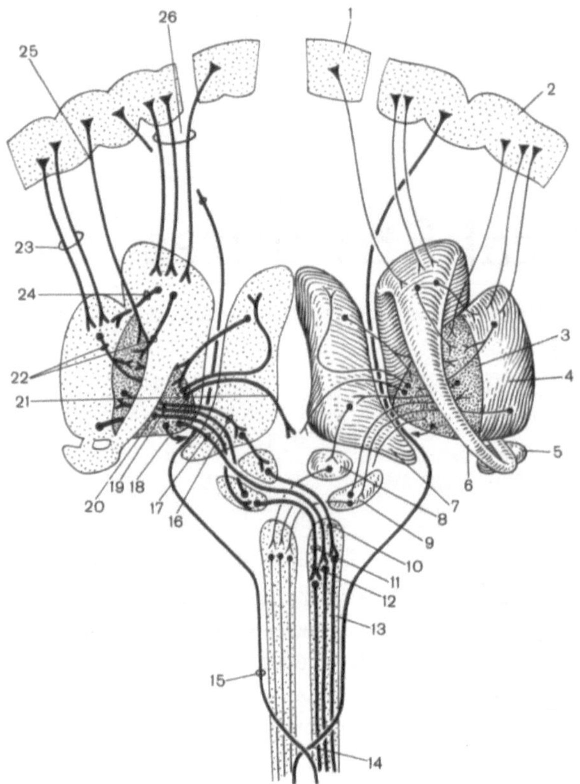

FIGURE 1. Scheme of the main cortical–subcortical neuronal circuits and interconnections of the basal ganglia (Cooper, 1956). 1, GC; 2, frontal cortex; 3, GP; 4, Put; 5, Am; 6, NC; 7, Th; 8, NR; 9, SN; 10, rubroreticular tract; 12, nigroreticular tract; 13, FR; 14, pyramidal decussation; 15, corticospinal tract; 16, pallidosubthalamic connections; 17, pallidoreticular tract; 18, AL; 19, nigropallidal tract; 20, strionigral tract; 21, pallido-hypothalamic tract; 22, striopallidal connections; 23, cortico-striatal tract (to Put); 24, striatal connections; 25, corticopallidal tract; 26, corticostriatal tract (to NC).

cortex and after the destruction of the basal ganglia. For instance, cats with basal ganglia ablated "seem to be like robots"—they monotonously walk around in circles for hours without changes in behavior—whereas cats without cerebral cortex perform various movements.

Klosovski *et al.* (1959) demonstrated that dogs readily tolerate extirpation of one Th, after which the behavioral changes and neurological deficit disappear rapidly. However, extirpation of both Th, although compatible with life, leads to profound behavioral disorders and reflex activity, which are only partially compensated.

Our present understanding of the role of the separate components of the basal ganglia in organizing motility is still far from complete. In the past few decades, the following concepts have evolved. The GP is an important central apparatus for the integration of both voluntary and reflex motor activity. The Str stands at a higher level of integration than GP and inhibits its functions. The Str is connected mainly with GP, whereas the latter has two-way ties with almost all cerebral structures. That is why the functional role of Str is implemented via GP, which allows one to speak of the striopallidal motor system.

It has been established experimentally that GP and SN exert an inhibiting effect on the Th nuclei (Veki *et al.*, 1977; Deniau *et al.*, 1978). Hassler *et al.* (1979) demonstrated that there is functional antagonism between the thalamocorticospinal and descending nigral systems.

In classical neurology, a lesion of GP is considered to give rise to the akinetic rigidity syndrome, whereas a lesion of Str results in the hypotonic hyperkinetic syndrome.

The extrapyramidal pathways participating in the regulation of motor functions consist of polysynaptic chains of neurons. In this respect, they are much more complicated than the pyramidal pathways, which consist of only two neurons. Such complex communications in the extrapyramidal system permit an important physiological role—instantaneous functional unification of many subcortical–stem structures in a variety of combinations for performing complicated motor actions and regulating muscular tone.

Recent anatomic–physiological investigations make it possible to identify the main pathways along which impulses circulate in the extrapyramidal system. These studies have established the principle of "neuron circles" in organizing the activity of that system. This principle, which was first proposed by the famous Russian physiologist Bernstein (1947), plays a vital role not only under physiological conditions but also in the pathogenesis of extrapyramidal disorders. These circles, which are made up of neuron chains, may be simple and consist of a small number of links (for example, Th–Str–Th) or more complicated, including many structures (Str–GP–Th–extrapyramidal zones of the cortex–Str). Since both stimulating and inhibiting impulses traverse these pathways, one may understand the complicated interactions among the nuclear structures in the extrapyramidal system. It has been established that these interactions are based on the feedback principle, which is so extensively employed in contemporary engineering.

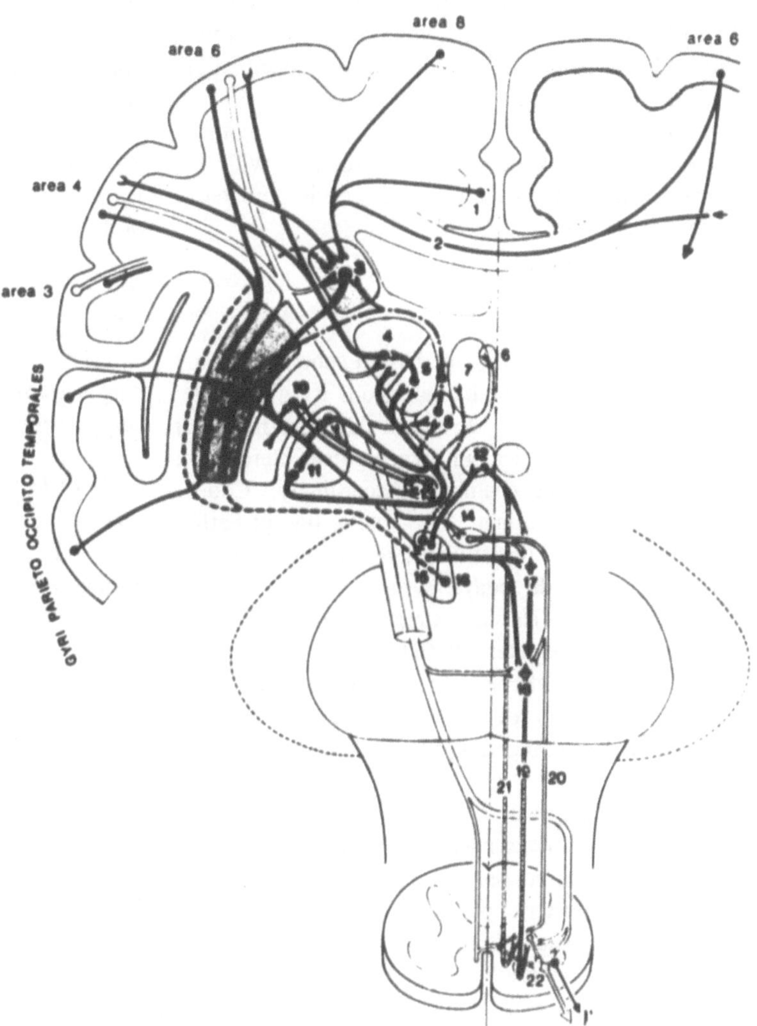

FIGURE 2. Interrelationships between the main subcortical, brainstem, and spinal structures and their connections with different areas of the cerebral cortex (Nieuwenhuys *et al.*, 1978). 1, Gyrus cinguli; 2, corpus callosum; 3, nucleus caudatus; 4, nucleus ventralis lateralis; 5, nucleus ventralis anterior; 6, nucleus habenulae lateralis; 7, nucleus medialis thalami; 8, nuclei intralaminares thalami; 9, putamen; 10, globus pallidus, pars lateralis; 11, globus pallidus, pars medialis; 12, colliculus superior; 13, nucleus subthalamicus; 14, nucleus ruber; 15, substantia nigra, pars compacta; 17, formatio reticularis mesencephali; 18, formatio reticularis pontis; 19, tractus reticulospinalis; 20, tractus rubrospinalis; 21, tractus tectospinalis; 22, cellulae motoriae cornus anterioris.

Irrespective of the ways the impulses circulate in the neuron pathways of the system, they must have an output to the peripheral effectors. These efferent pathways of the extrapyramidal system to the anterior horns of the spinal cord are also not fully resolved. It is noteworthy that the pyramidal tract, as numerous investigations have shown, is an important efferent link in the extrapyramidal system.

Hassler (1959a,b) describes two main routes from the basal ganglia to the anterior horns of the spinal cord: nigroreticulospinal and pallidorubroreticulospinal. He believes that a lesion in the former causes rigidity, akinesia, and tremor at rest, whereas a lesion in the latter causes various hyperkinesias (athetosis, chorea, etc.). Because the efferent pathways from Str, GP, and SN converge in mesencephalic and pontine FR, it has become clear that the reticulospinal tract (and not the rubrospinal tract, as had been assumed earlier) is the main efferent pathway of the extrapyramidal system. The fact that Monakow's (rubrospinal) bundle is a rudimentary system in man was emphasized by Burdenko and Klosovski (1937).

In conclusion, it should be stressed that notwithstanding the importance of the pathways and interrelationships among various cortical and subcortical structures in the pathogenesis of extrapyramidal disorders, pain syndromes, epilepsy, and so on, their study should not be overestimated. In view of the complicated structure of the brain, there is no reason to believe that even an exhaustive study of such pathways will establish the mechanisms of disseminating pathological impulses. The following analogy might illustrate this idea. At a large railway terminal we can see a multitude of crossing tracks; however, without a train timetable, we cannot tell on which track a specific train will arrive.

We begin our brief presentation of the stereotactic anatomy of deep-seated cerebral structures with Th, whose nuclei are among the main targets in stereotactic neurosurgery.

1.1. Thalamus

The thalamus (Th) is the largest subcortical mass of gray matter of the brain and attains its greatest development in mammals. In Greek the word *thalamos* means connubial couch; in archeological literature it refers to a secret chamber. Accordingly, Cooper (1965a,b) has stated that ". . . a long line of earlier investigators found this structure intriguing and mystifying, as we still do today."

Functionally, Th is an extremely important part of the brain and, as Walker (1966) has noted, a key to the understanding of all cortical functions. Because practically all impulses reaching the cerebral cortex pass through Th, this structure has been called the "gate to consciousness."

Anatomically, Th is part of the diencephalon. It consists of large ovoid masses of gray substance situated on both sides between CI and NC laterally and V_3 medially. The posterior parts of both Th are wider than the anterior parts, and the distance between them is less posteriorly than anteriorly.

Detailed study of the very complex morphology of Th (its nuclei and connections, its cytomyelo- and angioarchitectonics) has been the subject of many investigations. From the point of view of historical chronology, mention should be made of the works of Bechterev (1906), C. Vogt (1909), Walker (1938), Pines *et al.* (1939), Dzugaeva (1949), Papez (1942), Kurepina (1944), Zurabashvili (1957), Puzillo (1958), Hassler (1964), and others. Details of the morphology of Th are presented in monographs by Walker (1938) and Sager (1962) and in a chapter by Hassler in the stereotaxic atlas edited by Schaltenbrand and Bailey (1959) (Fig. 3).

The size of Th varies considerably: its length ranges from 26 to 37 mm (usually 30–31 mm), its height at the lateral border from 9 to 23 mm and at the medial border from 4 to 20 mm, and its width ranges from 12 to 26 mm at the middle, from 8 to 24 mm superiorly, and from 6 to 23 mm inferiorly (Rogulov, 1968). The Th is oval in shape on a frontal section through the middle of LI; in the majority of cases, its long diameter is situated horizontally but rarely vertically. The longitudinal axis of Th lies anteriorly–posteriorly and medially–laterally.

In man each Th is egg-shaped; one can recognize its anterior and posterior poles and four surfaces—medial, lateral, dorsal, and ventral. The anterior pole of Th is formed by the tuberculum anterius, adjacent to the head of NC. The thick posterior pole of Th (Pul) borders on the geniculate bodies. The medial surface of Th is the lateral wall of V_3, and the lateral is adjacent to CI. The lateral surface of Th is covered by a thin layer of gray substance (nucleus reticularis) separated from other Th nuclei by a thin layer of white substance (lamina medullaris externa). The ventral surface of Th

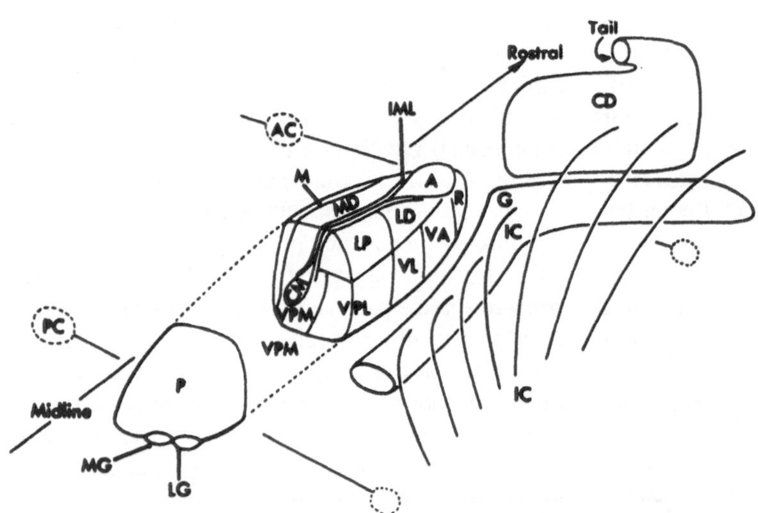

FIGURE 3. Thalamic nuclei and relationships of Th to LI, the head of NC, and CI. The right Th is shown from a posterolateral view in an "exploded" drawing in which Pul is depicted as detached by a frontal cut to show a frontal section of Th. LD, lateral dorsal nuclei; LP, lateral posterior nuclei; IML, internal medullary lamina; AC and PC, anterior and posterior commissures; Cd, head of NC; IC, CI; G, genu of CI (Willis and Grossman, 1973).

borders on Hypoth. Between the medial and ventral surfaces a hypothalamic sulcus extends from FM to Sylvian aqueduct, forming the border between Th and Hypoth. The superior (free) surface of Th, covered by a thin layer of white substance (stratum zonale), is located below F and CC. Externally, this surface is limited by NC and separated from it by stria terminalis.

In the anteroposterior direction, lamina medullaris interna passes through Th, dividing it into the lateral and medial groups of nuclei. The medial surfaces of both Th are connected by MI V_3, which is circular in cross section and varies in diameter considerably. The MI is located above the sulcus hypothalamicus.

The thalamic arteries have been extensively studied, both morphologically and angiographically. Arbitrarily, they may be divided into three groups—inferior, superior, and posterior. The inferior arteries, which supply the anterior and medial nuclei, are branches of the posterior communicating and basilar arteries. The superior arteries originate from the posterior choroidal arteries and supply the dorsal nuclei of Th. The posterior arteries, which are mainly branches of the posterior cerebral artery, supply the posterior and lateral nuclei. The arteries originating from the abovementioned sources form a network surrounding Th on all sides (Lazorthes and Salamon, 1971).

The nomenclature and definition of the numerous nuclei of Th are complicated and so far have not been fully agreed on. Approximately 100 large and small thalamic nuclei can be differentiated, some separated by thin strips of white matter. There are several classifications of thalamic nuclei (von Monakow, Vogt and Friedman, Le Gros Clark, Walker, Kurepina, Feremutsch and Simma, Hassler, and others), but the best are those of Walker (1938) and Hassler (1959b).

According to Walker, Th nuclei are divided into five groups: anterior, midline, internal, external, and posterior (Fig. 4). The anterior group consists of three nuclei—anterior ventralis, anterior dorsalis, and anterior medialis. The group midline is made up of five nuclei (paratenialis, anterior and posterior paraventriculares, medialis centralis, and central gray matter). The medial group consists of eight nuclei (dorsalis and ventralis mediales nuclei, centromedial nucleus of Luys, submedialis, lateralis, centralis, paracentralis, and parafascicularis). The posterior group consists of five nuclei, including the geniculate bodies. Kelly (1985) divided the thalamus into six nuclear groups:

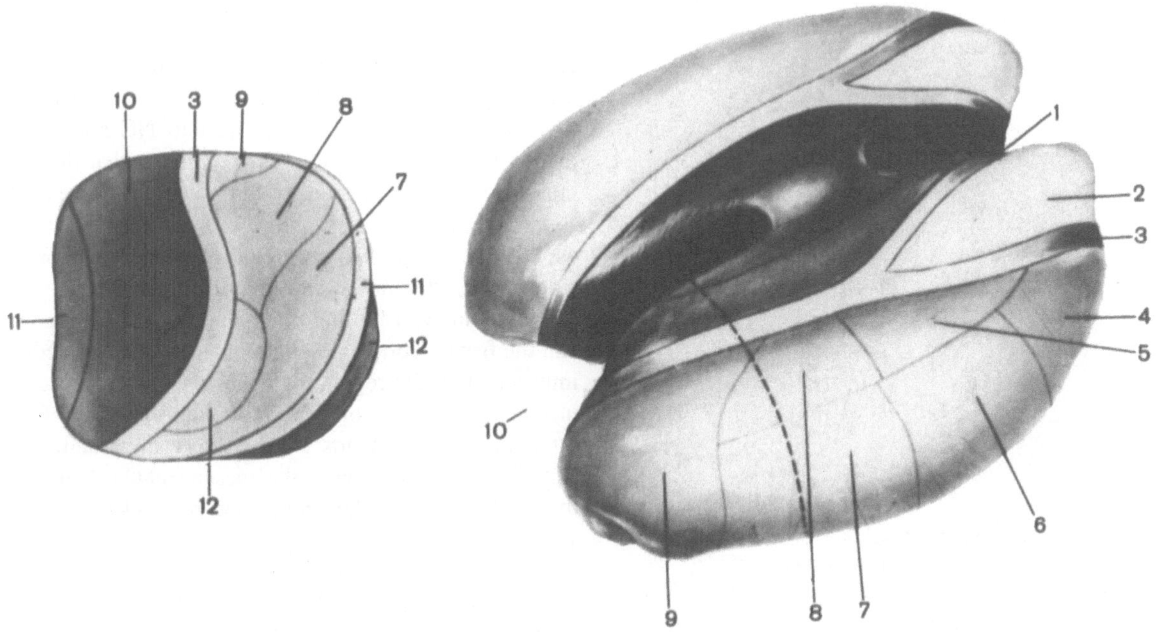

FIGURE 4. Simplified image of the main thalamic nuclei (Netter, 1975). Lateral group gray; medial, dark; anterior, white. Left frontal section along dotted line. 1, V_3; 2, nucleus anterior; 3, lamina medullaris interna; 4, nucleus ventralis anterior; 5, nucleus lateralis dorsalis; 6, VL; 7, VPL; 8, nucleus lateralis posterior; 9, Pul; 10, nucleus medialis; 11, lamina medullaris externa; 12, nucleus reticularis.

lateral (ventral tier and dorsal tier), medial, anterior, intralaminar, midline, and reticular.

Numerous investigations have been devoted to the afferent and efferent pathways linking Th with the spinal cord, brainstem, and other cerebral structures. Techniques that have been used to define these tracks include extirpation of Th, removal of various zones of the cortex with subsequent study of the retrograde degeneration of Th nuclei, stereotactic destruction of specific nuclei, electrostimulation and biopotential recordings from these nuclei with macro- and microelectrodes, and evoked potentials in Th and cortex. Figure 5 from the monograph by Willis and Grossman (1973) shows Th nuclei in a horizontal section and their main connections with other cerebral structures. It has to be noted that thalamocortical connections are, as a rule, two-way ties (Walker, 1966).

It is especially interesting to study connections between those Th nuclei referred to by Riechert as cortex-dependent and circumscribed cortical areas, especially those of the frontal cortex. These relationships are important in understanding the mechanism of rigidity, tremor, and other hyperkinesias. The frontothalamic paths (Puzillo, 1958; Hassler, 1959, 1966a; and others) consist of numerous bundles of fibers that connect the cortex of the frontal lobe, mostly the precentral region, to the medial, lateral, and (to a lesser extent) anterior Th nuclei (Fig. 6). The medial nuclear group is connected with the anterior part of the superior and middle frontal gyri and the triangular part of the inferior gyrus (area 45); the lateral group is connected with the superior and middle frontal gyri (areas 8 and 9); the anterior group is connected with the anterior part of the superior frontal gyrus (area 10). These frontothalamic fibers converge in compact bundles in CI and then, on the way to the cortex, spread out in a fanlike fashion. The bundles of fibers from the medial and anterior groups of Th nuclei proceed to the anterior limb of CI, and the bundles from the lateral group pass to the posterior limb.

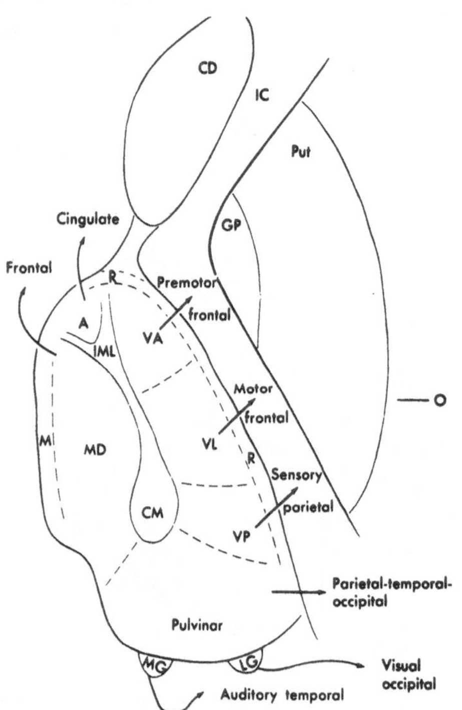

FIGURE 5. Cortical projections of specific thalamic nuclei; A, anterior; VA, ventral anterior; VL, ventral lateral; VP, ventral posterior; MD, medial dorsal groups; MG and LG, medial and lateral geniculate bodies. Nonspecific nuclei: CM, centromedian; M, midline group; R, reticular; O, intercommissural anterior–posterior zero point (Willis and Grossman, 1973).

At the point where CI passes into the corona radiata, the thalamic fibers split into a lateral part, which goes directly to the frontal cortex, and a medial part, which enters the subcallosal bundle to reach the corona radiata. This is important and explains the frequent bilateral effect of unilateral destruction of VL in extrapyramidal hyperkinesias.

A study of the retrograde degeneration of the lateral Th nuclei following focal cortical destruction has demonstrated that the projection of these nuclei is mainly to the region of the pre- and postcentral gyri. The intralaminar Th nuclei project to NC and CM, mainly to Put. Hassler (1959b) points out that the Th nuclei that receive afferent systems from GP serve as a source of a multitude of pathways to the cortical fields 4s, 6α, 6β, and possibly 8. In view of this, these zones may be included in the extrapyramidal system.

It is noteworthy that the connections between Th and the frontal cortex are bilateral and are undoubtedly important both for regulating voluntary movements and in the pathogenesis of extrapyramidal disorders. Descending thalamic paths are also very important, especially connections with the mesencephalic tegmentum (Johnson and Clemente, cited by Szekely *et al.*, 1962).

On the basis of their anatomic communications, the Th nuclei may be divided into three groups: (1) nuclei connecting Th with other subcortical structures; (2) specific (sensory) nuclei concerned with integration and transmission of sensory input from subcortical structures to the cortex; and (3) associative nuclei (mediodorsal, Pul) having projections only to cortical

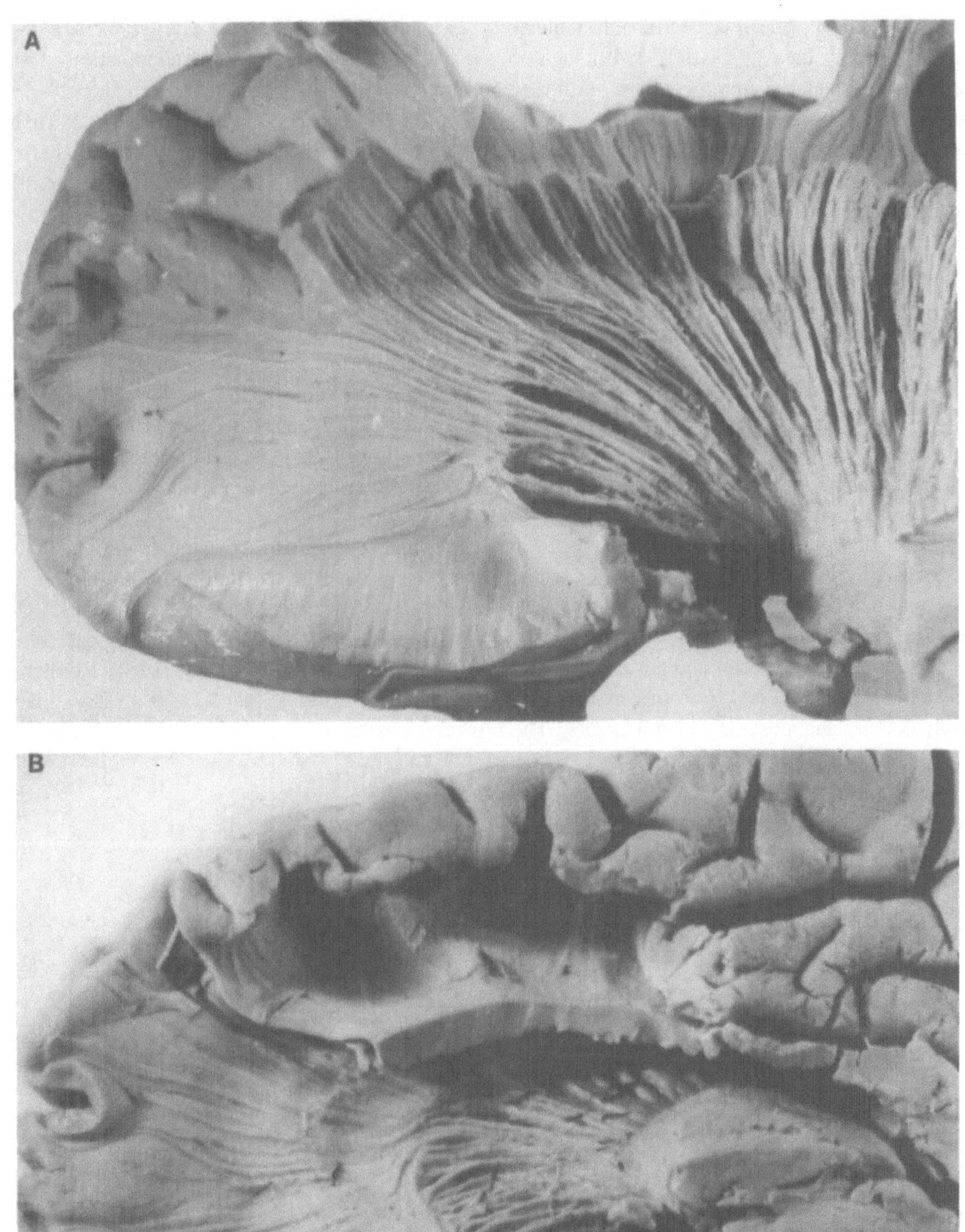

FIGURE 6. (A, B) Thalamocortical and corticothalamic pathways demonstrated with the aid of macrodissection of the pathways (Puzillo, 1958).

fields (Bousser, 1974). The nonspecific intralaminar Th nuclei or "sheath area" (Hassler, 1949) include CM, nuclei parafascicularis, intramedullaris, limitans, and reticularis, and others that receive afferent fibers via the spinoreticulothalamic pathways. This important nuclear group has no direct connections with the cortex: after extirpation of the cortical areas, these nuclei do not degenerate.

The lateral Th nuclei show a clearly defined somatotopic organization (Fig. 7). Below we present a short description of the main (but not all) thalamic nuclear groups.

1.1.1. Ventrolateral Nucleus

The classification of the nuclei of the ventrolateral part of the Th (VL complex) is of special significance for stereotactic surgery. The first classification of the nuclei, proposed by O. and C. Vogt in 1942, divided VL into three nuclei: oral, intermediary, and caudal, each having an external and an internal part. According to Walker, VL consists of anterior and posterior ventral nuclei (the latter subdivided into five other nuclei), and the ventral, dorsal, and posterior lateral nuclei. The VL, which measures on the average 9–10 × 8–9 mm, is one of the main targets of stereotactic operations in extrapyramidal disorders.

On frontal section the ventrolateral Th has the form of a convexoconcave lens—its medial and lateral edges form arcs with the convex facing laterally. The curvature of these arcs on the right and left sides is frequently asymmetrical. There is a general opinion that Pal, SN, Cer, and certain other motor-related subcortical structures converge upon VL and project to the motor cortex (Anasuma *et al.*, 1983a).

According to the classification by Hassler (1959a,b), VL corresponds to VO (nucleus ventralis oralis), which consists of four parts: nuclei ventralis oralis anterior (VOa), ventralis oralis posterior (VOp), ventralis intermedius (Vim), and ventralis oralis internus (VOi).

Three nuclei of this complex—VOa, VOp, and Vim—are of particular significance in stereotactic neurosurgery. It is noteworthy that VOa and VOp have substantial functional and morphological differences (Hassler, 1959b; Guiot *et al.*, 1961; Rümler *et al.*, 1972). It is presumed that VOa is a relay nucleus that receives afferent input from the medial part of GP as

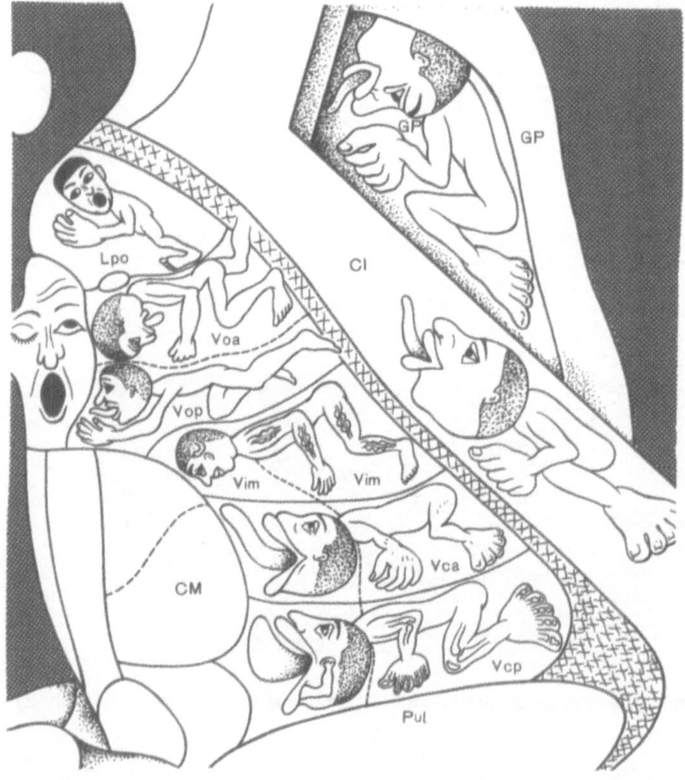

FIGURE 7. Scheme from the Hassler, Mundinger, and Riechert monograph (1979) illustrating somatotopic division of the six main thalamic nuclei, GP, and CI on a horizontal brain section. Relative sizes of homunculus parts reflect the representation of different parts of the body in corresponding subcortical structures.

well as from Nc and Put via Fth (H_1), which transverses CI. The main pathways from VOa project to the premotor cortex (field 6aα).

The VOp is the relay of the cerebellar systems coming from ND via Br Con and NR (dentatorubrothalamic tract). From VOp fibers pass to the motor cortex. The pathways from each ND, in the main, proceed to the contralateral VL, but part of the fibers also terminate in the homolateral nucleus (Walker, 1959). The VOp also receives afferent fibers via the Forel bundle and is the relay of the supravestibular tracts. The VOp projects to cortical field 4γ in the anterior part of the central gyrus (Hassler *et al.*, 1979).

Posterior to VOp and anterior to VPL lies a small nucleus, Vim, which is often included in the zone of destruction of the VL complex. The Vim projects to cortical zone 3a in the precentral gyrus. The vestibuloreticulothalamic tract is one of the main afferent paths to Vim. Situated medially to VOa and VOp is VOi (internal ventrooral nucleus), carrying the afferent pathways from the interstitial nucleus of Cajal and partially from Br Con. This nucleus receives vestibulocerebellar pathways and possibly projects to the premotor cortex (fields 8 and 6αβ). In general, the ventral lateral nuclear complex (VL) of Th is the subcortical relay to the motor cortex (Walker, 1938).

Neurons of the various VL nuclei differ in size and density. For example, the mean size of VOa and VOp neurons is 250 μm^2, those of VOi approximately 400 μm^2, and Vim 500–600 μm^2 (Hirai *et al.*, 1982).

It is necessary for a neurosurgeon to know the functionally important structures surrounding VL since any damage to them during an operation may lead to serious complications. In the caudal direction VL borders on VPL, damage to which may cause paresthesias on the contralateral side of the body. The ventral surface of VL is adjacent to NS, damage to which may result in hemiballism (Chapter 6, Section 11). Anteriorly, VL adjoins the posterior limb of CI in which the frontothalamic and frontopontine tracts pass. Any damage to them may cause psychic disorders (Taren *et al.*, 1969b). The part of CI where the corticospinal (pyramidal) tract passes lies lateral to VL. Any damage to this tract is accompanied by hemi(mono)paresis and, on the left, by speech impairment. Situated medially to VL are the dorsomedial Th nuclei and the mammillothalamic bundle, damage to which can also lead to psychic disorders.

In view of the individual variability of subcortical structures, especially VL, which is important for stereotactic calculations, studies have been carried out on the possible variations. It has been established that the distance from the midline to the lateral border of VL and thus to CI on a horizontal section of the brain is 13–15 mm at the level of VOa, 16–18 mm at the level of VOp, and 19–20 mm at the level of Vim (Guiot and Trujillo, 1969).

Table 1 presents the data of Spiegel (1965) on the relationship between the width of VL and the location of the lateral edge of the anterior horn of V at the level of CA. As can be seen in the table, this variability is quite significant.

Substantial individual variability of VL has been confirmed by the data of Rogulov (1968). The height of VL in a frontal section varies from 8 to 23 mm at its medial border and from 10 to 24 mm in its middle part. The width of this nucleus measures from 3 to 15 mm in its upper part, from 5 to 14 mm in the middle, and from 2 to 19 mm in its lower part.

1.1.2. Posterior Ventrolateral and Ventromedial Nuclei

These nuclei, VPL and VPM according to the classification of Walker or VC according to Hassler, are specific sensory or relay Th nuclei in which the

TABLE 1. Width of VL: Distance from Midline

Lateral edge of anterior horn at level of CA (mm)	Lateral edge of Th			Ventrolateral edge of Th		
	Min. (mm)	Mean (mm)	Max. (mm)	Min. (mm)	Mean (mm)	Max. (mm)
9–12	2	13.6	17	6	9.6	13
12–15	12	14.7	18	9	11.2	13
15.1–18	13	15.9	17.5	10	11.3	12
18.1–24	16	19	22	11	13.5	16
Entire group	9	14.7	22	8	10.8	16

lemniscal system terminates. Data on these structures are presented in Section 4.

1.1.3. Centrum Medianum

The centrum medianum (CM) is a comparatively small nucleus of Th lying medial to the VPL and somewhat anterior to the frontal plane drawn through CP. It has its greatest development in primates. It is related to the nonspecific Th nuclei because it has no direct projections to the cerebral cortex. There are two parts in CM: a magnocellular rostral portion and a parvocellular caudal portion. The functional significance of CM remains poorly understood. The center of CM is located 7–9 mm from the midplane.

This nucleus receives afferent inputs from GP. Passing through it are efferent pathways coming from the deep-seated Cer nuclei in VL via Br Con. According to Hassler, CM and the interlaminar nuclei project exclusively to both parts of Str, the parvocellular part of CM to Put and the magnocellular part to NC (Hassler, 1966b). These tracts pass through CI and the VL and VPL complexes. There is a separate tract from the emboliform nucleus of Cer to CM.

The CM is considered an important link in the medial system for the transmission and perception of pain (Section 4).

1.1.4. Pulvinar

This largest of the thalamic nuclei is composed of those nuclei forming the posterior part of Th. It consists of three parts—medial, lateral, and inferior—and has numerous efferent and afferent pathways linking it to many other structures of the CNS. After destruction of the dorsal area of Pul, there is degeneration of fibers, confirming its connection with certain zones of the cortex, primarily in the parietal and temporal lobes (Locke, 1960; Cooper *et al.*, 1974; Yoshii *et al.*, 1975). The corticotectal pathways pass through Pul to the midbrain. The Pul is also linked with the lateral geniculate body, the lateral and ventral groups of Th nuclei, and the intralaminar nuclei (Cooper *et al.*, 1974; Yoshii *et al.*, 1980). It was recently established that Pul has connections with the fastigial nucleus of Cer.

The Pul remains one of the least studied subcortical structures, and very little is known about its functional significance. It is believed that Pul participates in regulating motility and muscular tonus and that it influences the pain-conducting systems. These data serve as the basis for surgical destruction of Pul in certain CNS disorders (Chapter 9, Section 5.1.2).

1.2. Capsula Interna

The functionally most important pathway of the capsula interna (CI) is the corticospinal (pyramidal) tract, consisting mainly of thick myelinated fibers, originating from cortical pyramidal cells. At the level of the olives, the pyramidal tract in man contains more than a million fibers, 80% of which are myelinated (De Myer, 1959). This tract, in the form of a thin, compact bundle (4–5 mm in diameter), proceeds caudad into the posterior limb of CI laterally to VL (Guiot *et al.*, 1959; Hardy, 1966; Erlander *et al.*, 1975; Bertrand *et al.*, 1979). The corticospinal tract passing through CI and the cerebral peduncle sends many collaterals to SN and to the brainstem FR (corticoreticular tract) that have a close functional relationship to the extrapyramidal system.

The topography of the fiber bundles within the pyramidal tract is quite complex. The fibers concerned with the face, arm, and leg run separately. Most of the tract consists of fibers that innervate the facial muscles, forming a bundle that has the shape of a triangle in the horizontal plane and lies anteriorly amid VL, GP, and NC. Caudally and somewhat more laterally is the bundle for the arm, and still more posteriorly is that for the leg (Guiot *et al.*, 1968b; Bertrand *et al.*, 1976; Hardy *et al.*, 1979a,b). The central part of the pyramidal tract in CI is located posterior to a frontal plane passing through the middle of LI and at a distance of 20–22 mm from the midplane. The frontopontine tracts lie in front of the pyramidal tract in the anterior part of the posterior limb of CI. Functionally important two-way pathways connecting the orbital and prefrontal cortex with the medial nuclei of Th, SN, and the pontine nuclei travel in the anterior limb of CI. Fibers from the orbital cortex to Hypoth lie in the ventromedial part of CI.

Somatosensory pathways in CI proceed posterior to the motor fibers of the pyramidal tract.

1.3. Subthalamic Region

This relatively small deep-seated area of the brain has recently attracted attention, particularly of neurosurgeons, since its destruction has become a major stereotactic operation.

The subthalamus includes the ZI, fields H_1 and H_2 of Forel (CF), and part of the prelemniscus radiation. It lies ventral, medial, and caudal to the VL. The subthalamus was found to have the highest concentration of extrapyramidal tracts, which can be divided into ascending and descending systems. The former include the dentatothalamic and rubrothalamic tracts, which pass through Subth and terminate in VL. Here we should also include the reticulothalamic tract from FR of the pons and the midbrain.

The descending systems passing through CF include the pathways from the cerebral cortex (mainly from the frontal but also from the temporal and parietal cortex) as well as those from GP to NR and FR. In addition, Subth is connected with Hypoth and the mammillary bodies (tractus hypothalamotegmentalis dorsalis and tractus mammillotegmentalis).

In view of the need to perform stereotactic surgery, detailed studies have recently been done of the anatomic variants and relationships of the structures making up Subth, particularly CF and ZI (Lehman, 1972; Bubnov, 1975; Yukharev, 1979). These investigations have demonstrated that the localization of fields H_1 and H_2 is characterized by significant individual variability. Their length in the anteroposterior direction varies from 9 to 16 mm. They are largest from the middle point of LI to a point 5 mm anterior to CP. From this reference point, the posterior border of CF varies from 2 to 5 mm, and the anterior margin from 12 to 18 mm. The width of CF varies from 3 to 8 mm, and the distance from the midplane to its medial border is 2 to 8.5 mm, and that to the lateral margin is 6 to 16 mm.

The height of CF varies from 2.5 to 5 mm, and its position with respect to LI fluctuates from 1 mm above to 8 mm below the base line. It has also been shown that the size of CF depends on the length of LI. As the latter increases, the length of CF also increases (approximately by 2.2 mm), and it is displaced anterior to CP (anterior border by 2.4 mm; posterior by 0.8 mm). From the practical point of view, it is important to note that with an increase of the transverse diameter of V_3, CF is displaced laterally 2–3 mm (Yukharev, 1979).

The greater the length of LI, the more lateral is the lateral border of CF. At the level of the midpoint of LI, assuming its length to be 21–22 mm, the ventromedial angle of the nucleus is located 4–5 mm more lateral, and the midpoint of CF is 7 mm lateral. When the length of LI is 25–29 mm, the corresponding figures are 5.5–8.5 mm and 8.5–11 mm lateral (Lehman, 1972).

Yukharev (1979), on the basis of many series of 1-mm brain sections, identifies two parts of the CF: a pararubral part situated lateral to NR and related to the mesencephalon and a prerubral part located anterior to this nucleus and within the diencephalon. The border between these two parts is the frontal plane passing through the anterior pole of NR.

The pararubral part of CF is usually (in 78% of cases) trapezoidal in shape. It measures from 4 to 8 mm in length (average 5.79 mm), from 3 to 5 mm in width at the middle (average 4.64 mm), and from 3 to 5 mm in height (average 4.35 mm). As with CF, the size of the pararubral part generally correlates with the length of LI. When the length of this line is 22.9 mm or less, the length of that part (on average) is 5.1 mm, whereas when LI is 27 mm or more (on the average), it is 6.59 mm. The width of the pararubral part becomes greater as the width of Th increases.

The location of the pararubral part of CF is also subject to considerable individual variation. The medial and lateral borders of the central part of this structure are, respectively, 6–10 mm (average 8.12 mm) and 9–16 mm (average 12.14 mm) from the midplane. The superior border of this part is located from 1.5 mm above to 2 mm below LI (average 0.19 mm below), whereas the inferior border lies from 3 to 6 mm below (average 4.25 mm). The posterior border of the prerubral part lies from 2 to 5 mm (average 3.42 mm) in front of CP, whereas the anterior border is from 7 to 12 mm rostral (average 9.27 mm).

Since CF is situated above and medial to NS, damage to which may cause subthalamic hyperkinesia (Chapter 6, Section 11), one must define the "safety boundaries" for destruction of CF. Lehman (1972) estimated these boundaries in the following manner. With a long LI (27–29 mm at the midpoint), the safety boundaries for the center of CF would be 8 mm lateral to and 2 mm below LI, and those for the ventromedial angle 5 mm lateral to and 5 mm below. With a short LI (21–23 mm), at the same level, the center should be not more than 6 mm lateral to and 2.5 mm below LI, and those for the ventrolateral angle 4 mm lateral to and 5 mm below LI.

On the basis of the anatomic research of Yukharev (1979), destruction of CF should be an oval 8–9 mm long, 4 mm wide, and 3 mm high. The coordinates of the center of such a lesion should be 1.5 mm below LI, 9 mm lateral to the midplane, and 8–9.5 mm (depending on the length of LI) anterior to CP.

The location and variability of ZI and its interrelations with other subcortical structures have been studied in detail by Bubnov (1975) on 100 autopsy brains. The author found consideral variability in the location of this structure. Thus, its medial border is situated from 4 to 8 mm from the midplane, and its lateral margin from 16 to 22 mm. Only in the area from 8 to 16 mm from the midplane was ZI observed in all preparations. The location of this subcortical structure varies from 3 mm above to 6 mm below LI. Only in a small area 2 mm below LI was ZI present in all preparations.

The length of this structure anteroposteriorly varies from 11 to 18 mm. The ZI is always situated in front of CP. However, its posterior border is from 1 to 6 mm, and its anterior margin from 15 to 19 mm anterior to CP. In the area from 6 to 15 mm, ZI was present in all preparations. On the basis of these data, the coordinates for the center of the lesion should be 10 mm in front of CP, 1 mm below LI, and 11 mm lateral to the midplane. The optimal lesion should be 8 mm in length, 6 mm in width, and 4 mm in height (Bubnov, 1975).

1.4. Globus Pallidus

Globus pallidus or pallidum (GP) is an area of gray matter deeply seated in the cerebral hemispheres. Its name derives from its pale color, which arises from the large number of myelinated fibers interspersed between large nerve cells. The GP and the adjacent Put constitute NL; GP is the smaller part of NL and lies medial to and adjacent to Put from which it is separated by a thin layer of white substance, the lamina medullaris externa. Phylogenetically, GP is an older formation than Str. For instance, the pallidal complex is well developed in fish, whereas Str first appears only in birds. In man, Str forms by the end of the intrauterine period, later than GP. During the first months of an infant's life, the pallidal complex is the main cerebral structure governing movements, since myelinization of the striatal pathways is only completed by the fifth month of life. The motility of the newborn bears obvious "pallidal" features (Badaljan, 1975).

On both horizontal and frontal sections through the hemisphere, GP is shaped like an equilateral triangle, the apex of which points medially. The anterior pole of GP extends to CA. The CI lies medial and posterior to GP, and its ventral surface borders on the preoptic area.

Detailed information concerning the location of GP, the variability of its topographical relationships to CA and CP, and its distance from the midplane is presented in the stereotactic atlases of Spiegel and Wycis (1952b) and Schaltenbrand and Bailey (1959) as well as by Housepian and Carpenter (1957), Volkov and Yatsuk (1963), and Margorin (1970). The dimensions of this structure vary substantially, its length from 22 to 32 mm (in most cases 23–28 mm) and its width from 8 to 14 mm (most frequently from 10 to 14 mm). In a frontal section 20 mm anterior to CP, the height of GP varies from 10 to 15 mm, and its width from 12 to 15 mm. The distance from GP to the midplane of the brain is also not constant: anteriorly it ranges from 10 to 19 mm, and posteriorly from 21 to 31 mm (Margorin, 1970).

The location of GP with respect to CA varies considerably. Its rostral pole lies 3 to 12 mm anterior to it, and the caudal pole from 6 to 21 mm posterior (Margorin, 1970). The variability of the position of GP with respect to CP is approximately the same. Its anterior pole lies 25 to 34 mm from CP, and the posterior from 4 to 11 mm. The distance of CP to the medial part of GP varies from 18 to 26 mm, and that to its lateral part from 12 to 23 mm.

The GP consists of two parts, the medial and lateral segments, separated by a thin layer of white matter (lamina medullaris interna). The two segments of GP have different origins, the medial from the diencephalon and the lateral from the telencephalon (Kukujev, 1947). The lateral segment comprises 70%, and the medial 30%, of GP volume.

Microscopically, GP is composed of uniform large cells (40–50 μm) of triangular or fanlike shape and a multitude of bundles of well-myelinated fibers. The GP neurons have many dendrites, and their protoplasm contains a great deal of tigroid substance. To a great extent, they resemble anterior horn motor cells in the spinal cord and pyramidal cells of the motor cortex. A somatotopic localization has been established in GP, the oral being related to the head, the midpart to the arm and trunk, and the caudal segments to the leg.

The connections of GP with other cerebral structures are complicated and have not yet been fully analyzed. Grinstein (1946) emphasized that the "pallidum has direct and broad ties with subthalamic, mesencephalic and brainstem formations and with the cortex and thalamus."

Afferent pathways enter the GP from the cortex, Put, NC, Th, and NS, i.e., from structures directly related to motor regulation. The best-developed striopallidal connections are those in man, i.e., pathways passing to both GP segments from NS and Put. The SN is connected with the external segment of GP; moreover, the axons of its cells enter that nucleus directly through its lateral surface. Pathways from Th (from the medial and intralaminar groups of its nuclei) enter GP via CI. Pathways from the nonspecific Th nuclei (intralaminar nuclei, CM) proceed to the external segment of GP. There are two-way pallidosubthalamic connections: fibers originating in the external segment of GP pass to NS, whereas the tracts from that nucleus project to both parts of GP. There is also a powerful afferent input from the sensory systems (lemniscus medialis and the spinothalamic tract).

The direct afferent connections of GP with the cerebral cortex have not yet been extensively studied. Early investigations denied the existence of such pathways. However, recent investigations have discovered direct paths between the cortex and the striopallidal system (Riechert, 1980). In particular, it is presumed that corticopallidal fibers pass from the frontal cortex to GP as part of the corticospinal system tract.

Numerous efferent pathways of GP are significantly better developed and better studied. They form several bundles of fibers, the main ones being AL, FL, and FTh; pallidosubthalamic pathways are also known. The AL and FL originate in the medial segment of GP, which is the main efferent pathway from the entire striopallidal system. The AL, emerging from GP ventrally, circumscribes CI inferiorly and in field H_2 joins FL to pass medially. Both of these pathways terminate in VL, where they overlap with the tracts from SN. Therefore, practically all efferent fibers of the medial GP enter Th nuclei, where they terminate (Hassler *et al.*, 1979). Pathways from GP also extend to the prerubral area and CF, from which there is a short direct path to NR.

Both GP and NS are connected reciprocally. The efferent pathway from GP and NS passes in the medial segment as part of AL, penetrating CI. At the junction of AL and FL (field H_2), fibers form a bundle connecting GP with nuclei of the hypothalamic area (paraventricular nucleus, tuber cinereum). Fasciculus thalamicus (FT) in H_1 connects GP with the anterior part of VL (VOa). There are direct pathways from GP to CM, ZI, midbrain tegmentum, and FR of the brainstem. Part of the fibers from field H_2 proceed to NR of the other side via the subthalamic decussation. A separate bundle passes from GP to NS. There are also numerous connections between the two pallidal segments.

1.5. Striatum

Striatum (Str), the largest formation of the basal ganglia, is considered to be the central structure in the extrapyramidal system. The following morphological classification has been adopted for this structure. The corpus striatum consists of neoStr and paleoStr. The neoStr includes NC, Put, and possibly Cl; the paleoStr includes the medial and lateral parts of GP. The NL unites GP and Put. (Unless otherwise stated, Str refers to neoStr.) The two parts making up Str—NC and Put—have a common embryological origin from the lateral vesicle of the telencephalon.

Of all the basal ganglia, Str receives the largest number of afferent pathways. As Carpenter *et al.* (1976) pointed out, Str can be regarded as the receptive part of the corpus striatum, not only because it receives quantitatively the largest number of afferents but also because these afferents come from the most diverse sources. In the main, afferents come from the cerebral cortex and thalamic nuclei. The medial surface of the frontal lobe projects to the dorsal part of Str, the lateral surface to the lateral part, and the orbital surface to the medial part. Direct pathways from the temporal lobe enter the medial and lateral parts of Str.

The main efferent pathways issue from Str to both segments of GP and SN. The close functional relationships of Str with SN are implemented by various neurotransmitters—the strionigral pathways utilize γ-aminobutyric acid (GABA), and the nigrostriatal pathways dopamine. On the whole, Str is involved in many regulating and coordinating functions of the basal ganglia.

Following is a brief description of the two main parts of Str—NC and Put.

1.5.1. Nucleus Caudatus

Nucleus caudatus (NC), a C-shaped gray nucleus, is part of Str and is located along both sides of V, following its curvature. The nucleus consists of three parts—a head, a body, and a tail. The head forms the inferior wall of the anterior horn, into which it protrudes. Situated lateral to NC is Put, from which NS is separated by the anterior limb of CI, but inferiorly NC and Put unite in a common nuclear mass. The elongated body of NC forms the wall of V dorsal and lateral to Th and narrows into a tail behind Th, which enters the tectum of the temporal horn to terminate in Am. The NC neurons do not differ from Put neurons, and they are divided into large and small in a ratio of 20 to one. Both NC and Put receive numerous afferent pathways from many areas of the cortex, chiefly from the small pyramidal cells of the sensomotor zone as well as from the intralaminar nuclei of Th and from SN. Various parts of NC (lateral, central, and medial) project to different parts of SN.

Electrical and chemical excitation of NC experimentally inhibits behavioral activity, in particular motor reactions.

1.5.2. Putamen

Putamen (Put), a large nucleus of the basal ganglia that makes up the greater part of Str, is located between the external capsule and the lateral medullary lamina of GP medial to the insula. Ventrorostrally, Put unites with the head of NC. Putamen neurons are identical with NC neurons. The Put receives afferent connections from CN. There are grounds to believe that, functionally, Put has an inhibiting influence on GP. It is known that Put has a direct effect on motor activity (Denny-Brown *et al.*, 1945; Rozkanski and Lagutina, 1957) as well as on integrative cerebral functions (Laursen, 1963).

1.6. Substantia Nigra

Substantia nigra (SN) is a comparatively large nuclear formation in the brainstem situated ventral to NS and dorsal to the cerebral peduncle. This nucleus acquired its name from the macroscopically visible dark color of the nerve cells with large deposits of melanin pigment. The SN is subdivided into two parts: pars compacta, that area of the nucleus packed with large neurons containing melanin, and pars reticulata, with fewer neurons containing lipofuscin. The latter part is adjacent to the cerebral peduncle and cytoarchitectonically resembles GP.

The principal input to SN is from Str. There are direct projections from the head of NC into the rostral part of SN and fiber connections from Put to the caudal part of SN. The majority of strionigral paths terminate in the pars compacta. Unlike other main parts of the extrapyramidal system, SN receives no afferent input from FR and the sensory systems.

The efferent systems of SN include the nigrostriatal and nigrothalamic pathways as well as the pathway to the lateral segment of GP (Carpenter *et al.*, 1976). Originating from the large cells of the pars compacta, the nigrostriatal pathways pass through the ventromedial tegmentum and anterior limb of CI to terminate in NC and Put. Interruption of this dopaminergic pathway in animals substantially reduces the dopamine concentration in the homolateral Str. The nigrothalamic pathway projects mainly to VL and VA. Of all subcortical structures, SN has the highest dopamine concentration and is the main source of dopamine in Str. Efferent pathways from SN also pass to the anterior horns of the spinal cord, where they influence γ neurons (Hassler *et al.*, 1979).

The SN and Str have reciprocal two-way connections along which various neurotransmitters are transported. GABA, which inhibits dopamine synthesis in SN, is conducted from Str to SN. In the reverse direction, dopamine synthesized in SN passes from the pars compacta of SN to Put and NS. Consequently, reciprocal metabolic processes take place—GABA inhibits dopamine synthesis in SA, and dopamine inhibits acetylcholine synthesis in Str.

To date, direct pathways from the cerebral cortex to SN and from GP to SN have not been established. Destruction of the VL does not cause retrograde degeneration in SN. In all likelihood, there are no direct connections between these structures.

The dimensions of SN are quite variable. Anteroposteriorly, it measures 16 to 23 mm in length, and its width from the medial to the lateral margin varies from 15 to 20 mm. The height of the nucleus in its middle may be from 2 to 4 mm (Margorin, 1970).

1.7. Subthalamic Nucleus

The subthalamic nucleus (corpus Luysi) (NS) is a small lens-shaped deep-seated diencephalon structure lying medial to the base of CI. It consists of average-sized neurons. Afferent pathways of NS come from the lateral segment of CP and pass through CI. The existence of corticosubthalamic tracts has not been conclusively proven. Efferent pathways from NS project to SN as well as to the medial segment of GP and Str. Consequently, between NS and GP there are two-way connections similar to those between Str and SN. Pathways from NS to Th or the brainstem have not been established, nor has it been proven that NS receives afferent fibers. Partial NS destruction in experimental animals or in man by stereotactic operations causes severe hyperkinesias such as hemiballism or hemichorea on the contralateral side of the body (Chapter 2, Section 1.2).

The length of NS varies from 7 to 11 mm (average 8 mm), the width from 8 to 14 mm (average 9 mm), and the height from 4 to 8 mm (average 6.5 mm). The anterior border of the nucleus lies 13 to 18 mm anterior to CP, and the posterior border from 4 to 8 mm. The distance from the center of NS to the midline is 10.3 ± 1.7 mm; the medial border of the nucleus is located from 4 to 8 mm from the midline, and the lateral border from 10 to 15 mm (Bedrensky and Vasin, 1981).

1.8. Nucleus Ruber

Nucleus ruber (NR) is a comparatively large rounded nuclear mass in the brainstem at the level of the quadrigeminal plate. The NR is divided into magnocellular and parvocellular parts. Originating from the magnocellular part is the rubrotegmentospinal pathway proceeding to the spinal cord via the midbrain; from the parvocellular part the rubroolivary tract passes to the olive and then to Cer cortex. The afferent input to NR, in the main, is from Cer via Br Con. There are also fibers from GP and various cortical areas. No connections have been established between NR and SN (Hassler *et al.*, 1979).

The dimensions of NR are also variable (Yukharev, 1979): its length ranges from 7 to 11 mm (average 9.2 mm), its width at the central part from 6 to 10 mm (average 7.7 mm), and its height at its central part from 6 to 9.5 mm (average 7.6 mm). The distance of NR along the central plane from the midline to its medial border is from 0.5 to 2.5 mm (average 1.1 mm), and that to its lateral margin from 6.5 to 11.5 mm (average 9.1 mm).

Relative to LI, the superior border of the central plane of NR lies 1 to 3 mm (average 1.7 mm) and the inferior border from 8 to 11 mm (average 9.8 mm) below. The posterior border of NR is from 1 mm posterior to 3 mm anterior to it (average 1.1 mm), and the anterior border is from 7 to 12 mm (average 9.3 mm) anterior to CP. The longitudinal and horizontal dimensions of NR increase with an increase in LI and width of Th.

2. Limbic System

The limbic system (limbic lobe as first used by P. Broca) is localized on the medial surface of each cerebral hemisphere and consists of many structures, the detailed description of which does not come within the framework of this book. The limbic system or "Papez circuit" includes, first, the deep-seated structures of the temporal lobe—Am, Hipp, F—having close morphological and functional relationships with the anterior and dorsomedial nuclei of Th, Hypoth, and GC (Fig. 8). This system also includes a number of cerebral structures that are part of the rhinencephalon and

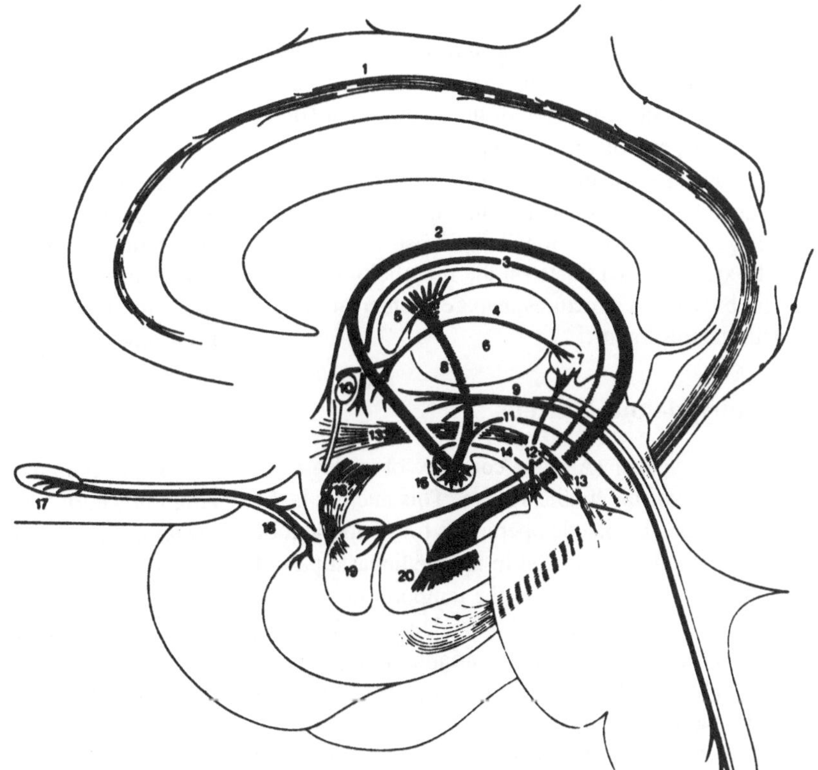

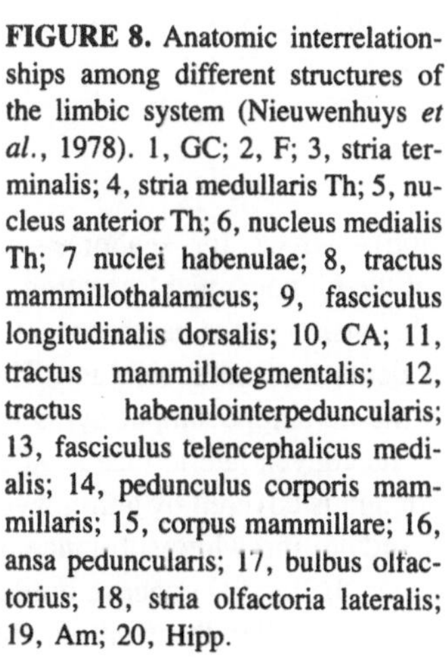

FIGURE 8. Anatomic interrelationships among different structures of the limbic system (Nieuwenhuys *et al.*, 1978). 1, GC; 2, F; 3, stria terminalis; 4, stria medullaris Th; 5, nucleus anterior Th; 6, nucleus medialis Th; 7 nuclei habenulae; 8, tractus mammillothalamicus; 9, fasciculus longitudinalis dorsalis; 10, CA; 11, tractus mammillotegmentalis; 12, tractus habenulointerpeduncularis; 13, fasciculus telencephalicus medialis; 14, pedunculus corporis mammillaris; 15, corpus mammillare; 16, ansa peduncularis; 17, bulbus olfactorius; 18, stria olfactoria lateralis; 19, Am; 20, Hipp.

phylogenetically older parts of the brain, being transitional from the brainstem to the hemispheres.

A number of these structures (substantia perforata anterior, stria olfactoria, indusium griseum, septum pellicudum) are collectively termed the paleocortex. The archicortex includes the uncus, gyri hippocampi, and several mediobasal structures. The boundaries of the limbic area and related structures have not been definitely established. Many authors include in this system not only Am and Hipp but also the anterior nuclei of Th, GC, Hypoth, mammillary bodies, septum pellucidum, certain midbrain structures, and the orbital and insular cortices.

Papez (1937) defined the neuronal circuit that is the central mechanism of emotions, including in it Hipp, GC, mammillary bodies, and anterior Th nuclei. It is difficult to separate the limbic system from FR of the brainstem and midbrain.

The main afferent pathway of the limbic system is the entorhinal cortex (parahippocampal gyrus). The connections between the limbic structures are very complex. Schematically, they may be divided into two large circuits. The first system (main limbic circuit) comes from the superior part of Hipp. Through the postcommissural F, this system projects to the mammillary bodies, from which the mammillothalamic tract proceeds to the anterior Th nuclei, which in turn project to various areas of the cingulate cortex. The second system comes from the inferior part of Hipp, giving rise to the precommissural F. These pathways project to the lateral nucleus of the septum, the lateral preoptic area, and the lateral Hypoth, and thence to the periaqueductal gray and to the FR nuclei of the midbrain and pons. The connecting link between these two systems is the internal hippocampal pathway system.

Next, we present data on the morphology of certain limbic system structures that are of interest to stereotactic neurosurgery.

2.1. Amygdaloid Nucleus

The amygdala (Am), a complex of at least six nuclei, is part of the limbic system. This nucleus is the main target of stereotactic operations for temporal lobe epilepsy. The Am is situated in the mediobasal part of the temporal lobe adjacent to the temporal horn of V. The shape and location of this nucleus, as for other subcortical structures, are quite variable. In the sagittal plane Am has the shape of a bean; in the frontal plane it is oval or almost square.

The Am has numerous connections, primarily with Hipp as well as with many cortical fields and deep-lying cerebral structures (Th nuclei and others). It has two-way connections with Hypoth, area preoptica, and mesencephalic tegmentum. Afferent pathways from tuberculum olfactorium and regio prepiriformis project to Am.

The variability of the dimensions and location of Am was thoroughly investigated by Sheljakin (1971, 1973). According to his data, the length of Am in sagittal sections 20–25 mm lateral to the midline ranges from 7 to 12 mm (average 8–9 mm), and its width varies from 16 to 24 mm (most frequently 19–21 mm). This nucleus is widest at the level of a section 5 mm posterior to the center of CA. The height of Am on sagittal sections 20 mm from the midline varies from 7 to 17 mm (most frequently 10–13 mm), and at a distance of 25 mm from 8 to 17 mm (most frequently 12–13 mm).

The angle between LI and a line passing through the largest diameter of Am varies considerably. This angle, opening anteriorly, at a section 20 mm from the midline is between 30° and 82° (most frequently 50° to 60°), whereas at a distance of 25 mm from the midline it is from 50° to 85° (usually 61° to 71°). At a section 20 mm from the midline, the center of Am is situated 12–20 mm (usually 15–16 mm) below LI, and in a section 25 mm lateral it is from 13 to 21 mm (most frequently 17–18 mm).

The distance of the center of Am from the midline in the above sections is from 19 to 28 mm (average 22–23 mm); the distance to the medial border of the nucleus is from 10 to 16 mm (usually 12–13 mm), and that to the lateral border from 27 to 37 mm (usually 32–33 mm). These data illustrate the great individual variability of the dimensions and location of Am, which must be taken into account in stereotactic operations for temporal lobe epilepsy and the aggressive syndrome.

2.2. Hippocampus

Hippocampus (Hipp)—Greek for seahorse—or Ammon's horn is part of the hippocampal formation in the temporal lobe, which consists of the parahippocampal gyrus, Hipp, gyrus dentatus, and subiculum. The Hipp is separated from the parahippocampal gyrus by the hippocampal sulcus. Its convex lateral face is adjacent to the temporal horn and is covered by a thin layer of white matter that extends to the white substance of the subicular area. Between the medial concave border of Hipp and the gyrus dentatus lies the fimbria, the continuation of the limb of F that anteriorly reaches the angle of Hipp gyrus. The anterior part of Hipp is enlarged and divided by sulci into separate tubercula. Posteriorly, Hipp curves medially and passes over to the posterior limb of F, where the two Hipp are connected by a thin commissure of white substance (commissura Hipp) that lies under the CC between the limbs of F.

Relative to CC and the hippocampal commissure, Hipp is divided into three parts: (1) precommissural, lying anteriorly from septum and extending from the olfactory triangle to the rostrum of CC; (2) supracommissural, lying in a thin layer above CC; and (3) the main part of Hipp, retrocommissural, which extends from the splenium of CC to Am (Stephan and Andy, 1982). On the basis of its morphology Lorente de No, in 1934, identified four zones in Hipp, from CA_1 to CA_4 in the mediolateral direction (from subiculum to gyrus dentatus). This classification is used today.

The fornix is the main efferent pathway of Hipp through which it receives an input from many structures of the diencephalon and midbrain. The afferent pathway that links Hipp with Hypoth comes from corpora mammillaria and passes to GC, to the entorhinal cortex, and, via the temporohippocampal fibers, to Hipp. This afferent pathway is part of the Papez circuit. There are also temporohippocampal fibers proceeding from the entorhinal cortex to segment CA_1 of Hipp (Cajal, 1911). The limbicohippocampal tract passes from the middle and posterior parts of CC to Hipp. Other connections of Hipp include pathways to the diencephalon, dorsomedial Th nuclei, and septal nuclei. The Hipp also has direct connections with the visual, acoustic, and olfactory cortices, as well as with the central and lateral parts of Am.

At present, it is assumed that Hipp is an important brain structure in the integration of memory. As many experimental and clinical investigations have revealed, Hipp participates in the appreciation of emotional and behavioral reactions.

The dimensions of Hipp, its interrelationships with other subcortical structures, and the degree and limits of its variability are extremely important for the stereotactic destruction of that structure in the treatment of temporal lobe epilepsy (Chapter 14, Section 6.5). These points were studied in detail on the basis of a large number of anatomic preparations by Sakare *et al.* (1982). According to their data, the length of Hipp in the sagittal section of the brain averages 37.9 mm. The width of Hipp in the frontal plane varies from 15.9 mm (anterior parts) to 12.1 mm (posterior parts), and its height from 11.1 mm (anterior parts) to 10.1 mm (posterior parts). All three dimensions of Hipp have wide variations.

In sections along the longitudinal axis of Hipp 20–25 mm lateral to the midline, the distance from its superior border to the intercommissural plane anteriorly varies from 8 to 11.6 mm, posteriorly from 2.3 to 6.6 mm, and from the inferior border to this plane 18.8–23.3 and 2.8–7.6, respectively. The distance from the middle of LI to the anterior pole of Hipp varies from 2.1 to 5.5 mm, and that to the posterior pole from 25.4 to 29.2 mm. The position of Hipp relative to the intercommissural plane is characterized by the angle formed by the longitudinal axis of Hipp and that plane. This angle, opening anteriorly, varies from 32.3° to 37.4°.

On frontal sections at the level of the anterior parts of Hipp, the distances from the medial and lateral borders and the center of Hipp to the midline are, respectively, 12.7–18.9 mm, 29.3–34.8 mm, and 21–26.8 mm; in sections through the posterior parts of Hipp these distances are 16.1–20.6 mm, 27.9–33.1 mm, and 22–26.8 mm, respectively. The angle between the longitudinal axis of Hipp and the midline varies from 4.6° to 13.7° (average 8.4°).

The distance from the pole of the temporal horn of V to the anterior part of Hipp—its hook—varies from 8 to 16 mm, and that to its center is from 16 to 32 mm.

2.3. Hypothalamic Region

The small area of the base of the brain located ventral to Th and adjoining the inferior part and floor of V_3 is called the hypothalamus (Hypoth). It contains many nerve centers of great functional significance. Hundreds of papers and scores of monographs have been devoted to the morphology, physiology, and pathology of Hypoth. Consequently, we present only the data on the morphology and pathways of Hypoth that are of significance for stereotactic neurosurgery.

The Hypoth is bounded superiorly by the sulcus hypothalamicus, inferiorly by the ventral surface of the diencephalon, anteriorly by CA, and posteriorly by the tractus mammillothalamicus.

Morphologically, Hypoth may be divided into three parts. Its anterior part—area preoptica or prothalamus—is situated above and rostral to the chiasm and posterior to CA and the lamina terminalis. The middle part includes the tuber cinereum, which in frontal sections has the shape of a triangle; its apex points toward the basal surface, and the base adjoins the anterior nuclei of Th, the CI, and the medial segment of GP. The tuber cinereum ends in the infundibulum, the apex of which forms the stalk of the hypophysis. The posterior portion of Hypoth contains the corpora mammillaria—two small hemispherical knobs lying posterior to the tuber cinereum.

Each of these three parts consists of many nuclei that are concerned with neurohumoral regulation of vegetative, visceral, endocrine, and metabolic functions, which, with the hypophysis, form a unified hypothalamohypophyseal system. According to contemporary data, Hypoth modulates certain types of behavior and participates in memory.

The cytoarchitectonic classification of Hartwig and Wahren (1982) divides the area preoptica into periventricular, suprachiasmal, supraoptic, and paraventricular nuclei and four principal preoptic nuclei (anterior, ventral, central, and lateral). The tuber cinereum includes the infundibular or arcuate, ventromedial, dorsomedial, lateral tuberal, perifornical, dorsal and lateral hypothalamic, tuberomammillary, and pallidothalamic nuclei. The mammillary body consists of three nuclei—medial, lateral, and intercalated. Cytoarchitectonically, these nuclei differ significantly from each other. Hartwig and Wahren (1982) identified nine types of neurons forming these nuclei. Some have a neurosecretory function and are connected mainly with the neurohypophysis.

Numerous fiber systems connect Hypoth with many cerebral structures. In addition to its efferent and afferent pathways, ipsilateral and contralateral associative tracts can be identified (Mosinger, 1950).

One of the main projections is the fornix (F), arising from Hipp and terminating in the lateral nuclei of the corpora mammillare. Emerging from this structure is another pathway—the mammillothalamic tract or Vicq d'Azyr bundle—passing above F and linking Hypoth with anterior Th nuclei, segmental nuclei of the midbrain, and periventricular gray. Lying adjacent to them is the mammillosubthalamic tract passing to Subth. Other pathways go from the posterior Hypoth to SN and NR.

The dorsal longitudinal bundle (Schutz' bundle) passes from Hypoth to the nuclei of the midbrain and medulla oblongata. This pathway, which transmits efferent impulses from Hypoth, sends fibers to the vagus nuclei and the sympathetic nuclei of the spinal cord, from which relays go to various internal organs to regulate their functions and maintain homeostasis.

Morphological investigations of the brain after posteromedial hypothalamotomy showed degeneration of neurons in VPL, parafascicular nuclei, GP, and FR of the brainstem. It has been shown that Hypoth has connections with VPL (Vopc), nonspecific, mainly dorsomedial, Th nuclei, and the somatosensory cortex. The posterior Hypoth receives considerable visceral sensory input (Sano *et al.*, 1975b).

3. Cerebellum

The cerebellum (Cer) consists of the vermis and two hemispheres, which in turn are composed of three main parts—anterior, posterior, and flocculonodular—separated by deep sulci. The flocculonodular part forms the ventral part of Cer.

The rostrocaudal cortex of Cer consists of from six to nine zones (Haines, 1981), which in all likelihood have various functions. The superior cerebellar peduncles (Br Con) connect Cer to the midbrain and adjacent subcortical structures, the middle cerebellar peduncles (brachium pontis) connect with the pons, and the inferior cerebellar peduncles pass to the medulla oblongata. The efferent fibers of Purkinje cells in Cer cortex, in the main, terminate in its deep-seated nuclei, nuclei fastigii, globosus, emboliformis, and dentatus, which are visible in cross sections of Cer.

According to current concepts, Cer is a very important part of the complex cerebral system concerned with integrating and coordinating locomotion. The Cer is the main organ coordinating motor activities (Hassler *et al.*, 1979). It may be regarded as an organ that receives information related to locomotion and modulates it in accordance with the desired goal. As many experimental studies have shown, the deep cerebellar nuclei exert a profound influence over the motor cortex.

The ability to perform these functions derives from the many afferent inputs it receives both from the periphery as well as from cerebral structures. The connections between Cer and the periphery (cutaneous, muscular, and joint receptors) are made through three systems that include the direct spinocerebellar, the spinoreticulocerebellar, and the spinoolivocerebellar tracts. These three systems pass through the middle and inferior peduncles to both the cortex and the deep-lying nuclei of Cer. The direct spinocerebellar pathways provide Cer with information about the state of segmental interneurons. In the Cer cortex the afferent systems give both "longitudinal" and "transverse" representation of the body. The Cer also has close connections with the extrapyramidal system (NC, Put,

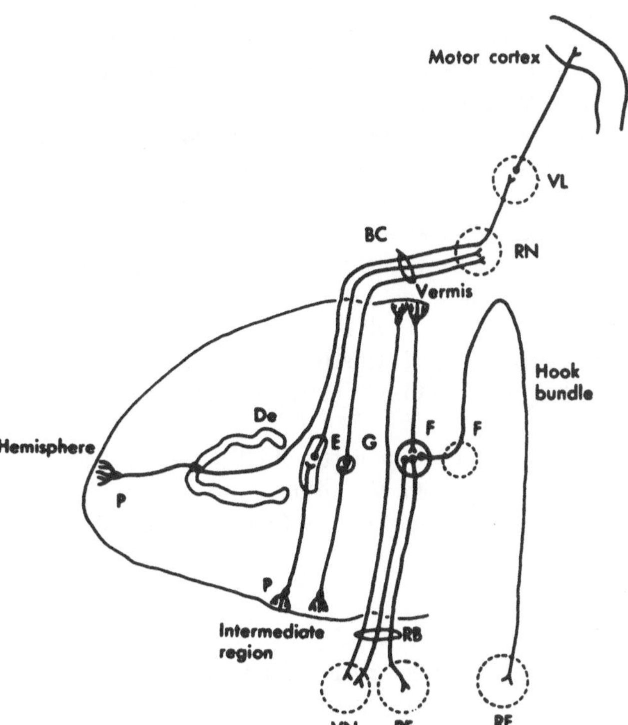

FIGURE 9. Cerebellar efferent pathways. The Purkinje cells (P) of the cerebellar hemisphere project to ND (De), forming the efferent system of the cerebellar cortex. The ND sends its axons through Br Con to the contralateral VL, which is connected to the motor cortex. The intermediate cerebellar cortex sends its output to the emboliform (E) and globose (G) nuclei, which in turn project to the contralateral NR (RN) through Br Con. The Purkinje cells of the vermis may either project directly to the vestibular nuclei (VN) or terminate in the fastigial nucleus (F), which projects in part ipsilaterally to VN and FR by way of the restiform body (RB) and in part contralaterally to the FR by way of the opposite Br Con (hook bundle) (Willis and Grossman, 1973).

SN, VL, and CM). From the olivo- and pontocerebellar pathways, Cer receives impulses from the cerebral cortex, NR, NC, GP, mesencephalic FR, and Pul.

The main efferent pathway of Cer (Fig. 9) is the dentatorubrothalamic tract. From the interstitial and dentate nuclei of Cer, impulses travel along the cerebellorubral pathways, through Br Con, into both parts of NR (parvocellular and magnocellular), and then to VL. The main part of this tract passes to the contralateral, and a small part to the homolateral, VL (VOp). From VL the cerebellar input passes to the sensomotor cortex; most goes to the cortex from ND, a smaller part from the interstitial, and the smallest part from the fastigial nucleus. The fibers arising in the dentate and interposed nuclei leave Cer via Br Con, decussate and distribute terminal ramifications contralaterally in NR, the nucleus reticularis tegmenti pontis, and the inferior olive (Anasuma *et al.*, 1983b).

There are various points of view regarding the influence of ND on the spinal systems regulating movement. Some authors consider that Cer nuclei provide the necessary level of activity for the spinal systems, particularly the γ systems (Granit, 1970, and others), whereas others assert that ND directly activates the cortical motor centers (Brooks *et al.*, 1972).

One of the deep-seated nuclei of Cer, namely ND, which is the object of destruction in certain extrapyramidal lesions, is of particular interest for stereotactic surgery. This nucleus, situated in the mediobasal parts of each Cer hemisphere, resembles a wavy or toothed oval-shaped disk, hence the name dentate.

Heimburger and Whitlock (1965), who first performed stereotactic dentatotomy, and then others (Zervas *et al.*, 1967; Slaughter and Nashold, 1968; Siegfried, 1971; Gortvai and Teruchkin, 1974; Koslowa, 1982), thoroughly investigated the spatial location of ND, its individual variability, and the relationship of this nucleus to the main stereotactic reference points—the line drawn tangentially to the floor of V_4 and one perpendicularly downward from Fm to that line.

Having investigated 100 cerebellar preparations, Koslowa has shown that the dimensions of ND vary within a wide range, its length from 13 to 23 mm (usually 15–21 mm), its width from 9 to 20 mm (usually 12–16 mm), and its height from 8 to 20 mm (usually 12–17 mm). As the height of V_4 increases, the length of ND grows by 1.29 mm, its height by 0.24 mm, and its width decreases by 0.6 mm.

The anterior border of ND lies from 2 to 12 mm rostral to the frontal plane passing through the apex of V_4, and the posterior border is from 8 to 19 mm behind that plane. The locations of the anterior and posterior borders of ND also vary with the height of V_4. For instance, if its height is greater than 11 mm, the anterior border lies 2.9 mm further anterior than when the height of V_4 is 5–7 mm.

The angle of inclination between the ND axis and the sagittal plane varies from 10° to 50° (average 28.7°). The center of ND lies 13.5 mm lateral to the sagittal plane, 4.4 mm below the horizontal plane, and 3.3 mm behind the frontal plane passing through the apex of V_4. If the height of V_4 is less than 8 mm, the coordinates of the center of the nucleus must be shifted 1 mm backward and downward in compensation, and with a height greater than 11 mm by 1 mm forward and upward.

The stereotactic coordinates of ND and its spatial interrelationships with V_4, required for stereotactic dentatotomy, are presented in Chapter 9 (Section 5.1.3).

4. Pain Structures and Pain-Conducting Pathways of the CNS

In all likelihood, René Descartes, in the first half of the 17th century, was the first to formulate the principle that any injury to the body triggers in the sensory nerves signals that pass to the brain. For more than a century of scientific neurophysiology, the pathways of conduction and mechanisms of pain have been studied in hundreds of thorough investigations, which have led to an understanding of its basic elements. However, in the past three decades, the concepts of classical neurology regarding the pain-conducting pathways and brain centers concerned with pain have undergone a radical reassessment and have become considerably more complex. The simple concept of conduction of pain along a three-neuron chain from the peripheral receptors to the spinal cord to the sensory nuclei of Th and the cerebral cortex has proved to be oversimplified. According to present-day concepts, the main pathways along which pain is relayed from the periphery to the cerebral cortex have a very complicated anatomic and functional organization (Bloedel and McGreery, 1975).

Hundreds of investigations have been devoted to the study of the anatomy and physiology of the nociceptive system of man (Fig. 10). The brief section in this chapter does not allow a full presentation of this complicated problem, and only a short discussion of the basic data of interest for functional and stereotactic neurosurgery is given. Additional information on pain and the pathogenesis of pain syndromes is presented in Chapter 13.

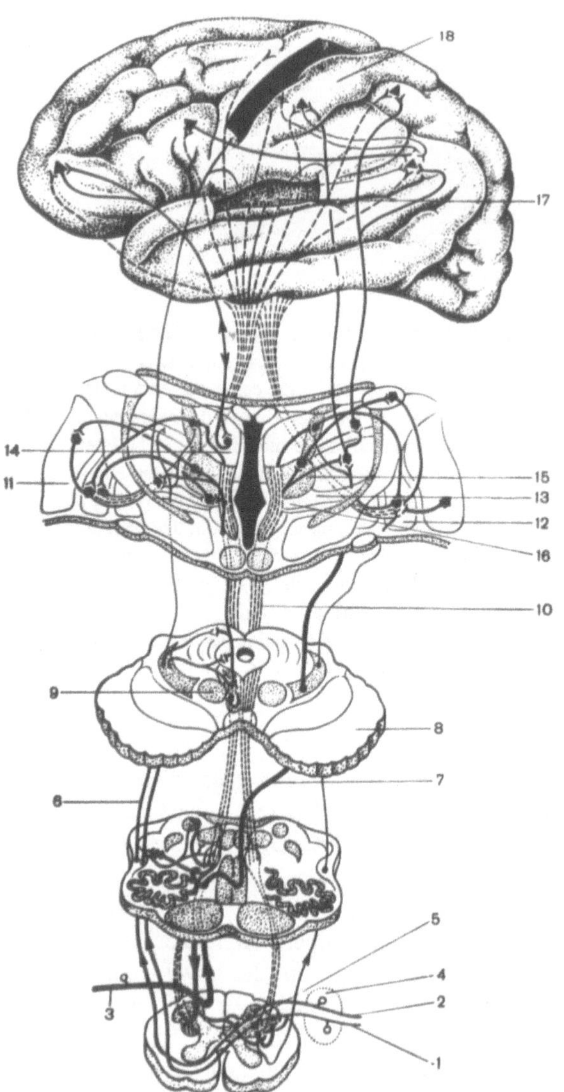

FIGURE 10. Scheme of pain-conducting tract and the brain and spinal centers related to pain (Hassler, 1969). Explanation in the text. 1, C fiber; 2, A-δ fiber; 3, A-β fiber; 4, spinal ganglion; 5, posterior horn of the spinal segment; 6, spinothalamic tract; 7, lemniscus medialis; 8, cerebral peduncle; 9, NR; 10, afferent reticular tracts; 11, Put; 12, GP; 13, VPL; 14, medial thalamic nuclei; 15, CM; 16, n. limitans; 17, second cortical sensory field (S-II); 18, gyrus centralis posterior (field 36).

Pain stimuli in the periphery are received by special receptors, nociceptors, the majority of which are polymodal since they react to different stimuli (mechanical, thermal, chemical, and others) (Zimmermann, 1981). It was established that the activation of cutaneous polymodal nociceptors depends on both direct effects of the stimulus and indirect effects mediated by the release of various biochemical substances. The nociceptors of both muscles and internal organs are stimulated by many endogenous chemical substances—substance P, norepinephrine, serotonin, bradykinin, prostaglandins, and others. However, it is accepted that there are two types of peripheral pain receptors, one reacting only to mechanical and the other to thermal or chemical stimuli. Nociceptors signal to the CNS not only the fact that they have been excited but also information about the intensity of the stimulus, which is coded in the frequency of afferent impulses (Zimmermann, 1981).

Pain from the skin and mucosa is transmitted by the axons of the intervertebral ganglia that make up the peripheral nerves. These axons are divided into A fibers (α, β, and δ) and C fibers, which differ in diameter (α, 12–22 μm; β, 8–14 μm; δ, 1–6 μm). According to the classical theory, each nociceptive fiber in the peripheral nerve carries its own specific modality: myelinated A-δ fibers transmit acute, sharp (or lightning), localized (or epicritic) pain, whereas the nonmyelinated C fibers have endings that are stimulated only by powerful irritants and carry diffuse, deep, burning (protopathic or slow) pain, which is a phylogenetically older sensation. According to another point of view, A-δ fibers mainly transmit mechanical or thermal pain stimuli. It has been established that repeated stimulation inhibits transmission of excitation along the A-δ fibers but intensifies excitation of C-fibers.

"Large" A-δ and A-β fibers have a lower excitation threshold than the "small" fibers. They transmit tactile and temperature impulses, and their stimulation usually does not cause pain (Hassler, 1972). However, if the intensity of the stimulation is increased sufficiently, these fibers may also transmit pain impulses (Casey, 1973a,b). Approximately half of the fibers entering the spinal cord in the posterior roots are nociceptive (Zimmermann, 1979).

The earlier supposition that pain sensation is also transmitted by the anterior roots was recently confirmed. It has been proven that a significant part of the nociceptive afferents pass into the spinal cord via the anterior roots (Applebaum *et al.*, 1977). These roots carry approximately a quarter of the afferent nonmyelinated fibers (Coggeshall *et al.*, 1975). Their presence explains the possibility of a relapse of pain after posterior rhizotomy.

Recent investigations have demonstrated that there is "rapid" pain, which lasts only during the period of the pain stimulus, and "slow" pain, which continues after the stimulus has stopped. These two types of pain have different morphological substrates. "Rapid" pain is conducted by thick myelinated fibers with a high conduction speed (25–50 m/sec). The nonmyelinated C fibers conducting "slow" pain are considerably thinner (less than 1 μm), and the speed of conduction through these fibers is much slower (about 1 m/sec). These data on "rapid" and "slow" systems of pain conduction are additional developments of the classical neurological concepts of epicritic and protopathic sensitivity.

Primary nociceptive impulses pass into the neurons of the spinal ganglia and then, via several rootlets into which each of the posterior roots is divided, enter the spinal cord. Each rootlet contains both thin nonmyelinated C fibers and thick myelinated A fibers, which form the posterior columns in the spinal cord. Before the rootlets enter the spinal cord, the C fibers lie laterally in each rootlet, and the A fibers lie medially. The first group of fibers conducts pain and temperature sensation, the second vibration and proprioceptive input.

Directly before it enters the spinal cord, the posterior root is surrounded by a so-called pial ring, which divides it into central and peripheral segments. A spatial organization is present only in the central segment: its middle contains large fibers of the γ system, the medial part has fibers of the leminiscal system, and the lateral part includes small fibers forming Lissauer's tract and activating the spinal neurons that give rise to the spinoreticulothalamic tract.

The large fibers conducting proprioception enter the posterior columns and also project onto spinoreticulothalamic neurons, exerting an inhibiting influence. The large fibers of the γ system terminate on the neurons of the anterior horns, forming part of the reflex arc of the myotatic reflexes (Chapter 2, Section 1.1).

The spinal posterior horn has a very complicated neuronal organization. It is divided into six Rexed laminae, in which the different sensory terminals end.

The substantia gelatinosa and Lissauer's tract play important roles in transmitting pain and possibly in the genesis of pain syndromes. Lissauer's tract is situated in the posterior part of the entrance zone of the posterior roots along the entire length of the spinal cord. Twenty-five percent of the fibers in this tract are of peripheral origin, and 75% originate from neurons of the substantia gelatinosa (Nashold and Ostdahl, 1979). After entering the spinal cord, the pain-conducting fibers form synapses with the neurons of the substantia gelatinosa, which contain a high concentration of opiatelike receptors. These neurons receiving pain afferents are localized mainly in laminae II and III of the posterior horns (Wall, 1969). The "gate mechanism" lies at this level. It has recently been established that the posterior horns have two types of nociceptive neurons—"specific" neurons, which respond to high-threshold afferent impulses from peripheral pain receptors, and "polymodal" afferents, which are stimulated both by pain impulses from A and C afferents and by nonpain impulses from low-threshold dermal mechanoreceptors (Zimmermann, 1979). The axon of the first pain-conducting neuron terminates in the posterior horn.

When the incoming nociceptive impulses to the neurons of substantia gelatinosa reach a critical level, the hypothetical "activation system" comes into action, including many supraspinal pain-controlling structures (nuclei of Th, Hypoth, limbic system, cerebral cortex), which give pain an emotional coloring.

The spinal cord has several systems conducting somatic and visceral sensation (Fig. 11). The first consists of the posterior columns, which conduct proprioception from the joints and vibration and fine sensibility. Fibers from the lower half of the body lie

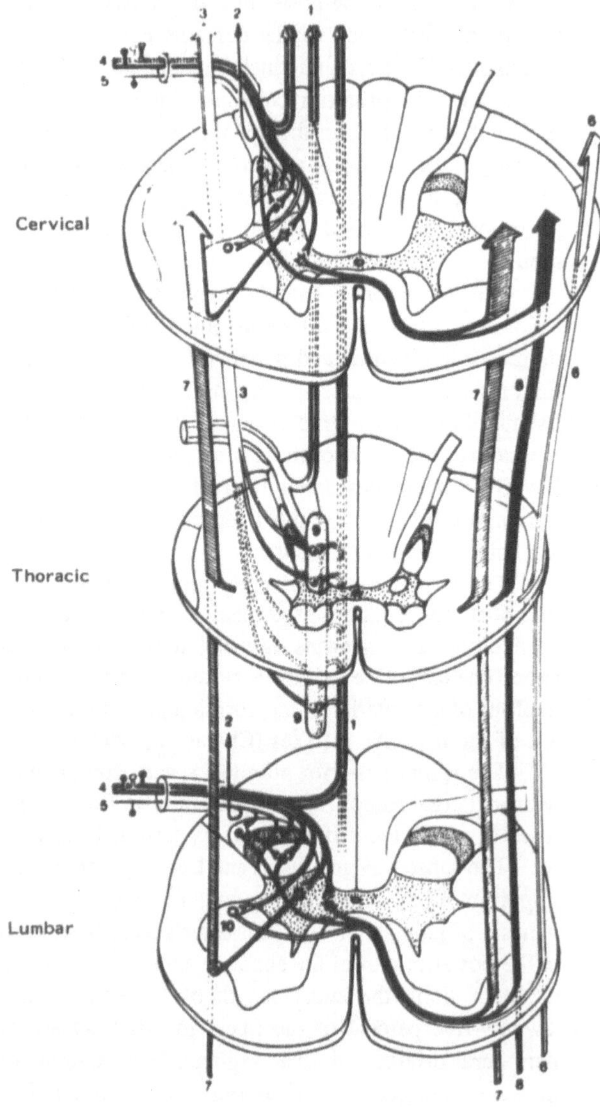

FIGURE 11. The origins and positions of the ascending pathways at three different levels of the spinal cord (Nieuwenhuys *et al.*, 1978). 1, Funiculus posterior; 2, fasciculus dorsolateralis; 3, tractus spinocerebellaris posterior; 4, A fibers of dorsal root; 5, C fiber of dorsal root; 6, tractus spinocerebellaris anterior; 7, tractus spinoreticularis; 8, tractus spinothalamicus; 9, nucleus thoracicus; 10, "border-cell."

medially in the posterior columns, forming the fasciculus gracilis, and those from the upper half lie laterally, forming the fasciculus cuneatus (funiculi Goli and Burdach). Both of these bundles terminate in the corresponding nuclei of the medulla oblongata. From these neurons, fibers, after crossing at the level of the olives, collect in bundles called the lemniscus medialis.

The second system is the spinothalamic tract, which transmits pain and temperature as well as afferent impulses evoked by mechanical squeezing of tissues from the entire contralateral half of the body (except the face). Presumably each of these types of sensation is transmitted by a separate group of fibers within the tract. It has been established that approximately 70% of the fibers in the spinothalamic tract are extralemniscal (Albe-Fessard *et al.*, 1975). There is another insufficiently studied spinocervical tract in man that is believed to "overlap" the functions of the two previous ones, e.g., transmitting pain and discriminative sensation (Willis and Grossman, 1973). All types of sensation from the face are conducted by the trigeminal system.

It has been proposed that there is another pain-conducting system, the archispinothalamic system, which has an anatomically distinct yet unidentified ascending pathway running near the central canal of the spinal cord (Gildenberg and Hirshberg, 1981a,b). Presumably this pathway is also concerned with the medial multisynaptic pain-conducting system.

The main pathway conducting pain sensitivity is the lateral neospinothalamic tract, consisting of posterior horn axons of neurons located ventral to the substantia gelatinosa. Fibers of this tract pass upwards for several segments, cross in the anterior commissure to the other half of the spinal cord, and proceed cephalad in the anterolateral column. Passing through the medulla oblongata and the brainstem, this pathway reaches VPL and terminates in its posteroventral part. Its somatotopic organization is described in Chapter 13.

The spinothalamic tract also has "rapidly" and "slowly" conducting fibers. It includes some uncrossed fibers for the ipsilateral transmission of pain. On the basis of investigations before and after commissurotomy, Šourek (1969) postulated that pain impulses are relayed not only in the spinothalamic pathways but also in the posterior columns of the spinal cord.

Electrostimulation of the spinothalamic tract causes sharp localized pain in the contralateral half of the body. Since the pain impulses in this tract are transmitted both by "rapid" and "slow" fibers, during stimulation of a peripheral nerve it is possible to identify two different biopotential waves in the spinothalamic tract.

It is noteworthy that whereas the pain-conducting pathways in the spinal cord are situated compactly and are easily identified in both morphological and electrophysiological investigations, after the spinothalamic tract enters the brainstem, a dichotomy occurs, making it substantially more difficult to identify these pathways. In the medulla oblongata, the spinothalamic tract lies slightly below the surface in the form of a compact bundle ventral to the inferior cerebellar peduncle and spinal trigeminal tract and dorsal to the inferior olivary nuclei. Section of this tract in spinothalamic medullary tractotomy is made at that level (Chapter 13, Section 7.1).

At the pons level, the spinothalamic tract is located lateral to the lemniscus medialis and medial to the middle cerebellar peduncle. In the midbrain this tract is situated in the dorsolateral part of the tegmentum, somewhat dorsal to the lemniscus medialis and lateral to the periaqueductal gray. At the mesencephalic level, the face fibers in the spinothalamic tract are medial, and those from the trunk and extremities are lateral. It is at this level that section of the spinothalamic tract is performed in open or stereotactic mesencephalotomy (Chapter 13, Section 5.2.5).

The lemniscus medialis is the main pathway for transmitting sensory input to the brain via the posterior columns of the spinal cord. The compact bundles of the lemniscus medialis also terminate in the specific Th nuclei (VPL), but some of the fibers end in the nonspecific intralaminar nuclei (Walker, 1966). In the spinal cord, fibers from the lower half of the body run in the ventral part of the tract, and fibers from the upper half run in the dorsal part. In the pons and midbrain, the tract changes from a vertical to a horizontal position. Fibers from the lower half of the body are situated laterally, and those from the superior half medially. At the level of the midbrain, between the periaqueductal gray and the trigeminothalamic tract, are the spinoreticular pathways, which conduct somatic and visceral sensation from the central zones of the head and the organs of the thoracic and abdominal cavities (Nashold, 1982).

For several decades, mainly after the classical work of Foerster, the lateral spinothalamic tract was considered to be the only pathway by which pain was relayed from the periphery to VPL. However, it has since been shown that other pathways conduct pain to the FR of the brainstem and intralaminar Th nuclei,

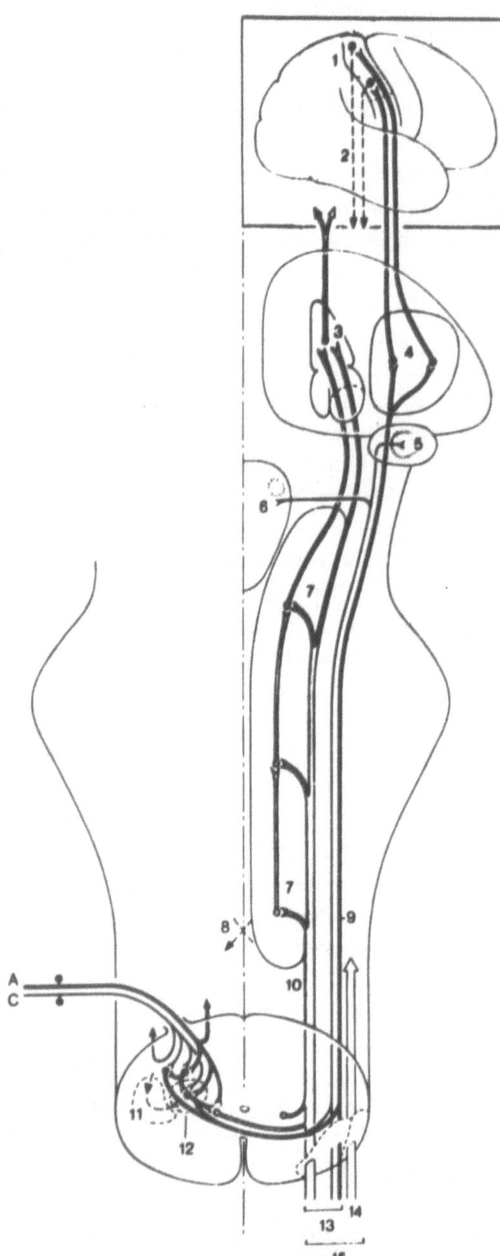

FIGURE 12. Neuronal connections of the spinothalamic (solid line) and spinoreticular tracts (Nieuwenhuys *et al.*, 1978). 1, Gyrus postcentralis; 2, tractus pyramidalis; 3, nuclei intralaminares thalami; 4, nucleus ventralis posterolateralis; 5, corpus geniculatum mediale; 6, griseum centrale mesencephali; 7, formatio reticularis medialis; 8, decussatio pyramidum; 9, tractus spinothalamicus; 10, tractus spinoreticularis; 11, tractus pyramidalis lateralis; 12, nucleus proprius; 13, fasciculus anterolateralis; 14, tractus spinocerebellaris anterior; 15, funiculus anterolateralis; A, A fiber; C, C fiber; A and C make up the radix dorsalis nervi spinalis.

including CM. This pathway was termed medial to distinguish it from the lateral (lemniscal) pathway (Fig. 12). The paleospinothalamic and spinoreticular tracts situated in the posterior Th medial to the neospinothalamic tract are related to the medial pain-conducting system. These tracts form numerous bilateral multisynaptic connections with FR of the pons and midbrain and pass to many other cerebral structures (CM, intralaminar nuclei, Hypoth, GC, Hipp, and others). Bilateral connections with the limbic system apparently determine the significant emotional component of pain.

The medial pathway forms the multisynaptic afferent system, which consists of a diffuse bilateral network. The specific projection systems of Th consist of monosynaptic pathways of thick fibers, whereas the nonspecific systems include polysynaptic pathways of thin fibers. The diffuse pain-conducting system of FR, especially rostral to the midbrain, according to both experimental and clinical data, is bilateral. To a certain extent, this explains the incomplete relief of pain and the occurrence of relapses after unilateral stereotactic destruction of nonspecific Th nuclei (Chapter 13, Section 5.2.1d).

Morphological investigations have demonstrated that the pain-conducting pathways to specific Th sensory nuclei sharply decrease in volume as they pass through the medulla oblongata, the brainstem, and the midbrain tegmentum. It has been established that in man, approximately only one-quarter of the fibers of the spinothalamic tract reach the sensory thalamic nuclei (Richardson and Akil, 1977). This is additional evidence for the existence of a second, extraspinothalamic pain-conducting pathway connecting with FR of the brainstem. The lateral spinothalamic tract is the main conductor of pain and thermal sensation, but there is now no doubt that diffuse and phylogenetically older spinoreticular pathways conduct the same modality.

In the mesencephalic FR, the spinoreticular tract merges into the reticulothalamic tract, which terminates in the nonspecific Th nuclei (CM, nuclei centralis lateralis, limitans, intralaminaris, parafascicularis, terminalis) and gray substance around the Sylvian aqueduct and V_3. Recent investigations have demonstrated the presence of nociceptive neurons in the medial Th, especially in a parafascicular nucleus (Reyes-Vazques and Dafney, 1983). The medial fibers (mainly CM and the intralaminar nuclei) also project to GP, where the afferent input connects with the subcortical efferent system.

The direct projection of the nonspecific pain nuclei of Th to the sensory cortex has not been established; however, it is presumed that such a projection does exist to the second sensory field (Fairman, 1966). Therefore, in addition to the classic spinothalamic pain system, there is another, more diffuse and complicated spinoreticulothalamic system.

The interrelationships between these two systems for relaying pain to the subcortical "pain centers" have not yet been adequately defined. Various suggestions have been made that the nociceptive input reaches FR only by the spinoreticular pathway, by collaterals from the spinothalamic pathway, or by both. Other connections are made by many collaterals of medial lemniscus into Fr. It is important to note that the lemniscal systems join with the spinothalamic pathway before entering VPL.

There is also a suggestion that the medial extralemniscal pain-conducting system is divided into two pathways in the mesencephalic FR. One ascending path terminates in the intralaminar nuclei; the other goes to the posterior Hypoth (Amano *et al.*, 1978). There is no doubt that these two systems are interrelated and are both under the regulating influence of other cortical and subcortical structures. It is considered that the medial system, consisting of nonmyelinated fibers, relays "slow," diffuse, poorly localized pain, whereas the lemniscal pathway, composed of thick myelinated fibers, conducts discrete "rapid" pain to VPL (Hassler, 1975). The interaction of these two systems serves as the neurophysiological basis of pain sensation.

The main part of the spinothalamic tract forming the lateral pain-conducting pathway terminates in the sensory (specific, relay) nuclei in the posterior areas of Th. According to the classification of Walker, these nuclei are called VPL and VPM (or Flechsig nucleus), and according to the classification of Hassler, Vc (n. ventrocaudalis). Hassler (1972a, 1976) defines four nuclei within Vc (VPL)—anterior (Vca), posterior (Vcp), parvocellular (Vcpc), and portal (Vc Por). The spinothalamic tract terminates mainly in Vcpc (in the basal part of Vc).These nuclei contain pain-sensitive neurons, the axons of which project to the postcentral gyrus of the cortex; moreover, Vca projects to field 1, Vcp to field 2, and Vcpc mainly to field 3b deep in the Rolandic sulcus (Hassler, 1960). Terminating in VPL are both (lateral and ventral) spinothalamic pathways; ending in VPM is the trigeminothalamic pathway conducting pain from the corresponding half of the face. It has been established that the spinothalamic tracts terminate not only in the contralateral but also in the homolateral VPL (Walker, 1938).

The somatotopic organization of VPL may be considered established, in particular by data obtained from stimulating that nucleus during stereotactic operations (Chapter 4, Section 7.1.2). The lateral part of that structure is concerned with pain and deep sensation from the leg and trunk, its midpart from the hand, and the medial from the face and head (Hassler, 1960; Bricolo, 1964; Hassler *et al.*, 1979; Hardy *et al.*, 1979a). The anterior and posterior parts of VPL (Vc) have separate somatotopic organizations: Vcp is concerned with the "joint homunculus," and Vca with the "tactile homunculus" (Hassler, 1972a). The fine somatotopic representation in VPL provides the anatomic basis for the sensory projections to somatosensory cortical areas I and II. The fibers from the thalamic sensory nuclei pass to the brain cortex through the postmedial part of the posterior limb of CI (Hardy *et al.*, 1979b).

Horsley (1909) presumed that sensory impulses from the parietal cortex and Th enter the motor zone of the cortex, which "processes" them into motor im-

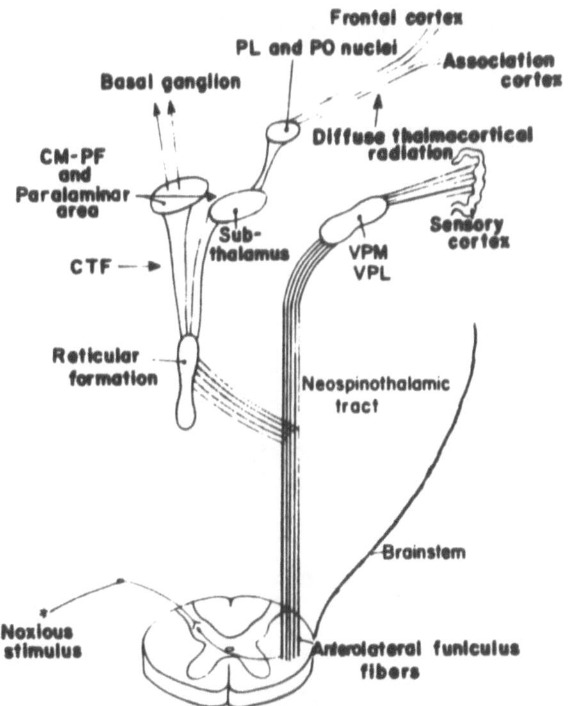

FIGURE 13. Diagram of the central pathways for pain (Richardson, 1974). Pl, pulvinar; CTF, central Th fasciculus.

pulses passing along the pyramidal tract. It was shown also that the peripheral afferent systems have direct access to the motor cortex via thalamic nuclei. Many neurons in the motor cortex have somatic sensory receptive fields, but the pathways which conduct the peripheral impulses are uncertain (Anasuma *et al.*, 1983a). However, Hassler (1975) assumes that only impulses of "rapid" pain enter the cortex from Th, whereas "slow" pain projects to the subcortical ganglia, in particular to the external segment of GP. Consequently, "slow" pain does not reach the cortical level but activates efferent subcortical structures. The cerebral cortex, where the final integration of the sensation of pain occurs, receives input not only from the main pain-conducting systems, but also from afferents modulating these sensations from nonspecific Th nuclei, the limbic system, the midbrain, Hypoth, as well as from other areas of the cortex.

In summary, one may conclude that in the human CNS there are two interrelated pain-conducting pathways to centers that, in modern literature, are acknowledged to be cortical and subcortical or, in respect to Th, as lateral and medial, respectively (Fig. 13). The former includes the spino- and trigeminothalamic pathways, posteroventral nuclei, and the somatosensory cortex. This system is concerned with conduction and perception of discrete localized pain.

The second pain-conducting system, directly connected with FR, includes the spinoreticular pathways and the medial, intralaminar, and caudal nuclei of Th. This polysynaptic system is phylogenetically older, and its pathways consist of nonmyelinated fibers. It is considered that the nonspecific system is concerned with the transmission of diffuse, poorly localized pain. An important feature distinguishing the medial system of pain from the lateral system is that stereotactic destruction within the framework of the medial system may not only eliminate chronic pain but leave pain sensitivity intact; i.e., it does not produce analgesia (Chapter 13).

In conclusion, it should be noted that certain recent investigations, which, however, cannot be considered final, have complicated to an even greater extent the existing concepts of the neurophysiology of pain and pain-conducting pathways. For example, it has been suggested that the afferent spinal systems conduct pain not selectively but along with polymodal sensation. It has been shown that the CNS does not have pain centers. Moreover, an even more unusual point of view has been advanced that the CNS in general has no specific pathways for pain (Dennis and Melzack, 1977) but that, depending on the setting one sensory system may convey tactile, proprioceptive, or pain sensation (King, 1979).

2

Extrapyramidal Mechanisms

The brain mechanism which performs involuntary, reflex movements has two additional parts. One of them inhibits the movement and the other increases it. . . .
I. M. Sechenov, Reflexes of the Brain, *1863*

We must identify individual reflex mechanisms wherever possible and yet strive to understand how these mechanisms are integrated into a functional whole.
Wilder Penfield

1. General Remarks

"Extrapyramidal lesions" and "lesions of the extrapyramidal system" are terms that have been used extensively in neurology for decades. They were proposed for those neurological motor disorders in which there are no clinical signs of pyramidal lesions or sensory disturbances. Such terms as "hyperkinesias" or "dyskinesias" are also frequently used in literature in reference to this group of syndromes. To a considerable extent, the term "dyskinesia" accurately reflects the clinical essence of this group of disorders, which are traditionally related to lesions of the extrapyramidal system. Ojemann and Ward (1973) emphasize that the term "dyskinesia" includes not only pathological involuntary movements but also related deviations of muscular tonicity such as rigidity, spasticity, changes in the activity (intensity) of voluntary movements, e.g., hypokinesia, and impaired associative movements.

It has generally been accepted that extrapyramidal lesions include a large number of separate nosological forms of CNS disorders—parkinsonism, cerebral palsy (double athetosis), dystonia musculorum deformans, torticollis spastica, hemihyperkinesias, essential tremor, myoclonus, and others. Each of these disorders is an independent state, but on the basis of at least three factors they may be grouped into the single "extrapyramidal" category in spite of their apparent differences in etiology, pathogenesis, and clinical manifestations.

The first factor is that motor disorders, particularly involuntary movements (hyperkinesias) and pathological changes in muscular tone (rigidity, spasticity) and regulation of voluntary movements (akinesia) predominate in the clinical pictures of all these disorders. The second factor favoring such a unification stems from the well-established opinion that these disorders are pathogenetically caused by lesions of the basal ganglia in the brain or (more generally) by lesions of the subcortical and brainstem nuclei and their projections (Fig. 14). Finally, the third factor that has become apparent during the past three decades is that one of the most effective methods for relieving these disorders is a stereotactic operation on deep-seated cerebral structures.

Extrapyramidal lesions are comparatively frequent and, as a rule, severe and often progressive disorders. Their management constitutes not only a medical but also a social problem, since there are millions of such cases throughout the world. If one adds the comparatively mild or "subclinical" forms of extrapyramidal pathology, then the number of such cases increases even further. A survey of 13,289 school children revealed hyperkinesias of various forms in 183 (1.38%) (Antonov and Shanko, 1975).

The loss of the ability to control precisely the movements of the body typical of a healthy person severely disables the majority of such patients, who periodically or constantly require assistance and care. Their grave suffering, both physical and mental, is aggravated by the chronic, progressive nature of extra-

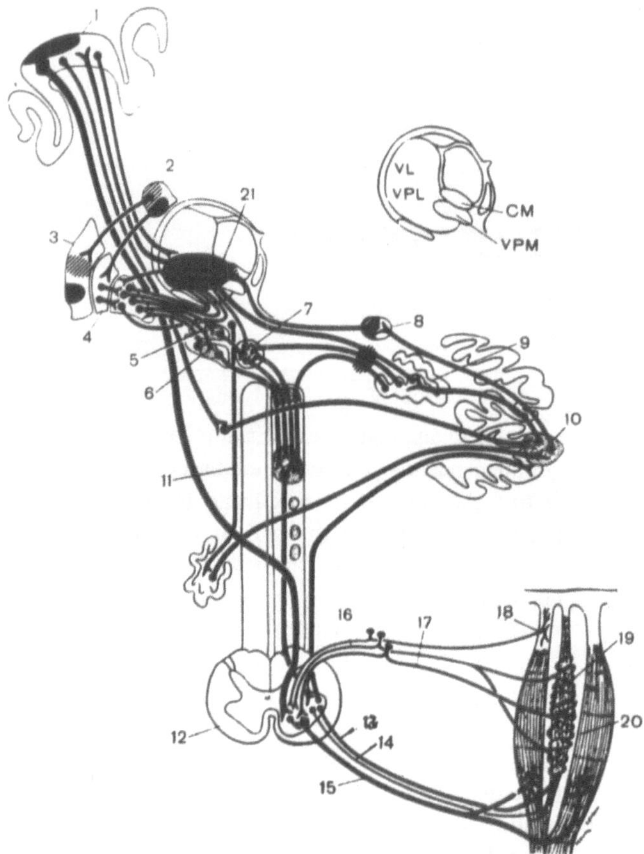

FIGURE 14. Anatomic pathways involved in conducting the pathological impulses that generate extrapyramidal dyskinesias (Cooper, 1965d). 1, Motor cortex; 2, NC; 3, Put; 4, CP; 5, NS; 6, NR; 7, SN; 8, nucleus emboliformis; 9, ND; 10, anterior lobe of Cer; 11, FR of the brainstem; 12, spinal segment; 13 and 14, γ-1 and γ-2 efferents; 15, α efferent; 16 and 17, afferents of groups I and II in the posterior roots; 18, Golgi body; 19, muscular spindle; 20, striated muscle; 21, localization of the stereotactic lesion. Right (above), scheme of Th nuclei.

pyramidal system disorders, which frequently last for years or even a lifetime without hope of spontaneous recovery.

The literature includes a number of classifications of hyperkinesias (Zucker, 1960; Schott and Lapras, 1961; Petelin, 1970; Biljik, 1971; Barbeau, 1975). These have various bases: the site of the lesion, whether generalized or localized, the character of the hyperkinesia, whether rhythmic or arrhythmic, etc. Since the pathophysiology of hyperkinesias has not yet been fully resolved, a classification is difficult. Such a classification should reflect contemporary knowledge of their pathophysiology, neuropathology, and biochemistry (including transmitters) and at the same time should incorporate the traditional pragmatic clinical approach. The author is of the opinion that a nosological principle must be the basis for a useful clinical classification of hyperkinesias.

The most complete and up-to-date classification of extrapyramidal lesions is the international classification adopted by the World Federation of Neurology. That classification was drawn up by the Research Committee on Extrapyramidal Diseases of the Federation under the chairmanship of Professor Lakke. The object of this committee was to establish a standardized, internationally acceptable clinical classification of disorders of the extrapyramidal system, a glossary of terms, and standard rating scales for abnormal involuntary movements and postures.

All extrapyramidal disorders are clinically classified into two groups: disorders of movement and disorders of posture and tone. The disorders of movement are further divided as follows:

1. Hypo- and akinesias
   a. Primary: Parkinson's disease, Shy–Drager syndrome, Huntington's chorea, progressive supranuclear palsy
   b. Secondary: toxic, traumatic, metabolic, infectious, space-occupying lesions, diffuse brain atrophy
2. Hyperkinesias
   a. Tremors
   b. Tics

c. Myoclonus
d. Chorea
e. Ballism
f. Athetosis
g. Akathisia

The disorders of posture and tone include the following:

1. Dystonia (primary and secondary)
2. Torsion spasm
3. Cogwheel phenomenon
4. Hypertonia
5. Hypotonia

It should be noted that these phenomena may occur in combination to form clinically defined syndromes.

1.1. Regulation of Motor Activity

It is not an objective of this book to review the physiology and pathophysiology of movements. This problem to which scores of monographs and thousands of scientific papers have been devoted has long ago become a separate chapter in neurophysiology, and possibly a separate science. Nevertheless, the author considers it necessary to discuss certain basic mechanisms that may prove useful to the neurosurgeon engaged in functional and stereotactic surgery. Three major factors have been identified as controlling voluntary movements: the force, the speed, and the displacement required.

Control of movements and muscular tone is achieved through integrated activity of complex functional systems of the CNS, which include a large number of interrelated and interacted structures of the brain and spinal cord—cortex, Str, GP, Th, NR, NS, SN, FR, nuclei and fiber pathways of Cer, and, finally, spinal segmental mechanisms. The extrapyramidal system consists of complex, polysynaptic chains of neurons with numerous interconnections. The complexity of this system reflects its important physiological role, which requires instantaneous functional integration of many cerebral structures in various combinations to execute motor acts and control muscular tone (Carew, 1985).

It has been established that the mechanisms triggering motor acts arise in the pyramidal system, whereas neuromuscular excitability and muscular tone to prepare the periphery are set by the extrapyramidal system. Strict synchronization of both systems ensures the coordinated activity of muscles so that motor performance is harmonious, precise, and smooth. In dyskinesias, each voluntary impulse initiating a movement evokes a burst of "extrapyramidal activity" that causes hyperkinesias and dystonic phenomena. The impression is that there is a kind of "short circuit" between the pyramidal and the extrapyramidal systems, leading to disintegration of the motor act.

In order to execute a coordinated movement, the CNS must fulfill three complex tasks: (1) spatial, to "select" the appropriate muscles needed to perform the movement; (2) temporal—each of these muscles must be activated or inactivated in a precise time relationship with other muscles; and (3) quantitative, the precise degree of excitation or inhibition of each muscle to achieve the desired movement.

In man, motor control is organized according to a hierarchic principle. Eccles (1980) defines the following levels that integrate voluntary movements: motor unit, spinal reflexes, brainstem reflexes, Cer vermis and cerebrocerebellar, Cer hemispheres, basal ganglia, and motor cortex. At the last level there is a mechanism for reverberating loops involving the associative cortex and subcortical structures (Roland *et al.*, 1980). At this level, "motor memory" is responsible for "preprogramming" movements.

The principal theory of motor control by the CNS requires a "peripheral control" genetically linked to the reflex concept of Sechenov, Pavlov, and Sherrington and elaborated by Bernstein (1947, 1966) and other investigators. According to the "peripheral theory," each coordinated movement consists of minute discrete phases, each based on the feedback principle, so that each phase is modulated by proprioceptive inputs. Each subsequent phase of the movement takes place only after the motor centers have received sensory information about the results of the preceding phase.

The theory of "central control" of movement has recently been enjoying wider acceptance and is being further substantiated by experimental data. Unlike the peripheral theory, the latter assumes that coordinated movements are determined by the CNS and controlled irrespective of feedback from the periphery. According to the central theory, all the information required to carry out a movement is stored in the neural structures executing it, and input from the periphery only supports the general level of activity of the central mechanisms controlling the movement.

The independence and interrelationships of the central and peripheral mechanisms controlling movement still remain unresolved and require further investigations. Among the numerous arguments in favor of the central control theory is the fact that certain refined movements (e.g., piano playing or high-speed typing) are so rapid that sensory feedback cannot play a significant role in their modulation (Evarts, 1971). Rapidly alternating movements about 10/sec are too fast to be under control of feedback loops because the time required for feedback from the muscles to spinal centers is about 0.1 sec (Eccles, 1980). Hence, these movements are more or less independent of sensory input. In pathological states, e.g., in ballistic movements, the speed of the hyperkinesia is also too great for modulation by feedback from the periphery (Eccles, 1980). At the same time, it is known that the CNS has powerful feedback channels, damage to which (e.g., section of posterior roots) leads to motor disturbances. The term "feedback" may designate various neurophysiological processes, but in reference to motor control, feedback signifies afferent signals from muscles being activated or already contracted. In the same action, there may also be "internal" feedbacks that precede the act—signals sent back from certain links of the neural chain involved in the motor activity.

About a decade ago, it was generally accepted that only descending influences originating in supraspinal structures provided facilitation or inhibition of the segmental and spinal circuits for motor control. Now, it is well documented that so-called long-loop reflexes starting in the spinal cord, relaying in various supraspinal structures, and descending to the cord play a very important role in motor control.

Coordinated activity at various levels of integration in the CNS is a very complicated and diverse process. It has been demonstrated that there is repeated recoding of information at each of these levels with numerous links between them. With lesions in higher centers, the lower integrative level of regulation (subcortical or spinal) takes over an "autonomous regimen" of work. This, however, does not mean that the pathogenesis of dyskinesia or muscular rigidity may be explained by the "exclusion" or "switching off" of supraspinal innervation. There are grounds to assume a more complicated mechanism (see below).

Spinal motor control is based on monosynaptic reflexes, one arc of which goes to the α-motoneurons from the muscle spindle receptors. The γ-motoneurons raise the sensitivity of the spindles to muscle extension and thus intensify monosynaptic reflexes, which may be inhibited by fibers from the antagonist muscle spindles. The axons of motoneurons give collaterals to Renshaw cells, which by feedback inhibit motoneurons. Another mechanism of inhibition is presynaptic inhibition, which reduces the excitability of primary afferents passing to the spinal cord via the posterior roots (Eccles, 1980).

In the literature, the anterior horns of the spinal cord have been compared to a programmed computer. They not only precisely carry out coded cerebral commands but also modulate them, depending on the changing conditions of the peripheral neuromuscular system.

1.2. Pathogenesis of Extrapyramidal Syndromes

The majority of theories on the pathogenesis of extrapyramidal syndromes are based to a greater or lesser extent on the theory of dissolution proposed by H. Jackson. Dissolution is the antithesis of evolution, and this theory presumes that higher motor centers of the CNS exert a permanent inhibiting action on lower ones and that if these higher centers are destroyed, the lower centers become disinhibited, resulting in the appearance of more primitive and uncontrolled activity. Such a theory is limited since it has been established that the higher motor centers not only inhibit but also activate the lower ones. Nor does it take into account the compensatory possibilities of various levels of the CNS. Thus, it is not based on a systematic approach to very complicated phenomena.

There is still no general theory of the pathogenesis of extrapyramidal disorders, although there have been repeated attempts to develop one. How is it possible to explain this in spite of the formidable advances in modern neurophysiology? Doubtlessly, the main reason lies in the complexity of normal movements and their neural systems of control. As was noted by the outstanding neurophysiologist Delgado (1969) the whole potential of modern-day cybernetics would not be sufficient to model a single step taken by man.

The investigations by Gelfand *et al.* (1964) on mathematical modeling of motor acts have demonstrated that the cerebral motor centers send to the periphery not direct and specific commands to individual muscles but commands for switching on "working matrices" located at segmental levels of the spinal cord and having considerable independence in activating the main elements of the motor act. Since there are no grounds to presume that these "signal matrix mechanisms" are damaged in dyskinesias, one may consider that the pathological supraspinal activity leads to incoordination of the matrices, causing disordered activation and ultimately disruption of the motor act. An example of this can be seen in the disturbed mechanism of reciprocal innervation in dystonia musculorum deformans—simultaneous contraction of antagonist muscles in the process of movement.

Cooper (1982) believes that all types of hyperkinesias are "sensory communication disturbances." This definition, emphasizing the role of the sensory component in extrapyramidal pathology, is nevertheless incomplete since it does not implicate the primary defect in the motor system of the basal ganglia and brainstem structures or biochemical metabolic disorders. Another question remains unclear: lesion of precisely which sensory system leads to the development of extrapyramidal hyperkinesias?

The study of extrapyramidal lesions has gradually led to the conclusion that certain premises of classical neurology with respect to the role of the functional and organic components in the pathogenesis of such disorders, including parkinsonism, need to be reexamined (Kandel, 1965). The fact that hyperkinesias may disappear completely after surgical destruction of certain subcortical structures is a convincing argument in favor of the concept of a functional (neurodynamic) factor in the genesis of this syndrome. One may presume that hyperkinesias and dystonias are based on specific lesions affecting a complex of cerebral structures, including the basal ganglia, brainstem, Cer, and cerebral cortex.

One of the main obstacles to understanding the pathogenesis of hyperkinesia and rigidity lies in the difficulty of producing an animal model. Unfortunately, no adequate model of the main extrapyramidal lesions in man has been developed. Numerous attempts to induce hyperkinesias in animals with lesions of GP, Th, NC, SN, and other structures, as a rule, have been unsatisfactory. When manifestations closely resembling extrapyramidal symptoms have been produced, they have proven unstable and were rapidly compensated.

Nevertheless, because the stereotactic method has made precise local destruction of subcortical structures possible, definite headway has been made recently in modeling human hyperkinesias. For example, partial stereotactic destruction of NS in monkeys leads to pronounced hyperkinesia in the contralateral limbs, resembling hemiballism or hemichorea (Carpenter, 1961). It is noteworthy that in those rare cases in which NS has been accidentally damaged during stereotactic operations on VL in man, hemiballism and subthalamic hyperkinesias have occurred (Chapter 6, Section 11).

One of the first experimental investigations in which local dystonic hyperkinesias were obtained was carried out by Foltz *et al.* (1959). They demonstrated that torticollis spastica develops in monkeys after destruction of the midbrain tegmentum (the medial part of FR, Br Con decussation, and posterior longitudinal fascicle). The authors concluded that torticollis results from destruction of the medial FR. In monkeys destruction of Put and GP induces transient dystonic hyperkinesias; however, if extirpation of the premotor cortex is added to the abovementioned lesions, then the symptoms persist for a long time (Dyakonova, 1968).

The second type of hyperkinesia that may be reproduced in animals is tremor. Damage to the ventral part of the tegmentum of pons and midbrain somewhat medial to SN in primates (and only in them) induces pronounced pathological tremor with a frequency of 4–7 Hz (Wycis *et al.*, 1957; Austin and Tsai, 1962; Poirier, 1972). It is assumed that this "tegmental tremor" lesion interrupts two pathways—ascending fibers to SN and the rubrospinal tract. However, since many pathways (tractus cerebellothalamicus, tractus nigrothalamicus, tractus rubrospinalis, tractus tegmentalis centralis, and others) pass through the ventrolateral part of the tegmentum, it is difficult to determine which of these tracts, when damaged, causes tremor. Rigidity, postural tremor, akinesia, and cogwheel phenomenon may be observed in a monkey after bilateral lesions involving the base of the midbrain and posterior Hypoth. These lesions destroyed many pathways including cerebellothalamic and nigrostriatal, but spared the rubrotegmentospinal pathway.

It is noteworthy that after such lesions in monkeys, there are fewer neurons in SN, resembling the state in parkinsonism, as well as a reduced concentration of dopamine in Str. Tremor may also be induced by destruction of ND of the cerebellum and its efferent pathways in Br Con (Carpenter, 1961; and others). In some cases, experimental tremor may be induced by stereotactic cryodestruction of NR (Ohye, 1982). A parkinsonismlike syndrome—tremor, rigidity, akinesia—has developed in rats after a microinjection of tetanus toxin in both NC (Kryzhanovsky and Aliev, 1978). These experimental models are approaching an adequate experimental model of parkinsonism.

The abovementioned experiments leading to the development of hyperkinesias have revealed a very interesting principle. In order to produce experimental hyperkinesia, it is necessary to destroy only a part of a subcortical structure (this is especially evident in lesions of NS), but in order to eliminate an already existing hyperkinesia permanently, it is necessary, as stereotactic surgery has shown, to destroy the subcortical structure completely.

A new stage in the development of animal models of parkinsonism was the selective destruction of the nigrostriatal dopamine system by stereotactic injection of 6-hydroxydopamine (6-OHDA) into SN of rats (Harik *et al.*, 1982). The nigrostriatal pathway rapidly and completely degenerates because of intracellular oxidation of 6-OHDA accumulated by SN neurons. Bilateral injection of 6-OHDA induces a syndrome closely resembling parkinsonism.

What then are the main pathophysiological mechanisms that determine the development of extrapyramidal phenomena? More than 100 years ago, Jackson, in his paper "On the anatomical, physiological and pathological investigation of epilepsies" (1873), expounded an outstanding hypothesis that all lesions of the CNS cause either "loss of function" or "overfunction." Based on this theoretical conclusion, Jackson considered negative (or minus) neurological symptoms to be produced by destruction of nervous tissue and positive (or plus) symptoms to be produced by hyperactivity. Examples of the first category are paralyses, anesthesia, astereognosis, aphasia, hemianopsia, and many other phenomena; examples of the second category include epilepsy, hyperkinesias, tremor, rigidity, chronic pain, nystagmus, and many vegetative phenomena. Jackson also proposed the "theory of dissolution," according to which damage to any cerebral structure "releases" structures at lower levels on which the affected structure normally has an inhibiting action.

Although this theory is still generally accepted, it is apparent that the genesis of neurological symptoms is much more complex. As examples, destruction of SN causes the appearance of the positive or plus symptoms of tremor and rigidity, whereas other sites of destruction—spinal cord, trigeminal nerve, or the brachial plexus—can cause permanent pain, which also must be regarded as a positive symptom. Examples of the reverse type included electrostimulation of certain subcortical structures, spinal cord, and peripheral nerves, which undoubtedly causes activation but not destruction of nervous tissue and induces negative or minus symptoms such as analgesia or hypotonicity.

Certainly, positive and negative symptoms are the final result of an "imbalance" of inhibitory and facilitating influences on the complex of low-level CNS structures.

On the basis of Jackson's theory, at the beginning of the 20th century there was a generally accepted opinion that rigidity and involuntary movements were caused by lesions of inhibitory mechanisms, causing the "release" of structures at a lower level. Wilson (1914) was the first to formulate the theory that the Str exerts an inhibiting effect on inferior motor centers, in particular, GP. Subsequently, this idea was elaborated on by Foerster (1921), who considered GP to be the main subcortical motor center, the source of constant motor and tonic stimulation of the muscular system in response to diverse afferent inputs to GP from the thalamic systems. The NeoStr, at a higher level than GP, has a constant inhibiting effect on it. Damage to GP leads to immobility, suppressed motility, and the development of muscular rigidity. Lesions of Str produce various forms of hyperkinesia. For decades, the Foerster hypothesis about Str being the "moderator" of GP was generally accepted.

Bernstein (1947) was of the opinion that the leading role in the genesis of dyskinesias was at the pallidothalamic level and offered a vivid description of "hyperkinetic dyssynergia":

> Hyperfunction at B level, arising as a result of disinhibiting painful processes, opens as it were, the doors to a phylogenetic menagerie, lying deeply dormant. And so from the depths of motility, monstrous, grotesque primordials come crawling out, without shape or substance, without sense or purpose, all kinds of torsion spasms, vestiges of ancient movements. . . .

Another hypothesis was proposed by Jacob (1932): degeneration of Str ends its inhibiting action on NR, so that the latter's "disinhibition" is responsible for the development of dystonia. Hassler (1959a) presumed that the pathogenesis of dystonia was principally that SN loses control over GP. Denny-Brown (1962) considers that subthalamic and mesencephalic nuclei are released from the control of GP.

The various theories of the pathogenesis of extrapyramidal states are based on the supposition that the clinical manifestations are triggered by a kind of "disinhibition phenomenon"; in other words, loss of higher motor control over lower centers releases the latter so that uncontrollable motor activity and muscular hypertonia develop. However, these authors disagree to a considerable extent about which particular structure is "released" from which subordinate control. Almost all subcortical ganglia—GP, thalamus, pallidum, NR, Subth, and FR—have at one time or another been assigned such roles. One cannot but note that the abovementioned concepts were speculative and were not substantiated by experimental or clinical findings. Moreover, these were proposed at a time when cerebral metabolic disturbances related to catecholamines and other neurotransmitter systems were unknown. If rigidity, as Foerster assumes, is based on "hyperactivity" of GP, which, without a doubt, is an important center of motor synergias, then electrostimulation of this structure should induce or intensify extrapyramidal symptoms. However, stimulation of this structure in both animal experiments and stereotactic operations on man has not confirmed this conclusion. Also, stimulation of NC and Put, which many authors expected to lead to hyperkinesia, did not cause either hyperkinesia or muscular rigidity. Stimulation of NC in cats produces an opposite effect—arrest reaction and reduced postural tone (Hassler and Dieckmann, 1967b). Moreover, stereotactic lesions of SN in animals do not produce hyperkinesias (Carpenter, 1961).

Nevertheless, there is some evidence in the literature that certain subcortical structures can activate dystonic phenomena and rigidity. Cooper (1965b,c) noted an intensification of dystonia after lesions of NC, Put, and ND. According to Narabayashi (1968), electrostimulation of a small zone posterior to VL increases hyperkinesia in dystonia musculorum deformans. There have been numerous descriptions of tremor intensification during electrostimulation of VL in stereotactic operations (Chapter 4, Section 7.1). It was also demonstrated that direct introduction of acetylcholine in GP during a stereotactic operation leads to an intensification of tremor in the contralateral limbs of parkinsonians. So far, it is difficult to assess these findings; however, one may assume that the basal ganglia contain neuronal structures that inhibit the generation (or circulation) of pathological impulses leading to hyperkinesias and disturbed muscular tone.

Although the "disinhibition" theory explains many facts, a number of important questions remain unanswered. First, there is no direct evidence indicating that hyperkinesia and rigidity are "generated" in GP, Put, SN, NR, FR, and Cer nuclei. Second, stimulation of these structures, both in animals and in stereotactic operations, generally does not lead to the appearance or intensification of rigidity or dyskinesia. Third, recordings of biopotentials during such operations do not reveal any clear-cut signs of elevated functional activity in deep cerebral structures examined. Fourth, lesions in individual subcortical structures, both in animals and in man, as a rule do not result in the development of extrapyramidal disorders. At the same time, the disappearance of rigidity and hyperkinesias is observed after stereotactic lesions of many subcortical structures (VL, GP, Subth, Pul, CM, and others). In one of our cases of severe generalized muscular dystonia, we demonstrated almost complete bilateral destruction of GP (Chapter 7, Section 3). In this case, is it possible to assert that rigidity was determined by "hyperactivity" of GP?

It is noteworthy that several authors are of the opinion that lesion of a single structure may lead to the appearance of different hyperkinesias (Laponogov, 1969). It is known that in several structures of the extrapyramidal system—GP, NR, SN, NS—there are similar groups of neurons, so-called motor cells. In view of this, the assumption was put forth that these cells form a special neuronal chain, damage to any link of which may lead to the development of hyperkinesia.

Another argument is the fact that stereotactic destruction of various subcortical ganglia can eliminate the same type of hyperkinesia. The logical conclusion to be drawn from these facts would be that in dyskinesias, a complex system of circulating pathological impulses implicate many subcortical, brainstem, and cerebellar structures. If an identical effect can be obtained by destruction of different cerebral structures, then in all likelihood the various lesions interrupt the pathways of the same "functional system" but at different levels of the CNS. As experience with stereotactic surgery for hyperkinesia and rigidity demonstrates, the "breaking" of the pathological sequence at any of its links leads to the disappearance of the primary phenomenon.

Consequently, one may assert that the pathogenesis of extrapyramidal disturbances can hardly be explained on the basis of fixed or limited lesions. The "localistic" concepts are now gradually giving way to systemic concepts involving complex networks of structures and pathways, primarily of the basal ganglia, brainstem, Cer, and cerebral cortex. In accord with this concept, Jasper (1966) considers that the bases of motor disorders cannot be found in isolated and separate disturbances of the pyramidal or extrapyramidal systems. Such disorders are caused by dysfunctions of various parts of the motor and even sensory pathways that disturb the functioning of the system as a whole.

The following sections of this chapter discuss the

main data on the pathogenesis of tremor, rigidity, spasticity, and akinesia and, in particular, the results of many years of studying muscular activity at our clinic in both healthy individuals and persons with various extrapyramidal lesions.*

In conclusion, brief mention should be made of the biochemical aspects of the pathogenesis of extrapyramidal phenomena. Discoveries in recent years have demonstrated that metabolic disorders involving catecholamines and possibly of other biochemical classes are of great significance. At present, it has been established that the main (but not the only) transmitter in the extrapyramidal system is dopamine and that the main dopaminergic neuronal pathway begins in the pars compacta of SN and ends in Str as beadlike neural terminals making a large number of synaptic contacts with cell bodies and dendrites of Str. Disturbances in the dopaminergic system play a very important role in the pathogenesis not only of parkinsonism but also of other extrapyramidal disorders. Biochemical disturbances in other extrapyramidal hyperkinesias in all likelihood are more complex than those in parkinsonism and are related to disruption not only of the dopaminergic but also of the GABAergic, cholinergic, and glutaminergic transmitter systems. Factual material on this problem is presented in Chapter 6, Section 6.

2. Tremor

2.1. Historical Note

Tremor is one of the few neurological symptoms known to medicine since ancient times. This is evident from the fact that the first classification of tremor was proposed in the second century A.D. by Galen, who distinguished two types of tremor: "palpitatio" ("trembling"), i.e., involuntary rhythmical movements at rest, and "tremor," the same movements occurring during voluntary movements. This classification by Galen, i.e., delimitation of tremor at rest and tremor of action, fully retains its significance today.

After 1500 years, Sylvius de la Boe in the 17th century modified Galen's terminology but did not change its essence. Instead of "palpitatio," he proposed the term "tremor coactus" to denote that it occurs both at rest and when the limbs are supported. "Tremor" was substituted for the term "motu tremolo," i.e., trembling that appears at the beginning of movement and disappears after its termination.

The outstanding contribution to the phenomenology of tremor was the classical monograph by Parkinson (1817), who identified "a shaking palsy" as a distinct disease. In that book, the reader can find a detailed and clear account of the tremor that characterizes the new disease. Parkinson described the development of the tremor from a mild, irregular twitching of the fingers to such a violent shaking that not only the bed of the patient but even the floor trembles.

Parkinson emphasized that tremor is usually present at rest and stops during purposeful movements for a short time ("less than a minute"). One of Parkinson's patients, an artist, was able to draw by making individual pencil strokes in spite of a powerful tremor since during such movements tremor was inhibited. Parkinson clearly differentiated the shaking palsy he described from a focal epileptic fit as well as from tremor induced by ". . . excessive consumption of alcoholic beverages, tea and coffee" and from ". . . tremor which is observed in old age."

Parkinson made the very interesting observation in one patient that after a stroke, the severe tremor disappeared in the paralyzed limbs. As motility returned to the limbs, the tremor reappeared.

Charcot (1888), although he accepted Galen's classification, proposed a more detailed and precise description of tremor at rest ("*tremblement de repos*") and intention tremor. In addition, Charcot described a "vibrating tremor" ("*tremblement vibratoire*") occurring in endocrine diseases (thyrotoxicosis) and intoxications (alcoholism and others). Charcot noted that in several cases of unquestionable parkinsonism, tremor was insignificant or absent; he used the term "*formes frustes*" to describe such cases. In the old literature, the term "paralysis agitans *sine agitatione*" was used to refer to complete absence of tremor in an otherwise typical clinical picture of parkinsonism. Charcot maintained that tremor of the head never occurs in Parkinson's disease: if the head is shaking, it is only because tremor of other parts of the body is "transmitted" to it. However, other authors at the beginning of this century (Westphal, Mendel, Wallenberg) demonstrated that "independent" head tremor may occur.

Déjerine described in detail the phenomenology of tremor in a chapter of his monograph *Semiologie du Système Nerveux* published in 1902. In this book the author presents a clear-cut definition of tremor that is applicable even today:

*The methodology and results of the investigations are presented in detail by Aizerman *et al.* (1974).

Tremor is characterized by spontaneous rhythmic oscillations of the body or parts of it around its position at rest. Tremor differs from other involuntary movements by the rhythmic nature of the oscillations in respect to the point of equilibrium. It is this peculiarity that distinguishes tremor from chorea, tic, ataxia, myoclonia.

Another eminent French neuropathologist André Thomas (1928) introduced the term "*tremblement d'attitude*" to designate tremor occurring while maintaining a posture and distinguished this type of tremor from intention tremor.

2.2. The Characteristics of Tremor

Rhythmic muscular contractions are a necessary and characteristic attribute of higher organisms (e.g., beating of the heart, respiratory movements, intestinal peristalsis). At times, one of these processes may "transmit" its rhythm to another, e.g., respiratory oscillations of arterial pressure. Complex locomotor acts (walking, running, and swimming) are also based on rhythmic contractions of many muscles.

It has long been known that tremor is a symptom of many diseases of both the CNS and other organs and systems. In certain diseases, tremor is the main or even the only manifestation. In such cases, it would be natural although not necessarily logical to presume that if the physician were able to eliminate the tremor, he would completely or largely cure the disease.

Modern literature, as well as that of the past, includes innumerable publications devoted to the study of tremor. Although this brief discussion is necessarily incomplete, it is useful since tremor is one of the major neurological manifestations that can be greatly ameliorated or eliminated by functional and stereotactic operations on various cerebral structures.

The scientific terminology related to tremor has yet to be clarified. In the literature quite a number of terms are used to designate the same phenomenon. In contemporary neurology, tremor is divided into three categories:

1. Tremor at rest (or static tremor) occurring in an externally supported limb with muscles relaxed.
2. Postural tremor (position tremor) occurring in an immobile limb actively maintaining posture against gravity.
3. Action tremor (kinetic tremor) occurring while performing purposeful movements. One variety of action tremor is intention tremor.

The International Classification of Extrapyramidal Diseases distinguishes seven kinds of tremor: physiological, rest, postural, intention, "rubral," "wing-beating," and flapping. It is assumed that each of these types of tremor has a different pathophysiological mechanism.

These terms characterize tremor on a clinical basis. However, tremors may also be classified etiologically as "alcohol tremor," "functional tremor," "hereditary tremor," or "familial tremor."

The most frequently encountered clinical type is tremor at rest. The majority of authors interpret the term "at rest" to mean that the patient is supine in bed or sitting in a chair with the limbs relaxed and supported.

Molina-Negro and Hardy (1971) proposed a new classification of tremor, differing substantially from the earlier ones. The authors consider "at rest" to mean during sleep. To designate the generally accepted "state at rest," they use the term "*la posture*" (postural), believing that in this case there is ". . . muscular tone maintained by afferentation and unconscious, nonvolitional influence," hence the term "postural." If the tremor develops when the patient performs a voluntary movement that involves central structures other than those maintaining posture, the tremor is called kinetic or action tremor.

A similar point of view was expressed earlier by Cooper (1966a–c), who also considered the term "tremor at rest" to be inaccurate. Since in the majority of parkinsonians there is muscular rigidity, tremor is observed not in relaxed but in "chronically contracted," "activated," or "hypertonic" muscles. These muscles relax only during sleep, when tremor disappears. Cooper (1966a) considers the term "intention tremor" also to be incorrect because the exciting factor may be an isometric contraction without intent to perform a voluntary movement. Intention tremor is most pronounced on approaching the goal (e.g., finger–nose test at the moment just before touching the target). Therefore, both tremor at rest and intention tremor should be considered tremor occurring on muscle contraction (postural) irrespective of an actual movement. Cooper concludes that both tremor at rest and intention tremor are based on identical pathophysiological mechanisms. One cannot but note that such a point of view overlooks the obvious difference between these two types of tremor. Molina-Negro and Hardy (1971) also believe that to group postural with intention tremor is a common error.

Essential or idiopathic tremor is considered to be

an independent disease, commonly of a hereditary nature. Although many authors classify tremor in the aged (senile tremor) as a separate category, it is now included as idiopathic tremor.

It is now generally acknowledged that no single pathological manifestation can be accurately interpreted without an understanding of normal function. The aim of this section is to discuss pathological tremor, particularly that of parkinsonism, the management of which is so important for stereotactic surgery. However, the pathogenesis and peculiarities of pathological tremor cannot be understood without a descriptive analysis of physiological tremor. For this, a brief discussion of the methods of recording tremors is necessary.

2.3. Graphic Recording of Tremor

Some means of objectively recording tremor is essential to study its pathogenesis. Improved recording methods have been developed since the first primitive pneumatic devices for transmitting oscillations caused by tremor to a recording system. At present, mechanograms (tremograms), which register the movements of an extremity, i.e., recording joint tremor, and EMGs, which record the biopotentials from one or several muscles, are commonly used. A mechanogram is an adequate and important method of assessing the presence and frequency of joint tremor; however, the recording of this tremor alone does not give the entire complex picture of tremorogenesis. Furthermore, during the recording one cannot always eliminate artifacts introduced by other movements of limbs or of the whole body.

Experience with objective recording of tremor has made it possible to formulate the main requirements of an ideal method of recording tremor:

1. Recording of tremor in one or, preferably, in three planes of space without changing the position of the limb (or its parts) or of the patient.
2. Recording all the main tremor parameters without distortion—frequency, amplitude, and direction of rapid or slow oscillation.
3. A highly sensitive system capable of precise calibration.
4. A multichannel recording system without inherent inertia (i.e., capable of recording rapid changes in the part of the body under investigation).
5. Convenience for the investigator and for the patient.

A tremogram is usually recorded with various sensors that transform mechanical movements into electrical signals. Several methods have been proposed: tensometric sensors with electronic amplifiers (Friedlander, 1956; Aizerman *et al.*, 1974), the piezoelectric unit with titanate–ceramic transducers (Rushworth, 1960; Findley *et al.*, 1981), miniaccelerometers (Marshall and Walsh, 1956; Cowell *et al.*, 1965; Jankovic and Frost, 1981), magnetic field recording between Helmholtz coils (Nashold *et al.*, 1966), and induction transformation of movements into electrical impulses (Teräväinen *et al.*, 1976). In addition, optical methods are also used, in particular, recording limb movements in a beam of light rays (Cooper, 1961), or cinematographic registration (Molina-Negro and Hardy, 1971).

Several of these methods require that during tremor recording the arm be held in specific positions, e.g., immobile on a special rest or in a beam of light. In view of this, tensometric sensors and accelerometers have advantages since when they are used, tremor may be registered with the limbs in any position and engaged in any movement. A significant advance in this area is the use of triaxial accelerometry, in which the vectors representing absolute values of acceleration are calculated by a computer (Jankovic and Frost, 1981). In recording of joint tremor using a piezo-electric sensor, one may single out two isolated variants—flexion–extension movements with or without a rotational component (Jurko *et al.*, 1963).

The method of recording joint tremor with tensometric sensors, which the author has used for more than 20 years, meets practically all the desired requirements and is simple and convenient to apply. Tensometric sensors, assembled according to different bridge patterns on goniometers, record reciprocal displacement of two articulated segments of a limb. Such sensors may be adapted to practically any joint. In practice, it is most convenient to record tremor in several wrist, elbow, or ankle joints, since it is important to compare the tremor on the two sides of the body. The goniometers with tensometric sensors that are fixed on these joints are not cumbersome for the patient, so recording of tremor may be continued for quite a long time (2–3 hr). This is an advantage for prolonged tremor recording during stereotactic operations. Signals from the tensometric sensors are fed through an amplifier to an oscillograph or electroencephalograph.

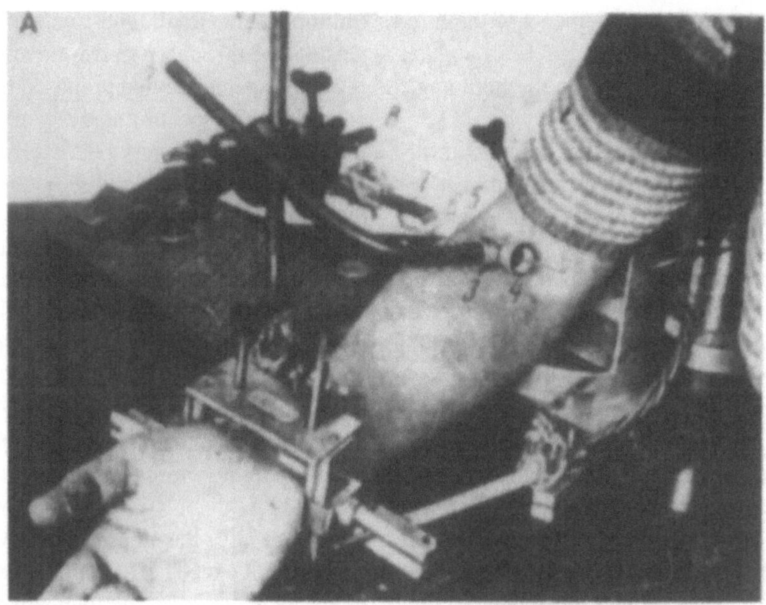

FIGURE 15. Different devices for recording tremor with tensosensors. (A) For recording arm tremor. (B) For recording leg tremor.

In recent years, we have developed a special experimental device (Fig. 15) for investigating angle changes (mechanograms) of various joints. To investigate the elbow joint, the forearm of the patient is fixed by a cuff to a long base that can make periodic swinging movements in the horizontal plane at different speeds. The displacement of the base is recorded using two types of sensors. A tensometric sensor connected to an amplifier records small displacements of the limb within a range of 0–10°, adequate for practically all types of tremors. For larger displacements of the forearm, i.e., active and passive movements, an angular displacement sensor records changes in the joint angle over a range of 0–90°.

It is the opinion of the author that the most informative method for studying any type of tremor is simultaneous recording of both joint and muscle tremor, i.e., tremogram and EMG. That is why simultaneously with the mechanogram, on the same paper, we record EMG with surface electrodes (distance between electrodes 1.5 cm) from several antagonist muscles of the elbow, wrist, ankle, and other joints. In some cases, for additional analysis, we record EMG from six muscles affecting the elbow joint during flexion and extension (Fig. 16). These investigations of tremor are performed both at rest with maximally relaxed muscles and the elbow joint on a support (tremor at rest) as well as when the forearm maintains a specific posture (postural tremor) or performs various movements (action tremor).

All investigations conducted both in healthy trained individuals and in parkinsonians may be divided into two groups (Aizerman *et al.*, 1974). The first group, arbitrarily called "static," included experiments with certain parameters—joint angle, angle of relaxation (or tension) of muscles, or degree of pressure on support—held constant.

The second group of studies, arbitrarily called "dynamic," included experiments in which the patient was required to change one of these parameters by flexing or extending the elbow joint at various speeds (minimal, maximal, and average), changing muscle tension, or through passive movements.

The special device proposed by the author and his coworkers for spectral analysis of tremor, mainly during stereotactic surgery, is described in Section 2.11.

It is necessary to discuss briefly a method for obtaining a quantitative evaluation of tremor—a technique developed by the author. For this purpose, we employed an integrator showing the summated activity of the biopotentials for a definite period of time. Usually, we integrated EMGs of the flexors of both wrists with the patient at rest. In this case, we integrated the total muscular activity for 30 sec, taking six to eight measurements and calculating the mean index for each hand in arbitrary units corresponding to microvolts per second.

Normally, the summated EMG activity of a muscle does not exceed 20–30 units. In parkinsonians, this index, depending on the intensity of tremor, increases many times, going as high as 500–700 units in the tremor form and 200–300 in the rigid form of the disease. After an effective operation on the basal ganglia, the EMG activity usually returns to normal.

Figure 17 shows the results of EMG integration in a parkinsonian patient with tremor more pronounced in the right hand. A bilateral effect was noted: postoperatively, tremor in the right limbs disappeared, and that in the left was noticeably diminished. Prior to surgery, the total activity was 659 units on the right and 177 units on the left. Three weeks after the left thalamotomy, it was 29 units on the right and 49 units

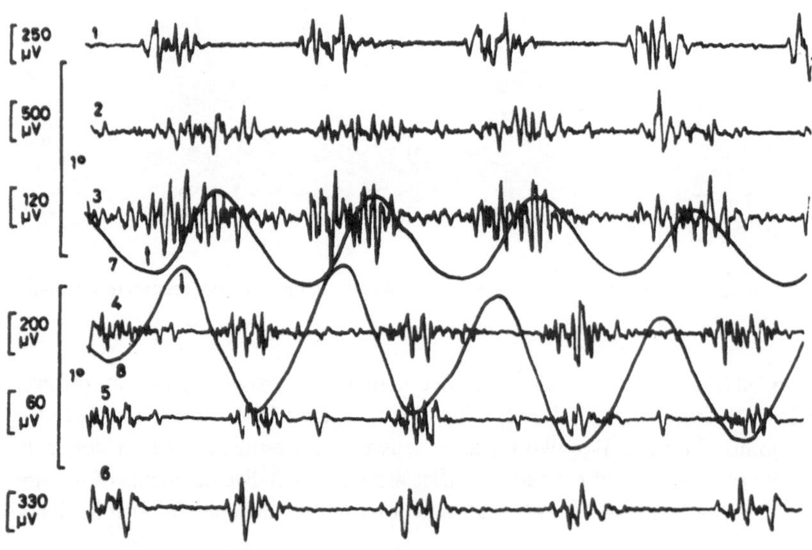

FIGURE 16. Simultaneous recording of EMGs of six muscles of the carporadial joint and mechanograms of active alternating movements in a parkinsonian patient. From above, EMG: 1, m. extensor carpi radialis brevis; 2, m. extensor carpi radialis longus; 3, m. extensor carpi ulnaris; 4, m. flexor carpi ulnaris; 5, m. palmaris longus; 6, m. flexor carpi radialis. Upper curve (7), wrist mechanogram at flexion–extension movement; lower curve (8), abduction–adduction movement. Flexion upward, extension downward.

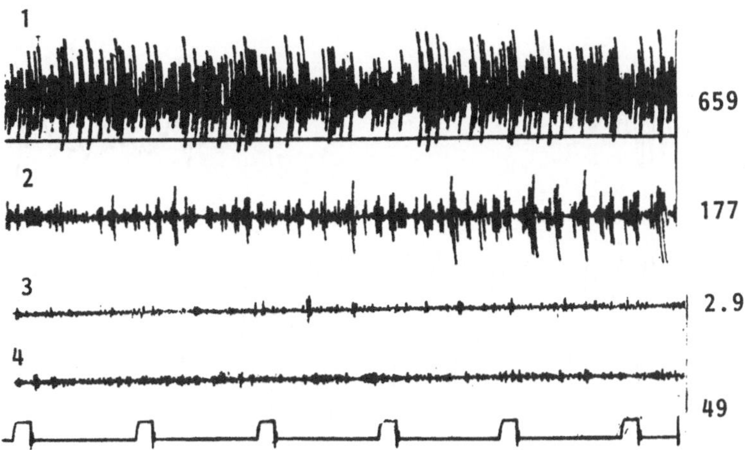

FIGURE 17. The quantitative evaluation of tremor by summing the bioelectric muscle activity with the aid of an integrator. The EMG recording is of tremor in both arms in a parkinsonian patient. Tremor was more pronounced in the right arm (1) than in the left (2). After left thalamotomy tremor disappeared in the right arm; its intensity diminished by more than 95% (3). Tremor in the left arm diminished about 75% (4). Time intervals, 1 sec (bottom trace).

on the left. The method allows one to obtain a quantitative estimate of tremor, to study its characteristics, and to assess objectively the results of surgical treatment.

2.4. Physiological Tremor

The human arm, when lying freely or holding a pose, seems to be immobile. However, if a tensometric sensor is attached to it, and its signals are recorded on continuous paper of an electronic device with sufficient amplification, then we can see a complex, irregular oscillating curve resembling that of instrumental noise. This so-called physiological tremor may be invisible to the naked eye but is present in all healthy individuals. At the same time it is important to keep in mind that under special circumstances (stress, emotion, shivering, epinephrine or other catecholamines), physiological tremor may greatly increase and become clearly visible.

Physiological tremor has been known for a very long time. It was first described by Schäfer over a hundred years ago. Approximately at the same time, Wolfenden and Williams (1888) considered pathological tremor at rest to result from "fusion" and subsequent amplification of separate oscillations of physiological tremor, whereas Herringham (1890) was of the opinion that the marked tremor at rest of parkinsonism masks invisible "physiological" tremor.

Physiological tremor is not perceived visually since its amplitude is very small, not exceeding several angular minutes. It can be seen only when its amplitude increases to approximately 0.5°. This means that even a tenfold increase in the amplitude of tremor cannot be observed either by the patient himself or by the examining physician. At rest, the amplitude of physiological tremor is extremely small, but holding a fixed posture (outstretched arms and so on) increases this amplitude substantially, so that a slight tremor can be visualized.

In characterizing any type of tremor, three parameters are of prime significance: amplitude, frequency, and form. Frequency of tremor is the number of oscillations per second. Physiological tremor (Fig. 18) is, as a rule, irregular. This means that its frequency varies all the time, and a tremogram is irregular. In physiological tremor, impulses from various motor units are, to a considerable extent, independent of each other. Synchronization is rarely observed and may be accidental rather than the result of some driving mechanism firing the motoneurons simultaneously.

Numerous publications on physiological tremor indicate that its frequency is 10–12 Hz. Our investigations have shown that in healthy individuals, physiological tremor signals from a tensometric sensor include peaks corresponding to frequencies of 1, 3, and 10 Hz (Kandel, 1981). Marsden (1978) noted that in the frequency spectrum of physiological tremor during muscular contraction there is a small peak of activity between 8 and 12 Hz. The frequencies of physiological tremor of both hands are usually identical, but if not, the difference is not great (Barlow, 1965). In tremograms of healthy individuals, one often also observes "superslow" oscillations, one every 2–5 sec (Dubenko and Nebotov, 1976).

Why and how does physiological tremor occur? So far, there is no answer to this question. No general theory on the mechanism of physiological tremor has been put forward. Marsden (1978) noted that ". . . the origins of physiological tremor are complex," and "the general conclusion is that many factors interact to produce such tremor." The author concluded that os-

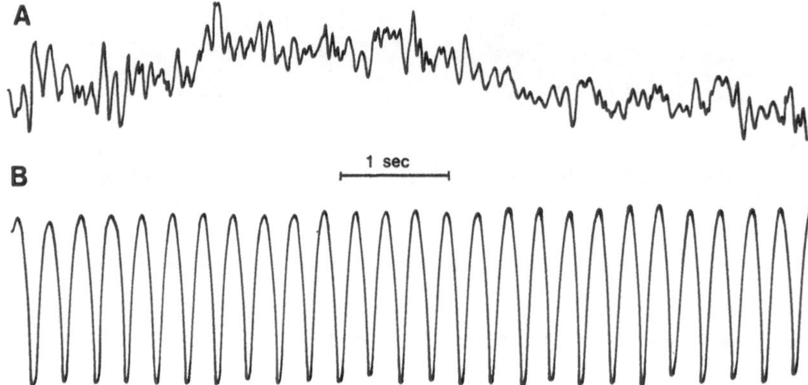

FIGURE 18. (A) Physiological tremor. Tremogram of the right carpal joint of a healthy subject. Completely irregular (''noise'') tracing of physiological tremor with mean frequency of 11 Hz. (B) Pathological tremor. Tremogram of the same joint in a parkinsonian patient shows a regular tremor with mean frequency of 4–5 Hz. Amplification of recording A is 30 times that of recording B.

cillation of a mechanical system, such as a human limb, is a function of both inherent mechanical properties of the system and the input into the system. Since it has been established that motoneurons of the anterior horns normally constantly send to the muscles weak desynchronized impulses that do not cause movement, a number of authors are of the opinion that physiological tremor is a direct result of that normal activity. Other authors tend to explain the appearance of physiological tremor by the existence of the stretch reflex, which is an important element of muscle control at the spinal level (Rack, 1978). It is assumed that the stretch reflex (contraction of a muscle in response to its elongation), which occurs by a feedback mechanism from γ receptors, forms a servo system ensuring, in particular, the maintenance of the position of a joint. The appearance of physiological tremor is explained as a result of inherent oscillations of the servo system arising from a ''delay in the closed contour of regulation'' (Matthews, 1962) or its instability (Halliday and Redfearn, 1956). The suggestion that physiological tremor reflects the α rhythm of the EEG has not been confirmed (Marsden, 1976).

As we have already noted, an important methodological factor in the investigation of tremor is simultaneous recording of joint tremor (mechanogram) and muscular tremor (electromyogram). Our analysis of physiological tremor simultaneously by mechanogram and EMG while the arm is held in a fixed position has shown that continuous noise activity is recorded in all the muscles involved in this act (Aizerman *et al.*, 1974). This activity (unlike oscillations of the joint angle) may be called muscular tremor. Since there are no periodic oscillations of this tremor, it is practically impossible to determine its frequency. During voluntary (upon command) tensing of a muscle as well as in maintaining a posture, irregular muscular tremor does not change, although its amplitude increases sharply, as noted by Sutton and Sykes (1967) and other authors.

We have direct experimental evidence that muscular tremor is a primary and joint tremor a secondary phenomenon, i.e., its direct result. If a joint is fixed with special clamps, creating so-called muscular isometry, then oscillations of the joint angle (articular tremor) will, naturally, no longer be present. But at the same time, muscular tremor, according to EMG data, does not change at all (Aizerman *et al.*, 1974). The same results are obtained with another test. If a healthy individual is given a command (in isometric conditions) to increase pressure on a hand lying on a rest, in the absence of joint tremor there is a sharp rise in the overall noise activity of muscles, but the muscular tremor will predominate in the muscles exerting pressure on the rest.

New data on the nature of physiological tremor have been obtained by us with a new technique, namely, training healthy individuals to hold a joint in a given position. Such training was performed using a special screen on which the trainee can see an illuminated blip oscillating synchronously with the signals coming from the sensors registering changes in the joint angle. The deviation of this blip from the center of the screen is proportional to displacements of the joint angle. The trainee was told to keep the blip within the screen. In other words, the trainee sees his physiological tremor in the muscles of one joint and gradually learns to ''hold'' it within the screen even as amplification is increased and, consequently, the oscillations of the illuminated blip increase. In this case, joint and muscular tremor were recorded with the aid of tensometric sensors and EMG. The maximal possible level of accuracy that was achieved by such training was 1–2 angular minutes.

As a result of this method of training, both joint and muscular tremor undergo substantial changes. The joint tremor synchronizes—i.e., the dominant frequency of bursts, approximately 10 Hz, stands out clearly, and the amplitude of other slower oscillations gradually diminishes. In other words, the tremogram of the joint becomes considerably more regular with a frequency of approximately 10 Hz.

A similar synchronization occurs in muscular tremor as monitored on EMG. After the subject attains maximum accuracy in holding a joint angle, one observes bursts with a 10-Hz frequency instead of continuous noise activity on the EMG. The better the training, the more clearly the 10-Hz bursts stand out from the continuous activity. This result has been established for the muscles of all joints investigated (wrist, elbow, ankle). It is noteworthy that such 10-Hz tremor occurs in each group of muscles, both flexors and extensors, but with opposite phases—bursts of one group come approximately halfway between periodic bursts of the other group.

On a voluntary increase of muscle tension in an untrained healthy individual, the amplitude of continuous activity increases sharply, whereas in a trained individual there is an increase only in bursts (i.e., 10-Hz tremor), and the low-frequency background between bursts does not increase. The EMG data have shown that in the process of training there is a gradual "stratification" of tremor into a high-frequency (10 Hz) component, the amplitude of which does not change significantly, and a low-frequency (approximately 3 Hz) component, which gradually decreases in amplitude.

A series of similar experiments with a significant load (5–6 kg) applied to the joint, which substantially changes proprioception, has shown no change in the burst pattern of the muscles. An increase in the load only increases the amplitude of bursts in the EMGs of all muscles of the joint. An analogous reaction—larger amplitudes of bursts—also occurs in an isometric condition that excludes joint tremor and also with voluntary pressing on a fixed support. Just as in an untrained individual, in this case there is only an increase in the asymmetry of muscular tremor related to the larger burst amplitude of the muscles exerting pressure.

Consequently, one can conclude that burst EMG activity (muscular tremor) while a joint angle is fixed is not in itself a pathological phenomenon. Tremor of precisely the same nature is observed under such conditions in a healthy but trained individual. This fact is important in understanding the mechanisms and the genesis of tremor. There is, however, a difference between muscular tremor in a healthy trained individual and that in a parkinsonian (see below). The amplitude of tremor in parkinsonism, quite naturally, is much greater, and as a result, the tremor is easily visualized. Furthermore, in healthy individuals the frequency of physiological tremor is always approximately 10 Hz, whereas in parkinsonians the tremor may have a frequency of either 5 or 10 Hz (see below).

The results of our investigations of both physiological and pathological tremor by spectral frequency analysis are presented in Section 2.8.

2.5. Pathological Tremor

Unlike physiological tremor, which is present in all healthy people but cannot be seen by the naked eye, pathological tremor is those rhythmic oscillations of the limbs, head, or whole body that can be detected visually. Tremor is a typical manifestation not only of parkinsonism but also of a number of other diseases of the nervous system (multiple sclerosis, essential tremor, etc.) as well as diseases indirectly related to damage to the nervous system (Basedow's disease, alcoholism, drug-induced tremor, etc.).

From the neurophysiological point of view, pathological tremor depends on alternating contraction of agonist and antagonist muscles as a result of synchronous alternating discharges of α-motoneurons of the spinal cord. As Rondot and Bathien (1978) stressed, the synchronized motor unit represents the basic phenomenon of tremor. Pathological tremor differs significantly from physiological tremor. The amplitude or intensity of pathological tremor is many tens of times greater than the amplitude of physiological tremor (Fig. 18). As a rule, it is considerably greater in the arms than in the legs. Amplitude is a very variable parameter of tremor, differing among patients and at times changing substantially and rapidly in a single patient. Moreover, the amplitude of tremor varies in different limbs (in both arms, in the arm and leg). The highest tremor amplitude is seen in the tremor and tremor–rigidity forms of parkinsonism. The amplitude of tremor in the rigidity and akinetic forms of the disease is insignificant.

Since joint tremor, as shown above, is doubtless determined by muscle tremor, it is apparent that the mechanogram of the oscillating movements of the limbs (tremogram) always depends totally on the bursts of all the muscles of the involved joint—the larger the amplitude of tremor bursts in EMG, the larger is the amplitude

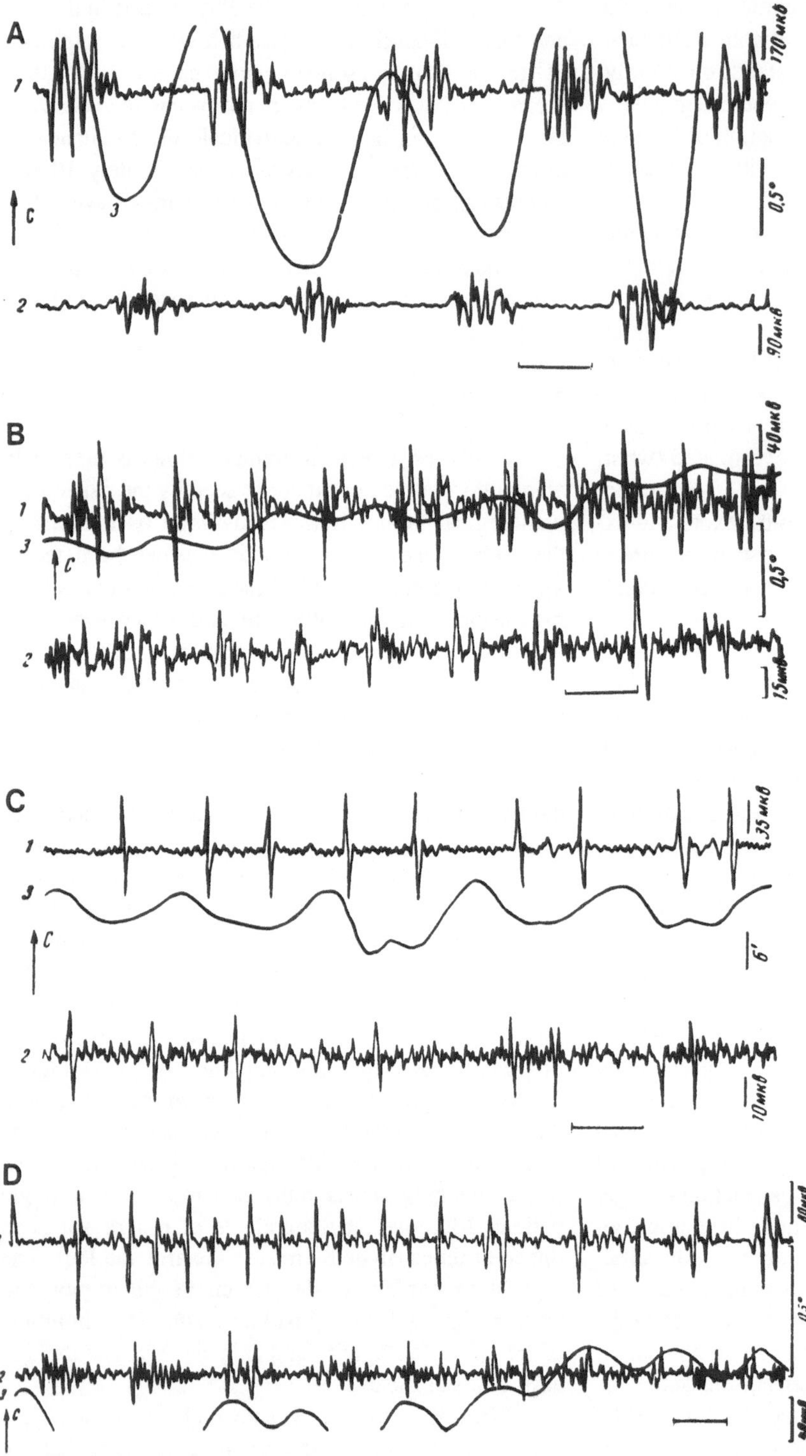

FIGURE 19. Electromyographic recordings in an experiment with a fixed joint angle in a parkinsonian patient. (A) The tremor with frequency of about 5 Hz has a high amplitude and practically no noise activity between bursts. (B) The tremor with frequency of about 10 Hz has a much smaller amplitude and pronounced background activity between bursts. (C) A rare case of "pure" 10–Hz bursting with intermediate amplitude. (D) Mixed tremor activity with different (5 and 10 Hz) frequencies in the same muscle. The periods of continuous activity are included in the record with bursts. Scalebars, 0.1 sec.

of the joint tremor. Moreover, this amplitude is directly determined by the phase relationships of the contractions of antagonist muscles. If, as is usually the case, they are in opposite phases, the tremogram will be rhythmical, regular, and of large amplitude. If contractions of a pair of antagonists are out of phase, then the regular pattern of the mechanogram will be disturbed and the amplitude of joint tremor decreased. This "floating phase" phenomenon can occur in both 5-Hz and 10-Hz tremor.

The frequency of tremor, i.e., the number of oscillations in 1 sec, is an important criterion for differentiating physiological and pathological tremor. According to data from mechanograms of tremor, the frequency of parkinsonian tremor is considerably lower than that of physiological tremor. The data of various authors on the frequency of pathological tremor seem to be quite similar. For instance, in quiet parkinsonians it varies within a range from 3.1 to 6.4 Hz (Lance *et al.*, 1963), from 3.8 to 8 Hz (Molina-Negro and Hardy, 1971), from 4.5 to 5.1 Hz (Van Manen, 1974), or from 4 to 5.3 Hz (Findley *et al.*, 1981), an average of 4–5 Hz. In the presence of active movements of the other arm or in emotional loads, the frequency of rest tremor increases by 1–2 Hz (Lance *et al.*, 1963). The frequency of postural tremor is greater than that of resting tremor by approximately 6–6.2 Hz (Findley *et al.*, 1981).

However, our investigation has shown that the incidence of pathological tremor depends on the method of examination. For example, pathological tremor of various amplitudes in the form of obvious bursts of muscular activity was found in the surface EMG of 96% of 120 parkinsonians investigated by the author. Depending on the frequency of these bursts, the tremors may be arbitrarily divided into three types: 5 Hz, 10 Hz, and mixed (Fig. 19). In the first group, the frequency of bursts is approximately 4–6 Hz, the bursts are clear, and there is practically no continuous activity; i.e., there is no "noise" between bursts. The amplitude of the bursts is high—many times higher than the amplitude of physiological tremor. In the most severe stage of the disease, even when flexion–extension movements are at maximum speed, the appearance of muscle tremor is not lost, and EMG activity comes in bursts.

The second variant is 10-Hz tremor in which the intervals between bursts are largely filled with "noise" of continuous background activity. "Pure" bursts of a 10-Hz frequency are observed rarely. A 10-Hz tremor resembles physiological tremor in a trained individual. The amplitude of this tremor is only half or one-third that of 5-Hz tremor, thereby confirming the inverse relationship of frequency and amplitude of tremor.

Finally, the third type of tremor is a combination of the two previously mentioned tremors. The frequency of bursts in the EMG of a single muscle of a joint is often unstable, with periodic bursts of 5 Hz, then 10 Hz, and back again. Intermittently, periods of continuous activity are seen in such a recording. As the disease advances, the continuous activity "recedes" more and more. The more severe the signs of parkinsonism, the greater and "purer" are the bursts.

How often are these three types of tremor encountered? According to our data, 5-Hz tremor was found in 23% of the cases, 10-Hz tremor in 34%, and mixed tremor in 39% of the cases; in 4% there was no tremor (Aizerman *et al.*, 1974).

If all the muscles about a joint contract at a frequency of 5 Hz, the joint should have the same frequency of oscillation, and the mechanogram would show a 5-Hz tremor. Since the amplitude of bursts of the surface EMG is greatest at this frequency, this tremor should form the clearest mechanogram. In the less common variants, all the muscles about a joint display 10-Hz burst activity, but the agonists and antagonists are out of phase. Accordingly, the mechanogram of the joint tremor will also have a frequency of 10 Hz.

The presence of different tremor frequencies in different muscles was first reported by Schwab and Cobb (1939) and has been confirmed by many investigators (Bishop *et al.*, 1948; Kandel, 1965; Alberts *et al.*, 1965b; and others). As investigations by the author have demonstrated, different muscles about a joint often contract at different frequencies (Fig. 20). For instance, the frequency of bursts may be approximately 5 Hz in one muscle and 10 Hz in its antagonist. Moreover, the frequency of bursts in any muscle may change from time to time. In such conditions, which are typical of parkinsonism, the frequency of tremor on the mechanogram will be 5 Hz because mechanical inertia of the joint will cut off all the higher frequencies. In other words, if the muscles about a joint have tremor frequencies of 10 Hz but one is 5 Hz, the frequency of joint tremor will be 5 Hz.

Since in most investigations joint tremor was recorded by mechanograms, it was inaccurately concluded that parkinsonian tremor always has a frequency of approximately 5 Hz. The frequency of "true" (i.e., muscular) tremor is very dynamic and changing. Since joint tremor is determined by muscle tremor, it is quite apparent that the mechanogram of limb movements reflecting joint tremor always depends on the

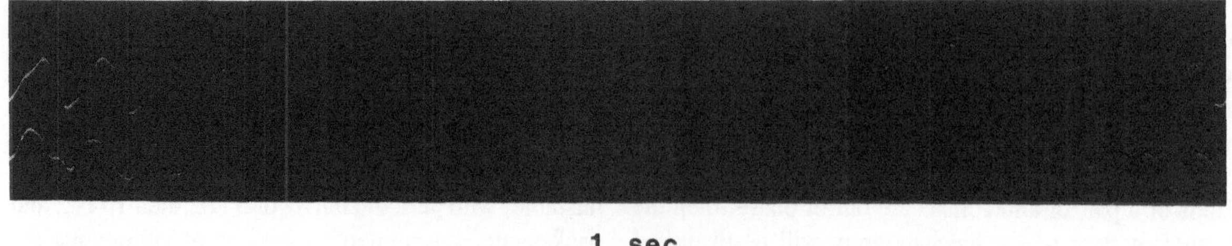

FIGURE 20. Simultaneous tremograms of the right (above) and left (below) carpal joints in a parkinsonian patient. The different tremor frequencies of both arms are clearly seen.

nature of the EMG of all muscles of the joint. In view of this, as the data of the author have shown, tremograms of various joints, even symmetrical joints, as a rule, have different frequencies or are out of phase, although mean frequencies are the same. However, the frequency of tremor may be higher on one side for some time and then be higher on the other. Accordingly, the ratio of frequencies may be changing all the time. Joint tremor and head tremor usually do not coincide.

An EMG investigation of tremor of antagonist muscles of one joint points to the relative independence of tremor in the two muscles. Burst activity in antagonists quite often reveals periods of spontaneous desynchronization in one muscle while clear-cut rhythmic activity is preserved in its antagonist (Fig. 21). The unloading test clearly reveals the absence of firm ties between burst activity of antagonists. The test involves the sudden release of a load attached to the forearm of a subject, in which case, the EMGs of the flexors and extensors of the forearm of a healthy individual show a 40- to 60-sec period of electrical silence in the loaded muscle 25–30 sec after unloading. On the background of silence in the flexors, a short burst of activity appears in the antagonists 35–50 sec after unloading (Fig. 22). Continuous activity is established in the forearm muscles after two or three such "reciprocal" bursts.

The author has conducted the unloading test in many parkinsonian patients. In each case, the test was carried out on both arms repeatedly with loads of 4–6 kg and repeated on subsequent days. It was found that in the majority of cases, changing the proprioceptive input significantly influences pathological tremor. In the majority of cases, unloading is followed by temporary desynchronization of EMG activity and the disappearance of tremor in the corresponding joint. In the majority of cases, when tremor returned, its phase had shifted.

Analysis of these data indicates the absence of a strict correlation between the frequency of pathological tremor and the form of parkinsonism. Nevertheless, certain generalities were evident. When tremor was predominant, its frequency was often approximately 5 Hz, whereas if pronounced extrapyramidal rigidity was prominent, both frequencies (5 and 10 Hz) were observed in most cases.

While investigating tremor in parkinsonians, we always noted the same direct relationship of joint tremor to muscle tremor as that described for physiological tremor. In order to exclude joint tremor, the elbow joint of the patient was firmly fixed in a special device. The EMG of the biceps and triceps revealed that muscle tremor does not change. This is easily verified by temporarily eliminating the isometric conditions—tremor in the arm begins immediately, while the EMG shows no changes.

If while the patient is holding his arm in some

FIGURE 21. Spontaneous desynchronization of bursting activity of m. triceps (above) with preservation of rhythmic bursts of tremor in the EMG of m. biceps in the same arm (below) in a parkinsonian patient.

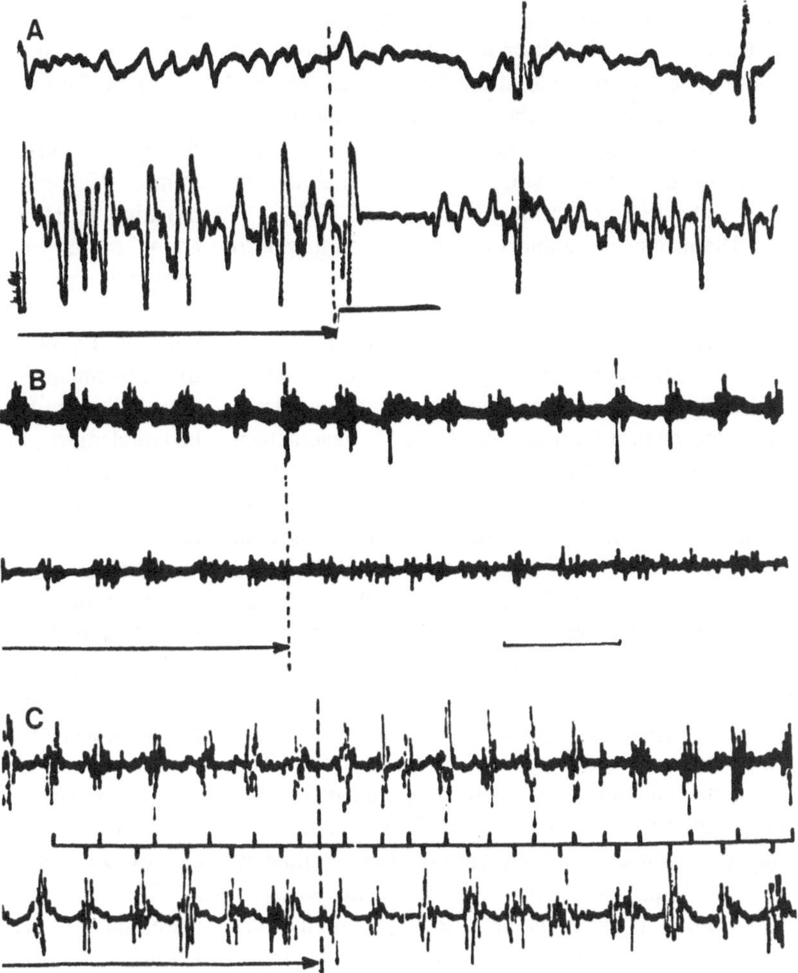

FIGURE 22. Change of tremor on EMG during unloading test. (A) Effect of sudden unloading on EMG of m. triceps (above) and m. biceps (below) in the healthy subject. "Silent period" in the electroactivity of m. biceps, which has held the load, is seen. Scale bar, 100 μsec. (B) Desynchronization of tremor bursts on EMG of m. biceps (below) in a parkinsonian patient following sudden unloading (vertical dotted line), and preservation of rhythmic bursts on EMG of m. triceps (above). Scale bar, 0.5 sec. (C) Different frequencies of rhythmic bursts of antagonist muscles in the parkinsonian patient after sudden unloading (vertical dotted line) (above, EMG of m. biceps; below, of m. triceps).

position he is given a command to tense the muscles of that arm maximally, burst activity increases to an even greater degree. In those cases in which the activity was continuous while the position was held, it changed to burstlike activity.

2.6. Action Tremor

Action tremor was first described in detail by De Jong (1926), who differentiated it from the generally known resting tremor. De Jong noted that action tremor is "a tremor [that] in contrast to resting tremor does not exist at rest and increases while moving." As the name itself indicates, this is a tremor visible to the naked eye, i.e., a pathological tremor appearing in an actively outstretched limb or during voluntary or passive movement. If one accepts this definition, then intention tremor should also be classified as an action tremor. In contrast to resting tremor, action tremor is independent of posture and not alternating but synchronous in antagonistic muscles.

When muscles are voluntarily contracted, resting tremor disappears because of desynchronization of the motor unit.

It is noteworthy that resting tremor usually disappears somewhat before an active movement begins, in the so-called initiation period. During EMG recording it is possible to observe a brief period of electrical silence before the appearance of action potentials (Lance *et al.*, 1963). Only rarely is resting tremor replaced by action tremor at the very beginning of a movement. If resting tremor is marked, it may persist (at the same frequency) during the active movement. In such cases, action tremor, naturally, is absent.

Action tremor is preserved all the time that muscular contraction continues. This tremor is often observed in parkinsonism but less commonly than in resting tremor. As a rule, action tremor is less pronounced on the more affected side of the body. However, it is not present in all cases, the reason for which still remains obscure. Action tremor has been noted in only half of the cases with resting tremor and in almost all cases without resting tremor (Lance *et al.*, 1963).

The majority of authors consider action tremor to have a considerably higher frequency than resting tremor, approximately 9–12 Hz (Angel *et al.*, 1969). Bishop *et al.* (1948) pointed out that at the beginning of an active movement, resting tremor in parkinsonians doubles in frequency. However, investigations carried out at our clinic (Aizerman *et al.*, 1974) have shown that action tremor, like physiological tremor, does not have any fixed frequency. During voluntary flexion and extension of the elbow joint, the intervals between bursts in the EMG of the two antagonists (biceps and triceps) vary considerably, depending on the phase and speed of the active movement. As the speed of these movements increases, the frequency of the action tremor steadily increases.

Two hypotheses regarding the genesis of action tremor have repeatedly appeared in the literature. The first is that this tremor is a pathological intensification (i.e., with large amplitude) of physiological tremor, since it is assumed that both types have approximately the same frequency (in this instance, the frequency of joint tremor) (Findley *et al.*, 1981). Naturally, the question arises whether action tremor and resting tremor have a common pathogenetic basis, since one may consider the former to transform into the latter with a change in frequency. Proceeding from this assumption, the second hypothesis claims that in the process of muscular contraction, resting tremor increases in frequency to become action tremor (Hoefer and Putnam, 1940; Bishop *et al.*, 1948). It may be said that neither of these hypotheses has withstood the test of time.

There are several convincing arguments against the hypothesis that action tremor in parkinsonism is an intensification of physiological tremor (Lance *et al.*, 1963). The frequency of action tremor is identical in muscles with short (facial muscles) and long (leg muscles) reflex arcs (Marshall and Walsh, 1956). The frequency of action tremor does not change, nor does this tremor disappear, after procaine block of muscles or intrathecal injection of procaine, although rigidity is temporarily abolished. After binding of a limb with a pressure cuff, rigidity as well as pain and deep sensation disappear, but action tremor persists. Since such procedures block the afferents group I fibers (and consequently the stretch reflex), action tremor cannot be dependent on proprioceptive feedback.

Transformation of one type of tremor into another does not occur. On the contrary, resting tremor usually disappears several seconds before the onset of action tremor. Whereas resting tremor is characterized by alternating bursts on EMG in the antagonist muscles, action tremor in these muscles, as a rule, is synchronous. It has also been noted that resting and action tremor are "inversely proportional"—the more pronounced the resting tremor, the less tremor during active movements, and vice versa. Of 23 cases with pronounced action tremor, six had no resting tremor, eight had moderate tremor, and only nine had intention tremor. On the other hand, of 17 cases with little action tremor, 15 had a pronounced resting tremor (Lance *et al.*, 1963).

Stereotactic operations on the basal ganglia, which eliminate resting tremor so effectively, have considerably less effect on action tremor. All these facts are evidence that these two types of tremor are not identical and have, in all likelihood, different pathophysiological mechanisms.

In many parkinsonians, we recorded EMG of antagonist muscles (biceps and triceps) and, simultaneously, mechanograms of the elbow joint in the process of voluntary flexion and extension of the forearm at different speeds. As the speed of these movements increased, the EMG picture changed significantly. During slow movements of the contracting muscles, action tremor was generally preserved. Then, as the speed of these movements increased, the intervals between bursts were filled with "noise," i.e., continuous activity. Finally, at a still higher speed, only continuous muscular activity appeared in the EMG.

As a general principle, this investigation showed that the transformation of burst activity into continuous activity, i.e., the disappearance of action tremor in parkinsonians, occurs at a higher speed of movement than in healthy subjects. As the disease grows in severity, the "threshold rate" at which burst activity disappears is "pushed back" further and further. Normally, this threshold differs significantly in muscles of various joints. According to our data, it is 8°/sec at the wrist joint and 30°/sec at the elbow joint. In parkinsonism this threshold rises to such a degree that it cannot be achieved even at a rate of about 100°/sec. Our experience has shown that even in the earliest stages of parkinsonism, when pathological tremor is still almost absent, there is an elevation in the threshold of continuous EMG activity.

In conclusion, one should attempt to answer an important question regarding action tremor: Is it a pathological phenomenon not found in healthy persons? Our study has shown that muscular action tremor occurs in healthy subjects who have undergone special training during both postural tests and voluntary movements. However, the amplitude of this tremor is substantially less than in parkinsonism (Aizerman *et al.*, 1974). Consequently, one may conclude that action tremor is not a tremor typical of parkinsonism, for it has been observed in voluntary movements of healthy subjects although with much smaller amplitude.

2.7. Tremor and Proprioceptive Input

The relationship of pathological tremor to proprioceptive input and, consequently, to the state of the γ system has been under discussion in literature for a long time. In spite of numerous studies, this question has not been finally resolved. In fact, the views of different investigators are quite contradictory. In earlier papers (Jung, 1941; Jung and Hassler, 1960), it was assumed that proprioceptive input had no influence on tremor. Back in 1930, Pollock and Davis sectioned the lower cervical and upper thoracic dorsal roots in a case of severe bilateral parkinsonism. After that operation, rigidity diminished temporarily, although tremor on the operated side continued, albeit at a lower frequency, and became still more intense. Subsequently, these data were confirmed by other researchers. In animal experiments, Ohye *et al.* (1975) noted that tremor was modified after rhizotomy but not arrested. Therefore, section of the posterior roots and, consequently, of the γ afferent fibers does not result in the disappearance of tremor.

Many investigators have carried out various peripheral procedures in order to clarify the pathogenesis of tremor. It was assumed that if such procedures significantly modified any parameter of tremor, it would imply a spinal role in the occurrence of tremor. In 1924, Walsh infiltrated the muscles of a trembling limb with procaine, which causes a selective blockade of the efferent fibers of the γ loop, after which rigidity and the cogwheel phenomenon, as well as the tendon reflexes, disappeared, but resting tremor and action tremor did not change (Lance *et al.*, 1963; Aronson, 1966). Since interruption of the proprioceptive reflex arc not only fails to eliminate tremor but even intensifies it, Hassler *et al.* (1979) are convinced that "tremor is not controlled from the periphery." However, when the motor points of muscles are anesthetized with procaine, tremor in the limbs disappears (Rondot and Bathien, 1978). In view of this, the authors assume that the spinal proprioceptive input must be activated in order for tremor to appear. A significant influence of the proprioceptive factor on tremor was noted by other investigators. This influence is confirmed by the unloading test, the results of which are noted above.

2.8. Interrelationship of Various Types of Tremor

The various types of pathological tremor described above have different interrelationships in the nervous system. Our investigations (Kandel *et al.*, 1974; Aizerman *et al.*, 1974; Kandel, 1982) with multichannel EMG recordings carried out on 72 parkinsonians have shown that these cases can be divided into three groups (Fig. 23). In the first group, tremor and, consequently, EMG activity at rest are absent, whereas if the arm is held in a certain position, burst activity appears with a frequency of 5 or 10 Hz (postural tremor). Slow active movements (flexion–extension) modulate this activity in amplitude and frequency (action tremor). As the speed of these movements is increased, they reach a threshold at which burst activity changes to continuous activity. In late stages of the disease, this threshold becomes so high that even at the highest possible rate of these movements burst activity remains without changing to continuous activity.

In the second group of parkinsonians, burst activity, usually with a frequency of 5 Hz, occurs at rest (resting tremor) and remains even while the subject holds a posture or performs active movements; however, the amplitude of the bursts is considerably greater than in the patients of the first group.

Finally, in the patients of the third group, at rest one observes not bursts but moderate continuous activity, which intensifies when a specific posture is held. During slow movements, one observes burst activity just as in patients of the first and second groups. As these movements are accelerated, continuous activity appears, but at a threshold considerably higher than in the first two groups.

On the basis of the data presented we draw the following conclusion. Burst electroactivity in muscles is not a pathognomonic symptom of parkinsonism, since it is also found in normal, healthy individuals. The main disturbance distinguishing this disease is the

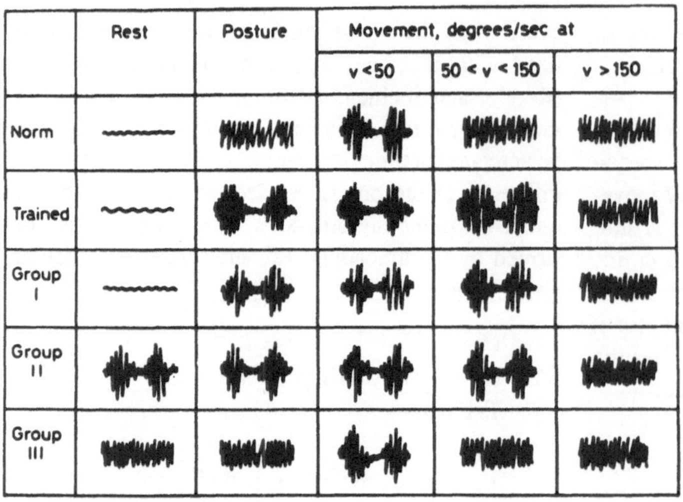

	Rest	Posture	Movement, degrees/sec at		
			v<50	50<v<150	v>150
Norm					
Trained					
Group I					
Group II					
Group III					

FIGURE 23. Bursts of continuous EMG activity in untrained and trained healthy subjects and three groups of parkinsonian patients during rest, posture maintenance, and active movements at different speeds.

"retreat" of continuous activity, leaving EMG bursts during voluntary movements. This burst activity at 5 or 10 Hz is a consistent sign of the tremor form of the disease. A pronounced increase in the amplitude of this activity progresses until resting tremor and postural tremor are transformed into visible joint tremor. The threshold rate of active movements at which continuous activity appears on EMG may serve as an index of the severity of parkinsonism.

2.9. Pathogenesis of Tremor

It would seem that such an obvious phenomenon as tremor, which has been studied for many decades, should have been fully explored, treated, and cured. All the more so because the main complaint, as it were, "lies on the surface" to be seen by all, including the patient himself, so that no analyses or special investigations are required for its diagnosis. Regretfully, even today we are still a long way from having a complete understanding of the nature and pathogenesis of tremor. In many cases, the etiology remains a mystery, the underlying mechanisms can be explained only in very general terms, and its treatment is not always effective.

We consider that at the same time, one cannot but note the very significant headway that has been made in our understanding of tremor over the past three decades. There can also be no doubt that much has been done to unfold the basic mechanisms of tremor, as the author has attempted to demonstrate in sections of this chapter. Nevertheless, one has to admit that there is still no generally accepted and satisfactory theory of the pathogenesis of tremor. Although the literature offers several such theories, none is able to explain fully the great diversity of experimental and clinical material. Indeed, the very fact that there are numerous theories on the pathogenesis of tremor indicates that this problem has not yet been solved.

Walker (1969) justly points out that any theory that attempts to explain the mechanism of pathological tremor must explain the following typical observations:

1. Alternating contractions of protagonists and antagonists.
2. Frequency and asynchrony of tremor in the arm and the leg.
3. The presence of tremor at rest and its disappearance during voluntary movements and in sleep.
4. Interrelationship with muscular rigidity.
5. Disappearance of tremor after destruction of the pyramidal tract, nuclei of Th, GP, and striothalamic pathways.

We consider the main unsolved questions concerning the pathogenesis of tremor to be the following:

1. What causes synchronization of discharges of spinal α-motoneurons and motor units, inducing alternating contractions of muscles?
2. What is the nature and localization of the mechanism generating rhythmic tremor? In other words, is there a so-called pacemaker of tremor, and where is it localized?
3. What is the tie between disturbed cerebral metabolism (especially metabolism of catecholamines in the nigrostrial system) and the appearance of tremor?

There are many reasons to consider the mechanisms underlying pathological tremor to be quite different from those of physiological tremor. It is also clear that synchronization of a multitude of sporadic, minute, practically invisible movements may lead to the appearance of pathological tremor amplified many times. A possible analogue is the laser, in which a countless number of asynchronous rays of light are transformed into a powerful quantum of energy.

2.10. Generator of Rhythmic Tremor

The data presented in the literature indicate that the majority of authors proceed from the assumption that there is a cerebral structure that generates rhythmic impulses that are transmitted to the spinal cord. It is presumed that the rhythm of tremor originates in the brain, just as the rhythm of respiration results from the generation of impulses in a bulbar center.

In view of this, it is necessary to discuss an important phenomenon discovered over two decades ago. In 1962 Albé-Fessard *et al.*, employing microelectrode recording, were the first to find during stereotactic operations on parkinsonians rhythmic neuronal discharges in certain Th nuclei (VL, VPL). This phenomenon aroused particular interest because the frequency of these discharges was the same as that of peripheral tremor. Subsequently, this fact was confirmed by many authors (Hardy, 1966; Jasper and Bertrand, 1966; Kandel, 1981; Lenz *et al.*, 1986). Later it was established that such rhythmic activity, synchronous with tremor, is present in certain neurons not only in Th nuclei, but also in GP, NC, and ZI. In contrast, Narabayashi (1982) considers that discharges or firing synchronous with tremor may be obtained in microelectrode recordings only from Vim and that in neighboring Th nuclei (VL, VA, VPL) such discharges are absent.

Three types of neurons were discovered: some that fired in a rhythm close to that of tremor (tremor cells) others that had bursts even when there was no visible tremor, and a third type of neuron that fired only during tremor, ceasing to discharge when the tremor stopped (Bertrand and Jasper, 1965). In functional tests, clenching the fingers into a fist, which stopped pathological tremor, caused the rhythmic discharges to disappear.

"Kinesthetic neurons" are activated only by muscle contraction or joint movement but do not react to touch or pin pricks on the skin. Only "kinesthetic neurons" were found in VL, whereas "tactile neurons" in this structure were absent (Bertrand *et al.*, 1969; Ohye and Narabayashi, 1979). In various Th nuclei, other investigators found neurons that were excited during initiation of a voluntary movement of one limb but not activated by passive movements of the same limb (Siegfried *et al.*, 1969).

Shortly after this phenomenon was described, it was thought that, at last, the long-sought tremor pacemaker had been found. One could imagine that the rhythm of tremor was generated in one of the Th nuclei. It was conceivable that thalamic discharges synchronous with tremor represented the pacemaker of the tremorogenic mechanism in Th and the cerebral cortex (Jasper and Bertrand, 1966). However, subsequently, doubts were cast on such a concept. It appeared in several cases that rhythmic discharges of neurons in subcortical structures occurred in the absence of visible tremor. Tremor discharges were found not in a single subcortical structure, as would have been presumed, but in many. Moreover, the neuronal discharges did not precede tremor oscillations but followed them with a mean latent period of 20 msec (Umbach and Ehrhardt, 1965). Further, it was established that there are two types of rhythmic activity in thalamic neurons, one with activity synchronous with the tremor rhythm and in phase with it, and the other at the same frequency but not in phase (Jasper and Bertrand, 1966). Only 68% of VPL neurons in patients with parkinsonian tremor and 31% of neurons in patients with cerebellar tremor have a discharge frequency identical to the frequency of EMG bursts. The rest of the neurons did not have such synchrony (Lenz *et al.*, 1985).

During voluntary movements, when peripheral tremor temporarily stops, rhythmic activity in the subcortical structures continues. In Vim and adjacent areas, there are neurons with both rhythmic and nonrhythmic discharges independent of peripheral tremor (Ohye, 1982). Moreover, cross-correlation analysis did not confirm the synchrony of the rhythmic bursts in Th nuclei and parkinsonian tremor (Walker, 1982). Analogous rhythmic activity in Th nuclei was recorded in cases of cerebral palsy or torsion dystonia in the absence of tremor (Ohye *et al.*, 1975). Alberts *et al.* (1965b) did not find any correlation between tremor and electric activity of the cerebral cortex, VL, or GP.

These observations do not seem to fit into the hypothesis that thalamic neurons are responsible for generating peripheral tremor. Therefore, it seems that the activity of neurons synchronous with tremor is not the cause but the result of tremor, i.e., the result of afferent impulses to Th from the contracting muscles.

Rondot and Bathien (1978), Hassler *et al.* (1979), and Narabayashi (1982) all conclude that neuronal bursts in VL synchronous with tremor do not imply that this nucleus is the source of tremorogenesis. The rhythms may be the result of ascending feedback proprioceptive impulses from the shaking limbs.

Confirmation of this conclusion was given recently by Lenz *et al.* (1986) by using a complicated autoregressive technique or transferring directed coherence functions relating thalamic and EMG recordings. The results of a study of 30 thalamic tremor cells by this technique demonstrate a significant role of sensory feedback in the generation of parkinsonian tremor. The authors suggest that the tremor is generated by oscillations of an unstable long-loop reflex arc.

On the basis of these data, one may conclude that in spite of scores of experimental and clinical investigations, including stimulation and recording of many subcortical structures during stereotactic operations, no structure in the brain has been identified that could be considered the pacemaker of tremor. Quite naturally, this led to the assumption that such a pacemaker is at the spinal cord level.

That the anatomic substrate of tremor is to be found in the cells of the anterior horns was first proposed by Jung (1941) but, as noted by Gybels (1963), was not supported by any experimental or clinical evidence. Jung considered tremor to be a primitive form of movement—an ''archaic relic'' analogous to the rhythmic swimming of fish—that was released by blockade of inhibiting impulses to the spinal cord. Subsequently, it was suggested that the tremor rhythm is generated in the ''interneuronal bulbospinal system'' (Jung and Hassler, 1960).

To study this possibility, a group of researchers including the author carried out a series of investigations, the results of which have been reported elsewhere (Gurfinkel *et al.*, 1965; Kandel, 1965, 1974a,b, 1981; Barlow, 1965a,b; Voronin and Kandel, 1971; Aizerman *et al.*, 1974; and others). As a result, a hypothesis on the generation of rhythmic tremor at spinal levels was propounded. According to this hypothesis, pathological tremor does not result from rhythmic impulses generated by a neuronal structure in the brain but from abnormal (nonrhythmic) influences of cortical, subcortical, and brainstem structures on the segmental mechanisms of the spinal cord. These influences activate segmental circuits so that rhythmic tremor is generated in the anterior horns of the spinal cord. In other words, generation of the rhythm of pathological tremor occurs at a spinal level, and supraspinal structures only modify the functional state of the segmental circuits of the spinal cord where the tremorogenic impulses occur.

The proposed ''spinal'' hypothesis implies that not tremor itself but the rhythm of tremor is generated at segmental levels, i.e., synchronization of (pathological) supraspinal impulses. Alberts *et al.* (1965b) also consider that the frequency of tremor is determined by spinal motoneurons. Another important argument speaks in favor of the ''spinal hypothesis'' of tremor genesis. As many neurophysiological experiments have shown, normal control of locomotion is accomplished by translation of a tonic descending message into rhythmic output in the spinal cord (Carew, 1985).

There can be no doubt that the primary pathology in parkinsonism and other extrapyramidal disorders is to be found in cerebral structures at brainstem or subcortical levels. Doubtless, the initial triggering of tremor is by neuronal structures and circuits at that level, from which nonrhythmic impulses pass to spinal centers. These supraspinal activities are not rhythmic, although thus far little is known about their nature or about the neural pathways carrying the impulses from the brain to the spinal level. Only at the spinal level is this nonrhythmic supraspinal activity transformed into a typical tremor rhythm.

The ''spinal hypothesis'' is based on a number of experimental and clinical observations. The first of these is that different frequencies and phases of tremor are seen in various parts of the body and even in different muscles around one joint. If there were a central pacemaker, one would expect the frequency of tremor in different parts of the body to be the same. If one were to presume the presence of two such central pacemakers (one each for the right and left sides), one would expect a stable relationship between the frequencies of tremor in left and right extremities. In order to explain the different tremor frequencies in joints and muscles by the central pacemaker theory, one would have to assume the presence in the brain in parkinsonism of hundreds of independent pacemakers setting the frequency of tremor for each muscle, since the number of such individual pacemaker should be at least equal to hundreds of biomechanically bound groups of muscles in each half of the body. It seems quite obvious that such an assumption is highly improbable. At the same time, these facts can be explained by the ''spinal hypothesis,'' which allows a wide range of independence for frequencies and phases of tremor in various parts of the body.

The second is the occurrence of spontaneous or

induced desynchronization of tremor in one antagonist muscle with the preservation of clear-cut rhythmic activity in the other muscle, which indicates that control of tremor in each muscle is independent. This is also confirmed by the shift in phase of tremor when it returned following the unloading test, which alters proprioceptive input. In resting tremor this procedure causes clear-cut desynchronization of electroactivity in the forearm flexor, although burst activity of the antagonist (triceps) does not change significantly. The fact that the tremor phase after unloading is "not remembered" suggests that the hypothetical pacemaker or tremor is not autonomous but is to a considerable degree dependent on afferent influences.

These facts also do not fit with the "central" hypothesis, which presumes that the generator of tremor is autonomous. Furthermore, if the rhythm is disrupted in some manner, its phase should be preserved on restoration of tremor. With the "spinal" hypothesis, the "remembering of the phase" is not only unnecessary but highly improbable. If generation of tremor occurs at the spinal level, tremor may be absent temporarily in one antagonist muscle while it is preserved in the second, and when tremor is restored, the phase relationships may be changed.

A third argument in favor of the "spinal hypothesis" is the latent period for changes in frequency and amplitude of tremor on stimulation of VL during stereotactic operations (Fig. 24). According to our data, this change in the tremor occurs in 500–900 msec, i.e., after three to five cycles of the tremor. If VL were, as many authors consider, the generator of pathological tremor, then electrostimulation of this nucleus should cause the tremor reaction to appear after the interval of time required for impulses to travel from VL to the limb muscles. Even if one assumes that the stimuli from VL do not have a direct route to the spinal segmental level but exert their influence through the motor cortex, the latent period should be much shorter than 0.5–0.9 sec.

Changes in pathological tremor during stimulation of VL after such a long period indicate that the influence of that structure on the segmental activity of the spinal cord is, in all likelihood, of an "adjusting" nature. This also favors the assumption that supraspinal influences only modify the state of the segmental spinal activity where the rhythm of pathological tremor is primarily generated.

Our investigation confirms the data of other researchers that activation of a single motor unit for muscle contraction, normally recorded by needle electrodes, requires on the average 7–14 msec. All motor units work asynchronously and independently of each other. In parkinsonism, the average time for stimulation of motor units is about the same, but they are firing synchronously in bursts that correspond with the rhythmic tremor seen on a mechanogram. Each burst consists of two to four impulses. The intervals between bursts are approximately 150–200 msec, and those between spikes within a burst are 30–60 msec.

As already noted, the specific pathophysiological mechanism of synchronization and, consequently, the appearance of characteristic rhythmic bursts at the spinal level remain obscure. We might assume that "elevated inertness" of the spinal Renshaw cells or the elevation of their threshold explains abnormal synchronization; however, the cause of this "inertness" has yet to be determined.

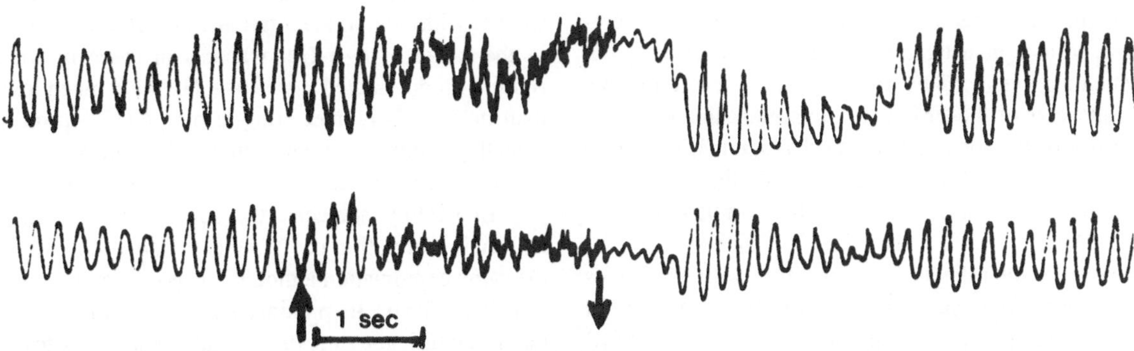

FIGURE 24. The influence of electrostimulation of the left VL (4 V, 50 Hz) on tremor during stereotactic operation for parkinsonism. Latent period of tremor reaction is 0.8 sec. Arrows indicate the period of stimulation.

In animal experiments, cooling of ND leads to abnormal synchronization and slowing of the burst rate (Meyer-Lohmann *et al.*, 1975), but there is no evidence that this structure is a generator of pathological tremor.

Because the presence of physiological tremor does not depend on synchronization (Freund and Dietz, 1978) one may assume that normally the cerebral structures provide an inhibiting mechanism preventing synchronization of spinal motoneurons, since the normal motor system is capable of long-term synchronization in special conditions (shivering, fatigue, stress) (Freund and Dietz, 1978). The authors have demonstrated that short periods of synchronization of motor units in physiological tremor is not random in most subjects. It is also possible that during a lapse of that inhibiting mechanism, no matter what the cause, synchronization of the incoming signals occurs, and consequently, tremor appears. However, if one accepts such an assumption, it is impossible to explain how destruction of VL and certain other subcortical structures eliminates tremor. There is also reason to consider that synchronization leading to the appearance of reciprocal alternating tremor results not from disinhibition but from stimulation of the spinal mechanisms.

2.11. Computerized Spectral Analysis of Tremor

Computerized spectral analysis of physiological and pathological tremor represents a new and promising approach that has been rapidly advancing in the past decade (Marsden, 1978; Freund and Dietz, 1978; Sato, 1982). As a rule, this study uses the integral EMG picked up by surface electrodes if the tremor spectrum is mainly in the 30- to 200-Hz range. However, as noted above, the main human joints such as the wrist, ankle, and elbow act as low-pass-band filters and cut off all frequencies above about 20 Hz. Therefore, tremor investigations in the frequency range at which the joint actually operates should record the low frequencies of the EMG. It is important that the comparison of the spectrums of mechanograms, reflecting the movements of the joint angles with the EMG spectrums, did not disclose any reliable correlations.

During the last decade, the author and Dr. Ivanova-Smolenskaya, in cooperation with Professors Aizerman, Andreyeva, and Khutorskaya (Institute of Control Sciences, Academy of Sciences of the USSR) have developed an improved method to investigate physiological and pathological tremor through analysis of the low-frequency range of the EMG record ''envelope'' (EEMG) (Fig. 25). It was shown that the mechanogram curve is very close to the envelope curve of the EMGs. We established that computerized analysis of EEMG spectra is much more informative than study of the EMG itself. The results of this study have recently been described (Andreyeva *et al.*, 1985, 1986; Kandel *et al.*, 1986; Ivanova-Smolenskaya *et al.*, 1986; Andreyeva and Khutorskaya, 1987).

Large groups of healthy subjects, parkinsonians, and patients with essential tremor were studied. A statistical estimation of muscular activity was made by recording numerous EEMG spectra from many muscles at different joints. For intraoperative monitoring of tremor, a special set-up was used (Fig. 26). Recording electrodes were brass plates with bits of palladium; their area was 0.8 cm^2, and the distance between electrode centers was 1.5–3 cm.

Every EMG went through a two-stage detector and a low-pass-band filter with a cut-off frequency above 20 Hz. Then the EEMG signals were fed directly into an analogue-to-digital converter, digitized, stored, and processed for power spectrum analysis (Fig. 27). Each examination produced at least 72 EEMG power spectra. Their arrays included many parameters (frequency, spectrum peaks, power spectra, histograms of spectral peak frequency distribution, power coefficient, and correlation coefficients) that characterized the tremor (Fig. 28). In processing, the data were analyzed in different ways, including effects of laterality and differences between upper and lower extremities.

One of the most important parameters is the spectrum peak, which refers to the frequency having the highest amplitude (greatest occurrence). The peak frequency changed constantly within the limits of ± 1 Hz. As spectral correlative analysis has shown, in parkinsonism with predominant tremor, the spectrum peak is in the range of 4–6 Hz, whereas in cases in whom rigidity predominated it was 6–8 Hz, both much lower than normal. It is interesting to note that even on the ''healthy'' side of patients with hemiparkinsonism, abnormal peaks not seen in routine EMG were recorded.

In healthy subjects, the spectral characteristics and their correlation coefficients varied widely during the same examination although the position of the joint remained fixed. In parkinsonism there was a statistically significant shift in the peak frequencies toward

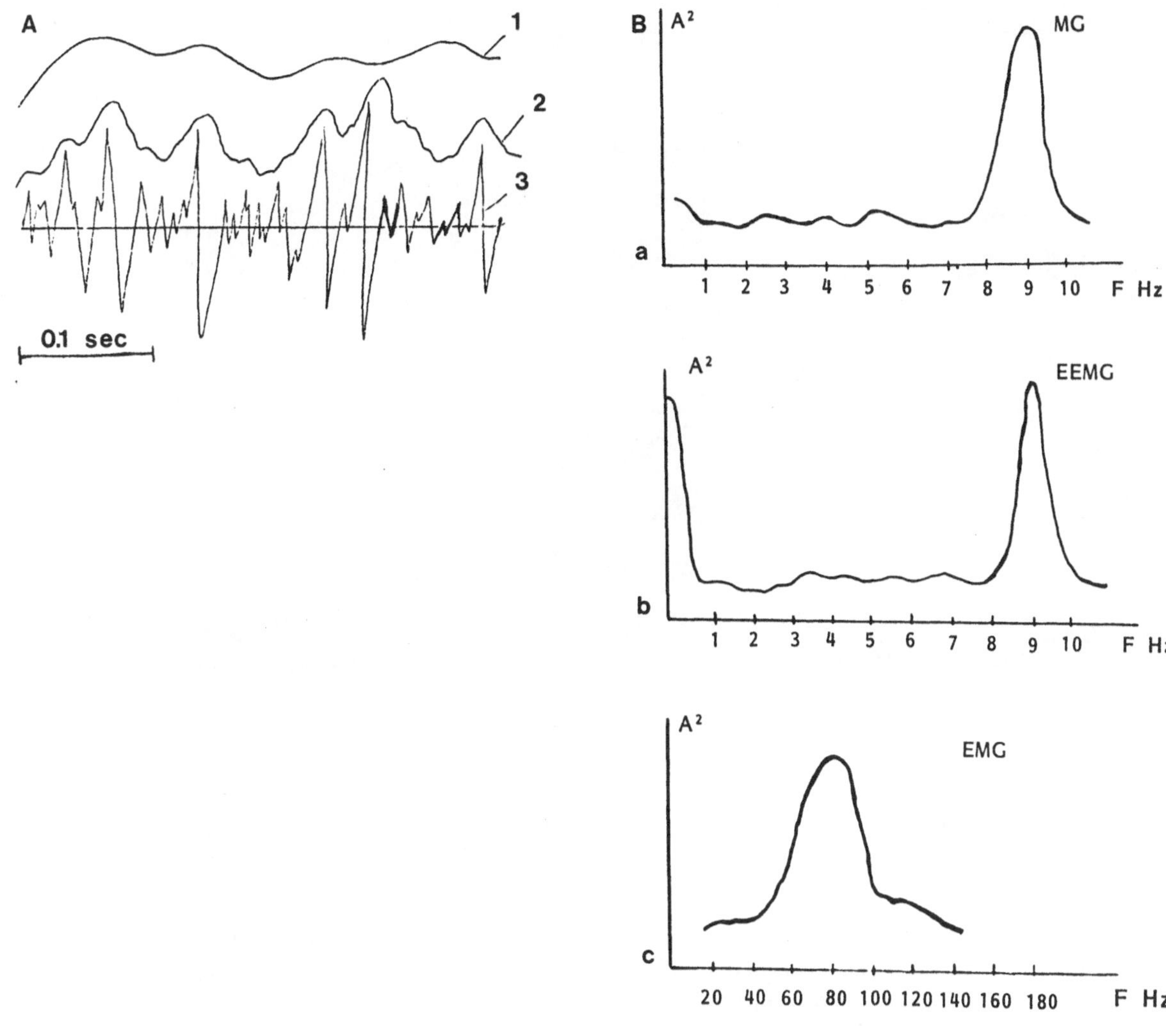

FIGURE 25. (A) Schematic image of tremogram (1), EMG record (2), and EMG "envelope" (EEMG) (3). (B) The spectra of frequencies of tremogram (a), EEMG (b), and EMG (c): the frequencies are shown on the abscissas, and the squares of the amplitudes at these frequencies on the ordinates. Note that the region of tremor frequency coincides with EEMG but not with EMG.

low (5–6 Hz) range, which is very rare in healthy subjects; the frequencies for the muscles on the same side of the body were largely coincident and had significant correlations. As in the case of healthy subjects, there was no correlation of peak frequencies for muscles of the right and left sides of the body, but in patients there was no difference in the frequency modes of the upper and lower extremities, such as those that characterize healthy subjects.

It appears that in normal persons there is no significant relationship between the activity of muscles in different joints on the same side, whereas in parkinsonian patients there is a definite correlation between the frequencies of muscle tremor in the extremities on the same side of the body with a decreased (approximately halved) principal frequency. In essential tremor, too, the "local control" present in a normal subject seems to be superseded by a signal from a central source that

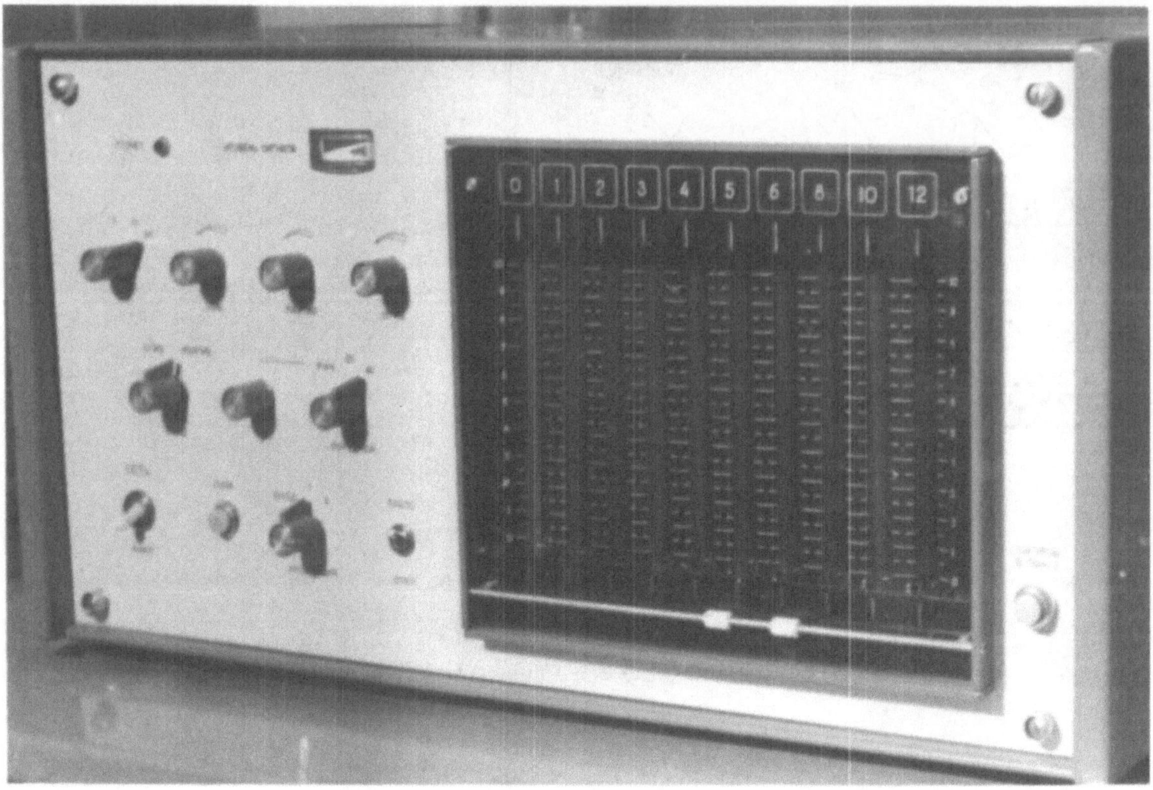

FIGURE 26. Our device for intraoperative monitoring of the frequency spectrum of tremor using tensosensors.

affects muscles of some joints symmetrically, whereas in parkinsonian patients a similar source affects different joints on the same side of the body. Furthermore, in cases of essential tremor, the "functioning of local sources" for lower and upper extremities differs from that of healthy subjects.

The processing of large arrays of EEMG spectra has revealed statistically significant differences among the three groups of subjects (healthy, parkinsonian and essential tremor patients). These differences are important for diagnosis and treatment of the diseases; they also give insight into the nature of motor control and its disorders. For example, the height of the spectrum peak and the constant spectrum component power at

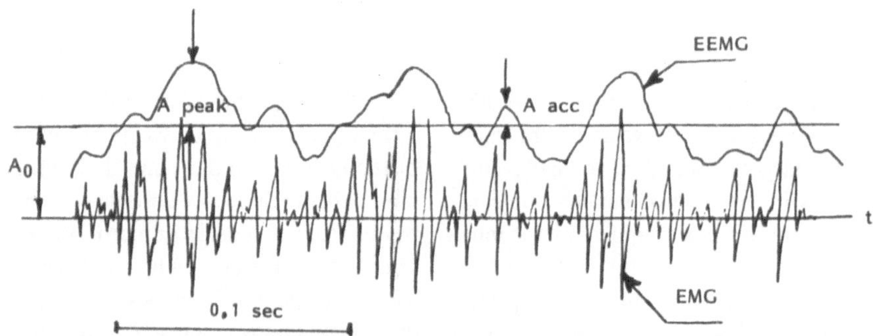

FIGURE 27. Envelope EMG signals fed into a computer. A_o, constant component; A_{peak}, amplitude of peaks; A_{acc}, additional peaks.

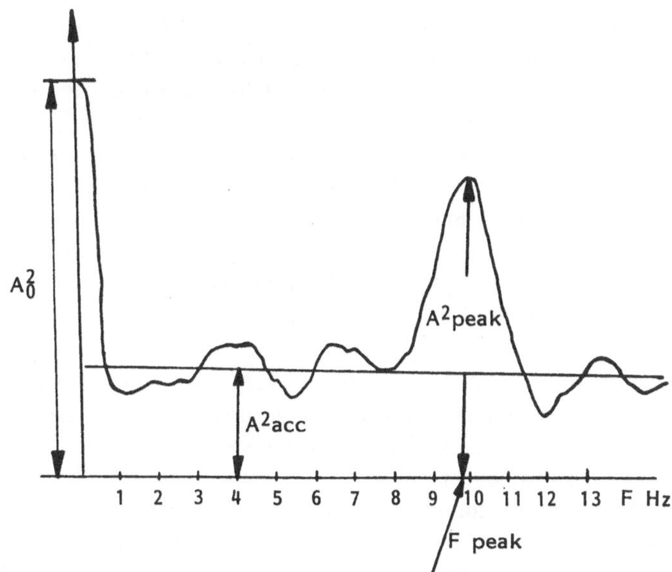

FIGURE 28. The main parameters analyzed by computer from EEMG and obtained spectrum. A_o^2, the square of constant component; F_{peak}, frequencies of peaks; A^2_{peak}, square of amplitude (or power) of peaks; A^2_{acc}, the mean power of additional peaks.

the wrist were, respectively, 120 and 1100 in healthy subjects, 1460 and 4800 in parkinsonians, and 310 and 1400 in essential tremor patients.

Further investigations with the aid of spectral analysis of tremor will, without a doubt, contribute greatly to an understanding of the pathogenesis of tremor.

2.12. Tremor and Cerebral Structures

A lesion of which cerebral structure leads to the appearance of tremor? In neurophysiology and neurology it is difficult to find a problem on which there are so many different points of view. Even today there is no satisfactory answer, although many other questions of a secondary nature depend on a clear solution. Is the development of tremor caused by a lesion of a single specific cerebral structure or several structures in different areas that form a single functional system? Can the existence of different kinds of tremor (postural, intention, and resting tremor and others) be explained by lesion of one structure, or must several different structures be involved? In which neuronal structure is the rhythm of pathological tremor generated? Why does tremor disappear completely on surgical destruction of several cerebral structures in which there are no morphological lesions of any known pathological process?

We here attempt to summarize the literature that has appeared on this problem; some data on modeling extrapyramidal phenomena, particularly tremor, are discussed in Section 1. There are grounds to assume that synchronization of spinal motoneurons and, consequently, motor units leads to the appearance of pathological tremor. On that basis, study of the pathogenesis of tremor requires an inquiry into the mechanisms and location of the synchronization process.

Nashold and Slaughter (1969) divide the existing theories of the pathogenesis of tremor into two main groups: epicentric and multicentric theories. The former presumes that a tremorogenic center is located in Th nuclei, possibly in VL. The latter considers tremor discharges of the lower motor neurons to arise not only from corticopetal impulses but also from disinhibition of many subcortical nuclei, the destruction of which eliminates the tremor.

Walker (1969) singles out four CNS structures in which tremor may be generated: the pyramidal tract, Th, FR of the brainstem, and the anterior horn of the spinal cord. In analyzing the participation of all these structures in tremorogenesis, one must agree with the viewpoint of Hassler *et al.* (1970), who consider that any theory on the pathogenesis of parkinsonian tremor must differentiate those cerebral structures damage to which causes the parkinsonism syndrome from those unaffected structures that determine the rhythm of tremor.

Throughout the entire history of neurosurgery, there has been no experimental or clinical evidence to suggest that tremor is generated in the cerebral cortex.

The fact that parkinsonism usually begins with trembling of one limb (sometimes even of one or two fingers) could implicate the cortical level in the genesis of parkinsonian tremor, but strict somatotopic organization is known to exist in many subcortical structures. Stimulation of the motor cortex or other cortical areas does not produce tremor or change its rhythm or amplitude.

Numerous attempts to find a rhythm synchronous with tremor in EEG or electrocorticogram (ECoG) in patients or animals have proved futile. Several authors have described θ rhythms in EEGs closely resembling parkinsonian tremor; however, our investigations showed this to be an artifact associated with the trembling head of a patient during EEG recording. Clinical data indicate that the higher cortical functions in parkinsonism, as a rule, remain intact, and morphological investigations do not reveal any specific changes in the cerebral cortex. Therefore, there is no reason to consider the "tremor generator" to be located in the cerebral cortex.

This, however, does not deny the possible important role of the cortical level in the genesis of the facilitating and inhibiting influences on neural "excitation cycles" that are of primary significance in the genesis of parkinsonian tremor (Alberts, 1969). The disappearance of all types of tremor in sleep is an indirect argument in favor of the participation of the cerebral cortex in tremorogenesis.

There can be no doubt that the basic cause of parkinsonian tremor is destruction of SN neurons. As Hassler *et al.* (1970) have emphasized, there has not been a single documented case of parkinsonism in which there was no loss of SN neurons. At the same time, the authors assume that the neurons pacing the rhythm of tremor cannot be SN neurons since "the damaged cells cannot generate impulses at a definite rhythm." Hence, tremor must be generated by another structure that is an antagonist of SN.

It is also presumed that SN normally desynchronizes the permanently weak spinal motoneuron activity that exists in the absence of any movement. In the event of damage to SN, the activity of these spinal neurons is synchronized, and pathological tremor appears (Hassler *et al.*, 1979). As a result of reduction of neurons in SN, there is an imbalance in the system of extrapyramidal control over motor function, which leads to synchronization of the steady outflow of impulses from the spinal motoneurons with a rhythm of 4–6 Hz. More specifically, tremor results from the elimination of the normally inhibiting influence of SN on GP and VL. However, this hypothesis does not explain why tremor is absent in the rigid and akinetic forms of parkinsonism.

Many authors consider tremor to be generated by Th nuclei, particularly by Vim (Guiot *et al.*, 1964; Cooper, 1966a; Narabayashi, 1969; Ohye and Narabayashi, 1979). This opinion is based on the fact that stereotactic destruction of this Th nucleus completely eliminates tremor. However, Hassler *et al.* (1979) report that with a totally intact Vim, tremor disappears completely after stereotactic destruction of VOa and VOp. Since in parkinsonism no morphological changes are found in VL and particularly in Vim, their relationship to the tremorogenic mechanism remains open. Narabayashi (1982) also believes that tremor and rigidity are related to cerebral structures: rigidity is related to VL, and tremor to Vim (as we have already noted, Narabayashi does not include Vim in VL). An analogous representation is said to exist in CF, the anterior part of which is concerned with rigidity and the caudal with tremor.

Jasper and Bertrand (1966), who do not directly assert that tremor is generated in the thalamic nuclei, use a more cautious wording: the "motor thalamus" has structures that may be viewed as "tremor energizers." Our data on the latent period of tremor following stimulation of VL also do not confirm that this structure is the generator of tremor.

Certain authors suggest more generalized systems as cerebral structures in which tremor may be generated—the dentatothalamic system (Hassler *et al.*, 1970; Struppler *et al.*, 1976), cerebellar pathways passing through Br Con to the midbrain tegmentum (Carrea and Mettler, 1955), the dorsomedial and ventral nuclei of Th and their projections to the promoter cortex (Gros *et al.*, 1966), and corticostrial and corticopallidal systems. In addition, nuclear structures such as GP (Denny-Brown, 1962), FR of the brainstem (Ward 1961), and Str (Rondot and Bathien, 1978) are considered to generate or facilitate the occurrence of tremor.

Levy (1967) attempts to ascribe the pathogenesis of tremor to defects in the general organization of motor activity. The functional interaction of three clearly defined systems—the corticopyramidal system, the cerebellar control of movements, and the motor mechanisms of the brainstem, principally FR, if disturbed, disrupts integration of the activity of these three systems, producing involuntary movements, in particular, tremor. However, this rational, global concept leaves unanswered the mechanism of the disruption and its relationship (if the theory applies to the resting tremor in parkinsonism) to destruction of neurons in SN.

Summing up these data, one must admit that it is still impossible to determine the brain structure responsible for tremor generation. It is also obvious that stereotactic lesions of the several subcortical structures that abolish tremor may not destroy the "generator" but only interrupt neuronal circuits involved in tremor.

One other important question regarding the pathogenesis of tremor still remains unsolved. Since there are valid reasons to assume that the tremorogenic mechanism resides in subcortical and brainstem structures, it is important to establish the pathways that project to the spinal centers and lead to synchronization of the activity of motoneurons in the anterior horns. Do these arrhythmic impulses go directly to the anterior horns, or do they first go to the motor cortex and then, as in other motor activity, pass in the corticospinal tract to the spinal motoneurons? As Evarts (1979) emphasizes, the outflow from the motor cortex is always the result of an inflow of sensory afferents from other parts of the cortex, numerous subcortical structures, and Cer. This afferent input reaches the motor cortex primarily via Th nuclei.

As was already noted, Parkinson assumed that tremor was transmitted through the pyramidal tract, since in one of his patients, after a stroke with hemiplegia, tremor disappeared, and as movements recovered, the tremor appeared again. Past experience with the surgical extirpation of the motor cortex and section of the pyramidal tract at any level leaves no doubt that an essential condition for the appearance of tremor is an intact pyramidal tract. It is thought that the pyramidal tract synchronizes the activity of spinal motoneurons, thus accentuating resting tremor. In addition, it is known that impulses from the motor cortex, descending in the pyramidal pathways, are responsible for initiating a movement and are simultaneously transmitted to the Cer cortex (Eccles, 1980). In view of this, the transmission of these impulses to Th nuclei via the dentatorubrothalamic pathway is possible. Levy (1967) considers tremorogenic impulses to pass along parapyramidal fibers located in CI medial to the pyramidal tract.

There can be no doubt that cerebral transmitter disorders play a definite role in the pathogenesis of tremor. Yet, only preliminary attempts have been made to "span the bridge" between disorders of catecholamine metabolism and tremor. From clinical observations, it is known that epinephrine induces or significantly intensifies tremor. In monkeys with lesions of the midbrain tegmentum and a postural ("tegmental") tremor, there is a substantial decrease in the level of dopamine and serotonin in Str on the same side (Poirier, 1972). If in this animal another destructive focus is made in VL, tremor disappears immediately, but the low striatal dopamine and serotonin levels remain unchanged. More detailed data on the role of biochemical disturbances in the pathogenesis of tremor are presented in Chapter 6.

3. Rigidity

The study of the pathogenesis of rigidity and spasticity must be based on an understanding of the mechanisms regulating normal muscle tone. Almost 150 years ago, Müller (1840) defined skeletal muscle tone as prolonged and tireless contraction of muscles ensuring maintenance of a certain posture of the body. It should be stated, however, that the mechanisms that maintain and regulate muscle tone are still not entirely understood.

3.1. Muscle Tone and Its Regulation

It was Sherrington who established that skeletal muscle tone was based on a proprioceptive (myotatic) monosynaptic reflex that is determined by the stretch reflex. This reflex was first discovered in decerebrated cats by Liddell and Sherrington (1924), who demonstrated that a stretched muscle contracts reflexly, thus counteracting stretch. In his definition Bechterev (1904) stressed the reflex nature of muscular tonus: "Muscular tonus is a constant support of a muscular tension since continuous input of afferent impulses from muscles to the centres from which they reflexively pass through efferent pathways again to the muscular system."

Tone is the reflex resistance of a muscle to stretching. It has been recently established that there are both short-latency monosynaptic and long-latency stretch reflexes, the arc of which includes supraspinal pathways (Marsden, 1978; Rothwell *et al.*, 1983).

The concept of the reflex nature of muscle tone is mirrored in the definition of Gurfinkel *et al.* (1965): "Tone is the state of the neuromuscular system characterized by the level of activity of the stretch reflex." The common observation of a contracted muscle and a muscle with loss of tone recalls the similarity of muscle tone and the activity of smooth musculature. This thought was formulated by Bernstein (1966): ". . . the work of the striated muscle, as a smooth muscle, proved to be that very same muscle tone which already was known very long ago to physiologists and clinicians, and which so stubbornly did not lend itself to precise definition."

The International Classification of Extrapyramidal Disorders (Section 1) includes the following traditional definition of muscle tone: "Tone is understood as a degree of resistance appreciated by an examiner during passive joint movements in a relaxed subject." Clinicians also usually define muscle tone as the mechanical resistance of a limb during passive movements of one of its joints. It is apparent, however, that this standard clinical definition simultaneously refers to two practically indivisible parameters—mechanical (viscoelastic) properties of muscle tissue and the reflex state maintaining muscle tone, i.e., the degree of muscle tension. One may judge the state of muscle tone at a given moment only by altering its length—either shortening or lengthening it.

Human motor activity is divided into phasic movements and posture. Since prolonged preservation of natural posture does not cause muscular exhaustion, it is presumed that this process is carried out by a special "energetically economical" form of muscular activity. The term "tone" is often interpreted as that form of muscular activity ensuring maintenance of posture. It is presumed that an important function of the striopallidal system is not only automating voluntary movements but also making them "economical," i.e., minimal consumption of muscular energy in movements.

Mammals have two types of muscles, "slow" and "fast," which perform both kinds of motor activity, phasic and postural (Buller *et al.*, 1959). The motor units of slow muscles tire very slowly, yet the muscular tension they develop is relatively weak. In contrast, the motor units of fast muscles produce a powerful and rapid contraction, but these muscles tire easily. This is explained by the fact that the fast muscles consume ATP—the main energy resource of the muscles—more rapidly.

There are large and small motor units, and as Evarts (1979) points out, a large motor unit can develop 200 times more tension than a small unit. Muscular contraction is initiated by the small units, and only if it must be strengthened are the large motor units involved.

In man, all muscles have approximately the same speed of contraction, and any limb muscle may function in both the phasic and postural activity. Unlike the earlier concepts, today it has been proven that many human muscles participate both in preserving postural tone and in performing rapid motor acts. Nevertheless, it may be considered an established fact that in postural activity, the dominant role is played by slow muscles, whereas phasic movements are carried out by fast muscle.

What is the physiological mechanism of tone? It seems that it is maintained both by proprioceptive impulses continuously arising in the muscle itself and by slight steady tonic activity of spinal α-motoneurons. Tone is regulated by a complex control system based on hierarchic principles: the basic mechanisms are at the segmental spinal level, and these are subordinate to the action of "multilevel" cerebral structures. A brief and simplified description of the regulation of muscular activity at segmental spinal levels follows.

Out of 10 million neurons in the spinal cord, motoneurons make up about 2%, but even these 200,000 neurons (Eccles, 1980) are apparently sufficient for organizing a multitude of diverse movements in man. For this purpose, information on these movements and the muscles performing them enter the spinal cord via the 600,000 afferents of the posterior roots. Each of the 200,000 motoneurons has 5500 synapses, which indicates the scope of the connections between the motor spinal centers and other CNS structures.

The motoneurons in the anterior horns of the spinal cord are divided into large and small α neurons and small γ neurons. Both types of neurons are involved in α–γ coactivation. The large α neurons innervate the fast limb muscles performing voluntary and reflex movements. The small α neurons send impulses to the slow tonic muscles ensuring postural reactions and maintaining the vertical position of the body (muscles of the torso, shoulder, and pelvic girdles). The α motoneurons activate the end plates through the release of acetylcholine.

The axon of each motoneuron innervates, on the average, approximately 100 muscle fibers, which constitute the motor unit. The density of muscle innervation directly relates to its functional load. For example, in ocular muscles working at the high speed necessary for rapid eye movements, one motoneuron innervates only three muscle fibers, whereas one to the biceps activates hundreds of muscle fibers (Evarts, 1979).

The brain controls the functioning of the motoneurons and their interaction by means of interneurons.

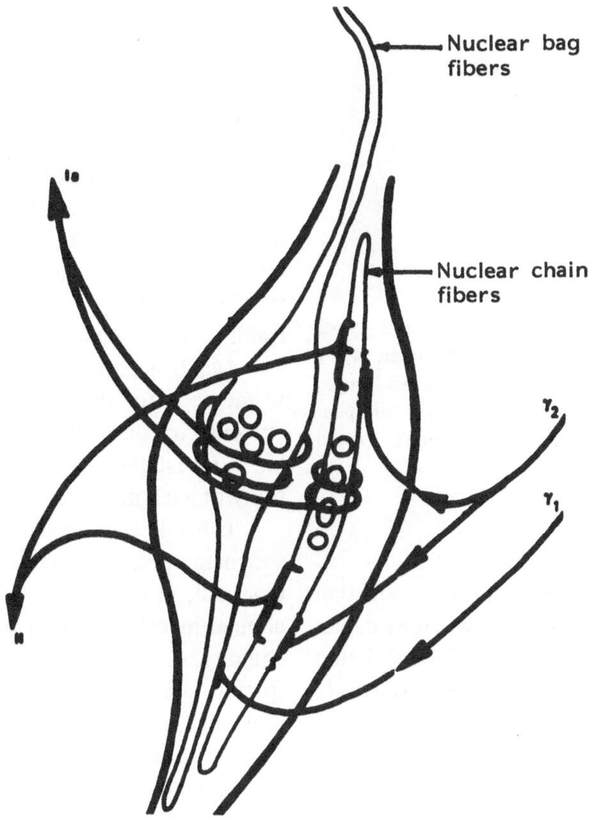

FIGURE 29. Two types of fibers in the muscle spindle: nuclear bag fibers and nuclear chain fibers.

Special interneurons of the anterior horns, Renshaw cells, receive collaterals from motoneurons and have an inhibiting effect on those motoneurons.

The γ neurons of the anterior horns are divided into static and dynamic. These neurons are not strictly motoneurons since they innervate only muscle spindles and do not directly cause muscular contraction. The discovery and thorough study of the γ system had profound significance for our understanding of the mechanisms regulating muscle tone. One of the main functions of the γ system is to regulate the sensitivity of the primary and secondary endings of muscle spindles.

Each fiber of the skeletal muscle has specific proprioceptors, muscle spindles, which are an important component of the stretch reflex that makes muscle length follow changes in spindle length. Intrafusal fibers making up the muscle spindle are divided into two categories (Fig. 29): nuclear bag fibers and annulospiral endings, which initiate Ia afferents, and nuclear chain fibers, which have "flower-spray" endings from which afferents of group II proceed to the spinal cord. The fine myelinated γ afferents from each muscle spindle pass through the posterior roots to the α motoneurons. Muscle spindles with γ afferents are highly sensitive sensors providing information about the physical state of the muscle fibers. As was shown by Granit (1970), every supraspinal excitation or inhibition of muscles affects the spindles.

The γ motoneurons innervating muscle spindles play an important part in motor activity and maintenance of muscle tone. During active contraction of muscles, the spindles become relaxed and, consequently, fire less frequently. During passive stretching of muscles, the spindle excitation is augmented, and their afferents excite the α motoneurons. Since the motoneurons innervating a muscle receive an afferent input from each muscle spindle, stretching synchronizes the discharge of motoneurons. During the stretching of a muscle, there are two types of responses from the afferents (Matthews, 1962). The first type (dynamic) is regulated by γ_1 motoneurons and depends on the speed of muscle stretching. The second type (static) depends on γ_2 motoneurons and on change of muscle length.

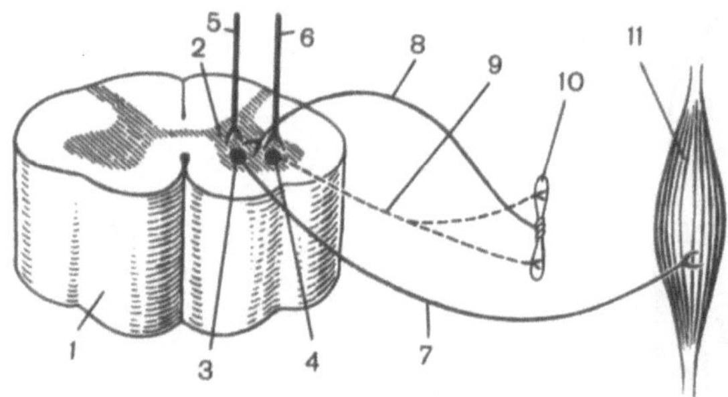

FIGURE 30. Simplified scheme of muscle innervation. 1, spinal cord segment; 2, anterior horn; 3, α motoneuron; 4, γ motoneuron; 5 and 6, supraspinal control of α and γ innervation; 7, efferent fiber of α motoneuron; 8, afferent fiber of annulospiral termination; 9, γ efferent fiber; 10, muscle spindle; 11, striated muscle.

Activation of α motoneurons leads to contraction of the muscle from which the afferent impulses came. Simultaneously, there is stimulation of the inhibiting interneurons, as a result of which the antagonist muscle relaxes. This is the basis of reciprocal innervation, which regulates the balance between antagonist muscles.

The muscle spindle, the γ efferents (γ_1 and γ_2), and the afferents passing to the α motoneuron together form the so-called γ loop, which is the second (after α motoneurons) innervating mechanism ensuring phasic and tonic muscle contractions (Fig. 30). Signals coming from the spindles to the motoneurons of the spinal cord are responsible for tonic muscle tension.

Recording of Muscle Tone

A record of muscle tone is important for studying its normal and pathological states. Conventional clinical investigation of normal and abnormal muscular tone is done by the physician on the basis of subjective assessments. In view of this, there have been numerous attempts to develop objective methods for recording and quantitatively evaluating muscular tone. Various devices for measuring muscular tone have been used; one of the first was proposed by McKinley and Berkwitz in 1928.

Clinical methods for measuring muscular tone may be divided into static and dynamic. Static methods involve measurement of "density" or "hardness" of muscles at rest or internal muscular pressure. Muscular tone may be measured in arbitrary units (myotones) with the Szirmai instrument, which is an elastometer measuring muscle resistance to pressure (Smirnov, 1976). Experience has shown that these methods are not precise, and, consequently, they are not widely used clinically.

The reinforcement test proposed by Schwab (1964) is suitable for clinical determination of even a slight degree of rigidity. As passive movements are made at the wrist joint of one arm, the patient is asked to clench the fingers of the other hand into a fist. If there is rigidity, resistance to passive movements increases.

Much more precise and informative are the dynamic methods for measuring muscular tone in either a stretched or contracted muscle. In such cases it is possible to investigate the reaction of the proprioceptive activity of each muscular fiber.

Gurfinkel and Safronov (1971) designed a tensodynamometer for quantitative evaluation of muscular tone and, in collaboration with the author, performed clinical investigations of pathological changes in tone (rigidity, spasticity) in various extrapyramidal disorders (parkinsonism, dystonia musculorum deformans, cerebral palsy, and others). This device (Fig. 31), similarly to other analogous apparatus, measures the resistance encountered in routine procedures used by the physician for investigating tone, such as passive swinging (sinusoidal) movements with a constant amplitude of the upper limb fixed horizontally on the base of the apparatus. The horizontal plane was selected so that the weight of the limb and the mobile piece of the apparatus would not affect the readings of the four tensodynamometric sensors mounted on the platform. The signals from these sensors, which are proportional to the resistance to muscle stretching, are fed into an oscillograph and recorded

FIGURE 31. Our instrument for measuring muscular tone as described in the text.

on paper. One thus obtains two superimposed tracings, one of oscillating movements of the platform with the limb attached to it and one of the mechanical resistance of the muscles of the joint being passively moved. Passive bending of the elbow joint at different speeds is recorded along with the angular joint displacement. The instrument has special devices to compensate for inertial and gravitational effects, thereby evaluating muscular tone in any dynamic range. This objective record indicates the muscle tone in units of torque (kilograms decimeters, kg dm).

The muscular tone thus measured correlates with the speed of the passive joint movement: at average speed, muscular tone is minimal, whereas at low and very high speeds it increases sharply.

According to the author's investigations in healthy subjects, the value of muscular tone varies from 0.5 to 1.5 kg dm, whereas in parkinsonism, this figure rises to 12–24 kg dm. Using this apparatus, we investigated muscular tone in many patients with rigidity before and after surgical treatment. Such a study has made it possible to measure objectively the muscular tone in afflicted muscles before and after an effective operation.

3.2. Pathogenesis of Extrapyramidal Rigidity

The above brief, and consequently, simplified discussion of muscle tone and its regulating mechanisms is a preamble to a most important clinical problem—the pathogenesis of rigidity, which is one of the main clinical manifestations of parkinsonism and other extrapyramidal disorders. Since the pathophysiological mechanisms of rigidity have not yet been fully disclosed, the present discussion focuses on three main aspects—experimental, biochemical, and clinical.

Rigidity has been known to clinicians for about a century. As was noted above, James Parkinson did not mention rigidity among the main symptoms of parkinsonism, although he did describe patients with the torso and neck bent forward. The first detailed description of rigidity in this disease was presented by Gowers in 1888. The International Classification (Section 1) defines rigidity as

> a form of hypertonia characterized by a constant uniform increase in resistance to passive movement, throughout the range of joint displacement, while the patient attempts to relax. This should be distinguished from spasticity.

It may be considered as established that extrapyramidal rigidity is based on hyperactivity of myotatic stretch reflexes, both phasic and tonic. Stretch reflexes are also divided according to their latent period into short and long-latency reflexes. The former are without a doubt monosynaptic, but the arc of the latter has not yet been established. It has been demonstrated that in rigidity both short- and long-latency reflexes are increased in the majority of cases (Berardelli *et al.*, 1983); however, in severe cases, there may be a marked increase only in long-latency stretch reflexes (Phillips *et al.*, 1959; Rushworth, 1969; Safronov, 1970). In the opinion of Hassler (1972b), Struppler (1973), and Narabayashi (1982c), rigidity in parkinsonism is a state of muscular tension caused by increased tonic stretch reflexes, whereas spasticity is caused by increased phasic stretch reflexes. This opinion is confirmed by data indicating that phasic stretch reflexes in parkinsonism are normal or only slightly increased or even diminished (Hassler, 1972a–c; Andrews *et al.*, 1972; McLennan, 1973).

The well-known Jendrassik maneuver leads to a large increase in afferent input from the stretched muscle spindles, which aggravates the stretch reflex and increases the tone of all body muscles as a result of burst activity in the γ system. Simultaneously, the deep tendon reflexes are hyperactive. When this maneuver is performed in parkinsonians, rigidity is intensified (Cooper, 1969b), but no increased tendon reflexes are observed (Hassler, 1972b).

Nevertheless, it is impossible to explain the pathogenesis of rigidity on the basis of elevated stretch reflexes, especially since the cause of this increase still remains obscure. The assertion that extrapyramidal hypertonia is based on dysfunction of myotatic reflexes is without a doubt valid, but this statement does not explain its pathogenesis.

The classical experiment by Sherrington (1898) involving transection of the brainstem in cats not only revealed a previously unknown type of rigidity, which the author called decerebrate rigidity, but also showed the prominent role of supraspinal structures in the regulation of muscular tone. In decerebrate rigidity, tone increases in the flexors but more markedly in the extensors. Deafferentation of rigid muscles by section of the posterior roots immediately eliminates rigidity, which seems to indicate that it is caused by an increased proprioceptive reflex, the receptor part of which is in the muscle itself. At the same time, it has been known for a long time that extrapyramidal rigidity differs substantially from decerebrate rigidity. In view of this, two types of rigidity have been defined. The first, α rigidity or "nonreflex rigidity," results from hyperactivity of α motoneurons and is independent of the γ system, since it does not disappear after posterior rhizotomy. It is assumed that α rigidity is the result of a disturbed balance between stimulating and inhibiting impulses from supraspinal centers to α motoneurons in favor of the stimulating input. An experimental model of α rigidity is decerebration, which results from ligation of all four arteries supplying the brain.

The second type of rigidity is reflex or γ rigidity or "plastic muscular tone," the classical model of which is decerebrate rigidity (Hassler, 1972b). Two other kinds of rigidity are also recognized: resting and activated (Webster, 1972). The author considers the former to be a function of Put and GP and the latter to be related to VL of the thalamus.

Steg (1964) considers extrapyramidal rigidity to be based on a primary increase in excitability of α motoneurons with an inactive γ system. It is noteworthy, however, that this point of view about the presence of α rigidity is not confirmed by the finding that excitability of spinal motoneurons in parkinsonism, as determined by the H reflex, is normal or even reduced (Bathien and Rondot, 1977).

The leading role of the γ system in the pathogenesis of extrapyramidal rigidity is acknowledged by most but not all investigators. Some authors consider rigidity to result from elevated γ activity (Asai *et al.*, 1960; Rushworth, 1960; Shimazu *et al.*, 1962). It has been established that injecting a muscle with procaine to cause a selective blockade of γ efferents leads to the disappearance of extrapyramidal rigidity. Other authors argue that in ridigity, unlike spasticity, there is hyporeactivity (inhibition) of the γ system (Steg, 1964; Hassler, 1966a) or an upset α–γ balance (Stern and Ward, 1962). Struppler (1973) is of the opinion that the disappearance of decerebrate rigidity after section of the posterior roots still does not offer absolute evidence indicating that rigidity is mediated exclusively through the γ loop. Since there are no objective quantitative methods for investigating the γ system in man, the solution of this problem is a matter for the future.

The activity of the γ system is regulated by an elaborate complex of cerebral structures that have both stimulating and inhibiting influences on it. There are different points of view regarding the influence of VL on the γ system. Certain authors believe that this nucleus activates it, but others consider its influence to be inhibitory.

A number of well-known clinical peculiarities distinguish extrapyramidal rigidity from pyramidal spasticity. In rigidity, muscular tone is constantly and uniformally elevated in both flexors and extensors, whereas in spasticity the tone in these muscles varies. In advanced parkinsonism, rigidity noticeably predominates in the flexors, producing a typical bent posture of the torso and upper limbs. Resistance to passive movements in rigidity is of a typical plastic or viscous nature. In rigidity, unlike spasticity, the cogwheel phenomenon is often present (Section 3.2), tendon reflexes are normal or reduced, and neither pathological reflexes nor clonus can be elicited. The "clasp-knife" phenomenon is typical of spasticity but absent in rigidity.

After amobarbital is injected into one carotid artery (Wada test), a transitory hemiplegia develops on the opposite side in parkinsonians, and rigidity disappears completely. In hemiplegic patients, spasticity does not change (Obrador *et al.*, 1961). This confirms that the main role in the genesis of rigidity is played by activating influences from the supraspinal level, whereas spasticity is a disinhibition syndrome of spinal genesis. Like tremor, rigidity disappears completely during sleep. Moreover, in approximately one-third of parkinsonians, in addition to pronounced rigidity, there are some signs of spasticity (Safronov, 1979). The combination of rigidity and spasticity (rigidospasticity) is quite typical of cerebral palsy (Chapter 9).

In his review, Hassler (1972b) proposes the following scheme for the pathogenesis of rigidity. Destruction of SN neurons leads to the "removal" of the dopaminergic inhibiting influence of Str on GP, which begins to have a facilitating influence on stretch reflexes. This influence comes in two ways: descending via FR of the midbrain and ascending via Th nuclei to the motor zone (6aα) of the cerebral cortex. Thus, rigidity in parkinsonism results from the loss of two factors: (1) the inhibiting influence of the nigrostrial system on the efferent neurons of GP and (2) the inhibiting influence of the nigrospinal system on interneurons and tonic stretch reflexes. As a result, these reflexes receive only the activating influence from three sources: (1) from the corticospinal pathway coming from cortical field 6aα, (2) from the pallidoreticulospinal pathway, and (3) from a reticulospinal pathway originating in the pontine reticular nucleus.

Our investigations of ridigity in parkinsonism (Aizerman *et al.*, 1974; Kandel, 1981) have shown that in passive movements under experimental conditions (Section 3.1.1), there are two changes in the EMG of the muscles involved. First, in the stretched muscle during a passive movement, irrespective of its previous electroactivity, there is continuous activity that intensifies as the speed of the passive movements increases. This activity levels off the bursts in the EMG at rest (resting tremor). The activity of the antagonistic muscle decreases sharply. The second type of change is the appearance of burst activity, the frequency of which increases somewhat as the speed of passive movements accelerates. Changes in the activity of the antagonist muscle are the same as in the first type. It is noteworthy that the first type of change in EMG is observed only in the early stage of the disease, whereas in severe cases, both types of change may occur, but in the second type, burst activity appears at comparatively slow rates of passive movements.

Therefore, increased electroactivity of the stretched muscles, irrespective of the nature of the activity, is a pathognomonic sign of ridigity, since this phenomenon is never found in healthy subjects. It indicates hyperactivity of tonic stretch reflexes. Further, general elevated muscle activity according to EMG data is seen in many (but not all) cases. The nature of the changes in activity during passive movements, whether continuous or burst activity, depends primarily on the speed of the movements. If, in the presence of tremor at any frequency, the bursts of antagonist muscles occur at irregular intervals and with some shift in phase, the muscle tension increases, and the patient's feeling of stiffness intensifies.

The biochemical aspects of the pathogenesis of rigidity are presented in other sections of this book, particularly in the section dealing with the pathogenesis of parkinsonism (Chapter 6, Section 6).

Cogwheel Rigidity

It has long been known that one of the typical signs of parkinsonism is the so-called cogwheel phenomenon, which can be felt by the physician during passive movement of a patient's limb, e.g., during flexing and extending of the arm at the elbow or wrist joint. When one attempts to flex the limb, the muscles yield jerkily, so that the examiner has the feeling of cogwheels moving on one another. It is generally acknowledged that the "cogwheel" sensation is the result of parkinsonian tremor "overlapping" muscular rigidity. However, certain authors doubt the validity of this view, since in some parkinsonians the cogwheel phenomenon occurs even in the absence of a typical resting tremor, even on EMG (Lance *et al.*, 1963). From clinical observations it is apparent that in the rigid form of parkinsonism the cogwheel phenomenon is observed considerably more frequently than in the tremor type. Moreover, the cogwheel phenomenon is always more pronounced in the wrist and elbow joints, and usually more in the arms than in the legs.

The cogwheel phenomenon is seen on EMG as relatively regular bursts occurring at a frequency from 6 to 9.5 Hz (Findley *et al.*, 1981). It seems that the frequency of the cogwheel phenomenon does not depend on the speed of passive movements in the joint (Lance *et al.*, 1963). In some cases there was no "overlapping" of tremor bursts on cogwheel bursts in EMG.

The cogwheel phenomenon has a spectrum of frequencies from rest tremor frequency to action tremor frequency. In view of this, Lance *et al.* (1963) feel that the cogwheel frequency depends on the "dominance" of one of these types of tremor. If tremor at rest is pronounced, then it "imposes" its rhythm on the cogwheel rigidity, and if action tremor is predominant, its frequency is imposed.

Our investigations (Aizerman *et al.*, 1974; Kandel, 1981) have shed new light on the mechanism. We showed that cogwheel rigidity most frequently occurs in rigidity as a reaction of stretching muscles by passive movements. During such movements there is one of two types of EMG activity, either continuous or burst, depending on the speed of the passive movements. The cogwheel phenomenon occurs only in the presence of burst activity, usually when the bursts have a frequency of approximately 5 Hz, since at a low frequency, the amplitude of the bursts is greatest. It is precisely in such cases that the physician's hand producing passive movements feels jerky resistance. If the amplitude of the bursts is small, these jerks cannot be felt, although the EMG quite clearly shows burst activity in the stretched muscles.

Since in pronounced rigidity there is a considerable increase in the "critical speed" of passive movements at which burst activity becomes continuous, the cogwheel phenomenon in these cases is more pronounced. Thus, cogwheel rigidity, like tremor, is caused by rhythmic bursts of activity in antagonist muscles as well as by decreasing continuous activity during alternative passive movements.

4. Spasticity

The increased muscle tone called spasticity is a well-studied neurological phenomenon that has caused great functional disability in patients. Like rigidity, spasticity in the strict sense is defined as increased muscle tone observed during passive movements. The components of spasticity are well known and distinguish it from extrapyramidal rigidity: "elastic" elevation of tone, "clasp-knife" phenomenon, synergia, synkinesia, spinal automatism, increased proprioceptive and decreased exteroceptive reflexes, the appearance of pathological reflexes, clonus, etc. On passive movement of a spastic limb, muscular resistance gradually increases, depending on the speed of the movement.

Spasticity is usually combined with reduced muscular strength in different ways (mono-, hemi-, and parapareses and -plegias). In the muscles opposing gravity, spasticity is usually more pronounced than in other muscles, e.g., greater in the flexors of upper limbs and extensors of lower limbs. This predominant localization of spasticity is manifest in the classic posture of hemiplegic patients. Spasticity is observed in a large number of neurological diseases—Bauer (1972) enumerates approximately 80 such diseases. Among the most common are cerebrovascular stroke, multiple sclerosis, craniocerebral trauma, cerebral palsy, and brain tumors.

Spasticity may be defined as a motor disorder resulting from disinhibition of the stretch reflex after an upper motoneuron lesion. Thus, spasticity is a disinhibition syndrome caused by disorders in the inhibitory mechanism of motor control.

Spasticity is most often the result of damage to the pyramidal tract (upper motoneuron syndrome) anywhere along its length from the cortex to the anterior horns of the spinal cord. The pathophysiological mechanisms of spasticity are very complicated. One of the main factors is hyperactivity of the γ system and motor units of the spinal cord as a result of inhibitory supraspinal influences (Granit *et al.*, 1957). Direct evidence of this can be seen in the disappearance of spasticity after section of the posterior roots or anesthetic block of γ fibers, but one cannot explain all aspects of the spasticity syndrome by this factor alone.

The second leading pathogenetic factor is a lowered threshold of excitation and hyperactivity of both α motoneurons and tonic stretch reflexes (Landau and Clare, 1964; Hassler, 1972b). This leads to the development of spasticity even in the absence of hyperactivity in the γ system (Bauer, 1972). One of the peculiarities of spasticity is the summation of excitation in the stretch reflex arc during movement (Herman *et al.*, 1974). Increased excitability of the stretch reflex in one muscle causes a similar increase in the antagonist muscle, which intensifies spasticity and impairs motor function through faulty reciprocal inhibition. The inhibitory influence of interneurons on the segmental motor apparatus is markedly diminished in spasticity. The typical predominance of spasticity in extensors is explained by the fact that they are innervated by tonic α motoneurons, which to a greater extent than phasic neurons are controlled by interneurons (Granit *et al.*, 1957). The inhibitory influence of these interneurons on afferents of the γ system disappears in spasticity.

In addition, Pierrot-Deseilligny (1983) singles out several other neurophysiological disorders associated with the development of spasticity: elevated polysynaptic excitability Ia, diminished autogenic inhibition Ib from Golgi tendon organs, decreased inhibition on the part of group II muscle afferents, reduced reciprocal inhibition Ia, decreased reverse inhibition, and reduced presynaptic inhibition of Ia terminals.

It has also been demonstrated that important roles in the pathogenesis of spasticity are played not only by lesions of the pyramidal tract but also by dysfunction of FR of the brainstem as well as many extrapyramidal and afferent systems. Fraioli and Guidetti (1977) distinguish two forms of spasticity: tonic and phasic. The former is characterized by intensification of postural stretch reflexes but not in voluntary movements. In phasic spasticity, stretch reflexes are also elevated, and spasticity increases greatly during movements. It is common knowledge that all types of spasticity are aggravated by any external irritant (light, sound, talking, etc.).

In the study of the diagnosis and pathogenesis of spasticity, it is important to examine the H reflex. Eliciting two H reflexes (H_1 and H_2) by a special technique enables one to distinguish clearly spasticity from rigidity (Ioku *et al.*, 1971).

In certain brain disorders, especially cerebral palsy, a combination of rigidity and spasticity is observed. This type of neurological syndrome is called "rigidospasticity." The author's investigations of the mechanisms of rigidospasticity and its dynamics after stereotactic dentatotomy are presented in Chapter 9.

Problems concerning the pathogenesis and therapy of spasticity were extensively discussed at the special international symposium in Vienna in 1971. The papers of that symposium (Birkmayer, 1972) contain much useful information on this problem.

The surgical management of spasticity in spinal cord lesions is discussed in Chapter 19.

5. Akinesia

Akinesia (bradykinesia) is the third (after tremor and rigidity) main symptom of parkinsonism and other extrapyramidal disorders. The majority of writers link akinesia with impaired mechanism for initiation of movement. Poirier (1972) defines akinesia as "a psychomotor disorder characterized by difficulties in initiating or executing a purposeful movement in the absence of paralysis." Hassler *et al.* (1979) consider akinesia to be manifested by the loss of involuntary, automatic, and expressive movements. The definition of the term "akinesia" proposed by the International Classification of Extrapyramidal Disorders (Section 1) is ". . . a disorder characterized by poverty and slowness of initiation and execution of willed and associated movements and difficulty in changing one motor pattern to another, in the absence of paralysis." Markham *et al.* (1966) include in the concept of akinesia not only difficulty in beginning movement and reduced speed and amplitude of movements but also the inability by tactile, proprioceptive, or visual sensations to recognize impaired postures caused by gravitational influences, and "brief paralysis" of movements often caused by external factors, in particular, sound.

Clinically, akinesia is manifested by a pronounced slowness and poverty of all voluntary movements, poor arm swinging, masking of face, "freezing" gait, and especially great difficulty in initiating a movement (impaired "starting function"). The pure akinetic syndrome without tremor and rigidity is encountered in not more than 8–10% of parkinsonians (Kandel, 1965). According to Barbeau (1972), the most frequent clinical manifestations of akinesia are prolonged initiation time in movements, defects in changes of the motor pattern, and a diminution of various associative movements. Barbeau characterized akinesia as "the most damaging of all the symptoms of parkinsonism."

An interesting phenomenon in the presence of akinesia is an especially marked deficit when patients are required to simultaneously perform two different movements with the same arm. Movement time for this complex task is much longer than for a simple movement (Wiesendanger, 1986).

Outwardly, akinesia seems to be the result of rigidity; however, it has long been noted that pronounced akinesia may occur when rigidity is insignificant or even absent. The rich experience in surgery for parkinsonism has demonstrated that after a stereotactic operation on the basal ganglia, rigidity, just like tremor, as a rule disappears completely, whereas akinesia usually remains or diminishes only a little. At the same time, pronounced rigidity may mask akinesia, which becomes clearly manifested after a stereotactic operation, when the rigidity disappears. Thus, these phenomena are independent of each other and have different pathophysiological mechanisms. At the same time, Hassler *et al.* (1979) consider that akinesia, like rigidity, is caused by loss of neurons in SN. Spiegel and Wycis (1967) noted the connection of akinesia with Hypoth disorder.

The pathogenesis of akinesia still remains obscure. Akinesia, in all likelihood, reflects a more generalized and severe lesion of the dopaminergic and other metabolic brain systems than that which induces tremor and rigidity. This is confirmed by the development of pronounced akinesia in mental patients after long-term use of phenothiazines, which block dopaminergic receptors. A direct correlation between severity of akinesia and the degree of destruction of dopaminergic neurons in SN, as well as with catecholamine deficit in Str, has been established (Hornykiewicz, 1976). In clinical investigations of akinesia, a sharp reduction in the concentration of dopamine and homovanillic acid in the brain and in CSF was discovered (Barbeau, 1972).

A distinct arrest reaction of all movements occurs in experimental lesion of NC. In animal experiments, pronounced akinesia may be induced by damaging the base at the upper brainstem and posterior Hypoth or medial Th as well as by injection of a cholinergic substance into Str.

Schwab *et al.* (1959) designed a special apparatus for the measurement of akinesia in parkinsonism. In the akinetic form of the disease, the stretch reflexes are inhibited, and γ rigidity is either absent or mildly pronounced, which is typical of this form of the disease. In akinesia the greatest changes in muscle tone were found in the performance of slow passive movements of a joint (100 sec and more). This is evidence that in akinesia there is hyperactivation of the static γ system (Safronov, 1974). Akinesia and tremor are in an "antagonistic relationship"—pronounced akinesia reduces not only the ability to execute voluntary movements but tremor as well (Molina-Negro and Hardy, 1971).

Our investigations (Aizerman *et al.*, 1974; Kandel, 1981) have shown that during passive movements in patients with the akinetic form of parkinsonism there is no inhibition of the activity of antagonists such as that observed normally. This impaired reciprocal innervation reduces the efforts of the muscles, making rapid movements impossible.

In the genesis of akinesia, an important role is played by damage to the afferent (sensory) part of the motor system, which leads to a "breakdown" of the mechanism that "switches it on." One may consider this "breakdown" to be of a functional rather than an organic nature. This is supported by clinical observations showing that under the influence of certain external factors, pronounced akinesia may suddenly disappear, usually for several minutes or even hours. This difficult-to-explain phenomenon was long ago described in neurological literature as "paradoxical kinesia."

L-Dopa significantly improves the clinical manifestations of akinesia, whereas stereotactic operations on the basal ganglia are not very effective (Chapter 6, Section 12).

3

Stereotactic Method

Science advances in spurts, depending on the achievements of methodology. With each new step in methodology, we seem to rise one step higher, from which a broader horizon opens up before us, revealing hitherto unseen objects.

I. P. Pavlov

1. Historical Note

A number of devices (electrode scalpels, myelotomes) for ensuring the relatively accurate introduction of instruments into desired brain and spinal cord structures of experimental animals were described in the second half of the last century. The first such devices (Dittmar, 1873; Woroschiloff, 1874) consisted of a simple guide and a miniature scalpel to make lesions in the motor and sensory pathways of the medulla oblongata and spinal cord of a rabbit.

Two British scientists, Horsley and Clarke, deserve the credit for introducing the stereotactic technique to neurophysiological experiments. In 1906 they devised an apparatus for stereotactic operations on animals, founded on basic anatomic principles. These investigators proposed the term "stereotaxis," which subsequently became firmly established in neurophysiology and neurosurgery (Clarke and Horsley, 1906).

However, 17 years before the publication by these authors, both Russian and European literature carried reports by the Professor of Anatomy at Moscow University, D. N. Zernov, who built a stereotactic apparatus, which he called an encephalometer, designed for both anatomic investigations and neurosurgical operations on the human brain. This apparatus was demonstrated on March 22, 1889 at a session of the Physico-Mathematical Society of Moscow University (Zernov, 1889) and described in the *Revue genérale de clinique et thérapeutique* and in *Manual on Descriptive Human Anatomy*. A year later, in a booklet published in both Russian and French, Zernov (Fig. 32) gave a detailed description of the encephalometer and its basic principles (Kandel and Schavinsky, 1972; Kandel, 1986).

The original apparatus of Zernov may be considered the prototype of a number of modern stereotactic devices. This instrument, made of aluminum, consisted of the following parts:

1. A basal ring (cr), which was fastened on the skull in the horizontal plane passing through nasion and inion.
2. An equator (e) fixed on the basal ring perpendicularly and graduated so that 0° lies above the sagittal suture of the skull.
3. A meridian (m), situated in the sagittal plane, which may be moved along the equator of the hemisphere.

The meridian was set at a certain number of degrees by adjusting a coupling on the meridian arch. The latter was divided into degrees, which are measured with the aid of two couplings on the arch. These couplings were radial with respect to the encephalometer sphere.

The device was attached to the skull in a fixed position with the aid of five bars. One was a plate applied to the upper edges of the orbits; another bar covered with solid rubber was fixed on the inion. Because of the variability of the form of the occipital bone, this bar was attached by a hinge. Two bars were introduced into the auditory meati to ensure fixation of the entire apparatus to the base of the skull. Since the auditory meati are not always symmetrical, the posi-

D. N. Zernov (1876–1946)

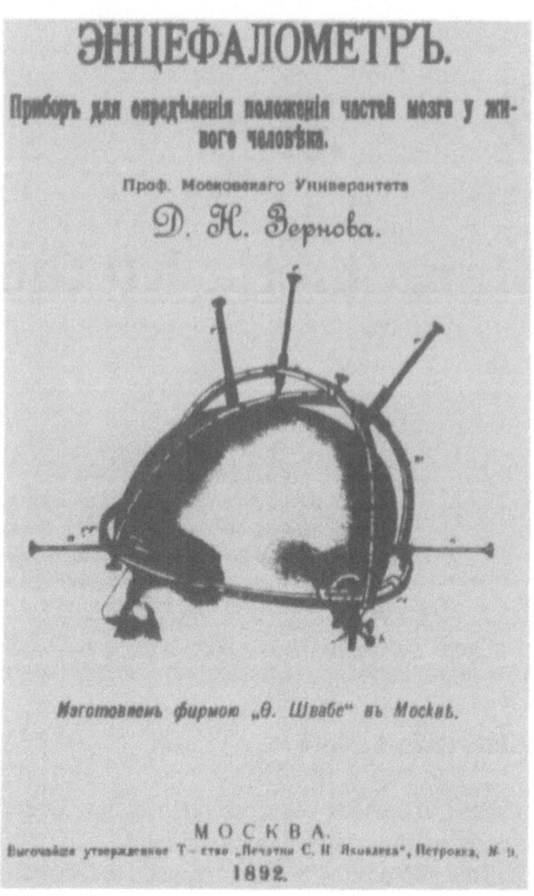

ЭНЦЕФАЛОМЕТРЪ.

Приборъ для опредѣленія положенія частей мозга у живого человѣка.

Проф. Московскаго Университета

Д. Н. Зернова.

Изготовленъ фирмою „Ѳ. Швабе" въ Москвѣ.

МОСКВА.

Высочайше утвержденное Т-ство „Печатни С. П. Яковлева", Петровка, № 9.

1892.

FIGURE 32. The booklet by D. N. Zernov with a description of the first stereotactic apparatus in the world.

tions of the ear bars were adjustable. The fifth, a midline bar, was fixed to the parietal part of the sagittal suture, thus stabilizing the whole apparatus.

In order to have the base of the apparatus midway between the nasion and the inion, the latter distance was measured with dividers. This value was subtracted from the known diameter of the meridian arch, and the remainder was divided by two to obtain the distance of the graduated pins of the frontal and occipital bars. Then the basal ring of the encephalometer was fixed.

To identify any point on the surface of the skull or the brain, the meridian was set up in such a manner that it passed over this point, and the tip of one of the pointers was set at this point. The spatial localization of the point was expressed in polar coordinates, i.e., the longitude was determined by the divisions on the equator, and the divisions on the meridian were used to determine the latitude of the point; then, based on these data, the target was found on the encephalometric map. This map was compiled by Zernov on the basis of anatomic investigations with the aid of the encephalometer and illustrated the mean positions of the cranial sutures and brain fissures as well as the range of their individual variability. Just like a geographic map, it had a graduated grid for locating the position of any point on the surface of the skull and brain. For determining any point on the brain surface, its coordinates were first found on the encephalometric map and then transferred using the meridian arch and the pointer to locate the site corresponding to the situation on the map. After that, the device was fixed to the skull, the pointer lowered to make contact with the bone, and the point thus obtained marked with ink.

If it was desired to determine what part of the brain surface corresponded to an injury or defect of the skull, the procedure was followed in reverse: the encephalometer was fastened on the head, after which the meridian and the pointer were set so that the pointer tip touched the border of the target area. Then the latitude and longitude of that point were noted on the divisions of the device, and these coordinates were transferred to the map. After several points on the margin of the target were plotted, lines were drawn joining them to produce an accurate projection of the area in question.

The encephalometer could also be used for determining the curvature of the skull. For this purpose the pointer had divisions that indicated the distance from the sphere of the apparatus. The resultant figures were plotted on a special diagram on which the external semicircle was equal to the radius of the sphere of the apparatus, with corresponding arcs each having a radius 2 mm shorter that the previous one. The points were plotted on each of the semicircles and joined by a line that showed the curvature of the skull.

The Zernov stereotactic device was first to employ a system of polar coordinates for determining the spatial localization not only of surface but of certain deep brain structures.

The encephalometer was employed successfully in a surgical clinic in 1889, as later reported by Zernov's pupil, Altukhov (1891). A middle-aged patient in very grave condition as the result of a skull fracture was admitted to Yauzki Hospital. The patient had developed Jacksonian epilepsy. The distinguished neurologist, Minor, proposed that a craniotomy be made exposing the left sulcus of Rolando. Prof. Zernov determined its localization using the encephalometer. A craniotomy was performed at this site, the dura mater incised, and a cerebral abscess exposed, from which a considerable quantity of pus was drained.

The encephalometer was used a second time in the same hospital in the autumn of 1889. After a cranial injury, a patient developed a right-sided hemiplegia with motor aphasia. It was decided to perform a craniotomy over the left sulcus of Rolando, the site of which was determined using the encephalometer. After incision of the dura mater, no changes on the brain surface were found, and the wound was closed. At autopsy an abscess the size of a walnut was found in the brain substance below the bone flap.

In 1890, a 13-year-old girl with Jacksonian epilepsy was admitted to the same hospital. She had had a craniocerebral injury and had a bony defect covered with soft tissues. Altukhov, with the help of the encephalometer, determined the cortical areas within the boundaries of the bony defect and transferred the contour to the encephalometric map for children compiled by the author. The boundaries of the defect on the map corresponded to the second left frontal gyrus. An operation performed by the prominent general surgeon Bobrov confirmed the accuracy of the localization indicated by the Zernov device. After surgical removal of the brain scar, the patient was discharged with significant improvement.

Prof. Zernov's pupil, Altukhov, carried out interesting investigations, which were described in his work *Encephalometric Investigations of the Brain in Accordance with Sex, Age and Cranial Index* (Altukhov, 1891). On the basis of anatomic studies conducted with Zernov's apparatus, Altukhov compiled detailed encephalometric maps for brachy- and dolichocephalic individuals, for men, women, and children, as well as a map of the mean positions of gyri and the main basal ganglia ("gray nodes") in dolichocephalics and brachycephalics (Fig. 33). Comparing his findings with the data of earlier investigators, Althukhov pointed out the advantages of the method he employed: first, in each case the conditions were identical (the position of the head in the sphere of the apparatus was always the same); second, in determining the position of the sulci, not only the distance to the nearest bone suture or some other reference point but also their relative positions with respect to each other were taken into account; third, any position of the pointer, as well as any point on the surface of the skull and brain, could be accurately designated in terms of stereotactic polar coordinates.

With the help of the encephalometer, Zernov and Altukhov determined the spatial position of the main basal ganglia, their projection on the cerebral cortex, as well as their dimensions in dolichocephalic and brachycephalic persons. In the former these nuclei are situated closer to the midline, and in the "flat sphere" (horizontal section of the brain) they have an elliptical shape, whereas in brachycephalics this shape is more circular. Tables were compiled on the mean positions of the basal ganglia of the brain. These authors admitted the need for further, more detailed investigations of the deep structures in the brain.

Altukhov (1891) wrote:

> At the moment, it is difficult to assess the practical significance of the data, but in my opinion one may be content with the fact that the conclusions drawn are unique, since the results of the encephalometric investigations are not arbitrary; i.e., they refer not only to a given skull but are always related to one and the same constant—the sphere of the encephalometer.

A most amazing prediction about the future development of stereotactic neurosurgery is reflected in the words of Altukhov that ". . . the encephalometer will, in the not-too-distant future, occupy a leading place among the most important and necessary instruments for surgeons."

In the years that followed, the encephalometer was used successfully in neurosurgical operations in several clinics in Moscow. In 1907 Prof. Rossolimo wrote: ". . . in our clinic not a single operation was performed without obtaining the best of results from the encephalometric method of Zernov." Rossolimo

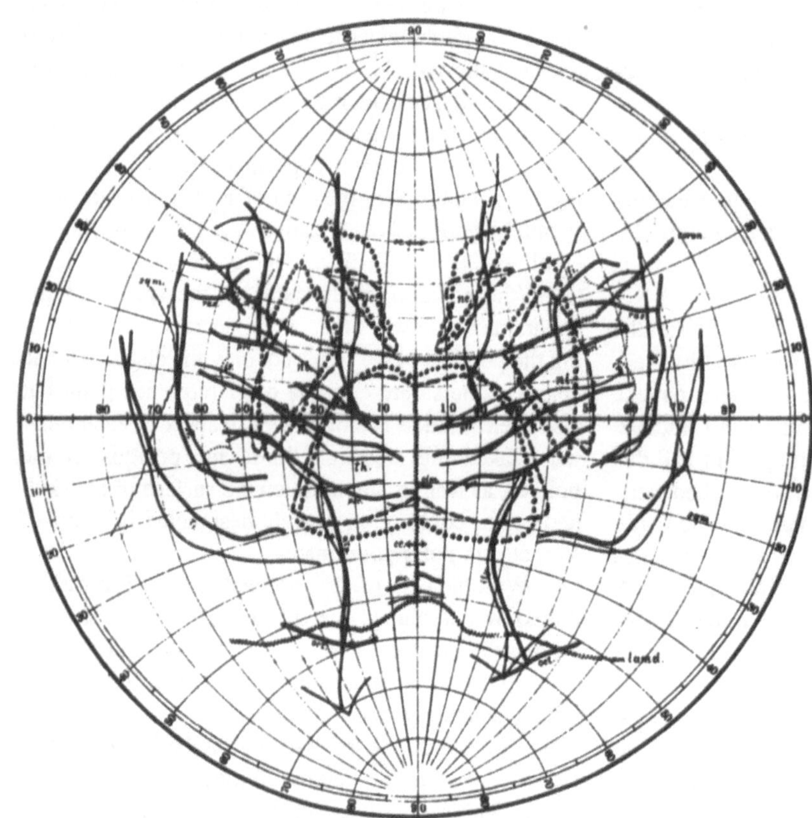

FIGURE 33. Map of the variability and average topography of cortical sulci and basal ganglia in dolichocephalic and brachycephalic heads (Altukhov, 1891).

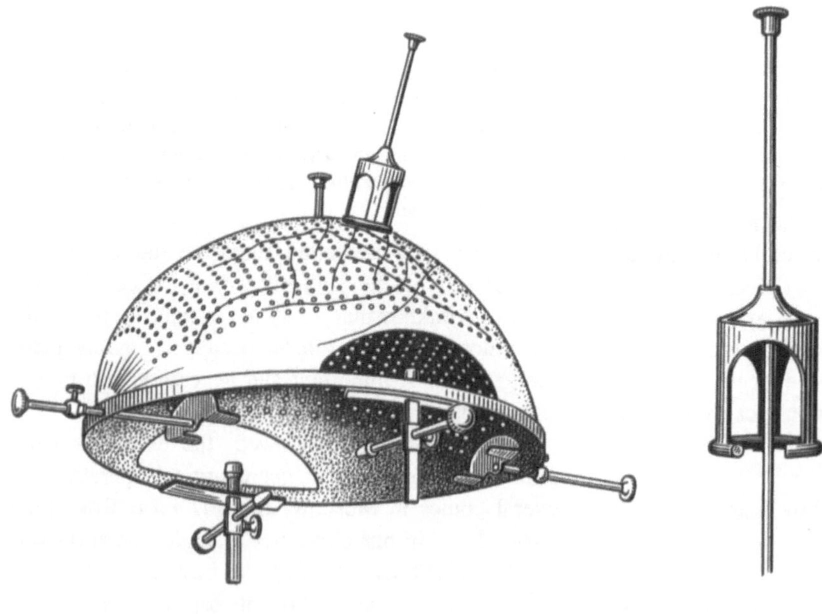

FIGURE 34. "Brain topograph": stereotactic apparatus invented by the Russian neurologist G. Ĭ. Rossolimo (1907).

modified the encephalometer and created a new device, which he named the "brain topograph" (Fig. 34). To the basal ring of Zernov's apparatus, Rossolimo attached an aluminium hemisphere on the surface of which there was a map of the cerebral sulci and gyri as well as the subcortical structures. Parallels and meridians were marked every 10°. On each square of this map there were from six to nine holes, depending on the width and interval between meridians. In order to determine the coordinates of some point on the brain surface, a hinged metallic pin tipped with an indelible pencil could be passed through each hole until it came into contact with the surface of the skull, thus marking the point. Such a brain topogram made it possible quickly to carry out encephalometric investigations not only in respect to sulci and gyri in the cerebral cortex but also relative to subcortical structures projecting on the cortex.

Without belittling the achievement of Horsley and Clarke in developing the stereotactic method, it would be fitting to emphasize that Zernov, Altukhov, and Rossolimo were its pioneers. They created the first stereotactic devices, were the first to emphasize the

Sir Victor Horsley (1856–1916)

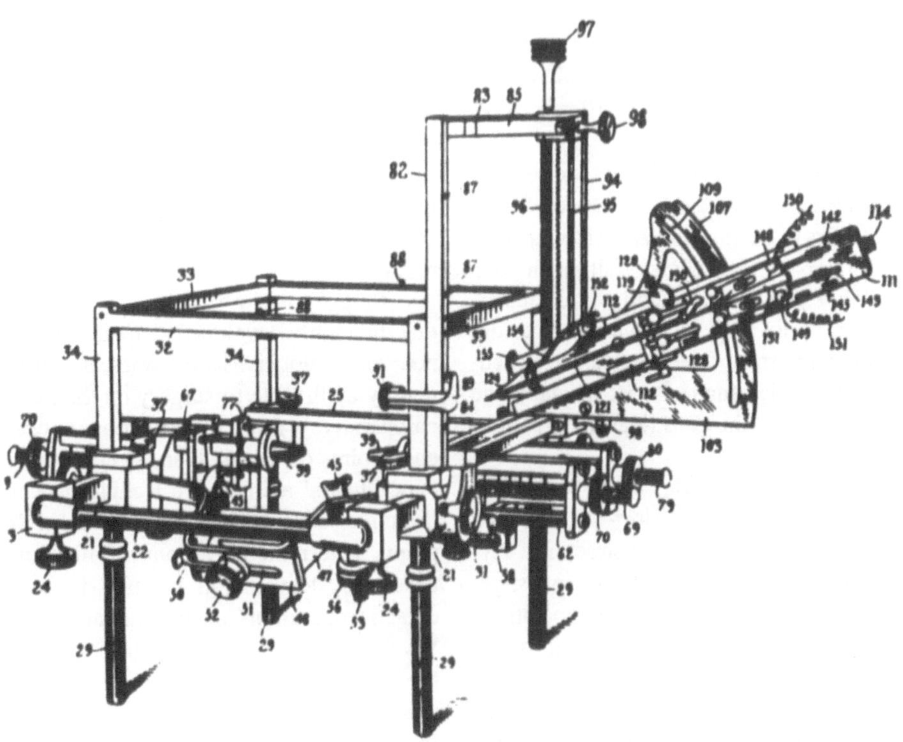

FIGURE 35. Original stereotactic apparatus invented by V. Horsley and R. N. Clarke for animal experiments (patent 22,455 of 1912).

great significance of this method for advancing neurosurgery, and successfully demonstrated its value by performing several operations on the human brain.

In 1906 the distinguished British neurosurgeon Sir Victor Horsley and his associate R. H. Clarke built the first stereotactic apparatus for neurophysiological experiments (Fig. 35). This apparatus was based on a three-dimensional system of coordinates—three planes perpendicular to each other in space, which were referred to external points on the animal skull. The frontal plane passed through the centers of the external auditory meati, and the horizontal (base line) through the same centers and the lower borders of the orbits, and the sagittal plane through the midplane of the skull. The point where these three planes intersected was arbitrarily considered the zero point.

The Horsley and Clarke stereotactic apparatus was attached to the head of the animal with five bars—two introduced into the external auditory meati, two in the lower borders of the orbits, and one an occipital bar.

The authors were the first to call this experimental method "stereotactic." The Zernov apparatus was apparently unknown to Horsley and Clarke, since they make no mention of it. Horsley and Clarke compiled the first stereotactic atlas of the brain of an experimental animal, and on that basis, they carried out a number of interesting investigations. In cats and monkeys, the zero horizontal plane was a plane passing 10 mm above the basal plane.

As so often happens to scientific discoveries that are considerably ahead of the science practice of their day, the unique works of Zernov and Altukhov, as well as the works of Horsley and Clarke (although to a lesser degree), went practically unnoticed and unappreciated for a long time. Only two decades later did the stereotactic method firmly enter the practice of neurophysiological laboratories and prove its special value in studying the functions of the CNS (Hess, 1932; Ranson and Magoun, 1933).

Many physiological laboratories in the world used numerous modifications of the Horsley–Clarke apparatus for intensive studies of subcortical structures, Hypoth, Cer, and brainstem. For the first time investigations with implanted electrodes to record biopotentials, stimulate, and focally destroy various deep brain structures were possible.

In 1918, Mussen proposed a stereotactic instrument, but it was never used in humans. In 1936, the German surgeon, Kirschner, constructed a stereotactic device for exact introduction of a needle into the Gasserian ganglion and its electrocoagulation in trigeminal neuralgia (Chapter 13, Section 7). In 1947, the neurologist Ernst Spiegel and neurosurgeon Henry Wycis in the United States performed the first modern stereotactic operation on deep structures of the human brain.

Ernst Spiegel (1895–1985)

Spiegel and Wycis were the first to demonstrate that the highly developed stereotactic technique used in neurophysiological laboratories at that time was quite readily applicable to surgery on the human brain. The comparatively reliable correlations between bone reference points on the skull and deep cerebral structures in experimental animals are lacking in man. Hence, the principle evolved that only intracerebral reference points determined by contrast medium in the ventricular system should be used for performing stereotactic operations on the human brain. Spiegel and Wycis deserve indisputable priority and lasting credit for elaborating the main prerequisites for the new method: the first modern stereotactic apparatus, the first stereotactic atlas of the human brain, and the first demonstration of the use of this method in the clinical treatment of many disorders of the CNS.

Henry Wycis (1911–1972)

Following Spiegel and Wycis, stereotactic operations on the basal ganglia soon began to be performed by other neurosurgeons—Leksell in Sweden, Riechert, Mundinger, and Umbach in the Federal Republic of Germany, Talairach, Guiot, Gros, and Pecker in France, Walker, Cooper, Andy, Tasker, Nashold, and Mark in the United States, Narabayashi and Sano in Japan, Gillingham and Walsh in Great Britain, Bertrand and Hardy in Canada, Krayenbühl and Siegfried in Switzerland, Obrador in Spain, Laitinen in Finland, Nadvornik in Czechoslovakia, Mempel in Poland, Kandel in the USSR, and others.

The history of stereotaxis of the spinal cord should be mentioned briefly. The prototype of the first stereotactic apparatus for experiments on the spinal cord of animals was built in the last century (1874) by the Russian physiologist M. Woroschiloff, who had worked in Ludwig's laboratory in Leipzig. This apparatus, which he called a "myelotome," was used for studying various pathways of the spinal cord in rabbits by their selective destruction.

The further historical development of the stereotactic method in neurosurgery is presented in Chapter 6.

2. General Principles

The stereotactic method or "stereotaxis" (from the Greek *stereos*, three-dimensional or spatial, and *taxis*, arranged, ordered) is a complex of techniques requiring special instrumentation and x-ray or CT and functional controls and using a system of spatial coordinates and special calculations to make it possible to introduce accurately a thin instrument (cannula, electrode) into a predetermined deep-seated structure of the brain or spinal cord in order to make a therapeutic lesion or take a biopsy. The essence of the stereotactic technique is to locate a cerebral target on the basis of an arbitrary system of coordinates incorporated in the stereotactic device.

Surgical stereotaxis is based on the calculation of precise spatial relationships between any structure deep in the brain and a number of points or landmarks that serve as intracerebral or (to a considerably lesser extent) cranial reference structures. Based on such calculations, it is possible stereotactically to reach surgically any structure situated in practically any area of the brain or spinal cord.

From the theoretical point of view, the determination of the center of a given deep-seated brain structure boils down to identifying its position in space. From analytical geometry, it is known that the position of a point may be determined with the aid of the Cartesian* system of rectangular or orthogonomal coordinates, reciprocally perpendicular planes (Fig. 36), or polar coordinates. In the rectangular system there are three coordinate planes (*XOY*, *YOZ*, and *ZOY*), each of which divides space into two parts. The lines of intersection of the coordinate planes form three axes of the coordinate system, abscissa, ordinate, and applicata, all of which intersect in one point called zero or the origin. The majority of countries have adopted the so-called European or right-handed system for the location of coordinate planes: the *X* axis is directed to the left from the origin, the *Y* axis forward (i.e., towards the person looking at the system), and the *Z* axis upwards. For the observer who is on axis *OZ*, movement of semiaxis *OX* to overlie semiaxis *OY* by the shortest route is counterclockwise, and the coordinate system is called right-handed. (This movement is clockwise in a left-handed system.) If all three coordinate axes are extended in the opposite directions, they form a system of three mutually perpendicular planes defining eight octants of space.

*Cartesius: Latinized name of the great French philosopher and mathematician René Descartes (1596–1650).

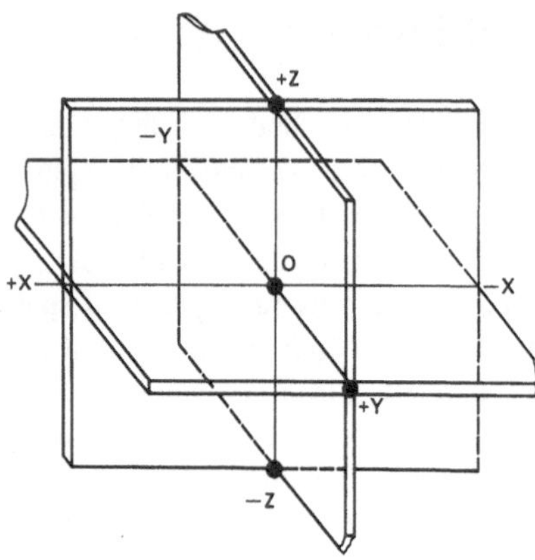

FIGURE 36. Complete system of coordinate axes and planes.

Different coordinates represent each point in space, so any three rectangular coordinates designate a point in space. The coordinates of any point inside the rectangular system are determined by its distances from the three coordinate planes, i.e., the lengths of the perpendiculars drawn from this point to the three planes. For points located on a coordinate plane, one of the three coordinates is zero.

In the polar or spherical (or radial) system of coordinates, the position of a point in space with respect to the coordinate axes is determined by its angular distances or polar coordinates. These coordinates are defined by three values: the length of the radius vector connecting the point with the origin, the angle between axis *OZ* and the radius vector, and the angle between axis *OX* and the projection of the point on plane *XOY*.

The earliest drawing of three coordinate axes of the skull that may coincide with the same axes of the brain was created by Leonardo da Vinci at the end of the fifteenth century (Fig. 37).

The reference points that are used for calculating a cerebral or spinal target point can be seen on x ray or CT scans or nuclear magnetic resonance (NMR) films. Since it is necessary to define two coordinates of the point, two x-ray films or scans, lateral and AP, are used to determine the coordinates of the point in the sagittal and frontal planes. The third coordinate (in the horizontal plane) can be calculated from the two coordinates already available. Thus, the planes of the two x-ray films form a Cartesian rectangular coordinate system (Fig. 38).

The following names for coordinate axes are used in stereotactic surgery: *X*, or lateral coordinate (frontal plane); *Y*, or anterior-posterior distance coordinate (sagittal plane); and *Z*, or vertical coordinate or height from the reference baseline (horizontal plane). The use

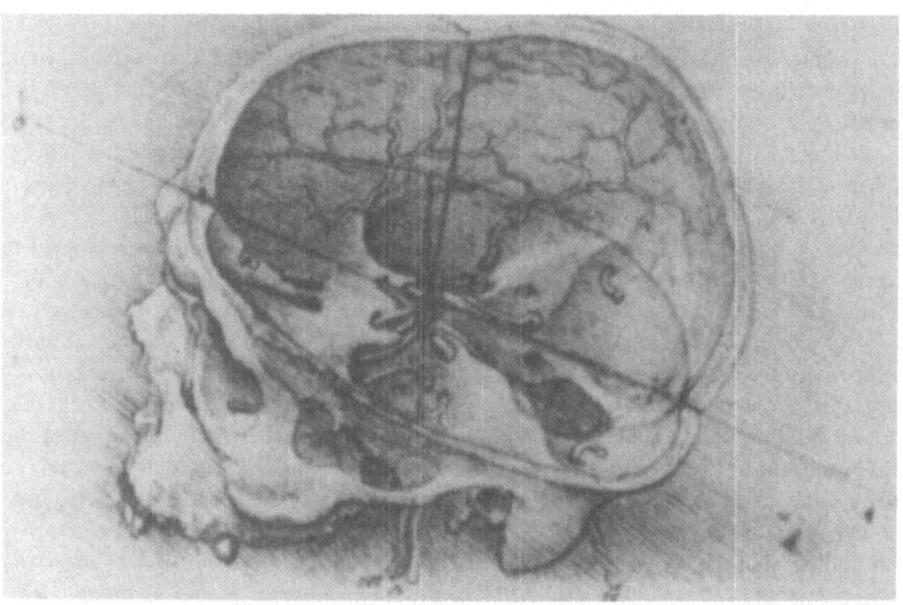

FIGURE 37. Leonardo da Vinci's drawing of three coordinate axes of the skull, which may coincide with the same axes of the brain (E. M. Todd, 1983).

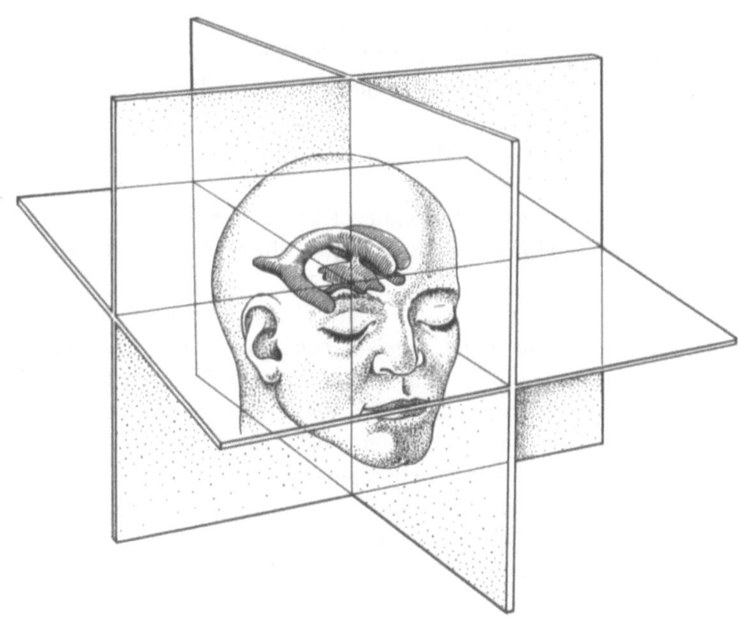

FIGURE 38. Schematic three-dimensional image of the site of the target point in the depth of the brain (Todd, 1972).

of computers, CT investigations, and NMR images for stereotactic calculations is described in Sections 6, 7, and 8.

X-ray investigations, including CT, not only are obligatory but probably represent the most complicated component of the stereotactic method. Such investigations require that a number of conditions be satisfied. The accuracy of stereotaxis, first and foremost, requires precise fulfillment of these conditions.

The anode of an x-ray tube is a spot source of energy, and that is why the beam of x rays always scatters. This phenomenon, which is called divergence, makes the image of the object on the film larger than the object itself (there are, however, exceptions to this rule, which are described below). It has also been established that an increase in the distance between the object and the source of the central ray, i.e., the ray at a right angle to the film, does not increase the degree of divergence (Spiegel and Wycis, 1952b). In the course of every stereotactic operation, divergence requires a twofold correction in all calculations: all dimensions on the films must be reduced in order to be brought to their true value, and the measurements made from the stereotaxic atlases (see below), which are to be transferred to the films, must be increased proportionally.

There are four ways of correcting divergence of x rays. The most effective method is to increase the distance between the x-ray tube and the patient's head. If this distance is more than 4 m, the beam of rays may be considered to be parallel, and the size of the object on the film may be considered true. However, even if the distance from the x-ray tube to the target point is equal to 4 m, the rays diverge, enlarging an object by 5% (Bertrand *et al.*, 1973). Increasing the focal distance, as proposed by Talairach, has been named teleroentgenography. Theoretically, if the focal distance is increased to infinity, the error in x-ray investigations cannot exceed 3 mm (Mundinger and Uhl, 1967). However, this method requires a very large operating room and a powerful x-ray apparatus. To overcome the difficulties connected with x-ray divergence, Schaltenbrand (1953) proposed a method of orthoroentgenography based on moving slots in front of the x-ray tube, thereby eliminating all nonparallel beams with the exception of those in the plane of the slots. This method is effective, but it did not find broad application in the practice of stereotaxis.

The third and most extensively used method involves the use of a metallic scale (a notched metal plate or two metal spheres separated a preset distance). These devices are fixed on the stereotactic apparatus or on the patient's skull parallel to the films and in a plane as close as possible to the plane of the intracranial target. By measuring on the film the distance between the notches, it is quite easy to determine the amount of divergence. The fourth method, from our experience, is the simplest and most convenient. From the known focal distance of the apparatus, which is constant for any given operating room, and the distance from the x-ray tube to the patient's skull, the stable "divergence coefficient" is determined using the simple formula:

$$M = P/(P - E)$$

where M is the x-ray enlargement coefficient, P is the focal distance, and E is the distance from object to x-ray film.

To obtain the true value for any distance on a roentgenogram done under standard conditions, the measurement made on the films should be divided by the constant x-ray enlargement coefficient, and to mark any given distance on the film, the true value should be multiplied by that coefficient. For rapid calculations of these values during an operation, we use a table that, within the range of 1 to 30 mm, gives the true distances and the distances with corrections for divergence.

It is obvious that images of cerebral structures closer to the x-ray tube have a greater degree of divergence than structures farther from the tube. Consequently, the enlargement of the right and left basal ganglia on a lateral roentgenogram will not be identical. However, since they are no more than 30–40 mm apart, if there is a sufficiently long distance between the x-ray tube and the skull, the difference will be only a fraction of a millimeter and can be ignored.

One of the prime objectives of x-ray investigations during any stereotactic operation is the ability to transform two-dimensional measurements on films of two projections into three-dimensional spatial coordinates for a given brain structure. If the central beam falls on the film at a right angle (orthogonal projection) and the films are in two mutually perpendicular planes, then from the projections of any point on these planes it is always possible to find the projection of this point on the third plane perpendicular to the other two. In other words, two orthogonal projections make it possible to

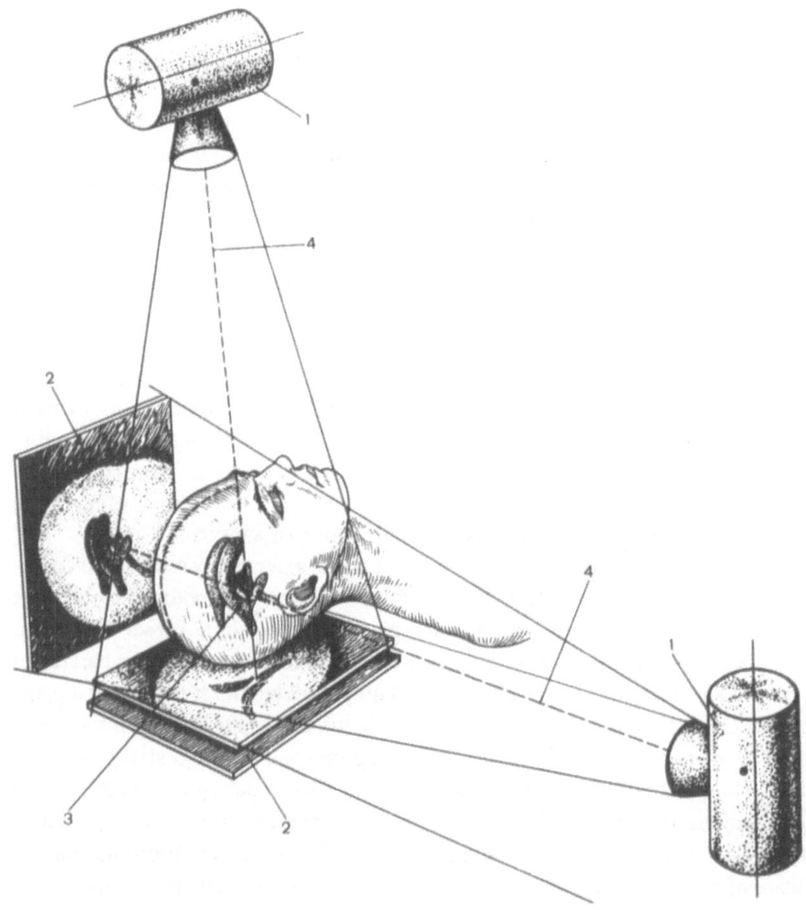

FIGURE 39. Scheme of the principle used for x-ray investigation in a stereotactic operation (Todd, 1972). 1, AP and lateral x-ray tubes; 2, corresponding cassettes; 3, cerebral target point; 4, AP and lateral central beams crossing at the target point.

determine the position of any point in three-dimensional space. Consequently, a necessary condition for x-ray control during an operation is that the target point must be at the intersection of two perpendicular central x-ray beams (Fig. 39).

These theoretical prerequisites make it possible to define the practical requirements for x-ray investigations during surgery. The operating room must be equipped with an x-ray apparatus having two tubes projecting in precisely perpendicular planes during the entire operation so that the central rays of both tubes intersect at one point.

A mandatory condition for accuracy in pinpointing the desired structure is that absolutely identical films (in both projections) coincide when superimposed. In the process of the operation, in order to obtain such films, the following requirements must be observed:

1. Constant distance between the anode of the x-ray tube and the center of the head, which is defined by the line connecting the external auditory meati and the midplane.
2. Constant distance between the center of the head and the film cassette.
3. Accurate projection of the central beam on the patient's skull so that this beam passes through the brain structure designated for destruction or stimulation. For this purpose, both x-ray tubes must have centering mounts with light crosshairs at the point of the central beam.
4. During filming in the lateral projection, the central beam must be perpendicular to the midplane of the head and to the plane of the cassettes, which, in turn, must be parallel to each other. In the AP projection, the central beam must also be perpendicular to the cassette and parallel to the plane passing through the upper borders of the orbits and the external auditory meati. It is desirable to control these factors on the screens of the electronic amplifiers.

An interesting modification of the stereotactic method was put forth by Fox and Green (1968a,b). In this method the AP and lateral x-ray tubes are connected to TV cameras whose central optical axes are parallel to the central rays of the tubes. With the aid of a specially oriented system of mirrors, both x-ray images are fed to two TV monitors. After conventional stereotactic calculations on the ventriculograms and determination of the correction angles between the target point and the electrode–cannula introduced through the burr hole, the oblique lines forming grids on both TV monitors are turned so that these lines are parallel to the electrode, which is then rotated in both projections to the correct angles. The TV grids are again turned so that their lines are parallel to the electrode, and the cannula–electrode is advanced to the target point deep in the brain. Since the scale ratios on the roentgenograms and on the TV monitors coincide exactly, the superimposition of the film onto the screen presents a visual image of all the intracerebral reference points. Without control films, the surgeon can see the electrode on the monitor as it advances into the depths of the brain and assess the accuracy with which it hits the target structure.

It must be emphasized once again that x-ray investigation will be accurate during a stereotactic operation only if the head is correctly and permanently fixed. Inaccurate fixation of the head in a stereotactic operation leads to displacements that must be corrected. During fixation of the head, the main objective is to avoid so-called rotational displacements in all three planes of the coordinate system. Most frequently, there is a slight turn of the head on its vertical axis, so-called rotational angulation, which can be seen on AP ventriculograms. In such cases, the vertical line drawn through the septum pellucidum does not coincide with the line drawn vertically through the middle of V_3 and the distances between temporal bones and lateral edges of the orbits are not equal.

A comparatively simple method has been suggested for calculating the degree of rotational angulation, which must be corrected (Hardy *et al.*, 1980a). On an AP roentgenogram, a vertical line is drawn through the middle of V_3, and on both sides parallel lines are drawn through the medial points of both orbits. In case of angulation, the midpoint between these two lines does not coincide with the midpoint of V_3. The horizontal distance between these two points is measured. Then on the lateral roentgenogram the distance between midpoint LI and the nasion is measured. In this manner, one obtains the lengths of two sides of a right-angled triangle, and the degree of angulation may be calculated.

It is quite apparent that if the head is not properly fixed in the stereotactic apparatus, i.e., if it is rotated to the right or to the left, the length of LI on the lateral film will be shorter than the true length because of nonorthogonal projection. Such a "shortening" of LI may introduce an error in the stereotactic measurements. Mathematical calculations have demonstrated that with a rotation of 12.5° LI is shortened by 0.7 mm

(Van Manen, 1967), with a rotation of 10°, by 0.44 mm (Hardy *et al.*, 1980a). A somewhat greater value for the shortening was obtained in model experiments on a skull with taut wires inside it: 1.0 mm with a rotation of 10° (Flamm and Van Buren, 1966). Since rotation within the range above is hardly possible in modern stereotactic apparatus, and a displacement of several degrees results in the shortening of LI by a fraction of a millimeter, this factor need not be taken into account.

2.1. Cranial Reference Points

As was noted earlier, the spatial location of any structure of the brain and spinal cord can be determined only with the aid of reference points, which may be either cranial (bone) or intracerebral. Let us first discuss the cranial landmarks, which have certain, although limited, significance in stereotaxy.

The variability of the shape and size of the human skull is well known and has been studied quite thoroughly. The ratio of the width of a skull to its length (cranial index) is one of the main indices making it possible to divide all humans into dolicho-, meso-, and brachycephalics: an index within the range of 70–74.9 is an indication of dolichocephaly, an index 75–79.9 of mesocephaly, and an index of 80–85 of brachycephaly (Gerlach, 1959). More than 70 reference points on the outer and inner surfaces of a skull are employed for characterizing its size and shape (nasion, inion, pterion, porion, lambda, bregma, and others) (Fig. 40).

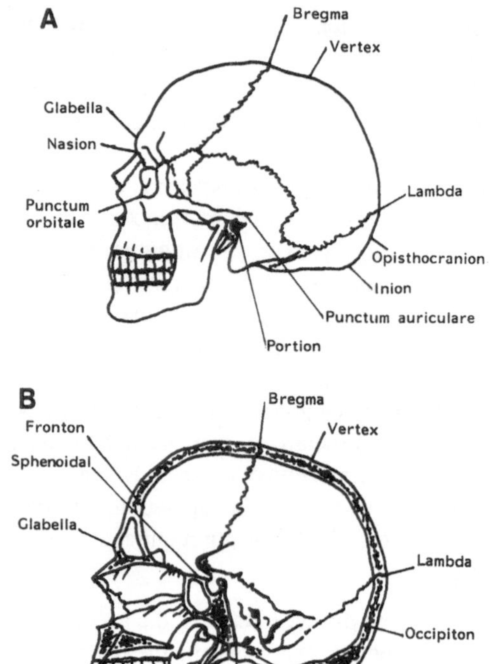

FIGURE 40. Reference points on the outer (A) and inner (B) surfaces of the skull.

For decades numerous investigations were carried out to study the spatial interrelationship between cranial reference points on the one hand and sulci, gyri, and deep structures of the brain on the other. This surface mapping, which is called craniocerebral topography or topometry, is of paramount significance for neurosurgery (Delmas and Pertuiset, 1959). For many years, neurosurgeons have been using the drawings of Krönlein, Egorov, and other authors to enable them, on the basis of cranial reference points, to plot the main sulci in the brain cortex.

The method of Egorov (1959) is based on two main lines, one along the circumference of the head ("equator") connecting the nasion and the inion and a second that connects these points in the midplane of the head. The second line should be measured and divided into four equal parts, thereby obtaining another three points (anteroposteriorly): F, P, and L. The line drawn from the superior external angle of the orbit to point L corresponds to the projection of the sylvian sulcus on the skull. The line drawn from point P downwards at an angle of 60° and intersecting the E "equator" at a right angle represents the projection of the precentral sulcus. The line corresponding to the fissure of Rolando runs parallel but 1.5 cm posterior to that line. However, it should be noted that these and other such schemes do not give the neurosurgeon the spatial location of deep subcortical and brainstem structures, which are the main targets in stereotactic surgery.

First, it is necessary to note the great variability of the main cranium dimensions on x-ray films in two projections. The author measured the dimensions of the skull on roentgenograms of 70 adult patients. The films were taken during surgery under standard conditions. The length of the skull was measured on the lateral projection (nasion–inion distance), and the width on the AP projection (the greatest distance across the external table of the skull in the parietotemporal areas). The range of variability in length was from 154 to 190 mm in males and from 144 to 184 mm in females; the range in width was from 132 to 160 mm in males and from 127 to 148 mm in females. These data confirm that the great variability of the size of the skull does not allow such measurements to be used for stereotactic calculations.

The interrelationship of cranial reference points and deep-seated structures in man (unlike in animals) is so variable that they cannot serve as the basis for determining the location of these structures (Mundinger and Potthof, 1961; Spiegel and Wycis, 1962; Kandel, 1965). As an example, the maximal variability in the distance between the vertical interauricular plane and the midpoint of LI is as great as 15 mm (Amador *et al.*, 1959), and that between the same plane and the center of GIP is 16 mm (Spiegel and Wycis, 1952b). The relationship between LI and the length and width of the skull, as well as between the height of Th and the height of the skull, are subject to such great variation that they cannot be of practical value in stereotactic calculations. A very striking example is presented by Amador *et al.* (1959): the distance from the interaural plane to the midpoint of LI in two skulls of the same size was 12.5 mm in one case and only 2.5 mm in the other.

The majority of neurosurgeons have abandoned the use of some "internal" cranial reference points that are visible on x ray of the skull: the anterior clinoid processus, the floor and other parts of the sella turcica, etc. A monograph by the author (Kandel, 1965) gave the results of measuring the distance from the center of VL, determined according to intracerebral reference points on 80 ventriculograms, to certain cranial structures located near the sagittal plane. It appeared that the distance from the center of VL to the end of the anterior clinoid processus varied from 25.5 to 35 mm, that to the end of dorsum sella from 22 to 31 mm, and that to the deepest point of sella from 30 to 39.5 mm. Investigations of the spatial interrelationships between the medial segment of GP and reference points of sella turcica also demonstrated that there are no constant correlations between them (Mundinger and Potthof, 1961). In our opinion, such a high individual variability precludes the use of intracranial reference points for stereotactic calculations.

Accurate determination of the center of VL on x-ray films, using cranial reference points, proved possible only in 68% of cases (Kamm and Austin, 1965). Such a low probability of accurately hitting these structures is unacceptable to the neurosurgeon.

In the early years of stereotaxis it was assumed that the basis for a "natural" coordinate system of the brain could be the horizontal planes formed by the cranial structures at the base of the skull, e.g., the Reid plane or the so-called Frankfurt plane, passing through the lowest border of the orbit and the middle of the external auditory meati. The use of these planes presumes strict symmetry of the skull, which is present in only 10% of cases; in 90% there is some asymmetry. The midplanes of the skull and the brain frequently do not coincide as a result of skull or brain asymmetry. Quite often, the internal auditory meati are located in different planes, and the interaural and midplanes, as well as Reid's plane, are not strictly perpendicular to each other. This nullifies the localizing value of these planes. All these data confirm that cranial reference points have no regular relationships to the intracerebral structures that are targets of stereotactic operations.

However, one should not conclude that cranial reference points are of no practical significance for the stereotactic method. They are important for determining the site of the burr hole in the skull bones. Cranial reference points are the goals of or are important in several stereotactic operations, such as the dorsum sella turcica in operations on the hypophysis (Chapter 15; Section 5), the foramen jugulare for destruction of the glossopharyngeal nerve, the oval foramen for destruction of the Gasserian ganglion, and dens of the second cervical vertebra in percutaneous tractotomy. These reference points ensure accuracy in hitting the target point in the operations mentioned. In rare cases, other reference points may be seen on conventional x-ray films (calcified pineal gland, other calcifications, as well as foreign bodies).

Reports have recently begun to appear again in literature about the use of cranial reference points for stereotactic calculations. According to Ohye *et al.* (1984a), two lines joining cranial reference points on lateral craniograms practically coincide: Twining's line connecting the tuberculum sellae and the internal occipital protuberance and the glabella–inion line (G–I). The authors found that CA lies at a distance of 20 ± 1.6 mm above Twining's line and that LI runs parallel to it directly caudal to CP. It is presumed that the length of LI is 23 mm, although this length may vary from 21 to 28 mm. We believe that this method using cranial reference points is not sufficiently accurate. If one combines the two deviations (distance from Twining's line to CA and length of LI) due to individual variability, then the inaccuracy comes to approximately 4–6 mm and even more, which is unacceptable in operations on the basal ganglia.

Another method of calculating coordinates of intracerebral structures is based on sophisticated computer software that enables reproduction of the entire course of the operation (Mundinger *et al.*, 1978a). The following is a brief sketch of the method. On 400 pneumoencephalograms made during operations for parkinsonism and other extrapyramidal lesions selected at random, four cranial reference points are marked on lateral films:

1. The small gap in the contour of the frontal bone above the frontal sinus.
2. The junction of the lamboidal suture with the sagittal plane.
3. The junction of the coronal suture with the same plane (bregma).
4. The tip of the clinoid process.

Once the distances between these points are measured on the films, the location of FM can be calculated by computer. In order to determine the line FM–CP, the authors use a line passing along the base of the anterior cranial fossa. They found that the tangent of the angle of that line and line FM–CP is directly related to the length of the line connecting the third and fourth points mentioned above. The coordinates of FM on the AP film are determined by echoencephalography.

Mundinger *et al.* (1978a,b) indicated that precise determination of reference points by this method proved possible in 88% of the cases. To assess this method, we employed the technique described above in 150 intraoperative ventriculograms. However, computer analysis of the data revealed quite significant differences between the location of points calculated according to cranial and intracerebral reference points. In view of this, we maintain that it is impossible to base stereotactic calculations on cranial reference points.

2.2. Intracerebral Reference Points

In every stereotactic operation it is necessary to determine the spatial localization of the subcortical structure that is to be destroyed. For this purpose, it is necessary to use the intracerebral reference points that may be determined on ventriculograms or pneumoencephalograms in two projections made under rigid conditions. The calculations of target points with computer technology (CT or NMR) without contrast material in the ventricles are presented in Sections 6, 7, and 8.

It is obvious that when a target point deep in the brain lies close to visible intracerebral reference points, its spatial location may be determined with greater accuracy than if situated more peripherally.

Before determining the intracerebral reference points, one must be certain that the ventriculograms were made with the skull correctly positioned. In this case, in the AP projection, the midline of V_3 is strictly symmetrical to both V and is directly below the septum pellucidum. In the lateral projection, it is necessary to see that the lines forming the superior contours of bodies of both V coincide and that the cranial reference points, mastoid processes, and the pyramid are superimposed. However, it is necessary to keep in mind that displacements of the central cerebral structures from the midplane in ventriculograms are much less than the displacements of "peripheral" cranial points.

In films of the AP projection (see Fig. 78), it is first necessary to note the shape, size, and degree of contrast filling of both V and V_3. In the shadow of the V, which resembles a butterfly, it is possible to see the body, the darker part of the shadow adjacent to the midline, and the anterior horns, the lighter and broader parts of the shadow extending laterally beyond the body of V. The convex line forming the margin between the body and the anterior horn outlines NC. The lateroventral contour of the medial part of body V is formed by both Th.

A thin line in the midplane and the septum pellucidum separate both V. This line begins below CC and descends to the upper point of the shadow of V_3, which usually has the shape of a narrow vertical fissure wider at its midpart by 3–4 mm. The diameter of V_3 is usually 14–15 mm. In hydrocephalus and in parkinsonism, the anterior part of V_3 acquires an oval or even spherical form. In these cases, the transverse diameter of V_3 increases from 3–4 to 9–10 and even 12–13 mm.

Ventral to the inferior part of V_3 it is possible to distinguish a narrow isthmus, the shadow of the Sylvian aqueduct, and below it, much wider in the form of an upright triangle, V_4 with its lateral recesses.

In the lateral projection (see Fig. 79), the overlapping elongated shadows of both Vs are seen, and it is easy to make out the body and the anterior, temporal, and posterior horns. The shape of V is dependent on those structures that are adjacent to it: CC, NC, Th, and Hipp. Important reference points for stereotactic calculations are the FMs linking both Vs with V_3. In the lateral projection both FMs overlap and appear as a narrow (3–4 mm) isthmus passing ventrally and posteriorly. The posterior border of FM corresponds to the anterior pole of Th. In pronounced hydrocephalus these foramina may be 6–8 mm in diameter.

Directly below FM at a distance of 2–3 mm and

slightly anterior to it in the contour of V_3, one can see a small (approximately 2–3 mm) semicircular notch, CA, the direct ventral continuation of which is the terminal lamina of V_3. In the anteroventral part of V_3 are two outpouches, recessus opticus and recessus infundibuli.

The posterior part forming the cupola of V_3 descends ventrally and posteriorly at approximately a right angle to join the narrow shadow of the Sylvian aqueduct. The second main reference point, CP, lies at the point of this transition. Directly above it are two caudal recesses of V_3, the small toothlike recessus pinealis and, above, the considerably larger recessus suprapinealis in the form of a horizontal strip 6–10 mm long and quite variable in shape.

The only reference point that often can be determined on an ordinary lateral x-ray is GlP. If it is calcified, one can locate CP 2 mm ventral and 4–5 mm anterior to the center of GlP. Above GlP there is sometimes a calcified commissura habenularum, a very small 2- to 3-mm-long strip in the form of a semicircle. The center of GlP is situated, on the average, 10.2 mm from the intraaural plane; however, this distance may vary from 4 to 20 mm.

A useful reference point on lateral ventriculograms is the massa intermedia (MI), which passes through the cavity of V_3. It has been established that the center of VL lies at the ventrocaudal pole of MI, which can often be seen in the center of V_3 in the form of a small (5 × 3 mm) oval-shaped structure. The distance between the centers of MI and CP varies from 13.5 to 16 mm (on the average 14.8 mm).

2.3. Linear Reference Points

An important advance in stereotactic calculations was the proposal in the 1950s to employ linear references, i.e., lines linking two reference points within the brain, as the basis for a cerebral coordinate system (Talairach *et al.*, 1952; Riechert and Mundinger, 1955; Cooper, 1955). The intercommissural line (LI) linking CA and CP (Fig. 41) was proposed as a "basal coordinate axis." In the following years, the majority of neurosurgeons began to consider this line as one of the chief guides for stereotactic calculations. The LI was used as the basis of the coordinate system of the stereotactic atlas of Schaltenbrand and Bailey. The midpoint of this line can be called "stereotactic zero." The LI that connects the centers of CA and CP differs very little from the Talairach line, which links the superior border of CA and the inferior border of CP.

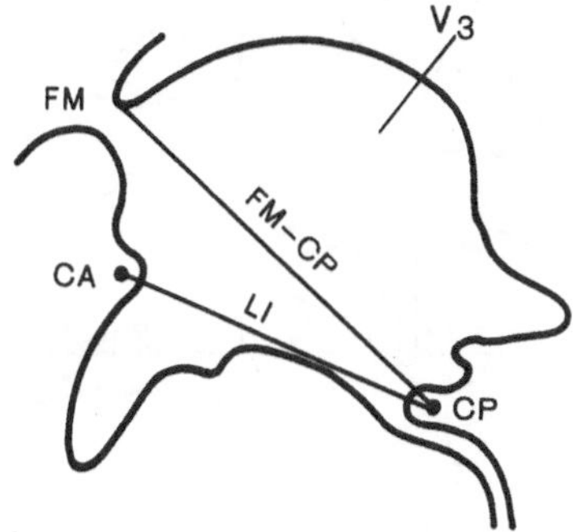

FIGURE 41. Schematic image of V_3 and two main reference lines (FM–CP and LI) on ventriculogram in lateral projection.

These lines are almost parallel, their divergence not exceeding 1.5°.

The advantage of LI is that its length is practically independent of the length and shape of the skull. The length of LI is not constant: according to various authors, it fluctuates from 20.5 to 28.5 mm. The average length of LI is given in the atlas by Schaltenbrand and Bailey as 23 mm, by Mundinger and Potthof (1961) as 24.6 ± 1.34 mm. The individual variations in the length of this line are less than those of other intracerebral reference lines. The LI is not parallel to the basal (Frankfurt) plane; the angle between them, open anteriorly, is approximately 15°.

As do other neurosurgeons (Riechert and Mundinger, 1959; Andrew and Watkins, 1969; Siegfried and Hood, 1983), in our stereotactic calculations, we use a different line. On a lateral ventriculogram this reference line connects the posterior border of FM with CP (Fig. 41). Although both of the lines (LI and FM–CP) terminate in one point (CP), they originate 3.5–4 mm apart vertically; i.e., line FM–CP lies above LI. The center of CA lies on the average 3.1 mm below and 2.8 mm in front of the anterior border of FM, and the angle between the lines of the posterior border of FM–CP and LI, when the former line is short (22–25 mm), is equal to 25 ± 1°, and when long (26–28 mm) is 21 ± 1° (Fig. 42) (Bedrensky and Vasin, 1981). The middle of FM–CP lies 2–2.5 mm above the midpoint of LI.

In our opinion, a significant advantage of line FM–CP is that it offers the possibility of more clearly determining the posterior border of FM (as compared to CA) on a lateral film. Furthermore, the line FM–CP is less variable than LI (Andrew and Watkins, 1969).

We (Kandel and Tsibulnikov, 1971) have presented detailed data on the reference line FM–CP. A study

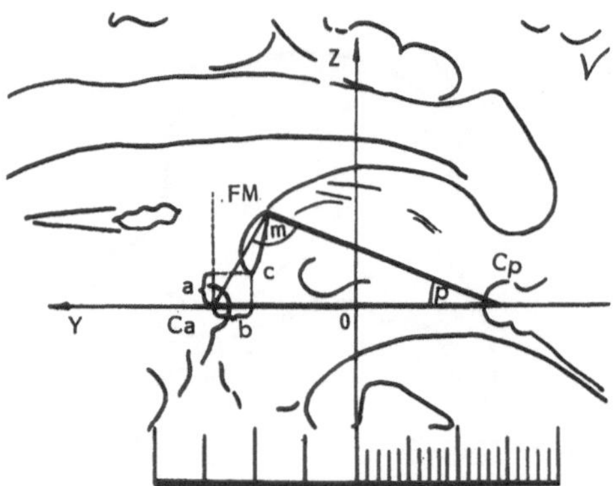

FIGURE 42. Relationship between the two main reference lines shown in Fig. 41. See text.

was made of the ventriculograms of 120 patients with extrapyramidal hyperkinesias obtained in the process of stereotactic operations. These patients, 72 males and 48 females, were 18 to 64 years of age. The average length of line FM–CP varied from 19.5 to 27 mm in males and from 18.9 to 27.4 mm in females. The average length of the line was 23.6 ± 3.6 mm and 22.97 ± 3.5 mm, respectively. In a standard brain that served as the basis for the stereotactic atlas in the monograph by Hassler *et al.* (1979), the length of the line between the posterior borders of FM and CP was 24.0 mm.

The length of the line FM–CP in various adult age groups, according to our data, does not vary significantly. Furthermore, in all our cases, we compared the length of this line and the degree of hydrocephalus, based on our method (Section 5). The small differences in the length of this line in hydrocephalus of degrees I, II, and III proved to be statistically insignificant, and, consequently, the length of this line does not depend on the degree of ventricular dilatation.

At times, in stereotactic calculations another linear reference is employed—the height of Th. It is determined by the length of the line drawn perpendicularly from the middle of LI to the highest point of the dorsal convexity of Th on the lateral roentgenogram. The height of Th is 14.6 ± 1.03 mm with variations from 12.1 to 16.8 mm (Mundinger and Potthof, 1961).

3. Individual Variability

A very timely and yet unresolved question related to the stereotactic method is that of determining the variability in size and location of subcortical structures. The morphological data presented in Chapter 1 indicate that the range of variability is significant. The high degree of variability of subcortical structures is well illustrated in the diagram of Van Buren and McCubbin (1962b) showing superimposed contours of sagittal sections (5 mm from the midplane) from 16 brains (Fig. 43). This diagram demonstrates that the difference in size of these structures frequently reaches 10–12 mm. In all likelihood it is impossible to find any two brains in sections of which the localization and dimensions of the basal ganglia coincide exactly in any plane (Fig. 44). Substantial variability is also characteristic of cerebral sulci and gyri and their relationship to subcortical structures (Fig. 45).

Another example of significant individual variability is given in the investigation of the volume of Str in 27 cerebral hemispheres (Zvorikin, 1982). For example, in extreme individual variants, the volumes of NC differed by a factor of 2, and Put by a factor of 3.7; moreover, these differences are not related to the dimensions of the cerebral hemispheres.

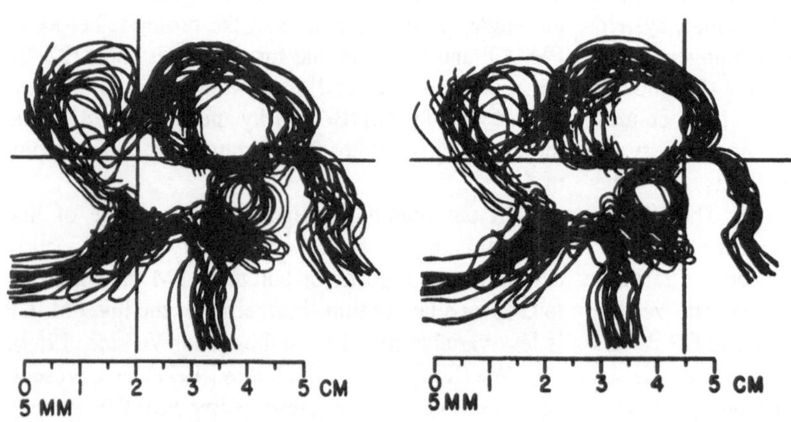

FIGURE 43. Superimposition of sections 5 mm from the midline of 16 brains. Left: Lateral projection of the center of CA and LI. Right: Lateral projection of CP and LI (Van Buren and McCubbin, 1962b).

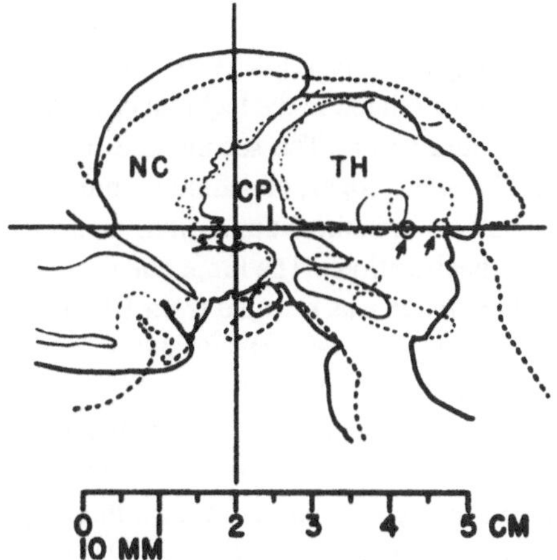

FIGURE 44. Superimposition of sagittal sections 10 mm from the midlines of brains weighing 1070 g (solid line) and 1785 g (dotted line). Two arrows, PC in each brain. Sections aligned by the lateral projection of the center of CA and LI (Van Buren and McCubbin, 1962b).

Information on the problem of variability is to be found in stereotactic atlases (see Section 4). It is necessary to emphasize that by stereotactic calculations, we can determine the location of some arbitrary structure within the cranium but not the center of that deep-seated surgical target. In the majority of cases, these points coincide, in which case the calculations are accurate. Difficulties arise in those comparatively rare cases in which an individual anatomic variation leads to a disparity between the true and measured data and serves as a source of error in calculations. The main difficulty in attempting to prevent this is that anatomic variability of subcortical structures does not lend itself to direct estimation. There can be no doubt that this factor is responsible for a certain percentage of ineffective stereotactic operations. Hence, one of the main and still incompletely resolved problems in stereotactic neurosurgery is to overcome the factor of individual variability (Brierly and Beck, 1959).

The most generally acknowledged practical method of correcting individual variability is the "relativity factor" (Riechert and Mundinger, 1959). This is a coefficient reflecting the relationship between the dimensions of any structure or, in general, any distance in a stereotaxic atlas, i.e., in a "standard brain," and those same dimensions obtained during an operation on a particular patient's brain. In principle, the "relativity factor" may be determined for each of the three axes of coordinates. In practice, it is usually applied only with respect to LI and is based on the ratio between the length of a "standard" LI or the distance between the posterior borders of FM and CP (according to our data, 23.3 mm) and the same dimension in the actual patient. In the majority of cases, this correction coefficient is not large (approximately 5–8%, i.e., 1–1.2 mm). However, if (in comparatively rare cases) the length of the line greatly differs (in either direction) from the "standard," then the value of the coefficient may be 1.5–2 mm. Obviously, the above coefficient is important since it substantially improves the accuracy of stereotactic calculations.

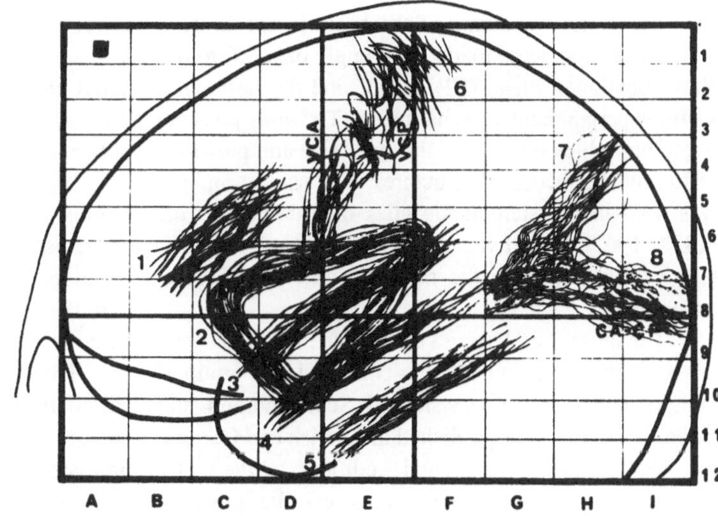

FIGURE 45. Statistical variability of the cortical sulci. 1, Inferior frontal sulcus; 2, insula; 3, Sylvian fissure; 4, superior temporal sulcus; 5, middle temporal sulcus; 6, central fissure; 7, parietooccipital sulcus; 8, calcarine sulci (Szikla, 1979).

4. Stereotactic Atlases

The morphological basis for the stereotactic method lies in the anatomic data on the shape, dimensions, and interrelationship of both surface and subcortical structures. The degree and range of variability of these data are given in stereotactic atlases of the human brain. They serve as the main source of information about numerous subcortical structures. These data are needed by the neurosurgeon for performing any stereotactic operation.

Each stereotactic atlas, and there are already several, is based on photographs of numerous sections of the "standard" or "medium" brain taken in various planes—sagittal, frontal, horizontal, and inclined planes.

The first stereotactic atlas (Spiegel and Wycis, 1952b) offers a detailed description of the stereotactic apparatus of the authors and the operating techniques and methods. This atlas is based on photographs of brain sections 5 mm thick made in three planes. The frontal sections were made parallel to the "mean guiding plane," corresponding to the line linking CP with the sulcus between the pons and the medulla oblongata. The GlP was taken as the center of the coordinate system of the brain. Subsequent experience demonstrated that this "stereotactic center" was not appropriate. The calcification of GlP, making it visible on ordinary roentgenograms, is observed in only 42% of cases (Kandel, 1965), and its localization based on CP visible on ventriculograms introduces additional inaccuracies to the stereotactic calculations. In addition, it has been established that reference points, in particular, GlP, are less accurate than linear references (Section 2.3).

The vertical sections of the brain in the Spiegel and Wycis atlas were made at an angle of +30° to the abovementioned plane (CP–pons), and the horizontal sections perpendicular to that plane. The atlas presents numerous tables on the variability of size and positions of the main subcortical structures. The same principle was used in the stereotactic atlas of Talairach *et al.* (1957, 1967), which was the first to employ a coordinate system based on the most accurate linear reference, LI.

The most complete and up-to-date manual on stereotaxy is the three-volume atlas compiled by a group of authors under the supervision of Schaltenbrand and Bailey. The atlas was published in English and German in 1959, and its second edition, edited by Schaltenbrand and Wahren, came off the press in 1977. The first volume of that atlas presents much data on morphology and physiology of subcortical structures, questions related to the techniques of stereotactic operations, and so on. The second and third volumes represent the atlas proper, containing scores of sections of a standard brain in three dimensions made according to precise specifications. The brain that came closest to the statistical mean was selected from 108 preparations and sectioned to serve as a standard on which all stereotactic operations might be based.

The main coordinate system on which the sections were taken is LI (to be more exact, the intercomissural plane), and the zero point of the coordinate system is the midpoint of LI, where a perpendicular was erected in the sagittal plane. Thus, LI and the perpendicular are two of the coordinate axes. The third axis passes through the zero point at right angles to the two previous coordinates. The frontal, sagittal, and horizontal sections of the brains in this atlas were made relative to these axes (planes). The atlas contains many charts based on a large number of preparations in which the coordinates of the main subcortical structures are averaged; it shows the range of variability of these structures and the "zones of reliably hitting them."

The stereotactic atlas of Talairach *et al.* (1957, 1967) was compiled especially for determining the stereotactic coordinates of deep structures in the temporal lobe, which are oriented with respect to the temporal horn of V.

Stereotactic atlases were also published by Delmas and Pertuiset (1959), Andrew and Watkins (1969), Van Buren and Borke (1972a), and Emmers and Tasker (1975).

The stereotactic atlas of subcortical structures, a supplement to the monograph of Hassler *et al.* (1979), is very informative and convenient for neurosurgeons. The coordinate system of this atlas is based on the above-described reference line FM–CP (Fig. 46). The brief commentaries for each section in that atlas contain much important information.

Valuable data on the topography and interrelationships of subcortical structures are also presented in the papers by Guiot *et al.* (1968a). Another formidable monograph, the atlas of Szikla *et al.* (1977a), presents the results of a detailed study of the stereotactic anatomy of the cerebral arteries.

The rapid development of stereotactic surgery stimulated the production of stereotactic atlases of the anatomic structures of the posterior fossa—Cer and the brainstem. The first coordinate system for stereotactic operations on these structures was proposed by Tal-

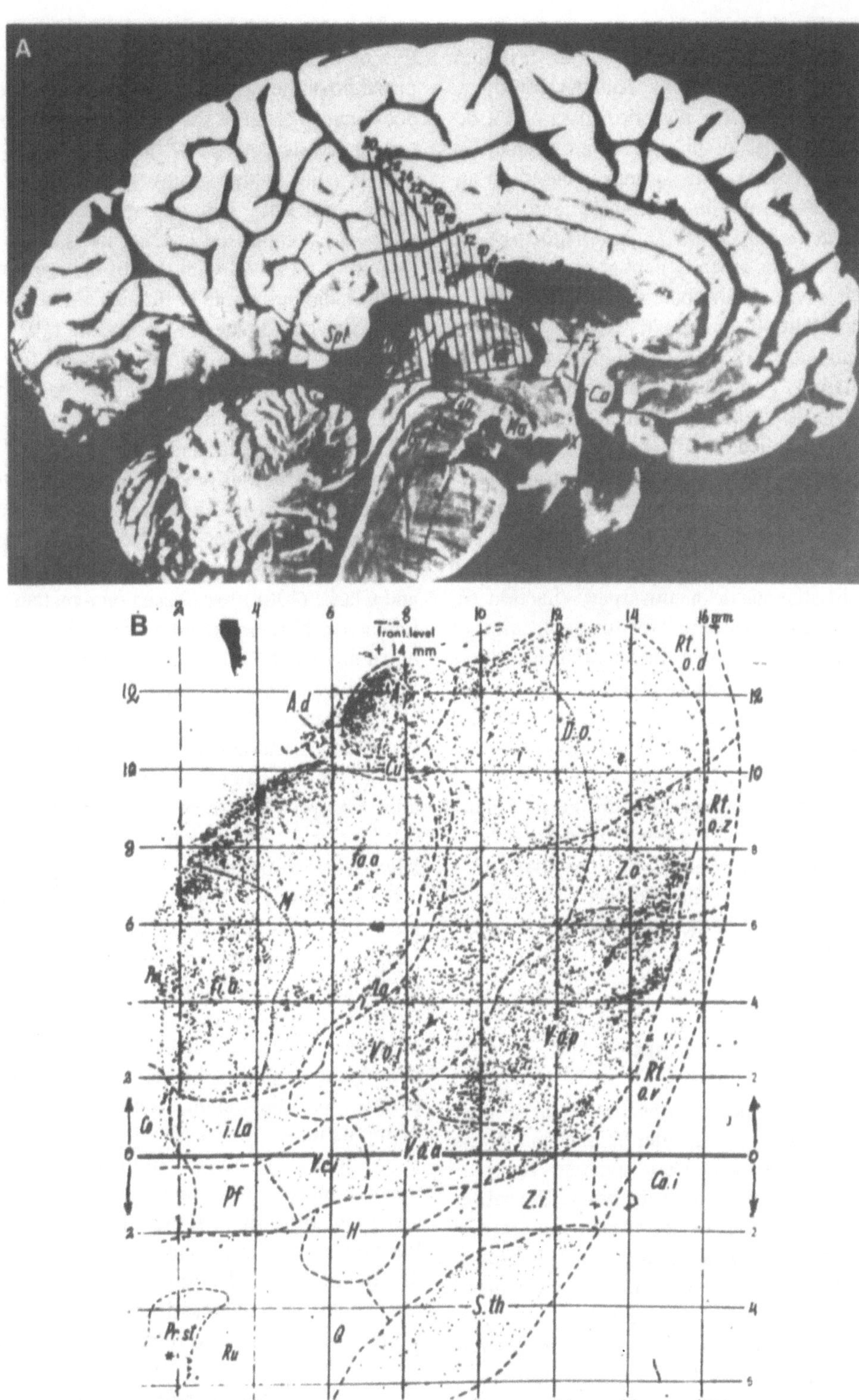

FIGURE 46. (A) Medial aspect of hemisphere with baseline of Th (line FM–CP). Perpendiculars to the line show the planes of frontal sections, each 2 mm from FM to CP; LI is also marked. (B) A frontal section 14 mm from FM (Hassler *et al.*, 1979).

airach (1957). The basis of this system is the fundus line of V_4 and the perpendicular to it from fastigium (Fm) (Chapter 9, Section 5.1.3). Another coordinate system for structures of the posterior fossa and a detailed stereotactic atlas of this area were created by Nadvornik *et al.* (1965). This system is based on another coordinate line, the line linking CP and Fm.

A *Stereotactic Atlas of the Human Brainstem and Cerebellar Nuclei* was recently published by Afshar *et al.* (1978). It gives details of careful studies of the individual variability of the structures.

Nadvornik *et al.* (1965) have compiled topometric maps of the spinal cord with an accurate coordinate system. A stereotactic atlas of spinal cord structures has also been compiled; it offers numerous sections with accurate topometric data on spinal neuronal structures and pathways as well as tables on their spatial variability (Zlatos and Cierny, 1975).

The use of computers, particularly computerized tomography in stereotactic neurosurgery (Section 6), has opened up new possibilities for the application of stereotactic atlases. Today there are stereotactic atlases in which images are generated by a computer on the basis of the data in its memory bank (Kall *et al.*, 1985). Images from these atlases may be superimposed on CT scans, and, what is more, the surgeon can compress or stretch these images, adapting them to the individual characteristics of the patient's brain. It is possible to get from the stereotactic atlas all the coordinates of a target in three dimensions. The reference points on CT scans may be reoriented in the same plane as is represented in the atlas.

5. Stereotactic Instruments

The stereotactic method was not used in neurosurgery until almost 40 years after Spiegel and Wycis created the first modern stereotactic apparatus for operations on subcortical structures of the brain. The subsequent rapid development of this technique gave a powerful impetus to the creation of new and improved stereotactic equipment, and this, in turn, stimulated further progress of stereotactic neurosurgery.

To date, more than a score of stereotactic apparatuses have been built. Quite often they differ from each other significantly in principle and complexity of design, manner of fixation to the skull bones, systems of coordinates, use of phantom devices, and so on. Both general-purpose and specific devices, for example, for operations on the hypophysis, the Gasserian ganglion, or the spinal cord, have been built. However, irrespective of design, each apparatus has retained the basic principle of the stereotactic method—integration of the coordinate system of the brain with the coordinate system of the stereotactic device.

In spite of the variety of systems of stereotactic instruments and, consequently, stereotactic calculations, the comparison ("tying up") of the coordinate system of the brain with that of the apparatus is based on the single principle of three-dimensional orthogonal and polar coordinates. As Abrakov (1975) noted, the seemingly wide variety of methods for performing such calculations often disguises the basic principle underlying all stereotaxis.

Both rectangular and polar coordinate systems are employed in various devices. Zernov's stereotactic apparatus, the first in the world, was based on the polar system of coordinates, whereas the first apparatus for experimental neurophysiology, developed by Horsley and Clarke (1908), was based on a rectangular system, which was also employed in designing the first modern apparatus by the pioneers of the stereotactic method, Spiegel and Wycis (Section 1).

The use of the rectangular coordinate system presupposes calculations in three planes of space situated at right angles to each other. Consequently, such an apparatus provides three advancing mutually perpendicular degrees of freedom. This method offers certain advantages: in particular, it makes it possible to determine accurately the enlargement of the object as a result of the parallax effect of the x rays as well as to install the apparatus repeatedly in exactly the same position if the operation is performed in two stages. Certain shortcomings of any apparatus based on a rectangular coordinate system are obvious; these include complexity of design, difficulties in fixation of the apparatus on the skull, and the time-consuming manipulations.

When using polar coordinates (Section 2) to locate a point in space, one determines the distance of that point from the origin and the angle between the line connecting these points and the polar axis. In order that the tip of the electrode–cannula accurately hit the target point, the stereotactic apparatus must allow three degrees of access—two rotational and one advancing. The rotational axes must be at right angles to each other and intersect at one point, and the forward movement of the instrument should proceed along its longitudinal axis, which passes through the point of intersection of the rotational axes. By simple formulas, the polar coordinates may be converted to rectangular. The most commonly used devices are based on a system of polar coordinates by which an electrode–cannula is introduced into the brain towards the target point. The specific angles are determined by lines drawn from the tip of the electrode–cannula, which has been advanced for a short distance into the brain, on ventriculograms in two projections.

From the practical point of view, all stereotactic devices used for brain operations may be divided into two groups:

1. Apparatus of comparatively simple design that is fixed in a small burr hole in the skull. The placement of such a device does not require preliminary x-ray control and is based on external cranial reference points.
2. Round or rectangular frames of more complicated design and large dimensions in which the patient's skull is fixed under x-ray control with the help of sharp pins that are screwed into the skull bones. The bases of the frames are used as a reference plane during the stereotactic operation.

The differences between the two types of apparatus may be summed up in the phrases "apparatus in the skull" and "skull in the apparatus."

The stereotactic devices of the first type offer a number of important advantages: they are simple and convenient, and their installation on the skull takes a short time. In recent years there has been a noticeable tendency to simplify the design of stereotactic devices because the complicated and cumbersome devices that require time for installation and for stereotactic calculations have not justified themselves in practice. However, at the same time, it is evident that the desire for simplicity and convenience in application should not be allowed to decrease the accuracy of the apparatus in reaching the subcortical target.

There is no need to describe even the majority of the existing pieces of equipment, which frequently differ very little from each other. Below, the principles of design and use of the main stereotactic devices are presented.

In light of the development of new stereotactic technology based on modern imaging techniques, many "classic" stereotactic devices seem to have become obsolete and old-fashioned. In spite of this, we present a brief description of the devices for three reasons. First, these apparatuses show clearly the development and evolution of the stereotactic method and its principles. Second, to this day, many neurosurgeons prefer to employ these reliable and precise devices in their routine work. There are also many neurosurgical clinics which do not have access to modern stereotactic CT-guided systems. Third, many "classic" devices are now modernized and are made compatible for use with CT and NMR imaging. Modifications of these instruments designed for use together with CT or NMR are described in Sections 7 and 8.

5.1. Stereotactic Frame of Spiegel and Wycis

Of the five models of stereotactic apparatus proposed by Spiegel and Wycis, we give a brief description of the last, which the authors named the "stereoencephalotome" (Spiegel *et al.*, 1956). The apparatus (Fig. 47), based on rectangular coordinates, consists of two main parts: a stereotactic device and a supporting target frame. The first part of the apparatus (the stereoencephalotome proper) consists of the basal ring, the electrode holder, and a measuring device. For fixing the stereoencephalotome to the skull there are graduated telescoping rods. Four metal pins are screwed into the skull bones. Mounted on these pins are the couplings of the vertical rods of the basal ring of the stereoencephalotome. The electrode holder has three degrees of freedom; i.e., it can move in three dimensions of space. The zero position and electrode-guiding correction angles are controlled by protractors in the first part of the apparatus.

The supporting target frame, the second part of the apparatus, may be considered a large cube formed by horizontal and vertical metal rods. Adjustment of the frame to the skull of a patient uses five fixation points: two pins that are introduced into the external auditory meati, two facial pins, and a pin to the occipital protuberance.

The basal ring of the first part of the apparatus is fixed on a thick glass plate, which forms the upper surface of the cube. On all four lateral surfaces of the cube there are vertical plastic plates with x-ray contrast markings; when the frame is in the correct position on the patient's skull, these opaque markings are superimposed on the films in both projections. These grooved markings are used for determining the two main planes: midplane and frontobiauricular, which is perpendicular to it.

The supporting target frame is set up in relation to the horizontal and midplane of the skull, whereas the stereoencephalotome is set up to correspond to the horizontal plane of the frame. Then pneumoencephalograms and roentgenograms in two projections

FIGURE 47. Stereotactic apparatus (stereoencephalotome) of Spiegel and Wycis (fifth model). Description in the text.

are made. The electrode is set at the zero position, corresponding to the intersection of the interaural and sagittal planes. After that, the support frame is removed, and stereotactic calculations are carried out.

On the film, in the lateral projection, the zero line, which is the continuation of the line of the electrode, is drawn, and the basal line is drawn perpendicularly to the former and passing through the center of GlP. On the stereotactic atlas, one determines the distance from the zero line to the frontal section in which the center of the subcortical target structure lies and at that distance draws a second vertical line on the film. The distance from the target point to the basal line is determined with the help of the atlas. The data are then transferred to the millimeter scale of the stereoencephalotome, thereby ensuring accurate insertion of the electrode towards the target structure. Similar but somewhat more complicated calculations are required for determining the trajectory of the electrode if the radial coordinate system is used.

The stereotactic apparatus of Spiegel and Wycis has a relatively high level of accuracy, although somewhat less than contemporary stereotactic devices. According to the inventors of this apparatus, in control experiments, the error in reaching the center of a preselected subcortical structure did not exceed 1 mm in 12 cases out of 20. In later designs, certain shortcomings of earlier models, such as the long time needed for fixation of the apparatus on the patient's skull, were overcome, but the apparatus remained complicated and cumbersome, and presently is not in use.

5.2. Stereotactic Frame of Riechert and Mundinger

The first model of this apparatus was designed in 1951 (Riechert and Wolff, 1951), and subsequently it was modified and improved (Riechert and Mundinger, 1955, 1959; Riechert and Spuler, 1982). The apparatus (Fig. 48) consists of three main parts, a basal ring, a target arch with electrode holder, and a phantom ring with a system of coordinates.

The massive metal basal ring is fixed on the skull of the patient under anesthesia with six sharp pins introduced into the bone at appropriate points. The basal ring is divided into 360°. The plane of the ring corresponds to the horizontal plane of the skull, the sagittal plane to the line 0°–180°, and the interauricular plane to the line 90°–270°. In taking x-ray films after pneumoencephalography, it is necessary that the plane of the basal ring be strictly perpendicular to the planes of both film cassettes and that the central beams in the two projections pass into the sagittal and interauricular planes, respectively. This is achieved with the help of a light centering device parallel to the central ray.

The stereotactic calculations in the two projections are determined by three coordinates: vertical (the distance from the target point along the perpendicular to the center of the basal ring), horizontal (from the same point to the interauricular plane), and frontal (from the same point to the midplane).

The data obtained are then transferred to the phantom device, which is an exact copy of the basal ring and target arch. After the metal pin fixed in the electrode holder comes into contact with the calculated point inside the basal ring, which represents the target point, it is possible to determine the angle and the depth of electrode insertion by the polar coordinates as well as by the burr hole site. Then the target arch with the electrode holder is attached to the basal ring on the patient's skull, and a small burr is used to drill a hole in the bone, through which the electrode is inserted into the brain to reach the target point accurately. For the destruction of subcortical structures, Riechert and Mundinger employ high-frequency electrocoagulation.

The use of the phantom gives this apparatus definite advantages over other devices by ensuring a high degree of accuracy (up to 1 mm) and general utility. Another advantage of phantom modeling is that no additional calculations are necessary. The apparatus allows introduction of the electrode to any point deep in the brain from practically any point on the skull surface. This apparatus is one of the best in stereotactic surgery.

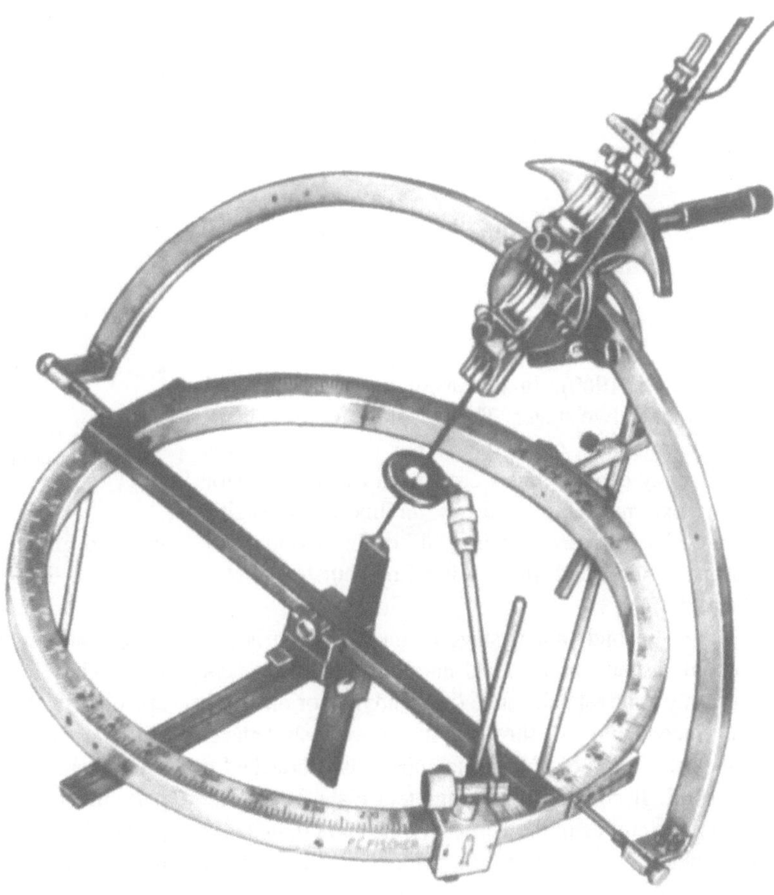

FIGURE 48. Stereotactic apparatus of Riechert and Mundinger. Description in the text.

Traugott Riechert (1905–1983)

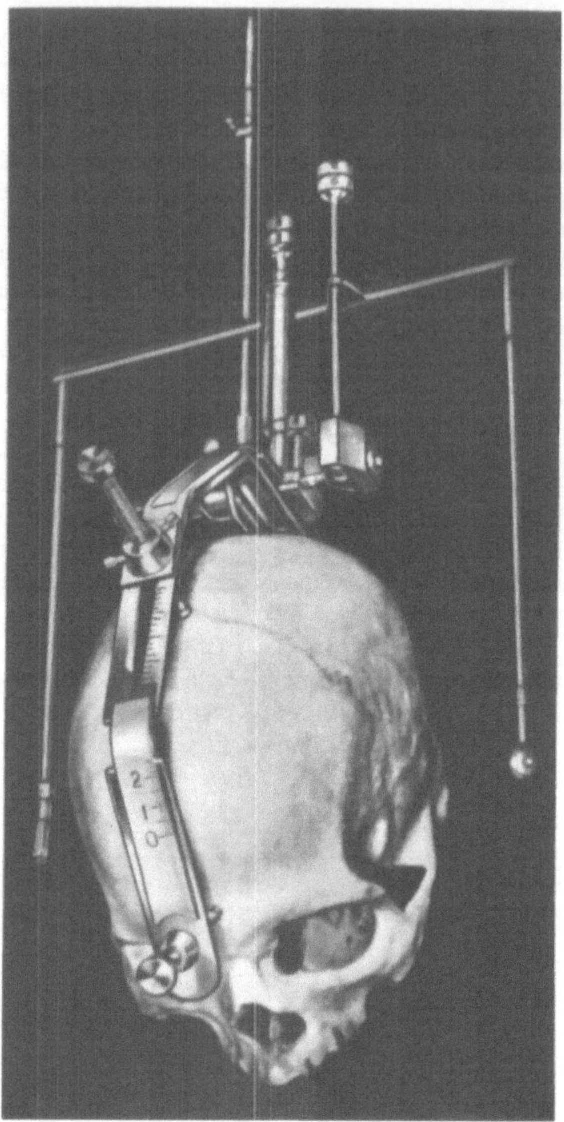

FIGURE 49. Stereotactic apparatus of Guiot and Gillingham. Description in the text.

5.3. Stereotactic Frame of Guiot and Gillingham

An original stereotactic apparatus was devised by Guiot in 1958 (Fig. 49) and subsequently modified by Gillingham (1960). In this method, the operation is performed in two stages. The first stage involves marking the midplane of the head. For this purpose, two small x-ray-opaque pins were laid in two narrow slots in the external table along the midline at the sagittal suture. Then a burr hole is made in the region of the coronal suture, and through it a fine catheter is introduced into V.

The second (main) stage of the operation is performed, usually under local anesthesia, in 2–3 days. An x-ray contrast substance (iophendylate or others) is introduced into V through the previously placed catheter. Then the stereotactic apparatus is attached to the skull. It consists of a long steel plate 3 cm wide with three curves in the vertical plane and graduated in millimeters. First, at certain points (lambda, coronal suture, and nasion) on the skull, three sharp pins are inserted along the midline as determined in the first stage of the operation. The stereotactic apparatus, which is set exactly in the sagittal plane from the nasion to inion with special attachments, is fixed to these pins. The electrode holder and the correcting device are placed at the posterior part of the plate. A metal structure in the form of a square inverted U is also fixed here on the main plate. Its horizontal arm is perpendicular to the main plate, and the two long metal pointers (trackers) are situated along both sides of the head parallel to each other, to the electrode guide, and to the plane of the main plate.

The distinguishing feature of this method is that the design of the apparatus allows the electrode to advance into the brain parallel to the midplane of the head. The second feature of this stereotactic method is the trajectory of the electrode, which traverses the posterior part of the brain (parietooccipital area) and V.

After the appropriate stereotactic calculations and the determination of the site of the target structure, the operation is performed under visual control with the aid of an electronic amplifier. Since the trajectory of the electrode is parallel to the midplane, it is necessary to control only the lateral projection. The desired subcortical structure in lateral projection is accurately superimposed on the two small holes at the ends of the trackers situated on both sides of the head. After such superimposition, the electrode that was introduced into the brain parallel to the trackers accurately hits the target, i.e., the center of the target structure. The depth of the electrode insertion is determined by fixing its tip in orfices of superimposed holes in both trackers.

In his initial work, Guiot used as the main cerebral reference the line connecting the posterior border of CA with the anterior border of the corpora mammillaria. This border is seen in the lateral projection as a "mammillary incisura" in the anteroinferior part of V_3. Subsequently, for the same purpose, the author began using LI as the main reference line. Guiot and Gillingham employ high-frequency diathermocoagulation for the destruction of subcortical structures.

This apparatus combines relative simplicity in design with a high level of accuracy. Yet, it should be noted that the introduction of the electrode from the posterior part of the brain may require it to traverse a considerably greater distance than if introduced by an anterior approach.

5.4. Stereotactic Frame of Leksell

Lars Leksell (1907–1986)

The apparatus proposed by Leksell in 1949 was one of the first to be built. It is based on the use of rectangular and polar coordinate systems. The main principle of the apparatus is such that the center of the arc always coincides with the brain target, which may be reached from any point on the skull. The apparatus consists of a metal coordinate frame in the form of a cube (Fig. 50). The rods forming this cube have horizontal and vertical radiopaque millimeter scales, which determine the three coordinate axes—sagittal, frontal, and horizontal. The second piece of the apparatus consists of a semicircular arch, also with a millimeter scale. The arc is fastened to the coordinate frame by two horizontal semiaxes fixed in the side rings that permit rotation. On this arch is an electrode guide that can be moved along both sides of the arch. The longitudinal axis of the stereotactic instrument passes through the geometric center of the arch.

The apparatus is fixed to the patient's head at three points by sharp stainless steel pins in such a way that the sides of the cube are parallel to the frontal and sagittal planes of the skull. The coordinate cube is rigidly attached by special rods to the short-distance x-ray tube and the cassette, which ensures their correct alignment in all planes (Fig. 51).

Pneumoencephalography, application of the apparatus, and filming are all done with the patient in a seated position. The coordinate cube is precisely oriented in the frontal and sagittal planes. With the patient in a supine position, a burr hole is made near the coronal suture. On the transparent "program map" showing the frontal and lateral images of the ventricular system, it is possible to calculate and mark the selected target point. Then the map is superimposed on the pneumoencephalograms to determine the three coordinates (X, Y, Z) on the scales of the frame.

Just as in other instruments built on the principle

FIGURE 50. Stereotactic apparatus of Leksell. Description in the text.

of moving the electrode in three dimensions, in Leksell's apparatus the scales for calculations on the basis of roentgenograms lie in other planes than the designated target point. In view of this, the method involves quite complicated mathematical corrections of the calculations using logarithms and a special diagram of spiral lines for correction of magnification. Target localization is calculated on data from conventional x-ray films either manually or with the Hewlett-Packard calculator with special stereotactic software. After the cal-

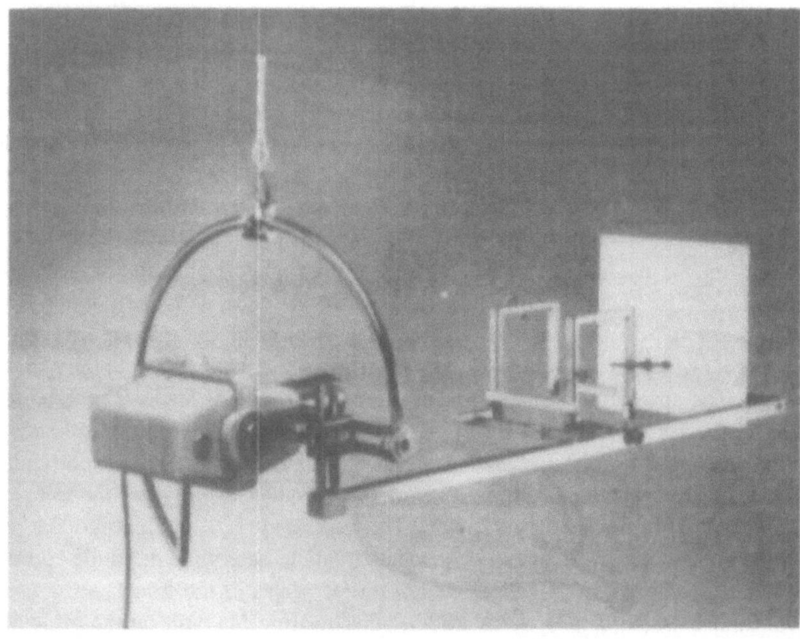

FIGURE 51. Portable x-ray tube combined with the Leksell stereotactic apparatus.

culated coordinates of the desired point are aligned with the radiopaque scales of the main frame, the appropriate angles are set on the mobile semicircular arch, which must be positioned so that its center exactly coincides with the target point. So arranged, with the arch in any position, the electrode hits the target. The required depth is precisely the radius of the arch. After these adjustments, the electrode or biopsy needle is advanced to the target point.

Leksell's apparatus is without a doubt one of the most accurate stereotactic devices. With it, an electrode or other stereotactic instrument can be introduced into any deep-seated structure in the brain from any point on the surface of the skull. Details on the techniques required to use this apparatus can be found in Leksell's (1971a) monograph, *Stereotaxis and Radiosurgery. An Operative System.*

Modifications of the Leksell apparatus for use with CT and NMR are described in Sections 7 and 8.

5.5. Stereotactic Frame of Talairach

This apparatus, based on the rectangular coordinate system, was devised by Talairach *et al.* in 1949 and then modified (Talairach *et al.*, 1958). It consists of a massive square aluminum alloy frame that is fixed on the patient's skull with four sharp pins (Fig. 52). This apparatus requires a long (approximately 4 m) distance between the x-ray tube and the object, as a result of which there is no need for a divergence correction. The main part of the apparatus consists of four square, hollow rectangles that are fixed on each side of the frame. On the two plates forming each rectangle there is a series of small holes spaced 1 mm apart. On the films in both projections, the holes of each of the two perforated plates must be superimposed on each other to confirm the proper orthogonal positioning.

In the lateral film it is necessary to select the hole of the rectangle that coincides with the target point deep in the brain. Then a graduated electrode is inserted into the selected hole and advanced accurately toward the target. The depth of penetration is controlled on the AP x ray with a similar grid. Talairach's apparatus makes use of the orthogonal approach to the target point in the brain—the electrode is introduced through the grid hole at a right angle to the stereotactic frame and parallel to the central x-ray beam. Talairach's second model includes a guiding device so that the electrode may be introduced not only in the horizontal but in any plane. A phantom device (''*table d'étude*'') is used for this purpose.

Although Talairach's apparatus is a general-purpose instrument, it is employed most frequently for operating in epilepsy. Talairach also proposed a special stereotactic apparatus for pituitary surgery (Fig. 53).

Recently, a modification of the Talairach apparatus was proposed by Scerrati *et al.* (1985). This model includes a target arch connected to the double grid to

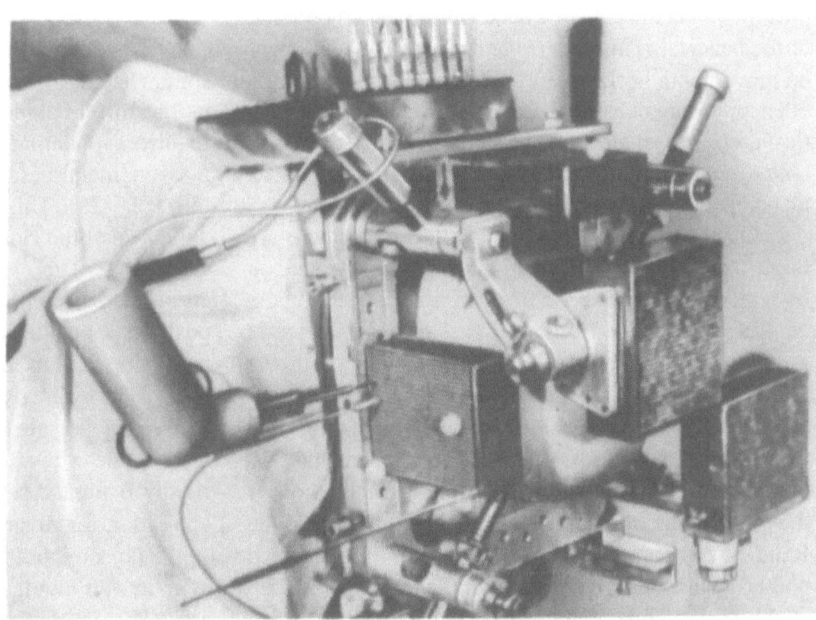

FIGURE 52. Stereotactic apparatus of Talairach combined with our cryosurgical device. Description in the text.

FIGURE 53. Stereotactic apparatus of Talairach for pituitary surgery.

allow the use of polar coordinates. A sliding bar and sliding carrier fixed to the frame allow the introduction of the electrode without using the double grid holes.

5.6. Stereotactic Frame of Rand and Wells

This apparatus consists of a massive square metal horizontal headrest with millimeter scales engraved on three sides (Fig. 54). A T-shaped arch with three sharp pins for fixing the patient's head is fastened on one side of the headrest. This arch (together with the head) may be moved both horizontally and vertically. On the two other opposite sides of the main frame there is a vertical arch with an electrode holder that can be moved in both directions along the arch and also in the horizontal plane.

On the opposite columns of this frame there are large round holes with centering crosses located on both sides of the head. The frontal axis connecting the centers of these apertures must pass through the target point inside the brain (the same principle as in the Guiot–Gillingham apparatus). In this case, the electrode guide, which is located in the same plane as the frontal axis, will ensure that the instrument hits the target point irrespective of the angle of inclination of the guide in the frontal plane. An original feature of the Rand–Wells apparatus is, therefore, "the fixed point," which must coincide with the target point deep in the brain. This is attained by the graduated movement of the patient's head in the sagittal and frontal planes and through a semiarch in the horizontal plane. In other words, in this apparatus it is not the direction of the electrode that is changed to hit the target point but the point is brought up to the fixed electrode.

The Rand–Wells apparatus has a high degree of accuracy and is a general-purpose instrument that can be used for any stereotactic operation.

5.7. Stereotactic Frame of Laitinen

This apparatus, developed in 1971, has a design somewhat similar to the Riechert–Mundinger apparatus but differs significantly from it in that it does not require a phantom device. The apparatus includes a massive metal basal ring and a semicircular arch fixed to it (Fig. 55). This is applied to the head of the sitting patient, and the ring is fastened symmetrically by four sharp pins. The electrode holder can move along the entire semicircumference of the arch. The patient is positioned on the table in a supine position, and the basal ring is connected to a holder built into the headrest. Then a burr hole is made near the coronal suture, the ventricles are tapped, and a ventriculogram is made. The x-ray tubes for lateral and AP films are rigidly fixed at a set distance from the head so that the correction for divergence is standardized.

The semicircular frame terminates in two hollow cylinders 50 mm in diameter. Films in both projections are taken in such a way that the central ray passes through small holes in both cylinders. Laitinen empha-

FIGURE 54. Stereotactic apparatus of Rand and Wells. Description in the text.

sizes that the accuracy of this device is within 0.25 mm, and a desired target point in the depths of the brain can be reached by the electrode from any direction. In addition to operations on the basal ganglia, the apparatus can be used for stereotactic operations on the posterior cranial fossa and the cervical spinal cord as well as for transnasal hypophysectomy.

Laitinen designed another guide for stereotactic destruction of the trigeminal and glossopharyngeal nerves (Chapter 13, Section 7).

5.8. Stereotactic Frame of Oliver–Bertrand–Tipal

This apparatus is a massive metal cube with millimeter scales on horizontal and vertical stanchions (Fig. 56). The cube is fixed on the patient's head by four pins. Two rings for centering lateral x-ray beams are attached on the lateral surfaces of the cube. These rings are centers of rotation for the metal frame, which consists of two rods and an arch and can be rotated in the sagittal plane. Two electrode holders with scales to mark the depth of penetration can move in the frontal

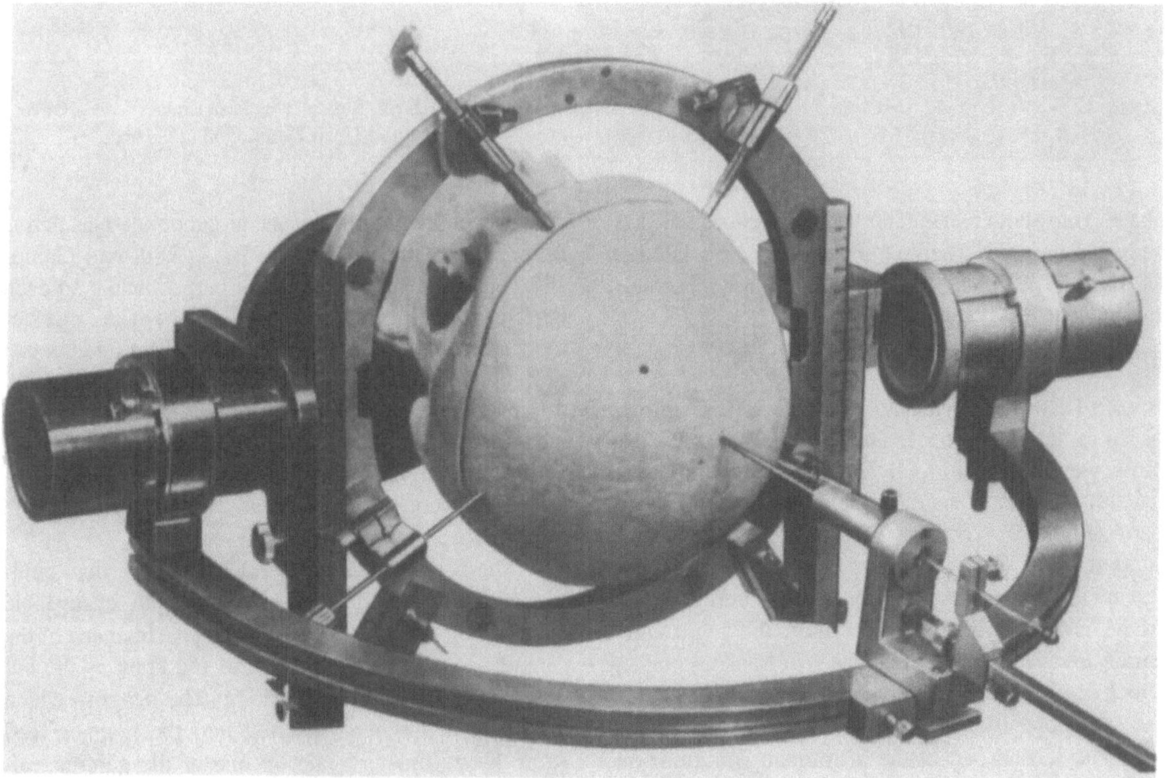

FIGURE 55. Stereotactic apparatus of Laitinen. Description in the text.

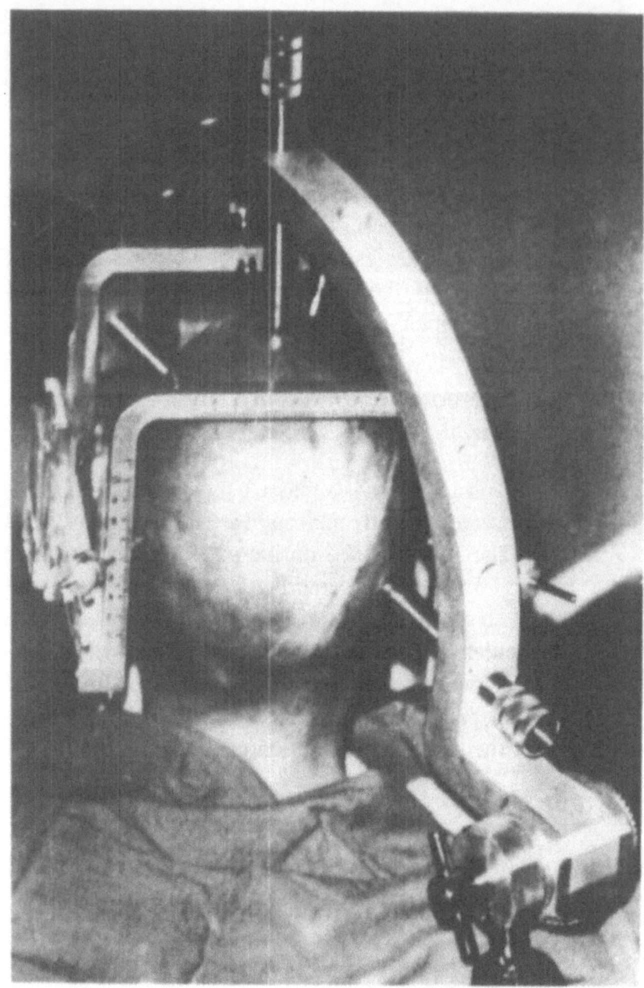

FIGURE 56. Stereotactic apparatus of Oliver, Bertrand, and Tipal. Description in the text.

plane. Thus, both rectangular and polar coordinate systems can be used. The instrument can be operated on the basis of CT data, digital subtraction angiograms, and nuclear magnetoencephalograms.

5.9. Stereotactic Frame of Kandel

In 1970 we designed a new stereotactic instrument based on the polar coordinate system. With this apparatus we have performed about 1400 operations to date. It is made of stainless steel and titanium in two parts (Fig. 57). The first part is a small rectangular support frame with a 30-mm-diameter hole in the center and three lugs that are fastened to the burr hole. The frame is firmly fixed by spreading the lugs with a spiral-threaded toothed ring. The gear transmission and a lock screw ensure reliable fixation in the burr hole for the entire operation.

The second part of the instrument is a guide and correcting device to ensure that the electrode or cannula is accurately aimed at the target point. This part is fastened onto the supporting frame with two clamps and consists of two protractors perpendicular to each other calibrated in degrees; the centers of the protractors correspond to 0°. The mobile frontal protractor can be moved relative to the fixed sagittal protractor. A roller device on the frontal protractor holds the cannula–electrode, which can move in two planes—sagittal and frontal. The necessary correction angles determined on ventriculograms can be transferred to the protractors.

The body of the apparatus has two vertical guiding columns and a transverse bar with changeable bushings of various diameters. The proximal end of the cannula–electrode that penetrates the brain as the bar descends is fixed in this bushing. The depth to which the cannula has been inserted into the brain is read from the millimeter scale on one of the guiding columns. For this purpose, when the tip of the cannula

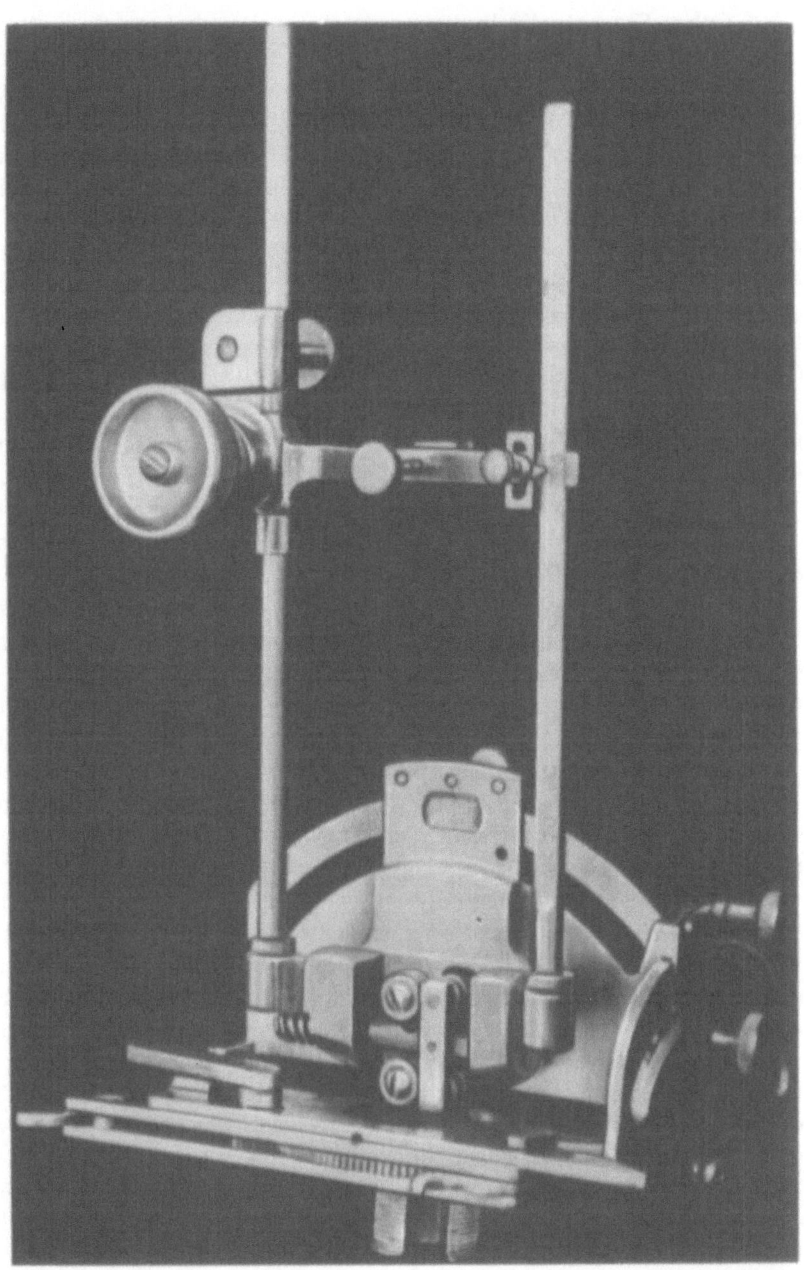

FIGURE 57. Our stereotactic apparatus. Description in the text.

makes contact with the cerebral cortex, the pointer is set at zero on the scale. The transverse bar is moved along the vertical columns with the aid of a rack-and-pinion gear, making it possible to introduce the instrument smoothly and evenly into the brain and, if need be, to advance the cannula in steps of 0.5 mm and lock it at any desired depth.

The roller device with spring clamps through which the cannula is inserted into the brain eliminates all play and also allows the use of cannulas with different diameters (from 1 to 4 mm).

The many years of work with this instrument have demonstrated its great accuracy in hitting a given subcortical structure (up to ± 0.5 mm), its portability, its rapid and reliable fixation on cranial bones, and its convenience for the neurosurgeon. The use of this apparatus in stereotactic operations is described in Chapter 4.

5.10. Other Stereotactic Instruments

In addition to those described above, stereotactic frames and devices have been devised by Bailey and Stein (1951), Monnier (1952), Wada (1953), Fairman (1959), Van Manen and Van Hoytema (1962), Cooper (1962a,b), Bradford (1962), Boctor (1962), Van Buren and McCubbin (1962b), Asenjo *et al.* (1966), Bertrand and Martinez (1966), Sugita and Murata (1966), Ray (1967), Todd and Wells (1969), Barcia-Salorio and Broseta (1976), Anichkov (1977), Huk and Baer (1980), Olivier and Bertrand (1982), Scerrati *et al.* (1984), Carol (1985), Olivier (1986), and others (CT-guided stereotactic apparatuses are described in Section 6).

6. Computer Techniques

In view of the rapid development of stereotactic surgery, the more complicated methods, and stringent requirements for accurately and reliably reaching a given cerebral structure, the use of computers for stereotactic purposes has been swiftly advancing in recent years. As Tasker (1987) concluded ". . . the most spectacular phase in the development of stereotaxis is the marriage with computer technology." Computers with varying degrees of complexity are extensively employed for both stereotactic calculations and verifying the position of the electrode (cannula) tip in the desired subcortical structures. It is noteworthy that a computer was first successfully employed by the Soviet investigators Belyaev *et al.* (1965) and Ivannikov (1969) at the research center headed by Bekhtereva for determining the stereotactic coordinates of target points in the brain. They used a computer for reciprocal conversion of coordinate systems.

Complicated computerized mathematical programs and software have been devised in subsequent years for determining and correcting the position of the electrode or cannula tip in the brain (Peluso and Gybels, 1972; Bertrand *et al.*, 1974a,b; Birg and Mundinger, 1975; Bertrand, 1982; Kelly, 1987). A large number of sections from the atlases of Schaltenbrand and Bailey or Van Buren and Borke are fed into a computer. By using analogue field plotters and a computer graphics terminal, it is possible to display various brain scans in any plane and to superimpose the inserted electrode–cannula in the brain as well as to calculate the correct angles for introducing the instrument. Any part of the instrument that is outside the plane of the scan can be seen as a dotted outline on the display (Bertrand, 1982). It is beginning to appear possible to employ computers to calculate the location of subcortical structures on the basis of cranial reference points without first visualizing the ventricular system (Birg *et al.*, 1977a,b; Mundinger *et al.*, 1978a; Bertrand, 1979; Laitinen, 1985).

Various models, in particular portable digital computers (Seitzer *et al.*, 1980), are successfully being used for stereotactic calculations. If 11 x-ray parameters derived from x-ray examination during surgery (eight in lateral projection and three in frontal) are fed into such a computer, it can produce the coordinates of the target point and the information needed to insert the electrode–cannula accurately (Dervin *et al.*, 1974). We use a minicomputer to calculate the depth to which the cannula tip must be inserted to reach the desired target point.

The use of a computer is most expedient for calculating several stereotactic targets, e.g., for implanting a large number of chronic electrodes. The on-line computer-graphic method also has been employed with success for evaluating the results of electrostimulation and recording of subcortical structures so that corrections can be made for stereotactic purposes (Saito and Ohye, 1974; Tasker *et al.*, 1977). Computer analysis also enables one to obtain histograms of the intervals between spikes of rhythmically discharging neurons in Th nuclei and to establish their correlation with peripheral tremor (Bertrand, 1982).

There have been successful attempts to put serial brain sections into the memory of a computer and have them reconstructed in three dimensions (Cahan and Trombke, 1975). This enables an automatic comparison of the coordinates of any brain structure with the coordinates of the structure in question obtained during the operation. A computer is used for obtaining on a color graphics terminal a three-dimensional image of any subcortical structure from a stereotactic atlas (Giorgi *et al.*, 1983). The three-dimensional display allows one to visualize the scans in nonorthogonal planes that contain the trajectory of the cannula–electrode. Color coding of diencephalic nuclei on inclined scans helps to identify various nuclei in a display that may differ substantially from that in a conventional stereotactic atlas (Giorgi *et al.*, 1983) (Fig. 58).

The main principle for the computer graphics technique proposed by Hardy *et al.* (1983) involves superimposing graphics of the corresponding cerebral scans on the on-line display of the trajectory of the stereotactic instrument. These scans from various ster-

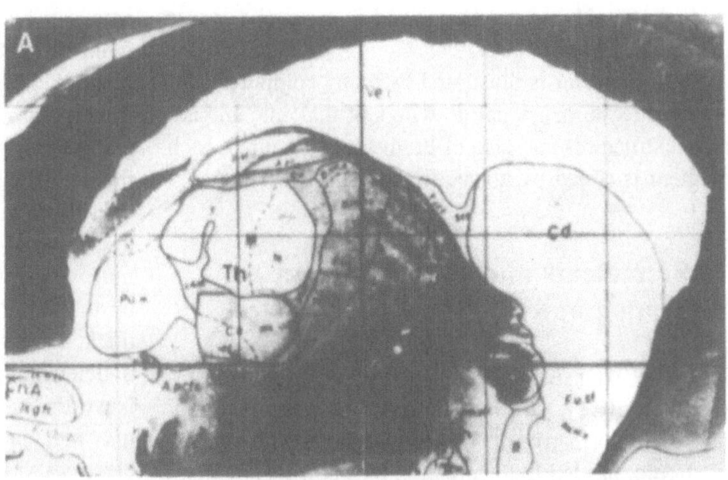

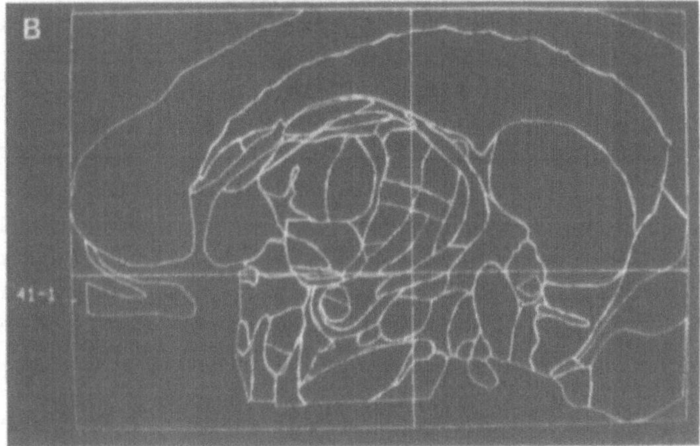

FIGURE 58. Three-dimensional processing of a stereotactic atlas (Giorgi *et al.*, 1983). (A) Brain sagittal section from the Schaltenbrand and Wahren stereotactic atlas. (B) Computer-generated image with digitization of the same section.

eotactic atlases are stored in the computer memory. The operator can retrieve from the computer any scan from a catalogue and, using the display, feed it into the graphics terminal. This display may be modified according to any parameter determined on a specific ventriculogram of a given patient.

Brown (1979) has proposed the use of three-dimensional computerized graphics to project a probe accurately to any cerebral structure in the future. A stereotactic atlas was incorporated into a computer program for visualizing the stereotactic probe relative to the atlas (Kelly, 1983).

A very impressive development in using a three-dimensional computer-based recording technique for stereotactic surgery was presented recently by Afshar (1987). The Andrew-Watkins (1969) and Afshar *et al.* (1978) two-dimensional stereotactic atlases were used for computer reconstructions of the thalamic nuclei and adjacent structures. Twenty-six separate structures were selected for computer digitization and three-dimensional reconstruction. Data were stored on discs and transferred to a mainframe computer. The *X* and *Y* coordinates from each brain section were scaled and rotated to a common origin and magnification (the *Z* coordinate is known because the distance between sections is constant: 1 mm).

A visual display unit with color graphics allows one to obtain the *X* and *Y* coordinates of the computer-stored structure. Each section of the atlas can be computer-generated as a hemisection.

The computer program written in FORTRAN allows the visualization of each image from different angles and also permits the image's rotation about any or all of the three orthogonal axes.

As Afshar stressed, once a specific target site has been chosen, adjacent structures can be constructed around the target point and viewed in any number of directions to access the optimal trajectory. Alternatively, once the trajectory angle is known to the reference plane, the relationship of structures to the electrode or cannula tip can be viewed as the probe passes to the calculated target.

This chapter is illustrated by many computer-generated three-dimensional drawings of thalamic and adjacent structures and also of brainstem structures; each of them is coded by a specific color.

7. Stereotaxis and Computerized Tomography

The extensive practical use of computerized tomography (CT), which represents an extremely valuable and informative method for diagnosing many CNS disorders, has led to the creation of a new technique in stereotactic operations on deep brain structures. Since CT investigations produce a three-dimensional reconstruction of brain anatomy and since the stereotactic method is based on three-dimensional coordinates of these structures, it seems natural to combine these two techniques in stereotactic neurosurgery. This combination has been called stereotactic computerized tomography (Bergström and Greitz, 1976).

It is now generally acknowledged that the use of the CT technique has made it possible not only to abandon ventriculography but to increase the accuracy of stereotactic operations and, in so doing, reduce the risk of complications. This technique allows one to use serial brain CT scans to determine accurately, with the aid of the computer of the same CT system, the stereotactic coordinates of any point in a deep-seated cerebral structure (Brown, 1979; Leksell and Jernberg, 1980; Colombo *et al.*, 1981; Bertrand, 1982). If a preprogrammed computer is used, the surgeon may put a cursor over a target point on a displayed CT scan. The point is then suspended within a three-dimensional computer matrix. After that, the computer calculates the stereotactic target coordinates (Kelly *et al.*, 1985).

Laitinen *et al.* (1985) have compared the accuracy of hitting the Th structures ND and Am with the aid of conventional ventriculograms and with CT. The mean difference in the two methods in determination of *X* and *Y* coordinates was not more than 0.6–0.7 mm, and the *Z* coordinate was the same in all cases. On CT scans, just as on conventional skull x rays, it is possible to determine the target points seen on the scans and to establish their coordinates. From these points, one can calculate the coordinates of any other point that cannot be seen on the scans.

It is comparatively easy to calculate the two coordinates for a target point on CT scans. The AP coordinate can be determined by the position of the target point in relation to the midline intersection of the AP plane with sections of the frontal and occipital bones. The lateral coordinate is determined by measuring the distance from the midplane to the target point. It is more difficult to calculate the vertical coordinate because it is necessary to take into account not only the thickness of the CT slices but also the plane of scanning. This problem in recent years has been eased by up-to-date scanners that produce a section only 1.5 mm thick. However, even in such circumstances, the error in determining the vertical coordinate may be 2–3 mm. Nevertheless, with the computer of these CT units, one may calculate the path of the stereotactic electrode (or electrodes) with an accuracy approaching 1 mm.

There are two principal methods of making stereotactic calculations on CT data. In the first, conventional CT is performed after various markers have been fastened on the head of the patient, for example, acrylic plates in which wires of various lengths, clearly visible on CT scans, are embedded (Lee *et al.*, 1978; Matsumoto *et al.*, 1985). Then the measurements for the coordinates of the target point made on CT scans before the operation are transferred to the roentgenograms of the head taken at the beginning of the operation according to strict stereotactic techniques (Gleason *et al.*, 1978; Kelly *et al.*, 1978; Russell and Brown, 1979; Kandel *et al.*, 1981; Gildenberg *et al.*, 1982; Rhodes *et al.*, 1982; Hardy and Koch, 1982; Gildenberg, 1983). As already mentioned, the plane of the horizontal CT scans must be exactly perpendicular to the subsequent x-ray films taken in two projections in the operating room as well as to the midplane of the head. Bone structures of the skull are used as reference points for transferring the desired data (Penn *et al.*, 1978b; Cail and Morris, 1979).

In transferring the CT data to conventional craniograms it is not obligatory to use a correction factor for x-ray divergence. One should measure at corresponding points the longitudinal and transverse diameters of the skull on photographs of CT scans and on ordinary films and then calculate the magnification factor.

The second method involves CT with a specially modified or CT-compatible stereotactic apparatus fixed on the head of the patient (Coloff *et al.*, 1973; Dervin *et al.*, 1974; Brown, 1979; Cail and Morris, 1979; Greitz *et al.*, 1980; Huk and Baer, 1980; Shelden *et al.*, 1980; Koslow *et al.*, 1981; Birg and Mundinger, 1982; Dubois *et al.*, 1982; Bertrand, 1982; Amano *et al.*, 1983; Hitchcock *et al.*, 1985; Carol, 1985; Iseki *et al.*, 1985; Gouda *et al.*, 1986). In this case, the precise spatial interrelationships between the apparatus and the target point are known (Fig. 59). If the stereotactic apparatus is not in the plane of the CT scan in which the target point lies, then by repeated scanning during the operation it is possible to determine the advance of the electrode to the target.

The modern CT scanners (for instance, 8800 General Electric) are supplied by an independent computer system, which stores all imaging information (CT, NMR, DA) for use in the operating room.

Integration of the stereotactic method and CT during an operation requires attention to a number of technical conditions: special transparent markers easily seen on CT scans must be installed on the stereotactic apparatus, which should be of materials that produce minimal artifacts on CT-scans, and special or standard computer software should be used for determining the target point (or points).

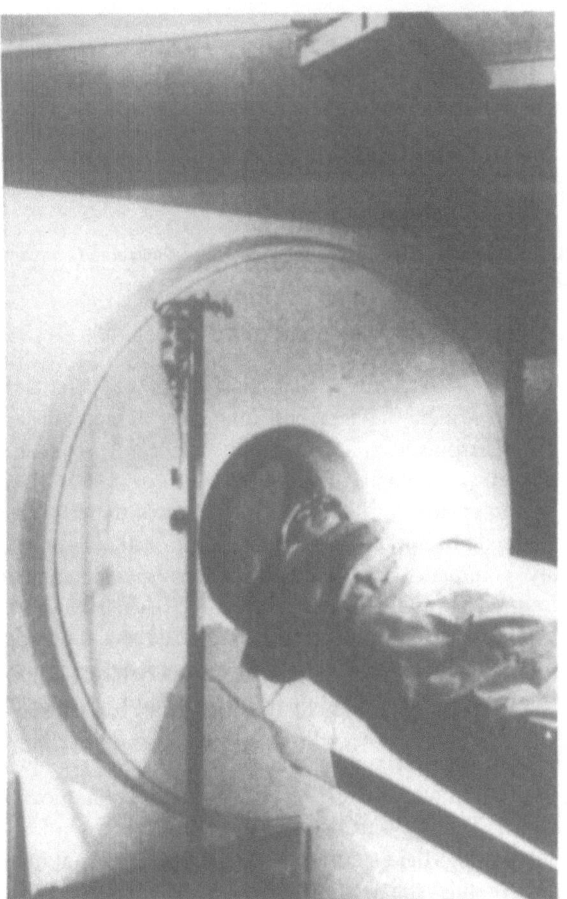

FIGURE 59. Patient with base ring of Riechert–Mundinger stereotactic apparatus during the CT investigation (Birg and Mundinger, 1982).

The majority of existing stereotactic apparatuses described in Section 5 are not suitable for CT-guided stereotaxy because the CT images are not satisfactory. This was the reason that modifications of the stereotactic apparatus, described in Section 5, were proposed in recent years. These instruments are made of materials that do not produce artifacts in CT studies (Plexiglas, acrylic, carbon fiber, nylon, aluminum, molybdenum disulfide). The reference systems on their frames are easily identified on CT scans, ensuring accuracy of the stereotactic calculations. They also permit intraoperative CT studies with the electrode deep in the brain. In CT investigation the plane of the zero CT scan should be parallel to the basal ring of the stereotactic apparatus. The target point and its coordinates are marked on the CT scans, and these data are fed into the computer, which calculates the parameters of the path of the cannula-electrode.

A number of systems for integrating stereotactic surgery with CT have been successfully employed recently (Perry system, Leksell system, Patil system, Brown–Roberts–Wells system, and others). The digital radiographic system for stereotactic calculations used by Fröder *et al.* (1983) consists of an x-ray image, an electronic amplifier, a TV camera with an output converting the image into digital form, a digital processor, and a display.

These systems and devices can be used in two ways, which Mundinger and Birg (1984) have termed indirect and direct methods. In the first, CT scans are made with the stereotactic apparatus (or its basal ring) fixed on the head, after which the patient is taken to the operating room and the stereotactic operation is performed. In the second, the operation is performed directly under CT control. In this case, the operating room must have a specially adapted CT scanner with a high resolving capacity (Lunsford, 1982; Fujita and Hosoda, 1984; Matsumoto *et al.*, 1985). Lunsford and Martinez (1984) operated on 48 patients with various brain tumors using a GE 8800 CT scanner (Fig. 60). Scanning was performed before, during, and directly after surgery. A fluoroscopic image intensifier connected to the scanner permits intraoperative angiography should the need arise. The technique described allows one to diagnose intracranial pathological processes in their early stages, prior to the development of a serious neurological deficit, with the aid of stereotactic biopsy to obtain an early and accurate histological

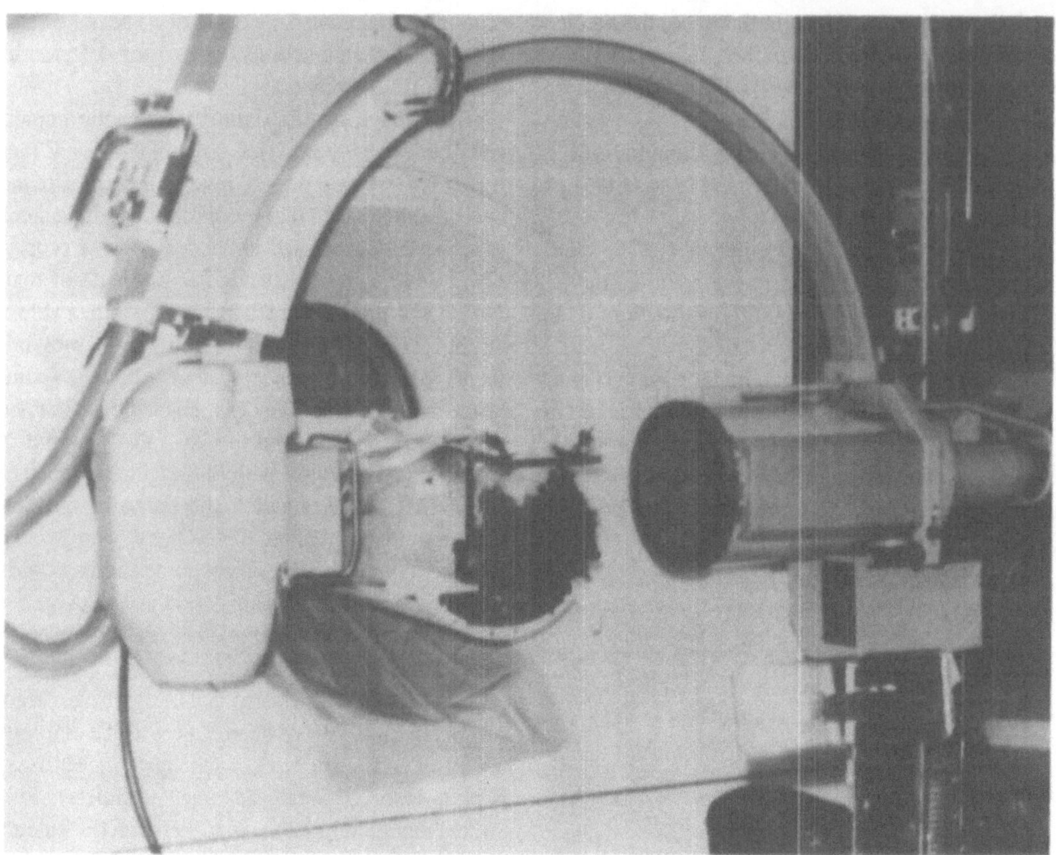

FIGURE 60. A CT scanner (GE 8800, General Electric Corp.) with fluoroscopic image intensifier in the operating room (Lunsford and Martinez, 1984).

diagnosis and immediately apply appropriate surgical treatment.

It should be noted, however, that the direct method is more complicated for technical and economic reasons. Even when the most up-to-date, rapid-action scanner is used, the duration of the operation is increased, and the location of the head of the patient in the scanner is inconvenient for the surgeon. Moreover, many stereotactic instruments are larger than the gantry of a computer tomograph. Nevertheless, there can be no doubt that intraoperative CT will find extensive application in the not-too-distant future.

A new model of the Leksell apparatus (CT model) has recently been developed allowing one to perform a stereotactic operation on the data of CT studies (Leksell and Jernberg, 1980). The coordinate frame of the apparatus, made of aluminum to reduce artifacts on the CT images, is installed directly on the table of the scanner, and the coordinates of the target point are determined on CT scans by comparison with fiducial markers. Fixation of the head in the apparatus is achieved with sharp carbon fiber pins, which are fastened in plastic and aluminum "sleeves." Then the Leksell stereotactic apparatus (Section 5.4), in its proper position, is attached to the CT scanner with a magnetic fixing rod. Two plastic disks with diagonal x-ray contrast stripes are attached to the stereotactic frame during CT investigations.

Law and Cocak (1982) proposed a simply designed adapter for the Leksell stereotactic apparatus, enabling CT investigations to be used for stereotactic calculations. The adapter is a three-sided polyethylene helmet that is fixed on the skull with nylon screws. Two aluminum wires serve as markers on the CT scans. The Leksell apparatus is put on top of the helmet, and the electrode is oriented in the usual manner for this apparatus (Section 5.4). Lunsford *et al.* (1983b) built a simple probe consisting of freely revolving circumferential arms, which are fastened to the main stereotactic frame of the Leksell apparatus. After the cannula (electrode) has reached the target point, it is fastened to the probe holder, and the semicircular arch of the stereotactic apparatus is removed. This permits CT scans to be done without artifacts (with the exception of those from the cannula itself).

Laitinen *et al.* (1985) also designed a simple adapter for CT-assisted stereotactic operations (Fig. 61). The adapter is an aluminum frame fixed on the patient's head by a nasion support and two earplugs. The adapter may be fitted to any existing stereotactic apparatus.

One of the most accurate methods is that of Birg and Mundinger (1973, 1982) in which the coordinates for the target point are determined visually directly on CT scans. The CT examination is performed after the basal ring of the Riechert–Mundinger apparatus (Section 5.2) has been fixed on the head of the patient. The zero point of the ring must coincide with the zero point of the scanner. The CT scans are made parallel to the basal ring. The thickness of the scans in the zone of interest does not exceed 1.5 mm. A computer is used for reconstruction of the sagittal and frontal sections, where FM, CP, the line connecting these points, and other reference points as well as the site of the burr hole are determined. This method may also be used for implantation of radioactive isotopes, stereotactic biopsy, and so forth.

The original skull-anchoring stereotactic frame was developed by Greitz *et al.* (1980). The system consists of a helmet made of a thermoplastic impregnated with photosensitive resin that hardens when exposed to ultraviolet light. The helmet is attached to the metal base ring by aluminum plates. A second base ring is attached to the CT-scanning table. Thus the patient's head, when placed in the helmet, remains in a constant relationship to the scanning table.

Shelden *et al.* (1982) designed a system consisting of the Riechert–Mundinger apparatus and a CT scanner. On the upper surface of the basal ring in that apparatus are a number of short graduated rods, the heights of which differ by 1 mm. This allows one to determine the three coordinate axes (X, Y, Z) on CT scans. A system very similar to this one is employed by Dubois *et al.* (1982).

Several methods for determining CT coordinates have been proposed in recent years (Bergström and

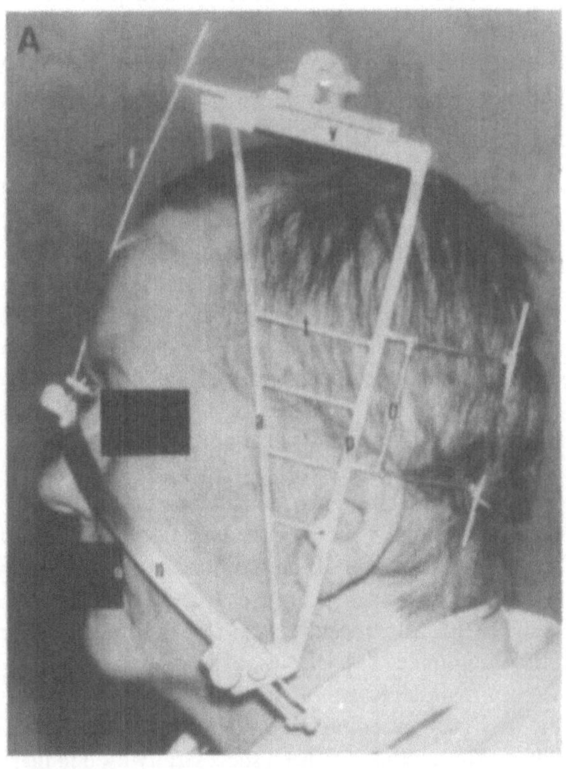

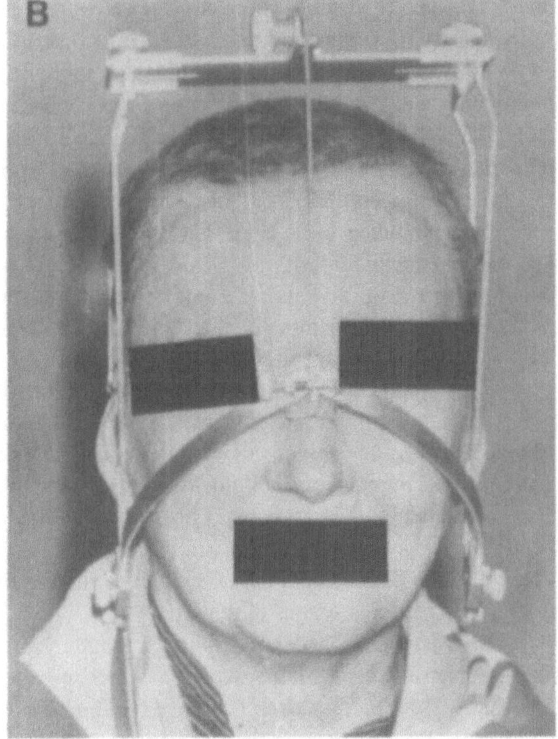

FIGURE 61. Laitinen's computed tomography adapter. (A) Lateral view; n, nasion support arm; a, anterior arm; p, posterior arm; v, vertex arm; t, transverse arm; f, frontal laterality indicator; o, posterior laterality indicator. (B) Anterior view.

Greitz, 1976; Gleason *et al.*, 1978; Mundinger *et al.*, 1978a; Penn *et al.*, 1978b; Wise and Gleason, 1979; Brown, 1979; Leksell and Jernberg, 1980; Kelly and Alker, 1981). The main problem is conversion of CT coordinates into coordinates of the stereotactic apparatus, and the overlaying of the systems. The accuracy of such stereotactic calculations depends directly on the resolving capacity of the CT scanner.

A number of mathematical methods have been elaborated for converting CT images into stereotactic coordinates. One of the complicated methods suggested by Schlegel *et al.* (1981), in brief, involves a computer that converts 20–30 CT images 4 mm thick into digits and records them on magnetic tape. Each of these images is distributed over a square grid consisting of approximately 65,000 dots. Thirty consecutive scans form a three-dimensional grid consisting of approximately two million dots. These data are put into the computer memory. The software then has the computer construct and produce images of any given section in any direction (Lunsford *et al.*, 1987). Although the required section usually does not exactly coincide with the three-dimensional grid, the discrepancy is small—the distance between the reconstructed plane and any neighboring point in the computer memory does not exceed 1.5 mm. The computer calculates the x-ray density (in Haunsfield units) of each reconstructed point by comparing the density of eight neighboring points and compares the dimensions of the CT scans and roentgenograms in both projections by linear interpolation. In this way, it is possible to reconstruct from CT images the coronal, sagittal, and radial planes, which are projected on a color TV monitor and displayed graphically. In both instances, images are produced of the skull and tumor contours in any given plane, which by superimposition are made to coincide with stereotactic roentgenograms.

The stereotactic coordinates of a target point (or points) inside a tumor are calculated on AP and lateral films by the routine stereotactic technique. The angles at which the stereotactic instrument is introduced are also calculated by the computer. This method has also been used for calculating the distribution of radiation doses in three-dimensional space after implantation of radioactive particles within the tumor.

The use of a Scout View computer program produces a lateral image of the skull on which parallel lines show the plane of each CT scan (Gildenberg *et al.*, 1982). This image may be compared to a conventional lateral roentgenogram made during the stereotactic operation; on it one can determine the plane of scanning and calculate the stereotactic coordinates of one or several target points. The vertical coordinates are marked off on the CT scan closest to the target point (''zero scan''). The cursor is set at the zero point, and the distance is measured from the midplane to the target point, the AP and lateral coordinates. The reference plane is the plane that crosses the established reference points at the base of the skull—frontal sinus, orbit tectum, sella turcica, sphenoid sinus, etc. Following the program, the computer establishes the distance between the reference plane and the zero scan. During the stereotactic operation, the zero plane and a parallel reference plane, taking into account the divergence correction, are plotted on the lateral roentgenogram.

One other method for performing a stereotactic operation with the use of intraoperative CT has been proposed (Fujita and Hosoda, 1984). When the head of the patient is fixed in the stereotactic apparatus, 2-mm-thick CT scans are taken parallel to LI (approximately 15–20 scans). The image of the sagittal plane is then reconstructed, and LI and the scan containing the target point are marked on it. Once the coordinates of that point are obtained, the electrode is then introduced into it, the accuracy being controlled by intraoperative CT.

Still another technique has been proposed for transferring the data of any diagnostic procedure (pneumoencephalography, angiography, gammagraphy, and others) to CT images and vice versa (Barcia-Salorio *et al.*, 1982). The method is applicable in any stereotactic operation, including biopsy, as well as in radiotherapy. The stereotactic apparatus is an aluminum frame with four sharp pins that are screwed into the skull bones under local anesthesia. For calculating the coordinates of the target point on the basis of CT data, three wooden scales are fixed on the frame: the anterior scale is square, and the two lateral are triangular, and all have metal wires in them. In CT investigations the five points corresponding to these wires are visible on the scans.

A special program fed into the computer allows one to determine by cursor the target point on the CT screen. On the basis of the positions of the five index points, the computer determines X and Z, the distances between the cursor and the corresponding axes. The third coordinate (Y) is calculated from the height of the CT scan, which is determined by the distance between the two lateral points indicated by the appropriate wires. This method has been successfully used in many stereotactic operations.

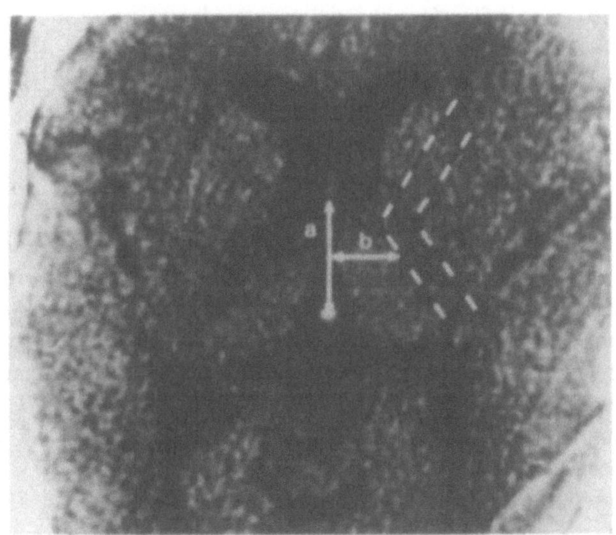

FIGURE 62. The distance from FM to GlP (line a) and the distance from the midpoint of this line to the medial border of CI (line b) on CT scan. The ratio of their lengths is 0.63 (Burchiel *et al.*, 1980).

It is possible to choose the optimal probe trajectory by using the computer cursor. If the trajectory to the target is unacceptable because it passes through functional zones, one may change the angle of the patient's head or the gantry tilt. The CT investigation is repeated until an optimal trajectory is found.

A simple technique not requiring additional equipment has been suggested for performing stereotactic calculations with CT (Burchiel *et al.*, 1980). Both FM and GlP are determined on one of the CT scans taken before the operation, and their centers are connected by a line on the scan (Fig. 62). Then one measures the distance from the midpoint of that line to the medial border of CI, and the ratio of the lengths of the two lines is determined. On the ventriculograms obtained during the operation, one measures the length of LI and multiplies that figure by the ratio obtained from CT data. The calculated figure is then compared to standard data from a stereotactic atlas. In determining the coordinates for the target point, one must take into account the ratio between the distance from the midplane in the patient and that in the stereotactic atlas. Since the midline of FM–GlP lies 5 mm above LI, the corresponding horizontal section in the atlas is used. This method can increase the safety of a stereotactic operation, especially in those cases in which, because of variability, the medial border of CI lies closer to the midplane than in normal cases.

For stereotactic biopsy of deep-seated tumors (Chapter 15, Section 1), clipping of aneurysms (Chapter 16), and removal of intracerebral hematomas (Chapter 17), we use our own method to transfer CT data to conventional skull roentgenograms. Then, by stereotactic calculations, we perform a biopsy in some deep-seated brain tumors (Kandel *et al.*, 1981).

Shelden *et al.* (1980) have described a new "microstereotactic" method for the localization and removal of small (up to 1 cm in diameter) pathological lesions (especially tumors) deep in the brain. Stereotactic calculations are performed by the computer according to special software with the aid of which one can see on the monitor screen three-dimensional colored images enlarged to given dimensions of the smallest lesions in any part of the brain (Shelden *et al.*, 1982). The operation is performed using a micromanipulator under control of the three-dimensional visual reconstruction of the lesion. Algorithms have been created to enable a computer to convert CT scans into three-dimensional pictures and interpret them quantitatively. Color coding improves visual perception of pathological lesions and other changes visible on CT scans.

A new achievement in stereotactic technique is the integration of CT imaging and the operating microscope (Roberts *et al.*, 1986). The sophisticated system provides for the reformatting and projection of CT data stored on the magnetic tape to the focal plane of the microscope in a common coordinate space. A set of fiducial markers are visible on the CT image and through the optics of the microscope, the position of which is determined by a three-dimensional referencing system. The system does not require a stereotactic frame. Using this system, the neurosurgeon sees the

CT information (e.g., a tumor boundary) superimposed upon the operating field in proper position, orientation, and scale.

The great number of computerized techniques, especially CT, for resolving problems of stereotactic surgery does not allow the author to go into the details of the mathematical and geometrical methods used in designing new devices. However, this review of the existing stereotactic CT techniques indicates that these new approaches are very effective. The use of computers surely enhances the accuracy of stereotactic calculations and accelerates and simplifies stereotactic operations, e.g., destruction of subcortical structures, implantation of subcortical electrodes, biopsy of deep-seated tumors or small pathological lesions, clipping of arterial or arteriovenous aneurysms, and evacuation of intracerebral hematomas. Nevertheless, there are some negative points of view concerning the value of CT data for stereotactic calculations. For instance, Fröder *et al.* (1983) consider that the determination of the coordinates of a target point on the basis of CT scans is not yet sufficiently accurate to replace calculations based on ventriculograms for stereotactic surgery.

There is still one other aspect that should be considered. If it is necessary to calculate the coordinates for several target points and the trajectories for introducing many electrodes into those points, then the use of CT techniques does considerably reduce the time needed for calculations. However, the majority of stereotactic operations have as their objective the destruction of only one subcortical structure. In these cases, the calculation of the coordinates for the target point on ventriculograms by the routine technique takes no more than 10–12 min. The use of computerized methods for this purpose will save only a few minutes of operating time, which is of no great practical significance.

The CT is also a valuable means of controlling the localization and dimensions of a stereotactically made lesion in deep cerebral structures and its effect on surrounding parts of the brain (Chapter 5).

Two modern CT- and NMR-guided stereotactic systems are described in the following sections.

7.1. Brown–Roberts–Wells Stereotactic System

One of the most modern stereotactic apparatuses is the Brown–Roberts–Wells (BRW) system, which we began using recently. This system has some important advantages. It may be used for conventional stereotaxy with determination of cerebral target point (or points) on ventriculography in both projections as well as without ventriculography by using modern imaging techniques: CT, NMR, and PET. In both cases the electrode trajectory is calculated by minicomputer and data processing is made in the operating room.

The apparatus (Fig. 63) consists of several main components. A nickle-plated aluminum head ring is rigidly fixed to the head at the beginning of the operation with four carbon fiber posts and screws. The head ring is positioned below the first CT scan, which does not interfere with the ring as well as with both x-ray films in routine (without CT) stereotaxy. During the CT-investigation the head ring remains in place.

The arc with another ring and millimeter scale (Fig. 64A) is attached to the head ring and also to a phantom base which is similar to the phantom of the Riechert–Mundinger device (see Section 5.2). The arc permits maximum flexibility of approach with great precision to any target in the brain from almost any angle, including lateral approaches to the temporal lobes, hypophysis, and posterior fossa structures. The massive floor stand to which the head ring is attached

FIGURE 63. Brown–Roberts–Wells stereotactic system (Radionics Co.).

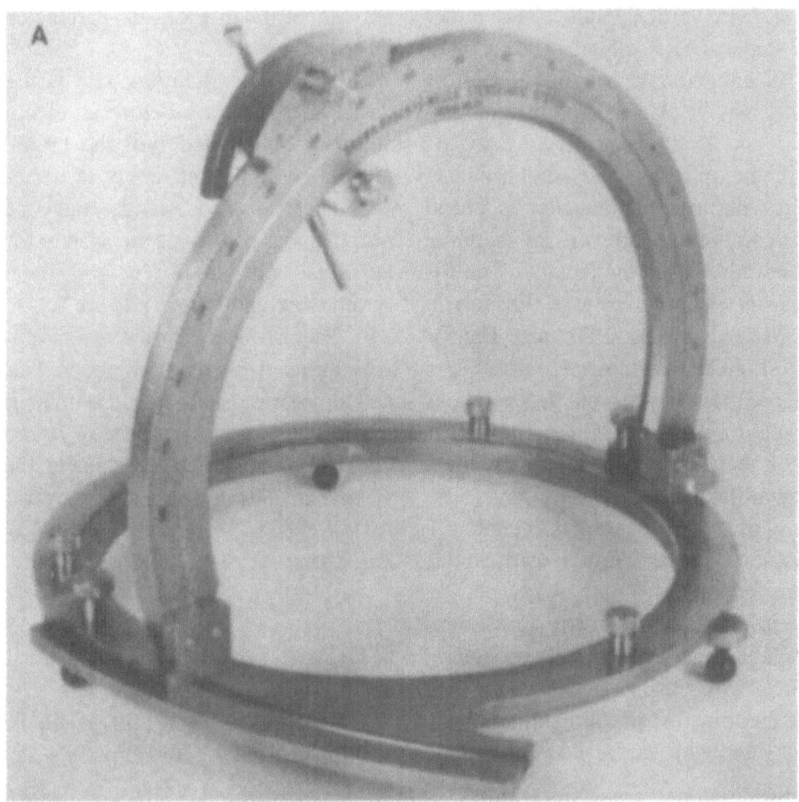

FIGURE 64. Basic components of the Brown–Roberts–Wells stereotactic system. (A) Arc system, which has four angular variables. (B) Localizer ring with carbon fiber rods for CT scan localization. See text for details.

allows movement in A-P, vertical, and lateral directions. Two cassette holders (AP and lateral) allow plain x-ray films to be taken. With a Mayfield headrest adaptor, the head ring may be clamped to the operating table.

Another part of the system—a localizing unit—consists of two rings with the same diameter as a head ring (Fig. 64B). It is composed of six vertical and three diagonal carbon fiber rods. When the localizing unit is attached to the head ring the rods serve as fiducials in three-dimensional space visible on CT scans. The localizer is used in CT-guided stereotactic procedures with any CT scanner without any special head orientation or using brackets or adaptors. The carbon fiber rods are shown up on the CT scans as nine index dots without causing any artifacts on the brain image. The coordinates of these dots and the target on the CT image are needed to determine the target position relative to the head ring.

In both procedures (CT-guided and ventriculographic) the obtained data are entered into a computer programmed with Epson HX-20, which calculates three target coordinates, the arc settings, and depth for a desired approach. The printer provides a permanent record of all parameters and calculated targets.

The BRW phantom base is used to confirm the target and approach before an electrode or biopsy instrument is inserted into the brain. This greatly increases certainly and safety of the procedure.

There is also a wide range of accessories that provide adaptability of the system to many stereotactic techniques (Talairach-type grid, the repeat fixation kit, stereotactic endoscope kit, etc.).

The BRW system is a universal device, which may be used for the conventional stereotactic surgery and also for the stereotactic biopsy, insertion of chronic deep electrodes, endoscopy, and so forth.

The first experiences using the BRW guidance system have been described in several papers (Heilbrun *et al.*, 1983, 1985; Levin, 1985; Hadley *et al.*, 1985; Appuzo *et al.*, 1987).

7.2. Patil Stereotactic System

Recently, Patil (Patil, 1983; Patil *et al.*, 1986) developed a new CT-compatible stereotactic system,

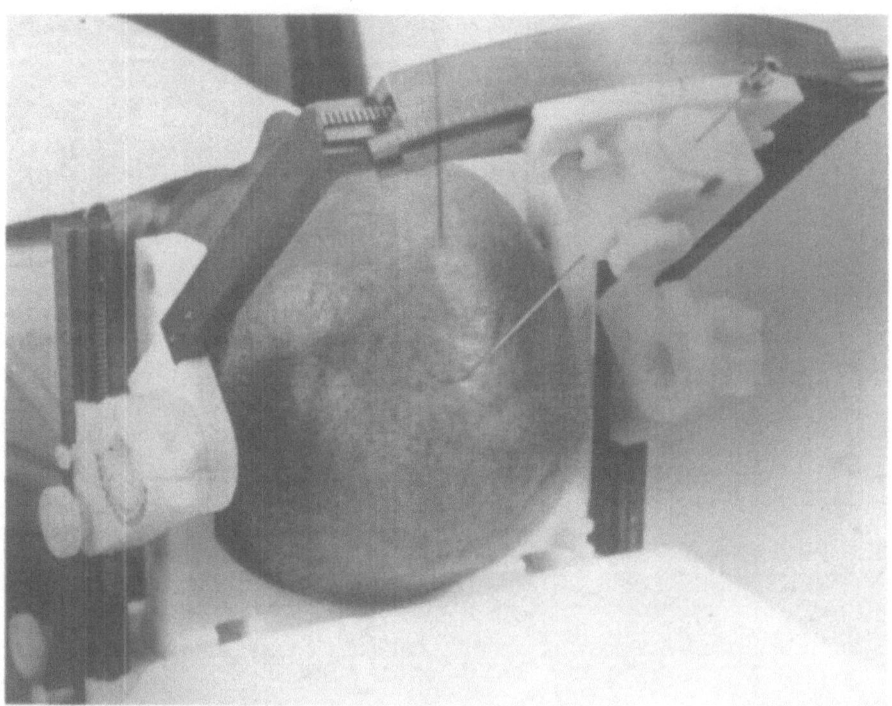

FIGURE 65. Patil CT-guided stereotactic system. Arc carrier with the probe holder is mounted on a movable circular attachment. General view of the system in place. See text.

which seems to be simple and effective (Fig. 65). This system, produced by Westco Medical Corp. (USA), utilizes the principle of placing the axis of rotation of the arc and the probe holder in the same CT plane as the intracerebral target. The use of laser-positioning light eliminates the need to calculate the *Z* coordinate. It is possible also to measure directly the *X* and *Y* coordinates from a single picture without the need for special computer programs or stereotactic calculations.

A plastic base platform fits on the CT table with a special attachment. Two aluminum vertical bars are mounted on either side of the platform. Each bar carries a movable circular attachment on which the arc carrier can be fitted. The arc is movably mounted on the arc carrier. The center of the arc lies on the line joining the centers of the circular attachments. The arc carrier rotates in the sagittal plane around an axis which coincides with the line.

The movable probe carrier is attached to the arc; the carrier allows the use of probes of various diameters. The plastic head holder, which is a separate part of the system, is movably mounted on the base platform. The head is fixed by four chromium-plated brass pins.

The centering edge of the groove in the middle of the base platform serves as the 0 reference point for the *X* coordinate. The top surface of the platform is the 0 starting point for the vertical movement (*Y* coordinate) of the circular attachment.

The CT scans in the region of the target are taken when the patient's head is fixed in the head holder. The adjustment of vertical bars are made without the arc and the arc carrier to avoid artifact. With the aid of the laser-positioning light the circular attachments are moved along the length of the platform until the light coincides with the vertical line in the middle of the circular attachments. The *X* and *Y* coordinates are measured from a single CT picture on which the target is visualized. Using the cursor of the scanner, the *X* and *Y* coordinates are measured according the above-mentioned distances. After adjustment of both coordinates is made by moving the circular attachment to the required height and the middle of the arc laterally, the arc carrier and also the probe holder are reattached and can be rotated to any desired angle so as to reach the target through any point on the skull. The burr hole can be placed before or after CT investigation.

The Patil system is used for stereotactic neurosurgery, brain biopsy, transsphenoidal hypophysectomy, radiosurgery, introducing radioactive sources, aspiration of brain abscesses, and so forth.

8. Stereotaxis and Nuclear Magnetic Resonance

A highly effective method for visualizing the human brain—tomography based on nuclear magnetic resonance (NMR)—has been used in functional and stereotactic neurosurgery during the last few years. Lars Leksell and colleagues (1985) considers that ". . . the use of NMR imaging in stereotaxy . . . will open up new areas of research in deep brain surgery."

In comparison to CT, the NMR technique has some important advantages that increase substantially its value in stereotactic neurosurgery. The great advantage of NMR in stereotactic surgery is the possibility of better contrast in comparison to CT between gray and white matter. This enables visualization of the intracerebral target without calculation of reference points (LI, etc.) (Turner *et al.*, 1986). Physical properties of NMR provide superior contrast resolution, clear imaging of the cerebral structures, and also rapid three-dimensional brain scanning at any angle. Moreover, the method allows direct target determination and clear imaging of a surgical lesion deep in the brain after the operation.

Because the NMR scanner creates a strong magnetic field, it is necessary to construct the stereotactic system from a nonferromagnetic material (aluminum, plastic, or brass). The visualization of reference points or markers using NMR stereotaxis may be achieved by coating them with mineral oil or by rods filled with a paramagnetic fluid.

Comparison of stereotactic coordinates obtained with NMR and CT has shown that the *X* and *Y* coordinates were virtually identical (the difference not more than 1–2 mm), but for the *Z* coordinate the difference in exceptional cases was 4 mm (Lunsford *et al.*, 1986).

The Leksell apparatus, which is described in detail in Section 5.4 and was adapted afterward by the author for application with CT investigations (Section 7), was recently used for NMR (Leksell *et al.*, 1985) (Fig. 66). For this purpose the coordinate-indicating disks were replaced by analogous plastic disks having vertical, horizontal, and diagonal tubes with a liquid indicator (vegetable oil) (Fig. 67). These tubes are temporarily attached to the frame during the NMR investigation. The adapter of the stereotactic frame for the NMR apparatus is made of fiberglass (instead of aluminum as for CT).

For NMR, the stereotactic frame is fixed on the patient's head with three or four pins made of fiberglass. Indicating disks are attached to the frame,

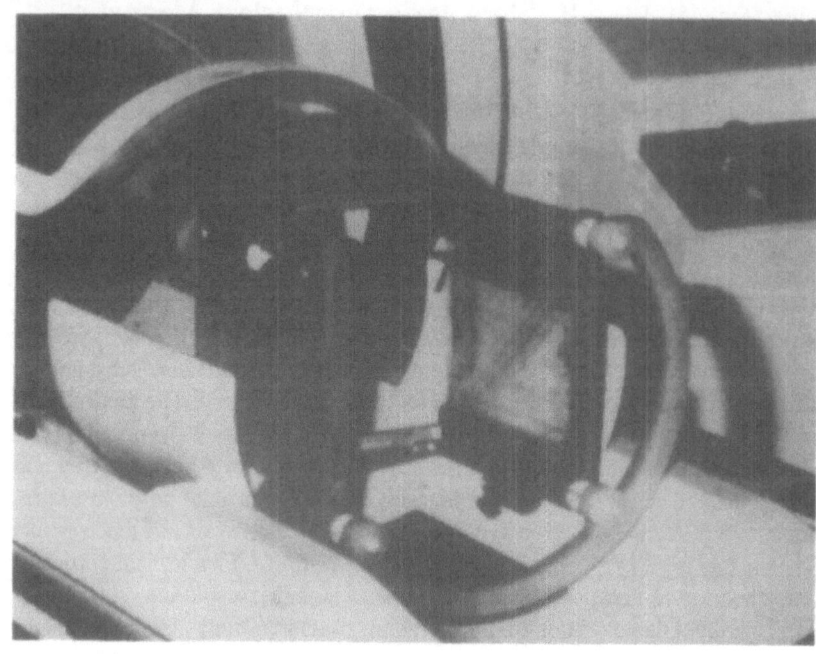

FIGURE 66. The NMR adapter and the coordinate frame in position in the RF coil of the Siemens Magneton (Leksell *et al.*, 1985).

which is connected to the RF coil of the NMR apparatus (Siemens Magneton) through an adapter. The required number of parallel slides are made in two perpendicular planes parallel to the coordinate frame basal plane and the frontal plane, respectively. If maximal accuracy is required to reach the target, the coordinates in the sagittal plane are also calculated. If necessary, accessory slides are made in the plane of the surgical approach, which allows imaging of the cerebral structures through which the electrode or other stereotactic

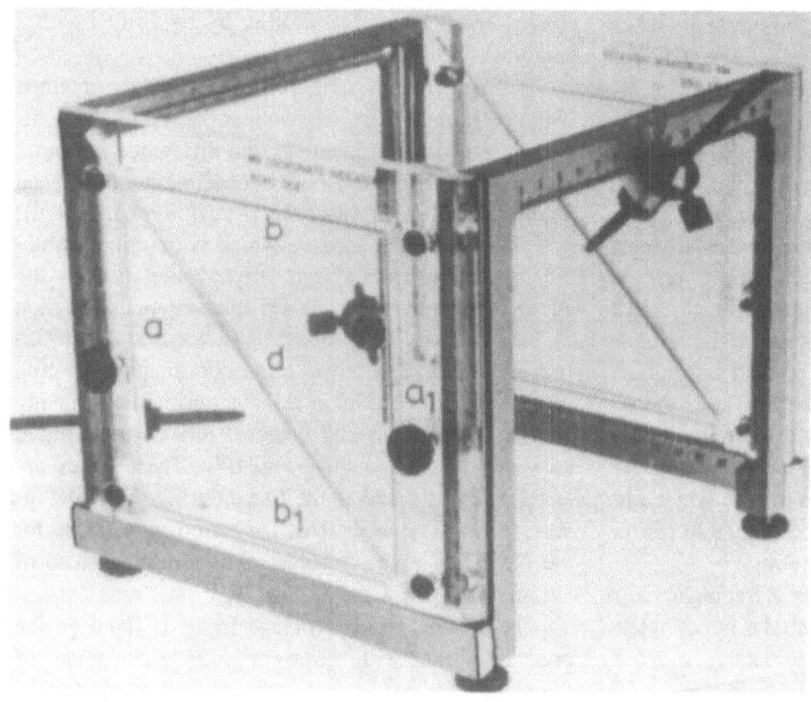

FIGURE 67. The coordinate disks for NMR with channels filled with oil. The channels a, a_1, b, b_1, and d appear as fiducials on the NMR scans (Leksell *et al.*, 1985).

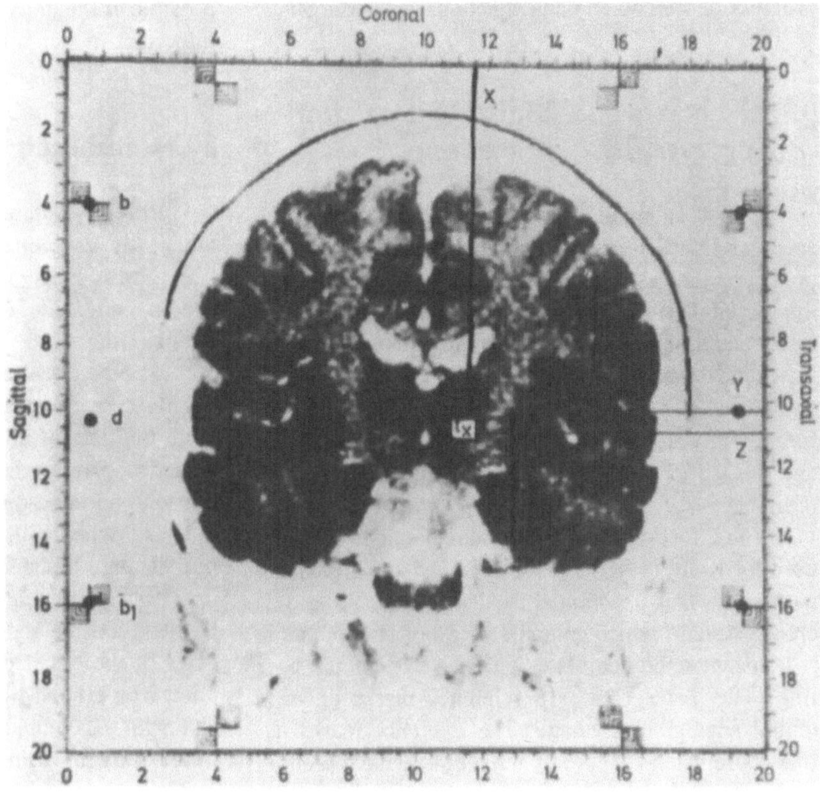

FIGURE 68. The NMR film superimposed on and aligned with the underlying coordinate scale. The cross lines through the target point (t) indicate the *X* and *Z* coordinates. The line intersecting the middle fiducial indicates the *Y* coordinate (Leksell *et al.*, 1985).

instruments pass. After that, on three scans in three dimensions, the target point and its *X*, *Y*, and *Z* coordinates are estimated as described for use with CT. These coordinates may be estimated on scan films or directly on the NMR screen. In the first case, the film is placed on the semitransparent disk with coordinate scales (Fig. 68). The four fiducials on the film should be correlated with the underlying scale so that all three coordinates of the target point can be determined. In the second case, the coordinate scales on the screen and the actual scan should be superimposed. The coordinates of the target point are obtained by placing the cursor on this point.

The software program from Siemens Medical Systems makes it possible to read the coordinates of the target point from the screen. As Leksell *et al.* (1985) stressed, by obtaining the coordinates from the screen directly, not only is the need to develop film eliminated but the surgeon can perform the necessary calculations directly in the operating room, some distance from the NMR equipment, with the aid of a separate satellite console. The possibility of storing all the data on a floppy disk makes it possible to obtain any desired image during the operation.

A control study of the accuracy of the Leksell stereotactic apparatus with NMR and CT was made using an agar-filled head phantom. The study showed that both imaging techniques produce an accuracy of better than 2 mm, but the *Z* coordinate measured by CT produces an error range of about 2–3 mm.

Other systems of NMR-assisted stereotactic devices were recently proposed, including those of Laitinen (1984), Olivier *et al.* (1985a), Peters *et al.* (1985), Lunsford *et al.* (1986), and others.

An NMR-compatible base ring is made of molybdenum disulfide and nine capillary tubes are filled with $CuSO_4$ solution. These tubes produce nine reference marks on each NMR slice (Kelly, 1987). In the Cosman adapter for NMR investigation in the BRW system, petroleum jelly is employed for contrast.

With the aid of a computer it is possible to combine CT and NMR data to get images of tumors in three-dimensional stereotactic space (Kelly *et al.*, 1985). There is every reason to consider that the application of NMR will add a new stage to the further progress of stereotactic and functional neurosurgery.

9. Stereotaxis and Digital Subtraction Angiography

Digital subtraction angiography (DSA) has been successfully used in recent years for determining target point coordinates. The advantage of the technique is the possibility of simultaneous subtraction of skull bones and the stereotactic frame. The use of DSA eliminates the danger of injuring cerebral vessels in case of an unusual electrode trajectory. This technique is used for the intracerebral introduction of multiple electrodes and also for biopsy of deep-seated brain tumors (Olivier *et al.*, 1985a).

The CT/NMR-compatible stereotactic headholder used by Kelly (1987) is also designed for digital angiography. This headholder is adapted to a special reference system, which contains arrays of nine points in radiolucent reference plates on each side of the base ring. The plates create 18 reference marks on each digital angiographic image. The reference marks define coordinate axes for any three-dimensional point in space for the correct annotation of that point on the angiographic image.

Birg *et al.* (1985) constructed a modification of the Riechert–Mundinger stereotactic device (Section 5.2) for use with an NMR scanner. The authors changed the metallic base ring to one of the same size made of a special plastic material. The ring has plastic screws with sharp pins made from quartz or steel, which produce only small local artifacts in the NMR scans. The coordinates of the target point are taken from small plastic reference tubes connected with the base ring using NMR scanner software.

In the new modification of the Patil system the coordinate markers located on the base platform are 5-mm diameter rods containing manganese chloride and are visible on the NMR pictures (Patil *et al.*, 1986). One rod in the midline groove serves as a marker for the X coordinate and another rod on the top surface of the base platform for the Y coordinate. The Z coordinate need not be measured due to the laser-positioning light and table indexing. The X and Y coordinates can be measured directly by the cursor of the NMR scanner. The determination of coordinates may be obtained for coronal and sagittal images.

The authors stress that NMR stereotaxis is valuable for biopsy of deep-seated gliomas that are poorly visible on CT scans. The NMR thin slices make it possible to see individual thalamic nuclei which is very important in stereotactic surgery.

10. Combined Visualization Techniques

Recently, attempts have been made to combine all the newly developed methods of brain visualization (CT, NMR, DSA) for use in stereotactic surgery. The use of composite data produces a scheme of intracranial volume.

Some new stereotactic devices (Olivier *et al.*, 1985a,b; Peters *et al.*, 1985; Kelly *et al.*, 1985) integrate the data of all three techniques during the stereotactic operation and display the integrated information on one three-dimensional image.

The tapes with CT, NMR, and digital angiography data are transferred to the computer system in the operating room. Each image may be displayed on the Ramtek console for calculation of stereotactic coordinates of the target point. The cross-correlation of the point with CT, NMR, and digital angiography data and also with the computer-stored stereotactic atlas is made by the surgeon using the cursor and trackball (Kelly, 1987).

11. Stereotaxis and Ultrasound

During recent years intraoperative ultrasonography has proven its great value in disclosing deep-seated pathological lesions inside the brain (Chandler *et al.*, 1982; Tsutsumi *et al.*, 1982; Sjölander *et al.*, 1983; Berger, 1986). The new generation of high-resolution ultrasound scanners has become an important tool for the neurosurgeon. It was shown that cerebral ventricles and cysts are hypoechoic and intracerebral lesions that are hypo- or hyperdense on CT are hyperechoic (Chandler *et al.*, 1982).

The increased use of ultrasonography in neurosurgery led to attempts to combine the method with stereotactic technique–ultrasound-guided stereotaxis (F.D. Brown *et al.*, 1984). The first steps of this new technique were connected with some difficulties. Because of the large diameter of the ultrasound probe, the approach to a deep lesion through a burr hole was impossible and a small craniotomy was essential. The Diasonics Company (USA) has developed new ultrasound transducers 16 mm in diameter (5.0 mHz and 7.5 mHz), which make it possible to obtain ultrasonic images through a burr hole.

Recently, Berger (1986) developed a skull-mounted apparatus for ultrasound-guided stereotactic biopsy through a burr hole using Diasonics equipment. The apparatus was used in a limited group of patients for biopsy and also for drainage of cysts and abscesses. The author believes that the advantages of this method over those guided by CT include less time required for the entire procedure (from 25 to 40 min), immediate confirmation of the target by imaging the echogenic needle track, assessment of cyst or abscess drainage, and detection of hemorrhage after biopsy.

The use the ultrasound as a method of destruction of brain tissue in a neurosurgical operation is described in Chapter 5. The combination of the stereotactic method with the laser technique is also described in Chapter 5.

4

Principles of Stereotactic Operations on Subcortical Structures

It is necessary to organize and bring the clinical observations of CNS diseases before, during, and after surgery to the exactness of a physiological experiment.

N. N. Burdenko

1. General Considerations

Stereotaxis, first of all, requires maximal accuracy. The success of any stereotactic operation depends on many factors but primarily on the accurate localization of the subcortical target. If the calculations are inaccurate, and the stereotactic instrument deviates from the target point only 2–3 mm, or if there are other technical defects, not only will the result be imperfect but very serious complications may arise. In the majority of stereotactic operations, the subcortical structure that is to be destroyed or stimulated lies close to other functionally important structures (pyramidal tract, chiasm, brainstem nuclei). That is why an error of only 2–3 mm in calculating the trajectory of the electrode–cannula can lead to very serious consequences. The three-dimensional anatomy of the target and adjacent structures is extremely important for the planning and calculation of the electrode trajectory.

The need for maximal accuracy in reaching a target structure has been confirmed by many neurosurgeons and by the author's extensive experience. There have been cases in which the introduction of the cannula into the brain to the calculated depth did not yield a clear-cut clinical effect but advancing the cannula for only 1–2 mm or (in accordance with the data of control films) changing its position slightly caused the main manifestations of the disorder, e.g., tremor, rigidity, or pain, to disappear completely.

The techniques and the various stages of different stereotactic operations are quite variable. Operations aimed at destruction of subcortical structures or their chronic stimulation differ significantly from each other. The techniques for destroying VL differ substantially from those used to make lesions in ND, Hipp, Am, hypophysis, or the spinothalamic tract in the cervical spinal cord. Moreover, the course of the operation depends on the type of stereotactic apparatus used, on the use of CT, and on many other factors. Since it is practically impossible to describe all the existing operations in a single chapter, this discussion covers only the main stages of the most common operations—operations on structures of the extrapyramidal and limbic systems. Other stereotactic operations, e.g., dentatomy, percutaneous cordotomy, hypophysectomy, or introduction of electrodes for stimulating subcortical structures, are described in their respective chapters.

Each stereotactic operation is based on certain principles and consists of several main stages. Arbitrarily, these stages may be compared to firing a gun at an invisible target; it is necessary to choose the target, to define visible reference points from which to calculate the coordinates of the invisible target, to adjust the gunsight accordingly, and then to fire at the target so as to hit the bull's eye.

The choice of target, i.e., the structure that is to be destroyed, is made prior to the operation. Since in different disorders these structures vary, the bases for making these choices are presented in the appropriate chapters. Below we have enumerated only the main stages of a kind of "generalized" operation. However,

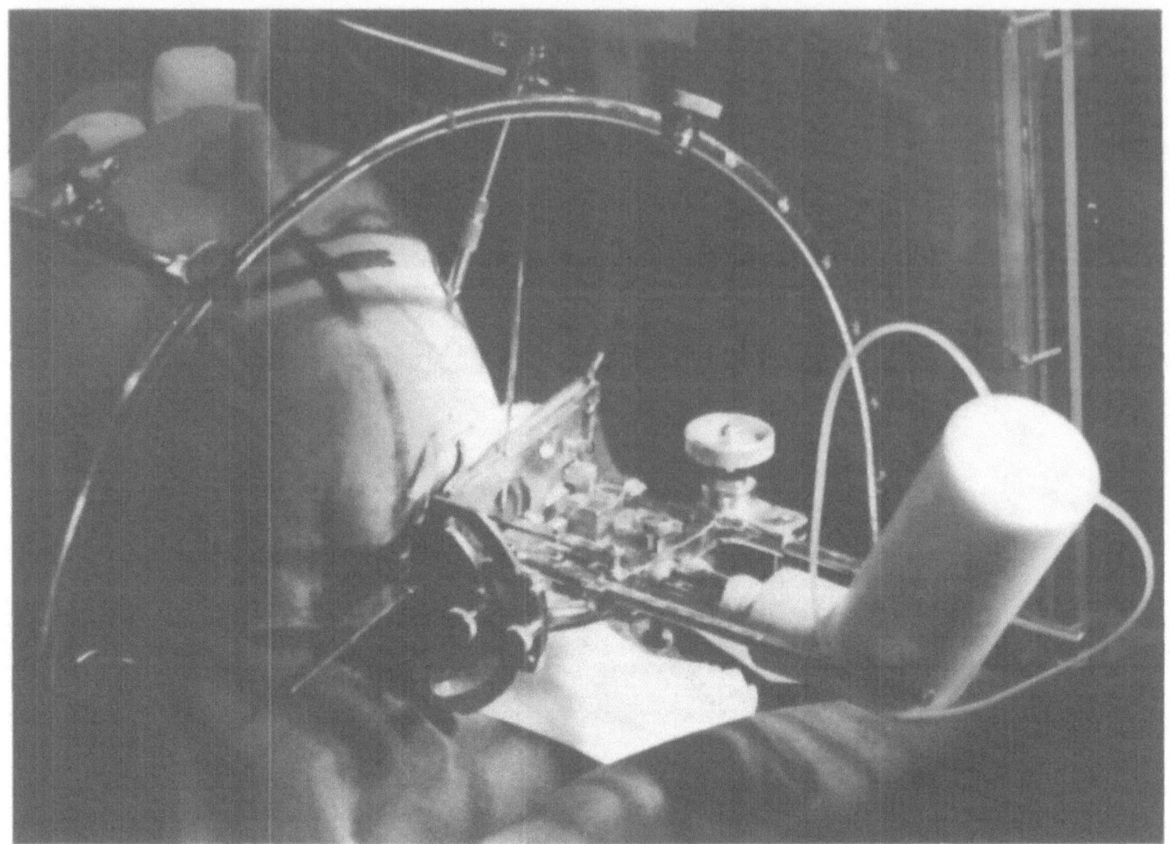

FIGURE 69. General view of the stereotactic operation.

it should be kept in mind that with different operating techniques, the sequence of these stages may be changed.

1. Anesthesia and drilling of burr hole.
2. Fixation of patient's head and installation of the stereotactic apparatus.
3. Contrast filling of the ventricular systems.*
4. Determination of intracerebral reference points on skull x ray in two projections, so that calculations can be made of the subcortical target on the basis of the data in stereotactic atlases.
5. Verification of the localization of the desired subcortical structure on the coordinate system of the stereotactic apparatus, and transfer of the data to the electrode.
6. Introduction of the cannula (electrode) into the desired subcortical structure under x-ray control or TV monitoring.
7. X-ray and functional control of the accuracy of the penetration into the structure.
8. Destruction of target structure and evaluation of the effect produced.
9. Extraction of cannula (electrode), removal of the apparatus, and closure of the surgical wound.

Naturally, we cannot describe all existing operative techniques but confine this discussion to the technique we have developed and used for more than 25 years. An overview of such an operation is presented in Fig. 69. Preparations for stereotactic procedures follow the usual neurosurgical routine. The operation is always performed in a single session.

2. Anesthesia

The position of the patient on the operating table depends on the technique and nature of the stereotactic procedure. For instance, in operations on the basal

*When stereotactic calculations are based on CT or NMR data, this stage is not required (see Chapter 3).

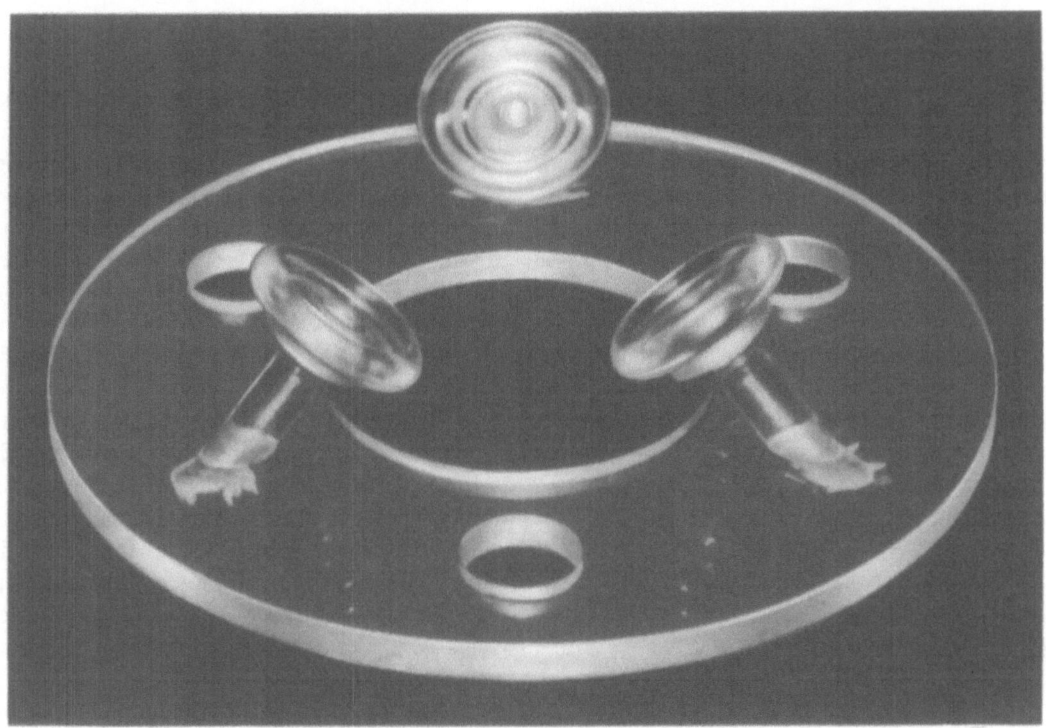

FIGURE 70. Our Plexiglas headrest for stereotactic operations on the basal ganglia.

ganglia, percutaneous cordotomy, and transnasal hypophysectomy, the patient lies in a supine position, whereas in stereotactic dentatotomy and percutaneous tractotomy, the prone position is more appropriate. Certain operations are performed with the patient in a sitting position. The use of CT has established that the position of the patient (supine, prone, or lateral) causes no displacement of cerebral structures.

In operations on the basal ganglia, the head of the patient is slightly raised and lies on a Plexiglas headrest of our own design (Fig. 70), which is not opaque to x rays. The center of the headrest has a large circular aperture in which the head is fixed by three rigid pins 5 cm in diameter; the head holder is connected by a ball-and-socket joint to stationary columns. The head of the patient is held firmly by the pins in the headrest so that its sagittal plane is perpendicular to the horizontal plane.

The operating field is prepared in the same manner as in other neurosurgical operations.

The types of anesthesia in functional and stereotactic operations differ. In certain conditions, for example, parkinsonism or pain syndromes, the method of choice is local anesthesia with mild premedication or neuroleptanalgesia. Local anesthesia is desirable because it allows the expected result of destruction of the subcortical structure to be observed directly on the operating table. The disappearance of tremor, rigidity, spasticity, or pain sensations is an important criterion for judging the proper placement of the lesion in the target structure. It is also important to record the effect of stimulation. It is quite natural that for these observations to be made, the patient must be fully conscious and able to reply to the surgeon's questions and carry out his instructions. The second important advantage of local anesthesia is the possibility of detecting immediately any complication and to treat it appropriately. Examples can be seen in the appearance of hemiparesis in thalamotomy or visual disorders in stereotactic destruction of a hypophyseal tumor.

Local anesthesia, however, has well-known serious shortcomings (insufficient analgesia, a feeling of fear). In view of this, we and other authors have, in recent years, frequently employed a comparatively new method of neuroleptanalgesia. After conventional premedication, initial narcosis, and intubation, the patient is given a mixture of nitrous oxide and oxygen. With this as a base, fentanyl and droperidol are periodically administered as necessary. In some cases, neuroleptanalgesia is used without intratracheal narcosis. The main advantage of this method is the ability to awaken the patient quickly and check the effect of the operation to verify the absence of possible complications (pareses, aphasia, etc.). Experience has shown that neuroleptanalgesia is the method of choice in many functional and stereotactic operations.

In those conditions associated with pronounced hyperkinesia as well as in excited patients, e.g., in mental disorders, local anesthesia is not used since movements obscure x-ray films even of the fixed head. In such cases, intratracheal narcosis using modern anesthetic agents with adequate premedication is required. When general anesthesia is used, the surgeon is unable to judge the result until the patient becomes fully conscious postoperatively.

3. Location of the Burr Hole

The site for the cranial burr hole determines the point of the cerebral cortex through which the cannula–electrode will be introduced to reach the subcortical target. In making this choice for the optimal and safest trajectory, it is rational to observe three conditions:

1. The point of introduction of the cannula in the brain must be at some distance from functionally important cortical zones (e.g., the anterior central gyrus).
2. On the way to the target structure, the cannula must not damage other important structures in the deep areas of the brain (central parts of CI). It is also desirable that it not pass through the cavity of V.
3. The path of the cannula from the cortex to the target structure should be as short as possible.

We pass the cannula to supratentorial subcortical structures through the cortex of the posterior part of the second frontal gyrus, a trajectory that meets the abovementioned conditions. We determine the site of this cortical area on the scalp using the skeletotopic scheme of Egorov (1959). Before the operation, a solution of brilliant green is used to draw a line A on the skin of the patient's head from the nasion to the inion along the midline of the skull. Then we measure the distance between these points, which varies considerably depending on the shape and size of the skull. Based on our data, this distance, on the average, is 36 cm in men (range of variability from 32 to 39 cm) and 34 cm in women (from 31.5 to 36 cm).

According to Egorov's scheme, the middle of this line corresponds to the vertex of the precentral sulcus. In order to introduce the cannula at a safe distance from the motor cortex, it is necessary to "step forward" 3.5–4 cm from this point. In practice, we take 12.5 cm from nasion when the distance between nasion and inion is 33–35 cm and 13.5 cm when this distance is 36–38 cm. Through this point on the sagittal line, a second line B is drawn in the transverse plane to symmetrical points at the tragi of both ears; it usually passes along or slightly in front of the coronal suture.

On line B, at a distance usually 3–3.5 cm to the left or right of line A, the point for the burr hole is selected; it is at the midpoint of the skin incision. If the distance of the burr hole from the sagittal line is short, the cannula may pass through V, but if it is large, the cannula traverses more of the central part of CI. The site of the burr hole can also be calculated using computer methods, in particular, computed tomography (Chapter 3).

The next stage of the operation is the making of the burr hole. Many neurosurgeons perform stereotactic operations using the Riechert–Mundinger apparatus (Chapter 3, Section 5.2) through a very small hole made with a fine drill in the skull, with the size of the hole corresponding to the diameter of the cannula–electrode. Such a technique simplifies the operation and reduces the operating time. However, our many years of experience have shown that not infrequently there are large veins on the surface of the cortical zone into which the electrode is to be inserted. If the instrument is introduced "blindly," such veins may be damaged, causing a subdural hematoma. In parkinsonism and other dyskinesias, there is frequently a significant thickening of the arachnoid membrane; if the dura mater is penetrated "blind," this thickened arachnoid may depress a considerable part of the cortex causing the rupture of a bridging vein at a distance. In order to avoid complications, the author is convinced that one should first make a comparatively small burr hole, permitting visual identification of cortical arteries and veins so as to avoid damaging them. Bertrand *et al.* (1973) also hold such a point of view concerning the need for opening of the dura mater.

After conventional preparation of the skin, a 3- to 3.5-cm vertical incision is made, and following hemostasis, the edges of the wound are opened with a retractor. Then a crown trephine 15 or 20 mm in diameter (Fig. 71A) is used to make the burr hole, and the bone "button" is preserved in physiological saline and re-

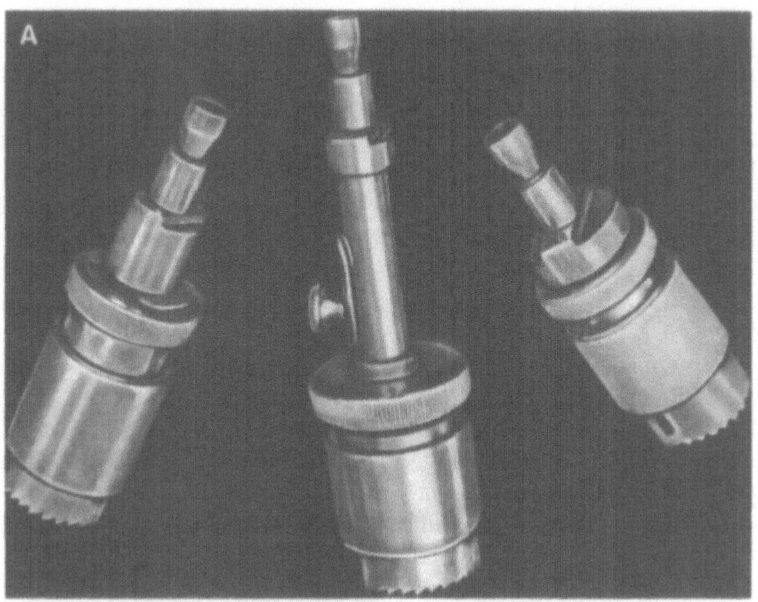

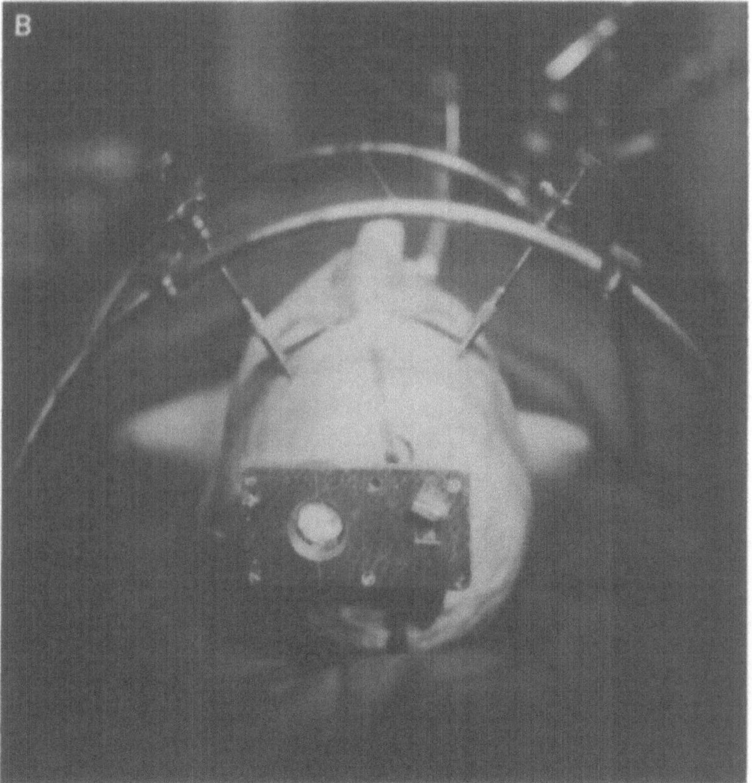

FIGURE 71. (A) A set of coronal burrs of different diameters for stereotactic operations. (B) The first part of stereotactic apparatus is fixed in the burr hole.

placed when the operation is over. The dura is opened with a crisscross incision, and its edges are pulled back. After that, the support frame of our stereotactic apparatus (Chapter 3, Section 5.9) is firmly fixed in the burr hole with the help of a screw key (Fig. 71B) that expands the three "legs" (pinions) of the support frame and fits them under the edge of the burr hole. This frame should be set up in approximately the correct position, i.e., in such a manner that the cannula would be oriented 1–2 cm behind the external auditory meatus in the sagittal plane and towards the internal angle of the orbit on the same side in the frontal plane.

The next stage of the operation involves rigid fixation of the patient's head. To the base of the headrest, in the frontal plane, we fasten two large metal semicircles joined by two beams that can be moved along these arches, which have a millimeter scale. To these beams are fixed two rods with sharp sterile tips. They are screwed into the soft tissue and external plate of the frontal bone symmetrically from both sides, thereby ensuring immobility of the patient's head throughout the entire operation (Fig. 72). If the operation is performed without narcosis, the sites of penetration are injected with a local anesthetic agent. In order to place the head in the proper position, we glue two small circular metal plates to symmetrical points on the tragi of both ears. With the head in the proper position on the screen of the electronic amplifier with TV monitoring, the centers of both circular plates should coincide in the lateral projection.

After that, the main part of our stereotactic apparatus (Chapter 3, Section 5.9), the guiding device with a fixed cannula for puncturing V, is attached to the supporting frame. This cannula, 12 cm long and 2.2 mm in diameter, is equipped with a mandrin and a small valve at the external tip. After coagulation of a small (2 × 2 mm) area of the cortical surface, V is punctured by advancing the cannula into the brain for approximately 40–45 mm. After CSF has been obtained, the small valve of the cannula is closed.

The next stage of the operation is to introduce contrast material into the cerebral ventricle.

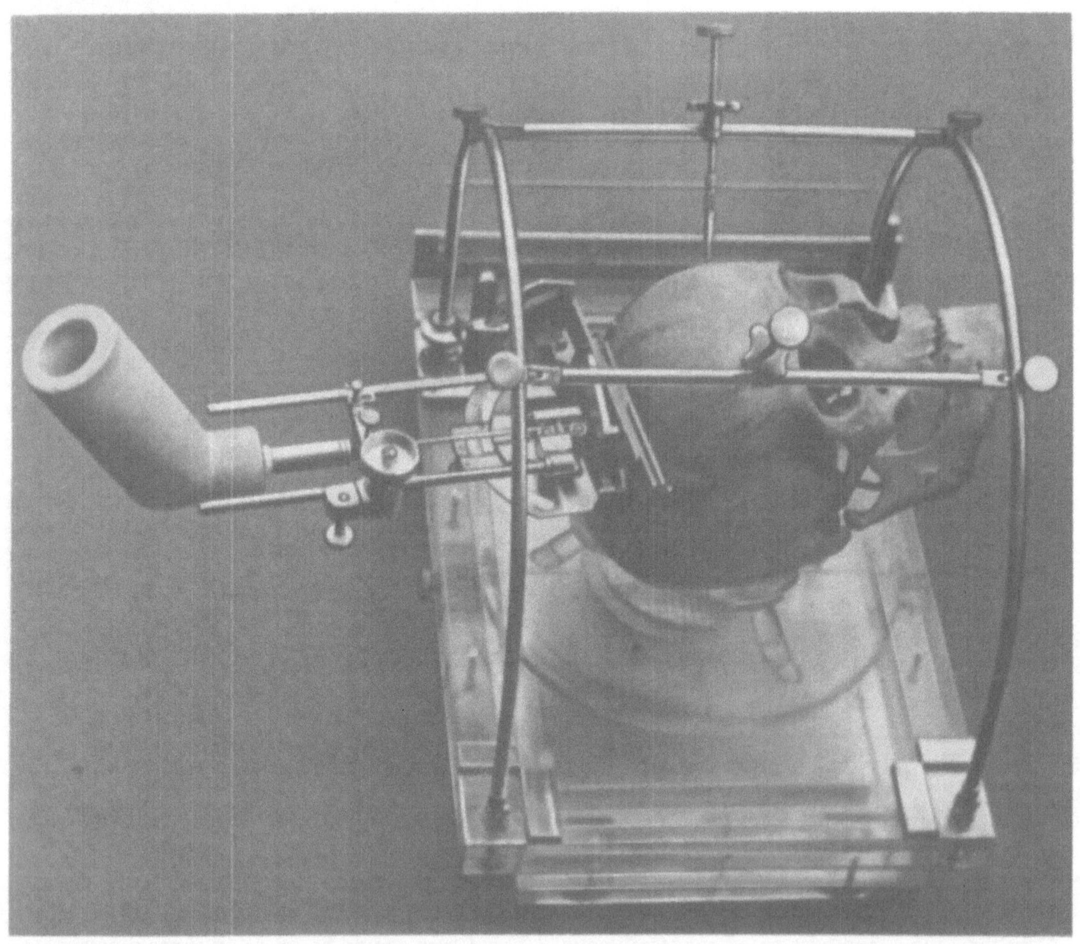

FIGURE 72. Fixation of the skull in the headrest with the aid of two sharp pins screwed into the bone.

4. Visualization of the Ventricular System

An outline of the ventricular system is required to determine the main intracerebral reference points on the x rays in two projections. If CT or NMR is not used for stereotactic calculations, a contrast medium is mandatory to outline the ventricles for each stereotactic operation. The accuracy of determining intracerebral reference points largely depends on the distinctness of the ventricular outline, which is essential for precisely determining the site of the desired subcortical structure.

Modern methods of cerebral visualization (CT, NMR) allow calculation of target points in deep cerebral structures without introducing contrast medium into the ventricular system. Detailed information on these methods is given in Chapter 3. Since these new methods are not available to many neurosurgeons, we must discuss in detail the traditional methods of stereotactic calculations based on ventricular visualization.

As our own experience and numerous data in the literature indicate, CT scanning prior to a stereotactic operation gives important general information about the state of the ventricular system and the degree of brain atrophy. In particular, in rare cases in which an extremely large hydrocephalus excludes the possibility of a stereotactic operation, preliminary CT scanning eliminates unnecessary ventriculography.

A study of angiograms and pneumoencephalograms to determine if it is possible to use angiography to locate target points in stereotactic operations has demonstrated the absence of reliable correlations between the course of cerebral vessels and intracerebral reference points.

Stereotactic surgery uses two known methods of ventricular visualization, pneumoencephalography (PEG) and ventriculography (VG), which are performed by introducing air or x-ray contrast media, respectively. The use of PEG initially demonstrated that this method ensured adequate visualization of the ventricular system but had serious shortcomings. After PEG there was often a worsening in the condition of the patient with cardiac disturbances, which often made it necessary to postpone the operation.

Potthof *et al.* (1972) pointed out that a transitory psychoorganic syndrome in parkinsonians, especially in the aged, was observed after pneumoencephalography more frequently than after the stereotactic operation itself. Riechert (1980) considers PEG an important risk factor in stereotactic surgery because it prolongs the time of operation. Subarachnoid spatial shadows may obscure intracerebral reference points, complicating or preventing stereotactic calculations. For example, air in the Sylvian fissure is often superimposed on the image of CA in a lateral film, making the visualization of V_3 unclear.

The use of PEG has still another drawback. Frequently with the patient in supine position there is a significant accumulation of air in the frontal area, and the deep-seated cerebral structures are displaced under the action of gravitational forces. This displacement, which can amount to 2.5–2.7 mm (Thulin *et al.*, 1972), reduces the accuracy of stereotactic calculations.

Many of the disadvantages of PEG are also inherent in pneumoventriculography. In view of this, an improved technique of air ventriculography (Siegfried, 1982) is employed. A needle or catheter is introduced into V and then advanced along the bregma–dorsum sellae line to FM. After 10 ml of CSF is withdrawn, 15 cm^3 of air is injected into V_3, which gives good visualization and reduces the side effects of conventional air ventriculography.

Many neurosurgeons, including the author, have resorted to so-called positive ventriculography, which is performed with a "heavy" contrast substance such as iophendylate (Myodil or Pantopaque, an iodine-containing substance in an oil base) that is considerably denser than CSF. However, in spite of its obvious advantages (high contrast), iophendylate also has serious shortcomings. Since it is heavier than ventricular fluid, this contrast medium does not fill the entire ventricular system but settles to the dependent region (posterior in the supine position). Iophendylate often breaks up into large globules, which rapidly flow to V_4 and then to the subarachnoid space of the spinal cord. In a number of cases, it has been necessary to manipulate the position of the patient's head in order to dislodge the iophendylate from the posterior horns to V_3. This is very undesirable and rarely successful. Emulsified iophendylate has only partially reduced these difficulties.

Although iophendylate is not very irritating to brain tissue and its meninges, its introduction into the ventricles in a volume of 5–6 ml frequently leads to undesirable side effects. In approximately 30% of the patients there is a worsening of their condition with fever and meningeal symptoms (Rostotskaya *et al.*, 1968). Another problem is the difficulty in extracting iophendylate from the CSF after the investigation. Since iophendylate is absorbed very slowly, it can still be seen in films 3 and more years after ventriculography.

These disadvantages of iophendylate prompted a search for a new, safer, less irritating and higher contrast medium. Since 1964, we have used for stereotactic operations a water-soluble x-ray contrast substance methylglucamine iothalamate (Conray) (Kandel and Plevako, 1966), which prior to that was used in neurosurgery only for cerebral angiography.

Conray is a 60% water solution of methylglucamine iothalamate and contains 28% iodine. It easily dissolves in CSF, has a similar density, is nontoxic, and is quickly absorbed into the bloodstream from the ventricular system. This substance is totally eliminated from the cerebral ventricles within 2–2.5 hr and is excreted by the kidneys in the course of 24 hr. Vast experience in the use of iothalamate for cerebral angiography has demonstrated that it ensures high-quality contrasting of cerebral vessels with minimal complications.

We have employed iothalamate in more than 1200 stereotactic operations as well as for diagnostic purposes in brain tumors and other CNS disorders. This experience has convinced us that iothalamate is the best contrast medium for stereotactic surgery (Kandel and Chebotaryova, 1972). As a rule, it gives excellent quality images of the entire ventricular system, especially V_3, the Sylvian aqueduct, V_4, and the subarachnoid space of the cervical spinal cord. In almost all ventriculograms, one can clearly see details of cerebral structures that are practically never visible on introduction of air, e.g., the pineal and suprapineal recesses of V_3 and MI. In the majority of cases, there is good visualization of FM, which enables us to determine with confidence the reference line (the posterior margin of CM–CP) that we use as the basis for our stereotactic calculations.

Introduced into the ventricles, iothalamate is tolerated well and does not cause vasomotor or vegetative reactions. Nausea, vomiting, and headaches were observed in only 7% of our patients. However, technical errors in the use of iothalamate may lead to serious complications. If the medium escapes to the cerebral cortex along the puncture track of the cannula, focal and even generalized epileptic fits may develop. In rare cases, the cannula may be erroneously advanced not into V but into the interhemispheric fissure, which also contains CSF. In such cases, the iothalamate will reach the cerebral cortex. To prevent this serious complication, we perform a simple test before we introduce the dye to exclude the possibility of the abovementioned complication. When there is any doubt that the cannula tip is in V, we introduce 4–5 ml of air. The filling of the anterior horn of V with air can be verified with an electronic amplifier with TV monitoring or by taking control films.

It is more expedient to introduce a slightly greater volume of contrast medium into V than the volume of CSF removed. A mixture of 3–5 ml of iothalamate and 2 ml of iophendylate dissolved in 5–6 ml of CSF or physiological saline is rapidly introduced into V; AP and lateral films made immediately after introducing the dye have the highest degree of contrast.

After both films have been taken, the valve on the outer end of the cannula is opened, and the CSF mixed with contrast medium is allowed to escape or is drawn off. After the cannula has been extracted, the puncture site should be thoroughly flushed with physiological saline to eliminate any trace of iothalamate from the cerebral cortex.

The rate of absorption of iothalamate from the ventricles may vary significantly. Our experience has shown that within 30–40 min after its introduction, the contrast material is reduced by approximately one-half and in 2–2.5 hr is almost totally gone from the ventricles. Delayed absorption of iothalamate from the ventricles was observed in only 8% of our cases, usually in elderly persons and patients with pronounced hydrocephalus. In such cases, a small amount of contrast medium was seen after 4–5 hr, especially in the posterior parts of V. As already noted, in the majority of patients it is sufficient to introduce from 3 to 5 ml of iothalamate; however, in pronounced hydrocephalus 6–8 and even 10 ml may be desirable.

A method of selective ventriculography of V_3 using a water-soluble contrast medium in stereotactic operations was proposed by Kim *et al.* (1970). Good visualization of the main reference points was achieved by introducing 0.5–1 ml of iothalamate dissolved in a similar volume of CSF through a soft catheter advanced to V_3 via FM. The authors emphasized that elderly patients with a labile cardiovascular system tolerate this technique considerably better than routine ventriculography (Siegfried and Braendli-Graber, 1980).

Although iothalamate gives good visualization of the entire ventricular system, its degree of contrast is less than that of iophendylate, especially in pronounced

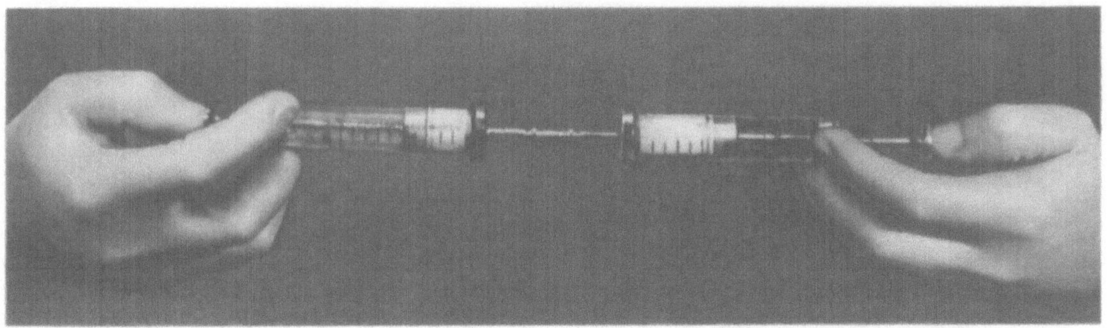

FIGURE 73. Preparation of iothalamate and iophendylate mixture for contrasting of the ventricles.

hydrocephalus. In order to obtain the advantages of both preparations, we have suggested ventriculography with a mixture of the two dyes (Kandel, 1978a).

As a rule, we employ an emulsified mixture of 2.5–3 ml of iophendylate, 3–4 ml of iothalamate, and 4–5 ml of ventricular fluid. If there are signs of pronounced hydrocephalus, the quantity of iothalamate may be increased to 5–8 ml. A simple procedure is employed for mixing the drugs. Both dyes with CSF are drawn into two syringes connected by a short metal tube. Then, alternate energetic pressing the plungers of the two syringes 10–12 times produces a well-emulsified mixture of both contrast substances directly before introduction into the ventricles (Fig. 73).

In recent years we have performed ventriculography with a mixture of iothalamate and iophendylate in 650 stereotactic operations (Fig. 74). In practically all cases we obtained excellent visualization of the details of the entire ventricular system (Fig. 75). The main advantage of this method is that iothalamate gives a very good contrast of the ventricles and the aqueduct while iophendylate intensifies the contrast of the im-

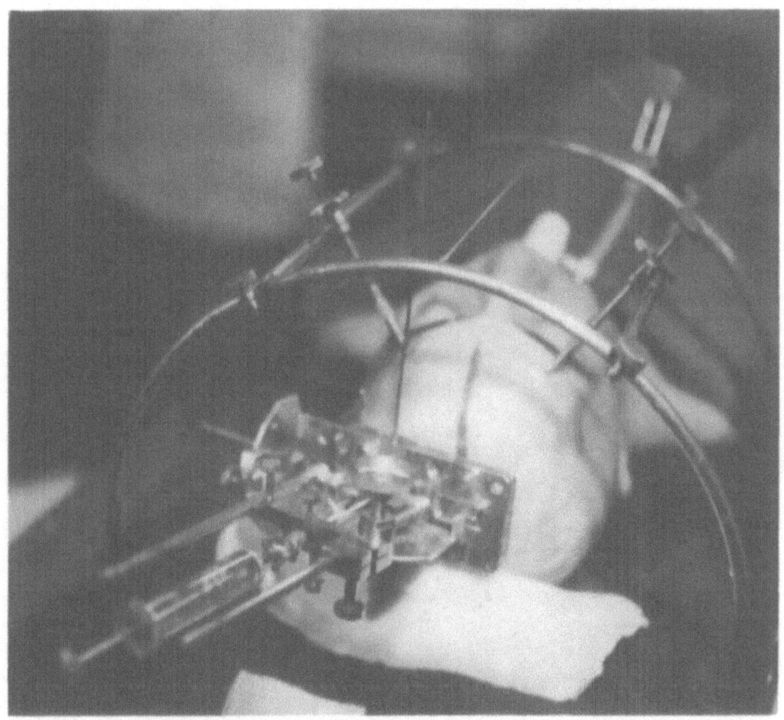

FIGURE 74. The head of the patient is fixed in the special device with the aid of sharp pins. Stereotactic apparatus is fixed in the burr hole. Puncture of V is performed, and contrast medium is introduced.

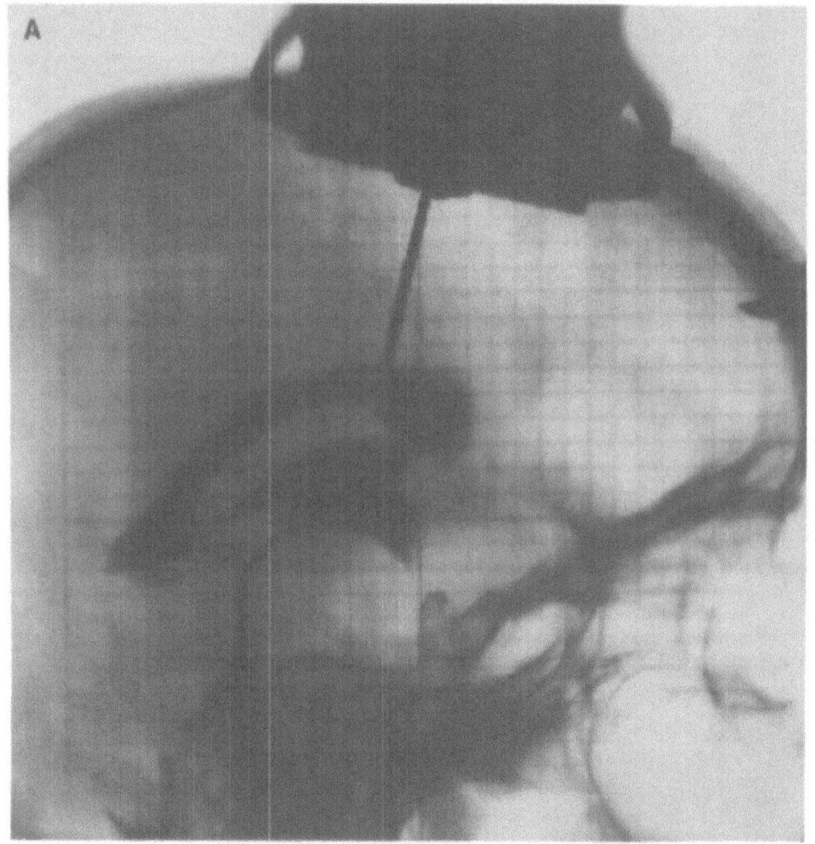

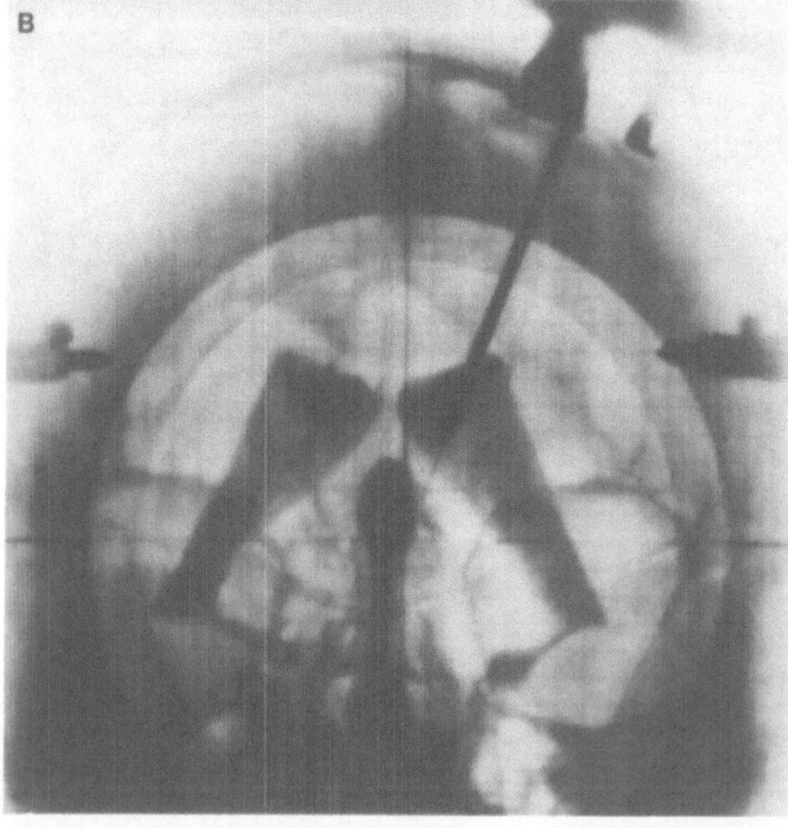

FIGURE 75. Ventriculograms after introduction of iothalamate–iophendylate mixture into ventricles. (A) Lateral projection. (B) AP projection.

ages, especially the posterior parts of V_3, the aqueduct, and V_4, which is very important for stereotactic calculations.

Ventriculography performed by this technique is tolerated well by patients. Since we have not observed a single complication directly related to this procedure, we believe that VG with iothalamate and iophendylate is one of the best contrasting techniques in stereotactic surgery if CT or NMR is not available for determining stereotactic coordinates.

5. Stereotactic Coordinates

Determining the spatial localization of the stereotactic target is the most complicated and important stage of the operation, one in which errors are apt to occur because of inaccuracies of x-ray techniques or individual variability of subcortical structures (Chapter 3, Section 3).

It should be noted that certain neurosurgeons, all having rich experience in performing stereotactic surgery on basal ganglia, use somewhat different calculations for stereotactic coordinates. For example, Laitinen (1985a,b), who collected questionnaires filled out by 16 leading neurosurgeons in various countries concerning choice of target for performing thalamotomy in parkinsonism, reported that the majority selected a target on LI approximately 5 mm posterior to the midpoint and 12–15 mm from the midplane. At the same time, it is noteworthy that the targets used by various neurosurgeons often varied by as much as 3–4 mm.

From ventriculograms made in two projections (Fig. 76) it is possible to make stereotactic calculations the ultimate objective of which are (1) to determine the main intracerebral reference points in AP and lateral films, (2) to calculate on the basis of these points the center of the subcortical target structure (or structures) to be destroyed, (3) to adjust the coordinate system of

FIGURE 76. Rapid electrochemical development of x-ray films during stereotactic operation by Picker–Polaroid device.

the stereotactic apparatus to the coordinates of the target structure and to determine the correction necessary to ensure that the electrode is centered on that structure, and (4) to control the accuracy of reaching the target.

It is convenient to perform such calculations on a glass table brightly illuminated from below or on an x-ray film viewer. Dividers, protractors, precision rulers, etc. are required. First, it is necessary to be sure that the films have been made properly: in the frontal projection the midplane must coincide with the midline of the headrest frame, and in the lateral projection the contours of the bony structures on both sides of the skull must coincide.

The first stage of the calculations is to determine the intracerebral reference points on both ventriculograms. Then the surgeon begins the most exacting part of the operation, i.e., determining the site of the subcortical structure into which the cannula (electrode) is to be inserted. These calculations must be compared with the data of a stereotactic atlas.

The calculations in ventriculograms of the lateral and AP projections for the localization of the two most frequent targets in extrapyramidal dyskinesias, VL and

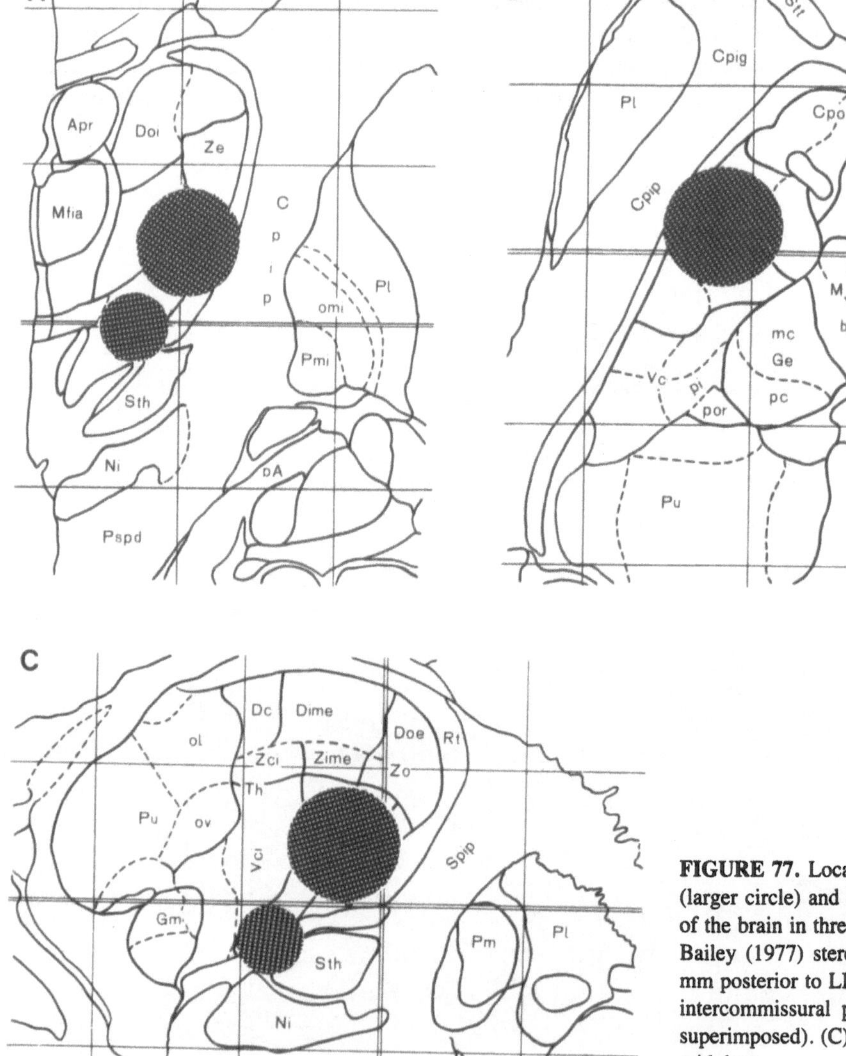

FIGURE 77. Localization of destroyed foci in VL (larger circle) and in Subth (smaller circle) on sections of the brain in three planes from the Schaltenbrand and Bailey (1977) stereotactic atlas. (A) Frontal section 4 mm posterior to LI midpoint. (B) Horizontal section in intercommissural plane (both foci in this section are superimposed). (C) Sagittal section 12 mm lateral to the midplane.

Subth (Fig. 77), are given in the following sections; calculations for locating other cerebral structures are presented in their respective chapters.

5.1. Anteroposterior Projection

After calculating the correction for divergence of x rays, one determines the midplane of the brain. For this purpose, a vertical line is drawn through the median fissure, along septum pellucidum, and continuing in the midline of V_3 (Fig. 78). Then one measures the vertical dimensions of the anterior part of V_3 and divides it into three parts. Through the junction of the middle and inferior third of V_3 a second line is drawn perpendicular to the first. The center of VL is located on this line at a distance of 12–14 mm from the midplane.

In hydrocephalus caused by the cerebral atrophy that develops frequently in various extrapyramidal disorders, the width of V_3 at its middle part is increased from 3–4 mm to 8–10 mm and even to 12 mm. In view of this, in the calculations an important question arises: is it necessary to take into account the enlarged V_3 (in the AP film) in determining the distance from the midplane to the center of the subcortical structure in question (VL, VPL, Subth)? There are two opposing points of view on this matter. On the basis of our personal experience, we believe that in pronounced hydrocephalus this distance should be increased by 1–3 mm to allow for the dilatation of V_3. This conclusion is backed by data on the variability of the distance between V_3 and the medial border of CI relative to the width of V_3. If V_3 is narrow (width 1–5 mm), the distance from the midline to CI is 16–20 mm, whereas if V_3 is wide (6–14 mm), this distance is 19–23 mm (Hardy *et al.*, 1979a).

Based on data from stimulation of many points in Th (Tasker and Emmers, 1969; Hardy *et al.*, 1979a,b), the spatial localization of VPL also depends, although not to a great degree, on the width of V_3. According to

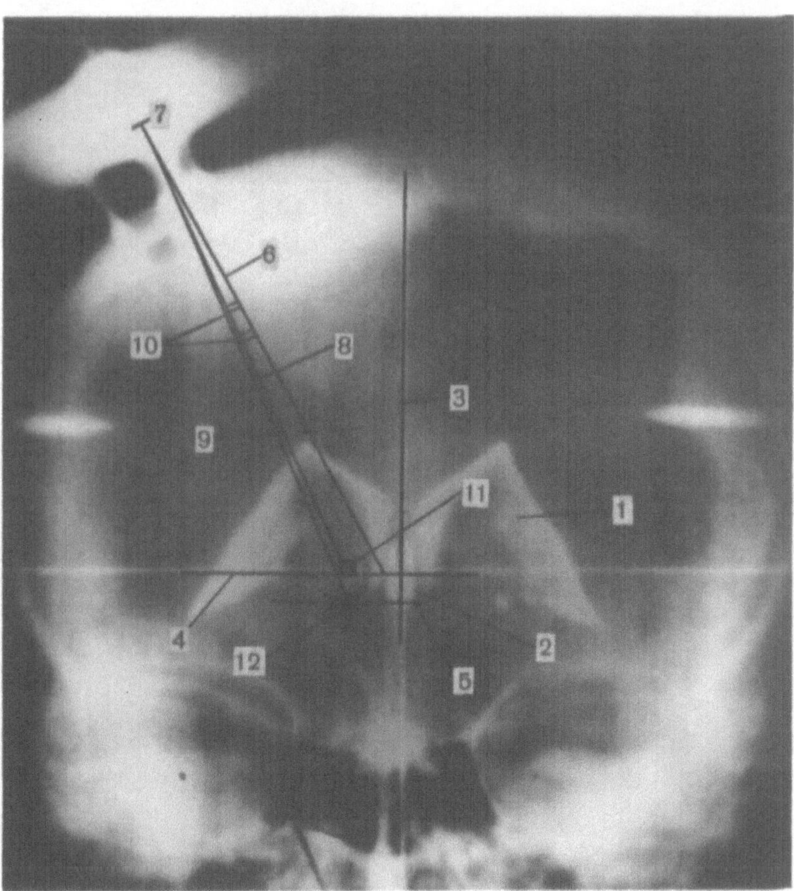

FIGURE 78. Stereotactic calculations for thalamic VL and Subth destruction. Ventriculogram in anteroposterior projection. 1, Posterior horn of V; 2, V_3; 3, midline plane of brain; 4, the line perpendicular to 3 on the border of the medial and inferior thirds of V_3; 5, line parallel to 4 drawn across inferior pole of V_3; 6, the zero line of the stereotactic apparatus; 7, "rotation center" of stereotactic apparatus during correction of the cannula direction; 8, the line connecting "rotation" and Subth centers; 9, the line connecting "rotation" and "VL center"; 10, correction angles of the cannula direction in the frontal plane; 11, the calculated VL center; 12, Subth center.

the anatomic investigations by Yukharev (1979), with dilated V_3, CF occupies a more lateral position, being displaced approximately 1.5 mm. However, other authors (Talairach *et al.*, 1957; Laitinen, 1966; Bertrand *et al.*, 1969) believe that dilatation of V_3 does not cause any significant lateral displacement of thalamic structures. If this view is correct, the localization of the center of a destructive focus relative to the midplane can be made irrespective of the degree of hydrocephalus.

The distance from the midplane to the target should be multiplied by the divergence coefficient, and the resultant distance should be marked off on the abovementioned horizontal line. This point is the center of VL.

The center of Subth is localized medial, ventral, and caudal to the center of VL. In order to determine it in the AP projection, one should draw another line below and parallel to the line passing through the center of VL. This line is drawn at the lower margin of V_3, and on it a point 10–11 mm from the midplane is marked. Thus, this point indicating the center of Subth lies deeper and medial to the center of VL.

5.2. Lateral Projection

In this projection (Fig. 79), we make use of two main reference points, the posterior borders of FM and CP, for the calculations. The former corresponds to the point where the curved inferior margin of V passes to the anterosuperior part of the cupola of V_3. The latter corresponds to the point where the caudal inferior

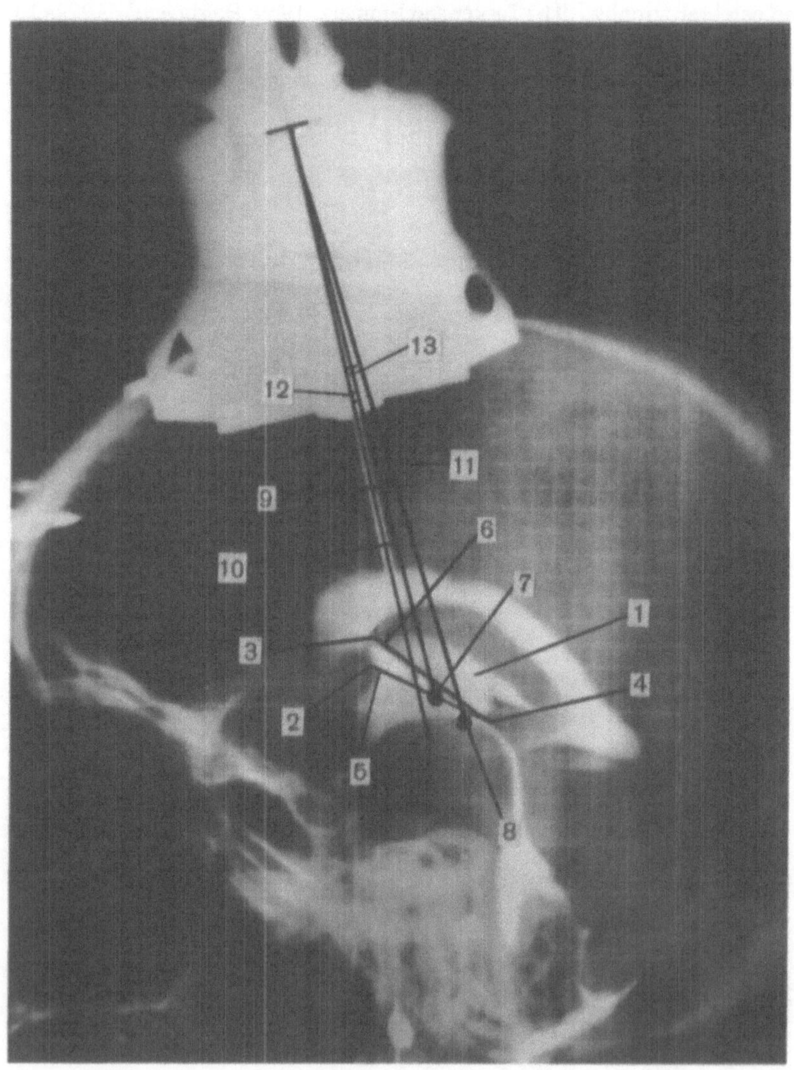

FIGURE 79. Stereotactic calculations from VL and Subth destruction on lateral ventriculograms. 1, V_3; 2, CA; 3, posterior edge of FM; 4, CP; 5, LI; 6, the line connecting posterior edge of FM and CP; 7, VL center; 8, Subth center; 9, the zero line of cannula direction; 10, the line from the "rotation center" to the VL center (13–14 mm on the line FM–CP and 2 mm inferior to that point); 11, the line connecting the "rotation center" with the Subth center; 12, the angle of correction of the cannula direction in the sagittal plane for VL; 13, the same angle for Subth.

border of V_3 (below the pineal recess) joins with the upper surface of the oral part of the Sylvian aqueduct. By connecting these two points on a lateral ventriculogram, one has the principal coordinate line FM–CP (its length and variability are discussed in Chapter 3, Section 2,3). This line is measured (corrected for divergence), and depending on its length, the "relativity factor" is calculated.

As has been noted, if muscular rigidity predominates in the clinical picture, VOa should be selected as the target for destruction, and if tremor or hyperkinesias are to be eliminated, then VOp and Vim should be destroyed. In accordance with this, on the line FM–CP, it is necessary to mark off 13, 14, or 15 mm from the posterior border of FM after adjusting this distance in accord with the divergence correction. The center of VL is the point located 1–2 mm below the designated mark on the line of the posterior border of FM–CP.

The center of Subth is located ventrocaudal to that of VL. In order to reach the center, it is necessary to make a mark 12–13 mm from the posterior margin of FM on the line FM–CP (after correction for divergence) and from that point to draw another line ventrally, perpendicular to the first. On this second line, mark off 6 mm and from that point another 7 mm caudally, so that this last point, which is the center of Subth, is 5–5.5 mm from the main line (posterior border of FM–CP). The focus of destruction should be an oval (in cross section) 3–4 mm wide and 6–7 mm long.

The next stage of the stereotactic calculations is to determine the angles of correction in both ventriculograms, i.e., the angle between the ventricular cannula and the trajectory line of the cryothermic cannula, which is required to reach the previously calculated target point accurately. Two more lines are drawn on each of the films to determine the angles of correction. The first ("zero") line passes along the axis of the cannula introduced into V. The "center of rotation" is then determined by the metallic pins of the stereotactic apparatus, visible in the films. The second line connects the "center of rotation" with the calculated target point. The angles between these lines are the angles of correction in the frontal and sagittal planes. Then the cannula is extracted from the brain, and the correction angles are transferred to both protractors of the stereotactic apparatus. This series of stereotactic calculations takes approximately 20–25 min.

Recently, in collaboration with our associates Peresedov and Goyhman, we have developed and put into practice an optimal system of calculations that requires only 6–8 min and enhances the accuracy of the calculations.

In essence, such optimization of calculations lies in the following. Any approach to finding the target point, i.e., the center of the subcortical structure to be destroyed, is based on the principle of proportional changes in all intracerebral dimensions as coordinates are transferred from the "standard brain" in the stereotactic atlas to the brain of the patient and then from the patient's brain to its contrast films. The proportionality implies, in particular, a similarity of triangles formed by two reference points (posterior border of FM and CP) and the center of the intended destructive focus (Fig. 41).

Calculations are simplified by the use of two specifically prepared stencils of transparent plastic, one for the lateral and the other for the AP projection. Each stencil has markings based on the principle of similar triangles. The stencils allow one to find the target point in both films in about a minute irrespective of the degree of x-ray divergence. There are two triangles, *ABC* and *abc*, having a common angle. The dimensions of the sides in the second triangle are equal to the corresponding dimensions in the Schaltenbrand and Bailey *Stereotactic Atlas:* apex *a*, the posterior border of FM; apex *b*, CP; and apex *c*, the target point. The triangle *ABC* indicates the position of the same points on the lateral ventriculogram of the patient.

The stencil is superimposed on the lateral x-ray picture so that the point CP coincides with point *B* on the stencil while the posterior margin of FM is on the line connecting these two reference points. In these conditions, the center of the focus will be found in the slot of the stencil coinciding with the point on the FM–CP scale. Through the slot, this point is marked on the film. The center of destruction in the AP projection is found in the same manner using the second stencil, built according to the same principle. The stencils may be used for x-ray films obtained with any divergence coefficients. It greatly reduces the time to determine the target point and enhances the accuracy of the calculations.

After fixing the electrode or cannula for local freezing in the stereotactic apparatus, we begin slowly to lower it into the brain towards the target. There are two ways of determining the depth of immersion, i.e., the distance from the surface of the cerebral cortex to the center of the subcortical target. This is a complicated problem since the planes in which the electrode or cannula moves are not parallel to the planes of the x-ray films in both projections. In order to resolve this problem, we employed a number of formulas obtained after processing the data in a computer. The results of the mathematical calculations are given in a table from which it is possible to determine swiftly the desired depth of penetration into the brain in each case.

In everyday practice, it is possible to use the following simplified but sufficiently accurate method. The cannula (electrode) is inserted into the brain to a depth approximately 3–4 mm less than the distance to the target point on the calculated VGs. Then, in the lateral film, which is closer to the orthogonal projection than is the AP film, one measures the distance from the active tip of the instrument to the target point with dividers, makes the correction for divergence, and then advances the cannula the required distance.

6. X-Ray Control

The main criterion for accuracy in reaching a given subcortical structure is x-ray control. For this purpose, after the cannula (electrode) is introduced to the calculated depth, control films are taken in the same two projections as those of the first films on which the stereotactic calculations were made. Then, with dividers, the main reference points and the center of the target structure are transferred from the original films to the control ones. In the majority of cases, the cannula has been introduced accurately, and the center of its active tip coincides with the target. In case of a difference exceeding 1 mm, which now occurs only rarely, it is necessary to correct the position of the cannula.

For this purpose, in both films, it is necessary to determine additional angles of correction (the angle between the axis of the cannula and the line from the center of rotation to the target point). After that, the cannula must be removed from the brain, the stereotactic apparatus corrected for the new angles, and the cannula reinserted into the brain. Repeat control films in this instance are mandatory.

The use of an electronic amplifier with TV monitoring in stereotactic operations substantially reduces the operating time. We employ two Delcalix monitors of the Dutch firm "Old Delpft." Two x-ray tubes are placed: one under the operating table (AP projection) and the other to the side (lateral projection). The electronic amplifier transforms the x rays into visible images on large TV screens. Each electronic amplifier has a video tape recorder to record the entire operation for subsequent playback. On the screens of the amplifiers it is possible to control the puncture of the ventricles, their filling with contrast substance, the proper direction of the cannula, and the (approximate) accuracy of the approach to the target. The intraoperative x-ray control is the principal means of ensuring that the target is being reached; however, it should be supplemented by numerous methods of functional control.

7. Functional Control

If the operation is performed under local anesthesia, the obvious criterion that the desired subcortical goal has been achieved is a positive clinical effect, which can be seen directly on the operating table. This result, as a rule, occurs 1–1.5 min after the cannula has reached the target point, i.e., before any attempt is made to destroy that structure. Our many years of experience, as well as data in the literature, have shown that a distinct clinical effect from slight mechanical damage to the target structure is a reliable prognostic sign of a good subsequent result from the operation.

The result naturally depends on the nature of the disorder: in parkinsonism it will be the disappearance of tremor and rigidity in contralateral limbs; in multiple sclerosis, only of tremor; in dystonia, of hyperkinesias; and in pain syndromes, of the sensation of pain. In operations performed under general anesthesia, the possibility of monitoring the clinical effect is obviously very limited, so various functional controls must be used.

7.1. Electrostimulation

Stereotactic surgery has given investigators a unique opportunity to perform electrostimulation of deep-seated cerebral structures in a conscious patient. This has made it possible to expand significantly and elaborate the information obtained earlier in neurophysiological animal experiments and to study those peculiarities characteristic only of the human brain.

The first results elicited by electrostimulation in the depth of the human brain in psychosurgical operations were published in 1955 by Sem-Jacobsen *et al.* of the Mayo Clinic. After their investigations, depth electrodes became used worldwide. To date, hundreds of such studies using advanced techniques have been published.

Electrostimulation is a very important component of stereotactic and functional operations at all levels of the nervous system—in operations on the cortex, subcortical structures, craniocerebral nerves, Cer, the spinal cord and its roots, peripheral nerves, and so on. The literature today includes scores of papers on electrostimulation of various subcortical structures: many nuclei of Th, Subth, CI, Hypoth, Put, NC, GP, GC, Am, Hipp, CC, deep-seated nuclei of Cer, and others.*

The results can be divided into three categories. The first includes data on the functional organization of subcortical structures, their interrelationships, rates of transmission of stimulation, pathogenesis of tremor, rigidity, and hyperkinesias, and so on, in other words, neurophysiological and pathophysiological investigations.

The second category, closely related to the first, concerns data of an applied nature. Electrostimulation of subcortical structures represents an important and informative stage in the majority of stereotactic operations. It is regarded as one of the main means of identifying subcortical structures, thus enabling the neurosurgeon to verify accurate placement of the instrument in the target structure or, on the contrary, to demonstrate that the electrode is in another, neighboring structure.

In the majority of operations on deep cerebral structures, electrostimulation should confirm that the electrode tip is not in CI, which is directly adjacent to the lateral margin of VL and the medial margin of the medial segment of GP. It should be noted, however, that the identification by stimulation of subcortical structures is only a relatively precise method. Since diffusion and current loops occur, the same sensory and motor effects may result from stimulation at different distances from CI.

Finally, the third and most rapidly developing technique is chronic electrostimulation of subcortical, cerebellar, and spinal structures and peripheral nerves. Chronic stimulation of these structures has recently been used extensively in the management of pain syndromes (Chapter 13) as well as epilepsy (Chapter 7), cerebral palsy (Chapter 9), and other disorders.

The effects of stimulation of deep cerebral structures may be divided into several main groups:

1. Motor (muscle reactions, muscular tension, tremor, convulsions in various parts of the body).
2. Sensory (paresthesias, pain, sensitivity disorders).
3. Ocular (eyeball movements, changes in the diameter of the pupils, opening and closing of eyes).
4. Vegetative (changes in arterial pressure, vascular tonicity, cardiac rhythm, respiration, perspiration, pallor).
5. Changes in higher nervous activity (general excitation, sleepiness, altered speech, emotional reactions such as euphoria, fear, rage, hallucinations, laughter, and so on).

The vast and diverse information that has been obtained from stimulation of deep cerebral structures has demonstrated a number of important principles, which should be discussed in detail. Stimulation of subcortical structures may produce irritation, activation, i.e., the appearance or intensification of certain functional activity, or inhibition of such functions, although the former occurs much more frequently than the latter. Moreover, opposite effects may be observed during stimulation of the same subcortical structure. For example, in stimulation of VL (see Section 7.1.1) one may observe both intensification and inhibition of tremor. This variability of effect arises in particular, from simultaneous excitation and inhibition as a result of polarization of nerve tissue.

The stimulation of certain structures can induce not only local, but generalized cerebral effects. For instance, stimulation of the Vicq d'Azyr bundle, the frontothalamic and frontopontine tracts proceeding into CI, may induce transient mental disorders (Guiot *et al.*, 1968b). With stimulation of VL, one sometimes observes a general activation of mental activity, excitation, motor activity, and so on (Bechtereva *et al.*, 1969). On the other hand, stimulation of other thalamic structures (intralaminary nuclei, Subth) may induce sleepiness (Bechtereva *et al.*, 1967).

The question of the existence of a dominant and nondominant Th cannot be considered resolved. Nevertheless, many investigators use these terms based on the assumption that the dominant hemisphere corresponds to the dominant Th. It is presumed that the dominant Th has a more complicated distribution of functions. In particular, stimulation of the dominant Th induces motor effects in the contralateral limbs and side of the face twice as frequently as stimulation of the nondominant Th (Schaltenbrand and Wahren, 1982).

The stability or reproducibility of effects of stimulation of different structures also varies considerably.

*The technique and effects of spinal cord stimulation are presented in Chapters 13 and 19.

It should be noted that the majority of investigations on stimulation of subcortical structures were conducted on patients being operated on under local anesthesia. Stimulation of these structures with the patient under general anesthesia usually causes wakefulness, and the EEG reveals a desynchronization.

Stimulation of certain subcortical structures produces stable and specific effects. Irritation of the pyramidal tract in CI gives rise to clonic or tonic contractions of the muscles of the opposite half of the body; stimulation of VPL produces various sensory effects; and stimulation of Hypoth evokes autonomic reactions. Stimulation of other structures induces less well-defined and less stable effects, sometimes of an opposite nature. For example, we have already noted that stimulation of VL in parkinsonism modifies tremor in only some patients; moreover, this may be either intensification or inhibition. Such inconsistent effects on tremor and other hyperkinesias were noted by several authors during stimulation of many subcortical structures—VL, GP, Subth, CF, CM, and others (Andy *et al.*, 1963; Spiegel *et al.*, 1964b; Mundinger, 1965c; Ohye *et al.*, 1975; Ganglberger *et al.*, 1970a,b; Smirnov, 1976). Finally, the stimulation of several deep cerebral structures does not produce any appreciable effects.

Numerous investigations, confirmed by our experience, have established that not only the occurrence of a response but also its nature depend on the parameters of the stimulating current. A change in frequency or voltage may not only intensify or diminish the response but may even reverse it. For example, at a stimulation frequency of 10 Hz, tremor did not change, but at 50 Hz, it was noticeably intensified, even at the same stimulus voltage and duration (Kandel, 1965). The relationship of tremor to the frequency of the stimulating current has been demonstrated in a number of papers (French *et al.*, 1962; Johansson and Laitinen, 1966; Hassler, 1974). Stimulation of GP, VOa, or VOp in parkinsonians with a 4-Hz current produced a slower rate of tremor, whereas higher frequencies (8, 25, 50 Hz) increased the rhythm of tremor. Stimulation of the other structures may affect the tremor differently (Narabayashi, 1968).

Motor effects such as discrete muscular contractions are best demonstrated with a low rate of stimulation, 5–8 Hz. As the frequency of the current is increased, these contractions merge into a continuous tonic spasm of the muscles. Stimulation with a high-frequency current results in motor effects involving a greater number of parts of the body than does low-frequency current (Andy, 1966). It has also been established that varying the current parameters used in stimulating cerebral structures transmitting pain may produce diametrically opposite effects. Stimulation at a frequency of 10–20 Hz induces analgesia, whereas a frequency of 60–100 Hz causes a sensation of pain (Richardson and Akil, 1974).

During low-frequency stimulation of PAG, there are usually no psychological reactions; however, an increase in the frequency of stimulation produces a pronounced sense of fear, confusion, and a feeling of warmth through the entire body (Amano *et al.*, 1982). It has been established that high-frequency (50–100 Hz) stimulation of NC causes intensified epileptic activity in Am and Hipp, whereas stimulation at a low frequency (4–6 Hz) sharply inhibits this activity (Chkhenkeli, 1981). Nashold (1982) noted that low-frequency stimulation of the midbrain does not cause any behavioral reactions or sensation of pain, which appear in high-frequency (up to 300 Hz) stimulation.

Walker (1982) considers the optimal parameters for stimulation to be biphasic impulses lasting for 1 msec at a frequency of 50–100 Hz. Certain authors prefer stimulation at a high frequency (200–300 Hz) (Toth and Tomka, 1968). Van Buren and Ratcheson (1973) believe that biphasic pulses cause less damage to nerve tissue than monophasic stimulation and that the duration of each pulse should not exceed 1 msec.

A thorough statistical analysis of changes in tremor during electrostimulation of deep cerebral structures in parkinsonian patients was conducted by Smirnov and Iovlev (1974). Stimulation was produced by bipolar rectangular impulses of 10 V, lasting for 1 msec in the form of a series of impulses at various frequencies. Their data reaffirmed that stimulation does not always (in fewer than half of the cases) cause changes in tremor. The changes (both activation and inhibition of tremor) depend primarily on the structure being stimulated. The greatest changes in tremor occurred during stimulation of VL (46%) and then, in descending order, of CM (34%), CI (22%), and VPL (21%). Changes in tremor during stimulation of GP and Put (13%) were considerably less frequent. The authors also confirmed that stimulation of all these structures may produce either intensification or inhibition of tremor; however, the changes are not the same but depend on the structure stimulated. There was a prevalence of tremor inhibition; it was observed 1.5–3 times more often than intensification.

As was demonstrated in several investigations (including those of the author), changes in tremor during stimulation of thalamic nuclei were, as a rule, bilateral and more pronounced in the arm than in the leg. It was also noted that stimulation of various nuclei, in certain cases, may produce changes (intensification or inhibition) in tremor on the contralateral side, whereas stimulation of other nuclei may affect tremor on the ipsilateral side of the body. Ipsilateral motor effects are observed on stimulation of the more medial parts of Th, and contralateral effects on stimulation of its lateral parts (Andy, 1966). Thus, these results and data in the literature indicate that electrostimulation of various subcortical structures during stereotactic operations did not identify any structure that could be considered the "central generator" of tremor.

A vast amount of information concerning physiological localization in Th and the midbrain has been collected in the monograph published by Tasker *et al.* (1982b). Their detailed analyses of computer-assisted display of data obtained by electrostimulation of about 10,000 sites along 835 trajectories during stereotactic surgery revealed a very complicated functional organization of the main subcortical structures.

There are two main techniques of electrostimulation. The first is during a stereotactic operation at a stage preceding the destruction of a target in subcortical structure. The second involves the introduction of a great number of thin electrodes into one or several subcortical structures; the electrodes are fixed to the skull and left in the brain for a long period of time (up to several months and even years).

After implantation of chronic depth electrodes, it is possible to stimulate subcortical structures and study their reactions over a period of time. This is of paramount significance in analyzing the neurophysiological characteristics of deep-lying structures. Chronic stimulation of various structures of the brain and the spinal cord has often been used in recent years for therapeutic purposes. The techniques of such therapeutic stimulation are described in detail in other chapters of this book.

Implantation of chronic electrodes in subcortical structures of the brain for neurophysiological animal experiments was first described in detail by Hess (1928). Intracerebral structures in man were first stimulated chronically by Pool in 1954 and then by Llewellyn and Heath (1962) and by Delgado and Hamlin (1962). Over the course of many years, implanted electrodes have been used by a Leningrad scientific team headed by Bechtereva (Bechtereva *et al.*, 1967, 1972a,b, 1977; Bondarchuk, 1971; Smirnov, 1976). These investigations gave diverse and important information not only about the functions of subcortical structures but also about the neurophysiological basis of emotion, memory, and other mental activities.

It is appropriate to comment on the different techniques used in electrostimulation in stereotactic surgery. The majority of neurosurgeons employ monopolar stimulation, although others feel that bipolar stimulation produces significantly less dissemination of current loops to surrounding brain tissue. Recently, bipolar stimulation has been used more frequently. Many researchers, including the author, employ a stylet electrode covered with a thin coat of insulating Teflon for stimulation (Fig. 80). As the graduated knob at the external end of the electrode is turned, a fine (0.3 mm) active tip made of a special alloy is extruded from the tube. This tip, protruding at an angle to the axis of the instrument, may be extended from 1 to 7 mm. Without removing the electrode from the brain but only revolving it around its axis, it is possible to stimulate a structure at some distance from the tube.

Electrodes of various designs, usually made of gold, platinum, palladium, stainless steel, iridium, and certain alloys, e.g., one of chromium, cobalt, and nickel are used for chronic stimulation of deep cerebral structures. Silver and copper, which were used earlier, have been abandoned because of their toxic effect on cerebral tissue. Electrodes composed of fine bundles of insulated wires (three to six or more) are stereotactically advanced into various subcortical structures. Multipolar platinum or stainless steel electrodes with Teflon insulation of the Schryver type are also frequently employed. Concentric bipolar electrodes, limiting the zone of current spread, and spring electrodes that change their rhythm to match the pulse and respiratory movement of the brain have been proposed (Pudenz *et al.*, 1975).

There is equipment with which one can both stimulate and destroy (radiofrequency thermocoagulation) subcortical structures (Chapter 5, Section 3.2).

It is preferable to stimulate with an anodal current from an impulse generator. If, on stimulation, a motor, sensory, or other effect is obtained, the stimulation should be repeated with the same voltage and then with higher and lower voltages in order to determine the threshold of excitation. Usually, we begin to stimulate with a low voltage (1–2 V) in an effort to detect the threshold for stimulation. Then the voltage may be gradually increased to 5–7 V. If current rather than voltage is being recorded, the minimal current stimulating subcortical structures is 5–6 mA (maximal, 40–50 mA). Most neurosurgeons, the author included, use

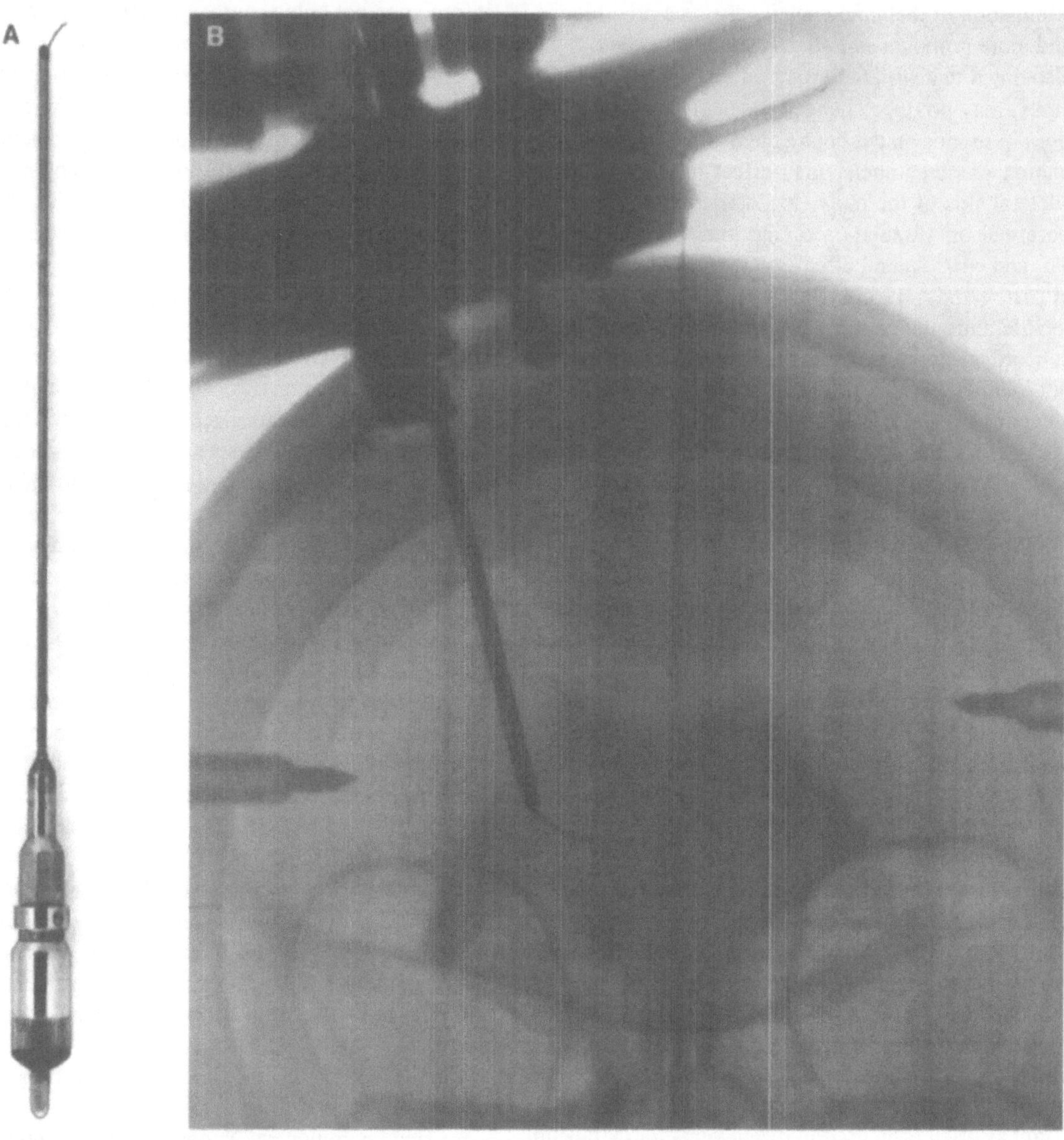

FIGURE 80. Electrostimulation of subcortical structures during stereotactic operation. (A) General view of the stylet electrode. (B) Intraoperative ventriculography. Active tip of the stylet electrode is extended medially.

a current with a frequency from 50 to 100 Hz, of rectangular form, and duration about 1 msec. It has been noted that pulses of longer duration damage nerve tissue. For certain structures more sensitive to stimulation, e.g., CI, a current with a lower frequency, 5–10 Hz, is recommended. Superhigh-frequency currents (500–700 Hz) have been used for stimulation (Andy, 1966). The total duration of each stimulation, which may be repeated several times, should not exceed 3–5 sec, since prolonged stimulation may induce an epileptic seizure.

During a stereotactic operation, stimulation is usually carried out at intervals as the electrode advances toward the target point. It is recommended that the stimulation be repeated every 2 mm so that the margin of a responsive deep structure will be detected early. The stimulation is continued until the response has passed a maximum and declines, which means that the electrode has passed beyond the target structure.

Preliminary studies have shown that noninvasive scalp electrostimulation of many points may produce responses from a subcortical structure located by computerized data (Rusinko *et al.*, 1985).

After these general remarks, more detailed data

on the effects of electrostimulation of various subcortical structures in man can be presented.

7.1.1. Ventrolateral Nucleus

Since the most frequent target for stereotactic destruction is VL, particular attention should be focused on the stimulation of this structure. Although VL stimulation does not produce any specific motor or sensory reactions such as occur following CI stimulation, changes in the tremor are observed much more frequently than after GP stimulation (Spiegel and Wycis, 1960; Hassler *et al.*, 1960; Kandel, 1965, 1981; Sugita and Doi, 1967; Tasker *et al.*, 1982b). These changes include modulation of the frequency or amplitude of tremor or both. The most common effect is an intensification (increase of amplitude) of the tremor, although arrest of the tremor is not infrequent. The rhythm of the stimulation is practically never precisely followed by the tremor. A very interesting fact is that latent tremor that becomes overt on stimulation has the same frequency as the usual pathological tremor of that patient, although the frequency of stimulation may vary over a wide range. An existing tremor becomes more intense on stimulation with a higher-frequency current (up to 100 Hz), whereas tremor inhibition requires a lower-frequency current.

The effect of stimulation on tremor also depends on the location of the electrode within VL. For example, electrostimulation of the anterior part of VL (VOa) reduces or arrests tremor, whereas stimulation of the posterior parts (VOp and Vim) increases the amplitude of tremor (Narabayashi, 1969; Vasin *et al.*, 1971; Ohye *et al.*, 1982a). These findings suggest that VL has two functional zones, one facilitatory and one inhibitory to tremor. It is thought that destruction of a zone that blocks tremor upon stimulation has a more pronounced and stable effect (Alberts *et al.*, 1965; Kandel, 1965, 1968; Vasin *et al.*, 1971; Van Buren and Ratcheson, 1973). However, there is an opposite point of view: if electrostimulation at one site intensifies or brings out latent tremor, it suggests that destruction of the subcortical structure stimulated will eliminate the tremor. Hassler *et al.* (1979) point out that stimulation of VOp with a frequency lower than the rhythm of tremor (4–6 Hz) slows it, whereas a current frequency higher than 6 Hz accelerates the rhythm. On stimulation of VL, the homolateral tremor becomes more regular and of increased amplitude.

Stimulation of VL often results in varied and even incongruous changes in the emotional–affective sphere, either sleepiness and flaccidity or a general increase in psychomotor activity, accelerated intellectual processes (Illinsky, 1970), euphoria, and dysphoria (Smirnov, 1976). Stimulation of the anterior part of VL in parkinsonian patients increases extrapyramidal rigidity (Narabayashi, 1968), whereas stimulation of Vim at times arrests dystonic hyperkinesias (Tasker *et al.*, 1982b). Stimulation of VOa and VOp with a current frequency of 25–50 Hz accelerates active movements of the limbs and impairs articulation (Hassler *et al.*, 1979) as well as facilitating the H-reflex (Laitinen and Ohno, 1970).

7.1.2. Sensory Nuclei

Stimulation of VPL with a low voltage (approximately 1 V) during stereotactic operations under local anesthesia induces paresthesias in the contralateral half of the body. With increased stimulus frequency and voltage, these paresthesias become stronger, barely tolerable, differing very little from pain.

Stimulation of sensory nuclei as well as CI (see Section 7.1.7) as a rule produces very discrete paresthesias. In our experience, if the electrode is advanced by only 1 mm from a point producing paresthesias, sensations quite frequently occur in another part of the body (Kandel, 1981). After stimulation with a current frequency of 1–10 Hz, which causes tolerable paresthesias, chronic central pain usually disappears for several minutes (Richardson and Akil, 1977). However, touch and pain irritation of the skin does not reveal any sensory loss. Stimulation of VPL also does not induce temperature sensations such as a feeling of warmth or cold.

During VPL stimulation, paresthesias often appear not only on the contralateral but also on the ipsilateral half of the body, which is an indication of bilateral pain representation in the sensory nuclei of Th.

The somatotopic organization of VPL is clearly evident during stimulation of different parts of that nucleus.

The torso and pelvic areas are represented in the superior part, and the limbs inferiorly. The individual parts of the body are represented in VPL not only along the vertical but also along the horizontal axes: in the lateral part of the nucleus is the leg, in the middle, the torso, and medially the arm and face (Albé Fessard *et al.*, 1967b; Vasin and Ratza, 1971b). The zonal representations of the body frequently overlap, especially those for the body and limbs.

During VPL stimulation, motor effects on the contralateral side of the body occur very rarely. Many authors consider that this results from the spread of current to CI and the pyramidal tract. However, Schaltenbrand and Wahren (1982) disagree with this point of view and consider that on stimulation of the posterior parts of Th, motor effects are induced by excitation of the thalamocortical afferent pathways.

Stimulation of VPM, where the trigeminothalamic pathway terminates, induces sensations in the head and face. The scalp, neck, face, tongue, and lips are consecutively represented in this nucleus from the dorsolateral to the ventromedial direction (Vasin and Ratza, 1971a,b). On stimulation of VPL with a low-frequency current (1–16 Hz), one may observe synchronization of the rhythm of the scalp EEG and, in the central cortex, the ECoG.

7.1.3. Pulvinar

Stimulation of Pul nuclei during surgery may produce paresthesias in the contralateral half of the body; however, their threshold is higher than in VPL (Richardson, 1974; Tasker *et al.*, 1982b). Ojemann and Fedio (1968) reported that stimulation of the left Pul, but not the right, results in short-term memory disturbances. Speech impairments occur on stimulation of the ventrolateral part of Pul. Excitation of the medial part results in the disappearance of spontaneous pain (Martin-Rodriguez and Obrador, 1975). Emotional changes are noted on chronic stimulation of the anterior part of Pul. Stimulation with a 50-Hz current produces both bursts of synchronous oscillations and spindles with periods of desynchronization of cortical rhythm on ECoG (Majorchik *et al.*, 1980).

7.1.4. Centrum Medianum and Intralaminar Nuclei

Stimulation of CM in patients with hyperkinesias does not cause pain. However, in such cases, one may frequently observe psychological reactions (fear, excitement), accompanied by vegetative changes (tachycardia, dyspnea). High-frequency stimulation of CM in patients with pain syndromes leads either to an aggravation or termination of spontaneous pain in both the ipsi- and contralateral halves of the body. Paresthesias in the contralateral half of the body occur quite frequently. On stimulation of the posterior part of the intralaminar nuclei, the paresthesias assume a burning quality (Sano, 1977).

Stimulation of CM with a 50-Hz current activates the EEG and ECoG, producing a desynchronization of the cortical rhythms and bursts of synchronous oscillations (10–19 Hz) and spindles (Majorchik *et al.*, 1980). Evoked potentials can be recorded from the ipsilateral cingular cortex (Sano, 1977).

7.1.5. Dorsomedial Nuclei

The effects of stimulating the dorsomedial nuclei have not been extensively studied. Several papers indicated that such stimulation during surgery induces mainly psychological reactions, in particular, excitability and anxiety (Richardson, 1974). In certain cases, the subjective sensations during stimulation of medial nuclei were unpleasant, and the patients requested termination of the procedure. In other cases, on the contrary, stimulation induced pleasant sensations. At times, patients laughed without cause and sometimes cried. Other authors (Tasker *et al.*, 1982b) have not observed any definite effects on stimulation of the dorsomedial nuclei.

7.1.6. Anterior Nuclei

Stimulation of the anterior nuclei of Th, in certain cases, induces turbulent affective reactions including a sensation of fear and even horror (Illinsky, 1970). These reactions disappeared immediately on termination of stimulation.

7.1.7. Capsula Interna

The effects of stimulating the posterior limb of CI are the result of excitation of descending pathways, of which the functionally most important is the corticospinal (pyramidal) tract, which passes through the posterior limb of CI (Chapter 1, Section 1.2).

Stimulation of CI is important in stereotactic operations on the basal ganglia. Since destruction of the pyramidal tract leads to grave consequences (hemiparesis, aphasia), the main object of stimulation is to establish that the active tip of the instrument is not within or close to the pyramidal tract. If this has occurred, it is necessary to correct the trajectory of the electrode by shifting it in a medial direction. If the electrode is in CI, stimulation of the posterior limb of CI causes typical so-called capsulary effects—clonic or, less frequently, tonic muscular contractions in the contralateral limbs and side of the face, usually at the rate of stimulation, but without any subjective sensations. For example, CI stimulation does not provoke the tingling sensation that is typical of VP stimulation.

Unlike Th nuclei, a motor response to stimulation of the posterior limb of CI is evoked not only by repetitive stimulation but also by single shocks or low-frequency (5–10 Hz), short-term (0.2 msec) shocks (Voronin and Kandel, 1971; Marossero *et al.*, 1972). The latent period of stimulation of CI is very brief, varying from 10 to 30 msec, whereas during stimulation of VL, this period is several times longer (Kandel, 1965; Vasin *et al.*, 1979). An increase in the frequency results in a proportional decrease in the threshold of excitation. The experience of the author, as well as of other investigators, seems to indicate that a weak current (0.5–2 V or 2–3 mA) should be used for stimulation of CI. It should be noted that motor effects from the posterior limb of CI may be obtained only on stimulation posterior to the frontal plane passing through the midpoint of LI (Bertrand *et al.*, 1965).

Depending on the current strength and distance from the electrode tip to the motor fibers of the pyramidal tract, there are either generalized or strictly local muscular contractions (at times, even twitching of a finger or of the corner of the mouth). If low-voltage (0.8–1 V) stimulation causes simultaneous contraction of the muscles of the face, arms, and legs, the electrode tip is in the center of the pyramidal tract. The basis of such a "generalization" still remains obscure. It may indicate anatomic rotation of the motor fibers within the pyramidal tract (Guiot *et al.*, 1959).

Stimulation with a 25- to 50-Hz current at times causes tetany or subtetany of certain muscular groups (Smirnov, 1976), which, in the opinion of some authors, indicates that the electrode is dangerously near the pyramidal tract.

In order to study the effect of CI stimulation on motor activity and tremor, we have stimulated CI during surgery in parkinsonians and dystonia musculorum deformans patients (Voronin and Kandel, 1971). With a laterally projecting monopolar electrode 0.2 mm in diameter, we delivered rectangular or sinusoidal stimuli, both single and of different frequencies (from 0.2 to 100 Hz), with a current strength of 1–1.5 mA. The tremogram was recorded with tensosensors placed on the wrist contralateral to the side of stimulation. Stimulation of CI with repetitive shocks with the same parameters in the same patient may produce different motor effects, e.g., only flexion or only extension of the joint or a combination of these movements. These data indicate that the spinal motor level has significant autonomy from supraspinal influences. Moreover, one may presume that the nature of the response is determined by the interaction of supraspinal excitation and spinal–muscular activity.

On stimulation with a sinusoidal current, the maximal amplitude of the motor reaction was achieved at a frequency of 5–7 Hz, i.e., at the frequency of pathological tremor in parkinsonians (Fig. 81). Our investigations have shown that depending on the current parameters, it is possible to calculate precisely the distance from the tip of the stimulating electrode to the pyramidal tract (Voronin and Kandel, 1971).

Stimulation of the posteromedial part of the posterior limb of CI induces sensory effects, usually a sensation of warmth in the contralateral limbs (Namba *et al.*, 1985). Stimulation of the anterior half of the posterior limb or the anterior limb of CI rarely produces motor effects (Hardy *et al.*, 1979a,b), but it may intensify any dyskinesias already present (Schaltenbrand and Wahren, 1982).

7.1.8. Subthalamic Region

Stimulation of Subth is of prime significance in destructive operations in this area, for it gives not only information of localizing value but also interesting data

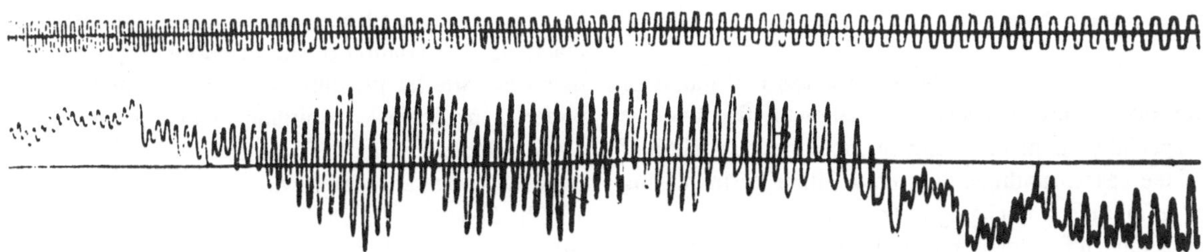

FIGURE 81. The variation of the intensity of tremor reaction on the frequency of stimulation with monopolar pulses of CI during a stereotactic operation in a parkinsonian patient. Upper curve: Gradual decrease of stimulating sinusoidal current frequency from 10 to 4.5 Hz. Lower curve: Mechanogram of the wrist angle recorded by tensosensors. Note the maximal reaction in the range of 5–7 Hz, which corresponds to the tremor frequency of the patient.

on the anatomic pathways related to that area and their functional importance.

It has been shown in a number of papers (Spiegel, 1965; Story *et al.*, 1966; Mundinger, 1968; Tasker, 1982b) as well as by our own experience that stimulation of Subth gives considerably more information than stimulation of VL or GP. In patients with parkinsonism, stimulation of Subth causes a partial or total arrest of tremor and sometimes synchronization or a change in amplitude. High-frequency (700-Hz) stimulation of the posterior nuclei of Th and Subth usually decreases the amplitude of the tremor on the contralateral side, whereas stimulation at a low frequency (5 Hz) increases the amplitude and disorganizes the rhythm of tremor. Disappearance of tremor during stimulation indicates that the tip of the instrument has reached the most "effective" zone, the destruction of which will result in complete relief of tremor. Stimulation of Subth with a weak current (2 V) at a frequency of 25–50 Hz as a rule causes rapid, asynchronous myoclonic twitching in the contralateral extremities and, at times, dystonic hyperkinesias.

In view of the small size of Subth, some stimulation effects are caused by excitation of the surrounding structures. It is believed that the myoclonic hyperkinesias are the result of stimulating NS. Since the nucleus and the root of the oculomotor nerve are adjacent to this area, stimulation often produces inward deviation of the ipsilateral eye. Less frequently, one observes convergence of both eyes or upward or downward movements. Irritation of the posterior longitudinal bundle causes mydriasis on the same side.

Stimulation of the hypothalamic tracts to Subth produces a variety of vegetative effects such as dilation of pupils, altered cardiac rhythm, fluctuations in arterial pressure, and nausea. At times, during irritation of Subth, it is possible to observe emotional reactions or sleepiness (Bechtereva, 1967).

7.1.9. Pallidum (Globus Pallidus)

Stimulation of GP does not cause any motor or sensory effects. However, excitation of this nucleus (especially the medial segment), in approximately half of the cases, modifies the tremor either by increasing its amplitude or changing its frequency. If tremor is not constant and is temporarily absent, stimulation often induces tremor, but always at a lower frequency than the frequency of the stimulating current (Spiegel and Wycis, 1960; Smirnov, 1976). Stimulation of GP usually increases the amplitude of parkinsonian tremor (Walker, 1957). In order to modify the tremor, the stimulation must be by a current of higher voltage and greater frequency than required for stimulation of VL to produce the same effect (Hassler and Riechert, 1959). So far, it is not known why stimulation of GP influences tremor in only approximately half of the cases.

Stimulation of GP in some cases retards speech (Ganglberger *et al.*, 1970b) and depresses and slows the rhythm of cortical potentials during ECoG recording. Van Buren and Ratcheson (1973) came to the conclusion that stimulation of GP does not give specific information useful for localization.

7.1.10. Nucleus Caudatus

Stimulation of NC is performed rarely. It has been noted that such stimulation causes impaired memory, confused speech, dizziness, involuntary smiling, as well as a decrease in the level of norepinephrine in CSF (Wood *et al.*, 1977). Inhibited motor activity, arrest of speech, and changes in consciousness have also been described (Van Buren, 1974).

7.1.11. Nucleus Ruber

The effects of stimulating NR have not been extensively studied, since its surgical destruction has no therapeutic indications. Komai *et al.* (1968) point out that low-frequency stimulation (below 10 Hz) in patients without parkinsonism causes irregular tremor in the contralateral arm as well as a sensation of fear. A higher-frequency current increases the amplitude of this tremor. In parkinsonians, stimulation of NR causes the tremor at rest to become irregular.

7.1.12. Hypothalamic Region

Electrostimulation of this region produces pronounced and varied autonomic, mainly sympathetic, reactions. Stimulation of the posterior Hypoth causes increased arterial pressure, tachycardia, unilateral or bilateral mydriasis, respiratory changes, most frequently hypernea, as well as EEG desynchronization with the appearance of δ waves (Spiegel and Wycis, 1962b; Sano *et al.*, 1970; Kalyanaraman, 1975). In such cases, arterial pressure increases by an average of 50 mm Hg, and pulse rate by 30 beats/min (Richardson, 1982a). Subjectively there are no sensory changes, but frequently the patient experiences a sensation of warmth or cold over the entire body, and sometimes a sensation of fear or even horror (Sano *et al.*, 1975b). Eye movements, in particular, deviation downward or sideways, are frequently observed on stimulation. Contraction of neck muscles with turning of the head has also been noted. Stimulation of the lateral nucleus of Hypoth induces a sensation of hunger (Quaade *et al.*, 1974). A general decline in pain sensation is also sometimes observed (Richardson, 1982a,b).

The results of posterior Hypoth stimulation enabled Sano (1974) to define the zone from which various autonomic reactions may be obtained. This zone, called the "ergotropic triangle of Hess," is bounded anteriorly by the mammillary body, the oral end of the Sylvian aqueduct, and the middle part of LI and circumscribes V_3 1–5 mm from its lateral wall. In lateral films this zone lies posterior and ventral to the junction of LI with a perpendicular drawn through its middle. The "ergotropic zone" coincides approximately with the so-called "sympathetic zone" described by Foerster (1935). According to others, Hypoth stimulation causes nystagmus and mixed sympathetic and parasympathetic pupillary and respiratory reactions (Kalyanaraman, 1975). Amano *et al.* (1978) reported that stimulation of the posterior Hypoth produced a significant rise of the HVA level in the fluid of V_3.

In central pain syndromes, Hypoth stimulation may intensify the pain (Mayanagi *et al.*, 1982) or may significantly (by 50–75%) diminish intractable pain (Richardson, 1982a).

7.1.13. Midbrain Tegmentum

Stimulation of the tegmental area is performed in pain-alleviating stereotactic operations (mesencephalotomy, Chapter 13, Section 5.2.5). The effects of stimulating this area, first described by Spiegel and Wycis in 1966, depend primarily on the location of the electrode in one of many functionally important pathways that are concentrated in that small isthmus of the brain. Stimulation of the spinothalamic tract in the midbrain tegmentum induces localized unpleasant pain and temperature sensations in the contralateral half of the body as well as moderate emotional reactions (Nashold *et al.*, 1969b). Stimulation of the more medial parts of the tegmentum (several millimeters from the Sylvian aqueduct) evokes a sensation of more diffuse pain accompanied by vegetative reactions and pronounced emotional disturbances such as a sense of fear. At times there may be an unpleasant sensation in the precordial or suprapubic region (Nashold, 1982).

Stimulation of the more ventral parts of this area through which oculomotor pathways pass results in abnormal movements of the ipsilateral eyeball, myosis, and nystagmus (Nashold, 1972; Tasker *et al.*, 1982b). Stimulation of the superior colliculus dorsal and lateral to the Sylvian aqueduct induces visual phosphenes in the contralateral field of vision (Nashold, 1972). Stimulation of the corticospinal (pyramidal) tract lying lateral and anterior to the oculomotor zone produces motor effects previously described. Excitation of lamina quadrigemina induces "the noise of a spinning top" in the contralateral ear (Spiegel and Wycis, 1962b).

7.1.14. Periaqueductal Gray Matter

Stimulation of PGM is an important means of managing pain syndromes and is described in Chapter 13.

7.1.15. Amygdala

Stimulation of Am with a 60-Hz, 4- to 8-V current causes various subjective sensations, epileptic attacks, and vegetative reactions and modifies the scalp EEG for several minutes (Laitinen and Toivakka, 1979). During such stimulation, there may be mydriasis and suppressed respiration for 30–40 sec (Narabayashi and Shima, 1973). But others authors report that stimulation of Am does not produce any specific reaction (Schaltenbrand and Wahren, 1982).

7.1.16. Hippocampus

Stimulation of Hipp in operations for temporal epilepsy results in vegetative and visceral changes that may include bilateral mydriasis and increased respiratory and cardiac rates. Sometimes stimulation of Hipp led to confusion and sleeplike states as well as to impaired consciousness, anxiety, oneiric reactions, and a hypomanic condition, which are typical of temporal epileptic attacks (Walker and Ribstein, 1957; Pampiglione and Falconer, 1960; Smirnov, 1976; Gurevich and Kartseva, 1974). In experimental animals, electrostimulation of Hipp causes stupor with significantly diminished reactivity to external stimuli.

7.1.17. Cerebellar Nuclei

Of the several nuclei of Cer, it is especially interesting to note the effects of stimulating ND, which is the target of destruction in stereotactic dentatomy (Chapter 9, Section 5.1.3). The following parameters of stimulation are recommended: rectangular pulses from 2 to 10 V, duration 2 msec, and frequency 25 or 50 Hz. The data on the effects of ND stimulation are contradictory: some authors observed motor responses (Hitchcock, 1975), and others only sensory effects (Nadvornik and Šramka, 1973). Stimulation of many points in ND has shown that the responses can be varied and even contradictory. At times, these differences may depend on whether the operation was performed under general or local anesthesia.

Motor reactions and changes in muscular tonicity in the ipsilateral extremities were noted on electrostimulation of the ventrooral parts of ND (Siegfried *et al.*, 1970; Nadvornik and Šramka, 1973; Vasin *et al.*, 1977). Other authors have described more diverse effects on the ipsilateral side such as increased muscular tonicity, a feeling of weakness in the muscles, deviation of the eyes, rotation of the head, and contraction of facial musculature (Nashold and Slaughter, 1969). In contrast with these data, Zervas (1970) did not note any definite effects of ND stimulation.

Sensory responses to stimulation of ND are inconsistent and were noted only in a few cases (Vasin *et al.*, 1977). The effect of stimulating this structure on tremor has not been sufficiently investigated. For instance, in two cases with intention tremor, stimulation of ND produced opposite effects: in one case there was activation, and in the other inhibition, of tremor (Nashold and Slaughter, 1969). According to other data, stimulation of ND modifies tremor (Toth *et al.*, 1974) and induces changes in ECoGs recorded from the central and precentral regions of the cerebral hemispheres.

The effect of chronic therapeutic stimulation of the Cer cortex and ND on muscular rigidity and spasticity is described in Chapter 9.

7.2. Electrosubcorticography

Electrosubcorticograms (ESubCoG) were first recorded in man by Meyers in 1940 and then by Williams and Parsons Smith in 1949 and Wycis *et al.* in 1949.

At the moment, there are two techniques for recording biopotentials of various cerebral structures: intraoperative recording and use of chronic implanted electrodes. Each of these techniques has its advantages and limitations. Recording during stereotactic surgery takes relatively little time and does not require leaving electrodes in the brain. However, an intraoperative investigation allows one to study the biopotentials of a limited number of subcortical structures. During such "acute" recording, there are frequently injury potentials, which may conceal important aspects of ESubCoG. If the operation is performed under general anesthesia or neuroleptic premedication, these drugs modify the neuronal electrical activity. With chronically implanted electrodes, the studies are conducted on a conscious patient who can carry out numerous functional tests, which are practically impossible to do during surgery. However, the implanted electrodes are the source of several problems and carry the potential risk of complications.

These considerations lead to the logical conclusion that each of these two techniques has its place in the diagnostic and surgical armamentarium. In certain clinical conditions (e.g., in epilepsy), both techniques may be used consecutively (Chapter 14), for as Walker (1982) emphasizes, the most valuable information may be gained from implanted electrodes.

Certain data on the results of recording biopotentials in different stereotactic and functional operations are presented in other chapters of this book.

Recording of subcortical biopotentials is one of the most important aids in stereotactic operations. Besides giving useful information about the functional characteristics and interrelationships of subcortical formations, this method in many cases makes it possible to identify certain nuclei by the nature of their electrical activity and consequently serves as a means of verifying the accuracy of electrode placement.

Many authors have demonstrated that EEG investigations with conventional laminar or needle scalp electrodes during stereotactic surgery rarely yield additional useful information regarding a subcortical structure (this does not apply to the use of evoked potentials—see below). Recordings of multichannel scalp EEGs (mono- and bipolar) during surgery in 90 patients suffering from various extrapyramidal disorders (parkinsonism, DMD, etc.) revealed that when the tip of the electrode reached a subcortical structure, in 60% of the cases there were no changes in the scalp EEG, and in the remainder only a nonspecific change, usually a decrease in frequency, in biopotentials (Kandel and Pokrovskaya, 1972). Slight but nonspecific changes in scalp EEGs in the process of freezing a subcortical structure (VL or Subth) occurred in half of the cases even though a complete or almost complete disappearance of tremor and rigidity had been obtained on the operating table.

After a stereotactic operation on the basal ganglia (especially after thalamotomy), in the majority of cases there is a transitory slowing of the α rhythm (Hassler, 1960; Ganglberger, 1961a,b). Our complete EEG investigations after stereotactic operations for different extrapyramidal disorders have confirmed that conclusion. In the majority of cases, the EEG remains essentially the same as prior to the operation. In two-thirds of the cases, slight changes, commonly a slower α rhythm and sometimes increased θ and δ activity, were observed during the first 3 weeks after surgery and then gradually disappeared. These data indicate that an operation on the basal ganglia does not appreciably alter the dynamics of biopotentials during or after surgery or cause a marked disturbance in the functional state of the brain. In those infrequent cases when such changes do occur, they are reversible.

Soon after the stereotactic technique was introduced into clinical practice, researchers began to study the biopotentials (ESubCoG) of the accessible subcortical structures. These investigations were intensified when chronically implanted electrodes began to be used, especially in stereotactic surgery of epilepsy (Chapter 14), since the epileptic focus (or foci) in deep-seated cerebral structures could be identified. Electrodes of various designs and made of different metals are used for recording ESubCoG. In particular, multipolar hollow-core flexible electrodes that cause little trauma have been devised (Comte *et al.*, 1983).

A special double-headed carrier device for intracerebral insertion of depth electrodes has been developed (Olivier, 1986). The device can be used with Olivier-Bertrand-Tipal or Leksell stereotactic apparatuses. Two heads mounted on the carrier allow one to perform a transcutaneous twist-drill trephination, and then the electrodes are inserted and anchored.

Without reviewing the numerous papers on ESubCoG, most published at the end of the 1950s and in the 1960s (Sem-Jacobsen *et al.*, 1955, 1956; Delgado and Hamlin, 1956; Grindel *et al.*, 1962; Bechtereva *et al.*, 1967; Cooper *et al.*, 1965a; Pagni and Marossero, 1965); Sem-Jacobsen, 1966), we note only that ESubCoG with conventional electrodes has given useful information about the functions of various subcortical structures, and their epileptogenic activity (see Chapter 14). These investigations have demonstrated the irregular rhythms of varying frequencies recorded from subcortical structures (nuclei of Th, Am, Hipp) and the practically flat tracing from the white substance.

We described low-amplitude asynchronous β activity (14–18 Hz) in combination with periods of α or slower rhythms from VL or GP (Kandel and Pokrovskaya, 1972). Many authors, ourselves included, have noted that the spontaneous electroactivities of the nuclei of Th, NC, and cerebral cortex are similar but not identical. The amplitude of biopotentials from the subcortical structures is approximately twice that of the scalp EEG. However, such characteristics have not proved very valuable for identifying subcortical structures as stereotactic targets, not even for reliably differentiating gray from white substance.

7.3. Microelectrode Recording

Stereotactic surgery has made it possible to study the biopotentials of subcortical structures by microelectrode recording, a technique that previously had been used only in experimental neurophysiology. The credit for introducing this technique to stereotactic neurosurgery belongs to Albé-Fessard *et al.* (1962), Guiot *et al.* (1962), Hardy and Bertrand (1965), Jasper (1966), and Bertrand and Jasper (1965). Microelectrode recording allows one to obtain useful information about the physiological state of small populations of nerve cells, even individual neurons, and to localize and identify subcortical structures. On this practical aspect—obtaining information of localizing value intraoperatively—we shall dwell in greater detail.

Microelectrode recording requires modern technical equipment. The number of neurons from which it is possible to record unit biopotentials during an operation or with the aid of chronically implanted electrodes depends primarily on the size of the electrode. Bipolar microelectrodes that record biopotentials from one or more neurons are made of very fine wire, the end of which is electrolytically tapered to a diameter of 10–20 μm and even 2–3 μm.

Semimicroelectrodes are made of various metals (steel, gold, platinum, tungsten) and have an active tip with a diameter of 20–60 μm. With very fine electrodes (20 μm), it is difficult to differentiate gray and white substance. Concentric multicontact bipolar semimicroelectrodes with a resistance of approximately 100 kΩ are usually used in stereotactic operations. The electrode is advanced into the brain in very short (several tenths of micrometers) steps by a hydraulic micromanipulator. The microelectrode should protrude 0.5 mm from the end of the guide tube to record electrical activity of the undamaged brain tissue. The microelectrodes are introduced via a thin steel guide, and recording is done with an electronic amplifer and stored on tape.

The author uses a semimicroelectrode that was built to our specifications by Shalnikov. It consists of three copper wires, 20 μm in diameter, each encased in a stainless steel casing 400 μm in diameter. The three encased wires are inserted into a common tube 1 mm in diameter. The casings are grounded. The distal end of the common tube is beveled, which increases slightly the distance between contacts. Three silver contacts on the ends of the copper wires protrude 20 μm from the slanted surface. The contacts are asymmetrically arranged with respect to each other.

The microelectrode activity seen as the needle passes along the trajectory to the target point may not only be watched on the oscillograph screen but also transformed into loud audio signals (neural noise) (Albé-Fessard *et al.*, 1966). A special audiomonitor reproduces and measures the noise level (Fukamachi *et al.*, 1977). The transformation of biopotentials from the cerebral structures into audio signals allows one not only to hear the neural noise but also to detect changes in the sound as the electrode passes from one cerebral structure to another. These events may be put into an on-line computer to transform them into graphic images (Ohye *et al.*, 1982a).

During the past two decades numerous surgeons have published papers discussing theoretical and practical uses of microelectrodes in stereotactic neurosurgery (Umbach and Erhardt, 1965, 1966; Albé-Fessard *et al.*, 1966; Gillingham, 1966a; Jasper, 1966; Hardy, 1966; Goto *et al.*, 1968; Bechtereva *et al.*, 1967, 1969; Bertrand *et al.*, 1967, 1969, 1973; Taren *et al.*, 1969a; Arutyunov *et al.*, 1970b; Kandel and Pokrovskaya, 1972; Velasco and Molina-Negro, 1973; Fukamachi *et al.*, 1973, 1977; Ohye and Narabayashi, 1979; Hardy *et al.*, 1980b; Ohye, 1982, 1987; Tasker *et al.*, 1987).

In the analysis of microelectrode recordings, two components are informative—background electroactivity (resting activity or noise) and neuronal spike activity. This method provides a reliable means of determining whether the electrode tip is in white or gray matter. Moreover, the boundary between them may be established as the electrode advances into the brain with a high degree of accuracy, within 0.5 mm. The different subcortical structures produce neural noise of varying intensity but always significantly higher than that from the white matter, in particular, from CI. Spike discharges, which are absent in the white matter, appear immediately after the electrode tip enters a subcortical nucleus.

Electric activity changes not only when the electrode tip passes from the gray matter to white and vice versa but also when it passes from a compact nucleus, e.g., the dorsomedial nucleus of Th, to a zone of rarefied neurons, e.g., lamella medialis Th. The intensity of the basic activity also varies with the size of the neurons: the larger the neurons, the stronger was their neural noise (Ohye, 1982). It is also assumed that the heightened activity of a subcortical structure is indicative of constant incoming sensory impulses (Guiot *et al.*, 1962).

Microelectrode recording may be used to delimit the superior and inferior Th boundaries and consequently to determine its height (Hardy, 1966; Bertrand *et al.*, 1973). As the electrode tip enters Th, the basic activity suddenly increases sharply, and it quickly declines as the electrode emerges from Th. On the basis of a microelectrode investigation, the superior border of Th (in adults) varies from 13 to 21 mm above LI, and the inferior from 1 mm above to 1 mm below that line (Fukamachi *et al.*, 1977). These figures correlate with morphological data. Bertrand *et al.* (1973) have also noted the possibility of precisely delimitating the inferior boundary of Th. However, difficulties arise in determining this boundary medially since the neural noise of Th passes into noise from Subth nuclei on its ventral surface (Ohye, 1982). This technique may also be used to delimit the anterior boundary of VPL: it lies 3 to 4 mm posterior to the perpendicular at the midpoint of LI (Jasper, 1966).

The microelectrode technique allows one to control the trajectory of the electrode when the target is VL or Subth. In the frontal approach to these structures, the electrode passes consecutively through the white matter of the semiovale, NC, and the dorsal nuclei of Th. The typical neural noise of comparatively low intensity appearing when the electrode tip passes from the white matter into NC serves as an objective criterion for judging whether the electrode trajectory is correct. Then this noise fades out, indicating that the electrode has emerged from NC, and appears again as the electrode enters the dorsal surface of Th. Subsequently, as the electrode passes from the dorsal to the ventral nuclei of Th at a point approximately 10 mm above LI, neural noise increases about 50%, and irregular spikes of large amplitude appear (Ohye and Narabayashi, 1979; Ohye, 1982), allowing precise delimitation of the boundaries between these two groups of nuclei (Fig. 82).

The microelectrode technique gives reliable identification of the sensory nuclei of Th and their boundary with VL, since the amplitude of spontaneous cellular activity is highest in VPL (Taren *et al,* 1968; Tasker *et al.*, 1982b). A low level of neural noise, moderate spontaneous bursts, and the absence of strong spike discharges are noted in VOa and VOp. In

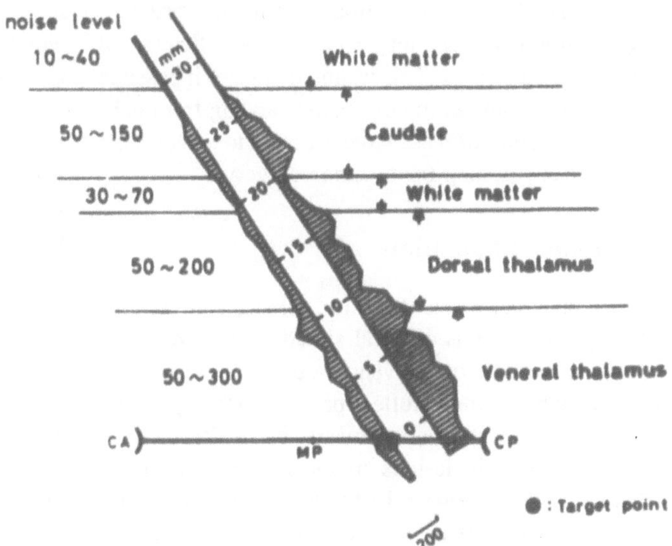

FIGURE 82. Neural noise level corresponding to each subcortical structure recorded by two electrodes during an operation for parkinsonism. The various values of the neural noise are expressed by the bar histogram at each point on the tracks. The histogram of noise level is indicated to the left, and the posterior track to the right. The numerical values between two tracks indicate the distance in millimeters from the target point. The lower thick line indicates LI; MP is its midpoint (Nakajima *et al.*, 1978).

contrast, recordings from Vim and VPL are characterized by high-intensity neural noise with large spikes (Nakajima *et al.*, 1978; Ohye, 1982). In recordings from VOa and VOp, the amplitude is less than approximately 75 μV, whereas from Vim it is more than 100 μV. Even more precise information about the boundaries of subcortical structures may be obtained by recording from two parallel microelectrodes 3–4 mm apart (Ohye, 1982). Data on the biopotentials of Am and Hipp are presented in the Chapter 14.

As noted at the beginning of this section, the use of microelectrodes has provided useful information about the functional organization of Th and other subcortical structures. Without dwelling in detail on these data, it should be noted that the microelectrode study by Jasper (1966) has shown that pain, touch, and temperature sensations as well as the joint and muscular afferent systems are represented separately in the sensory nuclei of Th. On movements of the extremities, the spikes in VL are reduced by 50–60%, and at times they disappear completely (Umbach and Ehrhardt, 1965).

The microelectrode method has made it possible to show that certain Th nuclei discharge in synchrony with tremor rhythm. This phenomenon is described in detail in the Chapter 2, Section 2.10.

Kinesthetic neurons and "neurons of voluntary movements" found in thalamic nuclei are also of great interest for neurophysiology (see below). However, it should be noted that changes in biopotentials related to motor activity cannot be recorded in all cases by the microelectrode technique. Such changes were found in only 13 out of 35 operations during which microelectrode investigation was carried out (Umbach and Ehrhardt, 1965).

Recording of biopotentials by microelectrodes during destruction of deep-seated Cer nuclei (dentatomy, see Chapter 9, Section 5.1.3) enables one to recognize the gray matter and consequently to distinguish the deep nuclei of Cer from the white matter. A practically flat EEG characterizes the white matter; however, when the electrode tip reaches ND, there is an abrupt increase in both frequency and amplitude of electroactivity (Siegfried *et al.*, 1970). As the electrode emerges from ND, the activity decreases instantaneously. At the same time, it is not possible to distinguish the dorsomedial from the ventrolateral part of the nucleus by biopotential recording, which diminishes the localizing significance of this technique.

Along with the majority of authors, we consider the possibilities of the microelectrode technique for identifying subcortical structures to be very important for stereotactic surgery. Bertrand *et al.* (1973) believe this method to be the most precise and useful of all neurophysiological techniques. On the other hand, Laitinen (1972) is of the opinion that recording of cellular activity gives no useful information for the localization of electrodes. Van Buren (1972) also believes that the microelectrode activity of VL and, in particular, Vim is not sufficiently distinct to identify these nuclei.

Nevertheless, one must conclude that despite its

limitations, the microelectrode technique provides useful information for intraoperative identification of subcortical structures. This technique makes it possible to overcome some of the difficulties arising from individual variability of these structures and to enhance the accuracy of reaching them during surgery.

7.4. Evoked Potentials

Evoked potentials (EP), i.e., changes in the biopotentials of various cerebral structures as a result of extraneous excitation by visual, auditory, somatosensory, or other stimuli, reflect perception and processing of sensory information by the brain. It has been established that in lesions of the nervous system at various levels—peripheral nerves, spinal cord, brainstem, subcortical structures, and the cortex—the EPs change, diminish, or disappear. After lesions of various neural structures, different components of the complex multiwave EP are reduced, or the latent period may increase. For example, late waves disappear in diffuse lesion of the brainstem, whereas the early components of EP disappear in lesions of the sensorimotor cortex. Evoked potential investigations may reveal subtle changes in sensory systems, even at subclinical levels.

Somatosensory evoked potentials (SSEP), in particular, trigeminal (TSEP), auditory (AEP), and visual (VEP), have been extensively employed in recent years in stereotactic surgery for more precise determination and identification of subcortical structures and lesions. Spiegel and Wycis (1952a,b) first used peripheral stimulation to identify centripetal pathways in stereotactic operations on the midbrain to alleviate pain. Subsequently, stereotactic surgeons at the Foch Hospital in Paris (Guiot *et al.*, 1964) developed and extensively employed EP investigations of tactile stimulation during the course of stereotactic operations to define the boundaries of the sensory nuclei of Th.

The EP technique may be applied in several ways. In stereotactic surgery, the main method is SSEP from the median or ulnar nerve on the hand or the peroneal nerve under the knee joint using short (0.1–0.2 msec) impulses of 60–80 V at a frequency of 1–10 Hz. Surface or needle electrodes of stainless steel are used for stimulation. Changes in background or evoked spike activity in the subcortical nuclei or white matter are monitored by mono- as well as bipolar recording. A monopolar electrode has a bare tip of 2–3 mm. A bipolar electrode with an interpolar distance of 2–3 mm is also used. The latent period of SSEP is approximately 10–20 (average 14–15) msec, depending on which nerve is stimulated and in which nucleus the EP is registered. One must bear in mind that general anesthesia changes the EP waveform.

Reliable results may be obtained only by EP averaging and the use of computers (Larson and Sances, 1968; Narabayashi, 1968; Giaquinto *et al.*, 1971). Usually, a computer averages from 100 to several thousand responses. This technique of averaging ensures a significant reduction in the variability of the response, since a computer filters values deviating greatly from the average as well as artifacts.

Somatosensory EPs are recorded only in the contralateral nuclei of Th (VL, VPL) (Fig. 83). They do not occur on the ipsilateral side. As our experience has shown the clearest SSEPs during an operation are found in VPL (Fig. 83B) and in the lemniscal zone, although they may vary in form and latency (Larson and Sances, 1966). These potentials may also be obtained from nuclei parafascicularis and limitans, CM, and Pul (Choi and Umbach, 1977). This confirms the significance of nonspecific Th nuclei in the transmission of sensory stimuli.

Many authors consider that recording of SSEP and VEP in Th nuclei is a reliable method for their identification during stereotactic operations (Choi and Umbach, 1977; Narabayashi, 1982a; Birk *et al.*, 1985). Registration of SSEPs during stimulation of the ulnar nerve enables one to localize precisely such an important structure as Vim. The SSEP may determine the posterior border of VL with an accuracy of up to 1 mm (Albe-Fessard *et al.*, 1967a; Sedan *et al.*, 1969). However, other authors believe that by the EP technique, it is possible to identify reliably only VPL and Vim (Walker, 1982). The localization of other important stereotactic structures (VL, CM) is only achieved in half of the cases.

Olfactory stimulation with strong odors leads to the appearance of EP in the medial part of Am, thus identifying this structure during amygdalotomy for temporal epilepsy (Chapter 14, Section 6.5).

Evoked potential investigation with tactile stimulation has confirmed the presence of somatotopic differentiation in VPL as well as in VOp and Vim (Hardy and Bertrand, 1965; Albé-Fessard *et al.*, 1967a). A very interesting observation was the demonstration of separate groups of neurons reacting specifically to tactile stimulation, squeezing of limb muscles, or passive limb movements (Jasper and Bertrand, 1966; Bertrand *et al.*, 1973).

Neurons of the posterior part of VPL are activated

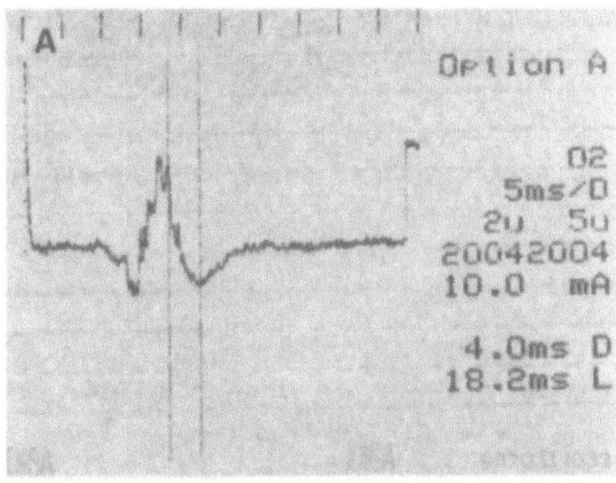

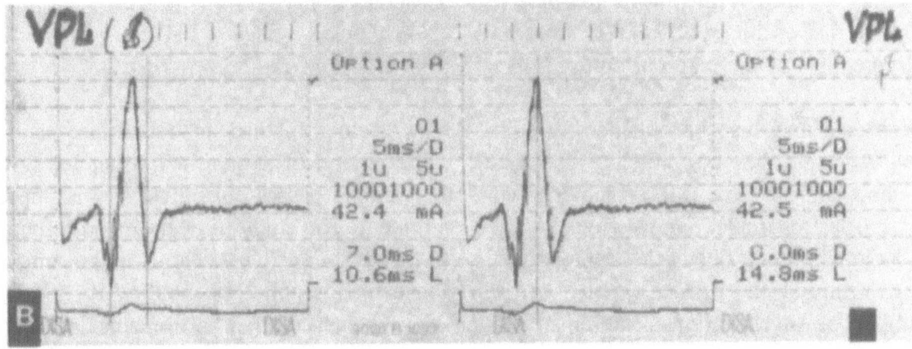

FIGURE 83. Somatosensory evoked potential (SSEP) recording on DISA apparatus during stereotactic thalamotomy. (A) The electrode is inserted into VL. The stimulation of the contralateral nervus medianus evokes (after a latent period of 17–18 msec) a positive–negative EP. Average of several hundred responses, each analyzed for 50 msec. (B) SSEP obtained from VPL. Stimulation of the same nerve. The complex of positive and negative peaks is much more pronounced than EP from VL.

by tactile stimulation of skin: some cells respond to pinching of the skin, and others to a light touch. In VL and the rostral part of VPL, many neurons respond by bursts of activity to gentle squeezing of the fingers or movement of joints on the contralateral side. Some of these cells discharge synchronously and rhythmically with the tremor, and when tremor ceased, their activity also ceased. A small number of cells were activated during passive movements but remained "silent" during voluntary performance of the same movements (Carreras *et al.*, 1967).

With a microelectrode technique, Ishijima *et al.* (1973) discovered two types of nociceptive neurons in the CM–parafascicular complex. These neurons fire in response to pinpricks in various parts of the body and have very broad receptive fields (e.g., the entire contralateral half of the body). The groups differed in their latent periods and duration of response.

Taren *et al.* (1968) noted that the destruction of an area directly in front of an EP zone activated by tactile stimulation of the contralateral thumb or corner of the mouth completely eliminated tremor of the arms. Destruction of an area in front of the zone responding to stimulation of the fourth and fifth fingers eliminated tremor in the legs.

This method of recording may be called "periphery–Th." Another method, "Th–cortex," is carried out during a stereotactic operation with implanted electrodes. Various subcortical structures are stimulated, and EPs are recorded in the cerebral cortex, usually in the central–parietal area, by either conventional scalp EEG or ECoG. Many interesting investigations have used this method to identify subcortical structures (Kelly *et al.*, 1965; Fairman and Perlmutter, 1965; Ganglberger *et al.*, 1970a; McSherry, 1982; Birk *et al.*, 1985; Sólyom *et al.*, 1985).

It was shown that the thalamic EP, recorded during stereotactic surgery, has the same peak latency as the cortical EP (Albé-Fessard *et al.*, 1985). It has also been established that stimulation of various thalamic nuclei induces EPs with comparatively long latent periods from different areas of the cerebral cortex. Since these discharges occur during stimulation of many deep-seated structures and in various parts of the cortex, their localizing value is relatively insignificant (Walker, 1982). Nevertheless, certain data can be useful in stereotactic operations. It has been noted that stimulation of VL includes two-phase EPs in the ipsilateral frontoparietal cortex that disappear after the stimulating electrode reaches the posterior border (Yoshida *et al.*, 1964). Other authors indicate that cortical SEPs evoked by stimulation of one VL are, as a rule, bilateral (Matsumoto *et al.*, 1973). Electrostimulation may be used to identify Vim since its stimulation, unlike that of other parts of VL, does not cause evoked potentials in the cerebral cortex (Narabayashi, 1969).

Informative data may also be obtained by stimulating VPL with a single stimulus, which gives rise to an EP with a short latent period in the ipsilateral sensorimotor cortex. Electrostimulation of various Pul nuclei results in large-amplitude EPs in the frontoparietal cerebral cortex (Yoshii *et al.*, 1975).

In a number of investigations, attempts were made to use the EP to evaluate both the immediate effect and ultimate prognosis of stereotactic operations for hyperkinesia. Struppler *et al.* (1972) report that this technique makes possible the accurate location of Subth. If the EPs disappear, both parkinsonian and essential tremor will cease immediately.

About a week after destruction of VL, the amplitude of EP diminishes. This EP depression correlates with the clinical result. If the outcome is good, there is usually a marked depression of the EP; however, there may also be a clinical improvement in cases in which the EP does not change appreciably (Upton *et al.*, 1982).

Monitoring of the visual EP has recently been used during removal of hypophyseal tumors either by the stereotactic method or via the microsurgical transsphenoidal approach. The EP to visual stimulation by 5-msec flashes of light provides useful information about the functional state of the optic nerves and pathways.

Monitoring of EPs in the cervical spinal cord has been used in operations on the spinal cord and its roots (Nash *et al.*, 1977). During removal of cerebellopontine angle tumors, useful information may be obtained by recording the AEP of the brainstem (Grundy and Jannetta, 1981). Another version of the EP technique, which may called "periphery–cortex," is used to identify the Rolandic fissure by recording EPs from various cortical zones during stimulation of peripheral nerves (Morell *et al.*, 1981).

In summary, one may conclude that the EP technique gives useful information both for human neurophysiology and for localizing CNS lesions, particularly in the brainstem. It represents an important supplementary method for identifying subcortical structures and assessing the value of both destructive and stimulation procedures. Although its value for determining the localization and identification of subcortical targets is limited, when taken with other localizing data, EPs are, without a doubt, useful for stereotactic neurosurgery.

7.5. Impedance Techniques

Another method to determine the function and identification of subcortical structures through which the cannula passes during surgery is impedance measurement. The study of deep-seated cerebral structures by this technique has been given the name stereoimpedoencephalography (Benabid *et al.*, 1979).

The measurement of impedance of cerebral tissue in neurosurgery has been used since the 1920s for localizing brain tumors (Grant, 1923). It was established long ago that the impedance of a normal brain (approximately 500–550 Ω) is twice that of tumor tissue (Gabibov *et al.*, 1972; Benabid *et al.*, 1979). This prompted numerous attempts to locate deep-seated brain tumors by impedance measurement (Pecker *et al.*, 1979a). For a number of reasons, in particular difficulties differentiating tumor from edematous brain tissue, this technique is not extensively applied.

The method is based on the measurement of the resistance of brain substance during the passage of a high-frequency current of sufficiently low intensity that does not stimulate or damage neural tissue. The impedance of normal brain tissue is determined mainly by the "concentration" of myelinated fibers. Since the electrical resistance of the three components of the brain is different—white substance has the highest impedance (600–700 Ω), gray substance an average impedance (200–300 Ω), and ventricular fluid the lowest impedance (less than 100 Ω). Thalamic nuclei have significantly less impedance than CI, the fibers of which are rich in myelin (Organ *et al.*, 1967).

The impedance value of cortex and subcortical white matter varies greatly; this variation is undoubtedly the disadvantage of the method. According to Bullard and Makachinal's data (1987), the cortex impedance ranged from 150–950 Ω (mean, 438 Ω) and subcortical values ranged from 200–1500 Ω (mean, 475 Ω). Edematous white matter has impedance of 25–100 Ω less than normal white matter.

The equipment for investigating impedance and phase angle consists of various electrodes, a radiofrequency generator producing a weak (several milliamperes) high-frequency sinusoidal current and an ohmmeter with an amplifier or impedance comparator. The permissible strength of the current passing through the brain for impedance measurement is 2 mA/cm^2. Impedance recording during stereotactic surgery is possible using the Radionics Lesion Generator (model RFG-3B).

Several types of electrodes are used: monopolar, bipolar coaxial, and four-electrode systems. The most reliable results are obtained with the latter, although the large number of electrodes somewhat increases the trauma of the procedure. Some authors use coaxial electrodes with the bare tip of the center electrode serving as one pole and the tube of the electrode as the second. Such a design allows measurement of impedance between adjacent contacts, i.e., within 4–5 mm. However, Laitinen *et al.* (1966) and Grechin and Borovikova (1975) prefer a monopolar electrode since in their experience changes in impedance and phase angle obtained with a coaxial electrode differ very little.

A monopolar electrode has a small bare tip 1–2 mm long, 1.15 mm in diameter, and an area of 1.8 mm^2. An electrode with a 3-mm tip has also been used (Fox and Green, 1969b). A plate fixed to the foot of the patient serves as the indifferent electrode; if convenient, this electrode may be inserted subdurally. The most informative data are obtained at frequencies of 10,000 and 100,000 Hz (Laitinen *et al.*, 1966).

Impedance measurements of deep-seated cerebral structures are made simultaneously with recording the phase angle, the changes in which as an electrode passes from one structure to another are more pronounced (Spiegel *et al.*, 1966; Laitinen *et al.*, 1966). Since the impedances of various subcortical structures differ, as the electrode advances into the brain, a profile can be obtained of the impedance changes along the trajectory of the electrode.

Several investigations of the use of impedance measurements for localizing subcortical structures have shown that the technique reliably differentiates white and gray matter and, consequently, the boundaries between them as well as the ventricular fluid (Dierssen and Marg, 1965; Laitinen *et al.*, 1966). This gives the surgeon information on the border of a subcortical nucleus (passing from the white substance to gray) and on the location of the ependymal wall and the ventricular cavity. In some cases, impedance measurements define the boundary between Th and Subth, although the method does not differentiate the individual Th nuclei (Grechin and Borovikova, 1975; Walker, 1982).

The direct correlations between impedence values and CT data have been established. These two physical parameters—impedance and CT density—are similar structural characteristics (Benabid *et al.*, 1979; Bullard and Makachinal, 1987).

There are grounds to believe that the impedance method may aid in localization of a stereotactic target. Moreover, in certain cases, variations in impedance may indicate that the electrode tip is close to a large vessel, the rhythmic pulsations of which produce impedance oscillations. The impedance method, used in conjunction with other methods of physiological control, is of definite value in stereotactic surgery.

7.6. Other Techniques

The H-reflex test, the equivalent of a myotatic reflex, is frequently examined in modern neurophysiology and is of definite localization significance. This test is a delicate indicator of the excitability of the segmental motoneuron pool and of the influence of supraspinal systems on it. The ratio of the maximal amplitudes of the H reflex and M reflex is an index of the excitability of the myotatic reflex. The H-reflex test is usually performed by electrostimulation of the tibial nerve in the popliteal fossa, with simultaneous recording of the EMG response of the gastrocnemius muscle.

Stimulation (usually biphasic rectangular impulses of 0.8-msec duration at a frequency of 0.05 Hz) is conducted with a cathode and an indifferent anode applied to the same leg. The intensity of stimulation should be 20–30% higher than the threshold value. As the voltage of the stimulating current is gradually raised, the amplitude of the H response increases to a level corresponding to the maximal excitability of the motoneuron pool. The voltage is then reduced in order to decrease the maximal H response by approximately one-half.

Stimulation of VL does not change the H reflex significantly. However, if the electrode tip is in or close to CI, the H response is noticeably greater (Vasin *et al.*, 1970). This may be useful for determining the boundary between these two structures during stereotactic surgery. Sugita *et al.* (1972) have shown that on the basis of changes in the H reflex on stimulation of VO, the superior boundary but not the inferior boundary of VO can be determined.

There is another technique of some localizing value in stereotactic neurosurgery. Polarographic determination of oxygen tension in deep-seated cerebral structures has revealed a statistically significant difference between white and gray matter (Grechin, 1973; Starikov, 1978). By installing a polarographic sensor on the electrode tip, it is possible to measure the change in Po_2 as the electrode tip passes from the white matter into Th nuclei. Shefer *et al.* (1970b), who successfully used this method in stereotactic operations for temporal epilepsy, demonstrated the possibility of differentiating the amygdala–hippocampal complex from the periventricular white matter of the temporal lobe.

In stereotactic operations as well as after implantation of chronic electrodes, it is possible to record local blood flow and temperature in subcortical structures with the aid of minithermistors and thermoresistors (Grechin, 1973).

7.7. Summary of Functional Controls

All the different methods of functional control help the surgeon to identify the numerous deep-seated cerebral structures and to implant the stereotactic instrument accurately. Moreover, these controls significantly increase the safety of stereotactic surgery by preventing damage to structures adjacent to the target point.

After all the x rays and functional controls have confirmed that the target structure has been accurately reached, it is possible to begin the main stage of the operation, the making of the lesion. Since the choice of the method of destruction is an important aspect of stereotactic neurosurgery and one not agreed on by all neurosurgeons, this subject is described in detail in Chapter 5.

Throughout the entire stereotactic procedure, it is essential to monitor carefully the condition of the patient. In addition to the usual cardiopulmonary controls, it is necessary (in operations under local anesthesia) to check the patient's consciousness, mental state, speech, size and uniformity of pupils, and superficial and deep tendon reflexes in order to detect pathological responses as early as possible. It is especially important to monitor muscular strength and sensation in the limbs to detect the first signs of complications such as hemiparesis. In pain-alleviating operations on deep-seated cerebral structures, particular attention should be paid to the subjective feelings of the patient and the clinical testing of pain and tactile and deep sensitivity.

After the process of destruction has been completed, the cannula (electrode) is removed from the brain, hemostasis is secured, and the stereotactic frame is removed. A small circle of thin inert dural substitute, which does not form cortical adhesions, is placed over the incised dura mater.

The "button" of bone that was removed at the beginning of the operation with a crown trephine is replaced. The aponeurosis and skin are tightly sutured with several interrupted sutures, and an aseptic bandage is applied, or the line of sutures is sprayed with plastic.

The duration of the operation by our technique for the destruction of one subcortical target is about 90 min, and for the destruction of two structures approximately 2 hr. All stages of the operation and stereotactic calculations are noted in a prescribed manner in the operative report.

5

The Production of Lesions in Stereotactic Operations

1. General Principles

The main objective and the ultimate aim of most stereotactic operations is the destruction of nuclei or pathways of the brain or spinal cord. In view of this, one of the most important problems of stereotactic surgery is the choice of the most effective, simplest, and safest method of destroying cerebral tissue. To a considerable degree, the reliability of a therapeutic operation and the incidence and gravity of postoperative complications depend on the lesion produced.

In the course of approximately 35 years of the development of stereotactic neurosurgery, several methods of destroying subcortical structures, based on the vast experience of experimental neurophysiology, have been proposed and tested clinically. Some did not prove satisfactory and were abandoned, but others are still being used and improved.

Stereotactic operations on subcortical structures are unique in that the manipulations are deep in the brain on small structures close to one another. These factors require that the method of destruction of cerebral tissue be precise. The ideal technique should meet the following requirements:

1. The possibility of inactivating a previously determined volume of cerebral tissue, i.e., making a lesion of a given size and shape. To attain this objective, the size of the focus must be directly related to the instrument of destruction.
2. The possibility of controlling the size of the lesion during the operation and, in particular, of terminating destruction immediately on appearance of symptoms of involvement of adjacent structures of the brain.
3. Minimal reaction of adjacent cerebral tissue to the lesion, which should be clearly delimited from adjoining structures of the brain. The less the reaction, the less danger there is of damaging adjacent structures.
4. The absence of general reactions or damage to the brain, i.e., minimal reaction of the cerebral tissue to the lesion, thus assuring few if any complications from the operation.
5. Almost total exclusion of the danger of vascular injury so as to reduce to a minimum the probability of intracerebral hemorrhage.
6. The possibility of reversing the inactivation of a structure before its destruction.
7. Procedures for destruction of cerebral tissue that are not unduly lengthy and that are convenient and simple.

The ideal method that meets all these requirements has yet to be found. This is evident from the fact that different methods of stereotactic destruction are used in neurosurgical clinics of various countries. Nevertheless, the cryosurgical method, because of its effectiveness, safety, and convenience, is, in our opinion, the best method of destroying the basal ganglia, deep Cer nuclei, and brain tumors. It meets virtually all of the requirements mentioned above. For some operations (destruction of the trigeminal ganglion or root, percutaneous chordotomy, destruction of the entrance zone of the posterior spinal roots), high-frequency electrocoagulation or thermocoagulation is undoubtedly the best.

The several methods of destruction used in modern stereotactic surgery may be divided into four main groups: physical, electrical, chemical, and mechanical. It should be stipulated, however, that such a classification is quite arbitrary. In view of this, the electrical methods are described separately, although, in essence, they belong to the category of physical methods. From the purely theoretical point of view, the use of heat and cold should be considered thermal methods of destruction. Therefore, the category of thermodestruction should include cryosurgery, thermocoagulation, high-frequency electrocoagulation, and induction thermocoagulation. In our description of lesion making, we therefore follow the accepted but not absolutely accurate classification in stereotactic neurosurgery.

To determine the optimal volume of destruction in a subcortical structure remains a complicated and unresolved problem. It is quite natural that the smaller the focus, the better: the less damage to the brain and to functionally important structures. However, there is a risk of incompletely destroying the target structure, which may result in only partial relief or a relapse of the tremor or rigidity. It may be said that the problem of the focus size is a search for "the golden mean."

In recent years, computed tomography has been used successfully for verifying the location of the destroyed focus after a stereotactic operation and for

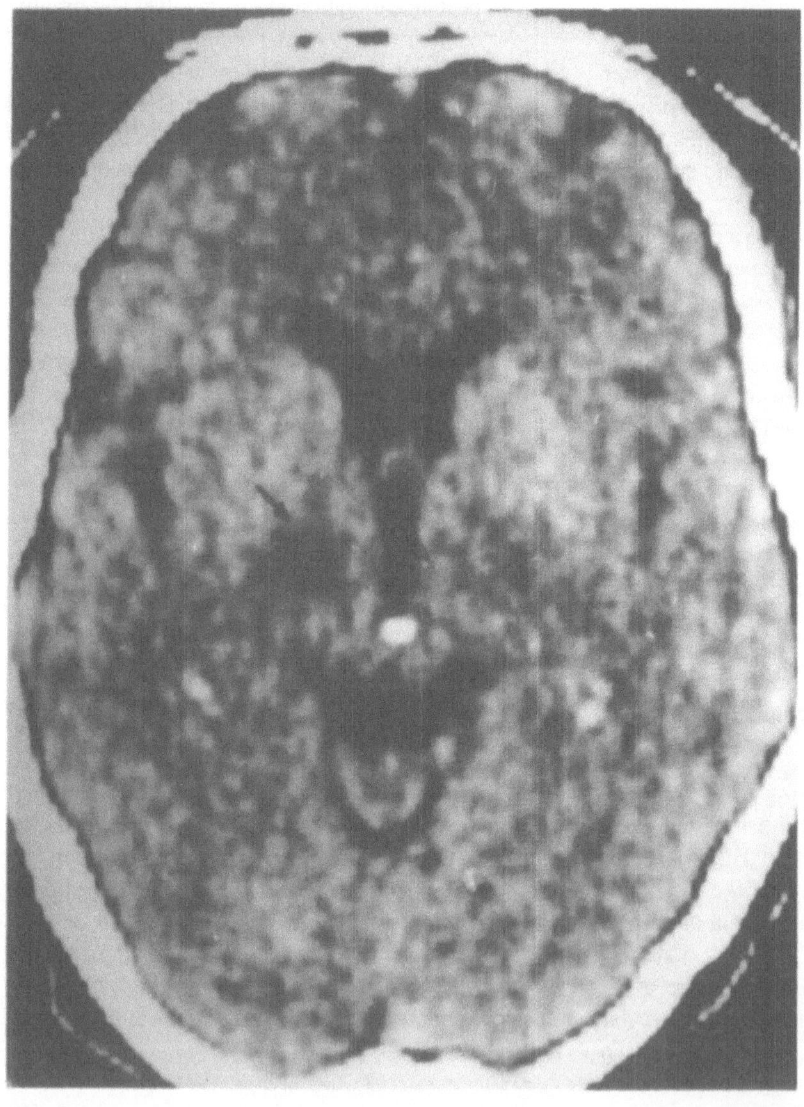

FIGURE 84. The cryogenic focus of destruction (arrow) in left VL 6 days after surgery. The perifocal edema of brain tissue is insignificant.

monitoring the results of the procedure. These investigations have shown that there are certain differences in the lesions made by various methods as shown by CT. For instance, after high-frequency electrocoagulation, the destroyed focus consists of a discrete 2 to 3-mm-diameter center of increased density typical of hemorrhagic necrosis. This is surrounded by a zone of decreased density with a diameter of up to 14 mm (Colombo *et al.*, 1981), which is presumed to be a zone of perifocal edema (Murayama *et al.*, 1979, 1982). Subsequently, the central area becomes a small zone of diminished density, several millimeters in diameter, which apparently represents the formation of a cyst.

According to our observation, the destroyed focus is visible on CT for as long as a year after the stereotactic operation. But others report that the thalamic focus is apparent in CT scans for only 3–4 weeks after the operation, after which it is no longer visible (Murayama *et al.*, 1982). However, some authors believe that dynamic CT scanning postoperatively does not give an accurate estimate of the volume of the lesion. A typical cryolesion on CT is presented in Fig. 84. Important information can be obtained from PET and NMR studies of thalamotomy.

2. Local Freezing

Cold both heals and kills. . . .
Hippocrates

The use of cold, along with curative herbs and bloodletting, represents one of the oldest therapeutic methods in human history. Cold compresses were used for treating fractured skulls and chest wounds in ancient Egypt in 2500 B.C. according to the Edwin Smith papyrus, which is the most ancient source of medical knowledge (cited after Tytus, 1968). In *The Iliad,* Homer also mentioned the use of cold washes for treating chest wounds.

Hippocrates, "the father of medicine," describes in detail the beneficial effect of cold to stop the bleeding of wounds and in traumatic edema. The famous Russian surgeon Pirogov extensively employed cold in the middle of the 19th century for treating wounds. He wrote: "Cold is certainly to be prescribed where a swollen, burning, and irritating wound is accompanied by parenchymatous (capillary) bleeding" (cited in Pirogov, 1961).

Cold is also one of the oldest anesthetic agents. Military physicians employed cold for surgical operations in the Middle Ages. Napoleon's chief surgeon, Larrey, noted that after wounded soldiers had been lying in the snow for a long time, it was possible to amputate injured extremities painlessly. More than 100 years later, this method was revived by the noted Russian surgeon Judin. In his *Notes on Military Surgery* during the Second World War, Judin (1943) gives a detailed description of anesthetizing an extremity by applying a tourniquet and cooling the limb. Judin noted a sharp decrease in the frequency of postoperative wound infections after cold anesthesia.

The local application of ether and later ethyl chloride to cool and anesthetize soft tissues was suggested about a century ago. At present, small surgical incisions are frequently performed under local anesthesia with ethyl chloride.

The investigations by Temple Fay at the end of the 1930s represent a landmark in the history of cryosurgery. He first attempted to alleviate pain in incurable cancer patients by irrigating the focal lesions with ice-cold water for several weeks (Fay and Henney, 1938). Subsequently, Fay employed local cooling in neurosurgery in craniocerebral injuries, encephalitides, and brain abscesses. He used a "cold bomb," which was inserted into the cavity after removal of brain tumors. Fay may also be considered the founder of general hypothermia, which he employed in 1938 to reduce the body temperature of a patient to 30°C for several days.

2.1. Biological Effects of Freezing

It is common knowledge that water is the main component of cellular structures of all living organisms from protozoa to mammals. That is why the freezing of water represents the theoretical basis for the action of low temperatures on biological tissue. At 0°C, water begins to form crystals of ice and gives off heat (80 cal/g). This heat, which is called the "latent heat of crystallization," must be removed or the water will not freeze.

The ice begins to form with the appearance of centers of crystallization. In the course of this process, the water temperature is close to 0°C, and after all the water transforms into ice, the temperature drops sharply. Films of the process of water transforming into ice have shown that the crystals formed have either rounded or sharp edges. These crystals freely deposit on those formed earlier, and that is why, after the appearance of crystallized foci, the process develops with the least expenditure of energy.

In certain conditions, water remains in a fluid state (i.e., does not transform into ice) at temperatures considerably below 0°C. This phenomenon is called supercooling. The process of ice formation occurs easily in a supercooled solution. If a crystal of ice or any other particle is introduced into such a solution, the process of crystallization begins almost instantaneously.

The number of separate ice crystals and crystallization centers depends directly on the speed of cooling. The more rapid the cooling, the fewer and smaller are the ice crystals formed. The size of the crystals also depends on the cooling speed. Slow cooling forms larger crystals and rapid cooling smaller crystals, since there are a greater number of crystallization centers.

Living tissue contains approximately 95% water. Numerous organic and inorganic substances are dissolved in both extra- and intracellular water of the living tissues. For this reason, ice formation in living tissues differs significantly from the freezing of water. The simplest model of the freezing of liquid in cells is the transformation into ice of a physiological solution with a freezing point of −0.55°C. If the physiological solution is supercooled to −21°C, most of the water turns into ice, and the NaCl concentration in the liquid portion is maximum. After all the water is frozen, the temperature may drop even more. Such dynamics of freezing are typical of all salt solutions (Rey, 1959; Luyet, 1960).

As many investigators point out, the mechanism by which low temperatures act on living tissues has not been fully resolved, although the main aspects of this process have been studied in some detail. The freezing of living tissue is not only a physical process but a significant biochemical transformation as well. Metabolic processes are terminated only when the temperature is below 0°C (Mazur, 1968).

The mobility of protoplasm, one of the main manifestations of the vital activity of the cell, depends particularly on the temperature of the medium surrounding the cell. As the temperature drops, the speed at which protoplasm moves decreases; moreover, if this occurs rapidly, mobility of protoplasm suddenly disappears, a phenomenon called "thermal shock," which is usually reversible (Meryman, 1966).

Protoplasm of a living cell freezes at a temperature close to 0°C, but because of supercooling, the cell may not be transformed into ice even at a considerably lower temperature. Nevertheless, there is a definite limit to supercooling: at a temperature of about −20°C every cell freezes totally (Mazur, 1968).

In cryosurgery it is important to study both the physical parameters of cooling and the biological principles involved in cellular destruction in living tissue. A mandatory condition in cryosurgery is the certainty that the entire volume of frozen tissue will undergo total and irreversible necrosis, thus excluding the possibility of the vital cellular activity returning on thawing. Many investigations have shown that without special protective measures the cells of mammals die when the temperature drops below −5 to −8°C. Gill and Frazer (1968) also emphasize that all living tissues, once frozen to a solid state *in situ,* perish. However, under certain conditions (even without employing cryoprotective substances such as glycerine), living cells, in particular, tumor cells, display a surprisingly high resistance to low temperatures.

The process of ice crystals forming in living tissue, as a rule, begins in the extracellular space, and only after the temperature drops considerably below 0°C do intracellular crystallization centers appear. Therefore, in order to transform extracellular fluid into ice, the temperature must be lowered to −5 to −10°C, and to freeze the intracellular fluid an even lower temperature is required, about −20°C.

With a relatively rapid drop, the temperature of living tissues may reach −20°C without the formation of ice crystals. The period between the drop of temperature below the freezing point and the beginning of crystallization is usually designated as the supercooling period. This phenomenon is clearly seen in temperature records made with a thermocouple during the freezing of the brain tissue (Section 2.8). Immediately after the beginning of the formation of ice crystals, the temperature rises to the freezing point, which for biological substances is −2 to −3°C.

Rinfret (1968) sums up the temperature–time relationship in the process of freezing a biological system (Fig. 85). The first phase is the lowering of the temperature in the liquid state of the system to the freezing point. The second phase involves a fall in temperature to below the freezing point (supercooling) with a subsequent return to that point, which indicates the beginning of freezing. The latent heat of crystallization is emitted during that rise. The third phase, the isothermal plateau, signifies the transition of the system to the solid state. And finally, the more or less rapid drop in temperature in the fourth stage indicates the completed transition to the ice phase in which the living cells are totally destroyed. An analysis of a similar temperature curve for local freezing deep in the brain during a stereotactic operation is presented in Section 2.8.

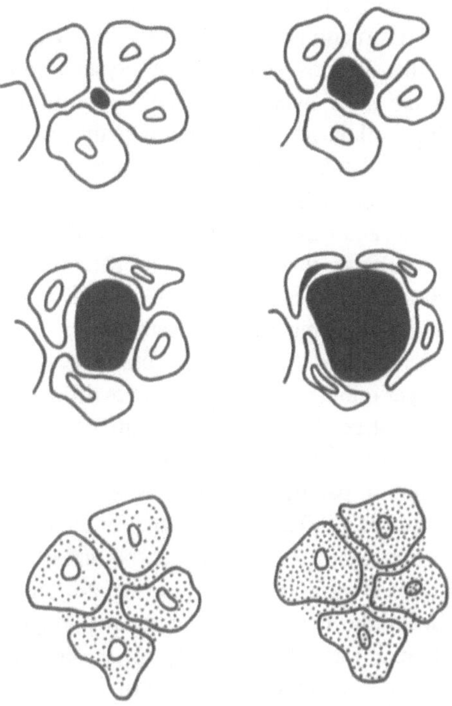

FIGURE 85. Rinfret's (1968) scheme illustrating two types of tissue freezing. Consecutive stages of development of ice crystals in extracellular space typical of relatively slow freezing. Besides mechanical compression, dehydration of the cells takes place as a result of intracellular fluid passing into the growing ice crystals.

So far, there is no entirely satisfactory explanation for the mechanism of the destruction of living cells and tissues by cold. Nevertheless, several hypotheses, supported by considerable factual material, have been put forth in explanation. Numerous investigations have established the main factors causing irreversible destruction of cells during their freezing and thawing:

1. Pronounced dehydration of cells in the process of ice formation extra- and intracellularly, leading to a sharp increase (''lethal concentration'') of electrolytes in cells.
2. Mechanical damage to cellular membranes by sharp ice crystals, as well as the compression of the cell bodies by these crystals.
3. Denaturation of cellular membrane phospholipids.
4. Termination of protoplasm mobility, so-called ''terminal shock,'' the nature of which is not yet clear.
5. Arrest of blood circulation in the frozen volume of tissue, leading to ischemic necrosis.

Lipoprotein complexes of cells containing much water are especially sensitive to cooling. Lysis of these complexes is also one of the causes of cell necrosis in freezing. Let us dwell in greater detail on the characteristics of each of the abovementioned factors.

In the formation of ice crystals, an osmotic pressure gradient develops, as a result of which water passes from the cells through the membranes into the pericellular space, where it freezes. This gradient increases in the supercooling state. It is noteworthy that the mass of ice occupies a volume 10% greater than the volume of water from which the ice forms.

Mazur (1968) calculated the intracellular and extracellular pressures occurring during the freezing process. This gives rise to a peculiar feedback-type process: when the water from the cell passes extracellularly and then freezes, the electrolyte concentration inside the cell increases as the amount of water in it decreases. This equalizes the intra- and extracellular pressures and slows the rate of dehydration until a kind of equilibrium is reached, which in turn prevents deep supercooling.

In the end, this process leads to a substantial increase in the intracellular concentration of electrolytes as well as of various toxic substances, which act on cellular metabolism, causing protein denaturation and sharp changes in pH. Rey (1962) calls this phenomenon osmotic shock.

The mechanical action of ice crystals on cellular structures depends to a considerable degree on the type and rate of freezing, which we discuss in the following paragraphs.

Sharp ice crystals, especially if intracellular, mechanically destroy cellular structures. Damage to cellular membranes is an important factor leading to tissue necrosis during freezing. Electron microscopy and biochemical studies have demonstrated that in the freezing process acute changes are observed in the phospholipid complexes of membranes: the two layers of the nuclear membrane separate, and the endoplasmatic net is ruptured, which leads to cell necrosis. Thus, ''thermal shock'' implies not only termination of protoplasm mobility but also ''lethal suppression of intracellular metabolism'' (Gill and Frazer, 1968).

When living tissue is cooled, the blood flow in it slows down sharply, and stasis occurs in vessels plugged with erythrocytes. In slow cooling, blood stasis plays an important role in the cryodestructive mechanism. However, if freezing is rapid, stasis does not develop to a sufficient degree, and cryonecrosis depends to a large extent on other factors. Termination of blood flow as a result of the freezing of large and small vessels also represents an important component in the formation of a cryonecrotic focus (Section 2.2). The hemostatic properties of cold, resulting in arrest of blood flow in frozen tissue, have led to its use for resection of certain parenchymatous organs (liver, kidneys, lungs, and others) having abundant vascularization.

One should examine those conditions that are necessary for success in achieving this destruction. The intensity of cell destruction in the freezing focus depends on the following four factors (Mazur, 1968): (1) rate of tissue cooling, (2) minimal temperature in freezing focus, (3) period of exposure to the low temperature, and (4) time and rate of thawing.

The literature frequently gives contradictory interpretations of the terms "rapid" and "slow" cooling. In view of this, we recommend the optimal classification of the rate of cooling presented by Losina-Losinsky (1972) (Table 2).

A significant difference can be seen in the quantity and size of the ice crystals produced by rapid and slow freezing. The number of crystals forming in living tissue during the freezing process is directly proportional to the rate of the process, whereas the size of the crystals is inversely proportional to the rate. Consequently, a higher rate of cooling produces a larger number of smaller crystals, whereas slow freezing produces fewer larger crystals. In rapid freezing (front of ice crystal formation advancing at 6 mm/min), minute crystals form, and in slow freezing (front of ice crystal formation advancing at 0.5 mm/min), the crystals are much larger (Meryman, 1956).

Numerous investigations have established that the biological action of rapid and slow freezing differs substantially. It is noteworthy that the influence of the rate of cooling on living cells and tissue is interesting not only from the theoretical point of view, but is also of practical significance. In designing cryosurgical instruments, the question arises: What rate of cooling—rapid or slow—is better for achieving total and irreversible destruction of cells? It has already been mentioned that in slow cooling ice crystals form only around the cells, whereas in rapid cooling they also form intracellularly, which undoubtedly promotes the destruction of cells through rupture of their membranes by sharp ice crystals. The formation of these crystals in cytoplasm, and especially inside the cell nuclei, inevitably results in their destruction. Moreover, higher concentrations of toxic substances and more acute changes in pH occur in rapid freezing.

TABLE 2. Classification of Rate of Cooling

Rate of cooling	Period of cooling	Rate of process
Very slow	Several hours	Up to 10°/hr
Slow	1 hr to 10 min	From 10°/hr to 10°/min
Rapid	10 min to 1 min	10–60°/min
Very rapid	60 sec to 10 sec	1–100°/sec
Super rapid	Less than 5 sec	More than 100°/sec

During rapid cooling at a rate of several dozen degrees per minute, the water does not diffuse from the cells into extracellular space. This intracellular water is at first supercooled, and then crystallization begins, proceeding in parallel with the formation of ice in extracellular spaces. This explains the formation of extracellular ice in slow cooling and extra- and intracellular ice in rapid cooling. However, during slow cooling the cells dehydrate considerably; e.g., if a single cell is cooled at a rate of 100°/min, its water content at −20°C will be approximately 50% of the norm, whereas at a rate of 1°/min, it will be only 4% (Mazur, 1968).

Rapid and superrapid freezing have recently been extensively employed in medicine for tissue conservation (lyophilization) because in superrapid cooling to a very low temperature (−120, −130°C), so-called amorphous ice, a uniform glasslike mass, forms in the cells and tissues rather that crystalline ice. However, the transformation of cellular protoplasm into a "glass-like state" without the formation of crystalline ice has not been definitely proven. Minute ice crystals are seen in the protoplasm of erythrocytes subjected to superrapid freezing in liquid nitrogen, but it is presumed that these minute intracellular crystals do not significantly damage cell constituents. Although Mazur (1968) terms these crystals "relatively harmless," it has been shown that cells may be destroyed as a result of cooling in the absence of intracellular ice formation (Gill and Frazer, 1968).

A number of investigations suggest that cells are destroyed not only in the process of freezing but also during thawing. In thawing there is a displacement ("migrational recrystallization" according to Rey's terminology) of ice crystals, which intensifies their destructive action on living cells. Moreover, when the temperature of tissue is increased, the high concentration of electrolytes has an especially potent destructive action on cells. These data seem to indicate that the results of cryodestruction depend to a considerable extent not only on the rate of cooling but also on the rate of heating. If heating proceeds slowly, the intracellular ice crystals continue to grow and, when they become large, damage intracellular structures. During very rapid heating, the crystals begin thawing earlier, and the survival rate of the cells may increase. Consequently, the thawing process is no less important for the destruction of tissue than freezing itself. Mazur (1968) proposes two ways to enhance the likelihood of irreversible cell destruction in the frozen zone: (1) increasing the diameter of the ice sphere beyond the destroyed zone so that the temperature in that zone is below −20°C and (2) thawing the ice sphere slowly (at a rate from 10 to 30°/min from the minimal temperature to total disappearance of ice).

In view of the factual material presented above, one may try to answer the question of the optimal rate of freezing and thawing for cryosurgery. Although it is quite difficult to give a final answer, there are sufficient grounds to consider that relatively rapid freezing, approximately 40–50°/min, is optimal. Substantially slower freezing (3–5°/min) seems to be inappropriate since intracellular ice formation does not occur. Similarly, it would hardly be rational to resort to superrapid freezing since amorphous ice forms. Unlike freezing, thawing must be slow, approximately 10–12°/min, to ensure destruction of cells. These optimal parameters for cooling and thawing serve as the basis for the design of our cryosurgical instruments.

The literature expresses the point of view that repeated freezing of the same part of the tissue (both healthy and pathological) "may increase confidence in the destruction of its cells" (Gage *et al.*, 1965; Cooper, 1965a). This viewpoint is based on the assumption that certain (especially tumor) cells may be highly resistant to freezing. Gill *et al.* (1968) investigated "singular" and "repeated" freezing in rat liver. These foci were subjected to different temperatures of the cryocannula (from −50° to −180°C) and different times of application (from 1.5 to 10 min). The authors report that at any temperature within the abovementioned limits, repeated freezing (up to five times) increases the volume of the destructive focus. For example, the diameter of a focus produced by two freezing–thawing cycles was 16 mm, whereas after five cycles it was 20 mm. It has also been established that freezing and thawing of living tissue increases its thermal conductivity by 10–20% and that repeated freezing–thawing cycles enhance the effect. Repeated local freezing is expedient, for example, for the destruction of tumors that are highly resistant to low temperatures.

2.2. Effect of Low Temperatures on Vessels

The development of the cryosurgical method and its ever broader application in various fields of medicine require a thorough study of the changes in the vessels of the cryodestructive focus since these changes may determine both specific operation complications, as well as the course of reparative processes in that focus. The most serious complication that may occur in the freezing process is bleeding from a vessel in the frozen focus as well as the formation of cryogenic thrombi in large vessles with subsequent ischemia of the zone beyond the lesion supplied by that vessel. The formation of aneurysmal dilation of vessel walls as a result of the cooling is also a possible complication.

Rosomoff (1959), investigating the effect of low temperatures on cerebral substance, noted vasodilation, enhanced permeability of vascular walls, and thrombosis of veins and capillaries. However, our experience has shown hemorrhagic complications after freezing to be extremely rare. Cooper (1962b) also emphasizes the absence of hemorrhagic complications. Goander *et al.* (1964), using this method in urological operations, incurred no hemorrhages. In gynecology, Cahan (1964) noted only an insignificant hemorrhage in the necrotic focus directly after freezing. The bleeding was quickly arrested, and no further hemorrhagic complications were observed. However, Miyazaki *et al.* (1963), in experimental cryogenic lesions deep in canine brain, found quite significant hemorrhage. Hemorrhagic complications, especially if large vessels are involved in the destroyed focus, have been reported by Coe and Ommaya (1964).

Cooper *et al.* (1965) consider that any biological tissue is destroyed after being subjected to a temperature of −20°C, the only exception being the muscular layer of large vessels, and they attributed the complications seen by Coe and Ommaya to imperfect technique.

Gage *et al.* (1967) found that large canine vessels frozen at a temperature of −120° to −160°C were resistant to cold. The structure and architectonics of collagen and elastic fibers proved to be little if at all disturbed, and the vascular wall remained intact. Muscular fibers underwent partial destruction but subsequently regenerated, and the vascular wall was fully restored within 6 weeks. This finding makes it practically safe to freeze segments of relatively large arteries. The authors also point to the absence of thrombosis in large vessels, which consequently retain their full functional capacity. There was not a single case of vascular dilation or aneurysm formation in vascular walls after freezing. No significant hemorrhages occurred in the tissues surrounding the frozen focus.

Walder (1968) studied the influence of local freezing deep in a feline brain at temperatures from −10° to −180°C and describes changes in the carotid artery and jugular vein of cats frozen at −150°C for 5–10 min. The muscular layer and intima in the frozen zone became thinner. The intima partially separated from the internal elastic membrane as a result of ice formation in the subendothelial layer and showed reactive proliferation. Changes in the adventitial, elastic, and muscular layers were insignificant. There was no thrombus formation in the vessels, and not a single case of hemorrhage was noted either during freezing or afterwards. A second series of experiments involved vessels damaged mechanically or by the introduction of mustard gas into the vascular wall. After the freezing of these vessels, no hemorrhage was observed after several months, but vascular thromboses were found in half of the cases.

Therefore, the published data demonstrate that changes in the cryogenic focus, in the perifocal region, and, in particular, in the vessels depend to a considerable degree on the temperature and the time of freezing. Small perivascular hemorrhages, as a rule, do not extend beyond the cryogenic lesion.

The nature of the vascular changes depends on the size and type of the vessel itself as well as on the state of the surrounding tissue. Of all the biological tissues, the most stable to the action of low temperatures is the wall of large vessels because of the great resistance of collagen and elastic fibers and relative stability of muscle.

In 1974 we studied the effect of freezing on arteries and veins of various sizes in rabbits. A cannula cooled by liquid nitrogen was employed to freeze the tissue. The temperature at the tip of this cannula approximated that of liquid nitrogen (−196°C). With the animal under urethane anesthesia, a cryocannula was inserted between the femoral artery and vein to touch both vessels. In other rabbits the femoral vessels were not isolated from surrounding tissues, and the cryocannula introduced between the artery and vein froze them together with the surrounding tissues. Thus, both large vessels approximately 2 mm in diameter and smaller vessels were frozen.

In order to study the action of cold on large vessels with thick musculoelastic layers in our next series of experiments, we introduced the cryocannula between the abdominal aorta and the inferior vena cava. The freezing time in the experiments varied from 3 to 20 min. Hence, in some cases only small arcs of the vessel wall were frozen at the point of contact with the cannula; in other cases, the vessel froze solid for 15–20 mm of its length with an ice thrombus temporarily stopping the blood flow. The rabbits were sacrificed from 24 hr to 3 weeks after the experiment, and the frozen areas were studied histologically.

At the beginning of the cryoprocess, one could observe the frozen vascular walls adhering to the active tip of the cannula. Subsequently, depending on the diameter of the vessel and the temperature of the cannula, the frozen zone gradually enlarged, extending from the point of direct contact with the cannula across the lumen to the opposite side of the vascular wall, thereby occluding the vessel with an "ice thrombus." Arterial pulsation distal to the thrombus ceased. As the freezing spread, it involved the surrounding tissues, forming an ice sphere around the main vessel. In order for the lumen of a large vessel to be occluded with an "ice thrombus," it was sometimes necessary to interrupt the inflow of blood briefly along the main artery. Depending on the freezing time, the ice sphere had a diameter up to 20–30 mm. The main vessels were frozen for a somewhat shorter distance.

After the freezing process was terminated, thawing began gradually. In the surrounding tissues, it began from the periphery, and in the main vessels from the upstream end of the vessel, earlier in the artery than in the vein. Through the comparatively thin vascular wall, it was possible to observe the return of blood flow and pulsations in the distal segment of the artery. In the majority of experiments, no gross changes could be seen in the vessels after thawing.

During the observation period after the operation, there were no signs of impaired blood circulation in the extremities. Not a single hematoma formed in the area of the operation. The postoperative period was uneventful in those animals in which the abdominal aorta and inferior vena cava were frozen. No intraperitoneal hemorrhage was noted, and at autopsy no blood was observed in the peritoneal cavity.

At various times after the isolated vessels (femoral arteries, veins, abdominal aorta, and inferior vena cava) had been frozen, specimens from the lesion as well as from proximal and distal segments were taken for histological investigation. In another series of experiments in which the main vessels were frozen *in situ*, all of the involved tissue was removed for investigation.

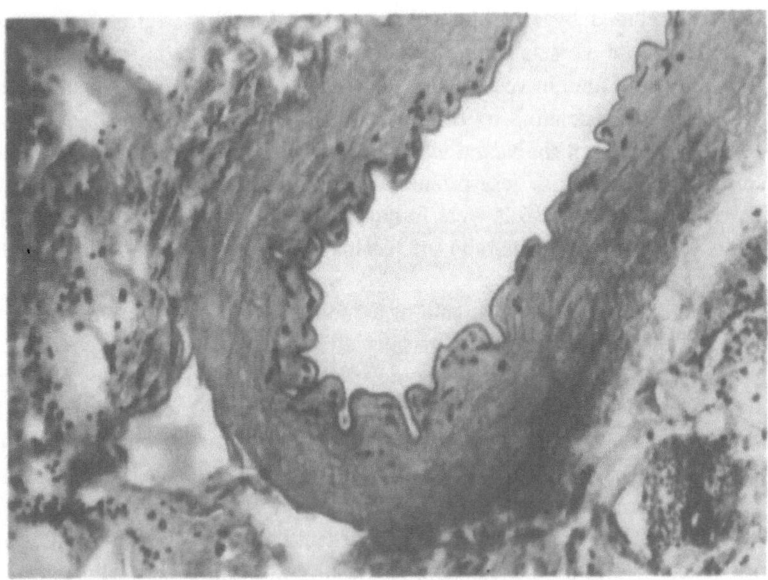

FIGURE 86. Necrosis of all layers of the wall of small artery after local freezing in an animal experiment. Hematoxylin–eosin. 7 × 20.

In the samples taken at various times, necrotic changes were quite pronounced in both the vessels and the surrounding tissues after 48–72 hr, whereas 1–3 weeks later reparative phenomena were more noticeable.

Microscopic examinations showed that the low temperature produced a necrotic focus involving all small vessels (arteries, veins, and capillaries) (Fig. 86). Their walls were homogeneous, cellular elements could not be differentiated, and nuclei were broken up. The muscle fibers surrounding the vessel were also necrotic. The larger necrotic foci in the muscular fibers were demarcated from the normal tissue by an inflammatory zone. Small perivascular hemorrhages were present among the necrotic muscle fibers. However, in no case were such hemorrhages found beyond the frozen focus. A positive Schick reaction was seen in the vascular walls as well as in the surrounding muscular fibers.

After 5 days, one could observe the growth of granulation tissue with large numbers of newly formed vessels, proliferating young fibroblasts, and macrophages around the foci of muscular tissue necrosis. Among the granulation tissue there were both regenerating as well as necrotic muscle fibers.

In the larger vessels the destruction was less marked than in smaller vessels; the larger the artery or vein, the less pronounced were the changes noted. The elastic fibers were fragmented and swollen, resembling short shreds with washed-out contours. The fibrous tissue was fragmented with interstices filled with plasma proteins.

In most of the main vessels, directly at the point of contact with the cryocannula, one could see the swollen and homogenized wall filled with acidic mucopolysaccharides and infiltrated with plasma proteins, in some cases involving only the subendothelial layers and in others the entire thickness of the wall, including the adventitia. Such plasma infiltrates were frequently accompanied by more or less pronounced segmental fibrinoid necrosis. As a rule, neutrophilic leukocytes filled the necrotic spaces of the wall. With special stains, used on the necrotic area of the vascular wall, the elastic fibers of the internal elastic membrane were seen to be preserved. The elastic fibers of the muscular layer were also unchanged, and only in the necrotic area was there fragmentation.

The elastic layers were preserved in practically all cases; however, in some the arterial internal elastic membrane was markedly swollen and lacked its normal distinct wavy appearance.

However, in none of the cryogenic foci, in the early period as well as after 3 weeks, was there a single aneurysmal protrusion in either small or large arteries.

Vascular thrombosis in small arteries, veins, and capillaries of the necrotic focus was common, but this was seen less frequently in larger arteries and veins. Usually, the vessels involved were dilated, their walls thin, and the internal elastic membrane straight. A positive Schick reaction was observed in the walls of vessels occluded by a thrombus. As a rule, thrombi occluded those vessels with necrotic walls. Parietal thrombi were found in vessels with segmental necrosis of the walls. Depending on the time after the thrombosis, the lumen of the vessel was in an early stage of clotting or in more or less pronounced organization. Thrombosed small vessels were found only within the limits of the frozen focus, and the lumina of the main arteries were not occluded.

Our experimental investigations demonstrated that as a result of freezing living tissues at a very low temperature from 7 to 20 min, the destruction of various vessels in the focus is not identical. Smaller vessels (arterioles, capillaries, and veins) are occluded by thrombi and have necrotic walls. Larger vessels, especially arteries, suffer to a lesser degree, and their walls retain their anatomic structure because of the resistance of elastic and collagen fibers to cold. The latter remain as a framework on which the vascular walls regenerate. That is why, in spite of pronounced morphological changes, the blood flow in these vessels is retained.

Insignificant petechial hemorrhages, which sometimes occur around the necrotic small veins and arteries, do not extend beyond the frozen focus and do not cause significant bleeding. The subsequent total restoration of blood flow in the main vessels of the frozen focus and the absence of hemorrhagic complications make it possible to use cryosurgery to destroy tissues near or even including large vessels.

In certain cases only cryosurgery can be used to destroy neoplasms when it is necessary to preserve the blood flow in a vessel close to a tumor. It should be considered, however, that the proximity of a main vessel may change the spherical shape of a focus, since not all the tissues of the target may be within the zone of freezing. That is why to destroy tissues with many large vessels, one should use a lower temperature and a large-diameter cannula, since the blood flow through a major vessel substantially interferes with the freezing of a target area.

The experimental data presented above led to the conclusion that the freezing method is relatively safe and without serious hemorrhagic complications either in or adjacent to the focus.

2.3. Experimental Cryoneurosurgery

Long before the cryosurgical method was used in neurosurgery, numerous experiments were done to study the local action of low temperatures on the central and peripheral nervous systems. The first reports on the action of cold on peripheral nerves were published at the end of the 19th century. It was demonstrated that the motor fibers of the sciatic nerve in a dog lose their conductivity when they are cooled to about +6°C. The sensory fibers are blocked at a lower temperature (+1°C). According to Denny-Brown *et al.* (1945), the conductivity of a peripheral nerve is totally blocked when it is cooled to +8°C for 30 min. Furthermore, it was established that nerve fibers of various structures have different resistances to cold. Thick myelinated fibers (A fibers) lose conductivity at a higher temperature than nonmyelinated C fibers. The experiments staged by Le Beau and Dondey (1964) demonstrated that synaptic transmission is blocked at temperatures below 20°C, whereas nerve conductivity is abolished at approximately 8°C.

This question was studied in detail by Siegfried *et al.* (1962), who demonstrated that there were three phases in the functional state of the nerve during cooling. In the first phase, from body temperature to +27°C, excitability of the nerve fibers increases. In the second phase, from +27°C to +20°C, excitability does not decrease or slowly returns to the initial level. In the third phase, from +20°C to 0°C, the conductivity of the nerve diminishes almost linearly. At a temperature somewhat above 0°C, excitability disappears, and if cooling is continued, it does not return after warming. The optic nerve is more resistant to cold than nerves in the extremities. That this nerve is totally blocked at 0°C is explained by its large number of small nonmyelinated fibers.

Experimental investigations on the effect of low temperatures on cerebral tissue began, quite naturally, with the cortex, which is quite accessible to freezing. The first experimental work was published back in 1883 by Openchowski, who cooled part of a canine cortex with a simple device to evaporate ether by a jet of warm air. On freezing the cortex solid, the author noted the development of contralateral paralysis and, in certain cases, epileptic convulsions. After the frozen parts of the cortex thawed, the paralysis disappeared. Openchowski was the first to observe that cooling the cortex produces a clearly demarcated focus of destruction without causing diffuse hemorrhage.

One of the first experimental investigations of the effect of cold on cerebral tissue was conducted by Speransky (1937). To produce experimental epilepsy, he froze various cortical zones of the dog employing a simple device in the form of a thin-walled metallic chamber, approximately 2 cm in diameter, through

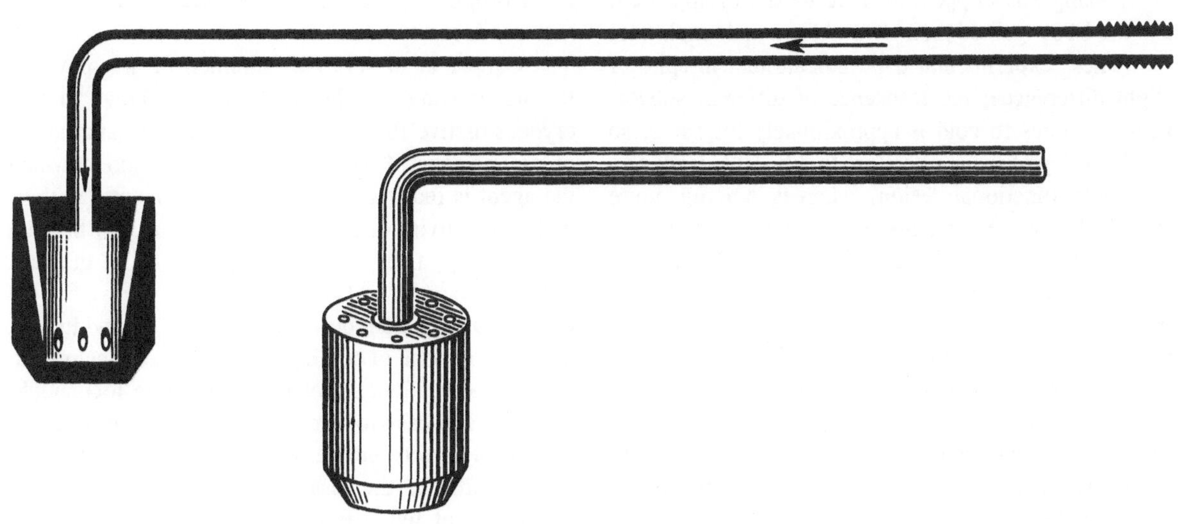

FIGURE 87. The simple device used by Speransky (1937) for local freezing of cortex in an animal experiment.

which carbon dioxide was passed (Fig. 87). After some time, at the freezing site (through the dura mater), a sharply demarcated cherry-colored spot corresponding precisely to the diameter of the freezing chamber appeared on the cortex. In sections, necrosis of cerebral substance could be seen to a depth of 2–4 mm.

Subsequently, the method of freezing the cerebral surface in animals was used by Hass and Taylor (1948), Balthazar (1957), Mazars *et al.* (1960), Ommaya and Baldwin (1963), and Cooper *et al.* (1965a). Experimental freezing of various cortical zones was attempted to develop a method for treating epilepsy caused by pathological cortical lesions.

After the construction of the first instruments able to produce strictly local cooling and freezing of structures not on the surface but deep inside the brain, the new method was used on a broader scale in neurophysiological investigations. One advantage of cold that caught the attention of investigators is the possibility of producing a temporary, reversible lesion of a deep structure of the brain, which is practically impossible with other methods. This technique enables one to study the effects of a temporary lesion on certain cortical and subcortical structures. In clinical conditions such a reversible lesion is frequently necessary before "final" destruction of a subcortical structure is carried out in stereotactic operations. This test is employed when for various reasons the surgeon is uncertain whether the tip of the instrument is in the target structure or if there is fear that cryodestruction may also involve functionally important cerebral structures.

The possibilities and limitations of making temporary subcortical lesions by cold in stereotactic operations were studied by a number of authors (Dondey *et al.*, 1962; Fasano *et al.*, 1964; Cooper *et al.*, 1965a; Kandel, 1965) (Table 3).

Cooper has reported similar results from cooling thalamic nuclei during stereotactic operations under local anesthesia. When VPL was cooled to approximately +3–5°C, the EPs in the sensory cortex were reduced sharply, and the patient experienced paresthesias on the contralateral side of the body. After

TABLE 3. Comparison of Results of Cooling

Author	Cooling		Result
	Structure	Degree	
Mark *et al.* (1961)	CN	+5°C	Pupils dilated
Siegfried *et al.* (1962)	CI	+5°C	Cortical responses blocked
Dondey *et al.* (1962)	VL	+20°C	Cortical EPs abolished
Cooper (1969a)	Cortex	−18°C	Somatosensory EPs abolished

the freezing was stopped, the paresthesias disappeared, and the EPs regained their normal form and amplitude.

These experimental data indicate that in spite of slight differences, the resistance of different subcortical structures to cold is approximately the same, so that lowering of the temperature to +5–10°C leads to a reversible functional lesion, whereas a temperature slightly below zero results in an irreversible loss of function because of necrosis of the cerebral structure.

Irrespective of the degree to which the temperature is lowered at the active tip of the cannula, and also irrespective of the diameter of the ice sphere around it, the temperature on the border between the ice and the cerebral tissue will always be approximately zero (to be more exact, −2 to −3°C). Because of the low thermal conductivity of ice, the temperature of cerebral tissue at a distance 2–2.5 mm from the surface of the ice sphere is only slightly below normal brain temperature. This provides the most important advantage of the cryosurgical method, a strictly limited focus of cryodestruction, for necrosis occurs in the cerebral tissue that is frozen and thawed without damage to adjacent cerebral tissue.

In addition to morphological investigations, data confirming this conclusion were obtained from our experiments (Kandel *et al.*, 1962). Trephination was performed in rabbits under light anesthesia. Electrostimulation of the motor cortex with 1–1.5 V evoked a motor reaction in one of the contralateral extremities. Then a cryocannula was introduced beneath that area (to a depth of 8–9 mm). Freezing was done in such a manner that the superior pole of the ice sphere emerged on the brain near the motor cortex. Electrostimulation of the intact cortex adjacent to the ice sphere established that at a distance of only 1–1.5 mm from the border of the frozen zone, the cerebral cortex retained its normal excitability to a Faradic current. This confirms the sharp localization of the cryodestructive focus.

2.4. Factors Determining the Size of the Cryogenic Lesion

In using the cryosurgical method it is very important to know the exact dimensions of the desired cryonecrotic focus in any tissue or organ, since the freezing, being deep within the tissue, cannot be controlled visually. This is a complicated problem, since besides the known physical parameters, the biological characteristics of various living tissues are important factors; these include the amount of the blood supply, the hydrophilic properties of the tissues, and the thermal conductivity.

Cooper *et al.* (1965a) summed up the physical factors relating to the form and volume of the cryodestructive focus as (1) flow intensity and nature of cooling agent, (2) temperature and pressure at which the agent is fed into the freezing instrument, (3) thermal conductivity of the cannula in tissue, (4) thermal conductivity of the metal in the active tip of the cannula, (5) diameter and surface area of the active tip, (6) temperature gradient between the active tip and its surface, and (7) rate of cooling. Since all these factors are known and can be controlled, the freezing foci should be stable and easy to reproduce. However, in freezing biological objects, besides the abovementioned factors, others influence the result: thermal conductivity and latent heat of living tissue, amount of blood flowing through the tissue, osmolarity, and so on. However, in cerebral tissue these factors change within such narrow limits that they are of no practical significance. The brain is known to have low thermal conductivity, which is practically uniform in its different areas. The specific cerebral blood flow is also stable and changes only in large vascular lesions of the brain. As a result, our experimental, morphological, and clinical experience as well as that of other authors demonstrate that the size of frozen tissue deep in the brain depends only on three factors: on the "power" of the cryosurgical instrument, on the quantity of liquid nitrogen injected into it, and on the duration of exposure. Since each of our instruments undergoes technical and biological (on rabbits and dogs) testing to determining its "power," we can determine the volume of cryodestruction simply by varying the quantity of liquid nitrogen used and the time of exposure. Numerous experiments have shown that under such conditions, the volume of the frozen sphere is quite constant.

Since establishing the size of the cryodestructive focus deep in the brain and its relationship to the quantity of liquid nitrogen used is very important for stereotactic operations, we have conducted a series of experiments to resolve this problem. The first involved focal freezing of a rabbit or dog brain with various quantities of liquid nitrogen. After extraction of the cannula from the brain, the ice sphere was measured when its diameter was maximum. As these experiments demonstrated, the assumption that the cryonecrotic focus in the brain corresponded to the diameter of the ice sphere proved correct. Moreover, with the same temperature at the tip of the cannula, the variability in the size of the focus was insignificant and, from the practical point of view, could be considered negligible. A second method of evaluating the lesion size involved measurements made during morphological investigations of cerebral sections. We had to take into account the shrinkage that occurred in the process of fixation.

We used another method to evaluate the size of the cryolesion in the brain of test animals (Kandel, 1981). Since lowering the brain temperature to 20–25°C impairs the blood–brain barrier, several substances (fluorescein, tubocurarine, penicillin, and others) can enter the brain. On the basis of these data, the following method was used to measure the size of the cryogenic focus. A cryocannula was introduced into the brain of the animal to a depth of 12–14 mm, and a cryothermal lesion was made. Just before the freezing, 4–5 ml of a 3% solution of trypan blue was injected intravenously. Because of the presence of the blood–brain barrier, the brain retained its normal color. However, in brain sections one could see a clearly defined circular frozen zone, the bright blue coloring of which contrasted sharply with the surrounding unstained normal tissue. This coloring resulted from local impairment of the blood–brain barrier in the frozen zone. The size of this colored area corresponded to the diameter of the ice sphere extracted from the brain after an analogous experiment.

The temperature dynamics were measured using six thermocouples, one of which was attached to the active tip of the cannula while the others were at distances of 1, 2, 3, 4, and 5 mm from the tip. If, in the freezing process, the temperature of the active tip of the cannula dropped by approximately 100°C, then at a distance of 2.5 mm the decline was 50°C, and at a distance of 5 mm only 25°C. Even when the temperature at the tip of the cannula was at its lowest, the temperature 5 mm from it did not go below 0°C, and at a distance of only 2 mm from the edge of the ice sphere, the temperature of the brain tissue remains at approximately +10°C, which is not enough to cause irreversible damage to the surrounding brain tissue. Our own data thus correlate very closely with that data of Cooper *et al.* (1966) and Walder (1968).

There are reasons to believe that the focus of destruction in stereotactic operations should be somewhat larger than the volume of the subcortical structure to be destroyed. In other words, the ice ball should extend 1–1.5 mm beyond the limits of the target. Such a conclusion is based on the fact that the temperatures in the center of the focus (near the active tip of the cannula) and at the periphery differ significantly. Mazur (1968) calculated the rates of freezing and thawing in the center of the focus and found them to be approximately the same, about 300°/min. Undoubtedly, the high rate ensures intracellular crystallization and, consequently, necrosis of practically all cells in the cryogenic focus.

However, the rate of cooling in the peripheral zone of the focus is approximately 30°/min. At the same time, the rate of peripheral thawing is even higher than that in the center (more than 300°/min). This means that the cells in the peripheral zone are frozen for approximately 30–60 sec, and the possibility of intracellular ice formation is less; a considerable number of the cells in that zone can probably survive. This is even more likely for nerve cells on the outer border of the ice sphere (Mazur, 1968).

Although these calculations seem convincing, experiments by Walder (1968) demonstrated effective coincidence of the borders of the 0° temperature front on the periphery of the focus and the necrotic borders determined by microscopic investigations. These data confirm that brain tissue exposed to a temperature below 0°C becomes necrotic. Nevertheless, it is apparent that between the frozen brain tissue, which subsequently undergoes necrosis, and the intact brain tissue with a temperature close to normal, there is an intermediate zone in which the cells are damaged reversibly.

2.5. Morphological Pathology of Cryogenic Brain Lesions

Use of the cryomethod required a thorough study of the dynamics of the pathological changes that take place in the frozen focus both on the surface of the brain and in deep-seated structures. An important question was the study of the pathological reactions occurring in the perifocal (with respect to the focus of cryodestruction) and adjacent zones. The first experimental studies, carried out before the cryomethod was used in neurosurgery, were on production of cryonecrotic foci in the cerebral cortex of animals.

The first morphological investigations of the frozen cortex were carried out by Hass and Taylor (1948) and then Balthazar (1957). The authors established that local cryofoci have clearly defined borders with adjacent undamaged brain tissue. The perifocal reaction was insignificant and did not lead to generalized brain edema or diffuse hemorrhages. A small cyst surrounded by normal brain tissue forms in the cortex of cats 1.5–2 months after freezing. Balthazar came to the conclusion that cold is "the most physiological method" for producing local destruction in the cerebral cortex. Hass and Taylor recommended freezing as "a harmless and palliative method" for destroying various living tissues.

The development of cryosurgical instruments for producing deep-seated brain lesions made it possible to carry out stereosurgical operations in animals and, consequently, to study the histopathology of the cryogenic foci.

After freezing a "brain cylinder" 2 cm in diameter in dogs, there was not a single case of hemorrhage, and all animals survived (Ries and Tytus, 1960). The cryonecrotic focus examined from 24 hr to 21 days after the lesion was produced was a well-defined "red infarct" without noticeable damage to the surrounding brain tissue.

In subsequent years the cryosurgical method began to develop rapidly and successfully. The effectiveness of the method for local destruction of normal brain tissue in stereotactic operations stimulated a search for an experimental model. A number of important questions concerning technique and effects of cold on brain tissue—optimal temperature, relationship of the size and nature of cryogenic focus to the quantity of cooling agent, evolution and morphological characteristics of the focus, perifocal and general reactions of the brain to the freezing focus, and several others—were repeatedly studied in experiments on various animals (Cooper *et al.*, 1965; Mazars *et al.*, 1966; Vikhert *et al.*, 1968; Walder, 1968). In parallel with the experimental investigations, a study was made of the pathological changes of cryodestructive foci in the basal ganglia on autopsy material (Cooper *et al.*, 1965; Kandel and Podgornaya, 1966; Kandel, 1974a) at various times after stereotactic operations.

The aim of our investigations was to solve two main problems. First, it was necessary to study the nature and dynamics of the destructive focus in deep brain structures, in particular, to determine whether the lesion was total and irreversible, a mandatory objective for employing the method in stereotactic surgery. Second, it was important to establish the nature and degree of both the general and perifocal brain reactions to a subcortical cryodestructive focus.

Under hexobarbital anesthesia in rabbits, a small burr hole was made over one cerebral hemisphere, and a cannula was introduced stereotactically into the thalamus for making a cryodestructive focus. The stereotactic atlas of Sawyer was used for determining the location of VL. The animals were sacrificed from 1 day to 6 months later. Serial sections of the hemispheres were made, covered with celloidin, and stained for microscopic examination. In these experiments, if the focus was located in the white matter or a subcortical ganglion, it was spherical. When the focus was placed in a cortical gyrus or adjacent to the ventricle, the lesion was somewhat irregular.

On the second or third day after the operation, a practically spherical focus of total necrosis measuring from 6×7 to 9×9 mm was found at the freezing site in the central part of the hemisphere. The necrotic zone was surrounded by a thin layer of granular spheres containing clumps of hemosiderin in their protoplasm. Adjacent to the zone of granular spheres, the tissue was slightly hyperemic with stasis in some capillaries and mild edema distending the perivascular spaces. There was some hyperplasia in the glial cells. The nerve cells around the necrotic focus were not changed.

On the fourth postoperative day, the frozen area was a mass of necrotic tissue with a few hemorrhagic areas. In the focus there were no structural neural elements, only a granulofibrous necrotic mass; surrounding it was a very narrow zone (0.15–0.3 mm) of granular spheres forming a clear-cut border around the focus. Around the border, the cerebral tissue was mildly plethoric with some pericellular and perivascular edema as well as proliferating glial elements. No inflammatory reactions, either diffuse or focal, were observed. Further from the destructive focus, the reaction rapidly disappeared.

Sections from brains of animals sacrificed after 5–7 days had no significant changes in the zone of cryonecrosis (Fig. 88). There was only a somewhat more pronounced perifocal proliferative reaction of the glial cells and a greater number of compound granular cells forming a thin wall (approximately 1.5 mm). Small vessels and loose, delicate connective tissue fibers bordered the layer of granular spheres. Leukocytes from the perifocal zone had migrated to the focal zone and engulfed necrotic debris. There was no evidence of diffuse edema. I.S. Cooper *et al.* (1965) have also emphasized that there were very insignificant changes in the brain tissue surrounding the destructive focus.

By day 10 to 14, the granular spheres filled the entire necrotic zone, which was clearly demarcated from the brain tissue (Fig. 89). As a result of intensive resorption, the necrotic mass was broken into separate areas, mainly in the center of the focus. Proliferation of glial elements and moderate hyperemia were present in the brain substance adjacent to the granular sphere

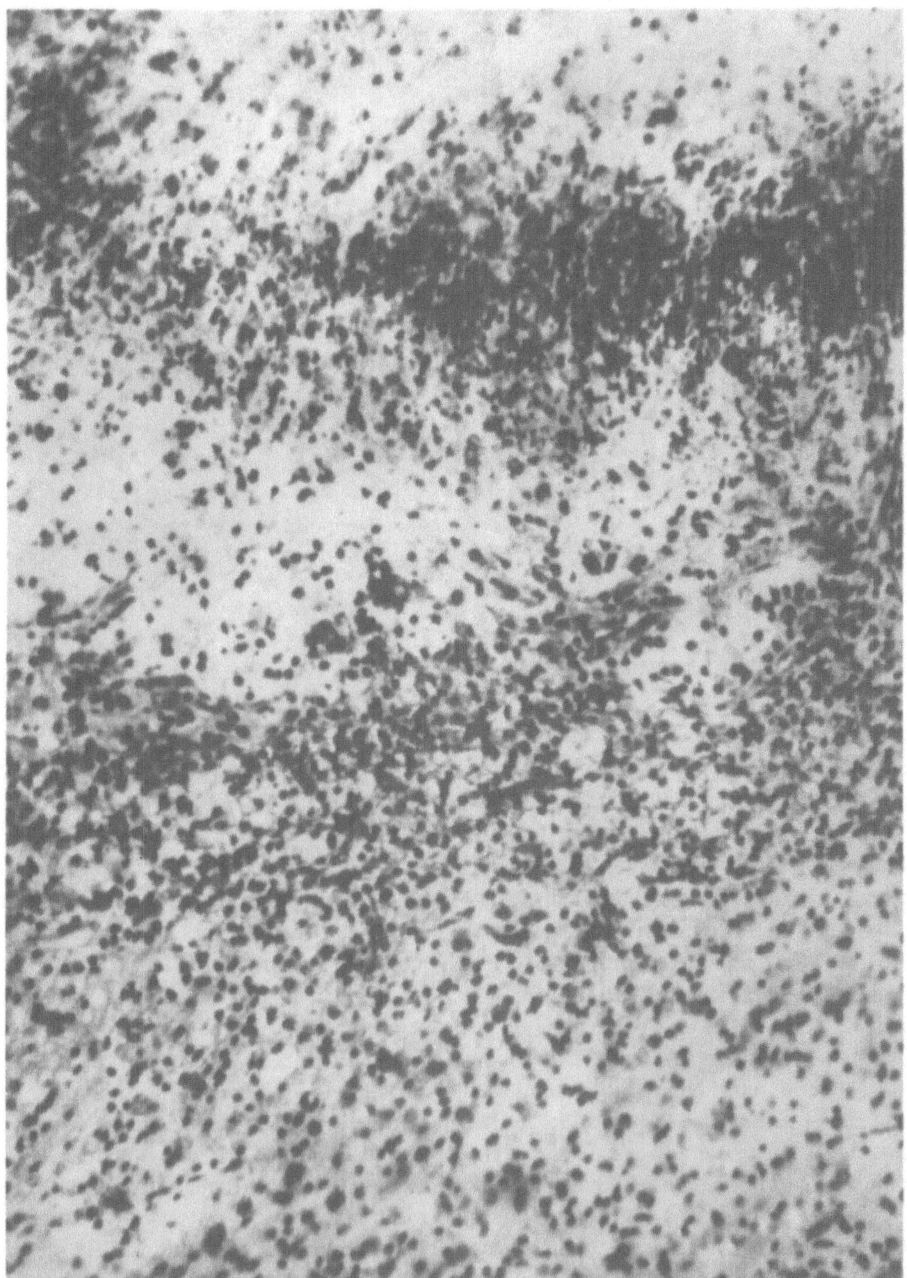

FIGURE 88. Histological picture of cryonecrotic lesion in rabbit subcortical white matter 8 days after freezing. From to ı to bottom: cryonecrotic zone, thin layer of granular spheres surrounding the damaged brain tissue with mild glial hyperplasia. Hematoxylin–eosin. 40 × 20.

zone. There was no evidence of stasis in the capillaries, only an insignificant pericellular and perivascular edema.

Further resorption of necrotic lesions resulting in cavitation occurred by the third week. The walls of these cavities had two layers: an internal layer consisting of a wide zone of compound granular cells and an external layer composed of a conglomerate of small vessels and intertwining connective tissue fibers. Mild hyperplasia of astrocytes was observed perifocally.

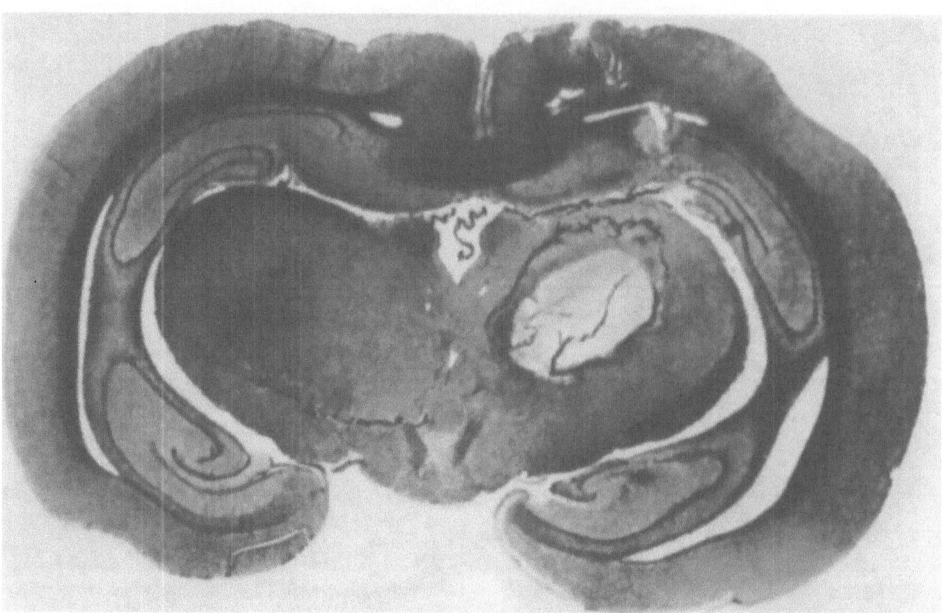

FIGURE 89. The focus of cryonecrosis on the frontal section of rabbit brain 12 days after local freezing (micromacrophotograph).

On the 22nd to 23rd day, in the center of the frozen zone, a cavity was beginning to form with thin walls composed of granular spheres and granular tissue. At 40 days, this cavity was larger, and its walls were formed by a narrow rim of granular cells, granular tissue, and a thin net of connective tissue fibers. Hyperplasia of the glial cells was considerably less pronounced, and the edema and hyperemia in the peripheral zone were absent.

After 2–2.5 months the walls of the cavity consisted of connective tissue fibers with interspersed areas of granular tissue and nests of hemosiderin. The glial cells in the cerebral tissue adjacent to the cavity walls were mildly hyperplastic. Three months after the freezing, the destructive focus was transformed into a cyst with a 0.5-mm-thick wall consisting of small vessels in a connective tissue matrix. There was no perifocal hyperplastic reaction of the astrocytes. Walder (1968) also found that by 3 months the cryogenic focus was transformed into a fluid-filled cyst. These foci have been studied in several autopsied cases. Approximately a year after the operation such a focus was a small (4 × 2 mm) gliomesodermal scar or cyst (Kandel, 1965; Rap and Mempel, 1971).

Summing up microscopic findings in the brain after experimental cryodestructive foci, one may conclude that the spherical frozen core of brain tissue (i.e., the part frozen solid and then thawed) undergoes irreversible necrosis. Subsequently, the necrotic tissue is resorbed, and a cystic cavity forms during the second month.

The insignificant reaction of the brain tissue around the focus is noteworthy. Vascular changes in the tissue surrounding the necrotic focus manifested by moderate local hyperemia were seen only during the first 10–12 days. No macroscopic hemorrhages in the necrotic zone or adjacent cerebral parenchyma were noted. Edema of the narrow zone of brain substance adjacent to the focus was minimal and disappeared completely in 2 weeks. Damage to nerve cells around the destroyed focus was not observed. The zone of moderate perifocal reaction did not exceed 2–3 mm in width in early (2–4 days) and 1–2 mm in later periods.

Electron microscopic studies of the frozen focus in the cat brain (Walder, 1968; Fraser, 1975) demonstrated that notable changes in brain tissue in the focal zone were present within a few minutes after freezing. The processes of the astrocytes around the neurons became unrecognizable, and the neuronal membranes were destroyed, increasing the intercellular space. Dendrites, axons, and synaptic structures seemed to be unchanged. The most pronounced changes immediately after freezing and thawing are found in the astrocytic membranes.

We carried out special investigations to estimate quantitatively the volume of the destroyed focus and its relationship to the quantity of liquid nitrogen used for freezing. The two-plane geometric method was used for determining the volume of destruction. The contours of the foci on histological sections were drawn to determine the size of the focus. The results were corrected by Simpson's formula.

These investigations demonstrated that in local freezing there is a positive correlation between the quantity of liquid nitrogen and the volume of the lesion: the more nitrogen in the reservoir of the instrument, the larger is the destructive focus.

In no case did we observe hemorrhagic changes in the perifocal zone during local freezing of deep cerebral structures. As mentioned above, a small hemorrhagic area in the focus was observed quite frequently, but the bloody infiltration was always within the focus and did not increase its size. Although Walder (1968) found microscopic hemorrhages in many cryodestructive foci deep in the brain of cats, macroscopic hemorrhages occurred in only 15% of the foci and did not extend beyond their limits or increase their size.

The "antihemorrhagic" effect of freezing stems from the peculiar action of low temperatures on small vessels (arterioles and capillaries) (Section 2.2). During the freezing process the blood flow slows down and then stops as a result of stasis in capillaries and small arterioles, which become packed with erythrocytes. The role of stasis in the destruction of brain tissue is especially pronounced during slow cooling.

However, among the numerous papers evaluating the cryosurgical method, only one (Coe and Ommaya, 1964) concludes that cryodestruction has no advantage over other methods of producing lesions in stereotactic surgery. According to their data, freezing may cause edema and hemorrhage beyond the frozen zone. It is difficult to explain the discrepant result. Cooper *et al.* (1965) believe that it may be related to technical defects in the cryogenic apparatus that the authors used in their experiments.

In addition to the morphological data presented above, histochemical studies of normal and neoplastic tissues after freezing and thawing have shown that enzymes and lactate were substantially elevated in all cases.

After approximately 1.5–2 months, the cryonecrotic focus transforms into a small cystic cavity or a barely noticeable gliomesodermal scar (Fig. 90).

2.6. Cryosurgical Equipment

Although freezing has long been used for the treatment of various skin lesions, the rapid advance of

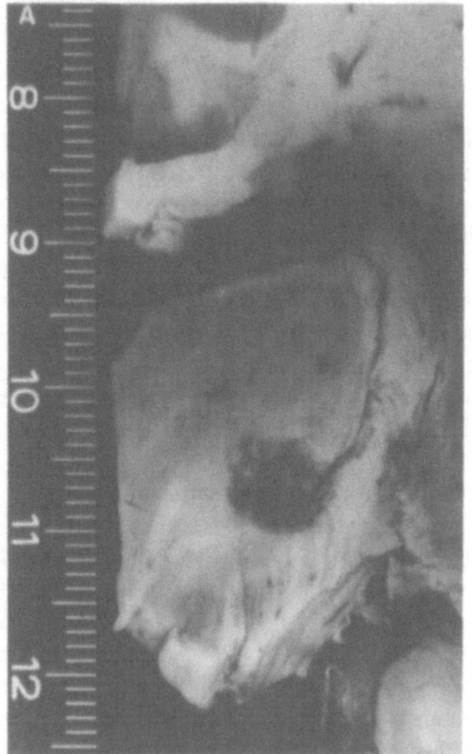

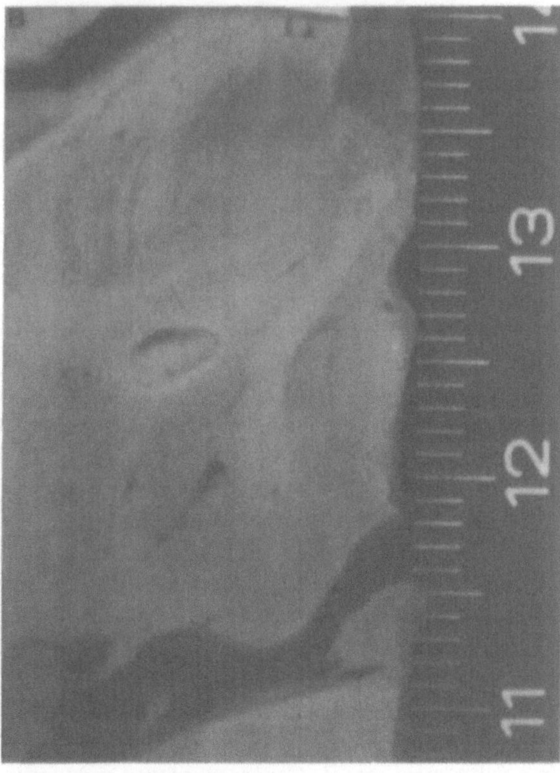

FIGURE 90. The foci of cryodestruction in VL at autopsy 3.5 (A) and 11 (B) months after stereotactic surgery.

modern cryosurgery became possible only as the result of technical progress in devising sophisticated instruments for both surface and deep cryogenic destruction.

The first attempts to build instruments for cryosurgical procedures on animals and man were made over a century ago. James Arnott, in 1845, was the first to make a simple instrument for local cooling and to employ it for treating cancer. A solution cooled by a mixture of ice and salt to a temperature of approximately −30°C was applied to various body surfaces such as uterus and breast to freeze cancer. In subsequent decades, various simple instruments for cooling skin and mucosa to destroy superficial pathological lesions were described.

The beginning of the 1960s may be regarded as initiating the modern era of cryosurgical instrumentation designed to destroy deeply seated tissues, especially brain structures. From the technical point of view, the aim was to construct a fine instrument that could be inserted into the brain and would have at its distal end a device capable of producing a temperature low enough to freeze a previously determined volume of brain tissue. At the same time, it was necessary to prevent damage to brain tissue adjacent to the shaft of the cannula.

It should be emphasized that cooling and then freezing a specific volume of deep-seated living tissue is a complicated task, since the circulating blood maintains a constant temperature of +37°C. The only way to resolve this problem is to use an instrument with high heat conductivity that can be cooled to a temperature below −20°C. Since the freezing is done deep in an organ to destroy pathological or normal tissue, in order to reach the target it is necessary to introduce a long, slender, and relatively harmless instrument. Consequently, the thinnest-walled tubes should be used for heat removal. In order to avoid damaging the tissue along the length of the cannula, thermal insulation is required except for the active tip where an extremely low temperature is to be created.

This complicated technical problem was first solved by Cooper (1961, 1962b), who designed an instrument for local freezing and used it for the first time in stereotactic operations on the basal ganglia. The early model circulated liquid nitrogen under pressure and removed the vapors with an aspirator. Thermal insulation of the cannula was provided by a copper coil through which an electric current was passed. In a subsequent model, Cooper employed vacuum insulation of the cannula except for its tip, which was a very important modification.

Approximately a year later, the author published his first results using a personally designed cryoinstrument, which is described below (Kandel *et al.*, 1962). Numerous subsequently built cryoprobes are being used successfully not only in neurosurgery but in many other areas of medicine as well (ophthalmology, gynecology, urology, oncology, general surgery, etc.).

The modern equipment used for cryosurgery can be divided into three groups depending on the means used to obtain low temperatures. The most widespread principle for the design of cryosurgical instruments is based on the change in the physical state (phase) of the cryogenic agent. Solid carbon dioxide or "dry ice," liquid nitrogen, and certain other substances may be used for this purpose. The temperature drops sharply on the transition of these refrigerants from the solid or liquid phase to gas.

Of all cooling agents, the one most often used in cryosurgery is liquid nitrogen, a powerful refrigerant that can produce a temperature of −196°C. It has a number of significant advantages: it is easily available, explosion-proof, incombustible, easy to transport, and relatively inexpensive. Nitrogen is stored in special Dewar flasks, which make losses to evaporation insignificant. The transformation of liquid nitrogen into gas absorbs a large quantity of heat (39 kcal per liter of nitrogen) and hence greatly reduces the temperature of the cryoprobe. It is generally agreed that liquid nitrogen is the most appropriate cooling agent for vacuum cryoinstruments.

The second principle by which low temperatures are produced is the Joule–Thomson effect (gas-throttling effect): cooling occurs on rapid expansion of a compressed gas (for example, nitrous oxide, carbon dioxide, or Freon passing through a choke valve). The depth of freezing varies with the degree of compression of the gas. This principle has found broad application in refrigerators but has been used rarely in cryosurgical instruments because it is impossible to get a sufficiently low temperature.

In recent years a number of medical specialties have used for various purposes cryogenic devices in which the cooling agent is nitrous oxide or carbon dioxide (Wright, 1971). The evaporation of N_2O causes a drop in temperature at the active tip of the instrument, but to no lower than −60 to −70°C. These cryocannulas have no vacuum insulation, which limits their use. Various models using N_2O for different purposes in medicine are manufactured by Spembly (UK), Frigitronics (USA), Dynatech (USA), and others.

TABLE 4. Properties of Cooling Agents

Cooling agent	Temperature of cooling agent under atmospheric pressure (lowest temperature at the active tip of cryoprobe)	Mode of transportation and storage
Freon 12	−30°	Steel cylinders filled with liquid at pressure of 10–15 atm.
Freon 22	−40°	
Carbon dioxide (dry ice)	−79.5°	Steel cylinders filled with liquid at pressure of 60–70 atm.
Liquid nitrogen[a]	−196°	Containers with vacuum insulation (Dewar flasks) filled with liquid at atm. pressure
Nitrous oxide	−89°	Steel cylinders under pressure of 60 atm.

[a]Liquid oxygen and other liquefied gases that closely resemble liquid nitrogen may also be used, but they present fire and explosion hazards. For the same reasons, the use of methane, ethane, and other combustible gases is undesirable. In the last two decades cryosurgical engineering has so greatly developed that a great number of cryogenic devices have been built and successfully used in medicine, especially neurosurgery.

The third principle is based on the Peltier effect, which is the opposite of the thermocouple effect. If an electric current is passed through a circuit consisting of two different metals, one will discharge heat, and the other will absorb it. The withdrawal of heat from one electrode will cool the other to approximately −25 to −30°C. Cryosurgical instruments based on the Peltier effect are rarely used in clinical practice because their cooling capacity is low.

Table 4 summarizes selected data on the most commonly used cooling systems.

2.7. The Author's Cryosurgical Instrument

At the end of 1961, at the Institute of Physics Problems, Academy of Sciences of the USSR, the laboratory supervised by Academician Shalnikov began working at my request on a new cryosurgical instrument with a high-vacuum shield for stereotactic operations on the basal ganglia. After technical trials and tests, we performed the first stereotactic operation with the new instrument in 1962. From the beginning, Shalnikov used as the guiding principle in constructing cryosurgical devices that they be as simple to use as possible. Over the next two decades, this cryosurgical device was modified by increasing its "power" (refrigeration capacity) and its reliability. It was made lighter, and the diameter of the cannula was reduced from 3 to 2 mm. At the same time, the "peak" temperature of the active tip was lowered from −60° to −180°C. Subsequently, we fabricated and tested seven models of the instrument for local freezing. During this period, more than 1900 cryosurgical operations were performed on deep-seated brain structures and brain tumors at the Burdenko Institute of Neurosurgery and at the Institute of Neurology.

We here describe the latest, improved, sixth and seventh models of the instrument* and its practical application for local freezing in stereotactic operations. This design, while retaining the advantages of previous models, has considerably expanded the possibilities of the instrument. The technical details of the design and use of earlier models were given in previous reports in collaboration with Shalnikov and others (Kandel *et al.*, 1962, 1970, 1973a, 1974, 1981).

The sixth model for local freezing (Fig. 91) is a portable, lightweight (80 g) instrument consisting of two parts. The first is a thin-walled stainless steel cryocannula 22 cm long with an external diameter of 2 mm, which is less than that of other cryocannulas (Frigitronics). Obviously, the smaller the diameter of the cannula, the less trauma there will be to the brain; however, further reduction of the diameter of the cryocannula to less than 2 mm seems inadvisable, since this would weaken the cannula below permissible levels.

Inside the 0.5-mm-thick cannula wall there is a thin (1 mm in diameter) hollow cold duct through which the liquid nitrogen reaches the active (noninsulated) tip of the cannula (Fig. 92). All joints of the metal tubes are fused by electric welding. The external end of the cold duct is connected to a fine polyethylene catheter that through a long plastic tube is attached to a conventional surgical aspirator.

The second part of the instrument is a removable thermostable 100-ml foam plastic reservoir that fits tightly onto the thicker peripheral end of the cannula. For freezing, the required volume of liquid nitrogen (temperature −196°C) is poured into the reservoir, passes through the cold duct to enter the active tip, cools it, evaporates, and is aspirated. In this way, the forward and reverse flows of the refrigerant pass through two concentric tubes in the cannula.

As in previous models, the body of the cannula is

*USSR patent No. 214746.

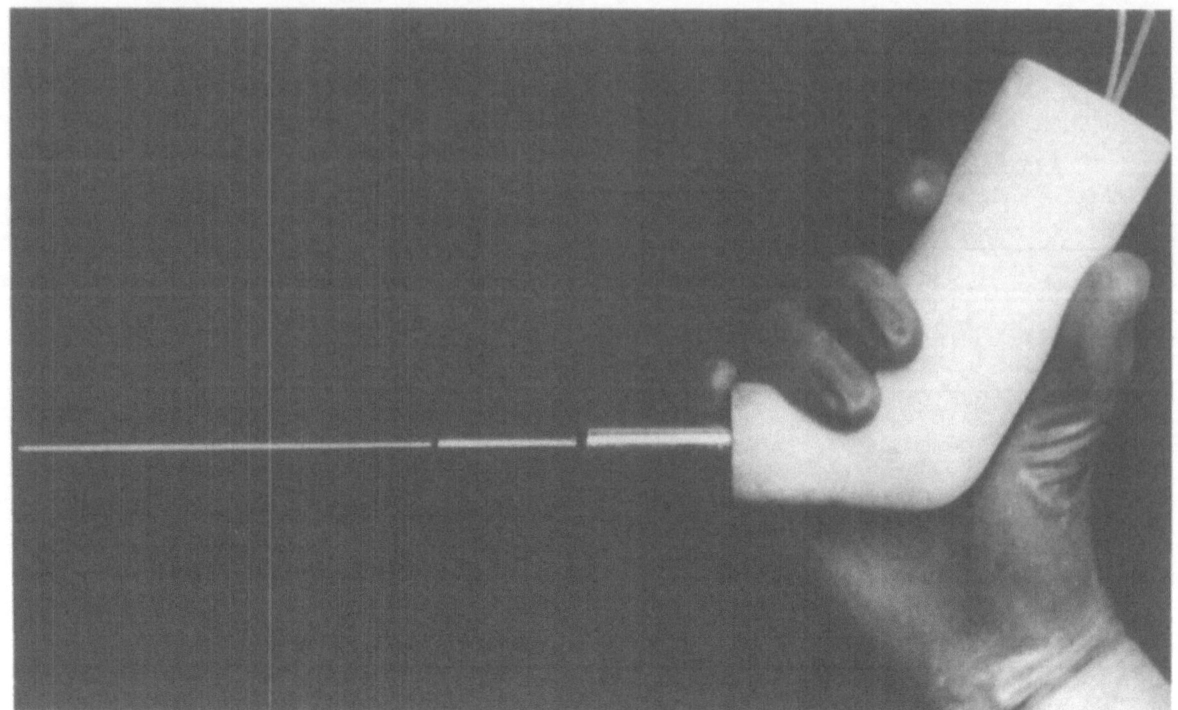

FIGURE 91. Our assembled neurosurgical device for freezing subcortical structures.

separated from the cold duct inside it (the distance between them being only 200 μm) by a vacuum layer (degree of vacuum 10^{-6}), ensuring a ''normal'' temperature of the entire external surface of the cannula with the exception of its 2 × 2-mm silver active tip. A deep vacuum is maintained with the aid of a charcoal collector that adsorbs any remaining gas. Direct contact between the cold duct and the wall of the cannula is prevented by a lining of thin quartz threads wound around the cold duct.

The metal part of the instrument is sterilized in alcohol for 5–6 hr and is connected to the reservoir just before the cannula is inserted into the brain. One must take care that alcohol does not enter the drainage tube of the cannula. After the liquid nitrogen is poured into the reservoir of the instrument, the temperature at the surface of the cannula tip gradually falls to −150° to −160°C. The body of the instrument, however, because of the vacuum thermoinsulation, remains at a temperature near that of the brain (the difference of a few degrees is of no significance).

A very thin thermocouple (copper–constantin) passes through the cold duct to the active tip of the probe. On moving paper the thermocouple records the drop in temperature and its subsequent return to the initial level. The entire cannula, with the exception of its active tip, is covered with a thin insulating coating of Bakelite varnish, which makes it possible to stimulate deep brain structures during the operation. The circulation of liquid nitrogen allows the ''power'' of the instrument to be increased several times and, as a result, reduces the freezing time. For instance, the destruction of VL or Subth takes approximately 90 sec.

Numerous experiments on rabbits have shown that filling the reservoir with 25 ml of liquid nitrogen produces an ice sphere with a diameter of 5–6 mm, 40 ml 8–9 mm, and 60 ml 13–14 mm. A large quantity of nitrogen (100–120 ml) can freeze a sphere about 30 mm in diameter, which makes it possible to use this instrument for the destruction of brain tumors. The exposure time in a typical case is approximately 9–10 min. For cryosurgery of brain tumors, we have designed a special instrument (Chapter 15, Section 2).

The seventh model differs from the sixth in that the active (noninsulated) tip of the cannula is located eccentrically on its axis, so the ice sphere forms asymmetrically with respect to the cannula tip. The formation of an eccentric ice sphere makes it possible to correct any slight inaccuracy in the position of the active tip of the probe. Moreover, it is possible to enlarge

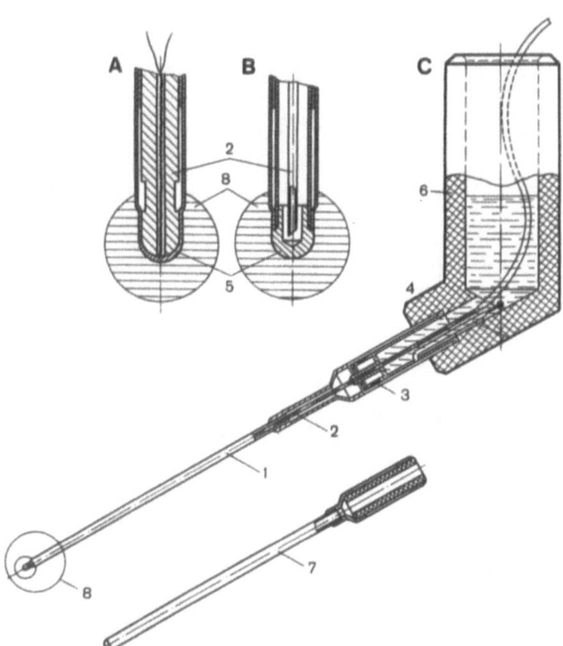

FIGURE 92. Scheme of construction (in section) of the sixth and seventh models of the author's device for local freezing of subcortical structures in stereotactic operations (Kandel and Shalnikov, 1970). (A) Magnified image of the active tip of the device. (B) Modification of the active tip permitting freezing excentrically about the active tip. (C) General view of the instrument. 1, Hollow cryoconductor; 2, capillary tube in the cryoconductor for liquid nitrogen circulation; 3, vacuum insulation; 4, reservoir of thermostable form plastic for fluid nitrogen; 5, bare active tip; 6, plastic tube to aspirator for gaseous nitrogen evacuation; 7, protective case; 8, ice ball on the active tip of the device.

the lesion eccentrically in the cerebral tissue without reintroducing the cannula. For this purpose, after the frozen sphere thaws, the cannula is rotated around its axis in the desired direction, and the freezing procedure is repeated. This is very convenient in operations such as cryohypophysectomy (Chapter 15, Section 3).

2.8. Technique for Producing Cryosurgical Lesions

The author's instrument for local freezing must be recalibrated after several (five or six) operations. For this purpose, the cannula is immersed in either gelatin or eggwhite heated to 37°C, both of which have a consistency similar to brain substance. After the required volume of liquid nitrogen has been poured into the instrument, one can visually time the formation of an ice sphere (Fig. 93). When the sphere reaches the

FIGURE 93. Series of photographs, from top to bottom, illustrating development and enlargement of the ice sphere on the "active tip" of the cannula of the freezing device inserted into gelatin. The time interval between successive films is 20 sec. On the last two photographs the beginning of melting of the ice sphere is clearly seen.

maximum diameter, it is extracted from the gelatin and measured with compasses. Control experiments have shown that the diameter of the ice sphere formed in these substances is only 1–1.5 mm greater than that in the brain.

The instrument should not be checked in water since convection prevents the formation of an ice sphere. However, it is possible to calibrate in water the power of new instruments for freezing. This is based on the precise determination of the quantity of cold in terms of calories being transmitted from the instrument to the water. In a small Dewar flask, 15 ml of water is poured, and its temperature is measured. Then the tip of the cannula is immersed into the water, and 40 ml of liquid nitrogen is poured into the reservoir. After the ice sphere has thawed, the cannula is removed, and the water temperature is measured again. The result, indicating the "power" of the cryoprobe, is expressed in grams of ice formed on the tip of the cannula. This value is determined by the formula $m = Q/q$, where $Q = m_1 (t_2 - t_1)$ (m is mass of ice in grams; m_1 is volume of water in Dewar flask; t_1 and t_2 are initial and final temperatures of water; q is latent heat of freezing, equal to 80).

The changing temperature in the freezing process during a stereotactic operation can be seen in Fig. 94. After the cannula is introduced into the brain, the temperature of the active tip rises from 22°C (room temperature) to 37°C (temperature of the brain). Eight seconds after 50 ml of liquid nitrogen is poured into the reservoir of the instrument, the temperature begins to fall at a rate of approximately 60°/min. In a matter of 30–40 sec, the temperature has dropped to zero. In several more seconds, the temperature at the tip of the cannula drops to −14°C (supercooling stage). At this temperature one observes a clear-cut peak (an increase of 6–8°C), which indicates that the supercooling stage is over and that ice crystallization has begun in the brain tissue surrounding the tip of the cannula. The formation of an ice sphere at the active tip of the cannula begins 40–45 sec after the nitrogen has been poured in and lasts (depending on its quantity) approximately 1.5 min. As the size of the ice sphere increases, the rate of fall of temperature is sharply diminished to approximately 1° in 4 sec.

The lowest temperature at the noninsulated tip of the probe of the latest design with 45 ml of nitrogen is −160 to −180°C. This temperature is retained for 70–80 sec. When the temperature stops dropping and the curve becomes an equilibrium between cooling (withdrawal of heat) and heating (by the blood) has been reached.

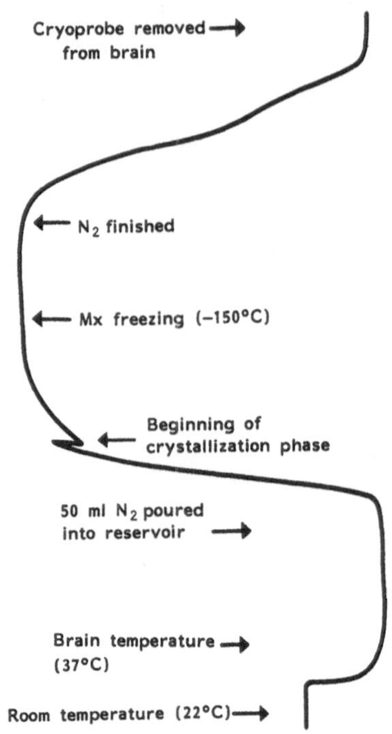

FIGURE 94. Dynamics of temperature (from the bottom upward) recorded with a thermocouple on the active tip of the cannula during freezing in a stereotactic operation. Explanation in the text.

Immediately after all the liquid nitrogen boils out in the reservoir of the instrument (total time about 2 min), the temperature at the active tip of the cannula at first gradually and then very rapidly begins to rise. In the course of 3–4 min, it reaches the temperature of the brain. By this time, the ice sphere at the active tip has melted, and the cannula can safely be removed from the brain.

In those comparatively rare cases in which the introduction of the cannula does not produce a clear-cut clinical effect, it is necessary temporarily to cool the subcortical structure. For this purpose, 8–10 ml of liquid nitrogen is poured into the reservoir to lower the temperature at the active tip of the probe to approximately 5°C. This inactivates a 4- to 5-mm zone of cerebral tissue (2–3 mm of tissue around the 2-mm cannula). If the test produces a definite positive clinical effect (usually in 1–1.5 min), the instrument is filled with sufficient nitrogen to make a therapeutic lesion. It is noteworthy that since it is possible to freeze a subcortical structure to a given temperature, and since it is known at what temperature a reversible functional inhibition of a specific deep structure occurs, by the cryogenic method one can objectively study the effects of such inhibition and, consequently, the functions of intracerebral structures.

Before filling the instrument with liquid nitrogen, it is necessary to connect the discharge plastic tube to a surgical aspirator. The pressure of the aspirator is adjusted for maximum suction. The required volume of nitrogen is poured from a Dewar flask into a measuring cup made of heat-resistant glass and thence into the reservoir of the cryoprobe (Fig. 95). The nitrogen should be poured slowly to reduce "boiling." Approximately 40–60 ml of nitrogen is required to create a cryodestructive lesion of appropriate diameter.

An ice sphere begins to form in 30–40 sec, as can be seen by the formation of a dense layer of rime, resembling snow, on the thickened external end of the cannula where it connects to the reservoir. The appearance of this "snow" is also evidence that the vacuum in the cannula has been preserved. Another indicator of the beginning of formation of the ice sphere and, consequently, of proper functioning of the instrument is the rapid loss of suction seen on the aspirator gauge.

In the latest models of our instrument, the boiling time for 40 ml of liquid nitrogen is 1.5 min, and that for 50 ml is 2 min. By this time, the clinical effect (under local anesthesia) must be quite evident. After the nitrogen has boiled off, the aspirator is switched off, and after 5–6 min, the time required for the complete thawing of the ice sphere, the cryodestruction can be considered complete, and the cannula is withdrawn from the brain.

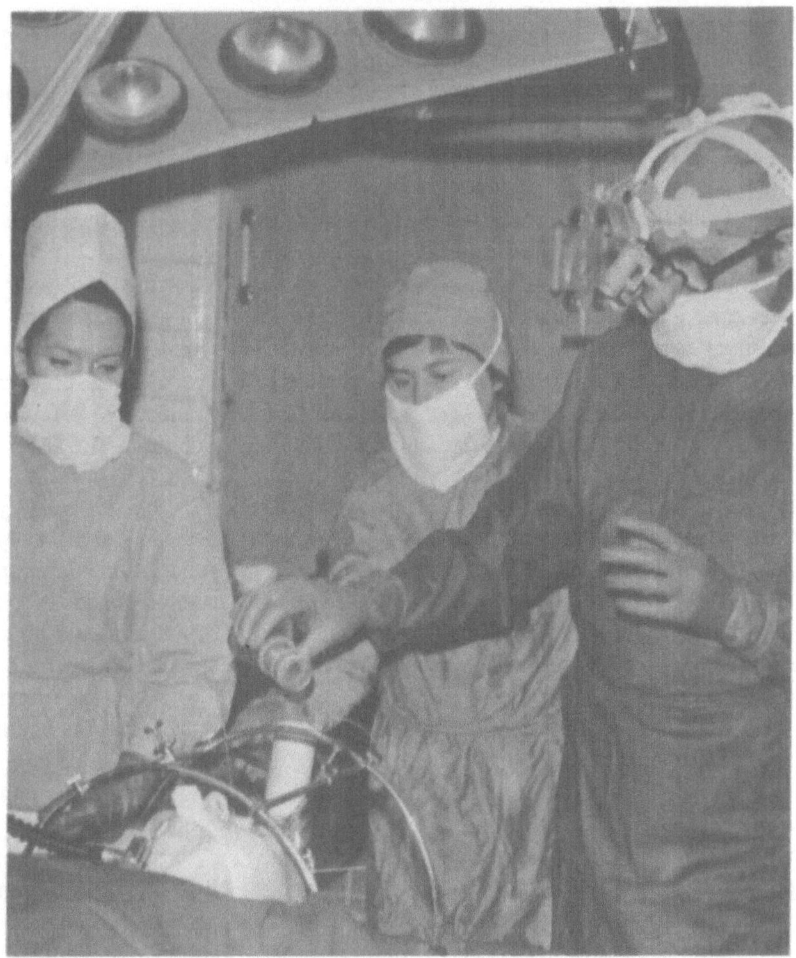

FIGURE 95. The liquid nitrogen is poured into the reservoir of the cryosurgical device.

If a second cryogenic lesion has been planned, after the appropriate calculations have been made, the cannula is again introduced, and the freezing repeated.

The results of stereotactic operations depend on many factors, including the method of destroying cerebral tissue. Obviously, to evaluate the results of surgical treatment on the basis of one factor, the type of lesion, is very difficult. Nevertheless, from an analysis of a large series of comparatively similar operations, one can draw some valid conclusions.

Cooper (1969a) gives a description of an interesting control experiment carried out in his clinic. The eminent neurologist Schwab (1966) reported the results of 50 consecutive stereotactic operations for parkinsonism performed by the cryosurgical method. In almost all cases there was a good therapeutic effect, such as abolition of tremor and rigidity in the contralateral extremities. There was no complications or mortality. But in the same period, in another clinic, Schwab observed a series of analogous operations in which thalamic nuclei were thermocoagulated. In this series, 6% of the operations had fatal outcomes.

In our monograph on the surgical treatment of parkinsonism (Kandel, 1965), we analyzed the results of our first 105 cryosurgical operations and noted a substantial reduction in postoperative complications and mortality as compared to the earlier method of destruction by alcohol. Hankinson (1969) also emphasizes that local freezing in stereotactic operations is much safer than other methods of destruction. In 400 operations performed by the thermocoagulation technique, mortality was 2.6%, and in 500 cryosurgical operations it was only 0.9%. There were also fewer postoperative complications, in particular, mental disorders. Hankinson notes that after cryosurgery was employed in his clinic, patients who previously, because of age or severity of the disease, were considered inoperable could be surgically treated.

Mempel *et al.* (1971), as well as other surgeons, present data on the great benefits and large margin of safety offered by cryosurgical destruction of subcortical structures.

We performed our first stereotactic operation using cryosurgery in 1962, and after that time it became our method of choice for operating on basal ganglia, Cer nuclei, and certain brain tumors. To the present day, we and our associates have performed approximately 1900 different cryosurgical operations. The more than 20 years of experience enables us to evaluate it objectively not only on the basis of experimental, morphological, and other data but also on the basis of clinical results. In the subsequent chapters of this book, we show that the results of stereotactic operations in parkinsonism, dystonia musculorum deformans, spasmodic torticollis, and other extrapyramidal lesions have improved significantly during the past two decades. This is not only because of improved techniques in performing these operations but also to a certain extent because of adoption of the cryomethod. The notable reduction in the frequency of postoperative complications and mortality, in our opinion, also resulted largely from the introduction of local freezing.

Summing up the data in literature and our many years of experience, we conclude that the cryosurgical method has been responsible for substantial progress in stereotactic surgery. In particular, it meets the requirements for an ideal method for destruction of tissue in deep brain structures.

3. Electrical Methods

3.1. Anodal Electrolysis

Electrolysis with a direct current to inactivate deep-lying structures of the brain in animals was first suggested by Simonov in 1866. At the end of the last century, electrolytic brain lesions in animals were studied using monopolar and bipolar electrodes. An important advance in the electrolytic method was made by Horsley and Clarke (1908), who showed that anodal lesions of the brain are more stable than cathodal, since in electrolysis fewer gas bubbles form on the anode. Subsequently, anodal electrolysis was extensively studied.

Spiegel and Wycis (1952b) were the first to employ anodal electrolysis for inactivating GP in man in stereotactic operations. They used a DC generator with a selenium rectifier and a special electrode consisting of a thin tube with a lateral hole through which a fine stainless wire could be extended at an angle of 120–130°. The experiments of Spiegel and Wycis demonstrated that the action of a 10-mA direct current for 1 min produces a brain lesion 4 mm long and 2–3 mm wide. The volume of destroyed cerebral substance was proportional to the quantity of electricity in coulombs.

Subsequent experiments showed that anodal electrolysis with a current of 5 mA or less for about 30 sec produced a more or less standard focus of destruction deep in the brain. When the strength of the current was increased, the size and shape of the area destroyed could not be controlled.

Although generators and electrodes like those described were successfully used for stereotactic operations, this equipment has a number of shortcomings. First, there is considerable fluctuations in the strength of the current activating the electrode. Second, stainless steel electrodes rapidly disintegrate in the course of electrolysis. After four to six coagulations, the active electrode must be replaced with a new one. Sweet and Mark (1953) came to the conclusion that the uncontrollability of anodal electrolysis and the occurrence of hemorrhages in the zone of destruction make this method unsuitable for clinical application. Electrolytic foci cause a noticeably greater reaction of the cerebral tissue than foci created by a high-frequency current. Nevertheless, anodal electrolysis has continued to be used. Mullan *et al.* (1965a,b) demonstrated that in order to obtain small, well-defined foci, it is necessary to use very fine electrodes and a weak current (2 mA and less) for a long period of time (about 30 min).

A modified apparatus for anodal electrolysis—a DC generator and electrodes of improved design—was made with our cooperation at the Moscow Scientific Research Institute of Surgical Equipment and Instruments. This apparatus was repeatedly tested both in animal experiments and in stereotactic operations on man. The generator consists of a transformer and a semiconductor rectifier that automatically compensates for fluctuations in line voltage. One of the requirements for stable electrolytic lesions is the absence of fluctuations in the resistance in the patient's body through which DC passes. Such fluctuations may occur for a number of reasons (the distance between the active anode and passive cathode, the formation of gas bubbles on the anode, changes in electroconductivity of brain substance during electrolysis, etc.). These fluctuations in resistance lead to corresponding changes in the current and, as a result, to undesirable changes in the size of the focus.

The design of the apparatus prevents fluctuations in the current entering the brain when the resistance changes during electrolysis. A special device automatically ensures a smooth rise or fall in the strength of the current. Before the coagulation, the parameters for electrolysis must be established. The strength of the current may be set with a meter from 1 to 15 mA. The duration of the current flow may be adjusted from 10 to 100 sec, after which it is automatically switched off.

The new design of the electrode permits three operations: electrolysis, electrostimulation, and recording of biopotentials. The electrode consists of a tube whose outer surface is covered with plastic insulation. At a distance of 2 mm from the tip of the electrode there is a lateral opening through which a thin wire is extended. After technical trials and experiments on animals, a special alloy that met most of the desired requirements was selected for the electrodes. By turning a knob calibrated in millimeters, it is possible to extend the active tip of the electrode into the brain in any direction from 1 to 7 mm.

In the insulating sheath of the electrode there is a metal contact for bipolar recording of biopotentials from subcortical structures; the second electrode is the active tip. Electrical stimulation may be carried out through the active tip of the electrode.

Experiments on dog brains demonstrated that the electricity required to produce desired foci varied from 100 to 800 mcoul. A comparative morphological study has established that a current of from 200 to 400 mcoul will produce an oval necrotic focus with a diameter of about 2 mm. However, with an 800-mcoul current (10 mA for 80 sec), the size of the lesion is increased three to four times (7–8 mm), and the size of the foci is more variable.

When 200–400 mcoul is used, the reaction around the necrotic zone extends 2 mm and has well-defined borders. With a current of 800 mcoul or more, the reaction is more pronounced and widespread.

We used anodal electrolysis in our first series (in the 1960s) of 52 stereotactic operations. After x-ray control of the position of the electrode, its active tip was extended for 3–5 mm in the calculated direction. Brief stimulation with a weak current verified the position of the electrode with respect to CI.

Usually, we made eight or nine small destructive foci using 250–300 mcoul (5 mA for 50–60 sec) for each electrolytic lesion. Then the active tip was withdrawn into the tube, the tube was rotated 45°, and the next set of lesions were made. The foci thus produced formed a semicircle. As a rule, we made these foci in three tiers separated by 2 mm. With this technique, we assumed that the electrolytic foci would fuse, producing total destruction of the subcortical structure.

Anodal electrolysis can be used to destroy tissue stereotactically and allows the surgeon to change the shape and size of the lesion and to terminate electrolysis in case signs of involvement of adjacent brain structures appear. One of the shortcomings of the method is that there is no certainty that the small foci

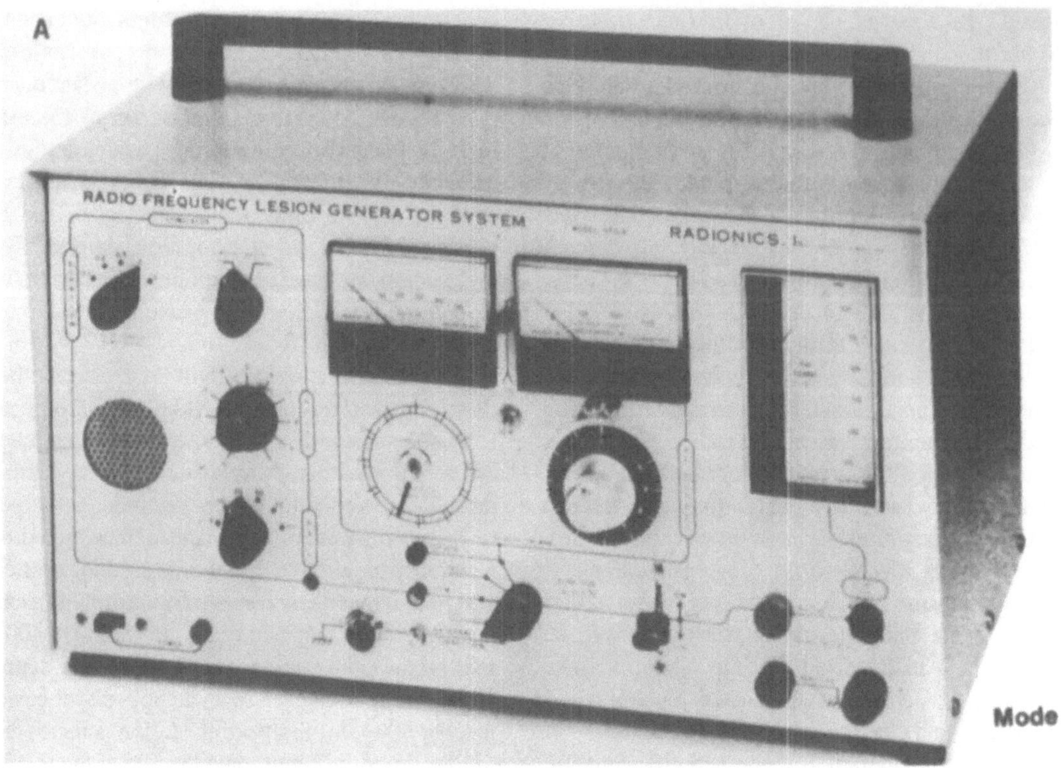

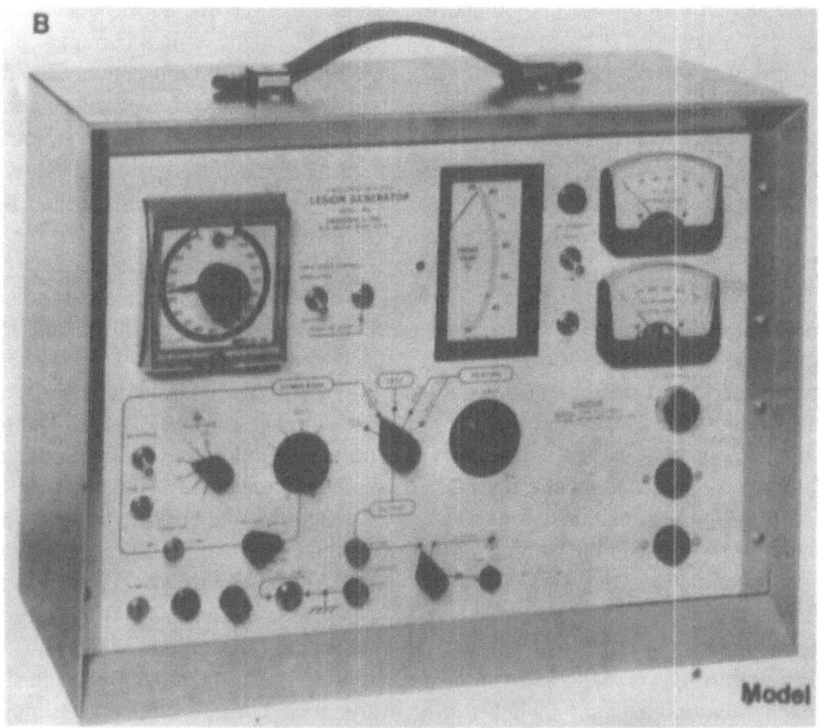

FIGURE 96. Two models of radiofrequency lesion generators (Radionics, USA).

will fuse and totally destroy the entire target structure. Moreover, the production of several foci takes considerable time, which prolongs the operation.

3.2. High-Frequency Coagulation

Many neurosurgeons use coagulation by a high-frequency sinusoidal current to destroy brain tissue in stereotactic surgery. In this method, first tested experimentally by Hess (1932), a high-frequency current is passed through cerebral tissue to produce a high temperature, which destroys nerve cells and fibers. Irreversible damage to nerve tissue occurs when it is heated to a temperature of over 60°C. In high-frequency electrocoagulation, the heating occurs not in the electrode but in the adjacent tissue, so the temperature of the electrode is always less than that of the surroundings. For example, with the tip of the electrode at a temperature of 62°C, the temperature at a distance of 3 mm from the tip of the electrode is 10°C higher, that at a distance of 4 mm is 6–9°C higher, and that at a distance of 6 mm is from 4° to 7°C higher (Watkins, 1965). The temperature at the tip of the electrode–cannula when high-frequency coagulation is used varies greatly depending on several factors: diameter of electrode tip, thermal conductivity of brain tissue, specific weight of blood and CSF in coagulated tissue, the presence of "scale" on the tip of the electrode, etc.

Numerous investigations have been done to calculate the current parameters under different conditions—tissue resistance, electrode surface area—and to calculate the relationships of these parameters to the size of the lesion. In particular, it was found that with high-frequency coagulation, there was a considerable difference in the amount of destruction of subcortical structures, depending on the conductivity and resistance (Meshcherski, 1961; Cosman *et al.*, 1983).

High-frequency electrocoagulation requires special generators. The customary devices used in neurosurgery are unsuitable for stereotactic operations since they do not generate pure sinusoidal oscillations. One of the better-known generators is the Wyss apparatus, which produces sine waves with a stable frequency within the range of 0.5–2 MHz. Usually in stereotactic operations, the following parameters are used: frequency 0.5–1.0 MHz, current 100–150 mA, exposure 30 sec. Such currents inactivate 200–250 mm^3 of brain tissue. The current for monopolar coagulation must be three times greater than for bipolar (Greenwood, 1955). However, Chehrazi and Collins (1981) demonstrated that both mono- and bipolar coagulation produce approximately the same size lesion.

Several modern generators of high-frequency currents to destroy brain structures are produced by Radionics (models RFS-SAV, RFG-3B5, 5s, 6, and 6s) (Fig. 96) and Fischer (Fig. 97). Besides producing the appropriate currents, these devices can monitor the temperature in the zone of destruction, stimulate subcortical and other structures, monitor impedance both digitally and acoustically, and automatically time biological phenomena.

The electrodes for high-frequency coagulation must meet certain specifications. Their size and shape are of great importance since they determine the configuation of the lesion (Dieckmann *et al.*, 1965). To heat the electrode rod in the process of coagulation requires good insulation; otherwise, considerable damage may be done to the brain. Bakelite varnish and

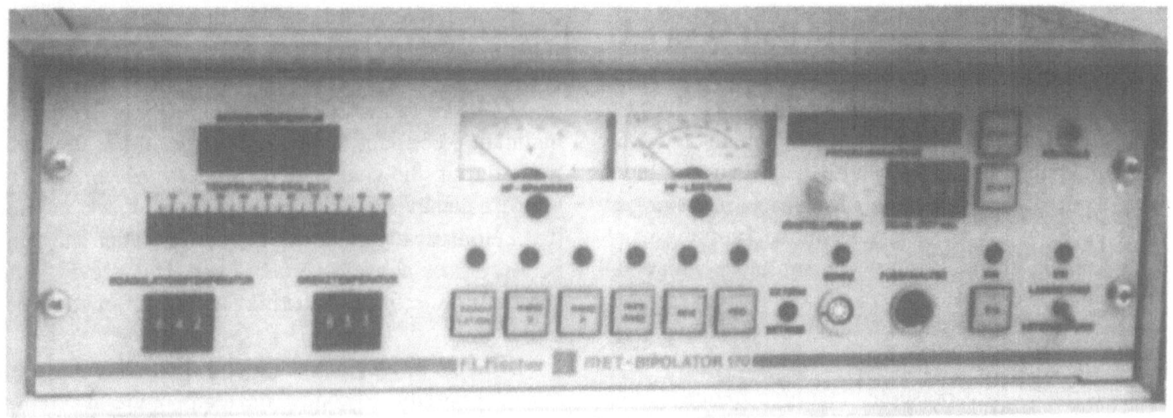

FIGURE 97. High-frequency lesion generator (F. L. Fischer MET GmbH, West Germany).

other dielectrics as well as a thin coating of glass are used for insulating the electrode. This insulation must be fully checked for leaks before each operation. Electrodes for high-frequency coagulation must be very thin since the smaller the surface area of the electrode, the higher the temperature will be in the surrounding brain tissue (Fox, 1970). It should be noted, however, that if the active electrode is too thin, the tissue at the tip of the electrode will "boil" with the formation of numerous gas bubbles, which make it impossible to increase the lesion to the desired size.

Fox (1970), in experiments on the brain of dogs demonstrated that with a 3-mm-long and 0.4-mm-thick active electrode tip, a strong current is not required: approximately 50 mA for 60 sec will produce a 5.5×5 mm lesion. In high-frequency coagulation, the size and shape of the focus depend not only on the parameters of the current and the length and shape of the electrode but also on the peculiarities of the tissue being coagulated. Increasing the strength of the current to 75 mA or higher leads to local "boiling" of tissue and the formation of scale around the electrode. Moreover, the current falls to zero. This occurs, however, only if the entire noninsulated tip of the electrode is nerve tissue. If only part of the electrode is immersed and its other part has contact with CSF, then part of the current is "shunted off," since the impedance of CSF is much lower than the impedance of nerve tissue. In this case, a current of 100–300 mA is required for coagulation.

The transformation of electric energy into heat during high-frequency coagulation is limited by the discharge of gas at the tip of the electrode, which prevents the further formation of heat. In such cases, further coagulation of the tissue stops in spite of the continued application of the high-frequency current. With a temperature of 80°C at the tip of the electrode, the enlargement of the destroyed focus stops 2 min after the current has been applied, even if the current is continued (Mundinger, 1973; Mundinger *et al.*, 1973).

The size of the lesion produced by high-frequency coagulation is also influenced by the presence of blood vessels in the focus. In view of this, foci made in gray substance, which has a greater blood supply, are always smaller than those in white substance with equal currents. Raising the temperature at the tip of the electrode to 75°C for 30–60 sec produces a lesion of approximately 9 mm^3 (Hitchcock and Teixeira, 1981).

Earlier, it was presumed that high-frequency currents would selectively destroy the small Aδ and C fibers while the large Aβ fibers to a considerable degree remain intact. However, recent experimental investigations using electron microscopy have shown that electrothermal coagulation, irrespective of the temperature at the tip of the electrode, destroys both small and large (myelinated) fibers to an equal degree (Smith *et al.*, 1981).

A high-frequency current can be used for a reversible or "physiological" inactivation of nerve tissue (Zervas, 1965). However, its use for this purpose involves a certain risk, since the temperature difference between the proauction of a reversible and irreversible lesion is relatively slight (Van Buren and Ratcheson, 1973).

The lesion after high-frequency coagulation is a central zone of necrosis circumscribed by a zone of incompletely destroyed cerebral tissue. According to a dynamic CT study, all subcortical lesions may be divided into two groups following high-frequency coagulation (Matsumoto *et al.*, 1985). Most lesions are low-density, but high-density lesions exist also.

Numerous experimental and autopsy studies have shown that high-frequency coagulation with a monopolar electrode produces circumscribed and predictable circular or oval-shaped foci. Less predictable foci result from use of a bipolar electrode, since coagulation sometimes occurs at each pole in spite of thermocontrol.

One of the main shortcomings of high-frequency coagulation is the formation of "scale" around the active tip of the electrode to which coagulated tissue sticks. It follows that on extraction of the electrode, it is quite possible that this "scale" might break off and cause mechanical damage with ensuing hemorrhage. Still another danger lies in the fact that local heating of a vascular wall may lead to the formation of an aneurysm with rupture and bleeding. In order to prevent such complications, the temperature of high-frequency coagulation should not exceed 60–70°C.

In order to avoid complications it is necessary to maintain a constant temperature at the tip of the electrode during the coagulation. Modern instruments for high-frequency coagulation have a feedback control that automatically shuts off the current when the temperature rises above a certain level. This prevents the cerebral tissue from "sticking" to the tip of the electrode.

The use of an electrode with a thermistor at the tip makes it possible to control the temperature in cerebral tissue and lessen the danger of hemorrhage during the destruction of subcortical structures (Fager, 1968). This has substantially decreased the variability in the size of the lesion. With thermostabilization, the fluctuations in the area of destruction proved to be 1.5 times less than without this control. Thus, the foci of inactivation produced by high-frequency coagulation with thermocontrol are stable and reproducible, making it possible to obtain predictable lesions. The dimensions of the foci correlate well with duration of the current and temperature at the tip of the electrode. Consequently, one may conclude that high-frequency coagulation with thermocontrol is one of the best methods for destruction of subcortical structures in stereotactic surgery.

3.3. Induction Heating Method

This method is based on a well-known physical phenomenon: a metallic conductor placed in an electromagnetic field heats as a result of electrical energy transforming into heat. This method was first tested experimentally by introducing small bits of wire deep into the brain and then heating them by induction (Carpenter and Whittier, 1952). When the head of the animal with implanted metal was placed in an electromagnetic field, a thermal focus, usually circular, was formed around the wire without significant damage to the surrounding cerebral tissue.

Clinically, this method, called telethermocoagulation, was first employed by Walker and Burton (1966), who introduced a small (8 × 1.6 mm) piece of stainless steel wire into a subcortical structure. Then the head of the patient was placed in a high-frequency electromagnetic field (over 200 kHz), and induction heating was repeated several times. Although the authors believed that the wire in subcortical structures did not shift since it was fixed by the surrounding destroyed cerebral tissue, the migration by a metal cylinder introduced into a subcortical structure is possible.

An interesting advantage of this method is the possibility of repeating the induction heating some time after the operation without leaving the electrode in the brain. However, this method does not produce foci of a definite size and shape; i.e., it is poorly controllable. Although the induction heating method with implanted pellets has not found widespread application in stereotactic neurosurgery, it is promising in those cases in which relapses may occur and repeated lesions are required, since several applications of induction heating may be given.

4. Radioactive Isotopes

After radioactive isotopes began to be used in neurosurgery approximately 30 years ago, scores of experimental and clinical investigations were carried out using various isotopes to create necrotic foci in the brain. More than ten different isotopes have been tested in clinical conditions.

The use of a radioactive isotope for the destruction of subcortical structures in stereotactic surgery depends, first of all, on the energy emitted by the isotope. Isotopes emitting α particles have not found any clinical application since they do not penetrate cerebral matter. On the other hand, γ rays display a high penetrating ability, creating large and poorly controlled destructive foci. However, in order to destroy cerebral gliomas, γ-radiating isotopes, for example, iridium-192, may be used.

The relatively safe application of isotopes was limited mainly to those generating β rays. Their advantage lies in their relatively short penetration of cerebral tissue, about 5–6 mm. The dimensions of the lesion depend on the activity of the isotope and the exposure time. The list of β radiants includes ^{90}Y and ^{109}Pd. ^{90}Y, which is most frequently employed, has a half-life of 2.5 days, and the energy of its β rays is 2.8 MeV. The half-life of ^{109}Pd is 13 hr.

^{90}Y is used in the form of ceramic or metallic granules. In experiments, we demonstrated that the volume of the cerebral destruction and the amount of perifocal reaction of the brain tissue are related to the activity of the ^{90}Y metallic granules. Granules with 1–3 mCi activity produced a necrotic focus in an animal brain not exceeding 8 mm with a mild perifocal reaction. Granules of 10–15 mCi gave a 20-mm zone of necrosis and a perifocal zone of severe degeneration and edema.

Radioactive isotopes were widely employed in stereotactic operations some 20 to 25 years ago (Riechert and Mundinger, 1960; Vikhert *et al.*, 1968; Talairach *et al.*, 1962). However, as more clinical experience was accumulated, the use of isotopes for making lesions gradually declined. At present, practically all neurosurgeons have abandoned this method of subcortical destruction for dyskinesias and pain.

Radioactive isotopes were not used in stereotactic surgery for hyperkinesias, epilepsy, or pain because of a number of drawbacks. One of the most serious is that the destruction of cerebral tissue and consequent development of a therapeutic effect proceed slowly. Approximately a week is necessary for the production of a necrotic focus. Hence, during the operation, the surgeon has no way of knowing how effective it will be. Another equally important drawback is that there is no possibility of terminating the destructive process if signs of damage to neighboring structures, especially CI, appear. Moreover, working with radioactive isotopes requires stringent radiation protection and observance of safety rules. Also, after stereotactic implantation of radioactive isotopes, postradiation edema of the brain may develop, causing increased intracranial pressure.

Although the use of radioactive isotopes in stereotactic neurosurgery for dyskinesias and pain was practically discontinued, in recent years there has been a return of interest in isotopes for stereotactic implantation in deep-seated inoperable malignant brain tumors (gliomas, germinomas, sarcomas, and others) as well as in pituitary tumors. Data on the use of interstitial irradiation for treating brain tumors are presented in Chapter 15.

5. Laser Techniques

Laser techniques in neurosurgery have been making rapid progress in recent years. It is common knowledge that lasers generate powerful, highly directional beams of monochromatic coherent electromagnetic energy capable of discrete local destruction of any tissue of the body.

A flexible light guide, conducting a laser beam, may be inserted into the thin tube of a conventional aspirator to direct it visually to a tumor or other cerebral pathological focus. The laser beam may be aimed at a target zone with the aid of a micromanipulator attached to an operating microscope. With the laser, the tissue of a tumor is turned into a necrotic mass that can easily be removed with an aspirator with practically no bleeding, since the laser instantaneously coagulates vessels. Various lasers—argon, ruby, heliocadmium, carbon dioxide, neodymium, YAG, and others—are now extensively used in medicine in general and in neurosurgery in particular. Carbon dioxide lasers have found the widest application since they coagulate tumor tissue without significant heating of the surrounding brain substance. More clinical experience is being accumulated in the use of lasers to destroy various cerebral pathological structures, especially deep-seated tumors (Heppner, 1978; Shelden *et al.*, 1980; Saunders *et al.*, 1980; Hara *et al.*, 1980; Kelly *et al.*, 1981, 1982; Takeuchi *et al.*, 1982; Perria *et al.*, 1983).

The first successful steps have been taken to combine stereotaxis and laser techniques. The stereotactic technique may be used to aim a laser beam at a deep-seated brain tumor and to vaporize it. Kelly and Alker (1981) employed a CO_2 laser in six cases to destroy tumors of Th, V_3, and other deep structures, with promising results. Kelly *et al.* (1982), using the same laser connected to the Todd–Wells stereotactic apparatus, removed a small deep-seated AVA supplied by branches of the anterior and middle cerebral arteries.

It is thought that the Nd–YAG laser has advantages over the CO_2 laser in removing tough and heavily vascularized tumors such as meningiomas, large acoustic neurinomas, and tumors at the base of the skull (Takeuchi *et al.*, 1982). A fiber light guide inserted into the tube of a conventional surgical aspirator should be kept 5 mm from the tumor and without direct contact with the blood or tumor. The Nd–YAG laser has also been used for destruction of the thick floor of sella turcica in transsphenoidal operations. Kosary *et al.* (1977), employing a CO_2 laser, bloodlessly removed a huge skull osteoma. The first successful attempts have been made to use argon or CO_2 lasers to destroy the dorsal roots entry zone in pain syndromes (see Chapter 13, Section 5.3.7) (Powers *et al.*, 1984b).

The possibility of carrying a laser beam to a pathological focus through a fiberoptic system opens up prospects for performing laser neurosurgical operations with the aid of an endoscope (coagulation of vascular plexuses, third ventriculoscopy, destruction of intraventricular tumors, etc.) It has been shown that removal of pathological foci with a laser does not cause significant perifocal edema of the brain.

A stereotactic device combined with a CO_2 laser was developed by Shelden and Jacques (Shelden *et al.*, 1980). It is produced by Coherent Company and called the Surgimetrix Stereotactic Neurosurgical System. A stereotactic ring is mounted on the patient's head. Stereotactic coordinates taken from CT scans are run through a preprogrammed computer which gives the X, Y, and Z figures; these are then applied to a phantom ring. The ring holds a probe delivery system that allows the neurosurgeon to choose the path of the electrode and to present all parameters of three-dimensional references of the lesion (tumor, cyst, and so forth).

The delivery system is then attached to the patient's ring. A hollow bullet-shaped probe with a tuliplike distal tip is introduced stereotactically into the tumor. The "tulip" is opened and the CO_2 laser is switched on. Precise manipulation of the laser beam vaporizes the tumor or other lesion. It is also possible to couple the system to the operating microscope.

At the present time, the laser technique is not used to destroy the selective "normal" structures lying deep within the brain. However, it is possible that such procedures may be done in the future when improved laser equipment is available for this purpose.

6. Remote Control Methods

These methods have a special place in the methods of local destruction of cerebral tissue. They involve the application of various types of energy "from the outside," i.e., penetrating the intact skull. In principle, they are ideal since, in general, they do not require any surgical procedures such as craniotomy. The first attempts to use remote control techniques were made in the 1960s.

6.1. γ Irradiation

In the treatment of parkinsonism, pain syndromes, and certain mental disorders, Leksell, after making stereotactic calculations, projected a powerful (280-kW) source of x-ray irradiation through the intact skull. The author called the method "stereotactic radiosurgery." In 1971 he described for the first time the "gamma gun." Its heavily shielded central body contained some 200 sources of ^{60}Co. With the head of the patient in the Leksell stereotactic apparatus (Chapter 3, Section 5.4), the coordinates of the target point are determined. Then the patient on the operating table is transferred to another room where his head is placed in a metallic helmet that contains the source of γ radiation and collimators. The operator uses a special device to transfer the coordinates of the target point to the collimator helmet. Then the surgeon selects the required number of collimators and the dose of radiation (maximum 20 krads). The γ irradiation focused on one point produces a radionecrotic lesion with a diameter up to 10×8 mm deep in the brain. Leksell has successfully used this method to destroy a few arterial aneurysms, small deep-seated AVM, pituitary tumors, craniopharyngiomas, and small acoustic neurinomas as well as to destroy deep brain nuclei to relieve parkinsonism, pain, and so on. About 1500 patients have been treated by this method (Leksell, 1987).

6.2. Proton Beam

Another promising method also belonging to the category of radiosurgery is the use of a proton beam ("proton scalpel") generated by a powerful synchrophasotron or cyclotron producing approximately 200 MeV.

A proton beam ensures absolutely bloodless neurosurgical operations. Since a proton beam cannot be precisely focused, in order to avoid damaging brain tissue lying in the trajectory of the beam, one of two methods may be employed. The first involves slow rotation of the head and body of the patient in two planes in such a way that the subcortical target is kept in the center of rotation and, consequently, receives the maximum dose of radiation. On the way to this structure, the proton beam passes through various parts of the brain, which receive only a fractional dose. The second method, which does not require rotation of the head, involves the use of a proton beam working on the principle of the Bragg peak. This peak occurs deep inside the irradiated substance at a point that can be calculated. The ionizing effect at the "terminal point" greatly exceeds the energy emitted along the path of the beam.

Considerable experience has already accumulated with the use of heavy particles for stereotactic destruction of intracerebral structures. This method is successfully being elaborated by Leksell in Sweden, Kjellberg *et al.* (1962, 1972) in Boston, and a group of Soviet scientists headed by Ruderman and Goldin at the Oncological Scientific Centre of the USSR Academy of Medical Sciences. This latter group has designed equipment for stereotactic proton hypophysectomy. The source of irradiation is the synchrotron at the Institute of Theoretical and Experimental Physics of the USSR Academy of Sciences, which produces an energy of 200 MeV.

In the first stage of the operation, the head is fixed in a special stereotactic apparatus, after which the target (hypophysis) is brought to the point where three axes intersect: the proton beam axis, the vertical axis of rotation of the instrument, and the longitudinal axis of the body of the patient. The correct alignment is confirmed by x-ray pictures. Then the patient is transferred to a rotating platform. An automatic device is activated to regulate the rotation of the patient and control the radiation dosage. Proton beam irradiation is performed during rotation of the head of the patient in two planes: around the vertical axis of the proton beam and the longitudinal axis of the body.

Good results were obtained in Boston by destroying the hypophysis with a proton beam on the basis of the Bragg peak in 97 cases of hemorrhagic diabetic retinopathy. Irradiation was produced by a powerful cyclotron at 160 MeV.

The advantages of this method are appealing, but its wide application is limited by the necessity of having such a sophisticated apparatus as a synchrophasotron, which is available only in large research centers in the field of physics.

6.3. Ultrasound

Ultrasound as a method of destruction is being more widely used for the removal of various brain tumors. In recent years efficient and convenient ultrasonic devices, for example, CUSA (Cavitron Ultrasonic Surgical Aspirator, Cooper Medical, USA) or Aloka (Japan), have been made commercially available. Isolated attempts in the past to use ultrasound to destroy subcortical structures in stereotactic surgery did not make much progress, and at the present time, ultrasound is rarely used for this purpose.

The term "ultrasound" refers to those acoustic waves with frequencies higher than the audible range of the human ear, i.e., over 20,000 Hz. Ultrasound is generated by various crystals that, when acted on by electrical energy, begin to oscillate with a resonant frequency.

Contemporary devices may generate very-high-frequency ultrasound. When focused by special lenses or reflectors, ultrasound will damage nerve tissue at the so-called "focal point." The mechanism by which ultrasound destroys cerebral tissue, although not yet precisely established, is considered to be related to a peculiar physical phenomenon called cavitation as well as (to a lesser degree) to the discharge of heat.

The possibility of destroying brain tissue by powerful focused ultrasound was repeatedly demonstrated in both experimental and clinical reports. It has been established that focused ultrasound can discretely destroy tissue deep in the brain without damaging structures in the path of the ultrasound (Fry and Fry, 1960; Fry and Myers, 1962; Zelondjev, 1968). Focused ultrasound may penetrate the brain to any previously calculated depth, so focused ultrasound, in principle, may prove to be one of the safest and most effective methods for inactivating brain tissue in stereotactic surgery.

One of the most important advantages of this method is that focused ultrasound destroys brain tissue but preserves vessels, which is important for preventing hemorrhagic complications. Moreover, a trackless lesion made by ultrasound does not cause perifocal brain edema (Fry *et al.*, 1981). In using ultrasound, there is a direct positive correlation between increased dosage and a larger lesion. Another promising peculiarity of this method is that more powerful ultrasound is required to destroy gray than white matter. In principle, this allows one to destroy pathways selectively without damaging nuclear structures. Another very interesting advantage is the possibility of producing a transitory, reversible effect on brain tissue without structural damage, making it possible to use focused ultrasound to study the functions of deep cerebral structures.

In spite of the advantages mentioned above, the use of ultrasound in stereotactic neurosurgery has a number of problems that are very difficult to overcome. Skull bones, which are of unequal thickness and contain air cavities, significantly distort and deflect ultrasonic rays. Then, bone absorbs a large amount of ultrasonic energy. In view of this, in order to focus precisely the ultrasonic beam on a target deep in the brain, the transmitter must be in direct liquid contact with the dura mater. This requires a large craniotomy (approximately 8–10 cm in diameter) to make it possible to focus the beam precisely. Attempts to minimize the size of craniotomy by decreasing the angle of dispersion of the focused beam did not improve the results, because the ultrasonic "focal point" increased in size significantly. Moreover, it is not possible to quantify the temperature at the lesion site.

At the end of the 1950s, several reports appeared on the use of ultrasound in stereotactic operations for parkinsonism, hyperkinesias, and epilepsy (Meyers *et al.*, 1959). Massive craniotomies were performed, and the operations lasted about 10 hr. Complicated and costly equipment was used, and many engineers and technicians were involved in the operation. Few such operations were done, and their long-term results were not published. Moreover, the aftereffects of ultrasound on the brain have not been sufficiently studied; certain investigators believe that the brain may be damaged by ultrasonic waves reflected from the base of the cranium. Even the methods of locating the ultrasonic lesion deep in the brain have not yet been elaborated fully. All these factors led to the discontinuance of these operations.

Ultrasound could become one of the best methods of destruction in stereotactic surgery if a method were developed for using it through an intact skull. For a number of years, Dr. Tyurina, under our supervision, has been conducting experimental and morphological studies with this objective in mind. Numerous experiments on cadavers and animals have demonstrated that with precise calculations and proper operating conditions, a designated lesion may be made deep in the brain through an intact skull. For this purpose, it is necessary to select for irradiation a part of the skull of approximately uniform (within a range of 10%) thickness and constant curvature. These parameters may be determined by CT or ultrasonic echography and by geometric methods. The center of skull curvature of the irradiated zone should lie on the central axis of the generator with an opening of optimal angle. The appropriate frequency of irradiation is calculated by a special formula. If all these conditions are observed, morphological control shows that irradiation through an intact skull causes local destructive foci (diameter 6–10 mm) deep in the brain of cadavers and animals, somewhat larger than calculated. The loss of acoustical energy during passage through bone is great, especially in old age. With considerable absorption of ultrasound by the skull, there is a danger of thermal damage not only to that bone but also to the adjacent part of the brain.

In summary, one may conclude that at the moment there is still no adequate technical advance to ensure an effective use of ultrasound as a destructive agent in stereotactic neurosurgery. Use of ultrasound in combination with stereotaxis is described in Chapter 3, Section 11.

7. Chemical Methods

In 1913 the Russian surgeon Razumovski was the first to inject alcohol into the cerebral cortex for Jacksonian epilepsy. In the years that followed, a number of surgeons made intracortical injections of alcohol in patients suffering from epilepsy or athetosis. The results obtained by these authors proved to be unsatisfactory, and the method was abandoned. At the beginning of stereotactic neurosurgery, Spiegel and Wycis (1950) again injected alcohol into subcortical structures of the brains of patients suffering from chorea and choreoathetosis with satisfactory results.

This was followed by reports by Cooper (1954, 1955) of the introduction of 96-proof (48%) alcohol into GP and VL, procedures he called chemopallidectomy and chemothalamectomy. Subsequent studies revealed that injections of pure alcohol into these nuclei close to functionally important structures frequently lead to serious complications (paresis and paralysis, temperature reactions, and mental confusion). In order to diminish the diffusion of alcohol, several authors employed various alcoholic mixtures (an emulsion of procaine in oil and wax, procaine with alcohol, a mixture of ethylcellulose and ethanol, alcohol and iophendylate).

We used the chemical method only in our initial work in stereotactic surgery (until 1962). The subcortical structure was injected with 0.3–0.4 ml of 96-proof alcohol mixed with a small volume of iophendylate to visualize the size and shape of the lesion radiographically.

Subsequent analysis of the data in the literature and of our own observations revealed a number of significant problems in using this method. In some cases, the introduction of alcohol caused a pronounced perifocal and general cerebral reaction. Then, the alcohol may spread through the intercellular channels to neighboring functionally important structures, damage to which may lead to serious complications. For these reasons, the chemical method for destruction of cerebral tissue was abandoned.

8. Mechanical Methods

The use of mechanical methods for inactivating brain tissue in neurosurgery began with the leucotome proposed by E. Moniz. The subcortical tissue was destroyed with a special leucotome consisting of a cannula with wire half-loop that could be advanced or retracted (Bertrand, 1958; Obrador and Dierssen, 1959). Rotating the tube around its axis after inserting it into the subcortical structure and extending the wire loop forms a spherical focus of destruction. The destroyed brain tissue produced by the mechanical method is usually irregularly shaped and often contains hemorrhagic areas resulting from rupture of small intracerebral vessels (Carpenter and Whittier, 1952).

The mechanical method is suitable for neurophysiological experiments on animals; however, in human stereotactic operations it has not been widely used. The "undercutting" of cerebral tissue by a loop creates the danger of rupturing cerebral vessels and producing intracranial bleeding. It is possible that "secondary" necrotic foci may develop some distance from the target structure as a result of damage to arteries supplying subcortical areas. The lack of quantitative control is also a major disadvantage. In view of these disadvantages, mechanical methods are rarely used in stereotactic surgery.

6

Parkinsonism

1. Historical Note

One hundred seventy years have passed since the English physician James Parkinson published the first detailed description of the disease that now bears his name (Fig. 98). Parkinson proposed that the disease, which he named "shaking palsy," was caused by an organic lesion of the medulla oblongata and the spinal cord.

James Parkinson's small monograph (66 pages), *An Essay on the Shaking Palsy* (1817), consisted of five chapters: (1) "Definition—history, illustrative cases," (2) "Pathognomonic symptoms examined—tremor coactus—scelotyrbe festinans," (3) "Shaking palsy distinguished from other diseases with which it may be confounded," (4) "Proximate cause—remote causes—illustrative cases," and (5) "Considerations respecting the means of cure." Today the first edition of that book is a great bibliographic rarity since fewer than ten copies exist in the world.

The book is based on long-term personal observations of six patients. The clinical picture was described briefly, but with great vividness and expression. The most typical symptom of the disease in the opinion of the author was tremor: "Involuntary tremulous motion, with lessened muscular power in part not in action and even when supported." Besides tremor, he described in detail other manifestations of the disease: stooping back, shuffling gait ("a propensity to bend the trunk forwards"), dysphagia, dysarthria, and hypersalivation. He also gave a clear-cut definition of propulsion (". . . to pass from a walking to a running pace"). The author justly emphasized that in this disease ". . . senses and intellects [are] uninjured." However, in his book Parkinson makes no mention of two important and typical symptoms of the disorder—rigidity and amimia.

Parkinson's views on the cause of the disease he described are only of historical interest. He considers the cause to be ". . . a disordered state of that part of the medulla which is contained in the cervical vertebrae" as a result of impaired blood flow "in the minute vessels" or "injury from sudden distortions." Quite possibly, this viewpoint stemmed from the fact that one of his patients had an injury of the cervical spinal cord before the onset of the disease. At the same time, he emphasizes that the nature of the morbid change and its localization in the medulla itself was still the subject of doubt and conjecture. Parkinson considered the "shaking palsy" an organic disorder of the CNS and recognized that specific therapy could not be advised until more was known of the disease. Nevertheless, he recommended symptomatic treatment, in particular, blood letting from the upper half of the neck and compounds for constipation.

In the course of the next century and a half, the study of parkinsonism had a long and complicated course. Only a few of the investigators who made significant contributions to the study of this problem are mentioned.

Approximately 80 years after the appearance of Parkinson's book, Charcot differentiated paralysis agitans from multiple sclerosis: "It would be incorrect to call paralysis a state in which there is a good retention of muscular strength." He also considered the term "agitans" inappropriate, since tremor may be insignificant or even absent in this disease, and proposed the name "Parkinson's disease" (Fig. 99). Ordenstein, a pupil of Charcot, for the first time treated this disease with belladonna alkaloids.

It is interesting to note that even at the beginning of the 20th century, many authors (Oppenheim and others) considered parkinsonism and disorders such as athetosis and chorea to be functional rather than

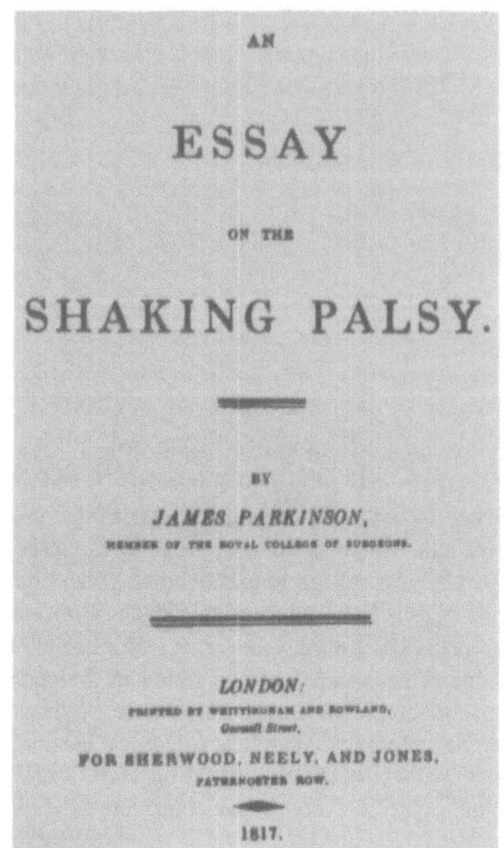

AN

ESSAY

ON THE

SHAKING PALSY.

BY

JAMES PARKINSON,

MEMBER OF THE ROYAL COLLEGE OF SURGEONS.

LONDON:

PRINTED BY WHITTINGHAM AND ROWLAND,

Goswell Street,

FOR SHERWOOD, NEELY, AND JONES,

PATERNOSTER ROW.

1817.

FIGURE 98. The cover of J. Parkinson's book, *An Essay on the Shaking Palsy* (1817).

FIGURE 99. Woman aged 62 described by Charcot in his *Lectures*. The onset of the illness was in 1868 after emotional stress. Tremor in right arm and retropulsion began in 1873. Paul Richer, who was Charcot's pupil, painted the patient in 1874 during her first treatment in Charcot's clinic at Salpetriere, Paris (A), and 3 years later (B). Note the typical parkinsonian posture of the patient.

organic disorders of the nervous system (''psychoneurosis''). Significant advances at the beginning of the 20th century were made by Foerster, Darkszewicz, Hunt, and other researchers, who confirmed the organic nature of the disease.

In 1917 von Economo published the first detailed description of encephalitis epidemica (encephalitis A). This new disease, undoubtedly of viral etiology, was recognized for the first time in 1915, during the First World War, at Verdun as well as in Austria. In the 1920s, Economo's encephalitis epidemica was pandemic in Europe with a very high mortality, up to 70%. Almost half the people who survived the disease subsequently developed clinical symptoms of Parkinson's disease (Margulis, 1940). This led to the belief that the chronic stage of epidemic encephalitis was parkinsonism. The outstanding study by Tretjakoff (1919) of its pathological anatomy demonstrated that the main morphological substrate of the disease was the destruction of the melanin-containing cells of SN.

In 1929 Critchley gave a detailed description of arteriosclerotic parkinsonism.

At the beginning of the 1950s, Spiegel and Wycis, for the first time, performed stereotactic operations on the basal ganglia of the brain, which was, in principle, a new approach to the treatment of parkinsonism.

The most important development in the treatment of parkinsonism was the use of L-dopa, which was proposed in 1961 by Birkmayer and Hornykiewicz and later elabotated by Cotzias (1968).

These have been the main historical milestones in the study of this disease. A great amount of literature has been dedicated to parkinsonism; moreover, the number of publications is steadily increasing. In Soviet literature there are two monographs devoted to this problem: the book by Dechterev published in 1927 and our monograph of 1965. The main problems of parkinsonism have been the subjects of seven international symposia held in Montreal (1959), Washington (1963), Edinburgh (1968), Zurich (1972), Vienna (1975), Quebec (1978), and Frankfurt (1981). Publications from these symposia contain a vast amount of information on the epidemiology, etiology, pathogenesis, clinical aspects, and treatment of parkinsonism.

2. Epidemiology

Parkinsonism is one of the most widespread chronic diseases of the CNS. Although at the present time typical Economo's encephalitis epidemica is encountered extremely rarely, abortive forms of the disease are often observed. The incidence of parkinsonism in recent years has been rising.

The incidence of parkinsonism in the United States 25 years ago was 0.6 per 1000 population (Kurland, 1958). Approximately 10 years later it increased to 1.1–1.15 per 1000 (Pollock and Hornabrook, 1966). The incidence of parkinsonism in European countries and North America is approximately 1 per 1000 population (Calne, 1977) or 1–4 per 1000 (Siegfried, 1978). This means that in the United States there are over one million Parkinsonian patients, and 40,000 new cases are diagnosed every year (Siegfried, 1978).

It is noteworthy that for reasons still unknown the incidence of the disorder varies substantially in different countries. Statistical data collected from literature on incidence in various countries range from 0.3 to 6.1 per 1000 (Siegfried, 1968). Surprisingly, the incidence of parkinsonism in Japan is approximately ten times higher than that in other countries.

Parkinsonism, which is a disease of elderly people, has become one of the main gerontological problems. Among people 50 years and older, the incidence of this disease increases about threefold. Approximately 2% of the population at the age of 70 and approximately 2.6% at the age of 80 suffer from parkinsonism (Kurland *et al.*, 1969). The study of the epidemiology of parkinsonism in the Soviet Union by Romenskaya (1976) revealed lower incidence figures: at the age of 40–49 years, 0.8 per 1000; 50–59 years, 1.9; and 70 years and older, 6.9 per 1000.

In the course of the past few decades the age of onset of the disease has clearly shifted towards the elderly. For example, in 1930 the average age at which the initial symptoms of parkinsonism appeared was 37 years, in 1940 it was 45 years, in 1950, 53 years, and in 1960, 61 years (Poscanzer *et al.*, 1969). Kaplan (1974) studied the age of parkinsonism cases in 110 hospitals of the United States in 1960 and 1970; the average age was 67.3 and 71.5 years, respectively.

Parkinsonism is less prevalent among blacks, possibly because of the increased melanin pigmentation (Kessler, 1978).

3. Mortality

Parkinsonism considerably reduces the longevity of man. Ten years after the onset of the disease, two-thirds of the patients have become gravely disabled or die. According to Canadian statistics, the mortality of parkinsonism per 100,000 population was 0.1 at age 25 to 44, 28.5 from 65 to 74, and 62.5 over 75 years. The overall mean annual mortality was 1.6 per 100,000 population, 84% of whom were people 65 years and older (De Jong, 1966).

Mortality among parkinsonism patients (from all causes) is three times higher than among the entire population of the same age and sex (Hoehn and Yahr, 1967). The longevity of patients with tremor as the initial symptom is somewhat greater than among those in whom the first symptom is rigidity or bradykinesia. The most frequent causes of death in such patients are bronchopneumonia and urinary tract infection. Such causes of death as trauma, peptic ulcer, and brain vascular lesions are much more frequent among parkinsonism cases than among the general population (Hoehn and Yahr, 1967). However, L-dopa treatment reduces mortality in parkinsonism (Zumstein and Siegfried, 1976).

4. Etiology

It has been generally accepted that parkinsonism can result from a number of conditions that are often difficult to identify. So far there are no trustworthy clinical or laboratory tests for determining the etiology of the disease, at least in the majority of cases.

On the basis of data in the literature and our own observations, the main causes of parkinsonism are the following:

1. Von Economo's epidemic (lethargic) encephalitis and other forms of encephalitis.
2. Cerebral arteriosclerosis.
3. Syphilis of the CNS.
4. Head injuries.
5. Intoxication (carbon monoxide, manganese, mercury, barbiturates). This group includes parkinsonism as a result of treatment with neuroleptic drugs and Rauwolfia derivates.
6. Brain tumors, mainly meningiomas.

Occasionally the parkinsonian syndrome is seen in grave forms of malaria, exanthematous fever, and recurrent typhus alimentary dystrophy.

The question of whether serious mental stress can cause Parkinson's syndrome remains open (Schwab and England, 1968; Vein *et al.*, 1981). In several of our patients this factor so clearly preceded the onset of the disease that it could not be ignored.

Data on etiologic factors in 450 of our cases are presented in Table 5.

In almost one-half of the cases it was impossible to establish the etiology of the disease. The diagnosis of "postencephalitic parkinsonism" can be considered reliably established in only one patient out of 10. In this respect our data are in accord with those in the literature. Mundinger and Riechert (1961), in a large group of patients, reported that in almost half of the cases the etiology of the parkinsonism remained unknown. Only in four out of 135 cases was the postencephalitic nature of parkinsonism established (Lieberman, 1974), and Hughes (1965) found encephalitis to precede only 9% of the cases. Hoehn and Yahr (1967) present the following etiologic data: primarily parkinsonism in 672 patients, postencephalitic parkinsonism in 96, and parkinsonism of other causes in 22 patients. In another series, idiopathic parkinsonism was noted in 86% of all cases (Alonso *et al.*, 1986).

TABLE 5. Prevalence of Etiologic Factors

Etiologic factor	Percentage of patients
Suspected encephalitis	21.4
Encephalitis	9.8
Various infections	7.7
Cerebral atherosclerosis	7.5
Head injuries	2.9
Intoxication	1.8
Etiology unknown	48.9

The discovery of slow viruses in recent years raised the possibility that these viruses might be the cause of parkinsonism. However, electron microscopy of brain structures in parkinsonism as well as tissue culture studies have not revealed any indications of a viral infection (Schwartz and Elizan, 1979). Repeated investigations of antibody levels to many different viruses in large groups of parkinsonian patients and in controls have not yielded any definite results, although the titer to herpes simplex virus was higher in parkinsonian patients (Marttila *et al.*, 1977; Marttila, 1980). One may consider that the viral etiology of idiopathic parkinsonism has yet to be proved.

In 1963 Poskanzer and Schwab proposed the "cohort hypothesis," which subsequently triggered lively discussions in the literature. The essence of that hypothesis is that all cases of parkinsonism are the consequence of von Economo's epidemic encephalitis at the beginning of the 1920s. It was presumed that encephalitis may have taken a subclinical course. The main argument of the authors is that since the pandemic of the 1920s, the average age at onset of disease has steadily increased, correlating with the number of years that have passed since the pandemic. If the "cohort hypothesis" is correct, then by the end of the twentieth century parkinsonism must totally disappear (with the exception of toxic parkinsonism) (Poskanzer *et al.*, 1969). We and other authors do not consider this hypothesis to be valid, primarily because parkinsonism, unfortunately, is not disappearing. Moreover, the majority of patients now 50–60 years old who suffer from typical idiopathic or primary parkinsonism were born after the pandemic of von Economo's epidemic encephalitis had ended.

Although almost 3% of our cases had a possible traumatic origin, the incidence of parkinsonism in another series of 1460 persons who suffered from head injuries was very low, 0.89% (Bogolepov *et al.*, 1970).

Drug-induced parkinsonism appeared about 25 years ago, after tranquilizers and neuroleptics entered psychiatric practice.

Another old hypothesis, which has long been known but which has not lost its significance, is the theory of selectively lowered resistance of the blood–brain barrier in the basal ganglia to various toxic influences. Parkinsonism caused by CO or manganese poisoning, hepatocerebral degeneration in copper metabolic disorders, and bilirubin lesions in subcortical structures related to Rh incompatibility have been cited as examples. However, so far it has not been possible to detect a toxic factor that causes lesions in SN neurons in parkinsonism.

The role of heredity in parkinsonism has been repeatedly discussed (Hassler, 1956; Petelin, 1970; Van Manen, 1974). A high percentage (23) of positive family anamnesis of Parkinson's disease was noted in the group of parkinsonian patients with the idiopathic form of the disease (Alonso *et al.*, 1986). An opposite point of view was first held by Charcot, who denied the significance of the hereditary factor. The early monograph by Dechterev (1927) presents convincing data on the absence of a hereditary predisposition to parkinsonism.

Earlier, we (Kandel, 1965) had expressed the view that the hereditary factor was insignificant. In only 5% of the patients on whom we operated were there cases of parkinsonism or other extrapyramidal lesions in their families or preceding generations.

To diagnose the etiology of parkinsonism on the basis of the clinical picture is often difficult or even impossible. A careful anamnesis still remains the best method of judging the etiology of this disease. In many cases the anamnesis combined with clinical investigations makes it possible to establish the cause of the disease with a considerable degree of certainty. However, in at least as many cases, it is impossible to establish the etiology. One frequently encounters the diagnosis "postencephalitic parkinsonism" or even"chronic encephalitis," both of which, like "the flu," are colloquial terms.

The clinical differentiation of Parkinson's disease and the parkinsonism syndrome of different etiology is difficult. In our opinion the diagnosis "parkinsonism of unknown etiology" is, in principle, more appropriate. Authors who establish the etiology of the disease in practically all cases rouse skepticism, for example, Hartman von Monakow's (1960) large series of cases in which the etiology of the disease was diagnosed in about 90% of the patients.

The distinction between "parkinsonism" and "Parkinson's disease" has no practical significance for a neurosurgeon in deciding on the indications for an operation. In contemporary literature these terms are commonly used as synonyms. Rarely the history and clinical picture may make it possible to establish an etiologic factor, in which case the term "parkinsonism" along with the appropriate attribute (postencephalitic, arteriosclerotic, etc.) is correct. However, in all other cases, the diagnosis "primary or idiopathic parkinsonism" or "parkinsonism of unknown etiology" is preferable to "Parkinson's disease" or "paralysis agitans."

Over several decades our understanding of the pathogenesis of parkinsonism underwent substantial changes. As time elapsed after the pandemic of lethargic encephalitis of 1914–1923, there was a noticeable decrease in the incidence of postencephalitic parkinsonism. On the basis of the archives of the Burdenko Institute of Neurosurgery for the period of 1935–1940, it appeared that 70% of the patients had, without a doubt, suffered from Economo's encephalitis; more than half were found to have had oculomotor crises and similar signs of encephalic disorders. In the majority of cases the first symptoms of the disease appeared before 30 years of age. However, even if epidemic encephalitis was the most frequent cause of parkinsonism 25 years ago, in recent years that etiologic factor has largely disappeared.

In order to determine the etiologic factors over the past two decades, we compared the frequencies of various etiologies in two approximately identical groups of patients operated on in 1958–1966 and in 1967–1977. The occurrence of confirmed encephalitis was reduced almost fivefold (from 16.2% to 3.4%), and head injuries 3.5-fold (from 4.3% to 1.3%). At the same time, the frequency of infections of possible etiologic significance more than doubled (from 4.9% to 11.2%), and the frequency of cerebral arteriosclerosis and intoxications did not change substantially. Primary parkinsonism, i.e., parkinsonism of unknown etiology, increased in frequency from 47.3 to 55.3%.

Many authors, however, have reported a noticeable increase in the frequency of arteriosclerotic parkinsonism and cases of parkinsonism of unknown etiology.

The differential diagnosis of idiopathic and arteriosclerotic parkinsonism offers no great difficulties when there is clear-cut clinical evidence of cerebral or general arteriosclerosis. Arteriosclerotic parkinsonism is characterized by peculiar psychic disorders: memory disorders, impaired intellect, emotional instability, and lability of affect. The disorder has its onset at an older age (after 65 years). The disease progresses swiftly in this form of parkinsonism. However, some authors doubt the existence of a specific arteriosclerotic parkinsonism and deny a cause–effect relationship between arteriosclerosis and parkinsonism (Eadie *et al.*, 1964; Siegfried, 1968; Pallis, 1971; Fahn, 1977). Indeed, one cannot but notice that in many hundreds of thousands of cerebral arteriosclerosis cases there are no signs of parkinsonism and, vice versa.

That there are no dependable criteria for the diagnosis of arteriosclerotic parkinsonism can be seen from the fact that its reported incidence varies greatly: Birkmayer *et al.* (1974) believe that its frequency is 2–4%, Selby (1967) 14%, Petelin and Koteneva (1975) 35%, and Vainshtock (1974) 59%. According to our observation (Section 4), arteriosclerosis can be considered an etiologic factor in only 7.5% of parkinsonism cases. On the basis of morphological studies, arteriosclerotic parkinsonism comprises less than 10% of all parkinsonism cases (Hassler *et al.*, 1979). It is difficult to deny the possibility that in many cases there is a coincidence of the two widespread diseases; so there was no significant difference in the frequency of clinical manifestations of arteriosclerosis in parkinsonism patients and in an age-matched control group (Marttila and Rinne, 1976).

Barbeau (1976) has proposed the theory that atrophy of the pigment-containing subcortical structures in parkinsonism, as well as in natural aging of the brain, is caused by insufficiency of special neuroendocrine cells of the hypothalamus (APUD cells). This interesting hypothesis requires further study.

Although a significant number of parkinsonian patients have antibodies to neural tissue in their blood, the possible role of autoimmune mechanisms in the etiology of the disorder is obscure.

It has been repeatedly pointed out in the literature that parkinsonism is observed with approximately the same frequency in both sexes; however, the idiopathic form is encountered more frequently in men. The higher morbidity in men is confirmed by the predominance of males operated on. For example, in our series 64% were men, and 36% women. In 1000 operations in Cooper's clinic, 67% were men, and 33% women (Waltz *et al.*, 1966). According to Hoehn and Yahr (1967), the average age at onset of the disorder is practically the same in men and women.

Summing up, one may come to the conclusion that the etiology of parkinsonism remains one of the most pressing problems of contemporary neurology.

5. Pathological Anatomy

Since parkinsonism, as a rule, is a disease of elderly people, morphological investigations frequently reveal nonspecific senile and arteriosclerotic changes in the brain. However, our main interest lies in those morphological changes that can be considered pathognomonic for this disease. Primarily, they include the degeneration and destruction of SN cells. These changes are so specific for parkinsonism of any etiology that many authors considered them the morphological substrate of the disease.

The assumption that parkinsonism might be the result of lesions in SN was first put forward by Blocq and Marinesco (1893) and Brissaud (1894), each of whom described a single autopsy case in which tuberculomas involving the SN led to the development of

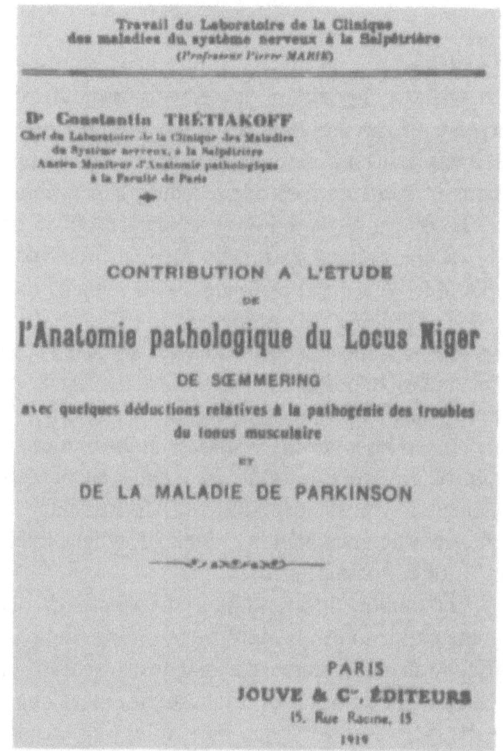

Travail du Laboratoire de la Clinique
des maladies du système nerveux à la Salpêtrière
(Professeur Pierre MARIE)

Dr Constantin TRÉTIAKOFF
Chef du Laboratoire de la Clinique des Maladies
du Système nerveux, à la Salpêtrière
Ancien Moniteur d'Anatomie pathologique
à la Faculté de Paris

CONTRIBUTION A L'ÉTUDE
DE
l'Anatomie pathologique du Locus Niger
DE SŒMMERING
avec quelques déductions relatives à la pathogénie des troubles
du tonus musculaire
ET
DE LA MALADIE DE PARKINSON

PARIS
JOUVE & Cie, ÉDITEURS
15, Rue Racine, 15
1919

FIGURE 100. The cover of Tretjakoff's (1919) book on the pathology of substantia nigra in parkinsonism.

hemiparkinsonism. Sixty years later a similar case was described by Hassler (1955).

Destructive changes in the cells of SN were first described by Tretjakoff (1919). In his brief but now classic monograph (Fig. 100), he presented a detailed description of the reduction in the number of nerve cells in SN and their lysis and phagocytosis by glial elements. Later authors confirmed the investigations of Tretjakoff and demonstrated that in parkinsonism 75–90% of the neurons are degenerated in the compact part of SN (Greenfield, 1955; Hassler, 1955; Pakkenberg, 1963). The cytoplasmic melanin lies extracellularly, and in the necrotic fields only shadows ("phantoms") are seen (Fig. 101). Accordingly, even macroscopically the SN seems to be colorless (Fig. 102).

Hassler (1955) demonstrated that in postencephalitic parkinsonism the cells of the entire SN are afflicted, whereas in shaking palsy the neuronal lesions are seen only in certain areas. The necrotic ganglionic cells are replaced by proliferating glial elements. Similar degenerative changes have been discovered in another pigment-containing nucleus of the brain, the locus ceruleus. Lesions of the GP, which are frequently described in earlier works, are less specific for parkinsonism and probably result from either encepha-

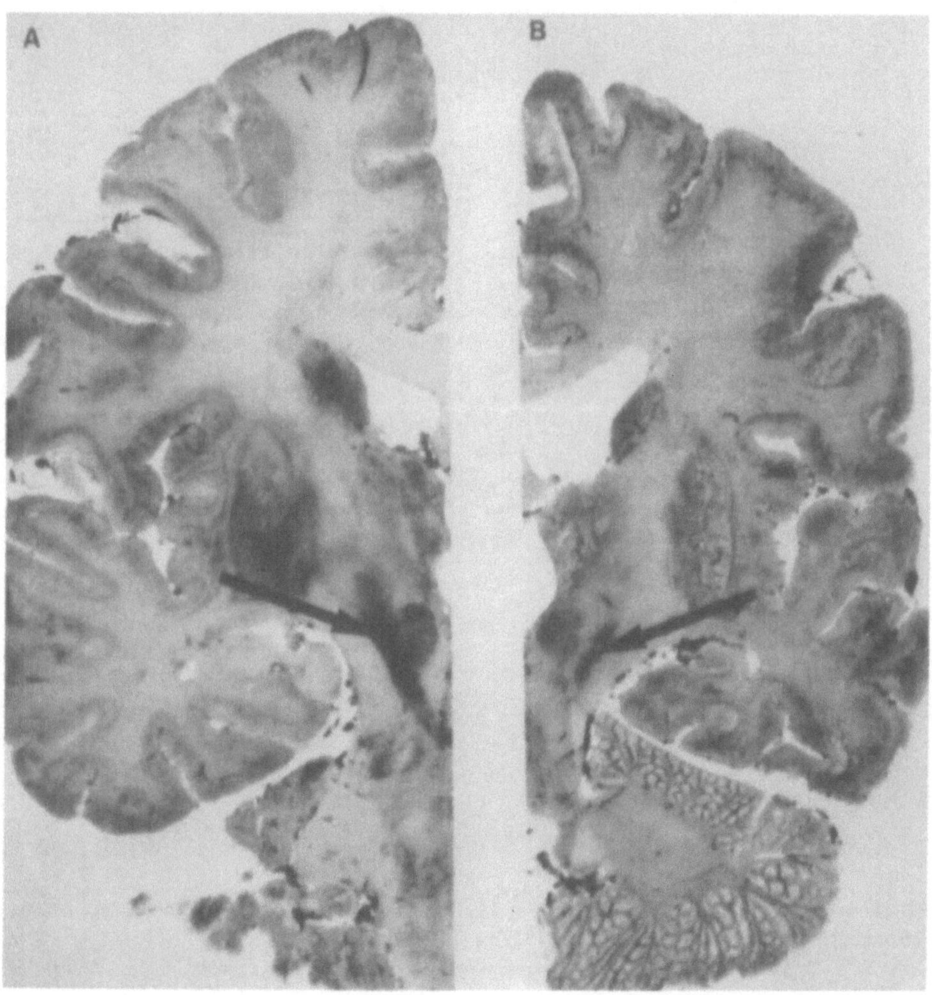

FIGURE 101. Microscopic changes in SN typical of parkinsonism. (A) Destruction of pigment-containing neurons. Melanin from the destroyed cells is eliminated by macrophagocytosis. (B) Lewy's inclusions in SN neuron (arrowheads) and destruction of the intracellular melanin (Morbus Parkinson, Sandoz, 1983).

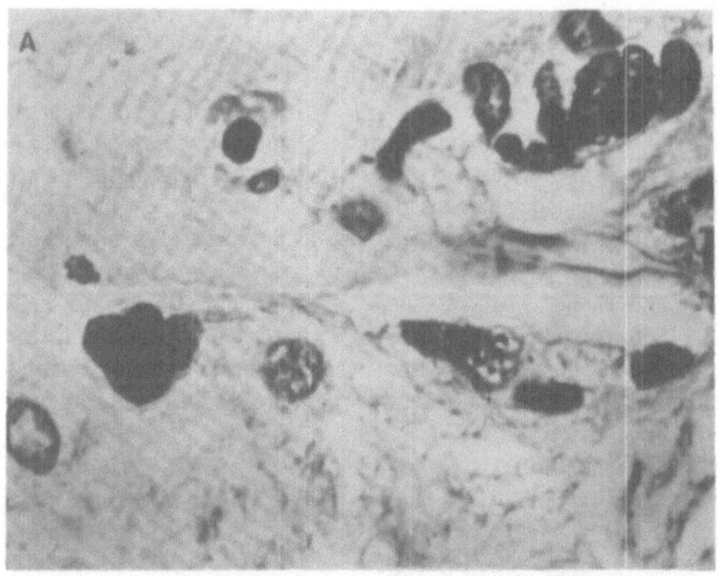

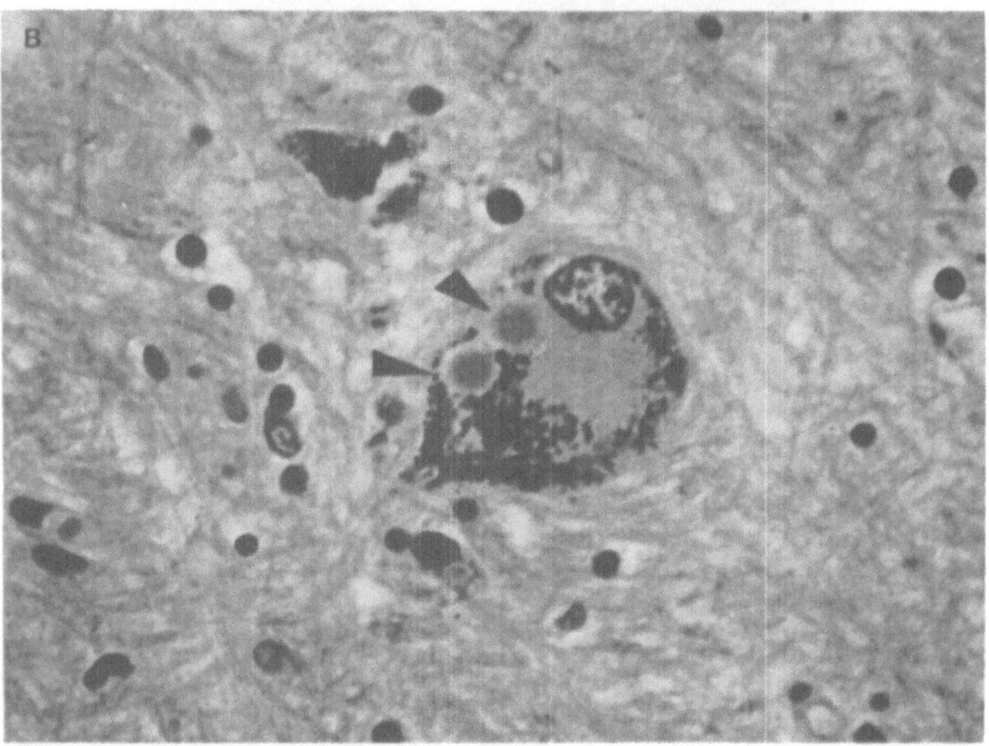

FIGURE 102. Substantia nigra destruction in parkinsonism. (A) Normal appearance of SN. (B) Marked degeneration of SN with loss of 70–80% of neurons in a parkinsonian case (Payne, 1969).

litis or senile involution. The only form of parkinsonism in which lesions in GP are prominent is that caused by CO intoxication (Hassler *et al.*, 1979).

The degenerative changes in SN specific for parkinsonism are associated with its partial destruction. A certain part of the ganglionic cells remain intact and therefore they can preserve their characteristic functions. It has also been established that even in cases of hemiparkinsonism, morphological investigations have shown bilateral lesions of SN (Davison, 1942). This may explain the steady progression of the disease with its subsequent "transition to the other side" and the development of bilateral symptomatology.

The pathogenesis of parkinsonism also includes other morphological changes of nerve cells that are almost pathognomonic for parkinsonism: Lewy bodies and Alzheimer neurofibrillary degeneration.

In brains of patients with idiopathic parkinsonism, the cytoplasm of the cells of SN, locus ceruleus, and dorsal nucleus of n. vagus (less frequently other cerebral structures) usually contains Lewy bodies, microscopic eosinophilic hyalin structures, the nature of which still remains uncertain. It is presumed that they are a product of neuronal degeneration. However, in postencephalitic parkinsonism, Lewy bodies are encountered extremely rarely. For example, Greenfield and Bosanguet (1953) found such structures in all cases of shaking palsy but in no case of parkinsonism of an encephalitic nature. These data were confirmed by Bethlem and Jager (1960). On the other hand, neurofibrillary changes in ganglion cells are found frequently in postencephalitic parkinsonism (Forno, 1966). On the basis of these cytological findings, it is possible to differentiate between idiopathic and postencephalitic parkinsonism (Hassler *et al.*, 1979).

Modern electron microscopic and histochemical investigations have given an insight into the pathomorphology of parkinsonism. Nevertheless, in the opinion of Lewis (1971), these investigations do not provide the key to understanding the nature of idiopathic parkinsonism. Actually, we still do not know the cause of necrosis of the nigral cells and whether this process is primary or secondary. As Lakke (1977) states, the cause of "spontaneous" destruction of the nerve cells in cerebral structures in parkinsonism still remains as much a mystery as at the time of James Parkinson. Nor have we fully uncovered the relationship between this process and disorders of the metabolism of cerebral catecholamines (see Section 6).

The second important pathomorphological change of parkinsonism can be called progressive diffuse brain atrophy manifested by hydrocephalus. Our monograph on parkinsonism (Kandel, 1965) and other investigations have stressed the frequent presence of more or less pronounced hydrocephalus in this disease. However, this factor has not received sufficient attention in the literature. In particular, no study has been made of the increasing hydrocephalus as the disease progresses, and no correlation has been established among the degree of hydrocephalus, the type of parkinsonism, and the duration of the disease. In this connection we have analyzed ventriculograms taken at random in the process of stereotactic operations on the basal ganglia in 105 parkinsonian patients. The etiology as well as the stage of the disease in these patients varied, and the age of the patients (20–65 years) and duration of the disease (3–20 years) covered wide ranges.

Ventriculography was performed by cannulating the anterior horn of one ventricle and injecting iothalamate or a mixture of iothalamate with iophendylate (see Chapter 4, Section 4). This process was monitored on TV screens in both anteroposterior and lateral projections.

The degree of dilation of the cerebral ventricles is the main criterion for judging the presence and severity of cerebral atrophy. In severe stages of the disease, many patients have not only hydrocephalus but also atrophy of the frontal lobes. In addition, it is usually possible to see dilated subarachnoid and cisternal spaces at the base of the brain. In order to obtain an objective evaluation of the degree of hydrocephalus, we made precise measurements with subsequent calculations of three indices previously shown (Kandel, 1965) to indicate the volume of the ventricular system

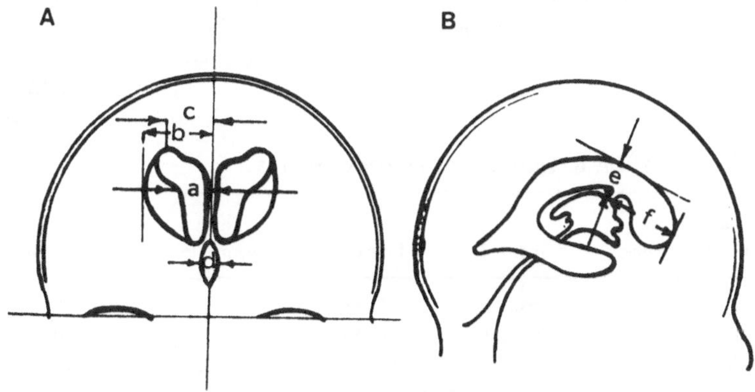

FIGURE 103. Our criteria for quantitative evaluation of hydrocephalus. (A) AP projections; a, width of anterior part of V body (distance from midplane to the "waist" of ventricle, formed by caput of NC); b, width of medial part of V body (from midplane to the apex of superior angle of the ventricle); c, width of anterior horn from midplane to the tip of the anterior horn; d, width of the anterior part of V_3. (B) Lateral projection; e, distance from posterior edge of FM to the tip of the anterior horn; f, distance from the same posterior edge to superior border of V body. Explanation in the text.

(Fig. 103). In all our measurements we used the appropriate correction factors for x-ray beam divergence.

In the anteroposterior x rays we determined:

1. The width of the anterior part of the body of V (distance from the midplane to the "waist" of V, formed by the head of NC).
2. The width of the middle part of the body of V (from midplane to the apex of its superior angle).
3. The width of the anterior horn of V (from the midplane to the most lateral point of the anterior horn).
4. The width of the anterior part of V_3.

On the lateral x rays we determined:

5. The distance from the posterior margin of FM to the most rostral point of the anterior horn.
6. The distance from the posterior margin of FM to the superior margin of the body of V.

Measurements 1, 2, and 3 were added to make index I; the sum of measurements 5 and 6 produced index II. Measurement 4 indicated the state of V_3. On the basis of this, the classification of hydrocephalus shown in Table 6 was established.

For controls, similar measurements were made on ventriculograms of 30 patients whose diseases were not accompanied by hydrocephalus (pain syndromes, spastic torticollis). After comparing the data on normal ventricles with those from 105 parkinsonian patients, we found that not one of the parkinsonian patients had ventricles of normal dimensions—all were noted to have hydrocephalus: degree I in 41.5% of patients, II in 40.2%, and III in 18.3%. It should be noted that hydrocephalus of degree III corresponds to a more than twofold increase in the volume of the ventricular system.

In certain serious cases hydrocephalus was extremely marked. For instance, in one 38-year-old female patient with pronounced rigidity in the severe, fourth stage of parkinsonism, the ventricles were so dilated that the distance from their walls to the cortex was less than 3 cm.

Our data have shown that hydrocephalus increases with the duration of the disease (Table 7). It was also clear that hydrocephalus correlates with the progression of the disease (Table 8).

From Table 8 it is evident that in stage II of parkinsonism, severe hydrocephalus (II and III) was observed in 39% of the patients, in stage III in 47%, and in stage IV in all patients.

Analysis of the data has revealed a correlation between the rate of progression of the disease and the severity of hydrocephalus (Table 9).

In "slow" progression of parkinsonism, pronounced hydrocephalus (II and III) was observed in 40% of the patients, in "average" in 57%, and in "swift" in 61% of the cases.

A corresponding analysis has shown the absence of any statistically reliable relationship between the etiology of the disease and the severity of hydrocephalus. However, the presence of extrapyramidal rigidity in the clinical picture is associated with considerably more pronounced hydrocephalus than in parkinsonism with tremor (Table 10).

Since the severity of hydrocephalus clearly correlates with both the duration and gravity of the disease,

TABLE 6. Indices for Hydrocephalus

Degree of hydrocephalus	Index I	Index II	Third ventricle
First	40–50	35–45	4–7
Second	51–60	46–55	8–11
Third	61–70	56–65	12–15

TABLE 7. Duration of Disease and Severity of Hydrocephalus

Duration of disease (years)	Severity of hydrocephalus (number of cases)		
	I	II	III
Up to 5	23	3	—
5–10	17	13	2
11–15	9	19	6
16–20	2	7	4

TABLE 8. Severity of Hydrocephalus in Various Stages of Parkinsonism

Stages of parkinsonism	Severity of hydrocephalus (number of cases)		
	I	II	III
II	27	16	1
III	24	18	3
IV	—	8	8

TABLE 9. Severity of Hydrocephalus Depending on Progression of Parkinsonism

Progression	Severity of hydrocephalus (number of cases) I	II	III
Slow	25	13	4
Average	17	18	5
Swift	9	11	3

it is obvious that as the disease progresses, hydrocephalus increases. To verify this assumption, a comparative analysis was made of the ventriculograms obtained in the first and repeated (on the other side) stereotactic operations on the basal ganglia. The interval between these two operations varied from 6 months to 5 years. Such a comparison demonstrated that in all cases the severity of hydrocephalus had increased.

The data in Table 10 confirm that as parkinsonism progresses, hydrocephalus gradually increases.

This gradual and steadfast progression of hydrocephalus can be explained only by increasing diffuse cerebral atrophy. The possibility that hydrocephalus results from CSF hyperproduction or a decrease in CSF resorption is ruled out by the finding that symptoms of increased intracranial pressure never occur in such cases.

The CT investigations in parkinsonism that we conducted in recent years fully confirm the data presented above. Atrophy of the brain was found in 60% of cases. In the majority of these cases hydrocephalus was combined with atrophy of the cerebral cortex. Similar data have been published by Becker *et al.* (1979), who carried out CT investigations in 173 parkinsonism cases. Atrophy of the brain was found in 51.4% of their cases and was quite pronounced in 20% of them.

TABLE 10. Severity of Hydrocephalus in Various Forms of Parkinsonism

Form of parkinsonism	Severity of hydrocephalus (number of cases) I	II	III
Tremor	10	5	1
Tremor–rigid	25	11	2
Rigid–tremor	13	15	4
Rigid	2	7	4
Akinetic	1	4	1

The cause of the diffuse cerebral atrophy still remains unknown. It is also unclear how a disorder of the cerebral metabolism of catecholamines leading to a dopamine deficit in subcortical structures causes a "trophic" disorder of cerebral tissue and, consequently, the development of diffuse cerebral atrophy. According to Schneider *et al.* (1977), there is no reason to consider this atrophy a consequence of cerebral arteriosclerosis.

This may even give rise to the question: Is not atrophy of the brain, manifest in hydrocephalus, the cause of parkinsonism and not the consequence? We believe that a negative reply should be given to this question. First, in the initial stages of the disease, hydrocephalus is not very pronounced. Second, with such a supposition one cannot explain why hydrocephalus is more severe in the rigid and less severe in the tremor form of the disease. Third, one should remember that in recent years a special nosological disorder has been well studied—the so-called "normal or low pressure hydrocephalus." The clinical manifestations of this disease considerably differ from parkinsonism.

Therefore, hydrocephalus is an important "secondary syndrome" in parkinsonism that aggravates the clinical manifestations of the disease. The severity of hydrocephalus may have an adverse influence not only on the progression and gravity of the disease but also on the effectiveness of surgical and medical treatment.

6. Pathogenesis

The pathogenesis of parkinsonism is a very complicated and still unresolved question which has been the subject of hundreds of investigations. One of the main difficulties has been the futility of all attempts to link the morphological substrate of the disease, destruction of the pigment-containing SN neurons, to the pathogenesis of the main clinical manifestations of parkinsonism (tremor, rigidity, akinesia, and others) (Chapter 2). The study of pathogenesis has traversed a long road mainly investigating the cause of the symptoms prominent in the clinical picture of this disease. In the study of this question, a valuable contribution has been made by stereotactic surgery.

During the past two decades, a very important scientific discovery was made that enabled one to claim that at the present time we have if not a complete explanation, in any case a foundation for a general theory of the pathogenesis of parkinsonism. It is important to stress that this achievement was not only of a theoretical nature, but also the scientific basis for a new and very effective method of treatment (see Section 7.9).

It was discovered that the basal ganglia of the brain, primarily Str and to a lesser degree SN, contain much more of the inhibitory neurotransmitter dopamine than all other structures of the CNS; 80% of the brain dopamine lies in Str (Hornykiewicz, 1971). There is also some dopamine in GP and a smaller amount in the cerebral cortex (Nyberg *et al.*, 1982).

The most intensive metabolism of dopamine takes place in Str, producing a high concentration of its final metabolite, homovanillic acid (Barbeau, 1972). This has led to the conclusion that dopamine (dihydroxyphenylethylamine), one of the basic links in the complex chain of cerebral metabolism of catecholamines, is the main motor transmitter in the extrapyramidal system.

It had already been established in experimental animals that reserpine causes a sharp drop in dopamine concentration in Str and, simultaneously, pronounced hypokinesia (Carlsson *et al.*, 1958). This suggested for the first time that there was a link between disorders in catecholamine metabolism and the pathogenesis of parkinsonism. This presumption was soon brilliantly confirmed: it was proven that in this disease, in the Str and SN, there is approximately a 90% reduction of dopamine (Fig. 104) as well of its main metabolite, homovanillic acid (Hornykiewicz, 1966a,b). The sharp reduction of dopamine in Str and SN in autopsy cases of parkinsonism, first observed by Ehringer and Hornykiewicz (1960), confirmed conclusively the important role of dopamine in the pathogenesis of this disease.

Scores of interesting experimental, biochemical, morphological, and clinical investigations published in recent years have yielded a vast amount of information on the biosynthesis and metabolism of catecholamines, serotonin, GABA, and other cerebral biogenic amines. Many links in these complicated biochemical processes have still not been fully studied, although there are already grounds to draw the basic conclusion that parkinsonism, to a considerable extent if not completely, is a disease of cerebral transmitter metabolism, giving rise to a severe dopamine deficiency in the basal ganglia.

At the same time other hyperkinesias (athetoid, choreiform, ballistic) occur not as a result of a deficit, as in parkinsonism, but because of excess dopamine in

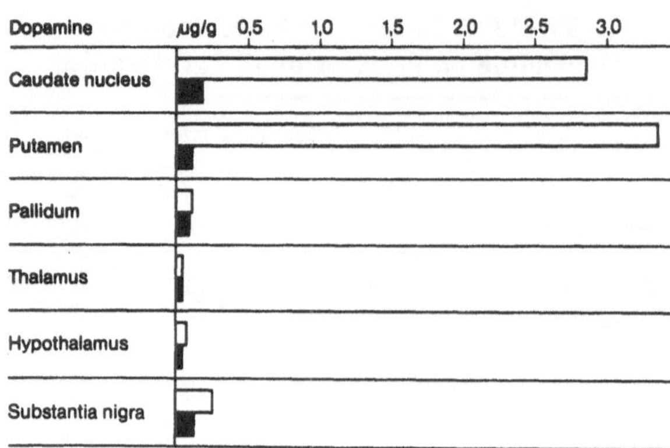

FIGURE 104. Concentration of dopamine in different cerebral structures in healthy persons (☐) and parkinsonian patients (■).

the basal ganglia. This is confirmed by the fact that prolonged intake of L-dopa very often leads to the appearance of these hyperkinesias (see below).

Cerebral metabolism of catecholamines and the basic data concerning its disorders have been outlined in detail in numerous works and monographs, which allows us to dwell on this question very briefly.

As noted above (Section 5), the main pathomorphological substrate of parkinsonism is the destruction of melanin-containing neurons in the SN and certain other structures of the basal ganglia. Melanin in the CNS (neuromelanin) is formed from catecholamines in a process regulated by a special hormone of the hypophysis. Among the catecholamines, the most important physiological role is played by dopamine, which (along with norepinephrine, acetylcholine, and serotonin) is the main cerebral synaptic transmitter in the extrapyramidal system.

It has been established that there are groups of neurons in SN producing dopamine from dopa. The dopamine is deposited in special vesicles inside the SN neurons. After reserpine is administered, these vesicles are depleted, and on introducing L-dopa into the organism, they are filled once again. These neurons have been named dopaminergic. Via the nigrostriatal pathways the dopamine enters Str neurons, where its concentration is highest. At this level the nigrostriatal dopaminergic pathway makes contacts with the GABAergic system as well as with the glutamatergic corticostriatal pathway. In normal conditions dopamine ensures activity of the dopaminergic system, which plays an important role in the regulation of motility. The local destruction of SN in monkeys causes a sharp decrease in the concentration of dopamine and norepinephrine in Str; however, this does not lead to the appearance of tremor.

The brain metabolism of catecholamines is an exceptionally complicated biochemical process, only the basic elements of which are known. In a simplified form, the pattern of this metabolism can be outlined as follows. The catecholamines are formed from the amino acids phenylalanine and tyrosine. The beginning of this multistage biosynthesis is the formation of dopa (dioxyphenylalanine) by hydroxylation of tyrosine under the influence of a tyrosine hydroxylase enzyme and its cofactor tetrahydropterin. This cofactor plays a very important role in regulating dopamine synthesis. The content of the hydroxylase cofactor in the CSF of parkinsonian patients is reduced by approximately 50% (Calne, 1983). It has also been established that tyrosine hydroxylase activity is sharply reduced in parkinsonism.

Another enzyme, dopa decarboxylase, transforms dopa into dopamine, from which norepinephrine is synthesized. Then, norepinephrine is transformed by other enzymes into epinephrine. Dopamine may be transformed in another way: under the action of monoamine oxidase it is subjected to oxidative deamination and decomposed to homovanillic acid.

These basic facts clarified the pathogenic role of necrotic SN neurons (Section 5), although the cause remains obscure. Direct correlations were found to exist among the severity of the degenerative changes in SN, the degree to which the dopamine concentration in Str decreased, and the severity of the parkinsonism (Schwarz and Fahn, 1970), in particular the rigidity and akinesia (Hassler *et al.*, 1979). The destruction of SN neurons, which are presynaptic, prevents the decarboxylation of dopa into dopamine.

Under normal metabolic conditions, dopamine entering Str from SN inhibits acetylcholine synthesis. The dopamine deficit in the basal ganglia upsets the normal balance between the main transmitter systems, dopaminergic and cholinergic, of the brain. In parkinsonism the deficit of dopamine acting on Str neurons increases their acetylcholine content, which leads to the appearance of the main symptoms of the disease (Hassler *et al.*, 1979). Significant changes occur in the dopaminergic receptors of Str, which presumably acquire denervation hypersensitivity and then, as the disease progresses, disintegrate (Wagner *et al.*, 1983).

Several hypotheses have been proposed about the cause or causes of the metabolic defect at the basis of parkinsonism: disturbed hydroxylation of tryptophan and tyrosine or their transport in the brain, disturbed synthesis of biogenic amines or decarboxylation of dopa, and so on. It has also been assumed that disturbances in amino acid metabolites play an important role, in particular, their transport at the cell membrane level (Lakke, 1977). However, these hypotheses all require experimental verification.

The dopamine deficit in subcortical structures is not only observed in biochemical investigations on autopsy material but has also been confirmed by positron emission tomography, which disclosed the depression of dopamine metabolism in parkinsonian patients (Leenders *et al.*, 1986). The authors supposed that the strio-nigral complex lost the capacity to maintain the constant dopamine concentration. The clinical determination of catecholamine levels in urine, blood, and lumbar and ventricular CSF is also important.

As a number of authors have demonstrated, the concentration of homovanillic acid in CSF, especially in ventricular fluid, where it is higher than in lumbar fluid, clearly reflects cerebral metabolism of dopamine and consequently may serve as an indicator of the functional state of the cerebral dopaminergic system. It has been found that the level of the main metabolite of dopamine, homovanillic acid, in the brain and CSF is considerably lower in parkinsonians than in healthy persons (Bernheimer and Hornykiewicz, 1965). Its concentration in ventricular fluid is only 36% of that in healthy subjects (Guldberg *et al.*, 1967). Its most pronounced decrease is usually observed when akinesia predominates; the decrease is smaller in tremor and rigidity. However, no correlation could be established between the level of homovanillic acid in CSF and the duration and severity of parkinsonism (Cunha *et al.*, 1983).

Our investigations (Barchatova *et al.*, 1981) have shown that the concentration of homovanillic acid in ventricular CSF of parkinsonian patients without L-dopa treatment was very low (average 226.8 ± 16.9 nM). In patients who received L-dopa (from 1.5–5 g/day) before the operation, this concentration increased about five times (average 1200 ± 179.9 nM). This finding confirms the increase of dopamine metabolism after introduction of its precursor, L-dopa.

In investigations of the urinary excretion of catecholamines (epinephrine, norepinephrine, dopa, and dopamine) and their metabolites, homovanillic and vanillylmandelic acids (Kandel *et al.*, 1973b), conducted jointly with Jadgarov, we confirmed that there is a significant drop in urinary excretion of dopa, dopamine, epinephrine, and norepinephrine in parkinsonism, although the excretion of the metabolites does not differ significantly from normal. These data indirectly confirm the catecholamine deficit in the basal ganglia in parkinsonism. A substantial correlation was found between reduced urinary excretion of dopamine and the severity of the main symptoms of parkinsonism (except tremor).

In two groups of people (healthy and parkinsonians), we studied the urinary excretion of catecholamines and homovanillic acid before and after a single oral dose of L-dopa (750 mg). The excretion of dopa and dopamine increased in both the healthy individuals and parkinsonians, but the increments were, respectively, 2.3 and 3.2 times greater in the former.

In the process of long-term L-dopa treatment of parkinsonism, we have also observed an increase in the sharply reduced urinary catecholamine excretion. According to our investigation (see above) and data from the literature, a similar increase in catecholamine concentration occurs in the blood and lumbar and ventricular CSF (Rinne *et al.*, 1971; Barchatova *et al.*, 1981). We observed a positive correlation between the duration of the treatment as well as the dosage of the drug and the increase in dopamine excretion. With a good clinical result of treatment, the dopamine excretion was almost twice as high as that in patients having an insignificant clinical improvement.

As our investigations (Kandel *et al.*, 1973; Barchatova and Kandel, 1984; Kandel and Yadgarov, 1984) have shown, after stereotactic operations on the basal ganglia, the reduced urinary excretion of catecholamines increases by approximately 40% and even higher in patients with a good clinical response. This observation confirms that the operation affects not only the functions of neuronal structures but also the disturbed biochemical component in parkinsonism. The destruction of VL and Subth may change the functional state of the related nigrostriatal system, which in turn leads to a normalization of catecholamine metabolism.

These recent investigations have led to the conclusion that parkinsonism is based on a primary biochemical defect in catecholamine metabolism in the brain. This biochemical defect undoubtedly lies at the basis of the neurological disorder, which previously was considered to be the main factor in its pathogenesis. Nevertheless, this question cannot be considered resolved. So far, there is no scientific proof of which defect is primary. The catecholamine metabolic disorders and positive results of long-term L-dopa treatment suggest that the neurological defect is secondary to the metabolic disorder. Nevertheless, a number of authors still consider the neurological defect to be primary (Birdsong and McKinney, 1974). They note that replacement therapy with L-dopa is often not effective and that the efficacy of such treatment gradually diminishes (Section 7.9). These authors believe that the progression of the disease, which cannot be arrested with L-dopa, is evidence of a primary neurological defect.

It is assumed that a balance among catecholamines, serotonin, and acetylcholine rather than the dopamine concentration is the predominant factor in the development of parkinsonism (Hassler *et al.*, 1979). When this balance is upset, cholinergic activity predominates. Other authors have also concluded that dopaminergic deficiency is not the only biochemical factor in the pathogenesis of parkinsonism (Hornykiewicz, 1976; Lakke, 1977). Certainly an important role is played by the complicated interrelationships among many neurotransmitter systems (dopamine, GABA, 5-HT, norepinephrine, serotonin, acetylcholine, and others).

Recent data have indicated that catecholamine metabolic disorders, in particular of dopamine, are not limited to the brain but are generalized and involve the entire organism. For example, an investigation by Barbeau (1975) demonstrated that in parkinsonism there is a substantial reduction in [^{14}C]dopamine absorption by blood thrombocytes irrespective of the severity or duration of the disease. In patients who had been treated with L-dopa, the decrease in dopamine absorption by thrombocytes was less pronounced than in patients who had not received such treatment or who had been treated with cholinolytic drugs.

A number of important questions concerning catecholamine metabolism in parkinsonism, in particular the dynamics of their metabolic changes in the process of drug and surgical treatment of parkinsonism, have still to be resolved. It remains unknown how disorders of catecholamine metabolism in subcortical structures, which represent the pathological basis of parkinsonism, lead to synchronous alternating discharges of spinal α-motoneurons producing tremor. Hassler *et al.* (1979) consider the main unresolved problem concerning the pathogenesis of parkinsonism to be the establishment of a link between the dopamine deficit in the nigrostriatal system and the activity of the thalamic nuclei, the destruction of which abolishes the main clinical manifestations of the disease.

In spite of the importance of the discovery of the role of catecholamine metabolic disorders in parkinsonism, the pathogenesis of the main clinical manifestations of this disease remains unknown. Further investigations must determine how cerebral catecholamine metabolic disorders are "translated" into the main symptoms of the disease—tremor, rigidity, and akinesia.

Another hypothesis of the pathogenesis of parkinsonism was recently advanced. It was established that in this disease there is reduced activity of the peroxidase enzyme associated with the dopaminergic system and the disintegration of free radicals. It is presumed that autointoxication with free radicals leads to the development of parkinsonism (Calne, 1983). This hypothesis sheds some light on the unexplained but statistically significant finding that the occurrence of parkinsonism is substantially lower in smokers. It is thought that the carbon monoxide entering the body in smoking neutralizes the accumulating free radicals.

7. Clinical Picture

The clinical picture of parkinsonism has been described fully and in great detail in monographs, handbooks, and hundreds of works (Davidenkov, 1922; Dechterev, 1927; Schwab *et al.*, 1959; Anosov, 1960; Denny-Brown, 1962; Doshay, 1960; Cooper, 1961; Kandel, 1965, 1975, 1981; Hoehn and Yahr, 1967; Selby, 1968; Siegfried, 1968; Calne, 1970; Pallis, 1971; Petelin and Koteneva, 1975; Ganev, 1976; Fahn, 1977; Stolyarova *et al.*, 1979; Hassler *et al.*, 1979; Ganev, 1976; Vein *et al.*, 1981; Birkmayer and Riederer, 1985). With such a vast amount of literature, there is no need to present the neurological manifestations of parkinsonism in detail. We dwell briefly only on those clinical manifestations of the disease that are important indications for surgery and are significant prognostic factors for surgical treatment; we also present data on the dynamics of certain symptoms of the disease after stereotactic operations on the basal ganglia.

7.1. The Beginning and Development of the Disease

The first manifestations of parkinsonism are of great interest since they are important indicators of the pathogenesis of the disease. The first symptom of the disease are typically either mild tremor or muscular rigidity, which initially is considered by the patient as "awkwardness," "heaviness," or "sluggishness" of the arm, leg, or both extremities.

The disease more often begins with tremor rather than rigidity. For example, tremor as the first symptom appeared in 56% of the patients, rigidity in 35.6%, and tremor and rigidity simultaneously in 8.4% (Kandel, 1965). Tremor was found as the initial symptom in 70% of the patients with primary parkinsonism disease and in more than half the patients with postencephalitic disease (Hoehn and Yahr, 1967).

The first symptom of parkinsonism to a considerable degree determines the future form of the disease. As a rule (to which, of course, there are exceptions), if the onset of the disease is manifested by tremor, the tremor or tremor–rigid form of the disease will develop. If the beginning is with rigidity, a rigid or rigid–tremor form of the disease will develop.

Tremor or rigidity in the majority of cases begins in one extremity. The initial involvement of an arm is observed, according to our data, 3.5 times more frequently than involvement of a leg (arm 71% of the cases, leg 20%, arm and leg simultaneously 9%). The cause of this sequence remains obscure. It is also difficult to explain the predominance of early involvement of the right extremities as compared to the left. According to our data, the disease began on the right side in 59% of cases, on the left side in 39.5%, and on both sides in 1.5%.

In the majority of patients the first indication of the disorder was tremor in the first three fingers or the entire hand with gradual spread to the whole arm. If the disease begins with tremor of the leg, a rare occurrence, the whole leg begins to shake. With the appearance of rigidity at the beginning of the disorder, the hand is involved first, and only gradually is the entire extremity affected.

In the majority of cases parkinsonism develops in a definite pattern. The involvement of one extremity (as a rule, an arm) spreads to the other extremity of the same side. Then the tremor or rigidity "passes over" to the arm and then the leg on the other side. Consequently, the progression of the disease in practically all patients passes through the stage of hemiparkinsonism, which undoubtedly is a favorable stage for surgical treatment. The involvement of the other side may be rapid or slow, occurring in a matter of months or years after the onset of the disease, but in the majority of cases it is inevitable in spite of drug treatment.

The steady progression of the disease is a characteristic feature of parkinsonism. Exceptions to this rule are encountered very rarely. Relatively stable clinical manifestations over a number of years have been noted in only 7% of our patients. Sigwald *et al.* (1959) found the disease to be stable or to have very slow progression in 90 out of 1700 (5%) patients in a follow-up of

10 years. Nonprogression of the disease was more often observed in young patients and did not depend on the etiology of the disorder. The reason for such a stationary condition remains unknown; it is unlikely to be related to drug treatment.

The rate of progression of the disease varies. In many patients, there is a very slow increase in tremor and rigidity in the extremities; other patients become bed-ridden invalids 2–3 years after the appearance of the first symptoms. So far, no reliable correlations have been found between progression of parkinsonism and sex, age at the onset of the disease, or other factors. If the initial symptom is tremor, the condition progresses less rapidly (Hoehn and Yahr, 1967). Post-encephalitic parkinsonism seems to progress somewhat more slowly than the idiopathic disease.

On the basis of the rate of development of parkinsonism, our cases may be divided into four groups. The group with rapid progression of the disease includes 38% of patients in whom the symptoms noticeably progressed and the state worsened every few months prior to surgery. In the group with an "average" rate (40% of the cases), the progression covered a period of 2–3 years. The third group included 15% of patients who had slowly progressing disease over the course of 5 years or more. The fourth group comprised 7% of the patients who had an arrest of the disease progression without any increase in the clinical symptoms. Consequently, in the majority of patients on whom we operated, the parkinsonism developed steadily and relatively swiftly.

7.2. Symptomatology

Any description of the neurological picture of parkinsonism must begin with its three main manifestations: tremor, muscular rigidity, and akinesia. However, these phenomena are described in detail in Chapter 2, and the clinical data on tremor and rigidity are presented in Sections 7.1, 7.6, and 7.7 of this chapter. Hence, in this section we outline other important aspects of the disease and their status after surgery.

7.2.1. Disorders of Posture and Gait

These disorders are very typical of parkinsonism. Quite often one glance at the characteristic gait of a patient is sufficient to diagnose the disease. No less typical is the posture of the patient, with the "stooping" back, somewhat lowered head, bent elbows, arms close to the sides, and slightly bent knees. Ordinary walking often requires a tremendous and at times painful effort. It is very slow and stiff; the legs literally do not lift from the floor (shuffling gait), and the steps are short and often uneven.

The absence of synkinesis of the arms (at first one, then both) during walking is one of the earliest signs of parkinsonism. A patient may move his arms in turn; however, the "automatic rhythm" of these movements is lost completely. As the disease progresses, other associated movements disappear too. We have repeatedly observed the total absence of synkinesis of the arms while walking in patients with insignificant tremor and rigidity. This disappearance of synkinesis is a very early manifestation of parkinsonism, the result of functional disturbances in the subcortical centers regulating associated movements.

A parkinsonian patient finds it especially difficult to turn while walking. Usually he must first stop, then gradually make the turn, and resume walking. Uncontrollable movements forward, backward, and less frequently to the side (pro-, retro-, and lateropulsion) are observed in approximately half of the patients. These movements result from the loss of the ability while walking to maintain a proper center of gravity, i.e., the correct positioning of the body weight over the area of support. The patient is compelled to run forward or backward, "catching up to his center of gravity."

A more or less pronounced improvement of gait, even its complete normalization, is one of the most striking effects of stereotactic operations on the basal ganglia. This improvement is difficult to evaluate quantitatively; nevertheless, we have noted an improved gait for long periods after the operation in 76% of patients. This improvement is related to a great extent to the severity of the gait disturbance prior to surgery.

The normalization of gait after the operation is primarily related to the relief of muscular rigidity. The gait becomes light, more rapid, especially in turning, the steps are longer, and there is a total or partial disappearance of pro- and retropulsion (Fig. 105). Along with the improvement in gait, there is a return of synkinesis of the arms during walking. This suggests that disturbed associated movements in parkinsonism are not caused by an organic lesion but result from inhibition of the centers controlling associated movements.

Physiotherapy promotes a rapid and full restoration of normal gait after effective surgery.

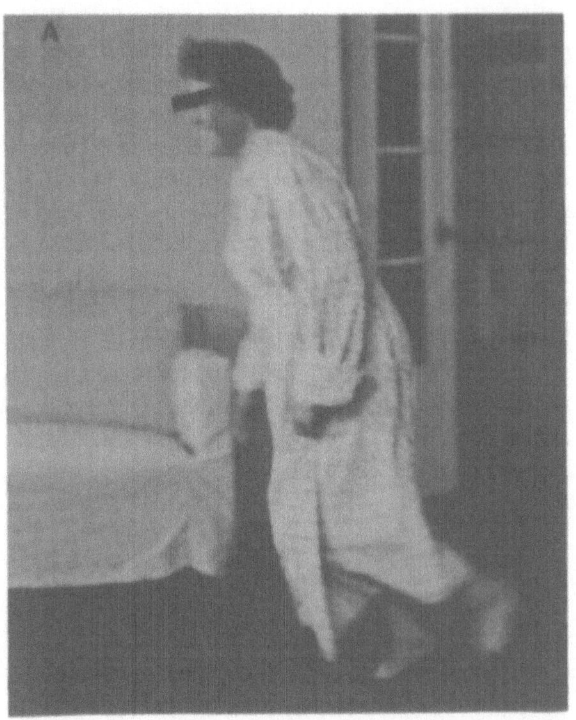

FIGURE 105. (A) Severe propulsion in 39-year-old parkinsonian patient. (B) Disappearance of propulsion and restoration of normal gait after surgery.

7.2.2. Speech Disorders

These disorders, present to various degrees, are also characteristic of parkinsonism. In the initial stages of the disease speech becomes quieter, monotonous, "dull," and lacking in emotional coloring. In stages III and IV, speech impairments are more severe: speech seems to "fade away," i.e., become quieter and less audible by the end of a phrase; quite often a patient "gets stuck" on a particular word or syllable and mumbles incoherently. In severe cases, speech becomes a barely audible indistinct whisper. In a number of our patients there was total aphonia. Sometimes there is no parallel between the degree of speech impairment and the severity of the disease—in spite of significant immobility, some patients retain a well-modulated speech.

The pathogenesis of speech disorders is complex and is determined by the interaction of a number of factors, both peripheral as well as central. Of peripheral factors, one should include rigidity of the laryngeal and respiratory muscles, tremor of the tongue and oral muscles. Disorders of the central mechanism of speech are determined by diffuse cerebral atrophy, cerebral arteriosclerosis, consequences of encephalitis, and so on.

Improvement of speech after surgery is less pronounced and occurs less commonly than, for example, the disappearance of tremor and rigidity. In the majority of cases there is no improvement in the greatly reduced speech volume after the operation. A clear-cut improvement of phonation was noted mainly in those cases in whom speech was impaired to only a mild degree prior to the operation. Nevertheless, in some patients speech improved after surgery, becoming louder, more discernible, and more "emotional."

7.2.3. Writing Disorders

The process of writing, which requires exceptionally fine motor skills, quite naturally is disturbed in the earliest stages of the disease. In parkinsonism with tremors, the handwriting becomes "shaky," "blurred," and quite often resembles a kind of "tremogram." In the rigid form of parkinsonism, the handwriting undergoes other changes—it becomes smaller, "fading" (i.e., the letters of one word become smaller), and, finally, totally illegible. It is possible to diagnose the form and stage of parkinsonism on the basis of a patient's handwriting. In stage IV of the disease, the ability to write is totally lost.

An improvement or normalization of handwriting is one of the striking results of effective surgery. After the disappearance of tremor and rigidity, the ability to write is restored rapidly even if the patient had been unable to write for many years. If the writing disorders were caused by akinesia, only a mild degree of improvement occurs after surgery.

7.2.4. Vegetative Disorders

These disorders represent frequent manifestations of parkinsonism, especially in the postencephalitic type. The most common vegetative disturbances are varying degrees of hyperfunction of the salivary, sebaceous, and sudoriferous glands. Salivation, one of the common symptoms of the disease, can often, but not always, be alleviated by drugs.

Increased perspiration is a common complaint of parkinsonians. In some (especially at night), there is such excessive sweating that the patient is wet with perspiration. Attacks of profuse perspiration may be associated with vegetative crises and are accompanied by increased tremor, hyperemia of the face, tachycardia, and emotional stress. Patients feel better in cooler surroundings and prefer to sleep uncovered. A glistening sebaceous face is typical of the outward appearance of many patients. In certain cases, the face and skin of the head are permanently covered with a layer of oily substance.

Some less frequent vegetative disorders are vasomotor lability, blushing, tachycardia, intense dermographism, protracted subnormal temperature, and a number of other disturbances. No clear-cut correlation has been found between the severity of vegetative disorders and stage of the disease. In some patients we observed pronounced edema of the legs without any heart or kidney pathology.

Our observations confirm data in the literature that vegetative disorders occur much more frequently and are noticeably more pronounced in postencephalitic than in idiopathic parkinsonism.

As parkinsonism progressed, a gradual decrease in body weight was observed in more than half of the cases. This may be explained on the basis of energy expended in tremor and rigidity and in some cases a result of reduced food intake because of chewing and swallowing disorders.

The pathogenesis of vegetative functional disorders in parkinsonism cannot be considered fully resolved. There are grounds to believe that these disorders result from lesions in Hypoth nuclei that are morphologically and functionally closely connected with the basal ganglia, primarily with GP. A more difficult question is whether the lesion in Hypoth is organic or functional. The development of parkinsonism after lethargic encephalitis, in which marked hypothalamic symptoms are observed, supports the first assumption. Nevertheless, we consider these disorders to be of a functional nature and to result from the action of other subcortical structures on Hypoth. After stereotactic operations on the basal ganglia, the vegetative disorders diminish considerably or even disappear completely in many cases, which would hardly be possible if the lesion in Hypoth nuclei were organic in nature.

7.2.5. Pain Syndrome

Pain is one of the frequent and excruciating symptoms. These pains are nearly constant and frequently more intense at night, depriving the patient of sleep. Quite often, the pain is localized in the legs, more often in the flexors and around large joints. A marked pain syndrome was observed in approximately one-third of our patients, and in half of these the pain was very intensive.

The main causes of pain are muscular rigidity, circulatory insufficiency in the muscles of the extremities, and contractures and deformities of the joints. Nevertheless, there is no direct correlation between the degree of rigidity and severity of pain.

The "pain form" of parkinsonism is one of the most favorable forms for surgical treatment. In the overwhelming majority of operated cases, the pain disappears completely, which favorably influences the patient's general condition.

7.2.6. Deformities of the Hands and Feet

Such deformities represent a relatively common finding in advanced stages of the disease. They develop gradually with progression of parkinsonism. In the initial stage, "dystonic contractures" of the hands and feet appear with hyperextension of the fingers of the hand at the interphalangeal joints and flexion of the toes with the big toes assuming valgus positions.

In severe cases, the dystonic positions become gross deformities of the hands and feet outwardly resembling rheumatoid deformities. These deformities, as a rule, are observed in the rigid–tremor and rigid forms of parkinsonism but are usually absent in the tremor form.

In the past it was presumed that such deformities and contractures were the result of secondary changes in the joints and ligaments apparatus and hence irreversible. This supposition was not confirmed. In many cases, we observed a rapid and total disappearance of these deformities, often immediately after surgery. Deformities of the hands and feet of many years' standing may disappear over the course of several hours after the destruction of VL and Subth, leading to the normalization of muscular tension. These observations leave no doubts that the deformities are caused by rigidity and, in particular, uneven hypertonicity in the minor muscles of the hands and feet; the joints and ligaments remain relatively intact even after such deformations have existed for many years.

The total disappearance of hand and foot deformities represents one of the most gratifying results of stereotactic operations on the basal ganglia.

7.2.7. Mental Disorders

Although Parkinson (1817) wrote "intellect being uninjured," a considerable number of patients, especially those with disease of long standing, display mental disorders varying from slight loss of memory and emotional lability to severe dementia. The most significant mental disorders are found in arteriosclerotic parkinsonism.

The etiopathogenesis of the psychoorganic syndrome is determined by the interaction of many factors. Besides changes with age, an important role is played by atrophy of the cerebral cortex and hydrocephalus which gradually progress with the advance of the disease. Mental disorders frequently occur or become aggravated due to prolonged L-dopa treatment.

7.2.8. Vestibular Disorders

One of the elements in a comprehensive evaluation of a parkinsonian's condition is an otoneurological investigation. In most of our 150 cases who had examinations both before and after surgery, we noted an absence of spontaneous nystagmus. Disorders of hearing, taste, olfaction, and optokinetic nystagmus were observed rarely.

It is worth dwelling in greater detail on induced nystagmus since the data obtained seem to be of interest and the question has received little coverage in the literature.

The standard caloric test was carried out: 60 ml of water at 25°C and 19°C was introduced consecutively into both ears while the induced nystagmus was observed. In almost 80% of the cases, vestibular excitability was diminished or completely absent. Depression of the vestibular response is usually observed on both sides; an asymmetric reaction is quite rare. In approximately 20% of the patients the vestibular excitability was within the normal range or slightly higher. During the caloric testing many patients displayed pronounced vegetative reactions (nausea, perspiration, reddening of the face, tachycardia, and so on). No clear-cut correlation was established between the stage of parkinsonism and the severity of otoneurological dysfunction.

The fact that there is a significant normalization of vestibular excitability in patients after surgery on the basal ganglia is quite interesting. Such an improvement was noted in 55% of the patients operated on (Kandel, 1965). It is noteworthy that after a unilateral operation an improvement in caloric reactions may occur on both sides. All these data indicate that destruction of VL or Subth removes the inhibiting influence of these (or other) subcortical structures on the vestibular nuclei.

7.2.9. Glucose Metabolic Disorders

We made a study of glucose metabolism in 62 parkinsonians before and after stereotactic operations on the basal ganglia. Since this question has not been addressed to any great extent in the literature, we will dwell briefly on the results obtained.

The blood glucose concentration was determined in parkinsonians 15, 30, 60, 90, 120, and 150 min after oral intake of 50 g of glucose. Analysis of the glycemic curves was based on the following indices: (1) initial level; (2) hyperglycemic coefficient; (3) carbohydrate absorption coefficient; (4) the time required for the glucose concentration to return to the initial level; and (5) the type of glycemic curve.

The resulting data showed that the majority of patients (78%) had pronounced glucose metabolic disorders, which were more frequent and greater when the clinical stage of the disease was more severe. More gross disorders were observed in the rigid form of the disease. The data imply that the pathogenesis of glucose metabolic disorders in parkinsonism is related to the pathological influence of the basal ganglia on Hypoth nuclei regulating carbohydrate metabolism.

After effective operations on the Th nuclei, there was a substantial normalization of carbohydrate metabolism in the majority of patients. One may presume that destruction of V1 and Subth reduces the pathological influences of these structures on Hypoth.

7.2.10. Respiratory Disorders

It has long been known that parkinsonian patients frequently suffer from various respiratory functional disorders—rapid or irregular breathing, reduced pulmonary ventilation, and so on. Attacks of respiratory distress are seen more frequently in postencephalitic parkinsonism. There are grounds to believe that this symptom is an indication of a previous encephalitis. The frequency and duration of these attacks can range from several minutes to 2–3 hr.

The chief causes of respiratory disorders in parkinsonism include restricted motility of thorax as a result of rigidity of respiratory muscles, hyperactivity of the parasympathetic system, reduced sensitivity of chemoceptors to CO_2, and lesions in brainstem structures.

From the point of view of surgical treatment of parkinsonism, it is very important to evaluate the state of respiratory functions. Respiratory complications after the operations, especially in elderly and weak patients, considerably increase the risk of surgical treatment and unfavorably influence its outcome. Riechert (1980) considers respiratory disorders the main postoperative complication and emphasizes the importance of making an objective preoperative investigation of pulmonary functions. In our clinic we made a detailed study of respiratory function in 32 patients with various forms of parkinsonism (Kandel, 1981). Of these cases (average age 57 years), four were in stage II of the disease, 15 in stage III, and 13 in stage IV. In the majority of the patients the investigations were repeated 2–3 weeks after a stereotactic operation on the basal ganglia.

On the basis of spirogram recordings the following parameters were calculated: respiratory frequency, respiratory volume, minute volume, vital capacity, Tiffeneau test, 1-sec forced expiration, maximal ventilation, oxygen consumption, respiratory coefficient, and oxygen consumption coefficient. To evaluate the respiratory response of CO_2, the bell chamber of the spirograph was filled with a hypercapnic mixture (6.8% CO_2 in oxygen), the CO_2 absorber was switched off, and the spirogram was recorded for 2.5 min to calculate the ventilation increase.

The O_2 and CO_2 content in the inhaled, exhaled, and alveolar air were determined with a gas chromatograph. On the basis of these data we calculated alveolar ventilation and the value of dead space by the Bohr formula.

With the aid of a gas analyzer (capnograph) working on the principle of selective absorption of infrared rays by carbon dioxide, the CO_2 concentration was monitored in various phases of the breathing cycle. Residual volume, functional residual capacity, dead space, and regularity of pulmonary ventilation were studied by the nitrogen clearance method in an open system using an azotograph and gasometer. From these data, the lung clearance index was calculated.

All values were related to standard conditions and compared to normal values of respiratory indices, which were calculated individually for each patient on the basis of anthropometric data and basal metabolism figures. In certain patients the gas content of arterial blood was determined using Clark and Severinghaus electrodes.

These investigations revealed that with a relatively normal rate of breathing, the minute volume is significantly increased (on average by 40%), and O_2 consumption is also substantially higher (on average by 39%). The functional and respiratory reserves in these patients were noticeably reduced. For example, the vital lung volume was decreased (on average) by 27%, maximal ventilation by 17%, the Tiffeneau test was 63% (75% being normal), and the ventilation response to breathing of the CO_2 mixture increased only 200% (300–400% being normal). Respiratory response to inhalation of the hypercapnic mixture showed that the ventilation volume in these patients was below normal.

All these respiratory changes probably result from the clinical effects of the disease (rigidity, tremor, vegetative hyperactivity, and so on), which cause a greater expenditure of energy for breathing, increased O_2 consumption and pulmonary ventilation, and reduced functional capacities and reserves for breathing.

Repeated investigations of respiration after an effective stereotactic operation demonstrated that the rate of breathing did not change substantially, but the increased minute respiratory volume dropped (on average) by 9%, i.e., toward the normal level. There was also a normalization of O_2 consumption, which diminished after the operation by an average of 24%. The vital lung capacity increased slightly, and the Tiffeneau test resulted in an average increase from 63% to 66.3%. There was a substantial (up to 350%) increase in ventilation on inhalation of the CO_2 mixture.

All the respiratory parameters normalized noticeably after a clinically effective operation and often returned to normal levels. The improved respiratory dynamics in parkinsonism may be explained by the clinical effect of the stereotactic operation—reduction or disappearance of tremor, rigidity, vegetative disorders, and so on.

7.2.11. Oculogyric Crises

In approximately 6% of our patients we observed oculogyric crises characterized by paroxysmal forced rotations of the eyeballs upwards or (less frequently) to one side. Not infrequently these crises were accompanied by blepharospasm. Besides oculogyric crises, other symptoms typical of postencephalitic parkinsonism were convergence paresis and disappearance of the accommodation reflex. The frequency and duration of these crises may differ; sometimes they are repeated several times per hour. This symptom is extremely distressful for the patient since it deprives him (usually for several minutes) of his vision. Oculogyric crises represent a pathognomonic sign of postencephalitic parkinsonism.

As a rule, this symptom disappears after a stereotactic operation on the basal ganglia. Kalyanaraman and Gillingham (1965) point out that the most effective operation for oculogyric crises is bilateral stereotactic destruction of the posterior limb of CI. We have repeatedly observed total disappearance of this phenomenon after unilateral or bilateral destruction of VL or Subth.

7.2.12. Sleep Disorders

A history of a prolonged period of marked sleepiness or (less frequently) insomnia represents a pathognomonic symptom of postencephalitic parkinsonism.

Investigations of the sleep–wake cycle in parkinsonians have demonstrated defective functioning of nonspecific integrative brain structures. The most typical sleep disorder is a pathological decrease in activating influences. The sleep mechanism suffers most in patients with an akinetic form of parkinsonism. Disorders were found in both "slow" and "rapid" sleep (Golubev, 1974).

7.3. Electroencephalography

Electroencephalography studies in parkinsonism have not revealed any changes that are specific for this disease. The most frequent alterations are a slowed α rhythm, reduced amplitudes of biopotentials, and diffuse, not sharply expressed δ waves (Ganglberger, 1961a,b; Kandel, 1965; Toth and Tomka, 1968). In almost half of the cases, no EEG deviations from normal were observed. The EEG reaction to hyperventilation also did not differ from normal in the majority of cases. No correlation was found between EEG changes and the stage, duration, or etiology of parkinsonism (Arkhipova and Illinsky, 1972). Scalp EEG investigations did not make it possible to differentiate between parkinsonism and Parkinson's disease (Grindel *et al.*, 1962).

After a stereotactic operation on the basal ganglia, the majority of patients showed slowing of the α rhythm over 2–3 weeks; the rest of the patients had no EEG changes (Ganglberger, 1961a,b; Kandel and Pokrovskaya, 1972). Slower alpha rhythm is, in all likelihood, the direct result of the destruction of thalamic nuclei which are known to participate in regulating electrogenesis of the cerebral cortex.

7.4. Electromyography

On the basis of our investigations using multichannel EMG, patients with different forms of parkinsonism may be divided into three groups (Kandel *et al.*, 1974). The first group includes patients in whom (1) resting EMG activity was absent, (2) postural tremor occurred while maintaining a pose with burst activity at 5 or 10 Hz, (3) during slow movements there was spike activity (action tremor), though modulated both in frequency and amplitude, and (4) during rapid movements up to the speed threshold, the EMG showed continuous instead of burst activity.

The second group included patients in whom burst activity, usually at 5 Hz, was observed at rest (rest tremor). The amplitude of this tremor in maintaining a posture and in active movements was much higher than in the first group.

In patients of the third group, there was weak continuous activity at rest that increased in maintaining a posture. Slow movements resulted in burst activity, just as in patients of the first and second groups. At a certain speed threshold this EMG activity became continuous (as in the first group), although the speed threshold was considerably higher than in the other groups.

7.5. Cerebral Blood Flow

Investigations have shown that cerebral blood flow (CBF) in parkinsonism is lower than in normals of comparable age. Many parkinsonians with marked hydrocephalus were found to have a substantial (up to 40%) decrease in CBF as measured by the Kety and Schmidt method. Employing a noninvasive technique (^{133}Xe inhalation), Bés *et al.* (1983) demonstrated that hemispheric CBF in parkinsonians was decreased by 10.6% compared to young individuals and by 6.7% compared to individuals approximately 70 years old. Reduced CBF may be explained by diffuse cerebral atrophy (Section 5), lesions in extracranial arteries, and cerebral arteriosclerosis.

7.6. Forms of Parkinsonism

The different forms of the parkinsonian syndrome have long been known. It has not been possible to establish a clear-cut correlation between the form of parkinsonism and its etiology. Nevertheless, in patients with tremor and tremor–rigidity, encephalitis (or suspected encephalitis) in the history was almost twice as frequent as in patients with the rigid–tremor and rigid forms (Kandel, 1965). As our experience has shown, the form of the disease is quite stable. In no case did we see a transition from one form to another. Without a doubt, different pathogenic mechanisms lie at the basis of different forms of parkinsonism. The identification and classification of these forms are of interest from the surgical point of view, since the results of operations on the basal ganglia depend on the type of parkinsonism. We believe that the most rational classification is into four types (Kandel, 1965).

7.6.1. Mixed Type

This class is characterized by presence of both tremor and muscular rigidity. In spite of the arbitrariness of comparing tremor and rigidity, in practice it is not difficult to determine which of these symptoms is predominant in the clinical picture of the disease. On this basis, it is expedient to divide the mixed type into tremor–rigidity and rigidity–tremor. This mixed form of parkinsonism is encountered most frequently—it was observed in 77.5% of our operated cases.

The mixed type is most frequently accompanied by the classical picture of parkinsonism. Tremor in the extremities is usually constant, of small amplitude and average intensity, of the "pill-rolling," "coin-counting" type. In purposeful movements, this tremor gradually diminishes, but in static tension states it noticeably intensifies. Tremor of the head, lips, and tongue occurs quite rarely. Muscular rigidity determines the typical posture of the patient and characteristic disorders of gait.

7.6.2. Tremor Type

This type is characterized by a constant marked tremor with an insignificant increase in muscular tension and little impairment of movement. This form is encountered less frequently—in our series it was observed in 9.5% of patients.

In this type there is a predominance of severe tremor of great amplitude and low frequency. Even at first glance, this "coarse" tremor, most pronounced in the distal segments of the extremities (especially the hands), differs noticeably from "fine" tremor, which is typical of the mixed form. The trembling quite often spreads to the head, tongue, and oral muscles, and at times there is a continuous violent tremor of the entire body of the patient.

In this form of the disease there is usually little or no general rigidity or bradykinesia. Speech and gait are affected to a lesser extent. Unlike other forms of parkinsonism, such patients find self-care relatively easy; however, fine and precise movements are almost or totally impossible because of the violent tremor (the patient cannot eat by himself, especially liquid food; he cannot bring a glass of water to his mouth or write).

A distinguishing feature of this form of parkinsonism is a high-amplitude violent tremor that can be inhibited for several seconds during purposeful movements. One of our patients had such a violent continuous tremor of his hands that it was difficult to believe his claim that he could shave himself with a straight razor. However, the patient would give his hand a powerful flick of the wrist, thereby inhibiting the tremor for several seconds during which he rapidly ran the razor over his face. Thus, a few flicks of his wrist allowed him to shave.

In static states such as while holding the arms extended, tremor is markedly intensified. Since rigidity is insignificant, there are usually no contractures or deformities of the small joints. Anatomic disorders are encountered less frequently than in other forms.

7.6.3. Rigid Type

The rigid type is distinguished by increased tone of the neck, torso, and limb muscles with a pronounced cogwheel rigidity. In the "pure" form, the rigid variant is encountered comparatively rarely—it was observed in 10% of our cases. In this form of parkinsonism, tremor in the extremities is insignificant and not constant.

Muscular rigidity disorganizes the activities of the patient. All movements become more difficult, which often creates the impression of paresis of feet and hands. In extreme cases of the disease, even passive movements are practically impossible, and the examiner has to exert great effort in order to flex the arms or legs of the patient. In still more severe cases, the extreme rigidity of the entire body leads to contractures in the joints, especially of the fingers and radiocarpal and talocrural joints. Gross deformities of the hands and feet develop, making movement impossible. The uneven distribution of tone leads to pronounced disorders in posture—kyphosis or scoliosis, or bending the torso forward while walking.

A peculiar feature of the rigid form is the akinetic component. In certain cases, it is even possible to speak of a rigid–akinetic form, which further accentuates the motor impairments.

Whereas in the tremor form of parkinsonism purposeful movements are usually impaired by a tremor that diminishes with movement, in the rigid (akinetic) form of parkinsonism the opposite occurs: a mild trembling in the hands appears or is intensified with any kind of action, for example, in the finger–nose test (action tremor).

The rigid type undoubtedly is the most severe and troublesome form of parkinsonism, causing the patient tremendous mental and physical suffering. In this form, to a great degree, the patient loses the ability to take care of himself and is soon totally immobile. The face becomes a fixed, frozen mask. Speech defects (barely audible whispers or even total aphonia) intensify the helplessness of the patient. The frequent pain in the rigid muscles (especially at night) aggravates the patient's suffering.

In rigid–tremor and rigid forms of the disease certain patients experience peculiar annoying feelings that they find difficult to describe in words. These, without a doubt, are not muscular pains or sensations of tightness (numbness) of the muscles, which often occur in parkinsonism, but rather a feeling of "motor anxiety." Such patients, if they cannot rise by themselves, ask to be turned in bed or helped from their chair and in 15–20 min again ask for help to be seated.

7.6.4. Akinetic Type

Akinesia is a peculiar motor disorder characterized by a marked decrease in the speed of active (purposeful) movements (Chapter 2, Section, 5). Charcot (1876a) noted that in a number of patients undoubtedly suffering from parkinsonism, trembling was insignificant or absent. He called these cases "*formes frustes.*" In earlier literature one often encounters the term "paralysis agitans sine agitacione" to describe cases in which there was total absence of tremor in parkinsonism.

The akinetic form of the disease is characterized by total absence of tremor and rigidity. Besides tremor, vegetative disorders, and muscular rigidity, akinesia is a very typical and prominent manifestation of parkinsonism. However, the severity of this phenomenon in different forms of the disease is quite variable. In the tremor form there is practically no akinesia. It is usually more apparent in the rigid–tremor form and, as a rule, is quite pronounced in the rigid form.

On examining a patient with a pure akinetic form, one's attention is drawn to the absence of tremor in the extremities although the appearance is typical of parkinsonism. Muscular tone is not markedly increased, the patient only very slowly and with great difficulty performs the simplest motions; he cannot look after himself, get up, sit down, or walk.

Results of surgical treatment demonstrate that rigidity and akinesia may exist independently of each other. After surgery, muscular rigidity quite often totally disappears, but the akinetic syndrome remains. Further proof arises from experience with the use of certain pharmacological preparations in psychiatric practice. After such treatment, the pure akinetic syndrome may develop without signs of rigidity.

It is quite easy to recognize akinesia if the muscular tone is not markedly increased but considerably more difficult to evaluate when rigidity is present. In the majority of cases the disappearance of rigidity after surgery has demonstrated that akinesia was insignificant, since the fine, rapid movements of the extremities are restored. However, in a number of cases even total eradication of rigidity did not lead to a significant improvement of motility, since the rigidity had masked pronounced akinesia.

The pure akinetic form of the disease occurs comparatively rarely; of 802 parkinsonians, absence of tremor was found in only 10%, and absence of rigidity in 10% (Hoehn and Yahr, 1967). The akinetic form was noted in 3% of the patients on whom we operated. In none of these patients was there a history of past encephalitis. Thus, an etiologic factor in the akinetic form of parkinsonism was rarely identified.

Although we are still far from fully understanding

the neural bases of tremor and muscular rigidity, considerable progress has been made (Chapter 2). Unfortunately, this cannot be said about akinesia, which remains one of the most mysterious phenomena in the pathology of the motor system. Nevertheless, the satisfactory results of surgical treatment for certain akinetic forms of parkinsonism have shown that akinesia is not always an irreversible defect of the central regulation of motility.

The classification we have adopted of four clinical forms of parkinsonism proved to be very practical for the purpose of selecting patients for surgical intervention and for assessing prognosis. A different type of classification for parkinsonians, suggested by Ojemann and Ward (1971), divided them into two groups, one with symptoms involving "midline structures" (torso muscles for maintaining postures and walking, amimia, tremor of tongue and lips) and the other with predominant involvement of the extremities. The authors considered the first group more suitable for treatment with L-dopa and the second to be candidates for stereotactic surgery.

7.7. Stages of Parkinsonism

The establishment of the stage of parkinsonism (i.e., the severity of the disease) in each patient is of great practical significance in assessing the indications for surgery and evaluating its efficacy. In our experience, determination of the severity of the disease on the basis of its main manifestations (tremor, rigidity, etc.) not only lacks objectivity, but in some cases is difficult. For example, a patient with severe but irregular tremor may, in certain circumstances, be able to cope with his work and fully care for himself, whereas an akinetic patient with insignificant muscular rigidity without tremor may be a helpless, bed-ridden invalid. One should not judge the stage of the disease by its duration. Certain patients become total invalids within 2–3 years after the onset of the disease, whereas others are ambulatory 15–20 years after the appearance of the first symptoms of parkinsonism.

It is possible to distinguish four stages of parkinsonism on the basis of the patient's functional abilities (Kandel, 1965). The classification of the disease depends on the abilities of the patient to perform his job and to carry out everyday activities.

Stage I corresponds to the insidious onset of an irregular and mild tremor in one arm or leg that is intensified by emotions and disappears when attention is distracted. The patient may inhibit the tremor voluntarily or by changing the position of the extremity. A mild impairment of one arm or leg is manifested by a slowness and awkwardness of movement. Associated movements of one arm during walking may be lost. There are no changes in gait, speech, or facies. The general state remains satisfactory. The patient is able to continue working and can look after his needs.

Stage II, as a rule, is the stage of hemiparkinsonism, but there may be early signs of involvement of the contralateral extremity. Tremor and rigidity on the affected side are pronounced. Tremor is almost continous and has the typical features of tremor at rest. It is no longer possible to inhibit tremor voluntarily. The gait is noticeably disturbed, and while walking the patient "pulls up" his leg. All movements are slow, especially in the affected extremities. The patient experiences a mild general impairment. After awakening in the morning, rigidity and tremor are less pronounced than later in the day. The voice becomes dull, vegetative symptoms appear, and there may be dull pain in the affected extremities.

The working capacity of the patient becomes limited. Individuals performing manual labor are compelled to switch to lighter work. Others performing mental labor usually retain their working capacity. There is a noticeable impairment of personal care, and all activities required in everyday life are performed slowly and with difficulty, often with the "healthy" arm.

Stage III of the disease is characterized by bilateral involvement and the typical clinical picture of parkinsonism. The habitus of the patient enables one to make the diagnosis immediately. The coarse or fine trembling of all extremities (especially in the tremor form) is continuous and stops only during sleep. The patient is unable to inhibit this exhausting tremor. Even during emotional activity or inattention, the intensity of the tremor changes very little. There is often a tremor of the head, lower jaw, and oral musculature. Rigidity is pronounced and involves not only the extremities but also the muscles of the torso and the neck (especially in the rigid form). Voluntary movements of the hand, feet, fingers, and toes are limited. Cogwheel rigidity is readily apparent.

The general impairment is quite significant, and all movements are noticeably slower. The gait is grossly disturbed consisting of small steps with practically no lifting of the feet from the floor, and the absence of synkinesia of the arms.

Usually the patient can walk a few dozen meters very slowly, but at times walking without assistance is impossible. The patient must even be helped to rise from a chair. In walking the patient experiences an irrepressible desire to move forwards, backwards, or to the side (pro-, retro-, and lateropulsion). The masklike face of the patient is almost devoid of expression; the voice is quiet, slow, monotonous, "dull," and without emotional modulation. Quite often toward the end of a sentence the speech "fades out," and it is impossible to understand the patient. Sometimes he seems to "get stuck" on one syllable and makes multiple repetitions.

There are often pronounced vegetative symptoms: salivation, profuse perspiration, and vasomotor lability. Many patients are troubled by constant dull pain in the rigid muscles, especially at night. Because of this the patient is compelled to change position frequently in bed. The patient's character and interests change: he retreats within himself, concentrates on his emotions, loses interest in his surroundings, stops reading, and avoids meeting unfamiliar people.

In this stage of the disease, working becomes impossible, and even personal care is limited. The life of the patient is restricted to his home or even to his room. If not constantly, he requires someone's assistance at least several times a day. Only with great difficulty can the patient rise slowly from a chair or sit down. He cannot turn from one side to the other in bed and finds it difficult to dress and wash without assistance.

Stage IV is characterized by practically total immobility. The bed-ridden patient, as a result of prolonged and irregular muscular hypertonus, has deformities of the fingers, hands, and feet as well as contractures of joints, especially the ankles and wrists. Speech becomes so faint that it is difficult to understand. In this stage, quite often, as a result of trophic and metabolic disturbances, the general health declines, and at times cachexia ensues, aggravated by difficulty in swallowing.

The masklike face and incoherent speech isolate the patient from the surrounding world (Fig. 106). Unable to express his thoughts, emotions, and feelings about his grave condition, the patient at this stage of the disease is confined not only to his room but to his bed and chair. All everyday activities (eating, washing, dressing, and so on) are lost, and the patient, unable to rise, sit down, or even turn in the bed, is totally incapacitated. The need for constant assistance and care is the most typical feature of this stage of the disease.

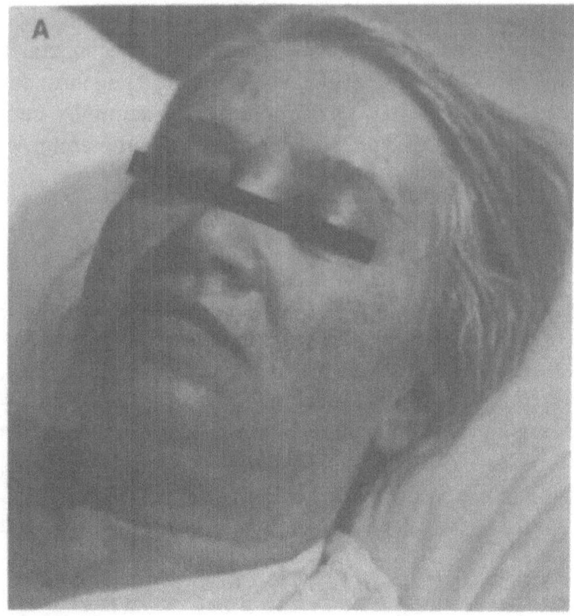

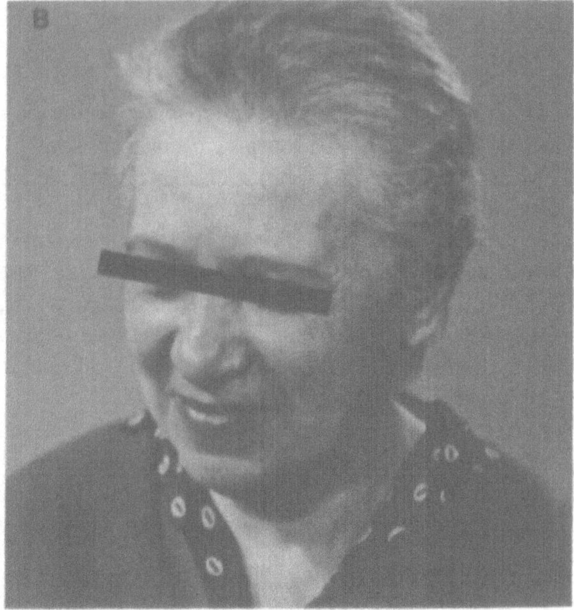

FIGURE 106. (A) Masked face in bedridden patient with rigid form of severe parkinsonism. (B) The same patient 2 years after bilateral thalamotomy.

This classification of parkinsonism in stages naturally does not attempt to include all the numerous variants of the clinical picture.

7.8. Diagnosis

Parkinsonism is one of the very few diseases of the nervous system the diagnosis of which usually does not present difficulty. For an experienced physician quite often it is sufficient to take a look at the patient. However, in the initial stage of the disease diagnostic difficulties may be encountered, and subsequent observations are necessary to confirm the diagnosis of parkinsonism.

In some cases it is necessary to distinguish early parkinsonism from essential tremor, pallidal degeneration, the tremor form of hepatocerebral degeneration, as well as certain other, rarer syndromes (the Shy–Drager syndrome, Creutzfeldt–Jacob disease, progressive supranuclear paralysis). Our experience has shown that if diagnostic difficulties are encountered in the initial stages of the disease, a study of the motor activity will serve as a reliable method to establish the diagnosis (Chapter 2).

Based on our experience (Kandel *et al.*, 1974), at least one of the following EMG phenomena must be present for the diagnosis of parkinsonism: (1) the presence of burst activity, (2) a change in frequency of bursts from 10 to 5–6 Hz, and (3) increased EMG amplitude. After successful stereotactic surgery, all of these phenomena disappear.

Differentiation between idiopathic (Parkinson's disease) and atherosclerotic parkinsonism is straightforward only when there are clear-cut clinical symptoms of cerebral and general atherosclerosis.

In atherosclerotic parkinsonism, the incidence of which has considerably increased in recent years, there are peculiar mental disturbances such as memory loss, decreased intellect, and emotional instability. An onset at an elderly age (after 60 years) is usually an indication of an atherosclerotic etiology.

The distinction between "parkinsonism" and "Parkinson's disease," although undoubtedly of definite interest from the etiologic point of view, is not of great practical significance for a neurosurgeon since it does not alter the indications for surgery. In modern literature these terms are usually used as synonyms. If the history and clinical findings allow one to establish the etiologic factor, the term "parkinsonism" prefixed by the causative factor (postencephalitic, arteriosclerotic, etc.) may be appropriate.

7.9. Medical Therapy

The treatment of parkinsonism with pharmacological agents has a history of over 100 years and a voluminous medical literature. We dwell on it briefly since in order to evaluate the indications for surgery, the neurosurgeon must know the results of drugs treatment.

7.9.1. Anticholinergic Agents

In 1867 Ordenstein and in 1887 Charcot proposed treating parkinsonism with belladonna alkaloids, the active component of which is atropine. Beginning with the 1940s and for several decades, the main trend in the medication of parkinsonism was the use of synthetic anticholinergic preparations (trihexyphenidyl, biperiden, etc.), sometimes in combination with antihistaminic, neuroleptic, and other drugs. The action of anticholinergics is to suppress central acetylcholine activity and to "neutralize" excess acetylcholine, which is a stimulating transmitter in Str, rather than inhibiting dopamine entering that structure. Undoubtedly, prolonged administration of these preparations gives a certain degree of relief to a considerable number of patients, and cholinolytic agents remain an important component in medical therapy of parkinsonism. However, over the course of many years patients gradually adapt to these preparations, substantially lowering their efficacy, and increased dosages are often accompanied by pronounced side effects. Moreover, patients may react adversely to these drugs. In addition, cholinolytics neither prevent nor retard the progression of the disease.

7.9.2. L-Dopa

The most significant development in the medication of parkinsonism is without a doubt the use of L-dopa (levorotatory isomer of dihydroxyphenylalanine). Treatment with this preparation is based on new concepts of the pathogenesis of the disease (Section 6). In the brain, L-dopa is rapidly transformed into dopamine, thus compensating for the deficit of this motor transmitter in the synapses of subcortical structures, primarily Str (Rinne *et al.*, 1971). That is why L-dopa treatment gives considerably better results than all the drugs used earlier.

The therapeutic value of L-dopa was discovered in 1961 simultaneously and independently by Birkmayer and Hornykiewicz in Vienna and by Barbeau in

Montreal. After their investigations, L-dopa therapy evolved through three main stages. In the first stage (at the end of the 1960s), treatment was conducted with small doses and resulted in a short-term effect. The second stage began in 1967, when Cotzias (1968) proposed treatment with large doses of L-dopa, which, as a rule, provided a significant and long-lasting therapeutic effect. The need for large doses of L-dopa is explained by the fact that a large part of the administered preparation is transformed into dopamine by the enzyme decarboxylase before it reaches the brain. The third stage was the combined use of L-dopa and an inhibitor of dopa decarboxylase, which substantially enhanced the efficacy of treatment.

A precursor of dopamine, L-dopa, unlike dopamine, penetrates the blood–brain barrier. One or two hours after intake of L-dopa, its concentration in plasma reaches its maximal level, and the high L-dopa plasma concentration returns to its initial level 4 hr later (Rinne, 1981). Dopaminergic neurons in subcortical structures capture the dopamine in the basal ganglia, thus restoring the disturbed balance between the cerebral dopaminergic and cholinergic systems.

At the present time scores of reports on the high efficacy of L-dopa in parkinsonism have been published (Yahr *et al.*, 1969; Anden *et al.*, 1970; Rinne *et al.*, 1971; Kandel *et al.*, 1973d; Birkmayer *et al.*, 1974; Barbeau, 1976; Hoehn *et al.*, 1976; Petelin *et al.*, 1977; Stolyarova *et al.*, 1979; Yahr, 1979; Vein *et al.*, 1981; Birkmayer and Riederer, 1985; Lakke *et al.*, 1980; Kandel and Yadgarov, 1984). It has been established that L-dopa is most effective in the akinetic form of parkinsonism, substantially reducing the muscular hypertonicity, but has no significant effect on tremor. There is a general opinion that this drug can control tremor only in exceptional cases. Further, L-dopa does not influence drug-induced parkinsonism because the responsible drugs block dopaminergic receptors in Str.

L-Dopa has substantially reduced the mortality attributed to Parkinson's disease. Whereas prior to L-dopa, the mortality was three times greater than that of the general population, it has dropped to 1.5 times after L-dopa treatment. The quality of life for the parkinsonian patient was improved (Yahr, 1976).

Severe decompensated dysfunctions of liver, kidneys, and heart as well as psychoses and psychoneuroses are contraindications to L-dopa treatment.

If there are no doubts as to the diagnosis, then as a rule each parkinsonian should be started with L-dopa. At the beginning of treatment there is usually a noticeable improvement. However, this therapy must be individualized and given under a physician's control. It is usually begun with small doses (250–500 mg per day), which are gradually increased. The maximum daily dose should always be established individually depending on the effect, tolerance to the drug, and the severity of side effects. The average daily dose is 3.5–4 g; however, in certain cases, it is increased to 6–8 g. The treatment is continued indefinitely, since after discontinuance the therapeutic effect disappears completely over approximately a fortnight.

In order to study catecholamine metabolism in healthy individuals and in parkinsonism, we investigated urinary catecholamine excretion in two groups (healthy persons and patients who had not been given L-dopa previously) before and after a single dose of L-dopa (750 mg). After such a dose the catecholamine concentration in the urine sharply increased, but to different degrees in healthy individuals and parkinsonians. In the former the excretion of dopa increased (on average) six times, that of dopamine 63 times, and that of norepinephrine 2.5 times; epinephrine excretion did not change. In parkinsonians dopa excretion also increased sixfold, but dopamine excretion increased only 28 times, and norepinephrine nine times. Since the initial levels of these substances were different in both groups, the absolute increments are significant. After a single dose, the absolute increase in urinary dopa excretion of healthy individuals was, on average, 192 mg/day. The absolute increment of dopamine in healthy individuals (average 11.4 mg/day) was 3.2 times greater than in the parkinsonian patients (average 3.5 mg/day).

After a single dose of L-dopa, the maximum urinary excretion of dopamine and norepinephrine occurred during the first 3 hr. There was also a marked increase in the excretion of the main metabolite of dopamine, homovanillic acid, but to an equal degree in both healthy individuals and patients. The excretion of the main metabolite of norepinephrine and epinephrine—vanillylmandelic acid—after administration of 750 mg of L-dopa did not change either in healthy individuals or in parkinsonians.

In the process of long-term L-dopa treatment, the renal excretion of catecholamines, which is considerably decreased in parkinsonians, sharply increases. For example, dopamine excretion increases approximately 40 times, dopa 15 times, and norepinephrine 2.5 times; only epinephrine excretion changes insignificantly (Kandel *et al.*, 1973d; Kandel, 1976a).

During L-dopa treatment with a decarboxylase inhibitor, dopamine excretion increased to a lesser degree, on average by almost 20 times, whereas dopa excretion greatly increased, approximately 90 times (six times higher than with L-dopa alone). Norepinephrine excretion increased fivefold (double that when L-dopa was used alone), and epinephrine excretion fourfold.

We compared data on urinary catecholamine excretion in untreated parkinsonians with the same results following long-term L-dopa treatment. It appears that when the clinical effect of treatment was good, renal excretion of dopamine was almost twice as high as when the effect was satisfactory or minimal. Dopa excretion, on the contrary, declined when there was a good clinical result. Excretion of epinephrine and norepinephrine did not change irrespective of the clinical effect. These facts permit certain preliminary conclusions. With a good clinical result, the brain dopaminergic structures are "saturated" with dopamine, and consequently dopamine excretion is elevated. In the absence of such an effect, the dopamine excretion tends to be decreased. These conclusions seem to confirm our data on the dependence of catecholamine excretion on the daily dose of L-dopa, since it was demonstrated that an increase in dosage leads to a proportional increase in the urine excretion of all four catecholamines.

Our data have also shown a positive correlation between the duration of L-dopa treatment and high dopamine excretion. For instance, compared with patients who received L-dopa for 1 month, dopamine excretion during treatment for 3 months increased 1.5 times, and threefold during treatment for 6 months. If the daily dose of L-dopa was increased during treatment, there was a statistically significant increase in the excretion of dopa, dopamine, and norepinephrine. In patients who received 2 g of L-dopa a day, the excretion was two to three times less than when the dose was 4 g per day (Kandel, 1976a).

Since an increase of catecholamine concentration was noted in both blood and CSF, one may assume that during treatment there is also an elevation of dopamine concentration in brain tissue. We have shown (Barchatova and Kandel, 1984) that the homovanillic acid concentration in CSF of parkinsonians who had not received L-dopa is only one-fifth that of patients receiving L-dopa (average 226 and 1200 nM, respectively). At the same time, no correlation was found between the level of dopamine metabolites in CSF and the clinical findings (Rinne *et al.*, 1971). One might have expected a correlation between the severity of parkinsonism symptoms and the L-dopa content of the plasma. However, it appears that there is no such correlation, nor does the L-dopa level in plasma correlate with the clinical improvement. Yet it does correlate with various side effects (vomiting, hyperkinesia, etc.) (Rinne, 1980).

As experience was gained in the treatment of parkinsonism with large doses of L-dopa, it became apparent that there were significant shortcomings. L-Dopa does not prevent progression of the disease since it is only a replacement and not a remedial therapy. Since dopa very rapidly decarboxylates in many organs, only an insignificant amount ever reaches the brain. This is the reason that there is no or only a short-term therapeutic effect in many patients.

The second important shortcoming of L-dopa lies in its various side effects, which are practically inevitable if it is used for a long time. The most frequent are the "on–off" and "wearing-off" phenomena, nausea, vomiting, mental impairments, hallucinations, memory loss, excitation, unstable postures, cardiac arrhythmias, sympathoadrenal crises, orthostatic hypotension, and anorexia. Several side effects may be observed simultaneously in many patients.

The frequency (up to 80%) of quite severe side effects in L-dopa treatment has been noted by most authors (Barbeau, 1975; 1976; Marsden, 1977, 1980; Stolyarova *et al.*, 1979; Yahr, 1979; Kandel, 1981; Pederzoli *et al.*, 1983). Special mention must be made of the extrapyramidal hyperkinesias resembling dystonia or choreoathetosis, which are extremely burdensome for the patient. Dystonic symptoms range from little jerks or orofacial-lingual movements to severe dystonic spasms of the extremities, torticollis, and bending of the trunk. Often there are myoclonic attacks of rapid intermittent twitching of the muscles of the extremities and the torso. Involuntary movements develop after a relatively long period of treatment in approximately 30–40% of cases. Frequently "peak dose dyskinesias" occur 1 to 1.5 hr after each dose; less often dyskinesias appear early in the morning or as the effect of the drug is wearing off (Shaw *et al.*, 1980). It is assumed that myoclonic twitchings and other hyperkinesias result from elevated serotonin activity triggered by L-dopa. The daily fluctuations of symptoms or "on–off" phenomena are caused by L-dopa accumulation during the course of treatment and gradually become more frequent. Rinne (1984) states that fluctuation of symptoms after 3–5 years of treatment was observed in 31% of patients, after 5–7 years in 50%, and after 10–12 years in 84% of patients.

The severity of side effects, particularly the "on–

off'' phenomenon, may be diminished somewhat by deprenyl, a MAO inhibitor (Birkmayer *et al.*, 1974; Rinne, 1982). In a number of cases, side effects may be lessened if the patient is not given L-dopa 2 days a week (weekly drug holiday) (Goetz *et al.*, 1983). If the side effects cannot be corrected by reducing the dosage, then L-dopa treatment should be discontinued.

The next important problem is adaptation to the drug, i.e., a gradual diminution of its therapeutic effect. To minimize this effect, L-dopa has been given in combination with a peripheral inhibitor of dopa decarboxylase (preparations include Sinemet®, Madopar®, and Nakom®) the enzyme that converts L-dopa into dopamine (Barbeau, 1972; Zumstein and Siegfried, 1976; Petelin *et al.*, 1977; Rinne, 1980, 1981; Kandel, 1981; Vein *et al.*, 1981). Because this enzyme is present in all tissues of the organism, a large part of the L-dopa decarboxylates in the body, and only 5% of it reaches the brain (Calne, 1977). The inhibitor suppresses dopa decarboxylase only in the periphery (heart, kidneys, liver), but L-dopa decarboxylation in the brain does not change appreciably because the inhibitor does not penetrate the blood–brain barrier.

Two inhibitors of peripheral (extracerebral) decarboxylation are currently used, benserazide (Ro 4-4602) and carbidopa, the former being more active. The ratio of L-dopa to carbidopa (Sinemet®) is 10 : 1 and to benserazide (Madopar®) 4 : 1. A combination of 150–200 mg of L-dopa and 50–100 mg of the inhibitor boosts the L-dopa plasma concentration to the same degree as would be produced by administration of 1000 mg of L-dopa alone (Rinne, 1980).

The suppression of L-dopa decarboxylation in the periphery elevates its concentration in the blood, thereby enhancing the penetration of the drug into the brain, where it is converted to dopamine. This makes it possible to reduce substantially (by more than half) the dose of L-dopa and diminish the frequency and severity of side effects resulting from catecholamine synthesis in the periphery.

The average daily dose of L-dopa with an inhibitor is 2–2.5 g. The common scheme is as follows: first week, 0.5 g per day; second, 1 g; third, 1.5 g; fourth and fifth, 2–2.5 g. After a good effect is obtained, the daily dose may be decreased. Even when L-dopa is combined with an inhibitor, the development of side effects may require that the treatment be discontinued. For example, of 178 parkinsonians, only 80 patients were continuing such treatment after 6 years (Shaw *et al.*, 1980).

Barbeau (1976), after many years of treating patients with the akinetic form of the disease, concluded that large doses of this preparation are most effective. A noticeable improvement was observed in only 53% of the cases, and L-dopa was unable to arrest the progression of the disease.

Several years later, after L-dopa preparations began to be used extensively for treating parkinsonism (approximately 1976–1978), many authors noted that the effect of this treatment was gradually diminishing. The decline in the efficacy after many years of L-dopa usage is explained by the gradual destruction of the presynaptic dopamine receptors in Str as the disease progresses. On the other hand, the morphological investigation of the postmortem NC in parkinsonian patients treated with L-dopa did not disclose the reduction of D_2 dopamine receptor density (Guttmann *et al.*, 1986). The authors believe that the diminished clinical effect of L-dopa after its long administration is not due to receptor drop-out but depends on other factors. Another possible explanation may be denervation or drug-induced hypersensibility of the synaptic apparatus participating in the genesis of dyskinesia. In view of this, a number of authors have recommended that L-dopa not be given in the initial stages of the disease, when the symptoms are mild or even moderate (Fahn and Bressman, 1984), and that its use be postponed until progression of the disease makes it absolutely necessary.

At the present time there seems to be no doubt that the effectiveness of L-dopa is lost completely some time after the beginning of treatment. There are some differences of opinion concerning the duration of that "effective period." Some authors consider that L-dopa loses its effect in 3–5 years (Kelly and Gillingham, 1980; Kelly, 1983), others an average of 8 years (Rinne, 1981). In the author's experience, the condition of the patient returns to the initial (pretreatment) level after 7–8 years. In rapid "malignant" progression of parkinsonism, the efficacy of L-dopa begins to decline within a year after the beginning of treatment, and in 4 years its efficacy is completely lost (Rinne, 1982). Accordingly, one may conclude that L-dopa does not affect the progress of the disease or the loss of neurons in SN. The "L-dopa era," about which so much was written at the beginning of the 1970s, proved to be very brief indeed.

A number of studies showed that L-dopa treatment prolongs the life expectancy of parkinsonians (Barbeau, 1976; Zumstein and Siegfried, 1976; Marttila and Rinne, 1976). However, Rinne (1982) demonstrated that in the first 4 years of treatment, the slight decline in mortality may be explained by the advanced stage of the disease in many patients. In subsequent years of treatment the decline in mortality is substantial, almost to the level of the general population. However, after 10 or more years of treatment, the mortality rapidly increases to the level of the first years of treatment. Shaw *et al.* (1980) found that if L-dopa was continued for a long (more than 2 years) time, the mortality was substantially reduced.

An attempt has been made to treat severe parkinsonism by repeated intraventricular administration of increasing doses of dopamine (Venna *et al.*, 1984). In an elderly patient suffering from parkinsonism for many years, in whom treatment with Sinemet® and bromocriptine caused grave mental disorders, dopamine (up to 16 mg) was injected over the course of several months into both cerebral ventricles via Ommaya reservoirs implanted in both anterior horns. Although this treatment lessened bradykinesia, such complications as marked mental disorders, myoclonias, and dyskinesias developed, and the dopamine administration was discontinued.

7.9.3. Dopamine Agonists

In 1887 Charcot proposed treating parkinsonism with ergotamine (an ergot alkaloid); more recently, a number of ergot derivatives have been used to treat this disease. Their mechanism of action differs significantly from that of L-dopa. They directly stimulate the dopaminergic receptors in Str; i.e., they are dopamine agonists. By acting on the postsynaptic link of the dopaminergic system in Str, which is undergoing progressive degeneration of the dopaminergic receptors, these drugs compensate for the endogenous dopamine deficit.

Compared to L-dopa, the dopamine agonist ergot alkaloid bromocriptine (2-Bromo-α-ergocryptine mesylate), which is a strong suppressor of prolactin secretion regardless of the cause, has a number of advantages: it is not transformed by decarboxylase, the level of which is decreased in parkinsonism, and it does not produce active or toxic metabolic products. The half-life of bromocriptine (5–6 hr) is considerably longer than that of Ldopa (2 hr), ensuring a longer therapeutic effect.

In recent years bromocriptine, most often in combination with L-dopa, has been extensively used in treating different forms of parkinsonism (Lieberman *et al.*, 1976a–c; Gerlach, 1976; Parkes *et al.*, 1976; Caraceni *et al.*, 1977; Lees *et al.*, 1978; Jansen, 1978; Fahn *et al.*, 1979, 1980; Yahr, 1979; Rinne, 1980; Schneider and Fischer, 1982; Jellinger, 1982). In many cases this preparation lessens tremor, rigidity, bradykinesia, and disorders of gait. Bromocriptine noticeably reduces the frequency and severity of the "on–off" phenomenon and rarely causes hyperkinesia. Bromocriptine also exerts a potent antidepressant action, which is desirable in the treatment of parkinsonism. The side effects of the drug are nausea and orthostatic hypotension, which may occur upon initiation of treatment.

Although bromocriptine has been shown to be less effective than L-dopa with an inhibitor, it can potentiate their action. Bromocriptine is indicated in those cases in which L-dopa produces no effect or in which an initial effect has diminished significantly or in which L-dopa causes pronounced side effects (Parkes *et al.*, 1976; Lieberman *et al.*, 1979). In a number of cases, it has been possible to reduce the dose of L-dopa or to completely replace it with bromocriptine. Moreover, bromocriptine has a therapeutic effect in approximately 20% of the patients who previously had received no medication.

Recently, there has been a tendency to treat parkinsonism with large doses of bromocriptine, up to 60 or even 100 mg per day. Bromocriptine causes side effects approximately as often as L-dopa, but involuntary movements and the "on–off" phenomenon occur less frequently. Orthostatic hypotension and mental disturbances are more frequent with bromocriptine, which also may cause diplopia, ergotism, and pedal edema in large doses (Calne, 1978).

Other dopamine agonists that also activate the dopaminergic receptors include lergotrile, pergolide, and lisuride. These preparations significantly reduce tremor in monkeys after destruction of the ventromedial portion of the midbrain tegmentum. Because the effect of lisuride does not depend on dopamine synthesis, the drug may be effective in cases of degeneration of dopaminergic receptors. The efficacy of lisuride is approximately the same as that of bromocriptine (Le Witt *et al.*, 1983).

Pergolide, a synthetic dopaminergic agonist (maximal daily dose approximately 6 mg), gives a definite therapeutic effect in cases of L-dopa incompatibility, but its effect declines noticeably in 6 months (Goetz *et al.*, 1983). "On–off" fluctuations diminish under the influence of pergolide. The efficacy of pergolide and bromocriptine is approximately the same

(Le Witt *et al.*, 1983), although some authors believe pergolide to be more effective (Lieberman *et al.*, 1978).

With an average daily dose of 50 mg per day, lergotrile reduces the L-dopa requirement by only 10%. In approximately two-thirds of the patients, treatment with this drug had to be discontinued because of side effects (Lieberman *et al.*, 1979).

The use of these dopaminergic preparations in managing parkinsonism has given promising results (Lieberman *et al.*, 1975; Rinne, 1982); however, their efficacy has not been determined. One other stimulator of dopamine receptors—piribedil (ET-495)—was tested in few groups of patients. Biochemical investigations have demonstrated that this preparation reduces the breakdown of dopamine and serotonin in the brain. However, its clinical efficacy was not very great (McLellan *et al.*, 1975).

Dopamine agonists have approximately the same side effects with the same incidence as L-dopa plus inhibitor (Rinne, 1982). In almost half of the patients, bromocriptine treatment had to be discontinued because of intolerance (Lieberman *et al.*, 1979). Since the side effects of dopamine agonists are the result of their peripheral dopaminergic activity, it is recommended that these drugs be prescribed together with peripheral dopamine antagonists. There are also indications that the prolonged intake of dopamine agonists depletes the dopaminergic receptors. Furthermore, these drugs may have a toxic effect on the liver.

Clinical tests have recently been carried out on a new preparation, L-*threo*-DOPS which compensates for the cerebral norepinephrine deficit in parkinsonism (Narabayashi, 1983a). In spite of the initial favorable effects from the use of these new drugs (bromocriptine, piribedil, and so forth), the long-term results have been quite limited.

7.9.4. Amantadine and Deprenyl

One medical treatment of parkinsonism is the use of the antiviral preparation amantadine (Midantan®, Symmetrel®, Verygit®). Its therapeutic effect was discovered accidentally by Schwab and England (1969), and in subsequent years it was thoroughly studied by many authors (Parkes *et al.*, 1970; Kadikov, 1973; Mankovsky *et al.*, 1974; Petelin *et al.*, 1977).

An important advantage of amantadine is that it is not toxic and practically causes no side effects even after many years of administration. Amantadine has practically no contraindications, although special caution is required in case of kidney impairment and mental disturbances. The average daily dose of the drug is 200–300 mg. It is recommended that amantadine be used in combination with L-dopa and cholinolytics. This drug undoubtedly has a therapeutic effect in rigid and akinetic forms of the disease and to a lesser extent in tremor. On the whole, the effectiveness of amantadine is considerably less than that of L-dopa.

Based on our observations, good results in prolonged amantadine treatment were obtained in one-third of the patients, satisfactory in one-third, and practically no effect in the rest. These observations suggest that if there is no effect after 2 weeks of this treatment, it may be discontinued. The fact that amantadine possesses a clinical action similar to that of L-dopa, and both preparations potentiate each other when combined, suggests that the mechanism of amantadine action is related to its effect on catecholamine synthesis in the basal ganglia. However, in our investigations (Kandel *et al.*, 1973b), urinary catecholamine excretion before and after amantadine treatment was not significantly different. Because of this, we conclude that the therapeutic action of the drug differs from that of L-dopa but remains obscure.

Some good results have been obtained in recent years in the treatment of parkinsonism with deprenyl, an MAO-B inhibitor, which is used as an additive agent with other medication.

The effectiveness of a new drug, budipine, was investigated recently by Jellinger and Bliesath (1987) in a double-blind trial of 31 parkinsonian patients. The additional administration of the drug with L-dopa with inhibitor gave good results on tremor and less pronounced effects on bradykinesia and rigidity.

7.9.5. Postoperative Medication

Postoperative treatment of parkinsonism, when a stereotactic operation has been performed because of poor response to L-dopa or undesirable side effects raises several important questions. Should L-dopa be continued postoperatively? How successful will such treatment be? How does the operation change the side effects of medication? The accumulated experience offers answers to these questions.

Our investigations (Kandel *et al.*, 1973d; Kandel and Yadgarov, 1984) have shown that renal catecholamine excretion (dopamine, dopa, epinephrine, norepinephrine) increases only slightly after stereotactic operations, which is an indirect indication that the dopamine deficit is still present in the basal ganglia. These data point to the advisability of continuing drug treatment after stereotactic operations. Consequently, medical therapy should be continued after the operation or started if it had not been employed preoperatively since it can potentiate and stabilize the beneficial effect. After the operation has eliminated tremor and rigidity, the akinetic component quite often remains. The latter responds well to L-dopa treatment; in fact, Narabayashi *et al.* (1973) regard L-dopa as synergistic to a stereotactic operation.

In view of the substantial clinical improvement that usually occurs after an operation, there are reasons to believe that the dosage of antiparkinsonism drugs can be considerably reduced. In our practice, we usually prescribed half the dose of L-dopa and cholinolytics that the patient had received before surgery. Subsequently, the dose is modified depending on the effect obtained. A similar policy has been recommended by Narabayashi (1983a).

In principle, postoperative L-dopa treatment is undoubtedly effective; however, there are different opinions about the degree of its efficacy. We studied a group of patients for whom L-dopa was prescribed for the first time after a unilateral or bilateral stereotactic operation. The majority of these cases had mixed (rigid–tremor or tremor–rigid) forms of the disease. A comparison of the results in this group with those of a similar but unoperated group showed that the effectiveness of L-dopa in the operated patients was noticeably less, although approximately half showed a clear-cut improvement (Kandel *et al.*, 1973a–c). There may be several explanations. The group of operated patients did not have the pure akinetic form of the disease, for which L-dopa is most effective. Moreover, a stereotactic operation on the basal ganglia modifies catecholamine metabolism in the brain and partly reduces the dopamine deficit in the basal ganglia, an observation indirectly confirmed by our data.

There are also other viewpoints on this issue. It has been assumed that the L-dopa efficacy is approximately the same in both operated and unoperated cases, but Laitinen (1972) considers L-dopa to be more effective in early operated cases.

Hyperkinesias such as chorea or dystonia exacerbated by prolonged L-dopa treatment entirely disappear postoperatively on the "operated" side of the body even though the drug is continued. If the operation is performed before the beginning of treatment, these hyperkinesias do not appear in the corresponding extremities (Laitinen, 1972; Van Buren *et al.*, 1973; Kandel *et al.*, 1973d; Cooper, 1977; Kandel, 1981; Kelly, 1983).

The data presented in this section emphasize the important role of contemporary drugs in the therapy of parkinsonism. The general principles of medical and surgical management of this disease and the criteria for selecting the optimum therapy are presented in the following section.

8. Indications and Contraindications for Surgery

Surgery is indicated for parkinsonians whom long-term medication has not benefited, for patients unable to tolerate such treatment, and when the steady progression of the disease jeopardizes working capacity or the ability of the patient to care for himself, provided there are no contraindications to surgery.

The following evaluation of the basic indications for surgery, naturally, is based on an analysis of the results of surgical treatment of parkinsonism.

8.1. Age

The youngest patient on whom we operated was 8 years old, and the oldest 78. The great majority of these patients (82.4%) were aged 40 to 60 years. Persons under 40 years of age comprised 8.5%, and those over 60 8.1% of cases.

Undoubtedly, there must be an "age limit" beyond which operations on the basal ganglia are not permissible. On the basis of our experience, we believe that the maximum age for surgery is 75 years, although this figure should not be considered absolute, and exceptions may be made in individual cases. Among the parkinsonians over the age of 65 years on whom we operated, the operation yielded a good or satisfactory result in the majority of cases, although the incidence of complications was higher. The chronological age of the patient is probably not as important as the physiological condition, the form and stage of the disease, and the state of the cardiovascular system.

Long experience with stereotactic operations in old parkinsonians has been accumulated by Riechert (1966b), who has operated on 503 patients between the ages of 61 and 70 years and on 31 patients between 70 to 75 years of age. Comparing this group with another group of parkinsonians of relatively young age (from

29 to 42 years), Riechert concluded that the main symptoms of the disease were approximately the same in both groups and that relief of symptoms postoperatively was also approximately the same in both the young and the old. However, in spite of the approximately similar degree of improvement after surgery, the patients of the younger group were able to improve their activities and working capacity to a considerably greater extent than the older group.

With improved techniques and better results of stereotactic operations in recent years, there has been a tendency to minimize the age restrictions. Siegfried and Zumstein (1976) have performed thalamotomies for tremor and pain on 91 patients aged from 70 to 90 years. In the majority of patients quite satisfactory results were obtained; however, there were more complications, and the mortality was higher. Therefore, operations on elderly and senile people, provided there are clear-cut indications, are possible and, without a doubt, effective. But at the same time, they have a greater risk of complications than operations on younger people.

8.2. Etiology

The etiology of the disease has no definite significance for the results of surgical treatment. The only exception is atherosclerotic parkinsonism (Section 4), which in many instances is an absolute contraindication to surgical treatment.

8.3. Form of the Disease

The tremor, mixed, and rigid forms of parkinsonism are all suitable for surgical treatment, although the results are not the same. In the akinetic form, indications for surgery are not so clear, particularly because L-dopa treatment is very effective in such cases. The majority of authors consider that if tremor and rigidity are absent, the operation is not indicated. On the other hand, we consider that in well-selected cases, the operation may give satisfactory results. Accordingly, the akinetic form of the disease is not an absolute contraindication for surgery, although in most cases medication is much more effective.

We operated on 85 patients in whom the diagnosis of arteriosclerotic parkinsonism was reasonably certain as well as on a number of others in whom an arteriosclerotic etiology of the disease was suspected. We believe that the indications for surgery in this condition should be quite strict, for almost half of these patients had speech and mental impairments in the postoperative period, and several died after surgery.

The significant risk of stereotactic operations in arteriosclerotic parkinsonism was also noted by Mundinger and Riechert (1961), who operated on 34 such patients, the majority of whom developed or aggravated mental disorders postoperatively.

8.4. Duration of Disease

Duration of the disease in itself is of no particular significance. We have operated on patients who have had the disease for 25–30 years, and good long-term results were obtained.

8.5. State of Internal Organs

Acute and chronic diseases of the liver and kidneys with associated functional disorders are absolute contraindications to surgery. Chronic lung conditions (emphysema, pneumoconiosis, and others), although not absolute contraindications, require serious appraisal, since they can lead to postoperative pulmonary complications. Diseases of the blood and severe diabetes are absolute contraindications to surgery.

The state of the cardiovascular system should be evaluated thoroughly. Since parkinsonism is a disease of the aged, cardiovascular disorders are observed quite often. If they are reasonably well compensated, such conditions do not contraindicate surgery, nor does pedal edema, which was present in approximately 5–6% of our cases. Apparently, this is a consequence of central vasomotor disturbances, since it disappears soon after the operation.

Frequent angina attacks, pronounced general and coronary arteriosclerosis, persistent hypertension, previous myocardial infarctions, cerebral hemorrhages, and ischemic attacks are all unquestionably contraindications to surgery.

We do not consider autonomic disorders contraindications to surgery, since, as a rule, they diminish or even disappear postoperatively. Nevertheless, after surgery it is necessary to take special measures to prevent complications, since in this period the vegetative control may become more impaired.

8.6. Brainstem Functions

Disorders in phonation, swallowing, or respiration indicating lesions in the medulla oblongata are very unfavorable prognostic factors. Moreover, they do not diminish as a result of surgery. As a rule, bulbar disorders should be considered contraindications to surgery.

8.7. Mental State

Pronounced mental disturbances are not typical of parkinsonism and are observed rather rarely even in far-advanced stages of the disease. Organic mental disorders are usually an indication of other conditions—cerebral arteriosclerosis, diffuse cerebral atrophy, schizophrenia, etc. At times parkinsonism may be combined with psychoses of an organic type (senile and presenile psychoses). This combination is an indication of severe diffuse cerebral lesions, in particular, cortical atrophy with marked dilatation of the ventricles. Typical psychic disorders can often be observed in arteriosclerotic parkinsonism (memory impairment, apathy, Korsakoff's syndrome).

Gross mental impairment (organic dementia) is an indisputable contraindication to surgery. Such patients have poor tolerance to surgery on the basal ganglia, and postoperatively the mental disorders may even be worse.

All types of schizophrenia, especially its chronic form, are absolute contraindications to surgery. In cases of mild memory impairment on a background of arteriosclerosis, the desirability of surgery must be resolved individually with consideration of the entire clinical picture. The advice of the psychologist or psychiatrist in such cases is of great value in aiding the neurosurgeon to make the correct decision.

8.8. Hydrocephalus

Hydrocephalus as a result of diffuse cerebral atrophy is observed in parkinsonism quite frequently (Section 5). We consider that even pronounced hydrocephalus without gross mental disorders is not an absolute contraindication to surgery, although the results in such cases may be less satisfactory. However, at times the hydrocephalus is of such a degree that a stereotactic operation is not technically feasible, since the stereotactic localization of VL or Subth is impossible. In such cases, Bouchard and Umbach (1972) suggested that an atrioventricular bypass be inserted to relieve the hydrocephalus prior to the surgery. Although these authors performed this operation in only four cases, they noted postoperatively not only a 20–30% reduction in size of the ventricles but also improvement of extrapyramidal symptoms.

8.9. Medication

Long-term medication with modern drugs is a mandatory prerequisite for a final decision for surgery (Section 7.9). Exceptions are justified only in those cases in which such treatment is of little or no use, if there is relentless progression of the disease, or in cases of drug intolerance. If the patient has not been treated with modern drugs (especially L-dopa), then it is necessary to try medical management for no less than 6–8 months; only if there is no improvement is it advisable to consider surgery.

Iatrogenic involuntary movements of various types developing after prolonged L-dopa therapy, if not relieved when the drug is stopped, constitute an indication for stereotactic surgery.

8.10. Stage of the Disease

The severity of the clinical manifestations represents one of the main criteria for surgery. Without a doubt, the more severe the disease, the greater is the danger of postoperative complications, and the smaller the chance for functional recovery. Nevertheless, we believe that there is no reason to operate too early when the risk of an operation is incommensurate with the initial symptoms of the disease. In stage I of the disease, when there can be effective medical treatment, there is no indication for surgery.

The second stage (the stage of hemiparkinsonism) is the most favorable for surgery because it is possible to restore full working capacity. All hemiparkinsonians on whom we operated have had very good and stable results, in fact, practically full recovery.

Determining the indications for surgery in stage III of the disease usually does not present any great difficulties. If medication is ineffective, there are no contraindications related to disease of internal organs, cerebral arteriosclerosis, or mental disorders, and the patient is not too old, then, as a rule, we recommend surgery. Most patients operated on in stage III of the disease have a significant improvement.

The decision is much more complicated when the patient is elderly, in stage IV of the disease, totally immobile and helpless, bed-ridden, and emaciated as a result of many years of suffering, with deformed hands and feet and severe vegetative disorders. Yet the patient's intellect may be preserved, and he may have no obvious brainstem involvement, cardiovascular disorders, or other contraindications to surgery. Both the patient and his relatives may urgently request surgery, regarding it as the only hope for salvation once medication becomes ineffective. In such cases, it is exceedingly difficult for the neurosurgeon to make a decision, since the case is on the borderline between operable and inoperable.

The aphorism of Burdenko that "the risk of an operation should not exceed the danger of the disease itself" is applicable to such cases. Even patients with the most severe forms of parkinsonism can live for many years, although with mental and physical suffering. It is easier for a surgeon to perform an operation than to refuse it, although the thought that "we have nothing to lose anyway" should not be allowed to justify an operation, either from the medical or the ethical point of view.

Without a doubt an operation in stage IV of the disease is less effective and more dangerous than one on a less severely afflicted patient. However, according to our data, the percentage of positive results after an operation in stage IV is not much lower than that in stage III. Of course, one must consider the relative evaluations of the postoperative result: for the patient in stage II to have a good result, he must return to his job, but a patient in stage IV has a good result if he can get up from his bed and walk around his room on his own.

A neurosurgeon with wide experience knows that even in the most hopeless case, it is sometimes possible to obtain an amazing result. That is why even in the most far-advanced cases of parkinsonism, the surgeon cannot say that there are no chances of obtaining a good clinical result from surgery. To reject surgery deprives the patient of his last hope and ultimately dooms him to a painful and helpless existence.

As mentioned above, we have operated mainly on severe cases: 45% of the operated patients were in stage III, and almost 35% had the most severe stage (IV) of parkinsonism. We intentionally broadened the indications for surgical intervention in stage IV of the disease, although we knew that this would have a negative effect on the overall results of treatment and would increase the rate of complications and even the mortality. Moreover, we recognized that surgical treatment of the most severe patients was an unresolved problem and hoped that the experience would enable progress to be made.

In summary, it may be said that the expansion of indications for surgery in the severe stages (III–IV) was considered justified only in the absence of serious contraindications.

8.11. Summary

The indications and contraindications for a stereotactic operation for parkinsonism are as follows.

The operation is indicated:

1. In patients up to 75 years of age.
2. In hemiparkinsonism (stage II).
3. In carefully selected cases in stages III and IV.
4. In tremor and tremor–rigid forms of the disease.
5. In patients unresponsive to or intolerant of drug treatment.
6. In the absence of contraindications.

The operation is contraindicated:

1. In patients over 75 years of age.
2. In the most severe and advanced stages of the disease.
3. In the akinetic form of the disease.
4. In patients responding to drug therapy.
5. In the presence of contraindications.

In atypical cases not meeting these criteria, the decision about surgery should be made on an individual basis.

8.12. Side of the Operation

Stereotactic operations for parkinsonism produce their main effect on the contralateral side of the body. Accordingly, the selection of the side for the operation depends on the predominant involvement of the left or right extremities. If the disturbance is of equal or near-equal degree on both sides, we perform a left-side operation to restore a functional right arm, which is more useful in everyday life, although the danger of complications after a dominant-hemisphere operation is somewhat greater. This explains the fact that we have performed operations on the left in 60% more cases than on the right side.

In some cases we chose the side for the operation on the basis of the presence of pain in the extremities of one side of the body. Because the pain as a rule disappeared after the operation, we performed the first operation on the side contralateral to the painful extremities.

9. Surgery

Surgical treatment of parkinsonism has had a long and complicated history that has reflected changing surgical concepts based on developing knowledge of the etiology and pathogenesis of the disease as well as advances in clinical neurophysiology.

9.1. Historical Note

The more than 60-year history of surgical management of parkinsonism has been characterized by a persistent search for more effective operative techniques. In the prestereotactic era, the 44 years from 1911 to 1955, 22 different surgical operations on the central and peripheral nervous system were introduced into clinical practice (Siegfried, 1980c). A retrospective review indicates that the early operations were often ill-founded, and, as a result, many were not effective. Nevertheless, even these yielded valuable insight into the pathogenesis of the disease (see Table 11). Figure 107 presents a scheme of the basic operations (including operations on the basal ganglia) proposed for the treatment of parkinsonism and other extrapyramidal lesions.

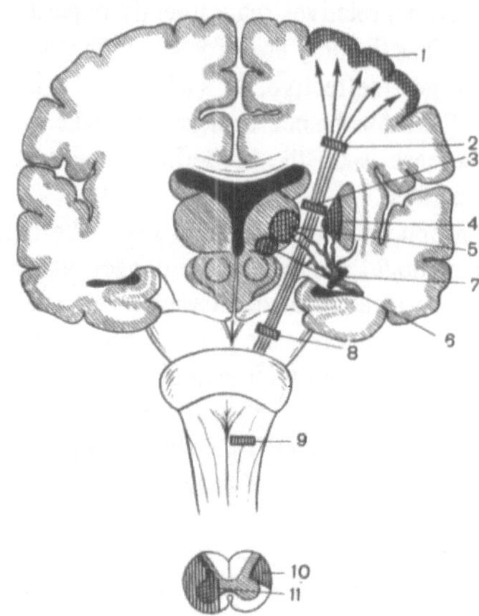

FIGURE 107. Scheme reflecting evolution of surgical operations for parkinsonism and other extrapyramidal disorders. 1, Extirpation of premotor and motor zones of cortex (Horsley, Klemme, Bucy); 2, section of pyramidal tract in the semioval center (pyramidotomy) (Polenov); 3, section of pyramidal tract in CI (Browder); 4, destruction of GP medial segment (Meyers, Spiegel and Wycis, Guiot, Cooper); 5, destruction of VL (Riechert, Hassler, Mundinger, Leksell, Cooper, Gillingham); 6, destruction of Subth (Spiegel and Wycis, Mundinger); 7, clipping of anterior choroidal artery (Cooper); 8, section of pyramidal tract in the brain peduncle (pedunculotomy) (Walker); 9, section of rubrospinal and tegmental tracts in medulla oblongata (bulbotomy) (Burdenko and Klosovski); 10, section of pyramidal tract in the dorsal part of spinal cord lateral column (Putnam, Puusepp, Mashanski); 11, section of lateral column of the spinal cord (Oliver).

TABLE 11. Brief History of Surgery for Dyskinesias

Cerebral (motor) cortex resections
- Motor cortex: Horsley (1890), Mish (1923), Eletski (1923), Nazarov (1927), Polenov (1928), Bucy and Buchanan (1932), Sachs (1935), Bucy and Case (1939), Klemme (1940, 1942), Putnam (1940)
- Frontal cortex: Baruk *et al.* (1953), Gros *et al.* (1955)

Subcortical procedures (internal capsule)
- Polenov (1928, 1937), Babchin (1934), Meyers (1940, 1942), Browder (1948)

Pallidotomy and anterior choroidal artery clipping
- Browder *et al.* (1953), Rand *et al.* (1956), Guiot (1952), Fenelon (1950)

Cerebral pedunculotomy
- Walker (1949a,b), Guiot and Pecker (1949), Wertheimer and Mansuy (1950), Hamby (1953), Borager (1955), Meyers (1956a,b), Pomme *et al.* (1958), Maspes and Pagni (1964)

Bulbotomy
- Burdenko and Klosovsky (1937, 1938), Ugrjumov *et al.* (1974)

Cerebellar dentatotomy
- Delmas-Marsalet and Van Bogaert (1935), Toth (1961)

Spinal cordotomy
- Puusepp (1930), Mashanski (1935), Putnam (1938, 1942), Ebin (1949), Oliver (1950), Hamby (1953)

Spinal radiculotomy
- Foerster (1908), Leriche (1912), Pollock and Davis (1930)

Sympathectomy
- Ganglionectomy: Brüning (1923), Wertheimer and Bonniot (1926)
- Periarterial carotid sympathectomy: Rosanov and Chugunov (1927), Alexander and Stolbun (1935), Gardner (1949)

For many decades the surgical treatment of parkinsonism and other hyperkinesias was based on the concept that the pyramidal system played the leading role in the genesis of involuntary movements. On this basis a number of operations were proposed to interrupt the pyramidal tract at different levels—from the motor cortex to the anterior horns of the spinal cord. It was presumed that these operations interrupted the main pathway along which the pathological impulses reached the periphery.

At the present time "pyramidal operations" have been relegated to the history of neurosurgery. They have been totally replaced by significantly more effective and safer stereotactic operations on the basal ganglia. The main shortcoming of "pyramidal operations," both at the cerebral as well as the spinal level, was that the disappearance of tremor or involuntary movements was accompanied by spastic paresis or paralysis of the contralateral extremities. Although voluntary movements might gradually return in some cases, the hyperkinesias also resumed. Moreover, operations on the pyramidal tract only relieved (but not always) one manifestation of parkinsonism, i.e., tremor; other symptoms remained unchanged.

9.1.1. Operations on the Cerebral Cortex

The pioneer of "pyramidal operations" was Sir Victor Horsley who in 1890 reported on the extirpation of the anterior central gyrus in a patient with athetosis. After the operation, Horsley noted a reduction of hyperkinesias in the affected extremities. Subsequently, operations at the cortical level were performed quite extensively. In the 1920s they were successfully carried out by many surgeons, in particular, the Russian surgeons Mish (1923), Eletski (1923), Nazarov (1927), and Polenov (1928). However, the shortcomings of the Horsley operation soon became evident. There was postoperative paralysis or severe paresis of the contralateral extremities, which usually receded gradually; however, relapses of involuntary movements were quite frequent. The cicatrix between the meninges and cerebral tissue in the area of the operation often led to subsequent epileptic attacks. Nor were the results of this operation improved by the proposal of Razumovski (1913) to introduce 2–3 ml of alcohol into the cortical motor centers (Koljubiakin, 1923; Polenov, 1928).

In following years it seemed that somewhat more encouraging results could be attained by the partial removal of the motor and premotor zone of the cortex (Brodmann areas 4 and 6), in which the involved extremities are represented as proposed by Bucy and Buchanen (1932), and then Sachs (1935). A noticeable improvement in severe hyperkinesias was observed after this operation in patients with athetosis or hemiballism (Bucy and Case, 1939; Putnam, 1940; and others). Although after this operation, the involved extremities were paretic, the authors believed that the patients were better than with the tremor. The tremor disappeared immediately after surgery, but, as a rule, relapsed. Muscular hypertonicity did not change postoperatively; it may be assumed that in these cases extrapyramidal rigidity was replaced by pyramidal spasticity. Subsequently, when parkinsonism developed on the other side of the body, the patient suffered from paralysis on one side and from tremor and rigidity on the other.

In 1940 and 1942 Klemme briefly reported on 100 operations involving extirpation of the premotor cortex (Brodmann area 6). After surgery, in almost half of the cases there was a lessening of tremor in the contralateral extremities without the development of paresis or paralyses. It should be noted that Klemme's data were not well documented and consequently doubts were raised regarding the value of these operations since their long-term results were not presented. Moreover, one cannot but notice the high rate of postoperative mortality (17%).

As more experience was gained, it became evident that extirpation of the motor cortex was an ineffective and risky operation that should be avoided.

9.1.2. Operations on Subcortical Pathways

At the beginning of this century there was the opinion that "the pyramidal motor pathway to the extremities could be cut at the under surface of the hemisphere near the point where this pathway enters CI. Accordingly, Polenov, for extrapyramidal hyperkinesias, instead of the Horsley operation, suggested "subcortical pyramidotomy," which involved cutting the pyramidal pathway just above the point it enters CI. Using a special instrument which he designed—a pyramidotome—Polenov (1928) performed this operation for the first time on a 20-year-old female suffering from right-sided hemiathetosis and Jacksonian epilepsy. After the pyramidotomy, the arm was paralyzed but subsequently partly recovered as the abnormal athetotic movements returned. However, it was soon apparent that pyramidotomy did not give a lasting effect. Moreover, as a rule, severe pareses or paralyses almost always occurred postoperatively.

The significant shortcomings of operations at the cortical level stimulated the search for methods of sectioning the pyramidal pathway at the level of the CI and cerebral peduncles. Destruction of the CI in order to eliminate tremor in parkinsonism was proposed by Browder in 1948. Employing a transventricular approach, he removed the head of NC to expose the fibers of the anterior limb of CI, and with the aid of a special bent hook, he gradually incised these fibers in the direction of the knee of the CI. In order to eliminate tremor completely, Browder recommended section of the CI to the point of producing severe paresis in the contralateral arm. He considered that only one out of 10 parkinsonians were suitable candidates for the operation, which should be performed on patients not older than 50 years of age.

The results of dissecting the pyramidal tract at CI were similar to those after extirpation of the motor cortex—a contralateral hemiplegia or hemiparesis. Moreover, in view of the cortical resection and the opening of the lateral ventricle, this operation was even more risky and complicated. After the Browder operation, frequently behavioral changes developed resembling the effect of prefrontal leukotomy. In view of these problems, this operation never gained general acceptance.

9.1.3. Frontal Leukotomy

The 1950s witnessed the publication of a very small series of observations concerning leukotomy for extrapyramidal disorders (Barük *et al.*, 1953; Gros *et al.*, 1955). The basis for this operation was the well-known fact that emotional factors intensify tremor in parkinsonism and abnormal movements in other hyperkinesias. Although some improvement was reported after leukotomy in a few cases, this operation for parkinsonism and dyskinesias should be considered theoretically unfounded and of little value.

9.1.4. Pedunculotomy

In 1949 Walker proposed a new operation which he called cerebral pedunculotomy. The operation involved section of the pyramidal tract at the level of a cerebral peduncle. Since pedunculotomy is the only "pyramidal" operation which is sometimes performed during the present day, its results should be discussed in somewhat greater detail. At first, Walker performed section of the lateral part of the cerebral peduncle which becomes accessible after lifting the temporal lobe. The first report by Walker (1952) of 4 operations in hemiparkinsonism was not very optimistic—hemiparesis or hemiplegia appeared in all the patients, while tremor and rigidity was only diminished, but not abolished.

In the next series of observations—9 operations for hemiballism, parkinsonism, and choreoathetosis—the results were better. The severity of hemiparesis which appeared postoperatively was less, while the abnormal movements were reduced to a greater degree. However, Meyers (1956a,b) did not confirm the efficacy of this operation in parkinsonism.

Guiot and Pecker (1949) proposed a modification of pedunculotomy which involved section of the central part of the cerebral peduncle through which the corticospinal (pyramidal) tract passed. Bucy (1951) reported on the results of three such operations: tremor was eliminated in a parkinsonian, but pronounced hemiparesis developed; in two patients with hemiballism there was a noticeable reduction in hyperkinesia with a slight lessening in muscular force. Satisfactory results from pedunculotomy in dystonia musculorum deformans and athetosis were noted by Wertheimer and Mansuy (1950), Hamby (1953), Pomme *et al.* (1958), and others.

One of the most impressive investigations of pedunculotomy was the work of Meyers (1956a,b), in which he analyzed the results of 17 bilateral operations in generalized dystonia and choreoathetosis. According to the data of the author, a "significant improvement" was observed in 13 cases. Meyers stressed that in order to obtain a better effect, section of a greater portion of the cerebral peduncle was necessary. Section of the medial portion of the cerebral peduncle through which the frontopontine pathways pass had no effect on abnormal movements, but did not cause hemiparesis (Meyers, 1951). Hamby (1953) also advocated bilateral pedunculotomy for dystonia, noting that neurological defects after this operation were noticeably less than after extirpation of the motor cortex. A very rare case of pedunculotomy in an 8-month-old infant with deforming muscular dystonia and hemiparesis was reported by Pomme *et al.* (1958).

Data which appeared in the literature later are contradictory as to the efficacy of pedunculotomy in extrapyramidal hyperkinesias. Two papers can serve as an example. The first reported a high frequency of serious complications and relapses postoperatively, while the authors of the second paper assessed the results of pedunculotomy much more optimistically.

Maspes and Pagni (1964) performed pedunculotomy by the subtemporal approach in 13 patients with dystonia and choreathetosis; a bilateral operation was performed in two of the cases. In all instances, flaccid hemiplegia or severe hemiparesis in the contralateral extremities developed postoperatively, as well as other serious complications (impairments of consciousness, epileptic attacks, and bulbar disorders). The long-term results were not considered hopeful.

Hyperkinesis reappeared to various degrees in five cases. Abnormal movements diminished in eight cases; however, hemiparesis remained.

Considerably more favorable results of pedunculotomy in various hyperkinesias were reported by Jane *et al.* (1968). They emphasize that the severity of hemiparesis after that operation was not great and was limited to the proximal muscular groups in the extremities.

At the present time, pedunculotomy is rarely performed in the treatment of parkinsonism and other hyperkinesias.

9.1.5. Bulbotomy

In 1935 Russian neurosurgeons Burdenko and Klosovski proposed and for the first time performed a new operation for extrapyramidal hyperkinesias, in particular in parkinsonism, which they called "bulbotomy." This operation was designed to section by the open method Monakow's rubrospinal tract and the central tegmental bundle at the level of the olives. The objective of the operation was to terminate involuntary movements and reduce muscular hypertonicity.

N. N. Burdenko (1876–1946)

Two years later Burdenko and Klosovski (1937) reported on the immediate results of bulbotomy in 11 patients with parkinsonism, bilateral athetosis, and torsion spasm. In five cases after bulbotomy there was some reduction in ridigity, while the other nine operations produced no effect; moreover, in one there was a lethal outcome.

From the historical point of view, bulbotomy was a new and progressive operation which was the first to show the possibilities of surgical intervention on medullary pathways. However, this operation proved to be ineffective, and at the present time, it is not performed.

9.1.6. Clipping of the Anterior Choroidal Artery

In 1952 Cooper, while performing pedunculotomy on a patient with postencephalitic parkinsonism, encountered such profuse hemorrhage that he had to clip an artery, which (as subsequently shown by carotid angiography) was the anterior choroidal, and close the wound. After this seemingly unsuccessful operation tremor and rigidity disappeared in the contralateral extremities without the development of hemiparesis. Thus, this unexpected complication gave Cooper the idea for a new operation in parkinsonism—clipping off the anterior choroidal artery to cause an ischemic infarct in the GP and Th. The anterior choroidal artery usually arises as the second branch of the intracranial carotid artery and supplies blood to GP, VL, SN, NC, NR, and other subcortical structures.

The first results of this operation in 52 parkinsonians was encouraging. In 65% of the cases, after the operation, the tremor at rest diminished or totally disappeared in the contralateral extremities, and in 75% of the cases muscular rigidity was decreased. A number of patients had noticeable improvement in gait and ability to take care of themselves. However, the operation had a 10% mortality and 8% incidence of hemiplegia. Other reports have appeared in the literature of a high percentage of complications and unstable results (Rand *et al.*, 1956). Difficulties concerning the surgical approach to this artery and its frequent anatomical variability were also emphasized. The rapid development of stereotactic surgery which began at that time led to the abandonment of that operation.

9.1.7. Operations on the Cerebellum

The first attempt to relieve the parkinsonism syndrome by destruction of ND was undertaken by Delma-Marsalet *et al.*, in 1935. In a few cases, the authors noted a reduction of muscular tonus; however, tremor became more pronounced. For a quarter of a century this operation was not practiced, but in 1961, Toth extirpated ND by the open approach in three parkinsonians. In two of them, tremor and rigidity diminished on the homolateral side.

At present, stereotactic dentatotomy is successfully employed for treatment of spasticity, mainly cerebral palsy.

Ludvig Puusepp (1875–1942)

9.1.8. Operations on the Spinal Cord

When operations involving section of the pyramidal tract at various levels were being developed, several procedures on the spinal cord were proposed. An operation on the spinal pathways was first performed by Puusepp (1930), who sectioned the posterior columns in one parkinsonian patient. The author noted that after the operation, which he called cordotomia posterior lateralis, there was a slight diminishing in rigidity without change in tremor.

In 1938, Putnam proposed section of the anterior column of the cervical spinal cord in choreoathetosis. Later, such operations were performed by Mashanski (1935), Ebin (1949), and Oliver (1950). Ebin also severed the opposite ventral pyramidal tract through the commissure. Mashanski (1935) transsected the anterior columns in 23 parkinsonians. Improvement was noted in only one-third of the cases, and the postoperative mortality was very high (22%). Hamby (1953) reported that the results of these operations were not good; an improvement was observed only in 23% of patients in whom the reduced tremor was present in more or less paralyzed extremities. In 42% of the cases, there was no improvement, in 29% the condition was worse, and 6% died. After the surgery the severe paresis in the extremities gradually diminished, but at the same time there was a relapse of the tremor.

Subsequently, Putnam (1942) proposed another variant of the spinal cord operation which he called pyramidotomy. The operation involved section of the corticospinal pathway in the lateral column at the level of the second cervical vertebra. Putnam reported that tremor disappeared postoperatively in 15 of 25 cases, but rigidity remained unchanged. Immediately after the operation, a hemiplegia was present on the side of the operation in all the patients, but with the passage of time, the motor power returned to a certain degree.

In 1950 Oliver reported on a new operation for parkinsonism which involved section of the entire lateral half of the spinal cord. However, in these cases it was difficult to evaluate the benefit since these patients had a paralysis of the limbs on the side of the operation with loss of pain and temperature sensation on the opposite side of the body.

The low success rate and high rate of complications made operations on the spinal cord pathways undesirable even in the "prestereotactic era." Today these operations are no longer performed for parkinsonism and hyperkinesias, although they are being successfully employed for the management of spasticity.

Even before operations on the spinal cord, operations on the spinal roots had been suggested. In 1908 Foerster proposed sectioning the posterior roots of the spinal cord to reduce extrapyramidal rigidity. In the 1930s this operation was performed by many surgeons. In the majority of papers, however, it was noted that there was a temporary and incomplete or no therapeutic effect. However, the Foerster operation has now been

Otfrid Foerster (1873–1941)

modified and is successfully being used to relieve spasticity, torticollis, and chronic pain syndromes.

In 1912 Leriche resected the V, VI, and VII posterior cervical roots in two parkinsonians without appreciable relief.

An extensive posterior rhizotomy was performed in a parkinsonian by Pollock and Davis (1930), who noted a significant reduction of rigidity in the arm on the side of the operation. However, postoperatively the tremor did not disappear although its rhythm became irregular. Due to the loss of sensation in the arm on the side of the operation, the movements were slow and awkward.

9.1.9. Operations on the Sympathetic System

Operative intervention on various parts of the sympathetic system was employed in the 1920s and 1930s for hyperkinesias, in particular, parkinsonism. In 1923, Brünning removed the superior cervical ganglion and performed a carotid periarterial sympathectomy, which, in the opinion of the author, improved the cerebral blood supply.

Extirpation of the entire cervical segment of the sympathetic trunk with section of the rami communicantes as proposed by Wertheimer and Bonniot (1926) was aimed at reducing muscular hypertonicity. However, the effect of the operation proved insignificant and brief.

Rosanov and Chugunov (1927), Alexandrova and Stolbun (1935), and Gardner (1949) reported a slight improvement in the condition of parkinsonians after carotid periarterial sympathectomy, and extirpation of the superior cervical and other sympathetic ganglia. Subsequent experience, however, demonstrated that these operations on the sympathetic system were ineffective in the treatment of extrapyramidal hyperkinesias. This type of surgical intervention, now discontinued, is of interest only from the historical point of view.

9.1.10. Operations on the Basal Ganglia

It would seem expedient to describe these operations at the end of the historical review of surgical treatment of parkinsonism since they represent the transitional stage to contemporary treatment of that disease—stereotactic surgery. Some 40 years ago when the first operations on the basal ganglia were performed, the idea doubtlessly seemed not only extremely bold, but also to involve a high risk. Neurosurgeons at that time considered the basal ganglia to be an "out-of-bounds zone," in respect to which the old aphorism "noli me tangere" was fully applicable. It was assumed that surgical intervention in this area inevitably led to a lethal outcome. Nevertheless, in the history of surgery, new progressive ideas have superceded traditional concepts and beliefs.

The trailblazer in basal ganglia surgery is the American neurosurgeon, Russell Meyers, who in 1939 was the first to extirpate by the open approach the head and anterior segment of NC in a woman with hemiparkinsonism, in whom a previous operation ("undercutting" of the premotor cortex) had not eliminated tremor (Meyers, 1940). This operation gave improvement for 4 years. However, because the operation did not give sufficient relief, it was necessary to find new surgical procedures. In subsequent years, Meyers (1942) tried several ways of sectioning or extirpating various parts of the basal ganglia. Approaching through the frontal lobe and anterior horn, Meyers tried destroying different subcortical structures alone and in various combinations: head of NC, anterior limb of CI, Put, GP, and AL. To some degree, each of these operations improved tremor and rigidity. An analysis of his results in 58 cases revealed that there was a significant improvement in 19%, some improvement in 43%, no improvement in 26%, and mortality in 12%.

As can be seen from these figures, the results were a long way from ideal. Nevertheless, Meyers established two important principles for the further development of stereotactic neurosurgery. He demonstrated that the most effective results were obtained after ansotomy, i.e., section of the pallidofugal pathways in AL. Second, he showed for the first time that ansotomy could reduce or even eliminate tremor, as well as rigidity, without producing paresis and other signs of pyramidal tract involvement, such as impairment of speech. However, this operation, which involves cutting of a cerebral hemisphere and exposure of the lateral ventricle, yielded a very high postoperative mortality (15.7%). In many cases (approximately 15%) there were postoperative epileptic fits, as well as grave mental disorders. Objectively assessing the high risk connected with open operations on the basal ganglia, Meyers did not recommend this operation for general use in neurosurgery.

Meyers' basic concept was further developed at the beginning of the 1950s by other neurosurgeons who also used the open approach to the basal ganglia (Fenelon, 1950, 1955; Guiot and Brion, 1952). In order to improve the technique of this approach to GP, Fenelon (1950) carried out a minor osteoplastic craniotomy and passed an electrode parallel to the sagittal plane of the frontal lobe. After the electrode was introduced into GP, it was electrocoagulated. Subsequently, Fenelon (1955) changed to the temporal approach to GP. After an osteoplastic craniotomy and exposure of the pole of the temporal lobe, he introduced a bent electrode through the inferior surface of the temporal pole to a depth of 40 mm and electrocoagulated GP. With this technique, Fenelon obtained satisfactory results in more than half of the parkinsonian cases.

Susequently, an open method for gaining access to the medial part of GP was employed by Guiot and Brion (1955), who reported on the first 40 operations involving destruction of that structure by a subfrontal approach. Satisfactory results were obtained in 42.5% of the cases; during follow-up, this figure dropped to 37%.

9.2. Summary

Such were the milestones in the initial stage of surgical treatment for parkinsonism by open operations on the basal ganglia. Evaluating the results of this stage retrospectively after almost four decades, one may conclude with a great degree of certainty that it played a very important role in the development of parkinsonian surgery, and provided a condition sine qua non for the transition to the contemporary stereotactic stage.

The main contribution of these early operations was the proof that destruction of a part of the basal ganglia may produce lasting relief of the effects of parkinsonism and a number of other extrapyramidal diseases without complicating motor or sensory disorders. Although experience has shown that the open operative approaches to the subcortical structures were unnecessarily complex, traumatic, and fraught with serious complications and some mortality, they did indicate a satisfactory approach to the problem with stereotactic techniques which was feasible. This knowledge opened the way for the development of stereotactic procedures.

10. Stereotactic Surgery

The present stage in the surgical treatment of parkinsonism should rightfully be called the stereotactic era. There can be no doubt that this stage represents one of the most striking and outstanding neurosurgical achievements of the present day. As stated in Chapter 3 (Section 1), the trail blazers were Spiegel and Wycis, who first used the new method to operate on the basal ganglia. The first six operations reported by these authors in 1954 involved stereotactic destruction of the medial segment of GP and AL. Encouraging results were obtained in four of the six patients in the form of a substantial reduction of tremor and rigidity in the contralateral extremities. Similar operations were subsequently performed by other neurosurgeons in a number of countries (Talairach, Riechert and Mundinger, Leksell, Narabayashi, Guiot, Cooper, Gillingham, Bertrand, Aronson and Walker, Obrador, Laitinen, Kandel, and others).

At the end of the 1950s we started to use stereotactic operations for parkinsonism in the Soviet Union. In 1960 and 1961 we published the results of the first 76 stereotactic operations. Good or satisfactory long-term results were obtained in 66% of the 42 operated cases. In subsequent years, the stereotactic technique of treating parkinsonism was adopted by neurosurgeons in a number of cities in the USSR.

An important advance in the development of surgery for parkinsonism was proposed by the Freiburg school (Riechert, 1953, 1959; Riechert and Mundinger, 1955, 1960) at the end of the 1950s. They changed the surgical target from the medial segment of GP to VL, which proved to be considerably more effective. In subsequent years, numerous papers were published on the immediate and the long-term results of the VL lesions (Riechert, 1966a; Laitinen, 1966; Mundinger and Zissner, 1966; Schwab, 1966; Cooper, 1969a,b, 1977; Scott *et al.*, 1970; Bertrand *et al.*, 1973; Hassler *et al.*, 1979). The vast experience of Riechert (1966a, 1980) and Mundinger (1969, 1973) in the surgical treatment of parkinsonism led them to an important conclusion about selective destruction of the anterior and posterior parts of VL, depending on the form of parkinsonism. It was shown that destruction of VOa and VOp is more effective when rigidity is predominant, whereas destruction of VOp and especially Vim gives better results when tremor predominates. Subsequently, this conclusion was repeatedly confirmed by many authors (Narabayashi, 1982a; Driollet *et al.*, 1973a; Kandel, 1975b, 1978a, 1981; Ohye *et al.*, 1982a).

The stereotactic coordinates given by various authors for the center of the lesion in VL that gives the best clinical results are not identical, although the differences in the various coordinates do not exceed 2–3 mm. Our data regarding the target are presented in Chapter 4 (Section 5). Laitinen (1966) considers that the most "effective" target point corresponds to a part of ventral VL with the following coordinates: 16.5 mm from CA along LI and 11.5 mm from the midplane. A deviation of even 2–3 mm from that target, in the opinion of Laitinen, reduces the effectiveness of the operation.

There can be no doubt that thalamotomy in parkinsonism produces a significant therapeutic effect that remains stable for many years. This has been confirmed in numerous papers, including our own data, which are presented below.

The extensive experience of surgical treatment of parkinsonism at the Cooper's Clinic and a statistical analysis of their results have been presented by Waltz *et al.* (1966). Computer analysis of 165 factors was done on 1001 parkinsonians. Long-term results of the destruction of VL and adjacent anterior VPL were studied in 203 patients 5–12 months after surgery. Bilateral operations were performed in 34.6% of the cases. Diminished tremor was observed in 90.6% of the patients, and extrapyramidal rigidity disappeared in 88.6%; 94.4% of the patients were better able to care for themselves.

Other authors presenting long-term results of such operations emphasize that there was marked improvement in 80–88% cases (Van Manen, 1969; Laitinen, 1972; Van Buren *et al.*, 1973; Siegfried, 1983). However, some authors point out that the substantial postoperative improvement gradually recedes in approximately 20% of the cases, mainly because of severity of the disease. Our data corroborate these conclusions (Section 12).

We began performing stereotactic destruction of VL in 1959. In our monograph *Parkinsonism and Its Surgical Treatment* (Kandel, 1964), we presented detailed data on 244 stereotactic operations on 187 parkinsonian patients. In the overwhelming majority of cases, the operation involved destruction of VL. In a long-term follow-up, it was shown that a stable and significant improvement was obtained in 74% of the cases. Surgical treatment proved to be more effective in the mixed and rigid types of the disease. In the tremor form, the percentage of good results was somewhat lower because of recurrence of tremor. Bilateral operations, although more effective, were more often followed by various complications. In the akinetic form of the disease, the operation was significantly less effective and was deemed indicated only in rare cases. A unilateral operation does not prevent the progression of parkinsonism on the other, unoperated side.

At the same time, neurosurgeons continued to search for a subcortical structure whose destruction would be even more effective in parkinsonism.

The next step in stereotactic surgery of this disease was the proposal by Spiegel and Wycis (1963) to destroy Subth. This operation, called a campotomy or subthalamotomy, was worked out in detail by Mundinger (1965a,c), who demonstrated that the long-term results of Subth destruction, especially with respect to relief of tremor, were noticeably better than those of destruction of VL or medial GP. For example, complete disappearance of tremor in the postoperative period was observed, respectively, in 82.6%, 41.7%, and 26.4% of cases. In a subsequent paper, Mundinger (1969) reported on the long-term results of 500 stereotactic subthalamotomies in parkinsonism. In 75% of the cases, tremor was eliminated completely. The author emphasizes that the greater the tremor before the operation, the more effective were the results. After destruction of VL, long-term follow-up showed tremor to remain completely absent in only 53% of the cases.

The successful results of subthalamotomy (Bertrand *et al.*, 1969; Velasco *et al.*, 1972) reaffirmed the important conclusion that complete and permanent elimination of tremor could be obtained without damaging the pyramidal tract. Whereas a VL lesion made near this tract often encroaches upon the adjacent

CI, the greater distance between Subth and the corticospinal pathway minimizes the chance of damaging it even slightly. It should be noted, however, that several authors with extensive surgical experience do not share Mundinger's view of the advantages of Subth compared to VL destruction. They consider the long-term results of these operations to be approximately the same (Story *et al.*, 1966; Hughes, 1969; Siegfried, 1974). Only one investigation (Fager, 1968) considers subthalamotomy for parkinsonism to be less effective and to cause more complications than VL thalamotomy. However, Fager's experience of only 13 subthalamotomies is obviously insufficient for drawing definite conclusions.

We later proposed performing a combined thalamosubthalamotomy in one stage, consecutively making cryogenic lesions in VL and Subth, and have done more than 350 such operations. Analysis of the long-term effects has demonstrated better results than following destruction of each of the structures separately (Kandel, 1972, 1978b, 1981), a conclusion subsequently confirmed by Gros *et al.* (1976a,b).

Stereotactic destruction of CM in parkinsonism was proposed by Markham *et al.* (1966). Their limited clinical experience did not allow them to draw any definite conclusions about the results of CM destruction. In any case, this operation has not seemed a promising application of stereotactic surgery for parkinsonism.

Stimulation of subcortical structures does not seem applicable to the treatment of parkinsonism. Toth and Vajda (1980) implanted electrodes in 18 patients for simultaneous stimulation of several subcortical structures (VL, GP, ND) as well as the motor cortex; later, microcoagulation was done in those structures. However, no long-term results were presented.

Since the indications for and technique of VL destruction have been described (Chapter 4, Section 5) we only present the details of subthalamotomy here.

10.1. Stereotactic Destruction of the Subthalamus

Stereotactic destruction of Subth structures (subthalamotomy, campotomy) is at present one of the principal operations used in the treatment of many extrapyramidal hyperkinesias—parkinsonism, cerebral palsy, dystonia musculorum deformans, spasmodic torticollis, hepatocerebral degeneration, Huntington's chorea, and multiple sclerosis.

From the anatomic connections of Subth (Chapter 1, Section 1.3), it is evident that this small, deep-seated nucleus is intimately related to many nuclei and pathways of the extrapyramidal system, in particular the dentatothalamic and dentorubral pathways.

Since the target points for Subth destruction, as first described by Spiegel and Wycis (1963), were Forel's fields H_1 and H_2 (CF), they termed this operation campotomy (from the Latin *campus* field). Subsequently, Mundinger (1965a,c) proposed, in addition to CF H_1 and H_2, the destruction of ZI and part of the prelemniscal radiation, a procedure more appropriately called subthalamotomy.

Since Subth is in close proximity to VL, somewhat ventral and dorsal to that nucleus, the stereotactic calculations for location of both structures are similar but not identical. Subthalamotomy involves a number of important pathways, which should be considered in greater detail. As already noted (Chapter 1, Section 1.3), the pallidofugal pathways leading to NR and FR, the rubrothalamic and dentatothalamic pathways to VL, and fiber tracts from the precentral frontal cortex pass through this zone.

The question of which of these pathways must be destroyed to eliminate tremor still remains open for discussion. Some authors believe that it is necessary to sever the pathways from the cerebral cortex to the spinal motor centers. Others consider it necessary only to block the efferent systems from subcortical structures to the brainstem and spinal cord. Mundinger (1965a,c) considers that the best results are obtained by the complete destruction of the dentatothalamic systems. In any case, to eliminate tremor, rigidity, and hyperkinesia, it is necessary to interrupt the entire system of fibers concentrated in Subth for a distance of several millimeters. Accordingly, the target in the area must be hit very accurately, within 1 mm.

Since NS is adjacent to CF and ZI, and since partial damage of NS could cause choreic or ballistic hyperkinesia (Section 11), the focus of destruction should not extend beyond the limits of Subth and thus should not be large (approximately $4 \times 4 \times 3$ mm), i.e., somewhat smaller than the usual foci in Th nuclei.

An important step in the operation is electrical stimulation to confirm the accuracy of hitting Subth. This stimulation produces complicated and diverse effects described in Chapter 4 (Section 7.1). In view of the small size of this area, the stimulating electrode must be a very fine stylet (0.6–0.8 mm) capable of being extended laterally in any direction. Stimulation is usually performed with a weak current of the following parameters: current 0.1–0.5 mA, frequency 50–100 Hz, and duration of stimulation 1–2 msec.

Morphological investigations have shown that the electrode to the Subth should be directed lateromedially and anteroposteriorly so that the danger of damaging CI and NS is minimal. Mundinger (1965a,c) destroyed Subth with a stylus electrode less than 1 mm in diameter, producing a necrotic lesion measuring 7–8 × 4.5–5 mm.

Stereotactic coordinates for hitting Subth are presented in Chapter 4 (Section 5).

It is evident that the best method for determining the most effective target point is to compare the location of the lesion with the clinical result. It is very important to allow as far as possible for the high individual variability of subcortical structures (Chapter 3, Section 3). In order to resolve this task, Mirkiyev, Pyatikop, and I analyzed the pertinent parameters on ventriculograms obtained on 60 parkinsonian patients, 16 with bilateral operations, with tremor and tremor–rigidity.

To analyze of the lateral ventriculograms and determine the optimal target point, we used the method of Velasco *et al.* (1972), dividing LI into ten parts and erecting on this base a grid of 100 squares (10 × 10). This made it possible to calculate the mean target and establish its coordinates irrespective of the length of LI.

The center of the lesion in both VL and Subth in each of the 76 operations was located on a square of the grid, and the points were then transferred to the composite scheme shown in Fig. 104. The squares on the grid were designated +1 to +5 along the vertical axis upwards from the reference line FM–CP and −1 to −5 downwards. Points along the horizontal axis were designated *A* to *Y*. In each square, symbols were entered to indicate the state of the tremor (absent, improved, no effect or relapse).

As one can see from Fig. 108, the results of this analysis proved very enlightening. A good result—total and lasting disappearance of tremor after VL destruction—may be obtained only by accurately hitting a small zone occupying square 1F and the anterior half of square 1G on the schema. The dimensions of that zone are approximately 4 × 4 mm, with the average length of the line from the posterior edge of FM–CP being 23.5 mm.

The effective Subth lesions lie in square 2I and the front quarter of square 2Y. The dimensions of this zone are smaller, 2.5 × 2.5 mm. It is important to note that even a small deviation from the center of these zones by 1.5–2 mm in any direction significantly lessens the relief of tremor. This once again emphasizes the necessity of taking individual variability into account and of striving for maximum accuracy in calculating the coordinates of the desired structure in stereotactic operations.

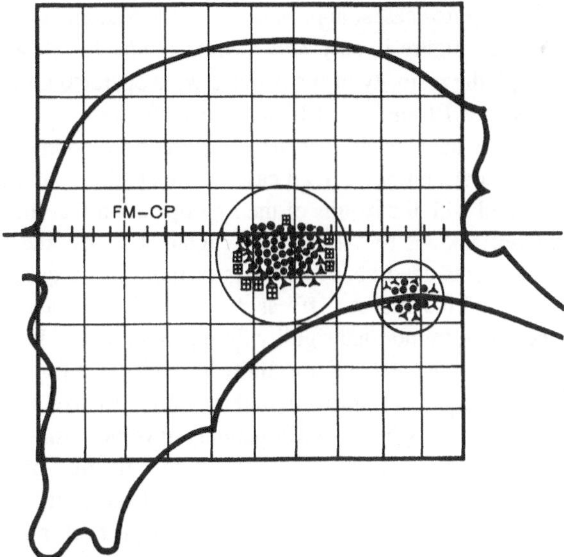

FIGURE 108. Dependence of clinical result on the site of the lesion in VL and Subth for tremor form of parkinsonism. The line from posterior edge of FM to CP is divided into ten parts, and a scale of 100 squares is erected. Small incisions on the indicated line correspond to 1 mm for its average length, 23 mm. Larger circle, VL; smaller, Subth. Results of stereotactic operations for tremor in 60 patients according to cryodestructive focus location are conditionally marked: black circles, complete disappearance of tremor; triangle, almost complete effect with persisting slight, inconstant tremor; a cross in a square, little effect or relapse of tremor.

10.2. Bilateral Operations

The advisability of a second operation on the other basal ganglia is a complicated and still not finally resolved question which obviously arises only when there is marked bilateral involvement.

Studies conducted in the 1960s demonstrated that the immediate results of an operation on the second side with respect to tremor and rigidity were the same as on the first side (Cooper, 1961; Riechert, 1962; Kandel, 1965; Bravo *et al.*, 1967). Nevertheless, an analysis of data in literature shows that even surgeons having great personal experience in stereotactic operations for parkinsonism perform bilateral operations in relatively few cases, approximately 15–25%, and even Cooper's figure is only 35% (Waltz *et al.*, 1966), although the majority of his patients were operated on in the stages III and IV of bilateral disease.

It is noteworthy that the question of a second operation does not arise in all cases. Some patients are so satisfied with the results of the first operation that they are not anxious to "tempt fate" again. Usually, these are patients aged about 60 years or more in advanced stages of the disease. If, as a result of a unilateral operation, tremor and rigidity disappear on one side of the body, the general disability diminishes substantially, the gait is improved, and the ability to take care of one's self is better, such patients have no desire to go through a second operation. Since the risk of a second operation is greater than that of the first, we believe that the surgeon should encourage such a rational decision.

However, only a comparatively small proportion of the patients come to such a conclusion—approximately half beg the surgeon to operate on the second side, especially if the first operation produced a good result. Nevertheless, in spite of the tempting possibility that the patient might get complete relief from the troublesome manifestations of the disease, neurosurgeons must be particularly cautious about performing a stereotactic operation on the second side. In the 1950s and 1960s, many surgeons assumed that one operation for bilateral parkinsonism was "half the job," but in later years this point of view was abandoned.

This reaction stemmed from the fact that bilateral operations involve considerably more risk than unilateral. The frequency and severity of complications such as dysarthria and mental impairments after the second operation are much greater.

At the present time, some neurosurgeons consider bilateral operations to be indicated only rarely. Others are more optimistic in their evaluation of the results and perform bilateral operations more frequently. For example, Bravo *et al.* (1967) present the long-term results of such operations in 300 cases. The tremor and rigidity on both sides were abolished in 61% of the patients, significantly reduced in 21%, slightly improved in 15%, and worse in 2%; postoperative mortality was 0.8%. In short, satisfactory long-term results were obtained in about two-thirds of the patients. More than half of the patients regained their working capacity or were able to work at home. These results remained stable; in other words, if the patient was able to work 3 months after the operation, he retained this ability for 4–5 or more years. Approximately the same results were reported by Riechert (1980) in 412 bilateral operations. However, his analysis of the long-term results showed that the frequency of relapses after bilateral operations was higher than after unilateral surgery. The eventual results of bilateral operations were 14% worse than the results of unilateral surgery, and the number of complications after bilateral surgery was twice as high. In other words, 76% of parkinsonians having bilateral surgery had the tremor eliminated or markedly diminished (Gillingham, 1966b).

It may be assumed that a good result of bilateral operations will be maintained for many years. Shichijo *et al.* (1983) examined 22 parkinsonians 10–15 years after bilateral thalamotomy and concluded that the results were quite satisfactory. The dyskinesias did not recur in any of these patients over many years of L-dopa treatment, and postoperative speech impairments developed in only one-third of these patients.

What time interval should be allowed to elapse between the first and second operations? In the 1960s there were a few reports of bilateral stereotactic operations performed in one stage or after a very brief interval (Cooper, 1959; Kalyanaraman and Ramamurthi, 1965); however, this practice was subsequently discontinued because of complications. At present, it is generally acknowledged that the interval between the two operations should be at least 6 months, and still better 1–2 years. Our experience fully confirms this opinion.

On the basis of our experience with bilateral operations on 120 parkinsonians, we feel that the advisability of an operation on the second side should be evaluated with great caution and strictly on an individual basis. In addition to the general indications for such an operation (Section 8), it is imperative that the result of the first operation be good and stable. If the first operation was executed technically correctly but proved to be ineffective, it would be unwise to recommend an operation on the opposite side. Moreover, even if the first operation had good long-term results but was followed by a grave postoperative period with serious complications, especially speech impairments, the surgeon should not operate on the second side. It is also necessary to take into consideration the wishes of a well-informed patient regarding the high risk of a second operation.

10.3. Clinicoanatomic Correlations

To understand the physiological basis of the effect of a stereotactic operation on various manifestations of extrapyramidal disease, it is important to make clinicoanatomic correlations, i.e., postmortem studies of the brain after stereotactic operations. These correlations are essential to determine not only the accuracy of hitting a given subcortical structure but, even more important, the correlation between the therapeutic effect of the operation and the structures actually destroyed. This is of immense significance for correctly determining the most effective target for the relief of various nosological forms, in particular, parkinsonism.

Beginning in the late 1950s, several such clinicomorphological investigations were published (Spiegel and Wycis, 1958; Cooper *et al.*, 1963; Krayenbühl *et al.*, 1964; Kandel and Podgornaya, 1966; Pagni, 1966; Markham *et al.*, 1966; Smith, 1967; Hassler, 1969; Hamien and Maloney, 1969; Hartmann von Monakow, 1962), and a well-illustrated book by Hassler, Riechert, and Mundinger was published in 1979.

Without going into the details of these investigations, we briefly summarize their findings. It was shown that comparatively small lesions within VL totally eliminated tremor, rigidity, and other parkinsonian manifestations. Total destruction of the basal segments of VOp was most effective. After partial destruction of this subnucleus, the disturbances tended to recur (Hassler *et al.*, 1970; Hartmann von Monakow, 1962). In destroying VOp, it is necessary to interrupt the pallidothalamic, rubrothalamic, and dentatothalamic pathways passing from ND via Br Con (Cooper *et al.*, 1963; Kandel and Podgornaya, 1966; Hassler *et al.*, 1970; Riechert, 1980). Accordingly, the effect of the operation on tremor is thought not to result from the destruction of VL itself but to arise from severance of its afferent pathways, especially FTh (Pagni *et al.*, 1966). Destruction of Vim is not necessary for the elimination of tremor since in many cases Vim was damaged only partially or even left intact (Hassler *et al.*, 1979).

It has also been shown that even total destruction of GP does not abolish tremor but only diminishes it by 40–60%. All authors agree that inclusion of the medial part of CI, adjacent to VL, in the lesion did not improve the results of the operation; in fact, it might even worsen them, and might cause complications (Riechert, 1972; Hassler *et al.*, 1979).

11. Prevention and Treatment of Postoperative Complications

The success of a stereotactic operation on the basal ganglia depends to a great extent on proper and well-organized care for the patients in the postoperative period. The prevention of complications both during and after the operation is an important task facing the surgeon and anesthesiologist.

These considerations are discussed in detail in this section because of the importance of this question and because these same complications may occur after operations for other extrapyramidal diseases. The principles of treatment of these complications are nearly the same irrespective of the primary diagnosis.

As experience has shown, complications during stereotactic operations in parkinsonism are observed comparatively rarely. For example, about 95% of our stereotactic operations proceeded smoothly in spite of the fact that the majority of the patients had a severe form of the disease and ventriculography was performed in all patients immediately before surgery. On the operating table, there was not a single instance of complications connected with impaired vital functions (shock, collapse, respiratory difficulties).

Among possible complications during the operation under local anesthesia, one must watch for the appearance of flaccidity, dulled mentation, or sleepiness, which may be manifested by slow or monosyllabic replies to questions or delayed responses to simple commands. If these impairments are not very pronounced, which is often the case, they should not cause the surgeon anxiety. As a rule, they disappear several hours after the operation. However, experience has shown that a marked impairment of the general condition, even without focal neurological signs or symptoms, may be the first indication of an intracerebral hemorrhage. As a therapeutic measure in these cases, we recommend a calcium chloride solution, aminocaproic acid, or vicasol as well as dehydrating agents.

Among the complications during surgery are local neurological disorders such as speech impairment in the form of either motor aphasia, if the operation is performed on the dominant hemisphere, or dysarthria. Signs of pyramidal tract involvement in the form of hemiparesis or hemiplegia may also develop during the operation.

These complications are an indication of the inaccurate placement of the electrode–cannula, which has damaged the pyramidal tract in CI. In such cases it is necessary to remove the cannula from the brain immediately. After the calculations are corrected, the cannula may again be introduced in the proper target point. For accurate hitting of VL or Subth, it is necessary, as a rule, to direct the trajectory of the cannula more medially. Therefore, if symptoms of partial damage to the pyramidal pathway appear it is necessary to make a correction and wait until the pyramidal symptoms totally disappear before making a lesion in the proper subcortical structure.

These adjustments are advisable only if the "local" complications are not very severe (slight speech impairment or mild paresis of the arm or leg). With the development of hemiplegia, which is an indication of gross damage to CI, or a focal neurological disturbance with a definite worsening of the general condition, the operation must be stopped immediately.

A lower facial paresis on the side opposite the surgery should not be considered a complication, since a mild paresis is observed in more than one-third of the cases. This occurrence is not significant and completely disappears in the course of several days. More than that, the appearance of slight facial asymmetry of the corners of the mouth is an indirect indication that VL has been hit accurately.

It is only necessary to describe briefly an uneventful postoperative course since no complications occur after the majority of operations. Nevertheless, in the first few days after surgery, it is necessary to take a few precautions to avoid possible complications, speed the recovery of function, and improve the general condition of the patient. On the basis of our experience, we recommend the following simple measures in the early postoperative period.

Although cerebral edema is a serious complication of many neurosurgical operations, after the operation on the basal ganglia it occurs rarely, and even when it does, its severity is usually not great. Accordingly, we usually do not prescribe dehydrating solutions intravenously after the operation. To prevent brain edema, we commonly give intramuscular injections of a diuretic, furosemide, once or twice a day in the first 2 or 3 days after the operation.

We consider the administration of ganglion-blocking preparations an important means of preventing and treating the main complications. Since in parkinsonians there is often marked stimulation of vegetative functions (increased salivation, perspiration, vasomotor lability, and high blood pressure), small doses of ganglion blockers are indicated in the first 2–3 days of the postoperative period. After stereotactic operations, as a rule, antibiotics were prescribed for 3–4 days; however, in almost 2000 stereotactic operations in our clinic, there were only two cases of meningitis.

One of the rare but dangerous complications of surgery on the basal ganglia is a disturbance of thermoregulation manifested by hyperthermia of central origin. In order to prevent this complication, we give small doses of amidopyrine *per os* or intramuscularly for 2–3 days. In many parkinsonians, because of rigidity of the respiratory musculature, breathing excursions are substantially restricted. In order to prevent pulmonary disorders, it is advisable to prescribe breathing exercises: 10–15 slow, deep inspirations every half hour. A regular supply of oxygen and cardiac stimulants (one or two injections per day) are prescribed for elderly or weak patients.

If the patient has tolerated the operation well and has no postoperative complications, then on the second or third day after the operation, and in recent years on the next day, we allow the patient to sit up and in 2–3 days to walk.

The majority of postoperative complications, which at present occur in only 5–6% of cases, are of a transitory nature; they pass off quite rapidly and do not affect the convalescence or the results of the surgical treatment.

Of the postoperative complications, one of the most serious is the previously mentioned hemiplegia on the side contralateral to the operation because of surgical damage to the pyramidal tract as a result of inaccurate stereotactic calculations or an atypical position of CI. As pointed out by Bertrand *et al.* (1973), when stimulation is employed during the operation, the risk of significant hemiparesis is practically negligible. In cases of mild hemiparesis, our experience has shown that normal movements are fully restored in the course of 1–3 weeks. In severe hemipareses and hemiplegias, the frequency of which does not exceed 1%, restoration of movement is usually incomplete. The same holds true for motor aphasia occurring simultaneously with hemiparesis after an operation on the dominant hemisphere. Treatment of hemiparesis and aphasia follows the generally accepted rules for rehabilitation.

The occurrence of dysarthria presents an important problem in the postoperative period. The frequency of this complication according to the literature and our data is 3–4%. However, in elderly patients with atherosclerotic parkinsonism as well as in those patients who had dysarthric problems prior to surgery, this complication occurs more frequently, up to 7–8% and after a bilateral operation 15% and higher. It is assumed that postoperative dysarthria is caused by damage to the corticobulbar pathways in CI rostral to VOa (Hassler *et al.*, 1979). Although dysarthria is observed as an isolated symptom, it is frequently associated with mental impairment, loss of memory, and pseudobulbar disturbances, especially swallowing disorders. As a rule, dysarthria gradually disappears, but in elderly patients and after bilateral operations, the recovery is slow and incomplete. The frequency of lasting dysarthria after unilateral operations is 1–2% and after bilateral operations 3–6%.

Mental impairments of various types are observed in 5–6% of cases, which is an improvement over the early period of stereotaxis, when these were observed in up to 10% (Laitinen, 1966). The disturbances range from mild confusion, depression, and apathy to severe psychotic syndromes with mutism, loss of bladder and bowel control, and sometimes delirium and hallucinations.

Just as in dysarthria, mental disorders are more frequently observed in elderly patients with vascular disease in whom psychogenic disturbances were present prior to the operation. Neverthless, this factor is not the only cause of mental disturbances, since in the majority of patients, even elderly persons, there are no such postoperative disorders. Data in the literature indicate that transitory mental impairment (disorientation, memory loss, etc.) is observed three times more frequently after VL destruction on the left side than after a similar operation on the right side (Kullberg, 1975). Based on our data, operations on the dominant hemisphere are also followed somewhat more frequently by mental disturbances, although the difference is not great.

All neurosurgeons with many years of experience in parkinsonian surgery (Cooper, 1963; Kandel, 1965, 1981; Dierssen and Obrador, 1967; Riechert, 1972, 1980; Siegfried, 1976) note that the frequency of complications after bilateral operations is at least twice as high as after unilateral operations (Section 10.2). In view of this danger, the indications for an operation on the second side should be rigidly observed and the decision made with great caution. For the same reason, the size of the lesion in the basal ganglia on the second side should be smaller than that on the first. Cooper (1969a), who carried out a great number of bilateral stereotactic operations, points out that after such operations protracted dysarthria occurred in 20% of the cases. The greater number of patients with speech impairment after bilateral than unilateral thalamotomy may be explained by the lack of thalamic compensation when both Th are damaged. It is presumed that the memory impairments that sometimes occur after bilateral stereotactic operations may be caused by involvement of the mammillothalamic pathways on both sides.

In the past it was thought that speech impairments and mental disturbances after bilateral operations were caused by the symmetrical lesions in both hemispheres. However, this theory was not confirmed since the number of complications occurring with symmetrical and asymmetric foci proved to be the same (Kandel, 1965; Dierssen and Obrador, 1967). Markham *et al.* (1966), on the basis of a morphological study on the brains of four patients after bilateral operations, also concluded that symmetrical foci in VL do not cause mental disturbances.

Intracerebral hemorrhages, occurring in the focus or near the focus of surgical destruction in the basal ganglia, represent one of the most dangerous complications after stereotactic operations. The majority of significant hermorrhages have a lethal outcome.

The incidence of hemorrhage is relatively low. In the initial period of the development of stereotactic surgery, they were observed more frequently (3.7%; Cooper, 1961). In recent years, this complication has been encountered extremely rarely. For example, in 500 recent stereotactic operations, we had only three cases of intracerebral hemorrhages (0.6%). Crevier (1974) surveyed 70 neurosurgical centers for the incidence of intracranial hematomas after stereotactic operations and reported that in 21,000 stereotactic operations, postoperative hematomas occurred in 0.8%. The mortality rate from these hematomas was 56%. The majority of these hematomas occurred along the trajectory of the electrode or cannula. The author emphasized that the frequency of hematomas did not depend on the method of destruction (electrocoagulation, freezing, leukotome, etc.).

Anatomic data indicate that along the path of the cannula (electrode) from the cortex to the basal ganglia there are no large vessels that, if injured, could lead to an intracerebral hematoma. This is true whether the cannula is inserted through the premotor cortex or is passed through the parietooccipital cortex. The occurrence of this rare complication may be attributed to certain specific factors, including cerebral atherosclerosis and arterial hypertension in elderly patients.

A rare but important factor predisposing to the development of intracerebral hemorrhages is a blood dyscrasia. Blood clotting (prothrombin factor and others) both before and after a stereotactic operation in parkinsonism is usually normal (Kandel, 1965).

In the majority of cases the cause of intracerebral

bleeding remains unknown. In the absence of arteriosclerosis, arterial hypertension, or blood dyscrasias, this complication is quite unexpected. However, the surgeon must always keep this possibility in mind and be prepared to treat it.

The first signs of intracranial hemorrhages may appear on the operating table but more frequently are seen in 8–10 hr. However, this complication may develop 3–5 days after the operation. The clinical picture of progressing intrahemispheric bleeding, to a considerable extent, is similar to the classical signs of a hemorrhagic stroke. The first symptoms are confusion and restlessness; then focal symptoms appear—pareses, aphasia, etc. A comatose state may indicate that the blood has ruptured into the ventricular system. The most rapid and accurate method of diagnosing this complication is computer tomography.

Treatment of this complication follows general neurosurgical principles (ice packs on the head, and aminocaproic acid and calcium chloride intravenously). If a lumbar puncture confirms the suspicion of blood in the ventricular system, a ventricular puncture is indicated to evacuate the blood.

Puncturing the operated area (i.e., the focus in the subcortical ganglia) for the purpose of aspirating the hematoma has been done only in rare cases. If there are clear signs indicating an intracerebral hemorrhage, it is necessary, without losing time, to perform a small craniotomy, incise the cortex and white matter, and remove the hematoma visually. In such cases, it is possible to use the stereotactic instrument to remove the hematoma (Chapter 18).

Prophylactic measures for preventing intracranial hemorrhage include a precise and careful performance of all surgical manipulations (in particular, slow and gentle introduction of the cannula into the brain) and moderate reduction of arterial pressure on the operating table and in the postoperative period with the help of ganglionic blockers.

One of the rare and striking complications of stereotactic operations on the basal ganglia is hemiballism (hemichorea), described in Chapter 10. Paradoxically, stereotactic operations can both cure and cause hemiballism. The cause is well known, although a number of questions are still unanswered. As noted in Chapter 10, hemiballism occurs as a result of operative damage to NS or its connections with GP. If NS is abnormally situated or if the stereotactic calculations are inaccurate, NS is likely to be damaged since it is located only several millimeters from the most frequent targets, VL and Subth. In rare cases described in the literature in which there has been some damage to NS, hemiballism has not appeared (Cooper, 1963; La Fia, 1969). This may be explained by the simultaneous destruction of the pallidofugal pathways in CF (Hassler *et al.*, 1979).

It is noteworthy that postoperative hemiballism is a rare complication. In the early period of surgical treatment of parkinsonism, it occurred somewhat more frequently and appeared after 1.7–2.5% (La Fia, 1969; Laitinen, 1972) or 4.6% of the operations (Gros *et al.*, 1964). At the present time, according to data in the literature, the frequency of this complication does not exceed 0.5–0.6% and, according to Hassler *et al.* (1979), only 0.2% in 3000 operations. In the course of almost 2000 stereotactic operations, mildly pronounced hyperkinesias in the corresponding extremities developed in eight cases (0.4%). These data give an indirect indication of the accuracy of the contemporary stereotactic method.

Postoperative hemiballism, its peculiarities, and its course have been described by several authors (Cooper, 1963; Kandel, 1965; Brion *et al.*, 1965; Hopf, 1968). The severity may vary from mild myoclonic twitches in the "operated" extremities to violent, sweeping, abnormal movements of these extremities characteristic of postapoplectic hemiballism.

In the majority of cases, these hyperkinesias appear in the first days after the operation and then gradually diminish and disappear. In all our patients with such complications, hemiballism disappeared in the course of 2–4 weeks after surgery without treatment. In extremely rare cases, however, abnormal movements continue and became a serious problem. Laitinen (1966) noted choreic movements in eight patients following 224 parkinsonian operations. In all cases, these movements soon disappeared. The only exception was one patient in whom progressive hemichorea continued for almost 2 years. The development of hemiballism long after surgery is quite rare, but in three patients described by Hughes (1965), hyperkinesias appeared 3 months after the operation and were permanent.

If postoperative hemiballism continues to an annoying degree, the only solution is to repeat the stereotactic operation. In such a case, the lateral segment of GP should be destroyed. Several such operations, described in the literature, have led to the total elimination of hemiballism, as in the two cases described by Tsubokawa and Moriyasu (1975a).

Among the rare postoperative complications, mention should be made of paresthesias resulting from damage to VPL and associated with hypotonia and unsteadiness. As a rule, these phenomena are transitory and disappear 1–2 weeks after surgery.

The postoperative mortality in parkinsonism depends primarily on the frequency and nature of complications related to various factors mentioned above—age and severity of the patient's condition, the type and stage of the disease, the operative technique, etc. The postoperative mortality in different series varies from 0.5% to 7%, reflecting, first and foremost, the severity of the disease in the operated patients. It is noteworthy that during the 30 years of stereotactic surgery for parkinsonism, postoperative mortality has been steadily declining. During the past decade, it has been 1.7% in our series.

12. Personal Observations

In the course of about 30 years, we operated on over 720 patients with parkinsonism in the age range from 8 to 76 years and performed approximately 850 stereotactic operations. Bilateral operations on the basal ganglia were done in more than 130 patients. The number of stereotactic operations in different periods has varied considerably. The greatest number of operations came in the period between 1958 and 1973. Then, with the introduction of L-dopa treatment and other dopaminergic drugs, the number of operations was reduced noticeably. Since L-dopa becomes almost ineffective in 5–7 years, during the past 6–8 years the number of stereotactic operations on the basal ganglia in parkinsonism has been gradually and steadily increasing. All of the operations (except at the beginning of our work) were performed using the technique described in detail in Chapters 4 and 5. The operation took less than 2 hr.

As in other extrapyramidal disorders, we considered the immediate result of the operation to be the patient's condition about 2 weeks after surgery, i.e., on discharge from the clinic. The long-term results were evaluated after a long period of time—up to 25 years from our first operation in 1958. These data were obtained from regular examinations of the operated patients in our outpatient unit as well as from analysis of the replies to detailed questionnaires. In many cases we used the reports of neurologists who examined the patients at their homes.

The evaluation of results of stereotactic operations in parkinsonism is a complicated problem. It is comparatively easy to rate the results of operations with respect to each of the main symptoms of the disease; for example, with respect to tremor, we could assess the effect in percentage improvement or according to a five-point scale. It is noteworthy that the data of various authors on long-term results differ substantially. For instance, the reported frequency of improvement in tremor after surgery (a reduction or total disappearance of tremor) varied from 65% to 99%.

It is considerably more difficult to give a general evaluation of the results of surgical treatment, since it is obvious that such terms as "excellent," "good," and "satisfactory" are subjective and the criteria used by different authors may differ greatly. Nevertheless, practically all neurosurgeons employ these terms for appraising their operative results. Several authors did a statistical analysis of several dozen indices. Although such a statistical approach has value, an isolated index or the sum of many (irrespective of their number) does not give a completely objective evaluation of the results of surgical treatment.

We believe that the evaluation of results should be based on the "functional" principle outlined above relative to the stage of the disease (Section 7.7). In other words, the degree of improvement after the operation should be determined by the patient's ability to look after himself and to carry on his work. In addition, it is necessary to take into account the relief or lessening of tremor, muscular rigidity, akinesia, vegetative disorders, etc. Such an evaluation of the results of treatment is impossible without consideration of the condition of the patient prior to the operation.

After these preliminary remarks, we discuss the criteria on which the long-term results have been classified into the following five headings: "good result," "considerable improvement," "slight improvement," "without change," and "worsening"* (Table 12).

Under the heading of "good result," we included patients without tremor and muscular rigidity in the contralateral extremities or, after bilateral operations, in all extremities. In such cases, there was a marked lessening of the general tension and, as a result of this, a good return of the functional activities that were limited prior to the operation, i.e., the ability to walk, sit, write, dress, and fully take care of themselves (Figs. 109, 110). Many patients who for years were confined to a chair or bed-ridden begin to walk freely and stand and sit down easily. They are able to write and perform fine movements with their fingers. In this group we have included patients who resumed work or are continuing to do their usual jobs after surgery.

*In patients who had bilateral operations, the evaluation is based on the result after the second operation.

TABLE 12. Results of Surgical Treatment (%)

Results	Good	Considerable improvement	Slight improvement	Unchanged	Worse
Immediate	66	22	5	4	3
Long-term	44	34	8	7	7

The majority of patients in this group had a marked lessening or disappearance of oral tremor, hypersalivation, attacks of profuse perspiration, muscular pain, as well as a considerable improvement in phonation, facial expression, mental state, and so on. Such good (in many cases, amazingly good) results continued for many years in 40% of the operated patients.

Under the heading "considerable improvement," we have included patients with significant postoperative improvement in all the above-mentioned parameters. However, the pathological signs did not disappear completely, and the functional recovery was less complete than in the first group. We also included in this group patients who had a good result after surgery but later, during a long period of observation, deteriorated, especially as a result of the progression of symptoms on the "unoperated" half of the body. All of the patients in this group were able to take care of themselves without need of outside help or assistance. Such "considerable improvement" was obtained in 38% of our cases.

Consequently, good or satisfactory long-term results were present in 78% of the operated patients. The long-term results in patients operated on during the past decade were somewhat better, 85%.

Under the heading "slight improvement" we in-

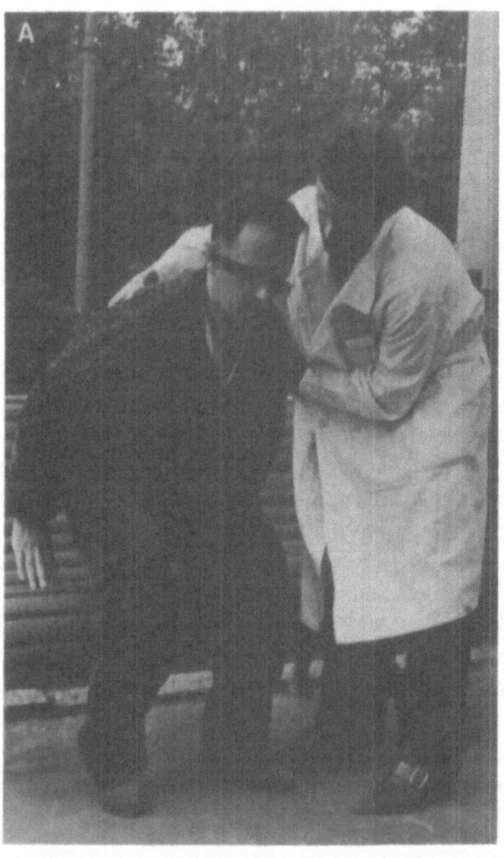

FIGURE 109. (A) The parkinsonian patient before surgery. (B) Three and a half years after stereotactic destruction of VL.

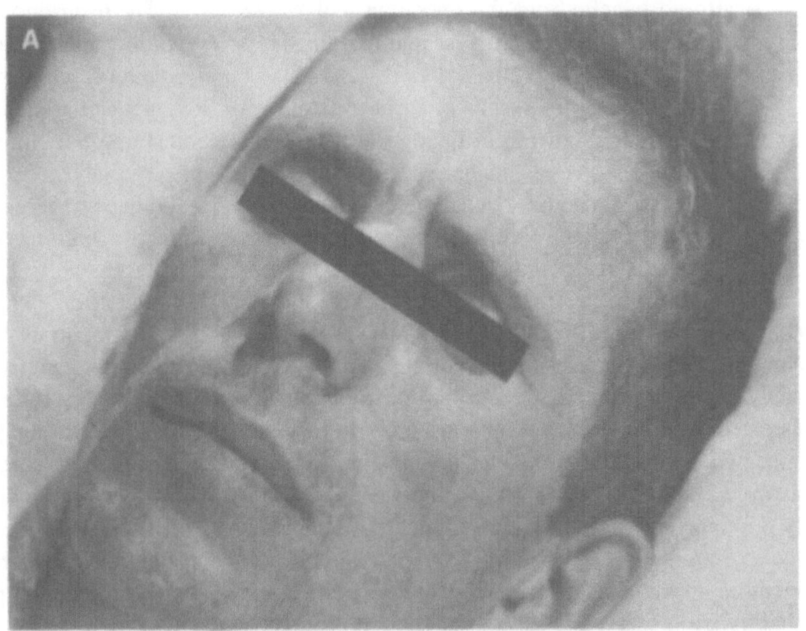

FIGURE 110. (A) Bedridden parkinsonian patient. (B) Restoration of self-care shortly after stereotactic thalamotomy.

cluded patients who did not show a marked postoperative improvement of their functional capabilities. Nevertheless, they had a lessening of tremor and rigidity as well as a return of the ability to perform a number of simple acts such as eating (usually with one hand), reading a book, and turning in bed. We have included in this group patients with partial relapses of symptoms while under observation or a further postoperative progression of the disease. Such a ''slight improvement'' was achieved in 8% of the operated patients.

The heading ''without change'' includes patients in whom the operation did not produce an effect or only a temporary improvement in the symptoms of parkinsonism (7% of the patients). The category of ''worse'' includes 7% of the operated patients who had practically no benefit from the operation and whose disease continued its rapid progress and those patients with disabling postoperative complications.

A comparison of the results of surgical treatment by various authors is always difficult, since there are differences in operative technique, severity of the disease, age of the patients, evaluation of results, and duration of observation. Nevertheless, our proportion of good and satisfactory long-lasting results is approximately the same as that of many other authors (Laponogov *et al.*, 1978; Nagaseki *et al.*, 1986; Matsumoto, 1986). For example, Bertrand *et al.* (1973), analyzing the results of stereotactic operations in 140 cases, reported excellent results in 26%—tremor disappeared completely, and patients were able to return to work. In 50% of cases, the results were satisfactory—tremor and rigidity were eliminated, but movements remained somewhat slow. A partial improvement in movements with mild residual tremor and rigidity was observed in 22% of cases. No postoperative changes occurred in one case, and intraventricular hemorrhage developed in one other case.

The long-term results of 1561 stereotactic operations for parkinsonism were briefly reported by Mundinger (1985). Tremor and rigidity were abolished or markedly improved in more than 80% of the cases. Operative mortality was only 0.4% and morbidity was less than 2%.

It is interesting to compare our early and long-term results, which are presented in Table 12.

From these data it is evident that in the course of time the benefits of surgery gradually diminish. For example, a significant positive effect was obtained at the time of discharge in 88% of the operated patients, but in the long-term observations, it was retained in 78%. This is explained mainly by early and late relapses of tremor as well as by the further progression of the disease. Return of rigidity occurred to varying degrees in 12% of our patients, which is consistent with the data in the literature. As has been noted by many authors confirmed by our own experience, a good therapeutic effect (total disappearance of tremor, rigidity, etc.) on one side does not prevent a gradual progression of the disease on the other, "unoperated" side.

Nevertheless, our data indicate that in the majority of patients obtaining considerable relief within 2–3 weeks after the operation, the improvement lasted many years. The frequency of complete relapse long after surgery, according to our data, does not exceed 5%.

The recurrence of tremor postoperatively in recent years has appreciably declined and at present is only 7% in our series. The improved results of surgical treatment of tremor were due to our new method of computing the elimination of tremor by computer frequency analysis (Chapter 2, Section 2.11). Previously, the frequency of recurrence of tremor was considerably higher: according to our data, 12%; according to Cooper (1969a), 11%; Riechert (1972), 15.8%; and Bertrand *et al.* (1973), 10%.

An important advantage of stereotactic operations on the basal ganglia for parkinsonism is the possibility not only of significantly improving motility and the general state of the patient but of arresting the progression of the disease even in rapidly advancing cases. In the majority of our patients, the good result on the "operated side" of the body was retained without change for many years. On the "unoperated side," as already noted, the process rarely stabilized and, as a rule, progressed steadily, although somewhat more slowly than before the operation. This observation often is an important factor in assessing the indications for an operation on the basal ganglia of the other side. On long-term follow-up (10–20 years), one may note that the progression of the disease has been arrested after surgery in many cases.

After unilateral stereotactic destruction of VL or Subth (or both structures simultaneously), tremor and rigidity are absent in the contralateral extremities. However, as our experience and the data in literature have shown, in approximately 15–20% of the cases there is a noticeable although less pronounced improvement in the homolateral extremities. One may assume that this effect is caused by surgical damage to the noncrossed descending corticospinal or bulbospinal tracts.

As our experience has shown, the results of surgical treatment in parkinsonism depend on the complicated interaction of many factors. One of the chief factors is the condition of the patient—whether or not he is a "good candidate" for an operation. The significance of many other factors, for example, accuracy of hitting the appropriate subcortical target, choice of this structure, and method of destruction, has been discussed.

The long-term results of surgery depend on such major factors as the type and stage of the disease. The dependence of the result of surgery on the type of parkinsonism (Section 7.6) is of great practical significance both in assessing indications for surgical treatment and for predicting results. A greater benefit on rigidity than on tremor has been noted by several authors and agrees with our findings.

The best long-term results in our own series were obtained in the rigid and rigid–tremor forms of the disease—respectively, 82% and 80% good and satisfactory results.* Somewhat less effective were operations in the tremor–rigid and tremor forms: 78% and 76% had good and satisfactory long-term results. This fall is the result to a large extent of the recurrence of some tremor after the operation. Nevertheless, even in the "pure" tremor form, the operation in the majority of cases causes a total disappearance of tremor for many years.

Indications for an operation in the akinetic form of parkinsonism are not agreed on, mainly because these cases respond well to treatment with L-dopa (Section 7.9). We agree with those authors who consider that patients without tremor and rigidity should not be operated on since thalamotomy does not reduce akinesia (Kelly and Gillingham, 1980). Prior to the appearance of L-dopa, we operated on approximately 20 patients with the "pure" akinetic form. The long-term results of these operations have shown that in almost half of these cases it was possible to obtain more or less satisfactory results.

We operated on about 70 patients with obvious atherosclerotic parkinsonism, although in many other patients operated on we suspected this etiology of the disease. The indications for surgery in this form of the disease should be greatly restricted. In almost half of these patients, speech and mental disorders of varying severity marred the postoperative period, and several patients died after surgery.

Fifteen years ago, the author (Kandel, 1972) proposed and began to perform the combined, consecutive destruction of two subcortical structures, VL and Subth. Our long-term results of such an operation proved it to be effective, a good result or a significant improvement being noted in 88% of the patients.

As our investigations have shown (Section 6), the comparatively slight increase in dopamine excretion after a stereotactic operation indirectly points to a persisting dopamine deficit in the basal ganglia. These data underscore the need to continue postoperative medication with L-dopa and other drugs.

*These figures do not include patients in the category of "slight improvement."

13. Concluding Remarks

The data presented in the previous sections allow one to give a well-founded reply to the question that enters the mind of every physician who must treat a parkinsonian: Which is preferable—medical treatment or surgery? In our opinion, this is not an either–or question. After many factors have been evaluated, treatment should be individualized. In principle, medical management is always preferable to surgery, providing, of course, the benefit is approximately the same. That is why we consider that treatment in all cases of parkinsonism, irrespective of the type and stage of the disease, should be started with L-dopa in combination with other drugs. This is especially true for the akinetic form of the disease, in which L-dopa is most effective and a stereotactic operation has little effect. It should be recalled, however, that this form is encountered rarely, in approximately 10% of the patients.

If there is no obvious intolerance to the drug, L-dopa treatment should be continued for several months, and only after that should the question of an operation be considered. However, there is also the point of view that preoperative treatment with L-dopa should be avoided as far as possible (Shichijo *et al.*, 1983).

As long as medication gives a consistent good result, there is no reason to consider a stereotactic operation. However, as stated previously (Section 7.9), L-dopa yields a good therapeutic effect in only about half of the cases, and its beneficial effect does not last longer than 5–7 years. Moreover, this drug does not reduce rigidity appreciably and has little effect on tremor. In contrast, an operation on the basal ganglia completely and permanently eliminates both rigidity and tremor in the majority of cases. Stellar and Cooper (1968) presented comparative results of treating parkinsonian tremor by stereotactic surgery and with L-dopa. After the operation there was an appreciable reduction or total disappearance of tremor in 85–90% of the cases. With L-dopa treatment (91 patients), there was no change in tremor in 57% of the cases, a reduction in 38%, and disappearance of tremor in only 5% of the cases.

In approximately half of the patients, L-dopa therapy is either ineffective or has to be discontinued because of intolerance or the development of complications. It should also be noted that prescribing L-dopa dooms a patient to many years, and possibly his entire life, of taking an expensive drug, since if the drug is stopped, the symptoms recur within a fortnight.

One may conclude that the introduction of L-dopa reduced to a certain extent but did not abolish surgery for parkinsonism. In the leading clinics of functional neurosurgery, the number of stereotactic operations in parkinsonism, which fell to a minimum in 1968–1975, once again began to increase rapidly beginning with 1976–1977. If the patient does not tolerate L-dopa, if tremor predominates in the clinical picture, or if, in spite of medication, the disease progresses, then surgery should be considered.

At the same time, the advantages of surgical treatment for parkinsonism should not be overestimated. Stereotactic operations do not offer a complete cure. However, in view of the severity of the disease, its progressive nature, and the frequent failure of other methods of therapy, it may be concluded that stereotactic operations are very effective and provide relief to the majority of patients. In three out of every four cases, we obtained good or satisfactory results by operation in spite of the fact that the majority of these cases were in severe stages (III and IV) of the disease.

The effectiveness of a stereotactic operation, the risk of which at present is minimal in our opinion, is superior to that of all other existing types of treatment. We fully agree with Siegfried (1980c), who stated that stereotactic destruction of a small part of VL or Subth eliminates tremor much faster and better than any medication and that the effect of such an operation lasts for the entire life of the patient. Our personal experience and the data in literature indicate that the benefits of surgery are considerably greater than those of L-dopa treatment. A good and lasting result after surgery was found in approximately eight out of ten cases compared with approximately 5 out of 10 with L-dopa treatment. In Tables 13 and 14 we show in brief the advantages and shortcomings of both medical and surgical treatment in parkinsonism.

TABLE 13. Advantages and Disadvantages of Medical Treatment of Parkinsonism

Advantages	Disadvantages
1. Significant improvement, especially in akinetic and (less frequently) rigid forms of the disease 2. Treatment possible at any age and in severe stages 3. Therapeutic effect lasts for several years	1. No therapeutic effect (in 25% of the cases) 2. Pronounced side effects requiring discontinuance of drug (in 20% of the cases) 3. Insignificant effect in respect to tremor 4. The drug must be taken for many years 5. Effect terminates 6–8 years after beginning of L-dopa treatment 6. Financial problems related to cost of drugs

TABLE 14. Advantages and Disadvantages of Surgical Treatment of Parkinsonism

Advantages	Disadvantages
1. Total disappearance of tremor and rigidity (in 80–85% of cases) 2. Results last indefinitely 3. Practical recovery with operation in stage of hemiparkinsonism 4. Smaller postoperative doses of drugs	1. Risk of serious postoperative complications (2–3% of cases) 2. Complications after bilateral operations (10–12% of the cases) 3. Low efficacy in akinetic form of disease 4. Age restrictions (over 70 years, inadvisable) 5. Various contraindications to surgery

These data seem to indicate that in many cases of parkinsonism the most beneficial therapy is combined medical and surgical treatment.

We conclude that surgery has an important place in the management of parkinsonism, especially since advances in stereotactic neurosurgery promise to yield even better results in the future.

To complete this chapter, we should note briefly that several scientific centers, first and foremost Stockholm, are developing an essentially new method for treating parkinsonism, which opens up unique prospects for the future. This involves the implantation in deep-seated brain structures of heterogenic embryonic tissue or the patient's own tissue capable of synthesizing dopamine. Experimental studies have shown that such tissue has no antigenic properties and, once implanted in the brain, is not rejected but grows rapidly, increasing in volume 10–15 times.

It has been shown experimentally that impaired motility in rats caused by lesion of the nigrostriatal complex with 6-hydroxydopamine (6-OHDA) diminished substantially after transplantation of embryonic SN tissue into Str (Perlow *et al.*, 1979; Freed *et al.*, 1983).

In a recently published neurosurgical paper,

Backlund *et al.* (1985) describes a preliminary investigation of the transplantation by stereotactic techniques of autologous adrenal medullary tissue into Str of two patients suffering from severe parkinsonism. The aim of this operation was to stimulate dopaminergic receptors of Str with monoamines of the transplanted tissue. It is presumed that the implanted cells will produce catecholamines to replace the deficit in the basal ganglia. The clinical effect after surgery was very moderate in both patients; in only one it was possible to decrease the dose of L-dopa.

About 30 such transplant operations were performed in different countries during the last few years, including about a dozen performed by a Mexican team in La Raza Hospital, Mexico City. Some surgical details are important. The routine approach to the adrenal gland was used and the removal of its medullary tissue, weighing about 1 g, was performed under the operating microscope. The cortical tissue of the gland should not be included in the transplant and must be removed carefully. The piece of medullary tissue was attached by a small clip to the surface of NC of the nondominant hemisphere.

The future will show to what extent this novel and interesting approach will have real clinical application.

7

Dystonia Musculorum Deformans

Since there is not as yet any drug that can reverse the symptoms of this disease, neurosurgeons became interested in the possibility of lessening the pain and muscle spasm by operating on delicate mechanisms within the brain.

I. S. Cooper

1. Historical Note

Dystonia musculorum deformans (torsion dystonia) (DMD) is a severe, chronic, and usually progressive disease of the CNS. This disease was identified from the large group of hyperkinesias as an independent nosological entity by Schwalbe (1908), although 20 years earlier Gowers (1888) had described a disease with symptoms quite similar to DMD (Eldridge, 1970).

In 1908, Schwalbe, in his doctoral thesis *Eine eigentümliche tonische Krampfform mit hysterischen Symptomen,* reported three patients, two brothers and a sister, who were studied for 30 years. Schwalbe concluded that their affliction was related to a previously unknown disease of a hereditary nature. This opinion was confirmed when two of three children of one of the affected brothers developed the same disease (Fig. 111).

Shortly, Ziehen (1911), from whose clinic Schwalbe's work was published, described in detail the symptoms of the new disease and confirmed its nosological independence on the basis of Schwalbe's three cases and his own two observations. In all five patients, the disease began in children of good health. Though the clinical picture of the disease was described in detail, Ziehen, as did Schwalbe, erroneously related the disease to the category of "hysterical neuroses" ("*tonischen Torsionneurose*").

A complete description of the clinical picture of DMD was presented by Oppenheim (1911), who was the first to identify correctly its organic nature. He believed that the disease was based on subtle destructive changes in the cerebral "tone-regulating centers." It is noteworthy that although Oppenheim did not support the view of Ziehen and Schwalbe regarding the hereditary nature of dystonia, Oppenheim's observations included two brothers suffering from this disease. Oppenheim was the first to remark that dystonic symptoms appear at the age of 8–14 years and increase during standing and walking. He noted the typical progressive course of the disease, which he considered similar to double athetosis; "transitive forms" between these diseases were described. Oppenheim called the disease "dystonia musculorum deformans," which, as the most descriptive term, is still used in current literature.

In the same year, a paper by Flatau and Sterling (1911) gave a description of an identical disease in two brothers 7 and 11 years old. The authors called the disease "*progressiver Torsionspasmus bei Kindern.*" Soon Fraenkel (1912) reported four cases of this disease and, by analogy with torticollis, termed it "tortipelvis," a name that was not accepted. At times, torsion dystonia, which was proposed by Mendel in 1919, has been used, but this term is less appropriate than DMD because the torsion-component is sometimes not the main symptom or is absent. In older publications the condition was also called the Ziehen–Oppenheim disease (proposed by Bernstein, 1912).

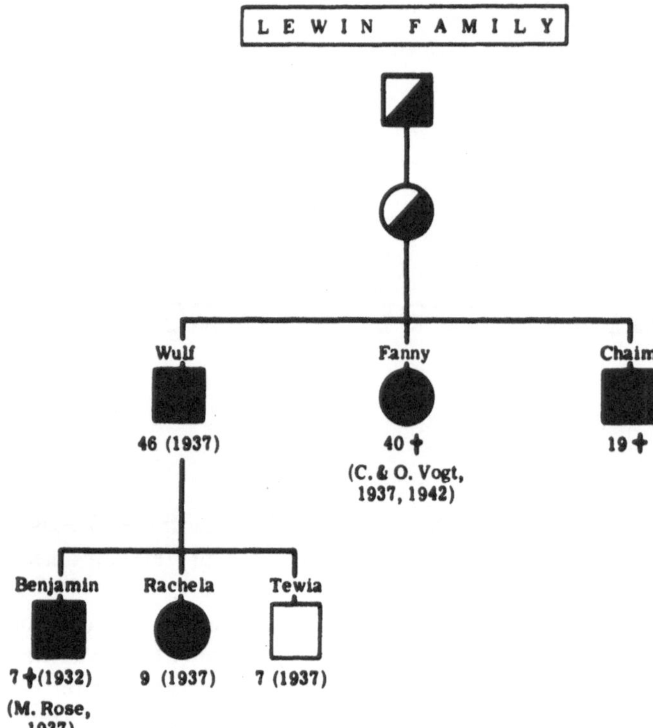

FIGURE 111. Pedigree of the Lewin family first described by Schwalbe (1908) for the first three generations (Zeman and Dyken, 1968). After Schwalbe, the family was described by Regensburg (1930). The pathological investigation of the brains of three members of the family was reported by O. and C. Vogt (1937) and Rosė (1937).

In the Russian literature DMD was described for the first time by Dzerjinski (1916) and Davidenkov (1919). Subsequently, Davidenkov and Zolotova (1921) reported three cases of DMD in one family. Davidenkov gave a clear definition of the term "dystonia": ". . . irregular distribution of the muscle tone, perverting normal posture or normal movement" (Davidenkov, 1957, p. 160). As Fahn and Eldridge (1976) emphasize, this term is applied not only to the peculiar type of involuntary movements but to the specific neurological disease. Dystonic movements can be considered inadequate muscle contractions producing motor activity.

As more reports accumulated in the literature, it was shown that DMD may be not only a "primary" but also a "secondary" state, i.e., may be the consequence of other CNS diseases (see Section 2). These reports raised serious doubts as to whether dystonia was an independent nosological condition. A number of authors have come to the conclusion that DMD is always a secondary disorder, i.e., symptomatic of some nervous system disturbance (Wimmer *et al.*, 1929).

Davidenkov in 1932 and again in 1957 and Herz in 1944a clearly stated that two forms of torsion dystonia exist independently. One is the disease *sui generis,* and the other a syndrome symptomatic of different diseases of the CNS. This principle, in our opinion, is correct even today.

There are reasons to suppose that DMD is not so rare a disease as it was considered earlier. After the first report by Schwalbe, about 1000 cases of the disease have been described in the world literature. Whereas 20–30 years ago only sporadic cases were reported in each paper, in recent years dozens of cases are reported every year.

2. Etiology

The etiology of DMD remains insufficiently studied. All cases of DMD may be divided into three groups based on the etiologic factor: (1) primary (idiopathic, sporadic); (2) secondary (symptomatic); and (3) hereditary–familial (Kandel and Vojtina, 1971). A more complex classification that includes approximately 30 proven and supposed factors and combinations of DMD with other diseases of CNS was proposed by Fahn and Eldridge (1976).

According to our classification, the majority of cases of the disease are assigned to the first group when neither history nor clinical or laboratory investigations disclose an etiologic factor (Zucker, 1960; Cooper, 1962a; Zeman and Dyken, 1967; Kandel and Vojtina, 1971). As far back as 1944, Herz remarked that the absence of a clear etiology, in many cases, is a reliable criterion to differentiate DMD from similar but known diseases of the extrapyramidal system. This principle is undoubtedly correct even now. In 105 cases out of 253 collected from the literature, the etiology of DMD was unknown (Zeman and Dyken, 1967). In 48 operated patients of the 62 who were analyzed in our monograph (1971a), the cause of the disease also remained unknown.

At the same time, the development of DMD was often reported to follow trauma to the extremities, freezing, psychic trauma, and so on. However, it is likely that such factors are not etiologic but only precipitating elements in the latent stage of dystonia.

Secondary (symptomatic) DMD, the dystonic syndrome described by many authors, is polyetiologic and can develop after other CNS diseases. Foerster (1921) and a number of other authors (Levy, 1922; Lwoff *et al.*, 1922) reported the development of dystonia after epidemic lethargic encephalitis or other encephalitides. There are many reliable reports of typical dystonia developing in athetosis, hypatocerebral degeneration (Jacob, 1932; Konovalov, 1948), Huntington's chorea, Hallervorden–Spatz disease (Marsden, 1976), cerebral rheumatism, brain tumors, and also after different infectious diseases (influenza, abdominal fever, brucellosis, malaria, pneumonia, etc.), intoxications (carbon monoxide, manganese, etc.), and head trauma.

It is known that the long-term treatment of certain mental diseases with tranquilizers and neuroleptics in some cases causes different extrapyramidal syndromes and, in particular, typical DMD. We operated on three patients who developed severe DMD after long treatment with neuroleptic drugs.

It should be noted that secondary dystonia is observed much less frequently than the idiopathic form. Only 10% of our patients had the syndrome of dystonia (Kandel and Vojtina, 1971).

The hereditary–familial form of DMD occurs next in frequency after idiopathic dystonia. In order to determine the genetic nature of any disease, a long-term follow-up of several generations is necessary. Detailed clinical investigations not only of the patients but of their healthy relatives in whom mild symptoms of the disease (*formes frustes*) can quite often be noticed are also very important. Such long-term and extensive investigations naturally encounter many difficulties and are seldom performed. In this connection it should be noted that DMD, unlike, for example, Huntington's chorea, the dominant hereditary nature of which was firmly established a long time ago, is a disease predominantly of children. This means that the majority of patients with dystonia are invalids their whole life, quite often bed-ridden, and thus, as a rule, do not have children. Consequently, it would hardly be possible to determine the transmission of the disease. Thus, the high rate of sporadic cases and comparative rarity of familial cases can not be a decisive argument against the hereditary nature of DMD.

Analysis of data from the literature shows that during the 70 years in which dystonia has been studied controversial points of view about the hereditary factor in this disease existed and continue to exist. Since Schwalbe's first report there is no doubt about the familial–hereditary nature of many cases of DMD, but the rate of confirmed familial cases varies according to the data of different authors and comprises only approximately 30–40% of all cases of DMD. Thus, Cooper (1976) indicated the hereditary factor in 28% of cases, Marsden (1976) in 22%, and we in 17%.

The question of whether dystonia is undoubtedly a genetically conditioned disease is still controversial. Though many authors deny the definitive significance of the hereditary factor in the etiology of the disease (Herz, 1944a; Wilson, 1954; Ribera and Cooper, 1960), it should be emphasized that Zeman and Dyken (1967) collected from the literature data about 117 familial–hereditary cases of dystonia and added 31 personal observations. Subsequently, Eldridge and Gottlieb (1976) collected from the literature descriptions of 768 families with cases of this disease, totaling 1179 patients. Approximately 60% of the patients had hereditary DMD.

Several illustrative examples follow. Wechsler and Brock (1922) observed a female patient whose brother, sister, father, aunt, and cousin suffered from dystonia but whose four brothers and sisters, mother, and grandparents had no signs of the disease.

Regensburg (1930) reported a unique family in which seven of eight children suffered with DMD, four of them in severe form. Their father and uncle also suffered with generalized dystonia. In all patients the disease had a progressive course.

Keyserling (1956) described a family in which the father and four of five sons, two of whom were twins, suffered with DMD. Zeman *et al.* (1960) observed an American family of 98 members, 12 of whom were affected by muscular dystonia. The genetic analysis showed that the disease was observed in four generations. Five patients had a severe generalized dystonia, two dystonia of mild degree, and five mainly local dystonia. The authors concluded that the disease was inherited as a dominant allele with high penetrance of the pathological gene. As a result of further genetic investigations, Zeman and Dyken (1967) described in detail five more families with dominant inheritance of DMD, and Hoefnagel *et al.* (1970) two families.

It has been established that genetic transmission of dystonia may occur both by dominant and recessive genes. Both types are encountered approximately at the same rate (Tabaddor *et al.*, 1978). Recessive inheritance includes observations in which the familial cases of dystonia were present in only one generation. From the point of genetics, the occurrence of the hereditary disease in children is quite possible if their parents were clinically completely healthy but one of them is a carrier of the pathological gene.

Larsson and Sjögren (1966) performed a detailed and careful genetic study of many generations of families with hereditary dystonia. After studying the whole population isolated in the north of Sweden, the authors found an unusually great number of patients with dystonia. The authors emphasized that this population is quite stable, lives in significant isolation, and includes many marriages between relatives. All these factors significantly facilitated the genealogical study (including the catamnesis data about expired patients), which covers the period from 1880 to 1960. The authors identified 121 patients with DMD, and 58 of them were still alive. Fifty-two of the patients were seen and examined clinically. The 121 patients (67 men and 54 women) belonged to several generations of 43 families that originated from three couples, whom the authors named "common ancestors." In 20 of these 43 families fathers suffered dystonia, in 13, mothers, and in one, both parents; in nine families the parents were healthy.

On the basis of their study, Larsson and Sjögren conclude that DMD is an independent hereditary disease that is caused by mutation of one gene with subsequent autosomal monohybrid transmission of the dominant type.

We observed many cases of hereditary dystonia, in particular two families in which each brother and sister suffered from DMD. In all four cases stereotactic operations on the basal ganglia were performed.

In discussing the role of the hereditary factor, one must note a point that has caught the attention of many investigators. The patients described in the early papers of Flatau and Sterling (1911), Oppenheim (1911), Mendel (1919), Rosenthal (1922), and others originated from the Jewish population of Poland, Galicia, and the western regions of Russia. The greater morbidity of dystonia in Jews, mainly those with the autosomal recessive inheritance, has been remarked in many papers (Eldridge and Gottlieb, 1976). On the basis of these data, Runge (1936) and Ryan (1950) concluded that Jews have a "special predisposition" to this disease. At the same time, dystonia was described many times in patients of different nationalities. Davidenkov (1932) pointed out that this disease is encountered in Frenchmen, Germans, Swedes, Brazilians, and others.

The existing reports of the disease mainly among Jews may be explained by the difficult living conditions of these people in Imperial Russia, when their isolation in regions or reservations led to increasing marriages among relatives. This last factor creates conditions under which recessive dystonia genes could manifest as the disease more often than with random matings.

The hereditary nature of DMD is also confirmed by the possibility of its development in twins. There are several such reports in literature. Kalinina (1966) observed two affected sisters—uniovular twins whose parents, however, were healthy. It is supposed that "gene penetrance" in this disease is not high, about 0.6, and the incidence of the pathological gene according to the Hardy–Weinberg law is approximately 1 : 200,000 (Zeman and Dyken, 1967).

An important argument in favor of the hereditary nature of dystonia is the presence of *formes frustes* of this disease, which are often encountered in families in which there are DMD patients. These forms, studied in detail by Zeman *et al.* (1960), were subsequently described by many authors.

In conclusion, despite many undoubted advances, the problem of DMD etiology has not been solved and requires further multidisciplinary investigations.

3. Pathology

Morphological investigations represent a reliable basis for understanding the pathogenesis and clinical manifestations in the majority of CNS diseases. However, DMD is an exception to this rule. The pathology of this disease has not yet been sufficiently studied. There are only about 30 authentic reported cases with adequate morphological studies of the brain.

It is necessary to emphasize that the main task of such an investigation is to discover changes in brain structures that are specific for DMD, e.g., changes that correlate with the clinical picture of the disease and are the basis of its pathogenesis. It is self-evident that changes in any brain structure could be specific only if they are observed in most or all autopsies. In this connection, extreme caution is required to interpret each observed change as specific for dystonia. It must also be noted that a necessary condition for retrospective analysis of cases reported in literature is that there be complete confidence in the diagnosis of DMD. In the first period of the study of the disease, many authors described other similar diseases, e.g., Wilson's disease and double athetosis with torsion and dystonic clinical signs. In this connection, many pathological cases in the literature of the 1920s through the 1940s described as "pure" DMD appeared after a retrospective study to be either the dystonic syndrome or another nosological form of extrapyramidal disease.

Furthermore, these patients usually die from different complications (cachexia, pneumonia, hyperthermia, etc.), which in themselves cause many morphological changes in the CNS. In this connection, there is the real danger of confusing the consequences of dystonia with its cause.

Richter (1923) carried out a morphological study of DMD and found significant changes in Put in the form of diffuse neuronal degeneration, though without changes in glia, vessels, or fibers. Changes in GP were insignificant. After Richter's observation, the pathogenesis of DMD in neurological literature for many years was connected with dysfunction of Str, and this disease was considered a "striatal syndrome." The following pathomorphological studies also disclosed selective damage to Put (Marinesco and Nicolesco, 1929; Davidson and Goodhart, 1938).

Herz in 1944 collected published reports of 26 pathological cases with the clinical diagnosis of torsion dystonia. The retrospective study showed hepatocerebral degeneration in four of them, epidemic encephalitis in four, and other diseases of the CNS in three. The majority of the remaining 15 cases were related to DMD, but as Herz remarked, the morphological data were often controversial and did not make it possible to draw any definite conclusions.

The investigation of the brain of a patient from the family L., the observation of which allowed Schwalbe (1908) to describe DMD as an independent disease, is very unusual and instructive. After many years, the famous neuromorphologists O. and C. Vogt investigated the same brain and reported their results in two papers. In the first (Vogt and Vogt, 1937), the authors described significant degeneration and loss of Put neurons, which seems to confirm Richter's data. Changes in other subcortical structures were insignificant and in the authors' opinion were not specific for DMD. The main conclusion of the authors, that Put is the locus morbi of DMD, is cited in many textbooks as a proven fact. But 5 years later, the Vogts (1942) published their second paper dealing with reinvestigation of the same brain; they rejected the conclusion of the first report and stated that the previously described changes in Put were secondary and that primary changes specific for DMD were connected with the destruction and cell loss in the parvocellular portion of CM.

Twenty-five years later, Zeman and Dyken (1967) again investigated the same brain, twice described by O. and C. Vogt, with application of modern morphological methods and drew a conclusion completely refuting the results of both publications. A careful comparison of slides of this brain and of control preparations showed that there was no loss of neurons either in Put or in CM. Zeman and Dyken (1967) were unable to get the brain of the patient, the son of one of the brothers of family L., for reinvestigation. This brain was studied in detail by Rose (1937), who agreed with the Vogts in finding primary atrophy of Put neurons. However, Zeman and Dyken studied the slide photographs made by Rose with magnifications of 100 and 450. A comparison with the control slides showed that there was no decrease of Put ganglion cells. The authors concluded that insignificant neuronal changes in both cases could not be considered specific for DMD.

In 1955, Meyers published a paper based on the largest series of pathological observations in world literature. In 15 cases of idiopathic DMD, the author disclosed a diffuse degenerative process, mainly in the striopallidal system. Hassler (1956) also remarked that the main pathological changes were localized in the striopallidum, although the most significant changes were observed not in Put but in the major and minor cells of GP. Less significant morphological changes were found in NC, NR, Subth, and also in ND and SN. In his following paper, Hassler (1966a) reported that in one case of dystonia significant degeneration and loss of neurons were apparent in CM.

Davidson and Goodhart (1938), in four typical cases of dystonia, observed status marmoratus of the basal ganglia, a pathological condition first described by O. and C. Vogt (1920). Similar data were obtained by Alexander (1942), who concluded that status marmoratus is the morphological basis of all juvenile cases of dystonia. This point of view was supported by other investigators (Denny-Brown, 1962; Johnson *et al.*, 1962).

Can status marmoratus be considered the pathological basis of dystonia? This conclusion is quite doubtful. It is obvious that Vogt and Vogt, as well as their pupil Rose, would have noted status marmoratus had it been present in their cases. Nor was this condition present in the pathological observations described by Schmitt and Scholz (1932), Dimitri (1935), De Lange (1945), and others. On the basis of a retrospective analysis, Zeman and Dyken (1967) question the three cases of Alexander (1942) on the basis that the status marmoratus in these cases was a consequence of a perinatal injury to the brain and that the cases should not be considered "pure" DMD. Zeman and Dyken concluded that partial sclerosis of Str was present in the first case of Davidson and Goodhart, and that in their other three cases, judging by the photographs, status marmoratus was not present at all. It seems that there is no reason to presume that the basis of dystonia is status dysmyelinisatus.

To sum up, all morphological studies of undoubted dystonia published in the literature can be divided into two categories. In the first group, which contains the majority of observations, no specific pathological changes that could be considered the cause of the disease were disclosed. A convincing study by Zeman and Dyken (1967) leads us to believe that in the cases of O. and C. Vogt and Rose no morphological changes were found sufficient to establish any clinicoanatomic correlation with DMD.

Two personal observations by Zeman and Dyken (1967), carefully studied by modern morphological methods, also ought to be included in this group. The brain preparations of both cases of hereditary dystonia were studied by several experienced pathologists, of whom only one knew the clinical diagnosis. In both cases no signs of neuronal degeneration were found in the brain structures. Several earlier and contemporary studies with negative results should be added to this group; the authors could find no cerebral morphological changes specific for DMD. Recent studies using CT also have not disclosed any changes in the brains of persons with this disease.

These observations should be contrasted to the second group, in which morphological changes were found in discrete brain structures that the authors considered primary and specific for the disease. In these cases neuronal degeneration was found in the striopallidal system, most often in Put and GP but also in CM.

In this connection, a morphological study of the brain of a 14-year-old patient with severe generalized DMD is interesting (the detailed description of this case was published previously, Kandel *et al.*, 1973c). Quite unexpectedly, the most significant pathological changes were disclosed in both GP, which were almost completely necrotic (Fig. 112A). These changes were not related to the routine stereotactic cryodestruction of VL performed in the patient. The second interesting finding was the presence in the GP of a small number of preserved astrocytes containing blue-black pigment granules giving a ferrum-positive reaction (Fig. 112B). We presume that some cases of typical DMD arise from lesions of the GP (pallidal degeneration). Changes present in the above case in NS and SN were much less significant.

Summing up the data on the pathological anatomy of this disease, one can conclude that the statement made by Zeman and Dyken in 1967 remains justified: "It would be more correct to assert that the pathological anatomy of deforming muscular dystonia remains unknown."

4. Pathogenesis

The pathogenesis seems to be the most complicated and the least studied of the problems in DMD. It must be admitted that recent advances in neurophysiology have not led to an understanding of the mechanism causing the dystonic phenomena and generalized motor disturbances. Analysis of the literature shows that all attempts to formulate a general theory of the pathogenesis of DMD have been unsuccessful.

It is believed that a lesion of the basal ganglia is the basis of this disease. Although the pathological picture is adequate for the understanding of the pathogenesis of most extrapyramidal diseases, the changes in DMD (see above) are not yet well enough established to indicate the brain structure involved and the nature of its malfunction to cause development of a dystonic picture. Hence, a satisfactory theory of the pathogenesis of dystonia is a matter for the future.

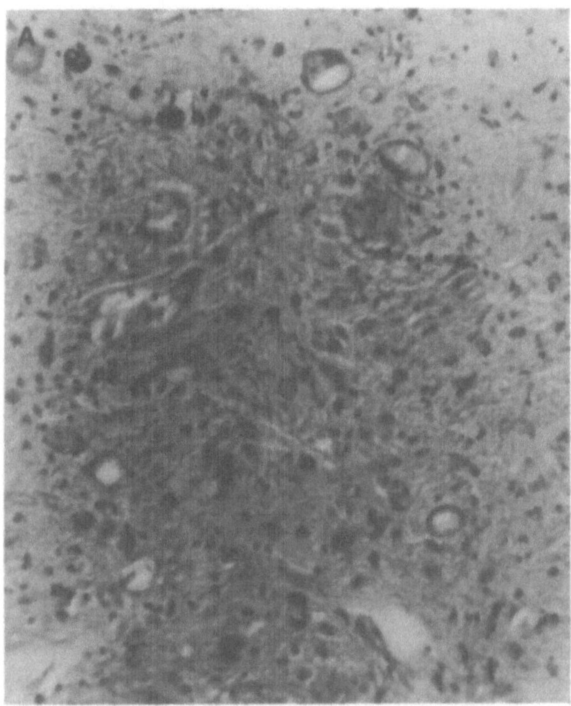

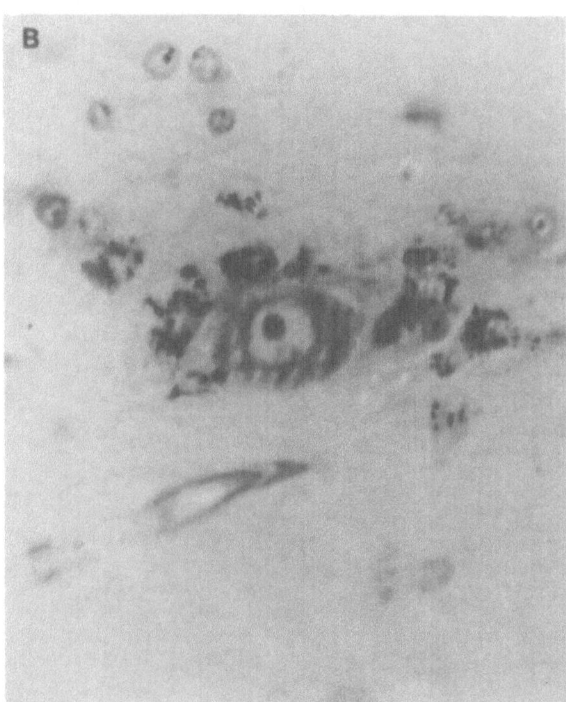

FIGURE 112. Pathological changes in GP in dystonia musculorum deformans. (A) Near-total destruction of GP. Nissl stain, 10 × 10. (B) Collection of iron-containing pigment in astrocytes surrounding abnormal GP neurons. Nissl stain, 40 × 40.

Stereotactic surgery for DMD not only appeared to be the most effective method of treatment but significantly improved our understanding of its pathogenesis. The surgical treatment of dystonia showed that irrespective of their pathogenesis, the symptoms are practically all fully reversible after a stereotactic operation. This fact indicates that dystonia depends not on a destructive lesion of a brain structure but on a "functional" disturbance of cerebral structures responsible for the regulation of motor activity (Fig. 113). The "reverse" fact is also evidence of this supposition—long-term administration of L-dopa quite often gives rise to hyperkinesias typical of DMD, and these are relieved after the drugs are discontinued (Chapter 6, Section 7.9.2).

The study of the pathogenesis of DMD and the results of stereotactic operations leave no doubt that the subcortical structures involved in dystonic phenomena have a pronounced somatotopic organization. This is evident from three observations. The first is that DMD begins in part of the extremity (palm, foot) and subsequently spreads to the whole extremity. Secondly, local forms of the disease occur, e.g., involvement of only the neck musculature in spasmodic torticollis (Chapter 8). Thirdly, in some cases, after stereotactic operations, the dystonia disappears in only one extremity and remains in another. All this implies that at any level of the CNS where pathological activity is generated, it may involve just one part of the body, possibly even one muscle group or an individual muscle.

5. Clinical Investigation

Because the EMG is the main clinical method for investigating the pathological state of muscles, EMG results in DMD have been described many times (Cooper, 1964; Podivinsky, 1964; Petelin, 1965; Kandel and Vojtina, 1971; Kandel, 1981). There is an almost continuous high-amplitude electroactivity (at times up to 200–300 μV) in the affected muscles, which is quite similar to the typical activity of voluntary muscle contraction. This activity has a frequency of about 80–90 Hz, which sometimes is replaced by a low-frequency component (12–18 Hz) and occasionally with periods of electrical silence in the affected

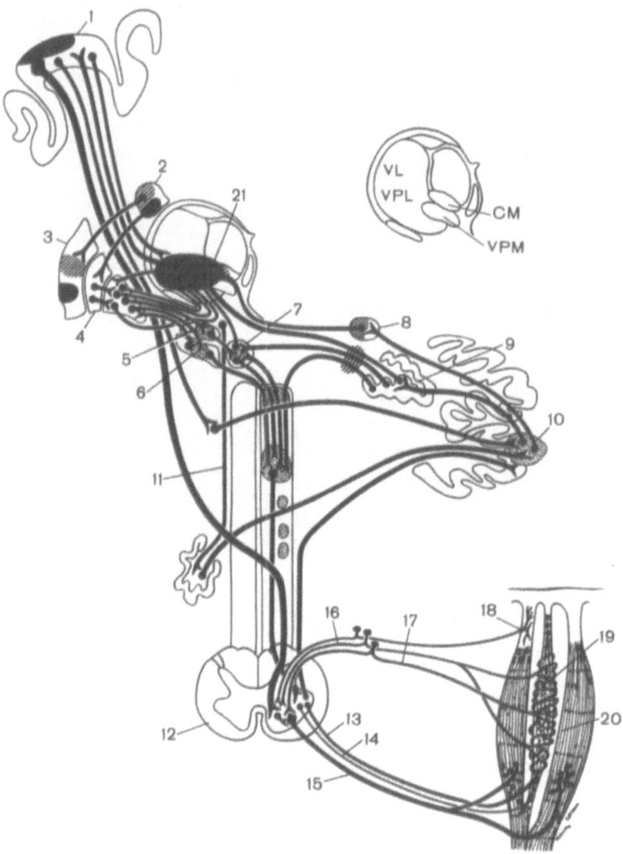

FIGURE 113. Scheme illustrating the anatomic structures of the brain involved in the pathogenesis of DMD and the tracts through which pathological activity passes to cause dystonic phenomena and intention tremor (Cooper, 1965c). 1, Motor cortex; 2, NC; 3, Put; 4, GP; 5, NS; 6, SN; 7, NR; 8, nucleus emboliformis; 9, ND; 10, anterior lobule of Cer; 11, the brainstem FR; 12, spinal cord segment; 13 and 14, γ-1 and γ-2 efferents; 15, α afferent fiber; 16 and 17, afferents of groups I and II in posterior root; 18, Golgi corpuscle; 19, muscle spindle; 20, striated muscle; 21, lesion site most effective for the relief of DMD. Inset upper right: Scheme of Th nuclei.

muscle, which does not differ from the normal EMG (Larsson and Sjögren, 1966). Affected muscles maintaining posture have continuous activity with repeated short bursts superimposed (Herz and Meyers, 1962).

The EMGs of antagonist muscles (for example, carpal flexors and extensors) clearly demonstrate one of the main pathophysiological characteristics of dystonia—simultaneous contraction of antagonist muscles on any attempt at voluntary movement (see below). Thus, the visually observed dystonic hyperkinesias and postures can be objectively confirmed by the EMG, but it must be noted that this test does not disclose the basis of the dystonic phenomena.

In Chapter 2, data were presented about the pathophysiological basis of muscle tone disturbances and about the role of cerebral structures in their pathogenesis. These data are supplemented with some facts regarding the pathogenesis of DMD.

The main "peripheral" mechanism of dystonia, which to a significant degree is involved in hyperkinesias and posture disturbances, is pathological spasm of antagonists or "oppositionism," which was noted by Wilson (1914) and studied in detail by several investigators (Davidenkov, 1957; Cooper, 1965d; Petelin, 1965; Kandel and Vojtina, 1971). Davidenkov (1957) defined this phenomenon as ". . . involuntary contraction of musculature opposing purposeful movement." In this case any voluntary movement is countered by simultaneous contraction of antagonists, which naturally distorts this movement or the posture and makes delicate coordination of different muscular groups impossible. In a certain sense, a "breakdown" of the reciprocal innervation mechanism lies at the basis of DMD.

However, it must be remarked that the principle of reciprocal innervation regulates mainly simple movements, which are controlled by structures at the segmental spinal level. A more complicated regulatory mechanism is characterized by complex motor, tonic, and postural motor activity, which was called coinnervation. In that case, efferent impulses not only activate the muscles carrying out the given movement but

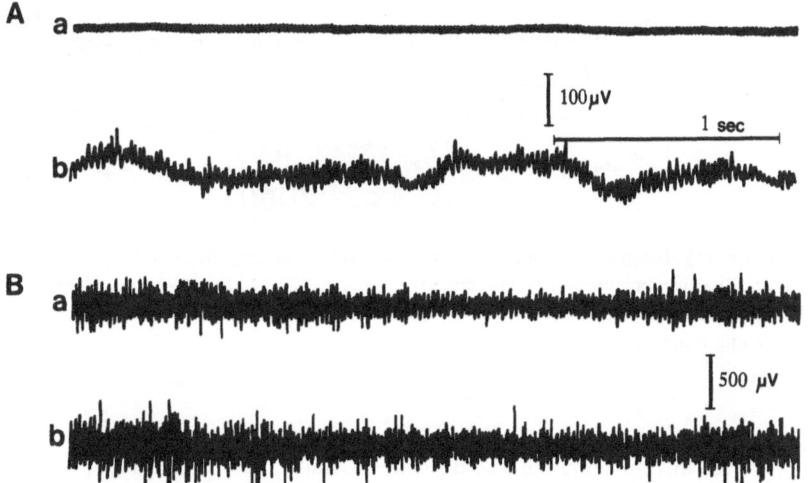

FIGURE 114. Typical spasm of antagonists in DMD patient. (A) EMG of m. biceps (a) and m. quadriceps (b) of the right hip. At rest there is marked spontaneous activity in m. quadriceps without activity in m. biceps. (B) The same recording during voluntary extension of the right leg. Note pronounced increase of electroactivity in both muscle antagonists.

radiate widely in other synergistic and antagonistic systems.

Simultaneous input of asynchronous impulses in agonists and antagonists leads to a disordered voluntary movement (for example, extension instead of flexion). This phenomenon is clearly seen in EMG recordings of antagonists (Fig. 114) disclosing disturbance of reciprocal innervation.

5.1. Westphal Phenomenon

Another pattern typical of DMD is the "shortening reflex," a phenomenon described by Westphal more than 100 years ago. In 1880 he first observed a "paradoxical muscular contraction" in some CNS diseases. This reflex later was called the Westphal phenomenon. According to the original description of the author, "the muscle, which is passively shortened when the points of its fixation are drawn together, contracts." Since such passive shortening, which is usually followed by relaxation of the muscle, causes exhaustion, the phenomenon is termed paradoxical. Consequently, unlike the above-described spasm of antagonists, this phenomenon results not from active contraction but from passive shortening of the muscle. The contraction involves not only the stretched muscle (stretch reflex) but also the muscle that is shortened. Westphal observed the phenomenon of passive shortening in many muscles, but particularly in the musculus tibialis anterior. If the foot is passively dorsiflexed, both it and plantar extensors contract tonically, fixing the foot in a position of dorsal flexion for some time.

The existence of the Westphal phenomenon, mainly in diseases of the extrapyramidal system, was confirmed by Foix and Thévenard (1922), who called this phenomenon the "posture reflex" (*réflexe de posture*). Foerster (1921) described the phenomenon as an "adaptive reflex" because in the author's opinion it adapted the muscle to its new length after passive shortening. Yanagisawa and Goto (1971) did a detailed EMG study of DMD and noted that the Westphal phenomenon was present in all patients but most often could be recorded from the anterior tibial muscle.

The Westphal phenomenon can be recorded electromyographically in approximately 30% of healthy subjects; however, the reflex is normally insignificant. This phenomenon is described in patients with cerebellar lesions, double athetosis, and parkinsonism. In many papers, the reflex nature of the phenomenon has been established, and for this reason it has been called the "shortening reflex." Though this reflex is sometimes present in normal subjects, when it is hyperactive, it must be considered pathological, reflecting a disturbance of the innervation mechanisms. The shortening reflex may be related to extrapyramidal rigidity, as it is present in many extrapyramidal lesions.

Since the pathogenesis and pathophysiological mechanism of the Westphal phenomenon have been insufficiently studied, we made a detailed investigation of this phenomenon in five patients with DMD aged from 11 to 22 years (Safronov and Kandel, 1975). All patients suffered to various degrees from the generalized form of the disease. Each patient was investigated repeatedly before and after stereotactic operations on the basal ganglia. The special apparatus for muscle tone investigation that was employed in this study is described in detail in Chapter 2 (Section 3).

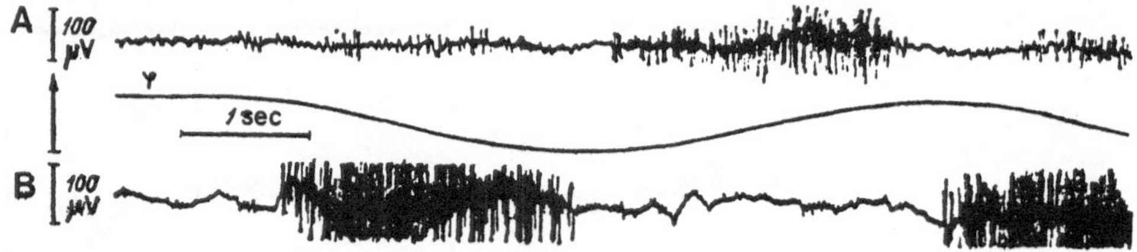

FIGURE 115. Electromyogram of m. biceps (A) and m. triceps (B) in DMD patient during passive movements of the arm on the monitor (mechanomyotomograph) described in the text. The middle curve shows flexion (upward) and extension (downward) of the forearm. Note the burst of electroactivity in the flexor during the passive flexion and in the extensor during the passive extension, the shortening reflex of Westphal.

With this apparatus it was possible to study the development of tonic spasm during the routine investigation of muscle tone. With repeated passive flexion and extension of the forearm, one could note the gradual increase in reflex activity in the flexor phase, after which the pronounced shortening reflex appeared (Fig. 115). The development of spasm may be divided into two periods: a period of relative inactivity with a slight increase in the shortening reflex and a second period in which high-amplitude activity increases sharply in the shortening muscles and the resistance curve reflects not resistance but "help" to the passive movement. At that moment, the amplitude of the tonic contraction is several times greater than normal. The tonic curve reaches its highest amplitude during flexion of the elbow joint (Fig. 116A).

5.2. Effect of Thalamotomy

In the same case, after stereotactic VL cryodestruction, a marked clinical improvement was noted—the hyperkinesias and severe dystonia in the corresponding extremities were abolished. Postoperatively the muscle tone was practically normal. The resistance curve during passive movements of the arm coincided with the curve of its movement, and the degree of resistance became the same as in healthy subjects (Fig. 116B).

Immediately after VL destruction, not only pathological spontaneous activity but also the Westphal phenomenon and spasm of the antagonists disappeared; this is an objective criterion to evaluate the results of the stereotactic operation. One more illustrative example follows.

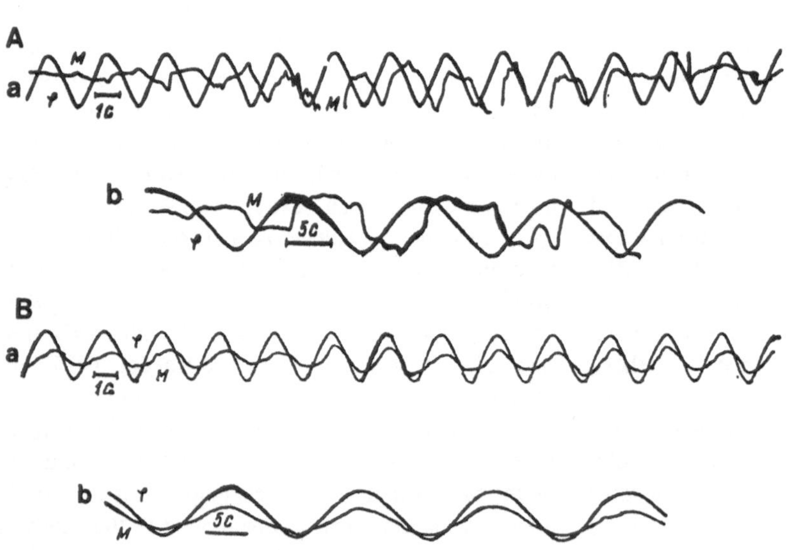

FIGURE 116. (A) Monitoring of mechanical resistance of the arm of a DMD patient at two speeds on the same device. M, The regular sinusoidal curve reflects repetitive activity of the monitoring device (flexion up, extension down); a, period 6 sec; b, 14 sec. Note the counterphase of curves M and ρ , which depends on reflex activation of the muscles in the phase of passive shortening. On the right is the resistance plotted in kgDm (see text). (B) The same monitoring in the same patient after stereotactic thalamotomy with a good clinical result. The same abbreviations as in A. The pronounced decrease of the shortening reflex after surgery normalized muscle tone. The mechanical resistance (right) decreased to normal.

In a 43-year-old patient with generalized DMD, the EMG study has shown that during the voluntary extension of the right leg there was a pronounced spasm of antagonists—the appearance of high-amplitude simultaneous activity in flexors and extensors of the leg (biceps and quadriceps of the thigh). After left cryothalamotomy, these dystonic signs in the right extremities were abolished completely.

Repeated EMGs of the same muscles were made 3 weeks after surgery and showed the return of normal reciprocal innervation. As Fig. 117 shows, in the voluntary extension of the leg, muscular activity occurred only in the extensors (quadriceps of the thigh), with minimal activity in the flexors. The disappearance of antagonistic spasm after the operation was readily demonstrated on the EMG during flexion–extension of the leg. The alternating activity is clearly visible in Fig. 117 during these movements.

Our investigations as well as Cooper's data (1965) confirm that spasm of the antagonists, which is the basis of the dystonic phenomenon, totally disappeared after a successful operation on Th nuclei, and the ability of antagonists to relax during voluntary contraction is restored completely.

Gurfinkel and Safronov (1971) supposed that afferents of group II generate the shortening reflex. The following arguments are used by the authors as confirmation of such an assumption: the shortening reflex has low-threshold activation, and the reflex is linked with the phase of movement and the asymmetry in the antagonist muscle groups, with more pronounced activity in extensors. All the characteristic features of the shortening reflex were noted in our investigations of DMD patients.

The fact that the shortening reflex is opposite to the normal myotatic stretch reflex allows one to suppose that the Westphal phenomenon reflects a disorder of the γ system. That the stretch reflex is absent in the antagonist of the muscle in which the shortening reflex occurs confirms this conclusion. However, the significance of the spindle afferents in the genesis of the Westphal phenomenon is not fully understood and requires future investigation.

Our study disclosed some new data about motor disturbances in DMD. The muscle spasm typical of the disease was found to have two stages: in the first, there is only a slow and relatively small increase of muscle tone, but in the second a very fast and sharp increase in the spasm occurs.

It may be presumed that the muscular spasm is preceded by summation of subthreshold efferent activity. There is reason to assume that this process takes place on the spinal neurons as a result of increasing intensity of the supraspinal efferent outflow. As is

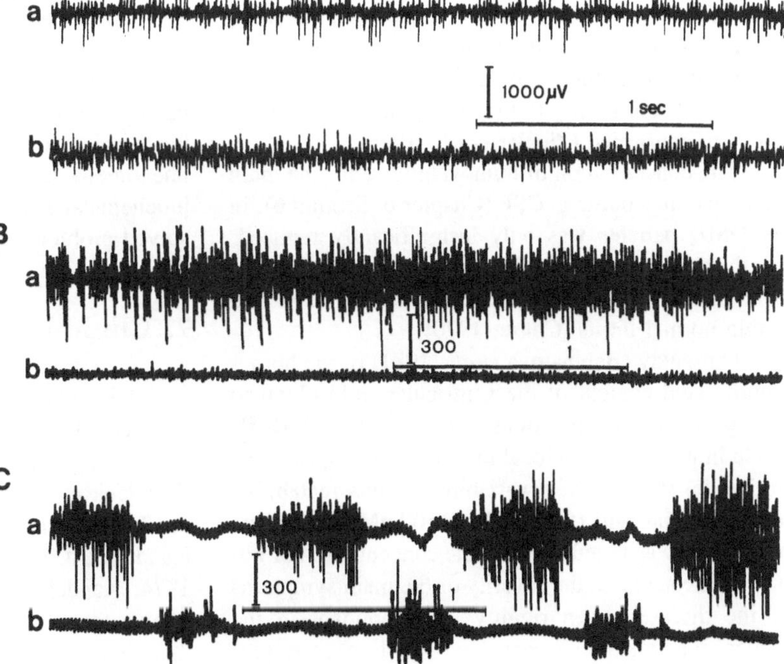

FIGURE 117. Disappearance of the "antagonist spasm" 3 weeks after effective stereotactic operation. (A) Preoperative EMG of the extensor (a) and flexor (b) of the leg. Marked "antagonist spasm" on voluntary extension of the right leg. (B) Disappearance of "antagonist spasm" 3 weeks after effective operation (the same designations). On voluntary extension of the leg, the burst of electroactivity occurs only in the m. quadriceps femori (extensor). (C) Restoration of normal reciprocity on voluntary flexion–extension of the leg. Clear alternation of activity bursts in flexor and extensor of the leg (the same designations).

known from clinical observations, any afferent stimulus (light, sound, pinprick, etc.) triggers an increase in muscular spasms and dystonic hyperkinesias.

6. Biochemical Pathogenesis

During the past two decades disturbances of cerebral metabolism, mainly of dopaminergic and other transmitter systems in the basal ganglia, have been discovered. The dominant role of these disturbances in the pathogenesis of several extrapyramidal diseases (parkinsonism, hepatolenticular degeneration, Huntington's chorea, and others) has been established. Similar biochemical investigations of DMD, unlike the above-mentioned diseases, have not yet disclosed the pathogenesis of DMD. This, in part, explains the ineffectiveness of drug treatment of this disease.

In spite of intensive investigations, it is still impossible to define the precise role of catecholamine and serotonin metabolic disturbances in the pathogenesis of DMD. There are data in the literature stating that the quantity of dopamine in the basal ganglia does not diminish in DMD as in parkinsonism but, on the contrary, increases (Barbeau *et al.*, 1963). However, the published data on this subject are controversial. Some authors have reported that urinary dopamine excretion in DMD is increased (Johnson *et al.*, 1962; O'Reilly *et al.*, 1965), but Zeman and Dyken (1968) did not confirm this in four patients in whom dopamine excretion was within normal limits. In some papers, an increased serum concentration of dopamine hydroxylase, which transforms dopamine into norepinephrine, was reported (Wooten *et al.*, 1973), but these findings have not been confirmed (Ziegler *et al.*, 1976).

The concentration of homovanillic acid, the main dopamine metabolite in CSF (Chapter 6, Section 6), in six DMD patients was only insignificantly reduced, and the concentration of 5-hydroxyindoleacetic acid (5-HIAA), the main metabolite of serotonin, remained within normal limits (Chase, 1970).

Our study (Barchatova *et al.*, 1981) of the homovanillic acid content of the ventricular fluid obtained during stereotactic operations in 24 patients with DMD (including ten with the local form of DMD, spasmodic torticollis) showed great variability of this metabolite in different patients (from 350 to 1100 nM). A definite correlation was noted between the concentration of homovanillic acid and the severity of the main symptoms of the disease: when rigidity was dominant in the clinical picture, the concentration was low, as in parkinsonism (average 392.9 ± 51.2 nM), but when hyperkinesias dominated, the concentration was markedly increased (average 1031.4 ± 89.3 nM). These data to a certain degree coincide with the results obtained by Tabaddor *et al.* (1978), who reported that low concentrations of homovanillic acid in the ventricular fluid are more typical of the local form of DMD beginning in adulthood, and an increased concentration in the ventricular fluid in children with rapidly progressing generalized DMD. The interpretation of these facts is not yet clear.

After stereotactic destruction of VL and Subth, our investigations showed a significant lowering of homovanillic acid concentration in the lumbar fluid. It was also established that in DMD patients a significant decrease in the tyrosine hydroxylase cofactor concentration in CSF occurs (this substance takes part in dopamine biosynthesis from tyrosine).

Other metabolic disturbances of amino acid metabolism (tyrosine, tryptophan, phenylalanine) are noted especially in hereditary DMD (Markova *et al.*, 1966; Barchatova, 1967). In the majority of patients with dystonia, an increase in the cholesterol level in blood was also noted.

The albumin content of blood in dystonia remains within the normal limits and does not change after surgery. Albumin and globulin fractions are also normal, but globulin subfractions have a tendency to increase sharply in the majority of cases (especially L_1- and L_2-globulins).

Unlike parkinsonism, the biochemical basis of DMD remains unknown, although one may presume that catecholamines play an important role in the pathogenesis of the disease. The mechanism by which biochemical disorders are transformed into dystonic phenomena is not clear. One may hope that future biochemical investigations will shed light on the unsolved problem of DMD pathogenesis.

7. Clinical Aspects

The clinical picture of DMD has been studied in great detail by many authors (Oppenheim, 1911; Mendel, 1911; Flatau and Sterling, 1911; Herz, 1944; Davidenkov, 1956; Cooper, 1962a, 1976; Zucker, 1963; Zeman and Dyken, 1967; Eldridge, 1970, 1976; Kandel and Vojtina, 1971; Mardsen and Harrison, 1974; Kandel, 1981; Patti *et al.*, 1985). It has been known that this disease more frequently affects men than women, but the difference is not great. Thus, among 253 cases of dystonia described in literature, 56.7% were men and 43.3% women (Zeman and Dyken, 1967).

The first symptoms of the disease most often appear at the age of 8 to 15 years, which suggests that DMD is a children's disease. There is a definite connection between the age of onset and the form of the disease. According to our data, generalized dystonia begins at an earlier age (average 10.2 years), whereas local forms of the disease begin at the average age of 19 years. It is accepted that the earlier DMD begins the more severe is the course of the disease (Cooper, 1969a; Kandel and Vojtina, 1971; Zeman, 1976). The data of Marsden (1976) are quite striking: in 86% of patients, when the disease began at the age of less than 10 years, generalized dystonia eventually developed; at the same time none of the patients with onset of the disease after 20 years of age (adult-onset dystonia) had the generalized form. Accordingly, two forms of the disease can be described, with onsets in childhood or adulthood.

In most cases the first symptoms appear in one leg, mostly as an inversion of the foot. Less often the disease begins with involvement of one hand, often as periodic spasms like a writer's cramp.

The main symptomatology of DMD consists of (1) hyperkinesias—involuntary movements or tonic spasms of the trunk, neck, and extremity muscles; and (2) muscular dystonia—uneven increase of tonus in different muscular groups leading to pathological postures.

Both main phenomena have been described by various names, for instance, "myostatic" and "myokinetic dystonia," or "movement dystonia" and "posture dystonia." Usually dystonic movements are mainly confined to the proximal parts of extremities and the trunk muscles. The specific and early symptom of the disease is scoliosis and lordosis of the lower thoracic and lumbar vertebral column, especially pronounced during walking. Typical of the more advanced stages of the disease are torsion flexion and rotation movements of the trunk (Fig. 118). The patient's back arches and rotates around its longitudinal axis. For this reason, the disease was called torsion dystonia.

Clinical observations show that in dystonia, postural muscle groups are more often affected, (so-called antigravity muscles). This is confirmed not only by the fact that postural disturbance is the main sign of dystonia but also that in a lying position at rest, when it is not necessary to maintain posture, dystonic signs as a rule alleviate or completely disappear. The extensors that maintain posture (mainly long dorsal muscles) are affected more than the flexors. In severe cases of dystonia, there is a strong tonic contraction of truncal extensor muscles leading to opisthotonus. The predominant spasm of extensors causes tensed straight legs with retroversion of the knee joints and "inversed" hands. Usually, the patients prefer to keep a totally flexed posture in which dystonic spasms decrease. There is no good explanation for these phenomena. The reason apparently is not only that extensors are stronger than flexors so the latter "overpull" during simultaneous contraction of antagonists. One may assume that the more intensive involvement of extensors reflects their specific physiological role to ensure postural reactions requiring prolonged tonic tension.

Hyperkinesias usually start as movement disturbances (awkward walking, abnormal posture), which produce bizarre body positions. The gait becomes awkward and unsteady. Some patients can walk better briskly than slowly, and the first steps usually require a greater effort than the following ones. Some of our patients preferred to run because the dystonic movements were less severe at that time. In severe stages of the disease, contractures develop so that the pathological postures of the extremities become fixed. During sleep, hyperkinesias and dystonic movements disappear.

There is also a peculiar local form of DMD: oromandibular or buccolingual dystonia or Madge syndrome, which is described in detail in the neurological literature (Delwaide and Desseilles, 1977). This dyskinesia involves the face, tongue, jaw bilaterally and sometimes spreads to the neck muscles. The etiology of the syndrome is obscure and it is called idiopathic. In recent years it was shown that similar orofacial hyperkinesias may develop after long-term L-dopa treatment of parkinsonism (Chapter 6, Section 7.9). According to the literature, this syndrome is observed in elderly patients.

In the neurosurgical literature we found only one report by Narabayashi *et al.* (1985) of two cases with satisfactory results after Vo-thalamotomy.

In recent years we operated on three men, aged 27, 49, and 54, with severe oromandibular dystonia. The cause of illness was unknown in these cases. Various medical treatments were unsuccessful. The syndrome was most disabling in the youngest patient, who, besides great difficulties in speech and facial grimacing, could not eat normally and consumed only fluid for more than two years. No neurological deficit or other abnormalities were noted in these patients.

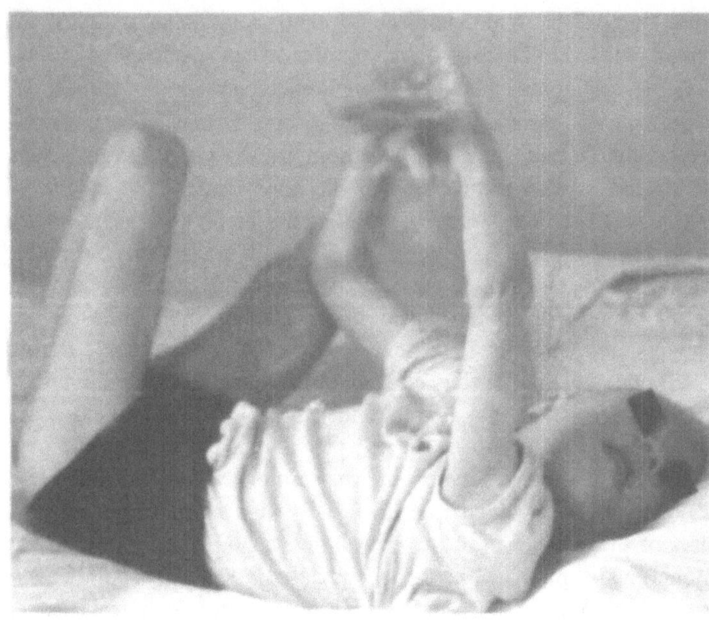

FIGURE 118. (A) Typical torsion of the body in severe juvenile generalized dystonia musculorum deformans. (B) The same boy 2 years after bilateral VL thalamotomy.

All three patients underwent stereotactic cryodestruction of VL. A marked improvement was achieved in all cases, but in two of them the effect was not stable and diminished gradually over a few weeks. These two patients were operated on a second time; then, the same operation was performed twice on the other side after 4 weeks and 9 months. Only a moderate effect was achieved after the bilateral operation. The marked improvement has been maintained in the third patient after 2.5 years follow-up.

Any sensory input (tactile, optic, auditory, etc.), as a rule, greatly intensifies the dystonic phenomena. Emotional and other stresses have a similar influence.

In some cases of DMD there is a severe pain syndrome with localization of pain in tensed extremities, back, and neck. As a rule the pain arises as a result of joint contracture and deformation. In severe stages of the disease various vegetative and trophic disturbances may develop. Only in rare cases are the so-called paradoxical kinesias described in the literature seen in which a patient with severe dystonia can quite easily carry out some complex motor acts such as dancing or playing the piano.

As a rule mental disorders do not accompany dystonia, as first noted by Oppenheim (1911). Moreover, in children with dystonia we and others have often noted an unusually high intellect.

The above-described phenomena are specific for generalized DMD. However, less frequently focal and abortive forms are also seen. It is important to emphasize that all these types of the disease are stable and, as a rule, do not become generalized except in cases in which a localized disorder is the initial symptom of the disease. Spasmodic torticollis, described in detail in the Chapter 8, is the most frequent focal form of dystonia. Hemidystonia is included in the group of hemihyperkinesias described in Chapter 10. Other focal forms are rare and are limited to one extremity.

Abortive forms of dystonia (*formes frustes*) are regarded as confirmation of the hereditary nature of the disease. Mild symptoms of dystonia are seen in these forms, which do not significantly handicap the patient or limit his working ability.

7.1. Stages of the Disease

Based on several criteria such as the ability of a patient to carry on everyday activity and to work and severity of clinical symptoms, especially hyperkinesias and rigidity, we distinguish four stages of generalized DMD (Kandel and Vojtina, 1971). In the first stage focal symptoms appear such as plantar flexion and inversion of the foot or writer's cramp. In this stage, the patient can stop the dystonic movement voluntarily or by changing the position of the extremity, so the working capacity is usually fully preserved.

In the second stage, the dystonia spreads to more muscular groups and is very difficult to correct by voluntary or passive movements. The ability to work in this stage is markedly limited.

The third stage is characterized by the development of the full clinical picture of the disease, with severe permanent hyperkinesias and muscular dystonia involving all or almost all muscular groups. In this stage, the gait is severely disturbed, the ability to work is lost completely, and self-care is extremely limited.

In the fourth stage, all manifestations of the disease are severe. As a result of a long-term and unbalanced increase in muscle tone, severe contracture and pronounced deformation of the vertebral column, thorax, and extremities develop. With these involuntary movements, patients suffer pain. In this stage they are severe invalids, bed-ridden, and require constant care. The general somatic state worsens significantly, and decubiti and cachexia often develop. Any intercurrent infection will likely prove fatal.

7.2. Diagnostic Studies

The various special diagnostic and laboratory investigations have not been of great assistance in the diagnosis of DMD.

Routine EEG examinations done by us together with Zhirmunskaya and Pokrovskaya in a large group of DMD patients did not disclose any significant abnormality in the majority of cases. Only in a few patients were significant disturbances of cortical bioelectrical activity (interhemispheric asymmetry, δ waves, etc.) marked. Photostimulation done in all cases did not reveal any significant change in brain activity. In only a few observations was there a slight increase of slow activity upon stimulation. Our findings have confirmed the opinion of other authors that there are slight EEG changes that do not have any significant diagnostic importance in DMD.

Angiography and CT scanning, based on our experience, have yielded no valuable data although some important abnormalities in the cerebral vascular system may be found in hemidystonia (see Chapter 10).

7.3. Differential Diagnosis

For the diagnosis of DMD in typical cases, anamnesis is very important if an etiologic factor is absent or the hereditary nature of the disease has not been confirmed. The development of the disease and its typical clinical picture with generalized or local tonic spasms of the trunk and extremity muscles allow one to diagnose DMD without difficulty. However, the disease does not always have a typical course. In the early stages and in some cases with an atypical course and uncommon neurological symptoms, it is difficult to differentiate dystonia from other organic or functional diseases that have a similar clinical picture.

Sometimes generalized DMD is difficult to distinguish from cerebral palsy, which is described in Chapter 9. Indeed, quite often slow involuntary athetoid movements and torsion dystonic phenomena as in cerebral palsy can be observed in dystonia. Differential diagnosis of these two diseases must be based on the fact that cerebral palsy is a "residual phenomenon" of a static pathological process. Its neurological picture is nearly always stable, and progression of symptoms, as a rule, is not observed. On the contrary, DMD is always a progressive disease. Unlike DMD, patients with cerebral palsy had birth trauma or a history of other brain lesions. A characteristic of cerebral palsy is the appearance of signs of the disorder in the first weeks or months of life, which is extremely rare in DMD.

Cerebral palsy differs from generalized DMD, as a rule, by initial manifestations of the disease not in one but in all extremities. With the exception of hemiathetosis, there is simultaneous involvement of the face, tongue, and neck. Hyperkinesia and athetosis are characterized by slow tonic muscle contractions simultaneously involving antagonists and agonists. Hyperkinesias in DMD are similar to athetosis but involve proximal muscles of extremities to a greater degree. In DMD the face usually is spared. The permanent grimace severely distorting the face is absent in double athetosis. In DMD, speech is affected much more rarely and to a lesser degree than in cerebral palsy.

Patients with cerebral palsy often have significant disturbances of mental functions, pronounced impairment of intellect, and, at times, severe dementia. In DMD, as a rule, there are no mental disturbances, and intellect is well preserved.

It is important to note the differential diagnostic features between DMD and hepatocerebral degeneration (HCD), which is described in Chapter 12. The following signs characterize both diseases: onset in childhood or at an early age; several members of one family affected; progressive generalized extrapyramidal rigidity; and severe hyperkinesia and contractures leaving the patient immobile. Hyperkinesias in HCD often have a torsion–dystonic character. The majority of these patients suffer dyspeptic and liver disturbances, allergic reactions, and hemorrhagic syndromes long before the onset of the first signs of extrapyramidal lesion. These symptoms are never present in DMD with such a course.

Although intellect is normally developed in DMD patients, it is often lowered in HCD, even to the point of dementia. Epileptic seizures, which are absent in DMD, are often observed in advanced stages of HCD. Laboratory investigations in HCD, as a rule, disclose typical metabolic changes (Chapter 12, Section 2). In DMD the Kaiser–Fleischer ring is absent.

Usually it is not difficult to distinguish DMD from Huntington's chorea—a progressive disease with typical choreiform hyperkinesias and pronounced dementia (Chapter 12, Section 3). Unlike DMD hyperkinesias, the chorealike twitches are more rapid, less "massive," and look more like purposeful movements. Both diseases may have members of several generations afflicted and show a progressive course. Huntington's chorea differs from DMD in that the hyperkinesias are random, the first signs of disease occur in adult age (usually from 25 to 45 years), and there are pronounced mental disturbances.

It is not difficult to distinguish DMD from parkinsonism. However, in extremely rare cases, when dystonia begins in patients of middle age with hyperkinesis in one extremity, parkinsonism may be suspected. A diagnostic difficulty sometimes occurs in adult DMD patients in the most severe stage of the disease, when the fixed dystonic postures resembles those of patients with the akinetic form of parkinsonism. Nevertheless, the anamnesis, development of the disease, type of tone disturbance, and hyperkinesias allow one to distinguish the diseases without great difficulties.

8. Indications and Contraindications for Surgical Treatment

At present there is no effective and reliable drug treatment for DMD. Curare-like, antihistaminic, and cholinolytic drugs seem to be ineffective. Several attempts in recent years to treat DMD with L-dopa have not given consistent results. According to several authors, this drug improves the rigid form of the disease (Melnitchuk and Sosnovskaya, 1973; Tkachev *et al.*, 1973; Markova *et al.*, 1978). Other authors completely deny any therapeutic value of L-dopa for DMD (Barrett *et al.*, 1970; Coleman, 1970; Patti *et al.*, 1985). About one half of DMD cases have been made worse by L-dopa treatment (Barret *et al.*, 1970). Evidence has been presented that L-dopa may aggravate the dystonic symptoms (Cooper, 1972). The reports of the effectiveness of carbamazepine (Tegretol®) in this disease (Isgreen *et al.*, 1976) have not been confirmed. One must conclude that in the absence of any other possibility of treatment, the only effective method is a stereotactic operation on deep brain structures.

Our long experience, as well as that of other neurosurgeons, confirms the opinion that the diagnosis of DMD, in principle, mandates an operation. Although this disease undoubtedly falls into the category of surgical diseases, surgical treatment is not indicated for all cases. Various contraindications may exist, but we believe that as a rule, generalized DMD is practically always progressive, and conservative treatment is ineffective. Accordingly, surgery is indicated in four patients out of five.

The perfection of stereotactic techniques and improvement of the results of surgery have led to a substantial broadening of the indications for operative treatment. If the disease is progressing so that the patient is unable to care for himself, one should not wait for further progression of the disease or postpone the operation for years.

As DMD patients are children or young people, the older age risk factor does not enter into the consideration for surgery. However, children younger than 6–8 years should not be operated on because they usually do not tolerant brain operations well. The long duration of the disease is not a contraindication to surgery, for our experience shows, even in patients who have had the disorder for 20–25 years, excellent results may be obtained.

The stage or, more exactly, the severity of the disease is a very important factor in determining whether a given operation should be undertaken. The majority of our cases were in the most severe stages

(III and IV) of the disease. Our experience shows that even in severe generalized DMD, good long-term results may be obtained. At the same time, patients in the most severe stage IV do not tolerate the operation well, and the risk of postoperative complications is high. Undoubtedly, each case must be decided on an individual basis. In proposing the operation even in the most severe stages, we proceed from the point of view that surgery is the only hope to alleviate torment and suffering. Refusal to operate dooms the patient at a young age to an unbearable existence, which sooner or later ends fatally.

We believe that broadening the indications for DMD surgery is quite justified, assuming of course, that contraindications (cachexia, severe somatic diseases, decubiti, etc.) are considered.

9. Surgical Treatment

Surgery for DMD as well as for other extrapyramidal lesions has followed a long and complicated path. However, practically all operations proposed and used several decades before the stereotactic era appeared to be ineffective, or the results were not comparable to the benefits of stereotactic operations on the basal ganglia. Almost all of these operations now are of historical value, so they are mentioned only briefly (for a more detailed description, see Chapter 6). Orthopedic operations on muscles, joints, and tendons, which included resection and transplantation of tendons, ligaments, and muscles as well as shortening of long bones, are considered ineffective. At present, teno- and myotomias for dystonia are seldom done, although they may be satisfactory for mild forms of the disease, when indications for stereotactic operation are optional, or sometimes as a supplement to these operations to correct residual deformation of joints and tendons.

At the present time, with the stereotactic procedures available, the advisability of major orthopedic operations is doubtful. We operated on several patients who had previously undergone operations for shortening of bones and elongation of tendons without effect. The results of stereotactic operations on the basal ganglia in these patients were worse than expected because the residua of the orthopedic procedures hindered the functional improvement of the affected extremities.

In the past many attempts were made to correct dystonic posturing by corsets or other devices. In old papers some benefit from these appliances in mild forms of the disease was noted. We, as well as other authors, feel that orthopedic corsets do not improve torsion dystonia. Various methods for fixation of the body and extremities also practically never achieve their purpose. Plaster of Paris bandages often increase hyperkinesias and can also cause severe pain in the immobilized extremities. As our observations show, dystonic patients, especially children, do not tolerate immobilization of the affected parts of the body.

Various resections of peripheral nerves (Stöffel operation) and spinal roots* as well as operations on spinal cord tracts and sympathetic ganglia have all been abandoned. At present, extirpation of premotor and motor cortical zones and dissection of the pyramidal tract in CI or in the cerebral peduncle (pedunculotomy) (Meyers, 1956a,b; Pomme *et al.*, 1958; Maspes and Pagni, 1964) are no longer advised for DMD. Before the stereotactic era, such operations in selected cases alleviated involuntary movements, but at the expense of spastic pareses or paralysis of contralateral extremities and sometimes focal epileptic seizures (Chapter 6, Sections 9.2, 9.3, and 9.5).

10. Stereotactic Operations

Stereotactic operations on the basal ganglia appear to be the most effective form of DMD surgery. The destruction of certain subcortical structures and their connections interrupts the "pathological chain" in which the neural activity responsible for producing hyperkinesias and muscular dystonia originates.

As in other hyperkinesias, the initial stereotactic surgery for DMD was destruction of the medial GP, but pallidotomy gave a beneficial effect in only approximately half of the cases (Cooper and Bravo, 1957). In 1960, along with other neurosurgeons, we began to operate on VL, the "main point" of the convergent paths connecting the extrapyramidal structures (Chapter 1, Section 1). One may conclude that to date VL is the structure of choice in DMD stereotactic surgery. In recent years, we have concentrated on destroying two targets in the basal ganglia in order to obtain the most effective result, i.e., combined destruction of VL and Subth (Chapter 6, Section 10). Other authors also recommend destruction of two or three thalamic structures—Vim, VPL, and sometimes also CM (Cooper, 1976).

The largest experience in the world with stereotactic operations on the basal ganglia for DMD has

*This statement does not apply to the surgical treatment of spasmodic torticollis (see Chapter 8) and spasticity (see Chapter 9).

Irving Cooper (1922–1985)

been accumulated by Cooper (1976), who, in 20 years, has operated on 226 patients. In 104 cases, unilateral, and in 122 cases, bilateral operations were performed. In 1976 the author summarized his long-term results of stereotactic destruction of Vim, VPL, and sometimes CM. Based on an average follow-up of about 8 years, he reported good or significant improvements in 69.7% of the cases, no change in 18.3%, and worsening (as a rule, the result of further progression of the disease) in 12%. His postoperative mortality was 2%. Dysphagia and dysphasia occurred after bilateral operations in 13% of cases. Postoperative pseudobulbar palsy was observed, particularly in patients who had displayed signs of oral and pharyngeal dystonia prior to the operation. Other authors have reported dysarthria after bilateral thalamotomy in more than half of the patients with different forms of DMD (Andrew *et al.*, 1983). Cooper (1976) noted that the result of surgical treatment was better in cases of hereditary DMD.

There are only two reports in literature of good results after stereotactic destruction of SN for DMD. Rand (1960) reported a unique case. In 1 year he performed six stereotactic operations destroying GP and VL but achieved only temporary improvement; relapses soon ensued. In the seventh operation performed by Rand, an open approach with electrocoagulation of SN through the lateral brainstem sulcus, the patient had no complications and obtained a significant improvement. Zapletal (1965) performed a similar operation in two cases of DMD. This operation (nigrotomy) appears to be associated with a high risk and has not been pursued.

Zervas (1977) observed a mild improvement in three of four patients with DMD after stereotactic destruction of ND. The same operation in two other patients did not give satisfactory results (Guidetti and Fraioli, 1977). Chronic stimulation of the deep brain structures appeared to be ineffective for DMD (Upton *et al.*, 1982).

Gildenberg (1978) reported one patient with DMD who showed significant improvement after chronic dorsal column stimulation of the cervical spinal cord (Chapter 13, Section 5.4), but Siegfried *et al.* (1981b) have found this method to be ineffective.

While preparing a DMD patient for basal ganglia surgery, a surgeon must choose the side on which to operate. If the disease is localized, the operation is done on the side contralateral to the affected extremities. For generalized dystonia, as a rule, a bilateral operation is performed in stages on the right and left basal ganglia with an interval of about 4 to 6 months.

The first operation is usually performed on the side contralateral to the more affected extremities. If the symptoms are equal on both sides, we usually perform the first operation on the left side to improve function of the right hand, which is more useful in everyday life. If pain is pronounced on one side, the first operation is performed on the corresponding side because pain is relieved after surgery in the majority of cases.

Because of the hyperkinesia, operations for generalized DMD are, as a rule, carried out under general anesthesia. In cases in which local anesthesia is employed, the therapeutic effect such as alleviation or full arrest of hyperkinesias and muscle rigidity is seen immediately on the operating table (Fig. 119).

11. Personal Observations

Over 25 years (1959–1984), we and our co-workers have operated on 188 patients with generalized

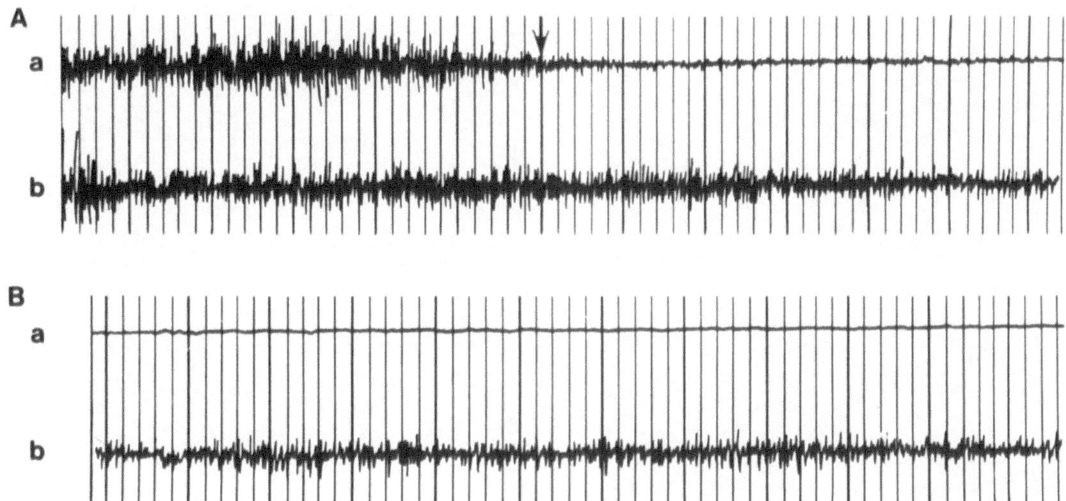

FIGURE 119. Electromyographic recording of forearm muscles in generalized DMD during a stereotactic operation under local anesthesia. (A) Cannula for local freezing is introduced into the right VL (arrow), after which dystonic signs in left extremities and typical spontaneous dystonic electrical activity in the left m. flexor digitorum superior disappeared (a). On the right side dystonic signs did not change, and electrical activity of the analogous muscle decreased slightly (b). (B) Electromyogram recorded at the end of the operation after local freezing of right VL. Spontaneous electrical muscular activity in the left digital flexor disappeared completely (a); that in the right decreased significantly (b).

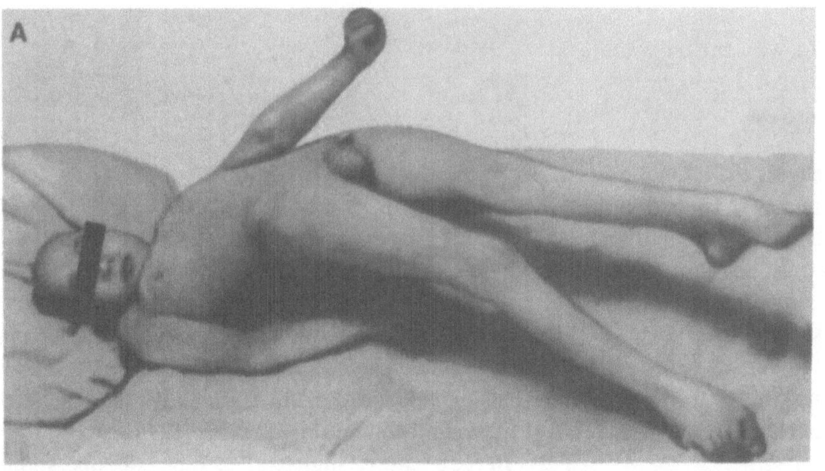

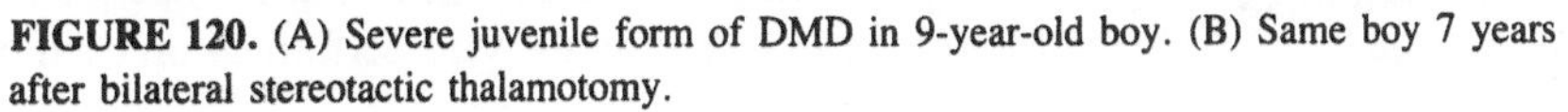
FIGURE 120. (A) Severe juvenile form of DMD in 9-year-old boy. (B) Same boy 7 years after bilateral stereotactic thalamotomy.

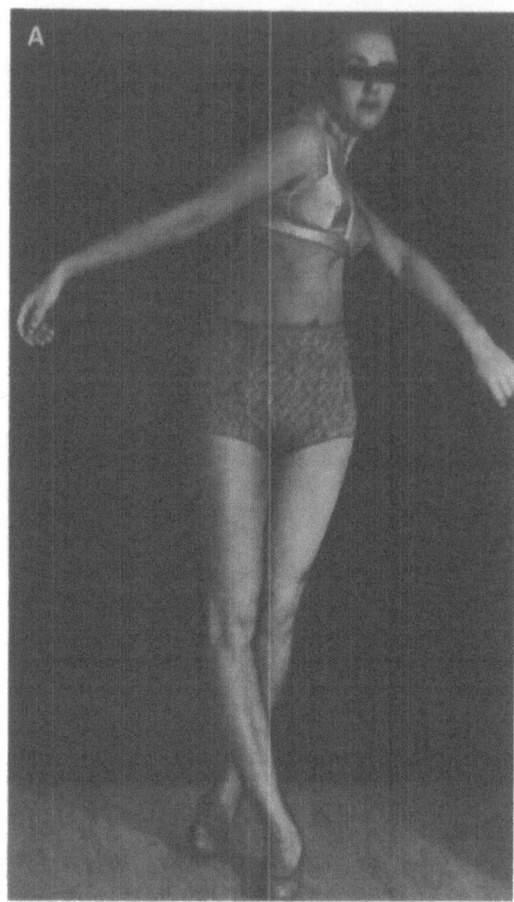

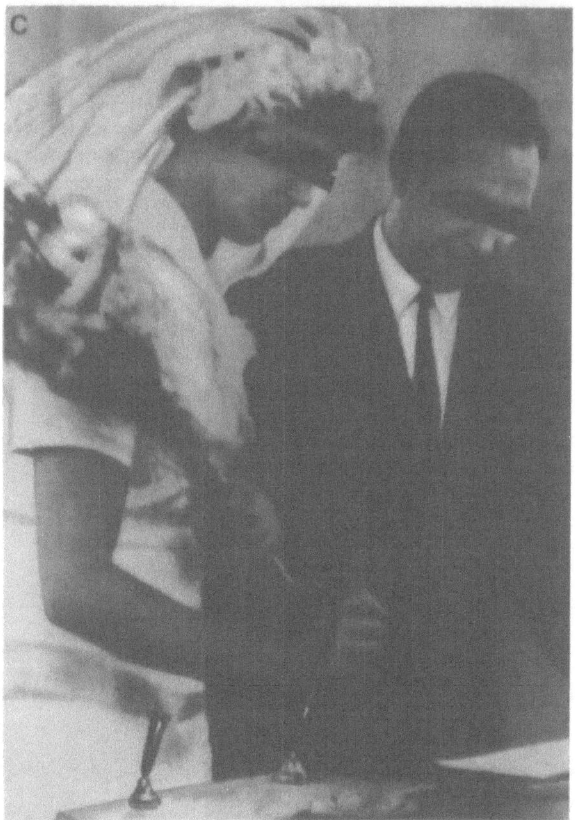

FIGURE 121. Generalized DMD in 24-year-old patient (A) before surgery and (B,C) 4 and 8 years after bilateral stereotactic operation on basal ganglia (thalamotomy).

form of DMD* (114 men and 74 women) on whom 272 stereotactic operations on the basal ganglia have been performed. The majority of these patients (109 cases) were from 20 to 50 years of age. The second group included children and teenagers (77). Only two patients were over 50 years (56 and 61). The duration of the disease, i.e., the time from onset up to the operation, was usually long. In only one-third of the patients was the duration of the disease less than 5 years. Two-thirds of them had suffered from 5 to 37 years, eight more than 25 years.

The majority (80%) of the patients at the time of operation were in the most severe stages (III and IV) of the disease according to our classification (Section 7.1). Over the years, all of these patients had received various therapeutic modalities, mainly pharmacological treatment, without marked benefit.

In 126 cases, unilateral stereotactic operations were performed; 62 patients had bilateral operations. More than two operations were performed on each of

*Patients operated on for spasmodic torticollis (Chapter 8) and for hemidystonia (Chapter 10) are not included.

FIGURE 122. (A) A 22-year-old patient with generalized DMD. (B) Two years after bilateral stereotactic operations on basal ganglia.

13 patients (three in eight patients, four in four, and five in one). The operations were performed on the right side in 60% of the cases and on the left side in 40%. Eighty-eight percent of the operations were performed under general anesthesia, and the rest under local anesthesia or neuroleptoanalgesia.

The majority of operations (172) consisted of stereotactic destruction of VL. In 74 cases, combined destruction of VL and Subth was carried out. In the rest, destruction of GP (early cases) or dentatotomy was performed. In 90% of operations, cryodestruction of the subcortical structures was done.

Complications after stereotactic surgey for DMD are relatively rare, less common than after surgery for parkinsonism and much less frequent than in cases of cerebral palsy. Complications of different degrees of severity and duration were observed in 10% of cases, but in the past 5 years the complication rate has decreased to 4.5%. Among the rare severe complications are mono- and hemiparesis, pseudobulbar syndromes with dysphagia and dysphonia (after bilateral operations), brainstem edema, and pulmonary involvement (pneumonia, atelectasis). In one case, severe meningitis developed after the operation. There were no postoperative intracranial hemorrhages.

The majority of these complications were transient and completely resolved 2 to 3 weeks after surgery. Long-lasting complications occurred in only five patients (hemiparesis in three and the severe pseudobulbar syndrome in two). Prophylactic medication and treatment of postoperative complications were carried out according to general neurosurgical practice (Chapter 6, Section 11).

After the 272 operations, five patients died (two from pneumonia and lung atelectasis, two from brainstem edema, and one from meningitis), an operative mortality of 1.8%. In recent years, there has been only one death in the last 90 operations (approximately 1%).

The long-term results constitute the main consideration for evaluation of the effectiveness of stereotactic surgery for DMD. Comparison of long-term and immediate results is undoubtedly also of interest. As noted above, the data from the literature indicate the great success of surgical treatment for this disease. Our monograph on DMD (Kandel and Vojtina, 1971), in which our experience in the first 102 stereotactic oper-

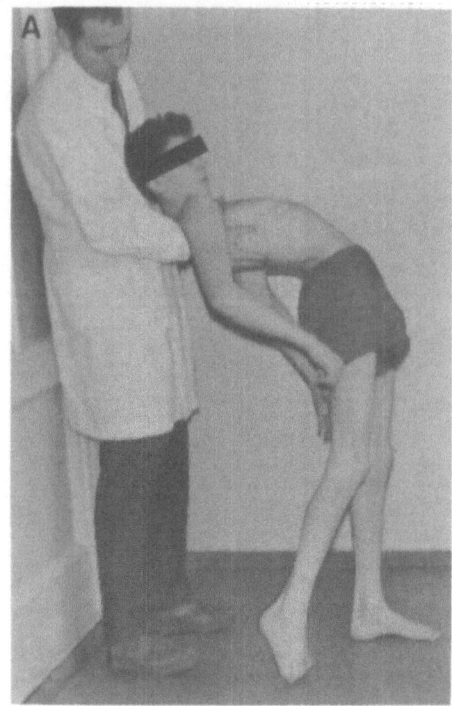

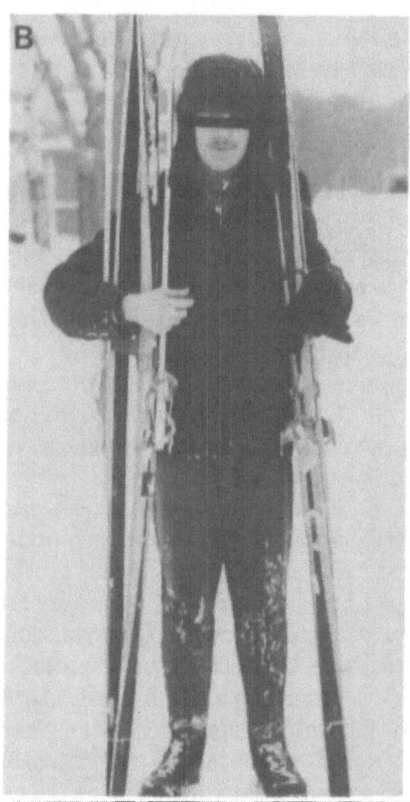

FIGURE 123. A 12-year-old boy with severe generalized DMD mainly in trunk muscles (A) before operating, (B) 3 years after two thalamosubthalamotomies (right and left), and (C) 12 years after surgery.

FIGURE 123. (*Continued*)

ations was summarized, revealed that in almost 70% of the patients good long-term results were obtained.

As for other conditions, the immediate results of operations were evaluated by comparison with the patients' preoperative conditions at the time of discharge from the clinic. The long-term results were assessed at some time between 4 months and 22 years after surgery. The results (as in determining the stages of the disease) were based on the patient's functional state, taking into account not only relief of hyperkinesia and dystonia but also the patient's ability to take care of himself or to engage in some kind of employment.

In the first or "good-result" group were those patients who showed practically complete recovery. Before operation they had been, as a rule, severe invalids, often bedridden with the generalized form of the disease. After bilateral operations, the hyperkinesia and dystonia in the trunk, neck, and extremities were abolished practically completely. These patients regained the ability to completely take care of themselves and to walk almost normally. They were able to attend school or to return to their previous jobs (Figs. 120–128). We also considered complete elimination of hy-

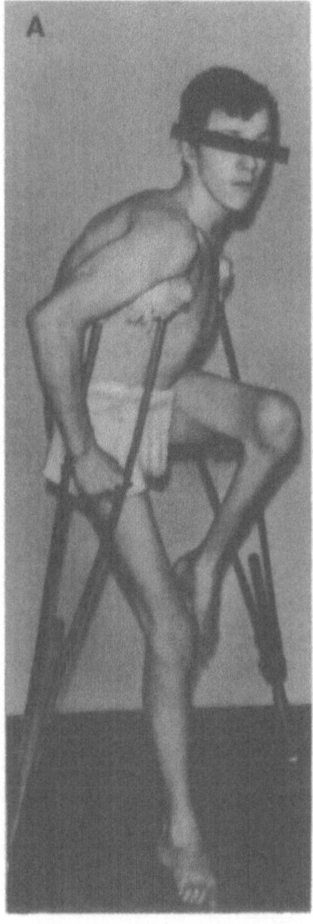

FIGURE 124. A 15-year-old patient with generalized DMD (A) before and (B) 3 years after bilateral stereotactic thalamotomy.

perkinesia and dystonia in the corresponding half of the body after unilateral operations as good results.

The second group included patients who, after the operation (or several operations), developed a stable "significant improvement" (60–70%) with marked alleviation of hyperkinesia and muscular dystonia. These patients were able to take care of themselves unassisted and to walk unassisted, although without normal gait.

The groups labeled "slight improvement" and "without change" need no comment. The "worsening" group included those patients who were not benefited by operation and who demonstrated further progression of the disease, or those with a change for the worse as a consequence of postoperative complications. The immediate and long-term results in these groups are presented in Table 15.

A comparison of the immediate and long-term results shows that they are quite similar. Accordingly, one may conclude that if a good clinical result is obtained after the operation, it remains stable for many years. Significant and steady improvement, which undoubtedly justified the operation, was noted in 76% of our patients.

In a number of cases at the time of discharge only mild improvement was apparent, but during the following several months the condition steadily improved. The reason for such a "delayed effect" of stereotactic surgery, first described by Cooper (1969a), remains conjectural.

A comparative study of the long-term results showed that the combined destruction of VL and Subth, which we have performed in recent years, is more effective than destruction of only VL. The frequency of "good" and "significant improvement" is 10% higher in the first than in the second group.

Practically all patients with generalized DMD required bilateral operations to achieve a complete therapeutic effect. For the local form of the disease, one operation on the side corresponding to the affected extremities proved sufficient.

A stereotactic operation on the second side should be performed not earlier than 4 to 6 months after the first. Such an interval is necessary to permit one to evaluate adequately the results of the first operation. The second operation is performed by the same technique as the first. If a good effect is obtained after the

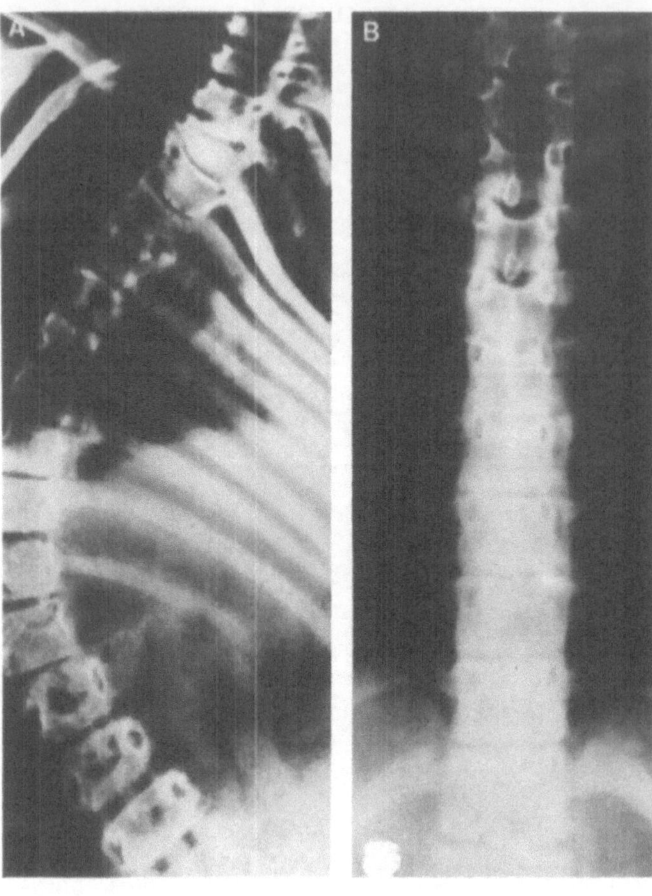

FIGURE 125. (A) Pronounced deformity of the vertebral column of a 16-year-old patient with generalized DMD. (B) Complete restoration of normal posture and absence of scoliosis after bilateral thalamotomy.

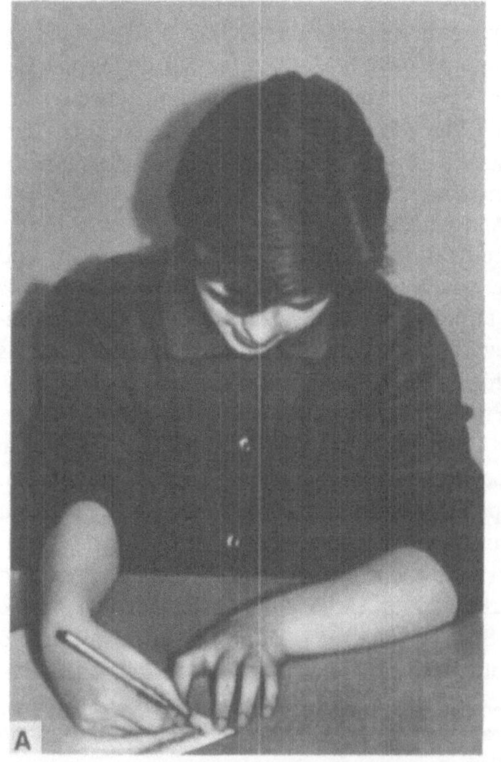

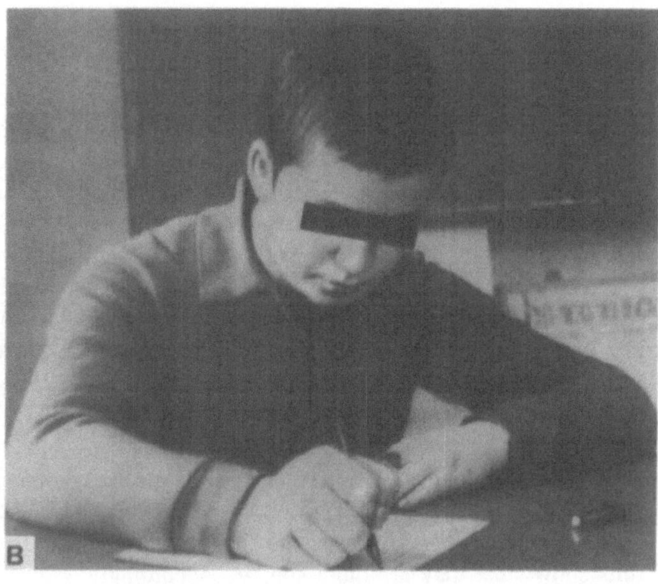

FIGURE 126. (A) Dystonic hand posture during writing in a 22-year-old patient with local DMD and (B) normal posture of the hand after stereotactic cryothalamotomy.

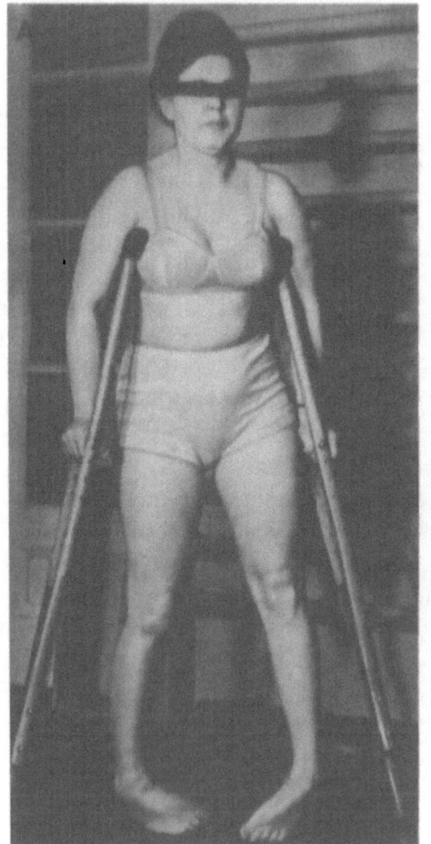

FIGURE 127. Dystonic posture of both legs in a 26-year-old DMD patient (A) before the operation and (B) soon after VL thalamotomy.

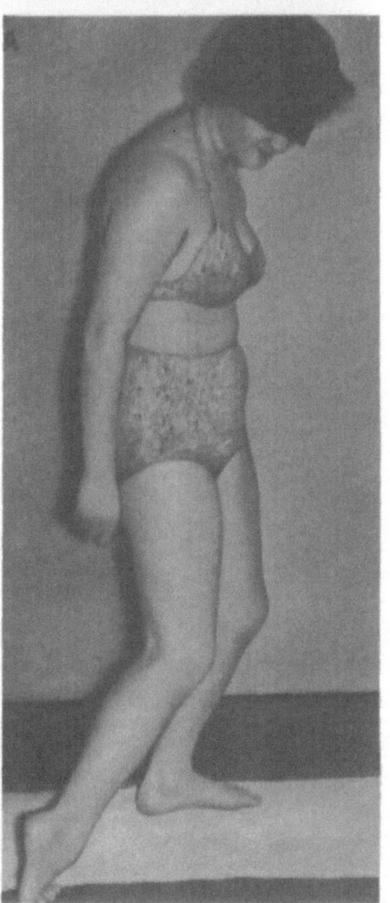

FIGURE 128. (A) Dystonic posture mainly in right leg in local DMD. (B) Normal posture and walking 3 months after left thalamotomy.

TABLE 15. Results of Surgical Treatment of DMD (%)

Results	Immediate	Long-term
Good	18	22
Significant improvement	50	54
Slight improvement	19	12
Unchanged	9	7
Worse	4	5

first lesion, the same stereotactic coordinates should be used on the other side. If the first operation gave only partial improvement, correction of the target for the second operation on the other side is advisable.

Indications for repeated operations on the basal ganglia of the same side may arise if the effect of the first operation was inadequate or if some time after a successful operation the DMD symptoms relapsed. In several of our patients, long after a successful operation, an unexpected and unaccountable dystonia recurred.

In the majority of cases the decision to perform a second operation is justified if the retrospective analysis of the first operative ventriculograms reveals a miss or near miss of the target structure. Since our stereotactic techniques have been perfected, repeated operations (on the same side) have not been necessary in recent years. In rare cases of relapse after a good effect, if only VL had been destroyed at the first operation, a lesion was made in Subth. In several of our cases, this procedure produced a good and stable effect.

One may thus conclude that modern stereotactic surgery is the most effective method for treating generalized and local forms of DMD.

8

Spasmodic Torticollis

There is no known cause nor morphological changes in this terrible disease in which the head is periodically or permanently twisted with great force to the side.

W. Dandy

1. Introduction

Spasmodic torticollis (ST) (or a wry neck in the old literature) is the name given to a dystonic and hyperkinetic state of the musculature of the neck leading to distorted, pathological posturing and forced turning or shifting of the head. This disease usually affects the lateral neck muscles and leads to an involuntary twisting of the head to the side. If there is a predominant dystonia of the posterior neck muscles bending the head forcibly back, the disease is called retrocollis; if the head is inclined forward, it is termed antecollis. The laterality of ST is traditionally defined by the side toward which the occiput is turned.

It can be considered an established fact that ST is a primary extrapyramidal lesion—a focal form described in the preceding chapter on DMD with isolated involvement of the neck muscles.* A number of convincing arguments may be presented in favor of this point of view. It has long been known that ST may be the first symptom of generalized DMD (Herz, 1944a,b; Davidenkov, 1960; Eldridge, 1970; Kandel and Vojtina, 1971). Generalized dystonia began with hyperkinesia of the neck musculature in 18% of our DMD patients. Secondly, after a stereotactic operation to relieve DMD, dystonia disappears not only in the trunk and extremities but also in the neck musculature (see Chapter 7). The point of view that torticollis is a focal form of DMD is supported by the majority of authors (Herz and Glaser, 1949; Ribera and Cooper, 1960; Hassler and Dieckmann, 1970a,c; Kandel, 1981, 1984; Van Hoof *et al.*, 1987).

However, ST as a general principle should not be regarded as a separate nosological disease. Nevertheless, ST is characterized by such a peculiar clinical picture and has such a typical development and course that this disease is customarily singled out as a separate nosological form so not to be lost in other focal forms of DMD.

Unlike generalized dystonia, first described by Schwalbe as a separate disease 80 years ago (Chapter 7, Section 1), torticollis has been known for several centuries. This is confirmed by the fact that operations to relieve this syndrome were performed in the 17th century (see below).

Spasmodic torticollis is undoubtedly a rare disease. The incidence of ST in the Canadian population is about 1 in 250,000 (Gauthier, 1986). Precise data on its incidence have not been reported in the literature; however, it may be presumed that ST is encountered just as frequently as generalized DMD. In several publications, authors analyze scores of their own observations. For example, Patterson and Little (1943) reported on 103 cases of ST, Arseni and Sandor (1960) on 50 patients, Sorensen and Hamby (1966) on 71,

*Here we are not considering ST caused by "peripheral" factors (congenital torticollis as a result of primary damage to neck muscles, torticollis in cervical spondylosis and intervertebral disk protrusion, atlantooccipital anomalies, tumors of the cervical spinal cord, and so on).

Cooper (1977) on 160, Hassler and Dieckmann (1982) on 92, and Kandel (1984) on 162 surgically treated patients.

It appears that this disorder commonly affects males. For example, in our patients, there were nearly three times more men than women. Other authors too have noted a significant predominance in the number of men.

2. Etiology

The etiology of ST, as well as of DMD in general, still remains a problem. However, although the organic nature of generalized dystonia has been finally established, some authors assume some torticollis cases to have a "functional" origin.

The psychogenic causation of ST was suggested long ago, since it is well known to clinicians that under the influence of emotions and other psychogenic factors, torticollis intensifies, and at other times, or under certain conditions, for example, by holding the head in a neutral position with only the finger touching the cheek, it temporarily disappears. In view of this, the term "torticollis mentalis" was suggested. "However it is proved," wrote Davidenkov (1960), "that the interpretation of this syndrome as a peculiar neurosis is obviously insufficient, and facts began to accumulate, indicating that at the basis of spasmodic torticollis there is something pointing to a greater degree to some kind of organic brain disease."

Nevertheless, in some papers (Sorensen and Hamby, 1966), the opinion is expressed that in a number of cases ST may result from "psychological causes," although among their operated patients, the authors made no such observations.

Contemporary concepts of the etiopathogenesis of extrapyramidal pathology emphasize the complex interrelationship of "functional" and "organic" factors. We, as do many other authors (Babchin, 1934; Davidenkov, 1960; Cooper, 1964; Podivinsky, 1968; Marsden, 1976; Hassler and Dieckmann, 1982) believe that ST is an organic lesion of the brain. The tremor in one hand often found in this disorder confirms the organic nature of ST.

Only in extremely rare cases can torticollis have a clear psychogenic nature. We can confirm this by one observation of an unusual beginning of the disorder. A breast tumor was discovered in a young woman, and it was decided to take a biopsy. Lying on the operating table, the patient, under great emotional stress, awaited the results of the biopsy, since she knew that in case of a malignant tumor, a radical mastectomy was inevitable. Turning her head, the patient saw the small skin incision, and thereafter felt her head being sharply "pulled" to the side. Although the tumor proved to be benign, and the operation was limited to a biopsy, the patient, whom we saw a year after that episode, had developed a typical ST.

Just as in DMD there is every reason to differentiate the "pure" form and the syndrome of ST. In the former, the etiology of the disease is unknown, which was the case in the majority of our operated patients; neither anamnesis nor clinical investigations revealed factors that could have caused the disease. The torticollis syndrome can develop on the background of many diseases and lesions of the CNS. In the past, one of the most frequent and well-known causes was von Economo's epidemic encephalitis; it was in the anamnesis of 20 out of 40 ST patients described by Abrakov (1952). At the present time this cause is extremely rare.

In rare cases it is possible to elicit a history of head and neck injuries prior to the beginning of the disorder. Walsh (1976) noted this factor in five out of 46 patients. Head injuries preceding the onset of ST were observed in only 6% of our operated cases. Because in many patients it is possible to demonstrate asymmetry of the skull and cerebral hemispheres in pneumograms, Hassler and Dieckmann (1970a,c) theorized that ST was pathogenically linked with birth traumas or skull injuries in early childhood. This interesting hypothesis requires more convincing proof.

Only in extremely rare cases can arteriovenous malformations in the deep cerebral structures be considered the etiologic factor of ST. Two such unique observations, verified angiographically, have been described by Lobo-Antunes *et al.* (1974).

In most cases, the etiology of ST still remains a mystery. As is known, the main and sometimes only means of establishing the etiologic factor is anamnesis. In the majority of patients operated on by us (63%), there were no data in the anamnesis that were important from the etiologic point of view. In the rest of the patients, the onset of the disorder was preceded (at different intervals) by influenza (20%), head injuries (6%), and mental trauma (7%). However, there was no certainty that these states played a genuine etiologic role and were not just inciting factors in the latent stage of ST.

The etiologic significance of heredity has rarely

been established. In the literature we managed to find few reports of a familial form of ST. In two cases, father and son were affected (Dzerjinski, 1916; Gilbert, 1977). Among more than 300 ST patients whom we observed, there was only one certain hereditary cases of this disorder. A mother had had ST for 30 years, and her 11-year-old son for 2 years. One may assume that unlike generalized DMD, the hereditary factor is much less significant in the etiology of ST, but according to Gauthier (1986), the incidence of familial cases of ST is about 5%.

In summary, one may state that the etiology of ST is unknown.

3. Pathology

The pathology of ST has been studied less than the pathology of generalized DMD. Since ST is not lethal, the literature includes only a few descriptions of pathological investigations of the brain.

Foerster (1933) believed that the cause of ST in his autopsied case was pronounced bilateral destruction of the large and small cells of Put producing a lacunar degeneration (*état criblé*). In another case of postencephalitic ST described by Grinker and Walker (1933), diffuse degenerative changes in the cortex, basal ganglia (especially in NC and Put), Purkinje cells, and ND were present. The authors stress that the diffuse nature of the changes does not allow a determination of the diseased structure responsible for ST.

Alpers and Dryer (1937) studied the brain of a 90-year-old male who had suffered from ST for 23 years. The authors found degeneration and replacement by glial elements of the large ganglion cells of Str and insignificant changes in the neurons of GP. Although the pathology observed in this case is more than can be attributed to age, the senility of the patient confuses the clinical picture.

Hassler and Dieckmann (1970a,c, 1982), in two cases of ST, found a reduction in the number of cells and fibers in Put as well as in the parvocellular part of the CM. However, a thorough morphological study by Zeman and Dyken (1967) and Tarlov (1970) failed to reveal microscopic changes in the brain structures related to ST.

Summing up the neuropathological data, one may conclude that the pathology of ST in the few cases studied has not been that of the "pure" form but the syndrome of torticollis with diffuse cerebral lesions. Nevertheless, on the basis of the pathological findings described above, one may infer that ST is related to lesions of Str. In those morphologically studied cases of DMD in which it was possible to identify the primary destructive changes, they were more often localized in Str, principally in Put (Chapter 7, Section 3). The question of why in some cases of generalized dystonia the lesion is in the Str and in others, only in ST remains open. Our observations confirm the data from the literature (Gauthier, 1986) that CT study does not disclose any abnormalities in ST cases (Van Hoof *et al.*, 1987).

4. Pathogenesis

The pathogenesis of ST cannot be considered resolved. The physiological basis of maintaining the head in a normal position and the voluntary cephalic movement in three planes of space is very complicated and only insured by coordinated activity of many muscles and muscular groups. Of the muscles participating in this complex function, the leading roles are played by sternocleidomastoid, trapezoid, and posterior muscles. As is known, the neck muscles have a more complicated innervation than other skeletal muscles. In movements of the head in various directions, the neck muscles act both as antagonists and as synergists (the coinnervation phenomenon).

The first theory of the pathogenesis of ST was proposed by Foerster (1911). He believed that GP was the source of constant motor and tonic impulses acting on the muscular system in response to diverse afferent stimulation of GP from the thalamic systems, whereas NeoStr has a constant inhibiting effect on the activity of GP. Since Str has a somatotopic organization, Foerster considered that in ST the part of Str that inhibits involuntary motor and tonic impulses to the neck musculature was affected. Moreover, he thought it possible that the development of hyperkinesia was influenced not only by switching off the inhibiting effect of Str on GP but also by the intensification of the inflow of excitation along afferent pathways of the thalamopallidal system.

Foerster's assumption concerning the role of Str as the cause of ST is valid even today. But if, as Foerster believed, dystonia is based on hyperactivity of GP, then it would be natural to assume that electrostimulation of that structure should trigger the appearance or intensification of the symptoms of the disease. However, numerous studies of the effects of stimulation of this structure, both in animal experiments and in human stereotactic operations (Chapter 4), did not confirm such a supposition.

Hassler and Dieckmann (1982) consider head turning to be initiated by pallidothalamic neuronal systems involving both segments of GP, AL, perforant fibers of CI, and VOa. The complexity and close interaction of these functional systems, in all likelihood, require the primary bilateral involvement of the neck muscles in ST.

In the course of the past two decades, it was shown that a more or less typical picture of ST can be obtained in animal experiments. This requires a localized stereotactic lesion in the medial part of FR of the midbrain tegmentum (Foltz *et al.*, 1959; Shimabukuro and Mori, 1969; Handa *et al.*, 1971; Mori *et al.*, 1985). Such a lesion destroys not only part of the mesencephalic FR but also the medial longitudinal fasciculus, the central tegmental fasciculus, and Br Con at the level of the magnocellular part of NR (Handa *et al.*, 1971). The ST induced in such experiments remained in the animals for several years and noticeably intensified in stress situations. However, the conclusion stemming from these experiments that ST is caused by lesions confined to the mesencephalic brainstem cannot be considered as proven, primarily because that concept is contrary to the data on the pathology of ST incriminating mainly Str and, particularly, Put.

Electrostimulation of Put in unanesthetized cats causes ipsilateral turning of the head. A similar inclination of the head is caused by stimulation of the interstitial Cajal nucleus, its afferent pathways to VOi, and efferent pathways leading to the posterior longitudinal fasciculus in the spinal cord (Hassler and Hess, 1954; Hassler and Dieckmann, 1982).

In view of this, it is assumed that ST occurs as a result of primary damage to Put, which in the normal state has an inhibiting effect on GP. When this action is terminated, greater activity of GP causes rotation of the head in the opposite direction (Hassler and Dieckmann, 1970a,c). Secondly, it may be considered as an established fact that an experimental model of ST may be produced by damaging other parts of the brain. A number of experiments confirm that rotating movements of the head and neck occur following damage to or electrostimulation of the deep-lying nuclei of Cer and its main efferent pathway—Br Con (Koella, 1955; Jung and Hassler, 1960).

These data led Cooper (1964) to propose the hypothesis that the causal factor in ST is damage to the dentato–rubrothalamic pathway, which disrupts the "sensory communications" leading from Cer to the motor and premotor cortex via Th. This hypothesis, in the opinion of the author, is the foundation for stereotactic destruction of VL in ST.

Mori *et al.* (1975) also produced a condition resembling ST in the cat. After unilateral stereotactic destruction of the paramedian zones of the dorsal portion of the cerebral peduncle, violent twisting movements of the head in various planes were observed. According to histological control, besides the lesion of the cerebral peduncles, the destruction also involved the nucleus of Cajal, its descending tract, and the medial longitudinal fasciculus. In other cases, destruction involved more lateral parts, including the medial tectospinal tract and FR. On the side of the destruction certain biochemical changes were observed, in particular, a decrease in the content of serotonin and catecholamines in a number of subcortical structures.

Fujita *et al.* (1971) demonstrated that the direction of the forced turning of the head depends on the localization of the electrolytic focus in the ventromedial part of the tegmentum of the midbrain; in more medial foci, there was ipsilateral turning of the head, and in more lateral foci, contralateral turning. There are also experimental data indicating that stimulation of NC causes turning of the head and trunk. After damage to NR in monkeys, Carpenter (1956) observed turning and inclination of the head to the side opposite the destroyed NR.

As one can see from the experimental data, the subcortical structure responsible for pathological head positions has not yet been established. In all likelihood, this complex pathology is associated with disorders in the functional system linking up a number of brainstem structures.

Sano *et al.* (1967), during stereotactic operations, stimulated various subcortical structures electrically and recorded EMGs of different muscles of the neck (sternocleidomastoid, splenius, trapezius). In operations for pain syndromes or mental disorders, stimulation of VL did not cause contraction of the abovementioned muscles, but stimulation of Hypoth or Subth as well as NR and the central tegmental tract caused their contraction on the homolateral side and turning of the head in the same direction.

Another picture was observed in the stimulation of the same structures in ST patients. Stimulation of VL produced discharges in the contralateral sternocleidomastoid muscle as well as in the homolateral splenius, whereas stimulation of Subth activated all homolateral muscles. Stimulation of the inferior olives caused a sharp contraction of the sternocleidomastoid and posterior muscles of the neck on the same side. Consequently, the contraction of neck muscles on the homolateral side in ST may be obtained by electrical stimulation of several subcortical structures—Subth, Hypoth, NR, the central tegmental tract, and the inferior olive. The authors believe that ST is a consequence of bilateral lesions of the brainstem.

In the literature, it is possible to find other points of view regarding the pathogenesis of ST. Certain authors link the development of the disease to damage to the vestibular system, which closely interacts with the tonic neck reflexes.

The claim that ST is a disorder of tonic neck reflexes or a release of truncal tonic reactions (Petelin, 1970) is, in all likelihood, correct; however, it explains neither the causes nor the mechanism of the disorder. Tarlov (1969), while stimulating vestibular nuclei in macaques, obtained various abnormal postures of the head and neck resembling ST.

It is presumed that an important role in head turning is played by the vestibulointerstitiothalamocortical path from the vestibular nuclei to the interstitial nucleus of Cajal, thence to VOi of contralateral Th, and then to cortical field 8 where the head- and eye-turning center is localized.

Of considerable significance in understanding the pathogenesis of ST are the EMG investigations of the affected neck muscles. The most typical feature is spontaneous high-amplitude spiking of these muscles at rest, which is never seen in normal, healthy muscles (Figs. 136, 137). As may be seen from our investigations, the characteristics of this pathological activity differ but clearly correlate with the clinical form of the disease. For instance, in the tonic form, continuous activity is observed, sometimes with short silent periods. The clonic form is characterized by burst activity, either as a regular tremor with a frequency of from 5 to 20 Hz or as high-amplitude bursts of irregular frequency. In the relatively frequent clonic–tonic ST, the EMG is a combination of both of these phenomena.

As our investigations have shown, in voluntary turning of the head, unlike involuntary jerking, EMG activity sharply intensifies, and the amplitude increases. However, voluntarily turning the head toward the wry-neck normalizes the EMG. When the patient is in a prone position, the spontaneous activity in EMG is minimal, whereas when in the vertical position, it is sharply intensified. Just as in DMD (Chapter 7, Section 7.2), the EMG of the neck muscles clearly shows a simultaneous contraction of antagonists, indicating a disordered reciprocal innervation.

On the basis of EMG findings in 20 cases of ST, Podivinsky (1960) distinguished three main forms of this disease: (1) extrapyramidal, (2) reflex, and (3) postparalytic (damaged cervical roots).

Summing up the data on the pathogenesis of ST, one may conclude that the basis of this syndrome is a disorder of the complicated physiological mechanism controlling movements of the head. The most important factor is the role played by Str, midbrain tegmentum, deep cerebellar nuclei, and cerebral cortex, the integration of which is still not fully understood.

5. Clinical Picture

The involuntary movements of the head and the pathological postures in ST are quite diverse and often change rapidly in the same patient. Just as in the generalized form of DMD, one can distinguish dystonic movements (hyperkinesias) and muscular dystonia. On this basis, ST may be classified according to the predominance of one of these phenomena, which are usually combined. The tonic form predominantes (70% of our patients), usually causing permanent contraction of the neck muscles and tilting of the head to the side (less frequently backward, and even less frequently forward).

The other, less common, clonic form of torticollis (30% of our patients) is characterized by rapid jerking of the head to the side, resembling myoclonus. Most frequently, these twitchings are arrhythmic, but sometimes they may resemble a rhythmic tremor as in parkinsonism. Quite often, one encounters a clonic–tonic form, which is a combination of the two preceding forms. Clonic jerking frequently occurs when the patient tries to return the head to the "zero" position.

Depending on the different groups of muscles involved in ST, it is customary to recognize three types in literature: "lateral ST," in which the muscles that turn and incline the head to the side participate; "backward ST," involving the muscles that throw the head back; and "forward ST," involving the muscles that incline the head forward. However, such a classification does not take into account the great diversity of head movements in ST. Accordingly, we consider it more expedient to recognize three basic forms of ST: (1) rotational or the so-called "pure" ST, i.e., twisting of the head on its vertical axis without deviation from this axis; (2) axial bending of the vertical axis of the head and neck to one side (antecollis, retrocollis, and laterocollis); (3) the most frequently encountered complex form with any combination of the first two types (Kandel, 1984).

It is easy to theorize that, in principle, six pathological postures of the head are possible: two versions (to the right and left) along the vertical axis of the neck and four inclinations (to the right or left shoulder, forward, and backward) with the bending of the neck. However, if we take into account that these postures are frequently combined with each other, then there are many more possible combinations.

In our series (162 operated patients), rotational ST was observed in 38% of the patients, to the left twice as often as to the right, for which there is no explanation. Pure inclination of the vertical head axis to either shoulder was found in 4%, and pure retrocollis and antecollis in 6% and 3% of the patients, respectively. In the remaining 49% of cases, combinations were observed: a turn along the vertical axis with throwing back of the head (the most frequent variant), the same turn with an inclination forward, a turn of the head with an inclination to the shoulder, a turn with an inclination to the shoulder and extension of the head, a turn with an inclination to the shoulder and forward incline, and retrocollis with an inclination to the shoulder.

Although in ST the head is usually forcibly turned to one side, in the majority of cases the muscles of both sides of the neck are involved in the pathological process, although the degree of their involvement varies considerably. In the majority of cases with rotational and combined form of ST, the severe contractions of the sternocleidomastoid and trapezius on one side and posterior deep cervical muscles (the rectus capitis posterior major, the obliquus inferior, and the splenius capitis) on the contralateral side are present. Usually, it is not possible to establish a definite relationship between the afflicted muscle and the form of ST (tonic or clonic).

Our EMG investigations (Kandel, 1981, 1984) showed that muscles were affected on half of the neck in only 8% of the patients (four of 50). Of the 46 patients with bilateral muscle involvement, 27 patients had four muscles involved, 14 had three, and five patients had two neck muscles affected. Since the involvement of the right and left sides of the neck is usually to different degrees, the head, as a rule, is turned to one side, depending on the lateral predominance of the dystonic manifestations. The fact that the neck muscles on both sides are involved in turning of the head to one side was noted by Herz and Hoefer (1949).

The injection of lidocaine in the affected muscles proved ineffective as a therapeutic method; however, the introduction of a local anesthetic in the posterior roots C_1–C_4 may be quite useful for EMG diagnosis of the nature and severity of lesions in individual or groups of muscles (Bertrand *et al.*, 1978).

Our observations, in agreement with those of other authors (Herz and Glaser, 1949; Abrakov, 1952; Cooper, 1965c,d), show that unlike DMD, this disease begins mainly at a mature age—from 30 to 45 years. The duration of the disorder before admission to our clinic was 1–3 years in 60% of the patients, 4–10 years in 30% of patients, and over 10 years in 10% of patients.

The disease usually begins in a healthy person with a feeling of an unpleasant, often painful, "pulling up" of the muscles on one half of the neck, but without turning of the head, an awkward feeling in the neck, or a desire to turn the head to the side or to straighten it. The patient often ignores this, and his attention is drawn to it by associates who notice the strange position of the head. This is followed by the gradual development of mild, involuntary, sporadic contractions of the neck muscles with usually prolonged intervals of freedom. The head is powerfully drawn to one side, less frequently forward or backward. At this stage, the patient is still voluntarily able to correct the deviation of the head; however, later it becomes more and more difficult, and, finally, impossible. The forced posture of the head then becomes permanent. The head is usually turned to the side opposite the tonically contracted and markedly hypertrophic sternocleidomastoid muscle.

The main clinical signs of ST in practically all patients develop comparatively rapidly, in approximately 1 year. If in some patients in the initial stages there are remissions after 1–3 months, then in the later stages the signs become stable, or the disease continues to progress slowly.

In certain cases, depending on the progression of the disease, the direction of the torticollis may change. For example, in one of our patients, for half a year the head was drawn to the right and then (approximately in a year) to the left, after which severe retrocollis developed; this did not remit over the course of many years. Another patient had a severe head turning to the left for 3 years; then, for 6 years he was practically healthy, after which his head began to turn and tilt to the right, a state that has lasted for more than 2 years.

The tonic contractions of the neck muscles and the turning of the head are markedly intensified by attempts to perform any purposeful movement, by holding the head in a vertical position, as well as in states of physical or emotional stress.

The severity of ST depends on the position of the body in space. As a rule, ST is mild or disappears in a prone position, reappears in a sitting position, intensifies during standing, and is maximal in walking. We operated on three patients in whom ST occurred only during walking. However, there may be exceptions to this rule. We observed two cases (one of whom we operated on) in whom ST occurred only in the sitting position. The hyperkinesias disappeared completely during sleep.

In the majority of patients there is constant pain (although of varying degree) in the tense neck muscles. This pain, as a rule, becomes a chronic complaint in ST and is intense in the most severe forms. There can be no doubt that the pain is the result of constant tonic tension of the cervical muscles. This is confirmed by the fact that stereotactic operations decrease or relieve this pain by eliminating tension on the neck muscles participating in the forced posture of the head.

Quite often in ST patients spine films will show changes in the cervical spine in the form of scoliosis, osteochondrosis, or spondylosis of varying severity. These changes most likely are secondary and cannot be considered the cause of ST.

Quite frequently in ST patients choreic tics, hyperkinesia, or intentional tremor are seen in one or, rarely, both arms. This fact, which was noted by Cooper (1965c,d) and Laitinen (1963), was confirmed by our findings. Hyperkinesia in one arm was present in 30% of our patients, a manifestation that indicates the extrapyramidal nature of ST.

The disease usually progresses slowly but steadily. A spontaneous remission or sharp decrease in ST is extremely rare. In more than 300 ST patients, we had only three with spontaneous remissions unassociated with some kind of treatment.

In severe forms of torticollis, the patient is unable to work and can take care of himself only with great difficulty. The turning to one side or retracting of the head prevents the patient from walking along a street, since he is unable to see obstacles in his way or moving vehicles. Most often the pathological turning of the head is the result of tonic tension in the hypertrophic and thick deep muscles of the neck on one side and the sternocleidomastoid muscle on the other side. The head, as a rule, is turned to the side opposite to the tense hypertrophied sternocleidomastoid muscle.

6. Stages of the Disorder

We can define four degrees of severity of ST (Kandel, 1984).

Stage I is mild ST. The patients can turn the head to either side independently and hold it in that position for a long time by dint of will. The torticollis generally develops when walking, under emotional stress, and during physical labor. In this stage, the patient retains his capacity to work.

Stage II is average severity. The patients, without the help of the hands, may hold the head in the normal position, but only for a short period of time. The ST appears not only during walking but also at rest. There are difficulties in dressing, and capacity to work is impaired.

Stage III is the severe stage of ST. The patient is able to hold his head in the normal position only with the help of his hands. The neck muscles are contracted and painful. During walking, the patient always holds the head with one or both hands. Such patients always have difficulties taking care of themselves, and their capacity to work is practically lost.

Stage IV is very severe ST. The patient is unable to place his head in the normal position even with the help of his hands. There are constant severe pains in the tense and hypertrophied neck muscles. The ST does not diminish at rest. There is total loss of working capacity, and the patient is unable to care for himself.

7. Other Manifestations of ST

Besides the main symptoms of ST, many patients are found also to have associated (although usually not pronounced) neurological abnormalities such as asymmetric nasolabial folds, slight deviation of the tongue, and hyperactive but asymmetric tendon reflexes.

Some patients have blepharospasm and involuntary contractions of the mimetic musculature, aggravated when an attempt is made to correct the abnormal posture of the head (see Fig. 135). Colbassani and Wood (1986) noted that approximately 10–20% of ST patients have associated motor abnormalities, particularly of the extrapyramidal variety.

Often one observes unusual phenomena in ST such as the disappearance of the head movements (partially or totally) during attention (reading an interesting book, watching TV programs, films, etc.) or during certain types of motor activity. One of our patients noted that in "unusual positions of the body"—skiing downhill, swimming, playing basketball or football, and in swift jumps—the abnormal head posture almost totally disappeared. The torticollis may stop when a person is moving; however, in some of our patients the ST occurred only in walking. In other cases, it diminished or even disappeared when the patient talked to a stranger, in particular, a physician.

It is interesting to note that voluntary turning of the head to the "other" side, as the movement is resisted (pressure of the cheek against the physician's hand), is executed more easily than when the movement is not opposed. In literature this phenomenon has been called counterpressure.

However, the most interesting and paradoxical phenomena often associated with ST are the "correcting gestures" or "geste antagonistique." If the patient places a finger or the palm of his hand on the rotated chin without any great effort and even with no intention of turning the head to a normal physiological position, the ST significantly diminishes or even disappears.

As we have noted many times, this phenomenon does not depend on which (right or left) hand the patient uses to touch either cheek. The following was observed in one of our patients: when the arms were raised, his head, which was forcibly twisted to one side, would return to the proper physiological position, and when he lowered his arms, his head would turn again to the side. This effect may be produced by light touch not only of the finger but of any other object. For example, several of our patients almost constantly held a rolled up newspaper to their cheek. In another patient, ST would diminish when a telephone receiver was placed to his ear. Sometimes a light touch of the fingers not to the chin but to the cheek, ear, or back of the head may relieve the ST. Certain patients stood or sat with their backs or occiputs pressing against the wall. One of our patients constantly bit his coat lapel, and another attached a clothespin to his ear lobe.

There were also still more unexpected and strange phenomena. One of our patients almost constantly carried a log on his shoulder, whereas another wore a cap with a lead plate weighing no less than 5 kg. Patterson (1943) observed cessation of ST when a patient put a hair on the cheek on the side of the muscular spasm. Monnier (1938) reported a patient whose condition was improved by whistling or rubbing his neck or even by turning on a green or blue light. These phenomena, which do not lend themselves easily to a logical explanation, in previous times led neurologists to suggest a "hysteric genesis" of the ST syndrome.

It should be noted, however, that such corrective gestures are effective only in the initial stages of the disorder. In severe cases, such a "small trick," as it was called by Podivinsky (1968), does not help, and the patients forcibly hold their heads with one or even both hands in an endeavor to correct the painful twisting of the head.

The phenomena described above have no acceptable explanation to date. As Cooper (1965c,d) stated, when a finger touched the chin slightly, thereby returning the head to the normal position, the nerve endings of this finger served as a proprioceptive "substitute servomechanism" that "informs" the postural centers of the brain about the position of the head in space. Such a hypothesis is interesting, but the author does not offer any valid arguments to support it. As we have already noted, the same effect is obtained when the chin is touched not with a finger but with some other object. Petelin (1965) considers these phenomena to be based on innervation connections with the tonic neck reflexes that redistribute muscle tone.

Podivinsky (1968), investigating these "correcting gestures" by EMG, showed that slight tactile irritation of the skin in the area of C_2–C_4 segments causes the hyperkinesis to diminish greatly or disappear on the afflicted side of the neck. Similar irritation of the other, "healthy," half of the neck results in a "burst" of involuntary muscular activity. In any case, the correcting gestures may supply specific proprioceptive input lacking in ST (Colbassani and Wood, 1986).

8. Differential Diagnosis

Differential diagnosis usually does not present any particular difficulties. An experienced physician will easily distinguish true ST from changes in the normal position of the head as a result of local pathological processes in the neck muscles or cervical spine. For instance, children may have a torticollis-type congenital anomaly on the basis of fibrous degeneration of the sternocleidomastoid muscle. A fixed position of the head may be the result of fractured cervical vertebrae, protrusions of the cervical intervertebral disks, or tumors in the cervical spine and spinal cord.

Abnormal postures of the head in such diseases can usually be differentiated visually from extrapyramidal ST. In more difficult cases, an analysis of the clinical pictures and the development of the disease as well as x rays and additional investigations will enable the physician to make the correct diagnosis. As we have already mentioned, ST may occur not only in DMD but also in parkinsonism, athetosis, Huntington's chorea, and other extrapyramidal disorders associated with these diseases. Nevertheless, even in such cases a differential diagnosis does not present any particular difficulties.

9. Surgical Treatment

Drug treatment, which is rarely successful in mild forms of the disease, is absolutely ineffective in severe ST. The use of cholinolytics (trihexyphenidyl, etc.), drugs acting on the affected muscles (injection of local anesthetics, alcohol, quinine solution, and others), and other methods produce either no results or only brief relief. L-Dopa also proved ineffective (Barrett *et al.*, 1970; Ansari *et al.*, 1972); treatment with L-dopa resulted in improvement in only one out of 17 ST patients (Shaw *et al.*, 1972). In a few papers the effectiveness of haloperidol and GABAergic drugs was noted (Gilbert, 1972; Colbassani and Wood, 1986). One of the new measures of conservative treatment is sensory feedback, which is described as effective in many cases (Korein *et al.*, 1976).

Surgical treatment is indicated in all cases of ST providing the following conditions are met: the extrapyramidal nature of the disorder has been reliably established; the disease has led to partial or total disability; the disorder has been present more than a year; conservative treatment has been ineffective; and there are no contraindications to surgery.

The first attempt to correct ST surgically dates back to the middle of the 17th century, when a German surgeon, Minnius (1641), dissected the sternocleidomastoid muscle.

The Russian surgeon Buyalsky (1850) was the first to cut the spinal accessory nerve for ST. Many years later, Morgan (1867) performed this operation as proposed by Romberg. In 1890 Collier clipped the accessory nerve with a silver wire.

9.1. Cervical Rhizotomy

Section of the roots of the cervical spinal cord for ST, first performed almost a century ago, still has an important place in the treatment of the disease today. Resection of the accessory nerves of the neck also has not been abandoned completely even though it is not very effective even for a short time. Because this operation is simple and safe, it may be recommended in mild cases.

In 1891, Keen first proposed unilateral section of the first three anterior cervical roots for ST; however, this operation was not very effective and caused severe atrophy of the neck muscles. After Foerster suggested posterior rhizotomy (Chapter 13, Section 5.3.6), Taylor (1915) performed the first unilateral section of four posterior cervical roots as well as resection of the accessory nerve for ST. Then McKenzie (1924) described an operation, first performed by Cushing, that consisted of subdural section of both anterior and posterior roots of the first three cervical segments and section of the accessory nerve.

In 1926, Foerster reported that section of five anterior and posterior roots on one side in ST yielded quite satisfactory results. Frasier (1930), in four ST cases, cut the first three posterior roots on both sides and at the same time sectioned the accessory nerves intradurally.

The operation was modified by Dandy (1928) and then by Olivecrona (1931), both of whom performed bilateral intradural section of the motor and sensory roots of the first three superior cervical segments as well as intradural section of the accessory nerve roots on both sides at the foramen magum. At a second stage, Dandy also performed Buyalsky's operation—section of the accessory nerve in the neck. Dandy noted a complete cessation of ST in 5 of 8 patients and improvement in 2. In the neurosurgical literature, this operation is termed the Foerster–Dandy operation (Fig. 129).

In the Soviet Union, this operation was first performed by Babchin in 1934, and it was repeated many times by other neurosurgeons in the years immediately following. In 1952, Abrakov reviewed 110 Foerster–Dandy operations in the literature. In 1960, Arseni and

Sandor published the results of 50 cases of which 41 were pure torticollis and nine were a fragment of DMD. Good results were obtained in the torticollis cases, but in dystonia the operation was considerably less effective. The postoperative mortality was high (12.6%).

One of the largest series of this operation was published by Sorensen and Hamby (1966). Of 71 ST cases, good results were obtained in 38, limited improvement in 22, and slight improvement in six. In four cases, the operation produced no effect, and one case had a lethal outcome.

Wycis and Gildenberg (1979) reported on the long-term results (follow-up from 1.5 to 22 years) of 26 cases after bilateral section of the three superior cervical roots and both accessory nerves. Even though the operation was restricted to anterior rhizotomy, the long-term results were satisfactory in 22 cases and some improvement in three cases. One patient died after the operation.

Sano *et al.* (1967) and Scoville and Bettis (1979) came to the conclusion that the Foerster–Dandy operation remains the operation of choice in ST even though the postoperative atrophy of the neck muscles is distressing for the patient.

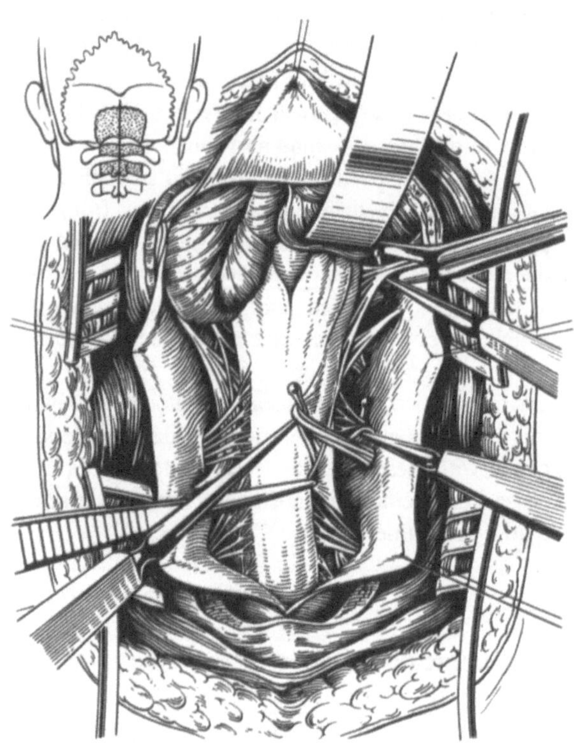

FIGURE 129. Foerster–Dandy operation for spasmodic torticollis: intradural cutting of the three superior anterior and posterior cervical roots and accessory nerves bilaterally.

Walter Dandy (1886–1946)

On the basis of the data from the literature, one may conclude that cervical rhizotomy is effective in ST. Even in severe cases, this operation completely or partially eliminates the hyperkinesis of the neck muscles. However, this operation has great risk. The extensive denervation of the neck muscles frequently (about 40–50%) leads to grave motor disturbances and serious complications (flaccid paralyses of neck muscles, limititation of voluntary head movements, dysphagia, sensory disorders, trauma to the vertebral artery, etc.). After cervical rhizotomy, 32% of 50 patients developed dysphagia, probably as a result of inability to raise the head at the appropriate step in the act of swallowing (Hamby and Schieffer, 1969). Epidural hematomas with acute compression of the cervical spinal cord after the rhizotomy were also described. In very rare cases, anterior rhizotomy results in another serious complication—ischemia of the cervical spinal cord caused by severing of the small arteries of the anterior roots. In 29 Foerster–Dandy operations, there were two lethal outcomes (Törmä and Troupp, 1958).

The advent of microsurgical techniques has made it possible to refine the operation on the cervical roots in ST. Bertrand *et al.* (1978), in this disorder, used a combined selective section of the upper cervical roots and a stereotactic operation on the basal ganglia (see below). Subsequently, they reported (Bertrand *et al.*, 1982) on 35 ST cases treated only by peripheral denervation. As the authors emphasize in this operation, unlike conventional rhizotomy, a selective denervation of the affected muscles is made with the aid of EMG. The number of muscles involved is usually greater than would be expected on the basis of the clinical data.

The operation is performed under mild narcosis without curare drugs so the surgeon can observe the effect of stimulation. An incision is made on the posterior surface of the neck with the patient lying on his side. Using an operating microscope, the surgeon exposes the cervical roots and their posterior branches from C_1 to C_7. By stimulation C_1 is identified extending from under the vertebral artery rostral to the posterior atlantal arc and the posterior branch of C_2. Then, based on the results of stimulation, the roots caudally to C_5–C_6 inclusive, and sometimes C_7, are sectioned. Then, the sternocleidomastoid muscle is denervated. In cases of retrocollis, the posterior branches to C_5 on both sides are also cut.

Using the technique described above, Bertrand *et al.* reported the following results: excellent, 34%; good, 54%; insignificant, 9%; and poor, 3% of the cases.

A few years ago a new, perspective mode of surgical treatment of ST was developed. Freckmann *et al.* (1981, 1983) described 11 patients in which vascular compression of the intracranial and intraspinal accessory nerve was disclosed at the operation. The compression was caused by different vessels: the vertebral artery, posterior inferior cerebellar artery, and spinal arteries. After microscopic inspection the nerve was dissected and a sponge was interposed between the nerve and one of the above-mentioned arteries. These operations of microvascular decompression have had very encouraging results, and are of importance also for understanding ST pathogenesis.

The recently proposed bilateral microsurgical decompression of the roots of the accessory nerves and resection of anastomoses between these roots and the posterior roots of the cervical spinal cord in 23 severe ST cases gave good results in ten and some improvement in another nine cases (Freckmann *et al.*, 1981, 1983).

Percutaneous posterior cervical rhizotomy was performed in two patients (Pagura, 1983) without producing changes in the involuntary movements but lessening the pain in the neck muscles.

Another approach to microsurgical rhizotomy is used by Colbassani and Wood (1986). A limited suboccipital craniectomy and laminectomy of C_1–C_3 are performed with the patient in the sitting position. After the upper three dentate ligaments are divided, microsurgical anterior rhizotomies are performed bilaterally with preservation of the radicular arteries. Then the patient is placed in the supine position and selective denervation of the involved sternocleidomastoid is made.

9.2. Stereotactic Operations

Occasional attempts have been made to destroy certain cerebral structures by an open approach in ST; however, the results were not satisfactory. For example, Sano *et al.* (1967) employed electrocoagulation of small zones (approximately 2 mm) in the inferior olive ipsilateral to the contracted sternocleidomastoid muscle in two cases. These operations, performed through an open suboccipital craniotomy, gave a marked improvement, but apparently the operation was not pursued. Stimulation of the inferior olives caused a pronounced tachypnea (up to 60 per min), which is an indication that this operation involves a certain risk.

The rapid development of stereotactic methods in neurosurgery and quite satisfactory results in the generalized form of DMD quite naturally encouraged the use of this method in ST. Stereotactic operations have substantially increased the understanding of the pathogenesis of ST and indicated the cerebral structures modifying it. Although there is still no firm theoretical foundation for these operations, the indications have been clearly defined, and the results have been critically evaluated.

The first stereotactic operation, pallidotomy, in ST was performed by Riechert in 1953. Several papers analyzing the results of stereotactic surgery in ST have been published in the interim. Cooper (1965d) reported on the surgical treatment of 75 cases. However, in this series the author included not only pure cases of torticollis but also cases in which ST was a component of generalized DMD. In a subsequent paper, Cooper (1977) summarized the largest experience in world literature with stereotactic treatment for ST. In 20 years, the author performed thalamotomy on 160 patients and in a long follow-up study noted a substantial improvement in 60% of them.

In an early paper, Laitinen (1963) reported five ST cases in which he performed unilateral destruction of VL or the medial segment of GP. Although improvement was noted in all cases with the exception of one, a relapse occurred after several months in all cases. At that time, the author concluded that stereotactic operations in ST offer no obvious advantages over cervical rhizotomy.

Subsequently, 52 ST patients underwent surgery at the Freiburg Clinic (Mundinger *et al.*, 1972). Others have operated on smaller series: Hassler and Dieckmann (1970a,c) operated on 16 patients; Caracalos (1972) on ten; Mori *et al.* (1975) on 17; Sano *et al.* (1972) on 12; and Driollet *et al.* (1975) on 13. In our monograph *Dystonia Musculorum Deformans* (Kandel and Vojtina, 1971), we presented the results of 20 stereotactic operations on VL in 16 ST cases; recently, the results of 152 operations were presented (Kandel, 1984). In 1976 Dieckmann reported the results of stereotactic operations on 70 patients on whom lesions were made in the Forel's thalamic bundle H_2 near VOa, in VOi, and in the interstitial–thalamic pathways to VOi. A good result was obtained in 47% of the cases, and an improvement occurred in another 29%.

In 1982 Hassler and Dieckmann analyzed the results of stereotactic operations on 92 patients who had had high-frequency coagulation of VOa and H_1 or VOi. In some cases, additional lesions of the internal part of Vim (connected with the vestibular tracts) were made. Electrostimulation of these structures prior to their destruction decreased the spontaneous electrical activity in various neck muscles.

The results of these operations were graded as follows: excellent and good in 67% of the cases, slight improvement in 23%, and no change in 10%. There was no postoperative mortality. The main complication (in 16% of the cases) was inability to perform fine movements in the extremities contralateral to the stereotactic operation. A second complication after destruction of Vim was ataxia in the contralateral arm. These complications disappeared in several months with active physical therapy.

An analysis of the results of these series of observations is both important and instructive. One of the main problems was to determine the most "effective" subcortical structure. Many authors, including ourselves, tested several structures as target points. As we have already noted, the first stereotactic operations in ST involved destruction of the medial segment of GP (Cooper, 1964). However, it was soon apparent that lesions of this target do not produce a consistent therapeutic effect. In the early 1960s, neurosurgeons working on the problem of ST changed to making a lesion of VL. For quite a long time, this nucleus was assumed to be the structure of choice for the most favorable results (Cooper, 1965c,d; Kandel and Vojtina, 1971). However, the ideal subcortical target is still the subject of discussion. Mundinger *et al.* (1972), who has had extensive experience in performing stereotactic operations for ST, considers subthalamotomy to be more effective than VL destruction. After lesions of Subth, usually on both sides, a marked clinical improvement was noted in 61% of the cases, whereas after destruction of VL it was observed in only 36%.

In their early work on ST, Hassler and Dieckmann (1970a) performed stereotactic destruction of VOi (nucleus ventrooralis internus), assuming that head movements were localized precisely in this thalamic nucleus. The stereotactic coordinates of VOi were 12 mm behind CA, 1 to 2 mm above LI, and 9 to 10 mm lateral to the midline. Because the long-term results of that operation were not satisfactory, the authors made lesions of both VOi and the interstitial nucleus of Cajal (Hassler and Dieckmann, 1970a, 1982). Although the long-term results after that operation were better, a lasting effect was obtained only in approximately half of the cases.

Bertrand (1976) noted good results in seven of ten cases after stereotactic destruction of VOi; however, four of the seven cases with a good outcome had bilateral operations. The author considers the results of VOi lesions to be better than those of his previous series of observations (30 cases) in which lesions were made caudal to VOi bilaterally in the majority of cases. Bertrand *et al.* (1978) later reported that bilateral destruction of VOi led to 10% complications in the form of pseudobulbar syndrome, whereas similar lesions in the posterior part of VL led to only 2%.

Another question that still remains to be resolved is the side of the operation. It may be performed on the side toward which the head is twisted or on the opposite side. Many authors (Cooper, 1965c,d; Andrew *et al.*, 1974) consider that for a lasting effect a bilateral operation is required.

Sano *et al.* (1972) proposed that a lesion be made of the mesencephalic interstitial nucleus of Cajal situated in the posteromedial part of Th for the relief of ST. The basis for such a procedure was the fact that electrostimulation of that nucleus intensifies cervical hyperkinesis. The coordinates of the Cajal nucleus according to Sano are: 7 to 9 mm behind the middle point of LI; 2 to 3 mm below that line; and 1 to 2 mm lateral to the midplane. Exactly the same coordinates are given by Hassler *et al.* (1979) in the stereotactic atlas. Sano's limited experience (12 operations) showed that in the majority of cases torticollis disappeared or was substantially diminished. The most pronounced therapeutic effect was noted when the posterior neck muscles were affected.

Andrew *et al.* (1974) performed stereotactic destruction of VO and CM in six cases with torticollis or retrocollis. Bilateral operation was performed in five cases, and unilateral in one. In three of the six cases, especially in the one case with a unilateral lesion, a good effect was obtained with the disappearance of hyperkinesias over the several years of follow-up. In the remaining three cases, the effect was temporary, lasting only several months. The authors note that in five cases a unilateral operation produced no effect. In their next study, Andrew *et al.* (1983) reported good results from bilateral thalamotomy in ten of 16 patients.

Driollet *et al.* (1975) reported the results of destruction of the VO nuclei of Th or Subth in 13 ST cases. In five cases, the operation was performed bilaterally. Subsequently, in four cases, a Dandy cervical rhizotomy was performed in addition. A good result was obtained in half of the operated patients.

Caracalos (1972) operated on ten patients. Of seven cases in which a unilateral operation was performed, substantial improvement was noted in only three. Three patients underwent bilateral surgery, and two were reoperated on the same side. In four patients in whom EMG revealed bilateral involvement, there was no effect, whereas in unilateral involvement, the results were good. In two of three cases, after a bilateral operation, the improvement was long-lasting. Mori *et al.* (1975) operated on 17 patients and noted that in cases of unilateral involvement of the neck muscles, and primarily unilateral discharges in EMGs, a good result was obtained by subthalamotomy on the opposite side. However, in cases of a bilateral involvement, even after bilateral thalamotomy, improvement was noted only rarely.

The fact that the results of stereotactic operations on different structures of the basal ganglia were not quite satisfactory stimulated surgeons to try combined surgical procedures. As was mentioned above, stereotactic operations combined with section of cervical roots and accessory nerves were performed (Bertrand *et al.*, 1978).

The limited early experience with stereotactic dentatotomy (Chapter 9, Section 5.1.3) showed that in ST this operation produced no effect (Siegfried *et al.*, 1970). Subsequently, however, a notable and lasting improvement in two cases was reported after bilateral dentatotomy (Siegfried and Verdie, 1977).

9.3. Stimulation of Spinal Cord

Chronic stimulation of the posterior columns of the cervical spinal cord has been investigated recently for the treatment of ST. Gildenberg (1977, 1978), who employed this method, obtained good results in four of six cases followed from 1 to 2.5 years. Moreover, he found that the best effect was produced by high-frequency stimulation (800–1000 Hz).

Waltz (1982) reported on the results of this procedure in 26 ST cases, and recently in 61 cases (Waltz, 1985). An improvement, to varying degrees, was obtained in 70% of cases followed up from 1 to 10 years. Dieckmann *et al.* (1985) employed bipolar cervical stimulation in 18 cases using high-frequency current (about 1000 Hz). The authors reported very good or good improvement of ST in 50%, but noted that the results were not as favorable as those of other neurosurgeons. The authors stressed that the therapeutic effect began only after months of continuous stimulation. The evaluation of the benefit of this new method in ST is for the future; presumably the first promising results will be confirmed by other authors.

Chronic stimulation of deep-seated cerebral structures in ST has yielded no effect (Upton *et al.*, 1982); however, low-frequency stimulation of Subth resulted in a substantial improvement in a limited series of cases (Mundinger *et al.*, 1977).

A summation of the published data allows one to draw several conclusions. It may be concluded that stereotactic operations on the basal ganglia in ST, on the whole, are undoubtedly effective. In approximately one-half of the cases, after the operation a substantial improvement is noted (almost to complete recovery), which can not be equaled by any other current method of treatment for this disorder. No less important is the fact that stereotactic operations are relatively safe and rarely cause serious complications. However, they are less effective in ST than in DMD and do not yield any tangible results in approximately 40 to 45% of the cases.

10. Personal Observations

We performed our first operation for ST in 1959. In the period inclusive of 1985 we have operated on 162 patients, 117 men and 45 women. Such a great difference confirms the higher incidence of this disease in men than women. These 162 patients underwent 206 stereotactic operations on different subcortical structures. The age of patients varied from 14 to 58 years; the majority were between 30 and 50 years of age.

The duration of the illness prior to admission was from 2 to 8 years in the majority of cases, but in some cases as long as 15 years. All those operated on were in the most severe stages (III–IV) of the disorder. This fact along with loss of working capacity and total lack of response to drug treatment for many years, were the main indications for surgery. Besides the usual clinical examination, EMGs were made of the neck muscles before and after the operation.

In more than half of the surgical cases (106 of 162), the etiology of ST was unknown. In the remainder, onset of the disorder was preceded by various factors that, with a certain degree of probability, can be considered etiologic, although it is impossible to prove a direct cause–effect relationship. A tonic-type ST was noted in 109 cases, and clonic in 53. The distribution of cases by the main forms of ST based on our classification has been presented above (Section 5). The standard cryosurgical method described in Chapter 4 was used in the majority of operations.

All operations were performed under endotracheal general anesthesia. It is noteworthy that in certain cases, the retrocollis was so pronounced that the bilateral tonic contraction of the posterior neck muscles did not disappear even under anesthesia. In three cases, because it was impossible to place the patient in a supine position, we operated with the patient in a semiprone position.

In the majority of operations (139), we performed stereotactic cryodestruction of VOa, VOp, and partially of VOi; in 48 operations, combined destruction of VL and Subth; in 17 operations, destruction of VOi; and in 16 operations, destruction of the nucleus of Cajal. A total of 105 patients had one operation, 42 patients had two operations (one on each side), three patients had three operations (two on one side and one on the other side), and one patient had four operations (two on each side).

The immediate results were evaluated when the patients were discharged from the clinic (10–12 days post-surgery), and long-term results were determined on the basis of repeated examinations and data from questionnaires.

We considered "good results" to be total or almost total disappearance of forced turning, jerking, inclining, or retracting of the head, total or practically total disappearance of tension and pain in the neck muscles, and the restoration of working capacity (Figs. 130–135). The pathological tension in the affected cervical muscles disappeared immediately after a successful operation. The muscles regain normal consistency and sometimes even become flaccid. There is a gradual decrease in the hypertrophy of the sternocleidomastoid muscle so that several months after the operation, it is hardly noticeable. Not only the forced twisting of the head but the severe blepharospasm that sometimes accompanied torticollis disappeared after the operation (Fig. 135).

An EMG study after an effective operation shows the practical disappearance of the preoperative spontaneous pathological electrical activity in one or several neck muscles (Figs. 136, 137).

We regarded as a "substantial improvement" incomplete but such marked reduction of all components of torticollis that after the operation, the patient could hold the head in a normal position for some time and turn it to both sides without use of the hands. Insignificant ST symptoms were manifested during walking or in times of excitement.

A "slight improvement" was considered when ST symptoms diminished but the pathological posture

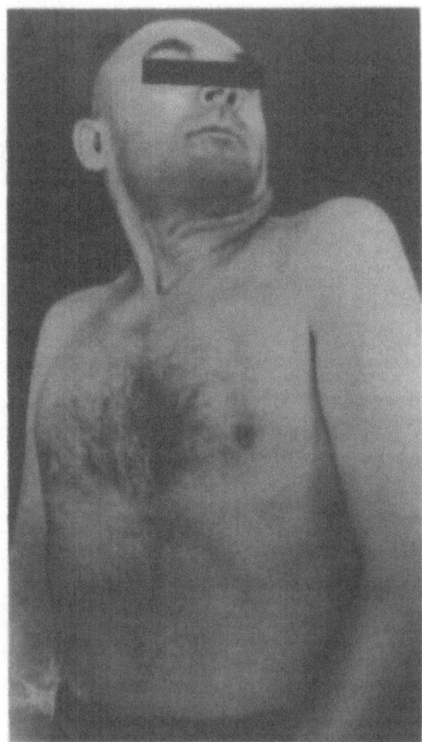

FIGURE 130. Rotational form of spasmodic torticollis in a 43-year-old patient (A) before operation and (B) 8 years after one-sided stereotactic VL thalamotomy.

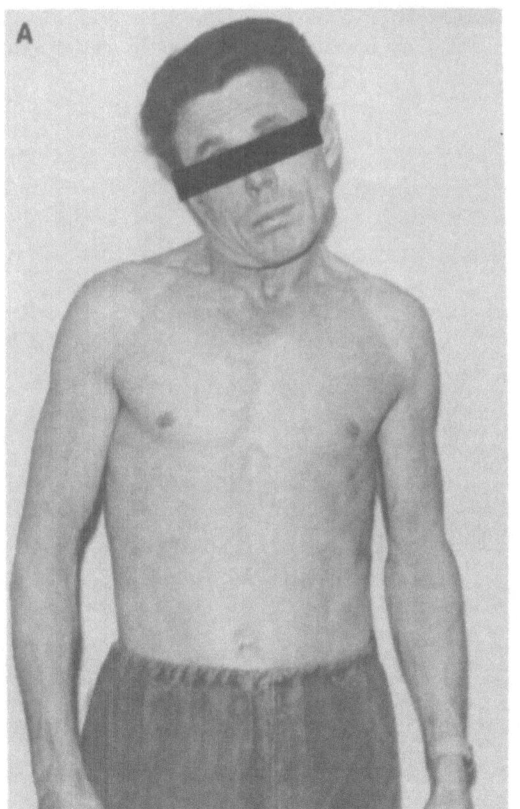

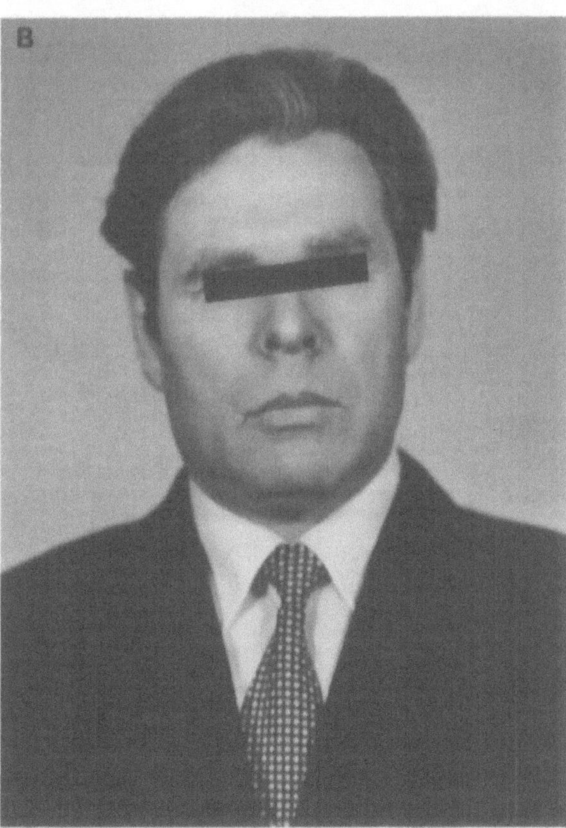

FIGURE 131. Axis form of spasmodic torticollis in 44-year-old patient (A) before operation and (B) 4 years after stereotactic destruction of left Voi and nucleus of Cajal.

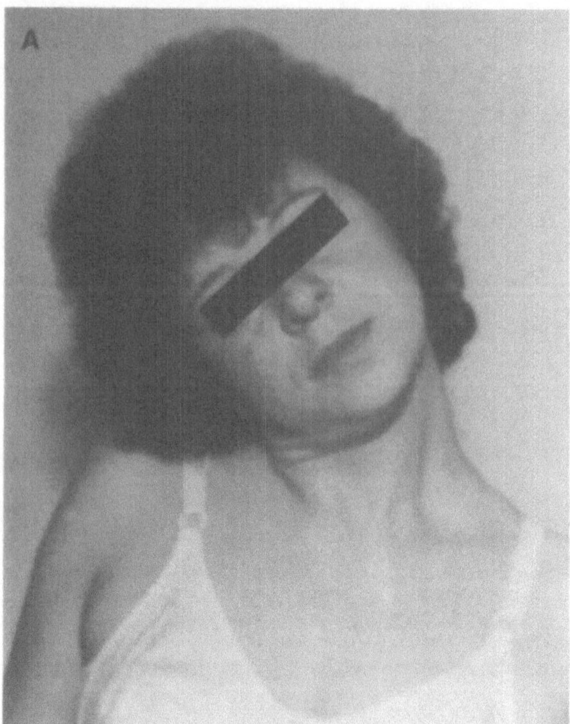

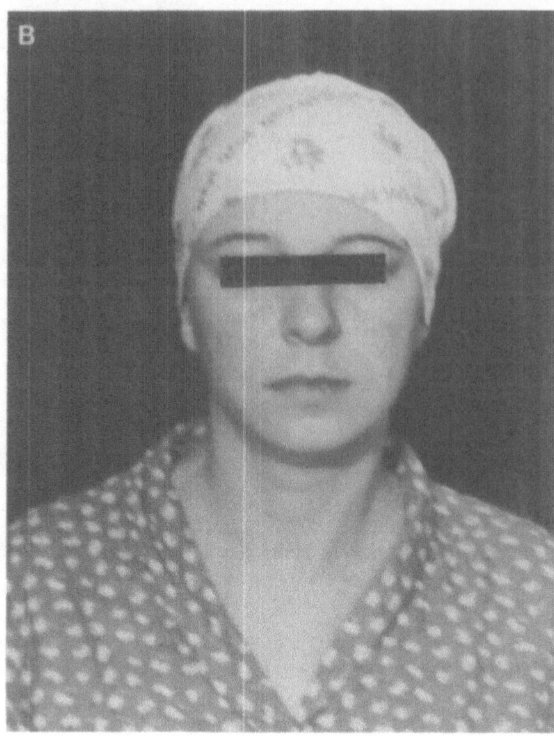

FIGURE 132. Axis form of spasmodic torticollis in 36-year-old patient (A) before operation and (B) 2 months after stereotactic destruction of left VL and nucleus of Cajal.

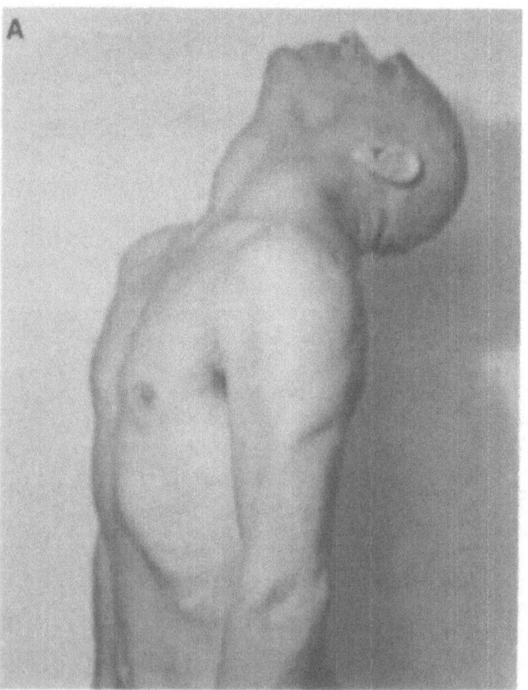

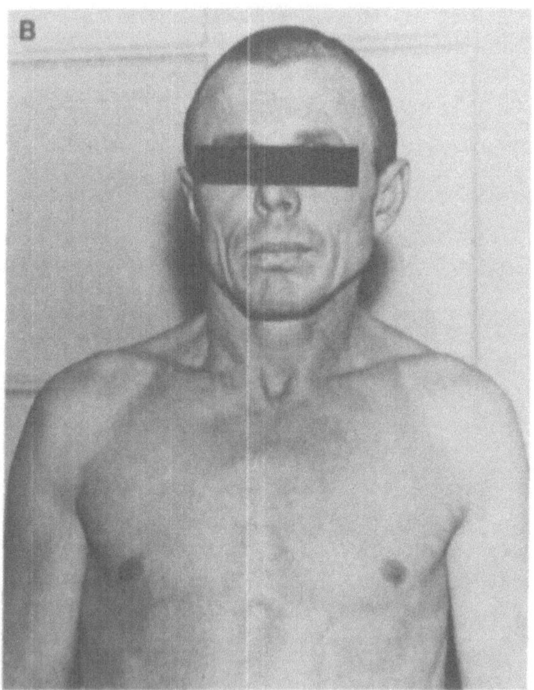

FIGURE 133. Severe retrocollis in a 42-year-old patient (A) before operation and (B) 3 years after second and 4 years after first stereotactic destruction of right VOp and Vim and left Voi and nucleus of Cajal.

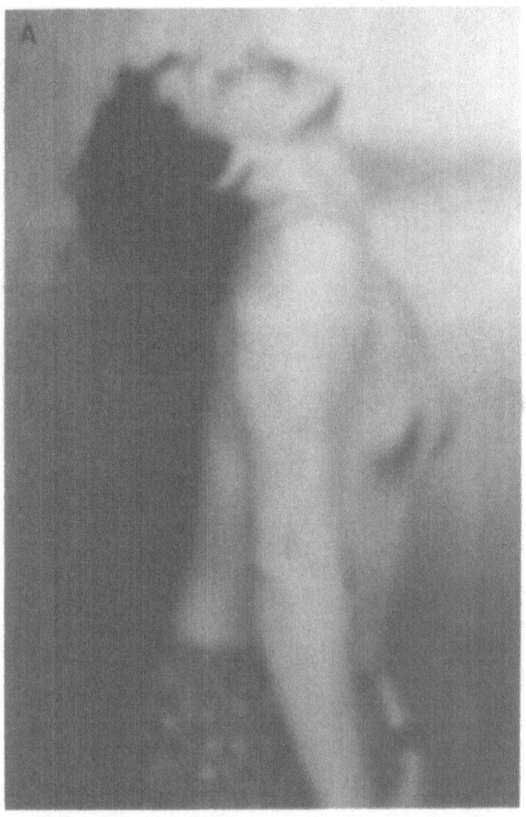

FIGURE 134. Severe retrocollis in 24-year-old patient present from the age of 11 years (A) before surgery and (B) about 2 years after stereotactic destruction of left VL and nucleus of Cajal and 9 months after right VL destruction.

of the head remained, although it was less pronounced than prior to surgery. Patients in this group found it easier to correct the position of the head at rest, although the torticollis returned during walking or the performance of active movements.

The average period of follow-up was 7 years, and in many cases it was 8 to 10 and even over 20 years.

Data on the immediate and long-term results of stereotactic operations (Table 16) indicate that more than half of the patients were practically well or substantially improved for many years. A comparison of the immediate and long-term results suggests that if a good immediate result was obtained, it persisted over the course of many years. If the effect was insignificant, with the passage of time, what little benefit there was gradually disappeared.

On the basis of our observations and those of other authors (Mundinger *et al.*, 1972), in approximately 20% of the cases an improvement occurs not immediately after the operation but 2 to 3 months later. We observed this delayed effect mainly in ST; in other extrapyramidal states, including DMD, the therapeutic effect, as a rule, is noted immediately after surgery. In some of our cases, a rapid and unexpected improvement developed 4 to 5 months later, but the cause of this is unclear.

Whether the result of the operation is related to the duration of the disease is not established, although it may be assumed that a prolonged rotation of the head will lead to secondary changes in the cervical spine

TABLE 16. Results of Surgical Treatment of ST (%)[a]

Results	Immediate	Long-term
Good	18	21
Significant improvement	34.5	32
Slight improvement	27.5	20
Unchanged or regression	10	27

[a]If the patient had two or more operations, a summation of the results of all operations is given.

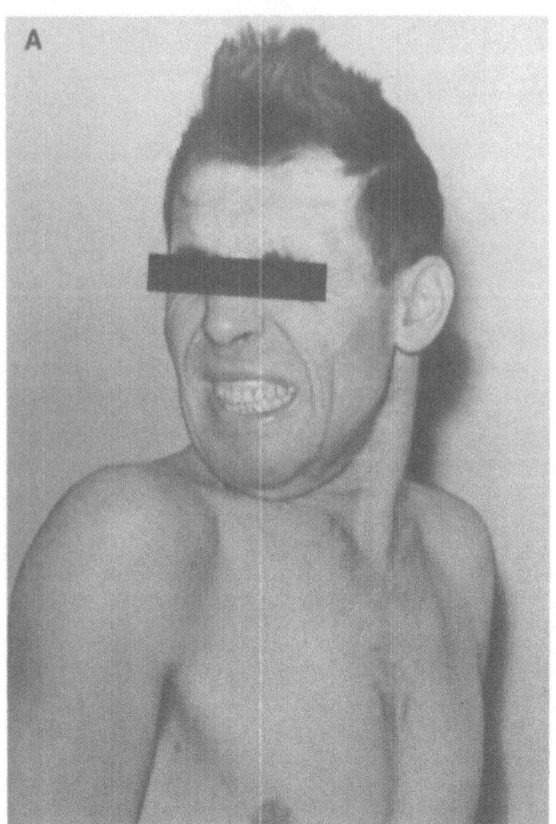

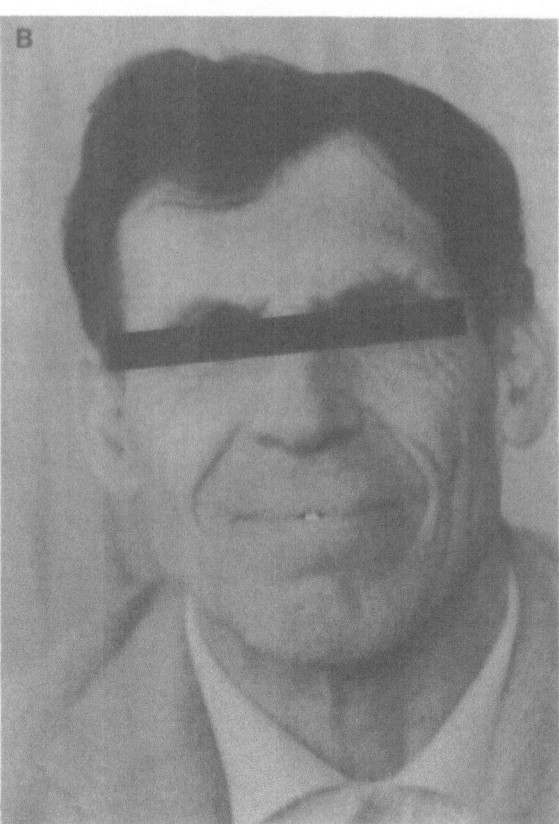

FIGURE 135. Rotational spasmodic torticollis with severe blepharospasm in a 46-year-old patient (A) before and (B) 9 years after bilateral VL thalamotomy.

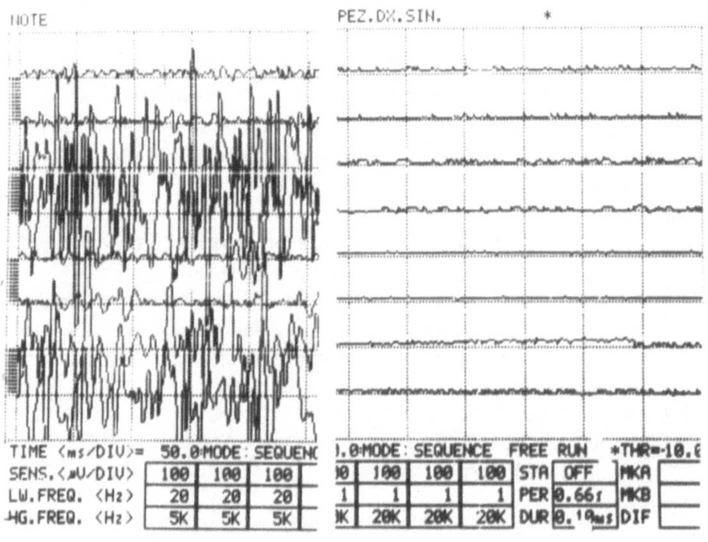

FIGURE 136. Electromyogram of the neck muscles in a 42-year-old patient with combined form of spasmodic torticollis. Left: Before operation. From above: I–II, right m. sternocleidomastoideus; III–IV, the same left muscle; V–VI, right m. splenius; VII–VIII, the same left muscle. Note the high-amplitude spontaneous electroactivity in left-side muscles. Right: Seven months after right-sided cryothalamotomy. Note the complete disappearance of spontaneous electroactivity in the neck muscles contralateral to thalamotomy.

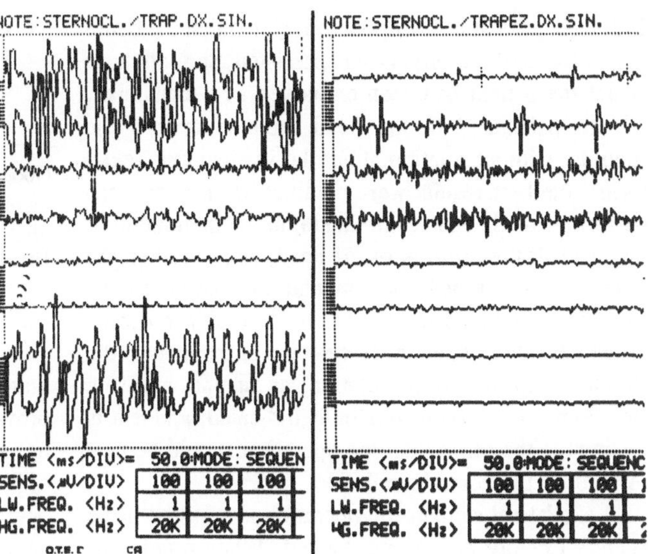

FIGURE 137. Electromyogram of neck muscles in 34-year-old patient with rotational form of left spasmodic torticollis. Left: Before operation. From above: I–II, right m. sternocleidomastoideus; III–IV, the same left muscle; V–VI, right m. trapezius; VII–VIII, the same left muscle. Note the pronounced continuous spontaneous electroactivity of right sternocleidomastoideus and left trapezius muscles. Right: Electromyogram 10 months after two-stage bilateral VL cryothalamotomy with a good clinical result. Note the virtual disappearance of spontaneous electroactivity in previously affected muscles.

(scoliosis, arthritis, etc.), which may impair the results of stereotactic operations.

One of the main unresolved questions in the surgery of ST is the side that should be operated on first.

Only a rotation of the head about the vertical axis (rotational form) is present in 40% of cases, so that the choice of the side for operation is frequently difficult, since in such a case, as a rule, the sternocleidomastoid muscle contralateral to the cephalic version is tense and hypertrophied, as are the posterior muscles on the opposite side. In this form of ST, the best results were obtained from operations performed on the side contralateral to ST. This conclusion is in agreement with the views of Hassler and Dieckmann (1982), who in the rotational form of ST also recommend destruction of fields H_1 and H_2 in the contralateral Subth.

On the basis of our experience, in the pure rotational form, we would recommend that the initial operation be done on the side contralateral to the ST; however, if a satisfactory result is not obtained, a second operation on the other side is indicated.

The following is an illustrative report of a lasting recovery from a rotational form of ST after unilateral stereotactic destruction of VL with more than a 20-year follow-up.

A 43-year-old male was admitted in August 1961 with complaints of an involuntary twisting of the head to the left and frequent jerking of the head to the same side for 3 years. At the age of 40, when he was otherwise in good health, he began to notice periodic involuntary drawing of the muscles of the left side of the neck and turning of the head to the left. As a result, the patient found it difficult to write; in fact, he was compelled to hold his head with the left hand when writing. Gradually, the involuntary movements intensified, twisting his head in a jerking fashion to the left. Prolonged drug treatment produced no beneficial effect. The patient's condition grew worse with intensified head twisting and involuntary protraction and elevation of the left half of the shoulder girdle (Fig. 126A). The forced twisting of the head was worse when walking. For 2 years the man was not able to work and had become an invalid.

Under general anesthesia the patient underwent a right chemothalamotomy. A mixture of 0.4 ml of alcohol and 0.2 ml of iophendylate was introduced into VL stereotactically.

The operation and the postoperative period were without complications. The hyperkinesia in the neck muscles was reduced considerably, as was the tension of the left neck muscles. For long periods of time, the patient was able to hold his head straight, but during emotional stress or in walking, there was a slight inclination of the head to the left. The forced movements of the shoulder girdle diminished substantially. In the following 3 to 4 months, there was a steady improvement so that half a year after the operation the patient was practically normal. A photograph taken 8 years after the operation shows the patient to be in good condition (Fig. 126B). A letter from the patient in

1981, 20 years after the operation, stated that the beneficial effects of the surgery had persisted, and, all these years, the patient had been employed as a technician.

In patients with the head inclined toward the shoulder without rotation about the vertical axis (axial form), the best results were obtained from operations performed on the side contralateral to the inclination of the head. If the position of the head, as is so often the case, is a combined rotation and inclination to the shoulder (combined form), it is much easier to decide on which side to operate. If the head is inclined to the shoulder opposite to the rotation, an operation on the side homolateral to the version is indicated, and if it is inclined to the same side, the operation should be performed on the contralateral side.

A bilateral stereotactic operation is required if the head is forcibly flexed or extended. When rotation of the head is combined with extension, the postoperative results might be different for individual components of this form of ST. If after a unilateral (on the opposite side to the inclination of the head) operation, the inclination, as a rule, disappeared with or without changes in the other components.

In 30% of our cases torticollis was combined with hyperkinesis or tremor in an arm on the side contra- or homolateral to the turning of the head. In these cases, we performed the first operation on the side opposite to the affected arm, since the side of the brachial hyperkinesia and the side of the neck muscles mainly affected always coincided. In the majority of cases a good effect was obtained on the torticollis; however, this result did not always correlate with the effect on the arm.

The quite high percentage of patients who showed only a slight improvement after one operation is explained by the high frequency of bilateral involvement of the neck muscles as noted previously (Section 5). In view of this, 42 patients underwent bilateral operations. In these cases, after the first operation, an improvement might not be noted. In all these patients, there were clear clinical and EMG indications of bilateral involvement of the neck muscles. Operations on the opposite side were usually performed 6 to 12 months after the first operation. Such an interval makes it possible to evaluate adequately the results of the first operation.

A comparison of the results of unilateral and bilateral operations, presented in Table 17, is of particular interest for the evaluation of stereotactic treatment of ST. As one can see from these data, significantly better results were obtained in bilateral than in unilateral operations (Table 17). A good result or a substantial improvement after the operation on one side was obtained in 38.5%, and after the operation on both sides, in 68% of patients. One may suggest that ineffective one-sided operations may be partly explained by multiple innervation of the sternomastoid muscle, not only by the accessory nerve, but from other sources (Hayward, 1986).

TABLE 17. Long-Term Results of Unilateral and Bilateral Stereotactic Operations for ST (%)

Results	Unilateral	Bilateral
Good	14.5	32
Significant improvement	24	36
Slight improvement	21.5	26
Unchanged	40	6

It is noteworthy that in bilateral operations, as mentioned above, a gradual improvement frequently occurred after the second operation in the course of a few months. Four patients underwent more than two operations on both sides. The end result of these operations was a significant lasting improvement in all cases.

The subcortical target giving the maximum clinical effect has not yet been found. As our experience has shown, the combined destruction of VL and Subth gives the best results in parkinsonism. However, our data do not confirm this for ST. An analysis of our results indicates that combined destruction of VL and Subth in ST offers no significant advantages over the destruction of only VL.

From analyzing the operations that produced the best results, it was shown that optimal results occurred if the focus of cryodestruction involved the anterior and posterior ventrolateral nuclei (VOa and VOp) and (partially) the internal ventrooral nucleus (VOi). As mentioned above, interest in the nucleus VOi increased recently when it was found that stimulation caused a feeling of the head turning and destruction led to good results in a number of ST cases. This nucleus is considered to be a thalamic relay in the vestibulointerstitio-thalamocortical chain, and in the opinion of certain authors, it is responsible for the rotational movements of the head. However, of 17 cases in which we performed unilateral destruction of VOi, a lasting good result was noted in only seven. In the remainder, even though the clinical effect was less satisfactory, the EMG showed an appreciable decrease in the pathological activity of the neck muscles on the side opposite to the operation. The insufficient effect in these cases may be explained by a bilateral involvement of the neck muscles.

In 16 cases we performed destruction of the interstitial nucleus of Cajal (in several cases in combination with destruction of VOi) and, partially, the fasciculus interstitiothalamicus. The following is an example of a complete and lasting effect after a cryodestruction of VL on one side and the nucleus interstitialis and VOi on the other.

Soon after suffering a cold, a 42-year-old man developed irregular and mild jerking of the head backward and to the shoulder. In 2 months these clonic jerks intensified and became constant. The muscles became tense and caused severe pain. In May 1978, stereotactic destruction of VOp and Vim was performed on the right side under general anesthesia. Immediately after the uncomplicated operation, there was only a moderate reduction in twitching and pain. This effect improved substantially 2 to 3 months after surgery. However, in another 4 months, tension in the muscles on the right side of the neck increased notably, as a result of which the head was retracted (Fig. 133A). When walking, and even at rest, the patient had to hold his head firmly with his hands. Consequently, one may assert that only very rarely is the clonic form of ST transformed into the tonic form or the rotational form into retrocollis.

A second operation was performed after a year—stereotactic cryodestruction of the nucleus interstitialis and VOi on the left side. Complete relief was noted immediately after the operation. The retrocollic position of the head disappeared, and the neck muscles were relaxed.

Three years after the second operation, the man is in good health and has no complaints. The posture and movements of the head are absolutely normal (Fig. 133B). He continues to work as a driver, as he did before developing the disorder.

A substantial improvement was noted in the majority of the other cases having destruction of the interstitial nucleus. One gets the impression that destruction of the interstitial nucleus is very effective when the posterior neck muscles are involved; however, our experience is too limited to allow a firm conclusion.

The selection of the proper "target structure" for the first and second operations is very important. It must be made after thorough evaluation of the clinical type of ST and the state of the neck musculature, including EMG data.

As our experience has shown, ST patients tolerate stereotactic operations much better than patients with other extrapyramidal disorders. Complications from our 206 operations were noted in only 28 cases (10.3%). They were of various kinds and in different degrees—mild paresis of the hand or mild hemiparesis (15 cases), marked dysarthria and disorders in swallowing (eight), impaired sensibility in the form of paresthesias or hypesthesias (four), and disturbed coordination (one case). These complications were noted with equal frequency after the first and second operations except for dysphagia and dysarthria, which occurred only after bilateral operations. In 24 cases, these complications were transitory and disappeared in the postoperative period or within 1 to 2 months. Significant lasting complications were noted in four cases (two hemiparesis; two gross dysarthria). In the last 10 years, the rate of postoperative complications has decreased from 10.3% to 4.5%. Progression of ST symptoms after the operation was not observed in any case.

In the 206 operations there was one fatal outcome, a patient who died after a second operation and was found to have status thymicolymphaticus.

11. Summary

In summary, one may conclude that surgical treatment is the only effective therapy for ST. Yet good and lasting results are obtained only in approximately 60 to 65% of operated patients. Nevertheless, the majority of these cases require an operation since pharmacological treatment, as a rule, is unsatisfactory. Only in mild cases of ST that do not affect capacity to work and in very rare psychogenic cases is surgical treatment not indicated.

At present there are two effective operations for ST, stereotactic destruction of the subcortical structures (VOa, VOp, VOi, Subth, and the interstitial nucleus) and rhizotomy of the cervical roots. The question of which operation is more effective has not yet been finally resolved. As other authors, we prefer stereotactic operations on the basal ganglia, which result in a lasting substantial improvement or recovery in more than half of the cases.

Further experience must be accumulated to resolve the question of the appropriate structure (or structures) for the target. Since in the majority of cases there is bilateral involvement of the neck muscles, bilateral stereotactic operations give almost twice as many good and lasting results as unilateral operations. The choice of the side for the first operation should be based on the type of ST, taking into account the group of muscles on one or both sides of the neck that are predominantly involved.

There are reasons to believe that the proper selection of cases for operation, further advances in the stereotactic technique, and possibly the discovery of the most appropriate subcortical structure for destruction will lead to even better long-term results of surgery for ST.

9

Cerebral Palsy

1. Introduction

Cerebral palsy (CP), a distinct nosological form sometimes called double athetosis or Little's disease, was described by Little in 1862 and subsequently studied by C. Vogt and Oppenheim (1911). However, it should be noted that certain authors consider athetosis as only one type of CP. The disease is regarded as a severe, usually nonprogressing disorder of the CNS characterized by hyperkinesias and muscular hypertonia. The traditional term "palsy" in the name of this disease is arbitrary since weakness, as a rule, is not prominent.

In 1964, the Little Club (a scientific society studying CP in Great Britain) proposed the following definition of the disease: "CP is a permanent, but liable to change impairment of movements and postures due to a nonprogressing defect or brain damage in early childhood." Symptoms of the disease caused by damage to the brain in its early stages are manifested at a very early age. Motor defects, as a rule, become noticeable when the child reaches the age of approximately 6 to 8 months.

Cerebral palsy cannot be regarded as a rare disease—its incidence is approximately 2.6 to 3.9 per 1000 newborn (Badaljan, 1975). Since there is no cure for this disease, the number of CP patients in many countries of the world runs into tens and even hundreds of thousands, making CP not only a medical but a social problem as well. Paradoxical as it may seem, one of the main causes of the increasing incidence of CP during recent decades has been the advances of medicine: with the modern therapeutic armamentarium (exchange blood transfusion, resuscitation measures, etc.) to save the lives of children with grave brain damage, children who previously were doomed are saved, but many of them develop CP.

2. Etiology

Cerebral palsy has many causes. One of the most frequent is hemolytic disease of newborn that is based on incompatibility of the blood of the mother and fetus. Disease of the fetus develops in one of 25 to 30 Rh-negative women (Badaljan, 1975). In an Rh incompatability, the antibodies developed in the mother induce isoimmunization and destroy the erythrocytes in the fetus. As a result, the brain of the fetus or newborn is damaged by the toxic action of bilirubin, the concentration of which increases in the blood many times (from a normal 3 to 15 μM to 350 to 500 μM). This gives rise to the so-called hemolytic jaundice of the newborn with bilirubin infiltration of subcortical structures—Str, Th, Cer nuclei, medulla oblongata nuclei—causing swelling and pericellular edema of their neurons as well as multiple small hemorrhages in the basal ganglia and brainstem. Similar hemorrhages occur in many internal organs. This condition is called kernicterus. However, kernicterus may also appear without an increase in the level of bilirubin in the blood.

Besides incompatibility of the blood of the mother and the fetus resulting from the Rh factor, there are a number of other common causes of CP—various diseases of the mother, toxemia of pregnancy, delayed labor, premature separation of the placenta, malposition of the fetus, and traumatic obstetric operations leading to impaired placental–uterine blood circulation—that may cause fetal hypoxia and ischemic brain damage with subsequent development of CP. There is evidence that abnormalities of pregnancy and parturition are the cause of CP in 70 to 80% of the cases. Various infectious, allergic, and other diseases in early childhood with a high fever and convulsions (meningitis, encephalitis, sepsis, pneumonia, chickenpox, and others) resulting in anoxic brain damage may also result in CP.

In some cases, CP appears to have been caused by damage to the cervical spinal cord during labor. At times, there is a combination of antenatal and perinatal pathological factors. Our personal experience does not confirm any hereditary factor in the etiology of CP, although the literature has quite a number of cases in which heredity has also been a factor (Mukherjee *et al.*, 1973).

3. Pathology and Pathogenesis

The morphological substrate of double athetosis is considered to be status marmoratus of the Str (Vogt and Vogt, 1919). The Str acquires a marblelike appearance caused by proliferation of fibrous astrocytes, which replace the destroyed neurons and gives an uneven appearance to the myelinated fibers. Hassler (1972c) emphasizes that status marmoratus often is not limited to Put and NC and sometimes may be seen in the GP, Th, and Cer nuclei, where there is also neuronal degeneration with subsequent gliosis. In other cases, the pathological substrate of double athetosis is demyelination and destruction of nerve cells in GP and NS (status dysmyelinisatus), which is usually the consequence of jaundice in newborns. In the more severe cases of CP, there is morphological evidence of diffuse brain damage—degenerated neurons in the cortex, atrophy of the white substance, and gliosis and cysts in the subcortical structures and brainstem.

Electron microscopic studies of biopsy material from VL, obtained in stereotactic operations of CP patients, reveals pronounced hyperplasia of the ribosomal apparatus of neurons, extensive changes in synaptic complexes, dystrophic changes in axons, and changes in the satellite glia (Romodanov *et al.*, 1983). As our experience has shown, ventriculograms made before stereotactic surgery often demonstrate hydrocephalus, which may be an indication of diffuse atrophy of the brain.

The pathogenesis of the main clinical manifestations of CP, spasticity, rigidity, and hyperkinesias, are presented in Chapter 2. Just as in DMD (Chapter 7), one of the main causes of motor disturbances in CP is the simultaneous contraction of antagonistic muscles.

At our clinic, a study was made of the descending supraspinal influence on the state of the segmental motor mechanism in CP patients (Artemjeva *et al.*, 1977; Kudinova *et al.*, 1976; Kandel, 1981). In order to evaluate its state, a study was made of the characteristics of the curves related to the impaired and the recovered H reflex induced by monopolar stimulation of the posterior tibial nerve in the popliteal fossa. The M and H responses of both anterior tibial and gastrocnemius muscles were recorded with bipolar surface electrodes separated by 2 cm.

In order to obtain an H reflex curve that demonstrates the dependence of the reflex amplitude on the strength of stimulation, the intensity of the excitation was changed. A number of parameters were analyzed on this curve: H reflex threshold, its maximum amplitude, and others.

A study was also made of the H reflex recovery, which demonstrated the dependence of the reflex amplitudes on duration of the interval between paired stimuli, which varied from 10 to 1000 msec. Without delving deeply into the results of this investigation, one may note that a number of pathological changes in the activity of the segmental apparatus were found in CP: impaired delayed inhibition, and increased excitability of low-threshold motoneurons in muscle antagonists.

We undertook similar investigations after bilateral stereotactic dentatotomy. The above-mentioned abnormal reactions became more normal after surgery; moreover, the degree to which these responses returned to normal correlated with the postoperative clinical decrease in rigidospasticity.

4. Clinical Aspects

The clinical picture, the course, and differential diagnosis of CP have been described in detail in handbooks on neurology and in numerous papers. Accordingly, these subjects are discussed only briefly here.

There exist approximately a score of CP classifications, e.g., spastic diplegia, spastic hemiplegia, and double hemiplegia. Based on clinical findings, atonic–astatic ("flaccid" form of CP described by Foerster in1909), hyperkinetic, and cerebellar syndromes may be distinguished. The classification of the American Academy of Cerebral Palsy (after Narabayashi, 1982c) is based on the main clinical manifestation of CP (spasticity, dystonia, rigidity, tremor, ataxia, atonia, etc.). In practice, however, mixed forms are frequently encountered, with a predominance of one or two of the above components. In spastic diplegia, the legs are more affected than the arms, and in double hemiplegia, both arms and legs are affected to approximately the same degree.

We dwell in more detail on the hyperkinetic or athetotic form of the disease, since this form is amenable to stereotactic surgery. The so-called hemiplegic form of CP is described in Chapter 10.

A characteristic peculiarity of CP is that the early signs of the disease are noted from the first few months of life. Impaired motor function is the basis of the clinical picture of the disease. In spite of the wide variety of these disorders, two basic phenomena are predominant, hyperkinesias and hypertonicity. With the exception of retarded mental development and cerebellar symptoms, which are not always present, the manifestations of the disease are consequences of the above-mentioned two phenomena.

Permanently increased muscular tone is usually a combination of rigidity and spasticity (rigidospasticity), usually with predominance of one or the other component. At the same time, in CP, one practically never observes hypertonicity of the hand flexors and foot extensors (Wernicke–Mann posture) typical of spastic hemiparesis. Siegfried and Hood (1983) emphasize that in cases of localized brain damage and preserved intellect in CP children, rigidity predominates over spasticity, whereas in diffuse brain damage with idiocy, spasticity predominates over rigidity. Such a conclusion is in agreement with our experience. An EMG of the affected muscles in CP reveals high-amplitude spontaneous activity without synchronization.

Hyperkinesias in cerebral palsy are usually extensive, "massive," involving the extremities and the neck and, to a lesser extent, the trunk muscles. These hyperkinesias intensify sharply when the individuals attempt to perform active movements or during excitement or heightened emotion. As a rule, slow "writhing" involuntary movements are observed, with flexion and extension regularly following each other. This leads to abnormal postures of the extremities, especially of the distal parts, in the form of hyperextension or hyperflexion. Hypertonicity of the adductor muscles in the thighs produces crossing of the legs. It is noteworthy that the nature and intensity of hyperkinesias in CP are highly variable. Besides the dystonic type of generalized hyperkinesia, one may frequently encounter quite mild choreoathetoid hyperkinesia without substantial rigidospasticity, in which case the patient retains the ability to walk, maintains satisfactory self-care, and at times maintains employment. Sometimes, one may observe CP cases in which the dominant symptom is a tremor less regular than in parkinsonism and not abolished by purposeful movements. Cerebral palsy is characterized by facial grimaces that result from hyperkinesias of the facial musculature and by speech impairments, most frequently in the form of dysarthria or explosive speech.

Severe muscular hyperkinesias combined with rigidospasticity quite frequently cause marked secondary changes in the locomotor system (joint contractures, deformation of extremities, scoliosis, etc.).

Very often CP is characterized by signs of corticospinal tract involvement—muscular spasticity, mild (and sometimes pronounced) pareses, increased tendon and cervical tonic reflexes, foot clonus, bilateral Babinski's signs, as well as pathological synkineses, oculomotor impairments, etc. Epileptic attacks of various types are not rare; however, petit mal is practically never observed. Epilepsy has been noted in 25% of CP cases (Davis *et al.*, 1982). Marked mental retardation (in approximately 40–50% of the cases) or debility, at times to the extent of idiocy, testify to severe diffuse brain damage.

Computed tomography has not aided in the understanding of the pathology and clinical manifestations of CP. The previously described atrophic changes in various cerebral structures, in particular, subcortical nuclei and frontal cortex, are found (Pederson *et al.*, 1982). At times, it is possible to observe porencephaly, signs of previous subependymal hemorrhages, multiple small cysts, or hydrocephalus (Koch *et al.*, 1980). In approximately one-third of CP cases, no significant cerebral pathology is observed.

It is necessary to emphasize the very wide variability in the severity of CP from mild twitching of the hands during the performance of purposeful movements to practically total disability. In average cases and severe forms of the disease with which the neurosurgeon has to deal, the habitus of the patient is so typical that an experienced physician can diagnose the case with a glance at the patient.

In the severe form of the disease with secondary muscular contractures, the patient is bed-ridden for life (Fig. 138).

In principle, we consider that there are indications for surgery in all cases of average and severe forms of the diseases, if there are no contraindications such as marked mental retardation (debility, idiocy), total absence of spontaneous speech, severe tetraparesis, atonic forms of the disease, the pseudobulbar syndrome, frequent epileptic fits, severe somatic diseases,

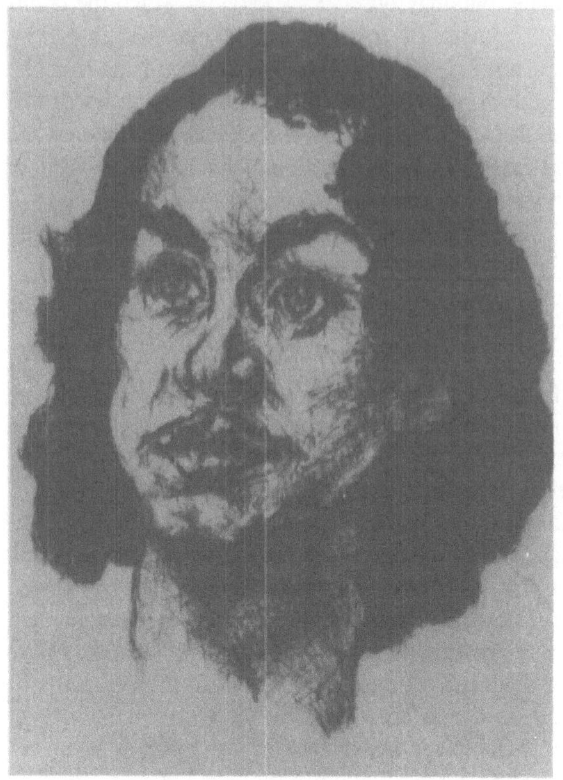

FIGURE 138. Portrait of Emperor Peter the Great, which was drawn by a 25-year-old-patient with severe cerebral palsy. Because voluntary movements of the extremities were impossible, the patient drew by holding the pencil in his mouth.

severe joint contractures, cachexia, and the presence of decubital ulcers.

5. Surgical Treatment

Since conservative treatment of CP is, as a rule, not effective, numerous attempts have been made in the past several decades to treat this disease surgically. However, these attempts did not give very favorable results. Various operations on bones, joints, tendons, and peripheral nerves also failed and were abandoned. Of all these operations, only two have given some benefit: section of the obturator nerves and cutting of the Achilles tendons. In certain cases, these simple operations may result in palliative relief.

Extirpation of the motor cortex by Horsley, first done in 1890, was the original brain operation designed to treat CP. Various modifications of this operation were made by many surgeons in subsequent years. However, in view of the few successes and the grave complications, extirpation of the motor or premotor cortex was discontinued and today is of interest only from the historical point of view. Section of the cerebral peduncle (pedunculotomy) and operations on the spinal cord to interrupt the pyramidal tract at various levels are not performed at the present time.

5.1. Stereotactic Operations

These operations represent the principal new ways of treating CP. Considerable experience in performing such surgery has been accumulated during the past three decades. The contraindications to these operations have been established.

Hassler (1972c) considers the main objective in the surgical treatment of athetosis to be the interruption of pathological impulses that originate in the partially damaged motor-regulating centers (Str, GP) before they reach the motor nuclei of the spinal cord.

5.1.1. Stereotactic Thalamotomy

Undisputed progress in the treatment of CP was noted after VL was chosen as the target point of stereotactic operations (Riechert, 1962; Cooper, 1964; Kandel, 1965; Gornall *et al.*, 1975; Narabayashi, 1967, 1977; Toth, 1972; Nesterov *et al.*, 1976; Broggi *et al.*, 1982; Zimbaljuk, 1984), and later Subth was used (Mundinger, 1969). Subsequently, a one-stage destruction of both structures was proposed (Kandel, 1972) and was often found to result in a significant diminution of extrapyramidal rigidity and hyperkinesia, thereby improving the patient's motor functions (walking, etc.).

If a marked benefit is obtained by the operation, then it is usually permanent, and recurrent disabilities are rare. It is noteworthy that a distinct improvement can be obtained in severe forms of CP in which none of the existing therapies offers any improvement. Nevertheless, it must be admitted on the basis of considerable experience that on the whole, the effectiveness of stereotactic thalamotomy for CP is definitely less than in other forms of extrapyramidal disease, in particular, DMD. First of all, this is because the operation has little influence on pyramidal spasticity. For this reason, neurosurgeons have sought new ways of treating this disease, especially by operations on other subcortical structures.

Stereotactic destruction of CM in CP proved ineffective (Ramamurthi and Davidson, 1975), although combined with lesions of other structures, this operation may give a significant improvement. Other authors recommend CM destruction in those cases of CP with severe hyperkinesia (Andy and Stephan, 1982) that are made worse by emotional stress (Galanda *et al.*, 1977). In an attempt to find a subcortical structure the destruction of which would improve the results in CP, Dierssen and Obrador (1967), in 26 patients, stereotactically destroyed the thalamoparietal pathways in CI, with quite satisfactory results in the majority of cases. However, these effects were not confirmed by other surgeons. Subsequently, a number of authors reported encouraging results from the stereotactic destruction of Pul (Cooper *et al.*, 1973) (see Section 5.1.2).

Hassler (1972c) considers that the partial improvement and relapses after VL destruction in CP occur because "additional" imputs that intensify athetotic movements enter the somatosensory cortex. After VL surgical destruction, this input may enter via the ventral nuclei at the base of Th. In order to eliminate all the afferent inputs, Hassler proposed destruction of several ventral nuclei, an operation known as "sagittal thalamotomy." According to the data of the author, this operation deafferenting Th yielded promising results in 31 patients with double athetosis. Long-term improvement of spasticity was noted in approximately 50% of the cases. Immediately after the operation, gait was more static; however, in several weeks, it became significantly better than prior to surgery.

As has already been discussed, the effectiveness of stereotactic thalamotomy (as well as other stereotactic operations) in CP is notably less than in other extrapyramidal conditions. If, in DMD, it is possible to restore a severely handicapped child to an almost normal state (unfortunately, not in all cases), then in CP it must be admitted that that goal is practically impossible. Of the scores of CP patients we operated on who were unable to walk from birth, there was not a single patient who could walk more or less normally after surgery.

This result is understandable. As mentioned in Chapter 7, children are stricken with DMD at the age of approximately 6 to 8 years, when the main motor stereotypes have already fully formed and reached a high level of development. That is why the elimination of muscular rigidity and hyperkinesias makes it possible relatively rapidly to restore motor acts, the cerebral mechanisms of which are not destroyed irreversibly in DMD. In CP the situation is different. These children, who were affected from birth, did not have a natural development of complex motor acts. They never were able to sit, walk, run, and so on. That is why even when, as a result of a stereotactic operation, there are favorable conditions for performing refined motor acts, the child's brain at the age of 10 to 15 years or older is unable to develop the mechanisms for motor functions that should have evolved naturally in the first years of life. For that reason, after a stereotactic operation for CP, it is of great importance to start a persistent and prolonged rehabilitation program, especially involving curative gymnastics and training of motor skills.

The fact that the results of thalamotomy were not very satisfactory stimulated the search for other subcortical structures the destruction of which would be more effective. Several stereotactic operations have been proposed for this purpose. We describe two in some detail since they have already passed clinical testing, and their effectiveness may be considered established. These operations, stereotactic pulvinotomy and dentatotomy, are used mainly for CP, although the literature presents several reports about their use for the relief of other extrapyramidal and pain syndromes (these data are presented in the corresponding chapters).

5.1.2. Stereotactic Destruction of Pulvinar (Pulvinotomy)

The pulvinar (Pul) is the largest part of Th and consists of several nuclei, the functional significance of which has not been established. Destruction of Pul nuclei is a relatively new stereotactic operation proposed by Cooper in 1973. He has described the technique of this operation using the cryosurgical method of destruction.

Pulvinotomy is performed both uni- and bilaterally in a one-stage operation. It should also be noted that Pul destruction is performed separately or in combination with the lesion in other Th nuclei (principally, VL). It is noteworthy that pulvinotomy is a relatively safe operation and without postoperative complications because there are no functionally important structures nearby. No mental, sensory, motor, or visual disturbances have been noticed after pulvinotomy.

In principle, the operative technique is the same as for destruction of VL or Subth (Chapter 4), the only difference being in the calculations of the stereotactic target. The coordinates for Pul are the following: 3 to 4 mm posterior to CP, 4 to 5 mm above the direct continuation of LI on lateral x ray, and 15 to 16 mm from the midline (lateral part of the nucleus) or 10 to 12 mm from that plane (medial part) on AP film.

No clear-cut indications for pulvinotomy have yet been established. At present, this operation has been clinically tested for two conditions, extrapyramidal syndromes and chronic pain syndromes. Data on its results in pain syndromes are presented in Chapter 13.

Since Pul is a relatively "new" structure in stereotactic surgery, several neurosurgeons have tried pulvinotomy in various extrapyramidal lesions, primarily for CP as well as for DMD, TS, and other hyperkinesias (Cooper, 1973; Martin-Rodrigues and Obrador, 1975; Galanda *et al.*, 1977; Guidetti and Fraioli, 1977; Broggi *et al.*, 1982). It has been established that this operation substantially reduces, although rarely dramatically, hyperkinesias, spasticity, and extrapyramidal rigidity. A significant improvement after pulvinotomy was obtained only in about half of the cases of extrapyramidal hyperkinesias with either unilateral or bilateral operations. The destruction of different parts of Pul diminished dystonic more than choreoathetoid hyperkinesias (Guidetti and Fraioli, 1977). At the same time, some authors saw no beneficial effect of pulvinotomy on the symptoms of CP (Broggi *et al.*, 1982). So far, it is difficult to make a positive statement regarding its effectiveness; however, in the opinion of the majority of authors, the results in CP are certainly encouraging. Lately, there has been a tendency to destroy Pul not separately but in combination with other subcortical structures.

Stereotactic destruction of ND is another operation that has been extensively employed in recent years for the treatment of CP. It is wise to dwell in greater detail on this relatively new and, in all likelihood, effective operation.

5.1.3. Stereotactic Dentatotomy

The first operation involving the destruction of deep Cer nuclei in parkinsonism through a small burr hole in the posterior cranial fossa was performed by Delmas-Marsalet and Van Bogaert in 1934. This operation ended fatally on the ninth day and was forgotten for many years. In 1961, Toth destroyed ND by the open method in three parkinsonians and noted diminished rigidity and tremor postoperatively.

After experimental investigations, R. C. Schneider *et al.* (1963) extirpated part of the cerebellar cortex in a CP patient and noted decreased spasticity and to a lesser degree improvement in athetoid movements. The authors did not report any postoperative intention tremor, ataxia, or dysequilibrium in that patient.

Advances in stereotactic surgery have made it possible to perform this operation stereotactically. The first such operation on the deep Cer nuclei was performed by Hassler and Riechert in 1960. After electocoagulation of the Fm nuclei, they obtained some diminution of spasticity in one CP patient. Heimburger and Whitlock (1965) were the first to perform stereotactic cryodestruction of ND, which plays an important role in regulation of muscular tone and of motor activity (Chapter 2); they obtained a substantial improvement in 11 patients, in particular, a reduction of spasticity in the extremities. Following Heimburger (1967, 1970a,b), other neurosurgeons performed stereotactic dentatotomy (Zervas *et al.*, 1967; Slaughter and Nashold, 1968; Siegfried *et al.*, 1970; Kanaka, 1972; Kandel, 1981; Fraioli and Guidetti, 1975b).

The exact mechanism by which ND destruction influences spasticity is still obscure, although it is generally acknowledged that the operation aims at eliminating the tonic influences of Cer on the muscular system. It is assumed that ND lesions block the tonic activity proceeding either to the motor cortex (Heimburger, 1970a,b) or to VL via NR (Siegfried, 1979).

For decades it was considered that destruction of cerebellar structures causes pronounced ataxia, intention tremor, and muscular hypotonicity. These beliefs, supported by clinical and experimental data (Carpenter and Stevens, 1957), retarded the development of stereotactic surgery of deep Cer nuclei for a long time. That is why the slight risk of dentatotomy was such an unexpected and distinct advantage. Even after bilateral dentatotomy there are practically no complications, and in particular, none of the cerebellar disturbances discussed above. Besides, the majority of authors destroy in a single procedure both ND, whereas a one-stage bilateral operation on Th nuclei involves the risk of complications.

Dentatotomy is usually performed under endotracheal anesthesia or under neuroleptanalgesia with the addition of local anesthesia in the area of the operation.

An important advantage of dentatotomy over many other stereotactic operations, especially thalamotomy, is the prone position of the patient. In view of this, it is possible to use the frame of the Riechert–Mundinger stereotactic apparatus turned upside down (Siegfried and Verdie, 1977). We have applied our own special headrest frame. The prone position is nec-

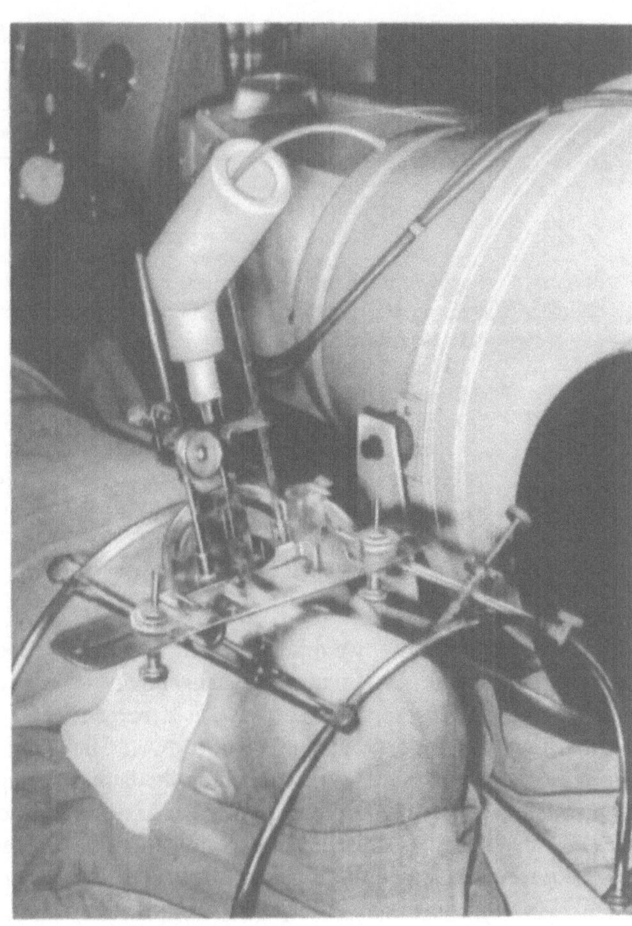

FIGURE 139. Stereotactic cryodentatotomy with the patient in a prone position. General view of the operation.

essary for the main stage of the operation (Fig. 139). Preliminary ventriculography is done with the patient in a supine position. Fractional pneumoencephalography is performed with the patient in a sitting position with the head slightly inclined forward (Siegfried, 1982). Since the main reference points of ND require a clear visualization of V_4, a direct picture is taken in the axial projection, i.e., with a 60° incline of the x-ray tube from the frontal plane. The lateral film is made in the usual manner. Extreme caution is required when turning the patient face down after the pictures have been taken.

After the patient's head is fixed in the special frame, one or two small burr holes are made in the posterior fossa at a distance of 3 cm lateral and 3 cm inferior to the external occipital prominence. The dura mater is coagulated, or a small incision is made in it. Then the stereotactic apparatus is set up.

The calculations for the localization of ND are not especially difficult provided that the main reference points are clearly visible on ventriculo- or penumoencephalograms. As usual, the calculations are made on pictures in both projections, using the data on spatial localization and individual variability of ND (Chapter 1, Section 3). It should be noted that the coordinates used by neurosurgeons for the center of the nucleus are somewhat different, since the borders of ND relative to the floor of V_4, Fm, and the midplane of the brain are not the same when based on data of anatomic studies by different authors (Table 18). The anatomical variability of the location of ND may be in some cases the reasons for failure after dentatotomy (Weigel and Mundinger, 1986).

In the calculations used by several authors for dentatotomy based on x-ray pictures, V_4 is the basis of the coordinate system required for determining the spatial localization of ND (Fig. 140). This main reference structure is convenient, first of all, because it is clearly visible in contrast studies and is situated near the target point. Furthermore, in spite of the significant individual variations in the dimensions of ND, the relationships with V_4 are relatively stable.

TABLE 18. Stereotactic Coordinates of Cerebellar Dentate Nucleus

Authors	Borders of ND (mean values, mm)					
	Lateral projection				AP projection: Lateral from midplane	
	Posterior from floor of V_4 (line AA_1)		With respect to line BB_1			
	Anterior border	Posterior border	Rostral	Caudal	Medial border	Lateral border
Heimburger and Whitlock (1965)	5	15	5	15	5	20
Zervas *et al.* (1967)	5	15	—	10	5	20
Slaughter and Nashold (1968)	6	23	3	12	5	20
Gortvai and Teruchkin (1974)	5	21	0–6	5–10	5	20

On the lateral x ray, we draw the main reference line (Fig. 141), line AA_1, that passes tangentially to the floor of V_4. The second line, BB_1, is a perpendicular dropped from the fastigium (Fm) to AA_1. The Fm must be very accurately determined on the lateral film, where it appears as the sharp tip of V_4 and resembles a "small tooth" situated more rostral than the posterior recessus, which forms a second but "smaller tooth" of lesser dimensions. Morphological investigations have shown that the dimensions of V_4 do not depend to a great extent on sex and age, although the length of the line BB_1, reflecting the height of V_4, i.e., the distance from its floor to Fm, varies from 10 to 14 mm (average 12.5 mm) (Siegfried, 1971).

As was noted previously (Chapter 1, Section 3), ND is oval, measuring (on average) 18 × 14 × 13 mm. The long diameter of this oval lies anteroposteriorly, dorsocaudally, and lateromedially. In view of this, the electrode should be introduced along the long axis of this nucleus. At a distance of 10–15 mm from the midplane, the angle of inclination of the long diameter of ND to line AA_1 is from 50 to 60°, whereas at a distance of 20 mm lateral, it is approximately 45°. With variability taken into account, the center of ND lies from 8 to 18 mm (average 11 mm) posterior to line AA_1 and 4 to 5 mm below line BB_1.

On an AP x ray, the main plane for stereotactic calculations is the midline dividing V_4 longitudinally into two equal parts, line CC_1. A second line, DD_1, perpendicular to the first passes through Fm and is 1 mm oral to and parallel to the line connecting both lateral recesses of V_4. The center of ND lies along this line 8 to 16 mm (average 14 mm) from the point of its intersection with the first line, i.e., from the midline.

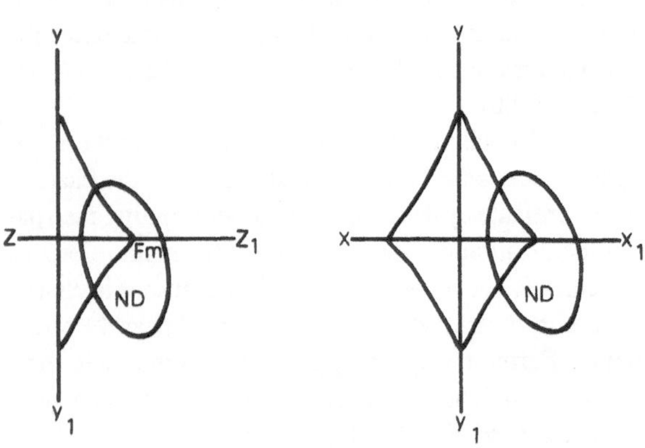

FIGURE 140. Simplified scheme of spatial location of ND in relation to the V_4 outlines. Left: Lateral projection. $Y–Y_1$, tangent to the V_4 floor; $Z–Z_1$, line drawn through Fm of V_4 perpendicular to $Y–Y_1$. Right: AP projection. $Y–Y_1$, midline of V_4; $X–X_1$, line drawn through Fm perpendicular to the midline.

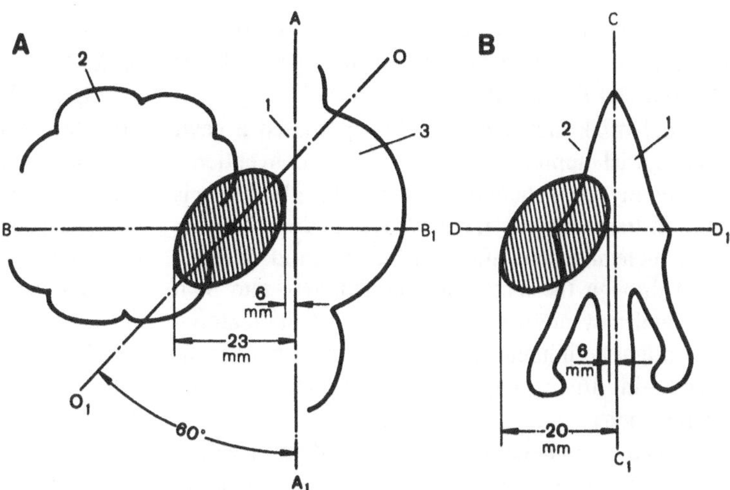

FIGURE 141. Stereotactic landmarks for dentatotomy. (A) Sagittal plane. AA_1, line tangential to V_4 floor; BB_1, line perpendicular to the previous line and drawn to Fm. The shaded oval is ND. Distances of medial and lateral limits of ND from the V_4 floor and the angle between major diameter of ND (O–O_1) and the V_4 floor are indicated. (B) Frontal plane. CC_1, the midline of V_4 floor; DD_1, the line perpendicular to the previous line passing through the apices of V_4 lateral recesses; 1, V_4; 2, Cer; 3, pons.

Other coordinate systems have been proposed for determining the location of ND. In particular, the line connecting CP with Fm has been suggested as the base reference line on a lateral film, and in the AP projection, the line perpendicular to the first and passing through Fm is used (Nadvornik *et al.*, 1965) (Fig. 142). This coordinate system has not been extensively used except by certain neurosurgeons (Vasin *et al.*, 1977).

A somatotopic representation in ND has been substantially proven by differential dentatotomy. After destruction of a part of this nucleus, the hypertonicity is diminished only in the lower or upper extremities (Nadvornik and Šramka, 1973). In view of this, it may be possible to destroy a specific part of ND when only one extremity or one part of the body is affected. For instance, the lower extremities are represented largely in the lateral part of ND, and the arms and trunk in the medial part (Heimburger, 1970b). A somewhat different organization was suggested by Nashold and Slaughter (1969), who used implanted electrodes, the extremities being represented in the lateral part of ND and the head and trunk in the medial part.

Two parts of ND—dorsomedial (magnocellular) and ventrolateral (parvocellular)—may be distinguished. According to Hassler (1950a,b) the ventrolateral part of the nucleus projects to VL, which may explain the better clinical results reported by several authors (Krayenbühl and Siegfried, 1972; Nadvornik and Šramka, 1973) from destruction of the ventrolateral part of ND. Siegfried (1971) considers this indicative of a direct connection between the ventrolateral ND and the contralateral VL. The coordinates of the center of ventrolateral ND are 1 to 3 mm posterior and 3 to 5 mm inferior to Fm and 14 to 16 mm from the midline in the AP projection. Other au-

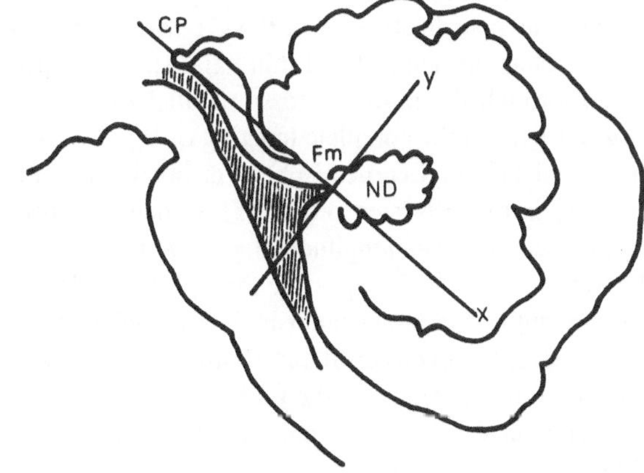

FIGURE 142. Schematic diagram of the coordinate system of ND in sagittal plane using intracerebral landmarks. Shadow area, aqueduct and V_4. Horizontal line connects CP and Fm; vertical line perpendicular to the previous line is drawn through Fm. X and Y, coordinate axes.

thors believe that better results are obtained from destruction of the lateral and intermediate parts of ND (Guidetti and Fraioli, 1977).

Nadvornik and Šramka (1973) proposed a new, transtentorial approach to ND. Unlike conventional dentatotomy, the patient lies on his back. A burr hole is made in the temporoparietooccipital area of the nondominant hemisphere, after which an electrode is introduced through the tentorium consecutively into both NDs, and they are destroyed. Fears that the tentorial veins might be injured by this approach were not confirmed—in 16 operations, these authors had no complications.

Electrostimulation as well as recording of biopotentials by microelectrodes in dentatotomy help to identify the location of ND (Chapter 4, Section 7.1.17). The introduction of a thin electrode with an active tip 40–50 μm makes it possible to detect ND by the sudden sharp increase in the amplitude and frequency of biopotentials when the electrode tip passes from the Cer white substance to the target nucleus.

At the present time, there are a number of important papers describing in detail the indications, technique, and results of stereotactic operations on deep cerebellar nuclei (Heimburger and Whitlock, 1965; Krayenbühl and Siegfried, 1969, 1972; Heimburger, 1970a,b, 1975b; Siegfried, 1971; Nesterov *et al.*, 1976; Guidetti and Fraioli, 1977; Vasin *et al.*, 1977) as well as our own investigations conducted in collaboration with other neurophysiologists (Kudinova *et al.*, 1976; Kandel, 1981).

The important advantage of CT-compatible stereotactic dentatotomy is the determination of the required target points directly from the CT scans without ventriculography or pneumoencephalography (Weigel and Mundinger, 1986). The base ring of the Riechert–Mundinger stereotactic frame (Chapter 3, Section 5.2) is fixed to the patient's head in a high position near the cortex. The coordinate axes of the CT and the stereotactic system are aligned and thin axial slices are produced through the posterior fossa parallel to the base ring. As a result the complete imaging of V_4, the aqueduct, and ND in Cer hemisphere is obtained. After that, V_4 is reconstructed on the CT screen in a midsagittal section in its longitudinal axis. A line perpendicular to the floor of V_4 passing through Fm is drawn. Subsequently, the target point is determined with the coordinates, 5 mm posterior and 10 mm inferior to Fm. The angle for insertion of the probe is 30°.

The plane of the entry point is determined on the oblique slide, and the distance of 14 mm from the midline next to the lateral recess is checked. The angle of the lateromedial approach of the probe is also 30°.

Using the resident software, the three coordinates of the target are determined, as well as ND depth and the trephination point in the occipital bone. Weigel and Mundinger (1986) recommend that the lesion be tubular-shaped with a length of 12 mm and a radius of 6–7 mm, thus reaching the entire ventrocaudal extent of ND.

To summarize, from the extensive clinical experience presented in the literature, it is possible to draw certain conclusions in spite of some contradictory results. There can be no doubt that after dentatotomy, there is a decrease and in certain cases even a total disappearance of muscular hypertonicity. However, this applies only to spasticity; the operation has no noticeable influence on extrapyramidal rigidity (Siegfried, 1971; Vasin *et al.*, 1977). Postoperatively, there is also a decrease in pathological postures and facial musculature spasms, and in a few cases there is improvement of speech.

As mentioned above, our investigations in collaboration with neurophysiologists demonstrated an objective decrease in spasticity of antagonist muscles after stereotactic dentatotomy, which was confirmed by spectromechanomyographic techniques that give an objective evaluation of muscular tone (Safronov *et al.*, 1978).

It is also noteworthy that even in clinically unilateral extrapyramidal disorders, better results were obtained from bilateral dentatotomy (Fraioli and Guidetti, 1975b). It has been noted by many authors and confirmed by our personal observations that clinical improvement is usually observed immediately after the operation and gradually enhances over the course of several months and at times up to 2 years (Vasin *et al.*, 1979).

A substantial decrease in muscular hypertonicity, especially the spastic components, after dentatotomy may be considered proven. However, the data regarding the side on which the tone diminishes after destruction of one ND are contradictory. Certain authors have observed a decrease in spasticity on the side homolateral to the operation (Nashold and Slaughter, 1969; Siegfried, 1971; Vasin *et al.*, 1977). Other authors (Heimburger, 1970a,b) noted an effect primarily on the contralateral side. Unlike other authors, Guidetti and Fraioli (1977) observed greater benefit after dentatotomy in dystonic and athetoid–dystonic syndromes, but in spasticity, the results were much more modest.

The majority of authors are of the opinion that

dentatotomy has no noticeable effect on hyperkinesias, but in those predominantly involving truncal muscles, dentatotomy is more effective than thalamotomy (Krayenbühl and Siegfried, 1972). Moderately satisfactory results after ND destruction were obtained in 46 of 64 patients (Heimburger, 1970a,b) and in 30 of 47 patients with various dyskinesias (Fraioli and Guidetti, 1975b). However, the authors admit that dentatotomy is not very effective in severe forms of CP with marked spasticity. It has also been established that destruction of the ventrolateral part of ND has no noticeable effect on tremor and cerebellar ataxia and is quite ineffective in TS.

An evaluation of the clinical results of dentatotomy naturally requires a comparison with the results of stereotactic destruction of Th nuclei. Such a comparison, based on the experience of 134 stereotactic operations (96 thalamotomies or subthalamotomies and 41 dentatotomies) in various extrapyramidal lesions, has been made by Krayenbühl and Siegfried (1972). They conclude that indications for both operations should be based on not the nosological form but the predominant extrapyramidal syndrome—either involuntary movement or muscular hypertonicity. Without a doubt, in hyperkinesia of the extremities, the best results are obtained from destruction of VL or Subth. However, in case of involuntary movements involving the more proximal and truncal muscles, that operation is less effective, and in such instances, there are indications for dentatotomy. Spasticity, typical of CP (double athetosis), does not diminish substantially after thalamotomy or subthalamotomy, and in such cases only dentatotomy may produce a therapeutic effect.

The choice of operation when there is a predominance of muscular hypertonicity depends on its nature, i.e., the presence of rigidity or spasticity. In rigidity, destruction of VL or Subth is preferable, whereas in "pure" spasticity or rigidospasticity, dentatotomy is more effective.

On the basis of this analysis, the main indications for dentatotomy are severe forms of CP as well as (to a lesser degree) certain forms of CP in which truncal muscles are primarily afflicted. A significant improvement directly after bilateral dentatotomy may be obtained in more than half of the CP patients (Fig. 143).

Long-term results (up to 10 years) have been presented by Siegfried and Verdie (1977) on 50 patients in whom 109 stereotactic operations on ND were performed. An objective evaluation has demonstrated that there was a clear diminution of spasticity in only one-third of the patients after the operation; approximately one-half of the patients were better able to look after themselves and be rehabilitated. According to the appraisal of Siegfried (1979), stereotactic dentatotomy produces a long and notable effect on spasticity in approximately 30 to 50% of the patients. Our limited experience with dentatotomy in CP patients allows us to speak of satisfactory results that, however, are far from what one would desire.

The effectiveness of dentatotomy should not be overestimated. As experience has accumulated, it appears that the long-term results are substantially worse than the immediate results. The percentage of good results obtained immediately after the operation decreased by almost one-half in 2 years (Siegfried and Verdie, 1977). In view of this, the initial enthusiasm for dentatotomy as a treatment for spasticity has diminished considerably in recent years, and the popularity of the operation has noticeably declined.

5.1.4. Stereotactic Pallidotomy

Just as in parkinsonism, the first stereotactic operations in CP involved destruction of the medial segment of GP, which gave promising results (Spiegel and Wycis, 1962a,b; Yasargil, 1962); however, follow-up reports were far from what one would desire (Mundinger and Riechert, 1963). At present, pallidotomy is not performed for CP.

5.1.5. Stereotactic Putamenotomy

Stereotactic destruction of Put in CP has no definite physiological foundation since there is no direct evidence that the Put plays a role in the pathogenesis of rigidity and spasticity in extrapyramidal lesions. Putamenotomy was proposed by Heimburger (1975a), who reported relief of hypertonicity, primarily in the distal musculature, in a small number of patients. Subsequently, this operation has been abandoned for CP.

5.1.6. Combined Operations

The desire to improve the results of surgical treatment of CP prompted repeated attempts to combine destruction of several subcortical structures in one operation or successive operations. The recent literature has a number of papers with promising results from such an approach (Balasubramaniam *et al.*, 1974; Hitchcock, 1975; Heimburger, 1975b; Kanaka and Balasubramaniam, 1975; Fraioli and Guidetti, 1975b;

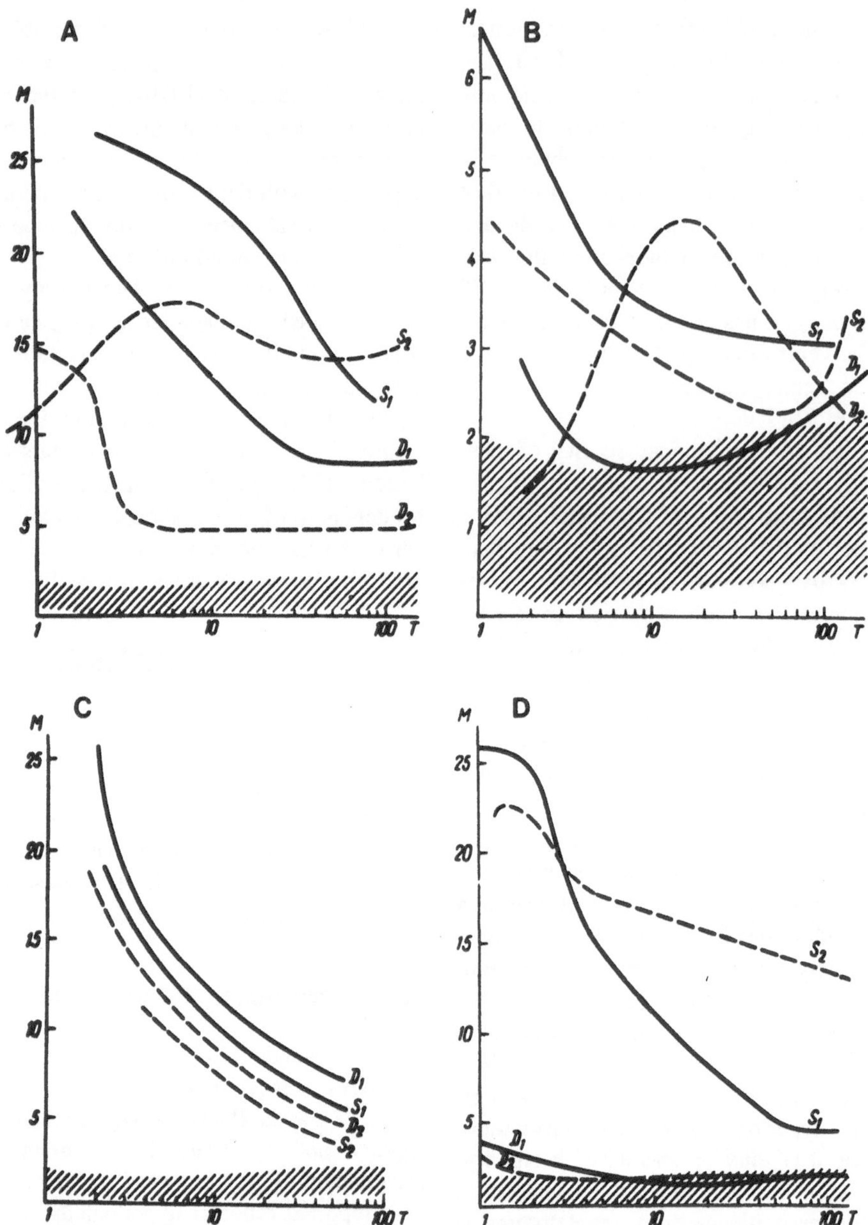

FIGURE 143. Spectromechanograms of four patients (A–D) with cerebral palsy before surgery (solid line) and after bilateral dentatotomy (dotted line). D_1 and D_2, right arm; S_1 and S_2, left arm. Note the marked decrease of muscle tonus after surgery, particularly in the high-speed zone. Abscissa, the cycle size (T) of sinusoidal movement in logarithmic expression. Ordinate, the value of muscle tonus in kgDm; shaded zone, normal value.

Ramamurthi and Davidson, 1975; Galanda *et al.*, 1977; Guidetti and Fraioli, 1977). However, such combined stereotactic operations make it difficult to assess the therapeutic effect of the destruction of each of the subcortical structures.

The "combinations" of structures are quite diverse. Destruction of several Th nuclei is the most frequent technique. For example, the clinical results in 94 CP patients operated by a stereotactic technique were analyzed by Balasubramaniam *et al.* (1974). These operations involved the consecutive destruction of VL, CM, and Vim. It is noteworthy that more than half of the children operated on were under the age of 10 years. On the whole, these operations gave encouraging results. For example, in 63 of 89 children with muscular hypertonicity, an improvement was noted in more than 50%. An improvement of approximately the same degree was noted in 38 of 69 patients with hyperkinesias. The authors consider that in those cases of CP that they term "sensory-induced hyperkinesias," stereotactic destruction of CM was more effective.

One should note the tendency to seek the advantages of dentatotomy and thalamotomy by combining these two operations, with the former being done on the side of affliction and the latter on the contralateral side. Kanaka and Balasubramaniam (1975) performed such a combined operation in 15 CP children, noting a significant improvement in eight of them.

Since dentatotomy frequently does not ensure an adequate therapeutic effect, Galanda *et al.* (1977) recommend another combined operation of one-stage stereotactic destruction of ND and Pul. Both "target points" may be reached through one burr hole in the parietooccipital area. Heimburger (1975b) has performed various stereotactic operations in CP children in different sequences, for example, dentatotomy, thalamotomy, a combination of both operations, putamenotomy, and so forth. Ramamurthi and Davidson (1975) destroy stereotactically CM and CF as well as ND and Am, whereas Guidetti and Fraioli (1977) combine destruction of VL and Pul.

Vasin *et al.* (1979) noted diminished spasticity and hyperkinesias in 42 CP patients who underwent combined operations—bilateral dentatotomy and destruction of one or more diencephalic structures (VL, ZI, Pul, CM, and others).

In view of the complexity and variety of combined stereotactic operations involving the destruction of several subcortical structures, it is not easy to evaluate their effectiveness. Nevertheless, there are grounds to believe that this form of CP treatment is quite promising and may improve the long-term results of surgery.

Along with the development of stereotactic techniques, efforts are being made to develop other ways to correct surgically the clinical manifestations of CP; in particular, operations on the spinal roots are being extensively advocated.

5.2. Selective Posterior Rhizotomy

Recently, one more functional operation aimed at reducing muscular spasticity in CP, section of the posterior spinal nerve roots, has been advocated.

The technique of posterior rhizotomy, which is usually performed under general anesthesia, is described in detail in Chapter 19. Here we discuss briefly only those data that are related specifically to CP. For this disease, bilateral section of the superior cervical posterior roots (C_1–C_3) with the patient in a sitting position has been proposed (Kottke, 1970). Unlike lumbar partial rhizotomy (Chapter 13, Section 5.3.6), complete posterior rhizotomy is usually done on the upper cervical roots.

After laminectomy of C_1–C_2, the dura mater is opened with a longitudinal incision. With a miniature hook the vessels are separated from the cervical posterior roots under the microscope, and all rootlets are sectioned on both sides, using a razor blade or microscissors. In 6 CP patients operated on by this method, there was a notable reduction of spasticity. It is believed that this results from restoring to normal the hyperactive tonic cervical reflexes. It is noteworthy that after the operation spasticity diminishes not only in the cervical muscles but in the trunk and extremities as well. Anatomic anomalies in which the C_1 posterior root is absent on one or both sides may cause confusion. Moreover, this root at times is not divided into rootlets at the cervical spinal cord. For example, root C_1 was absent bilaterally in six patients and unilaterally in four and was present bilaterally in only six patients (Guidetti and Fraioli, 1977).

Kottke's results have been confirmed by Heimburger *et al.* (1973), who performed this operation in 15 spastic patients with the quadriplegic form of CP. In addition to section of the posterior roots C_1–C_3, the superior rootlets of C_4 were sectioned. As the authors note, there was no "dramatic effect" after these operations, although in 13 of 15 patients there was a notable decrease in spasticity of the muscles of the neck, trunk, and extremities. The authors believe that cervical posterior rhizotomy is more effective in eliminating spasticity than dentatotomy and thalamotomy.

Guidetti and Fraioli (1977) performed section of the posterior cervical roots in 16 patients suffering from dystonic–athetoid syndromes with or without spasticity. A substantial improvement was noted in ten patients of the 16. These authors are planning to increase the number of posterior cervical roots sectioned. In certain patients they cut not only the posterior roots of C_1–C_3 but also the external branch of cranial nerve 11 on one side. Guidetti and Fraioli (1977) operated on another 24 CP patients with pronounced spasticity in whom they sectioned the posterior rootlets (radiculotomy) not at cervical level, but at the level of T_{12}–

S_1 or L_1–S_1. The authors point out that this group of patients was less severely afflicted than the group that underwent cervical rhizotomy. A notable or moderate improvement was observed postoperatively in 22 of the 24 patients—there was a decrease or disappearance of abnormal posture caused by spasticity in the lower extremities. Speech improved in two patients.

Fasano *et al.* (1977) consider posterior rhizotomy to be the most effective surgical method for treatment of spasticity. The spasticity relapsed after a successful operation in only 5% of cases. In a later paper (Fasano *et al.*, 1980), the authors reported the long-term results of posterior rhizotomy which they called "functional rhizotomy," in 80 CP patients with pronounced spasticity. In a 2- to 7-year follow-up, a significant improvement or total disappearance of spasticity was noted in more than 90% of the patients. There were no relapses of spasticity. Side effects were rare and transitory. In many cases, the authors observed a "suprasegmental effect" of rhizotomy, i.e., a diminution of spasticity in segments above the operative level. Section of the posterior roots will undoubtedly occupy an important place in the surgical arsenal for treatment of CP.

During the last few years, we performed 25 selective posterior rhizotomies: 18 operations for treatment of spasticity of the legs in patients with CP and 7 operations for pain due to avulsion of the brachial plexus (see Chapter 13, Section 5.3.7). The age of 18 patients with spasticity (10 males and 8 females) was 11 to 46 years. All patients except two were severely disabled and practically bedridden. Three previously underwent stereotactic thalamotomy without satisfactory effect.

After typical laminectomy of T_{11}–L_1 and separation of the lumbosacral roots under operating microscope, electrostimulation was performed to distinguish the "useful" and "harmful" spasticity (see above). The corresponding rootlets of each root were divided for three quarters of their diameter. As a rule the rhizotomy of L_3–L_5 and S_1 was made on both sides. There were no serious postoperative complications or morbidity.

The follow-up from a few months to 4 years has shown the substantial improvement in 14 of 18 patients. The marked decrease of spasticity allowed the patients to begin walking. Ten patients were able to walk without support. Mild but not useful improvement was noted in two other cases. There was no effect in one patient, and another patient died 1 year after operation from urinary sepsis.

5.3. Stimulation Methods

In 1973, Cooper *et al.* suggested a new method for treating muscular spasticity and rigidity, primarily in CP, by the chronic electrostimulation of Cer. At the end of the last century, Sherrington (1898) and Lowenthal and Horsley (1897) had demonstrated in an experiment that stimulation of Cer reduced experimentally induced rigidity. However, the specific mechanism by which chronic electrostimulation inhibits various extrapyramidal phenomena, particularly spasticity, has not been sufficiently studied. It is assumed that this stimulation inhibits the efferent cerebellar as well as the vestibular tracts, which in turn inhibit the activity of the cerebral and spinal motor systems via the cerebelloreticulospinal and cerebellothalamic systems. It is possible that stimulation antidromically activates the afferent pathways of Cer. It has also been suggested that stimulation of Cer cortex activates Purkinje cells, which inhibit the functions of the deep Cer nuclei (Cooper, 1973). There is yet another point of view that the clinical effect of Cer cortex stimulation does not depend on simultaneous inhibition of its deep nuclei. This opinion is based on the fact that a similar effect can be obtained by chronic stimulation of ND (Schvarcz *et al.*, 1981). It is possible that biochemical changes play a role in this process, since an elevated norepinephrine content has been found in CSF as a result of chronic electrostimulation (Wood *et al.*, 1977).

Chronic electrostimulation of Cer causes a reduction of both thalamic and cortical somatosensory EP in approximately half of the patients (Upton *et al.*, 1982). In the absence of such a response, stimulation is usually ineffective.

Chronic Cer electrostimulation is quite "young" in functional neurosurgery; however, by 1980 more than 1000 such operations had been performed (Penn *et al.*, 1980). This figure indicates that this method has had a comparatively large clinical trial, but nevertheless, its effectiveness has not yet been definitely assessed; in fact, there are contradictory opinions as to its value.

Chronic electrostimulation of Cer is performed with implanted electrodes on the cerebellar surface. Technically, this method is similar to that employed for stimulating the posterior spinal cord columns (Chapter 13, Section 5.4.2). Avery eight-contact electrodes with an area of 0.41–0.62 cm^2 for adults and 0.25–0.48 cm^2 for children have proven to be effective.

After a small bilateral suboccipital craniotomy made under general anesthesia, platinum plate electrodes are implanted on the surface of both Cer hemispheres. It is thought that in order to produce an optimal effect by chronic stimulation, the location of the electrodes on the cerebellar cortex is important (Penn *et al.*, 1978a,b; Haines, 1981); however, the optimal site for electrode implantation has not yet been established. Whereas Cooper *et al.* (1973) and Cooper (1978) used the anterior lobe of Cer for stimulation, other authors prefer to implant the electrodes on the rostral part of the posterior lobe (Bantli *et al.*, 1976; Penn and Etzel, 1977; Riklan *et al.*, 1977; Fasano *et al.*, 1982). A minireceiver implanted in the anterior thoracic wall or in the occipital area is connected to the electrodes. The transmitter is placed near the patient. Permanent stimulation starts 3 to 4 days after the operation.

A fully implanted lithium generator system producing a direct current (Neurolith Parasetter System, USA) has been used for chronic stimulation of Cer. After a period of this stimulation, spasticity, gait, and speech were improved in 70 to 80% of the patients (Davis *et al.*, 1980, 1981). The Cordis fully implanted programmed generator is also employed. New systems for Cer stimulation also provide for automatic switching of the current on and off at set intervals (usually 5, 7, or 10 min).

As Cooper, the pioneer of this method, has proposed, the stimulation of Cer is performed around the clock at short (several minutes) intervals using the following parameters: current 3 to 5 V, frequency 50 Hz, duration of impulse 0.5 msec. Other authors (Winkelmüller *et al.*, 1977; Ivan *et al.*, 1981; Fasano *et al.*, 1982) employ a current frequency of 100 to 200 Hz. Some authors recommend stimulation of Cer only at night (Fasano *et al.*, 1982).

Chronic electrostimulation does not cause any side effects or any neurological, psychological, or intellectual disorders even when the method is used over the course of many years (Davis *et al.*, 1976, 1980; Cooper *et al.*, 1982). Among the complications of this method that have been described are leakage of CSF along the wire into the subcutaneous pocket of the receiver (in 13.5% of the cases) and infection in the implantation area (in 5%). Quite frequently it is necessary to replace the implanted receiver or its leads because of breaks in the wiring; in the course of 5 years, 70% of the receivers had to be replaced, 17% because of breaks in the leads in the neck (Davis *et al.*, 1982).

As discussed, the literature offers conflicting evaluations regarding the effectiveness of the method. Cooper reported that 50 CP children treated by this method showed a significant decrease in spasticity and hyperkinesias and improvement in motor activity and speech (Cooper *et al.*, 1976c). In many children without mental retardation, the muscular tone became practically normal. Prolonged stimulation of the anterior Cer with implanted electrodes also produced a diminution of spasticity, rigidity, and muscle coactivation in the majority of CP patients (Davis *et al.*, 1977; Wong *et al.*, 1979). This effect occurred either immediately after the operation or after several days or months. The therapeutic effect on spasticity is more marked than that on hyperkinesia (Winkelmüller *et al.*, 1977).

An extensive study of the use of chronic electrostimulation was recently published by Davis *et al.* (1982), who implanted electrodes in 316 CP patients with spasticity. A marked improvement was noted in 64% of these patients. In the most severe cases, there was an improvement in 20 to 40%, athetosis diminished in 50%, and the range of movement increased by 20 to 30%. Speech and respiration also improved. It was easier to care for these patients (feeding, dressing, and so on).

Chronic stimulation of Cer cortex proved effective in decerebrate rigidity with opisthotonus in two children with severe posthypoxic damage to the CNS (Sukoff and Ragatz, 1980). Positive results from the employment of chronic stimulation have been reported also by other authors (Sedan and Lazorthes, 1978; Wong *et al.*, 1979; Dietz *et al.*, 1979).

It should be noted, however, that recently several papers have reported a negative assessment of the results of that method (Penn *et al.*, 1980; Whittaker, 1980; Manrique *et al.*, 1980; Gahm *et al.*, 1981). In a double-blind study, six experienced physicians were unable to differentiate the condition of eight CP patients with the electrostimulation apparatus switched on from that with it off (Whittaker, 1980). These authors consider the stimulation results to be placebo effect. Relatively poor therapeutic effects have also been published. Fasano *et al.* (1982), using chronic electrostimulation in 12 patients with spastic and involuntary movements, obtained a prolonged good effect in only 30% of them, no change in 30%, and a transient good effect in 40% immediately after operation, which gradually disappeared. A comparative study of tonicity and other motor disorders prior to the operation and over a long period after the electrode implantation did not reveal any significant improvement in the majority of patients. Fewer than half of the patients continued

Cer stimulation 2 years after electrode implantation, which in the opinion of the authors indicated a poor therapeutic result (Ivan and Ventureyra, 1982).

A chronic electrostimulation technique was developed that used not surface stimulation of Cer but stimulation of ND with platinum electrodes implanted in one or both of these nuclei (Schvarcz *et al.*, 1982). The methodology and stereotactic calculations were analogous to those in dentatotomy (Section 5.1.3). The period of percutaneous testing stimulation required to find optimal electrode combinations and current parameters continued for 8 to 12 weeks. After that the system was implanted. The maximal positive effect was obtained during stimulation for at least 2 hr with the following parameters: frequency 100 Hz, current 2 V, double stimuli lasting 0.25 msec. The limited experience of the authors (four patients with severe CP) produced promising results—improved motor function, decreased spasticity, and the performance of voluntary movements that prior to the operation had been impossible. Improvement in gait, equilibrium, and even speech was noted.

A few reports have appeared recently regarding the possibility of using chronic stimulation of the posterior columns of the spinal cord (Chapter 13, Section 5.4.2) in CP. This method resulted in an improvement in 84% of 90 patients (Waltz, 1981; Waltz *et al.*, 1981). However, according to other reports, the motor functions in CP patients were not helped by chronic stimulation of the posterior columns (Lazorthes *et al.*, 1981).

To summarize, the contradictory data currently in the literature do not permit a final conclusion concerning the effectiveness of chronic electrostimulation. Nevertheless, it should be emphasized that the positive reports of this method are based on a large clinical sample, whereas the negative results were obtained in a relatively small series of observations.

Recently, the first attempt was made to treat by stereotactic stimulation the so-called hypokinetic form of CP in which spasticity and rigidity are absent. Muscular tone in this form is low; because of muscular hypotonicity of the neck and truncal muscles, the child is unable to sit, stand, or hold his head, which "falls" forward or to the side. According to Šramka *et al.* (1982), in this form, chronic stimulation of the Br Con with implanted electrodes is effective. High-frequency impulses (200–400 Hz and 0.2–0.3 msec) are recommended for stimulation of Br Con (Galanda and Grác, 1983). Muscular tone is enhanced during stimulation, and the children are able to hold their heads in a physiological position, sit, and even walk.

6. Personal Observations

Over the course of many years, the author and his associates have performed a total of 232 stereotactic operations on 174 CP (double athetosis) patients.* There were 102 males and 72 females aged from 7 to 48 years. The majority of patients (116) were operated on between the ages of 10 and 20 years, and only 12 were under 10 years of age. In the majority of these patients, the disease had been discovered in the first months of life. In only 26 patients did the disease develop during the first years of life following infectious diseases with high fever.

The most frequent etiologic factor, established by anamnesis, was birth injury (45%). In second place were various infections of the perinatal period (20%). Hemolytic jaundice after birth was found less frequently than expected, in only 22% of the patients. The etiologic role of craniocerebral injury in the first months and years of life could be presumed in 3% of the cases. The course of the disease was not established for the remainder of the patients.

All the patients displayed, to varying degrees, the typical picture of CP previously described: greatly increased muscular tone in the form of rigidity or spasticity or both as well as generalized hyperkinesia involving the face, neck, trunk, and extremities. The majority of operated patients were invalids requiring total or almost total nursing care. They were bed- or chair-ridden and required constant help. The most severe stages (III and IV) of the disease were found in 128 patients out of 174. The main indication for an operation was hyperkinesia and increased muscular tone in the form of rigidity, spasticity, or both. The contraindications for surgery are presented above (Section 5).

The operations, as a rule, were performed under intratracheal anesthesia but sometimes under neuroleptanalgesia. One stereotactic operation was performed for the more afflicted extremities in 74 patients (including one-stage bilateral dentatotomies); two operations (right and left) with an interval of several months to several years were carried out on 66 patients, and three operations were done on the remainder.

The type of the operation performed has undergone a certain evolution over the course of a quarter of a century. The first eight operations in 1958–1959 involved destruction of the medial segment of GP. In subsequent years, the majority of operations (126) in-

*The surgical results in hemiathetosis and hemichoreoathetosis are presented separately in Chapter 10.

volved VL destruction. In 68 operations, a combined one-stage destruction of VL and Subth was performed. The rest of the cases were bilateral dentatotomy, Pul destruction, or selective rhizotomy.

In the initial stage of our work, a small number of operations were performed using alcohol injections or anodal electrolysis as the method of destruction. Since 1962, the cryosurgical method has been employed in all operations of this type.

There were no complications in 198 operations. After 34 operations, complications of varying severity occurred: pareses of contralateral extremities, pseudobulbar symptoms, postoperative pneumonia, etc. The majority of these complications were transitory.

Our experience and the data in literature indicate that severe CP patients tolerate operations on the basal ganglia (especially bilateral) less well than patients with other extrapyramidal lesions. This is evident from the high frequency of serious complications and postoperative mortality. One may say that the risk of a stereotactic operation in CP is greater than in parkinsonism, DMD, or hemihyperkinesias. According to our data, the frequency of serious complications after VL destruction was somewhat lower than after combined destruction of VL and Subth. In 232 operations there were seven deaths (3%) from various causes—pneumonia, hemorrhage from acute gastric ulcer, postoperative hemorrhage, and acute cardial insufficiency. During the last 10 years, the mortality rate decreased to 1.7%.

We evaluated the results of stereotactic operations in CP according to the generally accepted grades: "good," "significant improvement," "slight improvement," "no change," and "worse." The immediate results were determined at the time of discharge from the clinic, and the long-term results at various periods after the operation (6 months to more than 20 years). The evaluations were made on the basis of repeated examinations, answers to mailed questionnaires, or reports of neurologists who examined the patients. The results of the surgical treatment are presented in Table 19.

TABLE 19. Results of Stereotactic Operations in CP

Results	Immediate (%)	Long-term (%)
Good	16	12
Significant improvement	32	22
Slight improvement	37	44
Unchanged	12	16
Worse	3	6

The data in Table 19 indicate that, on the whole, the results of stereotactic surgery in CP are still far from what one would desire. A significant improvement immediately after surgery may be obtained in only approximately half of the cases. Unlike surgery for other extrapyramidal lesions, the long-term results of these operations are notably worse than the immediate results—only a third of the patients showed good or significant improvement over the course of many years.

However, this conclusion is based on our whole series, i.e., operations performed over more than 20 years. If one analyzes just the long-term results of operations done during the past 5 years, specifically combined destruction of VL and Subth, one sees that the figures are notably better than the overall long-term results—52% good or significant improvements compared to 34%. Nevertheless, as mentioned above, only one-half of the stereotactic operations in CP produce a significant effect.

Since cerebral palsy is always a severe and most often bilateral affliction of the CNS, bilateral operations are required in practically all cases (excluding hemihyperkinesias; see Chapter 10). Bilateral operations were performed in 66 of our patients. Our experience and that in the literature indicate that bilateral operations are more effective than unilateral; however, the incidence of complications in the former is notably greater. One of the main unsolved problems is the postoperative pseudobulbar syndrome with impaired swallowing and phonation in approximately 15% of the cases. In view of this, it is our opinion that bilateral operations on Th nuclei in CP should be restricted.

However, all of these observations do not mean that stereotactic operations, particularly thalamotomy, for CP are so ineffective that they should be rejected. Even a comparatively mild improvement in these patients represents a blessing both for them and for their relatives looking after them. If, after the operation, the patient is able to hold a spoon or turn the pages of a book, this is usually regarded as a great and joyous event. Since all other types of treatment are practically useless, stereotactic operations in severe forms of CP are justified.

10

Hemihyperkinesias

1. Introduction

The hemihyperkinesias include several nosological forms—hemiathetosis, hemichorea, hemichoreoathetosis, hemidystonia, and hemiballism. Such an arbitrary grouping of different extrapyramidal diseases is justified by their important clinical characteristic, namely, involvement of only one half of the body, which is indicative of the localization of the lesion in the basal ganglia of the contralateral cerebral hemisphere. On this basis, one might also include hemiparkinsonism in this group. However, we do not consider this appropriate since hemiparkinsonism (see Chapter 6) is only a stage of the disease, with inevitable later generalization. All of these hemihyperkinesias, as a rule, do not have a progressive course or a tendency to spread to the other half of the body. Accordingly, hemihyperkinesias can be considered residua of inflammatory, traumatic, vascular, degenerative, or other injury of the brain.

The literature does not have any statistical data on the incidence of each of the above-mentioned hyperkinesias. Based on our experience, hemiathetosis and hemichoreoathetosis are encountered frequently, hemidystonia occasionally, and hemiballism and hemichorea rarely.

2. Etiology and Pathogenesis

The etiologic bases of the hemihyperkinesias differ for different forms of this disease. The most frequent cause in our experience (Kandel, 1981, 1982) is encephalitis or meningoencephalitis, occurring in approximately 55% of cases in all forms of hyperkinesias. In our series of 86 patients, 49 gave such a history, including nine cases of tuberculous meningoencephalitis. In many patients (by history), there was a "lightning" onset of meningoencephalitis with a sudden high fever, disturbance of consciousness, and rapid (sometimes in 2–3 hr) development of hemiplegia. Lumbar puncture, if it was performed, showed pleocytosis in the CSF. After some time (usually several months), the movements of the paralyzed extremities gradually improved, but at the same time progressive hyperkinesias and muscle rigidity developed. Hemiathetosis and hemichoreoathetosis occur most often after meningoencephalitis but also after birth trauma or head injury in the postnatal period or in childhood. These frequent etiologic factors were present in 20% of our cases (Fig. 144).

The most frequent cause of hemiballism—a very rare and peculiar extrapyramidal disease—is a vascular lesion involving the subcortical structures of one cerebral hemisphere.

Hemiballism, first described 130 years ago by Todd (1853), was named that by Kahlbaum in 1867. Later, it became generally accepted as a distinct nosological form. Jacob (1928), on the basis of morphological data, first ascribed hemiballism to a lesion of NS (corpus Luysi). This conclusion was confirmed experimentally by Carpenter (1961), who produced hemiballism in monkeys by stereotactic lesion of NS. It is important to note that experimental hemiballism occurs only after partial (about 20% of volume) destruction of the above-mentioned structure. As clinical observations have shown, hemiballism in the majority of cases is caused by an acute vascular lesion (hemorrhage, infarction) in the region of NS (Bedwell, 1960). In the world literature, more than 150 cases of hemiballism or hemichorea have been reported as the result of hemorrhage in the NS. In these cases, hemiplegia developed initially, and after muscle power improved,

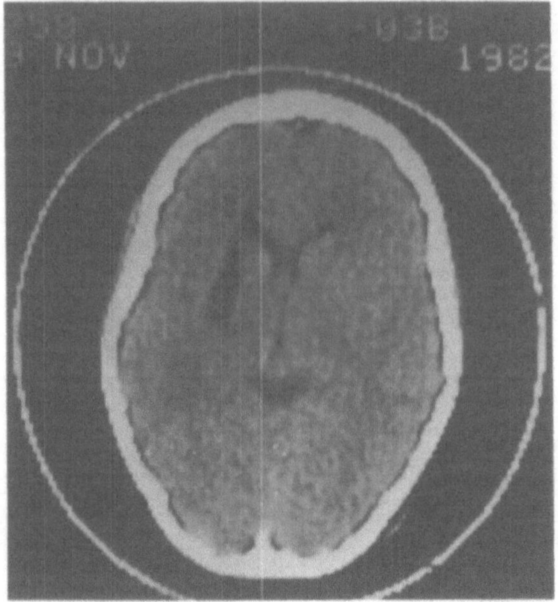

FIGURE 144. A CT scan of a 14-year-old girl witn right-sided hemichoreoathetosis developed after severe head injury. Note a low-density focus in the left NL.

massive involuntary movements appeared in the previously paralyzed extremities (posthemiplegic hemiballism). In many cases these clinical data have been confirmed by postmortem studies. In ten of 13 hemichorea cases, morphological investigations by Martin (1927) showed a hemorrhage and in two cases a tumor metastasis in the NS contralateral to the affected extremities.

Stereotactic neurosurgery occasionally sheds light on the pathogenesis of hemiballism. This hyperkinesia may develop as a rare complication when, during stereotactic destruction of VL, an unexpected partial lesion of NS was made (see Chapter 6, Section 11). However, the pathogenesis of hemiballism is not completely known. Several reports in the world literature confirm that hemiballism may develop without a lesion of NS (Martin, 1957; Cooper *et al.*, 1963), and several cases with morphologically proven lesions of this nucleus (Cooper, 1963) have not manifested hemiballism.

The question of the role of vascular factors in the etiopathogenesis not only of hemiballism but of other hemihyperkinesias as well is important from the clinical point of view. The role of these factors has not been studied to a sufficient degree, primarily because angiography and CT investigations are made very sel-

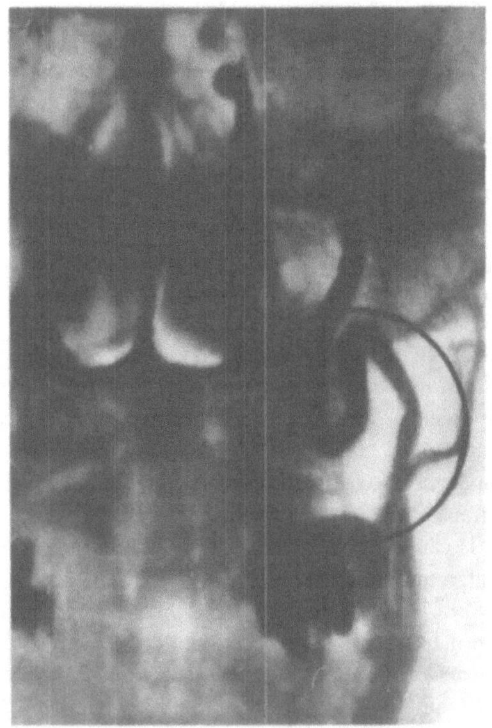

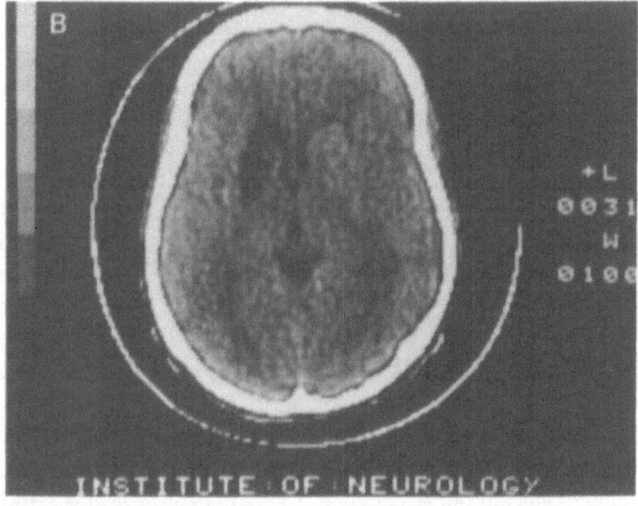

FIGURE 145. A 49-year-old women with left-sided hemiballism after an ischemic stroke in the right hemisphere. Bilateral kinking and tortuosity of the internal carotid arteries in the neck. (A) AP right angiogram. (B) Hypodense focus in right NC on CT scan.

dom in cases of hemihyperkinesias. As the literature shows, if hemihyperkinesias were not a consequence of an acute disturbance of brain circulation, vascular factors would be implicated extremely rarely. Occlusive disease of the carotid arteries in the neck or intracranially were observed most frequently. Radermecker *et al.* (1963) reported a 63-year-old patient with right-sided hemichorea in whom angiography disclosed an occlusive process in the intracranial part of the left internal carotid artery. Gönshirt *et al.* (1978) reported a case of transient hemiballism in the left extremities of a 65-year-old patient in whom angiography showed stenosis of the right internal carotid, vertebral, and basilar arteries. Margolin and Marsden (1982) reported two cases of hemihyperkinesias presumably caused by stenosis of the extracranial internal carotid artery on the side contralateral to the affected extremities.

Lobo-Antunes *et al.* (1974) found on angiography small arteriovenous malformations in the basal ganglia of the contralateral hemisphere in two patients with extrapyramidal hemihyperkinesias. Jellinger (1975) reported a unique case of a 60-year-old patient suffering from hemiballism for more than 20 years. The patient had an arteriovenous malformation situated in Th and NS. After rupture of the malformation and hemorrhage in the ventral parts of Th, the hemiballism disappared. As the author aptly remarked, a "spontaneous stereotactic operation" took place.

For a more detailed study of the etiopathogenesis of hemihyperkinesias, we (Kandel and Omorov, 1984) performed cerebral angiography and CT investigations in 26 patients with hemihyperkinesias (13 women and 13 men between 15 and 49 years of age). Twelve had involvement of the right and 14 of the left extremities; 12 had hemichoreoathetosis, eight hemidystonia, two hemiathetosis, two hemichorea, and two hemiballism. The disease had been present from 2 to 33 years. Only one of these patients had had an acute disturbance of the cerebral circulation. Angiography was performed on the side contralateral to the affected extremities. Axillary angiography was carried out on the right side (12 cases) to visualize the carotid and vertebral circulation. Carotid angiography was done on the left side (16 cases), and in one case additional left-side axillary angiography was carried out.

Various vascular abnormalities that could be considered major etiological factors in hemihyperkinesias were found in ten of the 26 patients. The most frequent vascular abnormality (in six patients) was kinking or coiling of the internal carotid artery. It is logical to suppose that the hemihyperkinesias in these cases were related to unilateral focal ischemia or destruction of subcortical structures as a result of circulatory disturbances in the internal carotid system. This supposition was confirmed in three of six patients with kinking of the internal carotid. In these cases, CT showed a small infarcted focus in the subcortical structures of the contralateral hemisphere. These were hypodense foci 2 × 1 cm in size in the region of NC and the anterior limb of CI. An illustrative case is presented below.

A 49-year-old patient developed a left-sided hemiballism several months after an acute stroke with left-sided hemiplegia. During the recovery of function of the left extremities, ballistic movements rapidly progressed. Right-sided axillary and left-sided carotid angiography disclosed kinking of the internal carotid arteries bilaterally at C_2 level (Fig. 145A). The CT showed a hypodensive focus in the region of the right NC without dilatation of the ventricular system (Fig. 145B).

In four patients occlusive vascular disease was present, in two cases stenosis of the internal carotid, in one case occlusion of the internal carotid bifurcation by a tuberculoma, and in the fourth case thrombosis of the middle cerebral artery. The cause of these occlusions in three of these young (17, 19, and 33 years) patients is unknown. In three of the four cases, CT disclosed intracerebral changes that could be responsible for hemihyperkinesia—hypodense foci of different size in the anterior parts of the NL and adjacent parts of the anterior limb of CI.

The next example illustrates the development of hemihyperkinesia after thrombosis of the middle cerebral artery, which produced a massive cortical and subcortical lesion in the cerebral hemisphere.

A 15-year-old patient with a history of birth trauma had had right-sided hemichoreoathetosis from early childhood. Carotid angiograms disclosed thrombosis of the left middle cerebral artery at the bifurcation of the internal carotid. The CT showed significant atrophy of the left cerebral hemisphere and a large hypodense area in the region of the inferior frontal, middle temporal, superior temporal, supramarginal, and angular gyri as well as the NL and posterior part of NC. The ventricular system was markedly deviated to the side of cortical atrophy.

The following case is a rare hemihyperkinesia with two pathological foci in the involved hemisphere, one in the basal ganglia and a second (calcified tuberculoma) in the basal cistern of the brain. One may assume that the hemihyperkinesia was caused by the

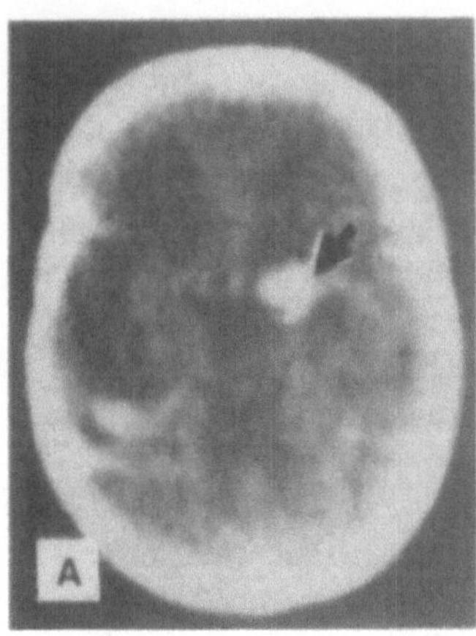

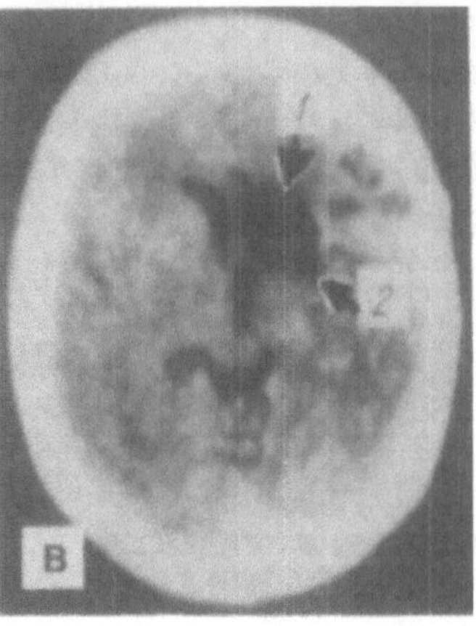

FIGURE 146. Occlusion of left internal carotid artery caused by tuberculoma after tuberculous meningoencephalitis developed in early childhood. (A) Calcification in the left suprasellar region on CT cisternography. (B) Large hypondense focus in the left NL and NC.

first focus, which developed as a result of middle cerebral artery thrombosis.

In early childhood, a 34-year-old women had experienced tuberculous meningoencephalitis. After 3 years, right-sided hemiathetosis developed gradually. Left carotid angiograms disclosed occlusion of the internal carotid artery as a result of compression by a tuberculoma. Cortical branches of the middle cerebral artery were filled by corticomeningeal anastomoses from the external carotid artery. Right axillary angiograms demonstrated an anterior trifurcation. On CT cisternography there was a calcification 2 × 1.2 cm in size in the left suprasellar cistern and an independent hypodense focus measuring 2 × 1.7 × 0.8 cm in the basal part of left NL and NC (Fig. 146).

Another vascular anomaly discovered on angiograms was a small AVM in the mediobasal region of the frontal lobe. This extremely rare case is presented below.

A 21-year-old patient had had right-sided hemichoreoathetosis since early childhood. Left carotid angiograms disclosed a small (2 × 2 × 0.3 cm) AVM in the mediobasal part of the left temporal lobe (Fig. 147). The anterior horn of the left lateral ventricle was markedly dilated.

Cerebral angiography and CT scans in patients with hemihyperkinesias gave a new insight into the etiopathogenesis of this extrapyramidal disease. As the above data indicate, approximately 30% of the cases were found to have cerebral vascular abnormalities with the aid of angiography. They probably are the etiologic factors of hemihyperkinesias. These vascular processes are varied and include kinking, deformation, stenosis, and thrombosis of the internal carotid and middle cerebral arteries and small AVMs.

3. Clinical Features

The clinical pictures of the different hemihyperkinesias have been well described in the literature. The clinical characteristics of athetosis (cerebral

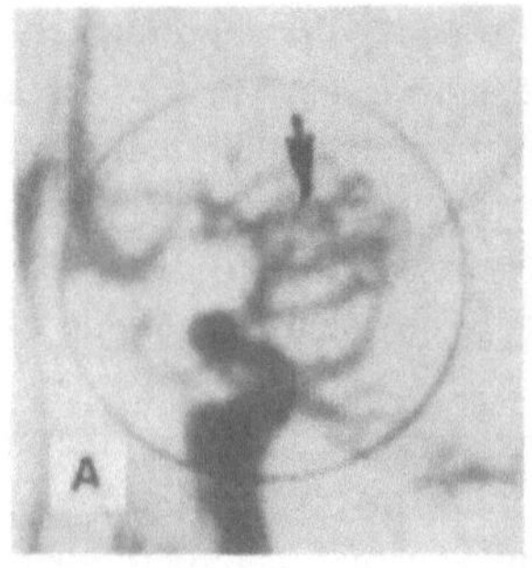

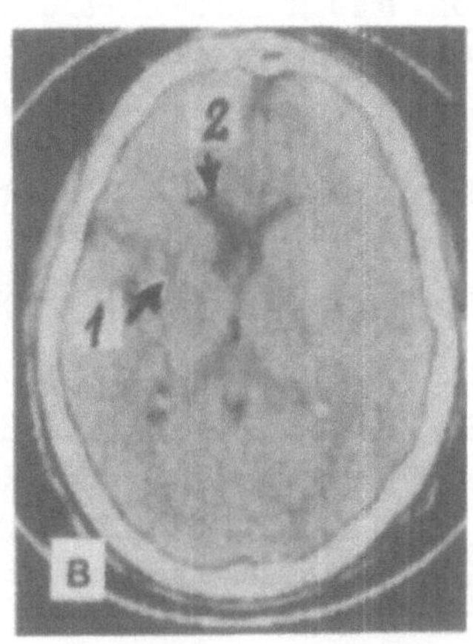

FIGURE 147. Small AVM supplied by the left internal carotid bifurcation in the mediobasal part of the temporal lobe in a 21-year-old patient with right-sided hemichoreoathetosis. (A) Left carotid angiogram. (B) CT scan with small AVM (1) and enlargement of the left anterior horn (2).

palsy) and torsion dystonia, presented in the corresponding Chapters 7 and 9, are directly applicable to unilateral forms of these conditions. The main clinical manifestations are hyperkinesias and extrapyramidal rigidity, often associated with a spastic component in the involved extremities.

The different kinds of hemihyperkinesias can be identified by the character of the involuntary movements. Hemiballism, which as a rule begins rapidly and soon involves the extremities of one half of the body, is characterized by swift, chaotic, arrhythmic involuntary movements of large amplitude in both extremities simultaneously or only in the arm or less frequently in the leg. Very rarely all four extremities (ballism) are involved. Hemiballism in its clinical manifestations is similar to hemichorea but differs from the latter in that there are more massive and chaotic hyperkinesias. Usually the facial and truncal muscles are not involved.

The use of the affected hand was greatly impaired or impossible in practically all of our patients with hemihyperkinesias. Many had to hold the affected hand with the healthy one most of the time or constantly to diminish the severity of hyperkinesia. Almost all patients have a mild to severe disturbance of walking because of the hyperkinesias in the leg. The muscle tone in the hyperkinetic extremities was markedly increased with extrapyramidal rigidity or sometimes elements of pyramidal spasticity. The involvement of the hand, particularly the severe hyperkinesia and muscle hypertonicity, as a rule, is more pronounced than the leg. In the majority of patients afflicted from childhood, there are atrophy and shortening of 3–6 cm of the affected extremities.

There is a significant increase of EMG activity in the affected extremities with a marked increase (in comparison with the healthy side) in the muscle biopotentials both at rest and during activity.

4. Stereotactic Surgery

The advent of stereotactic neurosurgery significantly increased interest in this form of extrapyramidal disease since drug treatment of hemihyperkinesias is, as a rule, unsuccessful. The various operations involving cutting of the pyramidal pathways at different levels that were done in the 1940s and 1950s gave no consistent or significant results. Meyers (1940) was the first to remove the head of NC and destroy GP and Put by an open transventricular approach in one case of hemiballism with good results.

Stereotactic operations on the basal ganglia appeared to be the only effective treatment for the different kinds of hemihyperkinesias. The number of papers on stereotactic treatment of hemihyperkinesias is quite limited. The first reports described one or two cases of hemiballism with good results after surgery (Talairach *et al.*, 1950; Martin and McCaul, 1959; Obrador and Dierssen, 1959; Wycis and Spiegel, 1962). Subsequent papers analyzed the results of stereotactic operations in four to eight patients with hemiballism (Andy, 1962; Gioino *et al.*, 1966). As a rule, the results of VL or Subth lesions appeared quite effective. In all 18 patients with hemiballism operated on by Siegfried (1974), the hyperkinesias were abolished completely. Only in one case did a relapse occur after several weeks; after a second operation, the hemiballism was abolished.

Destruction of the lateral segment of GP stereotactically has only been reported once in the literature. Tsubokawa and Moriyasu (1975a) performed lateral pallidotomy in two cases of hemiballism with a good clinical result for at least 2 years.

Other hemihyperkinesias (hemiathetosis, hemichoreoathetosis) have been treated by stereotactic procedures in only a few cases. Starikov and Staduchin (1970) reported on 17 patients with hemiathetosis; in the majority of cases, satisfactory results were obtained.

5. Personal Observations

The first stereotactic operation for hemihyperkinesis was performed by us in 1958. During the last 25 years, we have operated on 86 patients; the results have been reported previously (Kandel and Guluyev, 1977; Kandel, 1981, 1982). In this series, 56% were men and 44% women; their ages ranged from 10 to 67 years (75% were from 10 to 30 years). Clinically the 86 patients were divided into groups of hemichoreoathetosis (33%), hemiathetosis (31%), hemidystonia (23%), hemiballism (9%), and hemichorea (4%). The right and left sides were about equally often involved. In 85% of our cases the onset of the disease was before the age of 15 years. The duration of illness from onset until the stereotactic operation varied: 5% less than 3 years, 26% from 4 to 10 years, 17% from 11 to 15 years, 47% from 16 to 30 years, and 5% more than 30 years.

The most frequent etiologic factor in our series was encephalitis or meningoencephalitis in childhood, which occurred in 55% of the patients. Birth trauma

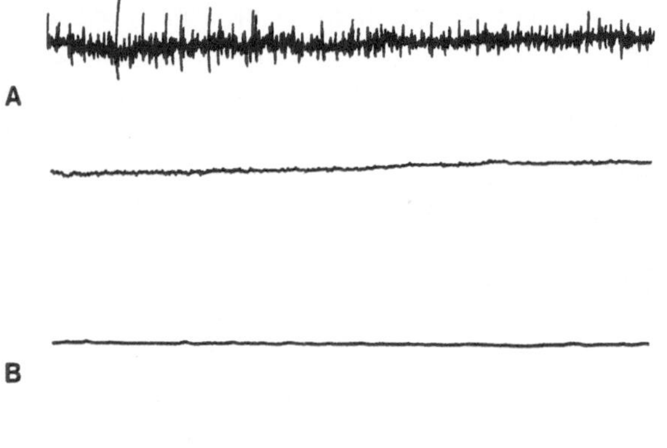

FIGURE 148. Hemidystonia in right extremities in 26-year-old woman. (A) Electromyographic recording before operation from superficial flexors of both forearms. There is pronounced spontaneous activity in the right arm (above) but no activity in the left (below). (B) The same recording after left stereotactic VL and Subth cryodestruction. Spontaneous activity in the right arm disappeared completely.

(including cerebral ischemia) and head injuries usually in childhood were responsible in 20%; vascular abnormalities, described above in detail, in 14%; and lesions of subcortical structures by tumors, tuberculomas, or calcification in 5%. In 6% of the operated patients, no etiologic factor could be determined.

In all patients, marked hyperkinesias and increased muscle tone were present in extremities of one half of the body. Walking was disturbed, and self-care limited. Many of the patients used the healthy hand to do everyday chores. Prolonged drug (including L-dopa) and other conservative treatments had been ineffective.

In 86 patients, 102 stereotactic operations were performed on the basal ganglia contralateral to the affected extremities. Fourteen patients were operated on twice, and two of them three times (mainly in the early cases), because of an unsatisfactory result from the first operation. At the second operation, a new lesion was made or the previous one enlarged.

The choice of anesthesia depended on the severity of the hyperkinesias. In rare cases in which the patient could lie on the operating table calmly, local anesthesia

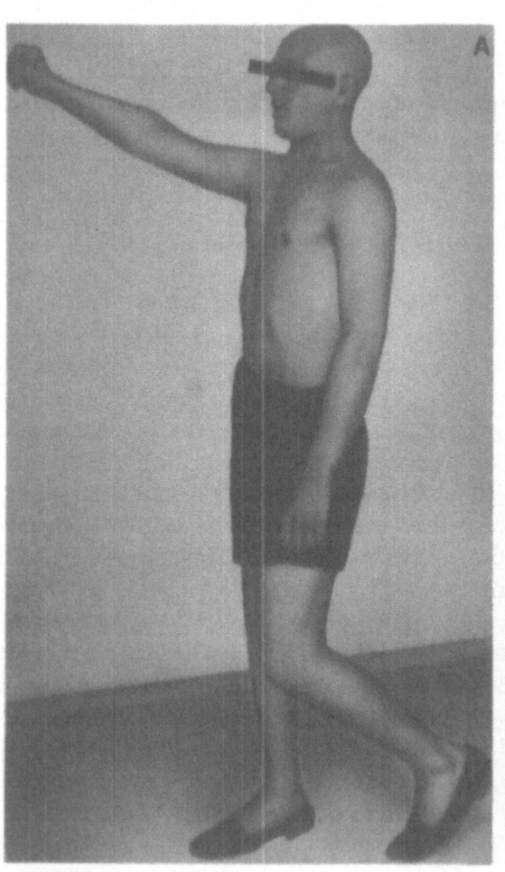

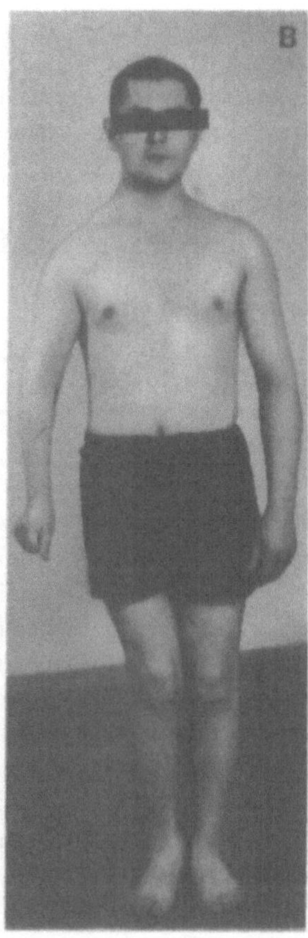

FIGURE 149. (A) A 23-year-old patient with right hand dystonia that developed in childhood. (B) The same patient 4½ years after left thalamotomy. Note the shortened right arm.

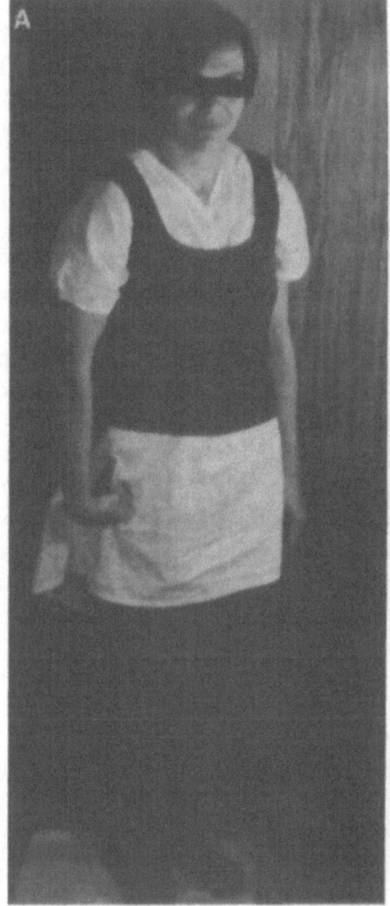

FIGURE 150. A 34-year-old patient with right-sided hemiathetosis (A) before and (B) 4 years after left stereotactic thalamosubthalamotomy.

with premedication was used, but in cases of pronounced hyperkinesias, intratracheal narcosis or neuroleptanalgesia was indicated.

As experience accumulated, the technique of stereotactic operations, choice of the surgical target, and methods of destruction were modified. Only the first 17 operations were performed after pneumoencephalography; in all subsequent operations ventriculography with iothalamate or an iothalamate and iophendylate mixture was used to outline the intracerebral stereotactic landmarks.

The selection of a subcortical target also slowly evolved. In the first five operations, the medial segment of GP was destroyed. For many years after that (62 cases), destruction of VL was the method of choice. During the last few years we have performed 35 combined operations making two lesions, a bigger one within VL and a smaller one in Subth. The main method for destruction of the target structure was freezing (Fig. 148).

There were no serious complications or mortality in the 102 operations. In eight patients a slight hemiparesis developed, which completely disappeared 2–6 weeks postoperatively.

In the postoperative period, intensive exercises and training of the affected extremities were encouraged, which undoubtedly promoted a more rapid functional improvement. This is confirmed by the fact that in some patients the functions of the extremities, especially of the hand, became significantly better not immediately after surgery but in the follow-up period.

Immediate results of the surgical treatment were evaluated at the time of discharge from the clinic (2–3 weeks after surgery), and long-term results after repeated examinations by neurologists who observed the patients and also on the basis of replies by the patients to detailed questionnaires. The follow-up period in these patients ranged from 2 to 18 years (average 6.6 years).

In the "good result" group we included patients who had almost complete relief of hyperkinesias in the affected extremities and normal muscle tone (Fig. 149–152). Subsequently, the patient was able to perform all everyday activities with the affected hand. Walking improved significantly, but some remaining difficulties caused not by hyperkinesia but by shortening of the arm or leg, muscle contractures, or joint changes. The majority of these patients were able to return gradually to labor activity.

The "significant improvement" group included patients with an improvement of 50–60%. The hyperkinesias were markedly diminished, the muscle tone became less pronounced, and walking less restricted. Their personal hygiene improved, they could perform household chores, and some patients began to do limited jobs. The "slight improvement" group included patients with an improvement of 20–30%. The results are presented in Table 20.

As the data in Table 20 indicate, during follow-up after surgery, good results or a significant improvement is noted in 74% of cases. No follow-up could be made in eight cases, and in four cases the results were not tabulated since the period of observation after surgery was short. The long-term results of surgical treatment are somewhat worse than the immediate results, although the difference is not great (74% and 80%, respectively, of good results or significant improvement). These data prove that the therapeutic effect of stereotactic operations as a rule remains stable for many years. Only six patients had a significant recurrence after an initial beneficial effect.

Twenty patients before the operation suffered from intensive pain in the affected extremities. After the operation the pain completely disappeared in 12 patients and was markedly reduced in eight.

It is important to analyze the clinical effects of

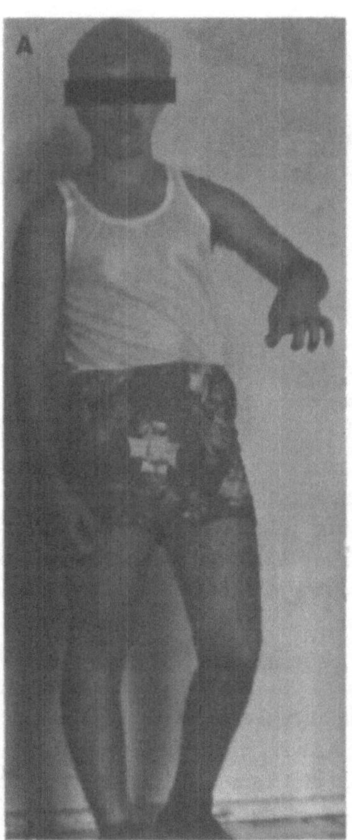

FIGURE 151. A 27-year-old patient with left hemichoreoathetosis (A) before and (B) 2 years after right cryothalamotomy.

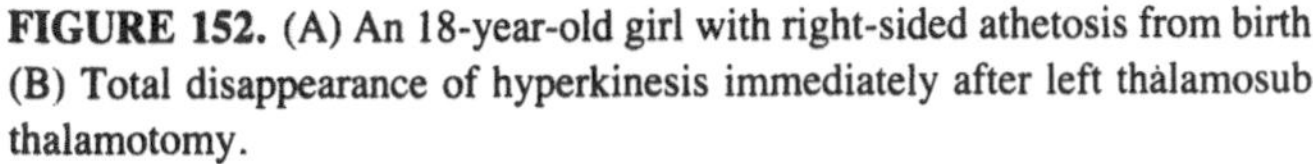

FIGURE 152. (A) An 18-year-old girl with right-sided athetosis from birth. (B) Total disappearance of hyperkinesis immediately after left thalamosubthalamotomy.

stereotactic intervention for hemihyperkinesias on the basis of the structure that was destroyed. For this analysis, we made random studies of ventriculograms of 63 operated patients, plotting the location of each subcortical lesion in three planes onto the corresponding plates of the Schaltenbrand and Bailey stereotactic atlas. The site of the lesion was correlated with the long-term results of the operations (Table 21).

As the data of Table 21 show, the combined destruction of VL and Subth is more effective than the destruction of only VL, but the difference is not great. In the first group, good results or a significant improvement during long-term follow-up were obtained in 72% of patients, and in the second group in 80%.

Correlation of long-term results with the clinical type of hemihyperkinesia showed that these operations are most effective for hemichoreoathetosis and hemidystonia. No correlation was found between etiologic factors and long-term results.

In some cases the focus of destruction in VL included a small part of CI lying laterally. However, the

TABLE 20. Immediate and Long-Term Results of Surgical Treatment of Hemihyperkinesias

Results	Immediate (*N*)	Long-term (*N*)
Good	36	30
Significant improvement	32	26
Slight improvement	15	11
No change	3	6
Worse	—	1
Unknown	—	12
Total	86	86

TABLE 21. Long-Term Results of Operations and Sites of the Cryogenic Focus

Effect of the operation	Ventrolateral nucleus (VL)(*N*)	VL + Subth (*N*)	Total
Good	15	11	26
Significant improvement	16	5	21
Slight improvement	10	2	12
No change	2	2	4
Total	43	20	63

destruction of these marginal fibers did not improve the clinical results but in several cases caused development of definite, although reversible, hemiparesis.

As our investigations with Barchatova and Omorov have shown, the level of homovanillic acid (the main metabolite of dopamine) in the lumbar CSF, which was elevated before the stereotactic operation for hemihyperkinesia, as a rule, significantly decreased (average from 250 nM to 100 nM). This points to a decrease of dopaminergic activity after stereotactic destruction of VL and Subth.

On the basis of our series, one may conclude that hemihyperkinesias, of which hemichoreoathetosis is the most common, are usually diseases of childhood. In the majority of cases, the etiologic factor is meningoencephalitis with subsequent hemiplegia. Hemihyperkinesias develop as motion returns to the paralyzed extremities but do not subsequently progress. Various vascular injuries in the brain as well as birth trauma often appear to be causes of hemihyperkinesias.

The only effective treatment of hemihyperkinesias is stereotactic surgery on the basal ganglia. In more than 70% of the patients, a significant therapeutic effect is obtained after surgery, and the effect remains stable for many years.

Complete or almost complete disappearance of hyperkinesias, muscle rigidity, and pain after surgery leads to a significant improvement of the functional activities of the affected extremities. The best clinical results were obtained after combined destruction of VL and Subth. Today, stereotactic operations for hemihyperkinesias are quite safe.

11

Multiple Sclerosis

1. Introduction

Almost 100 years ago Charcot described multiple sclerosis (MS) as a distinct nosological entity. Today, it is still one of the most important problems in neurology. Charcot not only pointed out the classical triad of symptoms pathognomonic for this disease (intention tremor, nystagmus, staccato speech) but also described other frequently occurring symptoms (diplopia, impaired vision, paresis of the extremities, vertigo, ataxia, muscular atrophy, etc.).

The incidence of MS in Europe, North America, and Australia is three to eight cases per 10,000 population; in Asia and Africa the incidence is several times lower.

The pathological anatomy, clinical picture, diagnosis, and course of MS have been described in detail in many papers and neurology texts, so it is unnecessary to repeat the details. But we would like to emphasize one finding that, in our opinion, is of importance in considering the pathogenesis of MS. Ventriculography performed before the operation in our series of cases showed in all 20 patients operated on a symmetrical hydrocephalus of varying degree. On the basis of a quantitative evaluation of hydrocephalus as described in Chapter 6, a first-stage hydrocephalus was present in two patients, second in 13, and third in five. This shows that MS is associated with diffuse cerebral atrophy of obscure etiology.

The most pressing unsolved problem in MS is its treatment. A persistent search for an adequate theory has not yet been fruitful, mostly because the etiology is still unknown. The most recent hypothesis implicates a viral etiology (herpes simplex, viruses of measles and parotitis, adenovirus, and others). However, immunologic factors have been given considerable attention as causes of MS, particularly a hereditary impairment of the immune system.

As regards stereotactic surgery in MS, we discuss only one of the main symptoms of this disease. The most serious and common manifestation is intention tremor, usually more pronounced in the hands. In many patients tremor predominates in the clinical picture and is the main cause of their total disability and of their inability to carry out even the simplest activities of everyday life. The principal characteristic of intention tremor, unlike the tremor in parkinsonism, is its absence at rest and its appearance on attempts to make purposeful movements. As a rule, it involves the extremities, head, and torso. In a prone position, the tremor is only moderate or even absent, but on any attempt to make a purposeful movement or to assume any kind of posture, a severe and large-amplitude rhythmic shaking develops, totally depriving the patient of the ability to make any purposeful movement.

In the beginning, tremor occurs in one extremity, progressively intensifies, and then spreads to other extremities. Gradually it involves the torso, head, jaws, and eventually the whole body. Frequently the patient is unable to move, not because of spastic paresis or paralysis but as a result of severe hyperkinesia. Consequently, one may speak of a specific hyperkinetic form of the disease, although other well-known MS symptoms are also present.

The pathogenesis of intention tremor is still not quite clear. Experimental postural tremor may occur after lesions of the medioventral area of the midbrain and ventral part of Br Con. Clinicians have long associated intention tremor with a lesion in Cer, particularly ND.

According to modern concepts (Cooper, 1969a,b), intention tremor results from impaired dentatorubral pathways, which pass through the superior cerebellar peduncle to NR and to nonspecific thalamic nuclei, which project to the cortex. Severe tremor in MS is called "myoclonia of action" and is believed to result from disinhibition (Mundinger and Kuhn, 1982). This syndrome is produced by a focus of demyelination in the Mollaret triangle (Hassler *et al.*, 1975; Riechert *et al.*, 1975). These authors believe that bursts of pathological impulses pass along dentatorubral and rubrothalamic pathways to VOp and ZI, to area 4 of the cortex, and then by the pyramidal tract to the anterior horns of the spinal cord. There is also impaired feedback from the muscular spindles (dysfunction of the γ system).

In respect to the pathogenesis of intention tremor and ataxia in MS, an interesting observation was made by Riechert *et al.* (1975), who thoroughly studied the brains of two patients who died some time after a stereotactic operation. They believe that tremor and ataxia may result from demyelination of various subcortical structures—NR, Put, NC, and SN. As a result, they consider impaired catecholamine metabolism to play a significant role in the pathogenesis of intention tremor. But in MS, in contrast to parkinsonism, there is not a deficit but an increase in the concentration of dopamine in Str.

Tremor in patients with cerebellar tumors has been studied by the tensometric technique with the forearm supported and the arm held horizontally in the air (Artaryan and Gurfinkel, 1971). The authors have shown that patients with cerebellar tumors develop significant changes in the physiological tremor with a marked increase in the amplitude of both high-frequency and slow waves. This is accompanied by "outbursts" scores of times higher than the normal amplitude of physiological tremor. The ratio of the amplitude of the high-frequency and slow oscillations also changes: normally this ratio is 1 : 4, but in case of cerebellar tumors it is 1 : 2. These changes may occur even in the absence of clinical intention tremor. The frequency of the tremor in patients with a lesion in Cer was the same as in patients with a physiological tremor. Yet the curve had areas of synchronization in the form of synchronous bursts at 6–8 Hz.

2. Surgical Treatment

2.1. Stereotactic Operations

It is generally conceded that no effective treatment for MS has yet been discovered. Conservative treatment gives practically negligible results, and for tremor it is useless.

Since stereotactic destruction of VL or Subth totally or partially abolishes tremor in parkinsonism and hyperkinesias caused by other lesions of the extrapyramidal system, one might think it reasonable to try to stop the intention tremor of MS by a similar operation. Another reason to consider stereotactic operations in MS was the assumption that some symptoms of certain chronic diseases of the CNS are caused by the partial impairment of the cerebral control. Complete (not partial) severing of an essential link in this mechanism might diminish or abolish the neurological dysfunction.

At the same time, it is evident that stereotactic operations on the basal ganglia in MS are not meant to arrest or even retard the progression of the demyelinization process. In this respect, surgical treatment of this disease is purely palliative and only is intended to abolish the tremor that makes patients complete invalids.

Because stereotactic operations are done for the hyperkinetic form of MS, a natural question arises as to whether this operation might create fresh foci of demyelinization. Riechert *et al.* (1975) concluded that stereotactic operations induce fresh demyelinization foci in one case out of ten. Mundinger and Kuhn (1982) did not note any unfavorable effect of such operations on the progress of the disease. Our observations fully confirm this conclusion.

In the literature there are several reports on stereotactic operations in MS. The first operation was performed by Cooper and Poloukhine (1960). In 1962 we performed the first stereotactic operation for tremor of MS in the Soviet Union and have reported the results of such surgical treatment (Kandel and Hondcarian, 1977, 1985).

Cooper (1967) reported good results of surgical treatment in 27 of 32 MS patients; the intention tremor either disappeared or was greatly reduced. There was one postoperative death, and two patients with spastic hemiparesis deteriorated. Obrador and Dierssen (1965) operated for intention tremor on various subcortical structures in 15 patients. The most effective procedure was the destruction of VL and Subth. The authors' conclusion differed from that generally accepted—they believed that the postoperative disappearance of intention tremor resulted from severance of pallidothalamic rather than cerebellar pathways.

Laitinen *et al.* (1974) performed thalamotomies in 26 MS patients. The immediate postoperative results were good in 73% (total disappearance of tremor in the opposite extremities); in 19% tremor was considerably reduced, and in 8% there was no effect. There were no postoperative deaths. There were various complications; for example, five patients developed transient hemiparesis. A follow-up (mean 3.3 years) showed that seven patients had died of the progressive disease, but the postoperative result was stable until death.

Augustinsson and von Essen (1977) stereotactically destroyed VL in 22 MS patients with marked clinical symptoms, in particular, severe intention tremor. The latter was significantly improved in all patients. The best results were obtained in patients in whom the process was relatively stable and not rapidly progressing.

One of the largest series of observations was reported by Mundinger and Kuhn (1982) on the immediate and long-term results of stereotactic operations in 84 MS patients. Good relief of tremor ("myoclonia of action") was obtained in 51% of the patients, and there was significant improvement in 19%. The good results at long-term follow-up were less than the immediate effects (70% and 81%, respectively). The authors believe that surgical intervention in MS is indicated only in the absence of other symptoms of impairment of the upper extremities (ataxia, paresis) and in the absence of an organic psychosis.

2.2. Operations on the Spinal Cord

In advanced stages of MS, mainly in the spinal form of the disease, severe spastic paraplegia sometimes develops in the legs with frequent painful muscle spasms. In such cases, drug therapy is rarely effective.

Functional operations at the spinal level, particularly selective spinal rhizotomy, although only palliative, significantly improve the state of these patients. The technique of these operations is described in Chapter 13. Sindou *et al.* (1982a) performed posterior rhizotomies on 15 MS patients with spastic paraplegia (painful spasms in the legs and flexion contractures). Postoperative follow-up (mean 3 years) showed a significant functional improvement. Of 11 bed-ridden patients, nine became able to sit and move about in a wheelchair. Almost all the patients reported a significant reduction of flexor contractures and painful spasms in the leg muscles. A considerable improvement in motor functions and sensibility in the legs, however, was of no practical importance.

Attempts lately undertaken to use stimulation in MS have been promising. Cook and Weinstein (1973) were the first to stimulate the posterior columns (Chapter 13, Section 5.4.2) in an MS patient with constant pain in the back. Besides alleviation of pain, there was an immediate and clear-cut regression of some neurological symptoms such as ataxia (Cook *et al.*, 1979, 1981). This was confirmed by Dooley and Sharkey (1981), who stimulated the posterior columns and reported a reduction of spasms and ataxia and an improvement of bladder function in 60% of patients with MS, Friedreich's ataxia, and other chronic diseases of the CNS. Waltz (1982) reported that 19 of 27 MS patients noted an improvement—decrease of spasticity, improved motor function, swallowing, gait, etc. Other authors report less optimistic results (Rosen and Barsoum, 1979). According to Ketalaer *et al.* (1979), dorsal column stimulation gave an objective improvement in only four patients out of 11.

Fredriksen *et al.* (1986) recently reported an informative study of epidural PCS in 49 MS patients aged 27 to 62 years. In the majority of cases, "moderate but meaningful" improvement was achieved, particularly in bladder function and walking ability. An improved quality of life was noted in a few patients. On the other hand, Rosen and Barsoum (1979) reported a failure of the PCS treatment in 9 cases and believed that the subjective improvement in a few cases was due to a placebo effect.

3. Personal Observations

Over the last few decades, we have operated on 22 patients with the hyperkinetic form of MS. In 17 patients we performed unilateral and in five patients bilateral operations on the basal ganglia (altogether 25 operations). The indication for surgery was bilateral intention tremor sufficiently severe to prevent the patient from caring for himself. There were 10 men and 12 women, mainly young and middle-aged (from 14 to 52 years).

Most of the patients operated on had suffered for a long time from this disease (from the onset of the first symptoms): 5 years or less in 12 patients, from five to ten years in seven patients, and from 10 to 20 years in three patients. The course of MS is known to vary. In 14 patients, it was rapid (2–3 years), in three slow, and in five "average."

Since MS is a relatively rare condition to be treated by stereotactic surgery, we began with the severest cases and only later broadened the indications somewhat. It should be emphasized that most patients (15 of 22) had been bed-ridden invalids for many years. The state of four patients was assessed as "average severity," and in three patients it was mildly incapacitating.

Almost all the patients had a variety of symptoms indicating lesions in different portions of the brain and spinal cord. The neurological picture varied in different patients, but practically all had the classical Charcot's triad. Most of them also had mild hemi- or parapareses, urinary disorders, pronounced truncal ataxia, absent abdominal reflexes, etc.

Of the 27 operations, 12 were performed under potentiated local anaesthesia. In the other cases, the tremor was so pronounced that endotracheal anesthesia had to be used.

The choice of a target point in MS is still unsettled. In 23 operations we made cryogenic lesions in VOp, and in four operations we made lesions in both VOp and Subth, which prevents a comparison of the results of lesions of these two structures. Since destruction of VOp, which is a "relay station" of the cerebellorubrothalamic pathways, leads to the elimination of all forms of pathological tremor, there is every reason to believe that it is the optimal target in MS.

We had no operative mortality in 22 patients but in five patients we noted a mild temporary increase in pyramidal signs, which disappeared within a month. The general impression is that even severely debilitated MS patients can tolerate surgery on the basal ganglia. Our experience, as well as the data from the literature, should allay fears regarding the risk of such operations. Even lumbar puncture was previously believed to be hazardous in MS patients.

The most impressive effect of this operation is the disappearance of severe tremor, usually noted at the moment when the end of the cannula reaches the target point. The immediate and long-term results in 22 patients are presented in Table 22.

During the follow-up period four patients died of the primary disease within a few years (they are not included in Table 22).

A successful surgical intervention significantly improved both the morale and the physical state of the patient, and disappearance of the incapacitating tremor markedly increased the ability to eat, dress, etc. Several patients who could walk unaided before the operation had an improved gait, although their ataxia was still present to a moderate degree. An unexpected result was the disappearance or reduction of a coarse spontaneous nystagmus in 14 patients. Staccato speech significantly improved in almost half of the cases.

TABLE 22. Results of Surgical Intervention for Tremor in MS.

Results	Immediate (*N*)	Long-term (*N*)
Total or almost total disappearance	14	9
Significant reduction	4	3
Unchanged	4	4
Unknown	—	2

Analyzing the long-term results, we have come to the conclusion in agreement with other authors that surgical treatment is much more effective in patients with a prolonged and slowly progressive (more "benign") disease. The results were notably worse or temporary in patients with a rapid course without remissions. This is also the opinion of other authors (Mundinger and Kuhn, 1982) who believe that it is better to operate on patients with a slowly progressive or stable course. However, we have no grounds to assume that surgical intervention has triggered an exacerbation or accelerated the course of the disease.

In conclusion, it should be emphasized that surgical treatment of MS is only palliative and is done to correct the one most annoying symptom, intention tremor. The operation does not influence the course of the primary pathological process and does not prevent or delay the inevitable fatal outcome. The data presented in Table 22 show that long-term results of stereotactic operations are less favorable than the immediate response because of progression of the disease. Nevertheless, we believe that at present these operations provide a unique opportunity to alleviate the incapacitating tremor of such patients. Moreover, these operations are tolerated comparatively well, and their effect lasts for many years.

12

Hereditary Degenerative Disorders

1. Introduction

In this chapter we present data on several relatively rare disorders of the CNS that are characterized by a progressive course and that lead to severe disability. Practically all of these disorders are hereditary, which makes them difficult to treat conservatively. Surgical treatment of certain manifestations of these disorders by the stereotactic method in many cases gives good and lasting results.

2. Hepatocerebral Dystrophy (Wilson's Disease)

2.1. Background

In 1912, Wilson described the clinical picture of a hitherto unknown disorder of the CNS, which he called "progressing lenticular degeneration." He described in detail the pathological anatomy of the new disease, which was characterized by lesions in the basal ganglia, principally NL, and stressed that it was always associated with liver cirrhosis. Hall (1921) proved that Wilson's lenticular degeneration and the pseudosclerosis described by Westphal (1883) and Strümpel (1898) were different forms of the same disease, which he named hepatolenticular degeneration.

Numerous investigations have been devoted to the etiology, pathogenesis, clinical picture, and course of this disorder. A significant contribution was made by Konovalov in 1960, who gave the disorder the name hepathocerebral dystrophy (HCD). He showed that the morphological changes in the brain were not limited to NL but at times were even more profound in other cerebral structures.

Based on contemporary data, HCD is a hereditary autosomal recessive disease caused by mutations of the gene responsible for the synthesis of ceruloplasmin, which is a complex glycoprotein of one of the blood plasma globulins responsible for copper metabolism. As a result of disturbances of ceruloplasmin synthesis in the liver, copper accumulates in the brain, mainly in the subcortical ganglia, and in other organs. This leads to chronic intoxication and the development of the clinical picture of this disorder.

In HCD the copper content of the liver is elevated, and there is an increased secretion of copper in the urine. The blood plasma has a significant reduction in the level of ceruloplasmin. As a rule, the clinical picture is that of chronic hepatitis, and morphologically annular cirrhosis of the liver is found.

The pathology of HCD is manifested by diffuse cytotoxic and angiotoxic changes in subcortical structures with degeneration of neurons, formation of Alzheimer glia of the second type, and pronounced changes in the walls of small vessels. Numerous investigators have described microscopic copper granules in NC, Put, GP, and commonly in the small vessels and capillaries. Computed tomographic scanning in HCD reveals a reduced density in the basal ganglia as well as atrophy of the cortex and cerebellum (Harik and Post, 1981; Kendall *et al.*, 1981). Along with the cerebral changes, there is progressive atrophic liver cirrhosis.

Konovalov (1960) described five forms of this disease. One of the most frequent and in all likelihood the gravest forms is the tremor and tremor–rigidity type. In this form the predominant feature is severe hyperkinesia manifested by pronounced progressing tremor of large amplitude involving the extremities, torso, and head. These peculiar movements of the extremities are rhythmic and of great amplitude (up to

50–60 cm), a feature that distinguishes them from hyperkinesias resulting from other extrapyramidal lesions. These movements, usually of the arms, resemble the flapping wings of a bird.

At rest, the hyperkinesias significantly diminish and sometimes even disappear completely. However, if any attempt at purposeful movement is made, for example, the finger–nose test or simply extension of the arms, a violent tremor occurs, which makes it impossible for the person independently to eat, dress, write, or carry out the other activities required in daily life. In this form of the disorder the patient, completely unable to care for him or herself, becomes a helpless, bed-ridden invalid.

The diagnosis in this phase of the disorder presents no special difficulties in view of the presence of such pathognomonic signs as a Kayser–Fleischer ring and typical changes in the content of copper and ceruloplasmin in the blood and urine. Hepatocerebral dystrophy is also characterized by gradual mental deterioration. The disease, as a rule, progresses steadily and within several years inevitably leads to death.

A number of new treatments have been proposed for this grave illness. Doubtless, the use of penicillamine, dimercaprol, and other drugs that restore the metabolism of copper to normal are quite effective and improve the neurological state. However, in some patients these drugs cause severe allergic reactions or do not produce the desired effect (Bondarchuk *et al.*, 1975).

2.2. Surgical Treatment

Since stereotactic operations on the basal ganglia have shown the possibility of totally eliminating all types of hyperkinesia and muscular rigidity, it was logical to attempt to use this method for the treatment of HCD. Stereotactic operations for this disorder are only palliative, for there is no reason to believe that these operations can influence the progressive course or the grave prognosis of this disease. The purpose of these operations is only to eliminate involuntary movements.

The literature offers little data on the results of operations for this disorder. Mundinger and Riechert (1963) reported that three patients with HCD were improved by surgery. Cooper (1965c) mentions a female patient with a grave dystonic form of HCD in whom unilateral thalamotomy produced a lasting bilateral improvement.

In 1968 we published the results of 12 stereotactic operations performed on eight patients with HCD (Kandel and Vojtina, 1968). Four patients had unilateral and four bilateral cryodestruction of VL. Six of the eight patients experienced relief of hyperkinesias.

Shefer *et al.* (1971) operated on 19 patients for HCD; six had bilateral operations involving destruction of VL and the field of Forel. Hyperkinesias and extrapyramidal rigidity were relieved or reduced in 12 patients. Bondarchuk *et al.* (1975) reported good results in two patients with HCD who underwent a combined treatment of prolonged (several months) stimulation of mesencephalic structures with implanted electrodes and penicillamine in small doses.

Our experience with stereotactic treatment of HCD consists of 22 operations on 18 patients—12 male and six female aged 22 to 35 years (average age 29 years). The disorder had been present from 4 to 12 years (average 7 years). All of these patients had the hyperkinetic form of the disorder. Hyperkinesias were of various degrees, but most were severe and all bilateral. Various biochemical investigations (copper level of blood and urine, protein fractions of blood, ceruloplasmin levels, fibrinogen, blood bilirubin, urinary amino acids, thymol determinations, and other tests revealed in all operated patients a significant disturbance of copper and protein metabolism as well as liver dysfunction. Eight of the patients had been jaundiced for some time.

In all of the patients the disease began with progressive tremor in the arms, most frequently of the right arm; in one patient the first symptom was head tremor. All had pronounced hyperkinesias of the arms and, to a considerably milder degree, the legs. Extrapyramidal rigidity was mild or absent. Because of the hyperkinesia, the patients were practically incapacitated, unable to eat, dress, or wash themselves. The majority had hyperkinesias of the torso and head, gait disorders, etc. In approximately half of the cases, there was mild intellectual deterioration typical of HCD, and in three of the cases, there was a pronounced dementia.

Fourteen patients underwent one operation on the side opposite the most affected extremities. Four patients were operated on bilaterally at intervals of 2 to 6 years. In most of the operations the target was VL (16 operations), and in six operations, VL and Subth. The operations were usually performed under local anesthesia. The hyperkinesias and rigidity in all cases were relieved on the operating table. There was one postoperative death, and in two cases there were lasting complications (moderate hemiparesis and pseudobulbar palsy).

Many years of observations have confirmed the

lasting significant therapeutic effect; in only a few cases was there a recrudesence. However, if the operation was unilateral, the hyperkinesias could progress on the homolateral side.

Based on a long-term follow-up, a good result was obtained in eight cases, a significant improvement in three, slight improvement in three, no change in two, and worsening in one. Several years after the operation, three patients died as a result of progression of the primary disease.

There is no evidence that a stereotactic operation can terminate or even retard the course of HCD. Our data fully confirm the observations of Shefer *et al.* (1971) that after these operations the increased copper content of the blood, typical of HCD patients, does not change. Neither is there any substantial change in the abnormal protein metabolism. A stereotactic operation will neither improve nor aggravate the basic metabolic disorder of HCD. Nevertheless, our limited experience has indicated a stabilization of the process so that patients lived for many years after the operation, and this suggests that in some cases VL thalamotomy may have a beneficial effect on the course of the disease. The following case illustrates this point.

A 26-year-old female patient was admitted to the Institute of Neurosurgery in 1963. Four years earlier she had developed a slight tremor in her right arm, and this tremor progressively worsened and involved all extremities, the arms more than the legs. Just as soon as the patient extended her arms, powerful involuntary flapping movements began, reaching an amplitude of approximately 50–60 cm (Fig. 153A). Her speech became impaired, and her gait disordered. The patient was unable to make purposeful movements and could not eat, dress herself, or write. For almost 4 years, she had been a grave invalid.

After a left cryothalamotomy, the involuntary movements in the right hand disappeared completely (Fig. 153B). An intraoperative EMG (Fig. 154) demonstrated that the right arm hyperkinesia stopped immediately after cryodestruction of VL. Ten months later, a similar operation was performed on the right side, after which the severe hyperkinesia in the left extremities (Fig. 153C) was immediately and totally relieved.

For more than 12 years after the two operations, their striking effect lasted. During this time the patient was able to look after herself without any help. She could walk, write, ski, and graduated from a college with good grades. In the course of the subsequent 8 years, there was some recurrence; slight hyperkinesias

FIGURE 153. A 29-year-old patient with hepatocerebral dystrophy. (A) Severe "beating wings" tremor of both hands. (B) Complete disappearance of tremor in right hand after stereotactic operation on the left basal ganglia (VL thalamotomy); tremor of left hand became less intense. (C) Disappearance of tremor in both arms after a similar second operation on the right side.

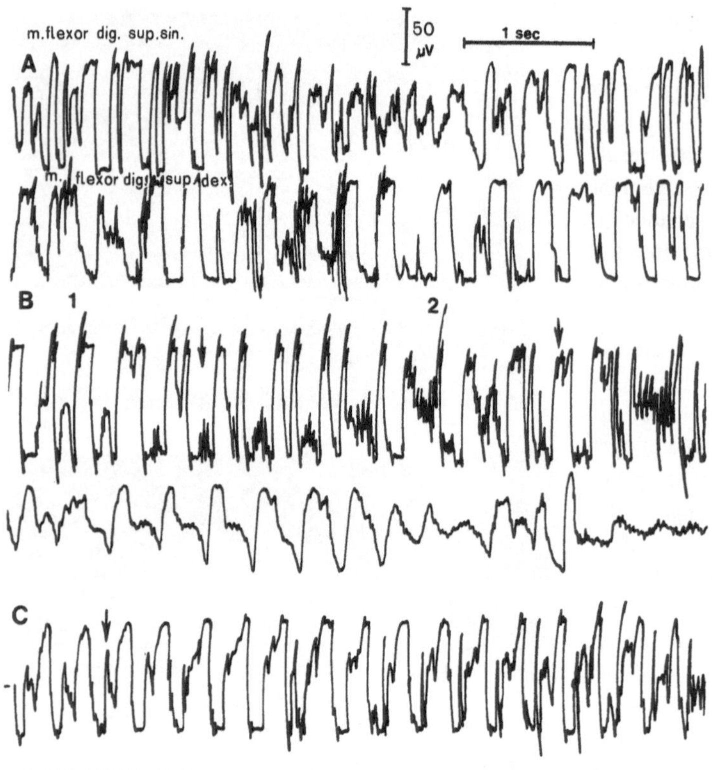

FIGURE 154. Three consecutive segments (A–C) of EMG records from both forearms of a patient with hepatocerebral dystrophy during stereotactic thalamotomy under local anesthesia. (A) Severe postural hyperkinesias in both arms at the beginning of the operation on the left basal ganglia (above, left arm; below, right arm). (B) After introduction of the cryoprobe in the left VL (1), the hyperkinesia in the right arm decreased substantially, and after the beginning of freezing (arrow), it practically disappeared. (C) After freezing of VL, the hyperkinesia in the right arm disappeared completely; in the left arm it did not change.

appeared in both arms, and her speech became impaired. However, even after 20 years her condition is not nearly as bad as it had been prior to the operations. She is able to look after herself and travel about and works as an engineer in a factory.

There can be no doubt that the effectiveness of penicillamine has lessened the need for surgical treatment of HCD. However, when this drug is ineffective or has side effects after long-term use, stereotactic operations on Th nuclei are certainly justified and timely. As a rule, they completely eliminate the severe hyperkinesias and allow the patient to resume normal activities. In spite of their palliative nature, stereotactic operations represent an effective treatment of HCD.

3. Huntington's Chorea

3.1. Background

This severe progressive hereditary disease of the CNS was first described by Huntington in 1872. Huntington described several patients in whom, at a mature age, choreatic twitching developed, followed by a progressive increase in the jerkings and later pronounced dementia. The hereditary nature of the disease distinguishes it from infantile infectious chorea as well as from a group of other conditions with choreatic movements, in particular, posthemiplegic chorea.

Huntington's chorea (HC) is a rare disease of the nervous system; its incidence varies, according to different authors, from 1.3 to 10 per 100,000 population.

Huntington's chorea has a number of typical clinical characteristics. Usually it begins in individuals near middle age, approximately 30 to 45 years, and its hereditary nature is evident from the frequent occurrence of similar cases in the family of the patients especially in the previous generations. Huntington's chorea is characterized by high penetration of a dominant gene, which means practically inevitable transmission of the disease to subsequent generations. However, this does not mean that HC is always hereditary, for the literature includes many "spontaneous" cases of the disease without a familial history.

From the morphological point of view, HC is characterized by degenerative, atrophic changes in the Str, with absence of small ganglion cells in NC and Put, which leads to their pronounced atrophy (Blinderman *et al.*, 1964). As the pathological process progresses, there is a symmetrical decrease in the size of both Str as seen on pneumoencephalograms. Degenerative neuronal changes in the cortex, GP, NS, and Purkinje cells have been frequently described. Cortical atrophy, mainly in basal frontal and temporal cortex, is present in more than one-half of the cases (Sax *et al.*, 1983).

Ventriculography before a stereotactic operation reveals pronounced hydrocephalus in the majority of patients, which is an indication of diffuse brain atrophy. For example, of 17 patients that we operated on, first-degree hydrocephalus was found in one, second-degree in ten, and third-degree in six. This symmetrical hydrocephalus was associated with significant atrophy of the NC.

The severity of hydrocephalus clearly correlated with the gravity of disease but not with its duration. Sax *et al.* (1983) also noted the high correlation among severity of choreic movements, mental disorders, and degree of NC atrophy. In several cases in which venticulography was repeated after an interval of several years, clear progression of hydrocephalus was noted. These data indicate that atrophy of the brain, particularly of the basal ganglia, is an important indication of the pathogenesis of the disorder.

Biochemical investigations, done mainly in the past decade, have revealed that the principal factor in the pathogenesis of HC is a still obscure metabolic disorder of the cerebral transmitters: dopamine, GABA, acetylcholine, glutamic acid, norepinephrine, serotonin, and certain neuropeptides (substance-P, angiotensin, somatostatin, enkephalins). Unlike parkinsonism, in HC the dopaminergic nigrostriatal pathway does not degenerate but is hyperactive and stimulates the dopamine receptors of Str neurons.

It has been established that in the basal ganglia in HC there is a significant reduction in GABA and the enzymes synthesizing GABA and acetylcholine. The GABAergic strionigral pathway undergoes pronounced degeneration. It is assumed that the GABA deficit in basal ganglia, mainly in Str, plays a leading role in the pathogenesis of involuntary movements (Bird and Iversen, 1974).

Clinically, HC is characterized by rapid choreiform involuntary movements in many muscle groups of the face, torso, neck, and limbs. These movements are practically continuous except during sleep, and the patient is in a constant state of motor tension, which intensifies during walking. Less frequently, HC assumes a rigid or akinetic–rigid form. Quite typical is the gradually progressing dementia with various mental aberrations.

Huntington's chorea usually begins in young or middle age, gradually but steadily progressing to physical and mental incompetence and then to an inevitable fatal outcome. It is an incurable disease—there is no effective means of treatment.

3.2. Surgical Treatment

Recently, attempts have been made to treat HC with stereotactic operations on the basal ganglia. These procedures, naturally, are only palliative, since they cannot arrest the degenerative process in the subcortical structures. However, as experience has shown, these operations eliminate or significantly decrease severe hyperkinesias. The surgical experience is very limited, since this is a rare disease with a poor prognosis. Reports of treatment of not more than eight patients have given encouraging but not stable results (Caracalos, 1972; Andrew, 1974). However, Gildenberg (1985) has expressed the opinion that an operation may aggravate the mental impairment.

In the course of more than 15 years, we have performed 21 stereotactic operations on 17 patients with HC, nine males and eight females aged 31 to 64 years (the majority were 35–45 years old). The disorder had been present from 6 to 15 years (mean 8 years). All the patients were in an advanced stage of the disease. Its main manifestations were the hyperkinetic syndrome and dementia. Severe hyperkinesias were present in 12 patients and less pronounced in five. Six patients were severely demented, six moderately, and five mildly. Unilateral stereotactic destruction of VL, or VL in combination with Subth, was performed in 13 patients, and bilateral destruction was done in four. There was no postoperative mortality. In two patients transitory hemiparesis developed. The result of the surgical treatment was graded according to the degree to which the choreic hyperkinesias decreased and personal care improved. A good result—practically total relief of hyperkinesias and improved gait—was obtained in seven cases. After bilateral operations the effect was noted on both sides. A significant improvement was observed in three patients, and slight improvement in two. Three patients were unchanged after the operation or showed subsequent progression of the disease; results are not known in two cases. Our experience has shown that in severe dementia the operation is not warranted.

Thalamotomy is considered to be the only treatment providing any relief to the patient with this severe disease. However, the mental impairment and the progression of the disease quite often nullifies the results of stereotactic surgery.

4. Essential Tremor

4.1. Background

This term is used for a chronic, slowly progressing CNS disease caused by pathological changes in the small neurons of Str. Essential tremor (ET), as a rule, is a monosymptomatic disease manifested only by tremor. Muscular rigidity, akinesia, hypomimia, static disorders, impaired gait, and other symptoms of parkinsonism are absent in ET. Only in the very early stages can there be difficulties in differential diagnosis. According to recent epidemiologic investigations, ET has a significantly higher incidence than parkinsonism. For example, Dupont (1980) sets the incidence of ET at 3.5–5.5% of the entire population over the age of 40. According to his data, among people 60–80 years old, the incidence of ET is 10–16 times higher than the incidence of parkinsonism. The incidence of ET in Finnish people is even higher, 9–10%, and in the age group 70–79 years it is 12.6% (Rautakorpi *et al.*, 1982).

This disease occurs more frequently in men than in women, without apparent cause, and progresses slowly. Unlike parkinsonism and senile tremor, ET usually begins at the age of 30–40 years, although it can occur at an early age—according to Riechert and Richter (1972), 40% of the patients show signs before age 20. There is no reason to consider ET a result or residual stage of some other cerebral disease.

There can be no doubt that heredity is an important factor in the etiology of ET (Davidenkov, 1932; Golubeva, 1962). In approximately 80% of patients a similar disorder has been found in other family members, and the distribution within families shows an autosomal dominant heredity pattern (Ivanova-Smolenskaya, 1978). In one study of 52 ET patients, a similar tremor was observed in 44 immediate relatives (parents or siblings) (Van Manen, 1974). Thus, ET appears to be a dominant hereditary disease with limited penetrance. Several authors prefer to call it familial tremor. In spite of the undoubted hereditary nature of the disease, there are also sporadic cases. According to Hassler (1956), ET is distinguished from parkinsonism by its pathological substrate, namely, *état precriblé* Str.

As a rule, ET begins simultaneously in both arms, thus differing from parkinsonian tremor, which generally appears first in one limb. At the onset of the disease, tremor usually occurs during active movements, thus resembling intention tremor (Brain and Walton, 1969; Cooper, 1969b). The main feature distinguishing ET from parkinsonian tremor is that the former is an action tremor or postural tremor, intensifying during purposeful movements. The frequency of the ET varies within a wide range from 6 to 12 Hz but most frequently is 8–9 Hz; i.e., the frequency of ET is higher than that in parkinsonism. With age, the frequency of ET diminishes (Marshall, 1962).

Unlike parkinsonism, ET, as a rule, is observed only in the arms. Frequently (in approximately one-third of the cases) there is pronounced tremor of the head, tongue, or lips, which is rare in parkinsonism. During emotional stress the tremor intensifies. Muscular tonus in the limbs is normal or reduced. The disease, as a rule, progresses slowly, sometimes with long remissions. In pronounced forms the patient loses his or her working capacity and experiences considerable difficulties in self-care.

Recently, good results in the management of this disease with β-adrenergic blocking drugs (propranolol, atenolol, obsidan, anaprilyn) were described in many reports (Teräväinen *et al.*, 1976; Ivanova-Smolenskaya, 1978). It is known that alcohol, though not a β-blocker, markedly diminishes ET.

4.2. Surgical Treatment

There are several reports in the literature of surgical treatment of ET by stereotactic operations with complete relief of tremor (Cooper, 1965b, 1969b; Spiegel and Wycis, 1962a,b; Obrador and Dierssen, 1966). Riechert (1980) reported a stable and good effect in 48 of 50 patients who had undergone stereotactic surgery.

Cooper (1965b) operated on 110 patients with intention and hereditary tremor of various etiologies. In 90% of these patients, tremor disappeared after stereotactic surgery. Obrador and Dierssen (1966) reported on 15 patients with intention tremor who underwent stereotactic destruction of various subcortical structures and concluded that the most effective results were obtained after destruction of VL and Subth. They considered the disappearance of ET as well as the tremor in SD after surgery to have resulted from interruption of pallidothalamic pathways.

In 1986, Mundinger *et al.* reported on their extensive experience in stereotactic treatment of ET. The authors operated on 105 patients, 18 of them bilaterally. There was no operative mortality or morbidity. Long-term follow-up has shown that 38% of the cases had complete relief, 39% showed clear improvement, 10% showed slight improvement, and 19% had not changed or had deteriorated.

We operated on eight patients with longstanding ET and performed unilateral or bilateral stereotactic cryodestruction of VL. In all these cases there was a good and lasting result with total or almost total disappearance of tremor. There were no complications or mortality after the operations. The following is an example.

For a period of 20 years before being admitted to the clinic, a 67-year-old patient had progressive tremor of the arms. There was no family history of similar afflictions. On admission, there was a pronounced medium-amplitude tremor in both arms, greater in the right, and a tremor of the head but none in the legs. Muscle tone was normal. The shaking interfered with eating and writing. The disease was gradually progressive. Prolonged drug treatment, including L-dopa, produced no effect. The blood pressure was 200/100 mm Hg.

A stereotactic left cryothalamotomy was performed under local anesthesia with ganglionic blockers used during the operation to lower the blood-pressure to 110/60. Immediately after the cannula was introduced into the VOp, the tremor totally ceased in the right arm. Three years after the operation, the patient still has no tremor in the right arm and writes and eats with the right hand normally. Tremor in the left arm has somewhat diminished.

The initial treatment of ET should be with β-adrenergic blockers; however, if there is no relief, a stereotactic operation on the basal ganglia, as a rule, gives good results with lasting elimination of tremor.

5. Hunt's Cerebellar Dyssynergia

5.1. Background

In 1914 Hunt described the disease that today bears his name and which he called "progressing cerebellar dyssynergia." The disease was characterized by three main symptoms, intention tremor, progressive ataxia, and scanning speech, which are indicative of a cerebellar lesion.

In his second paper (1921), Hunt described another six cases of this disease (among the patients were two brothers) and modified the name of the syndrome to "dyssynergia cerebellaris myoclonica." Hunt also added to its clinical picture, including myoclonus, epileptic fits, and dysarthria. In one autopsy case, Hunt reported primary atrophy of ND and degeneration of the superior cerebellar peduncle and dentatorubral pathways.

The disease begins at an early age with a very marked rhythmic high-amplitude tremor, which occurs during performance of any purposeful movement and disappears at rest. In addition, there is a pronounced ataxia. As the disease gradually develops, the patient becomes unable to walk and eventually totally disabled.

Subsequent writers (Critchley, 1962) elaborated on the Ramsay Hunt syndrome and described several similar cases. These had not only cerebellar but also spinal symptoms as well as epileptic fits. It is assumed that the morphological substrate of the syndrome is a lesion of the efferent pathways of ND; however, in one well-studied case of a typical Hunt syndrome, there were no morphological changes in cerebral or cerebellar structures (Choteau *et al.*, 1981).

The question of whether this syndrome is a distinct disorder remains open for discussion. Clinically, it is difficult to distinguish from myoclonic epilepsy and Freidreich's ataxia. Moreover, dentatorubral degeneration, typical of the Hunt syndrome, occurs in Creutzfeldt–Jacob disease, hypoxic encephalopathy, and others. Some authors consider it a hereditary disease transmitted by both dominant and recessive factors.

5.2. Surgical Treatment

There is no conservative treatment for this syndrome. In recent years a few attempts have been made to treat this condition by stereotactic destruction of the thalamic nuclei as well as by dentatotomy (Šramka *et al.*, 1972).

We have operated on three patients with the Ramsay Hunt syndrome. One such case is interesting for three reasons. First, it is unique in that the syndrome developed after lethargic encephalitis; second, after bilateral thalamotomy there was not only arrest of the gross intention tremor in both arms but also a considerable reduction in static ataxia, which has been assumed not to benefit by stereotactic treatment. And finally, the therapeutic effect obtained by the two operations was maintained over a period of more than 20 years.

A 31-year-old patient had had a severe head injury with loss of consciousness for 2½ hr at the age of 8 years. Subsequent development was normal. At the age of 17 he developed headaches and fever (up to 39°C). A tremor appeared in the right arm, and any attempt to sit threw him backwards. After several days the patient fell into a sound sleep for 7 days. He slowly recovered from the lethargic state, but for 2 months he was unable to speak, although he understood everything that was said to him. He had lost his equilibrium and, as a result, was unable to walk. Over the course of 6 months his speech gradually returned but was staccato and explosive.

Two years after the encephalitis, the tremor of the head became worse, and then the arms and legs became more involved. After progressing for 1 or 2 years, the tremor gradually stabilized. Repeated treatment in hospitals and clinics proved ineffective. Consequently, after 14 years the patient was a severe bedridden invalid. Because of gross ataxia, he was unable to stand or walk, and because of the acute tremor he could not take care of himself.

He was admitted to the Institute of Neurosurgery in April, 1961. His intellect was fully preserved, and he had adjusted well to his disease. He had no sensory disturbances, and his arm and leg strength was normal. His muscular tone was reduced, especially in the legs. He had a pronounced high-amplitude tremor in the arms, legs, head, and torso. When he was lying down, the tremor was not great, but when he was sitting or trying to make any purposeful movement, the tremor, especially in the arms, increased in amplitude, resembling the flapping of wings.

The second striking feature was severe ataxia of the torso and extremities, primarily in the legs, so that the patient was totally incapable of standing or walking. Unsupported, he immediately fell backwards. This clinical picture with the presence of profound intention tremor and severe ataxia was interpreted as Ramsay Hunt's cerebellar dyssynergia resulting from lethargic encephalitis.

At the first stereotactic operation in June 1961 0.3 ml of alcohol with 0.3 ml of iophendylate was introduced into the left VL. After the operation there was a significant decrease in the intention tremor of the right extremities and a marked reduction in ataxia. Two months later a similar operation was performed on the right side.

After the second operation, there was a significant decrease in tremor in the extremities, which became irregular. An improvement in speech and a marked reduction of ataxia in the arms and torso were noted. Gradually, the patient was able to sit up and later to walk independently with the help of two canes.

At a follow-up examination in 1982, it was apparent that the beneficial effect of the two stereotactic thalamotomies had lasted for more than 20 years. Intention tremor in the arms was insignificant, and the ataxia in the torso was not pronounced. The patient was able to walk with the help of two canes and could look after himself, sit up, dress, wash, and care for himself. However, the patient was unable to write legibly.

In spite of the limited surgical experience available, there is reason to believe that stereotactic operations on the Th nuclei are the only effective method of treating Hunt's cerebellar dyssynergia at present.

6. Myoclonus and Myoclonus Epilepsy

6.1. Background

Myoclonus (myoclonus epilepsy, Unverricht–Lundborg disease) is the name given to a chronic, progressive, and often hereditary disease of the nervous system producing rapid involuntary and arrhythmic contractions of separate muscular groups or individual muscles at a frequency of up to 40–50/min. These rapid muscular twitchings and jerks are distinguished from other hyperkinesias by the fact that they do not cause significant changes in the posture or position of the limbs. Myoclonus is noted equally at rest and during movements, thus differing from parkinsonian or intention tremor. If the myoclonic jerks are not associated with epileptic seizures, the disease is classified as myoclonus; however, if convulsions do occur, then it is called myoclonus epilepsy.

6.2. Myoclonus

According to the classification of Halliday (1967), myoclonus may be divided into three types: pyramidal, extrapyramidal, and spinal (or segmental). Whether myoclonus without epilepsy is a separate nosological form still remains an unsettled question. There are often difficulties in distinguishing myoclonus from chorea. When the muscular contractions in myoclonus are rhythmic, they acquire the features of tremor. Myoclonus combined with dementia is a characteristic manifestation of Creutzfeldt–Jacob disease as well as of Alzheimer's disease. Myoclonic jerks can also occur in other extrapyramidal lesions, for example, in HCD, ST, and certain hereditary diseases, as well as in encephalitis, posthypoxic encephalopathy, and vascular lesions of the brain.

Extrapyramidal myoclonus occurs only in the conscious state and is not associated with epileptic attacks (Dieckmann and Hassler, 1972). The EEG usually does not reveal epileptogenic activity. Voluntary movements as well as emotional stress intensify myoclonic jerks.

Hassler (1968) believes that the morphological substrate of extrapyramidal myoclonus is a lesion within the so-called Guillain–Mollaret triangle, which includes NR, the inferior olive, and ND as well as their main pathways. Uncoordinated or facilitated impulses stemming from that triangle in pathological conditions cause asynchronous activity of motoneurons (Dieckmann and Hassler, 1972). Besides the descending influences in the pathogenesis of myoclonus, a role is played by the ascending pathways to the motor cortex, in particular the dentatothalamic pathways and specific thalamocortical projections to 4γ area. The pathological mechanism responsible for the generation of myoclonic discharges may be connected with the accumulation of excess serotonin in FR, for Hassler (1969) produced experimental myoclonus by increasing the concentration of serotonin in subcortical structures.

6.3. Myoclonus Epilepsy

Unlike myoclonus, the nosological independence of myoclonus epilepsy does not raise any doubts. The disease, as a rule, occurs at a young age (up to 20 years), often as a hereditary–familial trait with autosomal recessive transmission. For example, 25 members of one family over eight generations suffered from essential myoclonus (Przuntek and Muhr, 1983).

Rare morphological investigations in myoclonus epilepsy reveal histological changes, primarily in Cer (loss of Purkinje cells, degenerative changes in ND, etc.). These findings suggest that this disease is based on genetically determined lesions of Cer.

The disease has a characteristic morphological substrate—special intracellular inclusions, mainly in Cer neurons, called myoclonic or Lafora bodies. Although these bodies are not found in all cases of myoclonus epilepsy, they are considered pathognomic for the disease. A study of Lafora bodies has demonstrated that they are related to glycoprotein (Austin *et al.*, 1968; Danner *et al.*, 1982). Whether these bodies are the cause of myoclonus epilepsy or their appearance results from metabolic disorders of nerve cells is a question that still remains open.

The pathogenesis, diagnosis, and clinical picture of myoclonus epilepsy have been described in a number of works (Gastaut, 1968). A detailed electrophysiological investigation was performed by Cabrini *et al.* (1972). Besides the constant myoclonic jerks, these patients have generalized epileptic fits and progressive mental deterioration. A conventional EEG reveals typical epileptic manifestations (spike–wave complexes). No correlation was found between the epileptic EEG activity and the myoclonic jerks. Recording with electrodes implanted in the sensory–motor cortex shows the presence of three-phase peaks, which are not seen on a conventional scalp EEG. A simultaneous recording of EMG and ECoG demonstrated that both spontaneous and myoclonic jerks were always preceded by cortical potentials. The latent period of the appearance of muscular contractions after cortical potentials was 8–9 msec for neck muscles and 30–35 msec for muscles of the lower extremities.

Besides widespread muscular twitchings and epileptic fits, this disease is characterized by comparatively rapid mental degradation.

6.4. Surgical Treatment

Conservative treatment of myoclonus and myoclonus epilepsy is seldom effective. This poor prognosis has been the stimulus for attempts to treat the condition by stereotactic surgery.

The first such treatment of myoclonus was described by Wycis and Spiegel (1969). A good effect was noted in three cases of this disease after campotomy and destruction of CF. Dieckmann and Hassler (1972) operated on 13 myoclonic patients, destroying the basal part of VOp. A follow-up of these patients after 3 months to 4.2 years showed clear good results. Both at rest and during active movements, myoclonus disappeared or was significantly diminished. In two patients there was also a decrease in homolateral myoclonus, which made a bilateral operation unnecessary. Although the results were good, in some cases they were marred by postoperative complications (ataxia and dyssynergia in the contralateral extremities and speech impairments). These complications disappeared relatively rapidly and remained stable in only two patients.

Good results of thalamotomy in myoclonus epilepsy were reported by Laitinen (1967). In four of six patients a notable improvement was observed after stereotactic destruction of CM (Ramamurthi and Davidson, 1975). In another case of this disease, Ramamurthi noted a good effect after bilateral combined destruction of CM and CF. In three patients who were operated on, quite satisfactory and lasting results were obtained. The myoclonus was substantially decreased, and epileptic fits became much less frequent, so that it was possible to reduce the dosage of anticonvulsants.

Other authors (Narabayashi, 1982a), however, believe that stereotactic operations for myoclonus epilepsy are not very effective. In cases with severe dementia as well as in rare palatal myoclonus, stereotactic operations are not indicated.

There is reason to believe that stereotactic operations on the basal ganglia, principally destruction of VL, Subth, and CM, are at present the only effective means of treating myoclonus and myoclonus epilepsy.

13

Pain

Alleviation of pain is divine labor.
Hippocrates

The neurosurgeon's battle against intractable pain seems more promising now than hitherto.
Lars Leksell

1. General Remarks

It would be rational to begin with the question, "What is pain?" Numerous definitions of pain have been suggested, a fact that in itself is an indication of the difficulties involved in formulating such a definition. The most suitable definition, proposed by Merskey (1978), was subsequently incorporated in the *Glossary* of the International Association for the Study of Pain: "Pain is an unpleasant and emotional experience associated with actual or potential tissue damage or described in terms of such damage." It is presumed that pain is an adaptive reaction of the organism acquired in the process of evolution and, consequently, a means of protection from harmful stimuli.

As Sano *et al.* (1975b) noted, the word "pain" has two meanings—one is the pain sensation, and the other the suffering. Pain is the most frequent cause of suffering, disability, and invalidism, afflicting millions of people throughout the world. Treatment of pain syndromes is an important and difficult task facing medicine in general and neurosurgery in particular. The physician quite often encounters a situation in which pain stops being a natural biological signal of alarm ("guardian of life and health"—Foerster), loses its protective sense, and becomes the main manifestation of a disease or even a separate disease *ipso facto*. In many instances, it is pain, and not the pathological process causing it, that is responsible for the grave state of the patient. Pain that is the source of much continuous suffering, disintegration of the personality, and chronic disability represents a very timely and to a great extent still unresolved problem.

The considerable broadening and increasing scientific investigations on problems of pain during the past two decades have led to the establishment of the International Association for the Study of Pain in 1973. The Association organized four world congresses on pain and publishes the journal *Pain*.

It is advisable to discuss briefly the classification of pain syndromes, which, undoubtedly, is important from a clinical point of view. Several such classifications have been proposed based on various factors. It is natural that all pain syndromes are divided into acute and chronic. The majority of authors separate as a distinct group somatic or nociceptive pain, occurring as the result of injuries, compression, inflammations, etc. of parts of the body as well as internal organs. In such types of pain there is usually no objective impairment of sensation, although, as Tsubokawa *et al.* (1982a,b) state, these pains may be excruciating.

Many investigators differentiate malignant or cancer pain, which has distinct characteristics, from other pain. And finally, practically all authors place chronic pain syndromes in a special group with various names such as central pain, denervation pain, dysesthetic pain (Tasker, 1982), deafferentation pain (Takeda, 1984; Hosobuchi, 1986), primary pain (Nashold and Friedman, 1972a,b), and neuropathic pain (Turnbull *et al.*, 1980). This category includes pain resulting from local lesions of the central or peripheral nervous systems at any level, associated with demonstrable impairment of various types of sensation.

Nashold *et al.* (1983) singled out three clinical

types of deafferentation pain: (1) dysesthesia following lesions peripheral to and not involving the spinal cord such as incomplete peripheral nerve injuries or viral lesions of the dorsal root ganglion; (2) dysesthesia with lesions of peripheral nerve and dorsal root plus spinal cord, such as those found in herpes zoster; and (3) dysesthetic pain with lesions of the spinal axis, brainstem, or thalamus. Examples of these types of pain include complete brachial plexus avulsion (Section 5.3.7), dysesthesia following direct spinal cord injury (Chapter 19), and a variety of insults to the brainstem and Th, the most prominent being the thalamic pain syndrome (Section 6.1). Central pain and its treatment are described in Section 6 of this chapter. A practical classification of pain has been proposed by Johansson *et al.* (1980)—neurogenic, somatogenic, and psychogenic pain. In practice, however, there are frequent cases when a patient displays the features of more than one of these types of pains, so that it becomes difficult to classify it in one of the above categories.

Since pain may occur as a result of a disorder of practically any organ, it is naturally not the purpose of this book to describe the multitude of pain syndromes. In this section we discuss those syndromes that have been included in the category of so-called chronic pain. Attention is focused mainly on neurosurgical management of pain and techniques that have been developed and improved in recent years.

Before describing the surgical management of pain, certain new theoretical concepts of the pathogenesis of pain are outlined. These concepts are based on the pain-transmitting structures of the CNS described in detail in Chapter 1. There is a tendency to divide the pathogenic factors involved in pain reactions into two groups: physiological (including anatomic, neurophysiological, and biochemical aspects) and psychological (including emotional, cultural, social factors, etc.).

The most fruitful methodological approach to understanding such a complex phenomenon as pain requires modern system analysis. In recent years attempts have been made to create a general theory of the pathogenesis of pain.

2. Gate Control Theory

The most recent theoretical concept of the morphological and neurophysiological basis of pain is the so-called gate control theory of pain proposed by psychologist Melzack and neuromorphologist and physiologist Wall in 1965. The gate theory soon became very popular, but at the same time, it has undergone significant transformations since it was first proposed by the authors. In 1978 Wall proposed a cascade gate pattern significantly more complicated than the one proposed by Melzack and Wall in 1965 (Fig. 155).

The gate theory is related, first of all, to the complex morphological organization of the posterior horns of the spinal cord, which can be divided into six Rexed laminae, each of which contains different sensory and, in particular, nociceptive terminals. It is assumed that gate control is localized in laminae II and III of the substantia gelantinosa, which receive all known types of afferent pathways from the skin, internal organs, and high-threshold muscular afferent pathways (Wall, 1978).

The basic principle of the gate theory is that at the level of the sensory input in the spinal cord there is a special control mechanism that regulates the transmission of incoming pain impulses to higher CNS structures; in other words, the neurons of the posterior horns allow certain sensory impulses to pass but, at the same time, act as a gating mechanism to others.

The morphological substrate of the gate is the aggregate of neurons of the substantia gelatinosa (Rexed's laminae II and III of the spinal cord), which lies beneath the apex of the dorsal gray matter and extends throughout the length of the spinal cord. All the peripheral afferent impulses coming into the spinal cord are distributed to different systems: posterior columns, substantia gelatinosa, and trigger of transmission neurons (T-neurons). However, before the afferent im-

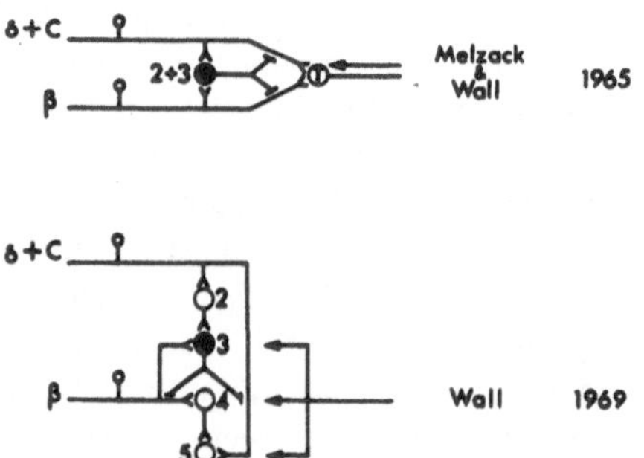

FIGURE 155. Two schemes of the gate control of pain from Melzack and Wall (1965) and Wall (1978). δ and C, small myelinated and unmyelinated afferents; β, large myelinated afferents; 2,3,4,5, cells in Rexed laminae of the same number; T, unspecified transmission cells; arrowheads, descending cerebral pathways. Black cells are inhibitory. Although the inhibition is shown as presynaptic, postsynaptic inhibition is also possible.

pulses reach the T-cells, they may be modified in other cells of substantia gelatinosa. Under normal conditions, this gate is sufficiently wide open for impulses to pass to the T-cells. If the intensity of the nociceptive impulses arriving from the periphery through small fibers in the posterior roots exceeds the threshold of the T-cells, discharges occur, which pass through the substantia gelatinosa into the ascending pathways to the subcortical structures and then to the sensory cerebral cortex. As a result of integrative activity of the hemispheres, this information is evaluated in the light of individual experience, emotional state, and other factors of higher nervous activity. Based on the feedback principle, commands from these higher centers are returned to substantia gelatinosa, which, depending on their nature, "opens" or "closes" the gate for pain afferents. It is presumed that depolarization of the neuronal elements closes the gate. If of high intensity, the impulses pass through the gate, resulting in the sensation of pain. Therefore, the substantia gelatinosa has an inhibiting action on the neurons of the posterior horns and on transmission of impulses along the posterior roots.

It has also been shown that the gating mechanism may either intensify or diminish the afferent impulses depending on the interaction of various fibers carrying sensory afferent impulses to the spinal cord. It is considered that at the level of the posterior horns there is a balance between small, nonmyelinated nociceptive C-fibers and the large, myelinated nonnociceptive A-fibers; the sensation of pain occurs when input along the small fibers predominates. It is also presumed that a function of the large fibers is inhibition of the small ones. If this function is eliminated, even a weak stimulus arriving via the small fibers is perceived as pain, which is quite typical of central pain (Section 6). According to the gate theory, increased activity of the low-threshold, large myelinated fibers inhibits the transmission of pain along the thin nonmyelinated fibers. It has been established that the perception of pain can be changed by acting on the gate with nonpain stimuli. For example, selective stimulation of the large myelinated fibers that do not conduct pain and temperature sensation can "close the gate" to impulses passing along the high-threshold, fine pain-transmitting fibers. The efficacy of therapeutic stimulation of the posterior columns of the spinal cord (Section 5.4.2) and peripheral nerves (Section 5.4.3) is a convincing argument in favor of this theory. The gate theory has served as the groundwork for the development of new methods of treating pain.

Nevertheless, it should be noted that certain authors do not consider the gate theory to be convincing (Mazars, 1976a). Indeed, on the basis of the theory, the pathogenesis of central pain cannot be satisfactorily explained.

In his book *The Puzzle of Pain,* Melzack (1973) notes that each new theory of pain during the past 100 years has stimulated further investigations. Undoubtedly, the gate theory has had a similar effect on the investigation of pain mechanisms.

These gating mechanisms at the segmental level of the spinal posterior horns play a very important role in the pathogenesis of pain syndromes. However, dysfunction of cerebral structures, in particular, the complex antinociceptive system of the brain (see below), is, in all likelihood, of predominant significance in the pathogenesis of chronic pain.

3. Antinociceptive System

It is essential to discuss briefly contemporary data concerning the inhibitory, antinociceptive, or pain-suppressing system the existence of which has long been known. However, discoveries in recent years have considerably expanded our understanding of the mechanisms of its functioning. Undoubtedly an impairment of the normal function of the antinociceptive system plays an important role in the pathogenesis of pain syndromes, especially chronic pain.

Any sensory input induces not only excitation but also inhibition of CNS structures. Moreover, any sensory information from the periphery is modulated by descending influences from cerebral structures. Historically, the antinociceptive system reverts to the central inhibition phenomenon, which was discovered by Sechenov in 1852. Physicians observed long ago that in strong emotional and stressful situations, a person may become insensitive to pain, or the threshold for its reception may be greatly increased. For example, 80% of soldiers wounded in battle do not complain of pain and do not request analgesics. The biographies of famous people relate how concentration of attention on some kind of activity deadens the feeling of pain. For example, Immanuel Kant, during attacks of gout, turned his attention to philosophical problems and could do without pain killers. Blaise Pascal overcame physical suffering by immersing himself in complicated mathematical calculations. Rachmaninoff wrote that any kind of pain disappeared when he began playing the piano. To stop pain, Stendahl listened to Hayden's *Mass*.

It has been established that a feeling of fear intensifies the reaction to pain, whereas emotions such as fury and aggressiveness sharply diminish these reactions. Such emotional stress changes the pain threshold and tolerance to pain (Waldman and Ignatov, 1976).

The anatomic substrate of this system, which has the ability to inhibit pain, was discovered quite recently after Mayer and Liebeskind (1974) established that stimulation of the periaqueductal gray matter (PAG) in animal experiments induces deep analgesia. The PAG, which is closely linked with the nucleus arcuatus Hypoth, has no direct connections with the spinal cord. In view of this, it is considered that PAG activates the nucleus raphe dorsalis magnus, which, lying along the descending pathways of the dorsolateral funiculus, inhibits the nociceptive neurons of the posterior horns (Fields *et al.*, 1977; Basbaum and Fields, 1978; Shah and Dostrovsky, 1980; Tsubokawa *et al.*, 1982a).

Stimulation of any part of that system, especially PAG and serotoninergic neurons of nuclei raphe and possibly the locus coeruleus, leads to the suppression of pain. The stimulation of PAG gives an analgesic effect not only peripherally, but also centrally. Thus, such stimulation inhibits the activity of the nociceptive neurons of Th nuclei (Emmers, 1981).

Apparently, this is not the only inhibitory system. Melzack (1975) considers that a similar mechanism is located in FR of the brainstem and that it inhibits the transmission of pain impulses at all synaptic levels of that system. It was shown in animal experiments that the stimulation of the Kölliker–Fuse nucleus in the dorsolateral pons or pontine parabrachial region results in potent inhibition of the responses of dorsal horn cells to both noxious and innoxious stimuli (Hodge *et al.*, 1986; Katayama *et al.*, 1986). There are good reasons to believe that the sensory cortex is also part of the antinociceptive system. For example, the stimulation of the secondary sensory cortical area activates the system and inhibits the nociceptive input into Th (Reshetnjak and Meizerov, 1981). The pathways inhibiting pain pass from the cortex to the posterior horns of the spinal cord via Th and FR (Adams *et al.*, 1974; Namba *et al.*, 1984). There is evidence that the antinociceptive system involves many subcortical and brainstem structures. That stimulation of many cerebral structures described in this chapter relieves pain supports such a conclusion.

The descending inhibitory system controls and modulates the intensity of the afferent inputs at many levels of the CNS. It has been established that the more this system is activated, the more intensive is the nociceptive input. As a result of damage to this system, normal sensory stimulation, which is always present, is perceived as a feeling of pain.

The antinociceptive system functions not only through neuronal activity but also by biochemical reactions, the mechanism of which was discovered and studied during the past decade. Goldstein *et al.* (1971) were the first to establish that there are specific opiate receptors in various CNS structures. Their concentration is highest in PAG, limbic structures, medial Th, midbrain, hypophysis, walls of V_4, and spinal substantia gelatinosa. Morphinelike substances—neuropeptides, endorphins, and enkephalins—form in the brain under normal conditions by enzymatic fission of their peptide precursors. Enkephalins have been found only within the CNS, mainly in structures related to perception of pain, whereas endorphins are also to be found in the hypophysis.

Endogenous opioid peptides have a wide range of actions on important physiological functions. They affect the pain perception threshold, regulate functions of the hypophysis and gastrointestinal tract, influence behavioral reactions, and even have a bearing on the pathogenesis of important mental disorders (depression, schizophrenia). Although it has not yet been established that there is a specific pain transmitter in the brain, it may be assumed that enkephalins are inhibitory transmitters and that endorphins regulate perception of pain, possibly at synaptic membrane level.

In animal experiments, acute stress elevates the level of opioid peptides (primarily β-endorphin) in the blood and cerebral tissue and, simultaneously, sharply raises the pain threshold (Akil *et al.*, 1976; Madden *et al.*, 1977). However, chronic action of the same irritants reduces the level of opioid peptides to normal. Clinical observations have shown that the endorphins in plasma and CSF in chronic pain are substantially reduced, which apparently is a factor in the genesis of the pain (Sicuteri *et al.*, 1978; Almay *et al.*, 1978; Salar *et al.*, 1982). Two different explanations of this phenomenon are proposed (Almay *et al.*, 1978). The first is that a general depression of the endorphinergic system caused by organic pain is followed by hypersensitivity to other nociceptive stimuli. The second theory is that inhibition of chronic pain causes an increased activity of the antinociception system, which, becoming saturated, is unable to protect the patient from other nociceptive impulses (Salar *et al.*, 1982). At the same time, on stimulation of PAG in clinical conditions, the concentration of opioid peptides, especially β-endorphin, in the CSF increases, while pain subsides significantly (Hosobuchi *et al.*, 1979).

Opioid peptides, especially β-endorphin, have high analgesic powers. For example, in terms of molar concentrations, β-endorphin is dozens of times more effective than morphine. Naloxone is a specific morphine antagonist preventing analgesia induced by endogenous opioid peptides. Apparently, this drug blocks the opioid receptors. However, naloxone does not prevent pain relief induced by stimulation of the peripheral nerve or spinal cord (Freeman *et al.*, 1983), and in many cases naloxone does not affect the analgesia induced by PAG stimulation (Meyerson *et al.*, 1985).

The opioid mechanism of the antinociceptive system, or, to be more precise, impairment of this mechanism, undoubtedly, plays an important role in the pathogenesis of pain. In all likelihood, only activation of this mechanism explains the efficacy of the new method of surgical management of pain by chronic stimulation of PAG and other cerebral structures (Section 5.4.1). However, increasing experimental and clinical data indicate that the biochemical mechanisms of pain are significantly more complicated. As was justly noted by Amano *et al.* (1982), endorphins represent only the tip of the iceberg.

In recent years, more data have accumulated indicating that besides opioid peptides in the brain there are also other yet unstudied biochemical systems modulating pain (Watkins and Mayer, 1982). Considerable attention has been focused on GABA as well as on serotonin as possible transmitters of pain (Reyes-Vazques and Dafny, 1983). Undoubtedly, catecholamines (dopamine, epinephrine, norepinephrine), histamine, prostaglandins, cholecystokinin, substance P, somatostatin, calcitonin, and others also participate in generating pain. In particular, it has been proven that peripheral afferent terminals contain substance P, somatostatin, and acid phosphatase, and substantia gelatinosa cells enkephalin and GABA.

Recent investigations have shown that substance P plays a very important role in the genesis of pain. The peptide was discovered in 1931, but its role in pain mechanism was disclosed only a few years ago. This neuropeptide is often called "substance of pain" because it is released in the focus of a tissue injury, inflammation, or vascular spasm. Tiny vesicles containing substance P are located in spinal ganglion neurons, in the spinal gray matter, and also in PAG. It is assumed that substance P release in the periphery depends on an axon reflex mechanism. The released substance is a source of powerful nociceptive impulses to the posterior horns of the spinal cord and from there to supraspinal structures (Wall, 1978; Zimmerman, 1981; Meyerson *et al.*, 1985).

Some investigators believe that there is a constant battle at different levels of the CNS between the two main pain-modulating neuropeptides: substance P, which activates nociceptive discharge, and endorphins, which inhibit it. It has been established that endorphins and morphine also inhibit substance P release.

Serotonergic neurons, which send axons in the posterolateral spinal cord funiculus to neurons at the segmental level, are primarily found in nucleus raphe magnus. The serotonin deficit occurring as a result of dorsal raphe nucleus destruction or the use of a serotonin synthesis inhibitor elevates pain sensibility. In view of this, certain authors consider serotonin to be the main transmitter in the descending antinociceptive system (Akil and Mayer, 1972; Messing and Lytle, 1977; King, 1980).

Analgesia induced by stimulation depends on the function of the descending pain inhibitory system passing from the raphe nucleus. It was shown that these pathways are mostly serotonergic (Basbaum and Fields, 1978).

Catecholamine-containing pathways from the brainstem to the spinal cord also play an important role in pain control mechanisms (Segal and Sandberg, 1977; Martin *et al.*, 1979). Analgesia induced by exogenous opiates or by PAG depends partially on catecholamine content (Taiwo *et al.*, 1985).

To sum up the data presented, one may conclude that pain, irrespective of the cause, becomes chronic not only because of the continuing action of its cause but also as a result of impaired function of the antinociceptive system. In this case, both the multilevel neural mechanism of pain suppression and the biochemical mechanism are impaired. There are reasons to believe that in chronic pain, the protective endogenous opioid mechanism does not function. This is confirmed by the fact that in patients with chronic pain the introduction of naloxone does not change either the intensity of spontaneous pain or the threshold of thermal pain (Lindblom and Tegner, 1979). Many other substances, mentioned above, participate in the pathogenesis of this pain.

4. Pathogenesis of Chronic Pain

There can be no doubt that the segmental mechanisms that transmit pain are constantly modulated by supraspinal actions of the multilevel complex system of cerebral structures. Although there is so far no convincing evidence of the existence of a cortical pain center, for decades it was assumed that pain becomes a conscious sensation only when it reaches the cerebral cortex. However, as experience has shown, resection of the somatosensory cortex (topectomy) does not prevent the perception of pain, though it may modify it. In view of this, there is the suggestion that appreciation of pain occurs not at cortical level but at the level of Th or even the brainstem. It has been demonstrated in animal experiments that stimulation of the sciatic nerve leads to rhythmic activity in VPL in the form of peaks of two types (Emmers, 1981). This suggests that the sensation of pain is coded in Th; moreover, one type of peak reflects intensity, and the other the modality of pain sensation. These spatial and temporal coding mechanisms are very important for pain appreciation.

Nevertheless, a very important role in the pathogenesis of pain syndromes is played by the higher cortical functions that are linked with sensation of pain. It is assumed that there are two "qualities of consciousness" that should be differentiated: simple "registration" of pain sensation (algesia) and a certain psychic equivalent of that sensibility connected with the agonizing perception of pain. In view of this, some authors differentiate "sensation of pain" and "suffering from pain." One example can be seen in herpes zoster or the state after resection of the trigeminal nerve root, when the patient experiences excruciating pain in anesthetic areas. Another argument is the fact that after prefrontal leukotomy, "suffering from pain" disappears, although "sensation of pain" remains.

Laitinen (1974) considers that the cerebral cortex, which is closely linked with the limbic system, "classifies" incoming sensory impulses into "dangerous" (pain), "harmless" (neutral), and others (for example, pleasurable). When the nervous pathways are damaged, the cortical neurons that do this "sorting" are unable to understand the new "language" of the changed input. This leads to the development of a peculiar state in which any sensory stimulus is perceived as a sensation of pain. Melzack (1971a,b) expresses a similar idea about "pain transformation" of afferent stimuli. It is quite probable that chronic pain syndromes occur under precisely such conditions.

One of the most interesting and convincing theories concerning the pathogenesis of chronic pain has been proposed and actively developed by Kryzhanovsky and co-workers (Kryzhanovsky, 1976a,b, 1979, 1980; Kryzhanovsky and Aliev, 1978; Kryzhanovsky *et al.*, 1974a,b)—the theory of "a generator of pathologically intensified excitation" in the CNS and, primarily, in the nociceptive system. Generators, which form under specific conditions, produce sensations of pain without any nociceptive activation from the periphery. This fact, in principle, was known long ago; however, an important merit of Kryzhanovsky's concept is that it is based on the author's experimental data.

The concept of "a pattern-generating mechanism of pain" proposed by Melzack and Loeser (1978) closely resembles this theory. This mechanism, which may occur at many levels of the spinal cord or brain, generated patterns of impulses in the absence of an inhibitory effect on the transmission of sensory inputs. Such a mechanism generating pathological pain may be triggered by any sensory stimulus.

A question that still remains open is: Why can the same lesions of the pain-transmitting pathways in some cases produce only anesthesia and a sense of numbness in the affected zone and in other cases lead to excruciating pain in the same zone? Many factors may be involved in explaining the still insufficiently studied pathogenesis of central, particularly deafferentation, pain. These factors may explain the hypersensibility of the pain-conducting pathways and their centers, damaged because of hypoxia, scarring, diminution of supraspinal inhibitory action, hyperactivity of synaptic mechanisms, or denervation hypersensitivity. In general, the neurophysiological basis of deafferentation or neurogenic pain has not been fully defined. Bettag (1966) considers that interruption of pain-conducting pathways activates "protopathic pain," which is conducted by fibers of the medial pain system) (spinoreticular pathways, nonspecific Th nuclei) (Chapter 1, Section 4). One of the most convincing arguments in favor of a central genesis of certain pain syndromes is the existence of phantom pains in the lower half of the body in paraplegics in whom total anatomic interruption of the spinal cord has been proven beyond doubt (Melzack and Loeser, 1978). Such pain remains even when the only possible pathway that could transmit impulses from the periphery, the sympathetic trunk, is blocked or interrupted surgically.

The deafferentation hypersensitivity phenomenon has been known for a long time and has been carefully investigated experimentally. Deafferented neurons in various CNS structures permanently or periodically generate spontaneous discharges ("firing") resembling discharges from an epileptic focus, which are not caused by any afferent impulses. It is assumed that these discharges are the cause of pain dysesthesia. Melzack and Loeser (1978) consider deafferentation pain to be the result of hyperactivity of neurons both in the spinal posterior horns and in the Th sensory nuclei ("pattern-generating mechanism") without any activation of peripheral pain pathways.

It is noteworthy that in central pain, loss of sensation is usually incomplete, which indicates only a partial lesion of the afferent pathways. In view of this, it is assumed that an important role in the pathogenesis of central pain is played by an imbalance of the ratio between different afferent fibers (A and C). There are grounds to believe that in chronic pain, the medial or "slow" pain transmission system is mainly involved in the pathological process (Chapter 1, Section 4). Another view holds that deafferentation pain occurs not as a result of excessive pathological input from the periphery but rather from insufficient sensory inputs from the afflicted parts of the body. This triggers the inhibitory pain mechanisms, producing subsequent hypersensitivity of the pain neurons in the cortex and Th.

When the main afferent sensory systems are damaged, the paleospinothalamic pathways and nonspecific thalamic structures play the main role in the pathogenesis of central pain, which is a "peculiar pathological form of sensation" (Pagni, 1976). Central pain occurs because of an "imbalance" between various pain-transmitting anatomic structures since part of this complicated system has been "put out of action." A similar supposition about the important role of impaired interaction of the lemniscal and extralemniscal systems in the genesis of chronic pain has been put forward by Winther (1973). Protopathic pain of central genesis occurs when the extralemniscal system is released from the inhibitory influence of the lemniscal system.

An increasing amount of information has been accumulated in recent years about the role of pathological changes at spinal level, primarily at the level of the posterior horns, where the gating mechanism is localized (see above), in the pathogenesis of chronic pain. It is highly probable that deafferentation pain is the result of spontaneous epileptiform activity of those spinal neurons to which the primary afferent fibers project (Loeser *et al.*, 1968; Powers *et al.*, 1984a). Wall (1980) noted that deafferentation pain implies "not only a loss of input, but actual degeneration so that spinal cord cells were free to act in a pathological way." Such activity can be caused by many factors: denervation hypersensitivity, switched-off inhibitory input, or nervous tissue irritation by cicatrices. The "peripheral" (spinal) theory is confirmed by the DREZ destruction operation (Section 5.3.7), which is very effective in certain types of pain.

For understanding the basic mechanisms of pain syndromes, it is interesting to note, as have many neurosurgeons when performing stereotactic operations, that electrostimulation of subcortical sensory pathways, including specific Th nuclei, under local anesthesia, has never produced a clear-cut sensation of pain in patients operated on for conditions other than pain syndromes. However, in patients with intractable pain, under the same conditions of electrostimulation, a sensation of pain is induced by stimulation at different levels—thalamic, capsular, subthalamic, and mesencephalic. This fact suggests that in subcortical stimulation during a stereotactic operation for pain syndromes, a painful response depends not only on the localization of the electrode in a particular structure being stimulated but also on the patient's interpretation of the sensory stimulus as painful.

The complex biochemical mechanisms (substance P, endorphins, serotonin, etc.) briefly described above play a very important role in the pathogenesis of chronic pain. In particular, the exhaustion of opioid production may be a possible explanation.

The data presented above naturally do not exhaust the possible explanations of the pathogenesis of pain, which currently has been recognized as the single, major problem of neurophysiology, neurology, and neurosurgery. Obviously, the basis of the pathogenesis of chronic pain requires further in-depth investigations.

5. Surgical Management of Pain

5.1. General Remarks

Management of pain syndromes is an old and very complicated problem of neurosurgery. For more than half a century, neurosurgeons have been diligently searching for the simplest, most effective, and safest operations to alleviate pain when conservative treatment offers no relief.

For decades such investigations were based on the supposition that if the pain-conducting pathways from the afflicted area are interrupted, the pain would disappear. Proceeding from this idea, many operations were proposed with the main objective of sectioning these pathways at various levels from the peripheral nerve to the cerebral cortex. Many of these operations proved effective and became firmly established in the practice of neurosurgery. However, years of experience have shown that section of the pain pathways does not alleviate pain in a significant number of cases; moreover, as the result of the operation, there is pronounced analgesia in the afflicted zone. However, even more severe deafferentation pain may develop in the denervated zone.

A partial explanation of this phenomenon was found comparatively recently. A study of the morphological substrate of the transmission and perception of pain has shown that the pathways and centers of the pain system are much more complicated and intricate than was previously assumed (see Chapter 1). The existence of two interacting systems of pain transmission, a nonspecific or medial and a specific or lateral tract, explained why the interruption of only one did not abolish pain. Another important aspect of that problem is "central" pain (Section 6), associated with hyperactivity of both cerebral and spinal pain centers. Peripheral operations in case of such pain naturally cannot provide relief. The existence of a multiplicity of afferent nociceptive pathways is the main course of ineffective surgical interventions for pain relief. Generally speaking, the limited knowledge of anatomic correlates and pathophysiological basic mechanisms of pain is one of the factors restricting the possibilities of surgical treatment.

Siegfried (1981) outlined the main requirements of any neurosurgical operation for the management of pain:

1. The procedure must be safe and not cause stress in the patient.
2. The rate of successful results must be high.
3. The rate of relapses and complications must be sufficiently low to be acceptable.

The rapid development of functional and stereotactic neurosurgery as well as mounting unsatisfactory results in the classical operations for pain alleviation stimulated an essentially new approach to the problem of stereotactic destruction at various levels of pain centers and pathways in deep-seated cerebral structures. The first stereotactic operation for intractable pain, mesencephalotomy, was performed by Spiegel and Wycis in 1947; the following year the same authors performed the first basal thalamotomy to interrupt the pain pathways where they enter Th (Spiegel and Wycis, 1948). The first destruction of VPL and CM was performed by Hecaen *et al.* (1949) and Talairach *et al.* (1949) and later by Monnier and Fischer in 1951.

It is noteworthy that functional and stereotactic surgery of pain, the fundamentals of which are presented below, not only proved to be an effective method of treatment, but also played an important role in studying the pathogenesis of pain.

The strategy for the management of intractable pain in any particular case is always a very difficult task. Neurosurgical treatment may be considered as a last resort, but it must not be done so late that severe drug addiction or marked changes of the patient's personality have developed. The first problem facing the neurosurgeon is to determine the indications and contraindications for any operation. These difficulties stem, first of all, from the fact that pain is a subjective sensation, which can be objectively evaluated only with great difficulty. Nevertheless, several successful approaches for assessing the intensity of pain objectively have been proposed in recent years (McGill's questionnaire, visual-analogue scale, multifactor test, etc.). The use of evoked potentials (EP) (Chapter 4, Section 7.4) represents a promising means to investigate this still unresolved problem. Recently, some investigators have managed to identify specific components of somatosensory EP in response to various nociceptive stimuli (Chapman and Wiesendanger, 1981) that disappear on stimulation of the posterior horns of the spinal cord. Possibly this method will allow an objective assessment of the efficacy of various operations for the alleviation of intractable pain.

The morphine saturation test (Hosobuchi, 1982) and the pentobarbital test (Plotkin, 1982) have been proposed as means of determining the effectiveness of operations to provide relief from pain. For this purpose, 1.5 mg of morphine in solution is injected intravenously every 60 sec until a total dosage of 30 mg has been given. Pain intensity is determined every minute on a ten-point scale. If residual pain remains after the test, a lapse of 20 min is allowed, after which 0.8 mg of nalaxone—an antagonist of morphine—is injected. In this test, somatogenic pain disappears completely in a matter of seconds after injection of 1.5–3.0 mg of morphine, the effect of which is removed by naloxone. The gradual lessening of pain and its recurrence after naloxone is an indication of organic peripheral pain and prognosticates good results after stimulation of deep cerebral structures, particularly PAG (Section

5.4.1). In case of deafferentation pain, the injection of morphine is not effective, and the success of PAG stimulation is doubtful.

In evaluating the indications for surgery for chronic pain, it is always necessary to consult an experienced psychiatrist, since it is known that pain (or, to be more precise, complaints of pain) may be a manifestation of many mental disorders (depression, hysteria, phobia, etc.).

It may be said that so far there is no single stereotactic and functional operation for the relief of pain. This is explained by the broad range of pain syndromes, the various locations and complexity of the above-described system of pain pathways and centers in the CNS, the wide variety of stereotactic operations, and insufficient evaluation of long-term results of destruction of various deep structures and pathways of the brain and spinal cord.

It would be no exaggeration to say that in no single area of stereotactic surgery are there so many unresolved problems as in the surgery of pain. Of these problems, two are most complicated: choice of the subcortical structure or structures, the inactivation of which is most effective, and the frequent postoperative relapses of pain, the cause of which until now has not had an adequate explanation.

Undoubtedly, in a considerable number of cases, modern functional operations yield significant and stable therapeutic relief of pain syndromes when all other means prove futile. Nevertheless, the problem of postoperative relapse of pain after a period of remission remains very real, since such relapses occur in approximately 30–40% of cases (Mundinger, 1974a; Voris and Whisler, 1975). In theory, there may be two main causes for pain relapses after destructive surgery on the nervous system for pain: alternative pain pathways "switch on," or sectioned nerve fibers regenerate. However, it is practically impossible to determine the cause of relapse in any specific case.

It must be emphasized that pain relapses may occur several years after any operation; consequently, a very long follow-up is required to make a final assessment of these operations. Moreover, it is necessary to take into account certain subjective elements when neurosurgeons assess the results of pain alleviation operations.

Hitchcock (1974) has proposed a convenient grading system for pain relief, consisting of five grades: (1) total disappearance of pain; (2) rare pain that disappears on administration of mild, nonnarcotic analgesics; (3) frequent pain that disappears on administration of the same analgesics; (4) constant pain that disappears on administration of potent narcotic analgesics; (5) constant pain that is not relieved completely by any analgesic.

In view of the great variety of surgical methods for the management of pain, their classification represents a difficult task. We consider it expedient to begin with data on all (both open and stereotactic) operations on either the brain or the spinal cord that involve section or destruction of nuclei and pathways related to pain. In view of the rapid development of functional neurosurgery, the stimulation techniques are outlined in Section 5.4 of this chapter, individually for the brain and the spinal cord.

Since central pain syndromes as well as trigeminal neuralgia are traditionally regarded as individual nosological forms, the surgical management of these disorders is presented in Section 6.

5.2. Operations on the Brain

5.2.1. Operations on Thalamic Nuclei

Destruction of various nuclei and pathways of Th represents one of the main trends in contemporary stereotactic surgery. A distinct advantage offered by operations on Th nuclei in intractable pain is that unlike operations at more peripheral levels, in Th there is clear-cut segregation of the different types of sensation—deep, tactile, temperature, and pain. Consequently, destruction at the thalamic level may interrupt only pain sensibility, leaving other modalities intact (D. E. Richardson, 1982a,b).

It is also noteworthy that in these operations, the pain pathways are interrupted relatively far from the motor and other conducting systems, so that these operations should not impair muscular strength. However, after operations on Th nuclei for pain, transitory impairment of consciousness is occasionally observed, presumably caused by damage to the nearby FR (D. E. Richardson, 1974; Hitchcock and Teixeira, 1981).

As discussed below, many specific and nonspecific Th nuclei are at present the targets of stereotactic destruction for intractable pain. However, the complex functional organization of these nuclei, their very number, significant anatomic variability, and the presence of dual pain systems all lead to considerable difficulties in identifying the most effective thalamic targets.

The extensive experience that has been gained in performing stereotactic operations on various Th nuclei has led to the unexpected and quite paradoxical conclusion that destruction of nonspecific thalamic structures is frequently more effective than destruction of specific sensory Th nuclei. This fact once again confirms the dominant role of the medial system (Chapter 1, Section 4) in the genesis of chronic pain. However, lesions within the medial system do not yield stable results, apparently, because of incomplete interruption of the pain pathways. As Sugita and Kobayashi (1982) indicate, these lesions involve only 10–20% of the diffuse pain system.

5.2.1a. Destruction of Sensory Nuclei. One of the first stereotactic operations on subcortical pain centers was destruction of Th specific relay sensory nuclei, the third neurons of pain sensation. In the first operations performed by Talairach *et al.* (1956) and Spiegel and Wycis (1962a), there was every reason to believe that the most effective pain relief operation would be obtained by destruction of the main "relay station" of pain—nuclei VPL and VPM, in which the spinothalamic and trigeminothalamic pathways terminate. As was described above, the entire sensory tract, in particular pain afferentation from peripheral receptors, terminates in VPL.

The stereotactic destruction of VPL differs from the operation on VL described above (Chapter 4, Section 5) mainly by the different calculations for the coordinates of the stereotactic target point. According to the Schaltenbrand and Bailey atlas, VPL lies caudal to VL and borders on its posterior part (Vim). The center of VPL has the following stereotactic coordinates: 4 mm anterior to CP, from 2 mm superior to 1 mm inferior to the line of the posterior border of FM–CP, and 15–17 mm from the midplane. Siegfried and Hood (1983) recommend somewhat different stereotactic coordinates of VPL: 3 mm anterior to CP, 1–2 mm below LI, and depending on the localization of pain, 13 mm lateral to the midplane in case of pain in the arms and 16 mm for pain in the legs. The stereotactic coordinates of the center of VPM are 3–4 mm anterior from CP, 2 mm below the line of the posterior border of FM-CP, and 8–9 mm lateral to the midplane.

An important stage of the operation is electrostimulation of VPL (Chapter 4, Section 7.1.2) to induce sensory responses referred to the contralateral half of the body. Recently, a detailed study has been made of the functional organization of the somatosensory Th nuclei, primarily VPL, on the basis of electrostimulation data obtained during stereotactic surgery (Tasker *et al.*, 1972; Hassler *et al.*, 1979; Hardy *et al.*, 1979b). Since somatotopic organization exists in VPL, it is desirable during the stimulation to define the zones of that nucleus that, when stimulated, intensify spontaneous pain. Primary destruction of these zones may result in a most significant therapeutic effect. Stereotactic destruction of VPL should be performed close to the base of Th, where lemniscus medialis terminates.

The technique and results of stereotactic destruction of VPL have been described in several papers (Riechert, 1960; Laponogov, 1965; Bettag, 1966; Siegfried, 1972a; Roth and Mark, 1973; D. E. Richardson, 1974; Watkins, 1975). From the experience of these operations, neurosurgeons encountered an unexpected and difficult-to-explain finding. Very often after VPL destruction there are significant impairments of pain and temperature sensibility in the contralateral half of the body. However, over the course of several days, the hypalgesia becomes less pronounced and rapidly disappears; in fact, prolonged hypalgesia was not observed in a single one of our cases after VPL destruction. However, at times, there was a notable diminution of tactile and deep sensibility.

The first results of VPL stereotactic destruction were received enthusiastically, since in the majority of cases, severe, intractable pain disappeared immediately after surgery. However, the analysis of long-term results was disenchanting. The hope for long-lasting relief of pain by VPL destruction was only partially justified. In spite of the fact that pain disappeared in the majority of cases immediately after the operation, in approximately half there was a relapse after a comparatively short period of time (usually within several months). The most likely explanation of these relapses is that the medial or spinoreticular pain system is left intact or that remaining collateral intrathalamic pain pathways may carry the nociceptive afferents to the cortical level.

Vasin and Ratza (1971) performed stereotactic VPL destruction in 25 patients with various pain syndromes (unrelated to malignant neoplasms). The results of unilateral destruction after periods from 5 months to 3.5 years were the following: practically total disappearance of pain in three cases, improvement in seven, and total relapse in two. After bilateral destruction, these figures were approximately the same. After destruction of VPL, relapses occurred in eight of 14 cases with chronic pain (Laponogov, 1965). As Obrador and Dierssen (1966) noted, destruction of sensory Th nuclei for central denervation pain is ineffective. Roth and

Mark (1973) note that the rate of recurrence has been reported to vary from 10 to 50%.

In the majority of our operated patients with thalamic pain syndrome in whom sensory nuclei were destroyed, relapses occurred within 3 weeks to 8 months.

In pituitary adenomas with acromegaly, as well as in basal arachnoiditis, severe head pain sometimes may develop. Such pain does not lend itself to any kind of therapy. In such cases, a good therapeutic effect after bilateral VPL stereotactic destruction has been described in a few cases (Vasin and Ratza, 1971).

Destruction of VPL very rarely causes serious complications (dysesthesia, sensory ataxia), although several cases of severe thalamic pain syndromes (Section 6.1) after surgery are described in the literature. In such cases agonizing hyperesthesia affected those parts of the body in which there had been no pain prior to the operation. The causes and pathogenesis of this quite fortunately rare complication are obscure. In one case, anatomically verified, central pain syndrome appeared after combined destruction of VPL and the intralaminar nuclei (Pagni, 1966).

In summation, it is noteworthy that in approximately half of the cases after VPL destruction pain of various types disappears, and this effect may last for many years. Without a doubt, such a result may be considered promising, considering that none of the existing methods of treatment relieves such patients. In one of our cases, there was permanent relief from severe phantom pain after VPL destruction (Section 6.2).

The limited success rate of operations on Th sensory nuclei prompted a search for new stereotactic methods of surgical management of pain.

5.2.1b. Destruction of Centromedian Nuclei. One of the important targets in stereotactic surgery for pain is CM, a small nucleus situated approximately in the center of Th and a part of the nonspecific afferent projection system passing to the cerebral cortex (Chapter 1, Section 4). Destruction of CM interrupts sensory impulses passing from Th to Str and GP. Recent investigations have shown that CM is an important link in the perception of pain; although some believe that a large part of the pain pathways of the medial system terminates in CM, others (Taub and Collins, 1973) state that there is no evidence from degeneration studies of the presence of direct spinal projections to CM.

Stereotactic destruction of CM for chronic pain was first performed by Talairach in 1949 and then by Monnier and Fischer (1951). Electrostimulation of CM during surgery does not induce any clear-cut sensations of pain. Patients who are fully conscious experience a difficult-to-explain feeling of tension, fear, internal excitation, and vague unpleasantness in the contralateral half of the body. At times, this is accompanied by increased pulse and respiratory rates, bilateral mydriasis, and at times dilation of the pupil on the side opposite to the stimulation.

Stimulation of CM in patients with intractable pain usually relieves the pain but also produces much the same paresthesias in the contralateral half of the body as stimulation of primary sensory nuclei. However, these paresthesias are less localized, and their excitation threshold is somewhat higher (1–3 V). Stimulation of CM during surgery is an important indicator of the prognosis; if the pain is eased, then destruction of CM leads to permanent relief of the pain syndrome.

The stereotactic coordinates of CM are the following: 7–9 mm posterior to the middle of LI, 1–2 mm above that line, and 6–8 mm from the midplane. After destruction of CM, pain disappears on the contralateral half of the body without an impairment of pain or other sensation. For bilateral or visceral pain, destruction of CM on both sides is recommended (Sugita *et al.*, 1972; D. E. Richardson, 1974); however, bilateral operations are more effective for unilateral pain (Hitchcock and Teixeira, 1981).

There have been several encouraging reports in the literature of stereotactic destruction of CM for pain of various types (Mark *et al.*, 1960; Bettag, 1966; Sugita *et al.*, 1972; Amano *et al.*, 1977; Hitchcock and Teixeira, 1981; Niizuma *et al.*, 1982). Sugita *et al.* (1972) performed this operation for pain caused by malignant tumors of the head, neck, abdomen, and pelvis. Simultaneously with CM lesions, the authors also destroyed the nucleus medialis fasciculosum (part of nucleus medialis dorsalis). A definite therapeutic effect directly after surgery was observed in approximately 80% of the cases. Subsequently, however, pain relapsed in many of the patients. Approximately the same results were obtained by Niizuma *et al.* (1981) from CM thalamotomy in 17 cases of thalamic pain syndrome. A significant or partial improvement was noted in only nine cases, and the pain relapsed in all cases after 2 to 7 months.

Unlike all other authors, Hitchcock and Teixeira (1981) reported good and lasting results from CM thalamotomy. Pain disappeared completely in all of their 19 patients after surgery, with only a few relapses. The authors believed that this success resulted from careful selection of patients.

The CM thalamotomy is indicated for cancer pain

resulting from damage to many different nerves. However, phantom pain, thalamic pain, and trigeminal neuralgia were not helped by CM destruction (Sugita *et al.*, 1972).

After unilateral CM destruction there is usually no intellectual or emotional disturbance, although Schaltenbrand and Wahren (1982) observed a partial Korsakoff's syndrome and speech impairment. Such disorders are observed more frequently after bilateral destruction of CM (Pagni, 1966).

In contemporary stereotactic surgery, isolated destruction of CM is rarely performed because of frequent relapses of the pain. Destruction of this nucleus is much more frequently combined with lesions of other subcortical structures, especially the parafascicular nuclei (see Section 5.2.1i).

5.2.1c. Destruction of Th Pulvinar. The pulvinar (Pul), the largest of the Th nuclei, consists of several divisions (Chapter 1, Section 1.1.4). Little is known about its functional significance. Morphologically, Pul has no direct relationship to pain perception, but it may be assumed that the nucleus is related to the complicated subcortical system modulating pain.

Some experience recently has been gained in the stereotactic destruction of Pul for intractable pain, an operation that has been named pulvinotomy. Since it is performed more frequently for hyperkinesias, the technique of the operation is described in Chapter 9. Pulvinotomy for pain is often combined with the destruction of other Th nuclei (Section 5.2.1). The question of which Pul nucleus should be the target for stereotactic destruction in pain has yet to be resolved. Adachi *et al.* (1971), D. E. Richardson (1974), Yoshii *et al.* (1980), and Yoshii and Fukuda (1979) consider that among the numerous Pul nuclei, destruction of supranucleus pulvinaris medialis is the most effective for relief of pain, whereas Laitinen (1977) prefers to destroy the anterior portion of Pul. In unilateral pain, pulvinotomy is performed on the side contralateral to the pain, whereas in bilateral pain, both sides are done in one stage or separately at various intervals. Bilateral operations are apparently more effective (Mizokami *et al.*, 1981). There is no neurological deficit after Pul destruction.

The clinical reports of pulvinotomy in pain syndromes are few in number and frequently contradictory. Cooper *et al.* (1973), who were the first to perform this operation, noted an improvement, especially in the necessity for narcotics, in all four patients on whom they operated. Martin-Rodrigues and Obrador (1975) reported the relief of severe pain syndrome in two cases after destruction of the ventrooral and ventrobasal portions of Pul. These patients had no sensory impairment postoperatively. The long-term follow-up has shown that the immediate results of pulvinotomy are significantly better than the long-term ones. A substantial improvement after destruction of medial or lateral Pul nuclei was obtained in only approximately half of the cases of other series (D. E. Richardson, 1974; Adachi *et al.*, 1971; Yoshii *et al.*, 1975). According to data by Laitinen (1977), pain was relieved completely or diminished substantially in 31 of 41 cases after pulvinotomy in various pain syndromes, but the effect in half of the cases lasted only 2–3 years. Pain relapses usually occurred within several weeks of the operation. The data of Fraioli and Guidetti (1975a) are even less optimistic—partial or total relapse of pain occurred within a few weeks or months in all five of their operated cases.

Yoshii *et al.* (1980) reported the results of pulvinotomy in 19 cases of different types of pain syndromes after several years; the results were excellent in four, good in four, and satisfactory in five, and in four there was no effect. The result of a bilateral operation was slightly better than a unilateral operation. Tasker (1982b) concludes that in intractable pain pulvinotomy is no more effective than VPL thalamotomy. According to our limited experience, a long-term satisfactory effect after pulvinotomy occurs in only about 40% of cases.

One may conclude that although Pul plays a definite role in pain perception, its destruction in pain syndromes gives a significant but variable therapeutic effect.

5.2.1d. Destruction of Intralaminar, Parafascicular, and Limitans Nuclei. These small Th nuclei are related to the nonspecific (medial) pain system (Chapter 1, Section 4). It is considered that the paleospinothalamic and spinoreticular tracts terminate in them. Certain authors include CM in the intralaminar nuclei. At least 25% of the neurons of nucleus parafascicularis and CM fire only in response to nociceptive skin stimuli (Ishijima *et al.*, 1973). This evoked activity is inhibited by morphine. These data served as the basis for stereotactic destruction of these nuclei for the relief of various pain syndromes (Mark *et al.*, 1963; Sano *et al.*, 1966). Isolated bilateral stereotactic destruction of the nucleus parafascicularis has also been described (Mark and Ervin, 1965).

The stereotactic coordinates of nucleus parafascicularis are 5 mm posterior to the midpoint of LI, 2 mm above this line, and 5 mm lateral to the midplane. The coordinates of nucleus limitans are 7 mm posterior to the middle of LI, 5 mm above that line, and 2–3 mm lateral to the midplane.

The destruction of these nuclei does not produce a sensory loss; postoperative complications are rare and not serious.

In approximately 60–80% of cases, especially in somatic pain, stereotactic destruction of these nuclei produces significant pain relief (Fairman, 1966; Sano, 1977); some authors believe that these nuclei are the optimal thalamic target (Takeda, 1984). At the same time, destruction of intralaminar nuclei for dysesthesias does not yield consistent results (Tasker, 1982a,b). Destruction of these nuclei along with other thalamic structures (Section 5.2.1i) is employed much more frequently than their isolated destruction.

5.2.1e. Destruction of Internal Medullary Lamina. The internal medullary lamina is a layer of myelinated fibers that divides Th into medial and lateral parts.

In 1966, Sano *et al.* proposed stereotactic destruction of the posterior half of the internal medullary lamina in pain syndromes and named this operation thalamolaminotomy. The authors believe that this operation interrupts the pathways between the specific and nonspecific sensory Th nuclei. In ten cases (three underwent bilateral operations) with pain of various types, effect was obtained in eight with follow-up from 1 month to 2.5 years.

The stereotactic coordinates of this structure are as follows (Poblete *et al.*, 1970): 7 mm posterior to FM on the line FM–PC, 2 mm above the line, and 7 mm lateral to the wall of V_3. Lately, this operation to alleviate pain has not been practiced extensively.

5.2.1f. Destruction of Dorsomedial Nuclei. This operation in combination with mesencephalotomy, performed by Spiegel and Wycis in 1947, was the first stereotactic operation carried out in the world. The DM nuclei are not on the pain pathways but are Th structures closely linked with the limbic system. Their destruction interrupts the pathways to the many areas of the frontal cortex as well as connections between the premotor and limbic cortex and Hypoth (Spiegel and Wycis, 1962a,b). It causes a peculiar "emotional indifference" to pain resembling the state produced by prefrontal leukotomy and other psychosurgical operations. Destruction of these nuclei is indicated mainly for relief of the "emotional–mental" components of pain, in particular in patients who have been taking narcotics for a long period. This operation, usually performed on both sides, was reported to give positive results in various pain syndromes in the 1960s (Mark *et al.*, 1963, 1965; Bettag, 1966). But later, Tasker (1982b) did not obtain a beneficial effect after destruction of dorsomedial nuclei for intractable pain.

In principle, the technique of dorsomedial nucleus destruction is similar to other operations on Th nuclei, in particular, VL. The stereotactic coordinates of the dorsal part of Th dorsomedial nuclei are the following: 7.5–8 mm posterior to CA, 12–13 mm above LI, and 4.5–5 mm lateral to the midplane (Roth and Mark, 1973).

This procedure is performed for the relief of pain comparatively rarely since there are frequent relapses as well as impairment of memory, mental functions, and other complications. More often, this operation is combined with such procedures as stereotactic destruction of other subcortical structures (Section 5.2.1i). Our experience of dorsomedial nuclei destruction in Tourette's syndrome is presented in Chapter 18.

5.2.1g. Destruction of Anterior Nuclei. The anterior nuclei of Th (nucleus anteroventralis) are a link with the limbic system, mainly with the mammillary body and Hipp. These nuclei are not part of the pain system of the brainstem or cerebrum. Their destruction, like psychosurgical operations, aims at eliminating the emotional component of pain. Only a few stereotactic operations on the anterior nuclei have been reported. A female patient with protracted hysterical pain obtained relief after unilateral destruction of the anterior Th nuclei (Andy, 1973).

5.2.1h. Basal Medial Thalamotomy. Spiegel and Wycis (1966) employed the term basal (or medial) thalamotomy to describe stereotactic destruction of extralemniscal reticulothalamic pathways passing from the midbrain to the nonspecific Th nuclei—CM, parafascicular, and interlaminar nuclei. The lesion was made in the medial portion of the base of Th caudal to CM in order to interrupt the medial pain pathways. There is no analgesia after these operations. Other authors use the term basal thalamotomy for the stereotactic destruction of both medial and lateral afferent pain pathways at their input into Th ventrally from CF.

The stereotactic coordinates for basal thalamotomy are the following: 1 mm anterior and 1 mm below CP (lateral projection), and 8–10 mm from the midplane (AP projection). With pain in the arms or face, the focus should be 1–2 mm more medial, and with pain in the lower half of the body, 1–2 mm more lateral.

Basal thalamotomy is performed on the side contralateral to the location of pain. In bilateral pain, the second operation may be performed after an interval of several weeks or months. Basal medial thalamotomy yields very effective results, especially in cancer pain, but they are transitory in approximately 80% of the cases (Spiegel *et al.*, 1966). Lasting relief of pain was obtained in no more than a third of the cases.

Hitchcock and Teixeira (1981) report total disappearance of pain in cancer in 83% of cases directly after the operation and in 66% after 6 months. In central pain, good results were obtained in 77% of the cases with a follow-up of 2 months to 8 years.

5.2.1i. Destruction of Multiple Th Nuclei. The great variety of stereotactic subcortical operations for pain syndromes indicates that so far the destruction of no specific subcortical structure has provided reliable relief in chronic pain. The difficulties in determining an optimal structure for destruction have stimulated the search for combined surgical techniques aimed at destroying two or more subcortical structures in a one-stage stereotactic operation. The combined destruction frequently involves several Th nuclei—VPL, CM, Pul, basal part of Th, and dorsomedial and intralaminar nuclei. Isolated attempts to combine destruction of various Th nuclei with the destruction of nonthalamic structures, for example, Hypoth or GC, have also been attempted. It is quite obvious, however, that in such circumstances it is difficult to assess objectively the efficacy of destruction of each structure.

The results of combined destruction of VPL and CM were first presented by Bettag (1966). In 23 of 28 operations, the pain syndrome disappeared completely and remained absent over several years of follow-up. The author operated on another 12 patients who had relapses after VPL destruction alone. Additional destruction of CM on the same side gave a good result in nine of the 12 patients. Bettag considers that combined destruction of VPL and CM produces a better effect than the destruction of one of these structures alone. For example, destruction of VPL alone gave a good result in 22 of 66 cases with intractable pain, whereas destruction of both structures (in 44 cases) gave 84% relief from 1 to 8 years (Bettag, 1975).

The results of combined destruction of VPL and CM reported by other neurosurgeons have not been so favorable. Cooper (1965b) noted a significant improvement in only half of 32 patients and total disappearance of pain for a period of more than 6 months in only one-third. In patients with a thalamic pain syndrome, destruction of VPL and CM had a good effect only in two of eight patients. Good results in pain of various kinds were obtained after combined destruction of VPL and the dorsomedial nucleus (Martins and Umbach, 1975). D. E. Richardson (1974), based on his experience, believes that combined uni- or bilateral destruction of CM and medial Pul gives the most complete and lasting relief of pain.

Combined destruction of CM and parafascicular nuclei (CM–Pf) is performed quite frequently and causes no sensory disturbances. Several neurosurgeons reported good results from this operation in pain as a result of various causes. It has been noted that bilateral destruction of CM–Pf is more effective than unilateral.

Leksell *et al.* (1972) destroyed by remote control CM–Pf in 25 cancer patients using γ radiation of 20–25 krad. Bilateral radiation was done in seven cases with disappearance of pain or a significant improvement noted in 10 of 25 cases. Using combined stereotactic destruction of CM–Pf in 22 patients with various pain syndromes, Mundinger and Becker (1975) obtained good long-term results in six cases, notable improvement in 11, and no effect in five. However, Amano *et al.* (1977), on the basis of their experience in 47 operations, considered that CM–Pf destruction is indicated in central pain but is not very effective in cancer pain.

Certain surgeons go even further and, in the process of a single stereotactic operation, destroy three to five thalamic structures involved in the transmission and perception of pain. For example, Choi and Umbach (1977) perform a "standard" combined destruction of Pul, CM, dorsomedial, parafascicular, and limitans nuclei in pain syndromes. In 37 cases with chronic pain, mainly in malignancy (bilateral operations were performed in six cases), this technique gave excellent or good results in 90% of the cases. Relapses were infrequent (18%). However, even combined operations do not always yield the expected effect. As an example, we describe one of our cases in which an improvement of thalamic pain was obtained after unilateral destruction of VPL and CM and bilateral cingulotomy.

Patient S., 57 years old, who had long been hypertensive, had an ischemic stroke in the left cerebral hemisphere. Four months later, a severe thalamic pain syndrome developed in the right limbs, which had a mild residual hemisparesis. Stereotactic cryodestruction of the left VPL improved searing pain in the right limbs immediately after the operation, but a relapse of the pain occurred after several weeks. Eighteen months later a second stereotactic operation was performed on the same side with cryodestruction of VPL as well as of CM. Postoperatively, although there was a significant lessening of pain in the right foot and shin, the searing pain in the right arm and right side of the trunk remained virtually unchanged. In view of the pronounced emotional component in the pain syndrome, 6 months later the patient underwent a third stereotactic operation, a bilateral cingulotomy. After that, the intensity of the pain diminished substantially, the emotional response decreased, and the patient become more calm. Nevertheless, the searing pain remained on the right side in the trunk and the arm but was less intense than before surgery.

One may conclude that in severe pain syndromes, combined stereotactic operations involving destruction of several Th nuclei are more effective. However, one must bear in mind that such operations are technically complicated and protracted.

5.2.1j. Destruction of Thalamocortical Pathways. The pathways relaying pain from Th nuclei to sensory and parts of frontal cortex are complicated and have not been adequately studied. They are difficult to identify during a stereotactic operation. Nevertheless, destruction of certain pathways between Th and the cortex has been attempted. The first report was made by Talairach *et al.* (1960), who performed stereotactic destruction of the pathways from Th to the second sensory field of the cortex in eight patients, with a significant improvement of pain noted in seven. Electrostimulation of these pathways during surgery induced various sensory effects as well as discomfort similar to the preoperative pain.

An analogous operation was performed in seven cases by Riechert (1966b). It is noteworthy that there was no impairment of sensibility after interruption of the thalamoparietal pain pathways. Astereognosis and spatial disturbances, which usually occur after removal of the parietal cortex, were also not observed.

Because the long-term results of these operations have not been published, one cannot consider the relief to be permanent. In the opinion of Mazars (1976a,b), destruction of the thalamoparietal pathways and removal of sensory cortex (so-called topectomy) have not been satisfactory.

Stereotactic destruction of pathways passing from Th nuclei to the mediobasal cortex of the frontal lobe also aims at interrupting thalamocortical pathways. The rationale of this operation differs significantly from the principle of the earlier procedure, since in this case the pain pathways are not interrupted. Thus, destruction of the thalamofrontal pathways closely resembles cingulotomy (Section 5.2.4). Romodanov and Mikhailovski (1965) performed this operation in four patients, noting an improvement in three. Encouraging results of this operation were also reported by Martinez *et al.* (1975).

At present this operation is rarely used. When the complicated topography of the thalamocortical pain-relaying pathways has been established more definitively, such procedures may be more effective.

5.2.2. Destruction of GP

Riechert (1980) reported that stereotactic destruction of GP resulted in an improvement of pain syndrome in one case; however, there has been no confirmation of the effectiveness of pallidotomy.

5.2.3. Destruction of Hypoth

Another pain-alleviating stereotactic operation aims to destroy the posterior or posteromedial nuclei of Hypoth. The operation, first performed by Sano (1962), has been named hypothalamotomy. Since it is also effective in certain mental disorders, the technique of the operation is described in Chapter 18.

The hypothalamus is not a component of the pain relay system; however, it is assumed to be related to structures facilitating the transmission and perception of pain in VPL, sensory cortex, and FR. The posterior Hypoth may be related to the pathogenesis of pain syndromes, since the medial part receives "slow" pain fibers and part of the paleospinothalamic tract.

Since pain reception is closely linked to stimulation of the sympathetic nervous system, a functional interrelationship between the ergotropic zone of Hypoth and the pain-relaying pathways is assumed. Experiments on cats have demonstrated that impulses from Aδ and C pain pathways enter the neurons in the posterior Hypoth (Sano, 1974). Morphologically, it has been established that the posterior Hypoth has ties both with specific sensory (VPL) and with nonspecific (Pf) nuclei of Th, which relay to the somatosensory cortex. It is possible that under normal conditions, the Hypoth contributes emotional coloring to pain and modifies those vegetative reactions that occur with pain (Sano *et al.*, 1975).

During hypothalamotomy under local anesthesia, painful stimulation of skin induces increased neuronal discharges in the posteromedial Hypoth. It is considered that destruction of this area in pain syndromes eliminates the diffuse protopathic pain without impairing epicritic sensation (Sano *et al.*, 1970b, 1975b).

Stimulation and coagulation of the posteromedial Hypoth substantially increases the β-endorphin level in ventricular fluid. Although the endorphin content returns to the initial level in several days, pain is relieved for a long time (Mayanagi *et al.*, 1982). It is also thought that the relief of pain after destruction of the posterior Hypoth can be explained on the basis of exhaustion of catecholamines in the subcortical structures, since stimulation of Hypoth increases HVA in the third-ventricle CSF (Amano *et al.*, 1979).

Nevertheless, these data do not explain the specific basis of the therapeutic effect of Hypoth destruction in pain syndromes. This remains more obscure because electrostimulation of the target during surgery does not induce sensory reactions. Patients without chronic pain frequently experience a sense of fear but no sensation of pain. Only in patients with intractable pain does stimulation cause a temporary cessation of pain (Mayanagi *et al.*, 1982). No impairment of sensibility, endocrinological functions, or mental activity are observed postoperatively.

Sano (1974) usually performs this operation on both sides at an interval of 7–10 days. The stereotactic coordinates in posteromedial hypothalamotomy are the following: 1–2 mm posterior to the midpoint of LI, 2–4 mm below that line, and 2–3 mm lateral to the lateral wall of V_3. Mayanagi *et al.* (1982) recommend a somewhat greater distance (5–6 mm) from the midplane. The destroyed focus in Hypoth should be no more than 4–5 mm in diameter.

The results of posteromedial hypothalamotomy have been analyzed by several investigations. Sano *et al.* (1975b) performed this operation in 18 patients with severe pain of various etiologies, mainly malignant tumors. The operations were done on one or both sides. After the operation, as a rule, there was a marked decrease in the severity of the pain. In chronic pain in the area of the trunk and limbs, better results were obtained after surgery on the side contralateral to the pain, and with pain in the area of the face, head, and neck, after surgery on the same side. Mayanagi *et al.* (1981) reported that in 20 patients with malignant and nonmalignant pain, the long-term relief was complete or almost complete in 60% of cases, partial in 35%, and nil in 5%.

Posterior hypothalamotomy is more effective in cancer pain than in pain of other origin (Fairman, 1972; Amano *et al.*, 1978; Mayanagi *et al.*, 1981, 1982).

Posterior hypothalamotomy is not an entirely safe operation. Šramka and Nadvornik (1975b) have described a hemorrhage in V_3 with FM occlusion after two such operations. An emergency craniotomy and evacuation of the hematoma saved the patients, and their neurological disorders gradually disappeared. Amano *et al.* (1978) consider hypothalamotomy to be contraindicated in patients in poor general condition, with mental disorders, and older than 70 years.

The long-term results of stereotactic hypothalamotomy have not yet been studied adequately. Nevertheless, one may consider this operation to be an effective method of managing pain syndromes, especially in malignant tumors.

5.2.4. Destruction of the Cingulate Gyrus

This operation also does not destroy pain pathways and centers but rather one of the structures in the limbic system that modifies the mental reaction. The emotional component plays a very important role in the pathogenesis and clinical manifestations of pain syndromes. Since emotional reactions are closely linked with the phylogenetically ancient limbic system and with the appreciation of pain, operations on structures of this system are most effective when there is a clearly defined emotional component. Surgical intervention in the limbic system often produces an effect resembling prefrontal leukotomy—the sensation of pain remains, but its agonizing emotional and psychological concomitants are absent.

Of the pain-alleviating stereotactic operations on the limbic system, it is first necessary to consider GC destruction, cingulotomy. Cingulotomy or cingulectomy by the open method was first performed by Ward in 1948 and then by Whitty *et al.* (1952), Livingston (1953), Le Beau (1954), and Sano (1954). The operation is successfully applied in certain mental disorders (Chapter 18, Section 2.1).

Morphologically, GC has two main attributes: first, extensive reciprocal connections with different noncortical zones, and second, tracts to and from the Th nuclei (Stephan and Andy, 1982). The GC is a polysynaptic pathway linking the medial frontal cortex with the anterior, dorsomedial, and intralaminar Th nuclei and also with Hipp. It has been established that damage to the GC may relieve certain pain syndromes, possibly explained by interruption of cingulostriatal fibers (Laitinen *et al.*, 1973).

Initially, cingulotomy was performed by an open approach, conventional osteoplastic craniotomy. Recently, as a rule, GC destruction has been done stereotactically, first by Foltz and White (1962) and later by Ballantine *et al.* (1967). Stereotactic cingulotomy has some important advantages in that the operation is relatively atraumatic, is tolerated well, and has minimal risk, since there are no functionally important structures nearby, damage to which could lead to serious complications.

Both GCs are located on the inferiomedial surfaces of the hemispheres above CC and extend for a considerable distance in the AP direction. Thus, there are several possible sites for destruction along the length of this gyrus. Certain authors consider the optimal target to be the middle part of the gyrus along the ventral margin. Other authors prefer destruction of the anterior (rostral) part of this gyrus, inferior and anterior to the knee of CC, subrostral cingulotomy (Laitinen *et al.*, 1973). Since at this site the GC is small, significantly less than posteriorly, a relatively small stereotactic destructive focus may totally interrupt the fiber tracts.

The main reference points for this operation are clearly visible in ventriculograms: the tip of the anterior horn of V, CC, and the midplane. It is our practice, with the patient under local anesthesia to make stereotactic lesions on both sides in the middle part of GC (Fig. 156). After ventriculography, stereotactic calculations are made using the previously mentioned reference points. Both GC are destroyed through one or two burr holes made at the level of the coronal suture and 2 cm lateral to the sagittal plane.

In the majority of cases, stimulation of GC induces no subjective sensation; however, in approximately one-third of cases, various emotional reactions are produced (Laitinen *et al.*, 1973).

For the destruction of the gyrus in its middle part, the stereotactic coordinates are the following: 15–20 mm posterior to the tip of anterior horn of V, 4–5 mm above the middle part of CC, and 5–6 mm lateral to the midplane. The lesion in the cingulum should be large, approximately 17–18 mm in diameter, and made on the two sides consecutively. For destruction of the rostral GC (rostral cingulotomy), the stereotactic coordinates are the tip of the anterior horn, 10–12 mm inferior and anterior to the knee of CC, and 5–6 mm from the midplane. Since the anterior part of GC is small, the lesion need be only approximately 8–10 mm in diameter. Some mental confusion may be observed very rarely in the first days after surgery.

One of the first papers demonstrating the effectiveness of cingulotomy was the report by Foltz and White (1962) on the results of different pain syndromes with drug addiction. In order to locate GC more accurately, combined conventional ventriculography with angiography may be used. They obtained satisfactory results after cingulotomy in four patients with intractable pain and three cases with phantom pain syndromes. Teuber *et al.* (1977) performed bilateral cingulotomy in 34 patients with pain or mental disorders, noting that this operation was more effective in chronic pain syndromes. In all likelihood, the largest experience with cingulotomy in intractable pain with drug addiction was presented by Brown (1977). In long-term follow-up of 43 operated cases, he reported that pain disappeared in 35% of the patients, and was greatly diminished in another 40%. Some authors have noted that the beneficial effect obtained directly after cingulotomy gradually diminished as time passed (Laitinen *et al.*, 1973; Hurt and Ballantine, 1974).

Other authors are more reserved in their assessment of the efficacy of cingulotomy in the management of severe pain syndromes. Wilson and Chang (1974), who performed bilateral anterior cingulotomy in 23 patients, the majority of whom suffered from pain caused by malignancy, reported a significant improvement of the pain in only 52% of the cases. Voris and Whisler (1975) also considered that cingulotomy seldom relieved pain; however, they admitted that the operation was quite effective in alleviating the feelings of fear, anxiety, and depression.

5.2.5. Operations on the Midbrain and Brainstem

5.2.5a. Mesencephalotomy. For several decades sporadic attempts have been made to manage intractable pain surgically by interrupting the spinothalamic pathways in the midbrain and brainstem. The operation involves destruction of pain-relaying pathways in the dorsal part of the mesencephalon at the level of the anterior part of the quadrigeminal plate (mesencephalic tractotomy or mesencephalotomy). The objective of the operation is destruction of the ascending spinothalamic and trigeminothalamic tracts at the tegmentum, where they converge beneath the inferior colliculus and lateral to PAG in the form of thin compact bundles at a distance of 3–4 mm from each other (Walker, 1940, 1972). The trigemino-

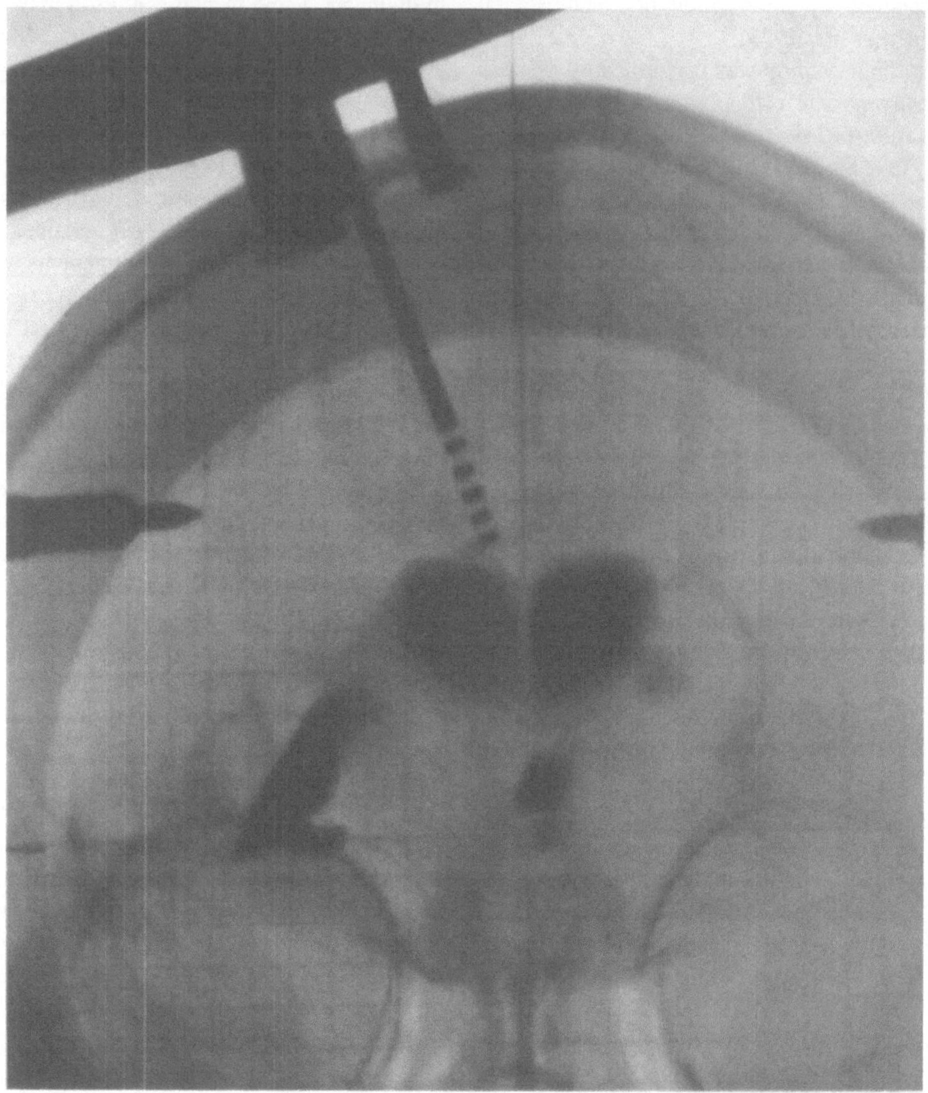

FIGURE 156. AP ventriculogram made during a right stereotactic cingulotomy. The multicontact electrode is introduced in the middle part of CC.

thalamic tract lies more medially and borders on PAG, whereas the spinothalamic tract lies laterally. Walker has shown a somatotopic arrangement of the fibers in the mesencephalic spinothalamic tract: the pain-conducting fibers representing the lower half of the body are located dorsolaterally, and those from the upper half ventromedially. There are reasons to consider that mesencephalotomy also involves the medially lying PAG, which plays a very important part in the pain inhibitory system (Section 3). Thus, we have a paradoxical result—both destruction and stimulation (Section 5.4.1e) of the same cerebral structure, in this case PAG, produce the same analgesic effect.

It is assumed that surgical destruction at the mesencephalic level also interrupts the diffuse spinoreticular system, thus interrupting both the lateral and medial pain-relaying systems (Walker, 1942a,b; Spiegel and Wycis, 1954; Nashold, 1972, 1982). Besides eliminating intractable pain, interruption of the spinoreticular pathways also reduces the emotional component of pain, thereby resembling cingulotomy (Section 5.2.4) and other operations on the limbic system.

Mesencephalotomy by the open method using an occipital craniotomy was first successfully performed by Walker in 1942. Section of the spinothalamic and trigeminothalamic tracts at the level of the superior quadrigeminal body leads to analgesia, thermoanesthesia, and pain relief in the contralateral half of the body but is a technically complicated and hazardous operation. After the first 29 such operations reported in the literature, the mortality was 24% (White and Sweet, 1955). In a number of cases there were oculomotor disorders and severe postoperative dysesthesias, which possibly were worse than the original pain. Although there have been few reports of its effectiveness (Drake and McKenzie, 1953; Zapletal, 1969), mesencephalotomy by the open approach should be considered an important historical landmark in the surgery of pain.

The "physiological permissibility" of this operation grew substantially after the introduction of the stereotactic method. The first stereotactic mesencephalotomy was performed by Spiegel and Wycis in 1947 on a woman with dysesthesia-type facial pain after several unsatisfactory operations on the trigeminal system. Two destructive foci were made, one lateral to the Sylvian aqueduct at the level of the superior colliculus involving the spinothalamic tract in the rostral part of the midbrain tegmentum at the base of CP, and the second in the dorsomedial Th nuclei. The relief of pain continued for the remaining 14 years of the patient's life.

In subsequent years, Spiegel and Wycis improved this operation. In patients with mental disorders, in particular, drug addicts, the authors destroyed not only the spinothalamic tract but also the dorsomedial Th nuclei. In 1966 the authors reported long-term results of stereotactic mesencephalotomy in 54 cases. Pain was relieved in 70% of the patients, but approximately half had relapses in the following months. Lasting relief from pain was noted in only one-third of the cases. Stereotactic mesencephalotomy was subsequently performed by many other neurosurgeons (Riechert, 1960; Mark *et al.*, 1960; Helfand *et al.*, 1965; Liberson *et al.*, 1970; Whisler and Voris, 1978; Seaber and Nashold, 1980; Amano *et al.*, 1980; Frank *et al.*, 1982; Shieff and Nashold, 1987) with encouraging results.

Several authors recommend mesencephalotomy on only one side, since bilateral operations involve greater risk. However, Nashold *et al.* (1969b) and Frank *et al.* (1982) do not consider bilateral operations exceedingly hazardous.

Stereotactic mesencephalotomy is usually performed under local anesthesia with mild premedication so that intraoperative observation may be made of the effects of stimulation in that area. Two approaches—posterior and anterior—are employed for introducing the probe into the midbrain target. Irrespective of the approach, the operation is performed on the side contralateral to the site of the pain. The posterior approach proposed by Spiegel and Wycis (1966) uses a burr hole in the parietooccipital area through which to pass a probe parallel to the midplane through CP.

The majority of surgeons now employ the anterior transfrontal approach with a burr hole near the coronal suture (Nashold, 1972, 1982; Amano *et al.*, 1980; Frank *et al.*, 1982). The trajectory of the probe passes 1.5 cm from the midplane via the posterior Th nuclei to the rostral midbrain in the area of CP and the aqueduct. Nashold (1982) considers the anterior approach as advantageous in that the probe enters the dorsal tegmentum in the rostrocaudal direction along the ascending sensory pathways. In order to avoid damaging the quadrigeminal plate as the probe advances toward the midbrain, Bosch (1983) recommends making the burr hole and introducing the probe 2–3 mm anterior to the coronal suture; in this procedure, it is very important to have a clear outline of CP and the rostral part of the Sylvian aqueduct in ventriculograms.

It has been established that in the lateral spinothalamic tract at the level of the dorsolateral portion of the midbrain tegmentum, there is a definite somatotopic arrangement—the facial area is represented more medially, and the trunk and limbs more laterally. The functional significance of this part of the brain makes it imperative that the stereotactic lesion in mesencephalotomy be small—no more than 3–4 mm in diameter. The optimal site of this lesion should correspond precisely with the site of the spinothalamic tract in the dorsomedial part of the midbrain.

The stereotactic coordinates of different authors for mesencephalic tractotomy vary within the following range: 2–5 mm caudal to CP, 2–5 mm below the plane of LI, and 6–8 mm from the midplane. Frank *et al.* (1982) emphasize that in order to prevent oculomotor damage, it is very important to avoid the collicular pathways of the quadrigeminal plate. For this reason, these authors consider that the focus should be made no less than 7–8 mm from the midplane. However, Whisler and Voris (1978) and Amano *et al.* (1979) select the target point 4–5 mm lateral to the midpoint of the aqueduct, and Bosch (1983) locates it at a distance 6.5–7 mm. In case of pain in the face and oral cavity, Nashold (1982) selects a target point 5 mm from the midplane, and in case of widespread pain affecting the neck, arm, and trunk, increases this distance to 7 mm (Fig. 157).

Stimulation and EP study in mesencephalotomy are important for accurate localization of the target. If the target in the spinothalamic tract is hit accurately, stimulation induces sensations such as localized pain and burning in the contralateral limbs (Nashold *et al.*, 1969a). In view of this, destruction is recommended only on obtaining clear-cut thermal sensations on stimulation, thereby avoiding damage to the lemniscus medialis (Frank *et al.*, 1982).

Stimulation of FR in mesencephalotomy usually produces severe pain in the contralateral half of the body (Nashold *et al.*, 1969a), whereas stimulation of PAG close to the aqueduct produces affective and emotional reactions (Nashold, 1982).

The appearance of homolateral oculomotor disturbances along with myosis during stimulation indicates that the tip of the probe is too deep in the ventral oculomotor pathways, whereas motor reactions are an indication that the corticospinal tract lateral and anterior to the pain-relaying pathways has been stimulated. Excitation of the lemniscus medialis induces sensations resembling passage of an electric current (Amano *et al.*, 1978).

Operations that have been performed correctly from the technical point of view result in a significant reduction or total disappearance of pain or temperature sensibility in the opposite half of the body. Hemianesthesia is a mandatory sign of an effective mesencephalotomy.

In contemporary stereotactic neurosurgery, mesencephalotomy is employed relatively rarely because it is technically difficult and often results in postoperative complications such as dysesthesia and oculomotor impairments. These include paralysis of upward gaze (Parinaud's syndrome) or down gaze (Foville's syndrome), bilateral ptosis, absence of pupillary reaction, miosis, diplopia, impaired convergence, as well as nystagmus. Oculomotor complications are explained by the fact that at the level of the rostral midbrain tegmentum, the spinothalamic tract lies in close proximity to the oculomotor nuclei and their pathways. Postoperative oculomotor disorders often recede, whereas severe dysesthesias, as a rule, persist. After stereotactic mesencephalotomy such dysesthesias occur much more rarely (in approximately 5% of the cases) than after an open operation. Dysesthesias considered to be caused by damage to the lemniscus medialis (Nashold *et al.*, 1979b) usually develop a long time after the operations, so they are not a problem in cancer patients with a short life expectancy.

Favorable results of mesencephalotomy for the relief of various types of pain have been reported by Nashold *et al.* (1969a, 1972). Of the 15 patients operated on for pain following traumatic avulsion of the brachial plexus, thalamic lesions, and other central syndromes, the pain totally disappeared in eight cases and partially in three. This operation is undoubtedly effective in cases of central pain. At the same time, the

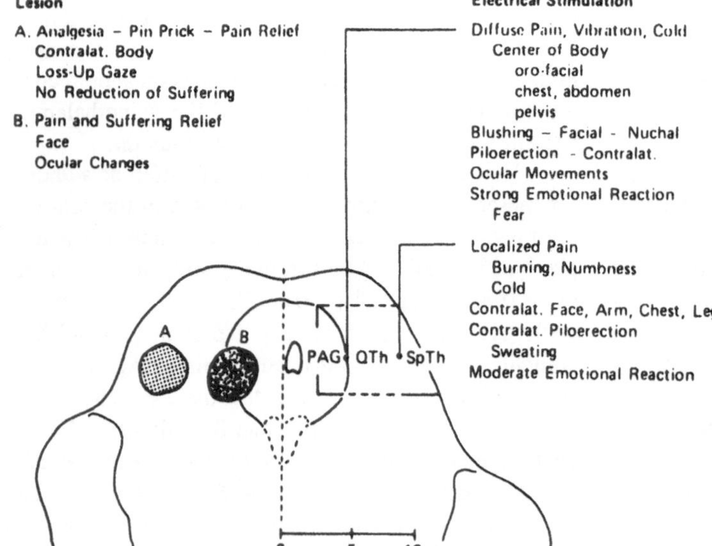

FIGURE 157. The effects of a stereotactic lesion and electrical stimulation in the human mesencephalon at the level of the superior colliculus (Nashold, 1982). PAG, periaqueductal gray; QTh, quintothalamic tract; SpTh, spinothalamic tract.

authors reported serious complications—postoperative paralysis of upward gaze in almost half of the cases, and residual severe dysesthesias in five cases. Similar complications after two operations were reported by Torvik (1959a,b).

In their next report, Seaber and Nashold (1980) discussed the relationship of oculomotor and pupillary complications to the site of the lesion. In 22 patients operated on for intractable pain, the destructive foci were located in the dorsomedial tegmentum of the midbrain; in a number of cases it was at the level of the CP, and in others, 5 mm posterior to CP and somewhat closer to the midline near the superior quadrigeminal colliculus. In a second group of patients, oculomotor and pupillary disorders, particularly upward gaze paralysis, were noted significantly less frequently after surgery. In all likelihood, the location of the second focus in the midbrain was more optimal.

After mesencephalotomy there are no respiratory disorders, which constitute one of the gravest complications in percutaneous cordotomy (Section 5.3.2). In view of this, unlike cordotomy, mesencephalotomy may be performed even in the presence of preexisting respiratory disorders. The postoperative mortality from this operation is 3–5%, which is approximately one-fifth to one-fourth that after open mesencephalotomy. In the series of 15 patients operated on by Nashold *et al.* (1969b), there was no fatal outcome.

Mesencephalotomy is considered most effective in pain caused by malignant tumors of the head, neck, and shoulder girdle (Voris and Whisler, 1975; Nashold, 1982; Frank *et al.*, 1982). Relief of pain for the duration of life should occur in 90–95% of patients (Whisler and Voris, 1978). A unilateral operation may produce bilateral pain relief.

In the management of pain syndromes, the destruction of the more rostral mesencephalon was also tried, an operation that has been named mesencephalic reticulotomy (Amano *et al.*, 1980). As the authors of this operation emphasize, it differs substantially from the classical mesencephalotomy since it aims to destroy not the lateral spinothalamic tract at midbrain level but the most medial part of FR, where the extralemniscal spinoreticular pain-relaying pathways terminate. This area is located on the border of PAG at the level between CP and the superior colliculus.

The stereotactic coordinates of the target point for mesencephalic reticulotomy are the following: 14 mm posterior to the midpoint of LI, 5 mm below that line, and 5 mm lateral to the center of the aqueduct. Recently Amano *et al.* (1985) reported the long-term results of this operation in 34 patients with intractable pain. Good results were noted in 68% of cases. Parinaud's syndrome developed after surgery in 26% of cases. Other neurological complications were few.

Frank *et al.* (1982) analyzed the results of rostral mesencephalotomy in 14 cases of thoracobrachial pain, frequently spreading to the neck and face, as a result of malignant neoplasms of the lung and breast. Mesencephalotomy was performed on both sides in three patients with bilateral pain. Directly after surgery, pain disappeared in all patients, and there was complete or practically complete analgesia in the contralateral half of the body. In a comparatively short follow-up (average 5 months) because of the primary disease, the pain recurred in only two cases. These patients underwent a second operation, which yielded a good result. In a subsequent report (Frank *et al.*, 1982), the authors gave the results of rostral mesencephalotomy in 40 cancer patients with follow-up from 6 months to 2 years. Directly after surgery, pain alleviation was noted in all cases, but the pain returned later in six cases. Anesthesia dolorosa developed in 15% of the patients, and transient oculomotor disorders in 20%. There was no postoperative mortality.

One may conclude that stereotactic mesencephalotomy is, without a doubt, effective in the treatment of certain cases of central pain, pain in the neck and head caused by malignancy, postherpetic pain in the face, and pain after trauma to the brachial plexus. Elimination of the pain may be attained in approximately half of these cases. However, the comparatively frequent complications have prevented its extensive application.

5.2.5b. Pontine Spinothalamic Tractotomy. Section of the spinothalamic tract at the pontine level by an open approach was first performed by Dogliotti in 1938. Although a few other attempts were made to perform open operations to destroy the sensory nuclei of the trigeminal nerve, this technically complicated and hazardous operation has not been widely practiced.

A similar procedure using a stereotactic technique (stereotactic pontine tractotomy) was proposed by Hitchcock in 1973. In favor of the operation, Hitchcock points out that at the pons level there is a dichotomy of the spinothalamic tract and the autonomic pathways. Hence, one may expect isolated destruction of the pain pathways at that level to induce cephalic analgesia without respiratory or bladder complications. The stereotactic coordinates of the brainstem structures were determined for this operation. The base of V_4, the Sylvian aqueduct, and Fm were selected as the reference points in ventriculograms in two projections.

In brief, that complicated operation is performed under local anesthesia with the patient in a sitting position. The stereotactic apparatus is attached at the burr hole in the occipital bone. With the map of the brainstem (Hitchcock, 1973; Hitchcock *et al.*, 1985), the target point is determined on ventriculograms made through another burr hole at the coronal suture. The stereotactic coordinates of the spinothalamic tract at the pontine level are the following: 5–6 mm anterior to the vertical plane of Fm perpendicularly to the base line of V_4 and 7–8 mm lateral to the midplane. The electrode is advanced to the target under fluoroscopic control. The induction of paresthesias at the site of the pain on stimulation with 0.1-msec rectangular impulses of 2–5 V and 60–100 Hz verifies the correct position. Because of the very small size of the pontine spinothalamic bundle, it must be "hit" very accurately, so that a lesion made by high-frequency coagulation does not exceed 3 mm in diameter. Analgesia in the painful area is present immediately.

In addition to the work of Hitchcock, Barbera *et al.* (1979) reported similar operations in five patients with malignant tumors in various locations. The authors consider intractable pain in the neck, upper limbs, and thorax to be the main indications for pontine tractotomy. Good results from pontine tractotomy were reported. There were no postoperative motor or oculomotor disorders such as occur after mesencephalotomy. However, the limited experience with pontine tractotomy does not allow one to draw final conclusions concerning its efficacy.

To conclude this section on operations on various brain structures for the relief of pain syndromes, we dwell briefly on one more pain-alleviating operation, destruction of the pituitary gland.

5.2.6. Hypophysectomy

Hypophysectomy is a special functional operation aimed at relieving pain in any part of the body. In literature, this procedure, pituitary neuroadenolysis, has also been done by chemical ablation of the hypophysis. It has long been known that after surgical destruction of a normal pituitary gland, one may frequently observe not only regression of the spread of metastases of hormone-dependent cancer (see Chapter 15) but also relief of pain. Later, it was shown that pain relief also occurred after hypophysectomy in hormone-independent cancer (Moricca, 1977). Moreover, quite frequently the pain disappears long before signs of regression of metastases are noted. This leads to the logical assumption that pituitary hormones play an important role in pain relief. After pituitary neuroadenolololysis there is a sharp increase in CSF adrenocorticotropin, TRH, vasopressin, and ACTH; however, the endorphin level does not change appreciably (Takeda *et al.*, 1983a,b). There is another point of view in the literature—pain relief is not related directly to the expected fall in the levels of known pituitary hormones (Ramirez and Levin, 1984). In view of this, it is assumed that the suppression of pain results not only from pituitary destruction but also from stimulation of the hypothalamo–hypophyseal axis (Takeda, 1984) or from a hypothalamic pain-suppressing capability triggered by hypophysectomy (Ramirez and Levin, 1984).

However, this operation may affect cerebral structures related to pain perception and not only endocrine regulation. From the other side, the operation does not change the normal sensitivity to pinprick or the appreciation of pain from acute injury (Miles, 1979). It has been established that the threshold of C-fiber excitation is elevated after this operation. The specific pain-suppressing mechanism after hypophysectomy remains obscure, although it is clear that the rationale of this operation is not only to eliminate pain but, by modifying the hormonal balance, to cause the tumor metastases to regress.

We perform the pituitary destruction under neuroleptanalgesia and, with our stereotactic apparatus, accurately introduce a cryoprobe transnasally and transsphenoidally into the sella turcica (see Chapter 15). Various methods are employed to destroy the adenohypophyseal tissue—high-frequency coagulation, cryodestruction, slow injection of a small quantity (1–1.5 ml) of alcohol (ethanol) or other lytic substances (Katz and Levin, 1977). It should be noted that the introduction of these latter substances into the pituitary has the risk of their spreading to the pituitary stalk or into the cerebral ventricle. When metrizamide was injected together with the lytic substance, ethanol, into the pituitary, in half of the cases, the contrast medium spread into V_3, V_4, and the Sylvian aqueduct (Takeda *et al.*, 1983a,c).

In the majority of cases, hypophysectomy leads to a significant alleviation of pain in malignant neoplasms. Such an effect was obtained in more than 90% of the cases by Levin *et al.* (1980), who injected pure alcohol into the pituitary. Hardy (1975a) performed transsphenoidal hypophysectomy in 160 patients with pain from metastases of breast cancer. Good and lasting pain relief was obtained in more than 90% of these patients. Takeda *et al.* (1983a,b, 1985), after hypophysectomy, noted pain relief in 80% of 102 terminally ill patients with intractable pain caused by diffuse cancer metastases. In hormone-dependent carcinomas the efficacy of the operation was greater than 95%. The authors stated that pain relief after hypophysectomy may be obtained in hormone-dependent as well as hormone-independent cancer, although 93% of patients had good pain relief in the former and only 70% in the latter. The clinical effect did not depend on regression of metastases.

Severe pain caused by bone metastases disappeared in almost 80% of patients after stereotactic hypophysectomy with high-frequency coagulation (Landolt and Siegfried, 1970).

Moricca (1977), who injected ethanol in the pituitary in 848 patients, has reported the largest experience in pituitary neuroadenolysis for the management of pain from cancer metastases. The procedure was repeated 3–4 times in many patients. Although the majority of patients underwent surgery for hormone-dependent cancer, 53 had other types of tumor (cancer of the lungs, urinary bladder, larynx, and others). Significant, prolonged relief from pain was observed in 708 cases (83%). Good alleviation of pain by the stereotactic destruction of the pituitary has been reported by many other surgeons (Katz and Levin, 1977; Tindall *et al.*, 1977; Lipton *et al.*, 1978; Kandel, 1981; Takeda *et al.*, 1983a,b, 1985). At the same time, certain authors have obtained less satisfactory results. Miles (1979) noted significant relief in only 42% of the patients operated on, and Lipton *et al.* (1978) reported total pain relief in one-third of the patients, partial pain relief in another third, and no effect in the remaining third of their series. Ramirez and Levin (1984) summarized and reviewed thirteen published series of patients who underwent surgical or chemical hypophysectomy for the relief of cancer pain. Both techniques gave similar results and produced pain relief in about 70–75% of the patients.

Levin *et al.* (1980) reported on three patients with intractable thalamic pain after stroke; after chemical hypophysectomy, the agonizing pain totally disappeared in all three patients who were followed up from 19 to 58 months.

Hypophysectomy as a rule is tolerated well by patients even in serious condition but may have various complications. Takeda *et al.* (1983a,c) noted transitory diabetes insipidus in half of 102 cases, visual field defects in 10%, and temporary ophthalmoplegia in 4% of the patients after surgery. If the needle is introduced too laterally, injury to the oculomotor or abducens nerves may occur. Very rarely, there were defects in the fields of vision, especially if 2.4 ml or more ethanol was injected.

Unquestionably, hypophysectomy is a very effective operation to eliminate pain from cancer metastases, especially bone metastases from mammary and prostatic cancer. However, this operation is not very effective in the treatment of pain from other causes.

5.3. Operations on the Spinal Cord

5.3.1. General Remarks

There are many operations on pathways and roots of the spinal cord for pain alleviation. Some of these procedures have already been employed in neurosurgery for several decades. We briefly discuss these classical operations performed by the open method, i.e., laminectomy with exposure of the spinal cord. Our primary focus is on new functional and stereotactic operations on the spinal cord.

For a number of reasons, stereotactic surgery on the spinal cord began to develop later than stereotactic surgery on the brain. Nevertheless, in the comparatively short period of 20 years, this new technique has achieved very tangible and striking success. A special stereotactic apparatus (Hitchcock, 1969a,b) (Fig. 158) has been devised for operations on the spinal cord by the Czechoslovakian firm Hirana (Fig. 159), by Radionics, and by others. The stereotactic devices for percutaneous cordotomy are described below (Section 5.3.2c).

Significant differences between stereotactic surgery on the brain and the spinal cord are obvious, yet the basic principles of the methods are identical. One of the most important considerations that must be taken into account and corrected for in stereotactic operations is the mobility of the spinal cord in the vertebral canal. It has been shown that the spinal cord becomes fixed in the vertebral canal only if the head is flexed, whereas stereotactic operations, in particular, percutaneous cordotomy, are more conveniently performed with the head in the "ordinary" position. Just as in cerebral stereotaxis, difficulties are encountered in stereotactic operations on the spinal cord because of the anatomic variations of the spinal pathways. At the level of C_2 segment near the cervicomedullary junction there is a

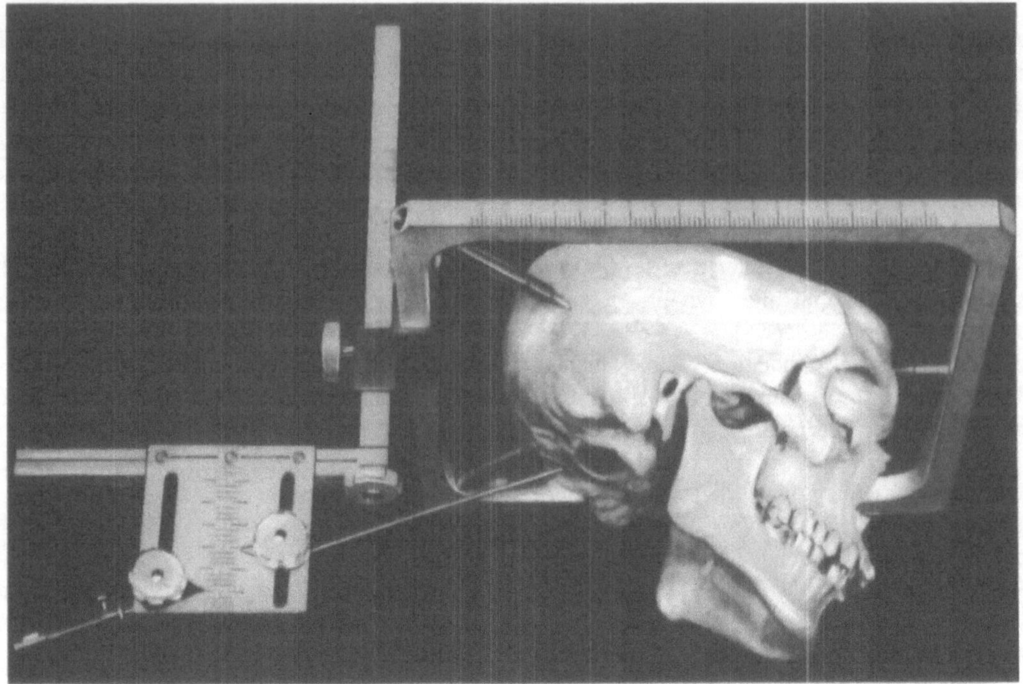

FIGURE 158. Stereotactic apparatus of Hitchcock for operations on the spinal cord. Description in the text.

significant variability of the size and position of the different tracts, which depends on the increased diameter of the spinal cord at C_2 level and also on the site of the decussation of the motor tracts.

In contemporary literature the term "stereotaxis of the spinal cord" refers to intervention at the craniospinal level by introducing an electrode into the occipital foramen between the base of the skull and the arch of the atlas. Certainly both percutaneous cordotomy by various approaches (Section 5.3.2) and myelotomy (Section 5.3.3) should be classed as stereotactic operations.

The AP diameter of the subarachnoid space at the upper segments of the cervical spinal cord is 20 mm; the diameter of the spinal cord is 10 mm, and the subarachnoid space is 10 mm (3 mm anterior and 7 mm posterior to the spinal cord) (Mullan *et al.*, 1965a). However, Iseki *et al.* (1982) reported that normally the AP projection of the spinal cord at C_1 level is 11 ± 0.3 mm, and the AP dimension of the subarachnoid space is 15.7 ± 0.9 mm. At C_2 level, these dimensions are, respectively, 10.4 ± 0.2 and 13.7 ± 0.5 mm. When the head is flexed, the vertebral canal (especially its posterior surface) is elongated, and the spinal cord adapts itself by "plastic deformation," changing its shape and thickness (Schvarcz, 1974). The anterior and posterior displacements of the cervical spinal cord in flexion and extension of the head differ at various levels. For example, at C_1, the displacement on the average is 3.3 mm, and at C_3, only 0.5 mm (Kawamura *et al.*, 1972). In view of such displacements, it is considered that fixation of the head inclined forward by a special stereotactic apparatus provides optimum position for fixation of the spinal cord.

Hitchcock (1969a) considers one advantage offered by the stereotactic approach to segments C_1–C_2 via the atlantoccipital membrane, as compared to percutaneous cordotomy at lower levels (Section 5.3.2), to be the greater accuracy in "hitting" the target point as a result of "immobilization" of the spinal cord.

In stereotactic operations on the spinal cord, one of the main reference points is the dens of the second cervical vertebra, which requires no contrast medium to visualize but only an AP film through the open mouth. The midline of its axis coincides with the midplane of the spinal cord (AP projection) with a strictly sagittal position of the head. Nevertheless, for each stereotactic operation on the spinal cord, it is necessary to employ contrast medium to define reference points that are not visible on conventional roentgenograms, so that the target point inside the spinal cord may be calculated.

Stereotactic operations on the cervical spinal cord (percutaneous cordotomy, medullary tractotomy, and

FIGURE 159. Chirana stereotactic apparatus for operations on the spinal cord (Czechoslovakia).

others), at the present time, have been the best developed and tested clinically; however, stereotactic procedures have also been used in operations on other parts of the spinal cord, for example, thoracic and lumbar myelotomy.

In such operations, just as in cerebral stereotactic surgery, it is important to control the site of the electrode, particularly by electrostimulation and impedance measurements. These methods are used to establish when the electrode, having passed through the subarachnoid space, enters the spinal cord and to identify the penetration of the pain pathways. Because of the small diameter of the spinal cord and the number and complex localization of its tracts, the size of the spinal cord lesion should be substantially less than that made in cerebral stereotactic operations. In principle, the diameter of these foci should not exceed 3–3.5 mm.

5.3.2. Cordotomy

The object of cordotomy is to interrupt pain-transmitting pathways in the spinal cord. As is known, the main tracts for pain and temperature sensation are the spinothalamic tracts traveling in the anterolateral quadrants of the spinal cord. The classical open cordotomy has been described many times, so we discuss the technique and results of that operation only briefly.

Many authors point out that cordotomy is effective only in organic, somatic pain and does not give relief in so-called functional pain. In order to differentiate such pain and, consequently, to select suitable patients for surgery, a preliminary epidural block with a 1–2% procaine solution is recommended. This test relieves organic pain for several hours but does not influence functional pain (Clark and Ervin, 1958).

The indications for cordotomy are quite variable. In first place, the intractable pain, mainly unilateral and localized below C_4–C_5 dermatomes, should result from malignant tumors, although the procedure has been successfully used in a wide range of benign pain of various etiologies (after injuries of the spine and its roots, in particular, after unsuccessful operations for intervertebral disk protrusions, adhesive arachnoiditis, intractable pain in the extremities of various types, in particular, malignant tumors, etc.). Nashold and Friedman (1972a,b) emphasize that cordotomy is especially helpful in thoracic and abdominal pain.

For pain caused by central lesions (thalamic and

phantom pain), cordotomy by any approach is completely ineffective. Neither is cordotomy indicated in head and face pain. For neuralgias of the peripheral nerves, cordotomy should only be a procedure of last resort (White, 1973).

5.3.2a. Somatotopic Organization of the Spinothalamic Tract. The spinothalamic tract was discovered about 100 years ago, but its function was unknown. In 1905, Spiller attributed the loss of pain and temperature sensation in the contralateral half of the body to a small tuberculoma in the anterolateral segment of the spinal cord.

The spinothalamic tract, a phylogenetically new nociceptive pathway to the brain, when stimulated, evokes a sharp, well-localized pain referred to a localized region of the body.

On the basis of anatomic investigations as well as stimulation in percutaneous cordotomy (see Section 5.3.2c), it has been established that in the spinothalamic tract there is a clear-cut "multilayer" representation of the parts of the body in the AP projection. Lying close to the surface of the spinal cord and slightly anterior to the dentate ligament in the posterolateral part of this tract are the pathways from the lower (sacral) dermatomes. In more central layers about the anterior horn are the lumbar, thoracic, and, finally, cervical dermatomes (Fig. 160). It has also been established that the pathways relaying cutaneous sensation lie closer to the surface in the spinothalamic tract, whereas the deeper parts of the tract contain fibers that transmit, respectively, temperature, deep sensibility to pain, and finally, visceral sensations (Poletti, 1982). The majority of the fibers in the tract represent the hand, whereas the fibers from the neck, torso, and lower limbs are represented by fewer fibers. At the cervical level, the spinothalamic tract is closer to the surface than at the thoracic level. The center of a cordotomy lesion with a diameter of about 3 mm should be located in the bundle of fibers from the hand.

The speed of conduction of impulses along the lateral spinothalamic tract, measured by the EP method, is 45–60 m/sec as contrasted with earlier data of 5–20 m/sec.

It is necessary to emphasize the great individual variability of the localization and somatotopic organization of the spinothalamic tract. A nondecussated spinothalamic tract, which occurs to various degrees, may explain relapses after cordotomy. In the literature are descriptions of extremely rare anomalies such as the absence of decussation of the spinothalamic pathways (Voris, 1951). In such a condition, cordotomy naturally induces analgesia in the homolateral half of the body.

The corticospinal tract, which must not be injured, lies only 1–1.5 mm posterior to the spinothalamic tract.

5.3.2b. Open Anterolateral Cordotomy. For the first time, a cordotomy after a conventional thoracic laminectomy for the management of intractable pain was performed by Spiller and Martin

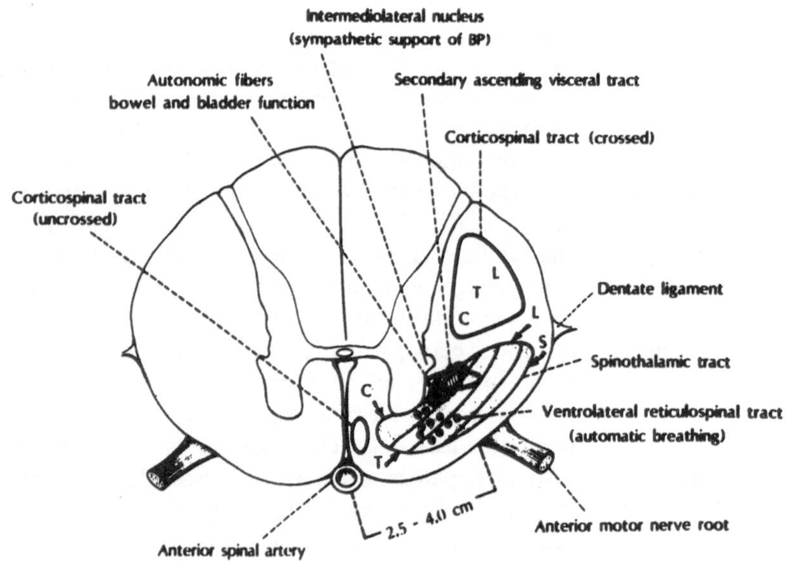

FIGURE 160. Somatotopic representation in the spinothalamic tract and its relation to the ventrolateral reticulospinal tract and corticospinal tract (Poletti, 1982). Explanation in the text. C, cervical; T, thoracic; L, lumbar; S, sacral.

(1912) and practically at the same time by Foerster (1913). Cordotomy may be performed at either the cervical or the thoracic level. Some authors recommend making the section at the cervical level even for pain in the legs and pelvic region (Ogle *et al.*, 1956; Graf, 1958; Mansuy *et al.*, 1976). After a technically correct interruption of the spinothalamic tract at any level, total analgesia should be produced below the level of operation on the contralateral half of the body.

In brief, the operative technique is as follows. The patient lies in a supine or, rarely, lateral position. In cordotomy at the high cervical level, the head of the patient must be somewhat below the level of the body in order to prevent air from entering the cerebral subarachnoid space (Schwartz, 1967). Cordotomy in the low cervical segments at the cervical enlargement is not recommended because of the risk of damaging the anterior spinal artery (Kahn, 1973).

The majority of authors perform this operation under general anesthesia or neuroleptanalgesia. Other authors (Schwartz, 1967) prefer local anesthesia, so as to be able to check the level of analgesia after the section of the spinal cord.

The usual laminectomy of the second and third vertebrae at high cervical or upper thoracic level, depending on the location of pain, is performed. If a unilateral cordotomy is to be carried out, hemilaminectomy may suffice. After the dura mater is opened by a sagittal, parasagittal, or transverse semicircular incision, the dentate ligaments are divided to permit turning the spinal cord. The cord is gently rotated approximately 45–60° by a needle holder or forceps grasping the dentate ligament, so that the anterior surface of the spinal cord is exposed laterally. If rotation of the spinal cord is difficult, which is often the case, one posterior root may be divided, preferably at the high cervical rather than the upper thoracic level. Kahn (1973) has noted that the cord can be rotated almost 90° so that the anterior spinal artery can actually be seen. The arachnoid membrane has to be removed from the surface of the cord and from the nearby posterior roots where the incision is to be made.

Then, in an avascular zone of the anterolateral quadrant, under a surgical microscope, a transverse incision is made beginning at the attachment of the dentate ligament to the cord. We, as do many other neurosurgeons, make the incision with a piece of a safety razor blade on a long handle or a special, slightly bent tractotome. In order to avoid rotating the spinal cord during incision, a special instrument with a mobile razor circumscribing an arch of 90° was used. Poletti (1982) performs the cordotomy using the Jacobsen instrument with a blunt tip bent at an angle of 45° to the axis of the instrument. Before making the incision of the cord substance, it is necessary to incise the pia just above the point where the instrument is to be introduced. It is recommended that the width of the spinal cord be measured in order to determine the optimal depth of the incision. Usually, the incision is made 4 mm long and to a depth of 4–5 mm in the ventral direction. To control the depth of incision, a distance 4 or 5 mm, depending on the width of the cord, is marked on the knife blade with bone wax. It is necessary to incise the entire anterolateral quadrant to the anterior spinal artery, which must be avoided.

The incision should be carried no more than 2 mm medial to the anterior root (Kahn, 1973). After the incision has been made, the patient is awakened so that the reaction to pin pricks and pinching of the skin may be determined. There should be no reaction to pain on the contralateral side of the body two to three segments below the incision. Many authors stress that the analgesia obtained after the incision must be complete. If there is only hypalgesia, even to a moderate degree, pain subsequently returns (White, 1963). Since there is a possibility of damaging the corticospinal tract, the strength and movements of the legs should be checked as the cuts are made. The dura mater is tightly sutured.

It is recommended that bilateral open cordotomy be made at two different levels with a distance of 6–8 mm between the incisions (Klar and Mletzko, 1960; Schwartz, 1967; Poletti, 1982). Some authors recommend bilateral cervical cordotomy even for unilateral pain (Mansuy *et al.*, 1976).

For more than 70 years, open cordotomy has been considered satisfactory, and even today, it is thought to be a reliable method for providing relief to patients with various pain syndromes, particularly in malignant lesions (Diemath *et al.*, 1961; White, 1963; Schwartz, 1967; French, 1974; Mansuy *et al.*, 1976). However, this operation is not devoid of complications, which limit its clinical application.

In the conventional technique of incising the anterolateral quadrant, the spinal cord must be rotated, which may displace its pathways, and as a result the incision may damage the corticospinal pathway lying posterior to the spinothalamic tract. Other complications, such as operative trauma, especially in seriously ill patients with malignant tumors, contribute, according to most authors, to a postoperative mortality rate of 10–15%. However, Mansuy *et al.* (1976) reported 3.2% mortality in 124 cordotomies for cancer pain.

There is also a relatively high frequency of complications, especially after bilateral operations, such as pareses of extremities, bladder and bowel disorders, dysesthesias, arterial hypotension, and sexual impairments. The frequency of weakness of the limbs after unilateral cordotomy is approximately 10%, and after bilateral cordotomy as high as 24% (Nathan, 1963). The mortality is also high after bilateral high (C_{1-2}) cordotomies.

One of the most serious complications after high cervical cordotomy, repeatedly described in the literature, is a peculiar type of respiratory paralysis (Ogle *et al.*, 1956; Belmusto *et al.*, 1963; Mullan and Hosobuchi, 1968). This complication results from damage to the descending pathway to the respiratory muscles at the level of the upper spinal cord, namely, the ventrolateral reticulospinal tract, which lies in the anterolateral quadrant slightly in front of the lateral spinothalamic tract (Hitchcock and Leece, 1967). Damage to this pathway during cordotomy causes respiratory insufficiency. An unusual paralysis of "automatic" (i.e., involuntary) respiration with the preservation of voluntary respiratory movements sometimes develops. This severe respiratory disorder during sleep ("sleep apnea") has been responsible for sudden death at night from respiratory paralysis, especially in the first 3 to 4 days after surgery. If the patient has even mild respiratory insufficiency before surgery (for example, as the result of a malignant lung tumor), then the risk of respiratory disorders after cordotomy increases sharply. In view of this, the respiratory capacity and O_2 content of blood before the operation as well as during the first few postoperative days must constantly be monitored.

A serious complication that is very difficult to eradicate is painful dysesthesia resembling thalamic pain below the level of the cordotomy, which develops late after surgery in approximately 6% of cases. A higher cordotomy in these cases is ineffective (Poletti, 1982).

The chief drawback of cordotomy by the open approach is the gradual recurrence, in the course of 1 or 2 years, of the original pain, presumably explained by the existence of other spinal pain-transmitting pathways (Kerr, 1975; Poletti, 1982). White (1963) performed cordotomy on 70 patients with various types of pain, mainly radicular, caused by a benign condition. Directly after the operation, the pain disappeared in more than 90% of the patients, but a year later, this figure dropped to approximately 50%.

In the first few weeks after cordotomy, the level of analgesia quite frequently drops by 3 to 6 segments, and after several months, by 6 to 9 segments, which to a considerable extent nullifies the initial positive postoperative result (Poletti, 1982). Many other surgeons have reported a comparatively high rate of pain recurrences long after cordotomy. Very pessimistic figures are given by Siegfried and Cetinalp (1981), who consider that 6 months after surgery the analgesia disappears in half of the patients, and after 3 years, in 80% of them. The effectiveness of open cordotomy also depends on the level of the cord section. Unilateral lumbar cordotomy results in good pain relief in 85% of cases, whereas high cervical cordotomy gave relief in only 50% (Nashold and Friedman, 1972a).

In view of the great number of relapses as well as various complications, in particular, painful dysesthesias, Poletti (1982) concludes that cordotomy is justified only in cancer patients with limited life expectancy. In several cases of recurrent pain after open cordotomy, percutaneous cordotomy (see Sections 5.3.2c–g) gave a good result (Tasker, 1982a).

In cases of pain in the area innervated by the brachial plexus, the arm and neck, a combined open high cervical cordotomy and section of the posterior roots C_2–C_4 have been recommended (White and Sweet, 1969).

Hardy *et al.* (1974) have developed a new approach for cordotomy. In this operation, section of the spinothalamic tract at the cervical level is performed by an open approach but, unlike conventional cordotomy, by an anterior approach to the cervical spinal cord. This approach, proposed by Cloward in 1958, is extensively used for removal of protruding cervical disks.

The operation is performed under endotracheal anesthesia. After removal of the intervertebral disk via an anterior cervical approach, a burr hole 18 mm in diameter is made in the vertebral body, the posterior wall of the body and the posterior longitudinal ligament are removed, and the dura is incised, exposing the anterior surface of the spinal cord. Then the operation is continued with a surgical microscope (×16). The anterior quadrants of the spinal cord, the anterior roots, the anterior spinal artery, and the dentate ligaments are clearly visible at the bottom of the narrow operative field. The spinothalamic tract is sectioned in an avascular zone (depth of incision, 4 mm) until the gray substance of the anterior horn, clearly visible under large magnification (×25), is exposed.

Microsurgical cordotomy by the anterior approach was performed by Hardy *et al.* (1974) in ten patients with pain caused by various cancers. After surgery, pain was relieved in all cases until the fatal outcome of the primary disease. This method may avoid the previously mentioned shortcomings of open cordotomy, which at present is performed rarely. But this traumatic operation in severely ill patients has prompted a search for safer and more effective procedures.

5.3.2c. Stereotactic Percutaneous Cordotomy. This operation has been extensively employed in recent years. As does the open operation, percutaneous cordotomy (PC) aims to interrupt the spinothalamic tract in the anterolateral quadrant of the spinal cord in the upper cervical region. Its main distinguishing feature is that it is performed by the percutaneous introduction of a thin electrode needle under x-ray control. The obvious advantage of such an operation, compared to open cordotomy, is, first of all, the minimal trauma, which reduces morbidity and mortality. At the same time, the precision of this technique substantially increases its effectiveness. In the majority of cases, no general anesthesia is required. If necessary, PC may be repeated. The complications resulting from open cordotomy (pareses, bladder dysfunction, dysesthesias, etc.) are observed much less frequently after PC.

Percutaneous cordotomy was proposed and first performed by Mullan *et al.* in 1963. Originally, these operations involved the manual introduction of a strontium needle in the spinothalamic tract between vertebrae C_1 and C_2 by a posterolateral approach. The tip of the needle containing radioactive strontium was inserted in the anterolateral quadrant of the ipsilateral half of the spinal cord. However, this technique had a number of serious drawbacks—inability to control the size of the lesion in the spinal cord, impossibility of arresting the development of the necrotic focus in case complications arose, radiation exposure of the patient, etc. Moreover, the use of strontium needles in PC posed a potential danger to the surgeon.

This technique was abandoned when Mullan *et al.* (1965a,b) introduced PC by anode electrolysis and Rosomoff *et al.* (1965) employed radiofrequency coagulation. Mullan inserted the needle into the spinal cord by a free-hand technique without any stereotactic device. Subsequently, the technique was modified and improved by several authors (Gildenberg *et al.*, 1967; Taren *et al.*, 1969a; Vihlein *et al.*, 1969; Tasker and Organ, 1973; Tasker *et al.*, 1974; Amano *et al.*, 1976; Kuhler, 1976). A special stereotactic apparatus was devised for this operation (Rosomoff *et al.*, 1965; Kandel and Pukanov, 1973; Iseki *et al.*, 1982) (Figs. 161, 162).

It should be emphasized that in spite of the seeming simplicity of this operation, it requires, on the part of the surgeon, in-depth knowledge of anatomic details, technical skills, and high attention to accuracy of stereotactic calculations.

The electrode for PC must meet special requirements. It should be a long (approximately 15 cm), thin stylet of stainless steel, insulated with Teflon® or polyethylene, with a thin (0.25–0.3 mm), free sharp tip of 2.5–3 mm. However, Mullan (1971) feels that the electrode tip must not be very sharp so that the surgeon is able to feel the penetration of the spinal cord. Special monopolar electrodes for PC with diameters of 0.25, 0.4, and 0.5 mm with a thermocouple at the tip are manufactured by Radionics and Owl Instruments (for example, Tasker electrode, Mullan–Portnoy electrode, Gildenberg electrode, Rosomoff electrode, etc.) (Fig. 163).

The electrode is introduced through an 18- or 20-gauge thin-walled spinal puncture needle. The active (noninsulated) tip of the electrode protrudes 2–8 mm from the needle. As an alternative to the straight electrode, a slightly curved one with an active tip allows the electrode to be inserted eccentrically in the spinal cord by 1–2 mm (Lin *et al.*, 1966; Gildenberg, 1972).

An electrode incorporating a temperature monitor at its tip to control the radiofrequency coagulation of the spinal cord has been devised (Levin and Cosman, 1980). This is the most reliable parameter, since neither the current strength nor the time of exposure but only the temperature at the tip of the electrode is the determinant of the amount of damage to the neural tissue. A thermocouple sensor 2–3 mm long and 0.25–0.5 mm in diameter is built into the sharp active tip of the electrode. The temperature and time parameters required to obtain various degrees of pain loss (from 75° to 85°C with exposure for 20–40 sec) have been established. The advantage of such an electrode is a more rapid destruction with preset parameters of temperature and exposure. The Levin cordotomy electrode is also manufactured by Radionics.

The majority of authors employ monopolar electrodes (Taren *et al.*, 1969a; Mullan, 1971; Tasker, 1982a). Bipolar concentric electrodes with a sharp tip, allowing more precise identification of the somatotopic representation within the spinothalamic tract, have also been devised (Amano *et al.*, 1976; Iseki *et al.*, 1982). The use of a bipolar electrode for stimulation in PC makes it possible to avoid a "shock" onset of the current and, consequently, allows more precise localization of the spinothalamic tract.

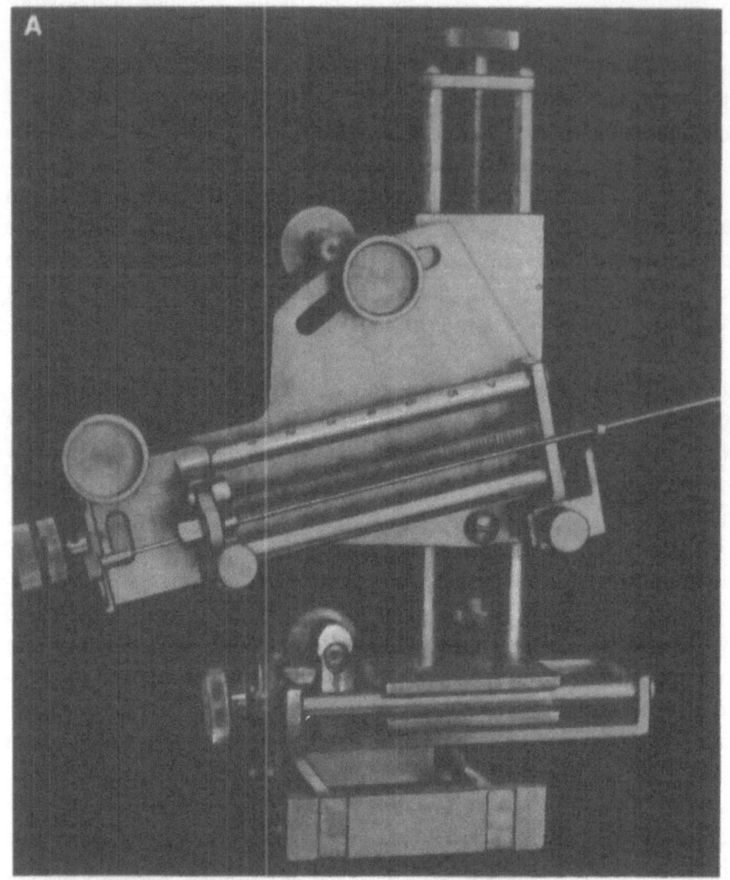

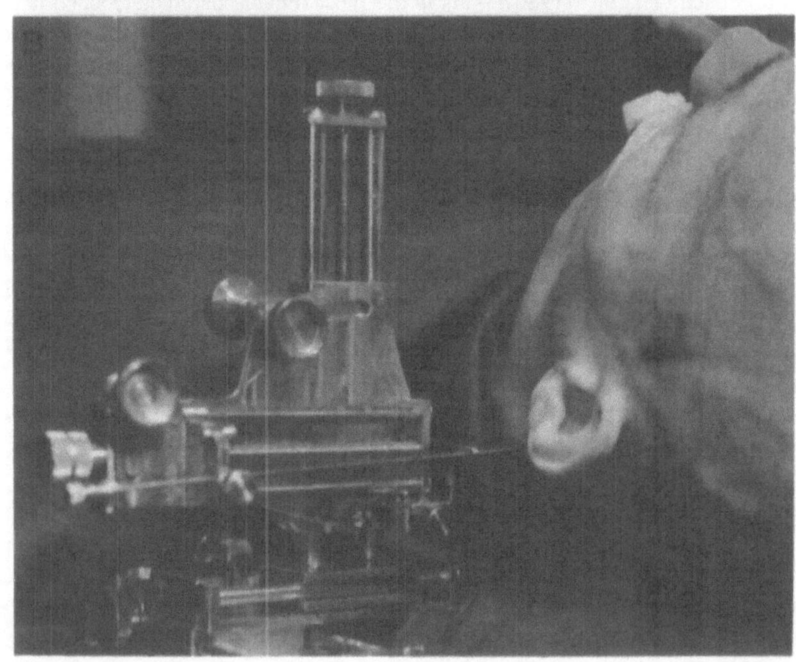

FIGURE 161. Our apparatus for percutaneous cordotomy. Explanation in the text. (A) General view of the device. (B) The device in place.

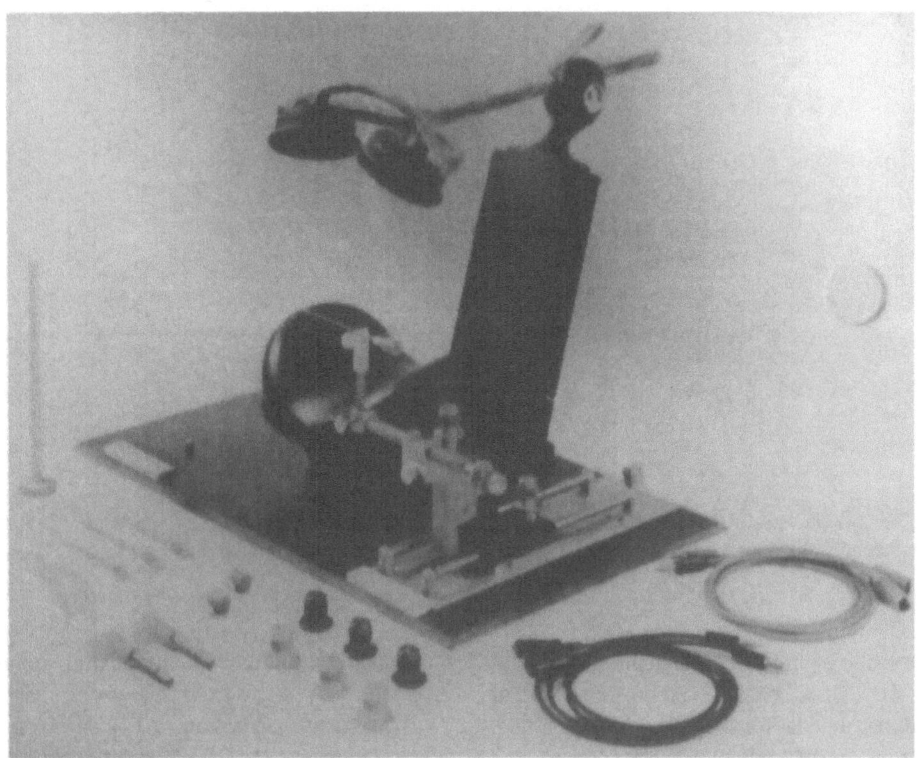

FIGURE 162. Rosomoff's headrest and stereotactic device for percutaneous cordotomy.

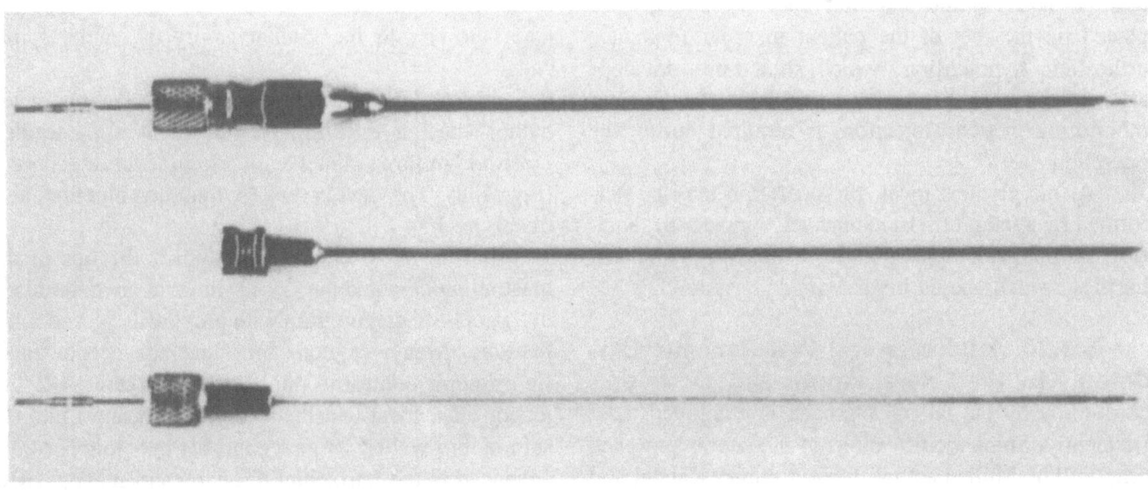

FIGURE 163. The Owl cordotomy electrode. A 0.4-mm electrolytically sharpened stainless steel wire projects 2 mm beyond its Teflon coating (bottom). A thin-walled 18-gauge LP needle (center), and the entire assembly (top) with 2 mm of the electrode projecting from the LP needle.

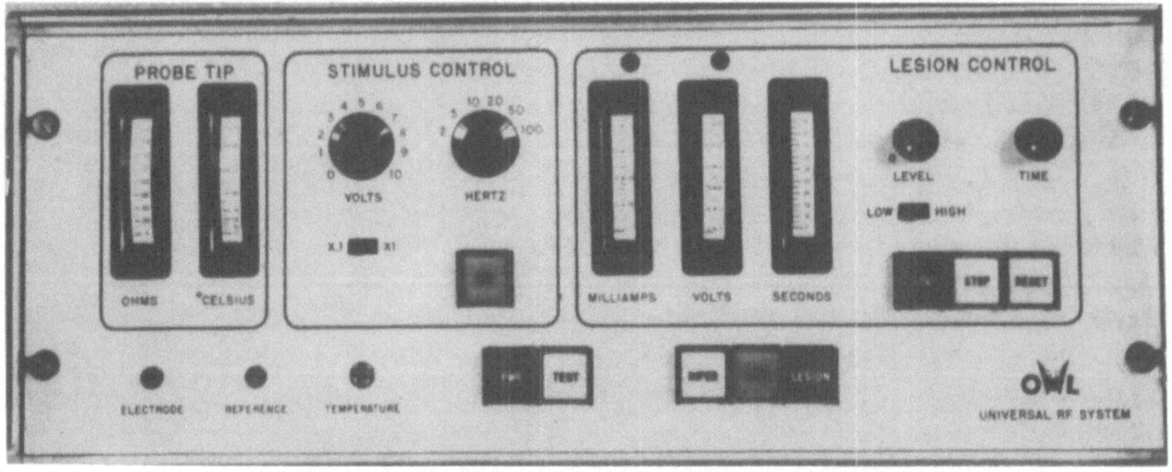

FIGURE 164. The Owl Universal RF System for measuring electrical impedance of CSF and the spinal cord. The system provides electrical stimulation at low and high frequencies and low and high outputs for making radiofrequency lesions.

For stereotactic operations on the spinal cord, and in particular, for PC, a special system consisting of electrodes and instruments for recording biopotentials, electrostimulation, measurement of impedance, and for radiofrequency destruction without changing electrodes has been manufactured (Fig. 164).

The operation is performed under local anesthesia or under neuroleptanalgesia, so that both the effects of stimulation and the sensory deficit may be determined. If it is necessary to sedate the patient, droperidol or diazepam is injected intravenously. During the operation, it is recommended that special bandages be placed on the feet of the patient in order to reduce orthostatic hypotension, which sometimes develops after interruption of the spinothalamic tract. X-ray monitoring in two projections is required during the operation.

At the cervical level, the operation may be performed by a lateral (most extensively employed), anterior, or posterior approach. Each has surgical peculiarities, which should be described separately.

5.3.2d. High Cervical Percutaneous Cordotomy by the Lateral Approach. Percutaneous cordotomy by the lateral approach is employed most frequently. Some modifications of the lateral approach proposed by Mullan are used in our clinic (Kandel and Pukanov, 1973). As a rule, the operation is performed under local anesthesia with mild premedication or neuroleptanalgesia; however, in case of severe pain, a short-lasting anesthetic (methohexital sodium) may be used.

During the operation, the head of the supine patient in a strictly "middle" position lies on a special headrest that we use for stereotactic operations on the basal ganglia (Chapter 4, Section 3). This headrest and its cassette holder allows one to take x-ray films of the cervical vertebral column in two projections. It is very convenient to control the position of the needle using an electronic amplifier with TV monitoring, also in two projections. The cervical vertebral column must be horizontal to prevent loss of x-ray contrast medium (see below) into the cranial cavity or caudal spinal canal.

In PC we employ a stereotactic apparatus of the author's design, enabling the movement of the needle–electrode in three planes with a high level of accuracy (Fig. 158). The device may be mounted on either side of the neck.

Before the needle is introduced, the tips of the mastoid process and the C_1–C_2 interval are determined in a lateral roentgenogram with previously placed x-ray markers. Many surgeons introduce the needle under fluoroscopic control in the lateral projection. After injection of a local anesthetic agent (procaine) into the skin, a thin-walled 18-gauge lumbar puncture needle is advanced in the horizontal direction subcutaneously to a point 1 cm below and 1 cm behind the tip of the mastoid process. The needle is advanced perpendicularly to the sagittal plane and enters the space between the C_1 and C_2 laminae. At a distance of approximately 6 cm from the skin, depending on the thickness of the neck, the dura mater is punctured.

Before penetration of the dura mater, a small quantity (1–2 ml) of procaine should be introduced through the needle, since, as a rule, at the moment of penetrating the yellow ligament and the dural sac, the patient experiences a momentary sharp pain. To insure that the needle is not in the subarachnoid space, an attempt to aspirate with a syringe should be made before the procaine injection.

In puncturing the dura mater, the surgeon experiences somewhat greater resistance than in conventional lumbar puncture. When the stylet is removed, a clear fluid rushes out. At times, the tip of the needle may be advanced into the spinal cord tissue, and the patient experiences a sharp pain, usually more severe than that felt from puncture of the dura mater.

In order to reach precisely the anterolateral quadrant of the spinal cord, the following reference points must be clearly visualized in the films: the dens of the second cervical vertebra, the anterior and posterior surfaces of the cervical spinal cord, and the dentate ligament, which indicates the frontal midplane of the spinal cord. We determine the reference points with three x-ray contrast substances—air, iothalamate, and iophendylate. At first, 5 cm^3 of air are introduced into the subarachnoid space. The air rises upward, outlining the superior surface of the spinal cord at the level of emergence of the anterior roots. In order to prevent air from entering intracranially, the head must not be raised higher than the body. A mixture of 1 ml of iophendylate and 4 ml of iothalamate is then injected into the subarachnoid space through the needle. Water-soluble contrast media (iothalamate, metrizamide, etc.) do not visualize the dentate ligament, but iophendylate, precipitating on both sides of the ligament, forms a horizontal "chain" in the lateral roentgenogram, clearly visualizing this very important reference point. This chain lies in the middle of the AP diameter of the vertebral canal. If the dentate ligaments are not visible, the needle has probably been introduced behind the desired point, and it is necessary to reintroduce it somewhat more anteriorly. As the iothalamate slowly settles, it outlines the inferior surface of the spinal cord. Metrizamide (Iseki *et al.*, 1982), tantalum powder (Smith, 1973), iophendylate (Ischia *et al.*, 1984), and iophendylate with air (Tasker *et al.*, 1974) are also used for contrasting the surface of the spinal cord in PC.

The reference points described above should be clearly seen in films in two projections (Fig. 165). The use of Polaroid™ film reduces the time for x-ray control. The lateral film is taken with the head in a strictly "sagittal" position; the AP film is best taken through the open mouth. This enables good visualization of the dens of the second cervical vertebra. Stereotactic calculations must be made on the films in order to define the spinothalamic tract. The point of entry of the electrode into the spinal cord (in lateral film) is 1 mm anterior to the dentate ligament. This point corresponds to the superficial fibers of the spinothalamic tract in which the sacral dermatomes are represented. Mullan (1971) recommends a point 1 mm anterior to the dentate ligament for leg pain and 2 mm anterior for arm pain. The target may also be determined by marking a point 2 mm anterior to the midpoint between the levels of air and iothalamate (i.e., the anterior and posterior surfaces of the spinal cord). This point is 10–11 mm from the posterior border of body C_2 (Ganz and Mullan, 1977; Ischia *et al.*, 1984a,b).

The target point in the spinothalamic tract should be calculated depending on the site of the pain. If pain is localized in the lower part of the body, the point in the lateral projection lies 5 mm anterior to the posterior surface of the spinal cord, and if the pain is at a thoracic level, the target is 8 mm. In the AP projection, corresponding target points lie 6 and 4 mm lateral to the midplane, as determined by a projection of the midline of the dens (Gildenberg *et al.*, 1967).

In PC the first passage of the needle (especially if done free-hand) is rarely on target. A slight correction (within 2 mm) may be made without removing the needle by shifting its shaft in the desired direction (Mullan, 1971). With our stereotactic apparatus, after the needle has penetrated the dura, films in two projections are made, and the required correction of the needle is determined. Then these data are transferred to the stereotactic apparatus, and the needle is introduced again. Then, the tip of the needle, as a rule, reaches the calculated point at the surface of the spinal cord. Then the thin electrode with 1–2 mm of exposed tip is inserted into the needle. The tip of the electrode, protruding 2–3 mm beyond the end of the needle, is advanced into the spinal cord (Fig. 162). To reach various parts of the spinothalamic tract the depth to which the electrode has penetrated is adjusted. To make a lesion of the more superficial part containing fibers from the legs and lower part of the body, the electrode must penetrate the spinal cord slightly less than 3 mm, and to destroy the

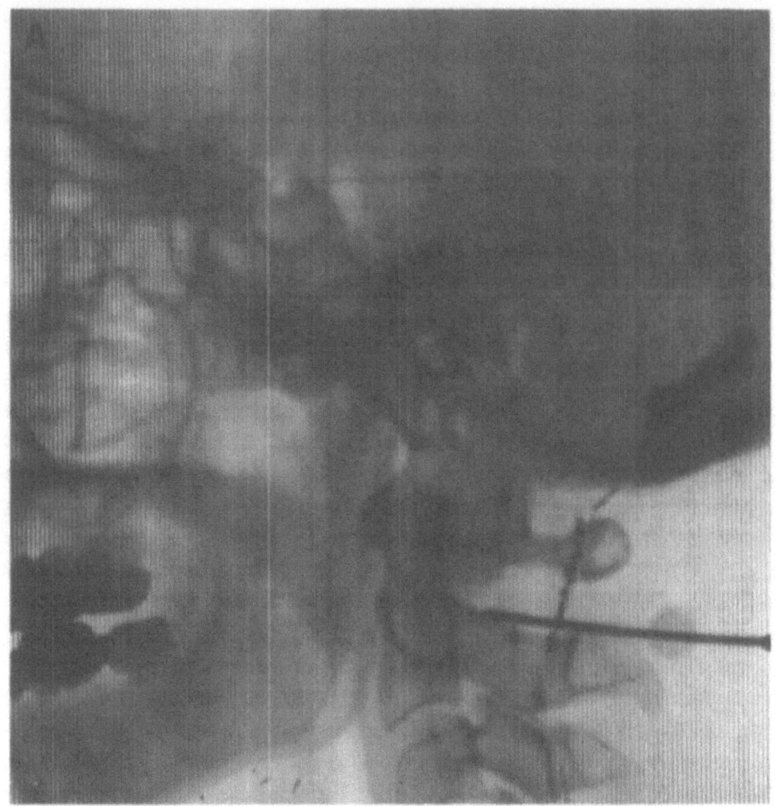

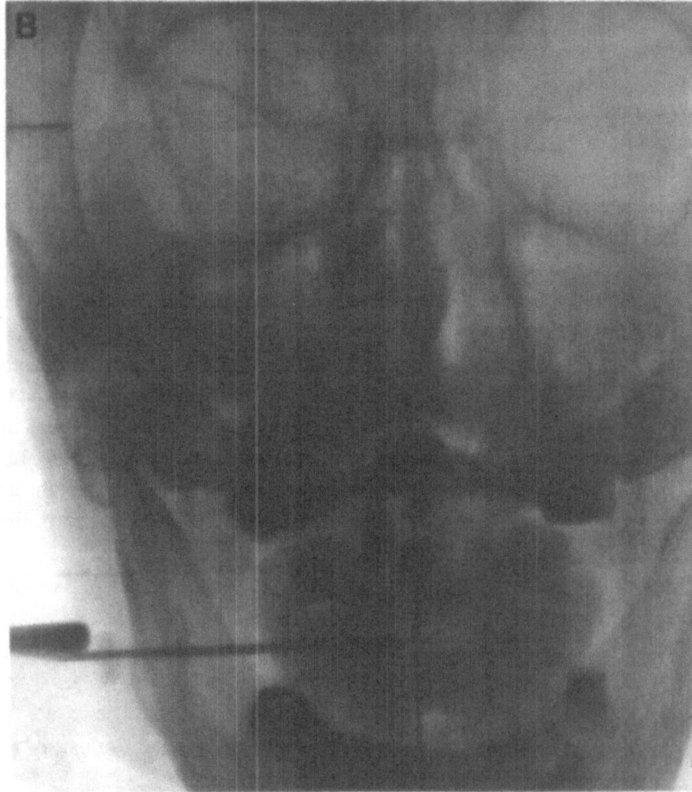

FIGURE 165. Percutaneous stereotactic cordotomy by lateral approach. After contrast medium outlines the spinal landmarks (see text), the protruding electrode is introduced into the spinothalamic tract in the anterolateral quadrant of the upper cervical spinal cord. (A) Lateral projection. (B) AP projection.

deeper fibers from the arm, 3.5 mm (Ganz and Mullan, 1977).

In many instances, the mechanical stimulation from the introduction of the electrode in the spinothalamic tract results in a marked reduction in spontaneous pain. Nevertheless, because the distribution of the fibers of the spinothalamic tract is quite variable, after the electrode has been advanced to the target point, the next procedure is a physiological verification of the site either by electrical stimulation or a measurement of impedance. The use of these methods is important because in rare cases there may be an anomalous anatomic localization of the dentate ligaments off the midfrontal plane, either in front of or behind it (Sweet, 1976). Stimulation allows one to determine the location of the electrode in the spinal cord and, consequently, to control the accuracy of "hitting" the target. The parameters of stimulation may vary; however, the majority of investigators, including the author, employ rectangular pulses of 1–3 msec, 1–3 V, and 1–3 mA. Ganz and Mullan (1977) believe that a current of up to 0.5 V is sufficient to induce paresthesias; if a more powerful current is required, the electrode is not in the spinothalamic tract, and a lesion should not be made. Various current frequencies are employed from 2 to 100 Hz, but the author prefers a frequency of 50–60 Hz.

A relatively simple test to determine the effect of PC has been proposed (Mayer *et al.*, 1975). If, on stimulation of the anterolateral quadrant of the spinal cord, the pain threshold on the contralateral half of the body is low, then the next radio frequency coagulation will cause total hemi-analgesia and relief of pain. If the threshold is high, the effect of the operation will be incomplete or, more often, nonexistent.

Depending on the position of the electrode, both sensory and motor responses may result from stimulation (Fig. 166). If the target point in the spinothalamic tract is stimulated by a current of 50–100 Hz, in the contralateral half of the body, especially in the distal part of the extremities, paresthesias, a sense of compression, an electric tingle, or feeling of warmth is perceived, or less frequently a sensation of cold, burning, or dull pain. Sensations of vibration or mild electric shock are sometimes experienced. Moreover, these induced sensory feelings should be localized mainly in the pain zone. If on stimulation, the pain involves large areas of one half of the body, the main criterion of "hitting" the target point should be the appearance of sensory feelings in the contralateral arm. Tasker *et al.* (1974) summarized the different effects of stimulation at 2 and 100 Hz when the electrode is precisely located in the spinothalamic tract. First, one may observe contralateral sensations and contraction of ipsilateral neck muscles, especially the trapezius, at thresholds about 2–3 mA. Furthermore, there are also contralateral sensations of tingling, warmth, or burning extending to the neck and hand at a threshold below 1 mA without tetanization.

Subsequent destruction of the target point should cause practically total loss of pain and temperature sensation in the entire contralateral half of the body below

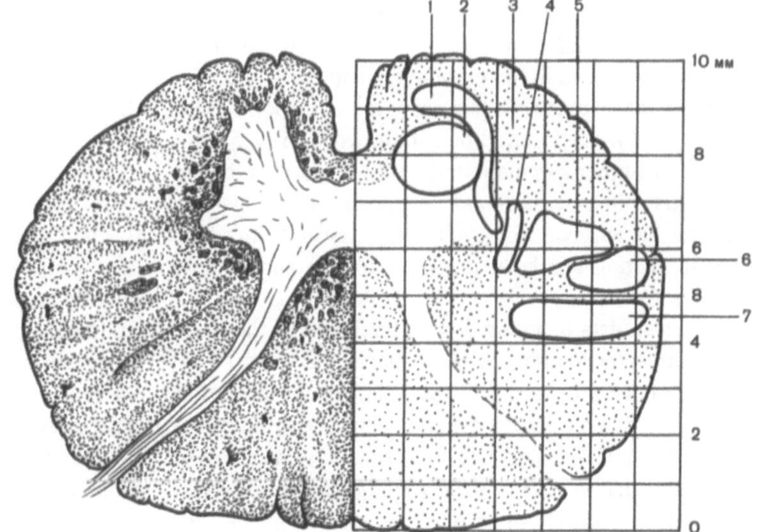

FIGURE 166. Scheme showing localization of tracts and "volume of representation" of different functions in the spinal cord based on electrostimulation data during percutaneous cordotomy (Mullan, 1971). The tracts giving sensory responses in contralateral extremities during stimulation should be the targets for destruction (anterolateral quadrant). 1, Respiration; 2, contraction of neck muscles; 3, sensibility in both hands; 4, neck muscles; 5 and 6, sensibility in contralateral extremities (5, hand; 6, foot); 7, movements in homolateral extremities; 8, dentate ligament.

C_6 and, in rare cases, below C_5 or C_4. Even when the electrode tip is accurately located in the spinothalamic tract, contraction of ipsilateral cervical muscles may occur on stimulation, indicating activation of local spinal reflexes (Tasker, 1982a).

If on stimulation various other effects or sensations in other locations are obtained, the electrode tip is not in the spinothalamic tract, and its position must be corrected. Contraction of muscles in the extremities on the side of the operation indicates stimulation of the more dorsally lying corticospinal pyramidal tract. The appearance of an ipsilateral sensory effect on stimulation indicates that the electrode is posterior to the crossed spinothalamic tract and in the ipsilateral tract. Sensory responses in or about the neck, paresthesias in both arms, and ipsilateral tetanic contractions of the cervical muscles at a low intensity of stimulation are usually an indication of direct stimulation of the anterior horn and, consequently, that the electrode tip is anterior to the spinothalamic tract (Tasker *et al.*, 1974; Ganz and Mullan, 1977). Severe pain in the back of the head, neck, and ear on the side of stimulation means that the electrode tip is not accurately placed but near the root C_2.

Ganz and Mullan (1977) mentioned one unusual effect of stimulation in PC: coincidental with paresthesias in the pain zone, simultaneous contractions of the muscles in the extremities on the same side occur, indicating stimulation of the corticospinal tract. The authors believe that this phenomenon results from "medullarization" of the cervical spinal cord, i.e., decussation of the corresponding corticospinal pathways at the upper cervical level. Since a lesion at this level may eliminate pain at the expense of paresis of the ipsilateral limbs, if stimulation does not induce a contralateral sensory response, or contractions of muscles are observed in the extremities and body, a lesion should not be made. In this case, the electrode should either be further inserted or withdrawn by 1–1.5 mm; then, by further stimulation, the proper location for the electrode tip can be found. A special apparatus enabling the surgeon to visualize the threshold of electrostimulation has been proposed (Nitter and Scholl, 1978).

From data resulting from stimulation of the spinothalamic tract with a bipolar electrode, Iseki *et al.* (1982) have proposed a "regression line" indicating the relationship between the site of the active tip of the electrode and the point of the highest dermatome in which analgesia is to be achieved. The authors consider that by using this line one can achieve greater accuracy in advancing the electrode to the target in the spinothalamic tract without using the dentate ligament as a reference point.

The second main localizing technique, i.e., method for controlling accuracy of reaching the target point in PC, is a special impedance measurement (Chapter 4, Section 7.5) made, for example, with Lesion Generator System Models RFG-3B, 5, and 5S and impedance monitor model IM-1 (Radionics). The impedances of the spinal cord and CSF differ considerably, the latter being considerably lower than that of the spinal cord. In view of this, a sharp, almost instantaneous increase of impedance occurs the moment the electrode tip comes into contact with the spinal cord (Fox and Green, 1969c; Gildenberg , 1969). Tasker *et al.* (1974) noted that impedance rises from the 400 Ω characteristic of CSF to 1200–1400 Ω of the spinal cord (Fig. 167). If impedance is 750–1000 Ω, the entire electrode tip is in the tissue of the spinal cord (Siegfried and Hood, 1983).

A special electrode allowing simultaneous determination of impedance and spontaneous electric activity of the spinal cord has been devised. Recording of biopotentials from "inside" the spinal cord reveals high-amplitude oscillations, which are typical of the anterior horns (Fujita and Cooper, 1976). Changes in impedance may be transformed into sound signals indicating the moment the electrode tip makes contact with and penetrates the spinal cord substance.

Summing up the data on the results of physiological control of hitting the target, Tasker (1982a) comes to the following conclusion. If stimulation with a current of 100 Hz and 0.5–1 V causes a contralateral sensory effect, usually a sensation of warmth, whereas stimulation with a current of 2 Hz and 3 V produces contraction of ipsilateral cervical muscles, then this is an indication with a reasonable degree of certainty that the electrode tip has accurately reached the spinothalamic tract.

If the x-ray and functional controls satisfy the surgeon that the spinothalamic tract has been accurately reached, the final stage of the operation—destruction of the tract—may be done. At present, the most frequently used method of making a lesion in PC is radiofrequency coagulation, and, less commonly, anodal electrolysis as proposed by Mullan *et al.* (1965a,b). Various types of generators (manufactured by Radionics, Owl Instrument, and others) with controls for regulating current parameters are used for radiofrequency coagulation. During the application, the strength of the current is gradually increased, taking

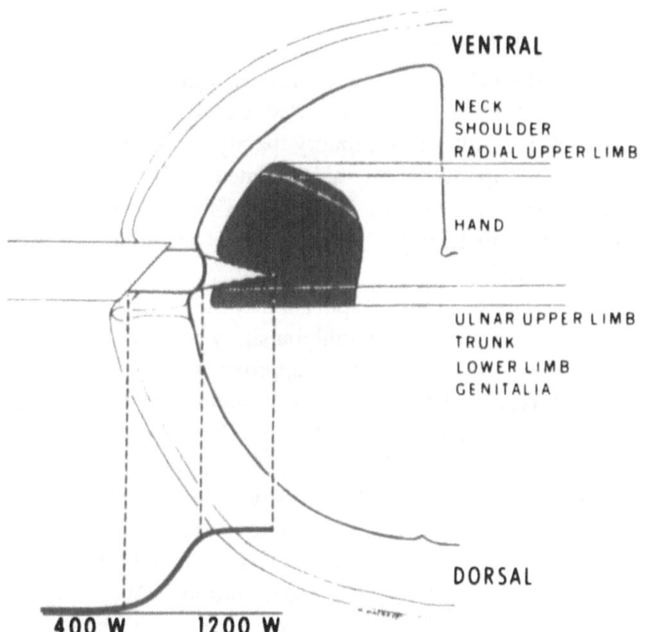

FIGURE 167. A scheme of percutaneous cordotomy (Tasker, 1982a). The electrode tip trajectory and location for introducing it into the lateral spinothalamic tract. The somatotopic representation within the tract is indicated. Note the sharp increase of impedance during the electrode penetration of the cord (left below).

into account the exposure time, since the size of the destructive focus should not exceed 3–3.5 mm.

If stimulation of the spinothalamic tract produces an appropriate reaction, as described above, one should begin making a small lesion with a short application (10–15 sec) of a high-frequency current of approximately 20–25 mA. If a moderate hypalgesia results in the pain zone, several consecutive and longer (up to 40–50 sec) high-frequency coagulations may be made by adding 5 mA each time, thereby enlarging the size of the lesion. The zone and degree of analgesia should gradually increase. The concluding two or three coagulations should be done with a current strength of 70–80 mA.

Mullan (1971) points out that the current required depends on the depth to which the electrode tip has been inserted in the spinal cord substance. If the tip is deep, then the current may be less intense, and vice versa. This results from the fact that the closer the electrode lies to the surface, the more it is cooled by CSF. The coagulation must be terminated immediately if paresis, even mild, appears in the arms or legs.

As previously stated, in PC it is necessary to achieve a clear-cut deep analgesia throughout the entire contralateral surface of the body below C_4–C_6, particularly in those dermatomes where there is pain. Testing on the operating table should be done after each coagulation, for it is necessary to demonstrate that pin pricks, pinching of the skin, and compression of the fingers or tendons of the contralateral half of the body do not cause pain. Nevertheless, as a number of authors have noted, and based on our limited personal experience, in rare cases a good therapeutic effect can be obtained even when there is no significant analgesia after a lesion of the spinothalamic tract is made. On the contrary, at times pain is not substantially diminished in spite of clear-cut analgesia.

After destruction of the spinothalamic tract, the patient frequently experiences burning pain in the neck, which disappears in a short time. Salar *et al.* (1982) noted an increase of the β-endorphin level in CSF after PC, which had been decreased before surgery as in the majority of cases with chronic pain.

For deafferentation pain, PC is not effective. For visceral pain, in particular, pain in the pelvic area, PC is not very satisfactory, nor is it recommended for pain spreading above the superior thoracic level (Mazars, 1976a).

By the lateral approach, PC has a significant risk of complications, the frequency of which varies considerably in the reports of different surgeons. As was already mentioned (Section 5.3.2b), respiratory disorders represent a relatively rare but very dangerous complication after high cervical (C_1–C_2) cordotomy by the open method, caused by damage to the descending respiratory pathways, particularly the reticulospinal tract passing in the ventral part of the anterolateral quadrant. This complication, although quite rare, also occurs after PC (Lema and Hitchcock, 1986). For example, Lin *et al.* (1966) reported that four patients died as a result of respiratory paralysis. In recent years, as experience has accumulated and surgical techniques improved, the frequency of this complication after unilateral PC is less—0.5% according to Tasker (1982a).

Among the other complications of unilateral PC, it is necessary to mention mild transitory paresis of the arms or legs, which occurs in approximately 15% of cases (Wepsic, 1976). Permanent paresis remained only in one case of 234 (Tasker, 1982a). One of the most serious but fortunately very rare complications of PC is burning dysesthesias occurring after a year or more in hypesthetic areas of the body. The frequency of this complication according to Ganz and Mullan (1977) is approximately 11%, and according to Wepsic (1976) about 5%. These dysesthesias are a very serious complication since they do not lend themselves to any kind of therapy. Among other complications, mention should be made of transitory bladder dysfunctions in approximately 5% of cases. Arterial hypotension occurs very rarely. The Claude Bernard–Horner syndrome is a comparatively frequent result of PC.

Postoperative mortality after unilateral operations varies greatly in the published series and depends primarily on the state of the patient prior to surgery. Tasker (1982a) reported a mortality in 234 operations of 0.5%; Koulousakis and Nittner (1982), after 143 PCs, reported that in the course of the first 2 weeks 12% of the patients with malignant tumors died, mainly from respiratory disorders.

Undoubtedly, the results of PC are good, but the degree and duration of relief of pain vary considerably in different reports. In assessing the results, one must take into account that in cancer patients a new pain caused by spread of the primary process may develop after surgery. Such pain should not be considered as representing a failure of PC. Moreover, because in the majority of cases PC is performed on patients with a progressing cancer, postoperative follow-up is limited in time.

The large series of cases reported by Tasker (1982) point to the effectiveness of PC. In practically all of the patients it was possible to identify and destroy the spinothalamic tract. Analgesia was obtained in the target area of the body in 92% of the cases. At the time of discharge, pain had disappeared completely in 88% of the patients, and in a subsequent follow-up in 71%. Mullan (1966) and Gildenberg (1982) also reported that pain disappeared in 80% of the patients after unilateral PC.

However, the majority of authors state that the long-term results of PC are not as good as the immediate results because of comparatively frequent relapses of pain. According to Kühner (1976), the immediate results were good in 77.5% of 52 PCs, satisfactory in 15%, and poor in 7.5% of the cases; however, the corresponding long-term results were 59%, 9%, and 32%. Ischia *et al.* (1984b) reported good results in unilateral cervical PC for pain resulting from a variety of metastatic lesions of the vertebral column; in 71% of 69 patients, pain disappeared or diminished considerably so that they could abandon or considerably decrease narcotics until their death.

5.3.2e. Bilateral High Percutaneous Cordotomy. It is necessary to dwell briefly on the problem of bilateral PC, the need for which quite often arises in cases of thoracic and abdominal malignancies and in bilateral pain from adhesive arachnoiditis. However, bilateral PCs are performed comparatively rarely. In a large series of cases (234 operations), Tasker performed bilateral PC in only 18% of the patients, Mullan (1966) in 13%, and Ischia *et al.* (1984a) in 7%. The majority of such operations were done in two stages, the second operation being performed not earlier than 1–2 weeks after the first. The frequency of postoperative respiratory disorders is considerably lower if the interval between the two operations is a month or more.

Bilateral high cervical PC results in total elimination of pain in approximately half of the cases; e.g., of 36 patients with cancer pain who had bilateral PCs, complete relief was obtained in 46%, partial in 12.5%, and no relief in 25% (Ischia *et al.*, 1984a). However, a bilateral operation has considerably more complications than a unilateral operation and should be considered high-risk surgery (Mullan and Hosobuchi, 1968; Rosomoff, 1969). The most dangerous complication is sleep-induced apnea described above (Section 5.3.2b). The frequency of this complication is substantially higher than after a unilateral operation and, according to the data of various authors, occurs in 4% (Rosomoff, 1969), 8% (Lipton *et al.*, 1974), or even 18% (Kühner, 1976). Some authors consider bilateral PC to be contraindicated if there are preoperative respiratory disorders. And finally, a number of authors have concluded that in view of the real danger of respiratory paralysis, bilateral operations above C_4, in general, are unwise (Gildenberg, 1974, 1976; Wepsic, 1976). On the basis of our limited experience, we also consider bilateral high PCs contraindicated.

The frequency of other complications after bilateral operations is also considerably higher. Serious bladder dysfunction has been noted six times more frequently than after unilateral PC (Tasker, 1982a). Pyramidal paresis has been observed in 17% of cases, whereas after a unilateral operation, it appeared in only 2.5% (Mullan, 1966). Reduced sympathetic tone frequently develops after the operation causing orthostatic hypotension and even fainting. This complication is variable, from 4% (Rosomoff, 1969) to 36–40% (Kühner, 1976; Ischia *et al.*, 1984a).

Mortality after bilateral cervical PC is approximately twice as high as after unilateral (9%, Kuhner, 1976; 12%, Rosomoff, 1969; 11.1%, Ischia *et al.*, 1984a). Some surgeons consider that in bilateral pain, an open bilateral cordotomy at a high thoracic level is preferable to bilateral cervical PC (Siegfried and Hood, 1983). These observations seem to favor the conclusion that bilateral PC should be performed only in exceptional cases, and if done, in two stages.

5.3.2f. Low Cervical Percutaneous Cordotomy by Anterior Approach. In order to avoid the above-mentioned complications in high PC by the lateral approach, PC at a low cervical level by the anterior approach has been developed, i.e., interruption of the spinothalamic tract at the level of C_6–C_7 or C_7–C_8 (Lin *et al.*, 1966; Gildenberg *et al.*, 1967; Gildenberg, 1976, 1982; Fox and Green, 1968c; Lipton *et al.*, 1974) (Fig. 168). At this level the lesion in the spinothalamic tract is located below the spinal origin of the phrenic nerve, and so it is possible to avoid disorders caused by damage to the efferent pathways innervating the respiratory muscles, in particular, the diaphragm. The main feature distinguishing the anterior from the lateral approach is that the spinothalamic tract is destroyed on the side contralateral to introduction of the electrode (i.e., on the side of the pain).

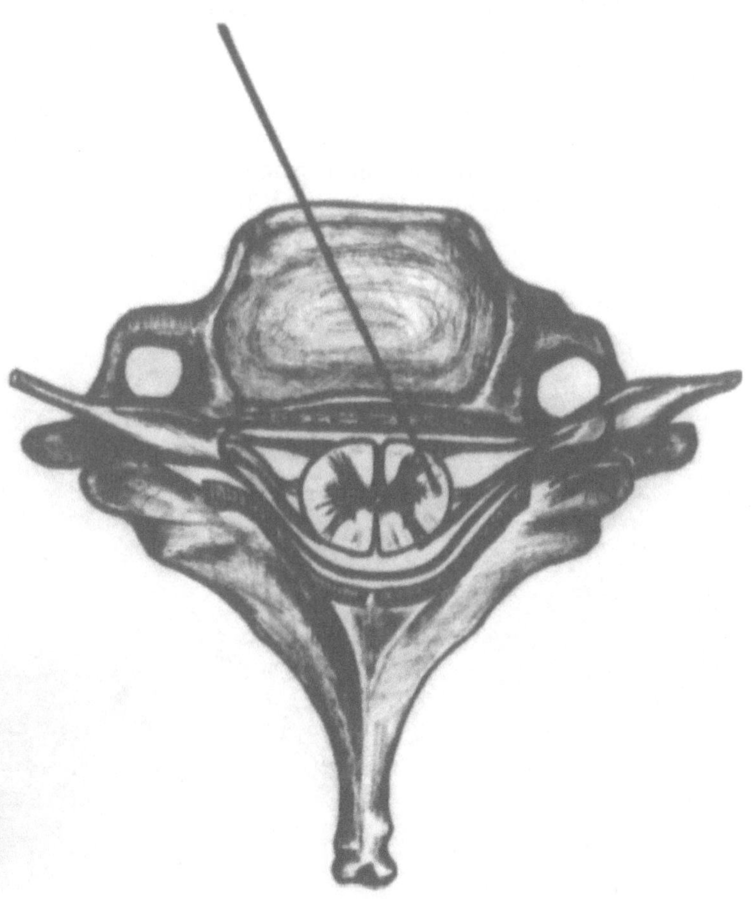

FIGURE 168. Scheme of percutaneous stereotactic cordotomy by anterior approach through an intervertebral disk of the lower cervical spine (Lipton *et al.*, 1974).

The operation by the anterior approach is performed under control of the image intensifier. The patient is placed in a supine position. Under local anesthesia, on the same side as the pain, an 18-gauge thin-walled lumbar puncture needle or a special needle for PC with stainless steel electrode extending 4 mm from the needle tip (the electrode is insulated with Teflon® except for 2 mm at the tip) is introduced in the anterior surface of the neck at C_6–C_7 between the esophagus and trachea medially and the neurovascular bundle laterally. The needle is directed slightly lateral to the sagittal plane toward the opposite side. The majority of surgeons use the free-hand technique for this procedure. In the anterior approach, the needle should be introduced only slightly deeper than the anterior margin of an intervertebral disk below C_4–C_5. Moreover, it is possible to change the direction as required based on the scout films, since the needle is not fixed in the disk. After taking control x rays and correcting according to the calculations, the needle is passed through the intervertebral disk, the dentate ligament, and the dura mater adjacent to the opposite anterolateral quadrant of the cervical spinal cord.

According to Gildenberg (1972), the desired target for the anterior approach depends on the area of the body to be relieved of pain. If pain is in the lower cervical dermatomes, the target is 3–4 mm lateral and just below the anterior surface of the spinal cord, or 8 mm anterior to the posterior wall of the cervical canal. With pain in the sacral region, the target is at the dentate ligament 8 mm lateral or 5 mm anterior to the posterior wall of the canal. Pain in intermediate regions requires lesions made between these two targets.

In order to reach the target point accurately, Gildenberg *et al.* (1968), Fox and Green (1968c), and Gildenberg (1969) recommend after every intraoperative x-ray control that a right-angled triangle be drawn on both films. The lower side of that triangle is the line from the tip of the needle to the target point on the AP film, and the vertical side is the same line on the lateral film. The angle formed by the vertical line and the hypotenuse is that by which the needle should be inclined from the vertical plane of the midline of the neck.

An important advance in anterior PC was the introduction of the stereotactic technique (Lipton *et al.*, 1974). The authors proposed a special stereotactic apparatus (Fig. 169) as well as a method for calculating the accurate introduction of the electrode (Fig. 170). The stereotactic device consists of an aluminum headrest and an overhead bow that permits two angular shifts of the needle at any selected point. It is important that the bow not obscure the path of the needle in either the lateral or AP projec-

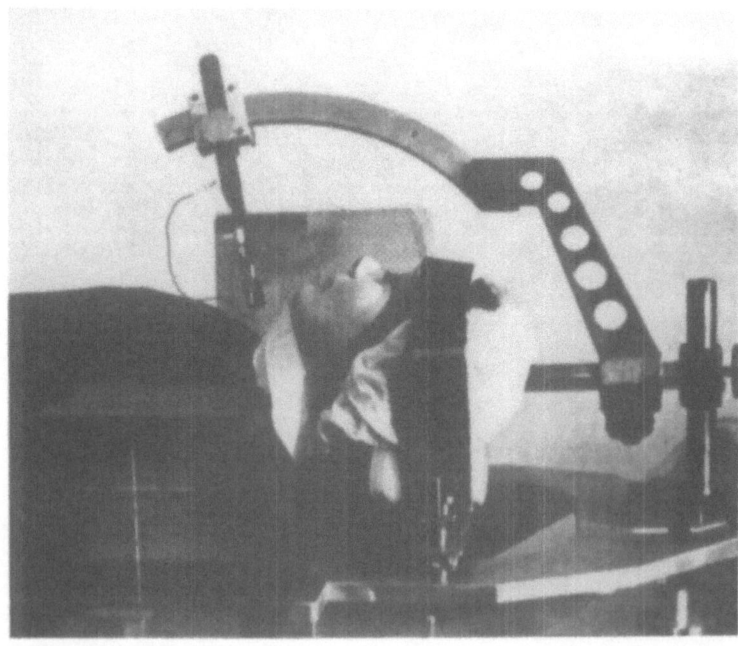

FIGURE 169. Lipton's stereotactic device for percutaneous cordotomy by the anterior approach.

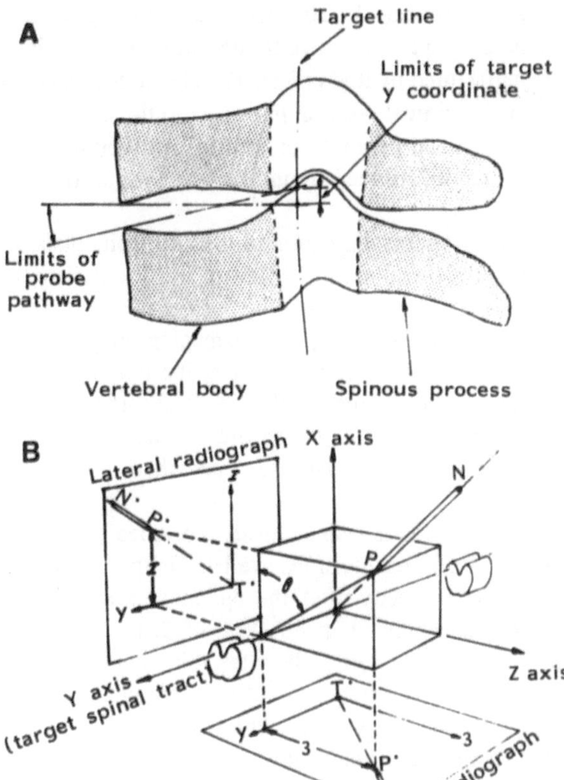

FIGURE 170. (A) Pathway limits through the disk space and permissible variation of the target *Y* coordinate. (B) Schematic drawing showing a section through the spinal cord. The angle is obtained from the coordinates *Z* and *X* on the lateral and anteroposterior radiographs, respectively (Lipton *et al.*, 1974).

tions, from which *X*, *Y*, and *Z* settings are calculated.

Fox and Green (1968c) employ for anterior PC an optical system with a grid of parallel lines on the object lens. The system is useful for correcting visually the direction of the needle.

Stereotactic calculations for the anterior approach to the spinal cord must be very precise, since the needle is fixed firmly in the interverbral disk, from which it is very difficult to change the trajectory. If the position of the electrode has to be corrected, the needle should be extracted to the anterior border of the interverbral disk, the angle of introduction changed as calculated, and the needle advanced once again. In elderly people with firm disks from osteochondrosis, it may be difficult to pass the needle through the disk.

In the anterior approach, one employs the same physiological controls to insure reaching the spinothalamic tract as in the lateral approach (stimulation, impedance measurements). During stimulation of the ligaments and the dura, i.e., before the needle enters the spinal cord, the patient experiences sharp pains in the back of the neck. The effects of stimulating the long pathways are the same as in the lateral approach (Section 5.3.2c). On reaching the spinothalamic tract, stimulation with a weak current (0.5 V) induces clear-cut paresthesias in an area on the opposite half of the body.

The clinical results of PC by the anterior approach have not been sufficiently studied for an assessment of that method. Lipton *et al.* (1974) have performed 18 such operations; however, only 12 were done completely. In six cases, it was impossible to pass through the intervertebral space. In 12 patients operated on for pain of malignant tumors, pain disappeared in six completely, in two partially, and in four, relapses occurred.

What is the comparative benefit of the lateral and anterior approaches described above? The majority of neurosurgeons feel that both approaches yield approximately similar results insofar as the relief of pain is concerned. In certain studies it has been stated that the anterior approach is less effective since relapses are more frequent; however, the rate of complications is lower.

Gildenberg *et al.* (1967) reported on the technique and results of 140 PCs in 92 patients. Some 112 operations were performed by the anterior approach at a low cervical level, and 28 operations by the lateral approach at a high cervical level. All the patients were operated on for intractable pain caused by malignant neoplasms. With the two approaches, approximately similar results were observed—pain was relieved in 84% and 89% of the patients, respectively. However, the low cervical approach, as was expected, had fewer postoperative respiratory disorders and other complications, in fact, no respiratory disorders were observed, whereas after cordotomy at high cervical level, three patients died of respiratory insufficiency.

Nevertheless, complications, usually of a transitory nature, occur after PC at a low cervical level. For example, transient weakness in the ipsilateral arm was observed in 12% of the patients, of the leg in 19%, and urinary retention in 4% of the patients. In his next paper, Gildenberg (1974) reported more than 300 operations by the anterior approach, 20%

of which were performed bilaterally. In 80% of the cases, satisfactory pain relief was achieved. In contrast with bilateral high PC, which is an extremely hazardous procedure, anterior PC may be performed on both sides for bilateral relief of pain. Alternatively, one anterior and one lateral high PC can be made if a high level of analgesia is needed on one side only (Lipton *et al.*, 1974).

It is concluded that anterior PC is a safer but technically somewhat more complicated procedure for pain relief, especially in patients with pulmonary dysfunction (Lipton *et al.*, 1974).

5.3.2g. Posterior Percutaneous Cordotomy. The posterior approach for PC is used rarely. In this approach the electrode is introduced with the aid of a stereotactic apparatus into high cervical segments of the spinal cord through the atlantooccipital membrane (Crue *et al.*, 1968; Todd *et al.*, 1969; Hitchcock, 1970) (Fig. 171). The technique of this operation is analogous to percutaneous trigeminal tractotomy described in the section on trigeminal neuralgia (Section 7).

Stereotactic cordotomy by the posterior approach is performed with the Hitchcock stereotactic apparatus (Fig. 158), which is attached to the cranium by three sharp pins—one at the midline of the forehead and one above each mastoid process. The operation is performed with the head in a fixed position, considerably inclined forward. The anterior surface of the spinal cord is brought as close as possible to the anterior wall of the vertebral canal, thus keeping the cord immobilized. A special electrode with a thermocouple sensor in the tip and a microdrive for posterior PC is available (manufactured by Radionics). The electrode must be in a strictly vertical position in respect to the midplane at C_1 level. The contours of the superior cervical segments are determined by introducing a small quantity of air.

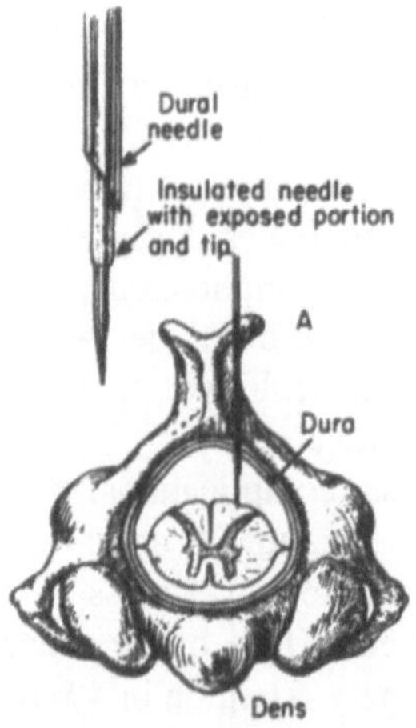

FIGURE 171. Scheme of percutaneous cordotomy by posterior approach (Todd, 1972). The special electrode is shown in upper left corner.

The stereotactic coordinates of the spinothalamic tract at that level are the following: 4–5 mm from the midline and 4–5 mm from the dens of the second cervical vertebra. The danger of damaging the posteroinferior cerebellar or vertebral artery has not been borne out in practice. Crue *et al.* (1968) consider that the posterior approach in PC ensures greater accuracy in reaching the spinothalamic tract than the lateral approach.

In 1976 Lorenz *et al.* made a detailed analysis of all the data on results of PC available in the literature at that time. Directly after the operation a satisfactory result was obtained in 80–96% of 2616 operations; however, in long-term follow-up, this figure decreased to 42–75%. The frequency of different complications also fluctuated in various clinics. Horner's syndrome and ataxia were noted most frequently. Other complications were less common: bladder dysfunction, 1.5–15%; transitory hemiparesis, 4–17%; permanent hemiparesis, 0–3% respiratory dysfunction, 0–4.6%. Postoperative mortality varied from 0 to 4.6%.

The analysis of such data indicates that PC is most frequently used and yields the best results in cancer pain (Gildenberg *et al.*, 1967; Tasker, 1982a). However, in benign pain syndromes (with the exception of deafferentation pain), this operation is, without a doubt, effective.

We have dwelt in detail on the description of techniques and results of PC since this operation is regarded as one of the best in the neurosurgeon's armamentarium of pain relief techniques. Lipton *et al.* (1974) considers PC to be the best of all the existing operations for the management of chronic pain. It is relatively simple, little or nontraumatic, reliable, and effective. Nevertheless, the wide use of neurostimulation has displaced PC, which, at present, is performed less frequently than previously.

5.3.3. Commissural Myelotomy (Commissurotomy)

One of the functional operations on the spinal cord in pain syndromes that has long been employed in neurosurgical practice is commissurotomy. This operation was first performed by Armour in 1926 and then by Putnam (1934) and Polenov (1937). The aim is to interrupt the decussating spinothalamic tract as it crosses in the anterior commissure of the spinal cord. The fibers relaying pain and temperature sensation pass to the other side of the spinal cord three segments above their entrance and then ascend to the sensory thalamic nuclei.

Commissurotomy offers two *a priori* important advantages over cordotomy (Section 5.3.2). The first is that by making one incision in the spinal cord it is possible to eliminate pain on both sides of the body. And the second is that this incision, made at a distance from functionally important long pathways of the spinal cord, sharply reduces the danger of damaging them.

The considerable experience of commissurotomy that has been accumulated in the course of several decades demonstrates the indisputable effectiveness of this operation in a number of pain syndromes. One of the main indications is bilateral pain caused by injury or other lesions of the peripheral nerves, roots, and spinal cord. This operation is particularly effective in injury of the vertebral column and in spinal arachnoiditis but has also been employed successfully in pain from inoperable tumors in the thoracic and abdominal cavities with invasion of nerve trunks and for pain caused by tumors of the pelvis and lower limbs.

For a successful operation, it is very important to determine the level and extent of the intersection of the anterior and posterior commissures in relation to the localization of the pain and disturbed sensation. For pain in any location, commissurotomy must involve three spinal segments above the superior borders of the affected dermatomes. It is necessary to take into account that the fibers of the spinothalamic tract proceeding from the lower extremities and the pelvic organs cross at the T_{10}–T_{12} segments of the spinal cord. The fibers from the upper extremities and torso pass over to the other side at a level several segments higher. In view of this, commissurotomy may be divided into superior, middle, and inferior, depending on the level of section of the anterior commissure of the spinal cord.

With pain in the legs, perineum, and pelvic organs, the entire lumbar enlargement is incised to the conus medullaris. The operation is usually performed under general anesthesia with the patient in the prone position or (less commonly) on his side. It is desirable to use a surgical microscope or magnifying lens. The proposed level of the laminectomy is marked on the x-ray film. The cranial border of the skin incision should be 5–6 cm above the upper segmental level of pain. As a rule, it is sufficient to resect the spinous process and arches of two or three vertebrae.

After the dura and arachnoid are opened and the spinal fluid is evacuated, the posterior median fissure is located by microscopic examination. Then it is necessary to displace or coagulate the vein that passes in or along the fissure. A longitudinal incision of the pia over the fissure is made with a microscalpel. Besides the mid- or paramidline posterior vein, another landmark for the midline is the thin arachnoid septum, which dips between the posterior columns and in which run fine arterial branches. If there are insurmountable difficulties in determining the midline and the operation is performed under local anesthesia, the posterior columns may be stimulated and the site determined from which paresthesias are produced in the sacral dermatomes.

Section of the anterior and posterior commissures may be accomplished by either a closed or open procedure. In the closed section, an ophthalmic scalpel or piece of safety razor is used to incise the spinal cord along the midline. However, a special myelotome (Cook and Kawakami, 1977) in the form of minispatula with sharp edges and blunt tip is preferable (Fig. 172). The length of this spatula (5.5 mm) is precisely the distance from the posterior longitudinal fissure to the dorsal end of the anterior longitudinal fissure, i.e., the optimal depth of the incision. After the myelotome has been inserted to this depth, the spinal cord is incised upwards and downwards for a

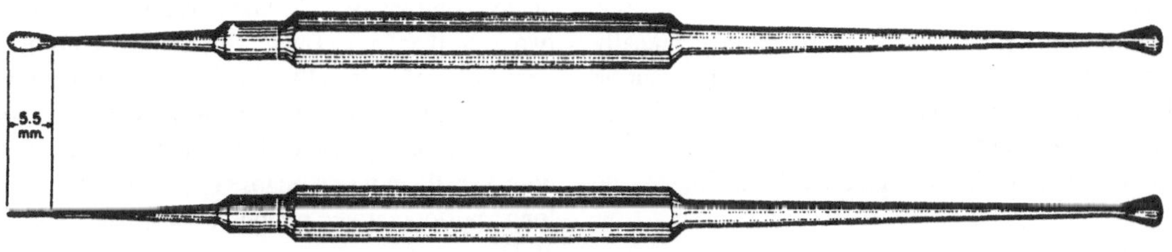

FIGURE 172. Myelotome proposed by Cook and Kawakami (1977). Description in the text.

distance of from 2.5 to 4 cm strictly in the sagittal plane. The use of this instrument avoids two main dangers of closed commissurotomy: first, the possibility of damaging the anterior spinal artery; second, the danger of deviating from the midplane and damaging the posterior columns.

The open procedure involves separation of the posterior columns of the cord with a small dissector or spatula, strictly in the midplane, and section of the commissure *ad oculus* or preferably with the aid of a magnifying lens or surgical microscope. A thin, sharp instrument (a special commissurotome or piece of safety razor in mosquito forceps) is advanced into the median fissure, and both commissures are divided longitudinally for a distance usually of 25–35 mm, depending on the extent of the pain. In certain cases, the commissurotomy may be made longer. For example, King (1977) incised the commissures for 65 mm in one case. The depth of the incision should be 6–7 mm so that the posterior part of the anterior sulcus is visible on the floor of the incision throughout its entire length and the minute branches of the anterior spinal artery that lie on the ventral surface of the spinal cord are not exposed on the lateral walls by the section of the commissures.

The open technique affords greater accuracy in section of the commissures, since the depth of the incision is controlled visually. However, the retraction by the spatulas may cause additional damage to the spinal cord, particularly the posterior columns, increasing the risk of complications and side effects. As a rule, there is no hemorrhage from the incised spinal cord. The surgical wound is closed in the conventional manner.

Spontaneous pain disappears or decreases immediately after commissurotomy, along with the appreciation of pain and temperature on both sides below the level of section. Directly after surgery, many patients complain of an unpleasant (but not painful) sensation in the hypesthetic* zone. Such sensations usually disappear in several days. Even after a technically correct operation, complications occur quite frequently, mainly transitory bladder disturbances and intestinal dysfunctions. Pareses of the legs after commissurotomy are rare.

Although commissurotomy is not a technically complicated operation, it requires particular caution and precision for its execution. First, the spinal cord must be dissected (as it were, into two halves) strictly in the midplane. Even the slightest deviation to the side may damage the spinal cord substance, particularly the posterior columns, leading to the development of complications.

The second precaution concerns the depth to which the cutting part of the instrument is inserted. The anterior and posterior commissures must be incised in such a way as not to damage the anterior spinal artery, lying in the anterior spinal fissure, with the tip of the instrument. It should be noted that there is an especially great danger of damaging Adamkiewicz' artery at the level of the low thoracic segments. Damage to this artery results in paraplegia caused by ischemic lesion of the caudal spinal cord. In view of this, it is strongly recommended that the cutting instrument not be inserted deeper than by two-thirds of the diameter of the spinal cord.

A successful attempt to perform commissurotomy using a CO_2 laser was recently reported (Fink, 1984).

Many authors have published good results from commissurotomy. In 1953, Wertheimer and Lecuire reported on their experience; over a 12-year period, they operated on 107 patients with various pain syndromes and achieved lasting good results in half of the cases. Šourek (1969) reported that incurable pain in the legs, pelvis, and abdomen disappeared after surgery in 25 cases. Commissurotomy has also been very effective in bilateral cancer metastases, but not in intrapelvic metastases (Cook and Kawakami, 1977). Lippert *et al.* (1974) also reported good relief of pain in 14 of 16 patients after commissurotomy. There was no mortality; complications were rare and mild, and a relapse occurred in only one of ten cases followed for several years. Sunder-Plassmann and Grunert (1976) obtained approximately the same results.

King (1977) performed commissurotomy in nine seriously ill patients (six with cancer pain) who were regularly taking large doses of narcotics. Directly after surgery and until death, the results were excellent in seven patients, who totally discontinued narcotics. In the remaining two patients, pain relapsed after several months.

Of 24 cases, most with pain from malignant tumors, Adams *et al.* (1977) reported that pain disappeared completely in 15 and partially in seven; only in two was there no effect. The authors consider commissurotomy especially effective in patients with pain in the rectum and perineum. In lumbar arachnoiditis, commissurotomy yielded a good immediate result; however, there was a relapse of pain in all cases after a long postoperative period (Cook and Kawakami, 1977).

*Touch is not lost.

Goedhart *et al.* (1984) performed commissurotomy in the conus medullaris from S_2 to S_5 in 10 cases with intractable malignant sacral pain, but only in a few cases was the pain relief satisfactory.

Complications after commissurotomy occur less frequently than after cordotomy and in the majority of cases are transitory. Among these complications, mention must be made of parapareses, bladder disorders, and dysesthesias. It is noteworthy that in certain cases spontaneous pain disappeared completely after surgery, although pain and temperature sensation decreased insignificantly or the hypesthetic zone was considerably smaller than the painful area (Lippert *et al.*, 1974; King, 1977). This fact leads to the assumption that disappearance of pain after commissurotomy results from interruption not only of the spinothalamic tracts but also of the extralemniscal pain pathways in the gray substance of the spinal cord (see Section 5.3.5). To summarize the literature, King (1977) considers that commissurotomy provides relief in 60 to 70% of patients suffering bilateral pain in the lower half of the body.

It is necessary to make a comparative assessment of commissurotomy and open cordotomy in the management of severe pain syndromes. In cases of unilateral pain, it is difficult to judge which of these operations is preferable. Their long-term results, judging by the data in the literature, are approximately the same. In case of bilateral pain, commissurotomy, in all likelihood, is preferable to cordotomy. Commissurotomy is more effective in pain of the lower limbs, abdomen, and pelvic areas. Complications in the form of parapareses and bowel and bladder dysfunctions occur less frequently after this operation. Finally, commissurotomy is performed in a single stage, whereas bilateral cordotomy is more frequently carried out in two stages.

Commisurotomy, although an effective operation for the relief of pain, is traumatic for seriously ill patients, and in contemporary functional neurosurgery is employed comparatively rarely.

5.3.4. Stereotactic Cervical Commissurotomy

This operation was proposed by Hitchcock in 1970. From the technical point of view, it is quite similar to PC by the posterior approach (Section 5.3.2g) and to stereotactic destruction of the descending tract of the trigeminal nerve (Section 7.3).

Although the experience of stereotactic commissurotomy is very limited, Hitchcock (1970), Nadvornik *et al.* (1974), and Schvarcz (1978) reported encouraging results in a small group of cases with pain syndromes of various origins.

5.3.5. Extralemniscal Myelotomy

After commissurotomy, quite often pain may be relieved without the development of analgesia in the corresponding part of the body. In particular, after stereotactic commissurotomy at the C_2 level (Hitchcock, 1972; Schvarcz, 1977b, 1978), pain may disappear from a considerable part of the body, although only a mild analgesia is produced in a limited area or not at all. To explain such an effect, it has been suggested that there is an archispinothalamic tract passing into the gray substance near the central canal of the spinal cord, the interruption of which results in the disappearance of pain without subsequent analgesia (Hitchcock, 1974). Section of this hypothetical tract may be done by both open and stereotactic procedures. The open operation, which Gildenberg and Hirshberg (1981b) call limited myelotomy, is performed at the thoracic level, i.e., above the level of segments corresponding to the pain zone. Visceral pain, in particular that resulting from malignant neoplasms of pelvic organs, is an indication for such an operation.

The operation is performed under anesthesia with the patient in the prone position. After laminectomy of T_9 or T_{10}, the midline of the spinal cord is identified with a surgical microscope, and a blunt microdissector is introduced into the posterior fissure of the cord to a depth of approximately 6 mm. The posterior commissure is then identified, and a longitudinal incision is made in the center of the spinal cord for 5–7 mm—approximately within the limits of one segment. The authors emphasize that since the aim is to interrupt the hypothetical pain-relaying pathways near the central canal, there is no need to cut the commissural pathways. The operation has practically no complications that are encountered after classical commissurotomy—paresis of the legs, deep sensory disturbances, etc. After this operation, a good effect was obtained in approximately 75% of cases with malignant tumors. A substantial decrease in pain was noted in eight of 11 patients with bilateral pain in the lower half of the body secondary to malignant tumor metastases (Gildenberg and Hirshberg, 1981a,b).

The same operation performed stereotactically has been named extralemniscal myelotomy (Hitchcock, 1972; Schvarcz, 1978). With a stereotactic apparatus, the needle electrode is introduced into the large cistern between the border of the major occipital foramen and the posterior arch C_1 and then into the midplane of the spinal cord between the posterior columns. Stimulation of the posterior columns induces a sensation in the upper half of the body. As the electrode advances deeper, this sensation passes to the lower half of the body and then disappears as the electrode is advanced so that its tip is situated anterior to the posterior columns. A lesion is made by a high-frequency current at this site. Although complete analgesia does not result from this operation, it relieves pain in any location, especially bilateral pain (Schvarcz, 1977), as well as postherpetic neuralgia (Gildenberg, 1982). Although this operation is not yet extensively employed, there can be no doubt that it is effective.

5.3.6. Posterior Rhizotomy

For many years one of the main surgical methods for the management of pain was total or (less frequently) partial resection of the spinal posterior roots, posterior rhizotomy. This operation by an extradural approach was first performed by Abbe in the United States and by Bennett in Great Britain independently in 1889 and then developed by Foerster in 1908 on the basis of Sherrington's (1898) classical experiments. Since this operation is now comparatively extensively used for spinal spasticity, it has been described in Chapter 19 on lesions of the spinal cord. Posterior rhizotomy for the spasticity of CP patients has also been discussed in Chapter 9.

Indications for posterior rhizotomy in pain syndromes are quite diverse—malignant tumors in the extremities, neck, and visceral organs, pain after unsuccessful operations on the intervertebral disks, occipital and intercostal neuralgia, posttraumatic and postoperative scars, peripheral neuralgia, especially that caused by painful scars after thoracotomy and bone fractures, coccygodynia, angina pectoris, and stomach, gallbladder, and renal disorders accompanied by intractable pain. In considering indications for rhizotomy in malignant tumors, it is necessary to take into account the general state of the patient, the possibilities of other therapeutic methods, as well as a life expectancy of at least 4–6 months (Barrash and Leavens, 1973).

An important decision facing the surgeon is the determination of the level and extent of the rhizotomy. This depends on the localization and distribution of the pain because an effective rhizotomy must denervate the affected zone. Accordingly, the rhizotomy may be unilateral or bilateral. White and Sweet (1969) give an informative table indicating which roots should be sectioned to relieve pain of various etiologies. Before surgery, it is recommended to introduce a local anesthetic in the spinal nerve under fluoroscopic control. A temporary disappearance of pain is an important criterion for prognosticating a positive effect by rhizotomy.

From two to six posterior roots on one or both sides are usually sectioned. It is desirable to divide not only the roots supplying the painful zone but also one or two roots above and below them. This is because each segment of the body is innervated by at least three posterior roots, as demonstrated by Foerster (1921). At least three or four posterior roots must be sectioned to produce a sensory loss in one dermatome.

The classical open posterior rhizotomy is performed under general anesthesia. For bilateral rhizotomy, a laminectomy, and for unilateral, a hemilaminectomy at the appropriate level is made considering the relationship of the vertebrae to the roots. Most authors intradurally section the posterior roots through a hemilaminectomy (White and Sweet, 1969; Loeser, 1972; Onofrio and Campa, 1972). After the dura mater has been opened, it is necessary not only to identify the posterior roots but also to distinguish them from the anterior roots, which, if the operation is performed at the level of the conus medullaris or cauda equina, may be difficult. Valuable information may be obtained by intraoperative EMG using needle electrodes, since electrostimulation of the anterior roots causes contraction of the innervated muscles.

According to White (1964) of 21 patients suffering from chronic pain caused by posttraumatic or surgical scars in the area of peripheral or intercostal nerves, a good result from open posterior rhizotomy was obtained in 12 cases. Seven of eight patients with root pain experienced a significant improvement after such an operation to relieve cervical or lumbar discopathies.

The operation is greatly enhanced by the microsurgical technique, permitting more exact identification of the roots, less traumatic sectioning, and greater ease in separating the thin radicular arteries that must be preserved. After the posterior roots that are to be sectioned are precisely identified, two or three dentate ligaments should be cut, and the spinal cord slightly rotated medially. Then the posterior roots are consecutively cut with a safety razor at the site of their entry into the spinal cord.

Scoville (1966) developed the open extradural preganglionic rhizotomy under local anesthesia for various types of pain in the lumbar region and legs, in particular, residual pain after removal of intervertebral disks. Half of the lamina is removed, the spinal ganglion is exposed, and the anterior root is separated in its dural sheath just proximal to the ganglion and sectioned. After a single rhizotomy (L_5 or S_1), pain was relieved in 50% of 12 cases, and after section of two roots, in two-thirds of the cases. After Scoville, a number of reports on extradural preganglionic rhizotomy have been made by White and Sweet (1969), Echols (1969), Barrash and Leavens (1973), and Strait and Hunter (1981). Echols obtained good results in 60% of cases, mainly after the failure of disk surgery. Our limited experience in extradural rhizotomy confirms the value of the operation especially if the pain area is not large.

Positive experience of open sacral rhizotomy in malignant tumors with intractable perineal pain has been reported by Crue and Todd (1964) and Felsööry and Crue (1976). In six of seven patients with cancerous invasion of the pelvis, sacral rhizotomy yielded a good effect. This operation was more effective and had fewer complications than injection of phenol in the dural sac (Kühner, 1976).

5.3.6a. Selective Posterior Rhizotomy. The idea of partial (or selective) rhizotomy was proposed and tried by Gros *et al.* (1967). The description of the operation is given in Chapter 19; here, we describe the technique of selective posterior rhizotomy developed especially for management of pain by Sindou *et al.* (1974a,b, 1983). The operation is based on the investigation by the author of the complex anatomic arrangement of the entrance of the posterior root into the spinal cord. Each posterior root is divided into three to ten rootlets, and each anterior root into three to five rootlets (Sindou *et al.*, 1983). After piercing the dura mater separately, both roots pass through the intervertebral foramen, where the posterior root forms the spinal ganglion. Just distal to the ganglion, the anterior and posterior roots join to form the spinal nerve.

Each rootlet of a given root has the same structure and the same proportion of myelinated and nonmyelinated fibers. Each rootlet consists of a peripheral portion, in which the nerve fiber with or without myelin is covered with Schwann cells, and a central segment, in which the fibers are sheathed by oligodendroglia. The junction of the two segments, called Tarlov's pial ring, is on average 1 mm from the junction of the rootlet and the spinal cord.

It is important that in the peripheral part of the rootlet, the fibers are not segregated according to size, but in the neighborhood of the pial ring, at the entrance to the spinal cord, the large and small fibers are anatomically separated according to their function. Thin or small nociceptive fibers (Aδ and C) are concentrated in the lateral part of each rootlet, large myotatic fibers are situated centrally, and large lemniscal fibers leading to the posterior columns run medially. In this operation, the small nociceptive fibers are separated from the large lemniscal fibers as each rootlet enters Lissauer's tract (Fig. 173). Cutting the small fibers in the lateral part of each rootlet relieves pain in that dermatome. Sindou called this procedure "selective posterior rhizotomy."

In order to avoid repeating the description of the technique of selective rhizotomy presented in detail in Chapter 19, only some additional data will be mentioned. According to Sindou *et al.* (1983, 1987), the operation is performed with the patient under general anesthesia in the sitting or prone position. For operation at the cervical level, the neck is flexed. The operation at the lumbosacral level is performed with the patient in the prone position. In unilateral pain, hemilaminectomy with conservation of the spinous process is sufficient.

After the dura mater has been opened, the roots of the cervical and thoracic cord can almost always be identified by checking their corresponding intervertebral foramina. The same technique, although more difficult, can be used for the lumbar region. For identification of the sacral roots, Sindou *et al.* (1974a,b) recommend measuring the filaments, taking into account that the S_2 and S_3 rootlets, which are responsible for sphincter and genital functions, are spread over a segment 20–35 mm from the coccygeal spinal cord–rootlet junction. After identification of the roots to be cut, it is very important to reawaken the patient to determine his responses to posterior roots stimulation, in particular to check the rootlets that, when stimulated, reproduce the pain. For this purpose, it is advisable to stimulate the rootlet of each posterior root separately in order to select the appropriate rootlets for dissection.

All manipulations on the rootlets are made under a microscope with magnification of ×10 to ×25. The arterial branches of each rootlet are separated and preserved. Before cutting, each rootlet is retracted by a small hook to expose the posterior root junction on the ventrolateral aspect of the spinal cord. After bipolar coagulation of the tiny pial vessels, an incision is made with a piece of razor blade ventrolaterally at the above-

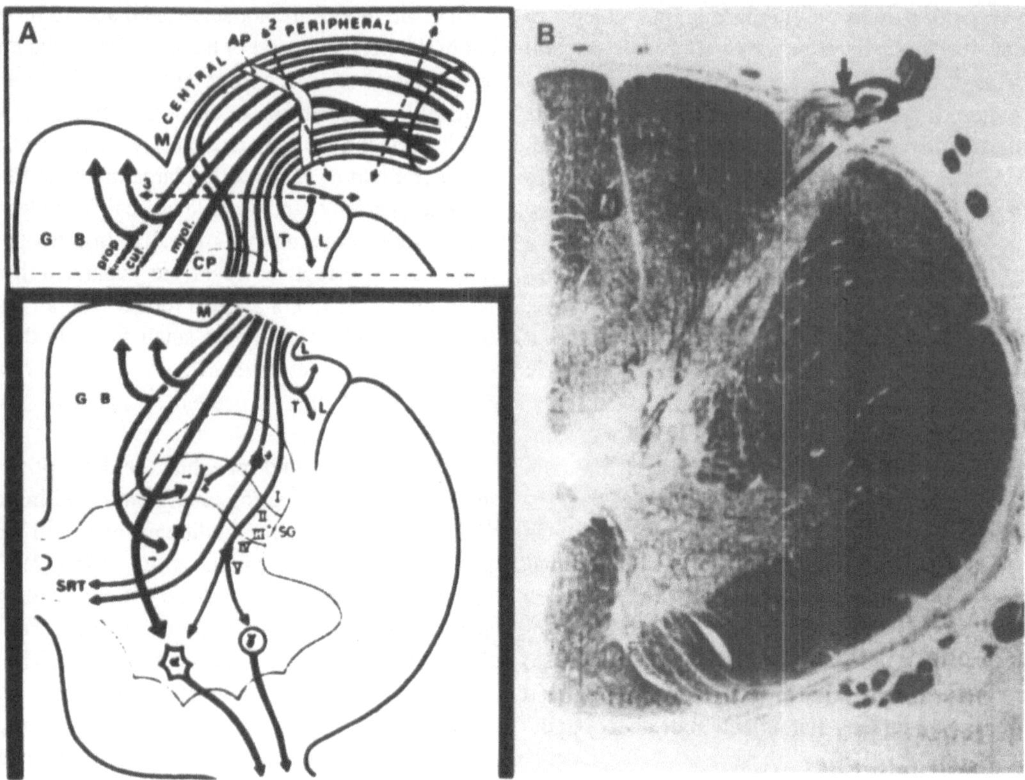

FIGURE 173. Organization of fibers at the posterior rootlet–spinal cord junction and posterior rootlet projections to the spinal cord (Sindou *et al.*, 1974a). (A) Top: Pial ring (AP) at the junction of a peripheral and a central segment of each rootlet. Peripherally the fibers are not topically organized but at AP the small fibers are situated on the rootlet surface, predominantly on its lateral side. In the central segment they regroup laterally to enter the tract of Lissauer (TL). Large myotatic fibers are located centrally, and lemniscal fibers medially. (A) Bottom: The small fibers terminate on the spinoreticulothalamic cells (SRT). The large myotatic fibers project into the anterior horn cells. (B) Rexed's lamination in the cervical spinal cord. The pial ring of the posterior rootlet (arrow). The solid line shows the incision for selective posterior rhizotomy, which extends for 1 mm into the rootlet–spinal cord junction.

mentioned junction (Figs. 174, 175). The incision into the spinal cord should be 1–2 mm deep and at an angle of 45° to the dorsal columns (Sindou *et al.*, 1974a,b, 1983). The procedure is repeated for each rootlet of one root. If necessary, selective rhizotomy should be performed bilaterally. If before surgery the bladder function of the patient is even partially preserved, root S_2 on the side where the pain is less should be left intact.

Several authors have noted tiny anastomoses among the four upper posterior cervical roots and the 11th cranial nerve (Vlahovitch *et al.*, 1976). They believe that these anastomoses may conduct nociceptive impulses from the cervical region and recommend that they be divided in addition to upper cervical posterior rhizotomy.

The clinical results of selective posterior rhizotomy in various pain syndromes depend on many factors but undoubtedly are better than those of total rhizotomy. Deafferentation pain does not develop after selective rhizotomy.

Good results from selective rhizotomy were obtained by Sindou and Lapras (1982b) in severe pain caused by lesion of the brachial plexus by a carcinoma of the upper lung (Pancoast–Tobias syndrome), and in carcinoma of the esophagus. The operation gave marked analgesic effects in all cases during the survival period (6 months on the average).

Selective lumbosacral rhizotomy resulted in significant relief of pain in 28 of 33 cancer cases. The results were not so satisfactory in painful nonmalignant diseases: in six patients suffering amputation pain, nerve injuries, or herpes zoster, long-lasting relief was observed in only three cases (Sindou *et al.*, 1983).

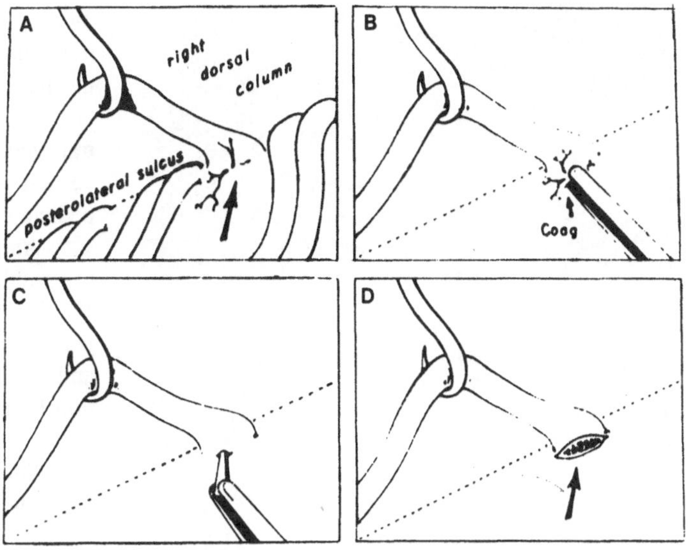

FIGURE 174. Technique of selective posterior rhizotomy under a surgical microscope according to Sindou *et al.* (1985a). (A) Lumbosacral rootlet retracted by a hook to make the ventrolateral region of the spinal cord–rootlet junction accessible (arrow). (B) Bipolar coagulation of the small vessels at the site of the incision. (C) Selective posterior rhizotomy by ventrolateral incision using a small piece of a razor blade. This incision is 1–2 mm deep and makes a 45° angle with the dorsal column. (D) Postoperative appearance of the incision (arrow).

Sacral and perineal pain is a difficult problem for surgical treatment. Most often this pain is caused by malignant tumors in that region, but benign chronic pain also exists. One of the most effective operations is a sacral rhizotomy, which is based on the sensory innervation of the region. It was established that S_2 root supplies the penis and scrotum in men, the labia and vagina in women, and part of the buttocks and bladder; S_3 supplies the perianal area, bladder, neck, and urethra; S_4 and S_5 supplies the coccyx and perianal area (Bohm *et al.*, 1956; cit. by Saris *et al.*, 1986).

Recently Saris *et al.* (1986) reported their impressive 15 years experience of sacrococcygeal rhizotomy for management of perineal pain. The operation was carried out in 28 patients with malignant pain (colorectal or urinary tract cancer) or coccydynia. A sacral laminectomy and bilateral cutting of S_1–S_5 or S_3–S_5 plus coccygeal roots were carried out. The authors concluded that sacral rhizotomy is an effective treatment for malignant perineal pain, but is ineffective for coccydynia and other benign perineal pain.

5.3.6b. Percutaneous Posterior Rhizotomy. A new and less traumatic technique, percutaneous posterior rhizotomy, has recently been developed and tested clinically (Uematsu *et al.*, 1974; Wepsic, 1976; Lazorthes *et al.*, 1976b; Verdie and Lazorthes, 1982; Pagura *et al.*, 1983; Nash, 1986). Although different approaches are used, usually an anterolateral or lateral route to the cervical roots (1 cm below the mastoid process) as in PC or posterolateral to the lumbar roots is preferred. The operation is performed under local anesthesia with mild premedication, but at the moment of painful high-frequency coagulation of the root, short-acting narcotics are often administered. X-ray control with an image amplifier is required during the operation. The technique of this operation at the cervical level resembles that of percutaneous cordotomy (Section 5.3.2d).

The position of the patient varies depending on the level of the operation: in cervical rhizotomy, a supine position is required; in thoracic rhizotomy, a prone position; in lumbar and sacral, a side position.

The introduction of the electrode into the intervertebral foramen is fraught with technical difficulties, especially at the thoracic level. In percutaneous rhizotomy, one must bear in mind that all posterior roots adjoin the inferomedial part of the vertebral pedicle. The cervical roots emerge from the intervertebral foramen perpendicularly to the sagittal plane of the vertebral column, the inferior cervical roots at an angle of 30–40°, and the inferior thoracic, lumbar, and first sacral roots at an angle of 25–35°. The superior thoracic roots emerge at an angle of 90° to the vertebral column so that the needle must be directed tangentially and may puncture the parietal pleura (Pagura *et al.*, 1983).

An insulated needle electrode with an exposed tip 3–5 mm long (Radionics and others) is used for percutaneous rhizotomy. It is introduced into the posterior root via the intervertebral foramen. For percutaneous rhizotomy at thoracic levels, the needle is introduced at a distance of 4–5 cm from the midline along the transverse process of the vertebra above the root to be ther-

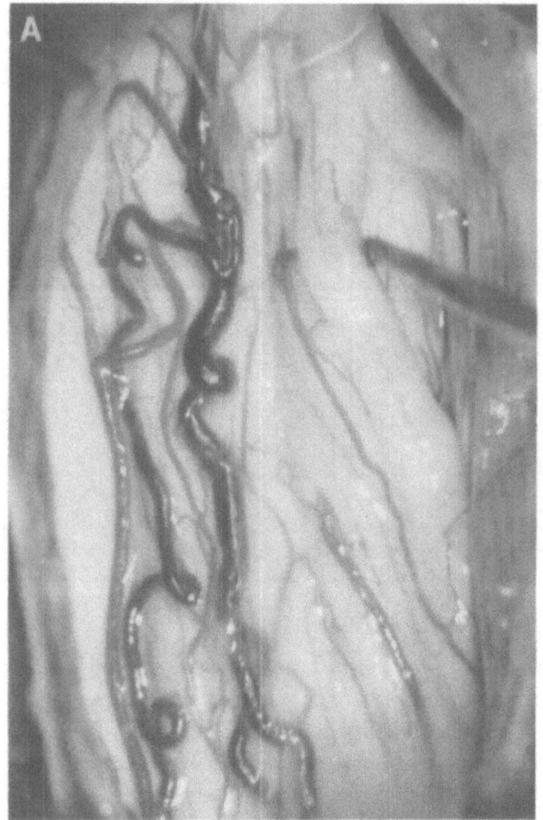

FIGURE 175. Selective posterior rhizotomy under a surgical microscope (×12). (A) Mobilization of one posterior rootlet and its elevation by microhook. (B) Cutting of the ventrolateral part of the rootlet with a microknife.

mocoagulated. Then the needle is advanced to a point 2 mm below the inferior edge of the vertebral pedicle. In the lumbar region, the needle is introduced at a distance of 7 cm from the spinous process of the vertebral transverse process lying below the target point and advanced upwards to the medial part of the vertebral pedicle (Pagura *et al.*, 1983).

An important aspect of the operation is electrostimulation with a weak current (approximately 1 V). The criterion for accurately reaching the root in percutaneous rhizotomy is contraction of the muscles and paresthesia in the painful dermatome on stimulation by a current first with a low (5–10 Hz) and then with a higher frequency (50–80 Hz).

Special caution is required in cervical rhizotomy since the posterior ganglion, especially at the level of C_1–C_3, is located only a few millimeters posterior to the vertebral artery. Accordingly, it is recommended that the head of the patient be slightly inclined so that the vertebral artery is displaced from the intervertebral foramen (Uematsu *et al.*, 1974). After the needle is introduced into the intervertebral foramen, a slight aspiration with a syringe is made to be sure that the needle is not in the subarachnoid space and has not penetrated the vertebral artery. If CSF is obtained on aspiration, the needle must be withdrawn 1–2 mm. After a sensory effect is obtained on stimulation, thermocoagulation (50–70°) is tested for 15 sec. Then it is repeated for 90–120 sec (Uematsu, 1982). Immediately after coagulation of the posterior roots, hypesthesia or total anesthesia appears in the corresponding dermatomes, and the patient notes a termination or marked decrease of pain. After rhizotomy at the thoracic level, x-ray films must be taken to exclude a possible pneumothorax.

At first, the main indication for percutaneous rhizotomy was well-localized monoradicular pain, mainly after disk surgery or as the result of intercostal neuralgia. In recent years, the operation has been used to relieve pain of such disorders as malignancies, postherpetic neuralgia, and postthoracotomy distress. Pagura *et al.* (1983) performed percutaneous posterior rhizotomy at various levels of the vertebral canal in 50 cases, for pain secondary to malignancy in 13 and for various benign processes, mainly root pain syndromes in 37. As a rule, two adjacent roots were coagulated. In 76% of the cases, the results were assessed as excellent or good with a mean follow-up of 9.3 months. Better results were noted in operations at the lumbosacral level, whereas in thoracic rhizotomy a good result was obtained only in half the cases. In the opinion of the author, this is explained by the anatomic peculiarities of thoracic roots mentioned above. A transitory motor deficit lasting only several days was noted postoperatively in two cases. Because of frequent failures, the procedures had to be repeated in 20%.

Wepsic (1976) noted dysesthesias after percutaneous rhizotomy, but in the series of Pagura *et al.* (1983), no such complications were observed.

Posterior rhizotomy is undoubtedly effective as a means of providing pain relief. However, modern techniques of selective or percutaneous rhizotomy are safer and less traumatic procedures and give longer-lasting pain relief than total posterior rhizotomy.

Unlike anterolateral cordotomy (Section 5.3.2b), anesthesia after rhizotomy never disappears. Complications occur rarely and, as a rule, are transitory.

5.3.7. Destruction of Entry Zone of Posterior Roots

The first verified case of traumatic avulsion of cervical roots from the spinal cord was described by Flaubert in 1872. A considerable increase in the number of brachial plexus traction injuries has been noted in recent decades. The most frequent causes are road accidents, especially those involving motorcycles.

Injuries to the brachial plexus may be divided into two main groups: (1) rupture of the primary and secondary trunks or postganglionic injuries and (2) avulsion of the brachial plexus or preganglionic injuries. Zorub *et al.* (1974), among 70 cases of brachial plexus injuries, described 21 cases of total avulsion with very little return of functions even after a long-term follow-up. According to data from the Burdenko Institute of Neurosurgery in Moscow, preganglionic injuries of the brachial plexus are encountered two to five times more often than postganglionic.

There is no need to dwell on the typical clinical picture of both pre- and postganglionic brachial plexus injuries. Nevertheless, it is necessary to emphasize that in approximately 20–30% of the cases with brachial plexus avulsion, severe pain develops in the denervated cervicobrachial area and paralyzed arm (Taylor, 1962; Nashold and Ostdahl, 1979). At times, such pain may occur with some paresis but not total paralysis of the arm. Yeoman (1968) reported severe pain in 32 of 46 cases with total rupture of the brachial plexus and in 22 of 40 cases in which movement and sensation were partially restored. Such pain, similar to the phantom pain syndrome (Section 6.2) and often pressing or burning, was usually so severe that the patient could not sleep at night. It became worse over a period of years. This pain, as a rule, is combined with phantom sensations in the paralyzed arm.

Severe pain in the area innervated by the injured roots usually does not appear immediately but develops 2–3 weeks after the avulsion and can appear even 2 years later (Thomas and Sheehy, 1983). This allows one to assume that pathophysiological changes take place in the spinal cord during this period that lead to the development of a pain syndrome (Nashold and Ostdahl, 1979).

The pathogenesis of chronic pain in brachial plexus avulsion still remains obscure, although the discussion of chronic pain in Section 4 probably applies. Nashold and Ostdahl (1979) consider the main role to be played by one or more of the following three factors: (1) hypersensitivity of the damaged neurons of the posterior horns in the dorsal roots entry zone (DREZ) as a result of sensory deafferentation; (2) damage to the spinothalamic and spinoreticular tracts; (3) local dysfunction of the DREZ neuron pool caused by activation or inhibition from Lissauer's tract. It is also assumed that avulsion of the dorsal roots eliminates the influence of the thick myelinated fibers on the neurons of the substantia gelatinosa. Increased activity of these deafferented neurons leads to the development of chronic pain.

The question of whether deafferentation pain after brachial plexus avulsion should be considered central (cerebral) or spinal pain still remains to be resolved. The permanent severe pain of total deafferentation of the arm seems to be related closely to phantom pain (Section 6.2), although the burning nature of the pain suggests causalgia. One of our patients with total avulsion of the brachial plexus had permanent severe pain in the paralyzed arm, resembling an electric current, which spread only to three fingers. Moreover, it seemed to the patient that his arm was hanging down, whereas in reality it was in a sling and flexed at a right angle at the elbow. Quite frequently, a shooting pain, like an electric current, radiated in the distribution of only one nerve (usually the median nerve). That coagulation of DREZ may eliminate or decrease the pain is an argument in favor of the spinal genesis of this syndrome.

Myelography in cases of brachial plexus avulsion visualizes the pronounced changes in the spinal cord, its membranes, and roots that may be seen at operation (arachnoid and dural scar adhesions, avulsed roots, and ruptured dural sleeves, cysts, and pseudomeningoceles). Myelography often demonstrates the emergence

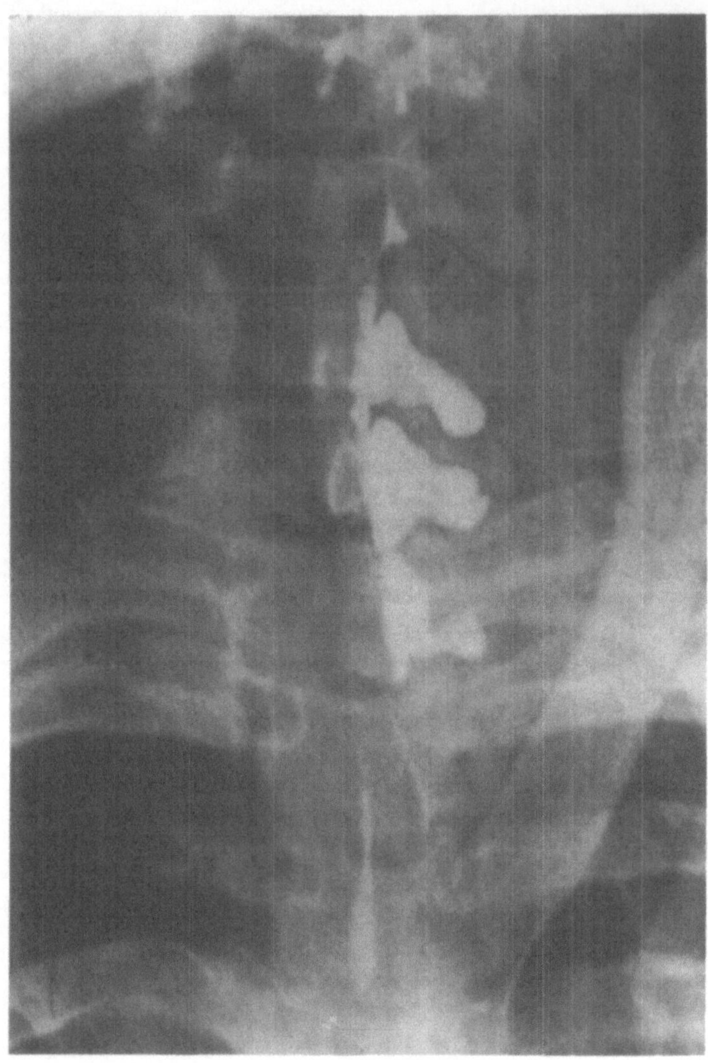

FIGURE 176. Cervical myelography of brachial plexus avulsion. Note the emergence of contrast medium beyond the dural sac and its spread along the avulsed roots.

of contrast substance outside the dural sac and along the torn roots (Fig. 176). As a result of avulsion of the roots, there is damage to the spinal cord, which, at that level, is atrophic. To differentiate pre- and postganglionic injuries of the brachial plexus, somatosensory EPs have been successfully employed of late.

Intractable pain after rupture of the brachial plexus does not lend itself to conservative therapy. Various operations employed to relieve this pain syndrome, e.g., sympathectomy, excision of the stellate ganglion, and posterior rhizotomy, were not very effective. Cervical cordotomy (Falconer, 1953) and mesencephalotomy (White and Sweet, 1969) may provide pain relief; however, the experience with these operations as described in the literature is limited. Neither transcutaneous nor posterior column stimulation provides sustained improvement (Thomas and Sheehy, 1983).

Stimulation of the spinal posterior columns in the cervical region has yielded contradictory results. Of seven cases, an excellent result was obtained in one; in two, the pain relapsed half a year after the operation; and in four cases stimulation produced no effect (Zorub *et al.*, 1974). Amputation of the paralyzed arm, frequently done at the request of the patient, does not relieve or even ease the pain, which becomes a phantom pain.

Nashold *et al.* (1976, 1983) proposed a new and promising operation for the management of this syndrome—coagulation of DREZ, including the substan-

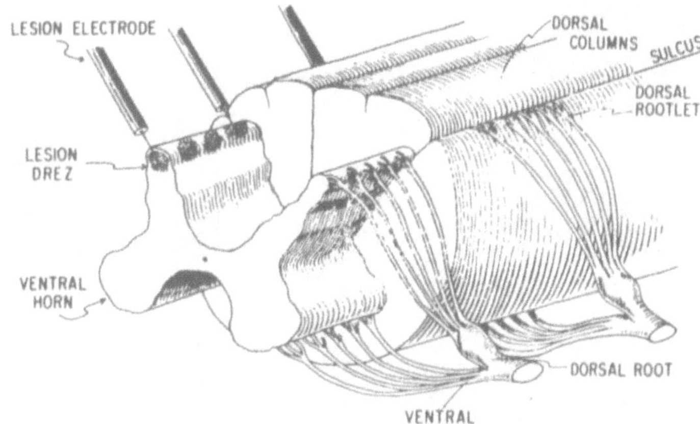

FIGURE 177. Anatomic relationship in radiofrequency coagulation of DREZ (Nashold and Ostdahl, 1979).

tia gelatinosa of Rolando and Lissauer's tract at the level of the damaged cervical roots (Fig. 177). It is assumed that in DREZ there is a complex interaction of stimulating and inhibiting circuits that modify incoming sensory impulses, but the precise mechanism by which this operation provides pain relief remains unclear. In all likelihood, it results from destruction of hypersensitive spinal neurons in DREZ or the interruption of primary nociceptive afferents. One might also consider the effect to be the result of interrupting ascending pain pathways (spinothalamic or spinoreticular tracts); however, on the basis of experimental data, this is improbable (Nashold and Bullitt, 1981).

It is also possible that the destruction of DREZ eliminates epileptiform activity in the supradjacent spinal cord segments. Such biopotential bursts have been observed both in animal experiments and in patients after posterior rhizotomy. These discharges may be the cause of central pain after a lesion of the spinal cord and its roots. In summary, one may assert that coagulation of DREZ "rebalances" the systems by which the neurons of the posterior horns interact with the input of pain impulses.

Coagulation of cervical DREZ is performed under anesthesia with the patient in a sitting position (Nashold and Ostdahl, 1979; Kandel *et al.*, 1987). After laminectomy of C_4–C_7 or C_5–T_1, the dura mater is opened. Then a microscope or magnifying lens is used to aid in separating the numerous arachnoid adhesions and cysts that develop after avulsion of cervical roots. On the side of the avulsion, the spinal cord is distorted, and the anterior and posterior roots are completely or almost completely absent (Fig. 178A).

As our experience has shown, this operation frequently involves difficulties in determining both the midline of the posterior surface of the spinal cord and the sulcus intermediolateralis adjacent to the dorsal column where the posterior roots bundles enter. Because of the presence of coarse arachnoid adhesions, small cysts, and fibrous scar tissues covering the damaged and atrophic half of the spinal cord, it is deformed to various degrees. On the inner surface of the dura mater, one may often see round defects several millimeters in diameter that resulted from the avulsion of the roots and their dural sleeves. In our experience, in these conditions, two reference points may be used to determine the lateral sulcus where the coagulation is carried out. First, on the corresponding side it is necessary to find two undamaged posterior roots, one above and one below the site of the scarred spinal cord where the posterior roots are absent. By joining the points where the undamaged roots enter the spinal cord with an imaginary line, one can determine the location of the lateral sulcus. By reference to the line along which the posterior roots of the other, undamaged side of the spinal cord lie, the scarred lateral sulcus on the damaged side may be plotted at the same distance from the midline. For the same aim, the intraoperative SEP, monitoring was successfully used.

The insulated steel electrode for coagulation of DREZ must be fine and very sharp in order to pass into the spinal cord substance through the scarred tissue and must have a 3-mm bare tip. Radionics produces a special electrode for making DREZ lesions. The electrode has a discrete sharpened tip 0.25 mm wide and 2 mm long with an indwelling thermocouple and shouldered insulation. Careful theoretical and technical specifications of the electrode were presented by Cosman *et al.* (1984). We employ a bipolar electrode with a 2.5-mm-long and very thin bar tip and an interelectrode distance

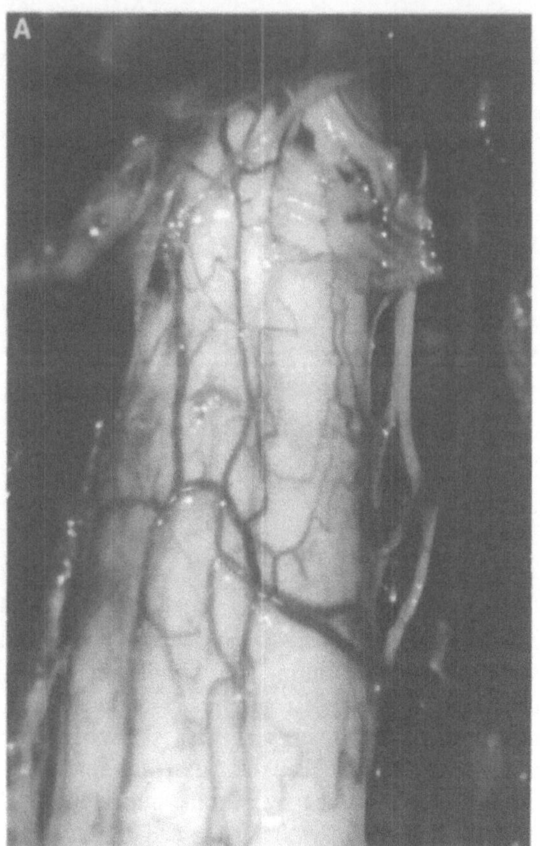

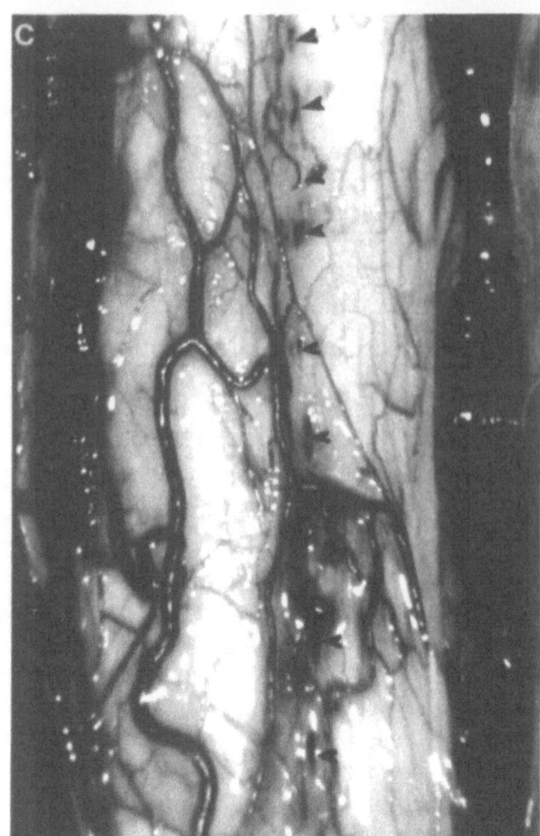

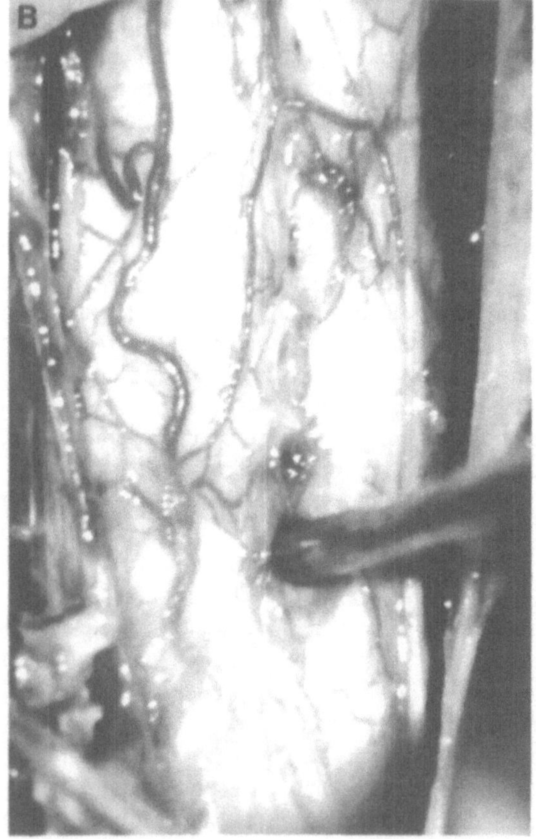

FIGURE 178. Stages of DREZ coagulation (photographs under surgical microscope, × 10–12). (A) Avulsion of the anterior and posterior cervical roots from the right side of the spinal cord. Note the preserved posterior root C_2 (upper right). There are no roots below. The surface of the cord is covered by scar tissue. (B) Coagulation of DREZ (second Rf lesion). (C) Chain of small radiofrequency lesions after completion of DREZ coagulation (arrowheads).

of 1.5 mm and produce the lesion with a high-frequency microcoagulator (manufactured by Esculap).

After the target zone is determined, the electrode is advanced to a depth of 2–3 mm in the sulcus intermediolateralis on the posterior surface of the spinal cord. Many points are coagulated (Fig. 178B) with a high-frequency current of 30–50 mA for 15 sec. These lesions produce a "chain" along the sulcus at a distance of 2–3 mm from each other. Altogether 10–20 lesions are made, depending on the extent of the damaged zone (Fig. 178C), i.e., slightly below the superior intact posterior root to the inferior intact root.

Friedman *et al.* (1984) have been making the lesions with a 0.25-mm thermistor, the active tip of which heats up to 75–80°C. The authors allow 15 sec for each coagulation and space the lesions 1 mm from each other, thus making approximately 25 lesions on a segment of the spinal cord. Hemorrhage does not occur.

Complications after DREZ coagulation are comparatively common, but in the majority of cases, they pass rapidly. Mild or moderately pronounced transitory paresis and disturbed sensation in the homolateral leg have been noted in approximately half of the operated cases. In certain rare cases, ataxia resulted from damage to the ipsilateral posterior column. Large doses of steroids for several days have been recommended to prevent complications after the operation (Nashold and Ostdahl, 1979; Friedman *et al.*, 1984).

Friedman *et al.* contend this new operation gives very good results in cases of severe pain after brachial plexus avulsion. In 11 of 18 cases, the pain decreased postoperatively by more than 75% over a comparatively long-term follow-up. However, complications developed in almost one-half of the patients—paresis in the homolateral leg or disturbed proprioceptive sensation in the same half of the body (Nashold and Ostdahl, 1979). In 1983 Nashold *et al.* reported that patients with brachial plexus avulsion who underwent the operation continue to experience pain relief for up to 5 years in 65% of cases. No new pain has been noted during the follow-up.

Thomas and Sheehy (1983) reported the results of DREZ destruction in 19 cases with partial or total avulsion of the brachial plexus. Significant relief from pain was noted in 16 cases that had been followed up to 2.5 years.

Another indication for this operation was recently established, namely, persistent burning pain in the legs, which develops in approximately 5–10% of paraplegics with a severely damaged spinal cord and cauda equina. Such pain, also of a deafferentation nature, usually involves the entire area below the spinal cord lesion. Usually this pain is classified as central pain, since numerous peripheral operations (sympathectomy, rhizotomy, cordotomy, posterior columns stimulation, etc.) are ineffective. Another concept holds that the sensation of pain arises in the spinal cord rostral to the lesion (Melzack and Loeser, 1978; Nashold and Bullitt, 1981). For relief of such pain, coagulation of DREZ in paraplegics is performed on both sides at the level of the lesion and for three segments rostrally.

Very satisfactory results of DREZ coagulation in paraplegics after various lesions of the spinal cord have been reported by Nashold and Bullitt (1981). In seven of 13 cases, persistent severe pain disappeared completely, and in the rest, it was diminished by more than 50% (follow-up from 5 to 38 months). A relapse of pain occurred in only one case a month after the operation; however, a repeated DREZ coagulation at a higher level gave lasting relief.

The first results from destruction of the DREZ with an argon or CO_2 laser was recently published (Powers *et al.*, 1984a). In 14 of 21 patients with pain from paraplegia, brachial plexus avulsion, or phantom limb, a satisfactory result was obtained (a decrease in pain by more than 50%). Levy *et al.* (1983) also used a CO_2 laser in three cases, one with avulsion of the brachial plexus and two with lesions of the spinal cord, and concluded that it offers several advantages—reduced operating time and lessened danger of damaging surrounding structures.

Destruction of DREZ proved effective in postherpetic neuralgia (Nashold, 1981). Friedman *et al.* (1984) reported on the results of the operation in 12 patients who had been suffering for many years from persistent burning postherpetic pain in thoracic dermatomes. In these cases, the authors destroyed the entry zone of the involved dermatomes as well as that of some dermatomes above and below the painful zone. Intraoperative recording of EPs directly from the spinal cord during stimulation of the intercostal nerves was employed for determining the location of DREZ. Total or almost total disappearance of pain was noted in eight of 12 patients over a period of from 6 to 21 months. Encouraging attempts have been made to use DREZ destruction on a broader scale, for pain caused by cancer metastases, pain after spinal cord damage, and phantom pain (Samii and Moringlane, 1983).

All these data seem to indicate that destruction of DREZ is an effective operation for the relief of different chronic pains caused mainly by lesions in the area of the posterior horns of the spinal cord.

Personal Observations. During the last 6 years we performed DREZ high-frequency coagulation in 24 patients (all men) with brachial plexus injuries and severe pain in the affected arm (Kandel *et al.*, 1987). The age of the patients varied from 22–60 years. All of them had a severe closed trauma in the shoulder region. The interval of time between trauma and admission to our clinic varied greatly from several months to 29 years. As a rule, pain appeared and progressed 2–3 weeks after trauma and only in few cases after 1–2 years. In the majority of patients we could identify two different types of pain. The first was permanent nonsharp pain that did not require the use of strong analgesics. This pain involved the dermatomes of the avulsed roots but had no defined borders. The second type of pain appeared on the background of the first pain. It was a sharp, burning, intractable pain like an electrical current. The paroxysms of such a pain may occur from 10 to 50 times a day. The duration of the paroxysm varied greatly from a few seconds to many minutes. The patients showed definite limits of pain, which corresponded with the damaged roots. Emotional tension, fatigue, change of weather, and many other reasons may provoke this pain. This second type of pain is mostly disabling. The patient adopted the antalgic pose with a hyperemic face and increased pulse and respiratory rate.

This sharp burning pain was very similar to causalgia and needed the administration of morphine or other strong analgesics, which often were not very effective.

Neurological deficit was usually severe with complete or near complete paralysis of the affected arm, atrophy of all muscles, absence of tendon reflexes, and decreased muscular tone. There was an anesthesia or (less often) hypoesthesia in the involved dermatomes and also a Horner sign on the affected side.

Positive contrast media myelography disclosed pseudomeningocele due to the rupture of dural sleeves. As a result, the invasion of the contrast medium along the ruptured roots was noted. Because of severe adhesions and disappearance of the subarachnoid space the contrast media was arrested on the level of the avulsion. Signs of atrophy of the cervical spine were also present.

Laminectomy or hemilaminectomy C_5–C_7 and sometimes also Th_1 was performed under general endotracheal anesthesia. The above-mentioned changes in cervical spinal cord were noted at the operation in all cases. Coagulation of DREZ was performed under the operating microscope as a chain of small bipolar lesions (0–16 lesions at each operation) were made, each of which penetrated the cord substance 2 mm.

In seven cases we combined DREZ coagulation with posterior selective rhizotomy (Section 5.3.6) of one intact root above and one intact root below the site on the avulsion. The rationale of that additional procedure was the spread of pain beyond the limits of interrupted roots.

During the follow up in 20 of 24 operated patients the results were considered either good (complete abolition of pain) or substantial improvement (mean follow-up: 2.5 years). Partial pain relief (about 50%) was noted in three cases and insignificant improvement in one patient. There was also one case of partial recurrence 1.5 years after surgery. There was a mild paresis with slight incoordination of the movements of the leg on the side of the operation in five cases. This complication, which reflects the influence of the lesions on the pyramidal tract and also on the ipsilateral spinocerebellar tract, disappeared in a few weeks, but a very light paresis remained in two cases. Some transient pain appeared in the healthy arm after surgery in two cases. It was considered a result of slight damage of the spinothalamic tract.

5.3.8. Administration of Opiates to the Spinal Cord

After specific opiate receptors, activated by Met-enkephalin, were discovered in the posterior horns of the spinal cord, mainly in substantia gelatinosa (Yaksh and Rudy, 1976; Lamotte *et al.*, 1976), and the inhibiting local action of opiates on nociceptive neurons was established, attempts were made to relieve intractable pain by intrathecal or epidural administration of local anesthetics and opiates, principally morphine. The latter activates opiate receptors in the substantial gelatinosa and inhibits secretion of substance P, suppressing the peripheral nociceptive input to the spinal cord (Poletti *et al.*, 1981). It has been established that opiates administered epidurally are absorbed by the CSF to a significantly greater degree than by the blood (Max *et al.*, 1981), and from the CSF they penetrate the posterior surface of the spinal cord and possibly the brainstem near V_4. Initially, this method was tried by anesthesiologists in the management of acute pain after various surgical operations on organs of the thoracic and abdominal cavities and on vessels of the extremities.

It has been established that repeated administration of anesthetics and opiates via a catheter into both the subarachnoid and epidural spaces causes analgesia that lasts for many hours and even a few days. An important advantage of this technique is that it can be used even in persons addicted to narcotics, who are common among cancer patients, especially as addiction was usually considered a contraindication to pain-relieving neurosurgical operations.

At the moment, preference is given to epidural administration of anesthetic agents. However, repeated morphine injections during the initial stage of the development of this method were associated with the risk of infection as well as suppressed respiration. These shortcomings were successfully surmounted by the development of implanted systems for constant infusion of given amounts of morphine (Fig. 179).

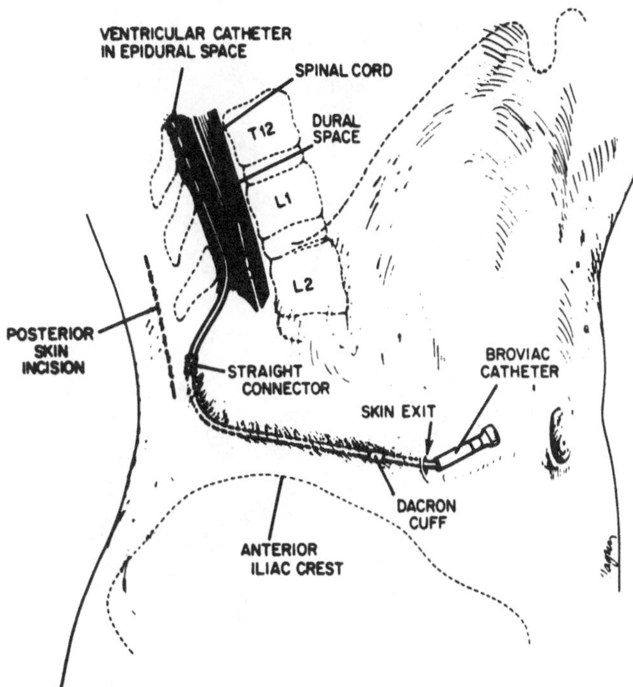

FIGURE 179. Diagram of a partially indwelling Broviac catheter system for long-term administration of morphine into the epidural space (Poletti *et al.*, 1982).

The first successful attempt to arrest radicular pain with a subcutaneously implanted system feeding a local anesthetic epidurally was made by Pilon and Baker (1976). For a period of almost 5 months, it was possible to arrest intractable pain totally in a patient with sacral metastases by repeated injections of an anesthetic agent into a subcutaneous reservoir connected with an epidural catheter. The motor power of the legs and the functions of pelvic organs were not disturbed.

Subsequently, Poletti and Schmidek (1982) improved the implanted system. A Silastic subcutaneous catheter, introduced epidurally in the posterior thoracic region, is connected via an "on–off" valve with a 20-ml reservoir implanted subcutaneously in the anterior abdominal wall. A Hakim valve was subsequently included in this system. Two milliliters of morphine are administered epidurally by transcutaneously pressing this valve 20 times (Fig. 180). The patient is able to perform this procedure independently at home.

Harbaugh *et al.* (1982) have also devised a system for epidural administration of morphine by use of an Infusaid™ pump (Model 100) with a 47-ml reservoir. Morphine enters the epidural Silastic catheter at a constant flow rate of 2–3 ml a day. The dose of morphine may be regulated by changing the concentration of the solution. After implantation, the pump may be refilled percutaneously approximately twice a month. This system has been used successfully in two cases of cancer pain. The Cordis reservoir with a catheter introduced via a Tohny needle has also been successfully employed (Cobb *et al.,* 1984). A new programmed Medtronic system, recently proposed, consists of a rotary pump, a batter, a high-frequency receiver, a transmitter, a microprocessor, and a reservoir with a bacterial filter. With this instrument, it is possible to program the quantity and time of morphine infusion, determine the quantity of solution in the reservoir, and switch the pump on and off. Other systems with telecontrol of the valve and computer are being tested (Poletti and Schmidek, 1982).

Saunders and Coombs (1982) have reported their first experience using a modified, fully implanted epidural system in 16 cases, 12 of which involved cancer pain. This system has a pump reservoir feeding morphine for several months at a constant flow rate of approximately 3 ml a day (10 mg of morphine in 1 ml). The system is designed for continuous use over a period of 5 years. When the correct technique was used, the authors did not observe any serious complications, in particular, respiratory depression.

Without dwelling on the details of the design of several implanted systems now being employed, we present, in brief, some data on the use of this method, mainly in chronic pain in the legs, abdominal cavity, and pelvis resulting from malignant tumors.

It has been established that infusion of morphine

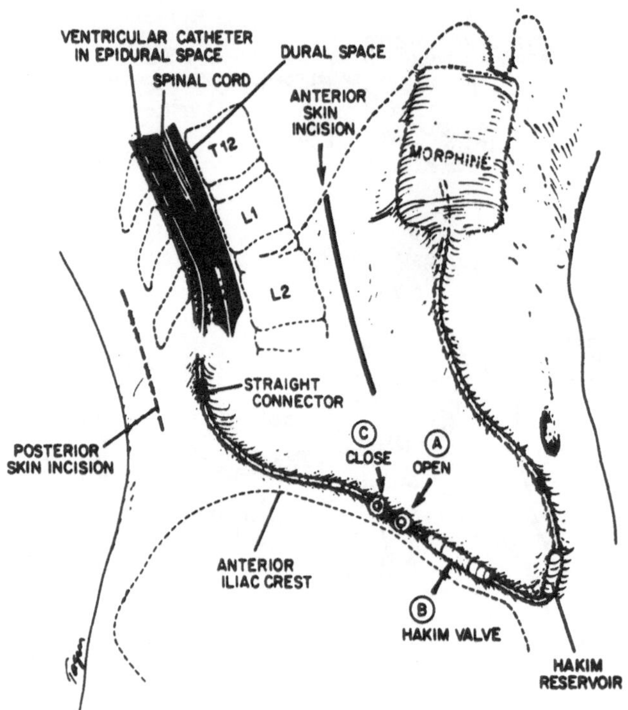

FIGURE 180. Diagram of a completely indwelling catheter system consisting of a morphine reservoir, shunt pump, and on–off valve for long-term administration of morphine into the epidural space. For the injection procedure, the on–off valve is opened (A), the Hakim valve is pumped 20 times (B), delivering 2 ml of morphine, and the on–off valve is then closed (C) (Poletti *et al.*, 1982).

and its derivatives in small doses epidurally at the lumbar level causes total analgesia in the lower half of the body, which usually spreads over 3–4 hr to produce analgesia in the upper extremities as well (Poletti and Schmidek, 1982). Epidural administration of narcotics is preferable to the intrathecal route because it causes fewer complications. Before permanent placement of the catheter, a single test injection of 1–2 mg of morphine intrathecally is recommended. The relief of pain for approximately a day indicates that the method will be effective (Cobb *et al.*, 1984).

Penn *et al.* (1984) employ another technique. With an external pump, 0.1 mg of morphine is repeatedly infused via an epidural catheter. The total daily dosage is 4.8 mg. If the analgesic effect is not sufficient, the dose is gradually increased to 14.4 mg per day. The test lasts 2–3 days, sometimes a week, during which period the dose of conventional narcotics is reduced by approximately 50%. The system is implanted permanently only if the pain is relieved by the test infusion of morphine more than 50%.

The daily dose of morphine required varies substantially depending on many factors; on the average it is 8–10 mg per day. Onofrio *et al.* (1981) reported that only 0.62 to 1.8 mg of morphine a day was required to eliminate pain in a case of sacral chordoma.

Cobb *et al.* (1984) implanted Ommaya or Cordis subcutaneous reservoirs connected to an intrathecal catheter in the lumbar region in ten cases of pelvic and sacral pain caused by cancer of the intestine, rectum, uterus, or prostate. Using small doses of morphine and other narcotics, the authors succeeded in controlling pain in the pelvis and sacrum over several months, but subsequently, the degree of relief diminished. For pain in the legs and neck as well as in cases of tumor invasion in the solar plexus, the method was less effective.

Penn *et al.* (1984) used a morphine infusion intrathecally in 14 cases of cancer pain. An excellent result was noted in eight, and a good result in five cases. These patients resumed care of themselves and were relieved of severe pain. Other authors have confirmed that this method is so effective in cancer pain that frequent narcotic injections may be discontinued (Chayen *et al.*, 1980; Lazorthes *et al.*, 1980; Zenz *et al.*, 1981; Greenberg *et al.*, 1982). Even in the terminal stages of malignant tumors, if repeated daily injections of large doses of morphine are replaced by two or three epidural infusions of only 2–3 mg, the physical condition and morale of these patients significantly improve.

Attempts have been made to administer narcotics

in the cervical region and even in the cisterna magna. However, some authors have had less encouraging clinical results. For example, Wang (1985) recently reported his experience with intrathecal morphine in 62 patients with intractable pain secondary to cancer of the pelvic organs. Forty-six patients who experienced pain relief from an initial test dose (0.5–2.0 mg) were further treated by low-dose morphine intrathecally with repeated single injections by external catheter or implanted pump. Only half were relieved of pain without side effects; however, there were no respiratory complications in this series.

There are grounds to assume that intrathecal or epidural infusion of morphine is effective not only in cancer pain, but also in pain of other etiologies, in particular, brachial plexus injuries (Penn *et al.*, 1984). Teddy *et al.* (1981) administered diamorphine epidurally by a catheter for postoperative pain after lumbar laminectomy for intervertebral diskopathy and obtained good results in practically all cases.

The technique may give rise to various side effects and complications, which occur less frequently in cancer than in other chronic pain cases (Saunders and Coombs, 1982). The most serious complication is apnea (Glynn, 1979), which may occur as morphine spreads to the cervical subarachnoid space and the brain ventricles. In view of this, monitoring of the respiratory rate and pupillary size is essential before and during the operation (Poletti and Schmidek, 1982). The patient must remain in a semisitting position for 12 hr (Lazorthes *et al.*, 1980). The dangerous respiratory disorder must be rapidly combatted by an intravenous injection of a morphine antagonist—naloxone in a dose of 0.4–0.6 mg—which at times must be repeated. Respiratory depression, fortunately, is a rare complication: in 14 cases of protracted morphine administration, it did not occur in a single instance (Penn *et al.*, 1984) or in eight cases described by Harbaugh *et al.* (1982).

Infectious complications are observed very rarely, especially when completely implanted systems are used. Nausea and vomiting occur frequently, usually several hours after morphine administration, but they can be relieved by naloxone (Poletti and Schmidek, 1982). Urinary suppression as well as allergic reactions such as pruritus are observed comparatively frequently and usually respond to naloxone. Sometimes CSF accumulates around the morphine reservoir used for intrathecal administration (Cobb *et al.*, 1984).

One problem arising from the long-term use of this method is the development of tachyphylaxis, which requires a gradual increase in the dosage of morphine. However, total tolerance to large doses develops rarely; Penn *et al.* (1984) observed it in only one of 14 cases.

The first reports on the analgesic effect of intraventricular administration of morphine have only appeared recently (Leavens *et al.*, 1982; Roquefeuil *et al.*, 1984). Small doses of morphine were administered via a ventricular catheter in the anterior horn connected to a subcutaneous reservoir. Roquefeuil *et al.* (1984) modified this method by connecting the catheter with a long plastic tube, which was connected to a collecting bag near the clavicle holding 200 ml of morphine solution. This system includes a Cordis valve to prevent reflux of ventricular fluid into the bag. The use of this method in eight cancer cases with severe pain gave a good and lasting analgesic effect. Small doses (0.5–1 mg) of morphine were circulated by intraventricular route in 38 patients with severe pain due to cancer of the cervicofacial region (Lenzi *et al.*, 1985). The intraventricular catheter was introduced through a burr hole in the precoronal region and was connected to a Cordis reservoir placed subcutaneously in the epicranium. The morphine was injected by pricking the reservoir 1–2 times daily or less frequently. Analgesia lasted 24 to 48 hours. The authors reported good results in the majority of cases. The quality of life improved greatly because of absence of pain and minimal side effects. A comparison of intrathecal and intraventricular administration of morphine reported by Lazorthes *et al.* (1985) has shown that the former is indicated for pain originating below the diaphragm, whereas the latter is preferable in cervicocranial pain which requires a much smaller dose of morphine (0.015 mg/day) than that used in the lumbar region (2 mg/day).

There is every reason to believe that this new method of treating severe pain will find broad clinical application, since it is a good alternative to systemic narcotics.

5.4. Stimulation Methods

5.4.1. Stimulation of Subcortical Structures

A comparatively new trend in the management of a number of CNS disorders, primarily intractable pain, is chronic stimulation of various subcortical and brainstem structures, now referred to as "deep brain stimulation." The first therapeutic stimulation of the human brain was carried out by Pool in 1954. The rapid progress in development of implanted heart pacemakers has been a potent stimulus to develop new techniques to use in the brain.

The main feature distinguishing this method from conventional stereotaxis is that the target structure is not destroyed but is subjected to prolonged stimulation with a weak current via an electrode or electrodes implanted in the brain. The possibility of not injuring cerebral structures is an obvious advantage of the method. There can be no doubt that stimulation techniques in pain syndromes are safer in principle than destructive procedures. That they are less risky is especially important in seriously ill and elderly patients. Two-stage operations for pain relief have also been performed in recent years. In the first stage, electrodes are implanted in deep-seated cerebral structures for a subsequent trial of stimulation. Then, depending on the results, the destruction of the corresponding structures may be performed.

Although the specific mechanisms of such pain-suppression techniques as electrostimulation of subcortical structures, spinal cord, and peripheral nerves still remain obscure, the techniques have indisputable experimental and physiological value. In 1969, Reynolds was the first to demonstrate in experiments on rats that stimulation of certain subcortical nuclei induced analgesia. Subsequent experimentation showed that this effect is most pronounced on stimulation of the periaqueductal or periventricular gray matter (PGM) and the dorsal raphe nucleus of the brainstem. Shortly after that, clinical reports appeared about the use of chronic electrostimulation of different subcortical structures for the management of persistent pain.

The efficacy of this method has been proven over the past 12–15 years, although the optimal subcortical structure for stimulation has yet to be determined. Positive clinical results have been obtained from the stimulation of many subcortical structures: VPL (Mazars, 1976a; Adams *et al.*, 1975; Sedan and Lazorthes, 1978; Siegfried, 1979; Turnbull *et al.*, 1980; Ray and Burton, 1980; Dieckmann and Witzmann, 1982; Tsubokawa *et al.*, 1982a,b; Young *et al.*, 1985; Kuroda *et al.*, 1985; Meyerson, 1985); VPM (Hosobuchi *et al.*, 1975; Siegfried, 1982; Roldan *et al.*, 1982); CM (Smirnov and Iovlev, 1974); medial Th (Adams, 1974); Pul (Smirnov, 1976); intralaminar nuclei (Thoden *et al.*, 1979); posterior limb of CI (Adams *et al.*, 1974; Namba *et al.*, 1984); lemniscus medialis (Mundinger, 1977; Mundinger and Salomao, 1980; Frank *et al.*, 1985); PAG (Meyerson *et al.*, 1978; Hosobuchi, 1979, 1983a,b, 1986; Richardson, 1981; Plotkin, 1982); Hypoth (Fairman, 1983); NC (Ervin *et al.*, 1966); and septal area (Gol, 1967; D. E. Richardson, 1982a,b; Schvarcz, 1985). Simultaneous chronic stimulation of two subcortical structures, e.g., Th nuclei and PGM, has also been tested (Frank *et al.*, 1985).

At the same time, a number of problems regarding the use of chronic stimulation remain to be solved. The indications for this technique have not been sufficiently well defined in respect to the cause and type of pain. The morphological and biochemical changes that may develop in the brain after prolonged stimulation of deep-seated structures have not yet been studied. Important ethical problems related to chronic stimulation of deep brain structures have yet to be resolved (Siegfried *et al.*, 1980). Technical problems such as displacement or breaking of electrodes, etc., arise frequently.

There are several hypotheses concerning suppression of chronic pain by stimulation of Th nuclei and CI. Melzack and Loeser (1978) consider that any damage to pain-relaying pathways causes hyperactivation of the posterior horn neurons and cerebral sensory nuclei, whereas stimulation of these nuclei inhibits this activation. Andy (1983) speculated that the pain-relieving effects of deep brain stimulation depend on altering the excitability state and/or the thalamic discharge patterns. Other authors believe that stimulation of Th neurons deprived of normal connections with peripheral pain-transmitting structures inhibits deafferentation pain. In particular, Mundinger and Salomao (1980) believe that deep brain stimulation alters the interrelationship between α–γ systems and C afferents, blocking nociceptive afferents to the brain.

The suppression of pain after stimulation of Th nuclei, brainstem, or spinal cord, as a rule, lasts several hours and even days after stimulation has been discontinued. This established fact leads to the logical conclusion that not only a neuronal but a biochemical mechanism lies at the basis of this phenomenon. One may assume that stimulation of certain deep cerebral structures releases endogenous opiate peptides, in particular, β-endorphin, inducing analgesia. If a patient is tolerant to morphine, stimulation of deep cerebral structures is not effective (Mayer and Hayes, 1975). At the same time, clinical and experimental data do not confirm the assumption that chronic pain relief by Th nuclei stimulation is the result of an opiate mechanism (Tsubokawa *et al.*, 1982a,b). The elevation of β-endorphin in CSF after deep brain stimulation may depend on the stress of the surgery (Meyerson *et al.*, 1985).

In all likelihood, not only opiates but other biochemicals, in particular, catecholamines, participate in the development of analgesia. The dopamine level in the ventricular CSF rises on stimulation of Hypoth (Fairman, 1981). It has been established experimentally that electrostimulation of VPL activates the nucleus raphe to the same degree as stimulation of PGM (Tsubokawa *et al.*, 1982a,b).

Even two decades ago, stimulation of any structure within the CNS with implanted electrodes was possible only with wires leading to an external current source. This created not only obvious inconveniences for the patient but also a constant risk of infection. The difficulty was surmounted after the development of subcutaneously implanted receivers that picked up signals from the transmitter through the skin. Special devices (manufactured by Medtronic, Biomedix, Neuromed, and other companies) that are totally or partially implanted have been developed in recent years for chronic stimulation of subcortical structures. Similar to the devices for spinal cord stimulation (Section 5.4.2), these systems consist of electrodes with leads, a receiver, a transmitter, an antenna, and a programmer. In the Medtronic system the electrode has four bare platinum contacts that are marked T_0 (the deepest) to T_3 (the uppermost). The surface of each is approximately 4 mm^2, and the distance between contacts is 2.5 mm.

For chronic stimulation of subcortical structures the Neuromed Company produces the MIND system (Multiprogrammable Implantable Neurostimulator Device), which consists of a four-contact electrode, a miniature implantable receiver, an antenna, and a transmitter. The latter produces impulses at a frequency from 20 to 175 Hz and duration from 50 to 500 μsec. It has four independent outputs, a common anode for monofocal stimulation, and a programmer for selecting all parameters of monophasic or biphasic stimulation. Avery Laboratories produces an eight-button plate electrode for deep-brain stimulation.

The Tesla Company of Czechoslovakia manufactures an implanted system for stimulating various cerebral structures. The advantage of this system is that the receiver is located not subcutaneously in the sub- or supraclavicular area but within a small cranial burr hole, making it possible to reduce the length of the implanted bipolar electrode to 7–10 cm. At the end of this 1.2-mm-diameter electrode, there are two platinum contacts 2 mm apart. In order to stimulate, the patient applies a disk-shaped plastic antenna to the scalp over the implanted receiver.

Electrodes of various metals (platinum, gold, tantalum, stainless steel, and different alloys) are used for stimulating deep cerebral structures. Multicontact electrodes of pure platinum are most commonly used at the present time. They consist of two to four intertwined wires. At the end of the electrode the wires are separated and made into small loops with a diameter of approximately 1 mm (Fig. 181). In this manner, each electrode has a contact point that is separated approximately 2 mm from another. The measurement of the resistance between the electrode and the neural tissue (McLellan *et al.*, 1981) confirms satisfactory contact. This resistance varies significantly, for example, 350 Ω for the Avery electrode and 750 Ω for Medtronic.

With the patient in a sitting or semisitting position, chronic electrodes are usually implanted through two small burr holes slightly anterior to the coronal suture and 2–3 cm from the midplane. The electrode is inserted stereotactically into the desired subcortical structure with a special Trend–Wells insertor. Then the electrode is fixed in the burr hole using rapidly setting plastic or cement. A special ring cap has been proposed to hold the electrode array, so that it may be transferred to another site or removed from the brain. The Richardson burr hole cap serves the same purpose.

After the electrode has been inserted, it is tested, usually by bipolar stimulation, which may take from several days to a few weeks. During the stimulation tests, in order to find the optimal combination of two contacts, pairs are taken consecutively, until the combination is found that, with the lowest voltage, suppresses pain maximally and induces definite paresthesias in the painful zone (Fig. 182). These sensations are neither painful nor unpleasant, although when the current is increased, the feeling becomes unpleasant, and if further increased painful. At the same time, various parameters such as voltage, current frequency, pulse duration, and time of stimulation are tested. If the current intensity during stimulation is below the threshold at which paresthesias appear in the part of the body where pain is felt, the pain is not suppressed (Turnbull *et al.*, 1980). These authors, as well as Boëthius *et al.* (1976), note that during stimulation of VPL the threshold of pain and tactile irritation of skin, as well as pain sensation on squeezing of muscles or tendons, do not change.

During the daily testing, if stimulation of VPL

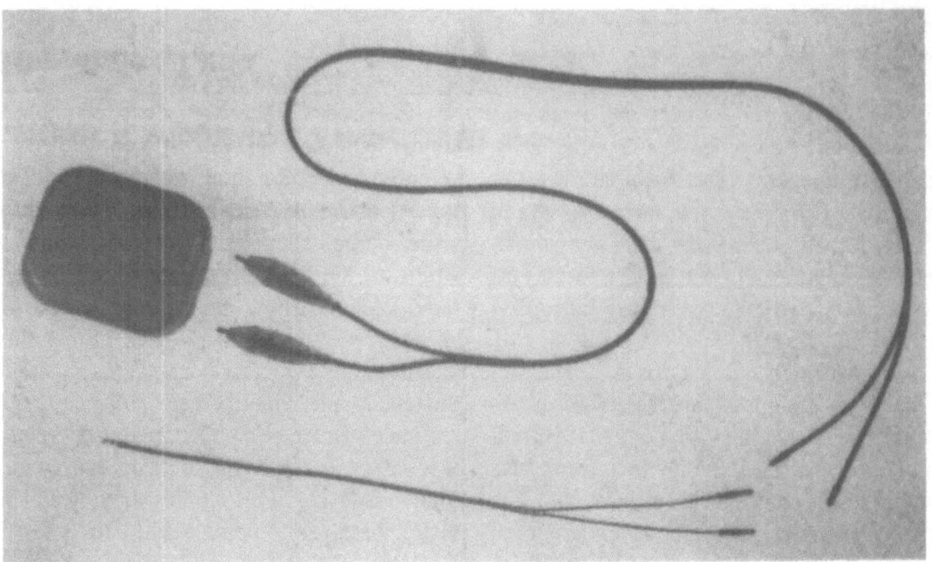

FIGURE 181. Thalamic stimulation system (model ST21, Biomedix).

causes the pain to disappear for at least several hours, the receiver may be implanted permanently.

Many authors stress the great individual variability of the effective parameters for stimulating subcortical structures. In chronic stimulation of VPM, the overall duration of such action during the course of a day varied from 4 to 60 min, whereas the frequency of the stimulating current ranged from 33 to 195 Hz (Siegfried, 1982). The optimal frequency for VPL stimulation varied from 25 to 100 Hz (Tsubokawa *et al.*, 1982a,b). A current 0.5–1 mΩ is usually used.

After the optimal combination of electrodes has been found, the receiver is implanted subcutaneously above or below the clavicle, and the electrodes are connected with the receiver under the skin of the scalp and neck. A disk-shaped antenna from the transmitter is glued externally to the skin over the implanted receiver. Then the patient, after receiving precise in-

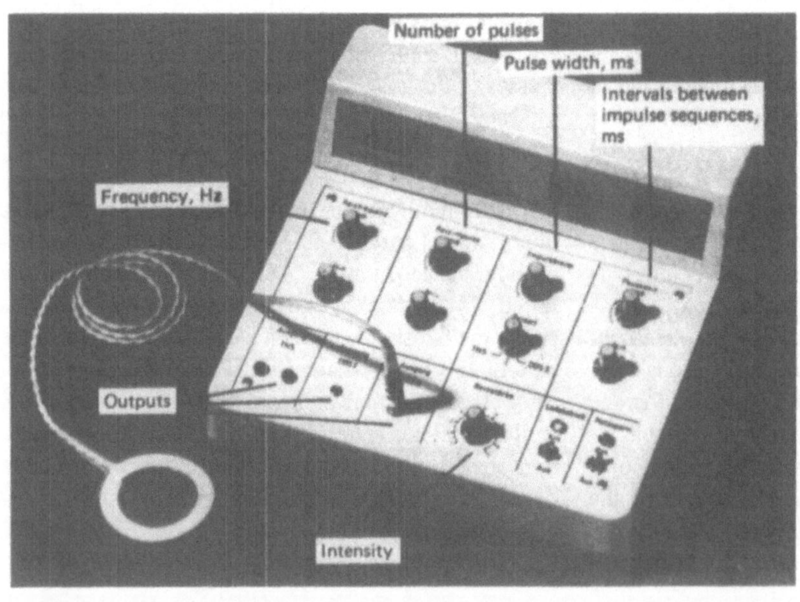

FIGURE 182. Stimulation and recording unit for deep brain stimulation with a programming unit with four parameters (frequency and impulse range, number of single impulses per sequence, and intervals between impulse sequences) which could be changed within physiologically feasible limits in order to establish a stimulation pattern most effective for the patient (Mundinger and Neumüller, 1982).

structions, can carry on the stimulation independently, usually several times a day.

5.4.1a. Thalamic Nuclei. Bechtereva *et al.* (1972b) reported favorable results from chronic stimulation of "thalamic and brain stem structures" in four cases (phantom pain syndrome, parkinsonism, Wilson's disease, complicated hyperkinesia). The implanted electrodes remained in the brain for a long period (up to 12 months). Bipolar rectangular impulses with a frequency of 50 Hz and duration up to 3 msec were used for electrostimulation, which was performed once or twice a week for 3 to 6 months. The authors attributed the therapeutic effect of electrostimulation "not so much to local as to general changes in brain activity."

Mazars (1975), one of the first to use this method, believes that chronic stimulation of VPL with implanted electrodes relieves pain by sensory deafferentation. If the pain is caused by excitation of nerve structures ("excessive afferentation"), this method is not effective. Mazars (1976a,b) reported the results of a long-term follow-up of 104 cases in which VPL alone or together with the lemniscus medialis was chronically stimulated with a weak current. The author considers this method very effective treatment in numerous pain syndromes (amputation stump pain, pain after avulsion of the brachial plexus, postherpetic neuralgia or peripheral nerves injury, and bulbar vascular lesions). In a case of severe pain after thoracotomy, periodic stimulation of VPL for 3 days led to relief of the pain for 12 years. However, electrostimulation produced no effect on thalamic pain.

Turnbull *et al.* (1980) stimulated VPL in 14 cases with benign pain, mainly caused by arachnoiditis developing as a result of repeated ineffective operations on lumbar intravertebral disks. As the experience of these authors has shown, even with prolonged (for months and years) stimulation of VPL, the patient does not acquire tolerance, so that an increase in voltage is not required.

Tsubokawa *et al.* (1976, 1982a,b) performed chronic stimulation of Th relay nuclei on six patients with pain of different origins. In five of them, the results were assessed as excellent or good. After beginning stimulation, the level of β-endorphin in the ventricular fluid increased almost twofold. It was also found that the analgesic effect was more pronounced and longer lasting if the stimulation was carried out after L-dopa loading. In their subsequent paper, Tsubokawa *et al.* (1985) reported the use of the method in 14 cases. Seven of them had somatic pain caused by malignancy, and the other seven had deafferentation pain. In both groups, there was good pain relief; moreover, the benefit was not related to the degree of increase of β-endorphin concentration in CSF. Recently, Hosobuchi (1986) reported new observations concerning the role of the dopaminergic system in pain relief obtained by thalamic stimulation. Sometimes there is a considerable loss of efficacy after a long period of stimulation. If these patients are placed on a regimen of L-dopa orally, the effectiveness of the stimulation is restored within a few days.

Richardson and Akil (1977) reported that chronic stimulation of the parafascicular nuclei in three pain cases (thalamic pain syndrome, phantom pain, lesion of the brachial plexus in mammary cancer) produced marked relief.

It is noteworthy that the published results on chronic electrostimulation of Th sensory nuclei have not been in agreement. For example, Mazars (1976a,b) reported good results in nine of 11 cases; in the other two cases, failure was the result of technical defects. However, Hosobuchi (1980) obtained satisfactory pain relief in only one of three cases. Of those undergoing chronic stimulation of thalamic sensory nuclei, 44% developed tolerance to the treatment (Young and Chambi, 1987).

It may be concluded that, according to the results of many investigators, the stimulation of sensory Th is mostly effective in treatment of neurogenic (or deafferentation) pain.

5.4.1b. Internal Capsule. Another subcortical structure the electrostimulation of which relieves chronic pain is CI. It is considered that such stimulation activates inhibiting pain pathways in the posterior limb of the CI leading to the parietal cortex from the sensory Th nuclei (Adams *et al.*, 1974; Hosobuchi, 1979). It has been established experimentally that CI stimulation suppresses nociceptive neuronal activity in VPM induced by irritation of dental pulp on the contralateral side. To a lesser extent, stimulation inhibits such activity in the posterior Th nuclei and CM (Nishimoto *et al.*, 1984).

Chronic stimulation of CI via implanted electrodes provides significant relief in various pain syndromes (Adams *et al.*, 1974; Namba *et al.*, 1984). Multicontact electrodes are inserted stereotactically in the target point at the following coordinates: 8–9 mm posterior to the midpoint of LI, 1–4 mm above or below that line, and 23–25 mm from the midplane.

After the electrode has been implanted, trial stimulations are conducted for several days. If an effect is obtained, the electrode is connected to the subcutaneously implanted receiver. Then the stimulation is carried on by the patient himself, who can vary the frequency of excitation depending on the effect.

For stimulation of the posterior limb of CI, the following parameters are recommended: biphase impulses of increasing voltage from 0.5 to 5 V, duration of impulse approximately 0.5 msec, frequency from 30 to 150 Hz. After each stimulation, pain disappears for a period of time (usually 1–2 hr).

Since it has been established that stimulation of sensory thalamic nuclei deprived of input of nociceptive impulses, as well as of CI, suppresses primarily deafferentation pain (Mazars *et al.*, 1974; Hosobuchi *et al.*, 1975; Boëthius *et al.*, 1976; Turnbull *et al.*, 1980; Plotkin, 1982), information on the results of this method is also presented in Section 6.

5.4.1c. Hypothalamic Nuclei. Experimental data (Watanabe *et al.*, 1980) and early clinical applications, although limited, have demonstrated that chronic stimulation of the posteromedial Hypoth provides pain relief (D. E. Richardson, 1982a; Mayanagi *et al.*, 1982). During stimulation of Hypoth (periventricular nuclei) in stereotactic operations for pain syndromes, Fairman (1983) found a close correlation between changes in dopamine content in the ventricular fluid and the results of the operation. If a good effect was obtained, the dopamine level increased substantially; if the results were poor, it decreased.

5.4.1d. Septal Area. Several attempts at chronic stimulation of another subcortical structure, the septal area, have been undertaken (Heath and Mickle, 1960; Gol, 1967; D. E. Richardson, 1982a; Schvarcz, 1985). Although these authors reported encouraging results, stimulation of that subcortical structure has found no broad application.

5.4.1e. Periaqueductal Gray Matter. One more method of chronic stimulation for pain relief has been developing rapidly and successfully, electrostimulation of the gray matter around the Sylvian aqueduct (PAG) and V_3. Since there are no direct pathways between PAG and the spinal cord, it is assumed that the analgesic effect from PAG stimulation is produced via the raphe spinal nucleus of the brainstem.

This method is based on two important considerations. First, the spinoreticular tract, the main ascending pathway of the medial pain-transmitting system, terminates in PAG as well as in the nonspecific Th nuclei. Second, reliable experimental data have been obtained indicating that electrostimulation of PAG, where opiate receptors are highly concentrated, causes deep analgesia in animal experiments because of inhibition of the neurons in the posterior horns of the spinal cord (Mayer and Liebeskind, 1974).

Electrical stimulation of the locus ceruleus, which is closed to the pontine parabrachial region, also inhibits nociceptive conduction at the level of the spinal posterior horns (Katayama *et al.*, 1986).

It has also been established that PAG is the site of action of the endogenous endorphin that suppress the perception of pain (Section 3). Experimentally prolonged (up to 30 min) stimulation of PAG causes excretion of endorphins and pronounced analgesia. This effect disappears after section of the spinal cord at any level. These observations confirm the presence of an antinociceptive system in the brain (Section 3). The role of endorphins in terminating pain has been proven by the fact that stimulation of PAG in patients with intractable pain increases the level of immunoreactive β-endorphin and enkephalin of ventricular fluid, particularly in V_3, to several times the normal amount (Akil *et al.*, 1978; Amano *et al.*, 1980, 1982; Tsubokawa *et al.*, 1982a,b, 1984). According to Amano *et al.* (1982), the β-endorphin level in 14 cases before stimulation was from 37 to 175 pg/ml, and after stimulation from 62 to 314 pg/ml. In all likelihood, this effect is specific for PAG or VPL, since stimulation of other structures, particularly CI or ZI, does not cause an increase in the ventricular endorphin concentration (Amano *et al.*, 1980; Turnbull *et al.*, 1980). These data confirm the supposition of stimulation-induced release of endogenous opioid substances.

It has been established that stimulation of PAG eliminates only the pain that can be relieved by morphine and, vice versa, that the analgesia induced by PAG stimulation is blocked by the specific opiate antagonist naloxone (Akil *et al.*, 1976; Tsubokawa *et al.*, 1984; Hosobuchi, 1986). The morphine saturation test is performed to differentiate somatic from functional pain; PAG stimulation is not indicated if the pain is not relieved by morphine.

After stimulation of dorsomedial nuclei of Th and NC during stereotactic thalamotomy, there is in the ventricular fluid a significant increase (2–3 times normal) in the concentration of methionine-enkephalin, one of the endogenous opiate peptides inducing analgesia (Furni *et al.*, 1980). However, the opiates cannot explain the protracted analgesic effect, since in several days the β-endorphin level in CSF drops to the initial level (Mayanagi *et al.*, 1981; Tsubokawa *et al.*, 1984; Young and Chambi, 1987). These data reaffirm the complex nature of the biochemical mechanism of the modulation of pain. There is an opinion that the increase in the level of β-endorphin in CSF is an artifact induced by the contrast medium used for ventriculography. Young and Chambi (1987) also believe that pain relief elicited by PAG does not depend on an endogenous opioid mechanism.

The stereotactic coordinates of PAG are the following: in lateral projection, 14 mm posterior to the midpoint of LI and 5 mm below the continuation of that line, i.e., 2 mm posterior to and 3–4 mm below CP; in AP projection, 3–4 mm lateral to the midpoint of the Sylvian aqueduct. Because of the very small target, the implantation of electrodes in PAG has to be made with utmost accuracy. Displacement of the electrode by 2–3 mm may lead to the absence of any effect (Meyerson *et al.*, 1985). Stimulation of the lateral part of PAG is performed more frequently somewhat medial to its border with the mesencephalic FR at the level of the superior colliculus and CP. A thin electrode of platinum or platinum–iridium alloy is implanted stereotactically in that area (Fig. 183). Electrodes may be inserted separately into the superior and inferior parts of PAG (Richardson, 1982a,b) or on one or both sides of the Sylvian aqueduct. It has been noted that bilateral stimulation is more effective (Hosobuchi, 1982). The electrode is temporarily fixed in the burr hole with a small piece of wax, passed under the skin, and brought out on the surface for stimulation.

The parameters for PAG stimulation vary from 1 to 8 V, from 0.5 to 1 mA, and from 10 to 120 Hz. Low frequencies usually do not cause any neurological disturbances. If the electrode is located close to the oculomotor nucleus, synchronous movements of the eyeballs occur on stimulation. High-frequency stimulation causes tachycardia, an increase in arterial pressure (by approximately 50% from the initial level), bilateral mydriasis, and horizontal nystagmus (Amano *et al.*, 1982). Stimulation of PAG with a frequency of 30–60 Hz induces a feeling of warmth and tingling in the painful areas on the opposite side of the body (Dieckmann and Witzmann, 1982).

The first test stimulus is made on the operating table directly after implantation of the electrode. For the next 1–2 weeks, control stimulations are continued in order to determine the effect on the patient. They are made using a special stimulator connected to the wires coming from the surface of the scalp. There are several ways to conduct stimulation trials. Usually, they are given five to six times a day, varying in duration from 15 to 60 min. After a few days, the number of stimulations is gradually reduced if a uniform effect has been obtained. Later, the stimulation is necessary every few days (Plotkin, 1982). Usually, a frequency of 60 to 120 Hz is used with a pulse duration of 0.1 to 1 msec; stimulation with high frequencies (more than 100 Hz) is more effective.

After the electrode and receiver have been implanted, the patient is given a stimulator and instructed in its use. After that, the patient selects the optimal frequency, voltage, and period of stimulation. Complications from PAG stimulation (dizziness, diplopia, etc.) are observed rarely and may be eliminated by modulating the parameters of stimulation (Thoden *et al.*, 1979).

In the majority of patients so treated, pain disappeared completely or partially (Richardson and Akil, 1977; Sedan and Lazorthes, 1978; Siegfried and Wieser, 1978; Ray and Burton, 1980; Young and Chambi, 1987). Stimulation of PAG is considered more effective than stimulation of the sensory nuclei of Th or the CI (Richardson, 1981). Stimulation of PAG led to the total or almost total relief of pain in 80% of 46 cases, whereas in stimulation of thalamic nuclei such an effect was observed in only 30% of the cases (Plotkin, 1982). However, the experience of Dieckmann and Witzmann (1982) has not confirmed such a conclusion. In two groups of patients, they found that stimulation of the sensory Th and PAG yielded approximately similar results, at least in deafferentation pain. Moreover, some authors consider that for cancer pain, stimulation of PAG is not very promising (Hosobuchi, 1982).

Analyzing the results of long-term follow-up of patients who underwent stimulation of PAG and the periventricular zone, Richardson (1982a,b) concludes that significant relief from pain was obtained in 60% of the cases. Dieckmann and Witzmann (1982) consider that stimulation of deep cerebral structures produces a considerable improvement only in deafferentation pain. Contrary to that point of view, Hosobuchi *et al.* (1979) believe that in central pain, stimulation of PAG is not effective. Hosobuchi (1986) presented the long-term results of PAG chronic stimulation. The author confirmed that in central deafferentation pain (thalamic pain, postherpetic neuralgia, etc.) the stimulation is ineffective, but in patients with pain of peripheral origin (mainly chronic lower back and leg pain) PAG stimulation gave pain relief in about 75% of cases.

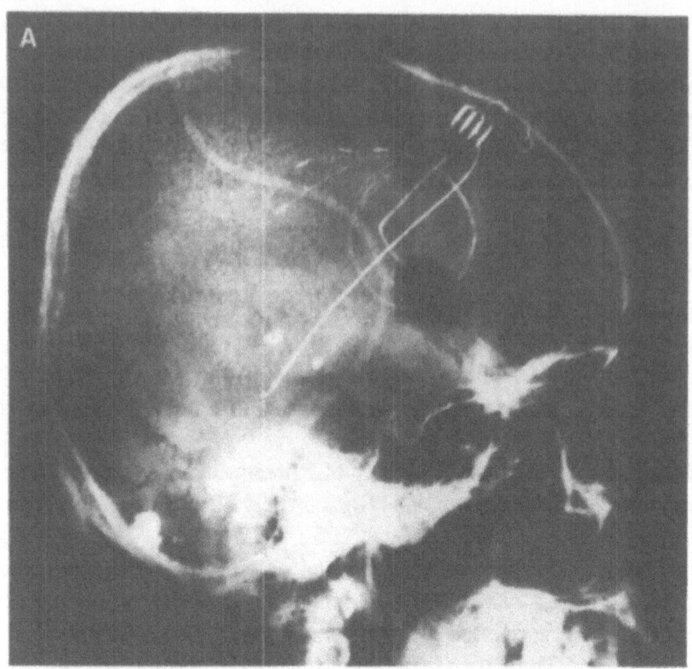

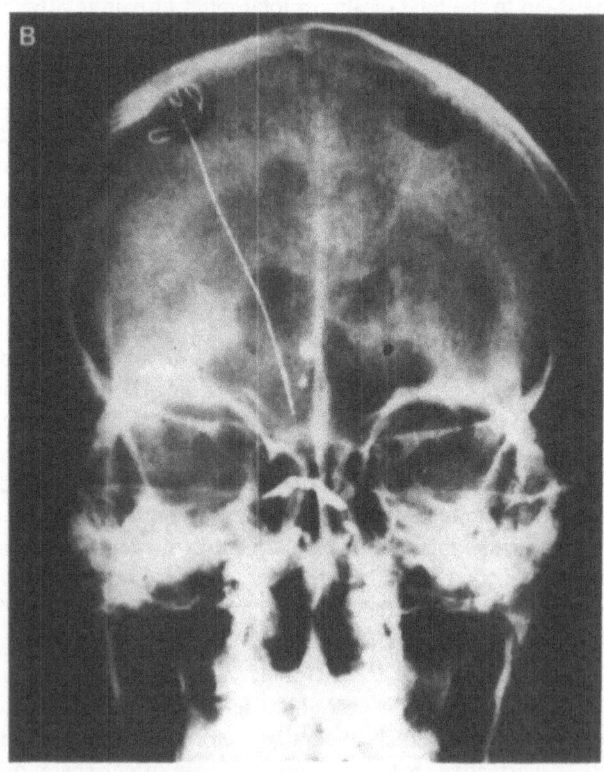

FIGURE 183. Implanted electrode in the periaqueductal gray matter (Amano *et al.*, 1982). (A) Lateral and (B) AP films.

Tolerance to stimulation or progressively less effective pain relief was noted in about one-third of the cases studied by Young and Chambi (1987). Summing up their data, it may be concluded that the efficacy of PAG stimulation may decrease gradually. A large series of 47 patients who underwent PAG stimulation was reported by Levy *et al.* (1984). Immediately after beginning the stimulation, good relief was noted only in 24 cases, but this effect was lost in 15 of them during long follow-up. It should be noted that certain authors deny any beneficial effect of PAG stimulation in either deafferentation or somatic pain (Amano *et al.*, 1982). Only nine out of 18 patients with malignant pain obtained good, stable, clinical relief (Meyerson, 1985).

Meyerson *et al.* (1978) have developed a technique for percutaneous PAG electrostimulation that in the majority of patients with severe pain from malignancy produced significant improvement.

The third European workshop on electric neurostimulation in 1979 came to the conclusion that chronic stimulation of PAG and the periventricular gray has passed the experimental stage and can be recommended for clinical application. It appears, without a doubt, to be one of the most effective stimulation methods in the surgical management of pain.

5.4.1f. Stimulation of Multiple Structures. There is now a tendency in pain syndromes to destroy several subcortical structures in order to increase the amount of sensory deafferentation. The same principle has been applied in stimulation. Dieckmann and Krainick (1979), in 29 cases with pain of various etiologies, implanted electrodes simultaneously in CM, the parafascicular nucleus, and PGM During a follow-up of 2 years, pain decreased by more than 50% in half of the cases.

Hosobuchi (1979, 1983a,b, 1986) concludes that chronic stimulation of PAG is effective in somatic pain, and stimulation of the sensory nuclei of Th in neurogenic deafferentation pain. Since both types of pain are frequently combined, we resorted to stimulation of both structures, either simultaneously or consecutively, using two sets of electrodes. In ten patients with pain of various etiologies, this method gave very satisfactory results. Hence, the structure to be stimulated remains a difficult problem. Richardson (1982a,b) considers that the best results are obtained from stimulation of structures close to raphe spinal nucleus, namely, PAG and the septal area.

As many authors have noted, the effect of stimulating subcortical structures with implanted electrodes diminishes with the passage of time. The reason is still obscure. It is assumed to be related to monoaminergic metabolism rather than to endorphins. This has been confirmed by King (1980), who has demonstrated that the use of monoamine precursors (L-dopa, L-tryptophan) prolongs the period of effective pain suppression by VPL stimulation.

Undoubtedly, the parameters for stimulation of deep cerebral structures are important for obtaining positive results. In order to attain maximum individualization, Mundinger and Neumüller (1982) proposed a programmed stimulation. A special device is used for constantly varying four parameters of stimulation (frequency, amplitude, number of impulses in a group, and the interval between groups). In this manner, an individual set of stimulation variants is chosen for each patient.

The long-term results of chronic stimulation of deep cerebral structures in pain syndromes vary substantially in different neurosurgical clinics. They depend on the criteria of patient selection, etiopathogenesis of pain, choice of subcortical target, method of stimulation, and other factors. In 1978, Sedan and Lazorthes reported the results of chronic stimulation of subcortical structures at several European neurosurgical clinics. An analysis of long-term results of more than 250 cases has revealed that positive results were obtained in 75% of the cases. At the previously mentioned 1979 European workshop on electric neurostimulation, Meyerson summarized the results in 324 cases from 13 clinics (quote by Siegfried *et al.*, 1980). These results were quite different; they revealed an improvement in the range of from 20% to 80% of operated cases. Such marked fluctuations in assessment of the effectiveness of the method prohibit us from passing a final judgment. Nevertheless, chronic stimulation of PAG seems to be a promising technique.

One may conclude that chronic electrostimulation of subcortical structures with implanted electrodes represents an effective method for the surgical management of various pain syndromes. There is reduction of pain on stimulation of several deep cerebral structures—VPL, VPM, CI, PAG, Hypoth, and septal zone. It is still difficult to say stimulation of which of these structures is most "effective" and for what type of pain. So far, it is impossible to explain logically the paradox that destruction and stimulation of the same structure (for example, VPL or Hypoth) yield the same effect, namely, relief of pain. Although a number of questions concerning the use of this technique have yet to be resolved, the therapeutic value of this promising method has been proven.

5.4.2. Stimulation of Posterior Columns of the Spinal Cord

One of the extensively employed and effective methods for the management of severe pain in recent years has been electrostimulation of the posterior columns of the spinal cord (PCS) using electrodes implanted sub- or epidurally at the corresponding spinal level. The term "spinal cord stimulation" has also been used recently for this technique; however, we prefer the term "posterior column stimulation," which is used in most of the literature. The idea of protracted stimulation of the spinal cord for the management of pain was proposed by Shealy in 1967 and applied in neurosurgical practice by Sweet and Wepsic (1968, 1974), Shealy *et al.* (1970), Nashold and Friedman (1972a), and Young *et al.* (1985).

The theoretical basis of the method is the "gate theory" of the pathogenesis of pain (Section 2). However, the specific mechanism of action of PCS on the perception of pain still remains obscure. In this respect, there are several hypotheses. There is no doubt that PCS is a kind of "substitute" for antinociceptive neuronal systems. The stimulation may lead to the inhibition of the "firing" of the deafferentated dorsal horn neurons (Privat *et al.*, 1976). There are also grounds to consider that such stimulation antidromically changes the functional state of the neurons in the posterior horns of the spinal cord, which "closes the gates" to pain afferents from the periphery (Hillman and Wall, 1969). It is also presumed that PCS inhibits the impulses along the fine pain-relaying fibers in the posterior horns by activation of the thick myelinated afferents passing in the posterior columns. Experimentally, it has been proven that PCS also suppresses the reaction of nociceptive neurons in the nonspecific Th nuclei (Cm, parafascicular nucleus). As the microelectrode technique has shown in PCS, these neurons cease to fire in response to stimulation of the contralateral sciatic nerve in cats (Nishimoto *et al.*, 1980). These data demonstrate that PCS modulates not only the activity of the descending pain systems but also the functional state of thalamic pain centers.

The suggestion was recently put forward that the pain relief from PCS results from activation of endogenous opiates that are present in substantia gelatinosa of the posterior horns. It is also assumed that PCS activates the efferents passing to PAG, the stimulation of which with implanted electrodes causes analgesia (Section 5.4.1e) (Sugita and Kobyashi, 1982). At the same time, certain authors deny that pain relief in PCS is connected with the activation of the opioid mechanism of pain control.

Spiegel (1982) presumes that PCS activates the input of cortical inhibiting impulses via the inferior cerebellar peduncle.

In animal experiments, PCS inhibits stretch reflexes, but only for a short period of time (Chapman and Wiesendanger, 1981). Somatosensory EPs do not change (Doerr *et al.*, 1978) during PCS, but there is inhibition of the elevated H reflex, which is substantially normalized (Illis *et al.*, 1976).

In PCS the pain sensation threshold does not change appreciably (Lindblom and Meyerson, 1975). However, if the stimulation is prolonged, the threshold rises, but only in the zone of paresthesias induced by stimulation (Doerr *et al.*, 1978). The tactile and vibration sensation threshold increases noticeably immediately after the beginning of PCS on both sides of the body. This effect, which is related to the action of PCS on the central inhibitory mechanisms of pain, lasts for a comparatively short period, whereas inhibition of pain continues (Lindblom and Meyerson, 1975).

Although PCS has been used for only about 15 years, the literature contains quite a number of important publications based on considerable clinical experience with a technique that is being constantly modified.

Posterior column stimulation has been tested in a wide range of pain syndromes of the most diverse etiology—in arachnoidites, trauma, and other lesions of the spinal cord, after operations on the spinal cord, in diseases and injuries of peripheral nerves, in phantom pain and causalgia, in trigeminal neuralgia, in pain caused by malignant neoplasms, in arthritis, and in certain diseases of internal organs. Nevertheless, it should be noted that many important methodological, technical, and other questions related to PCS have yet to be resolved.

The contraindications to PCS include pain of unknown etiology, pain in hysteria and psychoneuroses, pain in the legs after total interruption of the spinal cord, as well as narcotic addiction. It is also not suitable for headaches.

Two stages may be recognized in the development of the PCS technique. In the first stage, which lasted approximately 6–7 years, implantation of electrodes on the posterior surface of the spinal cord was done by a conventional open laminectomy, an approach rarely employed today.

Laminectomy with the removal of two or three posterior arches was performed four or five spinal cord segments above the superior dermatome involved in the pain. With pain in the thorax and arms, for instance, laminectomy was performed at the C_2–C_4 level, and with pain in the pelvis and legs, at the midthoracic level. The dura mater was opened with a longitudinal incision, after which the arachnoid membrane was divided, and then a flat electrode was applied to the posterior surface of the spinal cord. If the pain was localized to one half of the body, the electrode was placed slightly toward the "painful" side (Shealy *et al.*, 1970). The wire leading to the electrode was fixed by dural sutures, and the surgical wound was closed (Fig. 184).

It was soon established that subarachnoid implantation of the electrode posed two significant shortcomings: the formation of fibrous tissue around the electrode tip, which reduced the effect of the stimulation; and leakage of CSF if the dura mater were not tightly closed. The next step in improving this technique was subdural implantation of the electrode without opening the arachnoid membrane. However, even then, fibrous tissue formed around the electrode after approximately 1–2 years, but to a much smaller extent. Then certain surgeons placed the electrode endodurally, i.e., between two folds of the dura (Shelden *et al.*, 1975).

The second stage in the development of the method began with percutaneous epidural PCS. This technique simplified the operation considerably and made it practically safe (Hoppenstein, 1975). The advantage of the new method, which eliminated the need for a laminectomy as well as general anesthesia, lay not only in its simplicity and technical feasibility but also in the possibility of careful testing of the effect of stimulation.

Since the special devices for PCS are of great significance for the results of this operation, they deserve to be described in detail. There are two main systems for PCS—partially and totally implanted systems. A small 150-g transmitter measuring about 10 × 7 × 3 cm is powered by a 9-V battery. The current parameters can be regulated within the following range: voltage from 0 to 10 V, frequency from 1 to 120 Hz, impulse duration from 0.1 to 1 msec. The trans-

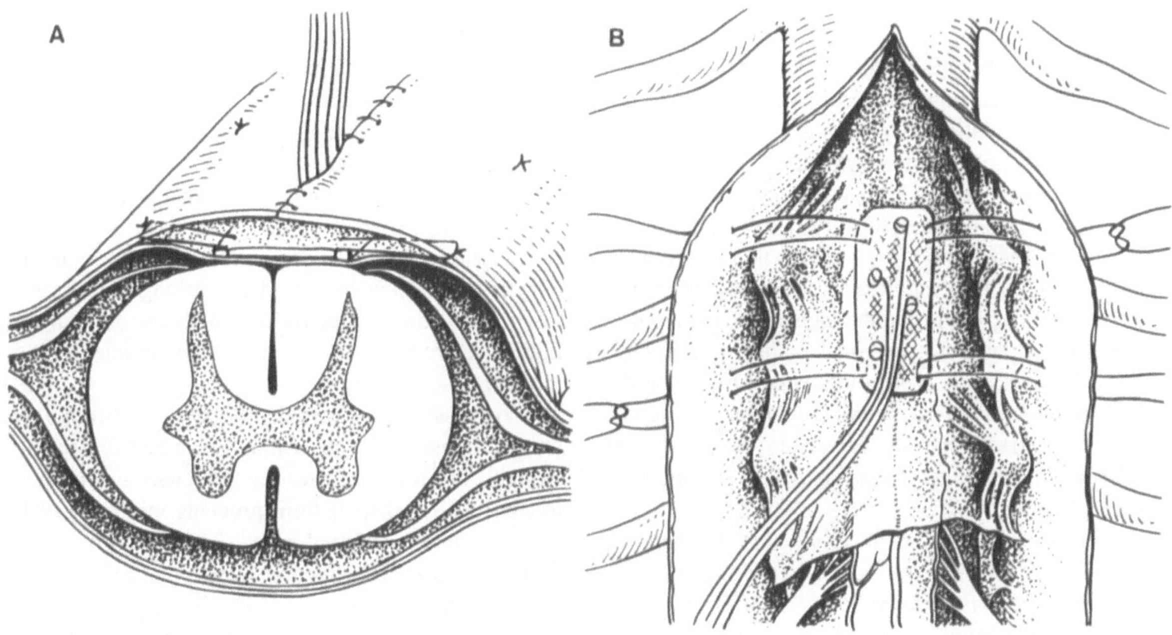

FIGURE 184. Subdural implantation of electrodes for posterior column stimulation (Nashold and Friedman, 1972a). (A) Location of electrodes with respect to spinal cord and fixation of leads in the dura mater incision. (B) Fixation of electrodes on the dorsal surface of the spinal cord.

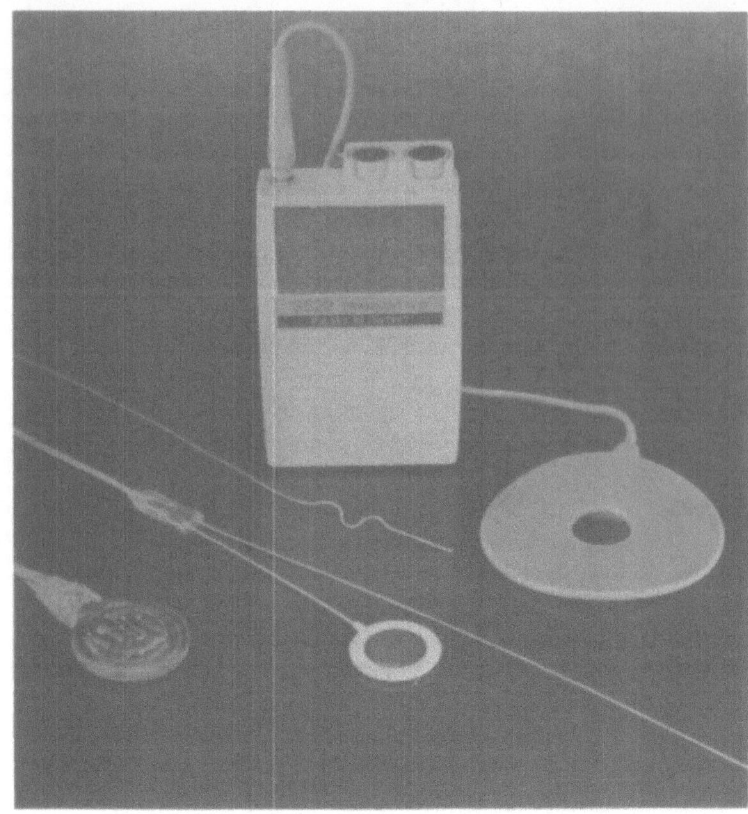

FIGURE 185. An arrangement for PCS: impulse generator, electrodes for the spinal cord, implantable electrode with antenna.

mitter may be checked with an oscilloscope, which shows the impulse amplitude and width as well as the intervals between impulses, from which data the frequency of stimulation may be determined.

Several implanted systems for PCS are now employed. The best known is the percutaneous spinal cord epidural electrical stimulation (PISCES) system manufactured by Medtronic (Fig. 185). The transmitter of this system has four regulators for controlling current parameters. Two are inside the transmitter and are used by the physician to set maximal voltage and impulse width. Such controls protect the patient from possible excess stimulation. Two external regulators for voltage and current frequency are adjustable by the patient, who is instructed by the physician before starting stimulation (see below).

The second part of the system is the antenna, which is a plastic circle connected to the transmitter by a flexible wire. The antenna is fixed to the skin of the thorax over the center of the subcutaneously implanted high-frequency receiver. Small adhesive disks, which should be replaced daily, are used to fix the antenna to the skin.

The electrodes for PCS are made of plastic-insulated, corrosion-proof nickel wire with a 3 to 4-mm bared tip. Both thick and thin electrodes are used. In the opinion of some authors, a thin electrode is preferable since it may be inserted through a smaller needle. However, other authors consider that a thicker electrode can be inserted more easily into the epidural space, with less chance of the electrode perforating the dura mater, which may lead to prolonged leakage of CSF. ''Floating'' electrodes (Medtronic) or Model E355 electrodes (Avery Labs) are frequently employed.

An S-shaped electrode (Sigma, Medtronic) has been proposed. On insertion, the electrode remains straight, but it bends after the guidewire has been extracted. The S-shaped form prevents migration of the electrode in the epidural space. There is also an electrode with a curved tip that allows one to change its position after insertion and so avoid obstructed areas of the epidural space. The electrode is introduced through an 18-gauge Tuohy needle. At the external end of the wire there is a connector that is attached to the appropriate instruments. The length of the wire varies, depending on the level of implantation. For the cervical and superior thoracic parts of the spinal cord, it is 40–45 cm; for the inferior thoracic, it is 75–80 cm.

In recent years there has been a definite tendency to employ bipolar stimulation instead of monopolar. It is assumed that during bipolar stimulation there is a concentration of current between the two electrodes, and consequently, one can use a lower intensity. At the same time, monopolar stimulation is technically more reliable. In this case, two electrodes are inserted a small distance apart. An effective method is to insert four electrodes simultaneously via a single epidural needle and position them at different levels of the spinal cord. Such a system enables testing of the effect of stimulation in 18 possible combinations of these electrodes as well as computer processing of the obtained results (Waltz, 1981).

An interesting new electrode model for PCS has been tested experimentally (Avery *et al.*, 1980). This electrode has a small "brush" of dacron fibers approximately 15 mm from the active tip. The authors consider that fibrous tissue should form around the dacron "brush," preventing it from migrating in the epidural space.

A miniature disk-shaped receiver picks up radio frequency waves from the transmitter and transforms them into monophasic rectangular impulses passing to the spinal cord. The receiver is encased in epoxide substance and covered with silicon rubber. A 38-cm-long wire is connected to the receiver.

The desired level of PCS is determined before the operation. In case of leg pain the electrodes are inserted at the midthoracic level, usually below T_7, and with pain in the thorax or arms, at the upper thoracic or cervical level. Insertion of the electrode by a puncture in the cisterna magna is not recommended, since difficulties may be encountered when the electrode is advanced caudally. Based on trial stimulations (see below), the electrodes should be positioned at the level of the spinal cord that induces paresthesias in the painful area.

The technique of implanting the PISCES system is as follows. With the patient prone on the fluoroscopic table, a Tuohy puncture needle is inserted under local anesthesia along the midline between the spinous process to the epidural space. The ease with which physiological saline is ejected from a syringe connected to the needle is an indication of accurate placement.

Under fluoroscopic control, the wire conductor is inserted into the epidural space via the needle and advanced upwards. Then the electrode is introduced to the desired level. If, during insertion, the surgeon encounters resistance, he should not try to force the electrode but should extract it a short distance and reintroduce it. Then, the second needle is inserted at the same level or somewhat below the first, and through it the second electrode is inserted in the same manner. A skin incision 6–8 cm long down to the subcutaneous fascia is made between the two needles.

The two wires emerging from the needles are connected to the trial stimulator. The needles are withdrawn from the epidural space to the subcutaneous fascia and then carefully removed. Fluoroscopic control should confirm that both electrodes have remained in place. For monopolar stimulation, it is necessary to have an indifferent electrode that is glued to the skin at some distance from the implanted electrodes. The wires are fixed to small subcutaneous disks ("anchors") and sutured to the fascia.

Another system for PCS is used by the Neuromed Company, the MSCS (Multiprogrammable Spinal Cord Stimulator). A thin polyethylene four-contact catheter electrode with removable stylet is introduced through a 16-gauge needle. The electrode has a variflex design, which prevents migration, improves its insertion accuracy, and prevents epidural adhesions. All four electrode contacts may be activated in any combination for monophasic or biphasic stimulation.

Stimulation trials are made with the patient on the operating table. In order to obtain clear-cut paresthesias in the painful area, the proper position for the electrodes in the epidural space must be found by continuing stimulation as one advances or withdraws the electrodes.

Then a small skin incision is made approximately 8–10 cm from the protruding wires. With a special instrument, a subcutaneous tunnel is made to the skin incision, and the electrodes are brought out. Both incisions are then sutured.

Stimulation trials are continued for approximately a week (Long *et al.*, 1981). The trials enable one to establish which stimulation, monopolar or bipolar, is more effective. Moreover, during this period various current parameters (voltage, pulse duration, and frequency) are tested in order to obtain an optimal effect. If stimulation induces clear-cut paresthesias in the painful area for several hours, followed by relief of pain, the chronic stimulation will be effective.

Analysis of both the stimulation threshold and

sensory reaction thresholds is of prime significance. The sensory reaction threshold is highly variable for each individual. If the current intensity is increased significantly above the threshold value, paresthesias become painful. Certain authors successfully employ computer analysis of threshold reactions to predict the efficacy of this method of treatment (Law and Miller, 1982).

If test stimuli with currents of varying voltage and frequency do not produce a clear-cut effect such as disappearance of pain, the electrodes should be removed. However, certain practitioners propose leaving the electrodes in for several days and continuing stimulation for 3–6 weeks in the hope of obtaining a delayed effect (Klingler and Kepplinger, 1982).

Test stimulation does not always give positive results. For example, based on the results of trials, implantation was done in fewer than half the cases (in 89 of 191) previously operated on for pain in one or both legs caused by arachnoiditis or after ineffective operations on intervertebral disks (Siegfried and Lazorthes, 1982).

The electrodes are implanted after the efficacy of PCS is confirmed and optimal current parameters are established. A small incision is made under the skin of the abdomen or in the subclavical area, and with a blunt dissector, a pocket is made in the subcutaneous tissue. A radio-frequency receiver is implanted in the pocket. Another subcutaneous tunnel is made from the receiver, and the wires are passed through the tunnel and connected to the receiver. Both incisions are sutured tightly.

After appropriate instructions by the physician, the patient himself performs stimulation at home using the transmitter in his pocket or attached to his clothes. As previously stated, the current parameters are determined on an individual basis. After numerous stimulations, the patients themselves usually find the most effective parameters. Rectangular impulses lasting 10–30 msec, 2–3 V, and a frequency of approximately 100 Hz are most frequently used. For PCS certain authors employ either a very high frequency, up to 500 or even 1400 Hz (Waltz, 1982), or very short duration, lasting 0–15 msec. The rate of stimulation also varies within a wide range. Some patients need practically constant stimulation, others require stimulation several times a day, and still others require it only when the pain is intense.

Since an objective assessment of the effect of PCS is practically impossible, the relief is usually graded, for example, 0 for total absence of pain, 4 for severe pain without changes, and 1, 2, and 3 for intermediate stages (Siegfried and Lazorthes, 1982).

Complications following implantation of electrodes occur quite frequently. One is the lateral displacement of the electrodes from the posterior surface of the spinal cord. The main cause of this complication is bending or rotation of the spinal column during movements. If the electrode has been lodged in the sleeve of the nerve root, stimulation causes pain in the corresponding dermatome, even with a low voltage.

The dislodgement of the electrode is one of the most frequent reasons for technical failure in PCS. For instance, more than half of 124 cases of epidural stimulation underwent reoperation because of electrode position changes (Sogabe *et al.*, 1985). The authors noted that the complication occurred most frequently within one month after the transplantation.

If, after a time, some sensations during stimulation lessen, or if the ordinary paresthesias develop into acute discomfort, then, as a rule, this signifies that the electrode has migrated from the posterior to the lateral surface of the spinal cord. Several methods for replacing the electrode to its proper position have been suggested. For example, it has been recommended to inject a small amount of physiological solution into the epidural space; in the majority of cases, the epidural electrode will return to its initial position (Kepplinger, 1983). Another complication is severe pain during stimulation, making it impossible to continue the procedure. This occurs rarely, in approximately 5% of the cases, and if such symptoms of spinal cord irritation appear, the electrodes must be removed. Very rarely there is leakage of CSF as a result of the electrode damaging the dura mater.

A certain percentage of ineffective operations, especially following prolonged use of PCS, are the result of various technical defects (displacement or damage of electrodes, poor contacts in the system, wire breakage, etc.). It is noteworthy that technical failures and malfunctions of the systems for neurostimulation, although still encountered, are much less frequent than at the end of the 1970s (Koeze *et al.*, 1984), when repositioning of the epidural electrode was required during the first month after its implantation in about one-half of the cases (Sogabe *et al.*, 1985).

The PCS method was further improved by the development of totally implanted systems that allow both mono- and bipolar stimulation. The Stimucord system manufactured by Cordis (Fig. 186) has a 70-g implanted stimulator analogous to a heart pacemaker and powered by a 6.4-V long-life lithium battery. The

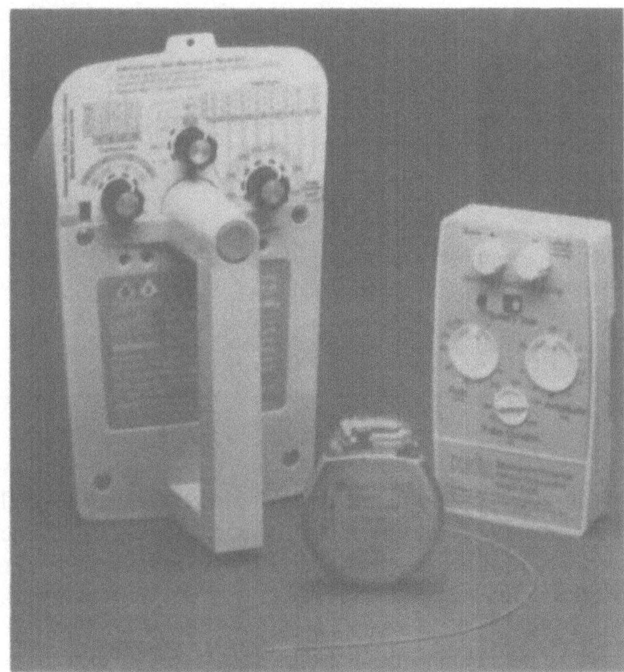

FIGURE 186. The Cordis Stimucord apparatus for spinal cord stimulation. Two models are manufactured for monopolar and bipolar stimulation. They are fully implantable systems with program control. Durable 6.4-V lithium–cuprite batteries are the source of energy. Limits of stimulation: current power, 0.8–85 mA; frequency, 10–100 Hz; stimulus duration, 75–315 msec. The patient can stop stimulation immediately by placing a magnet on the body over the implanted electrode.

stimulator is enclosed in an inert titanium case. The upper transparent epoxy part of the stimulator has a plug for connection to the wires leading to the spinal cord. The main advantage of this system is that there is no need for a transmitter and antenna. It has a separate programmer for setting the optimal current parameters. This device generates a pulsating magnetic field, which transmits the programmed parameters to the stimulator through the skin. The stimulator deciphers these signals and transforms them into impulses passing to the spinal cord. The current may be set from 0.8 to 8.5 mA, frequency from 10 to 100 Hz, and pulse duration from 75 to 315 msec. In order to transmit the desired parameters to the stimulator, the programmer is brought within 2–3 cm of the overlying skin. Stimulation may be stopped at any moment by applying a disk magnet to the skin at the implantation site of the stimulator.

The length of service of the stimulator depends on the lithium battery, which discharges at different rates with various current parameters. The average "service life" of a battery is approximately 3 years, after which the stimulator must be replaced.

Implantation of the epidural electrode and stimulator and subsequent testing are carried out as described above for partially implanted systems. The Cordis stimulator has a special aperture for fastening it to the subcutaneous tissue, which prevents its displacement during movements.

With these new PCS techniques, it has been shown that epidural stimulation with percutaneous electrodes is more effective than stimulation with electrodes applied after laminectomy. Clinical improvement after percutaneous stimulation proved to be 30–35% greater (Waltz, 1981).

A considerable number of studies summarizing both immediate and long-term results of treatment have been reported in literature. An analysis of these investigations shows that the data of different authors on the efficacy of PCS are often very variable. For example, Cook *et al.* (1981) reported a significant improvement almost in 90% of his cases, but other authors have not observed any effect in the majority of their cases.

As has already been mentioned, two periods may be singled out in the development of the PCS method. Analyzing the data in the literature, one may note a definite improvement in the results of treatment as the equipment for PCS has improved and as the criteria for selecting patients have become more specific. The first reports in the middle of the 1970s of the results with open implantation of electrodes gave a great number of ineffective operations and various complications. Recent papers (from the 1980s) indicate that the new totally implanted systems for PCS improved almost

TABLE 23. Results of Posterior Column Stimulation

Reference	Number of cases	Result		Complications
		Good	Satisfactory	
Shealy (1975)	80	20	32	48 transitory 15 replacement 5 removal
Nashold (1975)	30	25		15 recurrence of pain
Burton (1975)	75	44		
Long and Erickson (1975)	69	28	28	23 good and 11 satisfactory after 1.5 years
Hoppenstein (1975)	14	6	8	
Nielson *et al.* (1975)	130	54	33	33 good and 31 satisfactory after 6 years

twofold the success of the method. However, the reason for the varying efficacy of PCS in different cases remains obscure. In all likelihood, it is related to the differing role of "peripheral" and "central" factors in the pathogenesis of the pain syndromes. Some of these results are summarized in Table 23.

It has also been established that PCS induces a number of segmental and diffuse autonomic effects such as increased skin temperature, regional vasodilation, and an increase in peripheral blood flow (Cook *et al.*, 1976; Meglio *et al.*, 1980; Augustinsson *et al.*, 1982). Moreover, it was shown that PCS affects the central mechanisms of regulation of heart rate (Meglio *et al.*, 1986b). In spite of the fact that the mechanism of vascular effects is still not quite clear, there are sufficient grounds to employ PCS for the relief of pain of vascular genesis, in particular, arteriosclerotic and diabetic vascular lesions of the legs. Authors who have used this method report a significant therapeutic effect in 80 to 100% of cases (Augustinsson *et al.*, 1982; Meglio and Cioni, 1982). Stimulation in these cases resulted not only in the disappearance of chronic pain but, in many cases, in the healing of trophic ischemic ulcers, an effect that lasted for many months. In a number of cases, but not all, Doppler, plethysmographic, and rheographic investigations revealed an increased blood flow in the legs.

Posterior column stimulation proved effective in severe lesions of peripheral vessels producing pain and intermittent claudication (Tallis *et al.*, 1983). In six of ten cases of painful angiopathy after stimulation, pain in the legs at rest disappeared or decreased significantly, the walking distance before claudication increased fourfold, and the ergometer bicycle test revealed a 60% enhancement in these patients.

A cooperative investigation concerning the clinical use of spinal cord stimulation in peripheral arterial disease was presented recently by Spanish neurosurgeons (Broseta *et al.*, 1986). They carried out percutaneous epidural PCS on a low thoracic level in 37 patients with peripheral arterial disease of the lower limbs. The mean follow-up period of stimulation was about two years. Pain was relieved in the majority of cases and the distance walked before claudication occurred was significantly greater in 50%. Medium-sized cutaneous ulcers healed. Thermography revealed a rise in skin temperature in half of the cases. An improvement of blood circulation in the lower limbs, as measured by Doppler ultrasound, was also noted in half of the cases. The authors consider PCS a promising therapy of moderate peripheral arterial disease when direct arterial surgery is not possible or has been unsuccessful.

Siegfried (1983) reported a large series of cases with many years of follow-up after the operation. From 1973 to 1977 the author implanted electrodes via laminectomy and then switched to the modern method of PCS described above. In this second period, stimulation tests were made in 291 cases; however, permanent implantation of electrodes was carried out in only half of the cases, but about equally in persons with somatic and deafferentation pain.

In somatic pain, mainly residual back pain after ineffective operations on lumbar disks, the percentage of excellent and good results was 74% in the first year of follow-up and 71% after 4 years. In persons with deafferentation pain, the corresponding figures were 81% and 71%. However, in central pain syndromes (thalamic pain, phantom pain, and anesthesia dolorosa), PCS proved to be ineffective in practically all cases.

The difficulty of obtaining a long-lasting effect

from PCS has yet to be resolved. Within 3–5 years after initiating PCS, the number of cases relieved after electrode implantation has dropped to one-half (Krainick *et al.*, 1979; Winkelmüller, 1981) or almost to one-third (Long *et al.*, 1981). Observations made during reimplantation of electrodes reveal that the main cause for the late poor results is the tissue reaction to the electrode.

Contrary to all other authors, Mazars (1976a) considers that PCS is not at all effective and should be abandoned.

Summing up a large amount of data in the literature, one may conclude that a good result may be obtained from PCS for a number of years in 40–50% of the persons with severe pain of various etiologies (Riechert, 1973; Shealy and Maura, 1974; Pineda, 1975, 1977; Zumpano and Saunders, 1976; Urban and Nashold, 1978; Long, 1980; Krainick *et al.*, 1979, 1980; Kandel, 1986).

A new technique of combined PCS and stimulation of the paravertebral nerves for the relief of pain in a lower extremity was recently proposed (Urban and Nashold, 1982) (Fig. 187). In brief, the technique is as follows. With the aid of the method described above, the stimulating electrode is inserted percutaneously to the epidural space in the upper lumbar region, and then the electrode is shifted slightly laterally to the side of the pain. Then, 5 cm lateral to the spinal process nearest to the spinal nerve innervating the pain area but on the opposite side (the one without pain), a needle is advanced caudally and medially to the epidural space of the "pain" side. A second electrode is inserted through the needle and, under fluoroscopic control, advanced into the corresponding intervertebral foramen. The tip of the electrode is positioned at the proximal part of the spinal nerve on the anterolateral surface of the spinal cord. Then both electrodes are connected to the receiver, and trial stimulations following the above-described technique are made.

The authors emphasize that it is necessary to position the electrode where maximal paresthesias occur in the painful zone at minimum stimulating voltage. In the testing period bipolar stimulation is employed with a current of rising frequency between the epidural and peripheral electrodes. If a positive effect such as the appearance of paresthesia and then muscular contractions in the painful leg is induced, the receiver is implanted according to the technique described above. This was done in 16 of the 23 cases. In the remaining seven cases, paresthesias in the painful area could not be elicited, and the electrodes had to be withdrawn. The day after implantation of the electrodes, the pa-

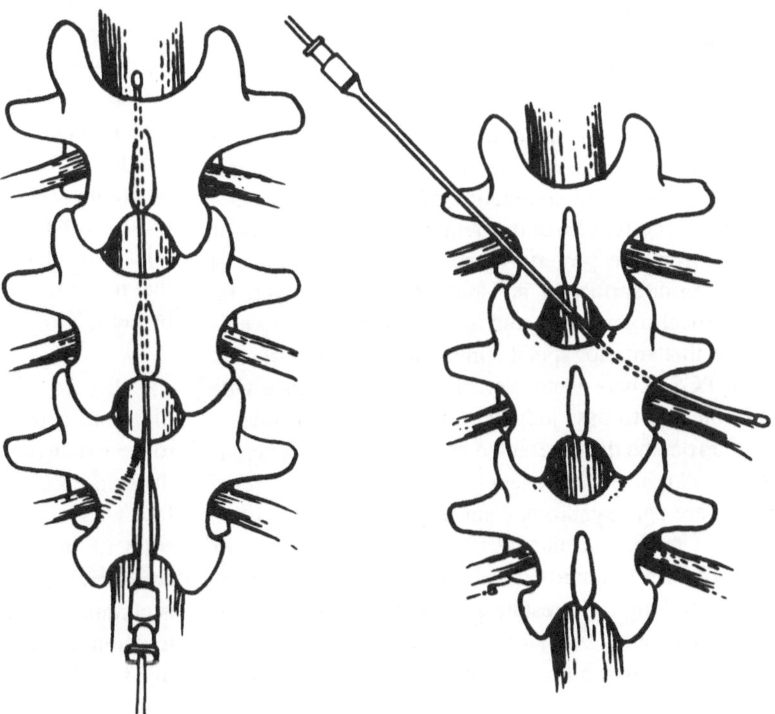

FIGURE 187. Scheme for combined PCS and stimulation of the paravertebral nerve (Urban and Nashold, 1982). Explanation in the text.

tients began self-stimulation. The parameters varied considerably, the frequency from 33 to 100 Hz, the voltage from 2 to 8.5 V, and the duration of stimulation approximately 0.2 msec.

In an 18-month follow-up, Urban and Nashold (1982) reported that in cases of pain in the loin and leg, phantom pain, damaged leg nerves, etc., 11 of 16 patients had from 25 to 99% relief; four had excellent results. There was practically no effect in five cases. The authors concluded that the method gave approximately the same percentage of good results as PCS, but there were practically no complications and considerably fewer diffuse paresthesias during movements. Further experience is needed to assess this method.

It may be considered that not only PCS but stimulation of the cauda equina may relieve pain in the lower half of the body (Richardson *et al.*, 1979; Blume *et al.*, 1982b).

From certain observations one may assume that pain can be suppressed by stimulation of other parts of the spinal cord than the posterior columns *per se*. In view of this, a number of authors have suggested that the term "spinal cord stimulation" be used rather than PCS. In some cases, stimulation of implanted electrodes on the ventral surface of the spinal cord or in the intervertebral foramen gave relief of pain (Larson *et al.*, 1975; Broseta *et al.*, 1982b). In view of this, percutaneous stimulation of the anterior surface of the spinal cord has been investigated (Lazorthes *et al.*, 1978); it induces similar analgesia without accompanying paresthesias.

Hoppenstein (1975) and Larson *et al.* (1975) have established that implantation of electrodes on both the anterior and posterior surfaces of the cervical spinal cord relieves pain; stimulation of the anterior surface requires a current 1/30 of that required to obtain the same effect from the posterior surface. The stimulation of the anterior surface of the spinal cord has, as yet, no theoretical foundation, and at present there is no evidence that anterior spinal stimulation is more effective than PCS. There is not yet sufficient experience with this method to pass judgment on its clinical value.

From the data presented above, one may conclude that PCS is an effective and safe technique for the relief of severe pain syndromes not responding to any other kind of therapy. Quite satisfactory and lasting results are obtained in approximately half of the cases. However, the long-term results are notably worse than the immediate ones. Further technical development should enhance the reliability and success of this procedure.

5.4.3. Stimulation of Peripheral Nerves

Electrical stimulation of the skin as a means of treating pain has been known for thousands of years. As far back as the seventh century, Paul of Aegina employed the shock of electric eels in the management of headaches. However, the treatment of intractable pain by percutaneous electrostimulation of peripheral nerves (PNS) appeared quite recently and is rapidly advancing.

There are several hypotheses concerning the mechanism of suppressing pain by PNS; however, the main ones may be called peripheral and central hypotheses. According to the former, the afferent input of the thin Aδ and C nociceptive fibers of the peripheral nerve causes a primary hyperpolarization of the first neuron, which is evidenced by the appearance of a positive antidromic potential in the posterior root. Another peripheral hypothesis presumes that PNS leads to a change in the level of excitability of the afferent fibers (Ignelzi and Nyquist, 1970); the analgesic effect is the result of blocking the small afferents caused by activation of the thick sensory nerve fibers (Matthews and Cadden, 1970; Campbell, 1981). The proponents of the central hypothesis consider that pain is abolished in PNS because of functional changes of the sensory structures in the CNS (Law *et al.*, 1980; Sweet and Law, 1983). Sanders *et al.* (1980) believe that PNS activates the neurons of PGM to release opioid peptides, particularly enkephalins, causing analgesia (Section 3). However, there is no correlation between the effect of PNS and the blood β-endorphin levels (O'Brien *et al.*, 1984). It was shown in experiments on monkeys that PNS evokes inhibition of the spinothalamic tract by activating other afferent fibers (Lee *et al.*, 1985), and this inhibitory effect is caused by release of endogenous opioids. Other authors consider that nerve stimulation modulates synaptic transmission in the CNS.

5.4.3a. Percutaneous Stimulation. This simple, nontraumatic, and perfectly safe method is likely to be extensively employed, although why PNS produces a good effect in certain patients and has no effect whatsoever in others still remains a mystery. In spite of the fact that the method is still "young," considerable experience in its clinical application has accumulated. The method is based on the principle of transmitting a pulsating current through electrodes applied to the skin. Numerous portable devices of very

FIGURE 188. Em-Set device (Finland) for transcutaneous peripheral nerve stimulation.

similar design have been proposed for PNS. They consist of a small (the size of a cigarette case) battery-powered generator of biphasic impulses of 0.4–0.8 msec. The first, negative phase of the impulses is longer than the second, positive phase. The frequency and intensity of the impulses may be varied from 1 to 100 Hz and from 0 to 80 mA. Two electrodes approximately 3 cm in diameter, covered with paste, are applied to the skin 3–4 cm apart and fastened with adhesive tape.*

FIGURE 189. Pulsar device (Spembly Medical, England) for transcutaneous peripheral nerve stimulation.

*Numerous instruments for PNS are available: Em-Set (Finland) (Fig. 188), Spembly (England) (Fig. 189), Medaid manufactured by Par Medex (Fig. 190), Soviet stimulators (Impulse, Delta-101, Elektronika-50, Iva MT, Axon-01), Feba (Yugoslavia), and Analgonic 1 (Czechoslovakia).

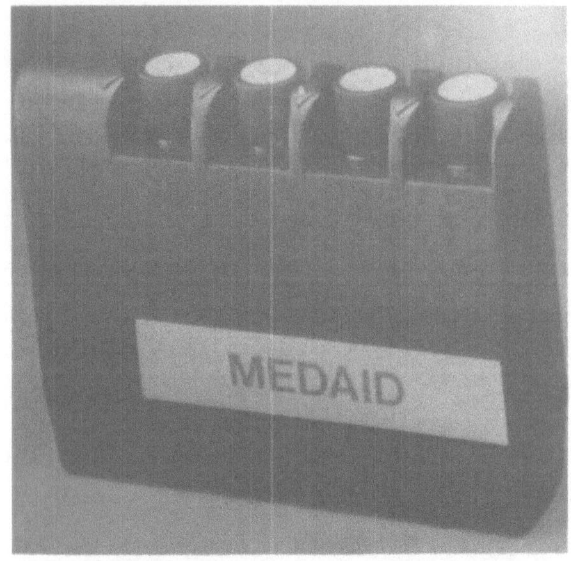

FIGURE 190. Medaid device (Par Medex LD, England) for transcutaneous peripheral nerve stimulation.

A current with a frequency of 20 to 100 Hz and an intensity of 15 to 80 mA is usually applied in PNS. It has been established that pain relief from PNS does not depend on the waveform of the impulses (Janko and Trontely, 1980). The intensity of the current is gradually increased until the patient feels a strong but quite bearable tingling sensation. The duration of a single stimulation is usually approximately 15–20 min but may be longer. When pain reappears, the stimulation may be repeated by the patient at home. The variability of the therapeutic effect of PNS is difficult to explain. In some patients, the relief lasts only while the stimulator is on, whereas in others, the pain disappears for many hours after stimulation.

Since this procedure is noninvasive and safe, its indications are very broad. In all likelihood, PNS is effective not only in peripheral pain (low back pain, pain caused by damaged peripheral nerves, posttraumatic arthritis, rheumatic pain, postoperative pain, coccygodynia, migraine, trigeminal neuralgia, chronic cervical syndromes, and spinal arachnoiditis) but also in central pain syndromes (phantom pain, causalgia, etc.). Of course, PNS is not to be used in patients with an implanted cardiac pacemaker. The electrodes require changing after approximately 300 hr of usage.

An interesting device called the Issal Sensimeter 1412 (Issal Surgical Instruments, Sweden) was constructed for quick and accurate measurement of cutaneous sensibility. The device allows one to measure the thresholds for perception of pain, maximal pain tolerance, and analgesia level during local and spinal anesthesia. These data may be very useful in the management of pain by PNS.

The proper site for the electrode still remains one of the unresolved questions in the clinical application of PNS. At the initial stage in the development of the method, it was considered mandatory to stimulate the nerves supplying the pain area (Long, 1974; Cauthen and Renner, 1975; Hiedl *et al.*, 1979). Later it was demonstrated that stimulation of a zone of anesthesia or area of diminished sensation is not very effective. In such cases it is recommended that the electrodes be applied, and PNS performed, in areas remote from the pain zones or even on the other half of the body (Laitinen, 1976). However, Shealy and Maura (1974) and Shealy (1975), who have had extensive experience with PNS, do not recommend stimulating remote zones since it is less effective. The area of the carotid sinus should not be stimulated.

The following technique is recommended at the present time. The electrodes are applied to the pain zone only if sensation is intact. In thalamic and phantom limb pain, as well as in postherpetic neuralgia, the electrodes are applied to the healthy shoulder and thigh in the zone remote from the former herpes zoster lesions. In sciatica and lumbago, the electrodes are applied paravertebrally on both sides.

After PNS, chronic pain caused by damaged peripheral nerves, spinal cord, and other disorders (Picaza *et al.*, 1975) was found to diminish by more than 50% in over half of the cases. Laitinen (1977) used PNS for 9 months in 46 cases with chronic pain of different etiologies (thalamic pain syndrome, postherpetic neuralgia, lumbar pain, cancer pain). As estimated on a special scale of intensity, pain in all groups of patients diminished by 39%, and the frequency of pain attacks and need for analgesics decreased by 46%. On the whole, the pain disappeared completely or almost completely in 41% of the cases, was treated satisfactorily in 37%, and not or only slightly relieved in 22%. In some cases with a good effect there was a gradual recurrence of pain.

The results of PNS obtained by Melzack (1975) are even better than those of Laitinen. In damaged peripheral nerve injuries, a good effect was obtained in 75% of the cases, and in phantom pain in 66%; however, the follow-up period was rather short. Less favorable but still good results were reported by Loeser *et al.* (1975). In a large group of cases (198 patients) with chronic pain, long-term complete relief was obtained in 12% of the cases, and short-term or partial improvement was achieved in another 68%.

A good analgesic effect was noted in 65% of 107 cases with chronic pain (Dildin *et al.*, 1984). In another series, significant relief from pain of various kinds (neurogenic, somatogenic, psychogenic) by PNS was obtained in half the cases (Johansson *et al.*, 1980). The authors consider the method to be more effective in neurogenic pain, especially if of peripheral origin and in the extremities. The results of PNS are less effective in chronic pain of the loin and neck: a greater than 60% diminution was noted in fewer than 40% of 89 cases (Corkill and Rosenblatt, 1981).

Peripheral nerve stimulation is effective in the management of pain after various surgical operations (Vanderark and McGrath, 1975; Rosenberg *et al.*, 1978), especially after laminectomy, although it cannot completely replace postoperative medication (Schuster and Infante, 1980).

The PNS method has produced encouraging results in headaches, although the relief produced varied with different types of pain. The best results were noted in Horton's syndrome and atypical facial neuralgia. The effect was somewhat less in trigeminal neuralgia and psychogenic headache, and in pain of muscle tension, the improvement was insignificant (Tessitore *et al.*, 1983).

In the literature, some reports have indicated considerably less relief than described above. Mercuri *et al.* (1980), in 100 cases of various types of pain, noted that in 54% of cases pain relief did not last more than 1 hr, whereas in another 23%, there was practically no effect. And PNS was ineffective in thalamic and facial pain (Bates and Nathan, 1980).

5.4.3b. Direct Stimulation. Besides percutaneous PNS, direct stimulation of nerves with implanted electrodes has been tried. The first such operation was performed by White and Sweet (1969), who applied platinum cuff electrodes on the ulnar and median nerves in a patient with posttraumatic pain in the arm. Chronic stimulation of these nerves resulted in almost total disappearance of pain. Then this method was successfully used in pain resulting from damaged peripheral nerves (Long, 1973; Nashold and Goldner, 1975) as well as in other pain syndromes (Picaza *et al.*, 1975). Just as in PCS, in order to determine the effect of this procedure before an operation, percutaneous stimulation is tried. The nerve is stimulated proximal and distal to the site of damage for approximately 30 min. Relief of pain and the occurrence of parethesias on percutaneous stimulation are good prognostic signs and indications for implantation of electrodes directly on the nerve. Stimulations of the nerve distal to the scar intensifies the pain, and PNS gives the best results when paresthesias occur locally in the painful area.

During the implantation of the electrodes for PNS, patient cooperation is essential, and these operations should therefore be performed under local anesthesia. After an incision is made in the plane of the damaged nerve, which is exposed for 3 cm proximal to the scar, thin monopolar and bipolar wire electrodes are fixed around the nerve. They are passed through a subcutaneous tunnel and connected to the receiver. The implanted cuff electrodes for PNS used initially had a number of shortcomings, since they could compress the nerve trunk (Nielson *et al.*, 1976) or, if too loose, slip around it.

In order to avoid these undesirable effects, Nashold *et al.* (1979, 1982) proposed multicontact electrodes made of four miniature platinum–iridium buttons, 2 × 1 mm in size, attached to small dacron disks approximately 12 mm in diameter and impregnated with Silastic. The electrodes are fixed to the epineurium with fine sutures (8–0). The wire from each disk, in a common cable, is connected to the radio-frequency stimulator. By stimulation one can determine the optimal combination of electrodes to produce paresthesias only in the painful area of the arm or leg. Often, even brief stimulation of the appropriate implanted electrodes eliminates pain for many hours.

A suitable site is then determined for implanting the miniature receiver (e.g., subclavically with arm pain or in the abdominal wall with leg pain), which is inserted into a pocket under the skin. A small disk antenna is pasted on the skin over the implanted receiver. The patient is then supplied with a generator connected to an antenna in order to permit self-stimulation.

An implantable peripheral nerve stimulator is manufactured by Avery Lab. Four platinum buttons are mounted in a cuff of silicon rubber, which is sutured around the damaged nerve. Thin wires encased in a common silicon rubber sheath lead from each button. This cable terminates in a male plug connected to the electrode position selector switch so that the effect of 21 different combinations may be tested.

After the optimal combination has been found,

the miniature silicon-rubber-encased receiver is implanted subcutaneously. The receiver transforms the energy from the generator into pulses of the same frequency and duration as the generator signals, which are conveyed along the wires to the stimulating buttons.

The transmitter, the main unit of the external system, is supplied by a 9-V battery and generates modulated high-frequency pulses (2.05 MHz). The amplitude and frequency of the transmitter pulses can be changed by the patient himself. Inside the receiver there are two controls regulating the amplitude and duration of each pulse, which may be set by the physician. The antenna relaying the pulses from the transmitter to the receiver is a hypoallergenic silicon-rubber disk enclosing a concentric wire. The antenna is pasted on the skin over the implanted receiver.

In arm pain, it is possible to apply the electrodes to the trunk of the brachial plexus, which innervates the pain zone; at the same time, the radio-frequency receiver is implanted in a subcutaneous pocket through the same incision.

Long (1973), Nashold and Goldner (1975), Picaza *et al.* (1977), Bohm (1978), and others have reported varying positive results from direct PNS at different times after injury to these nerves. It has been noted that the method is more effective when the electrodes are implanted at some distance from the nerve injury (Long *et al.*, 1981). In brachial plexus and phantom pain, the results were infrequently favorable, the pain disappearing only in rare cases. In cases of damaged lumbar roots resulting from intervertebral disk surgery, chronic stimulation of the sciatic nerve gave a good but transitory effect, the pain recurring in practically all cases within 6 months. It is noteworthy that pain was relieved completely or almost completely only when there was a positive effect from preliminary blockade of the nerve or appropriate sympathetic ganglia.

It has been noted that after prolonged periods of PNS with implanted electrodes (approximately 1 year), pain relief is long lasting, and stimulation may be given less frequently (Long *et al.*, 1981). Nashold and Goldner (1975) stated that one-third of the patients were able to reduce the frequency of stimulation after 6 months.

The largest experience with direct PNS has been summarized by Nashold *et al.* (1982), who have followed 35 cases from 4 to 9 years. The long-term results have not been as satisfactory as thought earlier. Of 19 cases in which electrodes were implanted on damaged nerves of the upper extremities, a significant reduction of pain was noted in 52%. Of 16 cases in which electrodes were applied to the sciatic nerve, the frequency of successful results was even less, 31%.

There is every reason to consider that PNS, both percutaneous and direct, is an effective means of treating many chronic pain syndromes, especially those caused by injury and other lesions of peripheral nerves and spinal roots.

6. Central Pain Syndromes

The term "central pain" has long been applied to chronic pain syndromes caused by various pathological processes within the nervous system. Such terms as "denervated," "deafferentation," or "neuropathic" pain are also frequently used to describe these conditions (Section 1).

In central pain, even ordinary tactile stimuli (e.g., when bed linen comes into contact with the skin) are felt as excruciatingly painful. Diffuse, vaguely localized burning sensations or feeling of pressure with a prolonged latent period between irritation and sensation characterize such pain. In the neurological literature, important features of these syndromes are reflected in the terms "hyperpathia" and "hyperesthesia."

Central pain syndromes caused by damage to a part of the nervous system may be divided into "spontaneous" and "hyperreactive" pain (hypersensitivity). As a rule, these phenomena dovetail with each other. The pathogenesis of chronic pain, as well as the significance of the antinociceptive system and opioid peptides in pain syndromes, are discussed in Sections 3 and 4 of this chapter.

Important evidence for the existence of constant irritative foci in subcortical structures in central pain syndromes was presented by Andy (1983). Using subcortical recording they found thalamic zones (mainly in nonspecific nuclei) of increased excitability and lowered threshold characterized by spontaneous spike and sharp wave bursts. The authors believe that these discharge patterns underlie intractable pain. They speculated that altering the excitability and/or thalamic discharge patterns by induced electrical stimulation may explain the beneficial effects of deep brain stimulation.

Conservative management of central pain essentially does not exist, if one disregards analgesics and narcotics. Surgical treatment is very difficult and uncertain. For decades, numerous attempts to interrupt the pain-transmitting pathways at various levels (peripheral nerves, spinal roots, spinal cord, medulla oblongata, and brainstem), as a rule, have not consistently helped central pain. The various operations for interrupting pain pathways may increase denervation hypersensibility of the cerebral pain centers and, in so doing, intensify rather than diminish the central pain.

Stereotactic neurosurgery has opened a new way for effective management of central pain syndromes—thalamic pain syndrome, phantom pain, causalgia, as well as postherpetic neuralgia and trigeminal anaesthesia dolorosa.

6.1. Thalamic Pain Syndrome

A most severe and excruciating central pain is the thalamic pain syndrome (TPS). This condition, Dejerine–Roussy syndrome, described in 1906, was named after the authors who first wrote about it. The pathophysiology of this syndrome is diverse and has been well studied on the basis of numerous clinical cases. It has long been known that TPS occurs most frequently after vascular lesions of deep-seated cerebral structures not exclusively involving Th nuclei. Although CT scanning has not yielded significant new data as to the causes of TPS, it has established that this syndrome develops more frequently than previously thought after minor hemorrhages in Th (Namba *et al.*, 1984). Previously, it was considered that ischemic foci were the most frequent causes of TPS. However, CT data have confirmed that the syndrome may occur even without pathologically dense foci in Th nuclei (Hirato *et al.*, 1984; Namba *et al.*, 1984) or when the focus is localized in other subcortical structures, e.g., in Put.

The pathological anatomy of TPS was investigated long ago. In 1943, Walker considered that lesions of the sensory Th nuclei were at the basis of this syndrome. Recent morphological investigations (Hassler, 1976) have shed light on the pathogenesis of TPS as well as on other central pain syndromes. Degeneration of the specific Th sensory nuclei related to "rapid pain" is found in such cases. Moreover, the medial system of conducting pain, nonspecific intralaminar and parafascicular nuclei, remains intact. Based on these data, it has been suggested that thalamic pain occurs as a result of the "release" of older subcortical pain centers from the inhibiting influence of the phylogenetically new specific thalamocortical pain systems. Hassler (1976) concludes that in thalamic, phantom, and other central pain syndromes, it is necessary to destroy stereotactically the nonspecific thalamic nuclei responsible for generating "slow" pain.

Another point of view relates to the fact that TPS may occur in lesions of nonspecific Th nuclei or reticular nucleus of FR with the specific VPL complex remaining intact (Cantor, 1973; Schott *et al.*, 1986). Some authors consider thalamic pain to be the result of a deficiency in epicritic sensory input in Th sensory nuclei (Bloedel and McGreery, 1975; Melzack and Loeser, 1978). And, last but not least, there is one hypothesis unlike the others. It is assumed that the pathogenesis of TPS is based not on the destruction but on the hyperexcitability of VPL neurons (Emmers, 1981), causing rhythmic activity induced by nonpain stimuli to develop.

The experience of stereotactic operations on Th nuclei indicates that after surgery TPS practically never occurs. In more than 2200 stereotactic operations, we observed such a postoperative complication in only two cases. Another such case has been described by Pagni (1966). As was demonstrated in Section 5.2.1, destruction of Th nuclei, both specific sensory and nonspecific, leads to the elimination of many pain syndromes, in particular TPS. In all likelihood, not only partial damage of thalamic nuclei but some other as yet undefined circumstance is required for the development of TPS. Accordingly, the pathogenesis of TPS is a problem awaiting solution.

The interval between ischemic stroke and the appearance of thalamic pain is usually several months or a year or two, although it can also be 2–3 years. Most frequently, pain and hyperpathia develop during the stage of sensory recovery after ischemic stroke.

Classical TPS consists of a severe "thalamic" pain in one half of the body with hemihyperesthesia, at times with mild hemiparesis, hemianopsia, and hemiataxia. A characteristic feature of thalamic pain is the diffuse, excruciating hyperpathia with spread of the pain to the entire half of the body with the exception of the face. On the background of constant pain, frequently there are paroxysms, either spontaneous or triggered by various (nonpainful) environmental factors. What is peculiar about such "spontaneous" pain is that it is not relieved by analgesics. Usually, TPS is not associated with severe hemiparesis; as Dejerine and Roussy noted in their classical description, the presence of severe paresis argues against the diagnosis of TPS.

Very typical of this syndrome are pronounced vegetative and trophic disturbances in the affected limbs such as edema, cyanosis, pallor, high or low temperature, changes in nails, and dry skin. Cerebellar disorders (adiadochokinesia, ataxia) are not rare, and, at times, athetotic hyperkinesias may be observed.

Conservative management of TPS involves analgesics and vascular drugs. In severe forms, the value of these drugs is very limited. Many patients gradually develop an addiction to narcotics with subsequent typical personality changes.

6.1.1. Stereotactic Operations

Thalamic pain syndrome was one of the first conditions to be alleviated by stereotactic operations on various Th nuclei. Destruction of VPL, most frequently performed (Section 5.2.1), gave immediate results that were, as a rule, favorable: the pain disappeared or markedly diminished, so that narcotics were eliminated or radically reduced (Bettag, 1966; Vasin and Ratsa, 1971b). However, the pain recurred in the majority of cases after a period of time.

Centrum medianum thalamotomy in TPS did not yield consistent positive results. Of 17 cases in which stereotactic destruction of CM was performed, pain almost disappeared in five and was improved in four more. However, the pain relapsed in all cases within 2–7 months after surgery. Nor did bilateral destruction of CM produce any better effect (Niizuma *et al.*, 1982). Similar results have been reported by Namba *et al.* (1984): an excellent or good result immediately after surgery in eight of nine cases; however, within 2 years, pain relapsed in all but one case.

Destruction of Pul for pain caused by cerebrovascular disease in 15 cases gave more favorable long-term results (follow-up from 3 to 10 years): in three cases, excellent, in four, good, in four, satisfactory, and in the last four, poor results (Yoshii *et al.*, 1980). However, Laitinen (1977), using a similar operation (anterior pulvinotomy) obtained less encouraging results in ten cases: a good effect in two cases, satisfactory in two, and poor in six (mean follow-up, 2 years).

Of 12 cases in which we operated for TPS after cerebral vascular disturbances, cryodestruction of various subcortical structures (VPL, CM, GC), produced significant and lasting relief of excruciating pain in only five cases, although immediately after the operation, all patients seemed better.

These preliminary results indicate that stereotactic hypophysectomy may relieve TPS (Section 5.2.6).

6.1.2. Stimulation Methods

In recent years there have been a few reports about the effect of stimulation in TPS and other deafferentation pain syndromes. Chronic stimulation of various subcortical structures (VPL, CI) by implanted electrodes produces a satisfactory and, to a certain degree, lasting therapeutic effect (Adams *et al.*, 1974; Hosobuchi, 1979; Tsubokawa *et al.*, 1982a,b). Hosobuchi and Adams (1975) obtained a good effect by simultaneous chronic stimulation of sensory nuclei and the posterior limb of CI. However, in only 14 of 29 cases with deafferentation pain did chronic stimulation of VPL and CI result in total disappearance of pain (Hosobuchi, 1979). In two patients with TPS after a stroke, good relief of pain was obtained from chronic stimulation of the CM–parafascicular complex (Andy, 1983).

Adams *et al.* (1974, 1977) obtained a significant therapeutic effect in four cases with TPS from organic brain disease by chronic stimulation of the posterior limb of CI. In seven cases with TPS, Namba *et al.* (1984) also employed chronic stimulation of the posterior limb of CI, reporting a good result in three cases, slight improvement in two cases, and a poor result in the remaining two (follow-up from 9 months to 2 years, 8 months).

The mechanism by which pain is alleviated, by chronic stimulation of Th nuclei and CI still remains obscure. It is possible that such stimulation compensates for an insufficient sensory input to specific Th nuclei. Another hypothesis is that such stimulation has a modulating influence on the antinociceptive system. Neither PCS nor PNS in TPS produces any effect (Bates and Nathan, 1980; Siegfried, 1983).

One must admit that the surgical management of TPS is still an unsolved problem. Nevertheless, the possibility of helping even a few of these incurable disorders, points to the need to continue with the existing operations while searching for a new surgical approach. In all likelihood, the most effective methods at present are chronic stimulation of subcortical structures (VPL, CI).

6.2. Phantom Pain Syndrome

Phantom sensation, i.e., the feeling that an amputated limb or part still exists, has been experienced by practically all persons after amputation except infants in whom the body scheme is not yet developed. In the majority of cases, such sensation gradually disappears;

however, in a number of cases, (about 10–20%) it develops into a burning, pressing, twisting, sharp, often intractable pain referred to the absent limb.* For the patient, such pain is the source of continuous suffering, preventing him from working and, quite often, even from sleeping normally. Frequently, superimposed on the constant pain are severe, aggravating spasms. Closely resembling the phantom pain syndrome (PPS) is severe, constant pain in an existing but totally denervated arm after complete traumatic avulsion of the brachial plexus (Section 5.3.7) and the pain in the lower body in patients with traumatic paraplegia (White, 1973).

The first brief description of phantom pain was presented by Ambroise Paré in 1551 (Keynes, 1952), and René Descartes almost a century later; the term "phantom pain" was first suggested by S. W. Mitchell *et al.* in 1864. In 1872, Mitchell described 90 veterans of the American Civil War with such pain, which the author considered "sensory hallucinations." The famous Russian surgeon Nikolai Pirogov described phantom pain in 1865 (cited in Pirogov, 1961). Even at that time, he pointed out that neuromas should not be regarded as the cause of this pain, since many amputees have neuromas that are not painful. This correct conclusion was not accepted then, and Charcot's opinion (1888) that phantom pain has a peripheral origin (neuromas, scars, ascending neuritis of the cutaneous nerves) predominated the literature for many years.

For several decades it was thought that the primary lesion in phantom pain was in the peripheral nerves. However, there can be no doubt that, unlike pain in a stump, it is not an irritative syndrome caused by a neuroma of a severed nerve (or nerves). As a rule, PPS is present soon after amputation, prior to neuroma formation, and resection of the neuroma very seldom eliminates the pain. In view of this, it has long been acknowledged that PPS belongs to the category of central pain. As already mentioned, phantom pain occurs in paraplegics with total interruption of the spinal cord. Finally, the fact that PPS disappears after surgical interruption of pain-relaying pathways and structures in the spinal cord and brainstem as well as after PCS points to the central origin of this pain.

To explain the mechanism of PPS, considerable importance has been attached to "breaking" the central antinociceptive system and loss of inhibiting input to deafferented neurons. As a result, normal afferent impulses are experienced as pain. There are still several other theories of the pathogenesis of phantom sensations. According to one, the main element in the pathogenesis of phantom pain is a deficit of sensory information from the missing limb to the complex body schema, which inhibits the pain centers of the brainstem FR (Melzack, 1971a,b; Majorchik *et al.*, 1980). Under such conditions, the central excitatory focus is no longer dependent on peripheral stimulation. To some extent, the second hypothesis contradicts the first in that there is excess afferent input from the severed nerve trunks of the amputated limb, resulting in hyperexcitability of the representation of the amputated limb in the cerebral body schema.

There is reason to consider the center of the body schema to be VPL, which integrates the awareness of the body parts in space (Miles and Lipton, 1978). The third hypothesis regards the main factor in the pathogenesis of phantom pain as deafferentation hypersensitivity of the neurons in the posterior horns and Th sensory nuclei (Section 4) (Melzack and Loeser, 1978; Kryzhanovsky, 1980), as well as the brainstem FR (Müke, 1974). Another recent hypothesis views PPS as a reafferentation syndrome in which the deafferented high-threshold neurons of the posterior horns are supplied again with primary low-threshold afferent fibers, both mono- and polysynaptically from adjacent and remote dermatomes (Howe, 1983).

Severe PPS, which frequently appears after amputation of limbs, does not usually respond to conservative treatment (White and Sweet, 1969). Many PPS patients gradually become addicted to narcotics.

6.2.1. Stereotactic Operations

It is difficult to enumerate all the surgical methods that have been tried in the past in the management of PPS; they include operations on peripheral nerves, resection of the obliterated arteries, section of spinal roots, cordotomy, commissurotomy, sympathectomy, mesencephalic tractotomy, excision of parts of the cerebral cortex, and leukotomy. As a rule, the long-term results of these operations have not been encouraging because of the high level of recurrences.

The many different surgical procedures that have been proposed for the management of PPS indicate that there is no absolutely successful operation. It may be said that the resections of nerves and vessels in the stump (excision of neuromas, reamputation, etc.) frequently performed in such cases are, as a rule, ineffec-

*Here we do not include postamputation stump pain, which is usually the result of peripheral factors and should not be referred to as central pain.

tive. This is attested to by the extensive experience of Soviet neurosurgeons during the Second World War and in the immediate postwar years. Operations on the sympathetic nervous system are also futile; the authoritative opinion of Grigorovich, in his monograph on the surgery of peripheral nerves (1981), was that not a single case of phantom pain had relief after sympathectomy. In a number of cases, an improvement was noted after anterolateral cervical cordotomy (Falconer, 1966; White and Sweet, 1969), however, the beneficial effect of this operation was not confirmed by other authors.

For some time, sporadic attempts have been made to remove individual areas of the cerebral cortex for various motor disorders. This operation, called topectomy, did not justify itself and has been practically abandoned, although isolated successful results of topectomy have been described in the literature. Török (1960) performed subpial resection of the motor and sensory areas of the cerebral cortex corresponding to the amputated left arm of a patient with phantom pain, and the pain disappeared postoperatively; however, the follow-up in this case was only 6 months.

The encouraging results obtained in the treatment of many pain syndromes by stereotactic techniques served as the basis for trying this method in PPS. However, fewer than 100 such operations have been described in the world literature, and only the immediate results have been published in approximately half of these cases. Usually, only the pain disappeared after stereotactic surgery, leaving intact the phantom sensation.

It is also noteworthy that these operations involved destruction of various cerebral structures, which makes it difficult to assess the effect of a given operation and the optimal stereotactic target. Sensory Th nuclei were destroyed in the majority of operations.

Cooper (1965b) reported that stereotactic thalamotomy in four cases of PPS yielded a good result in three. In two other similar cases, the PPS disappeared after combined destruction of VPL and CM. Single cases with good results from stereotactic destruction of VPL in PPS have been reported by Bettag and Yoshida (1960), Vasin and Ratsa (1971a,b), Roth and Mark (1973), and others. Thalamotomy in phantom pain gave a significant and lasting improvement in only 18% of the cases (Siegfried and Cetinalp, 1981).

The author (Kandel, 1971b) has reported a typical case of severe PPS that was totally cured by stereotactic destruction of VPL. For 23 years a 49 year-old patient suffered constantly from excruciating pressing and burning pain following a high amputation of his right leg after a war wound in 1944. In the course of several years, the patient underwent ten peripheral operations (repeated removal of neuromas, reamputation, sympathectomy, and others) without effect. For several years the patient took narcotics every day. In 1967, the author performed stereotactic cryodestruction of VPL in the patient [coordinates of the center of target: 18 mm posterior to FM, 5 mm above LI (length 24.5 mm), and 15.5 mm from the midplane]. The phantom pain stopped completely during surgery. The patient immediately abandoned narcotics. For six years after the operation (until his death from myocardial infarction), the phantom pain was completely relieved.

Laitinen (1977) performed a unilateral anterior pulvinotomy in ten cases of PPS. With an average follow-up of 2.5 years, a good result was noted in three cases, satisfactory in four, and poor in three.

A significant diminution of PPS was noted in four of six cases after stereotactic mesencephalic tractotomy in which a small destructive lesion was made in the medial part of the midbrain tegmentum (Section 5.2.5a) (Nashold *et al.*, 1969b). Electrostimulation of the lateral spinothalamic tract in the midbrain tegmentum intensified the phantom pain.

In two cases of severe PPS, a significant therapeutic effect was obtained after stereotactic destruction of the thalamoparietal pathways in the white substance of the of the parietal lobe (Section 5.2.1j) (Talairach *et al.*, 1959). However, this method was not pursued. Three patients with PPS had a good result after bilateral stereotactic cingulotomy (Laitinen *et al.*, 1973).

Müke (1974) performed percutaneous cordotomy in seven cases with phantom pain, with relief in four cases and a relapse in three others. The author recommends that this operation be done soon after the appearance of phantom pain, before "central irritation" sets in.

To conclude, mention should be made of a new method, unrelated to stereotactic surgery, for the management of PPS, destruction of DREZ (Section 5.3.6c). Powers *et al.* (1984b) reported that pain was completely relieved in two of three cases after this operation, whereas Saris *et al.* (1985) stated that six of nine patients were pain-free from 6 months to 4 years. In a few cases, after DREZ destruction, Samii and Moringlane (1983) reported pain relief.

Recently, we operated on a 34-year-old patient

with extremely severe PPS after disarticulation of the left arm as a result of a traffic accident. He required eight to ten injections of morphine every day for 4 years after trauma. Immediately after the DREZ operation on the cervical spinal cord, the pain completely disappeared, and the patient, pain-free, refused narcotics. The follow-up period lasted 2 years. It may be concluded that DREZ coagulation is an effective method of surgical treatment of the disease.

There were several attempts to alleviate phantom pain by direct PNS (Section 5.4.3b).

6.2.2. Stimulation Methods

Just as in other central pain syndromes, successful attempts have recently been made to use stimulation methods in the management of PPS. Mazars (1976a) considers that chronic bipolar electrostimulation of VPL with implanted electrodes is effective. He obtained an excellent result in 19 cases. Siegfried and Cetinalp (1981) obtained a good effect from stimulation of the Th sensory nuclei in 26 of 30 cases. However, stimulation of PGM or CI was not very effective. Smirnov (1976) noted that stimulation of Pul, CM, and the dorsomedial nuclei of Th arrested attacks of severe pain in patients with PPS. Nevertheless, chronic stimulation of several thalamic nuclei in four cases of PPS produced a good result in only two cases (Dieckmann and Krainick, 1979).

Posterior column stimulation (Section 5.4.2) has also been used in the management of phantom pain. The largest experience with PCS for treating phantom and stump pain is that of Krainick *et al.* (1975), who performed this operation in 73 cases. In both syndromes, the severity of the pain decreased by more than half directly after the operation in 65% of the cases. In their next series (Krainick *et al.*, 1980), the authors analyzed the long-term results (more than 5 years after beginning PCS) in the same group of patients. The number of cases with a positive effect from stimulation gradually decreased from 56% to 42%. It is noteworthy that in PCS the phantom sensations changed only in a few cases.

Nittner (1982) reported on the results of PCS in three cases of PPS after amputation of a lower limb. Because the usual site of stimulation in the midthoracic region produced unpleasant sensations over the heart and liver, the author considers the conus and epiconus to be the optimal site for the electrodes.

In two of four PPS cases, a significant improvement of pain was obtained by chronic combined PCS and stimulation of the spinal nerve (Section 5.4.2) (Urban and Nashold, 1982). According to Siegfried (1983), PCS in PPS is of practically no benefit.

Good results of PNS in PPS have been reported by Ray (1975). A significant effect was obtained in eight of ten cases. However, these results have not yet been confirmed by other authors.

The data presented indicate that stereotactic operations on deep-seated cerebral structures belonging to the pain-conducting and perception system and chronic stimulation of the spinal cord are, at the moment, the only effective methods for the management of PPS. Many questions of stereotactic treatment, especially the best target for destruction or stimulation, require further investigation.

6.3. Causalgia

Causalgia was first described by S. W. Mitchell *et al.* in 1864. The authors considered the disease to be an ascending neuritis with spread of the inflammatory process to the posterior roots and spinal cord. This severe pain develops after partial damage to peripheral nerves, most frequently the median or sciatic nerves as well as the brachial plexus, which contains a great number of sensory and sympathetic nerve fibers. Constant excruciating, burning pain and vasomotor disturbances in causalgia are most frequently localized in the wrist or foot but are usually not limited to the zone innervated by the injured nerve and radiate throughout the entire limb. This burning pain, as a rule, is intensified by emotional and physical factors (bright light, noise, changing temperature of the environment, etc.). The patient may experience relief by immersing the affected limb in running water or wrapping it in cold wet towels.

For many decades, there has been no doubt that causalgia is caused not by a pathological process in the injured nerve but mainly by "fixation of pain" in CNS structures. A damaged peripheral nerve serves only as a trigger for the development of complex pathological reactions in higher pain pathways and centers. That is why causalgia may be considered a form of central pain. Nevertheless, some authors emphasize the peripheral mechanisms in the pathogenesis of causalgia. In particular, it is assumed that in case of an injured peripheral nerve, impulses pass from certain nerve fibers (C-fibers) to other surrounding fibers, causing a kind of "short circuit." Pronounced sympathetic irritation, so typical of causalgia, is explained by "shunting" of sympathetic and somatic afferent nerve fibers.

Investigators today conclude that the causalgic syndrome is the result of a "retrograde neuronal reaction," during which there occurs a focus of pathological functional activity in the neurons of the spinal posterior horns (Sunderland, 1976). This focus is the source of pathological impulses transmitted to supraspinal pain centers, including the cerebral cortex.

Since causalgia has been described in detail in the literature, we shall not describe its clinical aspects, diagnostic criteria, and course. Conservative methods of treating causalgia do not exist. The formidable experience of the Second World War indicates that various operations on the injured nerve (extirpation of neuromas, "disentangling" nerves, etc.) are not effective or, as a rule, are followed by relapses of pain. The data in the literature concerning the efficacy of sympathectomy are contradictory, although there were numerous papers reporting good results from that operation (White, 1973). One may assume that sympathectomy provides relief from causalgic pain in acute cases but in chronic cases comparatively rarely; many patients become addicted to narcotic agents.

The vasomotor pain syndrome, called sympathetic dystrophy or Sudeck's atrophy, also developing after injury, closely resembles causalgia.

6.3.1. Stereotactic Operations

The only method that may provide hope of alleviating the suffering of these patients is stereotactic surgery of deep-seated structures of the brain. However, the data in the literature concerning such operations are limited, and a number of questions about the techniques of such treatment, in particular, the choice of the stereotactic target, still remain unanswered. Several successful operations on Th sensory nuclei have been described. Permanent disappearance of causalgic pain in the arm of one patient after destruction of VPL has been described by Shefer and Nesterov (1966). In one case of causalgia, stereotactic pulvinotomy resulted in relief of pain for 5 years (Yoshii *et al.*, 1980).

Single cases have been reported of a beneficial effect after stereotactic operations on pain pathways at the level of medulla oblongata and the midbrain. A severe causalgia, caused by injury to the radial and ulnar nerves, which persisted after several unsuccessful operations, was eliminated by medullary spinothalamic tractotomy (Section 5.2.5b) (Birkenfeld and Fisher, 1963). Percutaneous cordotomy (Section 5.3.2c) in causalgia has proved ineffective, although the first attempts to use a new operation, limited (extraleminiscus) myelotomy (Section 5.3.5), in this disorder have given encouraging results.

6.3.2. Stimulation Methods

Although the use of chronic stimulation in the management of causalgia is in its initial stage, the early results may be considered encouraging. Chronic stimulation of subcortical or spinal structures has been attempted. Turnbull *et al.* (1980) stimulated VPL in three cases of causalgia and succeeded in totally eliminating pain in all of them.

Andy (1983) described one case of causalgia after a gunshot wound of the right elbow in which many peripheral operations had failed, but chronic stimulation of the CM–parafascicular complex abolished pain completely.

In a case of causalgia in the arm, a good effect was obtained by electrostimulation of the lateral part of the midbrain tegmentum with a stereotactically introduced electrode (Nashold *et al.*, 1969b). Broseta *et al.* (1982b) employed chronic epidural PCS in 11 cases of causalgia present from 8 months to 6 years. An excellent result was noted in six cases, good in two, mild in one, and no effect was seen in two cases (mean follow-up 1 year). In five cases of causalgia reported by Nielson *et al.* (1975), PCS gave a good result with a follow-up of 2 years. Yamashita *et al.* (1981) reported a patient with causalgic pain for 7 years after damage to the median nerve whose pain completely disappeared after initiation of PCS with the PISCES system (Section 5.4.2).

To summarize the section on central pain syndromes, it should be emphasized that their management by functional and stereotactic surgery still remains a difficult and so far unresolved problem. The most promising operations are those aimed at destruction of the pathways conducting slow pain or long-term stimulation of certain subcorticotruncal and spinal cord structures.

7. Trigeminal Neuralgia

One of the many pain syndromes, trigeminal neuralgia (TN), first described by N. André in 1876, is traditionally defined as a separate nosological entity. Primarily, this results from the fact that statistically, TN is a comparatively common disorder that affects approximately five persons per 100,000 annually. The most frequent (approximately 80% of the cases) site of involvement is the second branch of the fifth nerve, less frequently the third branch, still less frequently both of these branches, and, very rarely (6%), the first branch (Menzel *et al.*, 1975; Brandt and Wittkamp, 1983). Trigeminal neuralgia occurs twice as often on the right (65%) as on the left side (35%), a finding not yet explained. The frequency of bilateral trigeminal neuralgia is 1–3%. Trigeminal neuralgia is observed more frequently in women (57.7%) than in men (42.3%) (Siegfried, 1983).

This disorder has a typical pathogenesis, clinical picture, and course. Recently, it has been demonstrated that the main cause of TN is compression of the root of the fifth nerve by an arteriole or, less commonly, a venous loop in the posterior fossa (see below). However, in approximately 20–30% of the cases, no compression of the root is seen, and such cases of TN must be considered essential or idiopathic.

The literature has considerable information on the pathogenesis of TN as well as other pain syndromes. An important role is considered to be played by changes in the opiate peptides inhibiting pain. However, a study has demonstrated no statistically reliable difference in the content of endorphin in the CSF of TN patients and a control group (Salar *et al.*, 1981), and no changes were found after either drug treatment or thermocoagulation of the Gasserian ganglion. In view of this, the authors conclude that there are no pathogenetic relationships between endorphins and TN.

The assumption that TN is an epileptic phenomenon has been discussed in the literature for many years. The paroxysmal nature of pain attacks, their "spontaneity," the presence of trigger zones, as well as the therapeutic effect of anticonvulsants all, to some extent, suggest such a hypothesis.

One of our own cases illustrates the unusual direct connection between TN and epileptic seizures. Both pathological conditions disappeared after percutaneous electrocoagulation of the Gasserian ganglion.

A 63-year-old patient had suffered for many years from frequent paroxysmal pain in the I-II branches innervating the left trigeminal nerve. Each attack of neuralgia was accompanied by a severe seizure of Jacksonian type in the right half of the face and in the right extremities. Sometimes the seizure became generalized with loss of consciousness. Just before admission the attacks of TN followed by focal epileptic fits were occurring very frequently—up to 20–30 per day. Intensive treatment by anticonvulsants gave little relief. CT study did not show any abnormalities. There were no other neurological signs or symptoms.

Percutaneous electrocoagulation of the Gasserian ganglion was performed under local anesthesia using the operative technique described in Section 7.2. Mild analgesia was achieved in the I-II trigeminal branches on the operating table. Immediately after the operation, the trigeminal attacks and also the Jacksonian fits disappeared completely. This effect has lasted during 1.5 years of follow-up.

For several decades TN has received the special attention of neurosurgeons. The clinical aspects, symptomatology, and course of the various forms of this disorder, particularly its most frequently encountered classical form of neuralgia (tic douloureux), have been described in scores of papers and monographs. This allows us to proceed directly to the problems of the surgical treatment without dwelling on the clinical aspects of the disorder.

First of all, it should be noted that in recent years new drugs have appeared—carbamazepine, phenytoin, clonazepam, and others—which undoubtedly are effective means of treating TN, although with the passage of time, the effectiveness of carbamazepine gradually declines. Moreover, carbamazepine causes various side effects in 69% of cases (Siegfried and Hood, 1983). In this respect, the figures presented by Van Loveren *et al.* (1982) are of interest. Although a beneficial effect from drug therapy was noted in the majority of 1000 cases of classical TN at the beginning, eventually 750 (75%) underwent operations by various techniques. To a certain extent, drugs reduce the indication for operative treatment; nevertheless, even today, surgery is the only hope for thousands of patients suffering from facial pain.

The search for the most effective and safest surgical treatment of TN has been in progress for two centuries. It is difficult to enumerate all the operations that have been proposed. Without dwelling on historical data, it should be noted that today many different operations are performed on the trigeminal nerve system for neuralgia. These operations may be divided into the following main groups:

1. Operations (dissection, alcohol injection) on the three peripheral branches of the fifth nerve.
2. Operations (dissection, compression, decompression, electrothermodestruction, glycerol injection) on the Gasserian ganglion and the sensory root of the fifth nerve.
3. Operations (section of the pathways of the tri-

geminal system) in medulla oblongata, midbrain, thalamus, and pain-relaying pathways from Th to the cerebral cortex.
4. Chronic stimulation of Gasserian ganglion and subcortical structures.
5. Microvascular decompression of the trigeminal nerve root.

Several surgical procedures on the sensory root and descending tract of the fifth nerve belong to the category of classical operations. The first intracranial operations in TN, section of the fifth nerve branches and resection of the Gasserian ganglion, were performed by Horsley *et al.* (1891), Krause (1892), and Hartley (1892) at the end of the last century. In the following decades, other operations were proposed: the Spiller–Frazier–Adson operation (section of the root by the extra- or intradural subtemporal approach), section of the trigeminal root (Dandy) by the posterior approach, and section of the descending root of the fifth nerve in medulla oblongata (Sjöqvist's operation) (Sjöqvist, 1937, 1938). Since these operations and their results have been described in detail in journals and manuals on neurosurgery, we shall concentrate on new surgical techniques for the management of TN, mainly stereotactic and functional operations.

7.1. Percutaneous Trigeminal Injection of Lytic and Other Substances

Dissection or alcohol injection of one of the three branches of the trigeminal nerve in TN has been employed for several decades. Since the relatively simple technique of these operations has been thoroughly described in neurosurgical literature, we shall not discuss them in detail.

The success rate of these operations is comparatively low; however, in mild forms of TN, dissection of a branch of the trigeminal nerve may eliminate pain for a long period of time. In the rarely encountered neuralgia of the first branch, resection of the supraorbital nerve often gives satisfactory results. In certain cases of neuralgia limited to the mandibular division, alcohol injection of the third branch is still used.

The technique of puncturing the oval foramen and the emerging third branch of the fifth nerve was proposed by Schlösser as far back as 1907. The third branch, after it emerges from the oval foramen at the base of the skull, divides into peripheral branches within 2–3 mm. In view of this, a block of this branch should be performed at the foramen ovale. Alcohol injection of the third branch is usually performed bv dentists. This procedure, done in the outpatient department, is obviously valuable in the initial stages of mandibular neuralgia. However, the relief from alcohol injection is often temporary, and recurrences of pain are frequent.

Another well-known technique is the infusion of alcohol into the Gasserian ganglion by percutaneous puncture through the oval foramen. This ganglion lies in Meckel's cave in the medial part of the pyramid. This method was proposed by Harris (1912) and Taptas (1911) and improved on by Härtel (1914). In 1940, Harris reported on 2500 cases of idiopathic TN in which absolute alcohol was injected into the trigeminal ganglion. After alcohol injection of the ganglion, the entire half of the face is anesthetized, but sensation gradually returns within a few years in approximately 40% of the cases, more frequently in the third division (Henderson, 1965).

The results of alcohol injection of the Gasserian ganglion have been described by many neurosurgeons, since this technique was extensively used in neurosurgical clinics for more than half a century. Much information on the high rate of relapses and various complications after this operation has accumulated. Umbach (1960) pointed out that the frequency of neuroparalytic keratitis after alcohol injection of the ganglion was 13%, and the frequency of relapses was approximately 30%. These complications led neurosurgeons to advise open intracranial operations with section of the root of the fifth nerve or its descending tract for TN.

However, other injection techniques were advocated.

The injection of hot distilled water into the Gasserian ganglion (Jaeger, 1959) was proposed. Livshitz (1973) reported good results from this method in 305 TN cases. The number of postoperative keratitic complications (5.3%) was approximately one-half of those after the surgical techniques described above, whereas the number of relapses (8.1%) were only one third. Yet, the hot water technique is not devoid of shortcomings, and the recurrences are still high, making it necessary to repeat the hot water injections in order to obtain complete relief.

The next step in the development of TN surgery was a proposal to inject the Gasserian ganglion with an agent less destructive than alcohol, namely, a 5% solution of phenol in glycerin, the high viscosity of which to a certain extent prevents it from spreading to surrounding structures (Vasin, 1973).

A serious drawback and a reason for frequent complications after injection of lytic solutions into the Gasserian ganglion is that they penetrate beyond Meckel's cavity into the subarachnoid space of the posterior fossa. To prevent this, Jefferson (1963) proposed that before injecting the solution, the head of the patient be sharply inclined so that the needle introduced into the Gasserian ganglion would be vertical.

Håkanson (1981) proposed another method to treat TN by injection of glycerol into the trigeminal cistern (Meckel's cave). The effect of glycerol was found accidently when the author injected tantalum dust in glycerol for trigeminal cisternography in a patient with a complicating paroxysmal trigeminal pain. Following this treatment, the patient's pain disappeared completely without facial sensory loss. This result encouraged the author to try the technique for trigeminal neuralgia per se.

The procedure is carried out with the patient awake in a sitting position after premedication. A 22-gauge needle is introduced through the skin into the foramen ovale. If CSF is obtained, a small quantity of metrizamide (less than 1 ml) is injected in Meckel's cavity under fluoroscopic control. The lateral x-ray film should confirm that the needle is in the cistern. After correct positioning of the needle, 0.2–0.4 ml of 100% glycerol (depending on the size of the cistern) with the addition of a small amount of tantalum dust for x-ray control is introduced into the cistern. After the injection, the patient is kept sitting in bed with the neck flexed to keep the glycerol in the cistern.

Håkanson (1982) reported on the results in his first 74 patients after 2 to 48 months. In 13 cases, trigeminal pain had recurred (18%), a recurrence rate two to three times greater than after radiofrequency coagulation of the trigeminal root. However, the procedure is much safer and involves less risk than radiofrequency lesion. After Håkanson's operation, there was no sensory loss, dysesthesia, or anesthesia dolorosa. Corneal sensation remained intact. About 60% of the patients complained of a slight numbness. Håkanson suggested that the glycerol preferentially destroyed the damaged large trigeminal demyelinated fibers.

Recently several authors have reported encouraging results using glycerol in trigeminal neuralgia (Lunsford and Bennett, 1984; Dieckmann *et al.*, 1985; Beck and Lobosky, 1985; Takusagawa *et al.*, 1985; Beck *et al.*, 1986).

Sweet and Poletti (1982, 1985) slightly modified Håkanson's technique in 25 patients, 24 of whom had trigeminal neuralgia and one who had constant pain. Their results led them to recommend a glycerol injection rather than a RF lesion for patients with TN who are willing to accept a greater risk of recurrent pain in return for the smaller likelihood of dysesthesias and severe loss of touch sensation.

Percutaneous retrogasserian glycerol rhizotomy was performed in 58 patients (Beck *et al.*, 1986). In 72% of the cases the results were good—complete cessation of pain. There was also marked improvement in 7% of the cases, but in 21% the operation was ineffective and in 11% recurrences took place after initial relief. The complications were rare and not severe.

7.2. Percutaneous Electrothermocoagulation

The next important stage in the surgical treatment of TN was an operation involving electrocoagulation of the Gasserian ganglion proposed by Kirschner (1932, 1933). For that procedure, the author designed a relatively simple device that may be regarded as the prototype of contemporary stereotactic apparatuses (Fig. 191). This device guided the needle to the Gasserian ganglion with relative accuracy through the foramen ovale. Then Kirschner destroyed the ganglion with a 350-mA diathermic current. In this technique, he followed certain other important principles of contemporary stereotaxis. He established the spatial relationships among the foramen ovale, the internal acoustic meatus, and the midplane. Correction for x-ray divergence was introduced for the first time. In the majority of cases, the author managed to reach the Gasserian ganglion on the first attempt. In 1942, Kirschner published his rich personal experience of 1113 operations of electrocoagulation of the Gasserian ganglion.

This operation rapidly won general acclaim and was extensively used in TN for several decades. Indeed, by its noninvasive character, simplicity, and relatively easy performance, this operation had advantages over the generally accepted open intracranial operations. Subsequently, however, more and more reports began to appear in the literature of potentially serious complications following this operation, including keratitis, blindness, deafness, other craniocerebral nerve palsies, primarily, the third, sixth, and eighth, thromboses, and ruptures of the internal carotid artery. It was thought that these complications were caused by difficult-to-control coagulation of the Gasserian ganglion.

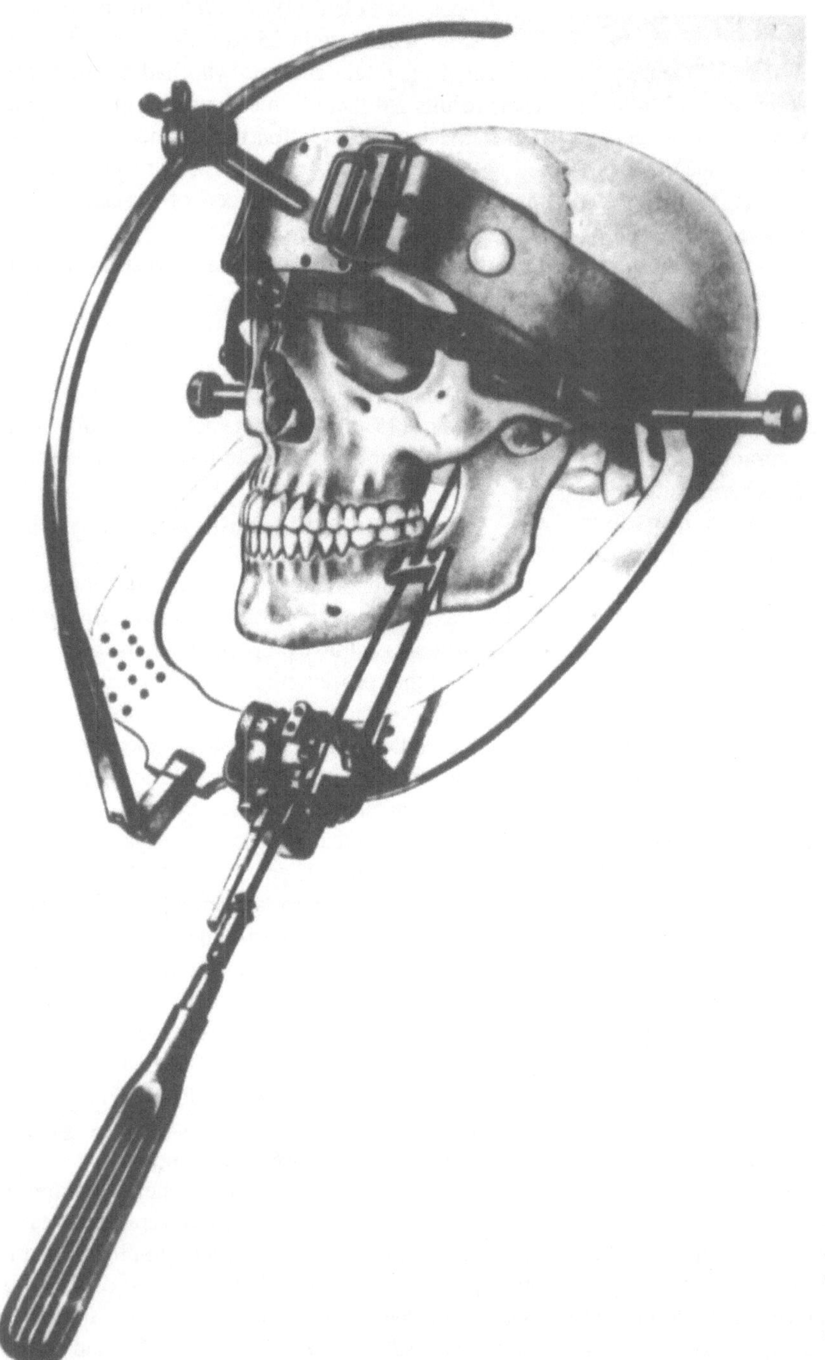

FIGURE 191. One of the first stereotactic devices: Kirschner's apparatus for precise introduction of the electrode into the Gasserian ganglion for its electrocoagulation.

The success rate of the method proposed by Kirschner was undoubtedly higher than that of alcohol injection of the Gasserian ganglion after Harris and Härtel. However, the number of complications and relapses after electrocoagulation was not much lower (Umbach, 1960). In view of this, there was a constant search for ways to improve this technique. One of the most effective modifications of the management of TN was percutaneous stereotactic destruction of the Gasserian ganglion or trigeminal root. The technique of this operation was elaborated by White and Sweet (1969), Sweet (1972), Schürmann *et al.* (1972), Sweet

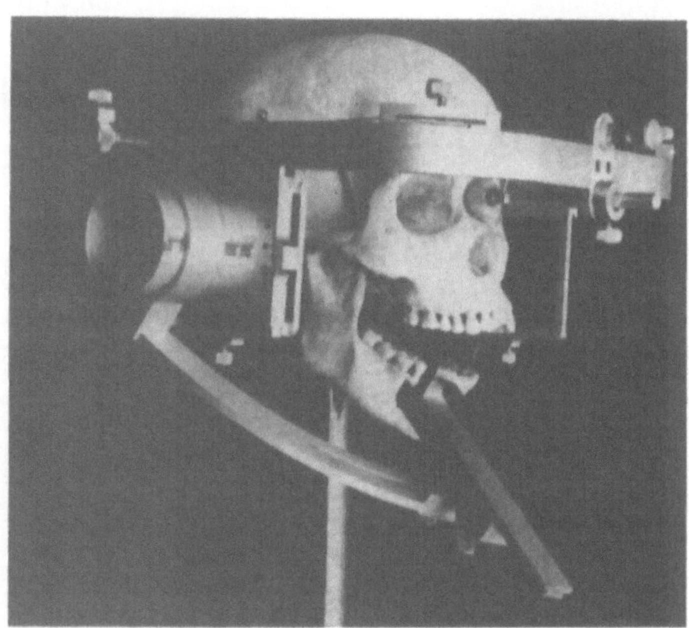

FIGURE 192. The trigeminus stereoguide with electrode introduced through the foramen ovale (Laitinen, 1984).

and Wepsic (1974), Nugent and Berry (1974), Siegfried (1977), and Tew and Keller (1977). The patient is given a local anesthesic agent, neuroleptanalgesia (Innovar®—a mixture of droperidol and phentanyl), or general anesthesia induced with short-acting preparations, e.g., brief intravenous anesthesia with methohexital. The needle is introduced percutaneously into the Gasserian ganglion or fifth nerve root via the foramen ovale through which the third branch of the trigeminal nerve emerges from the skull.

Certain surgeons use a stereotactic apparatus of some type for percutaneous electrocoagulation of the Gasserian ganglion. Laitinen (1984) devised a new stereotactic apparatus (Trigeminus Stereoguide) for destroying the trigeminal root through the foramen ovale or the ninth nerve through the foramen jugulare (Fig. 192). The apparatus is based on the following principle. The foramen ovale should be situated in the center of a spherical system consisting of a cylinder and a 90° arch, one end of which is fixed to the cylinder, and the other holds an electrode carrier. The location of the foramen ovale is determined by X and Y coordinates on lateral and axial roentgenograms of the patient's head, which is fixed in an oval aluminum frame by two ear pins and a third at the nasion. On the basis of calculations, the best trajectory for introducing the electrode either through the cheek or below the mandible is determined.

The stereotactic apparatus proposed by Waltregny (1982) allows one to destroy any one of the three branches of the fifth nerve (Fig. 193). The apparatuses of Lazorthes *et al.* (1976b) and of Modesti and Perl (1982) are similarly designed. However, with experience and training, many neurosurgeons prefer to introduce the electrode manually, as proposed by Härtel in 1914, considering the free-hand search a more reliable way to find the foramen ovale (Sweet and Wepsic, 1974; Tew *et al.*, 1978; Siegfried, 1983).

A thin-walled lumbar puncture needle with an external diameter of 1 mm is used for thermocoagulation of the trigeminal root. The outer end of the needle serves as the indifferent electrode, and the active electrode, a thin wire, has a slightly bent tip 2–3 mm long. Rotation of such an electrode within the needle enables the surgeon to find readily the nerve branch to be coagulated (Nugent, 1982; Tobler *et al.*, 1983). A hollow 20-gauge cannula with short bevel and sharp tip insulated except for the distal 10 mm has been used by Siegfried and Hood (1983). A special insulated cannula–probe (Type TIC, Radionics) with technical systems permitting both stimulation and radiofrequency coagulation of the root (Radionics Lesion Generator System, Fisher Generator, Owl Universal RF System, and others) has been devised for introduction into the Gasserian ganglion.

In brief, the technique of the operation is the following. The patient is in a half-sitting position. After infiltrating the skin with a local anesthetic agent 2.5–3 cm from the corner of the mouth on the side of the pain, the surgeon introduces the finger of his gloved

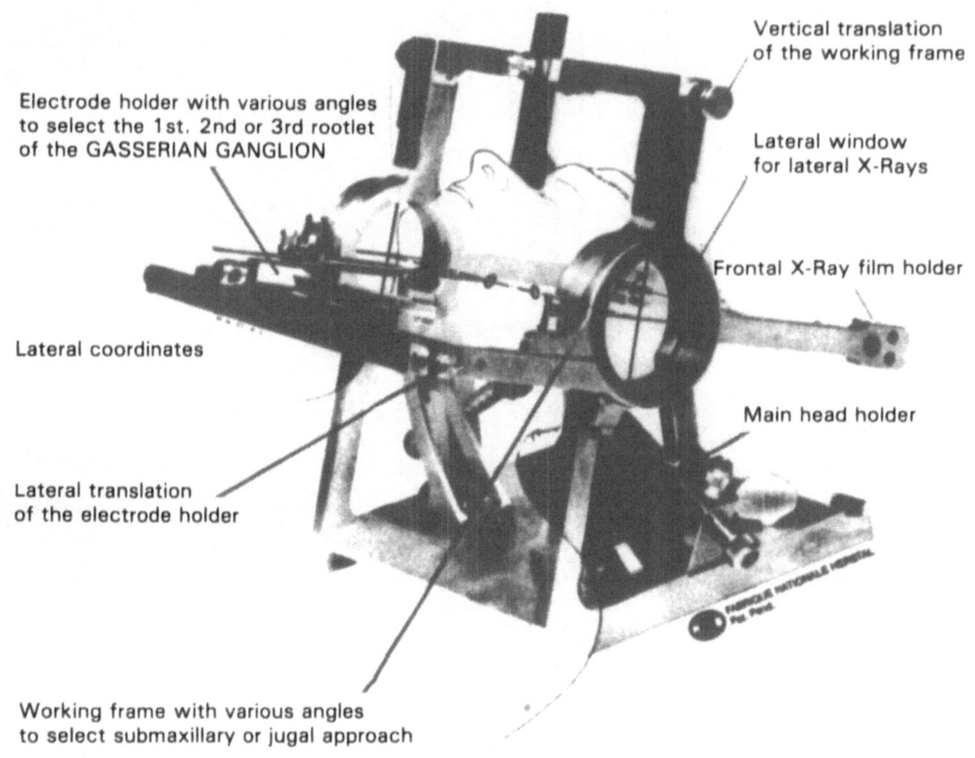

FIGURE 193. Stereotactic apparatus for trigeminal ganglionectomy (Waltregny, 1982).

hand into the mouth of the patient to locate the pterygopalatine fossa so as to guide the needle medially from the ascending ramus of the lower jaw (Fig. 194). The reference point for directing the needle is the intersection of the frontal plane 3 cm anterior to the auditory meatus and the sagittal plane of the center of the ipsilateral pupil. The needle must not penetrate the buccal mucous membrane, which would carry the risk of infecting the intracranial cavity. Moreover, it is important that the needle not deviate posteriorly or laterally, where it may pierce the internal carotid artery at its entrance to the canal of the petrosal bone. However,

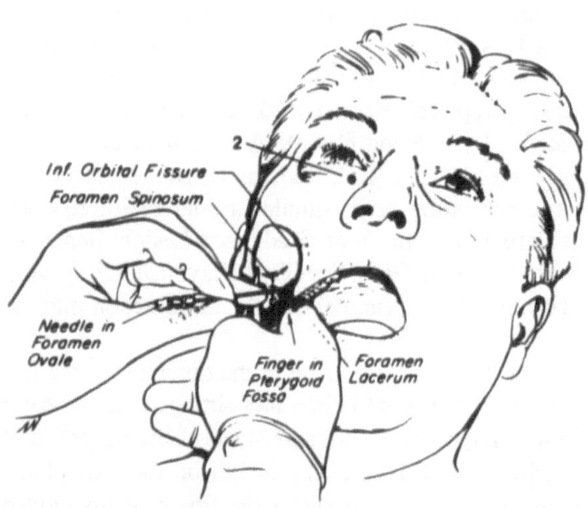

FIGURE 194. Free-hand needle placement into the foramen ovale. 2, Pupillary line (Tew *et al.*, 1978).

the danger of this complication should not be overemphasized, as was pointed out by Sweet and Wepsic (1974), who reported three cases in which the internal carotid artery was punctured without serious consequences.

The distance from the skin to the foramen ovale, and the depth to which the needle is inserted, is approximately 60–70 mm. It is recommended that one pass the needle through the middle or medial third of the foramen ovale (Sweet and Wepsic, 1974; Tew *et al.*, 1978). If this is done, then after the stylet is removed from the needle, CSF will flow out, which indicates that the target has been reached. However, there have been cases in which there was no question about the needle being accurately in Meckel's cave, but no fluid could be obtained. This may occur in basal arachnoiditis with adhesions (Vasin, 1973) as well as after previous operations on the root or injections of alcohol (Tew *et al.*, 1978). Contraction of the masseter muscle is also evidence of an accurate entry into the foramen ovale (Tew *et al.*, 1978).

It is essential for a successful operation to control the accurate insertion of the needle into the foramen ovale. This is done by taking x-ray films in two projections (Fig. 195) as well as observing the position of the needle on the image amplifier. It is recommended that AP film be used with the central ray passing from under the chin to the vertex (submentovertex x ray) (Sweet and Wepsic, 1974; Siegfried, 1977). The use of Polaroid x-ray film accelerates this procedure significantly.

The site of the foramen ovale is determined in films taken in two projections (Fig. 196). In the lateral film, the reference point is the junction of two lines: the clivus line and the line of the petrous ridge. These two lines form an almost perfect right angle. Anatomic investigations have confirmed that the foramen ovale lies at the apex of this angle (Nugent and Berry, 1974). In the AP film, the target point lies in the projection of the orbit 18 mm from the midpoint of the horizontal line passing through the middle of the internal acoustic meati and 7–8 mm medial to its lateral border (Nugent and Berry, 1974).

A simple and convenient method for fluoroscopic control of the position of the needle and the foramen ovale has been proposed by Tator and Rowed (1976). This method avoids the inconvenient position of the patient with a sharply inclined head required for taking an x-ray film with the central ray passing under the lower jaw. With this technique, the head of the patient lies in an oval depression of a special headrest fastened to the operating table. A mobile fluoroscope with an image amplifier is set at an angle of 45° to the plane passing through the lower border of the orbit and the acoustic meati. Then the head is gradually turned 35–50° to the affected side. On the image amplifier one can clearly see the foramen ovale, which is the reference point for introducing the needle.

There are several other techniques for checking the position of the needle relative to the foramen ovale. If the needle (irrespective of the mode of introduction) is pointing directly at the target point, then obviously

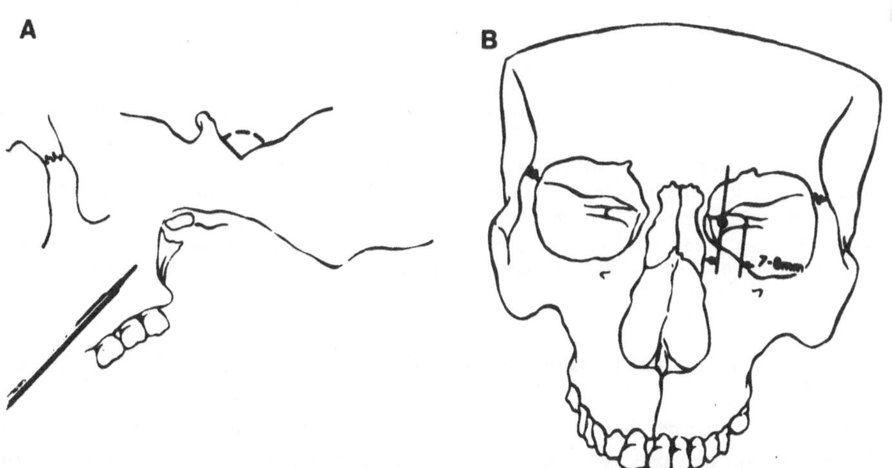

FIGURE 195. Percutaneous radiofrequency thermocoagulation of the trigeminal root for trigeminal neuralgia (Nugent and Berry, 1974). (A) Direction of the electrode to the junction of the lines along the angle the clivus and the petrous ridges (lateral view). (B) The target point lies approximately 7–8 mm medial to the lateral wall of the internal auditory meatus.

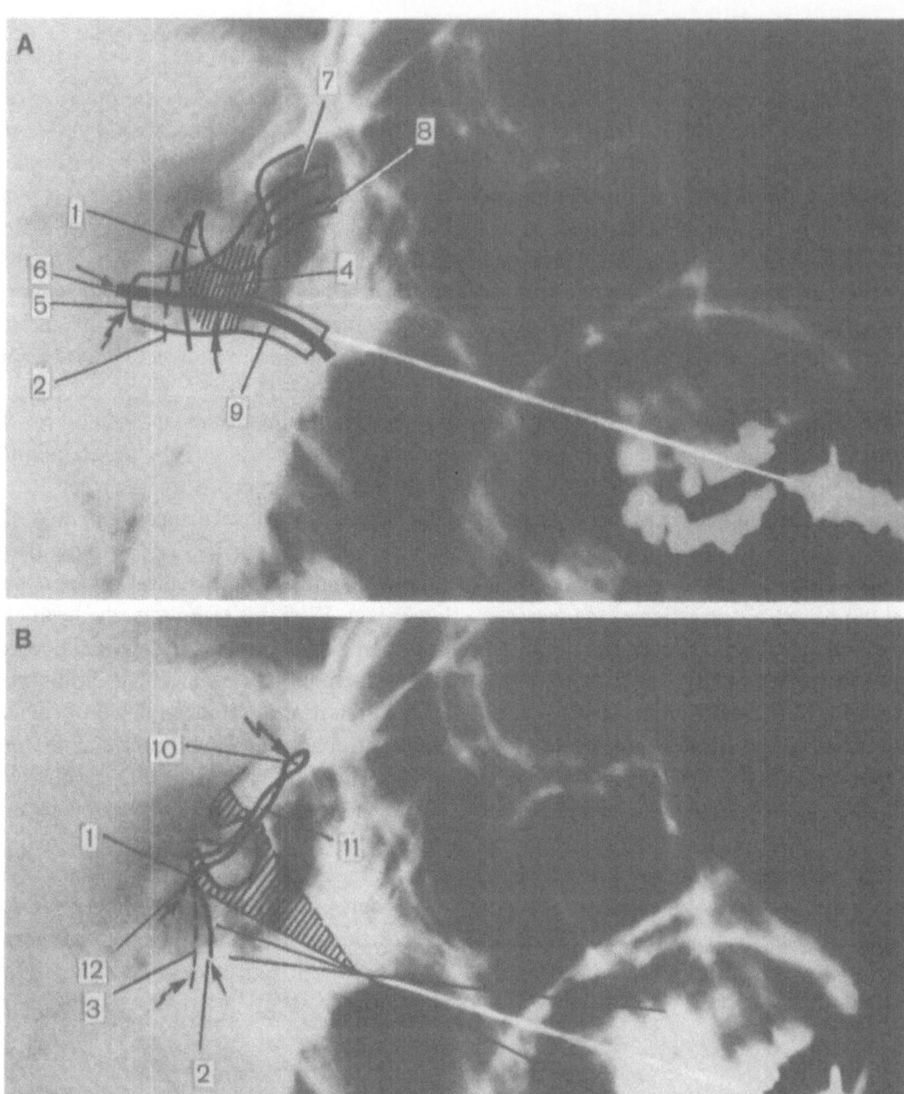

FIGURE 196. The main reference points on lateral films for percutaneous destruction of Gasserian ganglion (Onofrio, 1975). Correct direction of the needle–electrode. (A) Superimposition of Gasserian ganglion contours and three branches of the trigeminal nerve on lateral film. (B) Trajectory of the needle–electrode for destruction of each branch of the nerve. To avoid injury to other structures, the needle should not be introduced in the shaded zone. 1, Dorsum sella; 2, clivus; 3, petrosoclinoid ligament; 4, Gasserian ganglion; 5, sensory root; 6, motor root; 7,8,9, I,II,III branches of the trigeminal nerve; 10, fissura orbitalis superior; 11, III nerve; 12, IV nerve.

no problem should arise. If the needle in one of the films is seen to deviate from the trajectory, its position must be corrected. The "two-needle" technique is quite frequently used for this purpose. A second needle is introduced at an angle to the first, which serves as a guide in the film for determining the angle of correction. With two needles, it is possible to shift the tip of one with respect to the other by 1 mm (Vasin, 1973).

One must also bear in mind that a needle introduced somewhat below the foramen ovale (i.e., at a distance from the base of the skull), because of an axial projection, may falsely appear to be introduced correctly. To overcome this difficulty, Visca and Bernasconi (1972) proposed the following procedure. The head of the patient is turned 15° to the side of the needle, and a second axial film is taken. If the tip of the introduced needle is still at the foramen ovale, it must be correctly placed, but if the tip has deviated from the target, its direction must be corrected.

Difficulty in reaching the foramen ovale may be caused by a number of factors, in particular, by calcification of the stylopterygoid ligament, which often occurs in old age. If the needle fails to enter the foramen ovale, another pair of films must be taken, and the direction of the needle should be corrected. In rare cases, the foramen ovale is not visualized on the involved side but can be seen clearly on the opposite side. In such a case, a point symmetrical to the visible foramen with respect to the midplane is taken as the site of the foramen. After the needle has been introduced into the foramen ovale, it should be advanced to reach the trigeminal ganglion, taking into account that the distance from the foramen to the ganglion varies from 6 to 10 mm (average 8 mm) (Tew *et al.*, 1978).

An important criterion in judging that the Gasserian ganglion has been reached is the effect of weak stimulation. The patient should experience an unpleasant (but not painful) sensation such as itching and tingling in the distribution of one, two, or all three branches of the trigeminal nerve. By manipulating the needle, the electrode tip may be positioned so that the above-mentioned sensation occurs only in the site of the neuralgic pain with the lowest current intensity.

The effects of electrostimulation of the Gasserian ganglion or the trigeminal root for controlling the position of the electrode tip have been described by many authors (Livshitz, 1973; Nugent and Berry, 1974; Onofrio, 1975; Tew *et al.*, 1978). For stimulation, a current of 50–100 Hz frequency and a square-wave impulse of 1–5 msec is applied. The criterion for direct contact of the electrode with the root is considered to be paresthesias, tingling, or mild pain in one, two, or sometimes all three branches of the trigeminal nerve. These sensations are induced by a very weak current—from 0.05 to 0.2 V; if a higher voltage is required, the needle tip is not in the root and requires adjustment. Cobb and Fung (1983) found that the therapeutic effect was related to the parameters required to stimulate the nerve root: the lower the stimulation threshold, the better the results of the operation. During stimulation, besides paresthesias, there is a red blush of the face, which helps to identify that branch with which the electrode tip has made contact (Onofrio, 1975).

It is recommended that the electrostimulation be repeated after destruction of the Gasserian ganglion to verify the effect of the operation. An increase in the voltage required to induce the initial sensory paresthesias by at least five to six times is an indication of an adequate lesion.

Certain surgeons (Nugent and Berry, 1974; Tew *et al.*, 1978) prefer to perform a selective destruction of the nerve root rather than the Gasserian ganglion (Fig. 197), for patients appear to tolerate the effects of thermocoagulation of the root better than coagulation of the ganglion (Sindou and Keravel, 1979; Sindou *et al.*, 1987). In order to destroy only one of the three branches of the root, the electrode must be advanced into the root approximately 3–4 mm. It is important to take into account that along the course of the root to its entrance into the pons, the fibers rotate so that the

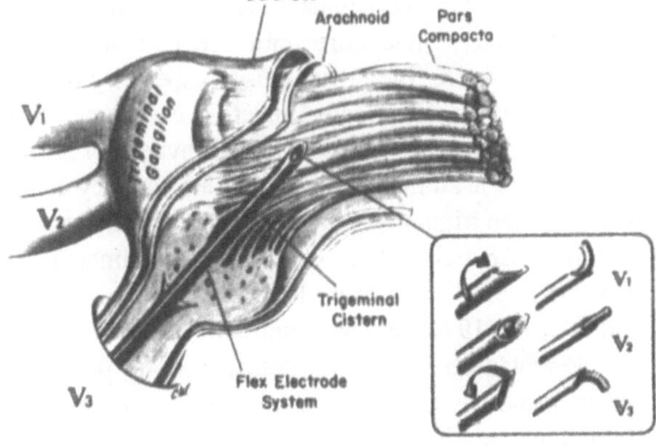

FIGURE 197. Introducing the electrode into the trigeminal rootlets. The curved electrode with its thermocouple is capable of producing lesions in any of the three divisions of the trigeminal root from a single position (right below) (Tew and Tobber, 1982).

fibers of the third branch distally lie lateral to the fibers of the first branch and proximally are lateral and caudal (Gudmundsson *et al.*, 1971). From the practical standpoint, the third branch lies closer to the surface (i.e., closer to the foramen ovale) in the root, the first branch is deep, and the second branch lies in between. The distance between branches is 5 mm (Tew *et al.*, 1978).

The next stage of the operation depends on the technique used. Either thermocoagulation or electrocoagulation of the Gasserian ganglion may be employed. A short-lasting general anesthetic such as methohexital at this stage is recommended (Siegfried and Hood, 1983). A stylet electrode having in its active tip a thermistor for controlling temperature is used for thermocoagulation. Destruction of the trigeminal nerve is achieved by raising the temperature at the electrode tip to 65–75° for 30 to 60 sec. Depending on the effect obtained, the thermocoagulation is performed one or more times.

The current parameters for radiofrequency coagulation are approximately the same as for PC—within the range of 50–110 mA and 12–20 V (Nugent and Berry, 1974) or 180–300 mA and 30–40 V (Siegfried and Hood, 1983). The coagulation is repeated several times. For the destruction of part of the ganglion or root, the usual time for applying the current is approximately 10 sec. Then the current is switched on two or three times for a few seconds. After the needle has been introduced into the Gasserian ganglion and during coagulation, one must regularly check pinprick and touch sensation on the ipsilateral half of the face.

As a result of destroying the Gasserian ganglion by any method, there should be a deep hypalgesia rather than analgesia in the corresponding facial zone. Nevertheless, total analgesia appeared in this zone in 20% of 800 operations (Nugent, 1982). If the appreciation of pain remains after the first thermocoagulation, Siegfried (1983) recommends repeating it with a slightly higher temperature until the patient feels a pinprick as touch. Tactile sensation, especially cornea sensation, as a rule, remains intact. It is assumed that analgesia without anesthesia, or with hypesthesia to a mild degree, is an indication that radiofrequency coagulation of the trigeminal nerve has interrupted conduction in the Aδ and C fibers; an effect on myelinated Aα and Aβ fibers occurs at a higher temperature (Sweet and Wepsic, 1974; Siegfried and Hood, 1983).

Percutaneous selective destruction of the Gasserian ganglion or trigeminal root is a safe and less traumatic operation easily tolerated by patients and so is especially indicated for elderly and debilitated patients. For that reason, it cannot in any way be compared to classical operations on the trigeminal nerve. After a percutaneous operation, an unpleasant numbness is quite frequently felt in the involved zone of the face. At the same time, serious complications are rare, the main one being neuroparalytic keratitis resulting from anesthesia of the cornea. The frequency of these complications, according to various authors, is not high—from 2 to 10%. Loss of the corneal reflex (in 800 radiofrequency coagulation operations) was noted only in 3% (and then in 2%) of the cases postoperatively (Nugent, 1982). Sweet and Wepsic (1974) found a postoperative frequency of corneal anesthesia in 10% of cases.

A dangerous but fortunately very rare complication of the percutaneous technique is puncture of the internal carotid artery with subsequent hemorrhage or the formation of a carotid–cavernous fistula (Sekhar *et al.*, 1979). A too deep introduction of the electrode—more than 5 mm below the clivus line—may damage the sixth, fourth, and third cranial nerves.

The period of hospitalization of a patient after this operation is usually only 2–3 days. The percutaneous technique offers a number of other important advantages. The possibility of performing it under neuroleptanalgesia enables one to test the effect on the operating table. Even when the sensation of pain disappears postoperatively, touch is preserved on the face.

A number of papers analyzing the results of percutaneous coagulation of the Gasserian ganglion or the trigeminal root have been published in recent years. In 1977 Siegfried analyzed the results of 500, and in 1981 of 1000 thermocoagulations of the Gasserian ganglion. It is interesting to compare these two papers to get an understanding of how surgical techniques were perfected in the short 4-year period. Of 500 patients aged 20 to 92 years, 416 had classical neuralgia, 35 had symptomatic, and 49 had atypical neuralgia. Complete relief in the most frequent classical form was obtained directly after the operation in 98% of the cases, and the frequency of early relapses was only 4.3%. These results were, without a doubt, considerably better than those obtained by any other surgical techniques.

In the group of atypical TN cases, the results were much less encouraging, with permanent relief of pain being obtained in only 22% of the cases. In view of this, the author concludes that in atypical facial pain, surgical treatment is not indicated.

In his next paper, the author submitted the results

of 1000 such operations. It is interesting that the oldest patient was 93 years old. Neuralgia of the second or second and third divisions of the trigeminal nerve was observed in two-thirds of the cases. The pain was on the right side in 61% of the cases. As the author emphasizes, in the 1000 operations, there was not a single instance of failure to reach the foramen ovale. In the first month after the operation in classical tic douloureux, pain disappeared in 98% of the patients, testifying to the exceptionally high success rate of the method. The frequency of pain relapses in the fortnight after the operation was 2% as compared to 4.3% in the previous paper. All the patients with relapse immediately underwent a second operation, which totally eliminated the pain. In this chapter, the author presents the long-term results of 135 cases followed up from 5 to 7 years. In this period, the relapses increased substantially to 20.7%. The majority of patients with relapses from insufficient loss of pain sensation in the neuralgic zone had a second operation with good results.

Sindou *et al.* (1987) reported the results of percutaneous radiofrequency rhizotomy of the V nerve after 609 operations. Excellent results were reported in 64% of cases, good results in 30%, no effect in 6% of cases, and recurrences in 7%.

The success of percutaneous destruction of the Gasserian ganglion and the minimal complications have been confirmed by many other authors (Onofrio, 1975; Menzel *et al.*, 1975; Sindou and Keravel, 1979; Nugent, 1982; Brandt and Wittcamp, 1983; Latchaw *et al.*, 1983; Frank *et al.*, 1985; Sonoda *et al.*, 1987). After thermocoagulation of the Gasserian ganglion, lasting pain relief occurred in 85% (Boggs, 1976), and after thermorhizotomy in 71% of cases followed up for 1 to 13 years (Brandt and Wittcamp, 1983). Percutaneous thermocoagulation of the trigeminal nerve root successfully relieved TN relapses after intracranial operations.

A large experience with radiofrequency coagulation of the Gasserian ganglion has been gained by Sweet and Wepsic (1974), who operated on 274 cases with facial pain, 214 of whom had typical TN. Directly after the operation, the pain disappeared in 90% of the cases without neurological complications or postoperative mortality. However, a follow-up of these cases from 2.5 to 6 years revealed relapses in 22% of the cases, which correlates fully with Siegfried's data presented above. Schürmann *et al.* (1972) performed electrocoagulation in 183 cases. Pain recurred in 24% of the patients requiring a second operation, after which relapses were recorded in only 7% of the cases. Similar results were reported in another study analyzing 65 TN cases in which electrocoagulation of the Gasserian ganglion was performed; pain remained in only one case (mean follow-up 1 year). Six patients with recurrences underwent a second operation, which eliminated the pain (Nugent and Berry, 1974).

These data lead one to conclude that relapses after percutaneous coagulation of the Gasserian ganglion are usually the result of an incomplete destruction. Relapses occur considerably more frequently in secondary (symptomatic) neuralgia than in idiopathic tic douloureux.

The large series of published cases indicates that destruction of the trigeminal root is approximately as effective as destruction of the Gasserian ganglion (Nugent and Berry, 1974). Tew and Tubber (1982) reported the results of percutaneous thermocoagulation rhizotomy of the trigeminal nerve in 700 typical TN cases with a follow-up of 20 or more years. Approximately one-third of these cases had undergone earlier ineffective surgical treatment. In 93% of the cases, coagulation of the trigeminal root gave an excellent or good result.

It may be concluded that percutaneous electrocoagulation of the Gasserian ganglion and of the trigeminal root gives approximately the same frequency of positive results, more than 90%. Based on a large clinical material (531 operations), Schürmann *et al.* (1972) compared the results of classical subtemporal retrogasserian resection of the trigeminal root and percutaneous electrocoagulation of the ganglion. The authors conclude that the long-term results definitely favor the second operation. There is every reason to agree with Siegfried (1977) that percutaneous destruction of the Gasserian ganglion is, at the moment, the method of choice for the treatment of TN.

In conclusion, it is necessary to mention briefly the use of percutaneous electrocoagulation in symptomatic TN. It has long been known that typical TN frequently occurs in multiple sclerosis, as first described by Oppenheim in 1923. It has been estimated that the frequency of trigeminal neuralgia in MS is 300 times greater than that in the general population (Cendrowsky, 1962). The combination of the two disorders occurs in 2% of the cases (Rushton *et al.*, 1981). Percutaneous thermocoagulation of the trigeminal ganglion, as described above, proved to be as effective a treatment of tic douloureux in MS as of idiopathic TN. Of 25 MS cases operated on by Siegfried and Lindenberger (1979), a long-term good result was observed in 96%. A lasting effect in MS patients was reported by other authors (Sweet, 1976; Brett *et al.*, 1982).

Good results were obtained after percutaneous thermocoagulation of the Gasserian ganglion in symptomatic TN caused by Paget's disease or by malignancy of the tongue, pharynx, jaws, or base of the skull (Siegfried, 1977, 1981).

7.3. Percutaneous Medullary Trigeminal Tractotomy

Trigeminal or medullary tractotomy aims to interrupt the descending tract of the trigeminal nerve as it passes in the medulla oblongata to the cervical cord. The second neuron or ventral (secondary) ascending tract of the fifth nerve relaying impulses to VPL originates in the nucleus of the descending tract (Taren and Kahn, 1962). Section of this tract by an open approach was first carried out by Sjöqvist in 1937. Since then, it has been extensively performed in neurosurgical clinics for the surgical management of TN.

We shall not dwell on the technique and results of this operation or its subsequent modifications (Kunc, 1965) since the Sjöqvist operation is the classic. Nevertheless, it is important to note the precision and accuracy of this operation, which involves making a small incision in the medulla oblongata at a level somewhat lower than obex to interrupt the descending tract of the fifth nerve. This tract has a specific topographical arrangement. The fibers of the first division of the fifth nerve lie ventrolaterally and descend to the C_2 segment. The fibers of the third division lie more posteromedially at the level of, or a bit higher than, the obex (Fox, 1973).

Precision is required in this operation because the incision above the obex may damage the restiform body, resulting in ataxia of the ipsilateral extremities, and section of the tract considerably lower than obex may damage the sensory fibers of the seventh, ninth, and tenth nerves.

In spite of the technical difficulties connected with the "closed" stereotactic operation on the medulla oblongata, Crue *et al.* (1970) and Hitchcock (1973) have described a technique for stereotactic percutaneous trigeminal tractotomy that was soon adopted by other surgeons (Fox, 1973, 1976a; Schvarcz, 1975, 1978; Grunert *et al.*, 1986).

The operation is performed under local analgesia or neuroleptanalgesia with the patient in a sitting or prone position. By using a specially made stereotactic device such as that Hitchcock designed for stereotactic operations on the spinal cord (Section 5.3), the spinal trigeminal tract may be approached. The head of the patient must be fully flexed on the chest in order to stabilize the cervical spinal cord. It is essential to have good visualization of V_4, aqueduct, obex, ventral and dorsal surfaces of the caudal medulla oblongata, and the upper part of the cervical spinal cord. This is achieved by introducing into the cisterna magna 1 ml of iophendylate emulsified in 1 ml of CSF (Fox, 1973) or a water-soluble contrast medium (iothalamate or dimeglumine iocarmate) (Schvarcz, 1978).

The main stages of the operation are similar to PC (Section 5.3.2) but with several important differences. The guide needle is introduced through the skin at the midline between the posterior border of the foramen magnum and the posterior arch of atlas. After passing through the atlantooccipital membrane and cisterna magna, the electrode should reach the target point. The descending tract can be destroyed at two levels, at the obex in medulla oblongata (Fox, 1973) and at the first cervical segment of the spinal cord (Hitchcock, 1969a,b; Schvarcz, 1975, 1978).

The most important stage of the operation is to find the target point and calculate its coordinates. An important reference point is the odontoid process, the midline of which, in the AP film, corresponds to the midline of the spinal cord. The descending trigeminal tract at the level of the first cervical segment is 3–5 mm ventral to the posterior surface of the spinal cord and 6–8 mm from the midplane. The stereotactic coordinates of the target point are the following: 6–8 mm from the midplane, slightly caudal to the obex, and 4 mm anterior to the floor of V_4 (Fox, 1976a) (Fig. 198).

The coordinates of each of the three divisions of the trigeminal nerve in the descending tract have also been established. The first branch is at a distance of 5 mm ventral to the posterior surface of the spinal cord and 7.5 mm lateral to its midplane; the third branch of the trigeminal nerve is 3 mm ventral and 6 mm lateral, and the second branch lies between the first and third branches (Schvarcz, 1977). Fox (1973) emphasizes that the closer the electrode tip is to obex, the more reliable is the position of the fibers of the mandibular branch of the fifth nerve as well as the seventh, ninth, and tenth nerves.

The introduction of the 0.5- to 0.6-mm electrode with a sharp bare tip through the needle is controlled by measuring impedance, making it possible to establish

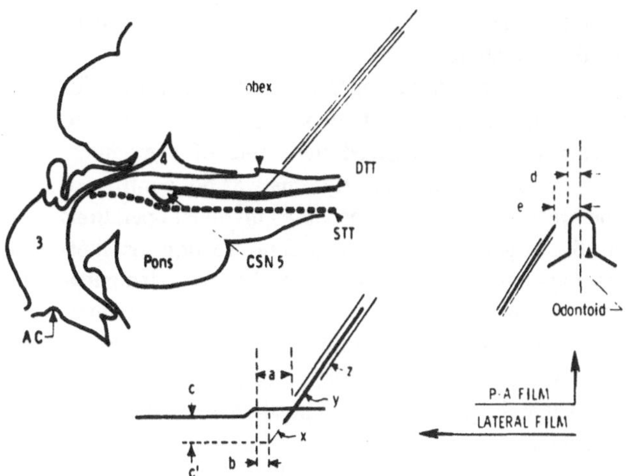

FIGURE 198. Placement of the electrode during percutaneous medullary trigeminal tractotomy (Fox, 1976). 3, V_3; 4, V_4; AC, anterior commissure; CSN 5, trigeminal nucleus; DTT, descending trigeminal tract; STT, spinothalamic tract; x,y,z, electrode, insulation, and guide needle, respectively; a, distance from the obex to guide needle tip; b, distance from the obex to electrode tip; c–c′, distance from the electrode tip to the plane of the floor of the fourth ventricle (c); d, distance from the odontoid midline to the electrode tip; e, distance from the odontoid midline to the guide needle tip.

the moment of contact with the spinal cord and the penetration of the electrode tip. That the descending tract has been reached is determined by monopolar electrostimulation with a weak current. A sensation in the distribution of one of the branches of the trigeminal nerve is confirmation (Crue *et al.*, 1970; Schvarcz, 1978). At the moment when the electrode reaches the descending tract, even without stimulation, the patient feels a sharp pain in the ipsilateral side of the face.

If paresthesias are felt in the ipsilateral half of the body during stimulation, the electrode tip is in the posterior columns of the spinal cord, and it must be replaced anteriorly and laterally. Similar sensations in the contralateral half of the body indicate that the electrode has been inserted in the spinothalamic tract deeper than the trigeminal descending tract, and in this case, the electrode tip must be withdrawn somewhat posteriorly. Somatosensory EPs are also used for verification of the trigeminal tract.

Radiofrequency electrocoagulation is usually employed for destruction of the descending tract. This is done fractionally, relying on induced facial sensations. A combined electrode has been designed that permits the recording of biopotentials, stimulation, and, finally, destruction of the descending tract (Schvarcz, 1975). The focus, which should include the spinal trigeminal nucleus and the descending tract of the fifth nerve, should be not more than 3 mm in diameter. Analgesia in one half of the face as well as in dermotomes C_2–C_3 appears immediately after the thermocoagulation. Since the descending fibers not only of the fifth but also of the seventh, ninth, and tenth nerves pass in the ipsilateral descending trigeminal tract, interruption of this tract causes analgesia in the face as well as in the oral cavity, rhinopharynx, and larynx. Touch sensation of the face as well as corneal sensation remain intact.

Percutaneous stereotactic medullary trigeminal tractotomy has been successfully employed not only in typical TN but also in other severe pain syndromes of the face (tumors of the face and neck, postherpetic neuralgia, anesthesia dolorosa). Fox (1973) reported a significant and lasting benefit from this operation in a small group of patients with malignant tumors of the head and neck.

Schvarcz (1975) reported on the results of 22 such operations (16 cases with pain syndromes of a benign nature and six cases with malignant neoplasms), employing the Hitchcock stereotactic apparatus. In the first group, in which the majority were cases of idiopathic TN, the results were excellent in all but one case. Similar favorable results were obtained in the second group of cases with malignant tumors. The follow-up was from 3 to 18 months, and the author emphasizes that no complication or prolonged side effect was observed. Touch sensation of the face and the corneal reflexes remained intact.

Postherpetic facial pain represents a peculiar trigeminal pain syndrome. Facial pain comprises about 15% of all cases of postherpetic neuralgia. Pain such as dysesthesias with hyperpathia is a serious and frequent complication, occurring in approximately 20% of the cases after facial herpes zoster (Siegfried, 1982). As a rule, such pain is localized to the innervation zone of the first division of the trigeminal nerve. Pain in the face occurs at various times after herpes zoster in those areas where sensation has been lost. Pain usually becomes very intense and rarely responds to conservative treatment. There can be no doubt that such pain is of the deafferentation type.

A great number of operations have been tried for the surgical management of postherpetic facial pain. Earlier papers emphasized the value of sympathectomy; however, subsequently, this was not confirmed. The greatest number of procedures involved operations on the trigeminal nerve including section or thermocoagulation of its major root or the Gasserian ganglion. All these operations inevitably ended in failure, although at times they eliminated hyperesthesia in the painful area (Siegfried, 1982). This convinced surgeons that the pathogenesis of postherpetic pain differs in principle from trigeminal neuralgia.

It is presumed that not only the nerves and sensory ganglia but also the entrance zone of the spinal roots and cells of the posterior horns are affected in the herpetic process. In view of this, it is assumed that in order to eliminate postherpetic facial neuralgia, it is necessary to destroy the spinal trigeminal nucleus and its descending tract. This assumption cannot be considered to be verified even though patients with postherpetic neuralgia experienced relief after stereotactic trigeminal tractotomy, first in three cases (Hitchcock and Schvarcz, 1972) and then in six cases in which the ophthalmic division of the fifth nerve was involved (Schvarcz, 1977a).

In a later paper, Schvarcz (1978) presented the results of stereotactic destruction of the spinal nucleus or descending tract of the fifth nerve in 100 cases (69 with benign facial pain and 31 with malignant tumors). A satisfactory postoperative effect was obtained in 87.5% of the cases with postherpetic pain, in 72% with other dysesthesias, and in 57% with anesthesia dolorosa. It should be noted, however, that Fox (1973) did not obtain such good results in a small series of cases with postherpetic neuralgia, in which improvement was only temporary.

Complications after trigeminal tractotomy are comparatively rare. Transitory ataxia as a result of injury to the dorsal spinal cerebellar tract at its input to the restiform body has been described (Fox, 1973).

Summing up, one may conclude that percutaneous trigeminal tractotomy is a more complicated and somewhat less effective operation than retrogasserian neurectomy for typical idiopathic TN. However, in the atypical (in particular, postherpetic) neuralgia and in secondary TN caused by malignancy, medullary tractotomy gives better results than percutaneous rhizotomy.

7.4. Stereotactic Midbrain Tractotomy

One of the effective operative methods in the treatment of certain cephalic pain syndromes is stereotactic midbrain tractotomy—mesencephalotomy (Section 5.2.5a), the open technique for which was elaborated by Walker (1942a). The aim of this operation is to interrupt the trigeminothalamic pathway at the midbrain level, producing a limited destruction of PAG behind CP. The so-called medial pain-transmitting system passes through this zone (Chapter 1, Section 4). At midbrain level (55 mm behind CP), the spinothalamic and trigeminothalamic pathways pass compactly at a distance of 1–2 mm from each other; the former lies medially and slightly below the latter. Walker (1942a) emphasized that after open mesencephalotomy analgesia involves half of the face; however, it is not always possible to produce analgesia in the oral cavity.

Midbrain tractotomy, the technique for which is described in Section 5.2.5a, is performed under local anesthesia on the side contralateral to the facial pain. In bilateral pain, it is possible, but because of the resulting deafness, inadvisable, in principle, to perform consecutive operations on both sides (White and Sweet, 1969).

There is comparatively little experience in performing stereotactic mesencephalic tractotomy for facial and head pain. The first stereotactic operation in the world was done by Spiegel and Wycis in 1947 on a patient suffering from severe facial pain, which was not relieved by retrogasserian section of the fifth nerve. The authors performed a combined operation—coagulation of the spinothalamic tract in the midbrain and destruction of the dorsomedial nucleus of Th. The authors reported in 1967(a) that there was practically total disappearance of pain for the entire period of follow-up (18 years). They emphasize that the dorsomedial nucleus was destroyed in order to interrupt the frontothalamic fibers, which play a definite role in the emotional component of pain sensations (Chapter 1, Section 1.1). To obtain a lasting effect, it is necessary to include the medial Th in the lesion. Orthner and Roeder (1966) reported encouraging results of stereotactic mesencephalotomy in 11 TN cases.

In malignant tumors of the head, neck, and throat, severe pain may frequently occur as a result of the involvement of several cranial nerves (fifth, ninth, tenth) and cervical roots, often on both sides. The usual pain operations (coagulation of the roots of the fifth and other nerves, tractotomy in medulla oblongata, and others) in these patients are not sufficiently effective.

In eight cases with severe pain in the head, neck, and rhinopharynx resulting from malignancy, Nashold (1972) performed unilateral stereotactic dorsomedial and mesencephalic tractotomy, making a lesion between the Sylvian aqueduct and the lateral spino- and trigeminothalamic tracts. Pain completely disappeared postoperatively in all cases (mean follow-up 9 months).

One of the shortcomings of this operation, which apparently is the main reason why it is not extensively used, lies in the troublesome postoperative dysesthesias, the pathogenesis of which remains unclear. It is noteworthy, however, that after a stereotactic operation, this complication is observed much less frequently than after open mesencephalotomy (Section 5.2.5a). According to Nashold (1972), dysesthesias were observed postoperatively in only one of 23 cases. Among the other complications, mention should be made of paresis of upward gaze and diplopia. Mesencephalotomy may be considered indicated only in atypical or secondary TN cases.

7.5. Destruction of Thalamic Nuclei

Singular attempts at stereotactic destruction of VPM, mainly in atypical TN, have been described in the literature. The stereotactic coordinates for destruction of VPM are the following: 9 mm caudal to the midpoint of LI or 3–4 mm anterior to CP, from 2 mm above to 2 mm below LI, and 8–10 mm lateral to the midplane. This operation did not give satisfactory results and is not practiced at present.

In a limited series of cases, it was found that in atypical facial pain, stereotactic pulvinotomy may yield a good result (Siegfried, 1977).

7.6. Stimulation Methods

Just as in pain at other locations, stimulation methods described in Section 6.3.2 have been intensively developed in recent years for the management of TN. In principle, the technical equipment and technique of these operations are the same as in other pain syndromes, yet they have a number of important peculiarities. Since, as described above, percutaneous thermocoagulation of the Gasserian ganglion or fifth nerve root is a highly effective and reliable method for the treatment of classical TN (tic douloureux), stimulation methods have been used mainly in so-called atypical neuralgia—in postherpetic facial pain, in anesthesia dolorosa after section of the fifth nerve, and in secondary neuralgias resulting from malignant tumors.

In ten cases with postherpetic neuralgia of the first division of the trigeminal nerve, Siegfried (1982) used chronic electrostimulation of VPM with implanted monopolar electrodes. Stimulation was performed several times a day for 5 min (on average) with a current of from 33 to 195 Hz. In a follow-up after 8 to 17 months, the results were very favorable: in five cases, excellent, in three, good, and in two cases, no effect.

In postherpetic TN, a significant effect from chronic stimulation of the lemniscus medialis was noted in several cases (Mundinger, 1977). However, a later study (Mundinger and Neumüller, 1982) reports that chronic stimulation of deep-seated cerebral structures in three cases of postherpetic and atypical trigeminal neuralgia produced no effect. Hosobuchi (1980) also noted that stimulating Th sensory nuclei was successful in only one of three cases.

Until recently, it was presumed that there was no effective treatment for anesthesia dolorosa, which occurs at times after section of the fifth nerve root or destruction of the trigeminal ganglion. One of the few successful attempts to treat this severe complication was made by Hosobuchi *et al.* (1973), who, by chronic stimulation of VPM, achieved considerable pain relief in four of five cases. In two cases of anesthesia dolorosa, the facial pain disappeared as a result of chronic stimulation of VPM (follow-up from 5 to 11 months) (Roldan *et al.*, 1982).

Meyerson and Hakansson (1978, 1986) published the first encouraging results of chronic stimulation of the Gasserian ganglion in TN. After an open subtemporal extradural approach, a bipolar electrode was applied directly to the ganglion and connected to a receiver implanted subcutaneously. Instead of pain, this stimulation induced a tingling sensation in one, two, or (more rarely) all three divisions of the trigeminal nerve. The technical difficulties of the open method prompted the development of a technique for percutaneous stimulation of the Gasserian ganglion by introducing an electrode through a needle. This method yielded good results in five TN cases (Meglio, 1983). Recently, the author (Meglio, 1984) reported the use of a system implanted in the Gasserian ganglion that consisted of a platinum–iridium monopolar electrode and a radiofrequency receiver. Good results were obtained in 12 cases of atypical TN. Limited benefits from percutaneous chronic stimulation of the trigeminal root in TN were also reported. A simplified, less invasive technique to implant a chronic stimulating electrode in the Gasserian ganglion and to bring the distal end of the electrode from the cheek to the neck was described by Spaziante *et al.* (1986).

Reports on the use of PCS (5.4.2) in the management of TN have also appeared in the literature. For this purpose, a bipolar electrode is introduced percutaneously into the subarachnoid space through the C_1–C_2 interspace for chronic stimulation. The present data on this method should be considered preliminary (Hoppenstein, 1975; Shelden *et al.*, 1975).

Reports on the therapeutic effects of stimulating skin trigger zones of the face in TN have been published recently. Such stimulations are performed several times a week using a 150- to 200-Hz current. Pain disappeared completely in almost one-half of the cases with primary TN and diminished notably in many others (Borromei *et al.*, 1982). Similar cutaneous stimulation for trigeminal pain produced a positive effect in only one-third of the cases (Nathan and Wall, 1974; Eriksson *et al.*, 1984).

7.7. Microvascular Decompression

In 1934, Dandy discovered that TN is frequently associated with various pathological processes in the posterior fossa. Of 215 TN cases that Dandy operated on by the posterior fossa approach, compression of the fifth nerve root by an artery was found in 66, by the petrosal vein in 30, by tumors in 12 cases, by arterial aneurysm in six cases, and by angioma in five cases. All together, a cause for the disorder was found in the posterior fossa in about 60% of the cases. It is noteworthy that in a subtemporal approach to the trigeminal nerve root, the pathological processes mentioned above could not be detected.

Decompression of the trigeminal root is a classical open operation for TN management proposed by Taarnhøj (1954). Recently (1982), the author published a 30-year follow-up review of 259 patients operated on by this technique. Of these patients, 228 were operated on through a subtemporal approach, and 31 by a posterior fossa route. There was a high rate of recurrence in about one-third of the operated cases, but the rate was much lower when the suboccipital approach was used than with the subtemporal approach.

In 1959 Gardner and Miklos proposed an operation involving vascular decompression of the trigeminal nerve root in cases of TN, and in 1962 he proposed an analogous operation on the facial nerve in the treatment of hemifacial spasm (Section 12). At first, Gardner performed the operation on 18 cases of TN in which, after surgery on the fifth nerve root by the subtemporal approach, the neuralgia had relapsed. Compression of the trigeminal nerve root by arterial loops was found in six cases, and other lesions in the pontine cistern that were undoubtedly the cause of the neuralgia in six cases; in the remaining six, no pathology was found. If the root was compressed, Gardner separated it from the arterial loop and interposed a piece of Gelfoam (quoted by Rand and Gardner, 1981). Section of the trigeminal root was performed only if it was impossible to separate it from the compressing artery.

The interest of neurosurgeons in this operation was rekindled by Jannetta, who, in a series of interesting studies (1967, 1976, 1977b, 1979a,b, 1981b), demonstrated that a principal cause of TN, which for many decades had been called "essential" or "idiopathic," was vascular compression of the nerve root and that the elimination of this compression leads to a permanent cure of the neuralgia. It is noteworthy that this cause is found much more frequently in recent years; moreover, compression of the root at the pons is of particular significance for the development of neuralgia.

Jannetta (1977b) found vascular compression of the fifth nerve in 88 TN cases, which involved compression or stretching of the nerve by the superior cerebellar artery in 82, by the anterior cerebellar artery in two, and by a vein in two. In a later series with almost twice as many cases, Jannetta (1979a), at operation, found vascular compression of the trigeminal root in 141 of 161 cases (88%). Compression of the root by an arterial loop was found in 100 cases, by a vein in 19 cases, and by both artery and vein in 22 cases. In 13 cases various tumors in the posterior fossa involved the fifth nerve root.

Apfelbaum (1977) found vascular compression of the root in 55 operated cases (artery in 35, vein in 13, and both artery and vein in seven). Oberbauer *et al.* (1982) noted compression of the nerve root by one of the cerebellar arteries in 83% and by the vein in 12% of the cases. Richards *et al.* (1983) found vascular compression of the nerve root in 46 of 52 operated TN cases. The most frequent cause of compression (25 cases) was a loop of the superior cerebellar artery. Neuralgia may be caused by compression of the descending spinal tract of the trigeminal nerve in medulla oblongata by a loop of the posterior inferior cerebellar artery (Sunder-Plassmann *et al.*, 1971). In very rare cases, TN may result from compression of the nerve by the elongated and dilated basilar artery (megabasilaris).

Anatomic investigations by Haines *et al.* (1980) have confirmed the argument in favor of vascular compression of the fifth nerve as the main cause of TN. At 20 autopsies of persons who had not suffered from TN, arterial compression or stretching of the nerve was found in only four cases and by the vein in another four. In 40 TN cases, surgical exploration of the posterior fossa revealed compression of the nerve by the artery in 31 cases and by the vein in another eight. However, contact between the fifth nerve root and the artery does not always indicate root compression. For instance, Van Loveren *et al.* (1982) found such contact in 82% of 50 operated cases; however, the authors consider that there was significant root compression in only 46% of the cases.

In subsequent years, microvascular decompression of the trigeminal nerve has been perfected by several neurosurgeons (Jannetta, 1976; Apfelbaum, 1977; Rhoton, 1978; Alksne, 1980; Waga *et al.*, 1979; Rand and Gardner, 1981; Ferguson *et al.*, 1981; Oberbauer *et al.*, 1982; Richards *et al.*, 1983; Barbe and Alksne, 1984; Szapiro *et al.*, 1985; Meglio *et al.*, 1986a; Kunze and Steiner, 1987; Sindou *et al.*, 1987) who operated with a surgical microscope. In order to approach the fifth nerve root, a small (3 × 3 cm) retromastoid craniotomy is made through a paramedian skin incision approximately 6 cm long. Lately, there has been a tendency to reduce the size of the craniotomy even more. Fukushima (1982) performs microvascular decompression of the fifth nerve through a bone opening only 1.5 cm in diameter. The semiprone position allows good exposure of the trigeminal nerve root from the pons to the border of the pyramid with minimal retraction of the cerebellar hemisphere and reduces the risk of air embolism. The bone is resected until the transverse and sigmoid sinuses are exposed. The dura mater is opened with a Y-like incision to the junction of these sinuses in the superolateral corner of the craniotomy. After retraction of the superolateral surface of the Cer hemisphere with a self-retaining retractor, the superior petrosal vein is exposed. The vein bridges from the superior cerebellar edge to the petrosal sinus at the junction of the tentorium and the petrous bone. This petrosal vein should be divided after coagulation and cut to expose the fifth nerve root, which freed from the arachnoid is now visible. Besides this subtentorial approach, successful attempts have been made to decompass the trigeminal root by a transtentorial approach (Klun, 1981).

As mentioned above, compression of the entry zone of the trigeminal root is most frequently caused by an enlarged caudal loop of the superior cerebellar or anterior inferior cerebellar arteries. As a rule, this loop is easily separated from the root, and a piece of Teflon, Dacron, or Gelfoam is inserted between them. In cases in which it is difficult to separate the root, one of the cerebellar arteries causing compression or irritation of the root is dissected.

This operation gives good long-term results, and in the majority of cases neuralgic attacks cease. The frequency of postoperative relapses is minimal (Jannetta, 1976, 1979a,b, 1981b; Rhoton, 1978). Without a doubt, the essential advantage of microvascular decompression is that after surgery, the function of the trigeminal nerve remains completely intact, so that anesthesia dolorosa does not develop. Nevertheless, complications may occur after this operation, although they are rare. Hematomas or cerebellar infarctions developed postoperatively in nine of 161 cases, and impaired function of the eighth nerve in 14 cases.

Several papers devoted to a comparison of the results of percutaneous high-frequency coagulation of the Gasserian ganglion (see below) and microvascular decompression in typical TN have recently been published. Such comparisons have led to contradictory conclusions. According to Apfelbaum (1977), the immediate relief of pain is approximately the same (88% and 96%, respectively); however, the frequency of TN relapses after the first method was 13%, and after the second only 5%.

A similar comparison of the results of percutaneous trigeminal rhizotomy and microvascular decompression has been presented by Ferguson *et al.* (1981). The first method was used in 55 cases with a mean follow-up of 30 months. In this group, 54% were cured, 42% relapsed, and pain continued after surgery in 4%. Comparable results after decompression (mean follow-up 28 months) were 71%, 17%, and 12%, respectively. Judging from these figures, vascular decompression yields significantly better results; however, it should be noted that the frequency of relapses after percutaneous rhizotomy in this report is substantially higher than in the series published by other authors.

Van Loveren *et al.* (1982), on the basis of many years of postoperative follow-up, also compared the results of percutaneous stereotactic rhizotomy in 700 cases and microvascular decompression of the trigeminal root in 50 cases. This comparison revealed that the results of both operations are approximately identical. After rhizotomy, the results were assessed as excellent in 61%, good in 13%, satisfactory in 5%, and poor in 1%, whereas relapses occurred in 20% of the cases. After microvascular decompression, an excellent result was reported in 68%, good in 12%, satisfactory in 4%, and relapses in 12% of the cases. On the basis of these data, the authors logically conclude that since percutaneous rhizotomy is less traumatic than exploration of the posterior fossa and can easily be repeated if pain should recur, stereotactic rhizotomy remains the operation of choice for TN.

The long-term results (average 43 months) in 37 TN cases after microvascular decompression were analyzed by Barbe and Alksne (1984). Trigeminal pain disappeared completely in 9% of the cases in which microvascular decompression was the first operation and only in 43% of the cases in which it was performed after section (in the majority of cases—percutaneous coagulation) of the fifth nerve root. This study also demonstrated the importance of the duration of the disorder to the results of the microvascular operation. In neuralgias lasting 9 years and more, a good result was obtained in only 42%, and in patients with a shorter duration, in 88% of the cases.

Swanson and Farhat (1982) proposed combining microvascular decompression with partial rhizotomy of the sensory root of the nerve. In 14 cases with classical tic douloureux, compression of the root by an artery was found in 12 and by a venous trunk in two cases. After separation of the root from the compressing vessel and insertion of a piece of Teflon between them, a small cut approximately 1 mm deep was made with microscissors in the posterolateral part of the sensory root without damaging the intermediate bundles. This incision was made 3 mm from the site of entry of the root into the brainstem. In a follow-up of 2 years, the neuralgia was completely relieved in all cases, without any disturbance of sensation or the occurrence of deafferentation pain.

In approximately 5% of TN cases, on exploring the posterior fossa, tumors of the cerebellopontine angle (neurinomas, cholesteatomas, meningiomas, and others) not diagnosed before the operation are discovered.

Since it is practically impossible to diagnose vascular compression before surgery even with the aid of angiography, serious difficulties arise in choosing the surgical treatment. Nevertheless, microvascular decompression is one of the best methods for the management of typical TN. In recent years many neurosurgeons have abandoned retrogasserian neurectomy and perform only microvascular decompression of the trigeminal root.

7.8. Microcompression of the Gasserian Ganglion

Compression of the trigeminal ganglion was known long ago. A thin instrument was passed through a small incision above the upper teeth into Meckel's cave. Then the Gasserian ganglion was compressed by rotating the instrument. Encouraging results using this method were reported by Ielasic (1959).

Recently, a simplified and less traumatic transcutaneous technique was used by Mullan and Lichter. This method called microcompression of the Gasserian ganglion was performed as follows. Under general anesthesia and with fluoroscopic control, a needle was introduced into Meckel's cavity. After that a thin Fogarty catheter W4 was introduced through the needle in the cavity, and the small balloon on its tip inflated with 0.7–0.8 ml of iothalamate. After a few minutes of trigeminal ganglion compression, the catheter was withdrawn.

The results of this relatively simple procedure are comparable with those of other trigeminal operations. For example, of 72 cases operated on by this technique, there were 15% recurrences, 6% significant dysesthesia, 10% mild dysesthesia, 3% significant hypesthesia, and 21% mild hyperesthesia (follow-up of 6 months to 6 years) (Mullan and Lichter, 1985).

Esposito *et al.* (1985) performed percutaneous microcompression of the Gasserian ganglion in 50 cases with essential TN. The procedures were done under neuroleptoanalgesia and compression time was 5 min. Eighteen patients did not receive any relief and underwent a second compression. A good effect was achieved in 7 of those cases, resulting in 39 of 50 cases of relief, a success rate of almost 80%.

Meglio *et al.* (1986a) used the technique in 92 cases affected by essential TN. Ninety percent of the patients had immediate pain relief, but at two years of follow-up only 60% of the cases were pain-free.

7.9. Endoscopic Dissection

Endoscopic visualization of the cerebellopontine angle has been developed in recent years. Through a small (3- to 4-cm) incision in the retroauricular area with the patient in prone position, a small (no more than 1.5 cm in diameter) burr hole is made, and the dura mater is opened in Trautmann's triangle, which is bordered by the sigmoid sinus laterally, the superior petrosal sinus rostrally, and the vertical semicircular canal medially (Oppel and Mulch, 1979). Then an endoscope with a diameter of approximately 4 mm is introduced into the incised dura mater and clearly visualizes many structures of the cerebellopontine angle (the fifth, seventh, eighth, ninth, tenth, and 11th nerves, jugular foramen, internal acoustic meatus, arteries, veins, etc.). During the endoscopic examination, one can distinguish the sensory and motor roots of the trigeminal nerve. Compression of the root by an arterial loop or vein is well visualized.

A section of the trigeminal nerve root may be performed through the endoscope. In one case of severe facial pain as the result of a carcinoma in the upper jaw, a good effect was obtained after endoscopic section of the sensory root of the trigeminal nerve as well as of the ninth and tenth nerve roots (Oppel *et al.*, 1981).

7.10. X-Ray Irradiation

In conclusion, it is necessary to mention briefly one other nontraditional method of treating TN, which has so far not been extensively employed and consequently cannot be fully assessed. Leksell (1971b) reported the long-term follow-up of stereotactic irradiation of the Gasserian ganglion and fifth nerve root using a large dose of x rays in two TN cases. The overall dose of irradiation by the rotation technique using the Leksell stereotactic apparatus was 1650–2200 r. Attacks of TN disappeared completely soon after the irradiation. Eighteen years after such a bloodless operation, both patients, already elderly, were considered totally cured and with retained sensation in the affected half of the face.

A linear accelerator with stereotactic centering of the beam of particles on the Gasserian ganglion has been used for treating TN (Kuroda *et al.*, 1972). The small number of cases has not made it possible to draw conclusions about the efficacy of this technique.

Therefore, it may be concluded that at present there are several methods for surgical management of TN. Depending on the form and nature of the disorder, the choice of operative technique must be strictly individualized. The most effective and reliable operations are percutaneous stereotactic destruction of the Gasserian ganglion or the trigeminal nerve root or microvascular decompression of the trigeminal root.

8. Neuralgia of the Glossopharyngeal Nerve

Glossopharyngeal neuralgia was first described by Weisenburg (1910) in a patient with a tumor of the cerebellopontine angle. The name "glossopharyngeal neuralgia" was proposed by Harris in 1921. White and Sweet (1969) suggested naming this disorder "vagoglossopharyngeal neuralgia" because of the important participation of the vagus in the pathogenesis of the disorder.

The disease is manifest by attacks of severe burning, shooting pain in the innervation zone of the ninth nerve in the external acoustic meatus, in the posterior half of the tongue, in the throat, larynx, soft palate, tonsil, and the area of the lower jaw. At times, the pain may irradiate to the face, upper jaw, orbit, neck, shoulder, or precordium. In pain paroxysms, ptosis of the upper eyelid on the involved side frequently occurs; attacks may be accompanied by syncope or epileptic states. Paroxysms of pain may frequently appear without visible cause but are usually provoked by chewing, swallowing food, especially cold liquids, or talking and are often accompanied by excessive salivation and lacrimation from the ipsilateral eye. Attacks usually last minutes and sometimes hours and are frequently repeated as many as several dozen times a day. At night they occur rarely.

Neuralgia of the glossopharyngeal nerve (NGN) is a rare disorder. Dandy (1927) described four cases of NGN in 450 cases of TN. Approximately the same ratio—1 : 1000—was noted in subsequent studies (Bohm and Strand, 1962; Rushton *et al.*, 1981). In the majority of cases, it is not possible to establish the etiology of NGN, and it is usually termed idiopathic or essential. At the same time, this disorder has been repeatedly described in tumors at the base of the skull or throat affecting the ninth nerve. At times, NGN may be caused by ossification of the stylohyoid ligament, which exerts mechanical pressure on the ninth nerve (Graf, 1959). Bilateral NGN has been reported in 12% of the cases (Rushton *et al.*, 1981). At times, NGN is combined with TN, approximately in 10% of the cases (Rushton *et al.*, 1981).

The disorder usually begins in people over the age of 40–45 years and affects men and women with approximately equal frequency. It is characterized by periods of aggravation and remission with attacks disappearing for several months and sometimes for many years (Rushton *et al.*, 1981). Conservative treatment of

NGN is not very effective, but in some cases carbamazepine provides relief.

The first attempt to treat NGN surgically was undertaken by Sicard and Robineau (1920), who made an extracranial section of the ninth nerve in three cases. Adson (1924) performed a similar operation in four cases. Pain disappeared directly after the operation, but relapses soon occurred in all cases. In 1927 Dandy, via suboccipital craniotomy, for the first time performed intracranial section of the ninth nerve and rostral fibers of the tenth nerve.

For several decades, surgical treatment of both primary (idiopathic) and secondary (in malignancy) glossopharyngeal neuralgia consisted of total or partial section of the ninth nerve. It was performed either by the open approach to the brainstem through the posterior fossa (Graf, 1959) or extracranially in the jugular foramen where the ninth nerve emerges from the skull base. Certain authors have recommended simultaneous section of the superior rootlets of the vagus (Robson and Bonica, 1950).

For precise identification of the ninth nerve in the posterior fossa, it is necessary to visualize the foramen magnum, jugular foramen, and internal auditory meatus. One must remember that the jugular foramen lies medial to and below the internal auditory meatus, and the ninth nerve is located below the seventh and eighth nerves as they pass out that foramen. In the majority of cases, ninth nerve section gives a good effect; however, there are pain relapses in a considerable number of cases. Swallowing disturbances are a comparatively frequent complication in such operations.

Another effective surgical treatment of NGN is selective tractotomy in the medulla oblongata analogous to the Sjöqvist operation in TN.

Since the pain fibers of the ninth and tenth nerves, in well-differentiated compact bundles, pass in the medulla oblongata between the column of Burdach and the spinal trigeminal tract, they may be interrupted by a small incision in this area. Kunc (1965) performed this operation by an open posterior fossa approach in 31 NGN cases, in 15 for idiopathic NGN, in seven for NGN combined with neuralgia of the third branch of the trigeminal nerve, and in nine for pain of tumors in this area. A good and lasting result was observed in the majority of patients with long-term follow-up.

Just as in TN (Section 7) and hemifacial spasm (Section 12), it has been established that NGN may be caused by compression of the ninth nerve and superior roots of the tenth nerve by the vertebral or posteroinferior cerebellar arteries. Laha and Jannetta (1977) found such vascular compression in six cases operated on for NGN. Decompression of the ninth and tenth nerves was performed under a microscope through a retromastoidal approach to the posterior fossa in three cases. A piece of Gelfoam was inserted between the artery and nerves. A partial relapse occurred in two cases after 6–12 months, and one patient died following surgery. An excellent result in severe NGN in a 75-year-old patient after microvascular decompression of the nerve has also been presented (Hamer, 1986).

A new technique of NGN destruction, percutaneous rhizotomy, an operation analogous to that for TN, has been proposed (Sweet, 1976; Lazorthes and Verdie, 1979; Tew and Tobler, 1982; Isamat *et al.*, 1981; Salar *et al.*, 1983a,b; Pagura *et al.*, 1983). The object of this operation, which is performed under x-ray control, is destruction of the trunk of the ninth nerve with a needle electrode introduced through the jugular foramen. This foramen consists of two parts—a medial (pars nervosa), through which the ninth nerve passes, and a lateral (pars venosa), in which the jugular vein is situated. For percutaneous coagulation of the ninth nerve, two approaches have been employed, an anterior, proposed by Lazorthes and Verdie (1979), and a lateral, proposed by Bonica (1954) (Fig. 199). Each of these approaches has its pros and cons.

Isamat *et al.* (1981) consider the lateral approach to the ninth nerve to be technically simpler than the anterior one. In the lateral approach, the trajectory of the electrode is shorter, and its position is more easily controlled on x rays. Also, orientation is easier if there is a tumor at the base of the skull causing NGN.

The technique of percutaneous dissection of this nerve by the anterior approach, in brief, is the following. A long needle is introduced through the skin of the cheek on the affected side at a distance 2.5–3 cm from the corner of the mouth, and under fluoroscopic control it is advanced toward the base of the skull in approximately the same direction as for the foramen ovale (Section 7.2). The needle should be at an angle of 12° to the vertical plane passing through the ipsilateral pupil and at an angle of 40° to the plane passing through the internal auditory meatus and the inferior border of the orbit (Siegfried and Hood, 1983; Giorgi and Broggi, 1984).

After the needle tip has reached the medial part of the jugular foramen lateral to the internal carotid artery, the correct placement of the needle is confirmed in control roentgenograms of the skull base and by electrostimulation with a frequency of 5 Hz and voltage of no more than 0.5 V. If the electrode tip is in the

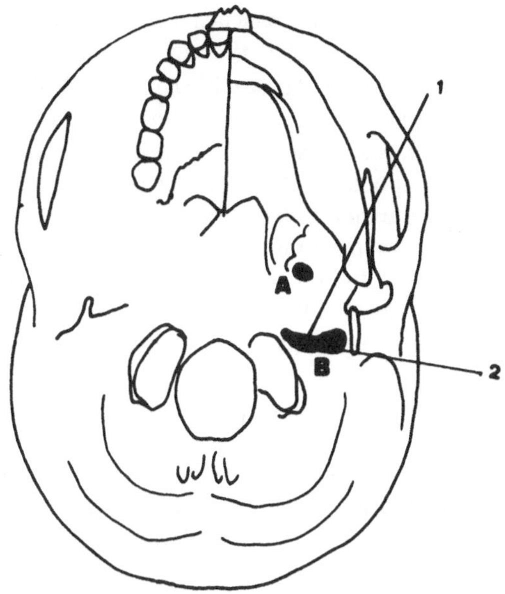

FIGURE 199. Basal skull landmarks for percutaneous destruction of the glossopharyngeal nerve (Salar *et al.*, 1983b). A, oval foramen; B, jugular foramen; 1, anterior approach; 2, lateral approach.

correct position, muscles innervated by the accessory nerve will contract. Then the frequency is increased to 50–75 Hz, causing dull pain and paresthesias in the innervation zone of the ninth nerve in the ipsilateral retropharyngeal area, external auditory meatus, and tonsil. An increase in the strength of the current causes contraction of laryngeal muscles.

During stimulation of the ninth nerve, there is an increase in arterial pressure and pulse rate; however, if the nearby vagus nerve is stimulated, a sharp decrease in the pulse rate by more than a half as well as a drop in arterial pressure by approximately 50% will occur. Such a vagal reaction is indicative of the relatively correct position of the electrode (Pagura *et al.*, 1983).

Because of the possibility of a sudden pronounced vagal reaction, which is observed frequently during both stimulation and coagulation of the ninth nerve, it is necessary to monitor arterial pressure and pulse rate during the operation. Should there be a vagal reaction during coagulation, it is recommended to discontinue it and administer a vagal blocking agent (atropine, 0.5–1 mg intravenously). Then the electrode should be withdrawn several millimeters, which should diminish substantially the vagal reaction (Isamat *et al.*, 1981).

Impedance measurements are another way to localize the target in this operation. When the electrode tip reaches the pars nervosa of the jugular foramen, the tissue impedance is approximately 700 Ω (Giorgi and Broggi, 1984). After radiofrequency coagulation of the ninth nerve at a temperature of 70–75°C, pain disappears completely in 1–2 min, and in the external auditory meatus there is a zone of hypalgesia.

Percutaneous destruction of the ninth nerve is undoubtedly an effective, reliable, and nontraumatic operation, although experience in its use is still limited. In practically all cases of NGN, after surgery, pain either disappears completely or only insignificant twinges remain. Percutaneous rhizotomy is effective in both primary and secondary NGN resulting from malignancy. Of eight cases with idiopathic NGN operated on by Isamat *et al.* (1981), the pain disappeared completely in seven during the operation. Two such operations for primary NGN were described by Tew and Keller (1977). Broggi and Siegfried (1979) obtained a good effect after percutaneous radiofrequency coagulation of the ninth nerve in facial pain from cancer, and Giorgi and Broggi (1984) reported good results in both idiopathic and secondary NGN. Pagura *et al.* (1983) in 15 cases relieved pain caused by cancer of the tongue, throat, tonsil, and pyriform sinus. Salar *et al.* (1983a,b) performed percutaneous selective thermocoagulation of the petrosal ganglion in the jugular foramen in eight NGN cases. Three had idiopathic neuralgia, and five had pain secondary to oropharyngeal cancer. Good results were obtained in all cases.

Rare complications such as dysphagia and dysphonia occur after percutaneous rhizotomy of the ninth nerve. Inaccurate direction of the needle may puncture the jugular vein; however, this is not considered a hazardous complication.

Therefore, there are three effective surgical techniques for the management of NGN: open or percutaneous section (coagulation) of the ninth nerve, section of the descending tract in the medulla oblongata, and microvascular decompression of the ninth nerve in the posterior fossa. Specific indications for each of these operations and their comparative success are still open questions that require a further accumulation of clinical experience.

9. Convulsive Tic

Convulsive tic (*tic convulsif*) is a rare and peculiar pain sydrome that is a combination of TN and hemifacial spasm (Section 12). This syndrome was first described by Cushing in 1920, who found three such patients among the 332 TN cases, and since then only 37 cases of this syndrome have been described in the world literature (Cook and Jannetta, 1984). Since facial pain is the dominant manifestation of convulsive tic, in the majority of cases attempts were made to section the fifth nerve root by the subtemporal or posterior approach. Tumors, arteriovenous aneurysms, or ectasia of the basilar or vertebral arteries were discovered in the majority of the cases. Hemifacial spasms did not diminish in most of the operated cases.

Recently, Cook and Jannetta (1984) reported the results of surgical treatment of 11 cases with convulsive tic in a group of almost 900 TN cases. Combined microvascular decompression of the fifth and seventh nerves by the technique described above (Section 7.7) was performed in these 11 cases. It appeared that compression was caused by different arteries: compression of the fifth nerve usually by the superior cerebellar artery, and of the seventh nerve most frequently by the anterior inferior cerebellar artery. Dilatation of the vertebral artery was found in two cases. A follow-up of these patients for an average of 6 years revealed that in eight of 11 cases, complete relief was obtained, i.e., elimination of both facial pain and hemifacial spasm.

10. Occipital Neuralgia

Pain in the cervicooccipital region is a frequent disorder that has been described in the literature many times. This pain may result from various causes, in particular, lesions of superior cervical disks, arthrosis of C_1–C_2 vertebrae, Arnold–Chiari malformation, fibrositis of the occipital muscles, and so forth. Pain is referred to the innervation zone of both occipital nerves, which corresponds to dermatomes C_1–C_2–C_3. This pain is often intense and paroxysmal and frequently radiates to adjacent areas.

A well-known and, as a rule, effective method in the management of occipital neuralgia is section of the major and minor occipital nerves. Recently, Oh *et al.* (1983) reported the results of this technically simple operation in 31 cases. In a long-term follow-up from 1 to 9 years, complete relief was noted in 84% of the cases.

Without dwelling on other studies, brief mention should be made of only one relatively new, simple, and effective technique of percutaneous radiofrequency thermocoagulation of the occipital nerves for pain in this area (Blume *et al.*, 1982a). A short needle is introduced on the side of the pain at a point 4 cm below the external occipital protuberance and 2 cm lateral to the midline. The tip of the needle should reach the occipital bone at the superior nuchal line. After the nerve has been reached, the stylet of the needle is replaced by a thermoelectrode, and the nerve is coagulated by raising the temperature to 90°C for 90–120 sec. The authors performed this simple operation in 450 cases with a follow-up of from 1 to 8 years. Excellent or good results were noted in 85% of the cases. The frequency of early pain relapses has recently dropped from 10% to 5%.

Ehni and Brenner (1984) described occipital neuralgia in elderly people, caused by degenerative changes in the lateral articulation of C_1–C_2. When steroids and local blocks with anesthetics did not provide relief, a long-lasting benefit was obtained by intradural rhizotomy of C_2.

11. Migrainous Neuralgia

Migrainous neuralgia, a pain syndrome that is frequently encountered, was first described by Harris in 1926. It is also called vasomotor migraine, cluster headache, the Horton syndrome, histamine cephalgia, or petrosal neuralgia. The disorder is characterized by severe attacks of sharp or pressing unilateral pain in the temporal and retroorbital areas. In severe forms, the pain attacks occur several times a week and last many (sometimes 10–12) hours. The pain is accompanied by autonomic manifestations such as lacrimation from the eye on the side of the pain and reddening and swelling of the mucosa of the nose and rhinopharynx. Just as in attacks of conventional migraine, there is sensitivity to noise and bright light. During attacks the superficial temporal artery dilates markedly, become tense, painful, and throbs. Rare cases of neuralgia have been encountered in which attacks of pain occur first on the right, then on the left side.

The pathogenesis of attacks in vasomotor migraine is still not entirely clear. It is presumed that the cephalalgic phase corresponds to the dilatation of the extracranial vessels with the release of vasopressor substances (serotonin and others) as well as spasm of the intracranial arteries (anterior and middle cerebral arteries), which has been confirmed angiographically (Masuzawa *et al.*, 1983). There is no valid explanation of the considerably high frequency (a ratio of 4 : 1) of cluster headaches in men and vasomotor migraine in women.

An attack of pain may be caused by various factors (cold, menstrual cycle, overfatigue, ingestion of alcohol, etc.). In many cases it is not possible to establish the cause of the attack. Spontaneous remissions for weeks and months are not infrequent. Chronic pain that is almost permanent with periodic exacerbations has been observed.

In many cases, drug treatment has been effective (pizotyline, ergotamine with caffeine, fonazine, as well as β-adrenoblockade, steroids, tranquilizers, antidepressants, and analgesics). However, such therapy frequently does not provide relief, and that is when surgical treatment is indicated.

One of the most effective operations is resection of a segment of the superficial temporal artery. A relatively long resection, not less than 3–4 cm, is recommended. This simple operation is performed under local anesthesia. After the operation, the attacks of pain often disappear for many years. The author has performed this operation in 16 cases with a severe form of the Horton syndrome. In 12 cases the painful attacks disappeared completely (follow-up from 1 to 16 years), in two cases the attacks occurred much less often, and in two cases the operation proved practically ineffective.

The following is an example illustrating the role of the superficial temporal artery in the genesis of this syndrome.

A 46-year-old man, a former boxing champion, had suffered for 6 years from frequent attacks of severe pain in the right temporoorbital region lasting from 18 to 20 hr. The pain was so intense that the patient made two attempts at suicide. During an attack the superficial temporal artery swelled to 4–5 mm in diameter and became tense and very painful to touch. After resection of approximately 3 cm of the artery in the right temporal area, the pain disappeared completely for 6.5 years. After that period the attacks of pain recurred, although with considerably less intensity. It appeared that during attacks the stump of the proximal part of the resected artery, easily palpable under the skin just above the zygomatic arch, became swollen and pulsated intensively. At a second operation, the stump of the artery to the zygomatic arch, approximately another 1.5 cm of the artery, was removed. The attacks of pain disappeared completely and have not recurred.

In the past, quite a number of different surgical procedures have been tried for the relief of severe headache: operations on the root and peripheral branches of the trigeminal nerve, including injections of alcohol and other lytic solutions, as well as x-ray irradiation. These procedures, as a rule, were ineffective.

Gardner *et al.* (1947) proposed for persistent unilateral headaches the resection of the great superficial petrosal nerve, which contains efferent preganglionic fibers innervating the mucosa of the nose, mouth, throat, and palate as well as the lacrimal glands. After resection of this nerve, Stowell (1963) obtained a significant improvement in 18 of 21 cases, and Watson (1983) had good results in three of four cases. However, other authors, have not had such good results. No convincing results were obtained from resection of the nervus intermedius. Combined intracranial resection of the large superficial petrosal nerve and the sensory root of the trigeminal nerve (Kunkel and Dohn, 1974) provided relief, but the operation has not been widely used. Maxwell (1982) reported good results from percutaneous radiofrequency coagulation of the fifth nerve root in eight cases of migrainous neuralgia. In the opinion of Watson *et al.* (1983), none of the many operations for cluster headaches guarantees a total and lasting therapeutic effect. We consider resection of the superficial temporal artery an effective method for treating the Horton syndrome; however, even this operation does not provide relief in all cases.

12. Hemifacial Spasm

This condition should not be included in the pain syndromes except for the very rare instances when it is combined with TN (Section 7). However, since microvascular decompression of the facial nerve has proved very effective in this disorder in recent years, it seems reasonable to dwell on this problem briefly.

Hemifacial spasm (HFS) is a distinct nosological disorder first described by Gowers (1888) and Brissaud (1894) and characterized by spontaneous, paroxysmal, often painful contractions of all or individual muscles of one side of the face, producing grimacing and closing of the eyes, often accompanied by ipsilateral hypersalivation and hyperlacrimation. Emotional excitation usually intensifies these spasms, distorting the face of the patient.

Hemifacial spasm is often a progressive disorder striking most often in people of middle age or old age, beginning with slight, periodic contractions of the orbicularis oculi muscles, spreading to the facial muscles, and eventually involving the platysma. Other craniocerebral nerves are usually not affected, although diminished hearing on the same side has been noted in some cases. The EMGs of the affected facial muscles in HFS reveal periodic, clonic, high-amplitude bursts or almost constant high-frequency discharges. Angiography and CT do not show any pathological changes in the brain substance and cerebral vessels (Fairholm *et al.*, 1983).

Although there is no doubt that HFS is an irritative syndrome caused by hyperactivity of a branch of the facial nerve, in previous literature there have been opinions that HFS is a psychogenic syndrome or extrapyramidal lesion. Gardner (1962) and then Jannetta (1970, 1976) were the first to demonstrate that the main cause of HFS is vascular compression of the facial nerve in the cerebellopontine angle. This was subsequently confirmed by many authors. A clear-cut compression of the entrance zone of the seventh nerve at the brainstem by the vertebral artery, the anterior or posterior inferior cerebellar, or cochlear arteries was discovered in 46 of 47 cases (Jannetta *et al.*, 1977) and in 73 of 74 cases (Iwakuma *et al.*, 1982).

The facial nerve may be compressed by the premeatal or postmeatal segment of the anterior inferior, posterior inferior cerebellar, or vertebral arteries (Rhoton, 1981). In 40 cases verified during surgery by Goya *et al.* (1983), 25 involved compression of the facial nerve by the anterior inferior, seven by the posterior inferior cerebellar artery, and eight by the vertebral artery. In 20 cases operated on by Fairholm *et al.* (1983), 12 had the nerve compressed by the anterior inferior cerebellar, one by the posterior inferior, and seven by the vertebral artery. Quite a number of cases of HFS caused by compression of the entrance zone of the seventh nerve by saccular aneurysms of the posterior inferior cerebellar and other arteries as well as AVM have been described in the literature (Gardner and Sava, 1962; Maroon *et al.*, 1978; Pierry and Cameron, 1979). In one postmortem examination of a patient with this disorder, the morphological investigation by Iwakuma *et al.* (1982) revealed compression of the facial nerve by a loop of the posterior inferior cerebellar artery. Fascicular demyelination of the nerve root and proliferation of Schwann cells were seen microscopically. There is an opinion that in compression of the seventh nerve, damage is done to the myelinated covering of its fibers with short-circuiting of the neural impulses. Subsequent EMG investigations demonstrated that this process is based on ephatic transmission of impulses and ectopic excitation (Nielsen, 1984).

Conservative treatment of HFS, particularly, with diazepam and carbamazepine, is practically ineffective. Previous blockades of the facial nerve with alcohol or phenol are not only very painful but usually ineffective, since relapses occur quite often.

Four methods of the surgical treatment of HFS are known at present: selective resection of the branches of the seventh nerve in the preauricular area or resection of the nerve trunk; partial resection of the facial nerve at its exit from foramen stylomastoideus (Scoville, 1969b); percutaneous coagulation of this nerve at the same site (Hori *et al.*, 1981); and microvascular decompression of that nerve in the posterior fossa. The technique of the latter operation, proposed and developed in detail by Jannetta (1970, 1977b), which is most frequently used at present, does not differ in principle from microvascular decompression of the trigeminal nerve root, described in Section 7.7, yet this operation has several important peculiarities. It is usually performed with the patient in a sitting position with dopplerographic control to detect any air embolism as soon as possible. The head of the patient is inclined toward the chest and turned 15–20° to the ipsilateral side. Certain surgeons, including the author, prefer to operate with the patient in a prone position.

A vertical incision in the skin several centimeters long is made 2 cm from the mastoid process. Then a craniotomy 4–5 cm in diameter is made posterior to the mastoid eminence, followed by the removal of air cells and closure with paraffin. It is necessary to open the sigmoid sinus for a better view of the seventh nerve. After CSF has been removed from the basal cistern and the cerebellar hemisphere in the region of the flocculus is retracted, the operation is continued under a microscope. The arachnoid membrane is resected, exposing the choroid plexus of the lateral recess of V_4 in the foramen of Luschka. As Jannetta *et al.* (1977) emphasize, it is very important to select the proper angle of approach and the correct position of the self-retaining retractor to expose the facial nerve lying beneath and medial to the eighth nerve. Then it is necessary to separate the seventh nerve root at its entrance zone from the compressing arterial loop. Compression of the nerve by the vein or by the artery and vein has been found in a few cases. When separating the compressing vessel, one must preserve the perforating arterioles passing to the brainstem.

After separation of the arterial loop, a piece of muscle, Teflon, or Ivalon sponge is placed between the loop and the brainstem to prevent compression of the roots of the seventh and eighth nerves. As Jannetta *et al.* (1977) indicate, it is necessary to place the Teflon precisely between the brainstem and the artery and not between the artery and the cranial nerves.

Decompression of the facial nerve by separating it from the compressing artery, without a doubt, in the majority of cases yields an excellent and lasting result with few postoperative complications or relapses (Jannetta, 1977b, 1981a; Fabinyi and Adams, 1978; Witzmann and Dieckmann, 1982; Fairholm *et al.*, 1983; Carlos *et al.*, 1986).

Jannetta (1977b) performed microvascular decompression on 45 cases and, with long-term follow-up, reported an excellent result in 38, good in two, slight improvement in three, and poor results in two cases. Postoperative paresis of the facial nerve developed in six cases; however, in all cases but one, it was temporary. Subsequently, Jannetta (1981a) reported on an exceptionally large series of 229 cases of microvascular decompression of the seventh nerve. The HFS disappeared completely in 93% of the cases postoperatively.

Of 40 cases operated on by Goya *et al.* (1983), an excellent result was obtained in 29, good in eight, fair in two, and poor in one case. Fairholm *et al.* (1983) observed that HFS ceased in 19 of 20 cases following microvascular decompression of the seventh nerve (the operation was repeated in two cases). Reduced hearing occurred postoperatively in four cases but returned rapidly in three. After an average follow-up of 18 months, there was not a single relapse of HFS. Carlos *et al.* (1986) reported good results in about 90% of operated cases. Impaired hearing is a relatively rare complication of microvascular decompression for HFS. In a large series of 143 consecutive patients, only 2.8% have had a significant hearing loss as a complication of facial nerve decompression (Møller and Møller, 1985).

Iwakuma *et al.* (1982) compared the results of the above three methods on the basis of 110 cases operated on for HFS that had been present from 3 months to 20 years. Partial section of the seventh nerve at the brainstem (Scoville, 1969b) was practically ineffective; relapses occurred in a short time in almost all cases. Resection of the peripheral branches and facial nerve trunk with coagulation of the stumps did not produce lasting results, for the spasms disappeared in only three of 20 cases.

The results of microvascular decompression of the seventh nerve were substantially better. Directly after the operation, HFS and facial synkinesis completely disappeared in 97% of the cases, and in a 3-year follow-up, a relapse occurred in only one of the 74 cases. Diminished hearing was noted postoperatively in 12 cases, which the authors attributed to damage to the acoustic nerve during surgical manipulations or to excess retraction of the cerebellar hemisphere.

It is noteworthy that HFS may cease after exploration of the posterior fossa even in cases in which vascular compression of the seventh nerve was not demonstrated at operation. Fabinyi and Adams (1978) found such pathology in only three of nine cases of HFS operated on; however, eight of these patients experienced relief after surgery. The authors attribute this to the separation of fibrous adhesions around the nerve and its wrapping with sponge.

A new technique for the management of HFS has lately been proposed, namely, percutaneous coagulation of the facial nerve by insertion of an electrode at its exit from the stylomastoid foramen (Hori *et al.*, 1981). This simple procedure, involving no risk whatsoever, according to the authors, proved to be very effective. Spasms of the facial musculature disappeared completely in 24 of 27 cases. With a follow-up of 1 year, there were only three relapses. In 60% of the cases, after coagulation of the seventh nerve, the facial muscles became weak; however, the authors state that this complication disappeared spontaneously in 1 to 4 months. Naturally, this raises the question of why there is such success after percutaneous electrocoagulation of the seventh nerve, whereas its open partial or total section is almost always followed by relapses. So far, this question remains unanswered. Possibly, the more distal destruction of the nerve by electrocoagulation may be the answer.

Consequently, today the most effective operation in the management of HFS is microvascular decompression of the seventh nerve. Percutaneous electrocoagulation of the facial nerve at the stylomastoid foramen is also, apparently, quite effective.

14

Epilepsy

1. General Remarks

This widespread, severe disease, which is difficult to treat, still remains one of the most important problems of neurology, psychiatry, and neurosurgery. Its importance and social significance become apparent from the single fact that 0.4–0.6% of the population of our planet suffer from epilepsy, i.e., tens of millions of people. Approximately one-third of these are afflicted with epilepsy from early childhood. For example, with a morbidity rate of approximately one person per 200 of the population, in the United States, there are more than a million epileptics. And since no fewer than 10% of them have focal epilepsy, at least 100,000 such patients might benefit from surgical treatment (Robb, 1975). According to other sources, in North America there are approximately a half million epileptics who are potential candidates for surgical treatment (Ojemann and Ward, 1975).

There are literally hundreds of drugs for the treatment of epilepsy, and their number increases with every passing year. Nevertheless, in spite of the considerable achievements of medical treatment, for a great number of patients anticonvulsant drugs are ineffective and do not control epileptic attacks. As is apparent, repetitive attacks may lead to total disability and mental impairment.

Thousands of scientific papers and scores of monographs have been devoted to epilepsy. A bibliography on epilepsy for the previous 25 years published in the United States in 1976 contains approximately 18,000 publications on this condition. A review of the present state of the problem would fill several thick volumes. Therefore, we consider it our task to present only a brief description of certain kinds of epilepsy and to summarize the basic data on the newest methods of surgical treatment for this disease, particularly stereotactic operations on deep-lying structures of the brain.

Questions pertaining to the clinical picture and diagnosis are dealt with briefly and only relative to the indications for surgical treatment. The description of the different types of epilepsy is based on the International Classification compiled by a special commission chaired by Gastaut. This classification is based on the anatomic, symptomatic, etiologic, and functional criteria with due consideration of data obtained from the surgical treatment of epilepsy. The classification was endorsed by the VIII World Congress of Neurology in Vienna in 1965. Minor changes were introduced to this classification at the XI Congress of the International League against Epilepsy held in New York in 1969.

2. Etiology

Epilepsy is a disease of many etiologies. Epileptic attacks may be attributed to various cerebral pathological processes—injuries, tumors, vascular, inflammatory, and generative processes, etc. Birth injuries occupy a prominent place in the etiology of epilepsy, mainly because of damage to the fetal brain as it moves through the maternal passages. Considerable significance is attached to other ante- and perinatal pathological factors, in particular, the frequent and grave complication of asphyxia that results from uteroplacental circulatory disorders. Damage to the newborn's head during the rendering of obstetric aid is a frequent cause of epilepsy. One should also mention secondary asphyxias of the postnatal period caused by lesions in the cardiovascular system, especially in premature infants, as well as the hypoxia of pneumonia. The convulsive propensity of the brain caused by asphyxial affliction may not appear immediately but only after a long period of time, when the effect of additional pathological factors acts on the nervous system.

In difficult deliveries, there is sometimes a rupture of the tentorium and a mushroomlike herniation of the temporal lobe or that part of the hippocampal gyrus where morphological changes are often found in epilepsy. The frequency of pathology in pregnancy and deliveries relating to the etiology of epilepsy varies according to the data of different authors from 6 to 44%.

In multifocal epilepsy, contrast and CT investigations often reveal gross morphological changes in the brain—hemiatrophy, porencephalic cysts, ventricular diverticula, aplasia of the temporal lobe, or absence of the septum pellucidum. In recent works, authors increasingly emphasize the role of a hereditary factor in the etiology of epilepsy creating an "epileptic diathesis" in the brain.

3. Pathogenesis

Ever since J. Hughlings Jackson in 1870 defined epilepsy as a "discharging lesion," many hundreds of investigations have been devoted to the pathogenesis of this disease. It is generally acknowledged today that epilepsy, as a disease, is the result of a complicated combination of two basic factors—epileptic "readiness" of the brain in general ("epileptization of the brain") and the development of one or several epileptogenic foci in various cortical and subcortical structures that generate periodic discharges of epileptic activity. The following stages in the development of epilepsy have been suggested: "epileptic neuron, epileptogenic focus, epileptogenic system, epileptic brain" (Stepanova and Grachev, 1976). The epileptogenic focus has certain functional interrelationships with the cortical and subcortical structures, forming a so-called epileptic system. The discovery and study of the kindling phenomenon (Goddard, 1967; Wada and Sato, 1974) demonstrated the complex interaction between different structures of the brain in the genesis and development of epileptic attacks.

The typical EEG spikes in epilepsy point to the simultaneous discharge of many thousands of neurons. At any moment the excitation that develops in a small population of neurons may reach the "critical mass," spread to other cerebral structures, and trigger an epileptic seizure.

The pathogenesis of epilepsy is quite complicated, and this is caused by the fact that, first and foremost, epileptic activity may occur in many structures of the brain and advance in various directions. The intensity of this activity also varies and is not always sufficient to trigger a "breakthrough" that will lead to an epileptic seizure. Certain authors explain the development of a primary, general, convulsive attack by the presence of an epileptogenic focus in the truncal and subcortical nuclei, and others in the brain cortex. There can be no doubt that the focal seizures are triggered by corticosubcortical structures, whereas a secondary generalization or primary generalized convulsive attack cannot occur without the participation of the brainstem and subcortical structures. The EEG investigations by Rossi *et al.* (1968), Pruvot *et al.* (1972), and Romodanov *et al.* (1974) have demonstrated in primary generalized attacks the possibility of bilateral and simultaneous convulsive activity in cortical and subcortical mechanisms. When this high-frequency discharge recruits bursting activity in neighboring neurons, a propagating seizure occurs (Ward, 1975). It is presumed that an epileptic seizure occurs only when the excitation from deep-lying structures reaches the cortex and disseminates throughout that region (Bancaud *et al.*, 1967).

There can be no doubt that an epileptic discharge from one focus can disseminate along different cerebral pathways, which explains the frequent polymorphism of epileptic attacks. At the same time, it has been established that there are structures and pathways that play leading roles in the generalization of epileptic discharges. First of all, these include CC as well as F, through which the discharges pass from Hipp and the diencephalic area, and through CA, which has pathways linking both temporal lobes.

Numerous investigations have established three main pathways for the spread of an epileptic paroxysm. The first is the irradiation of the excitation via intracortical zones. The second is a transmission through CC and other commissural fibers linking vast fields of both brain hemispheres. The appearance of so-called "mirrorlike" foci offers an example of the great significance of CC in the transfer of epileptic activity. The third pathway along which generalized convulsive activity propagates is via the subcortical and reticulocortical reciprocal pathways (Okudjava, 1969). Consequently, in both primary cortical foci as well as generalized convulsive attacks, the pathways of generalization are primarily of a "vertical" nature. A convulsive paroxysm that originates in the cortical neurons usually spreads to the brainstem and then generalizes and simultaneously involves both hemispheres of the brain. At the same time, it is presumed that the brain possesses inhibitory ("antiepileptic") mechanisms that block the emergence and generalization of epileptic attacks.

4. Epileptogenic Focus

In the almost 100 years since the classical works of Jackson in neurology, it has become customary to divide epilepsy into two main groups—focal or symptomatic, in which the focus is assumed to be a cerebral lesion both clinically and pathologically, and "essential" or "genuine" epilepsy, the cause of which cannot be established. However, the results of studies over the course of a quarter of a century tend to narrow the concept of "essential" or "genuine" epilepsy. Modern diagnostic techniques have made it possible to pinpoint an organic site of the disease in approximately half of the patients who had earlier been relegated to the category of genuine epilepsy.

There have been numerous investigations of the epileptogenic focus following various approaches: morphological, biochemical, EEG. There are several definitions of an epileptogenic focus, but we consider one of the most precise definitions to be that of Romodanov (1980): ". . . not simply a group of ganglionic cells capable of producing a convulsive focus, but a dynamic, constantly active pathological structurofunctional system."

Morphological and, in particular, electron microscopic investigations of the cortical epileptic foci, removed at operations, have revealed a substantial decrease of neurons and an increase of glial cells (Vaquero *et al.*, 1979).

A membrane potential disorder has been discovered in the epileptogenic focus, which may lead to spontaneous discharges of excitation. These potential changes depend on the uneven distribution of electrolytes, particularly the intracellular sodium ions. At the same time, there is an accumulation of potassium ions in the mitochondria of "epileptic" neurons, as a result of which the potassium–sodium balance is upset.

By using positron emission scanning, it was established that in the epileptogenic focus there are clearly defined changes in glucose metabolism; what is more, during an attack the metabolism increases, and in the interictal periods it decreases. Histochemical investigations have revealed in the epileptogenic focus a significant increase in the activity of various enzymes—glucose phosphatase, adenosine triphosphatase, and acid phosphatase (Vaquero *et al.*, 1979). Moreover, in the focus of epileptic activity there are changes in amino acid metabolism; in particular, glutamic and γ-aminobutyric acids diminish substantially.

Jasper (1962), on the basis of research done with Penfield and associates, believes that an epileptogenic focus consists of three concentric zones, each having its own EEG characteristics. In the center there is a core of brain tissue or lesion characterized by slow waves in the EEG. It is encircled by the focal epileptic zone, whereas on the periphery there is a zone of irritation that generates sharp waves of high amplitude.

On the basis of data obtained from recording of intraneuronic cortical biopotentials, it was concluded that the tonic and clonic phases of the epileptic paroxysm occur at the cortical level and differ from one another by the rhythm of their neuronic discharges and their organization (Okudjava, 1969). Investigations have demonstrated that a somatomotor attack is the result of excitation of the corticothalamic circuits connecting certain cortical zones with the corresponding specific thalamic nuclei (Gastaut, 1963).

Attempts have been made to determine the quantitative size of an epileptogenic focus, which may vary greatly. For example, on the basis of electrosubcorticographic data, it has been established that the minimal diameter of an epileptogenic focus in Th varies from 0.5 to 3 mm, and its volume varies from 0.3 to 20 mm^3 (Grachev and Stepanova, 1971). Interictal investigation of patients with complex partial epilepsy by positron emission tomography (PET) revealed areas of hypometabolism which have correlated with the epileptogenic focus (Mazziotta and Engel, 1984). The basis of this hypometabolic focus is still unclear.

5. Diagnostic Procedures

Talairach *et al.* (1958) were the first to state the most important task: establishment of clinical, electrophysiological, and anatomical correlations in every epileptic patient.

The most informative method for detecting and pinpointing the localization of an epileptogenic focus is electroencephalography. In the focus, typical high-amplitude discharges such as the "spike–wave" complex combined with slow δ and θ waves are recorded. These changes become more clearly defined on activating the epileptogenic focus (hyperventilation, photo- and phonostimulation). A pharmacological activation of the epileptic attack and the study of EEG changes induced by drugs are also important (Rossi *et al.*, 1974, 1978; Talairach *et al.*, 1974a,b; Wieser *et al.*, 1978; Ryabokon, 1980, Kambarova *et al.*, 1981). Many preparations are used for this purpose: chlorpromazine, pentylenetetrazole, propanidid, and methohexital. New methods of studying epilepsy are also being employed, including telemetric EEG and video recording of epileptic attacks.

Electrocorticography (ECoG) and electrosubcorticography (ESubCoG) are more valuable and informative methods for the diagnosis of all forms of epilepsy (see also Chapter 4, Section 7). Electrical activity of the human cortex was first recorded by Foerster and Altenburger in 1935. The first report of simultaneous recording of thalamic and cortical potentials in epileptic patients made by Spiegel and Wycis in 1950 was a first step in the study of subcortical epileptic EEG activity.

As the result of stereotactic placement of electrodes, it has become possible to perform stereo-EEG (see below), which makes it possible to investigate the epileptogenic focus in three dimensions. It was established that deep within the brain there may be isolated epileptogenic foci not showing any manifestations on scalp EEGs or even in records from the cerebral cortex. And bilateral diffuse firing in the EEG may be derived from a small epileptogenic focus situated deep within the brain.

Olivier and de Lotbiniere (1987) formulated three main reasons for using implanted depth electrodes with a computer detection program: (1) a "bitemporal focus" for determining which temporal lobe is clearly predominant in the onset and continuation of the seizure; (2) ambiguity in localization of the focus within the same cerebral hemisphere; and (3) multifocal or secondary generalized epilepsy. If all clinical or EEG investigations have shown a single predominant epileptogenic focus, the use of depth electrodes monitoring is unnecessary.

Angiography is rarely very informative from the point of view of localizing an epileptogenic focus. At the same time, pneumoencephalography in the majority of patients gives valuable information about the ventricular system and subarachnoid space. An external or internal hydrocephalus resulting from cerebral atrophy has been found quite often (Ryabokon, 1980).

The extensive use of CT in neurosurgical practice has considerably increased the possibilities of detecting and evaluating structural changes in the brain related to various forms of epilepsy. In CT investigations of epileptics, it has been recommended that the main scan not be performed at an angle of 25° to Reid's line as usual but parallel to that line. This lessens the interference of the bony calvarium on the structures of the middle cranial fossa, which is especially important when examining patients with temporal lobe epilepsy.

A CT examination in cases with epilepsy may indicate the side and the location of an epileptic focus caused by small tunors, AVM, or tiny glial cicatrices. By contrast enhancement (intravenous injection of iothalamate), it is possible to detect small foci of local gliosis and with the aid of CT cisternography with metrizamide to demonstrate medial temporal herniation indicative of mesial temporal sclerosis. However, even CT is not always sufficiently informative. For example, in half of 173 patients with generalized or temporal lobe epilepsy, CT did not reveal any organic cerebral changes (Ladurner *et al.*, 1979). A CT study of all forms of epilepsy reveals pathological changes in the brain in only 35–55% of the cases; however, in patients over 65 years of age, it is diagnostic in more than 85%. Even with the use of the new diagnostic techniques (CT, positron emission scanning), EEG in the interictal period remains the best method for detecting an epileptogenic focus.

In approximately one-third of epileptic patients, it is impossible to demonstrate an epileptogenic focus using all existing methods. These forms of epilepsy are characterized by bilateral synchronous, symmetrical discharges in the EEG. This suggests that the epileptogenic focus is localized subcortically or in the brainstem, from whence epileptic impulses radiate to the cortex of both cerebral hemispheres (Penfield and Jasper, 1959). On the basis of thorough studies of the electrophysiology of an epileptogenic focus, Jasper (1962) formulated concepts about "dominating," "dependent," and "mirrorlike" foci and studied their hierarchy. Interrelationships such as subordination, reciprocity, and domination may develop among epileptogenic foci in various cerebral structures.

Numerous experimental investigations have demonstrated that during attacks there is a sharp increase of cerebral blood flow (CBF), O_2 consumption, and glucose utilization. In view of this, numerous clinical investigations were conducted using modern techniques. Data concerning CBF in the interictal period proved to be contradictory: both increased and reduced CBFs were registered. A number of authors have presented convincing data that in cortical as well as deep-seated epileptogenic foci there is a decrease in local

CBF producing hypoxic cerebral tissue (Chkhenkeli and Bregvadze, 1979, 1980). These data have been confirmed by recent studies of positron emission tomography with radioactive oxyglucose and NH_3 and $^{15}O_2$. These investigations have shown that in the interval between attacks there is a significant total and regional decrease in CBF, O_2 consumption, and glucose metabolism. The greatest reduction of CBF was found in the temporal cortex as well as in Cer (by 35.7%).

Another method has been proposed for detecting and localizing epileptogenic foci by determining CBF with the aid of γ counter after intracarotid administration of ^{133}Xe (Hougaard *et al.*, 1976).

6. Surgical Treatment

6.1. General Remarks

Drug treatment of epilepsy in recent years has, without a doubt, made formidable advances. A great number of new and quite effective drugs have appeared. Treatment is controlled by measuring the concentration of these drugs in the blood. As a rule, a combination of several drugs is used in the treatment of epilepsy, quite often in doses bordering on toxic levels. Such treatment over the course of many years is harmless, but in approximately 25% of the patients there are various complications (drug encephalopathy, allergic reactions, toxic effects on internal organs, etc.). It is felt that no less than 20–30% of epileptics do not respond to medical treatment (resistant or refractory epilepsy) (Wilder, 1971; Robb, 1975). However, in more than half of the patients with grand mal and temporal lobe epilepsy, anticonvulsant drugs in any dosage do not effectively control attacks. In such cases, surgical treatment is indicated.

The majority of authors consider that surgery is indicated when attacks occur three to five times a month in spite of medical treatment. Since there is a comparatively rapid development of irreversible personality disorders in cases where the seizures are frequent, as well as a gradual formation of secondary and mirrorlike epileptogenic foci, if drug treatment proves ineffective over the course of several years, surgical treatment should not be postponed.

The majority of surgeons believe that drug treatment should be tried for approximately 3–4 years (Zemskaya, 1971; Walker, 1974; McNaughton and Rasmussen, 1975). However, there has been a tendency to reduce this period to 1–2 years (Thomalske, 1975; Zotov, 1977). The opinion is more often expressed in the literature that earlier surgical intervention is indicated in children (Shefer *et al.*, 1972; Falconer, 1972a,b).

Since a focus in the left hemisphere causes more disruption of cerebral activity than a right hemisphere one, it is considered that in "left-sided" epilepsy, an operation should be performed earlier.

The topical localization of the epileptogenic focus represents the most important task facing the neurosurgeon. In accordance with contemporary concepts, it is necessary to determine not only the location of the pacemaker of the epileptic discharge but also the site and interrelationships among the many cerebral structures that are involved in the epileptic system, in particular, those both facilitating and inhibiting the discharge.

6.1.1. Indications for Surgery

Indications for an operation irrespective of the type and etiology of epilepsy can be formulated as follows:

1. The presence of a relatively accurately localized epileptogenic focus, i.e., pinpointing the cerebral structure or structures causing the attacks.
2. A focal structure that lends itself to surgical destruction by some method.
3. The condition is progressive.
4. Ineffective long-term and systematic treatment with anticonvulsant drugs for 2–4 years without significant reduction of the frequency of attacks permitting the patient a normal life at work, home or school.
5. No absolute contraindications to operation.

The majority of authors consider that bilateral or multiple foci do constitute a contraindication, but some surgeons recently have extended, in selected cases, the indications to include multifocal epilepsy.

The surgical treatment of epilepsy was started a century ago by Sir Victor Horsley (1886), who was the first to remove a cortical cicatrix and surrounding brain tissue in the anterior central gyrus to relieve Jacksonian convulsions. Horsley believed that the surgical removal of the cortical focus of excitation would stop the attacks. He used electrical stimulation of the cortex for locating the epileptogenic focus. However, such operations performed by many surgeons after Horsley, as a rule, were followed by paralysis of the affected limbs.

Various surgical procedures have been proposed at different times in accordance with the then current understanding of epilepsy (for example, sympathectomy, excision of the carotid sinus nerve in the neck) and abandoned. It is noteworthy that the contemporary neurosurgical armamentarium of antiepileptic operations is indeed large. No less than a score of operation procedures are currently in vogue for the treatment of the disease.* In general, taking into account the great diversity in forms of epilepsy, it might be considered plausible for each of these forms to have its "own" therapeutic operation. Yet this concedes that the problem has not been resolved and that there is no generally acknowledged strategy for the surgical treatment of epilepsy.

The success of surgery for this disease and the noteworthy achievements attained in this direction during recent years cannot be doubted. Nevertheless, in approximately 20–30% of the cases, the therapeutic effect following surgery is unsatisfactory. This results from several factors: difficulty in locating the epileptogenic focus, multiple epileptogenic foci, in particular, the presence of secondary or mirror foci, incomplete destruction of these foci, etc. Relevant causes of surgical failure in partial epilepsies are connected with difficulties in delimitation of the epileptogenic zone and therefore of complete surgical extirpation of the zone (Rossi *et al.*, 1978).

It has been established that in the course of time secondary epileptogenic foci become autonomous; i.e., they can generate discharges that disseminate throughout the brain even after the primary focus has been eliminated. That is why, when there are secondary foci, the destruction of the primary focus does not abolish the attacks. At best, it may only be possible to lessen their frequency, as well as emotional and mental concomitants.

6.1.2. Types of Operative Procedures

Without dwelling on the historical aspects of the surgery, contemporary operations for epilepsy may be divided into three main groups.

The first group includes operations whose objective is to remove the epileptogenic focus, i.e., epileptogenic brain tissue. These are the most frequently performed operations, since they are the most effective. The operation involves extirpation of cicatrices, foci of sclerosis (gliosis) in brain tissue, small vascular malformations, various brain tumors, etc. Penfield has repeatedly emphasized that the main goal of the operation must be the fullest possible removal of the epileptogenic focus.

The second group includes those operations whose purpose is to sever the conducting pathways along which epileptic discharges are transmitted, i.e., the paths by which they reach often remote structures of the brain. In principle, operations on pathways along which epileptic discharges spread are less effective, since new pathways for dissemination are gradually formed. This is evident from the fact that attacks are abolished or sharply decreased in frequency after extirpation of the epileptogenic focus in 84% of patients, whereas after severing the propagating pathway or destroying the activating structures, the seizures are eliminated in only 55% of cases (Shefer *et al.*, 1974).

Finally, the third group includes several operations the aim of which is to "switch off" or eliminate the so-called activating cerebral centers, which are not the epileptogenic focus, but which intensify or generalize the discharges (Am, VL, Hypoth, and others). Since it has long been known that emotional tension and stress provoke epileptic attacks of all types, it is logical to assume that the destruction of structures subserving emotions, such as the limbic system, might also lead to a positive clinical effect.

Walker (1982) enumerated three main principles underlying stereotactic, functional operations for epilepsy:

1. Destruction of certain cerebral structures that would diminish the overall excitability of the hemispheres or, at least, of the cerebral cortex or would increase the inhibition of other structures.
2. Destruction of the epileptogenic ("critical") foci in the brain, which would eliminate the epileptic process.
3. Interruption of pathways through which discharges spread from the epileptogenic focus to other cerebral structures, which would abolish the clinical manifestations of the disease, particularly the epileptic seizures.

Severe disease of internal organs, impaired intellect causing debility, and old age are contraindications

*We shall not dwell on the various methods for cooling the brain, in particular, its ventricles, or on general hypothermia to arrest status epilepticus. In several publications the authors report encouraging results.

to surgery for any form of epilepsy. Walker (1982) believes that an IQ of 60 or less is a contraindication to operation in any form of epilepsy. It is noteworthy, however, that after operation, the intelligent quotient often becomes higher, even if prior to the operation it was substandard (Barcia-Salorio and Broseta, 1976). Most authors consider the presence of numerous foci of epileptic activity, especially in both hemispheres, to be a contraindication to surgical treatment.

6.1.3. Principles of Localization of Epileptic Foci

In contemporary neurosurgery for epilepsy, there are two clearly defined approaches: classical or open operations and stereotactic operations. It is our intention to dwell in detail on the second procedure, which at present is more commonly practiced.

The first stereotactic operations to destroy the intralaminar nuclei in petit mal epilepsy were performed by Spiegel *et al.* in 1951 (1951a) and in generalized epilepsy by Riechert and Wolff, also in 1951.

The stereotactic approach to the study and treatment of epilepsy has a number of undisputed and significant advantages. Stereotactic operations are much less traumatic, and therefore, the risk is less. Such operations involve the destruction of only relatively small deep-seated brain structures, i.e., a small nucleus. For example, stereotactic hippocampotomy for temporal lobe epilepsy involves the destruction of only approximately 5–8% of the amount of brain tissue that is removed in the classical anterior temporal lobectomy. Accordingly, there are few complications after stereotactic operations, such as neurological deficits, impaired memory, and mental functions. Yet, one encounters a dilemma in determining the optimal size of the lesion. Unquestionably, a smaller lesion makes the operation less traumatic and, consequently, complications less frequent. However, a smaller zone of destruction has not always been justified in practice. Rasmussen (1975) believes that small, discrete epileptogenic foci are encountered rarely. Much more frequently they consist of comparatively large areas of the brain. Consequently, the extirpation of a small epileptogenic focus may not substantially reduce the frequency of seizures.

It is necessary to emphasize an important factor in the majority of (but not all) stereotactic operations for epilepsy that distinguishes them from similar operations for dyskinesias and pain syndromes. The procedure is usually performed on a locus morbi, i.e., those deep-seated cerebral structures having significant morphological changes as the result of various pathological processes. However, in a number of cases, the location of the epileptogenic focus may not coincide with the site of the cerebral lesion. Walker (1982) points out that the presence of a spike focus in a deep-seated structure does not necessarily mean that the area is the site where the epileptic attack originates. The primary focus, situated at a distance, may produce propagated spikes in other related structures. Moreover, the firing of such a focus does not always lead to attacks. Nevertheless, a thorough investigation with implanted electrodes is the principal way to locate the focus. For example, in half the patients with frontal and temporal lobe epilepsy, scalp EEGs did not reveal the presence and location of the focus, which was discovered only by recording from implanted electrodes. It is noteworthy that it was in this half of the patients that subsequent stereotactic operations proved considerably more effective (Laws *et al.*, 1970).

Stereotactic operations require a very discrete and precise placement of many electrodes in previously designated cerebral structures. This opens up great opportunities both for therapy following diagnosis of the nature and location of an epileptogenic focus and for scientific research, since stereotactic operations make it possible to employ various neurophysiological techniques, including biopotential recordings from various cerebral structures, electrostimulation, measurements of impedance, local CBF, oxygen consumption, as well as direct examination of the tissue of the focus by biopsy. It is noteworthy that the development of stereotactic surgery has made a significant contribution not only to the treatment but also to the understanding of the etiology, pathogenesis, and mechanisms underlying the various forms of epilepsy. The first attempt to treat epilepsy by stereotactic operation was made by the pioneers of stereotaxy, Spiegel and Wycis, in 1950. The techniques and results of surgical treatment of the epilepsies by stereotaxy have been described in numerous works (Talairach *et al.*, 1958; Spiegel *et al.*, 1958; Zemskaya, 1961, 1972; Narabayashi and Uno, 1966; Heimburger *et al.*, 1966, 1978; Wycis *et al.*, 1966; Mullan *et al.*, 1967; Hori *et al.*, 1968; Shefer *et al.*, 1968; Jinnai and Mukawa, 1970; Belyaev, 1970; Romodanov *et al.*, 1971a,b, 1975; Mempel, 1971a,b; Jelsma *et al.*, 1973; Nashold *et al.*, 1972, 1973; Guidetti *et al.*, 1972; Bouchard *et al.*, 1975; Zemskaya *et al.*, 1975; Ojemann and Ward, 1975; Sigua and Chkhenkeli, 1976; Ganglburger, 1976; Skryabin *et al.*, 1976; Laitinen and Toivakka, 1979; Rossi, 1980;

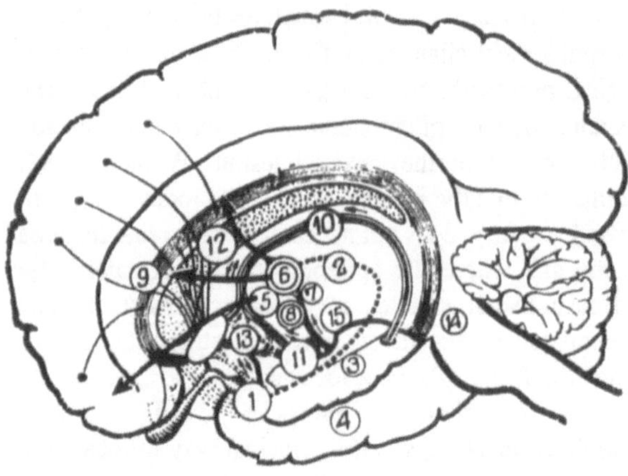

FIGURE 200. Scheme of main cerebral targets of stereotactic operations for epilepsy. 1, Am; 2, stria terminalis; 3, Hipp; 4, gyrus Hipp; 5, dorsomedial nucleus of Th; 6, anterior nucleus of Th; 7, CM; 8, VL; 9, GC; 10, F; 11, Hypoth; 12, CC; 13, CA; 14, mesencephalic FR; 15, CF.

Chkhenkeli, 1981, 1982; Ramamurthi and Kalyanaraman, 1982; Walker *et al.*, 1982; Sramka, 1985).

There are grounds to assert that stereotactic operations have significantly improved the surgical treatment of epilepsy. Yet, its rapid and successful development, as is always the case, has led to the emergence of new problems. For instance, the choice of the most "effective" subcortical structure for destruction is still an unsolved problem. This is evident from the large number of structures that are the targets for stereotactic operations in epilepsy: GP, VL, CM, CI, Subth, lamina medullaris interna, Am, Hipp, F, GC, CC, Hypoth, and others (Fig. 200). There are reasons to believe that in order to treat successfully the various forms of epilepsy, the manifestations of which are so diversified, an objective test must be used to select the most "effective" subcortical structure.

Henceforth, we shall discuss the surgical techniques used in only three forms of epilepsy—focal, generalized, and Kojevnikoff's (myoclonus epilepsy, described in Chapter 12).

6.2. Focal Epilepsy

Based on longstanding tradition, focal epilepsy includes those cases in which clinical, roentgenologic, or electrophysiological methods can demonstrate a localized organic brain lesion to be the cause of the attacks.

Almost a century ago, Jackson first expressed the idea that an epileptic attack begins in a hemispheric focus from which a pathological excitation spreads to adjacent cerebral structures. The result is a series of muscular contractions related to their representation in the cerebral cortex. As a result, convulsions may begin in the face, extend to the upper extremities or to involve the whole half of the body, eventually becoming a generalized fit. From experiments on monkeys as well as from clinical data, the preferential pathways along which an epileptic discharge spreads to involve the various cortical areas of both hemispheres as well as the many subcortical structures such as Put, VL, GP, and Am have been described.

In focal epilepsy, the focus is most frequently localized in the temporal lobe (28%) or on the convexity (22%) or the medial surface of the frontal lobe (8%); however, 21% of patients were found to have multifocal epilepsy (Zotov, 1977). Among the etiologic factors of focal epilepsy (with the exception of temporal lobe epilepsy), head injuries rank first, and brain tumors (mainly temporal and frontal meningiomas) second.

According to incidence, frontal lobe epilepsy ranks second after temporal lobe epilepsy (described in Section 6.5). However, the surgical treatment of this form of epilepsy is less effective than that of other localizations of the epileptic focus (Rasmussen, 1964, 1975; Talairach *et al.*, 1969, 1974a,b; Garsia Sola *et al.*, 1982). Over a period of 30 years, 250 patients with frontal lobe epilepsy underwent surgery at the Montreal Neurological Institute. In 75%, the attacks were caused by gliosis, cicatrices, or atrophic lesions of the frontal lobe and in the remaining 25% by brain tumors. Certain authors classify cingular epilepsy as a separate entity.

Frontal epileptic seizures are characterized by a number of phenomena that make the diagnosis certain (Rasmussen, 1964):

1. Rapid loss of consciousness with subsequent generalized convulsion without focal manifestations.

2. Rapid loss of consciousness with turning of the head and eyes to the opposite side followed by general convulsions.
3. Turning of the head and eyes in the opposite direction without loss of consciousness, possibly followed by a generalized seizure.
4. An epigastric aura alone or as the beginning of a generalized seizure.
5. Unusual sensations throughout the body ("heaviness," "weakness," etc.). As in the previous case, this may be the only manifestation of the seizure, or it may develop into a generalized convulsion.
6. Sudden disorders of thought ("forced thinking" after Penfield), which patients find difficult to explain.

Epileptogenic foci are very rarely localized in the temporal or occipital cortex; this appears in approximately 2–3% of the patients (Stepien *et al.*, 1980).

It has been demonstrated in recent years that CT investigations are very important in determining the indication for an operation in focal epilepsy and in choice of the surgical technique as well as in forecasting the results of surgical treatment. If CT reveals a more or less restricted lesion in the cerebral substance, the epileptogenous zone may not be confined or even localized in this area.

6.2.1. Surgical Procedures

After Horsley's first operation for focal epilepsy (extirpation of the motor cortex) in 1886, milestones in the surgery of epilepsy were the works of Foerster (1925), Foerster and Penfield (1930), and Polenov, who, in 1928 proposed a new operation termed subcortical pyramidotomy, section of the pyramidal tract with a special leukotome. After Foerster and Altenburger (1935) performed ECoG for the first time, this extremely valuable and informative technique was introduced into clinical practice by Penfield and his students (Penfield and Erickson, 1941; Penfield and Steelman, 1947; Penfield and Jasper, 1954). By this means, it was shown that the zone of epileptogenic brain tissue was, as a rule, much larger than the area of visible morphological changes.

The EP method (Chapter 4, Section 7.4) has lately been successfully used to determine the sensorimotor cortex in operations for focal epilepsy (Goldring and Gregorie, 1984).

We shall not discuss in detail traumatic epilepsy, which has been extensively studied; in such cases, the epileptogenic focus, as a rule, corresponds to the site of the former head injury. A subpial resection of the cerebral cortex in the epileptogenic zone, introduced by Penfield, has proven very effective in traumatic epilepsy. Walker (1949b) proposed subpial suction of the cortex through several orifices in the vascular arachnoid. Subsequently, this technique and several modifications were successfully used by neurosurgeons.

Cortical resection is a standard operation, which we shall only describe briefly. This operation is performed, as a rule, under potentiated local anesthesia, which allows observation of motor, sensory, and speech responses during electrostimulation and recording of the ECoG. The precise determination of the cortical areas is a very important stage of the operation and requires broad exposure of the cortex.

The determination of the epileptogenic focus and also of the sensorimotor area and speech zone (in the dominant hemisphere) is made by electrostimulation and ECoG. Many models of recording electrodes for exploring the cortex have been devised, usually held by a clamp to the margin of the craniotomy. Epidural recording, also useful, is made by strip electrodes made of Silastic and Teflon and containing four disks of stainless steel. After cortical resection, the EEG study has to be repeated to verify total extirpation of the epileptogenic focus.

On the whole, the results of surgical treatment for focal epilepsy by the classical open method may be considered very good. Rasmussen (1979a) summarized the largest series of focal nontumoral epilepsy operated on in the Montreal Neurological Institute from 1928 until 1974. Cortical resection was performed in 1407 patients. In 33% of them, the seizures were abolished completely, in 32%, the seizures became much less frequent, and in 34%, there was a moderate reduction in frequency. The extensive surgical experience accumulated at the Leningrad Institute of Neurosurgery indicates that positive results in long-term follow-up were obtained in 70% of the patients: in 31% there was a practically total termination of attacks, and in 39% the frequency diminished considerably (Ugrjumov *et al.*, 1969; Zotov, 1977). In another series of cortical resections of an epileptogenic focus, good results were obtained in 62–64% of the patients in follow-up periods from 1 to 12 years (Goldring and Gregorie, 1984). One of the chief reasons for ineffective operations in focal epilepsy is undoubtedly incomplete extirpation of the epileptogenic focus.

6.2.2. Stereotactic Procedures

Stereotactic operations for focal epilepsy (with the exception of temporal lobe epilepsy—see Section 6.5) are performed quite rarely. However, in those cases in which the epileptogenic focus is localized in a precisely established subcortical structure, such operations are possible and, in fact, expedient.

An interesting operating technique in multifocal epilepsy has been proposed by Zemskaya *et al.* (1975). After conventional craniotomy and the determination of the cortical epileptogenic focus with the aid of ECoG, an electrode is inserted stereotactically in one of the thalamic nuclei or in one of the mediobasal temporal structures projecting to the cortical focus. The subsequent surgical procedures are determined by the results of electrophysiological investigation. Based on these results, either the cortical focus is extirpated or the subcortical structure is destroyed or both are performed consecutively.

After experimental data indicated that the destruction of the anterior (VA) nuclei of Th inhibits epileptic attacks caused by penicillin, an attempt was made to treat cortical epilepsy by the stereotactic destruction of VA. The coordinates of that structure are 5 mm anterior to the midpoint of LI, 4 mm above that line, and 5 mm lateral to the midplane. Limited experience with these operations, however, has not yielded clear-cut results (Talairach, 1952; Ojemann and Ward, 1975; Mori *et al.*, 1982). In three patients with focal motor seizures, stereotactic destruction of the posterior limb of CI (posterior capsulotomy) (Jelsma *et al.*, 1973) produced a good clinical effect; however, permanent paresis of the hands occurred postoperatively in all the patients.

Talairach *et al.* (1974a,b) analyzed their own data as well as data from the world literature concerning the efficacy of surgical treatment of various forms of focal epilepsy. According to these data, the frequency of successful results in temporal epilepsy was 81%, in parietal, 68%, in perirolandic, 60%, and in frontal, 25%.

6.3. Generalized Epilepsy

According to the theory of Penfield and Jasper (1954), a "centrencephalic system" comprising part of FR and having two-way ties with both hemispheres of the brain is localized in the rostral brainstem. Paroxysmal activity in this system may cause general epileptic attacks, which Penfield considered a distinct form of epilepsy that he called centrencephalic epilepsy. Essentially, this form consists of primary generalized attacks (grand mal) without focal components and comprise 20% of all types of epilepsy in adults. Penfield believed that in these attacks the epileptogenic focus is localized in the brainstem or medial mesencephalic and diencephalic structures. Discharges from such a focus spread simultaneously to both hemispheres of the brain, causing a rapid loss of consciousness and general convulsions. Since no organic changes or pathological processes were found in these structures in this form of epilepsy, Penfield and Jasper assumed that centrencephalic epilepsy has a "functional" genesis.

This very interesting but not indisputable theory has triggered quite a long and continuing discussion. One objection to this theory was that centrencephalic fits are observed when there are organic lesions in the cerebral hemispheres as the result of tumors or operations on the frontal lobes. Another observation that contradicts this theory is that stimulation of various thalamic nuclei during stereotactic operation practically never induces epileptic attacks. On the basis of clinicopathological correlations, Williams (1965) raises doubts as to the occurrence of epileptic attacks in FR of the brainstem and presents formidable arguments indicating that thalamic nonspecific nuclei, in particular, medial and intralaminary, must participate in order to cause a generalized seizure.

In contradistinction to the classical theory of centrencephalic epilepsy, a number of authors believe that primary general attacks are triggered by paroxysmal activity in the cortex of the frontal lobe, particularly its medial and orbital surfaces. The absence of epileptic attacks in the frequently encountered truncal–subcortical lesions of vascular, inflammatory, traumatic, or other origin does not confirm the theory of centrencephalic epilepsy (Bancaud *et al.*, 1967; Zotov, 1976; Saradjishvili and Geladze, 1977; Karlov, 1978). In general, certain authors doubt the very existence of primary generalized attacks, believing that careful observation of all such patients will detect a focal phenomenon at the beginning of the seizure (Saradjishvili, 1980). Nevertheless, most researchers consider that clinical and EEG findings seem to indicate that epileptogenic foci in this form of the disease are most frequently localized in various deep-seated cerebral structures such as thalamic nuclei (Schott *et al.*, 1986) mesencephalic FR, or pons. At times, in the so-called paroxysmal phase of sleep, it is possible to observe a sharp increase in epileptic activity in deep-seated brain structures (Chkhenkeli, 1982).

The Wada test (intracarotid administration of sodium amobarbital), which causes a temporary functional "switching off" of one of the hemispheres, has been employed successfully for differentiating frontal and centrencephalic epilepsy. Typical changes in the somatosensory EP in centrencephalic seizures, in particular, bilateral negative waves with a latent period of 170–200 msec, have been described.

The surgical treatment of primary, generalized, genuine, or centrencephalic epilepsy still remains controversial. Several functional and stereotactic operations have been proposed for this purpose and tested in clinical conditions.

6.3.1. Callosotomy

It has been established that numerous commissures in the brain (CC, CA, CP, F, MI, and the hippocampal commissure) play a dominant role in the pathogenesis and generalization of an epileptic discharge. Numerous experimental investigations have revealed that the most important is the main cerebral commissure, the CC. V. Horsley (1886) demonstrated in animal experiments that the cutting of forebrain commissures prevents the spread of epileptic discharges. In 1940, Erickson confirmed that an epileptic discharge could disseminate from one cerebral hemisphere to another via CC. This prompted the idea of performing surgical dissection of CC in epilepsy. That operation was given the name "cerebral commissurotomy" or "callosotomy." It was first performed by Van Wagenen and Herren in ten patients in 1940. It is quite natural that such an operation is indicated only in those cases in which there are very frequent attacks and changes in the personality of the patient, when prolonged treatment with anticonvulsants is totally ineffective, and when there are contraindications to conventional operations aimed at destroying the epileptogenic focus such as the presence of numerous cortical foci or a focus located in a functionally important zone.

Callosotomy is performed under endotracheal anesthesia with the patient in a semisitting position. Wilson *et al.* (1978) recommended approaching CC by making two transverse cutaneous incisions, one parallel to and slightly anterior to the coronary suture, and the second in the posterotemporal area. A crown trephine is used to make two burr holes beneath these incisions, after which the dura mater is incised, the medial edge of the hemisphere is retracted, and the bridging veins are coagulated. Although dissection of the veins does not lead to significant complications (Bouvier *et al.*, 1983), in order to preserve them, it has been recommended that angiography be performed before surgery, and, in accordance with the data thus obtained, the burr hole or bone flap be placed so as to avoid large veins (Rayport *et al.*, 1983). Under an operating microscope, the CC fibers are dissected to the ventricular ependyma in both burr holes. Anteriorly the dissection includes the rostral part of CC, whereas the posterior edge of the section is the arachnoidal membrane covering the vein of Galen. In addition to dissection of CC, the subjacent hippocampal commissure is sectioned.

Rayport *et al.* (1983) perform callosotomy in two stages. In the first operation the anterior portion of CC (rostrum, genu, the anterior part of the corpus) is dissected, and at the second stage, the posterior portion of CC (splenium).

Patients tolerate callosotomy comparatively well. It rarely leads to serious complications such as impaired memory and behavior. The comparatively little experience in performing this operation indicates that in many cases satisfactory results have been obtained in general epilepsy. Nevertheless, since there may be mental disorders and other complications following this operation, certain authors recommend dissecting not the entire CC but only the anterior portion (Huck *et al.*, 1980). The presence of an epileptogenic focus in one of the frontal lobes with the dissemination of a discharge to the other hemisphere is considered the main indication for anterior callosotomy. Such an operation not only decreases the number of attacks but, in the majority of cases, leads to a substantial improvement of mental functions as shown by neuropsychological tests (Marino *et al.*, 1981; Bouvier *et al.*, 1983). Rayport *et al.* (1983) reported that callosotomy in nine patients with drug-resistant epilepsy did not result in a total arrest of the attacks, although their frequency and severity diminished noticeably.

Bouvier *et al.* (1983) are of the opinion that anterior callosotomy produces good results, but section of only the anterior portion of CC may not be effective. Wilson *et al.* (1978) describe only a temporary disappearance of general attacks after anterior callosotomy; a second operation with section of the posterior portion of CC produced a lasting termination of seizures.

Harbaugh *et al.* (1983) have summed up their many years of experience in performing open commissurotomy in 20 patients with various forms of epilepsy. As did Rayport *et al.* (1983), the authors recommend dissection of CC and the subjacent hippocampal commisure in two stages to diminish the operative trauma and the number of complications. This operation is more effective for the treatment of unilateral hemispheric epileptogenic foci.

Partial callosotomy may also be performed by the stereotactic method. Since closed stereotactic destruction of CC involves the risk of damaging one or both anterior cerebral arteries passing along its dorsal surface, the focus of destruction should be at least 10 mm from the medial plane and in the most anterior portions (Schaltenbrand and Wahren, 1982). A marked decrease in the frequency of seizures after selective partial callosotomy in 35 cases with multiform seizures was noted by Marino (1985).

In severe cases of general epilepsy that do not respond to any therapy, it is possible to disconnect the cerebral hemispheres. This operation involves section not only of CC along its entire length but also of CA, MI, the hippocampal commissure, and CP (Bogen and Vogel, 1965; Bogen *et al.*, 1969). It is regarded as an alternative to hemispherectomy. The experience with such operations is extremely limited; however, in certain cases a good result with total disappearance of epileptic attacks and even an improvement in the results of psychological tests has been obtained (Wilson *et al.*, 1975). At the same time, a so-called "disconnecting syndrome" may develop following these operations. All aspects of callosotomy are described in Reeves (1985).

We shall not dwell on a description of hemispherectomy, first proposed by Dandy for malignant gliomas (1928) and developed for treating epilepsy by Krynauw (1950) since this operation is not functional neurosurgery. One can only state that the literature gives numerous descriptions of good results in the most severe forms of epilepsy with pronounced mental disorders (Ralston, 1962; Belyaev and Shevchenko, 1972).

6.3.2. Stereotactic Procedures

There have been many important reports of the treatment of general epilepsy by the stereotactic method (Narabayashi and Mizutani, 1970; Bouchard and Kim, 1974). The group of Jinnai (Jinnai, 1966; Mukawa *et al.*, 1975; Jinnai and Mukawa, 1976; Jannai *et al.*, 1976), over the course of many years, has published the results of surgical treatment of severe forms of generalized epilepsy by bilateral stereotactic destruction of CF. Experimental investigations have demonstrated that such surgery cuts off the pathways whereby the epileptogenic discharge passes from the cortex to the brainstem as well as the striothalamic pathways. It has also been demonstrated experimentally that destruction of CF substantially raises the convulsive threshold of motor zones in the cortex.

Jinnai operated on 14 patients with genuine and 31 patients with symptomatic epilepsy. The center of the destructive focus had the following coordinates: 2 mm posterior to the middle point of LI, 4 mm below that line, and 8 mm lateral to the medial plane. The value of this work lies in the fact that the follow-up period of these patients ranges up to 10 years. The authors demonstrated that destruction of CF yields good long-term results; in more than two-thirds of the patients, the general seizures either ceased or became less frequent. Moreover, there was an improvement in the EEG findings. Bilateral operations were performed in many of the cases, which, as the authors point out, is necessary especially in genuine epilepsy.

Ramani *et al.* (1980) performed bilateral stereotactic destruction of CF in six patients with intractable epilepsy. Long-term (approximately 3 years on average) follow-up studies revealed that in four patients the generalized attacks became less frequent (by 50–75%). In seven of 11 patients, generalized attacks either ceased entirely or became much less frequent after stereotactic destruction of CF (Bojik, 1980).

Mullan *et al.* (1967) reported the results of unilateral stereotactic destruction of VL in nine patients with focal and general epilepsy; there was a marked improvement in six of them. The observations of other authors as well as our own correlate with the data on the efficacy of the stereotactic destruction of VL in epilepsy (Laitinen, 1967; Romodanov and Laponogov, 1971; Mundinger, 1976; Ugrjumov *et al.*, 1974; and others).

Posterior hypothalamotomy (Chapter 13, Section 5.2.3) in generalized epilepsy seems to be not effective. Sano *et al.* (1970b), who developed this operation, reported on 22 patients who underwent such surgery. Fits stopped in only one patient and became less frequent in six others.

The stereotactic destruction of another two subcortical structures, GC (Diemath *et al.*, 1966) and Put (Hori *et al.*, 1968), yielded hopeful results in a brief series of observations; however, in subsequent years, these operations have not been extensively used in the treatment of epilepsy.

In seven patients with generalized fits without any focal component, diagnosed as centrencephalic epilepsy, Jelsma *et al.* (1973) stereotactically destroyed the second quarter of the posterior limb of CI, including fasciculus thalamicus, VOa, and nucleus lateralis polaris. The authors named this operation anterior capsulotomy to distinguish it from posterior capsulotomy, whose objective it is to sever the corticospinal tract. In all cases, anterior capsulotomy proved ineffective. From several other works describing stereotactic destruction of various portions of CI, no conclusions are possible as to the efficacy of this procedure (Kalyanaraman and Ramamurthi, 1970).

Summing up, one may assume that functional operations on subcortical structures for the relief of generalized epilepsy are without a doubt effective, yet their value is still far from being proved.

6.3.3. Chronic Stimulation

A new trend in the treatment of generalized and sometimes focal epilepsy is the employment of stimulation by chronic implanted electrodes. Such attempts were prompted by numerous experimental investigations that showed that there are cerebral structures that inhibit the discharges of epileptogenic foci or prevent their dissemination to other cerebral structures (Okudjava, 1969). Today there are no longer doubts that such inhibiting structures are present in the human brain. First and foremost, they include the cortex and the deep-seated nuclei of Cer, NC, and quite possibly CM, intralaminary nuclei, and other structures (Kambarova, 1981; Šramka *et al.*, 1979, 1980, 1983; Cooper, 1976; Bechtereva, 1980; Chkhenkeli, 1982).

It has been established that stimulation via electrodes implanted in NC or ND sharply arrests the epileptic discharge both in the deep-seated nuclei of the temporal lobe and in the thalamic nuclei. This can be clearly seen by simultaneous recording of activity from Am or Hipp (Šramka *et al.*, 1980). In stereotactic operations for temporal lobe epilepsy, it was also shown that low frequency (4–6 Hz) stimulation of one NC (especially the ventral part of the head) inhibits epileptic activity in both Ams and interrupts a convulsive fit (Chkhenkeli, 1982). Even after brief stimulation, such an inhibiting effect may last for several hours. A similar effect was obtained by stimulating ND, for which a high frequency stimulating current is recommended. For NC, a low frequency current is advisable, since high frequency stimulation (50 Hz) of NC may intensify epileptic activity.

Several authors have recently reported promising results from chronic electrostimulation of ND, NC, Hipp, CM, and other structures (Bechtereva, 1980; Chkhenkeli, 1978b, 1982; Šramka, 1985; Šramka *et al.*, 1980). To date, only a few papers have appeared about the therapeutic action of stimulating an epileptogenic focus using implanted electrodes (Bechtereva *et al.*, 1977; Šramka, 1985; Šramka *et al.*, 1976, 1979). The mechanism of the antiepileptic action of stimulation remains unclear, and the therapeutic value of the method still cannot be considered proven. However, stimulation of the epileptogenic focus in status epilepticus may, paradoxically, eliminate that severe and dangerous condition (Bechtereva, 1980).

One of the new approaches to the treatment of various forms of epilepsy is chronic stimulation of Cer proposed by Cooper *et al.* (1973, 1976a). Since this operation is also used to treat spasticity in cerebral palsy, the technique is described in greater detail in Chapter 9. Numerous experimental and clinical data on the inhibiting influence of Cer on spinal cord motoneurons form the physiological basis for this operation. It is assumed that prolonged stimulation of Cer with implanted electrodes inhibits Purkinje cells, which activate epileptogenesis. Stimulation of the anterior portions of Cer depresses epileptic activity in Am and Hipp in temporal lobe epilepsy (Nashold *et al.*, 1975).

The technique of chronic stimulation of Cer is as follows: After making a small burr hole in the posterior cranial fossa, platinum electrodes are implanted on the cerebellar cortex. So far, however, it has not been established whether stimulation of the anterior or posterior surface of Cer is more effective. This operation, judging by the limited experience of Cooper *et al.* (1976a), proved effective in various types of epilepsy (psychomotor, myoclonus, grand mal). A pronounced improvement, even total elimination of seizures, was noted in ten of 15 patients in a follow-up period of 3 years. Another paper from the same clinic shows that chronic stimulation of Cer resulted in a considerable reduction of seizures in 18 of 29 incurable epileptics (Cooper, 1978). Out of 12 patients with grand mal, chronic stimulation of Cer led to a total disappearance of attacks in five patients and to less frequent attacks in six. In two of three patients with petit mal epilepsy, the seizures diminished in frequency (Davis *et al.*, 1982). The authors reported on the use of chronic cerebellar stimulation in 32 patients with seizures not responsive to medical treatment. Most of these patients were also afflicted with spastic cerebral palsy. The general results were as follows: seizures stopped in 57% of cases, became less frequent in 28%, and remained unchanged in 15%. It is, however, noteworthy that other authors do not confirm such a beneficial effect by chronic cerebellar stimulation (Strain *et al.*, 1979).

A valid evaluation of the efficacy of chronic stimulation of Cer for epilepsy will depend on more surgical experience and follow-up studies.

Various authors have expressed the viewpoint that stimulation methods are still in the experimental stage (Rossi, 1980). In spite of the good results mentioned above, certain investigators do not recommend prolonged electrostimulation of limbic structures or neocortex, since this could produce mirrorlike foci of epileptogenic activity (Livingston, 1975). Other authors deny the efficacy of chronic electrostimulation (Romodanov, 1980).

It is also known in that in epileptics even short-term electrostimulation of certain subcortical structures may induce an aura typical of their seizures and followed by an epileptic attack. For instance, stimulation of Am, Hipp, and the hippocampal gyrus in epileptics via implanted electrodes quite frequently leads to visual and other hallucinations (Mahl *et al.*, 1964; Weingarten *et al.*, 1977). All this gives grounds for apprehension that prolonged stimulation of various cerebral structures may increase the frequency of epileptic seizures.

It may be considered firmly established that at the present moment the main aim of a neurosurgeon in treating epilepsy and assuring the success of that treatment is to pinpoint the epileptogenic focus and destroy it by one of the available methods. Consequently, chronic stimulation of the focus should be regarded only as an accessory treatment that, first of all, helps to define the above-mentioned focus (or foci); only in specific cases can the focus be stimulated for therapeutic purposes. However, when it is impossible to discover the epileptogenic focus by all the present techniques, one may resort to chronic electrostimulation or to destruction of the pathways that disseminate the epileptic discharges or other cerebral structures that activate epileptic mechanisms in the brain.

6.4. Kojevnikoff's Epilepsy

The original form of epilepsy, described by A. Y. Kojevnikoff in 1895 and bearing his name, is caused by a chronic CNS lesion resulting from tick-borne encephalitis. This disorder was discovered in West Siberia and studied by Soviet scientists Zilber, Chumakov, Grashchenkov, and others in 1937–1940.

Encephalitis usually begins in the summer and only rarely in the autumn. The latent period after a tick bite is approximately 2 weeks. The acute period of tick-borne encephalitis is severe and also lasts a fortnight. The syndrome of Kojevnikoff's epilepsy (KE) usually develops 3–4 months after the acute period of tick-borne encephalitis. This disease most frequently occurs in children (Shubin, 1960; Zucker, 1963).

Pathological studies of the brains of deceased patients as well as surgical specimens of the motor cortex have revealed that in KE developing on the basis of tick-borne encephalitis, various structures are involved in the inflammatory process: first and foremost, the motor cortex of the cerebral hemispheres, the basal ganglia, Cer, and the brainstem. In the initial period of the disease (up to 6 months), the inflammation is acute and manifested by hyperemia of the cortex and meninges, lymphoid and plasma cell infiltration of the vascular walls, perivascular edema, and neuronal changes as in the acute form of Nissl's disease (edema and swelling of cells, uneven distribution of Nissl granules, nuclear ectopia, etc.). When the disease lasts more than 1 year, besides the acute inflammatory phenomena there may be signs of chronic inflammation such as thickening of the vascular walls and, in places, hyalinosis. On the background of perivascular and pericellular edema, it is possible to observe edematous and corrugated neurons of various shapes with a deformed nuclear membrane. According to other neuropathological reports, the changes were present only in the motor cortex and adjacent areas, resulting from various causes (most frequently encephalitis, less frequently infarction, injuries, and others) (Thomas *et al.*, 1977).

For decades the question of whether KE is a cortical or subcortical lesion has been discussed in the literature. Proceeding from the data of biopotential recordings from implanted electrodes, researchers came to the conclusion that KE has a cortical genesis (Bancaud *et al.*, 1970; Siegfried and Bernoulli, 1976). A quantitative analysis of the simultaneous recordings of stereo EEG and EMG as well as direct stimulation of the area confirms such a conclusion (Wieser *et al.*, 1978).

The primary epileptic focus, determined with the aid of EEG, ECoG, and ESubCoG, is localized in the motor cerebral cortex. Under the influence of stimulation from the focus, paroxysmal activity spreads to the deep-seated brain structures. If this secondary focus in the subcortical structures exists for a long time, it becomes "independent" and generates epileptic activity independently of the primary focus. On the basis of clinical and electrophysiological investigations, three forms of KE are defined: cortical, corticosubcortical, and subcortical (Nesterov, 1967).

According to current theory, KE has a combined corticosubcortical basis. Clinically, KE is usually manifested by an almost permanent myoclonic hyperkinesia, the convulsions of which are very polymorphic in localization, intensity, frequency of myoclonus, etc. Besides muscular jerks typical of KE, there are sometimes athetoid and choreic hyperkinesias.

Hyperkinesias in KE usually affect one half of the body. Bilateral myoclonias occur in only 10% of the cases (Nesterov, 1967). Myoclonic twitchings of various frequencies are usually nonrhythmic. Quite frequently they occur in series with pauses between them lasting from several seconds to several minutes. During purposeful movements, emotional stress, and auditory and visual stimulation, the hyperkinesias become more pronounced, but they diminish at rest and disappear in sleep.

During their evolution, the hyperkinesias may be stable over the course of many years, usually without any tendency toward generalization. Quite frequently more or less pronounced pareses develop as well as muscular contractures in those extremities affected by myoclonic twitching. As KE develops in childhood, one may frequently note trophic disorders in the afflicted extremities, which are stunted and are usually 3 to 4 cm shorter than the healthy extremities.

Generalized epileptic attacks represent the second typical clinical manifestation of this disease. As a rule, such seizures, although uncommon, begin with (in 19% of the cases, according to Nesterov, 1967), an intensification of myoclonic twitching with the subsequent spread of convulsions to other extremities, developing into a generalized attack with loss of consciousness.

6.4.1. Surgical Treatment

Medical treatment of this disease is not very effective. In view of this, over the course of several decades attempts were made to treat the disease by surgery: injection of alcohol into the cerebral cortex (Razumovski, 1913), extirpation of the motor and premotor cortex, and section of pyramidal pathways at different levels. All these attempts, however, were only partly successful, and the chief drawback of "pyramidal" operations was the paralysis connected with them. Indeed, if the myoclonic seizures did disappear, a spastic paresis of the corresponding extremities developed.

6.4.2. Stereotactic Procedures

The stereotactic method opened up new prospects in the treatment of KE. Yet, until now, the results of very few operations have been reported in the literature. We have operated on three patients with this disease. In two, the myoclonic convulsions practically disappeared, and the jerkings substantially diminished in the third patient.

Considerable clinical experience in the stereotactic treatment of KE has been published by Shefer and Nesterov (1965). In the ten patients they operated on, the disease was a result of tick-borne encephalitis. Myoclonic hyperkinesias in the face and extremities were combined with generalized epileptic seizures. After stereotactic destruction of VL, the hyperkinesias disappeared in half of the cases and diminished in the rest. Approximately the same results were obtained with respect to generalized epileptic attacks.

One may conclude that at the present time, a stereotactic operation on thalamic nuclei represents the only effective method of treating KE.

6.5. Temporal Lobe Epilepsy

Over the past 30 years, the greatest success in the surgical treatment of epilepsy has been achieved, first and foremost, in temporal lobe epilepsy (TLE). An extremely valuable contribution to the investigation of the pathogenesis, clinical manifestations, and surgical treatment of this disease was made by the famous Canadian neurosurgeon and neurologist Penfield and his school, as well as by Walker and Falconer.

As the name itself indicates, TLE is associated with lesions in the structures of the temporal lobe and consequently should be classified as a focal epilepsy. At the same time, this disorder has such a clear-cut morphological substrate and such typical peculiarities of its clinical manifestation that, in 1958, an international symposium decided that it should be designated as a separate nosological form. An important factor in this decision was the fact that TLE occurs relatively frequently. According to the data of many authors, TLE patients make up approximately half of all epilepsies. In Great Britain, where approximately one per

1000 of the population (Falconer, 1965) has the focal forms of epilepsy, the epileptogenic focus occurred most frequently in the temporal lobe (50–60%).

6.5.1. Etiology and Pathogenesis

The etiology of TLE is quite diverse. The most predominant etiologic factors include prenatal and natal disorders, various contagious diseases, mainly in childhood, and head injuries. However, in approximately one-third of the patients, the etiologic factor is not clinically established (Stepien *et al.*, 1980).

In the majority of cases, TLE is caused by an organic lesion deep in the mediobasal structures of the temporal lobe, particularly Am and Hipp. In the majority of patients, the temporal lobe is affected on both sides. The most frequent cause of TLE, based both on data from autopsies and morphological investigations of the temporal lobe removed at operations, is sclerosis of the medial part of the temporal lobe, including Am and Hipp (incisural sclerosis after Penfield or mesial sclerosis after Falconer). Such sclerosis of the deep-seated temporal nuclei was found in half of the patients who underwent temporal lobectomy. Various other pathological processes (cicatrices, infarctions, small tumors, etc.) were observed in the other half (Falconer and Taylor, 1968). The most pronounced morphological changes are observed in Hipp: disintegration of pyramidal cells, hyperplastic astrocytes and microglial elements, and changes in the vessels. It is presumed that sclerosis of Hipp is the result of a birth injury caused by damage by the edge of the tentorium during delivery or by occlusion of the choroid artery supplying hippocampal gyrus in the notch of the tentorium. Falconer (1965) considers that sclerosis in the deep-seated structures of the temporal lobe stems from "anoxic episodes" in early childhood.

The epileptogenic focus in the deep-seated structures of the temporal lobe has a higher rate of metabolism than the adjacent cerebral tissue, which is an indication of excessive O_2 consumption that can be determined with implanted polarographic electrodes (Shefer *et al.*, 1970b). This had led to the assumption that incisural sclerosis is a secondary phenomenon caused by the high metabolism of the neurons in the epileptogenic focus during seizures. The accompanying hypoxia leads to necrosis of the cellular elements of Am and Hipp.

Besides incisural sclerosis in removed specimens, quite frequently one may see minute, almost microscopic "cryptogenic" glial tumors. In the majority of cases, these structures are more malformations than true blastomas (Falconer, 1965). And last but not least, in approximately a quarter of the operated cases, no morphological changes accountable for the temporal seizures were found in the excised temporal lobe.

Quite frequently the epileptogenic focus in the temporal lobe, as time passes, leads to the formation of secondary foci in the mesencephalic FR irrespective of the primary focus (Chkhenkeli, 1981). This may account for the approximately 30% of patients with a clinical picture of TLE who have an epileptogenic focus or foci in other parts (lobes) of the brain (Zotov, 1977). Pneumoencephalograms or CT in TLE frequently reveal an enlarged, displaced, and deformed ventricular system, especially the temporal horns, the cisterns, and the subarachnoid spaces.

The etiology of the condition in 1100 TLE patients who underwent surgery at the Montreal Neurological Institute during the past 50 years was as follows: birth injuries, 24%; inflammatory and cicatricial processes, 15%; brain tumors and other mass lesions, 15%; postnatal head injuries, 12%; other causes, 12%; unknown etiology, 22% (Rasmussen, 1979b).

It has been established that the Am is the pacemaker of epileptogenic activity in the homolateral Hipp (Chkhenkeli, 1978b). Yet the occurrence of epileptic attacks confined to Am or to Hipp is observed rarely. Stereo-EEG data indicate that both structures are involved more or less simultaneously in the epileptic discharge (Wieser and Yasargil, 1982).

6.5.2. Clinical Aspects and Diagnostic Procedures

This disorder manifests itself at a younger age, usually between 10 and 20 years, than some forms of epilepsy. For example, in 62% of the cases, the attacks occurred below the age of 15 years (Belyaev, 1970), and in 75%, below the age of 20 years (Stepien *et al.*, 1980). Falconer (1972b) considers this form of epilepsy to be a childhood disorder which should be diagnosed at an early age.

Although generalized convulsive attacks (grand mal epilepsy) are not commonly observed in TLE, nevertheless, when the epileptogenic focus is localized in deep-seated structures of the temporal lobe, typical primary generalized attacks may occur, complicating the diagnosis of TLE (Saradjishvili, 1980). Six main types of attacks in TLE may be recognized:

1. Sensory (auditory, gustatory, olfactory hallucinations).

2. Affective (dysphoria, states of fury, aggressiveness, fear, alarm, ecstasy, suicidal thoughts).
3. Autonomic (abdominal, cardiac).
4. Dysmnesic (amnesia, illusions, hallucinations, reminiscences).
5. Automatisms or psychomotor attacks (pharyngooral, simple and complex locomotor acts).
6. Twilight states of consciousness (confusion, etc).

In TLE, these types of attacks may occur singly or together in various combinations.

The aura is particularly important for the diagnosis of TLE. The precursors of an attack are, as a rule, polymorphic but clearly defined. They may be the same as observed in other forms of focal epilepsy. Still, there are certain auras that are typical of TLE: drowsiness, olfactory, auditory, and gustatory hallucinations, depersonalization, *déjà vu* states, and some others. Commonly (in approximately three of four patients), autonomic auras are present.

So-called psychomotor attacks, occurring in the majority of cases, are most typical of TLE. In such an attack, on the background of altered consciousness, the patient performs complicated locomotor acts devoid of any purpose. In such cases, stereotype movements, usually short in duration, are quite frequent. Typical convulsive seizures occur in approximately half of the cases after the psychomotor manifestations.

Proceeding from stereo-EEG data recorded from deep-seated structures and the clinical picture, certain authors single out five types of psychomotor attacks caused by lesions in various mediobasal structures, namely, mesiobasal limbic cortex, temporal pole, posterotemporal neocortex, opercular (insular) cortex, and frontobasal cingular cortex (Wieser *et al.*, 1980; Wieser and Yasargil, 1982).

Besides psychomotor seizures, TLE may produce typical emotional and personality disorders with excitation, euphoria, aggressiveness, violent emotional outbursts, fits of fear, anger, dysphoria, etc. Paroxysmal mental disturbances in TLE are often equivalent to an epileptic attack. Changes in character and behavior as well as impaired memory are observed in the majority of cases and serve as the most obvious clinical manifestations of the disease.

When the epileptogenic focus is localized in the left temporal lobe, the attack often involves auditory hallucinations, amnestic phenomena, or complicated motor automatisms, and when localized in the right temporal lobe, affective paroxysms and *déjà vu* phenomena. Impaired memory in TLE is also more pronounced with left-sided epileptogenic foci.

The majority of TLE cases displayed pronounced hyposexuality, although in certain cases, the opposite phenomenon was observed. After resection of the temporal lobe, sexual activity not infrequently became normalized (Walker, 1972).

In TLE, formidable difficulties often arise both in respect to diagnosing the affected side, i.e., the right or left temporal lobe, and in differentiating unilateral from bitemporal epilepsy. It is also necessary to emphasize the danger of overspecifying TLE, because the clinical picture typical of TLE may also be observed in foci beyond the temporal lobe.

The EEG represents one of the main methods for diagnosing TLE. This investigation should be conducted not only "statically" but also with methods of activating pathological rhythms (phono- and photostimulation, hyperventilation, administration of drugs provoking epileptic activity, etc.) as well as telemetric and sphenoidal EEG recordings (Rasmussen, 1983; Manzano *et al.*, 1986). Another effective method for diagnosing TLE, as well as other forms of epilepsy, is EEG recording in light sleep, when pathological changes in the temporal EEG appear much more frequently than when the patient is awake.

Temporal lobe epilepsy is characterized by typical EEG changes (peak–wave complexes) recorded from the mediobasal temporal areas from one or both sides. Discharges from Hipp through the limbic circle can spread to GC and be recorded by scalp EEG (Grindel and Bragina, 1974). However, many authors point out that conventional scalp EEG does not allow one to determine precisely the site and the borders of an epileptogenic focus or even the side where it is located. That is why the most informative method of diagnosing TLE and determining the focus today is to record the biopotentials of the deep-seated structures of the temporal lobe with multilead implanted electrodes (stereo-EEG) (Fig. 201). Approximately ten electrodes with a diameter less than 1 mm each having about ten contacts are implanted in the Am, Hipp, and hippocampal gyrus through several 2- to 3-mm burr holes in the temporal bone (Fig. 202). The biopotentials are usually recorded over a long period of time, sometimes several weeks. From these recordings one selects the contacts that give the most valuable information and continues to study the recordings.

In stereo-EEG, it is possible to determine the af-

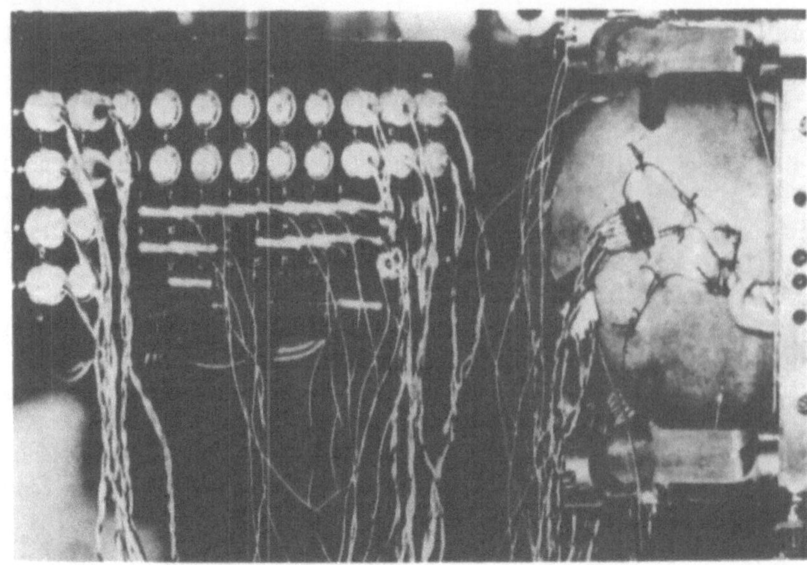

FIGURE 201. Stereo-EEG localization of epileptogenic area within the brain. Flexible multilead electrodes are inserted by the stereotactic technique. The patient's head is held by the stereotactic frame. Electrodes are attached to the skin (Talairach *et al.*, 1974a).

flicted side by counting the epileptic discharges on both sides; on the side of the focus their number is nine to ten times greater (Falconer, 1965). It is important to record stereo-EEG during a spontaneous seizure. Siegfried and Hood (1983) believe that such recordings give more diagnostic information than EEG recordings in the interictal period or in seizures induced by electrostimulation or by pharmacological agents.

6.5.3. Surgical Treatment

The treatment of TLE remains a complicated problem. Compared to other forms of epilepsy, TLE is

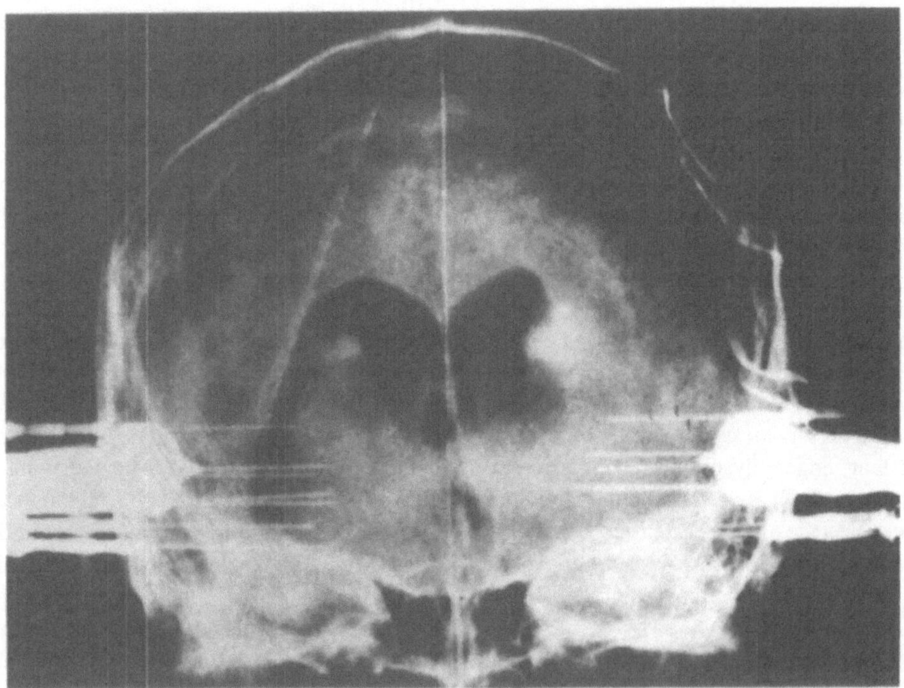

FIGURE 202. Multiple electrodes are stereotactically inserted bilaterally in the deep structures of both temporal lobes (AP film; Todd, 1972).

frequently resistant to medical therapy. In approximately one-third of the cases, long-term anticonvulsant therapy has no effect. In such cases, providing the conditions described in Section 6.1 are met, surgery is indicated. It has been established that a long period of seizures prior to the operation (over ten years) substantially reduces the outlook (Jensen, 1976). This, without a doubt, is a weighty argument in favor of early operation. The main contraindications to operation are pronounced debility with an IQ below 60 (Falconer, 1965) and an epileptic psychosis.

Over the course of almost three decades, the main and most effective operation for this disorder has been partial resection of the temporal lobe, temporal lobectomy.

6.5.3a. Temporal Lobectomy. The technique of performing this operation was thoroughly elaborated by Penfield and Baldwin (1953), who proposed subpial resection of the temporal cortex including Am and Hipp. We shall dwell only briefly on the classical operation, which is described in detail in literature. It involves a resection of the anterior half of the temporal lobe (to the inferior horn), including its deep-seated structures and sometimes the insular area.

Electrocorticography (ECoG), a mandatory component of such operations, makes it possible to detect the epileptogenic focus, establish its boundaries, and, after resection of the temporal lobe, control the effect of its removal. For this purpose, EEG recordings are made during the operation both from the surface of the temporal lobe and from implanted electrodes in its medial structures. An EEG investigation of the activity of the mediobasal portions of the temporal lobe with the help of implanted electrodes is the sole method of determining the affected side when this proves impossible on the basis of clinical data.

We shall not describe the technique of the standard temporal lobectomy. The majority of neurosurgeons perform the so-called standard resection of the temporal lobe with slight individual variations on the basis of electrographic data. One of the pioneers in the surgical treatment of TLE, Falconer (1965), and other neurosurgeons preferred to perform this operation under local anesthesia. With ECoG it is possible to pinpoint the location of the focus as well as to induce the aura or seizure typical for the given patient by electrostimulation of the deep-seated structures.

After conventional osteoplastic craniotomy, the cortex is incised along the superior temporal gyrus, and then the incision is continued downwards to the floor of the middle cranial fossa, and the inferior horn is opened up. This is followed by removal of a "block" of brain tissue approximately 5–6 × 3–4 cm in size, which includes the pole of the temporal lobe and its deep-seated structures, Am and the oral part of Hipp. This classical resection involves the removal of approximately 30–35 cm^3 of brain tissue, although the volume of Am and Hipp, the destruction of which is the object of the operation, comprises only 10% of the volume of the excised temporal lobe (Šramka and Nadvornik, 1975a). Niemeyer (1958) proposed removal of Am and Hipp by a transventricular approach via the inferior horn V_1; however, this operation was employed extensively only in the recent period when the microsurgical method was developed (see below).

Electrocorticograms made immediately after temporal lobectomy usually show the disappearance of epileptic activity. In those cases in which spiking is observed on the border of the excised temporal lobe, additional suction removal of the adjacent firing cortex, including the insular cortex, is usually performed. However, as Falconer (1965) and other authors have demonstrated, even if there is spike activity around the resected zone, seizures may cease almost completely after the operation. From this, the conclusion was drawn that, as a rule, there is no need to remove the insular cortex. If the epileptogenic focus is confined to the cortex of the temporal lobe, which occurs very rarely, subpial resection of the cortical focus may be sufficient (Rasmussen, 1980, 1983).

Classical temporal lobectomy is undoubtedly an effective operation. In the majority of cases the psychomotor attacks are abolished or their frequency is significantly reduced. The long-term results of these operations reported by different authors vary considerably. Nevertheless, if one assesses these results in general, in 30–40% of the cases the attacks cease after the operation, and they are considerably diminished in frequency in another 20% (Romodanov *et al.*, 1971b; Stepien *et al.*, 1980). An improvement was observed in 74% of the 125 TLE cases operated on by the open method (Shefer *et al.*, 1970a). Approximately similar results were reported by Van Buren *et al.* (1975): of 143 TLE cases, attacks totally ceased after surgery in 21% of the cases and considerably improved in 46%. Temporal lobectomy for medically refractory seizures in 50 children yielded the following results: 54% of the cases were seizure-free, 24% had only occasional seizures without loss of consciousness, 10% had fewer seizures, and 12% were unchanged. The Wechsler Intelligence Scale had shown practically no changes after surgery (Meyer *et al.*, 1986).

Jensen (1975, 1976) collected and analyzed a vast

amount of literature throughout the world on the results of temporal lobectomy. This statistical review summarizes 2282 operations performed from 1928 to 1973. According to the summary, a total arrest of TLE attacks varied within the range of 27.8 to 61.8%, the average being 43.6%. However, apparently some 15 to 20 years ago, temporal lobectomy reached the limit of its possibilities in the treatment of TLE. Moreover, in the process of accumulation of clinical experience and analysis of follow-up results, certain shortcomings of temporal lobectomy became evident, namely, significant trauma to the brain, at times causing intellectual disorders and other complications, and the relatively frequent late recurrence of attacks.

Memory impairment of various degrees after removal of Hipp represents a serious problem in temporal lobectomy. According to Walker (1957), this is observed in 10–15% of the operated patients. Other authors point to a considerably higher frequency of memory disorders after temporal lobectomy. After the removal of the greater portion of Hipp, these impairments were more pronounced and recovered only to a small degree. Investigations of the Penfield school demonstrated that memory impairments occur much more frequently in bilateral destruction of Hipp. This gives grounds to assume that in unilateral destruction of Hipp, its function is compensated by Hipp on the other side. Pribram (1960) noted the peculiar nature of the memory impairments in these patients: they seem to lose the ability to draw on their memories, whereas formal tests for memory did not yield any gross deviations from the norm. A pronounced Korsakoff's syndrome develops in certain cases (Bein and Boreiko, 1972).

Another frequent complication of this operation is homonymous hemianopsia (Meerson and Tetz, 1975), total or partial, in 40% of the cases according to a statistic review by Jensen (1975). The resection of the posterior temporal foci in the dominant hemisphere runs the risk of postoperative speech disturbances.

After resection of both temporal lobes, it is often possible to observe pronounced mental and behavioral disorders, visual agnosia, and hypersexuality (Klüver–Bucy syndrome). The greater the volume of resection, the more pronounced are these complications. Opening the temporal horn, inevitable in temporal lobectomy, may subsequently cause cerebrospinal fluid disorders. Hemipareses develop in approximately 5% of the cases after the operation.

Late relapses of attacks are possible after resection of the temporal lobe with early good results. Recurrences within a period of 3 years developed in 30% of operated patients, and from 3 to 9 years in another 6% (Skryabin *et al.*, 1976). Unilateral classical lobectomy produces practically no effect in mutilfocal epilepsy, in particular, when there are epileptogenic foci in both temporal lobes, which occurs in approximately 30% of patients. At the same time, bilateral temporal lobectomy is strongly contraindicated because of the possibility of severe memory impairment.

Recently, Wieser and Yasargil (1982) proposed an original and, in all likelihood, promising operation involving selective microsurgical removal of Am and Hipp by an open method (amygdalohippocampectomy). Unlike the large craniotomy that was used earlier, a minor frontosphenotemporal craniotomy (the pterygoid approach of Yasargil) is performed. After opening of the arachnoidal membrane covering the Sylvian fissure and evacuation of the CSF from the basal cisterns, the frontal and temporal lobes are spread apart with retractors, exposing the internal carotid, the anterior and middle cerebral arteries, and their branches. Then an incision 15–20 mm long is made in the mediobasal portion of the superior temporal gyrus laterally from M_1, between the polar–temporal and anterior temporal arteries. After the temporal horn is opened, the amygdaloid complex is identified and removed. Then the branches of the posterior temporal artery are coagulated, the optic tract is separated, and Hipp is isolated along its entire length and is removed *en bloc* to the external geniculate body and the beginning of the peduncle.

This operation differs from the classical lobectomy by the considerably smaller volume of resected brain tissue. Only the epileptogenic structures, Am and Hipp, are removed in the operation, and the dorsal, lateral, and basal parts of the temporal lobe remain intact. The authors performed this operation on 27 patients with frequent drug-resistant psychomotor attacks. The long-term results (average 2-year follow-up) were better than in standard temporal lobectomy, since the seizures completely disappeared in 22 of 27 cases.

An alternative to open operations in TLE is stereotactic surgery, which has rapidly been advancing over the past three decades. Indications for selective stereotactic destruction of the deep-seated temporal structures (Am and Hipp) became better founded after it was proven that even when these structures are not the source of epileptogenic discharges, their destruction interrupts the pathways for the dissemination of discharges from the temporal cortex to Th and Hypoth (Walker, 1982).

6.5.3b. Stereotactic Procedures. Stereotactic operations in TLE were initiated by Talairach in 1955 and by Narabayashi in 1958. These researchers explored the epileptogenic sites using electrophysiological methods. In order to locate the epileptogenic focus deep in the temporal lobe and not accessible to conventional scalp EEG, the authors implanted a great number of multicontact electrodes (up to 200 contacts) in the brain. As was mentioned above, this investigation, called stereo-EEG, makes it possible to study the spread of discharges from the epileptogenic focus to other brain structures in three dimensions. Stereo-EEG is required for resolving such questions as indications for the operation, choice of subcortical structures for destruction, surgical approach, and even for the prognosis after surgery. Subsequently, Talairach *et al.* (1958) summarized their extensive experience in a well-known study.

The majority of neurosurgeons today prefer the stereotactic method of treating TLE to the classical temporal lobectomy. We have already discussed the principal advantages of the stereotactic approach in the treatment of all forms of epilepsy (Section 6.1). These are especially true in respect to TLE, since the stereotactic destruction of the deep-seated temporal nuclei is less traumatic than temporal lobe resection. The chief indications for a stereotactic operation in TLE are typical temporal seizures or grand mal with a clearly defined temporal component, the presence of bitemporal or multiple unilateral foci, and the presence of pronounced psychic and emotional disorders (Shefer *et al.*, 1974; Narabayashi, 1979). Another indication is attacks accompanied by unilateral hyperkineses; in such cases, the simultaneous destruction of Am or Hipp and VL eliminates not only seizures but the hyperkinesia as well (Skryabin *et al.*, 1976).

6.5.3c. Stereotactic Amygdalotomy and Hippocampotomy. The objective of stereotactic operations in TLE is the destruction of one or both of the deep-seated nuclei in the temporal lobe, Am or Hipp. Their stereotactic anatomy is presented in Chapter 1. As is known, these nuclei have a low excitation threshold, which is one of the reasons for their usual participation in an epileptic discharge. Yet the destruction of which structure is more effective is a question that has still to be resolved. There are various points of view. The majority of authors favor amygdalotomy (Schaltenbrand *et al.*, 1966; Mempel, 1971a, 1972; Stepien *et al.*, 1972; Shefer *et al.*, 1972; Vaernet, 1972; Hitchcock *et al.*, 1973; Siegfried and Ben-Shmuel, 1973; Andy *et al.*, 1975; Skryabin *et al.*, 1976; Siegfried, 1977; Small *et al.*, 1977; Heimburger *et al.*, 1978; Laitinen and Toivakka, 1979; Mempel *et al.*, 1979; Romodanov, 1980). If there are mental and personality disorders, some neurosurgeons believe that amygdalotomy is more effective. However, in recent years it has been established that among the structures of the temporal lobe, the Hipp has the lowest excitation threshold, and epileptic activity on ESubCoG in TLE predominates significantly in Hipp. In view of this, many neurosurgeons consider destruction of Hipp a more effective operation (Šramka and Nadvornik, 1975b). It is also presumed that stereotactic hippocampotomy not only eliminates the epileptic focus but also severs the pathways along which the epileptic discharge passes. However, some authors believe that the benefits of amygdalotomy and hippocampotomy are approximately equal (Narabayashi and Mizutani, 1970; Balasubramaniam and Kanaka, 1975).

Stereotactic operations for TE are usually performed under local anesthesia, which makes it possible to retain speech contact with the patient and observe the effect of stimulation of the deep-seated structures. Moderate premedication is often employed (analgesics, anticonvulsants, antihistaminic drugs). The burr hole is located somewhat anterior to the coronal suture and 2–3 cm lateral to the midline; Talairach uses another approach via the temporal squama.

The main stereotactic reference point is the temporal horn, which is located close to the structures designated for destruction—Am, Hipp, and perihippocampal cortex. In lateral and, especially, in anteroposterior projections of x-ray films, the temporal horn has a complicated configuration. Moreover, it is necessary to take into account the great individual variability of the temporal horn. Its length varies from 21 to 36 mm; moreover, the difference in the length of the right and left horns may be as much as 7 mm, whereas the width varies from 8 to 14 mm with the two sides differing as much as 3 mm (Chkhenkeli, 1982).

The amygdala is localized on the anterosuperior wall of the temporal horn, protruding into the ventricle. In view of this, a good contrast visualization of the temporal horn is essential for the success of the operation.

The center of Am has the following stereotactic coordinates: 5 mm anterior to and 5 mm above the tip of the pole of the temporal horn. According to other data, Am lies 6–7 mm above the basal line drawn along the base of the temporal horn and 2–3 mm behind a perpendicular to that line drawn through the apex of the inferior horn. Difficulties are often encountered in respect to the stereotactic calculations for the location of Am because of the enlargement of the scarred temporal horn. Asymmetry of the temporal horns, in particular, enlargement of one as the result of an adherent scar, may serve as a differential diagnostic indication of the affected side. Moreover, this enlargement requires consideration in those cases in which the coordinates of the temporal horn on the contralateral side are used to calculate the position of structures on the scarred side.

In case of a marked deformation of the temporal horn, it is recommended to use LI (Chapter 3, Section 2.3) as the main reference line. Stereotactic calculations of Am with respect to LI have the advantage that the position of both commissures is more stable than the temporal horn (Talairach *et al.*, 1957). However, in this case, it is necessary to take into account that the individual variability of Am in respect to LI is greater than that in respect to the temporal horn.

The center of the anteromedial part of Am has the following coordinates in respect to LI: 2.5–3 mm posterior to CA, 15 mm below LI, and 23–25 mm laterally to the midplane (Heimburger *et al.*, 1978). Somewhat different figures are given by Narabayashi (1982b). According to his data, the center of the medial part of Am is located at a distance of 17–20 mm laterally to the midplane, whereas the center of its lateral part is at a distance of 22–28 mm. Andy and Stephan (1982) give almost the same data of 18 mm and 20–26 mm. The optimal diameter of the zone of destruction in amygdalotomy is approximately 9–10 mm.

In bilateral TLE the operation is performed on both sides, usually in one but at times in two stages. The amygdala complex consists of several nuclei (Chapter 1, Section 2.1), in view of which, it is necessary to choose the part the destruction of which will be most effective. In the opinion of the majority of neurosurgeons, the optimal goal is the destruction of the medial part of Am along with striae terminales (Narabayashi and Shima, 1973; Mempel *et al.*, 1979). Some surgeons, however, prefer destruction of the basolateral part of Am, which has more connections with Hipp. Other authors believe that partial destruction of Am does not exclude preservation of epileptic activity in the remaining part and accordingly recommend total destruction of this structure (Skryabin *et al.*, 1976).

The main method of locating an epileptogenic focus, and consequently an imperative preliminary stage of a stereotactic operation, is protracted multiple ESubCoG after implantation of several multicontact electrodes in the deep structures of the temporal lobe (Figs. 199 and 200). An electrographic study of the deep temporal structures, the most informative diagnostic procedure, is usually started 2–3 days after the electrodes have been implanted, and their localization is confirmed by conventional craniograms. The results of 2–4 weeks of recording determine, to a considerable degree, subsequent surgical procedures. The registration of local, stable epileptic activity from one or several subcortical structures in the temporal lobe is considered to be the main criterion for their destruction (Talairach *et al.*, 1958; Shefer *et al.*, 1974; Bouchard *et al.*, 1975; Walker, 1982). Laws *et al.* (1970) emphasize that the scalp EEG made it possible to detect the subcortical epileptogenic focus in only half of the TLE patients. In the other half of the patients, indications for an operation were established only after analyzing the deep recordings from structures of the temporal lobe.

A comprehensive electrophysiological investigation is required for differentiating the medial and lateral parts of Am. Records of spontaneous electrical activity as the electrode advances toward Am give a nearly flat recording, but at the moment the tip of the electrode enters that nucleus, the ESubCoG displays pronounced high-frequency activity ("damaging activity"), which is followed by clear-cut spontaneous spiking with a frequency of 30–40 Hz and higher, typical of epilepsy. Spontaneous paroxysmal activity is most often observed in the lateral part of Am (Umbach *et al.*, 1973); this activity can be seen rarely in scalp EEG and ECoG.

An effective method of identifying the medial part of Am is olfactory stimulation (for example, the odor of ether), which induces potentials only in that part of the nucleus (Narabayashi and Shima, 1973). Electrostimulation of Am prior to its destruction gives a variety of effects (Chapter 4, Section 7.1.15). In cases of TLE, as a rule, stimulation of Am induces an aura or psychomotor seizure, whereas this is observed quite rarely on similar stimulation of Hipp (Falconer, 1965). With chronic implanted electrodes in the deep-seated temporal nuclei, telemetric stimulation is possible while recording biopotentials. After amygdalotomy, spiking discharges in the scalp EEG present prior to the operation usually disappear. However, even if the clinical result is good and the seizures disappear, these discharges may be preserved (Andy and Stephan, 1982).

What is especially noteworthy is that after unilateral or even bilateral stereotactic amygdalotomy, there are practically no impairments of speech, gnosis, or other complications (Heimburger *et al.*, 1966; Narabayashi, 1982b). Memory disturbances are very rare and are usually transitory in nature. After a bilateral operation, it is quite often possible to observe olfactory impairments to varying degrees and lasting a long period of time. After such an operation, there is a considerably greater improvement in psychological functions than after temporal lobectomy.

In view of the anatomic peculiarities of Hipp (Fig. 203), stereotactic calculations for its destruction are more complicated than similar calculations for lesions of Am. For such stereotactic calculations, as well as for destruction of Am, two systems of coordinates are employed. The first, which can be called thalamic, is based on the reference line LI and is routinely used in all stereotactic operations for extrapyramidal lesions. The second system, elaborated by Talairach *et al.* (1957) especially for deep-seated temporal nuclei (temporal system), is based on the main temporal line drawn along the base of the temporal horn, which is situated close to the nuclei of the temporal lobe. The temporal system served as the basis for the popular stereotactic atlas of Talairach *et al.* (1957), which was especially designed to depict the structures of the temporal lobe. The basic dimensions and variability range of Hipp are presented in Chapter 1.

The correlations between the two systems are marked by significant individual variability. For example, the value of the angle between LI and the basic temporal line, according to data of Talairach *et al.* (1958) obtained from anatomic investigations, fluctuates between 23.5° and 37.5°. Similar calculations on ventriculograms of 40 patients demonstrated an even greater variability of that angle—from 17° to 52° (Chkhenkeli *et al.*, 1979).

A serious shortcoming of the thalamic system of coordinates is the remoteness of LI from the deep-seated structures of the temporal lobe, which increases the inaccuracies related to individual variability. Nevertheless, often even gross morphological changes in the temporal lobe such as deformation or aplasia do not cause any changes in the length or position of LI. Heimburger *et al.* (1978), who have great experience in TE surgery, prefer the thalamic system.

Although both systems of coordinates are used in practical stereotactic surgery, the majority of neurosurgeons prefer the more reliable temporal system of coordinates. Yet, even this system has its shortcomings. For example, great difficulties are encountered when employing it in cases in which a cicatrical process causes deformation of the temporal horns.

Talairach *et al.* (1958, 1962) perform hippocampotomy via the lateral temporal approach after careful studies of recordings from Hipp and adjacent structures. Numerous electrophysiological investigations have demonstrated that with biopotential recordings from Hipp it is possible to register both rapid (20–30 Hz) as well as slow (4–6 Hz), rhythms of an irregular and unstable nature with synchronous and asynchronous periods. Biphasic slow waves with periods of electrical silence are seen in marked sclerosis of Hipp. Pathological activity is most often recorded from the anterior and medial part of Hipp.

The technique of making stereotactic calculations after Talairach is essentially as follows. A coordinate grid is drawn on a lateral ventriculogram. The base of the grid is the line AB or the basal line drawn along the base of the temporal horn. Perpendicular to this line, two other lines are drawn, CD, passing through the apex of the temporal horn, and EG, parallel to CD, drawn through the beginning of the ventricular triangle. The distance between lines CD and EG is approximately 40 mm. The coordinates of all the mediobasal nuclei and cortical areas of the temporal lobe are plotted on this grid, which shows the projections of all the parts of Hipp: area 1, uncus; 2, medial parts; 3, caudal parts. The diagram depicts the coordinates and the true dimensions in millimeters on the basal line AB as well as the points of its intersections with lines CD and EG. The uncus of Hipp is at a distance of 8–16 mm from the pole of the temporal horn, its medial portion at a distance of 16–32 mm, and the caudal parts occupy the remainder of Hipp along line EG.

The distances of the various parts of Hipp from the midplane differ and are also subject to substantial variability. The uncus lies at a distance of 22–27 mm, and its medial and posterior parts are at a distance of 28–30 mm. The center of Hipp lies 3 mm above the basal line and 13–15 mm posterior to line CD; the center of the hippocampal gyrus is 2 mm below the basal line and 15–20 mm posterior to CD.

On the basis of anatomic data, the following maximum dimensions for destructive lesions in various parts of Hipp are recommended: anterior part, 10 mm; medial, 8 mm; posterior, 6 mm.

Since Hipp extends for a considerable distance along the temporal horn (Fig. 201), the most rational trajectory of the electrode is along its long axis. This makes it possible to destroy Hipp completely as the electrode gradually advances or selectively to destroy

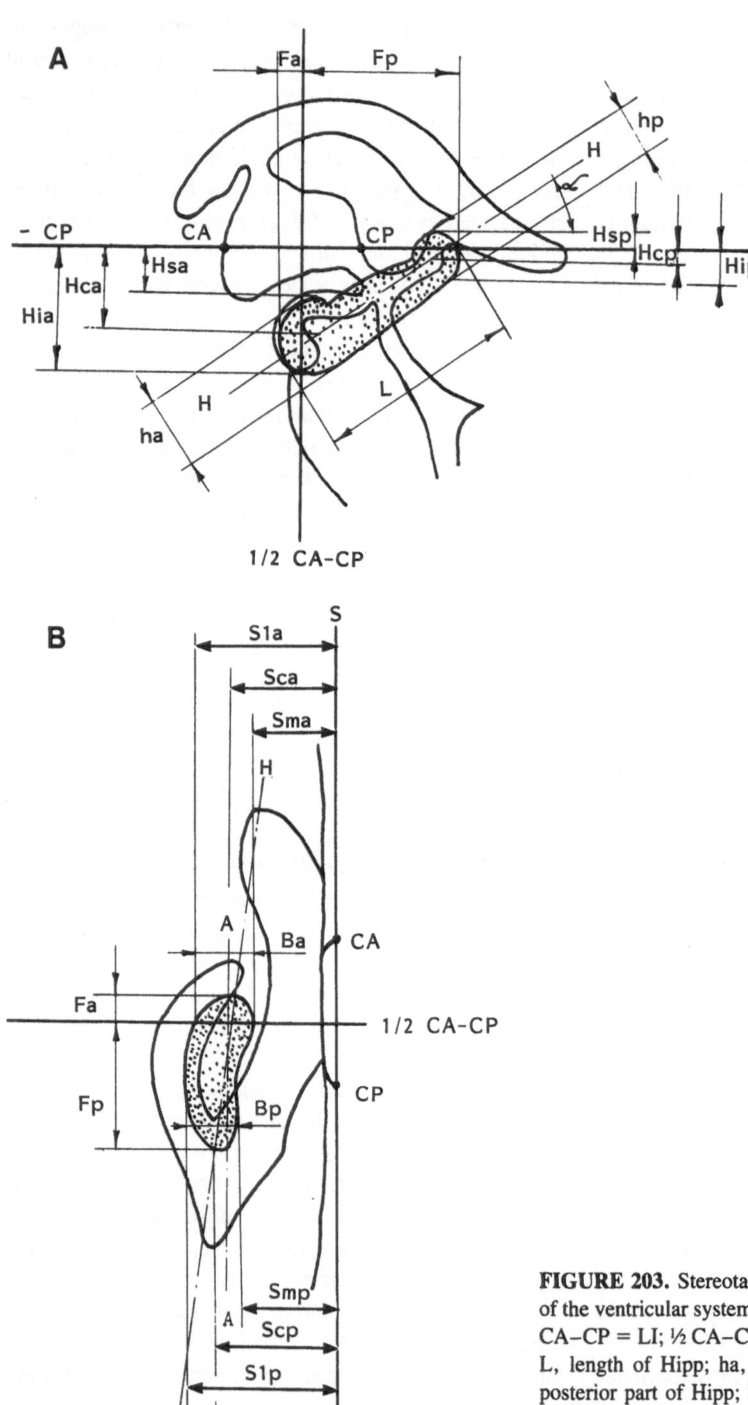

FIGURE 203. Stereotactic anatomy of Hipp projected on the contours of the ventricular system. (A) Sagittal section 20 mm lateral to midplane. CA–CP = LI; ½ CA–CP is middle of LI; HH, longitudinal axis of Hipp; L, length of Hipp; ha, height of anterior part of Hipp; hp, height of posterior part of Hipp; Fa, part of Hipp anteriorly; Fp, posterior to the perpendicular to the midpoint of LI. (B) Hippocampus in horizontal section. SS, midsagittal plane; A–A, parasagittal plane 20 mm lateral to the midsagittal plane; Ba, width of anterior part; Bp, width of posterior part of Hipp.

those parts in which pathological activity is registered. For this purpose, it was suggested that the electrode be introduced into Hipp via the ventricular triangle along the medial wall of the temporal horn; however, this technique has not been extensively used. The idea of introducing the electrode via the long axis of Hipp was developed Nadvornik *et al.* (1975) and Šramka and Nadvornik (1975a), who described the operation as stereotactic longitudinal hippocampotomy. In the opinion of these authors, the approach that they have suggested eliminates a serious defect in the Talairach technique (see above) in which, for the total destruction of Hipp, it is necessary to introduce several electrodes from various points on the lateral surface of the temporal lobe. Instead, the authors introduce only one electrode via the parietooccipital area. For this purpose, a burr hole is made at a point situated 55 mm above the external occipital protuberance and 40 mm lateral to the midline. In accordance with the stereotactic calculations, the trajectory of the electrode coincides with the longitudinal axis of Hipp, so that it is totally destroyed as the electrode advances orally (Fig. 204).

The operation proceeds as follows. A stereotactic apparatus with an electrode is installed in the burr hole. On a lateral roentgenogram, two points are plotted, between which the Hipp is to be destroyed throughout its entire length. The first point is 5 mm posterior to the anterior pole of the temporal horn, and the second 30 mm from the first along the axis of the temporal horn. The line connecting these two points is the trajectory of the electrode and forms an angle 23° with LI. The electrode should be introduced at this angle in the sagittal plane.

Corresponding to the calculations on the anteroposterior film, the electrode is introduced at an angle of 30° to the midplane. The above-mentioned two points, indicating the oral and caudal borders of Hipp, lie at distances of 23 and 30 mm, respectively, from the midplane. Therefore, the anterior border of destruction corresponds to the uncus of Hipp, and the posterior border is the point where it bends toward CC. In the longitudinal approach to Hipp, it is possible to damage the vascular plexus of the temporal horn, which is located medially and superiorly; however, this is highly improbable.

In spite of the fact that the trajectory of the electrode in this technique is considerably longer than in the temporal approach, the authors consider it less traumatic to the temporal structures. The results of this operation proved favorable: a marked improvement was noted in 10 of 13 epileptics (Sramka and Nadvornik, 1975a). Stereotactic destruction of Hipp, as a rule, does not lead to impairment of intellectual functions, emotional disturbances, or gnostic defects, and memory impairment is observed rarely and transiently.

The rich experience gained by many neurosurgeons clearly indicates that treatment of TLE by resorting to stereotactic destruction of Am or Hipp is very effective (Talairach *et al.*, 1958, 1974b; Heimburger *et al.*, 1966, 1978; Balasubramaniam and Kanaka, 1975; Nashold *et al.*, 1973; Ugrjumov *et al.*, 1974; Mundinger *et al.*, 1976, 1981; Shefer *et al.*, 1974; Skryabin *et al.*, 1976; Chkhenkeli, 1978a; Narabayashi, 1979; Mempel *et al.*, 1979; Romodanov, 1980; Walker, 1982). The most striking result of the operation was the total elimination of various temporal seizures in approximately 25% of the cases and a considerable reduction in their frequency in another 25%. Only a slight improvement was noted in approximately 30%. The best long-term surgical results were obtained by Talairach *et al.* (1974a,b) who report improvement to various degrees in 86.5% of the patients.

Long-term follow-up (average 6 years) has established that after amygdalotomy, a substantial improvement of TLE attacks was observed in half of the pa-

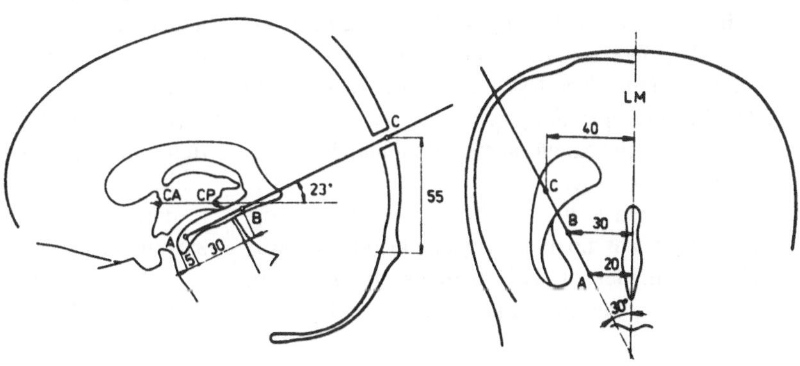

FIGURE 204. Schematic illustration of the electrode track to Hipp by the posterior longitudinal approach (Šramka, 1985). Left: Lateral view. Distances in millimeters. A–B, axis of Hipp; CA–CP = LI; C, the point of electrode insertion. Right: AP view. LM: midplane.

tients who underwent surgery, in a third of the patients with behavioral disorders, and in half of the patients with both of these forms (Heimburger *et al.*, 1978).

In another series, in 87% of 46 TLE patients with pathological behavior, amygdalotomy produced a stable improvement in the form of reduced attacks, a lessening of "emotional tension" and higher motor activity, as well as a normalization of aggressive behavior (Luczywek and Mempel, 1980). Unilateral amygdalotomy produced a marked improvment in 15 of 27 patients with a single epileptogenic focus and in six of ten patients with bilateral foci. A bilateral operation produced a good result in four of eight patients with bilateral foci (Vaernet, 1972).

Skryabin *et al.* (1976) present the results of a comprehensive evaluation of stereotactic treatment of TLE that included three components: fits, mental status, and social readaptation. In the group of patients with unilateral temporal foci, a clinical improvement was obtained in 73%, and in the group with bitemporal foci in combination with pronounced mental changes, in 44%. However, it was noted that emotional disorders regressed more slowly than epileptic attacks.

A no less important effect of this operation was the significant improvement in the mental and emotional states of these patients, which was noted by all authors: the disappearance or great reduction of "tension," anxiety, aggressiveness, turbulent emotional outbursts, etc. The results of stereotactic amygdalotomy in 60 TLE patients, demonstrated that in approximately half there was a marked reduction not only in temporal attacks but also in aggressiveness, a passion for violence, and hyperactivity. In another 40% of the patients, the reduction of these psychopathological manifestations was not marked. Several years after the operation, the results became somewhat worse. Narabayashi (1977, 1979) points out that the long-term results of these operations are stable with elimination of the attacks and normalization of EEG in more than half of the patients.

After stereotactic lesions in Am, Talairach and Szikla (1965) observed a disappearance of seizures in nine of 14 patients. The lessening in the frequency of attacks after the operation, as a rule, was associated with a reduction in behavioral and emotional disorders. Numerous psychological tests have demonstrated a significant improvement in mental functions after the operation. There was no improvement in only approximately one-quarter of TLE patients. It should be noted, however, that the authors subsequently evaluated the results of this technique less optimistically (Talairach *et al.*, 1974a,b).

Of 45 patients who underwent uni- or bilateral stereotactic electrocoagulation of Am, psychomotor attacks totally disappeared in eight patients, they were reduced by more than 50% in 17, and they diminished slightly or remained unchanged in 20 (Vaernet, 1972). Nashold *et al.* (1973) performed uni- or bilateral stereotactic amygdalotomy in eight patients with psychomotor attacks, and a substantial reduction in frequency and intensity of attacks was noted in four.

The effectiveness of amygdalotomy is emphasized by Ramamurthi and Kalyanaraman (1982). Of 21 patients with unilateral foci, attacks stopped completely in ten and diminished in another seven patients; of ten patients with bilateral foci, attacks ceased in four cases and diminished in another three. The results of treatment were better in those cases in which the epileptogenic foci were determined by scalp EEG prior to the operation.

Long-term follow-up demonstrated a significant improvement in three of four patients with complex partial attacks and aggressive outbursts who underwent unilateral stereotactic amygdalotomy.

An interesting observation by Narabayashi (1977) indicated that after stereotactic amygdalotomy, patients demonstrated higher sensitivity to anticonvulsant drugs, so that attacks could be controlled with lesser doses of drugs.

It is noteworthy that certain authors have reported less effective results of stereotactic operations in TLE than presented above. For example, Adams and Rutkin (1969) destroyed temporal nuclei in 25 patients with relief of seizures in only one case in an 18-month follow-up.

6.5.3d. Stereotactic Amygdalohippocampotomy. In order to improve the results of isolated stereotactic destruction of Am or Hipp, several authors destroyed both structures in a one-stage operation (Talairach and Szikla, 1965; Nashold *et al.*, 1973; Ugrjumov *et al.*, 1974; Sigua and Chkhenkeli, 1976; Nadvornik *et al.*, 1977). Destruction of Am and the rostral part (uncus) of Hipp may be performed via the anterior (temporal) approach with a single pass of the electrode without changing its direction. If Am and the middle or posterior part of Hipp must be destroyed, it is necessary to make a second introduction of the electrode.

As Mempel *et al.* (1979) emphasize, bilateral medial amygdalotomy combined with unilateral anterior hippocampotomy is more effective in TLE patients with marked emotional and behavioral disorders. The operation has no unfavorable effect on general mental activity; in fact, IQ, memory, and learning in some cases improved after the operation.

Because Am and Hipp are well delineated by NMR, a new technique of stereotactic resection of Am and Hipp with the aid of NMR was proposed recently in medically intractable complex partial seizure of temporal lobe origin (Kelly *et al.*, 1986, 1987). NMR scanning was performed with the patient's head fixed in the stereotactic frame with a feducial localization system. The configuration of Am and Hipp from NMR imaging was digitized into a three-dimensional computer image matrix. Through a burr hole in the temporo-occipital region and using the computer-assisted coordinates, Am and Hipp were resected by a computer-monitored carbon dioxide laser beam. The authors reported that 16 of the 17 patients experienced cessation or dramatic reduction of the temporal lobe seizures. The postoperative visual field defects were nondisabling. Transient minor speech problems after left-sided operation were observed in two cases.

It may be assumed that the long-term results of a combined operation of Am and Hipp are somewhat better than the destruction of each of these structures separately.

6.5.3e. Stereotactic Fornicotomy and Anterior Commissurotomy. There is still another stereotactic operation that is performed in TLE, fornicotomy or destruction of the fornix (F), often combined with destruction of CA (anterior commissurotomy). This operation is usually used after a primary operation, stereotactic destruction of the temporal structures (Am, Hipp), has been ineffective. Bilateral stereotactic destruction of F in TLE was proposed by Hassler and Riechert (1957) on the premise that it would interrupt the main efferent pathway from Am and Hipp, along which the epileptic discharge generalizes within the limbic system. Fornicotomy may be performed on one or simultaneously on two sides (Umbach and Riechert, 1964; Umbach, 1966; Bouchard, 1971; Mundinger *et al.*, 1976). It is especially indicated in children with severe mental disorders (auto- and heteroaggressiveness, psychomotor excitation) (Barsia-Salorio and Broseta, 1976). This operation is often combined with the destruction of other subcortical structures.

The stereotactic coordinates of F are determined from the stereotactic atlas. Destruction of F, which is quite long, is performed at the posterior edge of CA, immediately in front of FM. During the stereotactic operation, which is usually conducted under local anesthesia, it is very important to verify the interruption of F by electrostimulation. The current must be weak so as not to cause a generalized seizure. The appearance of consciousness disturbances, drowsiness, hallucinations (Umbach, 1966), as well as vegetative reactions (pallor, tachycardia, mydriasis, etc.) during electrostimulation indicates that the target point is being stimulated.

Fornicotomy is an operation involving little trauma and usually does not cause any serious complications (Riechert, 1980). Long-term results analyzed in several papers demonstrate that it is effective in certain forms of epilepsy (Umbach, 1966; Bouchard *et al.*, 1975). Attacks of TLE terminated completely in approximately a third of the cases after the operation (Umbach and Riechert, 1964). The frequency of grand mal epilepsy diminished and vegetative and psychic disorders reduced in the majority of the patients. Schaltenbrand *et al.* (1966) consider the fornicotomy more effective than destruction of Am, but Sugita *et al.* (1971) noted less favorable results of fornicotomy—an improvement, expressed to various degrees, was noted in only approximately one-half of TLE patients in 2 to 6 years of follow-up. It should be noted that although fornicotomy has not been used extensively in TLE surgery, it is indicated in those cases in which amygdalotomy or hippocampotomy did not yield a satisfactory result (Umbach, 1966; Bouchard *et al.*, 1975).

6.5.3f. Stereotactic Hypothalamotomy. Few papers in the literature discuss posteromedial hypothalamotomy for the treatment of TLE. This operation, which was proposed and elaborated on by Sano, is performed mainly for treating pain syndromes and certain mental disorders and is therefore described in detail in Chapters 13 and 18. Sano *et al.* (1970b) perform this operation in TLE on both sides, whereas Skryabin *et al.* (1976) operates on one side. Stereotactic destruction of the posteromedial Hypoth has given promising results (Balasubramaniam and Kanaka, 1975); however, this operation has not found extensive application in surgery for epilepsy.

6.5.3g. Stereotactic Destruction of Anterior Thalamic Nuclei. Several attempts have been undertaken to destroy NA as a treatment for TLE

(Mullan *et al.*, 1967; Bouchard and Umbach, 1972), but without clear-cut positive results.

6.5.3h. Combined Operations. The data presented in this chapter indicate that although stereotactic surgery represents an important advance in the treatment of various types of epilepsy, the problem is still far from being resolved. Stereotactic destruction of certain deep-seated brain structures (Am, Hipp, VL, F, etc.) does not ensure the termination of seizures in all cases. In view of this, in epilepsy, as well as in certain other brain disorders, there is a tendency to combine stereotactic operations, i.e., one-stage or stage-by-stage destruction of several subcortical structures. We have already discussed briefly amygdalohippocampotomy (Section 6.5.3d). In the severe cases, patients with mental disturbances, epileptic status, and diffuse atrophy of the brain, a number of neurosurgeons perform other combined operations, for instance, bilateral amygdalotomy or amygdalohippocampotomy with unilateral destruction of F and CA, or sometimes different combinations of destruction of CM, VL, GC, and anterior, ventral, or dorsomedial Th nuclei (Adams and Rutkin, 1969; Romodanov and Laponogov, 1971; Bouchard *et al.*, 1975; Koshino *et al.*, 1975; Skryabin *et al.*, 1976; Romodanov, 1980). After such operations, in the severe group of patients, in the majority of cases, the results were somewhat better than after destruction of only Am or Hipp.

Long-term results of stereotactic destruction of several structures—lamella medialis Th, F, and Am—in TLE are described by Mundinger *et al.* (1975) with a mean follow-up of approximately 6 years. In 24% of the patients, seizures terminated completely, and in another 24% there was a substantial reduction.

In another paper Mundinger *et al.* (1981) compare two series of personal observations on TLE patients. These groups differed in the number of subcortical structures that were stereotactically destroyed in a one-stage operation. In the first series (33 patients), Am, F, and CA were destroyed; in the second (12 patients), in addition to these three structures, ZI and Pul were also destroyed. In long-term follow-up (mean approximately 6 years), a substantial improvement was noted in 63.6% of the patients in the first series and in 74.9% of the patients in the second. These data show that the destruction of a greater number of subcortical structures may be more effective.

Destruction of the above-mentioned structures in various combinations together with unilateral destruction of the anterior ventral Th nucleus is indicated in centrencephalic epilepsy with mental disorders and characteristic EEG changes. In this form of epilepsy, which responds to surgical treatment poorly, good or satisfactory results were obtained in 16 of 19 cases (Bouchard *et al.*, 1975).

Another form of combined treatment that may be effective for bilateral epileptogenic foci is open destruction of the temporal lobe on one side and stereotactic destruction of deep-seated temporal structures on the other side (Umbach *et al.*, 1973; Shefer *et al.*, 1974; Sigua and Chkhenkeli, 1976). Stereotactic amygdalotomy on the other side may also be effective if there is a recurrence of seizures after unilateral temporal lobectomy.

There are grounds to believe that combined operations are more effective, although naturally such operations are more complicated and longer in duration. This is evident from an interesting study based on a vast amount of clinical material published by Romodanov *et al.* (1974). The authors analyzed the results of many years of a joint study by the neurosurgical clinics in Kiev and Warsaw. In all, 440 classical and 151 stereotactic (destruction of Am, CF, and Hipp) operations and 48 combined operations—initially, resection of the temporal lobe, followed by stereotactic destruction of CF—were carried out over the course of 15 years. Good results after classical operations were noted in 55% of the patients, after stereotactic procedures in 39%, and after combined operations in 60%, and improvement, correspondingly, was seen in 31, 42, and 25%.

Mundinger *et al.* (1981) summed up the results of stereotactic destruction of Am or Am in combination with other structures (F, Subth, Pul, etc.) in TLE, published by 11 authors from 1964 to 1979. The overall number of patients was 257, and absence or reduction of seizures was noted in 65% of cases.

The combination of fornicotomy with amygdalotomy and medial thalamotomy (Umbach *et al.*, 1973) is very effective in psychomotor seizures with vegetative disorders. Chkhenkeli (1982) employed more than 20 different combinations of stereotactic destruction of subcortical structures (Am, Hipp, F, GC, CM, VL, CF, striae terminales) unilaterally or bilaterally in various forms of epilepsy. He concluded that single small lesions do not ensure a stable therapeutic effect. Such an effect can be obtained only if massive destruction of several subcortical structures, especially Am on both sides and Hipp on the side homolateral to the dominant epileptogenic focus, is achieved. Additionally, it is desirable to destroy those structures that propagate the epileptic discharge.

Skryabin *et al.* (1976) in TLE recommend supplementing stereotactic destruction of Am with lesions of CA and F; they also noted a good result from unilateral hypothalamotomy after ineffective bilateral amygdalotomy.

Spiegel and Wycis (1962b) first performed the combined destruction of temporal and thalamic nuclei in epilepsy. Subsequently, it was demonstrated that the combined destruction of Am on the side of the epileptogenic focus and VL or Subth on the contralateral side is effective in relieving the combination of TLE and athetoid hemihyperkineses (Skryabin *et al.*, 1976).

In performing operations that destroy several deep-seated brain structures, there may be concern that the greater trauma of these operations may have a higher risk. However, the data of a number of authors indicate that combined operations do not increase the frequency of postoperative complications (Romodanov and Laponogov, 1971; Chkhenkeli, 1982).

The greatest experience in the surgical treatment of TLE by one unit, namely the Montreal Neurological Institute, was summarized by a leading expert in the field, T. Rasmussen (1983). Thirty-seven percent of the 894 evaluated patients with nontumoral TLE were seizure-free and 26% of patients had a marked reduction of seizures (63% of favorable results).

Even if the seizures stop completely after the operation, protracted medicinal treatment is recommended, but with only one or two drugs and in smaller doses than prior to the surgery. If there is a total absence of seizures for 1.5–2 years, the antiepileptic drugs may be discontinued.

15

Brain Tumors

The neurosurgeon should not only know that the subject has a cerebral tumor, where the tumor is, but also what the nature of the tumor is, since the operability and future of the patient depend on it. . . . Nothing can be of greater value to a neurosurgeon than the ability immediately to visualize from the gross appearances of a tumor what will be its histological nature. Too great emphasis cannot be laid on this.

H. Cushing

1. Introduction

In principle, surgery of brain tumors is not included in stereotactic or functional neurosurgery. However, in some cases stereotactic technique may be employed for biopsy, destruction, or removal of brain tumors.

Among the many brain tumors, which differ in histostructure, localization, diffusion, volume, and so forth, stereotactic treatment was first used quite effectively in pituitary tumors. However, stereotactic destruction of other types of brain tumors by various techniques has lately found greater application, mainly for biopsy or in those cases where conventional radical removal of the tumor is impossible.

The value of employing stereotactic techniques in neurooncology is reflected in the fundamental and well-illustrated monograph of Pecker *et al.* (1979a). The use of stereotaxis in combination with neuroroentgenologic and electrophysiological methods in 282 cases of brain tumors in various locations, and especially the general application of stereotactic biopsy, made it possible to select the most appropriate approach and therapy in each case.

2. Stereotactic Biopsy

The great importance of precise histological diagnosis of brain tumors for determining the indications and contraindications for surgery, for planning surgical tactics, and for prognosis is well known.

Even the modern imaging techniques (CT, NMR) permit no more than a probable histological diagnosis of intracranial tumors. In many instances, a precise histological verification is possible only by a biopsy of the tumor. Taking a biopsy is especially important in cases in which difficulties arise in determining whether deep-seated gliomas are operable and, if so, how radical their removal should be, or when there is uncertainty in regard to the preoperative histological diagnosis, which occurs quite frequently. In this respect, Broggi and Franzini (1981) noted that in 10 of 25 patients the histological diagnosis of the brain tumor obtained by stereotactic biopsy was different than was assumed on the basis of clinical data, including CT. In those cases in which a diagnosis is not possible on the basis of CT data, a biopsy may be of decisive importance in making differential diagnosis between brain tumors and certain other intracranial processes (encephalitides, gliosis, etc.) for which surgical treatment is not indicated. On the basis of CT data, at times it is practically impossible to differentiate a small tumor deep in the brain from a small infarction, although this is very important in deciding on the appropriate treatment.

Biopsy is also indispensable in certain rare clinical cases presenting peculiar diagnostic difficulties, for example, when two tumors of different histological structure are present in the same brain: benign and malignant gliomas, metastasis and glioma. Biopsy data are of decisive importance in another rare and especially difficult clinical situation. When there is a malignant tumor in one of the internal organs and CT or another technique demonstrates a brain tumor, which may not be metastatic but a meningioma. Radiation therapy without a preliminary biopsy of a tumor also cannot be recommended at present.

Free-hand biopsy of the brain, which has been practiced since the time of Cushing and Dandy, has some obvious shortcomings, the main one being the difficulty of accurately locating the pathological focus deep in the brain. Naturally, the smaller the focus, the more difficult it is to locate. Free-hand biopsy usually requires needling the brain, which is quite traumatic and increases the risk of such complications as cerebral hemorrhages or brain edema. A biopsy taken from the zone near the brain tumor may lead to an erroneous conclusion of an inflammatory or degenerative process.

Free-hand biopsy through a burr hole, as practiced in neurosurgery during the course of several decades, is a dangerous procedure. Certain authors report about 30% mortality after such a biopsy in malignant tumors. "Blind" needling of the brain in various directions in order to locate the tumor or cyst today should be condemned as not advisable in modern neurosurgery.

Stereotactic biopsy (SB) of brain tumors, first performed by the pioneers of stereotaxis, Spiegel and Wycis, represents essentially a new diagnostic method. As long ago as 1947, Spiegel proposed stereotactic aspiration of intracerebral cysts. However, it was only during the past decade, after the introduction of CT, that this method has come to be extensively employed in clinical neurosurgery (Conway, 1973; Waltregny, 1976; Shetter *et al.*, 1977; Cohadon *et al.*, 1977; Edner, 1979; Rushworth, 1980; Kandel *et al.*, 1981; Ostertag *et al.*, 1980; Colombo *et al.*, 1982; Greenblatt *et al.*, 1982; MacKay *et al.*, 1982; Lunsford and Martinez, 1984; Apuzzo *et al.*, 1987). Stereotactic biopsy has a number of a priori indisputable advantages over conventional intraoperative biopsy. First, it is less traumatic and consequently safer, especially in high-risk patients. There is every reason to believe that SB should be performed in the majority of cases of intracerebral and other tumors.

Before performing SB, as far as possible, it is advisable to establish the localization of the tumor and the approximate position of its core or central part, which may be done by angiography, ventriculography, scintigraphy, stereo-EEG, impedansometry, and other diagnostic methods (computed tomography is discussed in the next section). On angiograms one may determine whether there are large veins in the intended entry zone of the biopsy probe as well as arterial branches in the path of the probe. In highly vascularized tumors (hemangiomas, cavernous angiomas, and others), SB is indicated with great caution because of the risk of hemorrhage. For this reason, angiography is desirable in considering the risks of biopsy. Combined stereotactic CT with angiography is also very useful (Gahbauer *et al.*, 1983).

In principle, SB does not carry a high risk and is comparatively well tolerated, even by patients in a grave condition. The rate of complications and mortality after SB is only a fraction of that after conventional biopsy. Many authors recommend the use of steroids before and after the procedure in order to reduce the risk of brain edema. Nevertheless, in some cases SB also causes minor hemorrhages in the parenchyma of the tumor and at times intensifies perifocal and generalized brain edema. In one of the largest series of cases described in the literature, 503 cases of deep-seated brain tumors, two-thirds of them in the area of V_3 (Ostertag *et al.*, 1980), mortality after SB was 2.3%, and transient aggravation of symptoms 3%. No mortality after SB was reported in other large series of cases (Scarabin *et al.*, 1978; Edner, 1975; Bouvier *et al.*, 1983). Edner (1981) reported 11 serious complications after 345 SBs (subarachnoid and intraventricular hemorrhages, general brain edema). The rate of transient, rapidly subsiding complications (increased neurological deficit) is about 7–9% (Daumas-Duport *et al.*, 1982; Lobato *et al.*, 1982). Shetter *et al.* (1977) compared the results of SB and free-hand biopsy performed by the same neurosurgeons in two groups of patients. In the first group there were 9% mild complications and one fatality, whereas in free-hand biopsy, 20% mild and 10% severe complications occurred.

Stereotactic biopsy is performed under local anesthesia as well as under endotracheal anesthesia (Pecker *et al.*, 1979a; Kandel *et al.*, 1981; Bouvier *et al.*, 1983). Practically any of the stereotactic apparatuses described in Chapter 3 may be used for SB. Some newly developed devices are also commonly used (Greitz *et al.*, 1980; Brown, 1981) (BRW-System), most of which are anchored to the skull. After the stereotactic apparatus is installed and the appropriate stereotactic calculations are done, a special thin instrument is introduced into the central part of the tumor to take a biopsy. Several such instruments have been designed. An original instrument connected to a stereotactic apparatus has been proposed by Backlund (1971) (Fig. 205). This instrument consists of a fine cannula with two stylets, the first serving only for in-

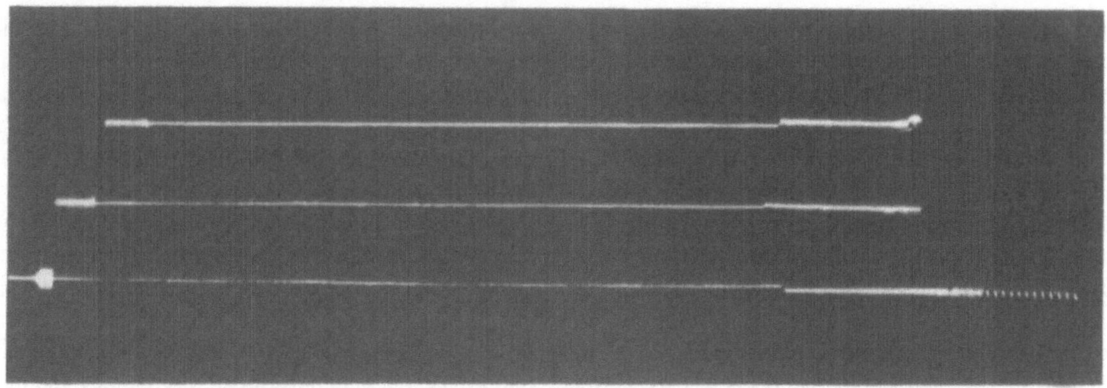

FIGURE 205. Backlund's (1971) cannula for stereotactic biopsy of train tumors. Explanation in the text.

troducing the cannula into the tumor, after which it is replaced by the other with a threaded tip. The end of this stylet, which protrudes out of the cannula for 10 mm, is screwed into the tumor, and the cannula advanced. With its sharp edges, the cannula cuts a thin cylinder of tissue (10 × 1.2 mm) with a volume of approximately 15 mm^3, which is fixed in the thread. After removal from the instrument, this tissue is forwarded for histological examination. No less than 3 mm^3 of tumor tissue is required for a reliable pathological diagnosis.

Another needle design for biopsy has been proposed by Kalyanaraman and Gillingham (1964). This needle has a groove that, when the needle is turned 360° along its axis, cuts a piece of the tumor or cerebral substance weighing approximately 15–20 mg. The authors performed 150 biopsies without complications. Miniforceps with a diameter of 1–2 mm protruding from a metallic tube are also employed (Sugita *et al.*, 1975; Waltregny, 1976; Cohadon *et al.*, 1977; Ostertag *et al.*, 1980; Apuzzo *et al.*, 1987). We use an instrument for biopsy resembling the one described in Section 5 on pituitary tumors. There are also other instruments designed for biopsy (needles of Cone, Gildenberg, Hankinson, Sedan; micro-alligator forceps by F. L. Fischer AG; flexible bronchoscopy forceps, Olympus Co.).

Sugita *et al.* (1972) first introduced into the biopsy zone a thin (2.2 mm in diameter) fibroscope in order to be certain there were no large vessels there. Certain needle designs offer the possibility of leaving radiopaque markers (clips and so on) at the biopsy site.

Many investigators, including the authors, recommend taking several biopsies, i.e., removing pieces of tissue along the trajectory (or trajectories) of the instrument every 5–10 mm as it advances from the periphery to the center of the tumor (serial biopsy). This is necessary, first of all, because the histostructure of the center and periphery of the tumor may differ significantly, and secondly, because the central part of the tumor may be necrotic, and a histological diagnosis in this case will be impossible to make (Apuzzo and Sábshin, 1983; Heilbrun *et al.*, 1983). To obtain more reliable information about the histostructure of the tumor, it is recommended that biopsies along two instrument tracts be made (Daumas-Duport *et al.*, 1982). It is also recommended to leave a clip in the tumor as a marker for brachytherapy (Sugita *et al.*, 1975).

The use of intraoperative neurophysiological methods (impedance and stereo-EEG monitoring), which permit one to confirm the exact location of the probe tip, improves the effectiveness and results of SB.

Laitinen and Toivakka (1972) performed SB in 18 patients with a preliminary diagnosis of deep-seated brain tumor. The authors demonstrated that ESubCoG recording in the process of biopsy with a multipolar electrode, in the majority of cases, allowed one to determine clearly both the presence and the borders of the intracerebral tumor.

Another method that confirms the presence of a deep-seated tumor is continuous impedance monitoring before SB. The improvement of impedance equipment in the last two decades has resulted in more successful use of impedance techniques in stereotactic surgery and especially in stereotactic biopsy and destruction of brain tumors. When the active tip of the instrument is introduced into the tumor, the impedance sharply diminishes approximately from 10,000–16,000 to 400–600 Ω (Tasker, 1985). Based on the calculations, the cannula, with an inner electrode for monitoring impedance, is advanced to the first point inside the tumor through the stereotactic apparatus. The tip of the recording electrode extends 2 mm beyond the inner end of the cannula. As the biopsy instrument is advanced, the impedance is read; the impedance in ohms is plotted along the abscissa, and the depth from the surface of the cortex in millimeters is plotted on the ordinate. The impedance of tumor tissue differs markedly from the impedance of the cerebral cortex, but according to Broggi and Franzini (1981), it may be either considerably higher or lower than the cortical impedance. Depending on the impedance monitoring, biopsies are taken from previously selected points along the entire course from the surface to the deepest part of the tumor.

Impedance monitoring was very useful in SB of different brain tumors and other mass lesions (Tasker, 1985). The impedance drops in a striking fashion when the electrode passes the border between the surrounding brain tissue and a wide variety of brain lesions. Using the CT-guided Leksell frame and impedance measurement the target was attained in 26 cases and was missed only in one procedure. Bullard and Makachinal (1987) also recommended this technique.

The importance of impedance or neural noise monitoring during stereotactic biopsy of brain tumors was shown in several reports (Organ *et al.*, 1967; Becker *et al.*, 1970; Benabid *et al.*, 1979; Broggi and Franzini, 1981; Tasker, 1985; Bullard and Makachinal, 1987; Yoshida and Kuramoto, 1987).

There are two methods for determining the pathological nature of biopsy material—the rapid smear preparation and the longer conventional method. In the former, pieces of tissue, immediately after they are obtained, are placed on glass slides and delicately smeared over the glass. Then, the thin layer of tissue is stained with several drops of Loeffler's methylene blue, studied under the microscope, and photographed. In the conventional cryostat method, pieces of tissue are imbedded in paraffin, sectioned, stained, and studied microscopically. From the experience of 150 SBs in brain tumors, Daumas-Duport *et al.* (1982, 1983) emphasize the need to distinguish two components of the biopsy: the tumor tissue proper and infiltration of brain tissue by tumor cells. In the second case, not only tumor cells but elements of edematous cerebral tissue may be seen in the preparation. It is apparent that the conventional method is more accurate; however, the agreement of histological diagnoses obtained by the two methods is quite high—95% (Ostertag *et al.*, 1980).

Many neurosurgeons and pathologists have also noted that the smear technique is reliable and has distinct advantages, especially when only small biopsy specimens are available and rapid diagnosis is required during operations (Hitchcock *et al.*, 1986). The smear technique is described in detail in a special monograph written by Adams *et al.* (1981). However, other authors consider that only 70% of the diagnoses made on the basis of the smear preparation are correct.

Recently a third technique was tested for SB—the "imprint" or "touch" preparation. Hitchcock *et al.* (1986) gave a good evaluation of the technique, which enabled them to make a rapid diagnosis in 100 cases of brain lesion biopsy. The authors stressed that the method gave excellent details of cell morphology and tumor architecture.

In those cases in which the tumor is found to be obviously inoperable, and the surgeon decides to resort to cryodestruction or to introduce radioactive isotopes, the important advantage of the rapid smear preparation is apparent. The data on the histostructure of the tumor obtained in several minutes allows the surgeon to perform the desired procedure immediately.

Just as in any biopsy, a small fragment of tumor tissue obtained by SB may not be sufficiently representative to reflect the histostructure of the tumor as a whole. In order to determine the "reliability" of stereotactic biopsy, de Divitius *et al.* (1983) compared the histological and cytologic diagnoses in biopsy and in surgical or autopsy material. In 64 cases, the diagnoses made on the two types of material coincided in 81.5%. Approximately the same frequency of different histological diagnoses in biopsy and in surgical specimens of gliomas, 21%, was obtained by Lobato *et al.* (1982) in 109 SB. A precise histological diagnosis was established by SB in 88% of 70 biopsies by Bouvier *et al.* (1983).

Sedan *et al.* (1981) summarized the results of 770 SBs of malignant gliomas from three neurosurgical clinics in France (Bordeaux, Rennes, Marseilles). The preoperative diagnosis was defined more precisely and the histology of the tumor was established in 41.5% of the cases, and in 23% of the cases the SB diagnosis was the same as the preoperative diagnosis. In 28% the diagnosis was changed, and in 7.5% of the cases no diagnosis could be made.

Conway (1973) performed SB in 31 patients with various deep-seated brain tumors (pituitary tumors, gliomas, pinealomas, etc.). A precise histological diagnosis was made in 25 cases. In certain cases Conway takes a column of tumor tissue up to 60 mm long for biopsy. A successful stereotactic biopsy of these tumors has been reported by Pecker *et al.* (1975) and Frank *et al.* (1985).

It is well known that the diagnostic problems and the choice of treatment are very difficult in deep-seated tumors around the III ventricle, especially in tumors of the pineal region.

Stereotactic biopsy in deep-seated tumors may be performed to introduce fine silver wires into the tumor to serve as markers for subsequent radiation therapy (Sugita *et al.*, 1975).

As an illustration, one of our cases is described in brief. In a 25-year-old female patient, CT confirmed the clinical data on the presence of a large inoperable tumor of the septum pellucidum and V_3 with a big cyst protruding in V_3 and blocking one FM. Stereotactic puncture was performed after making the appropriate calculations, the cystic fluid was withdrawn, and a biopsy was taken from a deep tumor, which proved to be an astrocytoma. This procedure resulted in a temporary improvement in the condition of the patient.

Stereotactic aspiration of colloid cysts of V_3 have yielded good long-term results in five cases (Bosch *et al.*, 1978). The authors believe that this relatively simple and safe operation offers significant advantages over the technically complicated open, direct removal of colloid cysts, which has a 15–20% postoperative mortality. Recently three other cases of stereotactic complete aspiration of colloid cysts of V_3 were presented (Rivas and Lobato, 1985). The operations were performed with the aid of a CT-adapted stereotactic device of Leksell-Jernberg. The authors noted the different viscosities of these cysts; hyperdense cysts on CT have a denser colloid substance than hypo- or isodense cysts. Donauer *et al.* (1986) performed stereotactic aspiration in ten cases of colloid cysts of the third ventricle and concluded that this technique is a method of choice for simultaneously diagnosing and treating colloid cysts. The introduction of ^{90}Y implants into cystic craniopharyngiomas after SB of these tumors has also been performed (Gahbauer *et al.*, 1983).

If the aspiration of a mass lesion in the region of V_3 or the midbrain after SB is not possible, it is recommended that ventriculostomy be performed during the same operation (Apuzzo, 1985).

The biopsy of the cyst wall is very useful for exact pathological verification of the lesion, but is associated with the danger of intracavitary hemorrhage. After such a biopsy the installation of a permanent drainage system is recommended, such as the Rickham reservoir with catheter (Apuzzo *et al.*, 1987). CT control for several days after the procedure is essential.

In brain abscesses, SB is not only a diagnostic but also a therapeutic procedure, permitting aspiration of the contents of abscesses and the introduction of antibiotics into the cavities (Moran *et al.*, 1979).

The introduction of computerized tomography in neurosurgical practice greatly stimulated the broader application of SB.

3. Stereotactic Biopsy with Computed Tomography

Computed tomographic investigations opened up previously inaccessible opportunities for diagnosing even small pathological foci deep in the brain. Today with CT it is possible to identify a brain tumor and other mass lesions in the earliest stages of their development, frequently at a "subclinical level." Nevertheless, in spite of the unique diagnostic possibilities offered by CT, quite often it does not allow one to differentiate certain pathological processes, especially of small volume, e.g., tumors and infarction foci in the brain as well as tumors of different histological types.

Since the data on the use of CT in stereotactic surgery are presented in detail in Chapter 3, we only dwell briefly on the recently published works regarding the use of CT for improving and modifying SB techniques (Laitinen, 1976; Cohadon *et al.*, 1977; Conway, 1977; Kelly *et al.*, 1978; Benabid *et al.*, 1979; Pecker *et al.*, 1979a,b; Piskun *et al.*, 1979; Bosch, 1980; Kandel *et al.*, 1981; Broggi and Franzini, 1981; Yeates *et al.*, 1982; Greenblatt *et al.*, 1982; MacKay *et al.*, 1982; Apuzzo and Sabshin, 1983; Bouvier *et al.*, 1983; Ohye *et al.*, 1984a; Horner and Potts, 1984; Frank *et al.*, 1985; Poza *et al.*, 1985; Sahni *et al.*, 1985; Heikkinen, 1986; Hood *et al.*, 1986).

Before evaluating the published results, two remarks must be made. Authors express the accuracy or rate of success of biopsy in percentages, but the basis for the figûres may be different. The majority of authors consider a success to be an exact histological diagnosis by obtaining characteristic tissue samples (for instance, the grade of astrocytoma), but some authors consider a biopsy successful if it yields a diagnosis of the pathological processes or other useful information (for instance, differentiation between tumor and inflammation or disclosing a cyst inside a tumor).

The second remark concerns the need to distinguish between "case success rate" and "procedure success rate," as was stressed by Greenblatt *et al.* (1982). That is important because about 10% of all biopsies are performed more than once, primarily because of failure of the first procedure. The "case rate" is the percentage of successful diagnoses, in this particular case, independent of the number of the procedures. The "procedure rate" is the percentage of successful results for each procedure. Since in many papers this difference is not reported, it is difficult to give general figures separately. Performing SB on the basis of CT data has greatly reduced the risk of complications. For example, there was not a single complication after SB in 94 cases with deep-seated parasellar tumors (Edner, 1981). With SB combined with CT, it was possible to make the correct histological diagnosis in 96% of various intracranial pathological processes (Lunsford and Martinez, 1984). It is noteworthy that in 26% of the cases, SB established a diagnosis that, on clinical data, including CT, prior to biopsy, had not been suspected or had been considered improbable. There was no mortality after SB, and complications occurred in only 5.9% of the cases. After the histological diagnosis was established, radiation therapy or chemotherapy was prescribed for the majority of patients with malignant tumors. In a series of 261 stereotactic biopsies, only two patients had neurological complications and one patient died as a result of postoperative hematomas (Kelly, 1987).

Lunsford *et al.* (1986) reported the use of both NMR and CT for stereotactic biopsy in three cases of deep mass lesions. The Leksell stereotactic apparatus, compatible for CT and NMR investigation (Chapter 3), was employed directly in the operating room. The authors concluded that NMR imaging provided better resolution than CT, especially with coronal and sagittal reconstructions.

Modern stereotactic techniques combined with CT investigations make it possible to biopsy safely mass lesions even in the brainstem (Sahni *et al.*, 1985). The instrument is introduced supratentorially and is passed through the incisura without penetrating the tentorium. Coffey and Lunsford (1985) used this technique for SB of tumors and hematomas in the midbrain and pons in 12 cases. In some, the lesion was stereotactically aspirated. Depending on the location of the lesion, the authors used transfrontal or transcerebellar approaches through burr holes.

The several SB techniques based on CT data that are used at the present time may be divided into two groups (see also Chapter 3). In the first, a conventional CT is made with the patient's head securely fixed in a stereotactic apparatus or fiberglass helmet. The radiopaque markers of various length, corresponding to the CT scan, are fastened on the head. Colombo *et al.* (1982) place three 1-mm lead spheres encased in plastic subcutaneously for this purpose.

The markers are changed for scans at different levels. Horner and Potts (1984) divide the marker system into two groups—discontinuous and continuously changing markers. The former are the same size on all CT scans within a certain interval (Shelden *et al.*, 1982). The latter change depending on scan level (Leksell and Jernberg, 1980; Brown–Robert–Wells System; etc.). Then, after a certain period of time, SB is performed in the operating room by transferring the data on the CT scans to roentgenograms in two projections by the conventional stereotactic technique.

In the second type, the head of the patient is also fixed in the stereotactic apparatus, and the biopsy is made directly in the CT scanner on the basis of the information obtained without transferring it to conventional roentgenograms. For this technique large-gantry scanners are more appropriate.

An original technique using the biopsy needle as a reference point was proposed by Huk and Baer (1980). A special device with the needle in the sagittal plane is fixed on the head of the scanning table without a patient, and CT scans are obtained. Then the same scans are made with a patient in position, and the two series are superimposed. The needle image is used as a zero point, and the angles of insertion of the biopsy instrument are calculated.

Modern SB techniques include the cursor function and external laser beam incorporated into the CT scans (Greitz *et al.*, 1980). The laser beam is used to align each CT scan with the stereotactic device.

The accuracy of the coordinates determined by the above-described techniques is high—about 1 mm for X and Z coordinates and about 3 mm for Y coordinates.

During the past 6 years we have been using the following simplified technique (Kandel *et al.*, 1981): CT is performed prior to the biopsy using a CT scanner 1010 (EMI, Great Britain). The head of the patient is fixed securely in a special device for the entire period of scanning to ensure the compatibility of the tomographic scans. Moreover, we do not observe the conventional condition that the initial plane of scanning coincide with the orbitomeatal plane; it may be at an angle to that plane or somewhat above or below it. However, it is obligatory that the head of the patient during subsequent biopsy be secured so that the initial plane of scanning (and consequently all the following planes parallel to the initial one) is perpendicular to both the lateral and AP skull roentgenograms.

Then a line is drawn on the head of the patient

along the initial plane of scanning, and small metallic markers are glued to this line. When the head is fixed in the stereotactic frame for the biopsy, this manipulation under electronic amplifier control allows one to position the plane with the markers perpendicular to the AP and lateral roentgenograms. The central x-ray beam should be in the plane of the initial scan. Then on a Polaroid photograph of the CT scan that shows an enlarged image of the tumor, the point (X) (or points) of the desired biopsy is determined. The coordinates of that target point (Fig. 206) are (1) the distance XX_1 from the target to the sagittal line on the skull and (2) the distance FX_1 from the midpoint of the external plate of the frontal bone to the projection of the target point on the midline.

After a suitable period of time (it may be several hours or 1–2 days), SB is performed. The stereotactic apparatus designed by the author (Chapter 3, Section 5.9) is installed in the burr hole, usually near the coronal suture. The head of the patient is fastened in the stereotactic frame so that the initial scanning plane is perpendicular to the AP and lateral films (the technique is described above). Then x rays are taken in both projections. After that, on each film a line is drawn through the markers to designate the initial scanning plane, and, parallel to it, several more lines corresponding to the planes of the CT scans on which the tumor is seen. The distance between the lines corresponds to the known thickness of each scan, allowing for the x-ray divergence. On each of these lines dots are made to define the borders of the tumor. By connecting these dots, one obtains a contour map of the tumor on roentgenograms in both projections (Fig. 207). Then on the corresponding lines one marks off the target point (or points) inside the tumor from where the biopsy is to be taken.

Since in the photos of the scans all real images are reduced, it is necessary to multiply the distances XX_1 and FX_1 on CT scans by coefficient constant for the given CT scanner (on our scanner, 4.3) as well as to

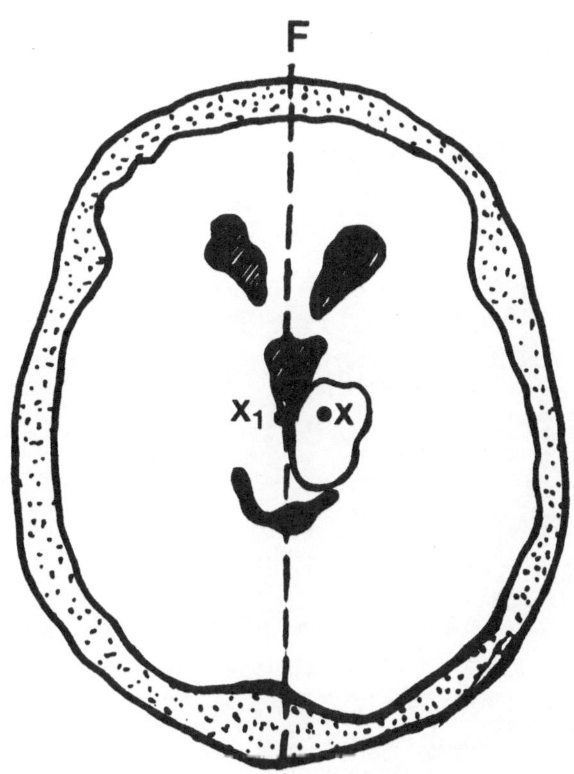

FIGURE 206. Scheme for calculation of the target point according to CT scan in a deep-seated brain tumor for stereotactic biopsy.

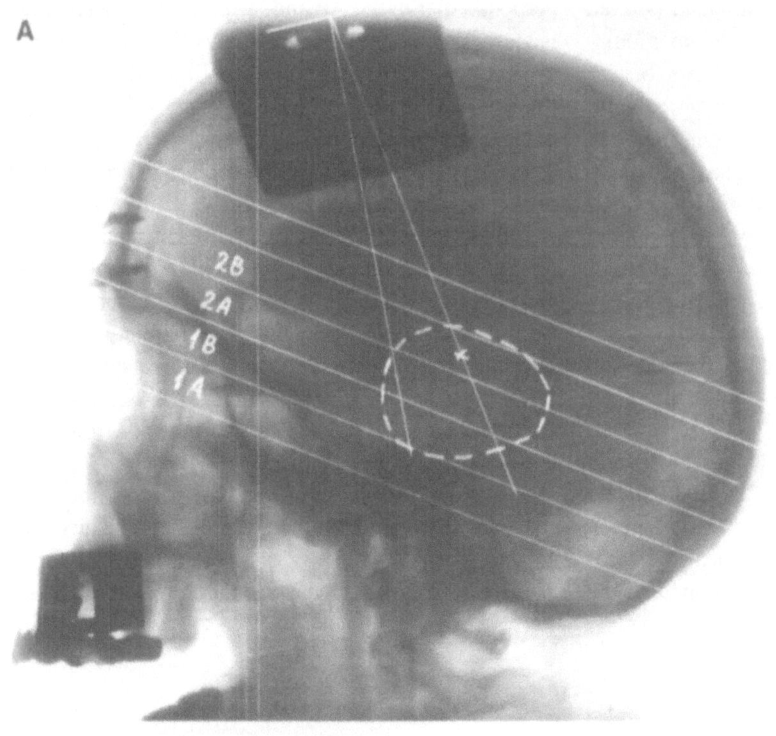

FIGURE 207. The planes of CT scans and contour of the intracerebral tumor transferred to the stereotactic x-ray films in both projections during the operation. The selected target point for the stereotactic biopsy is marked by small crosses on (A) lateral and (B) AP views.

correct for divergence. In this manner, the target point (or points) for the subsequent biopsy is demarcated on the AP and lateral roentgenograms. Then following the technique described in Chapter 4, the trocar is introduced to the target for biopsy; pieces of tissue are taken along the trajectory: two or three biopsies before the target point and two or three beyond it (serial biopsy), but all within the tumor contour drawn on both roentgenograms.

Since it greatly increases the possibilities of making an accurate diagnosis, the SB method based on CT data assists in defining the indications for and extent of surgery. In addition, it clarifies the most desirable treatment, particularly the advisability of stereotactic cryodestruction, RF coagulation (see below), or stereotactic implantation of radioactive isotopes. The following are three examples.

A 50-year-old patient was admitted with complaints of headache, weakness of the left extremities, and double vision. He showed rotary nystagmus, left-sided hemiparesis with hypertonus, and a constant tremor of the left hand. Ventriculography and angiography a year before admission had not revealed any clear-cut signs of tumor, although the cerebral ventricles were enlarged, and the Sylvian aqueduct was partially occluded. The CT scans demonstrated a relatively large tumor in the right Th (Fig. 208) that seemed to be inoperable. After the CT data were transferred to the films in both projections, a SB of the tumor under local analgesia revealed an astrocyte. A cryocannula (Chapter 5, Section 2) was used to make two cryodestructive foci, each 20 mm in diameter, in the tumor.

The condition of the patient improved substantially after surgery: headaches and nystagmus disappeared, left-sided hemiparesis improved significantly, tremor of the left hand disappeared almost completely, and gait improved considerably. However, practically all symptoms recurred after 1 year. A second operation involving cryodestruction of the tumor by the same technique was performed, and again a significant improvement was noted: the headaches and hand tremor disappeared, and gait improved. The patient tolerated both operations well. The follow-up after the second operation was 2.5 years.

Even when CT is employed, there are clinical situations in which it is very difficult to differentiate between an intracerebral hemorrhage and a glial tumor. In the case described below, SB made possible the diagnosis of a hemorrhage in a subcortical glioma.

A 57-year-old patient was admitted to our clinic after the rapid development of focal neurological symptoms including paralysis of the left arm, paresis of the left leg, left-side hemihypesthesia, central paresis of the seventh and 12th cranial nerves, and meningeal irritation. History revealed hypertension and three previous similar vascular episodes that had subsequently regressed, and CT investigations showed both a hemorrhagic focus and a glial tumor in the area of the left NL and adjacent white substance of the right hemisphere.

From the SB of the focus, 8 ml of dark blood with a small quantity of detritus was aspirated. Histological examination of the biopsy revealed an astrocytoma with signs of dedifferentiation. After craniotomy and incision of the cortex, about 30 ml of dark clotted blood and considerable tumor tissue at a depth of about 6 cm were removed. Subsequent histological investigations confirmed the diagnosis of astrocytoma. There was no significant change in the condition of the patient.

Difficulties often arise in differentiating between a glioma and subcortical encephalitis. In such cases SB is invaluable in making the correct diagnosis.

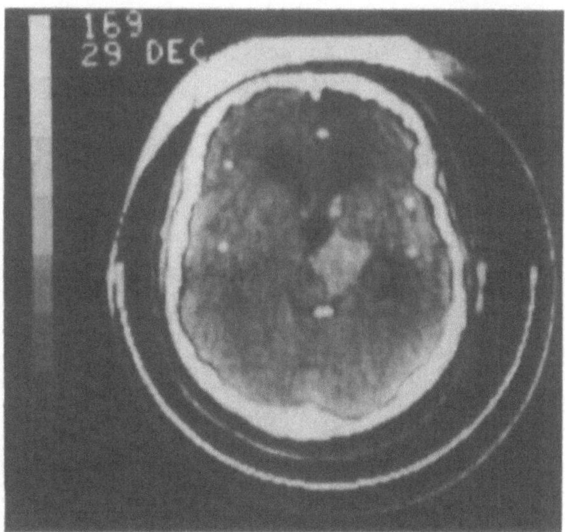

FIGURE 208. Inoperable astrocytoma of the right Th extending into V_3. Preoperative CT scan. A hyperdense zone occupying the whole right Th is seen; marked obstructive hydrocephalus.

A 39-year-old female was admitted to our clinic with complaints of headache, weakness, and numbness in the right extremities. Neurological status revealed mild right-sided hemiparesis with anisoreflexia and meningeal irritation. Investigations by CT after intravenous contrast enhancement revealed a focus of diminished density in the white substance in the area of the junction of the parietal, occipital, and temporal lobes of the left hemisphere. A focus of accumulated contrasting substance irregular in shape and nonhomogeneous in structure was found above this focus in the parietal lobe. These CT data did not allow us to exclude a glioma. A worsening in the condition of the patient and negative neurological dynamics supported this presumptive diagnosis. Stereotactic biopsy of the focus revealed a demyelinization with proliferation of all glial elements, primarily astrocytes, with the formation of globoid cells, small glial nodules, and intranuclear inclusions. The patient was transferred to the neurological department with the diagnosis of subacute progressing leukoencephalitis.

SB may be an effective treatment in the case of deep-seated intracerebral cysts. One of our observations with complete aspiration of the intrathalamic cysts is presented in Figure 209.

Nauta *et al.* (1984) proposed an interesting technique that allows visualization of the large cerebral vessels on CT scans used for SB. After routine AG, the catheter is inserted into the carotid or vertebral artery, and during CT scanning a single small bolus (4 ml) of contrast medium is injected into the artery. On the CT images, the arterial and venous phases are clearly visible, providing important information for SB.

In recent years, CT has been used to detect pathological processes in the brain at an early stage of their development, for example, gliomas or metastases less than 1 cm in size or small arteriovenous malformations. The most modern CT systems visualize pathological cerebral lesions as small as 2–3 mm. This requires an essentially new surgical technology for their removal. By classical, open operation, a surgeon can no longer reliably detect and remove such a small lesion with minimal damage to the cerebral tissue surrounding the focus.

Today it is not enough to determine and transfer the target points inside the tumor from the CT scan to stereotactic roentgenograms. One must establish and visually reconstruct the volume and form of the entire tumor as well as its relationship to large blood vessels. Directly after stereotactic biopsy, it is recommended that CT scanning of the brain be repeated to exclude possible hemorrhages into the tumor (Hahn *et al.*, 1979; Apuzzo *et al.*, 1987). In 1985 the results of the collective investigation of SB presented at the meeting in Marseilles were published. Twelve French neurosurgical clinics analyzed 3052 SB of brain tumors. The exact diagnosis of the tumor histology were obtained in 84% of cases. The mortality rate was only 0.6%, transient complications were present in 4.5% of cases, and stable neurological deficits were observed in 1% of cases. The introduction of radioactive sources into tumors and cysts was performed in many cases after SB (Benabid *et al.*, 1985).

The preliminary selection and subsequent identification of parts of the tumor for biopsy opens up broad prospects for a further comprehensive study of the morphology, histogenesis, and biochemistry of glial tumors at various stages of their growth. In addition to the above-mentioned diagnostic advantages, this method will permit more adequate means of comprehensively treating intracerebral glial tumors, depending on their dimensions, localization, histological structure, stage of growth, etc.

4. Stereotactic Destruction

Treatment of intracerebral glial tumors, as is known, still remains one of the main unresolved problems in neurosurgery. More than half of these tumors, because of their wide dissemination and localization in the deep parts of the brain, are considered unsuitable for radical removal. In view of this, neurosurgeons have been seeking new methods of effectively destroying inoperable gliomas.

4.1. Radioactive Isotopes

We shall dwell very briefly on the introduction of radioactive isotopes into brain tumors (interstitial brachy therapy) by the stereotactic method since several special monographs and numerous articles in the literature have been devoted to this question.

For evaluation of the results of SB, we collected 6850 cases from 36 papers in the literature. The great majority of cases were deep-seated brain tumors. The location of tumors and other processes where SB was carried out were in diminishing order: cerebral hemispheres, basal ganglion, thalamus and hypothalamus, pituitary, pineal region, ventricular system, cerebellum, and midbrain. The majority of cases (about

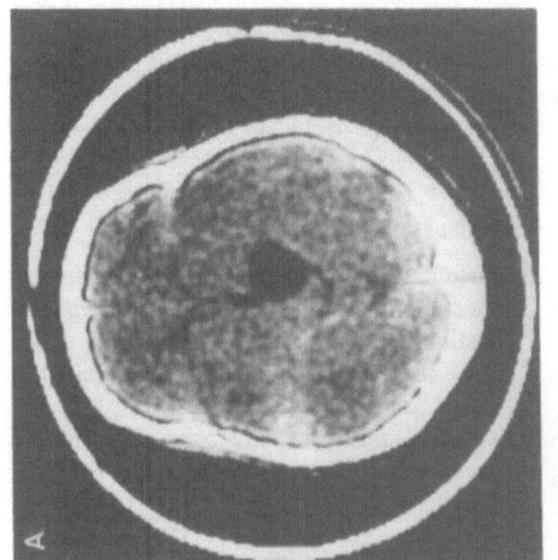

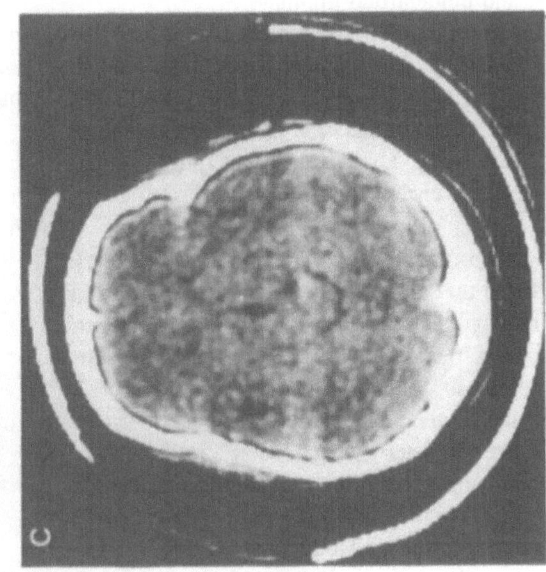

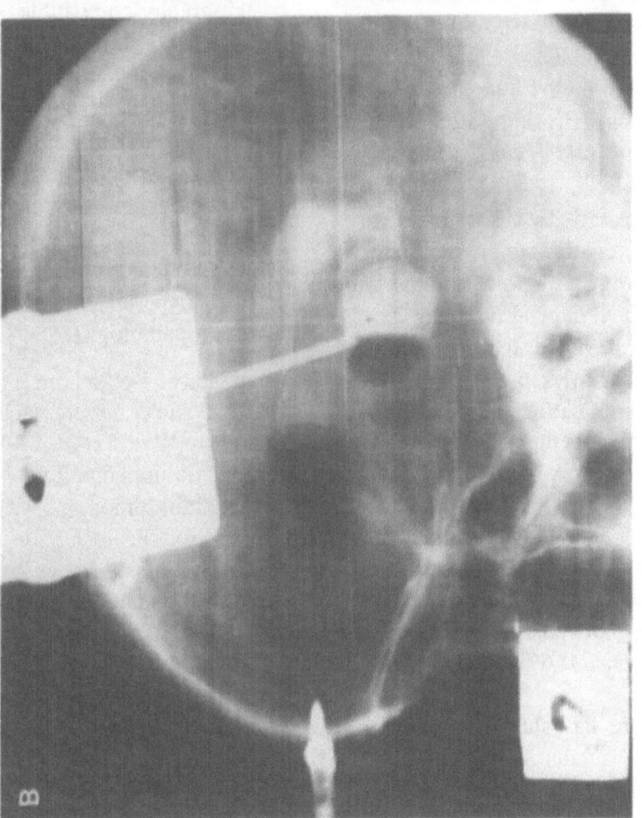

FIGURE 209. Stereotactic aspiration of the intrathalamic cyst in a 42-year-old patient. (A) CT scan before surgery. (B) X ray during the stereotactic aspiration of the cyst. After removal of the colloid fluid, the cyst cavity was filled with contrast medium. (C) CT scan two weeks after surgery. The cyst is not visualized.

65%) were various gliomas (in diminishing order also): different grade astrocytomas, glioblastoma multiforme, oligodendrogliomas, ependymomas, plexus papillomas, and unidentified gliomas.

In the group of nonglial tumors (about 25%) the majority consisted of craniopharyngiomas, pituitary adenomas, metastases, epidermoids, germinomas, meningiomas, neurinomas, and granulomas. The remaining 10% were abscesses, hematomas, gliosis, arachnoid cysts, encephalitis, necrotic tissue, and so on.

Keeping in mind the above-mentioned remarks we calculated that the overall rate of success of SBs is 92%. The "procedure success rate" is less than the "case success rate" and may be evaluated as about 85%. The rate of obtaining the exact histology of the lesion is much higher, especially in the past few years, than the rate of determining "useful information." Stereotactic biopsy may be very successful even in intrinsic tumors of the midbrain, pons, and medulla, as was recently reported by Hood *et al.* (1986).

The rate of successful biopsies depends on many factors, which will be summarized below. For instance, the method is more reliable for large tumors than for small ones, and more reliable for highly malignant astrocytomas than for unusual primary tumors or metastases.

As a rule, stereotactic biopsy is not a very dangerous procedure and is comparatively easily tolerated by patients even in a grave condition. At the same time, each SB is connected with the risk of complications and adverse effects, which depend on many factors. As Pecker *et al.* (1979a,b) noted, the importance of the information gathered justifies taking the risk.

The rate of different complications and the mortality rate in the collected 6850 SBs depended on many factors.

Summing up the mortality rate, which varies according to the published series, one may conclude that the overall mortality rate is about 1.5%. Two of the most successful series were a collected French series of more than 3000 cases and Apuzzo's series of 500 cases, with mortality rates 0.6% and 0.2%, respectively.

Analysis of the literature has shown that intracranial hematomas (epidural, subdural, intracerebral, and intratumoral) are the main reason for serious, frequently irreversible complications. The overall rate of hematomas after SBs is 2%. In the majority of cases the reasons for the bleeding remain uncertain.

In second place in the list of complications is the development of neurological deficits or aggravation of the preexisting deficits. The complication depends on such factors as the malignancy of the tumor, its size, location in functionally important cerebral areas, and the increase of brain edema. The overall rate of serious complications is 8% (including hematomas) with permanent disability about 4%.

Summing up the results of the SBs published in the world literature and seen in our personal experience, one may conclude that these results depend on the following eight factors:

1. Nature of the pathological process (tumor, nontumoral cyst, abscess, hematoma, inflammation, etc.).
2. Site, size, and histology of the brain tumor or other mass lesions.
3. Choice of the target point (or points) inside the mass lesion; the use of a single or serial biopsy.
4. Techniques of the biopsy (free-hand; stereotactic technique; CT-, NMR-, or digital angiography-compatible technique; biopsy instruments, impedance monitoring; fibroscope).
5. Histological (staining) techniques (smear, frozen/paraffin).
6. General condition of the patient.
7. Development of complications and adverse effects.
8. Correct interpretation of the results based on the experience of the neurosurgeon and neuropathologist.

Radioactive ^{32}P, a pure β emitter, was first introduced into cystic craniopharyngiomas by the stereotactic method by Wycis *et al.* (1954). An early paper on the stereotactic implantation of granules of radioactive ^{198}Au in a brain tumor was written by Talairach *et al.* (1955).

The last three decades have witnessed the successful development of stereotactic implantation of various radioactive isotopes in diffuse, deep-seated, often semimalignant or malignant tumors that cannot be removed by conventional surgery and are resistant to external irradiation (Mundinger *et al.*, 1959; Blagoveshchenskaya *et al.*, 1968; Mundinger and Hoefer, 1974; Kelly *et al.*, 1978; Frank *et al.*, 1985; Riechert, 1980; Lunsford *et al.*, 1985; Apuzzo *et al.*, 1987; Kelly, 1987; Dyck *et al.*, 1987).

Permanent or temporary interstitial irradiation of deep-seated brain tumors is used. For long-term irradiation, the radioactive sources are implanted into the tumor permanently and cannot be removed (Mundinger *et al.*, 1978b). For temporary brachytherapy different techniques are utilized, particularly a double-catheter after-loading system (Gutin and Dormandy, 1982; Gutin *et al.*, 1984; Kelly, 1987). The catheter is introduced stereotactically into the selected target in the tumor and is fixed in the skull bone with CT control. The inner catheter containing a predetermined number of radioactive sources with high activity is inserted for a certain period of time. After that, both catheters are removed. Interstitial implantation of radioactive isotopes offers a number of valuable advantages. It enables one to create an effective dose field in a limited volume of tissue, leaving the surrounding cerebral structures intact. Another advantage is the continuous emission of the radiation, ensuring a more potent destructive action on tumor tissue during the protracted irradiation.

Introduction of different radioisotopes into cystic brain tumors, mainly into craniopharyngiomas, is continuing. Good results of intracavitary irradiation using ^{32}P colloidal chronic phosphate in eight cases of craniopharyngiomas and in two cases of cystic astrocytomas of the third ventricle were reported recently (Lunsford *et al.*, 1985).

Musolino *et al.* (1985) reported the favorable results of stereotactic treatment for expanding cystic craniopharyngiomas by endocavitary introduction of β-radiation sources (^{189}Re, ^{198}Au, ^{90}Y). The authors noted the cessation of fluid formation and progressive shrinkage of the expansive cysts in all 18 cystic craniopharyngiomas. Follow-up study ranging from 1 to 6 years revealed cyst obliteration in 75% of the cases. There were no side effects or complications during the follow-up. About 3000 rads depth-dose is considered as the safer dose for endocavitary irradiation of craniopharyngiomas.

An extensive experience in interstitial radiation of more than 1000 brain tumors, mainly in deep regions, has been accumulated by Mundinger and Weigel, (1983). These operations, which are mostly palliative, consist of stereotactically implanting in the tumor tissue an isotope that then causes radiation necrosis of the surrounding tissue. This method is frequently used in malignant tumors at the base of the skull, in inoperable tumors of Th, midbrain, and area of V_3, cerebral trunk, as well as in pituitary tumors (Section 5).

The stereotactic implantation of radionuclide sources into brain tumors for interstitial therapy should be done using a computer. It is very important to calculate the isodose envelope in such a way that maximum radiation is delivered to the tumor without significant damage to the surrounding brain tissue. The computer displays the slices of the three-dimensional tumor volume on the video monitor. Using the cursor, the neurosurgeon may choose the point (or points) for isotope seed implantation. The computer then calculates the coordinates of the implantation and also displays the aggregate isodoses for each slice (Kelly *et al.*, 1985).

Today there is a wide choice of radioactive isotopes, differing both in half-life and type of emission. Short-acting isotopes such as radioactive yttrium, iridium, and palladium are most frequently used for the destruction of brain tumors; less often, the long-acting isotopes of cobalt and tantalum are used. The majority of these isotopes produce a mixed α and β emission. In the past, radioactive gold was implanted in gliomas; however, with more effective isotopes being developed in recent years, the use of radioactive gold has been practically abandoned. Now ^{198}Au is used only for implantation in some cysts of craniopharyngiomas. At the present time, a ^{90}Y implant is used for this purpose (Gahbauer *et al.*, 1983).

Yttrium is a pure β emitter with a small radius of destructive action (Chapter 5, Section 4). We reported the first results of the stereotactic implantation of ^{90}Y granules in inoperable tumors at the base of the skull (Kandel *et al.*, 1960, 1963). According to biopsy reports, all these tumors were malignant (cancer, sarcoma, chordoblastoma) and growing into the nasopharynx. The ^{90}Y granules with an activity from 5 to 15 mCi, 0.8 mm in diameter and 1 mm long, were implanted transnasally into the tumor under visual TV monitoring. A set of cannula probes of our design (Lyass *et al.*, 1963) was used for implantation. The granules of yttrium were placed in the tumor tissue at a distance equal to the diameter of the necrosis (approximately 2 cm).

Since all the patients had incurable malignant tumors, the results of treatment could be considered satisfactory. In follow-up periods from 10 months to 5 years, of six cases, only one patient died. The rest displayed a clear-cut clinical improvement with regression of symptoms, and one patient worked for 5 years (Blagoveshchenskaya *et al.*, 1968). Therefore, even in the most hopeless cases, the implantation of ^{90}Y in a tumor at the base of the skull can significantly prolong the life of the patient.

The heavy α-radiating ^{90}Co is used in the form of needles 2 mm in diameter, advanced stereotactically into the tumor or the part of it that remains after conventional tumor removal (Riechert, 1980). For implantation in deep-seated brain tumors of average and large size, preference is given to a milder α radiant, ^{192}Ir, having lower radiation intensity (0.61 MeV) and longer half-life (74 days). Before the isotope is stereotactically implanted in the brain tumor, complicated dosimetric calculations of the ionizing radiation are carried out, using special phantoms for this purpose.

The localization and size of the tumor are determined by CT as well as ventriculography, angiography, scintigraphy, and other diagnostic methods. Then a small wire of ^{192}Ir 0.5 mm in diameter and 3–5 mm long is introduced stereotactically to the central parts of the glioma. After that, the central parts of the tumor may receive a large dose of radiation, up to 15,000 rad (Kelly *et al.*, 1978; Apuzzo *et al.*, 1987; Dyck *et al.*, 1987).

In another variant of this method, a metallic tip containing a large dose (up to 120 Ci) of ^{192}Ir is advanced stereotactically into the center of the tumor. A dose equal to 6000–12,000 rad is delivered to the center of the tumor, and the surface receives approximately 3500 rad. The smaller dose on the surface is quite adequate, since the rapidly growing cells in the periphery of the tumor are more sensitive to radiation. Depending on the specific conditions, in certain cases, a conventional open operation is first performed in order to reduce the volume of the intracerebral tumor and to mark its remaining part with clips and tantalum powder. Then wires of ^{192}Ir are introduced into the remaining tumor.

For many years the neurosurgical clinic in Frieburg has been developing the technique of stereotactic implantation of small-gauge (0.3–0.5 mm) wires of ^{192}Ir (an alloy of 70% platinum and 30% iridium), most frequently into malignant inoperable gliomas of Th, midbrain, and superior brainstem. The great activity of the isotope makes it possible to create a very high isodose in the tumor. Good results in multiform glioblastomas are obtained only with a focal dose of no more than 10,000 rad. Reports from that clinic (Mundinger and Metzel, 1970; Mundinger and Hoefer, 1974; Mundinger *et al.*, 1978b; Mundinger and Weigel, 1983) have analyzed the long-term results in these cases. It is emphasized that the majority had malignant gliomas resistant to conventional distant γ therapy. Many of these patients have been living for 3–5 years, displaying a substantial clinical improvement and regression of neurological symptoms.

A new low-energy radionuclide, ^{125}I in the form of a powder absorbed on Dowex and put in small titanium granules, was recently proposed for stereotactic implantation into a brain tumor (Mundinger and Weigel, 1983). This isotope, concentrated within a radius of about 2.5 cm, possesses mild radiation and, as the authors state, causes minimal radiation of the brain tissue surrounding the tumor.

^{125}I sources were implanted stereotactically in seven pineal tumors with good results in five cases (Frank *et al.*, 1985).

Colloid solutions of many radioactive sources are used for injection into intracerebral cysts associated with gliomas or craniopharyngiomas. These colloids act on the secreting walls of the cysts and prevent their reaccumulation.

One of the main complications of interstitial radiation of intracranial tumors is radionecrosis of brain tissue with the formation of necrotic foci and cysts. Since this complication develops slowly and generally a long time after implantation of the isotope and is accompanied by an increasing neurological deficit with symptoms of high intracranial pressure, it is quite often erroneously mistaken for recurrence of a glioma. At times, it is necessary to resort to an open operation to remove large foci of radiation necrosis that develop in the tumor after massive radiation.

The rapid development of intraoperative ultrasound examination of the brain with new high-resolution systems creates an acceptable alternative to CT-guided SB (Chandler *et al.*, 1982). Recently, Berger (1986) designed a special apparatus for ultrasound-guided biopsy. This device permits one to overcome the main disadvantage of the method—the need for craniotomy. A skull-mounted apparatus with ultrasound probes 16 mm in diameter with 5.0 and 7.5 MHz rates permits ultrasound-guided SB through a burr hole.

The device has been successfully used in 19 patients for tumor biopsy as well as aspiration of cysts and abscesses. The lesions were of mixed echogenicity, and only the cysts were hypoechoic.

4.2. Cryodestruction

A very promising technique for the destruction and removal of various brain tumors is cryodestruction, the principles of which were presented in detail in Chapter 5 (Section 2). Numerous experimental and clinical investigations have demonstrated that in the majority of cases, freezing of brain tissues leads to its total necrosis. However, certain types of tumors display very high resistance to low temperatures. Our ex-

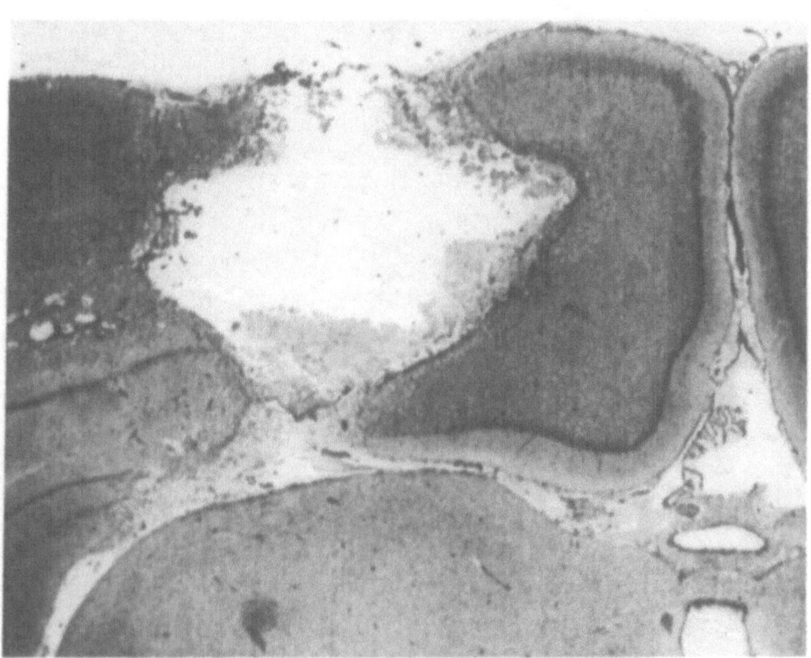

FIGURE 210. Frontal section of rabbit brain 50 days after local freezing of a transplanted human malignant brain tumor. The cavity has replaced the destroyed tumor. Hematoxylin–eosin, ×10.

perimental investigations with Yablonovskaya in local cryodestruction of heterotransplanted glial tumors in rabbits revealed that total destruction may be achieved in the early stages of the development of tumors (Fig. 210). But later, cryodestruction inactivates only the central part of the tumor, leaving infiltrations continuing to grow in the peripheral parts (Fig. 211).

For the cryodestruction of brain tumors since 1966, we have used a special device (Fig. 212), designed and built jointly with Shalnikov. This lightweight (140-g) instrument consists of a plastic reservoir for liquid nitrogen and three interchangeable cannulas with diameters of 4, 5, and 6 mm. Depending on the amount of liquid nitrogen poured into the reservoir and time of exposure, at the active tip of the cryoprobe, an ice sphere with a diameter up to 50–55 mm develops, which is quite sufficient for the destruction of small and medium-sized brain tumors.

We use the cryomethod for the destruction of brain tumors in three ways (Kandel and Beijesin, 1970; Kandel, 1981). Crydestruction, i.e., destruction of the tumor *in situ* without its subsequent removal (Fig. 213A), is indicated in deep-seated intracerebral gliomas of the basal ganglia and midbrain, when conventional radical removal is impossible. Before the operation, various diagnostic methods (angiography, CT, etc.) are employed to pinpoint precisely the margins of the tumor. The contours of the tumor are transferred to stereotactic roentgenograms in two projections, as described in the preceding section. Then a burr hole is made in the skull overlying the tumor projection and avoiding the functionally important parts of the cerebral cortex. Based on the stereotactic calculations, SB is taken. Then a cryocannula of the appropriate diameter is introduced into the central part of the tumor stereotactically, and the previously calculated volume of tumor tissue is frozen. Having determined the contour of the tumor from the previous diagnostic examinations, after the first cryodestruction it is possible to advance the cryoprobe to other parts of the tumor and repeat the freezing to achieve the maximum destruction of the neoplasm.

Cryoextirpation or total removal of a frozen tumor (Fig. 213B) has been used primarily in benign tumors (meningiomas), although it may also be used in gliomas located on the brain surface. After conventional osteoplastic craniotomy, the tumor is located and partially exposed. Then, as in the previous technique, the cryoprobe is advanced into the center of the tumor, which is then frozen. In its solid frozen form, it can be separated quickly and easily from the surrounding cerebral tissue.

Cryoextirpation combined with cryodestruction, the combination of the two techniques described above (Fig. 213C), is indicated in large infiltrating tumors, the total removal of which is impossible in view of

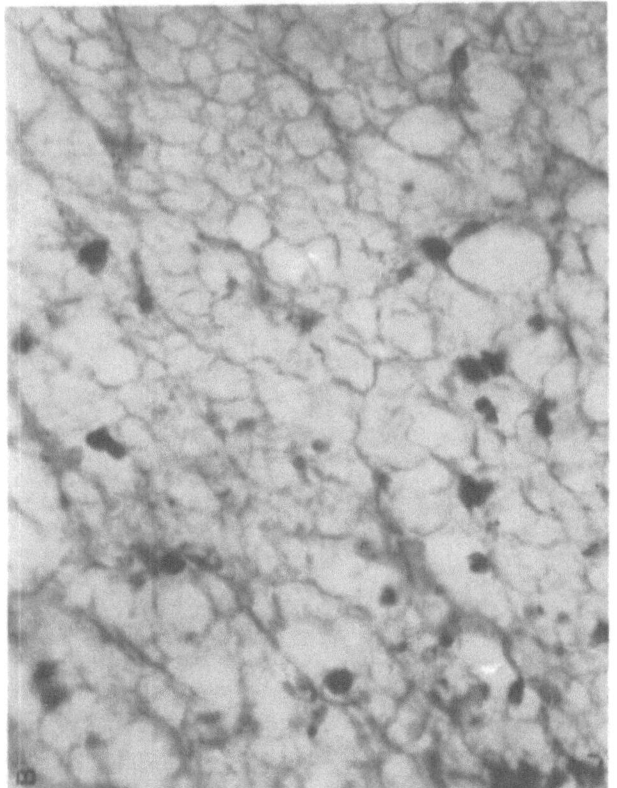

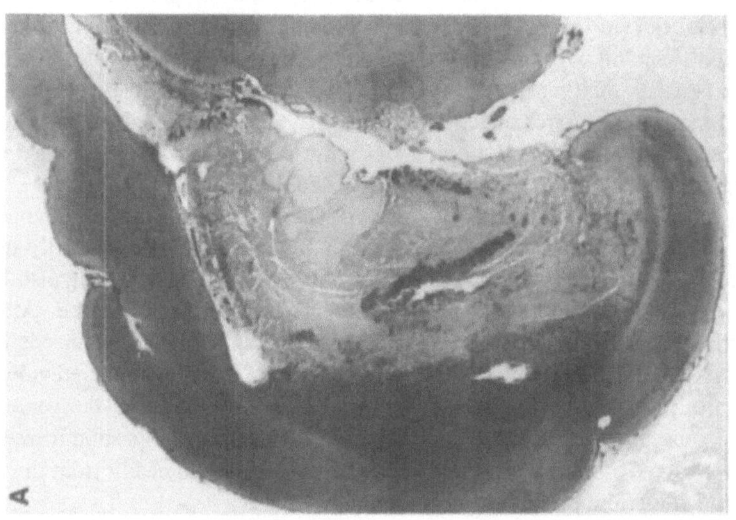

FIGURE 211. Cryodestruction of the central part of transplanted human brain tumor in rabbit brain. (A) Frontal section of the brain 3 weeks after cryodestruction. The focus of total destruction of the central part of the tumor is clearly seen. Hematoxylin–eosin, ×10. (B) Microscopic view of the zone of complete destruction of the tumor, ×200. (C) The zone of dystrophic neoplastic tissue, ×200. (D) The zone of the infiltrating tumor, ×140.

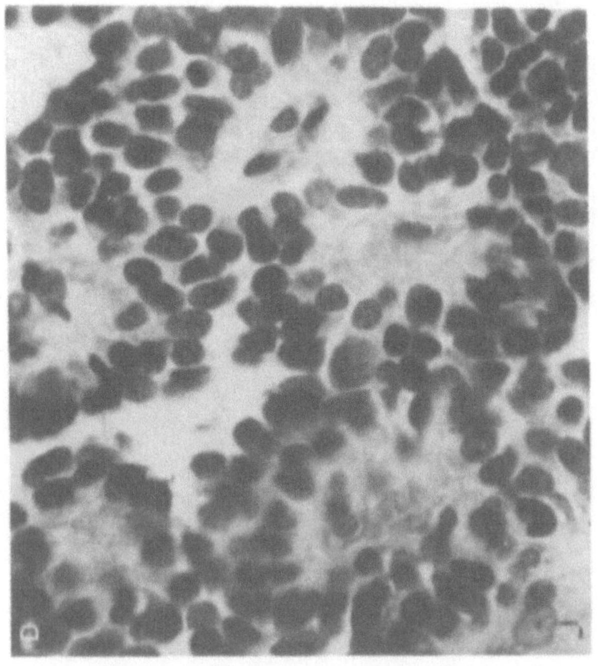

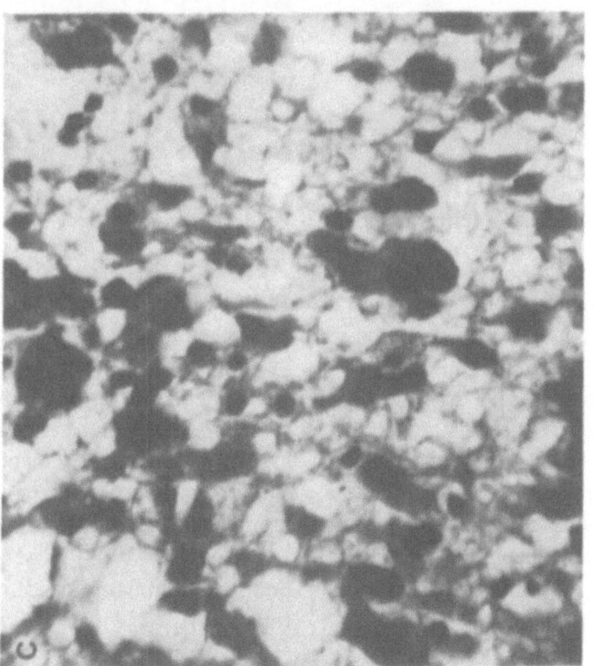

FIGURE 211. *(Continued)*

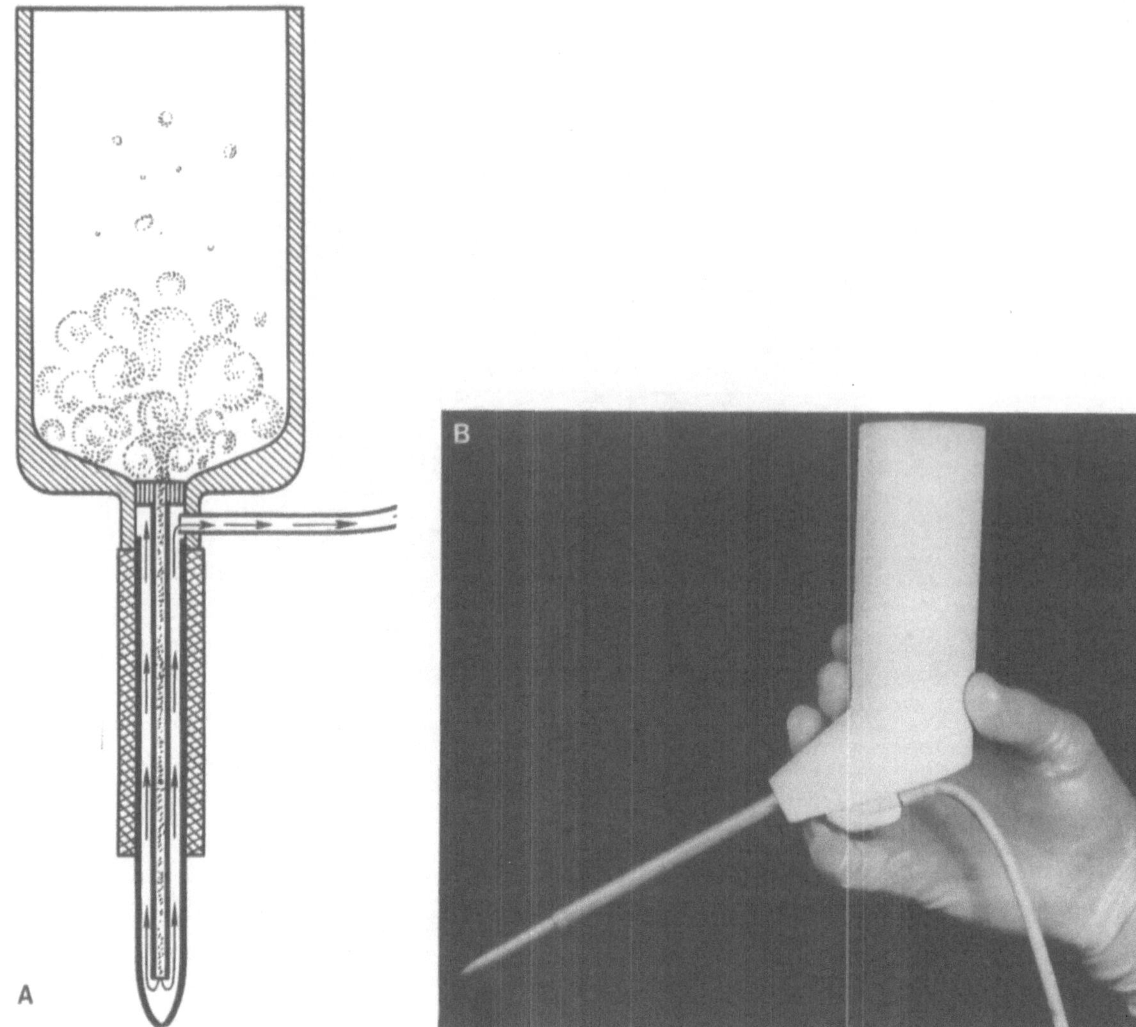

FIGURE 212. (A) Schematic diagram of our cryogenic device for brain tumor freezing. (B) General view of the device.

their spread into the mediobasal formations, basal ganglia, and brainstem. In such cases, it is possible to freeze and remove the greater part of the tumor, whereas the remaining unremovable deep-seated part may be frozen and, after thawing, left *in situ*. In rare cases, this method may also be used for meningiomas that are intimately attached to anatomic structures and that, if damaged, could lead to serious complications (e.g., with the sagittal sinus). The same technique can also be used for basal meningiomas that have a patent main cerebral artery passing through them.

Over many years, we have accumulated experience in the surgical destruction of more than 120 supratentorial brain tumors at various sites and of different histologies. This has been described in our earlier publications (Kandel and Beijesin, 1970; Kandel, 1974a,c, 1981). The use of cryodestruction in convextity and falx meningiomas as well as in meningiomas at the base and on the tentorium proved to be an effective supplement to conventional surgery (Fig. 214). All these tumors (with the exception of two) were removed totally, and in many years of follow-up, there was not a single recurrence. There was also no postoperative mortality.

Conway (1973) froze pineal tumors in two children with clinical improvement in both cases and with follow-ups of 2 and 5 years. Romodanov *et al.* (1971a,b) proposed combining cryosurgery of malignant cerebral gliomas with administration of chemotherapeutic agents directly into the tumor tissue as a further development of this method. With CT, it is possible to control visually the process of freezing, a

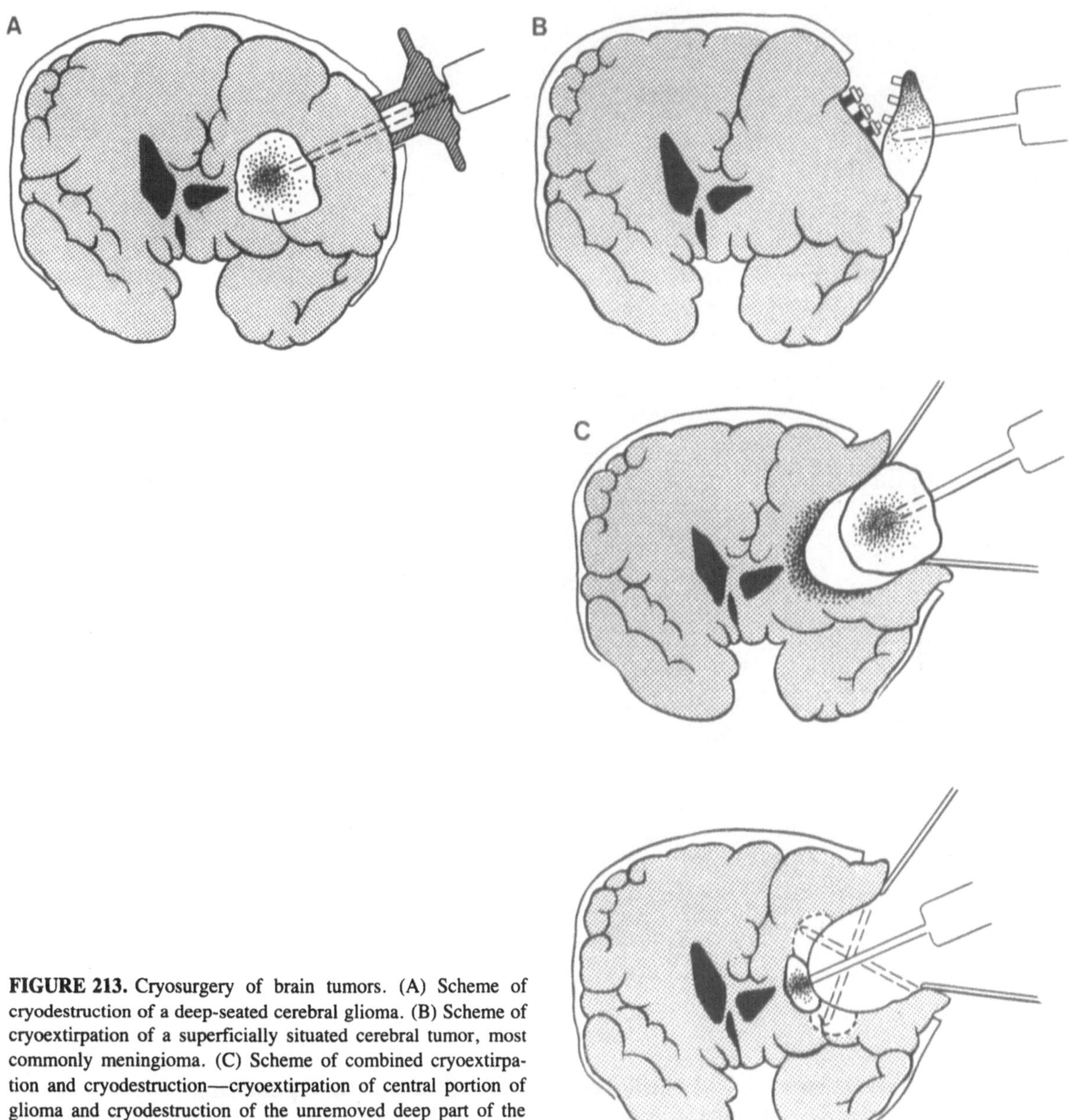

FIGURE 213. Cryosurgery of brain tumors. (A) Scheme of cryodestruction of a deep-seated cerebral glioma. (B) Scheme of cryoextirpation of a superficially situated cerebral tumor, most commonly meningioma. (C) Scheme of combined cryoextirpation and cryodestruction—cryoextirpation of central portion of glioma and cryodestruction of the unremoved deep part of the tumor by overlapping freezing lesions.

technique used for freezing small deep-seated tumors. With CT control it is possible to see how the ice sphere encases the tumor.

At this time, a comparison of the results of cryodestruction and the conventional method of removing gliomas is not feasible. There can be no doubt that cryodestruction is palliative, since, as a rule, it is impossible to determine the precise borders between the tumor and the surrounding cerebral tissue. Nevertheless, in a number of our cases after cryodestruction of malignant gliomas, relapses were observed only after long periods of time, and the life-span was longer than that reported after conventional therapy. These encouraging data stimulated further development of stereotactic cryosurgery in neurooncology.

5. Pituitary Tumors

Surgery of pituitary tumors traditionally represents one of the main problems in neurosurgery. That

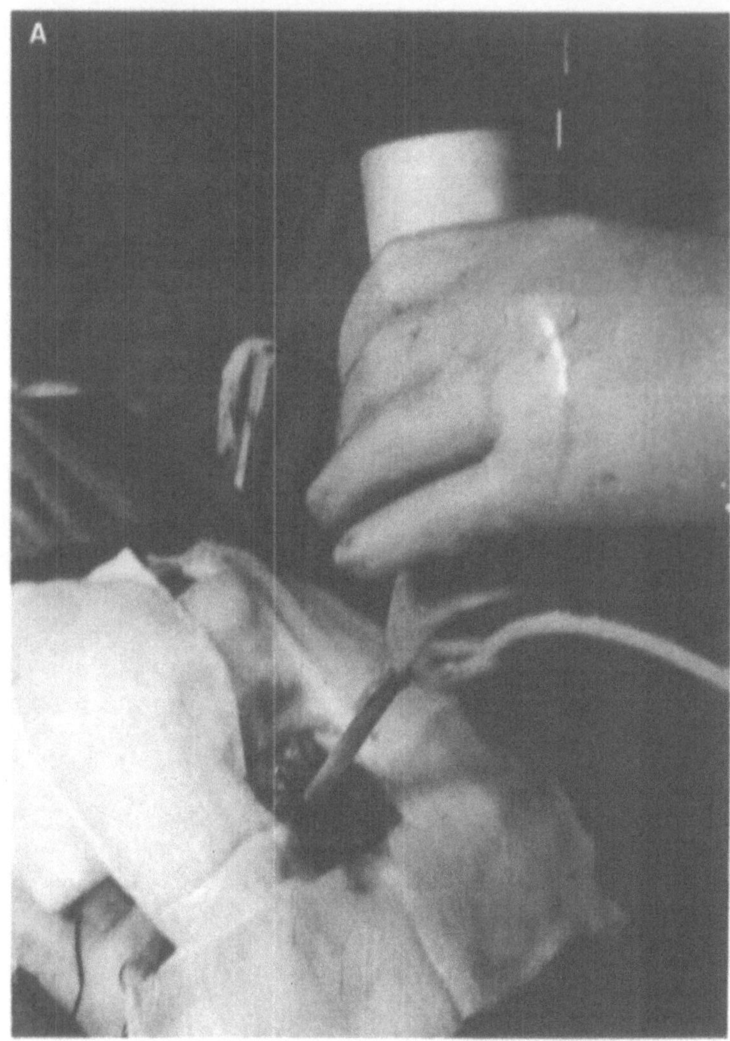

FIGURE 214. (A) Convexity meningioma frozen solid. (B,C) The frozen tumor is removed.

this problem has not yet been resolved is indicated by the fact that at present several surgical methods are used to treat these tumors: open extirpation by the intracranial approach, extirpation by the transnasal–transsphenoidal approach using the microsurgical technique, stereotactic destruction of these tumors by various methods, as well as nonsurgical management (x-ray and γ radiation, proton beam, bromocriptine treatment, etc.).

Without commenting in great deatil on this major problem, to which at least a dozen monographs and hundreds of papers have been devoted, we present data from the literature and our own experience with stereotactic surgery of pituitary adenomas.

5.1. Stereotactic Operations

The advances made by stereotactic neurosurgery stimulated the application of this method to pituitary tumors. These operations may be divided into two groups: operations on pituitary tumors for the purpose of their destruction, and operations to destroy a normal adenohypophysis in hormone-dependent cancer and certain endocrine disorders. The second group, naturally, does not belong to the category of operations on pituitary tumors; however, the stereotactic method in both cases has very much in common. Accordingly, the technique of destroying a normal hypophysis is presented briefly at the end of this section.

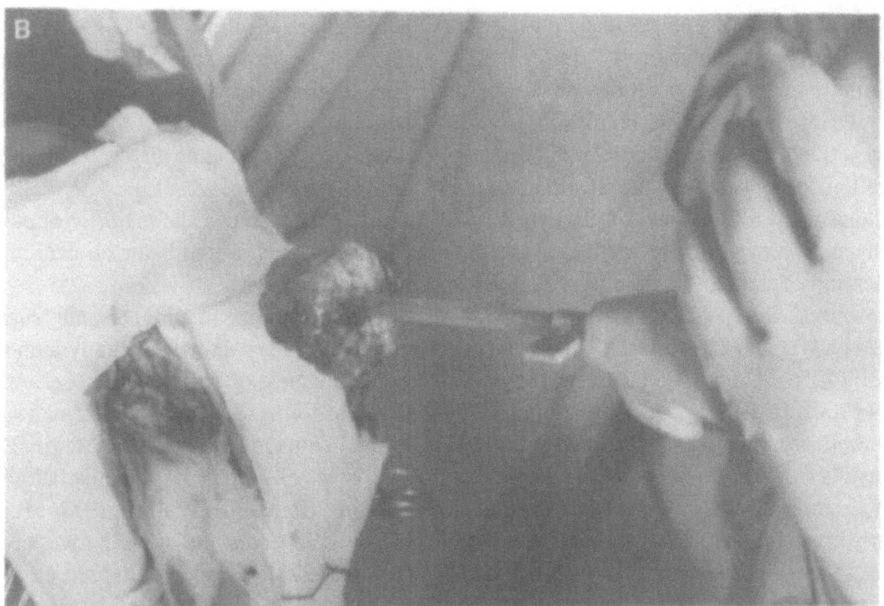

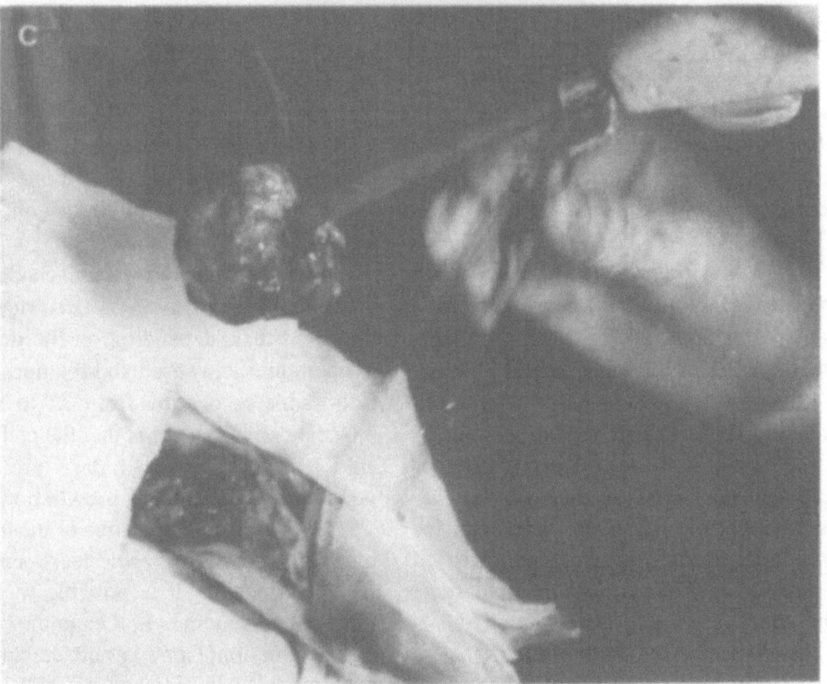

FIGURE 214. *(Continued)*

It should be noted that the pituitary is a very convenient object for stereotactic operations. The sella turcica is so located that it is clearly visible on conventional cranial roentgenograms and serves as a reliable reference point that does not require contrast media to visualize the ventricular system. The sella is seen even more clearly when pituitary tumors enlarge the sella turcica substantially.

Stereotactic operations on the pituitary require high accuracy and great caution to avoid damaging adjacent important functional structures (chiasma, oculomotor nerves, cavernous sinuses, internal carotid arteries, hypothalamic area). Their damage during surgery may cause serious complications. Consequently, when performing stereotactic operations on the pituitary, irrespective of the technique employed, strictly localized destructive foci are required. Another important consideration for a successful surgical outcome is to preserve intact the posterior lobe of the hypophysis (neurohypophysis), damage to which could lead to serious complications (diabetes insipidus and others).

The first stereotactic operations on the hypophysis were performed by Wycis *et al.* (1954), Riechert (1955), Talairach *et al.* (1956), Guiot (1958a), and Mundinger and Riechert (1960). In 1961 at the International Symposium on Stereoencephalotomy, Talairach presented remarkable results from transnasal stereotactic implantation of ^{90}Y granules both in a pituitary tumor, in particular, one producing acromegaly, and in a normal hypophysis for severe diabetes, metastases of malignant tumors, Cushing's syndrome, and malignant exophthalmos. In the USSR, the first stereotactic operations for implanting ^{90}Y granules in a pituitary tumor at the base of the skull were performed by the author in 1960.

Two main approaches were used in stereotactic pituitary operations—transfrontal (Wycis *et al.*, 1954; Cooper, 1963; Mundinger and Riechert, 1962, 1967; Poblete and Zamboni, 1974) and transnasal–transsphenoidal (Talairach *et al.*, 1956; Guiot, 1958; Hardy, 1969, 1975; Kandel *et al.*, 1973a; Babchin *et al.*, 1967; Zervas and Hamlin, 1974; Seymour *et al.*, 1978). The transfrontal approach in which the stereotactic instrument passes through the frontal lobe before penetrating the pituitary tumor is now rarely used since it is frequently the cause of serious complications. During the past decade the transnasal–transsphenoidal stereotactic approach has been generally accepted both for removal of a pituitary tumor and for operations on a normal hypophysis. Such an approach allows one to perform all the surgical manipulations under the diaphragm of the sella turcica, which reduces to a minimum the chances of damaging important supra- and parasellar structures. In this approach there is a potential risk of infection entering from the nasal cavity and sphenoidal sinus as well as the development of postoperative CSF rhinorrhea. However, the use of modern antibiotics and careful surgical techniques have made it possible to minimize these complications and to cope successfully with them.

Various methods of destruction, described in detail in Chapter 5, are employed for destroying pituitary adenomas as well as a normal hypophysis.

Stereotactic destruction of the pituitary by implanting radioisotopes undoubtedly gives satisfactory results. For example, Mundinger (1965b, 1974c) reported on the long-term results of transsphenoidal destruction of pituitary adenomas by stereotactic implantation into the tumor of ^{192}Ir in doses that varied depending on the size of the tumor, from 0.47 to 2.25 mCi. In 73 patients, the majority of cases followed up for many years, a significant clinical improvement was noted, particularly in vision.

Radioisotope techniques are successfully being employed today; however, many authors have noted a high incidence of various, often serious, postoperative complications (CSF rhinorrhea and subsequent meningitis, damage to the chiasma, etc.). The frequency of rhinorrhea after transnasal implantation of isotopes varies from 10 to 23% (Walker, 1967b; Rand *et al.*, 1968; Hardy, 1969), and the mortality was approximately 7% (Lyass, 1971). This led to the search for safer methods of destroying the pituitary in stereotactic operations. One of the most promising was cryodestruction, extensively used in stereotactic operations on the basal ganglia (Chapter 5, Section 2). The cryosurgical method that we have been employing in brain tumors since 1968 is, in our opinion and in the opinion of other authors (Conway *et al.*, 1970), one of the best methods for the destruction of pituitary tumors, although other techniques may be satisfactory.

The experimental work of Saglam *et al.* (1972) on cryohypophysectomy in *Macaca* rhesus monkeys demonstrated that, depending on the degree to which the temperature is lowered and the duration of freezing, it is possible to destroy from 72 to 96% of the hypophyseal cells. It appears that the cells most sensitive to cold, in descending order, are those producing gonadotropic hormones, growth hormone, and ACTH. The cells of the anterior lobe of the hypophysis producing TSH proved to be the least sensitive to freezing.

By freezing, it is possible to destroy practically the entire adenoma. For example, in 16 cases, after cryodestruction, Conway and Garcia (1970) found total destruction of the hypophysis in 12 and only the vestiges of adenohypophyseal tissue in the sella turcica in the remainder.

Cryodestruction does not involve a great risk to the cranial nerves since they are protected from the cold by the "thermal barrier" formed by CSF, cavernous sinuses, and carotid arteries. "An internal thermal barrier" formed by the arteries passing through the intermediate lobe of the pituitary body to a certain extent protects the neurohypophysis from cooling. Moreover, the resistance of the hypophyseal capsule and the bones of the skull base to cryonecrosis is so great that the defect at the bottom of the sella heals normally after transnasal stereotactic surgery. The orifice at the base of the sella after cryohypophysectomy is closed by fibrous connective tissue within the first week and by a thick scar at the end of the first month. With this in mind, we as well as other authors consider that the orifice in the sellar floor may be left open without danger of nasal CSF leakage developing (Conway and Garcia, 1970; Norell *et al.*, 1970; Kandel *et al.*, 1970, 1973b; Kandel, 1981).

Transnasal stereotactic cryohypophysectomy is indicated both in hormone-active and inactive pituitary adenomas, provided the tumor is principally located within the sella. With caution, the cryosurgical method may be used in cases of limited suprasellar extension. It is most expedient to use this method for large tumors destroying the sella floor and growing into the sphenoidal sinus. If the adenoma is large and extends far beyond the sella superior and laterally, causing visual disturbances, a transnasal stereotactic operation is not indicated. To destroy a suprasellar extension by this method is not only technically difficult but dangerous, since edema of the remaining tumor may cause acute compression of the chiasma and blindness (Mundinger and Riechert, 1967; Kandel, 1981).

One of our cases involved a large endosellar tumor with supraseller extension for approximately 12 mm. Besides pronounced acromegaly, the patient had bitemporal hemianopia and slight reduction of vision. Stereotactic destruction of the adenoma was without complications; however, approximately 8 hr after the operation, a sharp decrease of vision to 0.1 in both eyes was noted. That evening, a classical subfrontal operation was performed to remove the suprasellar part of the tumor, which had been compressing the optic nerves and the chiasma. In the course of 1 week, the vision was restored to 0.8 in both eyes, and the hemianopia disappeared.

As a practical rule, cryodestruction of an adenoma by the transnasal approach may be possible if the tumor protrudes above the sella not more than 8–10 mm. In case of a larger suprasellar growth of the tumor, extirpation by the classical intracranial approach or by a microsurgical transnasal–transsphenoidal approach is indicated.

The techniques of stereotactic cryohypophysectomy employed in our clinic, as well as the results of surgical treatment, have been described in part in our earlier works on this topic (Kandel *et al.*, 1970, 1973a; Kandel, 1981).

5.2. Preoperative Preparations

Stereotactic cryohypophysectomy requires a thorough preoperative examination and special preparation of the patient. On conventional films and tomograms it is necessary to make a careful study of the sella and adjacent bony structures and to obtain an estimate of the size and primary direction of extension of the tumor. Particular attention should be paid to the degree of enlargement and the shape of the sella, the degree of destruction of the anterior and posterior clinoid processes, as well as the sella floor, the thickness of the bone at the base, the volume of the sphenoidal sinus, etc. As a basis for comparison, one should take into account that the sella in lateral projections normally varies from 90 to 120 mm^2 (Schinz *et al.*, 1952). Roentgenograms of nasal sinuses must exclude sinusitis, which is a contraindication to operation.

The preoperative x-ray examinations that we practice include CT scanning, cisternography with metrizamide or pneumography of the basal cisterns, cerebral angiography, as well as venography of cavernous sinuses.

Recently, we have been using enhanced CT scanning (EMI-Scanner 1010) with a scan thickness of 10 or 5 mm on which to base the indications for surgery and the choice of technique for hypophyseal adenomas. Investigations by CT, especially frontal scans, yield much information about the size, shape, localization, and spread of the adenoma. In the majority of cases, adenomas have an increased x-ray density; however, in rare cases the density does not differ from that of the cerebral substance even after contrast enhancement (isodensive tumors). Particular attention should be paid to clear visualization of the chiasmatic cistern. Deformation of the cistern by a suprasellar extension of the tumor is important for diagnosis. In case of a parasellar growth, it is possible to observe a unilateral filling defect in this cistern. In a number of cases we performed CT cisternography after endolumbar injection of contrast substance, which gives more information than conventional CT.

It should be noted, however, that CT enables one to diagnose reliably only adenomas with extrasellar extension. Intrasellar tumors are less often visible on CT scans; nor does the chiasmal cistern show significant changes. An expanded diaphragma sella may frequently be seen in CT scans. Since the majority of tumors we operated on had no suprasellar extension, CT did not always give additional data of surgical importance. It should be noted, however, that the most modern scanner models allow one to diagnose small (approximately 5-mm) endosellar adenomas (Faria and Tindall, 1982). In those cases in which CT confirms a substantial extrasellar growth of the tumor, transnasal stereotactic surgery is contraindicated. In view of the above limitations, we frequently resort to pneumocisternography or contrast CT cisternography, which allows one to visualize and assess the state of the chiasmal, interpeduncular, and pontine cisterns situated above the diaphragma sella (Fig. 215). Such a study makes it possible to confirm or to exclude a suprasellar extension of the tumor and to determine its superior border. The latter is very important in planning the surgical approach.

A valuable advantage of pneumocisternography is the possibility of demonstrating the presence, size, and localization of subdiaphragmal arachnoidal recesses, which extend into the cavity of the sella in approximately 20% of the cases. Their detection is very important because if they are damaged during surgery, a CSF rhinorrhea and possibly a postoperative meningitis may occur. In view of this possible complication, calculations for the cryofoci are carefully made (see below) so as not to damage these recesses. Riechert (1980) also stresses the importance of cisternography in stereotactic operations on the pituitary.

Preoperative cerebral angiography is performed to verify the diagnosis of pituitary adenoma and to assess the extent of its supra- or parasellar extension and the presence of large vessels in the tumor tissue. In case of suprasellar growth of the adenoma, such typical changes can be observed on the angiograms as upward displacement of A_1 dislocation of the carotid siphon. Moreover, arteriography should totally exclude an arterial aneurysm simulating a pituitary tumor or meningioma of the tuberculum sella.

The second important goal of angiography is to establish the exact distance between the siphons of the internal carotid arteries on both sides of the sella. It is necessary to know this distance, which normally is 14 mm (Bergland *et al.*, 1968), in order to perform the operation safely; moreover, this distance must be taken into account in the stereotactic calculations. Anatomic investigations have shown that in very rare cases the internal carotid siphons may lie very near the midplane, separated from each other by only 5–6 mm. In such cases there is real danger of cryogenic or mechanical damage to the carotid arteries during surgery.

Both of the above-mentioned goals may be successfully achieved by angiography, which allows simultaneous visualization of a maximal number of cerebral arteries. For this purpose, in our clinic we have

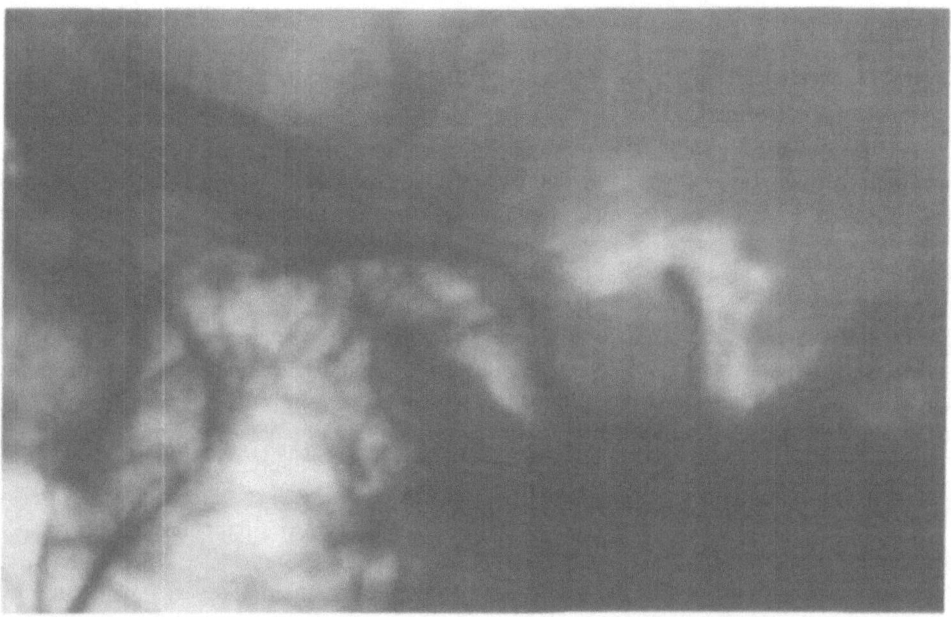

FIGURE 215. Pneumocisternography performed before cryohypophysectomy. The sella turcica is markedly enlarged. Note the absence of suprasellar growth of the pituitary adenoma.

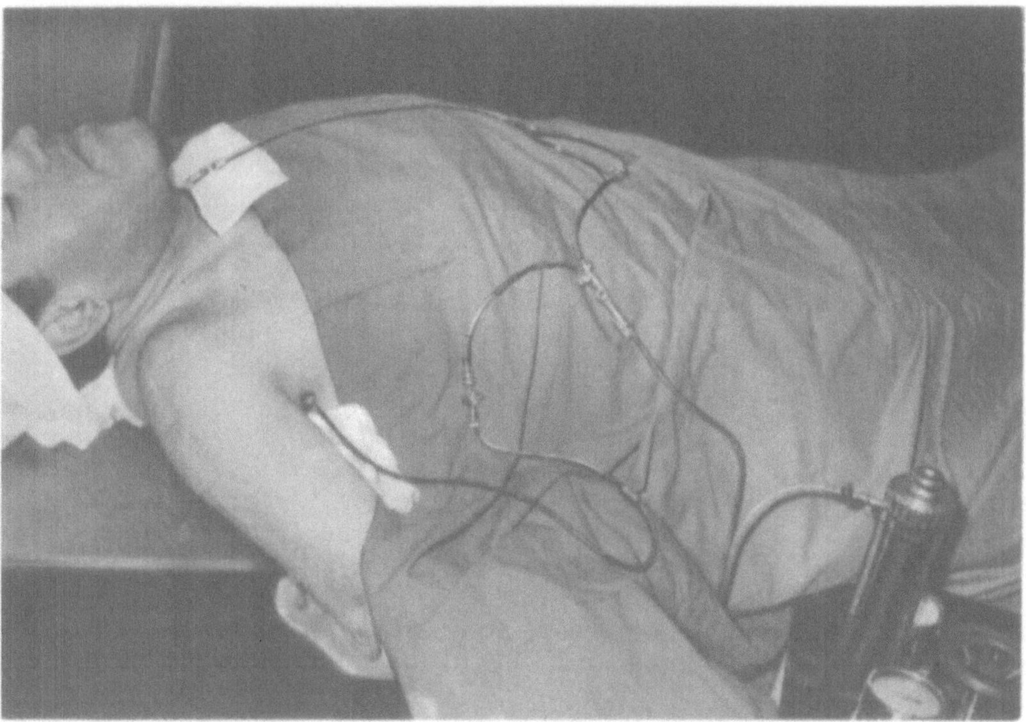

FIGURE 216. Simultaneous axillary and carotid angiography in a patient with a pituitary tumor. Needles are introduced in the right axillary and left carotid arteries. The Y-shaped tube connects the needles with the automatic AG syringe.

proposed and have routinely performed seriatim bilateral carotid–axillary (carotid on the left and axillary on the right) angiography (Fig. 216). This technique yields more information than bilateral carotid angiography and is simpler and easier to perform than total cerebral angiography. To puncture the axillary artery, we use a needle with an internal diameter of 1.6 mm, and for the common carotid artery, a needle with a diameter of 0.8 mm.

Through a Y-tube both needles are connected by catheters (diameter 4 mm) with an automatic AG syringe. The length of the catheter for the carotid artery is 40 cm, and for the axillary, 20 cm. The catheter to the axillary artery is filled with contrast substance, while the catheter to the carotid artery remains filled with a physiological salt solution. When contrast medium is injected, the dye in the right axillary artery enters the intracranial vessels before that injected in the carotid artery, thereby filling at the same time both vessels to the cranial cavity and obtaining simultaneous filling of all the cerebral vessels.

Then, with the automatic syringe, 50–60 ml of contrast medium is injected with a pressure of 5.5–6.0 atm (depending on the arterial pressure of the patient). The contrast drug simultaneously fills the right carotid and vertebral arteries in the neck, all their intracranial branches, as well as the left carotid artery and its intracranial branches (Fig. 217). In this way, the angiograms visualize all the intracranial vessels with the exception of a small intracranial segment of the left vertebral artery. As a rule, the investigations are performed under local anesthesia. If necessary, the contrast medium may be introduced a second time.

The object of basal venosinusography, which may be included in the x-ray investigation prior to cryohypophysectomy, is to visualize both cavernous sinuses. This allows one to determine the dimensions of the transverse diameter of the sella as well as to assess the parasellar extension of the pituitary tumor. In a number of cases, it is possible to visualize both intracavernous sinuses on the axial films, thereby confirming the longitudinal diameter of the sella. Good contrasting of the cavernous sinuses allows one to avoid damaging them during surgery.

While preparing for a transnasal stereotactic operation on the hypophysis, the hormone profile of the

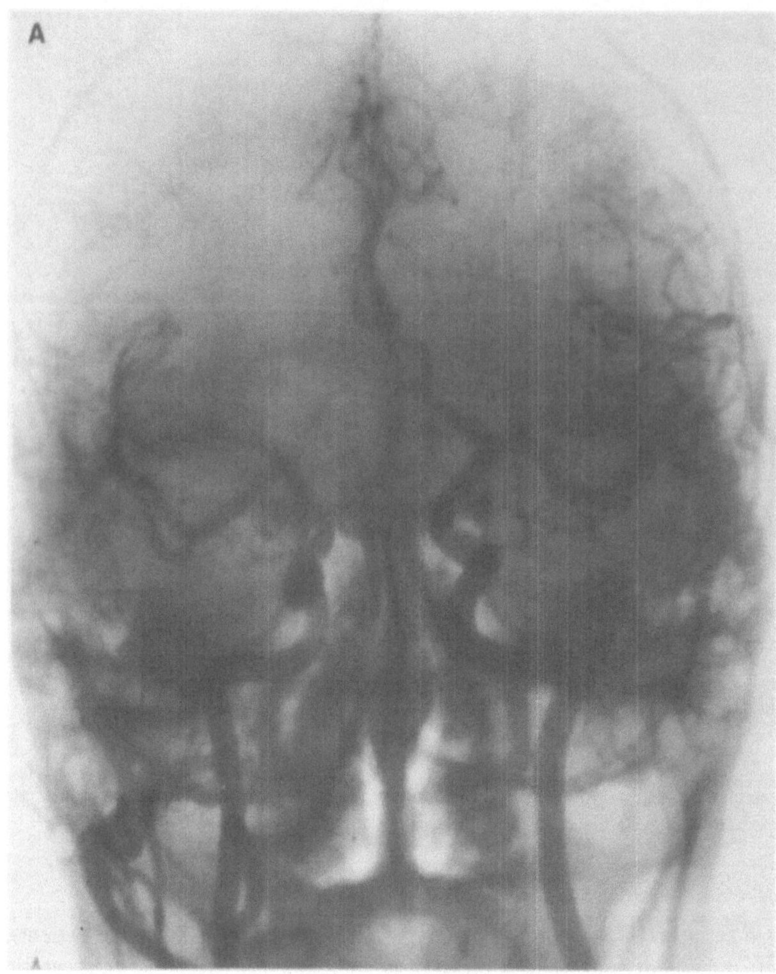

FIGURE 217. Simultaneous axillary (right) and carotid (left) angiography in pituitary tumors. (A) Angiogram in AP projection. Filling of the right carotid, right vertebrobasilar, and left internal carotid arteries. Significant elevation of left A_1 segment is evidence of a marked suprasellar extension of the pituitary adenoma. The right A_1 segment does not fill. (B) The same angiography in another case of pituitary adenoma. The absence of elevation of the A_1 segments indicates the absence of suprasellar extension of the adenoma.

patient must be thoroughly studied. In the first place, this confirms the clinical data on the degree of hormonal activity of the hypophysis adenoma, and second, it offers the possibility of assessing the efficacy of the operation by comparing pre- and postoperative data. Of the various endocrinological investigations that directly or indirectly reflect the functional activity of the pituitary and related endocrine glands, the most important determinations are the somatotropic hormone (STH) or human growth hormone (HGH) of the anterior lobe of the pituitary in blood serum by radioimmunoassay, prolactin level in serum, andrenocorticotropic hormone (ACTH) in blood, thyrotropin (TSH) and gonadotropin (GH) levels in urine, level of 17-ketosteroids (KS) and hydroxycorticosteroids, summary 17-oxyketosteroids (OKS) and their separate fractions in daily urine and blood and after injection of insulin (insulin tolerance test), assimilation of radioactive iodine by the thyroid gland, content of general and free thyroxin in plasma, thyrotropin reaction to releasing factor, determination of a number of biochemical indices such as inorganic phosphorus and calcium in serum, sugar, and cholesterol in blood, and glucose tolerance test.

An increase in the concentration of STH, prolactin, and other hormones is an indication of hormonal activity of the pituitary adenoma, whereas the normalization of these indices postoperatively is evidence of effective destruction of the tumor. The level of KS and OKS indicates the presence and degree of hypocorticoidism, which sometimes requires the prescription of steroid therapy.

In the biopsy material obtained during the operation (see below), using the immunocytochemical method for hormonal granules in the cytoplasm of tumor cells, one may test for certain hormones: CTH, prolactin, ACTH, thyrotropin, and follicle-stimulating hormones.

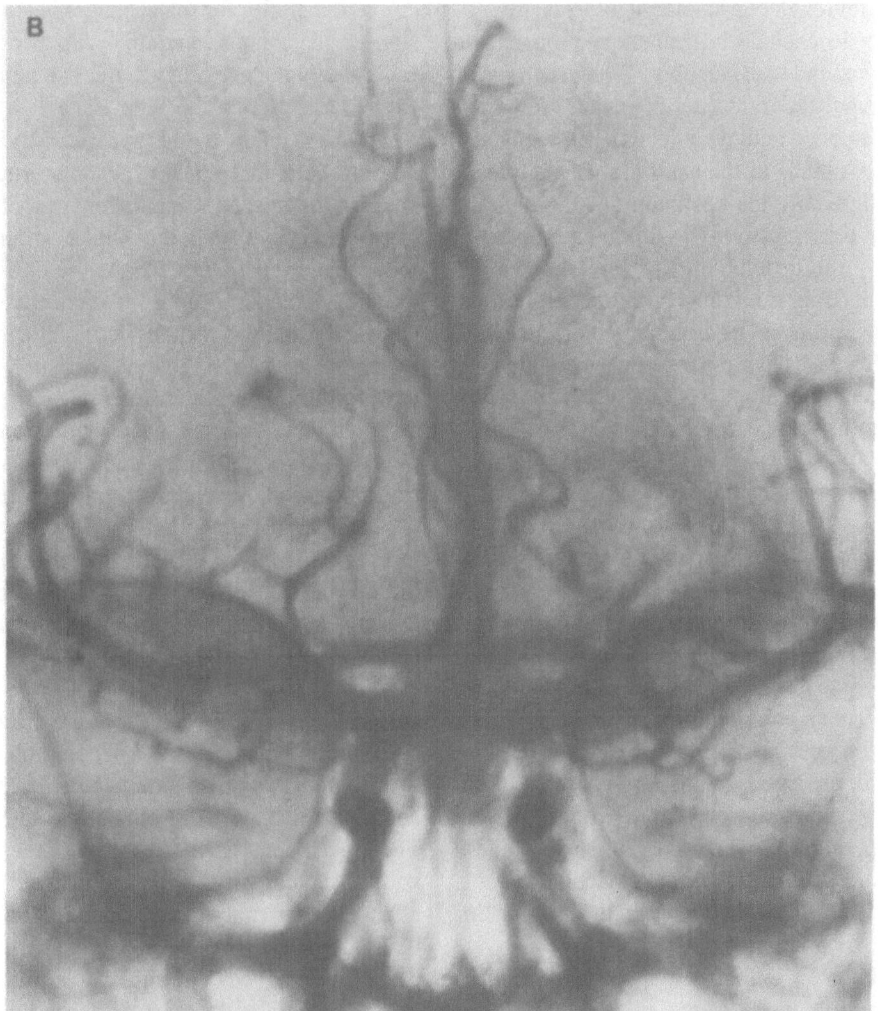

FIGURE 217. *(Continued)*

The recent developments of clinical and immunohistochemical methods of investigating the hormonal profile, of electron microscopy of tumor tissue, and of staining of cells by the immunoperoxidase technique have led to a modern and better-founded classification of pituitary adenomas. The previous classification terminology was based on routine histopathological stains (eosinophil, basophil, chromophobe, mixed adenomas), whereas the modern classification divides pituitary adenomas on the basis of the hormones they produce, recognizing that one adenoma may produce two or more hormones (Landolt, 1978; Scanarini and Mingrino, 1980). By this criterion, one may recognize prolactinomas, somatotropic, corticotropic, thyrotropic, and gonadotropic adenomas, as well as adenomas with combinations of these hormones. It should be noted, however, that this classification, as well as the previous one, does not correlate precisely with the clinical endocrinological syndromes (acromegaly, gigantism, amenorrhea–galactorrhea, Cushing's syndrome, Nelson's syndrome, the Forbes–Albright syndrome, etc.). For example, hyperprolactinemia may be observed both in hormonally active (prolactin-secreting) as well as in inactive pituitary adenomas. That is why some authors prefer to differentiate hormonally active and hormonally inactive adenomas not according to their histological characteristics but on the basis of clinical data, including the level of specific hormones in the blood (Wilson and Dempsey, 1978).

When preparing patients for surgery, it is important to make a thorough otorhinolaryngological investigation in order to exclude inflammatory processes in the rhinopharynx and nasal sinuses. When such disorders are observed, the transnasal operation on the hypophysis is postponed until the infection has been totally eliminated. Cultures of the nasal and faucial mucus should be made before the operation to determine the microflora and their sensitivity to different antibiotics. Depending on the results of this investigation, a powder is made from a mixture of the antibiotics and sprayed into the nasal cavity from an atomizer several times a day for 3–4 days prior to the operation. Directly before the operation, the rhinopharynx is processed with nitrofurazone.

5.3. Surgical Equipment

In stereotactic operations on the pituitary, we employ an apparatus we have specially designed for transnasal stereotactic operations on the pituitary. This device ensures reliable fixation of the patient's head, and the appropriate insertion of the surgical instrument to the desired sellar point with an accuracy of 0.5–1 mm. In addition, our stereotactic apparatus provides the possibility of electrocoagulation, tumor biopsy, and biopotential recording, as well as x-ray control at each stage of the operation. To a considerable extent, these requirements are also met by the stereotactic apparatus used by Talairach and Riechert, especially designed for operations on the pituitary.

As already noted, our apparatus, made of stainless steel and titanium, consists of two parts. The first is a metallic platform with a round aperture in the middle. The second piece of the apparatus is fixed to the platform, insuring accurate advance of the surgical instrument. This part consists of two mutually perpendicular protractors graduated in degrees that have index pins to allow an accurate setting on both protractors of the correction angles for the cannula. The double-base design of the main protractor, with two rotation axes, ensures a high degree of accuracy in guiding the trocar to the target within the sella.

On the body of the apparatus there are vertical guiding columns along which our trocar moves advancing the freezing cannula into the sella turcica. The trocar is fixed in a roller with a spring clamp on a movable terminal, allowing the use of trocars of different diameters. The depth to which the trocar is introduced into the nasal cavity is monitored by the millimeter scale on the guiding column. The trocar is advanced using a rack and pinion gear with an end pin. The surgeon turns the gear so that the sharp trocar pierces the wall of the sphenoidal sinus and the sellar floor, which is quite thin in pituitary tumors. During destruction of a normal pituitary, when the bone structures are quite thick, a sterile hammer may be used for hitting the trocar with light, frequent blows and advancing it smoothly by 1–2 mm; it can be fixed securely at any given depth.

The 200-mm trocars that we designed for cryohypophysectomy (Fig. 218) are made of a stainless steel and titanium alloy. Depending on the operation, we use trocars with an external diameter of 3 or 4 mm and a corresponding 2.2- and 3-mm internal diameter. The trocar consists of two parts: A cyclindrical rod with sharp cone tip, which is inserted into a thicker tube. This external thick-walled tube 130 mm long has beveled edges to match the beginning of the sharp tip of the internal rod. The tube has a round aperture for easy extraction of the trocar by the movable terminal of the stereotactic apparatus. Such a design ensures a strong and reliable trocar. Various instruments can be introduced through it into the sella turcica for biopsy, recording of biopotentials, and cryodestruction of pituitary tumors.

A biopsy is an essential stage of the operation.

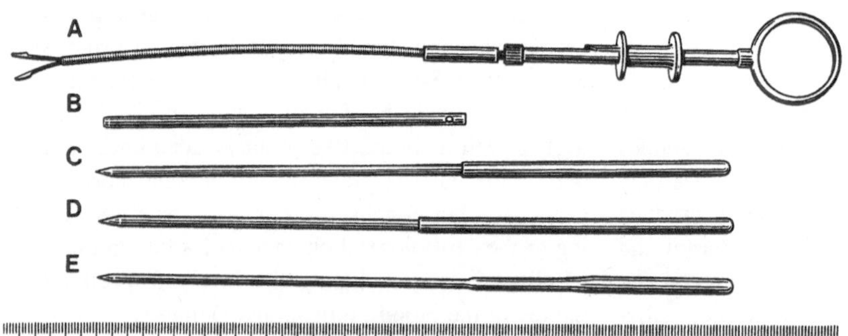

FIGURE 218. Instrumentation for transnasal – transsphenoidal operations on the hypophysis. (A) Instrument for biopsy of pituitary tumor; (B) external portion carrier of trocar with an open end for removal; (C) internal part of trocar; (D) assembled trocar with 4-mm external diameter; (E) assembled trocar with 3-mm external diameter.

Various methods are employed in stereotactic operations (Section 3), depending on the consistency of the adenoma tissue, such as aspiration, removal of fragments with a special dissector or a miniature tumor forceps, and so forth.

It is expedient to begin a biopsy with an attempt to aspirate the contents of the sella. In many cases, the tissue of a pituitary tumor has a soft or even semiliquid consistency, and using a syringe one can frequently aspirate a sufficient amount of tissue for histological investigations. For aspiration biopsy, we use a cannula with the tip beveled at an angle of 60°, which is advanced into the pituitary through the external tube of the trocar. Its tip protrudes from the tube by 2 mm. Several syringe aspirations are made, and then the cannula is extracted and washed out with a physiological solution. Pieces of tumor tissue are usually found in it. This method of obtaining a biopsy is simple and relatively safe but is ineffective if the neoplastic tissue is firm. In such cases, we use miniature biting forceps (Fig. 219), which, closed, are introduced through the trocar to the needed depth; a biopsy is then taken under electronic amplifier with TV control.

An apparatus designed by the author is used for local freezing (Chapter 5, Section 2.7). Its compact size and light weight offer a number of advantages for the surgeon. In operations on the pituitary it is important to perform cryodestruction eccentrically from the active tip of the cannula. This provision has been incorporated in the seventh model of our cryosurgical instrument in which the small uninsulated part of the cannula is located not at its tip but on the lateral surface. When liquid nitrogen is poured into the reservoir of the instrument, the ice sphere forms eccentrically at the tip of such a cannula, i.e., mainly on one side of the cannula. Thus, by turning the cannula around its axis, several partially overlapping freezing foci can be made without repeated introduction of the cannula. By pouring in liquid nitrogen each time, it is possible to create a cryodestructive focus of a desired size and shape. The cryogenic cannula is introduced into the sellar cavity through the trocar, which serves as the guide, and protrudes 2.5 mm from the tip of the trocar.

The second part of the device is a light reservoir for liquid nitrogen made of heat-resistant plastic. The reservoir that we use in stereotactic cryohypophysectomy differs from the reservoir for stereotactic operations on subcortical ganglia by its greater capacity as

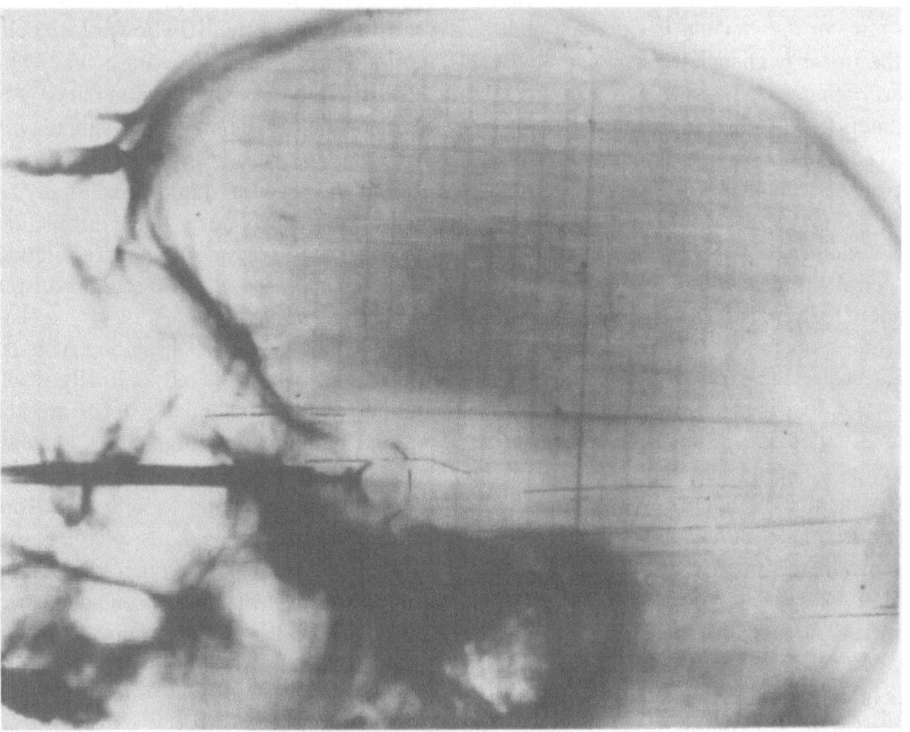

FIGURE 219. Stereotactic biopsy of pituitary tumor by miniature forceps, shown before a small piece of adenomatous tissue is taken.

well as its shape: the reservoir for cryohypophysectomy is linear with a rectangular cross section. The two parts of the device are sterilized separately and are connected before the cannula is introduced into the sellar cavity, the reservoir being fitted on the external, wider part of the cannula for freezing.

The instrument for cryohypophysectomy is used as follows. The two parts of the cryoinstrument are connected and inserted in the stereotactic apparatus. The end of the cannula is put into the trocar, which has been introduced into the seller cavity, and the cannula is advanced smoothly until its active tip protrudes 2.5 mm from the trocar. The insertion of the cannula is monitored on the TV screen and by x-ray films, after which the aspirator connected to the instrument is switched on, noting the vacuum level on the manometer. The required volume of liquid nitrogen is poured from a Dewar flask into the reservoir of the instrument.

A volume of 80 ml of nitrogen and an exposure of 2–3.5 min are required to obtain a 14- to 15-mm-diameter eccentric ice sphere at the tip of the cannula. In 1.5 min after the aspirator is switched on, there is a significant drop in the vacuum level as measured by the manometer on the aspirator, and a layer of frost resembling snow appears on the plastic tube leading to the aspirator. After the freezing–thawing cycle, the tip of the cryogenic instrument, together with the trocar, is lowered 3–4 mm, or the cannula is extracted and the direction of the trocar is changed.

Thus, overlapping cryodestructive foci may be made in the anterior two-thirds to three-quarters of the sellar cavity. The cannula may be withdrawn or shifted after thawing of the ice sphere for 5–6 min, when the temperature at its tip reaches +20°C.

The small diameter of the cryogenic cannula (2 mm), which permits it to pass through a 3-mm-diameter trocar, undoubtedly makes the procedure less traumatic. In transnasal operations on the pituitary, this is very important, since the size of the operative bone defect on the sellar floor depends on the diameter of the trocar. The smaller this defect, the less are the dangers of postoperative CSF rhinorrhea and infection of the cranial cavity. A number of authors believe that defects in the sellar floor created during the operation with a diameter of more than 3 mm should be closed with a metallic, silicon, or muscle "cork" to prevent CSF leakage. However, the use of a muscle "cork" is not very reliable, whereas metallic and silicon plugs pose a serious obstacle if a repeated operation is required.

We do not fill the aperture in the sellar floor, and nevertheless, we have had no protracted CSF rhinorrhea or meningitis in any case. This may be explained by the fact that before the operation all patients underwent pneumocisternography, which makes it possible to visualize the position of the recesseses of the basal cisterns and to avoid damaging them (Section 5.2). Besides that, cryodestruction does not prevent normal healing of the defect in the sellar floor.

5.4. Technique of Cryohypophysectomy

Adequate anesthesia plays an important role in stereotactic cryohypophysectomy. It is necessary to eliminate the elements of pain and fear for the patient in order to enhance the possibility of keeping in contact with him. This is necessary for timely detection of possible cold damage to the cranial nerves lying lateral to the sella in the cavernous sinus as well as to the chiasma, which, lying superiorly, is in serious potential danger during cryohypophysectomy. In order to check the functions of these nerves during the main stages of the operation, the surgeon should have the patient perform certain tasks that will check vision and the function of the oculomotor nerves (Fig. 220).

Paresis of one or more cranial nerves may be a complication of cryohypophysectomy and, according to Rand *et al.* (1968), has been noted in 10% of the operations. In our series, the incidence of this complication was substantially lower (see below). The initial indications of such a paresis are, as a rule, reversible and point not to destruction but only cooling of the fibers of the nerve below approximately +15°C. If, in this case, freezing is immediately stopped, the functions of the nerve are rapidly and totally restored. Such complications may occur in the destruction of a normal pituitary, but in cryodestruction of pituitary adenomas, this danger is less, since the tumor displaces the craniocerebral nerves.

The choice of anesthesia in cryohypophysectomy still cannot be considered as finally resolved. Certain authors employ potentiated local anesthesia with or without endotracheal intubation, whereas others consider general inhalation anesthesia necessary. The author uses the following technique. Before application of the stereotactic apparatus, pledgets soaked in a 1% solution of tetracaine hydrochloride to which a few drops of epinephrine have been added are inserted into the cavity of the rhinopharynx. This is followed by local anesthesia injected at the two points where the pins are to be set, fixing the head of the patient in the stereotactic frame.

Barbiturates and short-term muscle relaxants are

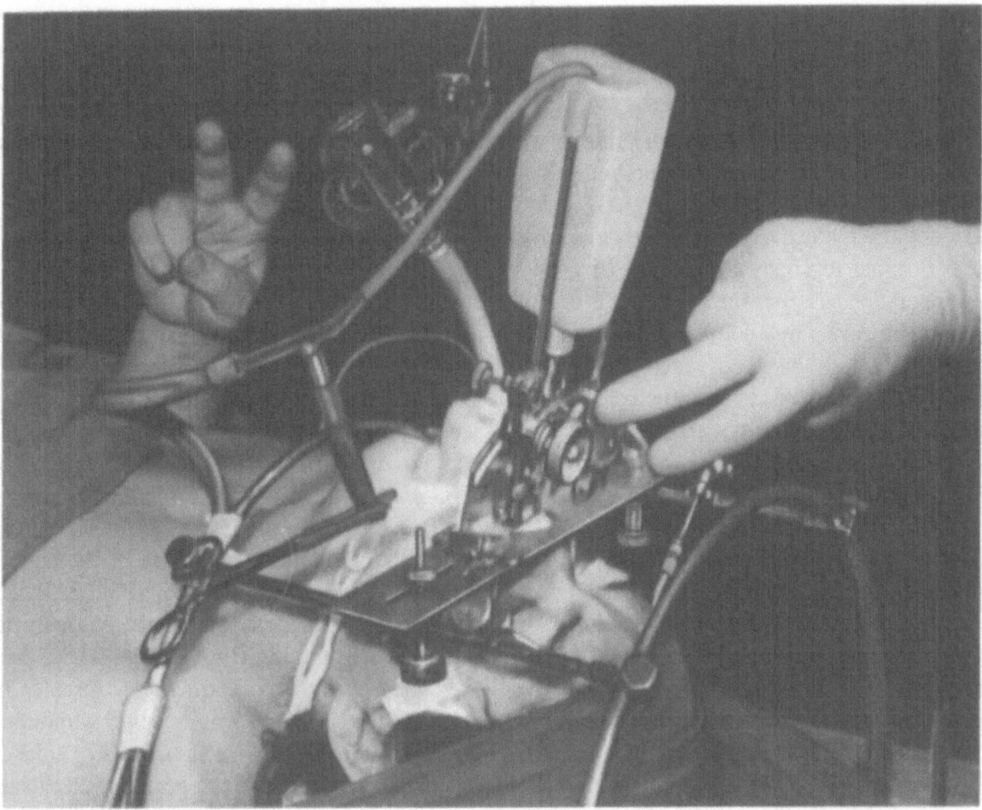

FIGURE 220. Transnasal stereotactic cryohypophysectomy. Control of patient's vision and eye movements at the moment of freezing. The patient, under neuroleptanalgesia, correctly indicated how many of the surgeon's fingers she could see.

administered intravenously for the initial anesthesia. We consider endotracheal intubation mandatory in order to avoid aspiration of blood from the nose. Then we employ neuroleptanalgesia combined with inhalation of nitrous oxide in small doses.

Before beginning the operation, it is necessary to review again the craniograms in two projections, the CT scans, the pneumocisternograms, and angiograms that pinpoint the site, direction of growth, and mass of pituitary tumor. In the lateral roentgenogram taken according to stereotactic rules, the vertical line bounds the anterior two thirds of the sella, which corresponds to the posterior border of the tumor, since the intermediate and posterior parts of the pituitary, damage to which is undesirable, are localized behind that line.

The air line of the basal cisterns on the pneumograms serves as the superior border of cryodestruction. The sellar floor and clinoid processes form the anterior and inferior borders of the zone to be destroyed. With the "total focus" of cryodestruction in the lateral projection determined in this manner, the AP roentgenogram must be analyzed. The superior border of the focus to be destroyed is also the air shadow in the basal cisterns, and the lateral borders are formed by the medial surfaces of the siphons of the internal carotid arteries at the sites of their closest proximity as well as the cavernous sinuses. These reference points are visible on arterio- and venograms. The sellar floor forms the inferior border. The prime directions of the tumor growth to the right or left, as well as the width of the nasal passages are determined on the roentgenograms. If the tumor is evenly distributed in both directions, the trocar is introduced through the wider nasal cavity.

Once the total volume of cryodestruction has been determined, one must plan the number of cryofoci required for radical destruction of the pituitary adenoma. When the cryogenic apparatus of the author is used, the diameter of each spherical cryodestructive focus is approximately 15 mm, and multiple foci must partly overlap. In this instance, one must decide if it is possible to cryodestroy completely through one nostril, or whether it will be necessary to introduce the cannula consecutively from both sides and create cryofoci in the tumor on both sides of the midline, which is necessary when the sella is large.

Depending on the size of the pituitary tumor, we usually create from two to six overlapping cryodestructive foci. In case of relatively small adenomas, two to three foci are required; however, if they are large, five to six and even more foci may be needed. Most frequently, the operation was performed through one of the two nostrils; however, in a number of patients, the freezing cannula was introduced consecutively through both nostrils. The centers of cryodestructive foci are marked in both roentgenograms, and subsequently, they serve as reference points for stereotactic calculations during operations.

The patient lies in a supine position on the operating table. A cushion is placed under the shoulders of the patient, and the head is slightly retracted and lies on a special headrest described in Chapter 4, Section 3, as in conventional stereotactic surgery. All three fixation rods of the rest must be firmly pressed to the head, the sagittal plane of which must be strictly perpendicular to the horizontal plane. Rigid fixation of the head is achieved by two sharp pins, which are screwed into the frontal bone symmetrically from both sides.

After the head has been fixed in the required position, the stereotactic apparatus is set up. The metallic rectangular platform with the aperture in the center is fixed by screws to the rods connecting both arches of the headrest. The platform, movable in three planes, is set up so that the center hole is over one of the nostrils.

The second part of our stereotactic apparatus, the guiding device with the trocar of the required diameter, is fastened to the platform. Based on the external reference points, the trocar is guided in the correct initial direction to the middle part of the sellar floor in the lateral projection and to the sagittal plane in the AP projection. Then the trocar is smoothly advanced through the inferior nasal passage until its sharp tip touches the floor of the sphenoid sinus. The depth from the external nares varies from 50 to 70 mm. X-ray Polaroid films are taken in both projections and developed in a Picker-Polaroid camera. This camera, with electrochemical development, produces a dry film in 30 sec. The film is then used for calculations to determine the precise direction of the trocar and the distance required to reach the first target point in the sella.

A line along the axis of the trocar and a line from the center of its rotation to the target point are drawn on the roentgenograms. These two lines indicate the angle of correction. If the trocar is correctly aimed at the target point, these two lines coincide; i.e., the angle of correction is zero degrees. Then the distance from the tip of the trocar to the target point is determined, and this value is reduced in accordance with the x-ray divergence (Section 5.3).

After the correction angle has been determined, the trocar is removed from the sphenoid sinus, and the angles are transferred to the protractors of the stereotactic apparatus. The trocar is again inserted to the required depth (Fig. 221). If a control roentgenogram shows that its tip is at the target point (Fig. 222), the internal part is removed, and the tumor is biopsied according to the technique described above.

For the first cryodestructive focus, one should choose a point that is the least likely to damage important structures surrounding the adenohypophysis. Most often this is along the midplane of the anteroinferior part of the sella, near its floor (Fig. 223). Here it is possible to take a biopsy relatively freely and then begin cryodestruction (Fig. 224).

The second cryodestruction is usually made by inserting the trocar 4–5 mm deeper without changing direction. Since this means further enlarging the cryodestructive focus to the posterior part of the pituitary, it is necessary to check carefully that the position of the active tip of the cannula is correct. This is done by x-ray control and by recording biopotentials from the site of planned cryodestruction. For this purpose, a special electrode is introduced into the trocar.

Adenohypophyseal tissue does not contain nervous elements generating biocurrents and so is characterized by bioelectric silence. The posterior lobe (neurohypophysis) produces biocurrents visible on EEG in the form of spikes with a frequency of 8–12 Hz and an amplitude of 50–80 μV. Cryodestruction is inadvisable in an area producing spikes even with a lesser frequency and amplitude than indicated above. In this case, the electrode should be withdrawn until the bioelectric activity has totally disappeared, and only then should freezing be performed. After two or sometimes three cryodestructive foci have been made near the sellar floor in the midplane, one may begin destruction of the superior parts of the adenoma by changing the direction of the trocar and reinserting as described above.

During the freezing process, it is necessary to observe the patient (see above) in order to check the reactions of the optic and oculomotor nerves. At the first signs of these functions being disturbed, which may occur by the end of the freezing cycle, this process

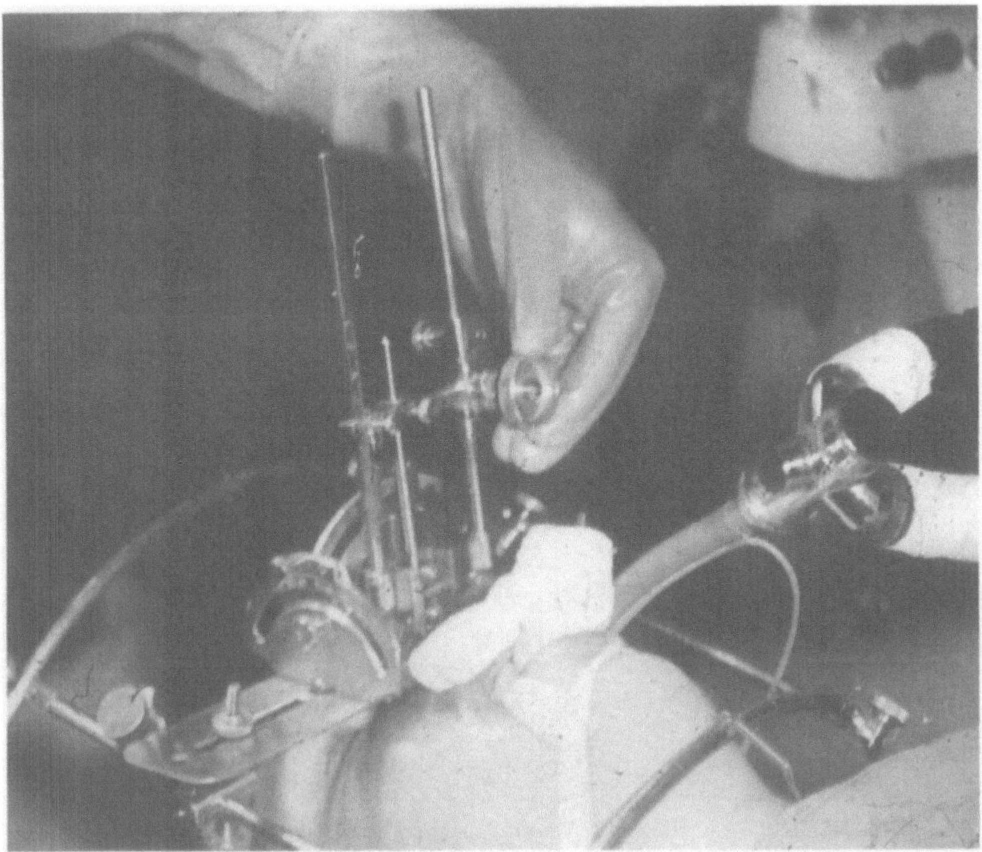

FIGURE 221. Transnasal stereotactic cryohypophysectomy. After stereotactic calculations, the sharp trocar is inserted through the floor of the sella turcica.

should be terminated by switching off the aspirator and injecting warm water into the reservoir with liquid nitrogen.

If the pituitary tumor is so large that it cannot be totally destroyed from one side of the nasal cavity, the operation is continued through the other side. For this purpose, the trocar is extracted, and the stereotactic apparatus is shifted so that the aperture in the center of the platform is under the other half of the nasal cavity. Then the manipulations are repeated in the same sequence as described above. In the majority of cases, however, all the planned cryodestructive foci could be made through one side of the nasal cavity.

After the last cryodestruction and thawing, the cryoinstrument is carefully extracted. Moderate hemorrhage, which sometimes occurs, is stopped by repeatedly inserting hydrogen peroxide tampons in the nasal passages. Then, the nasal and pharyngeal cavities are thoroughly inspected using a laryngoscope to verify that all hemorrhage has stopped completely. If epistaxis continues, measures are taken to arrest it by anterior and posterior tamponade, and only then is the patient extubated. We performed anterior tamponade in 20% of our patients, but in not a single instance was a posterior tamponade required.

The duration of the operation depends on the anatomic configuration of the sellar area and on the size of the total cryodestructive focus. It usually ranges from 2 to 3 hr.

5.5. Postoperative Care

After the operation the patients remain in a semisitting position for 2 days and are given dehydrating drugs, antibiotics, and coagulants.

The tampons are removed from the nasal cavity the day after the operation if there are no signs of bleeding. Should bleeding continue, the tamponade

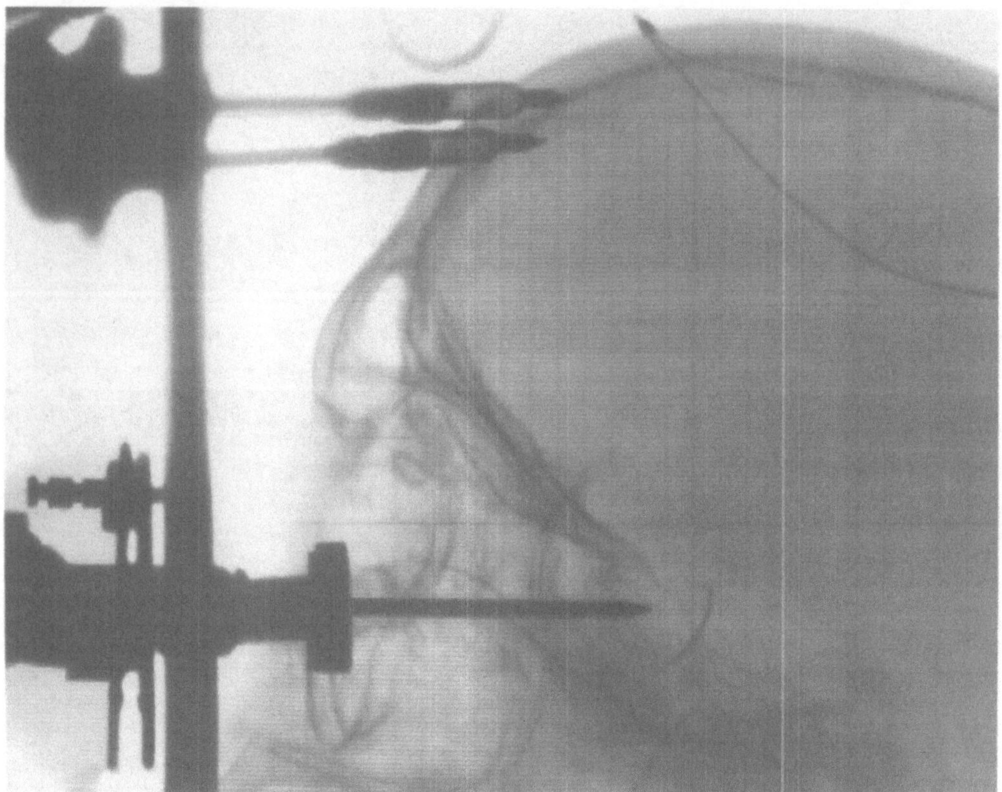

FIGURE 222. Sharp trocar is stereotactically inserted to the calculated target point in the sella turcica.

may be repeated; however, in our series of cases, in not a single instance was this required. Following surgery, some neurosurgeons prescribe hormone therapy with corticosteroids for preventive purposes in all patients (Mazars *et al.*, 1966; Conway, 1970), whereas others prescribe hormone therapy only if there are signs of hypopituitarism (Norell *et al.*, 1970). We used low-dose hormone therapy in the postoperative period in only a few patients when there were clinical indications. On the third or fourth day the patients are allowed to get up if there are no complications.

A number of complications that may occur after transnasal operations on adenomas have been described in the literature. Damage to the internal carotid artery with subsequent intracranial bleeding or the formation of a carotid–cavernous fistula is the most serious of the complications to result from incorrect placement of the trocar. This may result in arterial hermorrhage into the nasal cavities or intracranial hemorrhage. These well-known symptoms develop rapidly. Should a carotid–cavernous fistula develop, a unilateral pulsating exophathalmos may result. If ligature of the common or internal carotid arteries in the neck does not arrest the hemorrhage, emergency trephination and clipping of the internal carotid artery intracranially are indicated. Damage to the wall of the cavernous sinus may result from misinterpretation of the data from basal venosinusography. None of these complications occurred in our series of cases.

Damage to the cranial nerves, previously mentioned, may result from misdirection of the trocar or cold damage during cryodestruction. This complication is identified by checking the function of the cranial nerves during surgery (Section 5.4) and requires instant termination of the freezing process. If the function does not return within the next 20–25 min, the operation must be terminated. This complication was not observed in our series of cases.

Damage to the basal cisterns with subsequent CSF rhinorrhea may result from inaccurate placement of the instrument or misinterpretation of the data from pneumocisternography. If this occurs, during surgery or in the postoperative period, cerebrospinal fluid may leak from the nasal cavity, intensifying on increase of the intracranial pressure. Such CSF rhinorrhea is treated by dehydration, repeated lumbar punctures, or lumbar

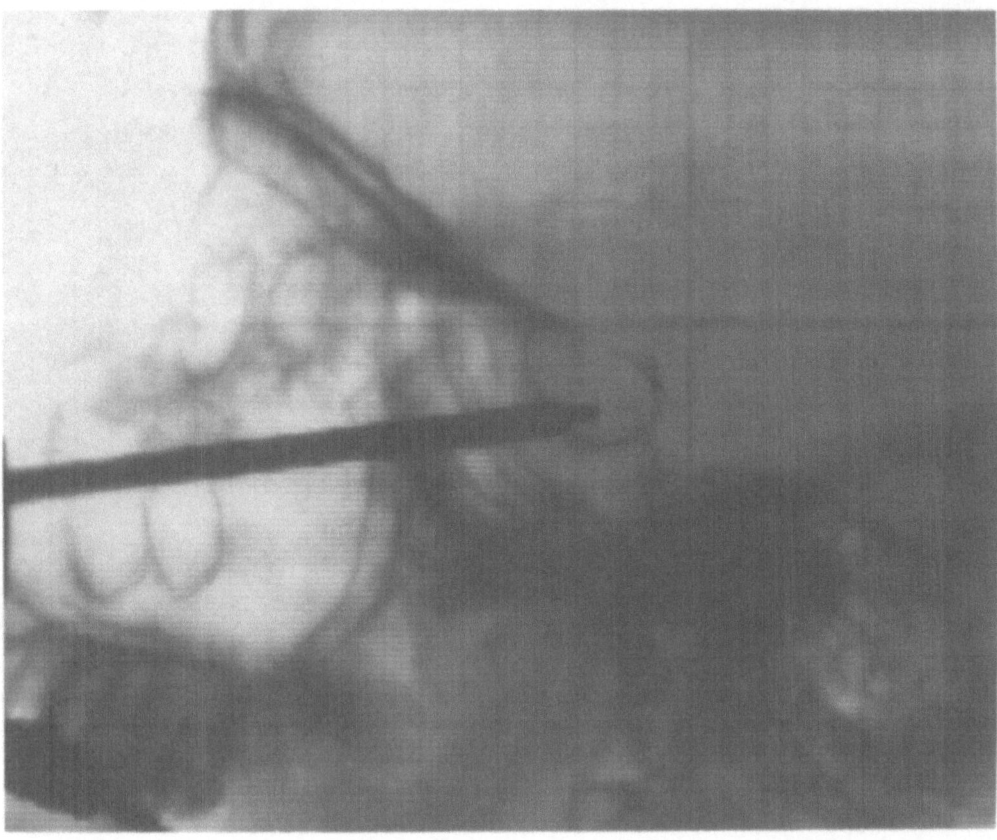

FIGURE 223. The cryocannula has been introduced into the lower part of the pituitary adenoma and the freezing is in progress.

drainage for 1–2 days, and meningitis is prevented by antibiotics. If conservative treatment is not effective, closure of the defect in the sellar floor by a muscle, plastic, or metallic stopper is recommended. The frequency of this complication in cryohypophysectomy is 1–4% (Blaezel and Lazarus, 1965; Harrison *et al.*, 1970). In our operative cases, there was no protracted nasal liquorrhea or meningitis postoperatively.

Complications caused by damage to the posterior lobe of the pituitary develop as a result of extension of the cryodestructive focus posteriorly. However, some authors consider that even almost total destruction of the posterior lobe of the pituitary causes only a transitory and not pronounced diabetes insipidus (Conway and Garcia, 1970). This rare complication appears in the postoperative period in the form of a syndrome of panhypopituitarism and diabetes (apathy, arterial hypotension, extreme thirst, reduced appetite, amenorrhea) as well as reduced metabolism and hormonal levels in blood (STH, gonadotropins, ACTH), causing a drop in the levels of 17-OKS and 17-KS in urine and blood. The treatment is replacement hormone therapy. There was no pronounced syndrome of postoperative panhypopituitarism in our cases.

Diabetes insipidus to varying degrees was observed in approximately 40% of the cases postoperatively but responded well to treatment. Extreme thirst was observed in only six of our cases (10%), who drank up to 3–4 liters of liquid per day. This requires short-term (for about 1 week) treatment with vasopressin in small doses, after which the diabetes insipidus disappears completely. Mortality after cryohypophysectomy is low and varies depending on the condition of the operated patients. Of 23 patients with pituitary adenoma operated stereotactically, two died (Rand *et al.*, 1968).

Data in the literature indicate that cryohypophysectomy was performed mainly in cases of pituitary adenoma, most frequently involving acromegaly, Cushing's disease, or Nelson's syndrome (Adams and Seymour, 1968; Conway *et al.*, 1969; Rand *et al.*, 1969; Norell *et al.*, 1970; Kandel *et al.*, 1970; Har-

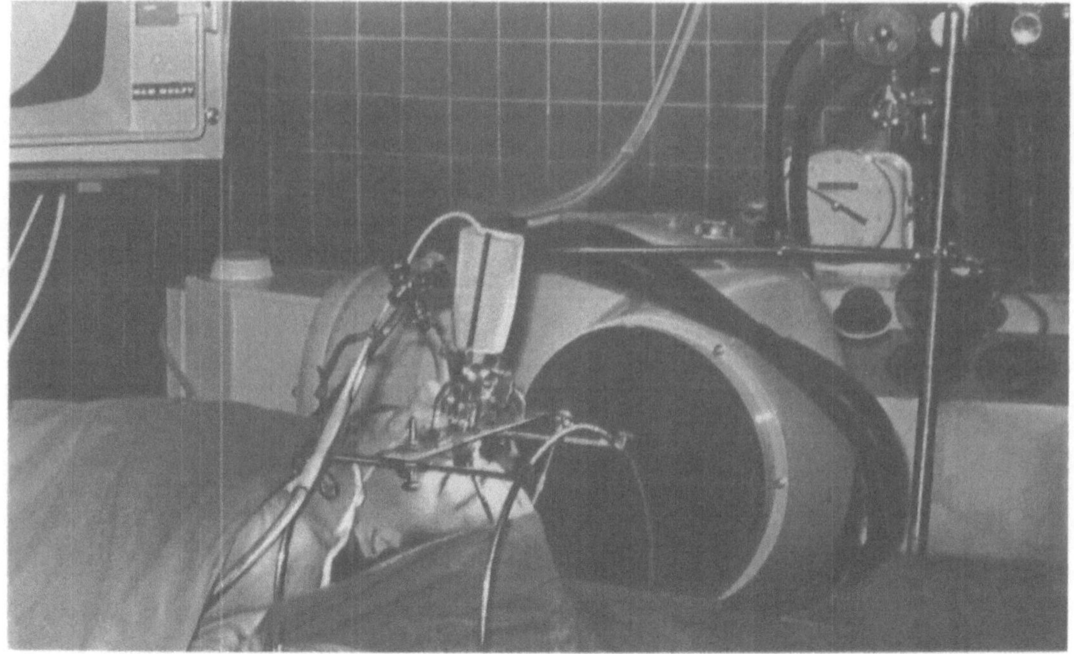

FIGURE 224. Transnasal stereotactic cryohypophysectomy: general view of the operation. The head of the patient is fixed in our stereotactic headrest. The stereotactic apparatus is fixed in the burr hole, and the cannula of the freezing device is introduced into the sella cavity. Liquid nitrogen is poured into the plastic reservoir. The nitrogen vapor is evacuated via the plastic tube, which is connected to a conventional surgical aspirator.

rison *et al.*, 1970; Ditullio and Rand, 1977; Seymour *et al.*, 1978). At present, it is generally acknowledged that acromegaly is an indication for removal of the pituitary adenoma, since the mortality without surgery is two to three times higher than in operated patients. Without surgery, 50% of these patients die before the age of 50, and 89% by 60 years (Evans *et al.*, 1966). Summing up the data in the literature, one may conclude that a significant clinical improvement is noted in 80–85% of the patients following surgery.

5.6. Personal Observations

In recent years we have performed stereotactic transsphenoidal cryohypophysectomy in 62 patients with pituitary adenomas, 34 women and 28 men aged 11 to 59 years. The duration of the disease ranged from 1 to 14 years (average 4.5 years). The dominant clinical syndrome in 42 patients was acromegaly; in nine it was galactorrhea–amenorrhea, in five adiposogenital dystrophy, in four gigantism, and in two patients, mild endocrine symptomatology. In 59 patients the histological structure of the tumor was established during surgery by biopsy (according to the old classification): 42 patients, eosinophil adenoma; ten, mixed; four, chromophobe; three, basophil.

The main group was made up of patients with pronounced acromegaly. Of 42 patients in this group, an eosinophilic adenoma was verified in 35 by biopsy during the operation, in six, a mixed tumor was identified, and in one patient, the histological structure of the tumor was not established. In approximately half of the patients, the main symptoms of the disease developed over the course of 2 years; in the second half, they developed over the course of 5–10 years. Prior to the operation 26 patients in this group had undergone conventional γ therapy (5000–6500 rad per course), which had produced a temporary beneficial effect in only half of them, followed by a further aggravation.

Almost all of the patients with acromegaly experienced headaches, in half of them severe and accompanied by nausea, dizziness, photophobia, and lacrimation. Practically all of the patients complained of general weakness, profuse perspiration, fatigue, impaired memory, and reduced working capacity. Acromegaly—enlargement of the face, hands, and feet to various degrees—was present in all patients. Almost all displayed prognathism, separation of teeth, and fine plication of the facial skin. The skin was grayish-yellow, especially marked on the face with its enlarged features. The shoe size of these patients increased from the onset of the disorder by 1.5 to 6 sizes. Practically all the women with acromegaly had amenorrhea and hypertrichosis, primarily on the shins, whereas men experienced diminished libido and potency.

Various hormones in blood serum (Section 5.2) as well as biochemical substances were investigated in the majority of patients with acromegaly before and after surgery. A very substantial (maximum eightfold) increase in the level of the growth hormones (STH) was present in practically all patients; however, in several, this increase was moderate. In about one-third of the patients, prolactin in blood was markedly elevated. Investigations of 17-KS and 17-OKS did not reveal any significant deviation from normal. Several patients were found to have changes in the glucose tolerance curve. The basal metabolism was increased. The content of cholesterol as well as Ca^{2+} in blood in acromegaly rarely exceeded the normal level, although hypercalciuria was noted in several cases. A significant rise in the level of inorganic phosphorus in blood was noted in one-third of the patients. Diabetes mellitus was present in one-third of the patients.

In all the patients with acromegaly, skull films revealed a typical enlargement of the sella, which was present to a marked degree in the majority of patients. In 34 of the 42 patients, the adenoma was only intrasellar, which was confirmed by contrast x-ray investigations, CT scanning, and clinically by the absence of visual disorders. In the rest of the patients, moderate suprasellar extension of the tumor was demonstrated clinically and roentgenologically. A distinguishing feature in the majority of patients was the extension of the tumor into the sphenoidal sinus, which was sharply reduced in size. Not only was the size of the sella enlarged in these patients, but its floor was totally destroyed because of tumor invasion into the sphenoidal sinus. Ophthalmological investigations often revealed moderate bilateral exophthalmos and lacrimation accompanied by headaches and photophobia. One-third of the patients were found to have changes in the field of vision, concentric narrowing or bitemporal hemianopia. Visual acuity was reduced in 17 patients (from 0.8 to 0.3).

The patients with acromegaly usually did not reveal significant disorders in their neurological status. Only a few patients were found to have paresis of convergence, bilateral Marinesco's sign, asymmetry of the nasolabial folds, or mild anisoreflexia.

In the literature, we did not find any data on the use of transnasal–transsphenoidal stereotactic cryohypophysectomy in pituitary adenomas in children. Consequently, we consider it of interest to report on three operated patients, an 11-year-old girl and a 13-year-old boy with the syndrome of acromegalic gigantism and an 18-year-old youth with the adiposogenital syndrome, in whom the disease had developed at age 16. In the first patient, intraoperative biopsy revealed an eosinophil, in the second patient, a mixed adenoma, and in the third, a basophil adenoma. The clinical picture of these patients varied considerably. In the girl and the boy, pronounced gigantism (height 187 and 179 cm, respectively) was combined with acromegaly and pubertas praecox. In all three patients x-ray investigations revealed a substantial enlargement of the sella, deepening of its floor, straightening and thinning of the dorsum sellae, and narrowing of the sphenoidal sinus. Angiography and pneumocisternography in the girl demonstrated moderate suprasellar extension of the tumor. In the youth with a basophil pituitary adenoma, the predominating endocrinological syndrome was adiposogenital dystrophy; he also had bitemporal hemianopia and reduced vision in one eye.

Transnasal stereotactic cryohypophysectomy, the technique of which is described above, was performed in all 62 cases. All of the patients (except two) tolerated the operation well. None of the complications described above (nasal bleeding, damaged carotid artery, CSF rhinorrhea, meningitis, or death) occurred. For several days four patients had polydipsia, which was controlled by small doses of antidiuretic hormone.

Serious complications developed in two patients. In one of those described above, cryohypophysectomy was followed by a sharp reduction of vision, which required urgent removal of the suprasellar part of the tumor by the classical subfrontal approach. Thrombosis of the left middle cerebral artery developed in the second patient 3 days post-surgery. Pronounced spastic hemisparesis persisted in this patient.

All the other patients noted a subjective improvement directly after the operation, which in the majority of cases was confirmed by objective data, in particular, repeated biochemical and hormonal investigations. The clinical improvement was manifested by a relatively rapid diminution of external signs of acromegaly as well as arrest of further growth in gigantism. Quite frequently there was a significant decrease of edema, as a result of which the nose, eyelids, lips, and fingers became smaller, and the hands and feet reduced in size (Fig. 225). After the operation some of the patients

FIGURE 225. Girl with gigantism and acromegaly 8 years after stereotactic cryohypophysectomy.

took a smaller shoe size, and they were again able to wear rings, previously too tight, on their fingers. As a rule, the yellowish-gray color of the face that made the patients look much older diminished or disappeared. Menses were restored in several of the women following surgery.

Headaches, suffered to various degrees prior to surgery by all the patients, terminated completely or diminished significantly on discharge. At the same time, lacrimation and photophobia reduced substantially or disappeared completely. Practically all the patients with reduced vision before surgery had restored or improved eyesight, and in two cases, the failing vision stabilized. In the majority of patients with hemianopia or concentrically narrowed fields of vision before surgery, the visual fields improved considerably after the operation.

In parallel with the clinical improvement, there was a rapid postoperative return to normal of hormonal and biochemical tests. The STH level in blood, which had been very high prior to surgery, returned to near normal a month following surgery in all 42 patients with acromegaly except five. In two patients with gigantism, it remained markedly elevated even though it was reduced to almost one-third of the preoperative level.

Investigations of excretion of 17-KS and total and free 17-OKS in urine before and after surgery revealed a mild diminution postoperatively. However, this reduction was not accompanied by clinical manifestations and did not require replacement therapy.

Postoperatively, the majority of patients demonstrated a reduction in the previously high level of inorganic phosphorus in serum and in some cases of calcium in urine as well as a tendency to decrease cholesterol in the blood and a normalization of glucose metabolism.

In all patients with the galactorrhea–amenorrhea syndrome, the duration of the disease varied from 2 to 6 years. Enlargement of the sella turcica in these patients was less pronounced than in acromegaly; however, the so-called microprolactinomas (less than 1 cm in size) were not found in this group. There was no significant suprasellar extension of the tumor in any case. All patients displayed a very high concentration of prolactin in serum with elevated STH in three cases. Before surgery, four women had been treated with bromocriptine, with only a transitory improvement.

All these patients tolerated cryohypophysectomy without any complications. A normal menstrual cycle was restored and galactorrhea disappeared postoperatively in five women, and galactorrhea was markedly diminished in another two cases. Repeated postoperative investigations of blood prolactin showed a return to normal in four cases, a significant reduction but still elevated level in three others, and in two cases with a very high preoperative prolactin level, a decrease by half, but still high level.

A distinctive clinical syndrome was observed in a small group of patients with adiposogenital dystrophy, all of whom histologically had basophil tumors or chromophobe pituitary adenomas with moderate suprasellar extension of the tumor. Two patients, after the operation, had rhinorrhea (3–5 days) relieved by repeated lumbar punctures. There were no other complications. A significant postoperative clinical improvement was noted in all the patients.

In our entire series there were two serious complications related to the surgery, but there was no postoperative mortality.

The follow-up of these patients varied from several months to 15 years (in the majority of cases from 4 to 12 years). The good postoperative effect in practically all cases remained throughout the entire period in 50 of 59 cases (long-term results in three cases remained unknown), i.e., in 85%. In ten patients there were various signs of recurrence of the adenoma at different times.

A typical illustrative case is presented. A 41-year-old female patient was admitted to our clinic in April, 1971 with complaints of enlarged extremities, especially the hands and feet, enlarged nose and eyebrows, raspy voice, and constant headaches. She had first noticed enlargement of the feet and then the hands in 1959. Since 1968, she had developed headaches, general weakness, and progressive enlargement of facial structures and feet. Amenorrhea appeared over the next 2 years, and shoe size increased by three units.

On examination she had pronounced acromegaly, enlarged nose and lips, separation of teeth, macroglossia, and enlarged and expanded feet and hands. Although she had moderately pronounced bilateral exophthalmos, her vision was not reduced, and her fields of vision were normal. Craniograms showed significant enlargement of the sella with a depressed floor, thinned dorsum sellae, and a small sphenoidal sinus.

There were no indications of suprasellar growth of the tumor from the data of single-stage bilateral carotid–axillary angiography and pneumocisternography. The STH content of the blood was 69 ng/ml. The combination of typical endocrine and roentgenologic signs with the progressing course of the disease was the basis for diagnosing an intracellar pituitary tumor.

A stereotactic transnasal cryodestruction of the pituitary adenoma was performed. The histological diagnosis was mixed pituitary adenoma. The postoperative course was uneventful. Immediately after surgery, the headaches disappeared, acromegaly diminished, and normal menstrual cycle returned. The STH level decreased to 8 ng/ml.

Thirteen years after the operation, the patient's condition is quite satisfactory. She has no headaches and no weakness or increased fatigability. She holds a full-time job as a bookkeeper.

Six patients underwent a second operation because of continuing growth or relapse of the tumor in 1.5 to 6 years after the first operation. A massive cryodestruction of the tumor was repeated in three of these cases, whereas the other three were operated on by the open subfrontal approach with microsurgical removal of the suprasellar part of the tumor. Two other patients were reoperated soon after the first operation. In one patient, for technical reasons, only a biopsy of the tumor had been taken. In the second patient, cryodestruction was done in two stages because of the gigantic size of the adenoma (approximately $7 \times 5 \times 4.5$ cm), in which nine large cryodestructive foci were made.

It is noteworthy that the frequency of relapses or continued growth of the tumor after stereotactic cryodestruction in our series was somewhat higher (approximately 15%) than that reported by other authors who removed adenomas by the transsphenoidal microsurgical approach (Hardy, 1975a, 1983; Faria and Tindall, 1982; Baskin *et al.*, 1982). In the majority of cases this may be explained by the large size of the adenomas that invaded the sphenoidal sinus. In these cases, total cryodestruction as well as a complete open removal of the adenomas was not technically feasible. At the same time, we consider that stereotactic cryohypophysectomy is a simpler and less traumatic operation with fewer complications than other surgical techniques used for the treatment of pituitary adenoma. The main indication for the stereotactic cryodestruction of these tumors is the presence of a large STH-secreting adenoma growing into the sphenoidal sinus and producing acromegalic syndrome. On this basis as well as from the clinical experience of the author and data from the literature, one may reasonably conclude that transnasal stereotactic cryodestruction in the management of pituitary adenoma is highly effective and relatively safe.

6. Stereotactic Destruction of the Normal Hypophysis

In concluding this chapter, we discuss briefly the stereotactic destruction of a normal pituitary. The data on hypophysectomy in the management of chronic pain caused by malignant neoplasms are presented in Chapter 13, Section 5.2.6.

The use of hypophysectomy began to develop in the early 1950s (Johnson *et al.*, 1958; Talairach *et al.*, 1962; Hartog *et al.*, 1965; Mundinger and Riechert, 1967; Blagoveshchenskaya *et al.*, 1968). In the initial period, the pituitary was removed by the classical open approach, and later by the transsphenoidal stereotactic technique.

A vast amount of literature as well as our personal experience have demonstrated that this operation is without a doubt effective in the treatment of metastases of hormone-dependent cancer as well as in diabetic retinopathy and Cushing's disease. Cancer of the mammary gland, prostate, uterus, thyroid, adenocarcinoma of the kidneys, chorionepithelioma, and others are hormone-dependent tumors.

The technique of stereotactic destruction of the normal pituitary does not differ substantially from the operation described above for pituitary adenoma. However, it should be emphasized that since the normal sella is only a fraction of its size in pituitary tumors, the stereotactic calculations should be more precise, and the total volume of destruction, naturally, should be considerably less (Fig. 226). The destruction of a normal pituitary involves greater technical difficulties in passage of the instrument through the undamaged sellar floor than in penetrating a sella containing a pituitary tumor.

The results of the different techniques for destruction of a normal pituitary, cryogenic, radioisotope methods, high-frequency thermocoagulation, introduction of lytic substances, and others, have been described in the literature.

The following data from the literature illustrate the efficacy of this operation and coincide with our limited experience. A large experience in stereotactic high-frequency thermocoagulation of a normal pituitary (approximately 300 operations) for diabetic retinopathy and metastases of breast cancer has been acquired by Zervas and Hamlin (1974). Using a transnasal approach, they make a large number (10–20) of small foci in the pituitary with the active tip of the electrode heated to 80°C. A relatively infrequent complication (in 15 cases) was CSF rhinorrhea, with subsequent meningitis in five cases. In metastases from breast cancer, hypophysectomy produced objective remission in 63 of 186 patients, and pain disappeared for a period of 3–6 months in another 41 cases.

In diabetic retinopathy, stabilization or improvement of vision postoperatively was established in 80% of the cases (Zervas and Hamlin, 1978). In severe forms of diabetic retinopathy, especially in young people, hypophysectomy will prevent progression of the process for many years and also reduce by approximately one-half the required dose of insulin.

Isotopes of yttrium (^{90}Y), iridium (^{192}Ir), and iodine (^{125}I) have been used successfully for the destruction of a normal pituitary (Mundinger, 1974c). According to the data of this author, in practically 60% of the patients with metastases from breast cancer, there was lasting remission, and intractible pain disappeared in the majority of cases. Other authors report that destruction of the adenohypophysis promotes termination or regression of the pathological process in 35% of patients with hormone-dependent cancer (Moricca, 1977).

Poblete and Zamboni (1974) reported good results from stereotactic implantation of ^{90}Y in a normal pituitary by the transfrontal approach in 300 patients, the majority of whom had metastases from breast cancer. Eight or more yttrium granules with an overall dose of 15 mCi were implanted in the pituitary. Complications (rhinorrhea, impaired fields of vision) were observed comparatively rarely. Two of 13 patients with diabetic

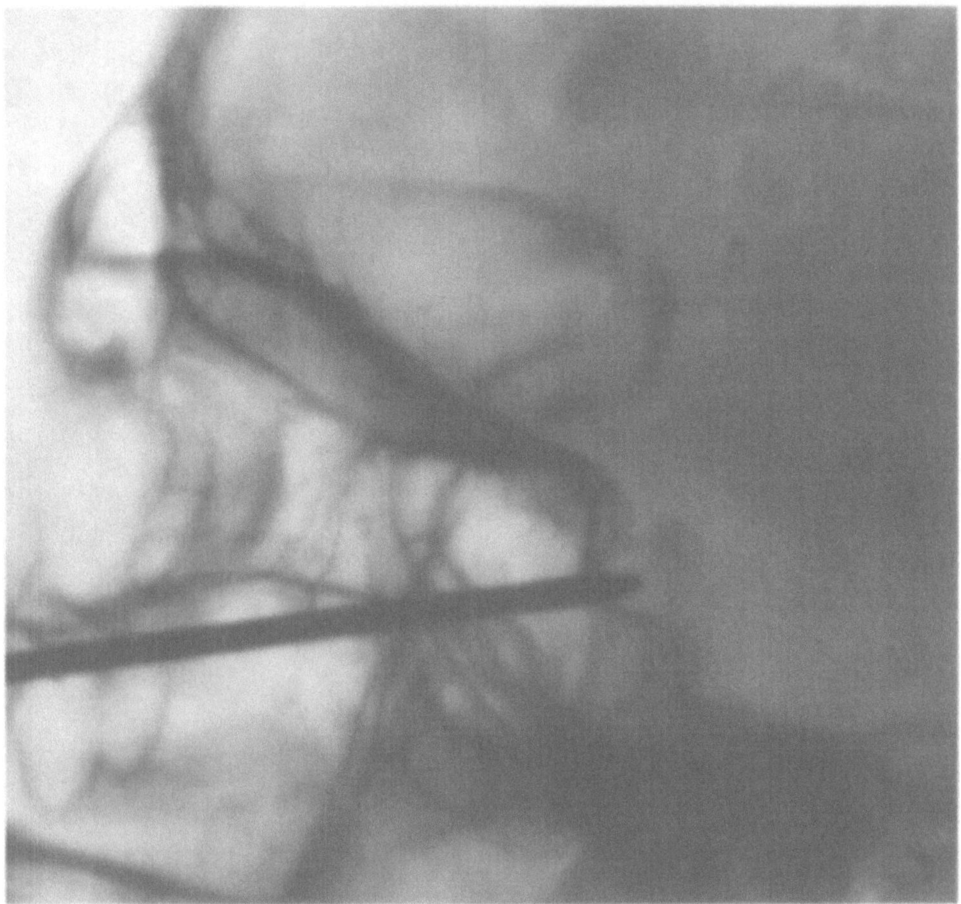

FIGURE 226. Stereotactic cryodestruction of the normal hypophysis in a case of disseminated breast cancer. The active tip of the cryocannula is introduced into the normal sella turcica.

retinopathy died from hemorrhages during surgery. The authors considered this to be the result of pronounced vascular disorders in diabetes and, in view of this, rejected the transfrontal approach for retinopathy.

Another indication for hypophysectomy is malignant exophthalmos, which in the majority of cases is caused by hyperthyroidism. This severe illness is accompanied by corneal ulceration and secondary glaucoma and ultimately leads to loss of vision. A number of authors have reported good results from stereotactic destruction of the pituitary in this disease. For example, in 11 of 12 patients operated on by Sedan and Harter (1966), there was a rapid disappearance or significant diminution of exophthalmos after implantation of ^{90}Y or ^{198}Au in the pituitary.

16

Cerebral Arterial Aneurysms and Arteriovenous Malformations

Obliteration of vascular lesions has been simplified with sterotaxy.

L. Leksell

1. Introduction

The use of the stereotactic method in different cerebral vascular disorders is a relatively new and promising achievement.

It is well known that the surgical treatment of cerebral arterial aneurysms and arteriovenous malformations (AVM) is one of the most important and pressing problems of contemporary neurosurgery. Although there have been developments during the past two to three decades, the great advances in that field have not fully resolved the problems, which are briefly described below.

2. Arterial Aneurysms

In the case of cerebral arterial aneurysms, a direct attack and clipping of the aneurysmal neck *ad oculus* is the best mode of treatment. The development of the microsurgical technique and its wide use in clinical practice (Drake, 1968, 1976; Yasargil, 1969; Rand, 1969; Guidetti, 1973; Yasargil and Fox, 1975; Zlotnick *et al.*, 1976; Pertuiset, 1979; Pia, 1979) have resulted in a remarkable improvement in technique and greater safety. Microsurgery has led to a significant decrease in postoperative morbidity and mortality. For example, the use of the microsurgical technique for 8 years in 505 patients with arterial aneurysms has reduced the mortality to 4%, and during the last 4 years, in 374 patients to 1.9% (Yasargil and Fox, 1975). Using microsurgical techniques, the mortality rate after aneurysm neck occlusion decreased to 3.7%; the occurrence of ruptured aneurysm during the operation decreased to 8%, and the erroneous severing of functionally important vessels decreased to 0.7% (Zlotnick *et al.*, 1976). In his well-known monograph, "Microneurosurgery" (1984), Yasargil presented the following results of surgical treatment of 1012 arterial aneurysms: good, 85.6%; fair, 5.6%; poor, 3.1%; and postoperative mortality, 5.7%.

In spite of that remarkable progress, the classical open operation by direct approach to an aneurysm still remains one of the most technically difficult procedures in brain surgery. The serious condition of the patient in the acute stage of subarachnoid hemorrhage as a consequence of arterial spasm, brain edema, cerebral infarction, or intracerebral hematoma, involvement of important brain structures, etc., in many cases makes the radical operation of aneurysm neck clipping too risky or technically impossible. In spite of a marked reduction in postoperative complications in recent years, the danger of rupture from surrounding dense adhesions about the aneurysm and the possibility of injury or spasm of nearby arteries as the consequence of brain retraction remain considerable.

During the last decade, many improved clip models were proposed and used in practice, but slipping of the clip after aneurysmal neck clipping also remains an unsolved problem. The rate of this complication, which leads to refilling of an aneurysm, is different in the experience of several authors: 17.6% (Steven, 1966); 13% (Drake and Allcock, 1973); 10% (Konovalov, 1973).

It is not necessary to describe the accepted and well-known operations for arterial aneurysm in this chapter. We shall present only briefly the new techniques that have been developed in the past 10 years. The search for these methods was stimulated by the difficulties and shortcomings of the existing classical operations for arterial aneurysms.

Below we briefly present four new nontraditional techniques; the first two require the use of stereotactic methods.

1. Stereotactic electrolytic thrombosis.
2. Stereotactic magnetic thrombosis.
3. Thrombosis by the balloon catheter.
4. Thrombosis by the use of coagulants and slowed blood flow.

2.1. Stereotactic Electrolytic Thrombosis

Very rare cases of spontaneous thrombosis of arterial aneurysms described in the literature led to the idea of artificial electrolytic thrombosis. For this purpose, Mullan *et al.* (1964) developed an atraumatic approach to aneurysms by means of a cannula introduced stereotactically through a burr hole. A stainless steel electrode with a sharp active tip less than 1 mm in diameter and 4–7 mm long (depending on the size of the aneurysm) was inserted within the aneurysm cavity, and a positive direct electric current was applied. Of the first 12 patients operated on by Mullan *et al.* (1965c), complete thrombosis occurred in only two internal carotid aneurysms. Hemiparesis developed during the procedure in five cases as a result of the passage of thrombi into the parent vessel.

Later, Mullan (1974) used a stainless steel, copper-coated electrode, which created a denser clot in the aneurysmal sac. Encouraging results were obtained in the majority of 61 patients, but four of them died immediately after the operation because of intracranial bleeding or thrombosis of the parent vessel, and two other patients, who had been in grave condition preoperatively, died in the postoperative period. In the following 2 months, eight patients died from recurrent bleeding of incompletely obliterated aneurysms. The authors consider that the changes in blood circulation after partial thrombosis increase the intraaneurysmal tension and the tendency to rupture.

Four patients with anterior communicating aneurysms were operated on by Samotokin and Khilko (1973), who stereotactically introduced thin electrodes into the aneurysms and applied anodal electrolysis for 1 to 3 hr. The operation was combined with an intravenous infusion of coagulants. In all cases the volume of the aneurysm, although reduced by 30 to 40%, was not completely thrombosed. The mortality rate after electrolytic thrombosis was 10% because of escape of the thrombi from the aneurysm into a parent vessel (Alksne, 1972).

Electrolytic thrombosis associated with a risk of complications does not provide reliable sidetracking of an aneurysm and is now used very seldomly.

2.2. Stereotactic Magnetic Thrombosis

Based on the idea of Mullan *et al.* (1964) of the possibility of stereotactically introducing a needle electrode into an aneurysmal sac through a burr hole, Alksne (1970) developed the technique of stereotactic magnetic intraaneurysmal thrombosis. The prototype was as follows. Under general anesthesia, a relatively thick magnetic probe 6 mm in diameter made from an alloy (Alnico-5) was introduced through a burr hole near the coronal suture using a Wells apparatus. The probe was inserted adjacent (within 1 mm) to the aneurysmal sac. Through a catheter in the internal carotid artery, a suspension of carbonyl iron in a 25% solution of human serum albumin was introduced. The suspension consisted of iron particles 4–5 mm in diameter. The magnetic probe remains in the brain for 4 days.

Under the influence of the magnetic field, a quantity of iron particles accumulated in the aneurysmal cavity, causing thrombosis and excluding the aneurysm from circulation. The results were controlled by intraoperative angiography. In several cases complete thrombosis occurred, but in the majority of patients, it was only partial because not enough iron particles entered the sac. There was a risk of thrombosis of the parent arteries because the iron particles were washed into them (Alksne, 1970).

For these reasons, the technique was modified by a second variant—the insertion of ferrosilicone or an iron–acrylic mixture directly into the sac (Alksne, 1972). As in the first variant, the stereotactically inserted magnetic probe was moved to within 1–2 mm of the aneurysmal sac. Through the probe a thin needle was inserted to puncture the aneurysmal wall. The moment of penetration was monitored on an arterial pressure recording made by a transducer connected to the needle. Then the sac was filled with the iron suspension, which was held within the sac by the magnetic probe. After this operation, the probe remained in the brain for several days. Alksne (1972) reported on his experience in 41 operations. In some cases successful thrombosis was achieved, but 13 patients died after the operation; five of them had been in grave condition preoperatively. The complications were connected to the introduction of the solution around an aneurysm or to embolization of cerebral vessels, especially in case of larger aneurysms (greater than 1 cm) or a wide aneurysmal neck.

To improve this technique further, Alksne and Smith (1977) used a suspension of iron powder in methylmethacrylate. This mixture not only lessened the fragmentation of the thrombi but reduced significantly the polymerization time. As a result, the magnetic probe was left in the brain for 1 hr instead of several days.

Rand and Mosso (1972) used a ferrosilicone mixture for stereotactic thrombosis in a case of a multichambered anterior communicating artery aneurysm. A dense thrombus was created within a few minutes after the aneurysm puncture and insertion of the mixture. After withdrawal of the needle, a ferrosilicone deposit was left around the aneurysm; because of the rapid hardening of the silicone, the risk of cerebral emboli was diminished.

The technique was further improved by transsphenoidal insertion of the magnetic probe for anterior communicating artery aneurysms. A probe of the same diameter was introduced by the sublabial route into the sphenoidal sinus and afterwards intracranially through a small drill hole in planum sphenoidale. The probe tip was located extradurally about 1 mm from the aneurysm, the wall of which was punctured and an iron suspension introduced into the sac. This technique, in 22 cases of anterior communicating artery aneurysms, substantially improved the results over those of the previous series of cases (Alksne and Smith, 1980). In 17 of 22 patients, complete aneurysmal thrombosis occurred with only three serious complications (hemiplegia, rhinorrhea). There was no postoperative mortality.

The method of stereotactic magnetic thrombosis is undoubtedly effective and, in all likelihood, will be improved in the future.

It is known that electric and magnetic fields will cause a thrombus to form on a vascular wall because of the static potential between the intima and blood elements. This phenomenon was the basis of an original method to produce thrombosis of a cerebral aneurysm using a stereotactically oriented external field of a powerful magnet in which a patient's head is placed (Kikut and Kadish, 1973, 1976; Kikut, 1976). This creates a magnetic field that may cause intraaneurysmal thrombosis. The authors use a special apparatus to orient the power magnetic fields stereotactically. The method was used in 13 patients with berry aneurysms (Kikut and Kadish, 1973, 1976), and decreased the size of the aneurysm by one-third to two-thirds in 11 cases. In two giant aneurysms, not even partial thrombosis was obtained.

Because no intracranial operation is performed, the safety and absence of surgical complications are advantages of the method, but its effectiveness remains to be proved. Only many years of observation will reveal if only partial thrombosis of an aneurysm will prevent future rupture.

2.3. Balloon Catheter Thrombosis

At the Burdenko Institute of Neurosurgery in Moscow, Serbinenko developed a new method of catheterization and occlusion of cerebral arterial aneurysms. Special plastic catheters 0.5–1.5 mm in diameter with a detachable balloon on the tip are inserted into the feeding artery of the aneurysm and aneurysmal sac after puncture of the carotid or vertebral artery under control of an electronic amplifier. The author divides the method into deconstructive and reconstructive operations, the former consisting of permanent occlusion of the parent artery, and the latter preserving the blood flow through the artery.

In his first report Serbinenko (1971d) described the obliteration operation (carotid balloon occlusion at the level of an aneurysm, at the level of the syphon, or in the neck) in 14 patients with aneurysms at different locations (supraclinoid, eight; subclinoid, three; intracavernous, three; in the neck, one). Later, several authors reported the results of similar operations for subclinoid or other unclippable aneurysms (Zozulya and Shcheglov, 1976; Zlotnik and Sekach, 1976; Weil *et al.*, 1987) and middle cerebral artery aneurysms (Zozulya and Shcheglov, 1976).

In the first report of the new obliteration technique, the operation was successful (Serbinenko, 1974) in only four cases; in four other cases the balloon catheter occlusion of an aneurysmal cavity was a failure.

By 1977 the Burdenko Institute of Neurosurgery had employed endarterial surgery in 25 patients with internal carotid artery aneurysms. In 21 cases obliteration operations were performed, and in only four cases, reconstructive operations. Zubkov (1974) reported the results of the balloon catheter technique in 24 cases of arterial aneurysms, in only nine of which was complete occlusion successful. In their next report (Zubkov and Matzko, 1982), the number of operated cases had increased to 64. Six patients died after surgery from different complications (rupture of the aneurysm or thrombosis of major arteries).

Shcheglov (1979) has the largest experience in using the balloon catheter technique for the treatment of arterial aneurysms. He operated on 65 patients, including seven cases of anterior communicating aneurysms, and was able to exclude 57 aneurysms from the general circulation.

According to Serbinenko, substantial technical difficulties in obliterating arterial aneurysms are encountered if the angle between the aneurysmal neck and the parent vessel axis is near 90° or if the neck is long and thin so that it is impossible to introduce the catheter. The same opinion was expressed by Zubkov (1973a), who considered the operation to be indicated only if there is a wide aneurysmal neck for catheter insertion.

Fox *et al.* (1987) used the detachable balloon for proximal artery occlusion in unclippable carotid and basilar aneurysms. The permanent morbidity was only 1.5%. Extra-intracranial bypass was also performed in about one half of the cases. All carotid aneurysms below the ophthalmic artery were thrombosed, but thrombosis was achieved in only one half of the supraclinoid and basilar aneurysms.

The balloon catheter technique is delicate, effective, and less traumatic, but the complications are not inconsiderable. It has been used extensively in the last few years, particularly for AVMs (see Section 3). In arterial aneurysms, the method has some limitations but does yield excellent results in selected cases.

2.4. Thrombosis by Coagulants and Slowing of Cerebral Blood Flow

A technique was developed at the neurosurgical clinic in Leningrad (Khilko, 1966; Samotokin and Khilko, 1968, 1973) for aneurysmal thrombosis by injecting large ϵ-aminocaproic acid at the same time that the carotid artery in the neck was partially compressed by a fascial cuff and controlled hypotension was maintained by ganglion-blocking drugs.

The method was used in sub- and supraclinoid aneurysms with encouraging results. The most favorable condition for thrombosis is an aneurysm with a long narrow neck that comes off at a sharp angle to the feeding artery. In this case, complete thrombosis was achieved in seven of ten carotid artery aneurysms (Samotokin and Khilko, 1973). Wide-necked and large aneurysms are not suitable for this technique. Attempts to thrombose anterior communicating aneurysms were unsuccessful.

Zubkov (1974) used a modification of artificial thrombosis by coagulants combined with arterial hypotension and decreased blood flow. In two cases coagulants were introduced intraarterially by catheterization of the internal carotid and middle cerebral arteries to the neck of the aneurysm. However, only partial thrombosis was obtained. A second attempt caused thrombosis of the middle cerebral artery. The author concluded that it was difficult to confine the thrombosis to the aneurysm. The thrombi are not stable, and pieces break off and embolize to cerebral vessels. Samotokin and Khilko (1973) pointed out the danger of thrombi washing off an aneurysm and into the cerebral circulation.

Complete thrombosis of an arterial aneurysm using coagulants and slowing the cerebral blood flow is possible only in about one-half of the cases; thrombosis of an anterior communicating artery aneurysm is practically impossible. The risk of thrombosis of parent vessels is relatively high.

3. Arteriovenous Malformations

Total extirpation of brain arteriovenous malformations (AVM) is a most satisfactory method of treatment. Only a radical operation will relieve a patient from the continuous threat to life from subarachnoid or parenchymatous hemorrhages and epileptic seizures that may lead to a major neurological deficit and mental deterioration.

The failure of nonsurgical treatment of cerebral AVMs is clearly reflected in a very interesting investigation of the natural history of AVMs, described in two papers by Troupp *et al.* In the first report (Troupp *et al.*, 1970), the authors analyzed the results of a long-term follow-up of 137 patients with AVMs treated nonsurgically between 1942 and 1967. Forty percent of these patients were found to be in good condition, 21% fair, 24% disabled, and 10% died from the AVMs and 5% from other reasons. In the second paper, published in 1977 (cited by Drake, 1979), Troupp described the results of follow-up in 102 patients from the same group. Only 20% of the patients were in good condition, 10% fair, 20% disabled, and 18% died as a result of the AVMs and 7% from other causes. Twenty-five percent were lost to follow-up. As these figures show, about half of the patients without surgical treatment became disabled or died.

In the last 15 years, the microsurgical technique has significantly increased the possibilities of a successful radical removal of AVMs and has improved the results of surgical treatment. Microsurgery made it possible to remove AVMs in hard to-reach locations such as the brainstem, paraventricular regions, or posterior fossa, which in past years were considered to be inoperable. Yet, there remain many unsolved problems in the surgery of AVMs.

In Table 24 we have summarized a number of large series of AVMs reported in the literature, including two reports of the world literature. These data indicate that only one-half of the hemispheric malformations were totally extirpated. In the other half, radical extirpation was impossible because of the large size or location of AVMs in deep or functionally important regions of the brain.

As the data of Table 24 indicate, one-half of patients harboring AVMs are doomed to gradual deterioration as a consequence of epileptic fits and increasing neurogical deficits. The patients live constantly under the threat of intracranial hemorrhage, which often ends fatally. When a direct attack is too dangerous, as it is in about 50% of these patients, the problem of palliative surgical treatment arises.

Palliative operations in AVMs that cannot be removed totally have been carried out for several decades. The main procedure is ligation or clipping of the arteries supplying the AVM. But in papers published in the 1950s, some neurosurgeons expressed the opinion that such operations were not very effective. Indeed, in many cases after the clipping of arteries, an increase in the diameter of other small feeding vessels was noted. In the last decade, the above-mentioned opinion has been reconsidered. The experience with artificial embolization of the AVMs (see below) has shown that exclusion of afferent vessels with the help of many emboli causes a substantial decrease in the volume of blood supply and, consequently, the size of the AVM. Moreover, this resulted in a decrease of the seizures and the risk of intracranial bleeding.

It is practically impossible to exclude a giant deep-seated AVM from the cerebral circulation. The decrease of its blood supply and, as a result, its volume represents the only rational treatment at the present

TABLE 24. The Operability of AVMs According to Data from the Literature

Authors	Year of publication	Number of cases	Total extirpation	Percentage of total extirpation
Olivecrona and Ladenheim	1957	125	81	65
Houdart and le Besnerais	1963	150	36	24
Pool and Potts	1965	523[a]	187	36
Pernett and Nishioka	1966	453[a]	119	26
Tönnis and Walter	1967	215	118	55
Moody and Poppen	1970	105	51	49
Filatov	1972	240	62	26
Morello and Borghi	1973	154	88	57
Mingrino	1978	196	98	50
Pellettieri	1980	166	119	72
Guidetti and Delitala	1980	145	92	63
Albert	1982	178	124	70
Debrun *et al.*	1982	46	16	35

[a]Collected statistics.

time. Many neurosurgeons now recommend the open clipping of afferent arteries in inoperable AVMs of basal ganglia, brainstem, and posterior fossa (Pertuiset *et al.*, 1976; Fox and Al-Mefty, 1977).

We shall not describe the indications for total extirpation of the AVMs, the technique of the open operation, or the surgical results because the problem is outside the field of functional neurosurgery. We shall only briefly describe some new techniques aimed at the total or partial exclusion of the AVM from the cerebral circulation using stereotactic or endovascular techniques. The more promising techniques are as follows:

1. Artificial embolization.
2. Combined stereotactic and open operations.
3. Stereotactic or open cryothrombosis.
4. Thrombosis by detachable balloon technique.

3.1. Artificial Embolization

In 1960 Luessenhop and Spence proposed artificial embolization of AVMs and later published the results of the new technique (Luessenhop and Prosper, 1975; Luessenhop and Mujica, 1981). Silastic-coated steel spheres were used for embolization. The spheres were introduced in the external or common carotid artery, from where they entered the internal carotid artery. The emboli may be introduced directly into the vertebral artery. The embolization is accomplished by a catheter introduced into the femoral artery. Many neurosurgeons have used this method (Sanchez *et al.*, 1969; Kandel and Nikolaenko, 1969; Khilko, 1969, 1976; Djindjian, 1972; Kritcheff *et al.*, 1972; Kusske and Kelly, 1972; Zubkov, 1974; Seeger, 1975; Luessenhop and Prosper, 1975; Fox and Al-Mefty, 1977; Stein and Wolpert, 1977; Mullan *et al.*, 1979; Sano *et al.*, 1980) and reported very encouraging results.

The rationale of the method is that the feeding arteries of the AVMs are, as a rule, of larger diameter than normal cerebral arteries. Because the blood flow in the aneurysmal arteries is much stronger, the emboli enter the hypertrophic vessels and plug their lumens, thus excluding them from the circulation and thereby diminishing the volume of the AVM. It is important to note that afferent vessels of AVMs do not participate in the blood supply of normal brain tissue so that their exclusion does not cause ischemic lesions.

The main purpose of artificial embolization of AVMs is to occlude as many feeding arteries as possible and yet leave intact the arteries supplying normal brain.

Emboli made from different materials have been used. We have injected emboli of biologically inert silicon mixed with tantalum powder to make them visible on plain x rays. Silastic sponge emboli are also used. Mullan *et al.* (1979) suggested cutting the sponge cylinders, 5 cm long and about 3 mm in diameter, into small pieces (1–2 mm) for introduction into the carotid or vertebral arteries. During each procedure, which might be repeated several times depending on the results, the authors introduced several dozen such emboli.

Silastic spheres impregnated with barium (Luessenhop and Mujica, 1981) and a rapidly solidifying biopaste (aron-alpha) (Sano *et al.*, 1980) are also used.

The best method for evaluating the effectiveness of embolization is angiography during and after the operation. For the same purpose, the CBF may also be used (Kandel *et al.*, 1971). Objective evidence of the effect of embolization are a substantial decrease of a very high CBF, mainly in the affected hemisphere, and a drop in high oxygenation of cerebral venous blood, an increase in the general cerebral vascular resistance, and $AVDO_2$ (difference of oxygen between the carotid artery and the jugular vein).

A considerable experience in artificial embolization, mainly on inoperative AVMs, has been collected during the last two decades. Fifteen years after beginning this technique, Luessenhof and Prosper (1975) reported the results of 94 embolizations in 55 patients with AVMs. Follow-up was from 2 months to 14 years (mean 4.5 years). If intracranial hemorrhage had occurred before embolization, it recurred after the procedure in about one-half of the cases. Epileptic seizures were diminished substantially in one-half of the patients who had had fits before the operation. A mild neurological deficit appeared after embolization in only four cases.

Khilko (1974) described the results (after up to 8 years) in inoperable AVMs in 40 patients. The complete obliteration of the lesion occurred in three cases, a marked decrease in volume occurred in 23, and an insignificant decrease occurred in 17 cases. A neurological deficit developed in five cases because the emboli entered the normal brain arteries. Sano *et al.* (1975a) embolized 21 AVMs; after the operation, four patients died, and three had neurological complications. In two of ten cases operated on by Tzonos *et al.* (1975), such complications also developed.

As the majority of authors have noted, the diameter of feeding arteries of the AVMs is one of the decisive factors in the indications and contraindications to embolization. The procedure may be successful and not too risky if the diameter of the feeding artery is substantially larger than that of normal cerebral arteries. The angle between the feeding branch and main arterial trunk is also very important. Embolization is indicated in only 65% of AVMs (Khilko, 1970) or in about 50% (Kandel and Nikolaenko, 1969). In the rest of the cases, there were contraindications because of peculiarities of the structure and hemodynamics of the AVMs.

As experience has shown, embolization is effective mainly in AVMs of the middle cerebral artery and the vertebral–basilar system. For AVMs supplied by the anterior cerebral artery, embolization is not only ineffective but is associated with a high risk, because the main stream of the carotid artery carries the emboli to the middle cerebral artery.

Serious complications are possible if emboli lodge in arteries supplying normal brain tissue or in the proximal part of main arteries. This risk increases after introduction of a large quantity of particles producing a decreased blood flow through the AVM. In one of our cases, the introduction of eight silicon–tantalum emboli excluded the main part of an AVM. The ninth embolus stopped at the bifurcation of the internal carotid artery causing a cerebral infarction with hemiplegia. Subsequently, the motor power partially recovered. Several authors reported that particles lodged in normal cerebral arteries in more than one-half of all embolization procedures. Debrun *et al.* (1982) also believe that the risk of particles lodging in normal cortical vessels as a result of the procedure "is extremely high."

If relatively big arteriovenous fistulas are present in the vascular tree, as is frequently the case, the artificial emboli enter the cerebral venous system and then the lung parenchyma. However, experience has shown that such emboli in the lungs do not cause harmful consequences.

The further development of the artificial embolization technique came with the successful combination of selective catherization of the arteries feeding the AVMs with the immediate introduction of rapidly polymerizing plastics (Pevsner and Doppman, 1980; Kerber, 1980; Debrun *et al.*, 1982). Isobutyl-2-cyanoacrylate (Bicrylate), used for embolization, is polymerized rapidly in the vascular conglomerate and "switches off" the AVM partially or sometimes completely. Successful attempts at intraoperative embolization of the AVMs have been made by introducing a silicon polymer into the cortical arteries (Sano *et al.*, 1975a), Gelfoam particles impregnated with iophendylate (Drake, 1979), a mixture of 50% bicrylate and iophendylate, which polymerizes in 3 sec (Cromwell and Harris, 1980), or a mixture of isobutyl cyanoacrylate, iodized oil, and tantalum powder in different proportions (Picard *et al.*, 1984).

Debrun *et al.* (1982) have analyzed their extensive clinical results of bicrylate AVM embolization. The plastic material was introduced through a balloon catheter or directly into the feeding arteries intraoperatively. Of 39 patients, complete exclusion of the AVM was noted in 30 cases, and partial exclusion in nine cases. Various, mainly transient, complications developed in about one-half of the patients, but there was no postoperative mortality.

Summing up the data accumulated over the course of many years on artificial embolization, one can conclude that the technique is undoubtedly effective. However, the method is applicable only in about 50–60% of cases with inoperable AVMs. The main potential hazard of the method is embolization of arteries feeding normal brain tissue.

3.2. Combined Stereotactic and Open Operations

Attempts were made long ago to combine the advantages of the classical open approach to deep-seated AVMs with the stereotactic technique. It is well-known that even if the exact location of an AVM has been established by AG data, the detection of a small AVM is not easy. Quite frequently, the search for an AVM during an open operation is associated with substantial trauma to the brain tissue. The rationale of combined operations is to disclose the AVM by the stereotactic technique and then to remove it by an open operation. The first attempts at such combined procedures were made more than a quarter of a century ago (Riechert, 1955) but were not widely accepted.

In a combined operation, the basal ring of the Riechert–Mundinger apparatus is fixed on the patient's head, and the usual stereotactic calculations based on the pneumoencephalograms are made. After an ordinary bone flap has been performed, the metallic arch with stereotactic coordinates is transferred from the phantom device to the basal ring. A thin probe is introduced and directed towards the AVM. After the probe reaches the AVM, a cortical incision is made for a routine approach to the lesion by dividing the brain tissue with spatulas, using the probe as a guide. The feeding vessels are clipped, coagulated, and the AVM removed *ad oculus*.

In 1960 Guiot *et al.* used a stereotactic technique to approach a small AVM in the lateral ventricle. After making a limited bone flap and a narrow (25-mm) channel in the brain, they removed the malformation completely. Riechert and Mundinger (1964) reported four cases of deep-seated AVMs successfully removed in this way. Later, Riechert (1980) described the use of the combined technique in 18 patients to remove small AVMs in the basal ganglia and the third ventricle. In 16 cases, the lesions were totally excised, and in two cases the feeding arteries were ligated.

Matsumoto *et al.* (1971) used the combined method in two deep-seated paraventricular AVMs. After stereotactic introduction of the needle and its replacement by a Teflon catheter to the malformation, the needle track was stained with a blue dye. After craniotomy and a small cortical resection using the stained channel as a marker, the AVMs were found and extirpated.

One successful stereotactic electrocoagulation of a single vessel feeding a small AVM in the paraventricular area of an 8-year-old child also has been reported (Cahan and Rand, 1973). A bipolar electrode was introduced near the vessel through a burr hole. During coagulation, the authors rotated the electrode to prevent its "sticking" to the vessel.

3.3. Cryogenic Thrombosis

The cryogenic method has been used recently for AVM treatment. It is known that small vessels in the cryolesion are thrombosed and destroyed, but large vessels remain intact because of the high resistance of collagen and elastic fibers to cold (Chapter 5, Section 2.2). After thawing, the blood flow through the vessels is restored (Walder *et al.*, 1970; Kandel, 1974b). These data laid the groundwork for development of the cryogenic technique of AVM thrombosis.

In 1975 Walder reported the results of the cryomethod in 35 operations in 27 patients for AVMs in different locations. After an ordinary craniotomy, a cryoprobe 2–6 mm in diameter was introduced by hand into the region of the aneurysm under angiographic control. Using liquid nitrogen as a freezing agent, a temperature of about −100°C was obtained at the tip of the cryoprobe for 5 min. As a result, few lesions were required except in very large AVMs, in which 20–25 lesions were made. Total destruction of the AVMs was achieved in one-half of the cases (14 cases of 27). Recurrent bleeding developed in three patients in whom the thrombosis was not complete, and one patient died. The author explained the incomplete thrombosis by the rapid blood flow in the AVM, which interfered with the hypothermic effect.

Pedachenko and Orlov (1977) used the stereotactic technique for cryothrombosis of AVMs in seven cases. Thrombosis could not be obtained in two large AVMs with a multichannel blood supply in spite of many attempts at freezing for 2 hr. The AVMs diminished in size in two patients, and in three small AVMs (less than 20 mm in diameter) a complete thrombosis was achieved. Subsequently, Pedachenko and Orlov (1977) reported 21 cryothrombosis operations on 15 AVMs; total thrombosis was obtained in only seven small lesions.

The technique of open or stereotactic cryothrombosis may be used in both operable and inoperable AVMs, but complete thrombosis is obtained in less than half of the cases, mainly in small AVMs.

3.4. Balloon Catheter Occlusion

Balloon catheter occlusion of the feeding vessels, developed by Serbinenko (1971a, 1974, 1976; Serbinenko *et al.*, 1978), represents an important advance in the surgical management of AVMs not suitable for radical operations. In recent years, the technique has been used successfully by many other neurosurgeons (Zubkov, 1974; Zozulya and Shcheglov, 1976; Zlotnik and Sekach, 1976; Romodanov *et al.*, 1979; Lysachev *et al.*, 1982; Lasjaunias *et al.*, 1986).

In 1976 Serbinenko reported the results of the technique in 47 patients with AVMs, mainly inoperable because of their multichannel blood supply or inaccessible location. In these cases, 120 large arteries feeding the AVMs were occluded by a detachable balloon catheter.

In their next paper, Serbinenko *et al.* (1978) described balloon occlusion of 253 afferent vessels in 80 AVMs and reported the results from 1 to 6 years after surgery. The authors came to the conclusion that the success of the operation depended on the reduction of blood flow in the malformation. The best results were noted in patients with epileptic seizures. In spite of the success of this technique, one must admit that it is not absolutely safe. In two cases, large feeding branches of the middle cerebral artery were ruptured by the balloon

(Serbinenko, 1976). Zubkov (1974) reported that five of 12 cases of AVMs died after the procedure, three as the result of CBF disturbances after the intravascular operation from brain infarcts.

At present, this method of occluding afferent vessels in radically inoperable AVMs is considered one of the most promising in the treatment of these cerebral vascular lesions.

These data seem to indicate encouraging results from the new, more effective, and safer surgical methods for treating cerebral aneurysms and AVMs. The techniques cannot be used in all cases, and their limitations and shortcomings stimulate the search for better methods.

4. Stereotactic Clipping of Aneurysms

There is no doubt that direct *ad oculus* clipping of the neck of an arterial aneurysm or total extirpation of an AVM is the most radical and reliable method of surgical treatment. However, there are reasons to believe that stereotactic clipping of vascular lesions is not only possible but advisable in cases in which a classical operation is impossible or excessively dangerous. This situation led us to propose a new approach to the problem and to develop a "pure" stereotactic operation. The primary aim of this method was to achieve a higher degree of safety in the surgical management of aneurysms and AVMs.

In 1971, we devised a new method of sterotactic clipping of arterial aneurysms and AVMs through a burr hole; in 1973, we used the technique in clinical practice for the first time (Kandel and Peresedov, 1975, 1976).

The technique has many *a priori* advantages compared to classical open operations. The stereotactic method allows one to reach the target point in deep parts of the brain with minimal trauma to the brain tissue and without manipulation, which can prompt brain edema and other complications. This method does not require retracting the brain and vessels of the circle of Willis, hence substantially reducing the risk of developing or increasing vasospasm. There is no necessity to free the aneurysm from adhesions, which frequently leads to rupture of the vessel.

The development of the new technique required the creation of an essentially new device, which provides for the application of a clip to the vessel or the aneurysmal neck with the help of a stereotactic apparatus. The method was used clinically after testing the equipment in many technical and animal experiments.

4.1. Technical Equipment

The equipment for the new operative technique consists of two main components: our stereotactic apparatus, described in Chapter 3, Section 5.9, and a special device for clipping. The device must permit passage of a clip to the neck of the aneurysm or vessel stereotactically through a burr hole, opening of the clip near the vessel, followed by its closure, the possibility (in case of complications or wrong clipping) of removing the clip to put it on another part of the vessel, and of modifying the compression of the clip during the process. Moreover, all manipulations of the clip inside the brain are controlled by the extracranial parts of the device.

The device* was constructed and has been used by us in clinical practice since 1973. It consists of several parts (Fig. 227). The principal part of the device is a thin-walled stainless steel tube 17 cm long with an outer diameter of 3.2 mm. A conical narrowing inside the tip of the tube opens the clip. At the outer end of the tube there is a special structure that consists of a rectangular shackle graduated in millimeters with two nuts. This controls the opening and closing of the clip and disconnects it from the device after it has been applied to the vessel or aneurysm. The device is connected (Fig. 228) to our stereotactic apparatus (Chapter 3, Section 5.9).

There are two interchangeable metal pivots that are inserted consecutively into the tube. The pivots fit snugly inside the tube. The first (guide) pivot has a short thin terminal segment that protrudes from the end of the tube for several millimeters. The protruding segment from the tube tip is equal to the distance from the end of the tube to the ends of the clip blades when they are fully opened. With this pivot installed, the tube is introduced stereotactically to the target point in the brain. Once the target has been reached, the guide pivot is removed, and a second (working) pivot with a special clip attached to its tip is inserted into the tube. The outer end of the second pivot has a scale graduated in millimeters indicating the distance between the clip blades as they open inside the brain.

The active principle of the device, consecutive

*USSR Patent No. 452,336, U.S. Patent No. 4,241,734, as well as patents in West Germany, Canada, Great Britain, France, and Japan.

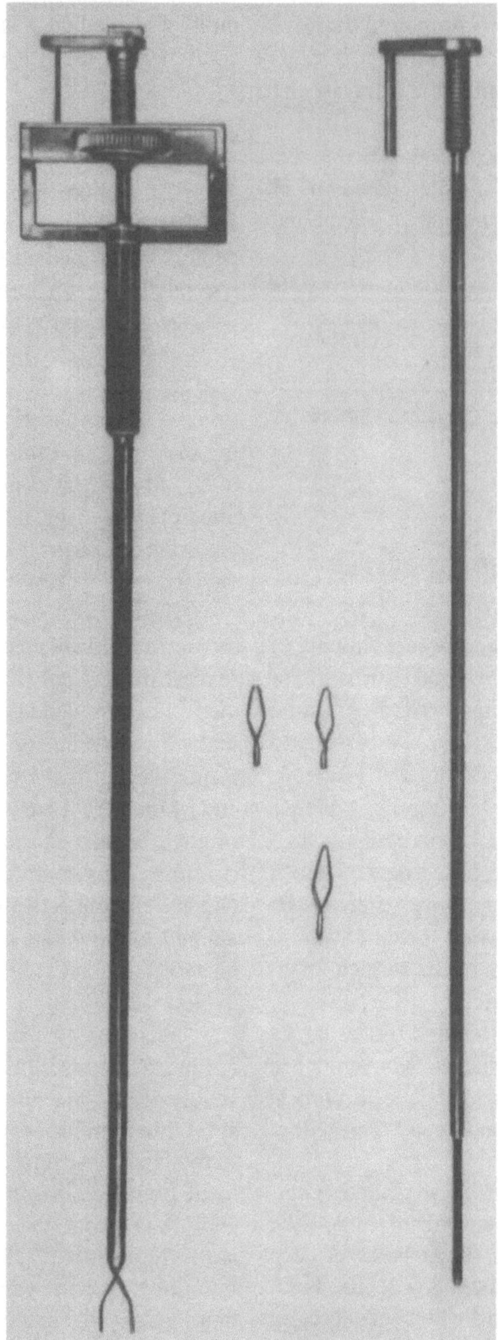

FIGURE 227. The author's device for stereotactic clipping of cerebral aneurysms and AVMs. Explanation in the text.

introduction of both pivots and phases of clip opening and closing, are shown schematically in Fig. 229).

4.2. Clips

To date, about 40 different neurosurgical clips (Fox, 1976b) are used for permanent or temporary clipping of aneurysmal necks and brain vessels.

Special removable stainless steel clips of different sizes have been constructed for use with our device (Fig. 230). The length of the clips varies from 10 to 17 mm, and their weight from 80 to 140 mg. They are of the crossing-spring type with parallel blades that can be used to clip vessels from 1 to 7 mm in diameter. It is

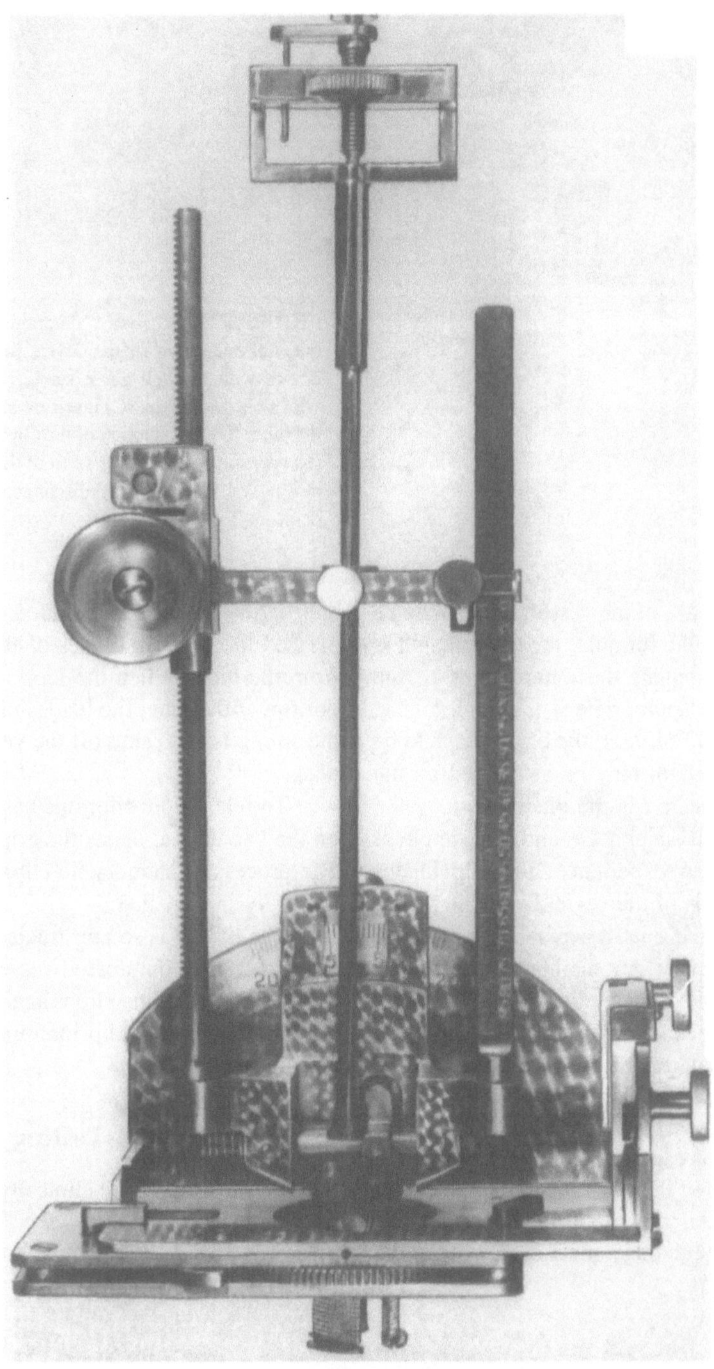

FIGURE 228. Device for stereotactic clipping of aneurysms and brain vessels attached to our stereotactic apparatus.

important to note that the blades of the clip passed through the 3-mm tube may be opened to about 8 mm without loss of the spring property. The clips are made of biologically inert, highly elastic, and anticorrosive steel consisting of 18% chromium, 9% nickel, and 0.5% titanium.

Our clips are flat springs with preliminary tension of 30–35 g/mm^2. The compressing force of the clip (40–45 g/mm^2) is constant because it depends on the clip tension and not on the surgeon's hand. The parallel blades of the clip prevent slipping. The length of the clip is also important, for reliable clipping requires that the blades completely occlude the vessel. Hence, the length of the blade should equal not the diameter but

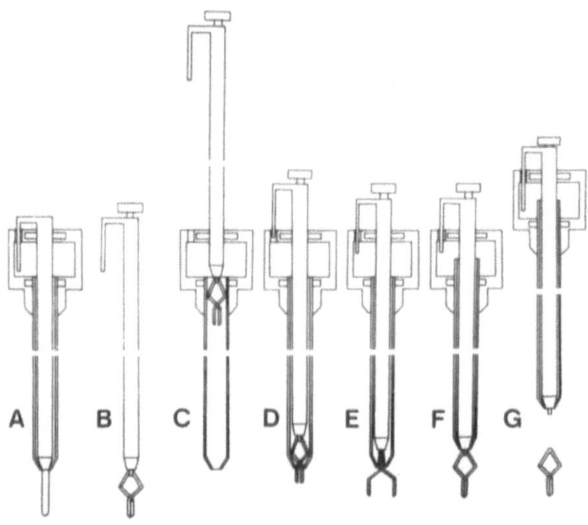

FIGURE 229. Schematic diagrams of the principle and sequential operations of the device for stereotactic clipping: (A) device with inserted guide stock; (B) working stock supplied with a clip on the tip; (C) insertion of the stock with the clip into the tube; (D) introduction of the clip into the conical narrowing; (E) opening of the clip; (F) exit of the clip from the cannula and closing of the clip; (G) disconnection of the clip from the device.

one-half the circumference of the vessel, which may be calculated as πr. From the formula, the blade length is equal to πr, where r equals the outer radius of the vessel. The formula is appropriate if the angle of the blades to the vessel is 90°; if less, the blades have to be a little longer. Their length may be calculated by the formula $l = r/\sin\alpha$, where l is the blade length, r the radius of vessel or aneurysmal neck, and α the angle of the approach. From our experience, the clip blades have to exceed the walls of the vessel by about 1 mm.

The clips are made in such a way that after closure the blade tips are in contact, but the blades are slightly separated. The surgeon selects a clip before the operation that is suitable to enclose the feeding artery or aneurysmal neck, as determined angiographically. The opened clip should have a gap sufficiently wide to exceed the outer diameter of the vessel or the neck by 1.5 to 2 mm. The clip is easily attached to the tip of the pivot before it is inserted into the tube. The fulcrum or loop of the clip is fixed in the grip of the effector pivot. As the clip passes through the conical narrowing at the end of the tube, the shoulders of the loop are compressed to open the blades of the clip as they emerge from the tube. When the loop moves past the narrow portion of the tube, the blades of the clip are closed by the spring loop to shut off the vessel or the aneurysmal neck.

To release the clip, one has only to press a button on the handle that opens the grip and releases the clip. The process of stereotactic clipping is presented schematically in Fig. 231.

The device is so constructed to permit removal of the clip on the compressed vessel. By rotating the button on the control the clip is again fixed in the grip, and on withdrawing the clip loop into the tube, the blades are opened.

4.3. Experimental Testing

Before being used clinically, the new method was tried both *in vitro* and *in vivo*. The reliability of clipping was tested on a number of thin-walled silicone

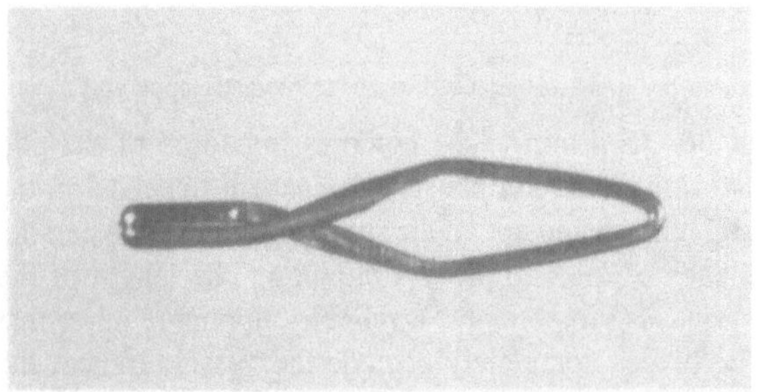

FIGURE 230. Middle-sized clip for stereotactic clipping (×2.5).

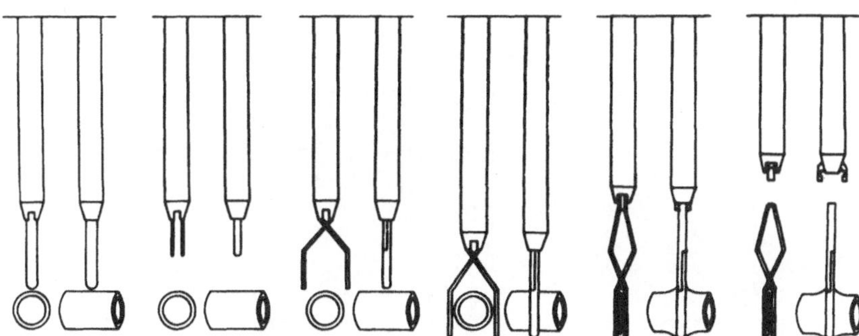

FIGURE 231. Schematic diagram of the consecutive stages and relationships of the clip and the vessel during stereotactic clipping.

tubes. After occluding the tube, the clips tolerated an increase of pressure in the system of more than 300 mm Hg. In dogs, clipping the carotid or femoral arteries completely obliterated the vessels. Histological studies demonstrated the absence of damage to the vascular walls.

In another series of experiments, the clips were introduced into the brains of dogs for periods of several hours to 6 months. Morphological investigations revealed minimal focal reaction in the brain tissue.

4.4. Preoperative Calculations

After an exact diagnosis of an aneurysm or AVM has been made, preoperative calculations are made on angiograms in both anteroposterior and lateral projections. The aim is (1) to select the point of clipping (target point) of the feeding artery (or arteries) or the aneurysmal neck; (2) to locate the point for the burr hole in the skull; and (3) to determine the appropriate approach for the clipping.

The target point must be selected on the basis of several factors. In aneurysms, the target point is the middle of the neck, which has to be relatively long and narrow. The diameter of the neck at the target point is carefully measured so that the size of the clip opening can be determined. Sometimes it is difficult to determine the target point because the neck is poorly visualized in one or both projections. In such cases oblique angiograms are helpful. The selection of the target point for clipping an aneurysm of the anterior communicating artery is discussed below.

In the case of AVMs, the selection of the target point is a more difficult task. We have experience with clipping of anterior and middle cerebral arteries and their branches. With clipping of the posterior cerebral artery, which is rarely possible technically, we have had no experience. The target in AVMs may be at any point along the feeding artery or arteries. As a rule, however, a point is selected distal to the last normal branch of the artery.

It is logical that the clipping of the distal part of the artery close to the vascular conglomerate is safer and more effective because the proximal branches supplying blood to normal brain tissue remain patent. But in many instances, clipping a feeder close to the AVM may be dangerous because of the possibility of damaging angiomatous tissue or large draining veins coming from the vascular knot. Because of this factor, we prefer to clip the arteries quite far (about 4.6 and in rare cases even 8 cm) from the nidus. Our experience has shown that clipping the hypertrophic feeding vessel some distance proximal to the AVM yields good results without any signs of a neurological deficit. This is explained by the fact that these arteries supply only the AVM and do not irrigate normal brain tissue.

Before each operation, the trajectory of the tube through the subcortical structures is analyzed with the aid of the Schaltenbrand and Bailey stereotactic atlas. It is necessary to know in which brain structures the clip blades will be opened near the aneurysmal neck or vessel. By such an analysis, damage to functionally important structures may be avoided.

The main reference line (LI) is determined by bony landmarks because there is no need for precision as in routine stereotactic operations. The calculations are made on angiograms or plain films (Fig. 232). On a lateral angiogram, the point of the angle between the inner horizontal and vertical plates of the frontal bone is connected to the anterior tip of posterior clinoid processes (line A). A second line (B) perpendicular to the first passes through the middle of sella turcica. Based on our calculations, the second line passes through the CA and the anterior edge of FM in 92.5% of cases. The distance from the first line to CA is 22.5 ± 2.2 mm; the second line and LI are almost at a right angle, 87.5 ± 5.3°.

Proceeding from these data, one marks the point

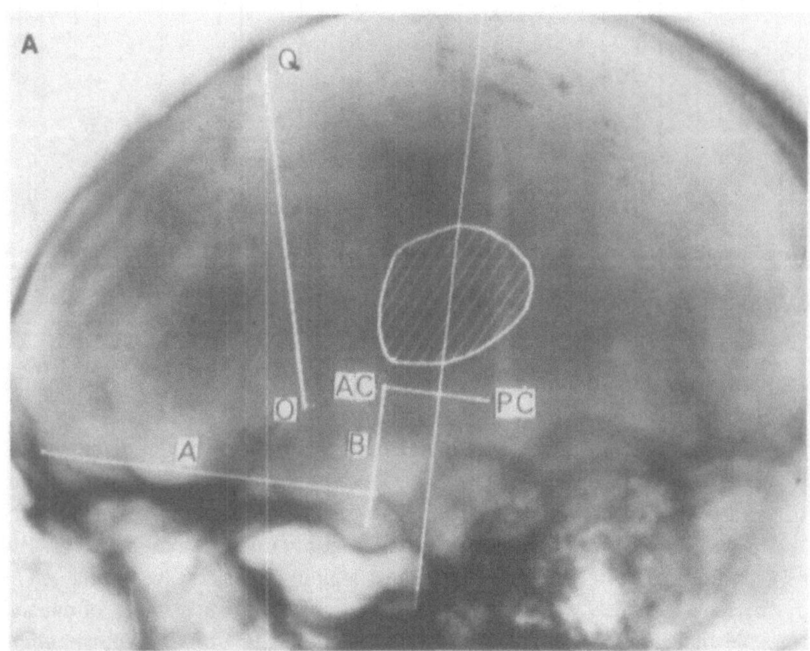

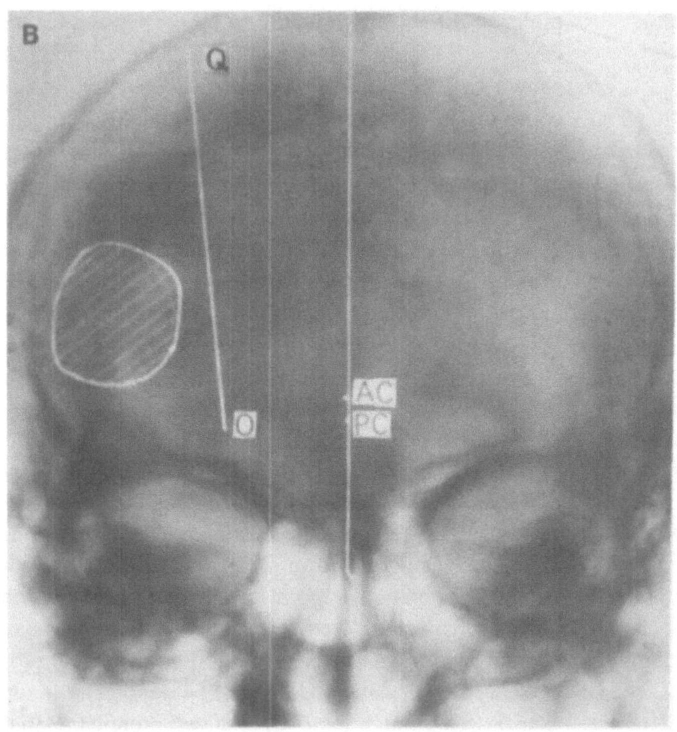

FIGURE 232. Outline of AVM transferred from angiograms to lateral (A) and AP (B) skull x rays. The AC–PC (LI) line is calculated with the use of a stereotactic atlas to determine the subcortical structures in the trajectory of the clipping device: A, line connecting the apex of the angle between vertical and horizontal plates of the internal surface of the frontal bone and superior tip of the clinoid process; B, line perpendicular to line A drawn through the middle of the sella; AC–PC, intercommissural line; O, target point; Q, site of burr hole; QO, the calculated trajectory of the clipping device to the feeding artery of the AVM.

CA and draws a perpendicular in the direction of CP. Because the mean length of LI from our data is 23.5 mm, one marks that distance on the perpendicular, pinpointing CP. The perpendicular to the middle of LI is the second reference line of the stereotactic atlas. Then, the distance is determined between the above-mentioned lines and the line connecting the center of the burr hole and the target point based on the atlas sections. These data are transferred to the sections, and the brain structures along the trajectory of the cannula are identified. We use the same method for determining the brain structures that may be involved in reaching the AVM. In this case, it is necessary to determine the distance from the reference lines to the borders of the AVM on the angiograms.

The determination of the amount a clip may open is an important factor in preoperative calculations. As is well known, the dense adhesions that firmly bind the aneurysm and surrounding brain tissue and vessels are produced about the aneurysmal sac and neck after a subarachnoid hemorrhage. Disturbing these adhesions may cause the aneurysm to rupture, although in our series of cases, there were no such complications. To be safe, it is imperative to choose a clip of the correct size. When opened, the blades must not touch the aneurysmal sac; in fact, there should be some space between the blades and the sac (about 0.5–0.7 mm). However, if the blades are opened too widely, nearby vessels that should not be clipped may be disturbed. For instance, if the aneurysmal neck has an outer diameter of 3 mm, the clip should be opened 4.5 mm. We determine the inner diameter of the neck or vessel by measuring on angiograms using a magnifying glass graduated with 0.1-mm marks. Correction of the x-ray divergence must be made to obtain the true inner diameter. If the neck wall has a thickness of about 0.5 mm, 1 mm must be added to the measured inner diameter.

After choosing the target point, two decisions must be made: determination of the site of the burr hole and the position of the plane of clip opening. For this purpose we transfer the target point and vessel axis to the drawing board and make certain geometrical calculations that we and co-workers Peresedov and Goihman have developed. The aim of the calculation is to determine the common location of such geometric elements as the axis of the artery or aneurysmal neck to be clipped, the plane of clip opening, and the axis of the clipping device.

In making preoperative calculations, it is useful to know the following terms: (1) target point (the center of the vessel or aneurysmal neck); (2) line of clipping (a line perpendicular to the axis of the vessel or the neck); (3) line of approach (the axis of the clipping device); (4) angle of approach (the angle between 2 and 3 above); (5) plane of approach (the plane containing 2 and 3 crossing in 1); (6) plane of clipping (the plane contained by 3 perpendicular to 5); (7) axis of the burr hole (a line perpendicular to the center of the burr hole); and (8) angle of inclination (the angle between 7 and 3).

The location of the plane of clipping (or the plane of the open clip blades) is very important because the plane and axis of the vessel should be nearly perpendicular. This relationship creates the optimal condition for clipping and practically excludes slipping of the clip.

In choosing the center of the burr hole (Fig. 233), which determines the location of the line of approach, one must consider the following factors: (1) the line of approach through which the tube will pass should not involve functionally important areas of the brain; (2) the best angle of approach is 90°, but it must not be less than 45°. Should it be less than that critical value, the clip may slip. In that case it is necessary to change the location of the burr hole; (3) the angle of inclination, which depends on the construction of the stereotactic apparatus (in the case of our apparatus, the angle must not be more than 20°).

4.5. Operative Technique

The operation is carried out with the patient under general anesthesia or neuroleptanalgesia. Routine premedication is used. The patient's head is placed in the supine position on the stereotactic head holder and fixed firmly with two sharp pins.

A sagittal line, the point at which it crosses the coronal suture, the preselected site of the burr hole, and the plane of clip opening are drawn on the patient's skull. After beginning the anesthesia, a catheter is inserted percutaneously into the carotid artery on the side of the aneurysm for intraoperative angiography (Fig. 234).

A skin incision 4 to 5 cm long is made through the mark for the burr hole, preferably near the coronal suture. A burr hole 25 mm in diameter is made with a crown trephine. The dura mater is incised. Our stereotactic apparatus (Chapter 3, Section 5.9) with the clipping device attached is fixed in the burr hole according to the preselected plane of clipping. The guide pivot is inserted into the device. After coagulation of the cortical point, the tube is introduced about 1 cm into the brain in the approximate direction of the target point.

The next stage of the operation is to make the stereotactic calculations. Anteroposterior and lateral plain films are taken, and the preselected target point is transferred from the preoperative angiograms to the films made on the operating table. Ordinary stereotactic calculations are made for the correct orientation of the clipping device. The calculations must be very accurate because the effectiveness of the clipping obviously depends on the accuracy with which the target point can be reached. Corrections for the angles in both

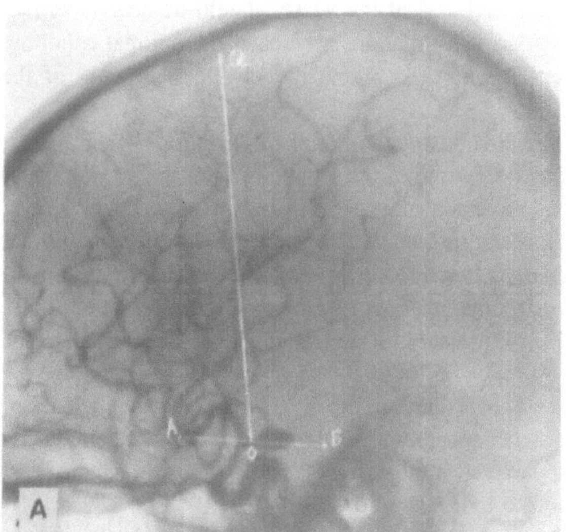
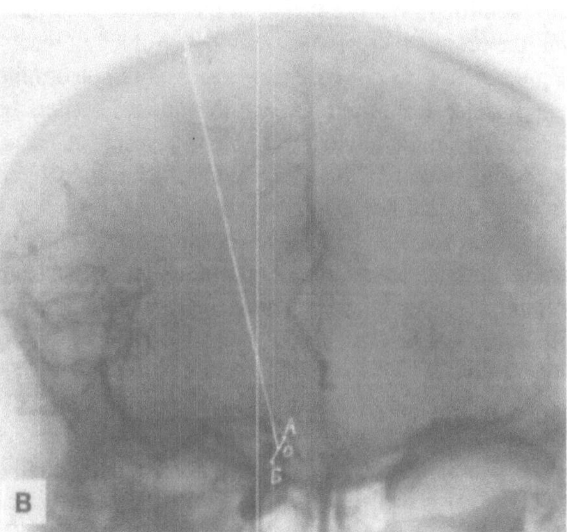

FIGURE 233. Preoperative calculations of the site of the burr hole on lateral (A) and AP (B) angiograms. AB, axis of the arterial aneurysm; O, target point, the site of the neck to be clipped; Q, site of the burr hole; QO, trajectory of the clipping device.

projections are calculated and then transferred to the protractors of the stereotactic apparatus.

Under the control of an image amplifier with a television monitor, the tube is introduced into the brain in the correct direction to the calculated depth until the tip of the guide pivot touches the target point. The guide pivot is now pulled out and replaced by the operative pivot, which carries a clip of the appropriate size that is introduced into the tip of the tube (Fig. 235).

Anteroposterior and lateral control angiograms are made. The target point on the preoperative angiograms is then transferred to the newly obtained angiograms. If the angiograms confirm that the tip of the tube is in the correct place, the clipping device is oriented in the required plane for clip opening, as calculated before the operation. By turning a special ring on the coupling on the outer end of the device, the clip is pushed out of the tube and is opened in the immediate

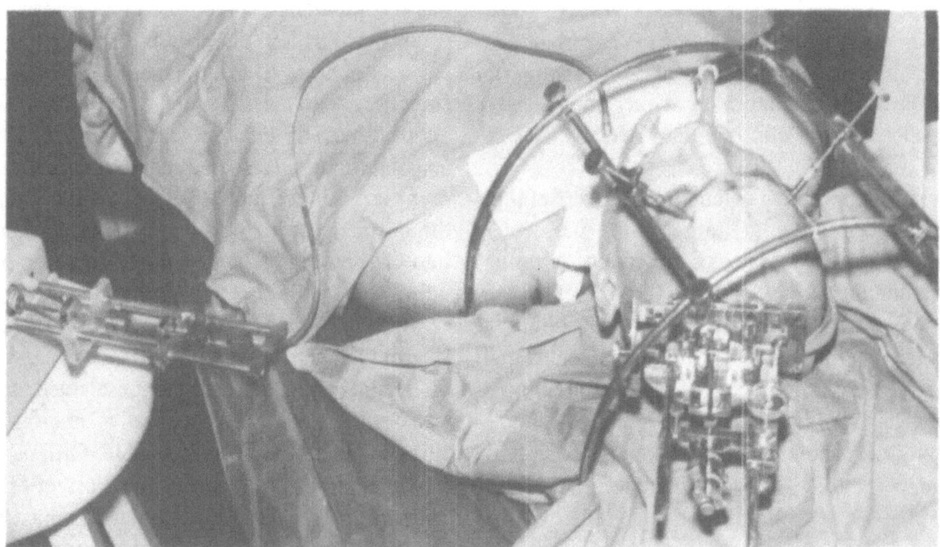

FIGURE 234. The catheter is inserted into the carotid artery for intraoperative angiography during the stereotactic clipping of cerebral aneurysms.

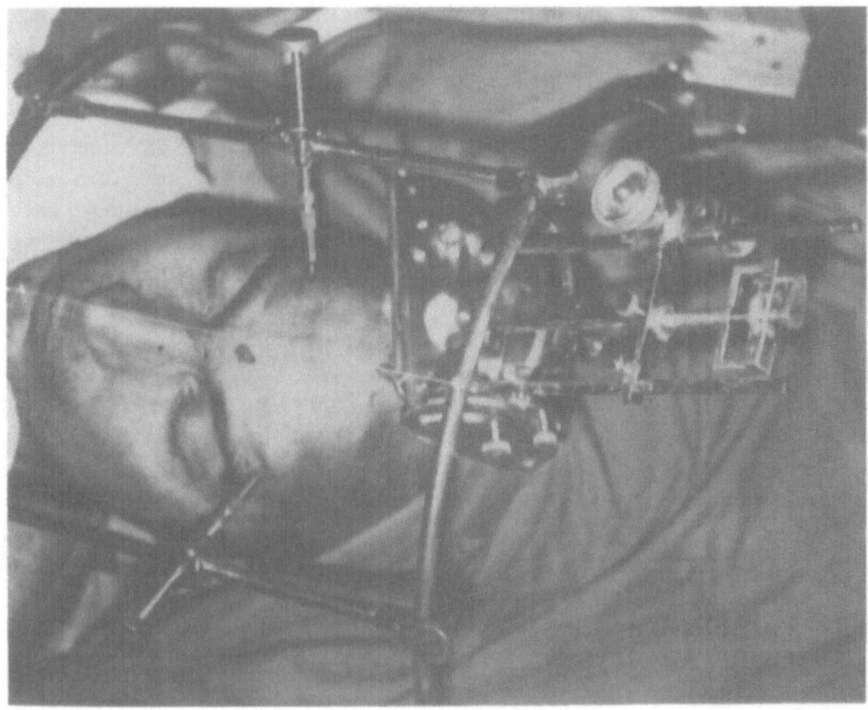

FIGURE 235. The general arrangement of the operative field during the main stage of stereotactic clipping.

vicinity of the aneurysmal neck or the artery. The clip is opened by being compressed in the tapered end of the tube. The distance between the clip blades in millimeters is checked on the scale on the outer end of the device. After the clip blades are opened to the required degree under control of the TV monitor, angiograms are made again. If these confirm the correct position of the clip, it is carefully and slowly advanced a little deeper and applied to the vessel or aneurysmal neck.

At this stage of the operation, the blood pressure is lowered to 50 to 60 mm Hg by an intravenous injection of trimethophan. Further turning of the ring extrudes the clip past the tapered end of the tube and closes it by the spring action. Another angiogram should verify effective clipping.

Immediately after clipping, the patient is awakened so that the function of the contralateral extremities and speech can be checked. If this function is satisfactory, neuroleptoanalgesia is repeated, and the blood pressure is increased gradually. The clip then is released from the pivot by pushing a button on the outer end (f the operative pivot. Under television control, the tube is withdrawn from the brain.

The stereotactic apparatus is released and removed. The bone plug is replaced, and the wound is sutured. The time of the clipping procedure averages between 1.5 and 2 hr. If it is necessary to clip two or more vessels, the entire procedure may be repeated again.

The construction of the device provides for the opening and removal of a clip if the patient's condition is aggravated (for example, development of hemiparesis) or the vessel is not occluded, such as a partial or incorrect clipping. The clip may be opened and withdrawn into the tube by turning the ring at the end of the device in the opposite direction turned to apply the clip.

4.6. Indications and Contraindications for Stereotactic Clipping

The possibilities of the new method have not yet been fully realized because the indications and timing of stereotactic clipping of arterial aneurysms and AVMs have not been completely established. The indications for this alternative treatment depend in each case on the evaluation of numerous factors: the location and size of the aneurysm, the size of the neck, the patient's condition, coexisting diseases, the presence of an arterial spasm, and the time interval from the onset of the subarachnoid hemorrhage.

In general, this technique is advisable in carefully

selected cases of arterial aneurysms and in cases of giant or deep-seated AVMs, when direct attack may be dangerous or technically impossible. As a rule, the stereotactic clipping of the AVMs is palliative, but in selected cases of small AVMs fed by a single artery, this technique may be the method of choice.

Below we shall present only the main factors in connection with indications for stereotactic clipping.

In arterial aneurysms, stereotactic clipping is indicated for supraclinoid aneurysms, for aneurysms of the internal carotid bifurcation and the anterior cerebral artery, and for clipping of the dominant anterior cerebral artery (A_1) for anterior communicating artery aneurysms and the trunk for aneurysm of the anterior cerebral artery. The possibility of stereotactic clipping of arterial aneurysms in other locations has not been established.

In our experience, stereotactic clipping neither caused nor aggravated an arterial spasm. This confirms the possibility of clipping arterial aneurysms stereotactically even in the presence of marked arterial spasm and on patients in poor condition. This is a tentative conclusion and may be altered as further experience is accumulated.

The absence of a neck to the sac and giant aneurysms are contraindications to this method.

In AVMs, stereotactic clipping is indicated for AVMs of any size supplied by the anterior and middle cerebral arteries and their branches. The clipping of the posterior cerebral artery supplying AVMs and of aneurysms of the vertebral–basilar system is apparently inadvisable. The technique is also indicated for AVMs supplied by one or several hypertrophied branches of the above-mentioned arteries. In the situation of so-called "scattered" blood supply by many small feeding arteries, the method is not indicated.

4.7. Clinical Results

To date we have performed 58 stereotactic clipping operations on 55 patients: 24 operations for arterial aneurysms and 34 operations for AVMs. Indications for surgery, the technique, and the results are described in previous reports (Kandel and Peresedov, 1975, 1976, 1977a,b, 1980, 1987b; Peresedov and Kandel, 1983).

4.7.1. Arterial Aneurysms

Twenty-three patients with arterial aneurysms, (nine males and 14 females) were aged 22 to 53 years. Before the operation nine patients had had one and 14 had had two or more subarachnoid hemorrhages. By four-vessel angiography, supraclinoid aneurysms were

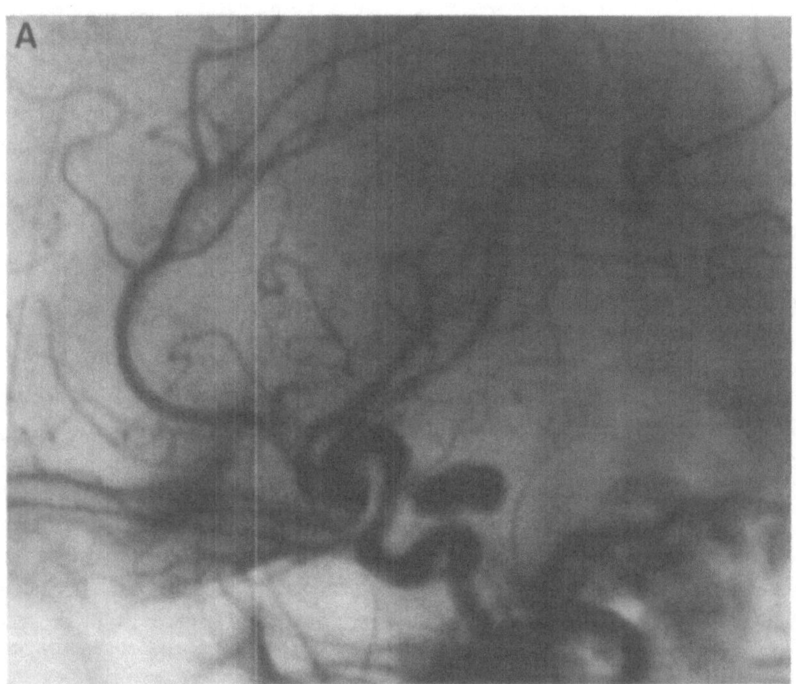

FIGURE 236. Supraclinoid arterial aneurysm in a 49-year-old woman, that caused three severe subarachnoid hemorrhages. Stereotactic clipping of the aneurysm neck is performed. Angiograms in both projections: (A) before surgery; (B,C) during the operation; (D,E) 2 weeks after the operation. The aneurysmal neck is clipped; the aneurysm is not filled.

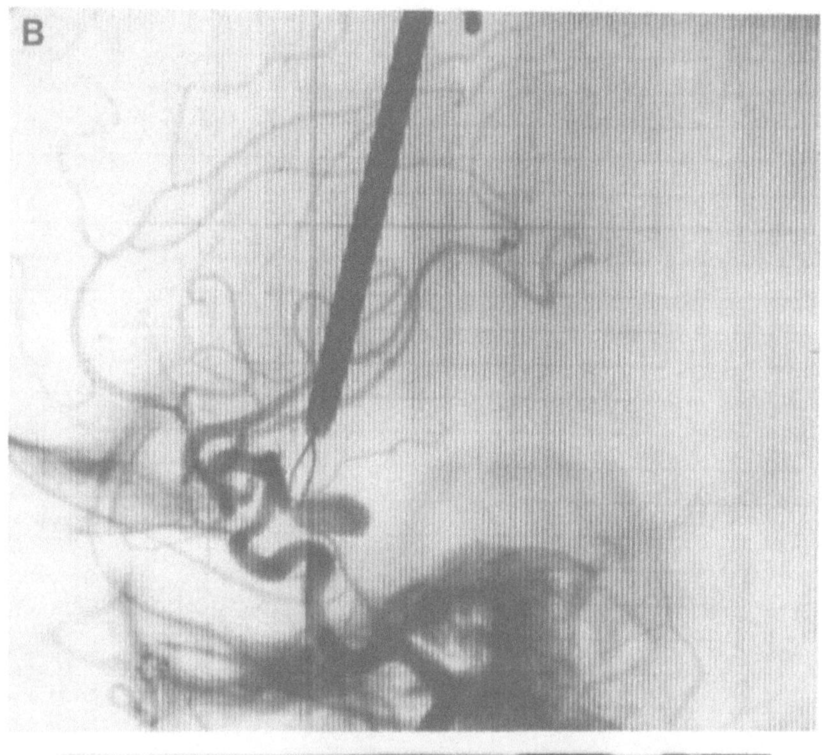

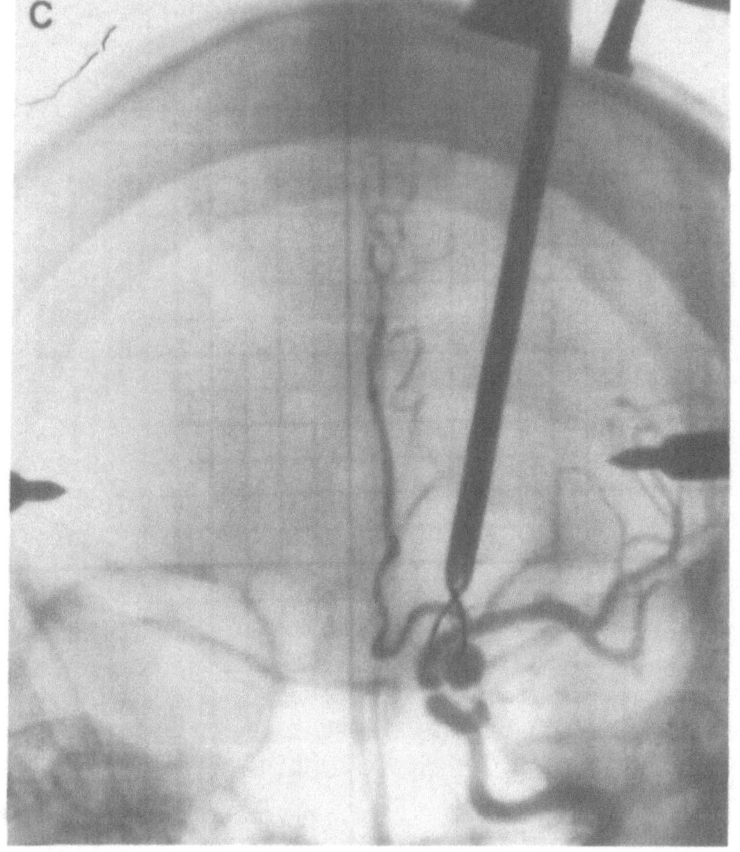

FIGURE 236. (*Continued*)

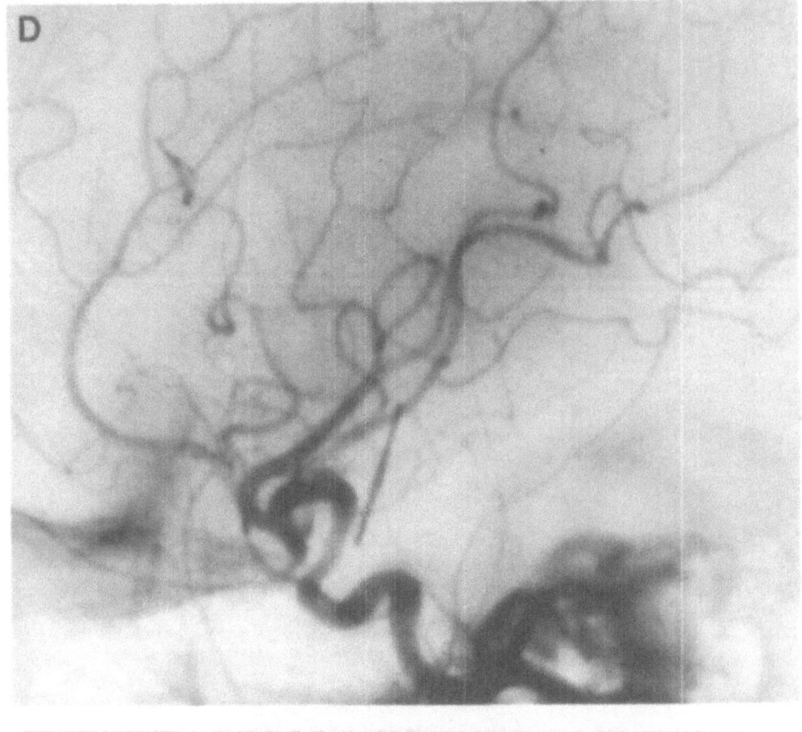

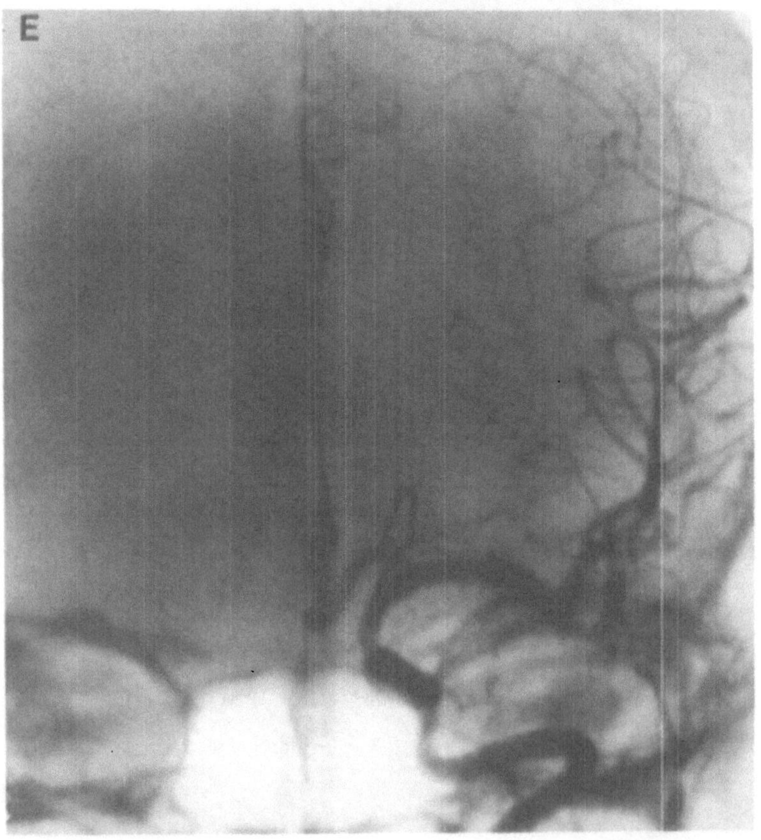

FIGURE 236. (*Continued*)

disclosed in 16 cases (nine on the right and seven on the left side), aneurysms of the anterior communicating artery in five cases, aneurysms of the internal carotid bifurcation in one case, and aneurysm of the anterior cerebral artery at the origin of frontopolar artery in one. The volume of the aneurysms varied from 250 to 900 mm^3.

The patients were operated on at different times after their subarachnoid hemorrhages—from 3 days to 11 months. Their preoperative conditions varied from good, without any complaints (14 patients) to deep sopor (5). Four patients were in fair condition.

Twelve out of 23 patients had no neurological symptoms. In the remaining 11 patients there were various signs such as marked meningeal syndrome, third nerve palsy, local pain in the frontoorbital region, hemiparesis or hemiplegias, disturbances of sensibility, papilledema, and so forth.

In the majority of cases (14 of 23), spasm of neighboring arteries was seen on the angiograms. In seven cases, including five patients in grave condition, a severe diffuse arterial spasm (more than 50%) occurred. In other cases, the spasm was mild and involved mainly the supraclinoid segment of the internal carotid artery.

The 16 supraclinoid aneurysms, which are the most common of all intracranial aneurysms, were of different sizes. In most cases their neck and sac were directed posteriorly, ventrally, and laterally. The diameter of the neck varied from 2.2 to 3.5 mm. The narrow neck of the majority of the supraclinoid aneurysms makes them particularly suitable for stereotactic clipping.

All supraclinoid aneurysms were successfully clipped. Below are two illustrative cases.

A 49-year-old woman was admitted to our clinic after two severe subarachnoid hemorrhages 17 months and 1 month previously. On hospitalization, her general condition was satisfactory. She had paralysis of the left third nerve, slight hemiparesis of the right side, and signs of mild sensorimotor aphasia. Angiograms demonstrated a bilobed supraclinoid saccular aneurysm (14 × 7 mm in lateral projection) arising from the left internal carotid artery at its junction with the posterior communicating artery (Fig. 236A). The neck of the aneurysm was narrow. There was a mild spasm of the left carotid artery.

Stereotactic clipping of the aneurysmal neck was carried out under general anesthesia. A trephine opening was made 3 cm posterior to the coronal suture and 3 cm to the left of the midline. The stereotactic apparatus with the clipping devices was introduced in the opening, and a catheter was percutaneously introduced into the left common carotid. The target point (aneurysmal neck) was transferred from angiograms to the AP and lateral plain films, and stereotactic calculations were made. The tip of the first pivot touched the target point at 76 mm from the cortex. It was replaced by a second pivot with a compressed clip on the tip. The clip was placed on the aneurysmal neck, and its correct position was confirmed by control angiography (Fig. 236B,C). During the clipping, the arterial pressure was lowered to 70 mm Hg.

The patient was awakened for several minutes in order to check movements in the contralateral limbs; no changes were noted. The clip was closed slowly by means of the external ring of the clipping device. An additional control angiogram showed that the aneurysm neck had been clipped and that there was no filling of the aneurysm. The clip then was released, and the device was withdrawn. The bone button was replaced, and the wound sutured. The operation, performed under continuous television monitoring, took 1 hr 40 min. There were no postoperative complications and no neurological changes. The patient was able to walk 3 days after surgery. Her aphasia and third nerve paralysis disappeared completely shortly after the operation.

Control angiograms made 2 weeks after surgery demonstrated the elimination of the aneurysm (Fig. 236D,E). The lumen of the internal carotid artery at the level of the clip was not changed. Nine years postoperatively, the patient exhibits no neurological deficit, and her general condition is good. She is working full time.

One week after a mild head injury, a 38-year-old woman suddenly developed a severe headache and vomiting but no disturbance of consciousness. Two weeks later, the severe headaches returned; these were centered mainly in the left frontoorbital region. The same day full paralysis of the left third nerve developed.

Angiograms made immediately after admission to our clinic showed an aneurysm on the internal carotid artery at its junction with the posterior communicating artery. The size of the aneurysm as measured in lateral angiograms was 11 × 7 mm. Stereotactic clipping of the neck of the aneurysm, which was located at a depth of 77 mm from the cortex, was performed (Fig. 237). The arterial pressure was lowered to 60 mm Hg during the clipping.

The postoperative course was uneventful. The patient was up and walking 4 days after the operation. A control angiogram confirmed the successful clipping of

A

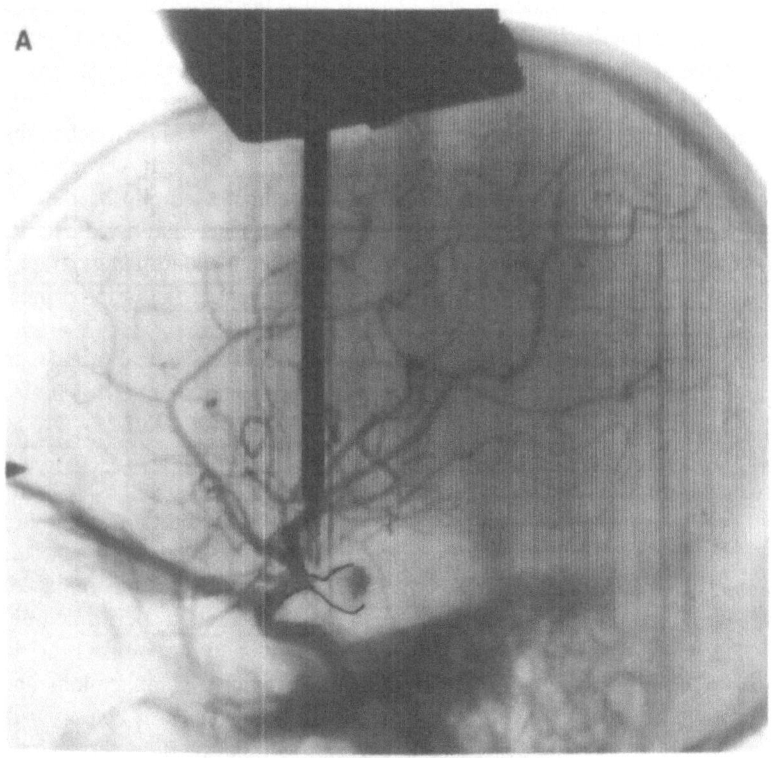

B

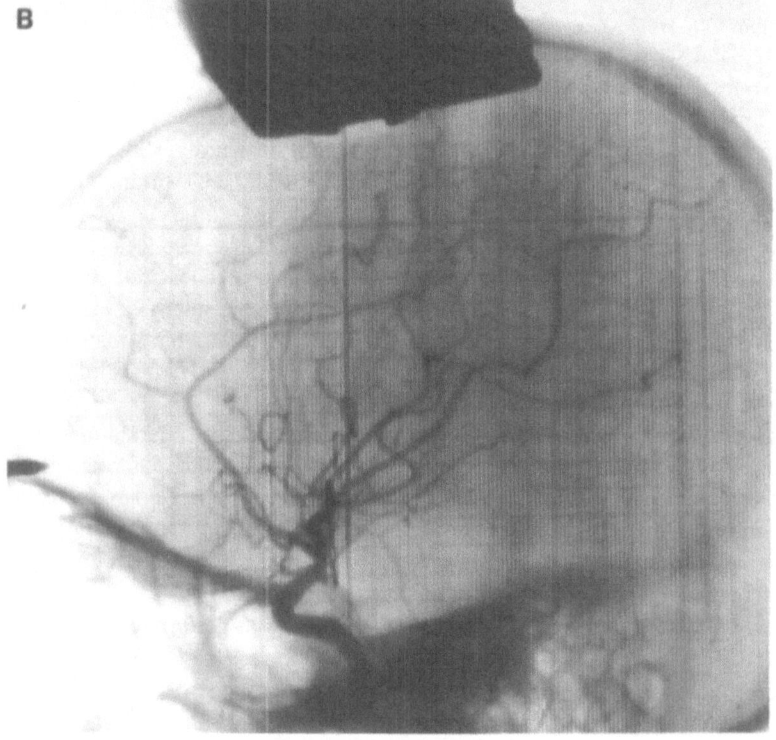

FIGURE 237. Supraclinoid arterial aneurysm in a 38-year-old woman on intraoperative lateral angiograms. (A) Immediately before stereotactic clipping; the clip is opened above the aneurysmal neck. (B) The neck is clipped and the stereotactic device is withdrawn from the brain. (C) Control angiogram after surgery showing that the aneurysm does not fill.

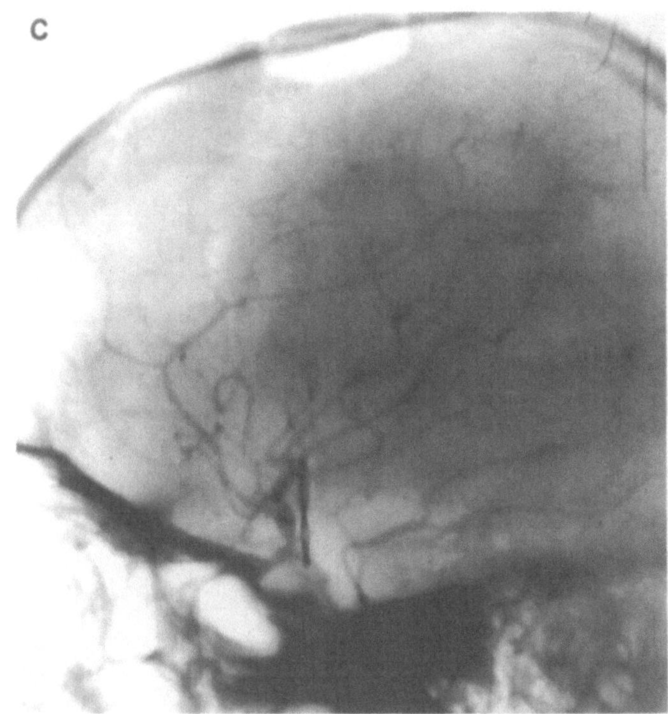

FIGURE 237. *(Continued)*

the aneurysm. Mild spasm of the middle cerebral artery disappeared after the operation. A 4-year follow-up has shown complete recovery of third nerve function. There was no rebleeding during the follow-up period. The patient works part-time.

Aneurysms of the internal carotid bifurcation are an important surgical problem. The approach and direct attack on these aneurysms are associated with many technical difficulties. The following example illustrates the possibility of successful stereotactic clipping of such an aneurysm.

A 48-year-old man suddenly developed severe headaches but no impairment of consciousness. A lumbar puncture was not done. About 1 month later, sharp headaches and pain in the right eye recurred. After 5 days, left-side hemiplegia developed. The patient was admitted to our clinic in serious condition. There was hemiplegia with decreased muscular tone and pathological reflexes.

Angiography disclosed a saccular aneurysm of the internal carotid bifurcation with marked spasm of the internal carotid and its main branches (Fig. 238A). The neck of the aneurysm was 2.5 mm in diameter. After careful preoperative calculations, stereotactic clipping of the aneurysm was performed under general anesthesia (Fig. 238B). The depth of the target point was 77 mm.

There were no complications after the operation. Movements of left extremities gradually returned, and the patient began to walk 3 weeks after surgery. A follow-up examination 1 year later showed a slight hemiparesis. The patient was in good general condition.

In spite of remarkable improvements in the results of a direct attack on aneurysms of the anterior communicating artery by the microsurgical approach, both the technical difficulties and complications remain serious unsolved problems. We believe that because of the special anatomic relationships in that region, stereotactic clipping of the neck of such an aneurysm is impossible or may only be done in exceptionally rare cases in the future.

Several neurosurgeons have proposed the open clipping of the dominant anterior cerebral artery, an operation that may be done by the stereotactic method. It is known that many anterior communicating aneurysms are filled only from one anterior cerebral artery, which in such cases is called dominant. Apparently there are several reasons for this phenomenon: hypoplasia of the other anterior artery, the presence of anterior trifucation, a sharp angle between the axis of the aneurysmal sac and the anterior cerebral artery, etc. In such cases, clipping the A_1 segment of the dominant artery, which supplies the aneurysm, leads to a sharp

A

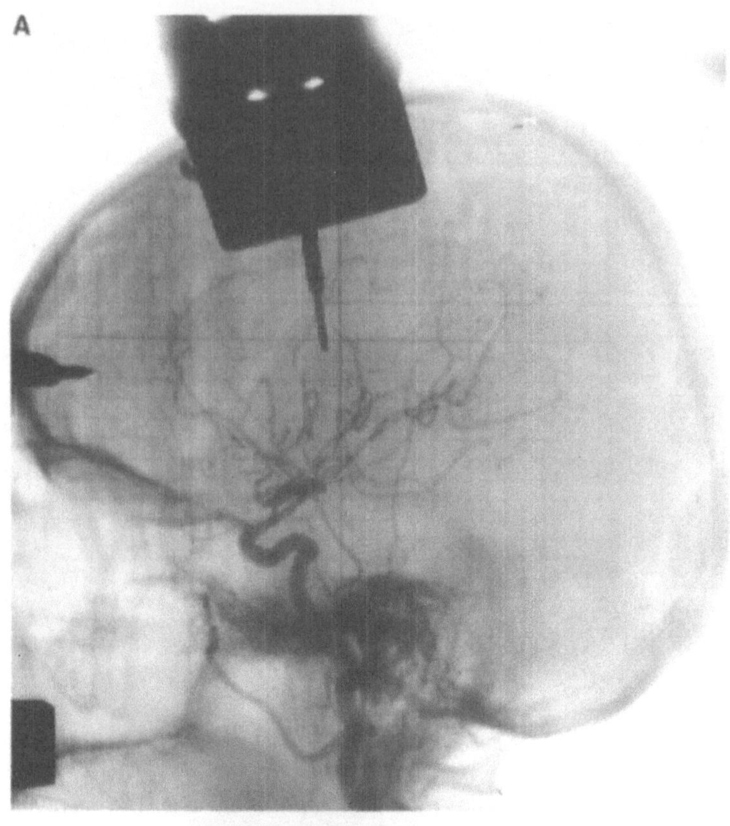

B

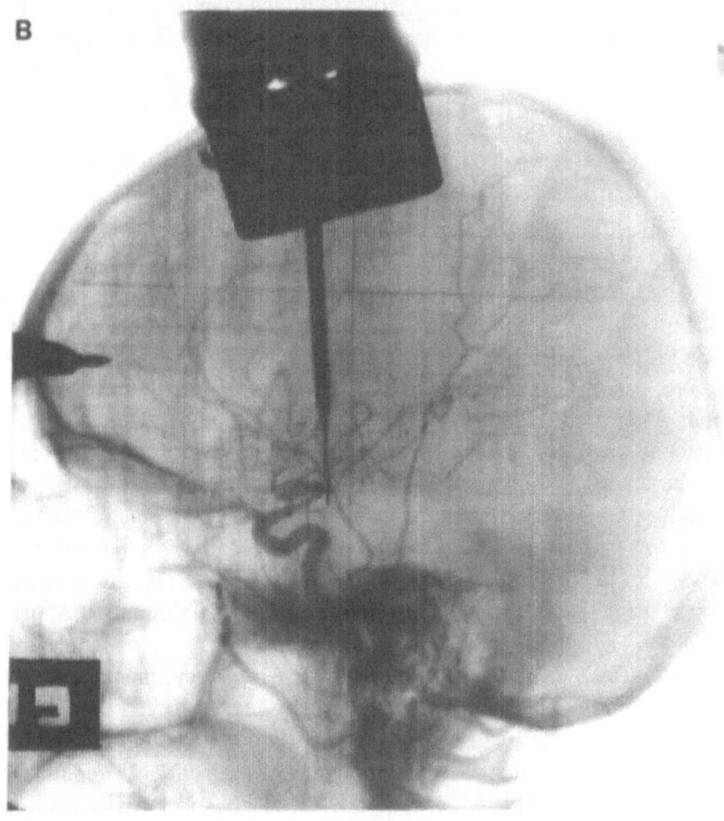

FIGURE 238. Small saccular aneurysm of the internal carotid bifurcation in a 48-year-old patient. Marked spasm of the carotid syphon and its main branches. Intraoperative angiograms before (A) and after (B) stereotactic clipping of the aneurysmal neck.

decrease of pressure in the aneurysm and subsequently its thrombosis.

This open operation was first performed by Logue (1956), who published a large series of observations in 66 cases. Postoperative mortality was 15%, and the rate of rebleeding 6% (Logue *et al.*, 1968). In a subsequent paper, Durity and Logue (1971) described the long-term results of the operation in 111 patients with aneurysms of the anterior communicating artery. The postoperative mortality in this serious operation increased to 26%. Follow-up angiography has shown that 23% of the aneurysms were obliterated completely, about 30% were diminished, and in 7% of the cases, the size of the aneurysms increased. Of 82 patients followed by the authors for an average of 8.5 years, only two died from recurrent hemorrhage.

Zlotnik *et al.* (1971) performed this operation in nine cases of anterior communicating aneurysms. Control angiography disclosed that there was no filling of the aneurysms in seven cases, and in the remaining two cases, the aneurysm had decreased in size.

Hugenholtz and Morley (1972) reported on a series of 23 patients with follow-up from 3 to 10 years after the same operation of proximal occlusion of A_1. The postoperative mortality was 13%. There was no recurrent bleeding in any of the surviving cases.

In 68 operations the dominant anterior cerebral artery was clipped by Hockley (1975). The mortality rate was 10.3%, and two more patients died from recurrent hemorrhage.

An important point in the operative technique is the selection of the segment of the anterior cerebral artery for clipping, because the clip may damage functionally important branches such as Hübner's artery, the diencephalic arteries, or the aneurysm itself. Scott (1973) recommends clipping the most proximal part of the anterior cerebral artery at a distance of 2–3 mm, considers it less advisable at 4–5 mm, and only in exceptionally rare cases should it be clipped 10 mm from the internal carotid bifurcation.

The above-mentioned data from the literature show that open clipping of the dominant anterior cerebral artery is an effective method to control anterior communicating aneurysms, especially if clipping of the neck is technically impossible; however, the postoperative mortality is relatively high (about 12–15%).

We have stereotactically clipped the dominant anterior cerebral artery in five cases of anterior communicating aneurysms. One patient was operated on twice.

A 44-year-old man began to suffer from sudden headaches and repeated vomiting and on one occasion lost consciousness for 3 hr. Initially, he showed signs of meningeal irritation: slight pyramidal signs on the right side and bloody spinal fluid. After admission to our clinic 3 weeks later, he complained of a headache, but his general condition was good.

Right-sided axillary and carotid angiography disclosed a medium-sized saccular aneurysm of the anterior communicating artery (Fig. 239A). Left carotid angiography showed no filling of the aneurysm. The aneurysmal neck was not clearly visible. Since stereotactic clipping of the neck seemed technically impossible, the dominant right anterior cerebral artery was clipped. A burr hole was made 1 cm anterior to the coronal suture and 3 cm to the right of the midline. The distance from the cortex to the target point was 62 mm. The clip was placed on the artery under television control.

The postoperative course was free of complications. A subsequent left carotid angiogram showed that the right anterior cerebral artery (A_1) had been clipped, and the aneurysm no longer filled (Fig. 239B). A control examination 3 years postoperatively found the patient in excellent condition, working full-time, and having no complaints.

Our limited experience led us to the conclusion that stereotactic clipping of the dominant A_1 is technically possible. However, evaluation of the results requires more observations. Control angiography in our four cases (in one case, it was not done for technical reasons) has shown that in three cases the aneurysms were not filled, and in one case, it filled from the other side, although such filling had not occurred before the operation.

Stereotactic clipping was performed in one "peripheral" aneurysm of the frontopolar artery. Clipping of A_2 segment of the anterior cerebral artery resulted in excluding the aneurysm from the circulation. This was confirmed angiographically. A mild hemiparesis developed in this case postoperatively but resolved 2 weeks later.

Three patients of 23 (13%) died after stereotactic clipping of arterial aneurysms. Two patients in grave condition with supraclinoid aneurysms were operated on and expired on the fourth and fifth day following surgery from pulmonary thromboembolism. The third patient died on the third postoperative day after successful clipping of a supraclinoid aneurysm from an unexpected rupture of a second AVM in the same hemisphere; this AVM had not been visualized in preoperative angiograms.

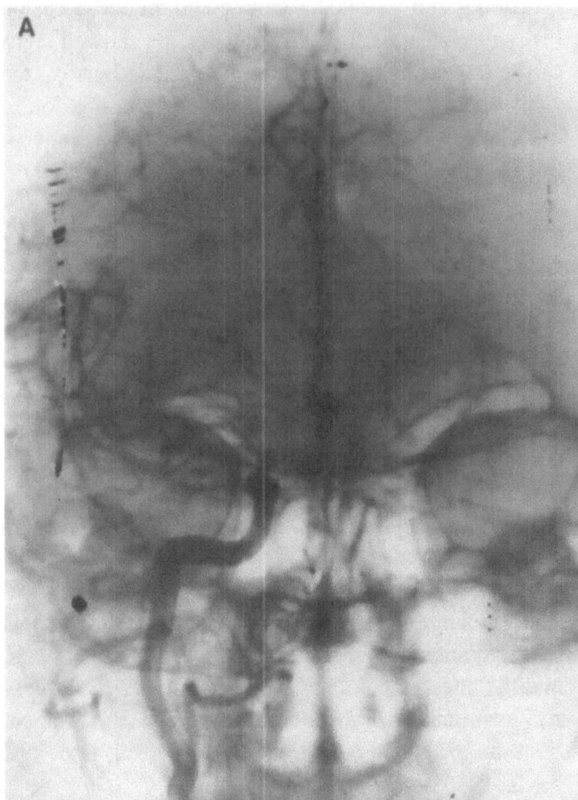

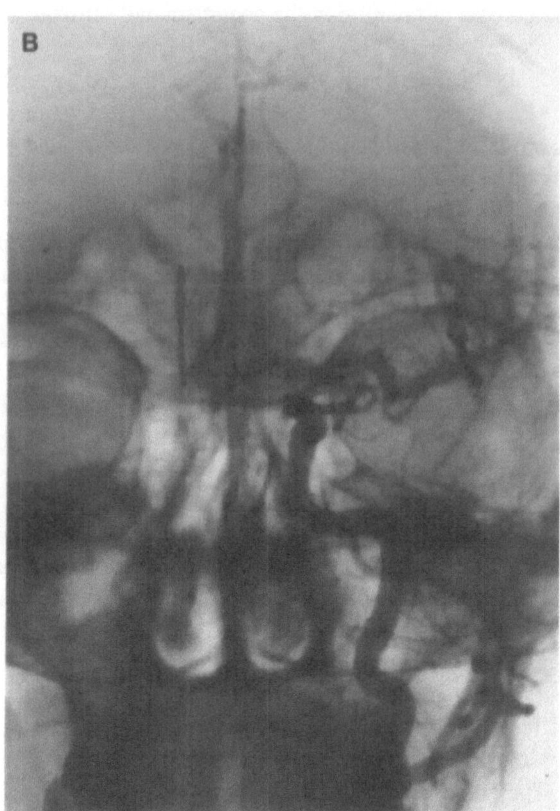

FIGURE 239. Aneurysm of the anterior communicating artery in a 44-year-old man. (A) The aneurysm filled only in the right-sided angiogram. There is a spasm of the internal carotid and A_1. The aneurysm is not filled from the left side although the left A_1 is visualized. (B) Left carotid angiogram (AP view) after stereotactic clipping of right A_1 shows filling of both anterior cerebral arteries, but the aneurysm did not fill.

Angiographic control in all cases was done during the operation, about 1 month after surgery, and (in 14 cases) in the period from 1 to 5 years later. It is interesting to note that the contrast medium may remain in the aneurysmal sac immediately after clipping and is sometimes visible on the plain films (Fig. 240). Control angiography following surgery has shown that all clipped aneurysms were excluded from blood circulation.

Long-term follow-up in 18 patients (two cases were lost to follow-up) from 6 months to 12 years confirmed the absence of repeated subarachnoid hemorrhages in all cases.

Our limited experience with stereotactic clipping led to the following conclusion. The operation is technically possible and effective in selected cases of supraclinoid aneurysms, aneurysms of the internal carotid bifurcation, and those of the A_2 segment of the anterior cerebral artery. The clipping of dominant A_1 in anterior communicating artery aneurysms is also possible, but its practicality needs confirmation.

Stereotactic clipping of arterial aneurysms does not increase preexisting arterial spasm or cause spasm if it was absent before surgery. There was no rupture of the aneurysms during or after surgery, and brain edema did not develop.

It should be emphasized that stereotactic clipping is an atraumatic operation that the patients endure easily. The majority of patients were ambulatory 2 to 3 days after surgery.

4.7.2. Arteriovenous Malformations

We have performed 34 stereotactic clipping operations in 32 patients with AVMs, 19 men and 13 women aged 13 to 52 years. The period of time from the onset of the illness to the operation varied greatly, from several months to 30 years. Intracranial hemorrhage

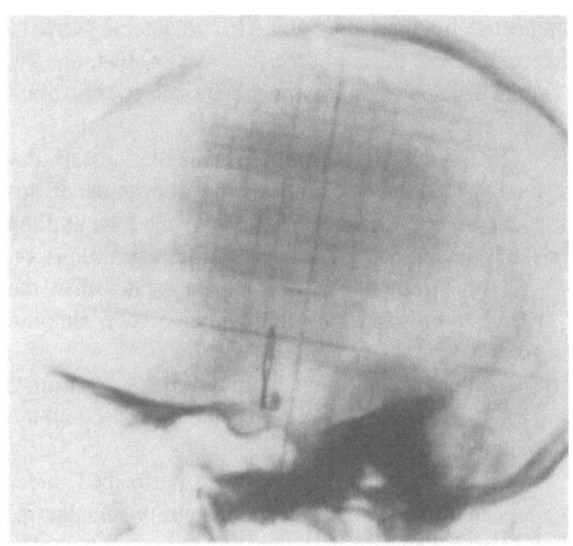

FIGURE 240. X-ray film 2 hr after stereotactic clipping of the supraclinoid aneurysm. Note the contrast medium that remains in the sac after clipping.

was the main clinical manifestation in 18 patients, epileptic fits in nine, and severe headache in five (two with pulsating bruit in the head).

Fifteen patients were admitted in good general condition, 14 in a mildly severe state, and three patients in bad condition. In the majority of patients, hemiparesis of various degrees was noted; meningeal signs, motor aphasia, and memory disturbances were present in some cases. The neurological deficit was, as a rule, the consequence of previous subarachnoid or parenchymatous hemorrhages.

The angiography was first performed on the side of the supposed AVM. If a lesion was found, angiography of the other side was performed. The contribution of contralateral arteries to the AVM blood supply was investigated in all cases. Selective angiography was sometimes used to determine the specific vessels involved in the anomaly.

The volume of the AVMs as determined by the Tönnis method varied widely from 8.4 to 198 cm^3; AVMs of small and middle size (up to 20 cm^3) were present in seven cases, large ones (up to 100 cm^3) were present in 17, and giant angiomas (more than 100 cm^3) were present in eight cases. Fourteen AVMs were located in the right hemisphere, and ten in the left, five in the region V and V_3, and three in the middle frontobasal regions.

In all cases the operability of the AVMs, i.e., the possibility of its total extirpation, was determined. For that purpose not only the volume and sources of blood supply of the AVM were taken into account but also its relationship to functionally important brain structures such as the sensorimotor cortical region, subcortical nuclei, mesencephalic and diencephalic structures, and lateral and third ventricules. The Schaltenbrand and Bailey stereotactic atlas was used for the evaluation of these factors (Section 4.4). This method made it possible to determine the brain structures involved by the AVM, which is of paramount importance in deciding the advisability of radical removal.

The large and giant AVM cases involved to some degree the central gyri, one or two cerebral lobes, parts of the mesencephalon, thalamic nuclei, Subth, CI, etc. The majority had multichannel blood supplies from several arterial systems. In half of the cases the supply was from arteries of both hemispheres. On the basis of the analysis, one may conclude that 25 AVMs undoubtedly could not be completely extirpated, in four cases the total removal was apparently possible but only at great risk, and in only three AVMs would it have been possible to perform complete excision. Therefore, stereotactic clipping is, first and foremost, a method of treatment for radically inoperable AVMs. As we show below, in certain, selected cases, this method can ensure total occlusion of an AVM from the circulation, and in such cases, it may be regarded as a radical method of treatment.

The selection of the target point, i.e., the part of the feeding artery that should be clipped as well as the location of the burr hole, has been described. Because this location determines the path of the cannula, it is necessary to choose a trajectory that will pass at a safe distance from the AVM and its draining veins. The target point and the centerpoint of the burr hole may be connected by lines on both AP and lateral films to determine, with the aid of the stereotactic atlas, that the cannula will not damage functionally important brain structures.

The preoperative geometric calculations as described in Section 4.4 showed that the angle of approach (the angle between the axis of the tube and the axis of the clipped vessel) in all our operations on AVMs was from 58° to 88°, much greater than the "critical level" at which the danger of clip slipping exists.

The diameters of the feeding arteries were from 2 to 6 mm. The size of the clip selected was based on the diameter of the vessel.

All operations on AVMs were performed under general intratracheal anesthesia or neuroleptanalgesia. The arterial pressure was lowered before clipping to 50–60 mm Hg. In case of neuroleptanalgesia, the patient was awakened to test the function of the contralateral extremities and speech. The depth of the clipped arteries from the cortical surface varied from 30 to 83 mm. The operating time was from 2 to 3 hr.

There were no serious or lasting complications in the 32 operations and no mortality. Only once did the clip slip from the artery; however, 1 week later the procedure was repeated, and the clip was set correctly. No increase of neurological deficit was noted postoperatively in any case. All patients withstood the operation well and began to walk 3–4 days after surgery.

It is well known that total and specific cerebral blood flow (CBF) in the presence of the AVMs increased substantially. As our previous study has shown (Kandel and Nikolaenko, 1971), the high total CBF in hemispheric AVMs correlated well with the size of the aneurysm and was from 16 to 300% greater than normal figures. At the same time, the cerebral O_2 consumption remained normal (3.25 ± 0.12 ml/100 g per min).

In five patients we investigated total and specific CBF before and after the stereotactic clipping of AVMs of different sizes. The Kety and Schmidt method with our chromatographic modification (Kandel and Nikolaenko, 1971) was used. The data are presented in Tables 25 and 26.

As these data have shown, the total cerebral CBF increased by 33–222%, and the specific CBF by 50–290%. In the case of large and giant AVMs, the CBF substantially increased in the other hemisphere as well. After surgery the high CBF in two patients fell to a nearly normal figure, and in three patients it diminished to 1.5 times normal. One may conclude that the CBF study is an objective criterion of the effectiveness of stereotactic clipping of the feeding vessels of an AVM.

Control angiography was performed in all cases at the end of the operation, on average at 3 weeks following surgery (excluding three cases), and in 18 patients at follow-up (from 1 to 7 years). These data have shown that in five patients the AVMs are excluded completely, in 20 patients the volume of the AVMs was diminished to a varying degree, and in seven cases the volume did not change because of filling from other channels. There were no cases of increase in the aneurysm volume.

A favorable clinical result was present in all patients at follow-up (from 6 months to 10 years). There was no repeat or primary intracranial hemorrhage. Generalized or focal epileptic seizures disappeared in seven patients and became much less frequent in two. Motor and sensory disturbances that were the consequence of intracerebral hemorrhages disappeared gradually. Hemiparesis persisted to various degrees in five

TABLE 25. Total CBF (ml/min)Before and After Stereotactic Clipping of the Feeding Artery of AVMs

Case number	Before clipping		After clipping	
	tCBF	Percentage of normal	tCBF	Percentage of normal
1	2422.7	322	1894.9	253
2	1015	135	708.4	95
3	1647.8	220	1066.1	142
4	1945.3	259	788.2	105
5	980.7	131	807.8	108

TABLE 26. Cerebral Blood Flow (ml/100 g per min) Before and After Stereotactic Clipping of the Feeding Artery of AVMs

	Before clipping		After clipping	
Case number	In the hemisphere with AVM	In the other hemisphere	In the hemisphere with AVM	In the other hemisphere
1	210.4	136.7	167.1	103.6
2	81.4	63.6	51.4	49.8
3	125	110.4	82.2	70.1
4	165.7	114.2	62.5	50.1
5	102	38.1	80.4	35.1

cases. The severe and pulsating headaches disappeared immediately after surgery in all four patients. Speech disturbances were restored completely in one patient and became minimal in two others.

Several illustrative examples of long-term results after stereotactic clipping of AVMs of different size and location are given. Arteriovenous malformations fed by the middle cerebral artery are very common. When the AVM is large and is located in a functionally important cerebral area, radical extirpation may be dangerous and difficult. The following case illustrates successful stereotactic clipping of the main trunk of the middle cerebral artery proximal to an AVM.

A 21-year-old woman suddenly developed severe headaches, increased weakness in the right extremities, and then lost consciousness for 20 hr. On admission to our clinic, paralysis of the right arm was marked, and paresis of the right leg, severe motor aphasia, marked papilledema, and bloody CSF were present.

Angiography demonstrated a large AVM located in the parietofrontal region and supplied by hypertrophied branches of the left middle cerebral artery (Fig. 241A). The AVM measured 4 × 3 × 2.5 cm. Its radical removal from the deep central part of the left hemisphere was undoubtedly connected with a high risk of serious functional disturbances.

Stereotactic clipping of the main trunk of the middle cerebral artery was performed (Fig. 241B). At a distance from the surface of 58 mm, a clip was put on M_1, the diameter of which was 3.5 mm at the target point. A control angiogram disclosed complete elimination of the AVM from the circulation. The postoperative course was uncomplicated. The severe neurological deficit disappeared in 1 month. About 5 years after the operation, the patient was in good condition without any complaints. There was no recurrent bleeding.

The next case also shows that the clipping of a proximal branch of the middle cerebral artery did not cause any neurological deficit.

A 25-year-old patient had a history of two severe subarachnoid and parenchymatous hemorrhages with consequent pronounced right-sided hemiparesis. For many years he had been experiencing as many as three generalized epileptic seizures a month. Four-vessel angiography indicated a giant AVM (volume 153 cm^3), occupying practically the entire left temporal lobe. The malformation was supplied by many branches of the left middle cerebral artery and to a lesser extent by the posterior cerebral artery (Fig. 242A). It was decided that total extirpation of such a large and deep lesion was technically impossible.

Stereotactic clipping of the main feeding branch of the middle cerebral artery near the internal carotid bifurcation was performed and confirmed angiographically.

Intraoperative angiography showed total nonfilling of the AVM (Fig. 242B). There were no complications. The CBF diminished by two-thirds in the affected hemisphere and by one-half in the other hemisphere, where CBF became almost normal.

Six years after surgery the general condition of the patient was good. The neurological deficit diminished substantially. The patient cared for himself and walked independently. During follow-up only two mild epileptic fits had occurred. There were no intracranial hemorrhages.

Arteriovenous malformations supplied mainly by

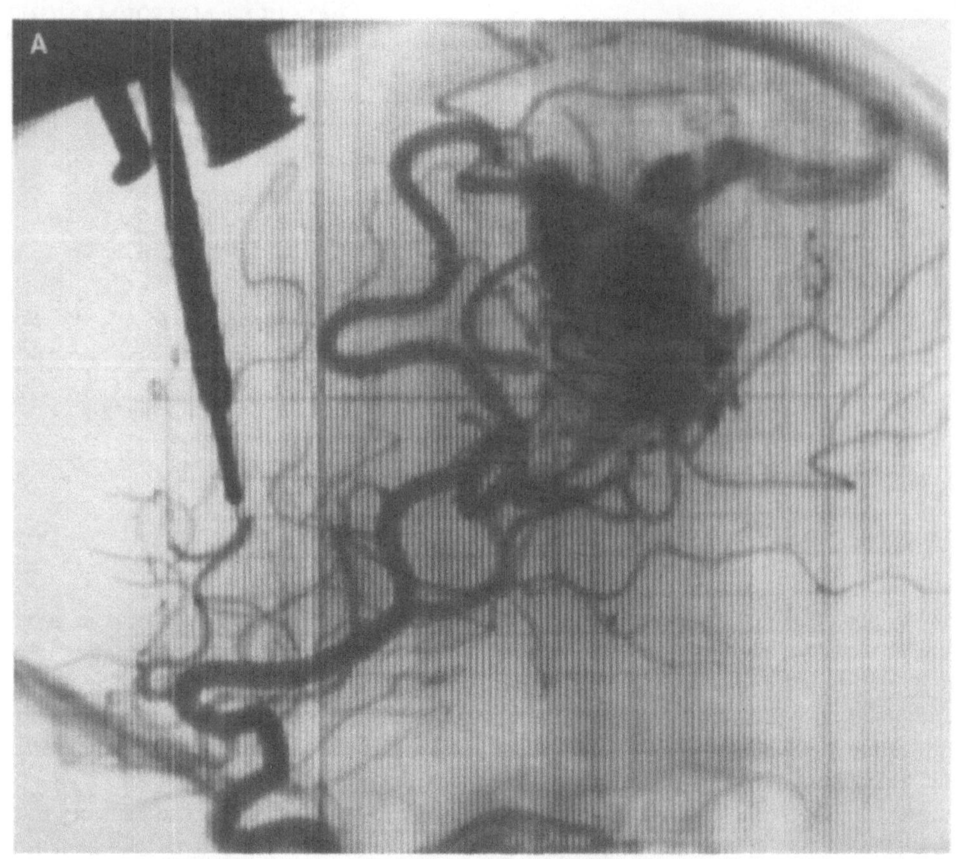

A

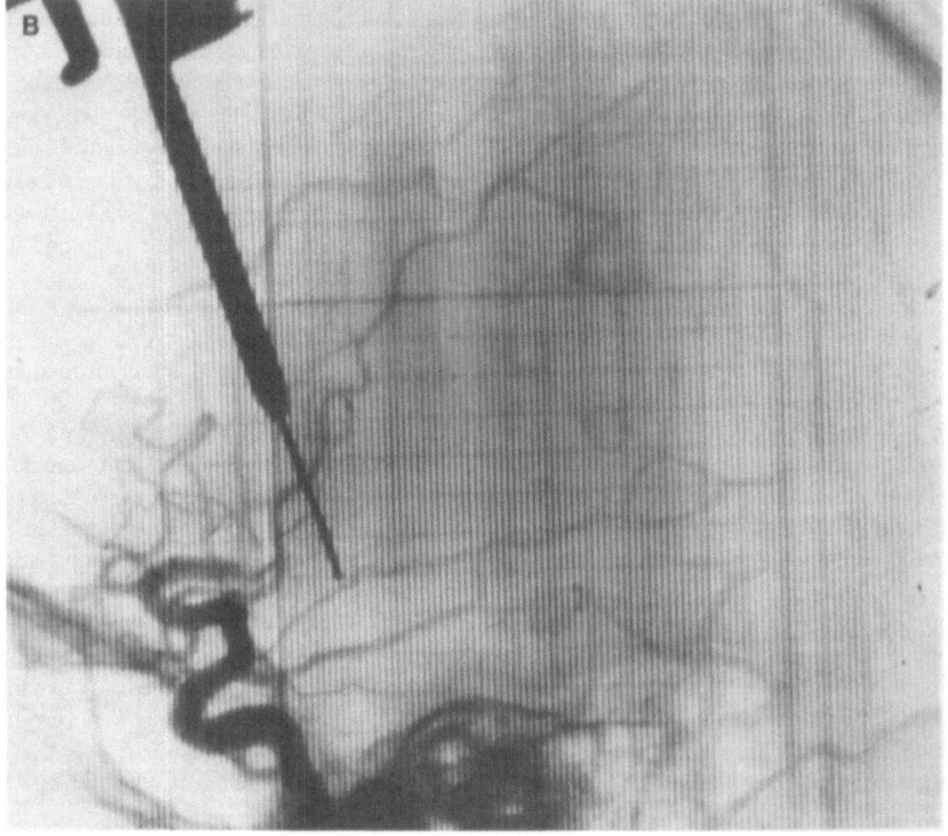

B

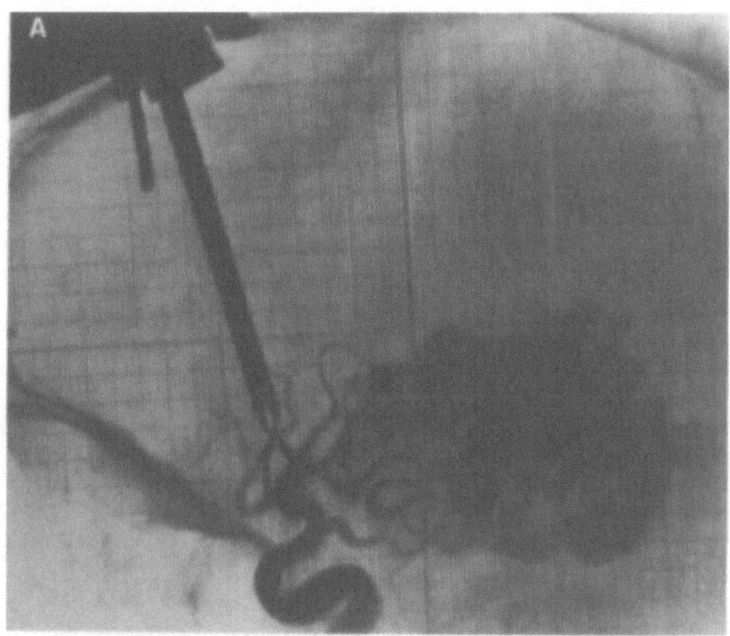

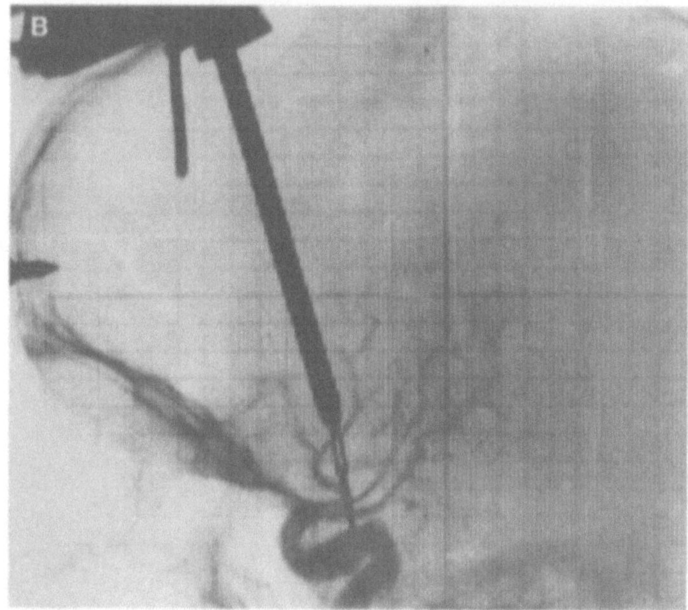

FIGURE 242. A giant AVM occupying the entire left temporal lobe and supplied mainly by many branches of the left middle cerebral artery. Stereotactic clipping of the main trunk of the artery in its proximal segment. Intraoperative angiograms (A) before and (B) after clipping. Deep branches of the middle cerebral artery remain intact.

the pericallosal artery are frequent in neurosurgical practice. That artery may be the sole feeder of the AVM, or other arteries may contribute to the blood supply. These AVMs are frequently located deep in the CC and are often only partially accessible to radical extirpation. The following case illustrates the possibility of clipping the pericallosal artery, which supplied an inoperable AVM of CC and basal ganglia.

A 13-year-old boy had suffered intraventricular hemorrhages over a 4-year period before admission.

FIGURE 241. A large AVM supplied by markedly enlarged branches of the left middle cerebral artery. Stereotactic clipping of the main trunk of this artery led to complete exclusion of the AVM with no increase in the neurological deficit. Intraoperative lateral angiograms (A) before and (B) after clipping.

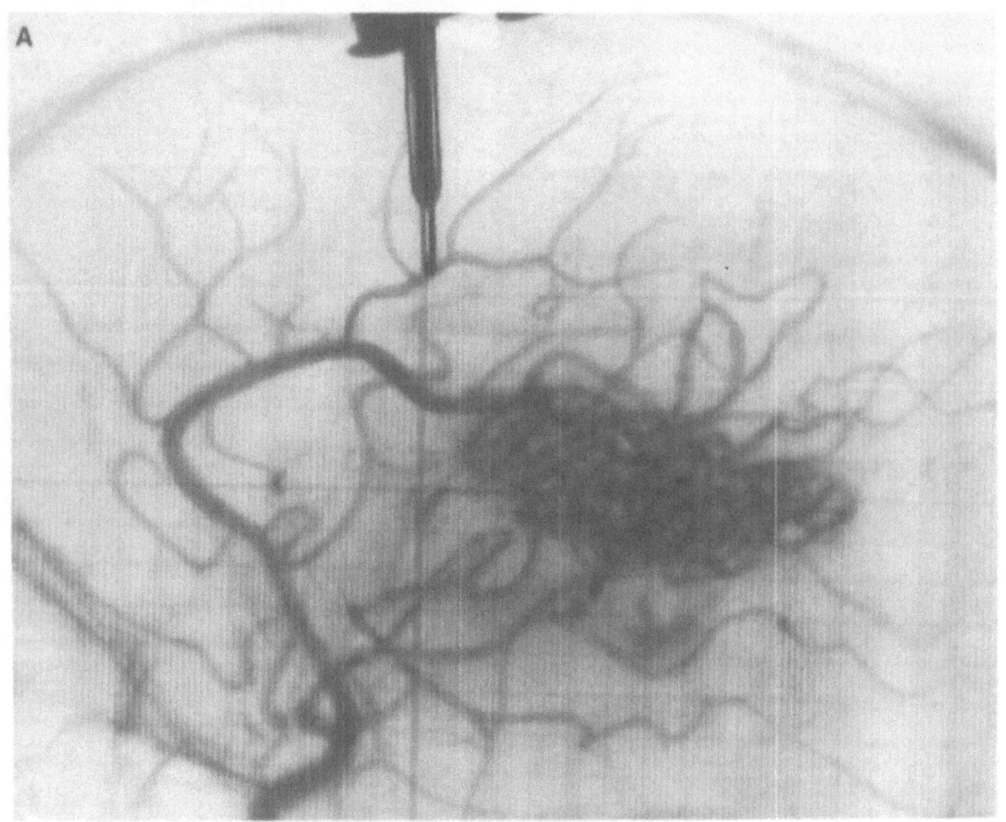
A

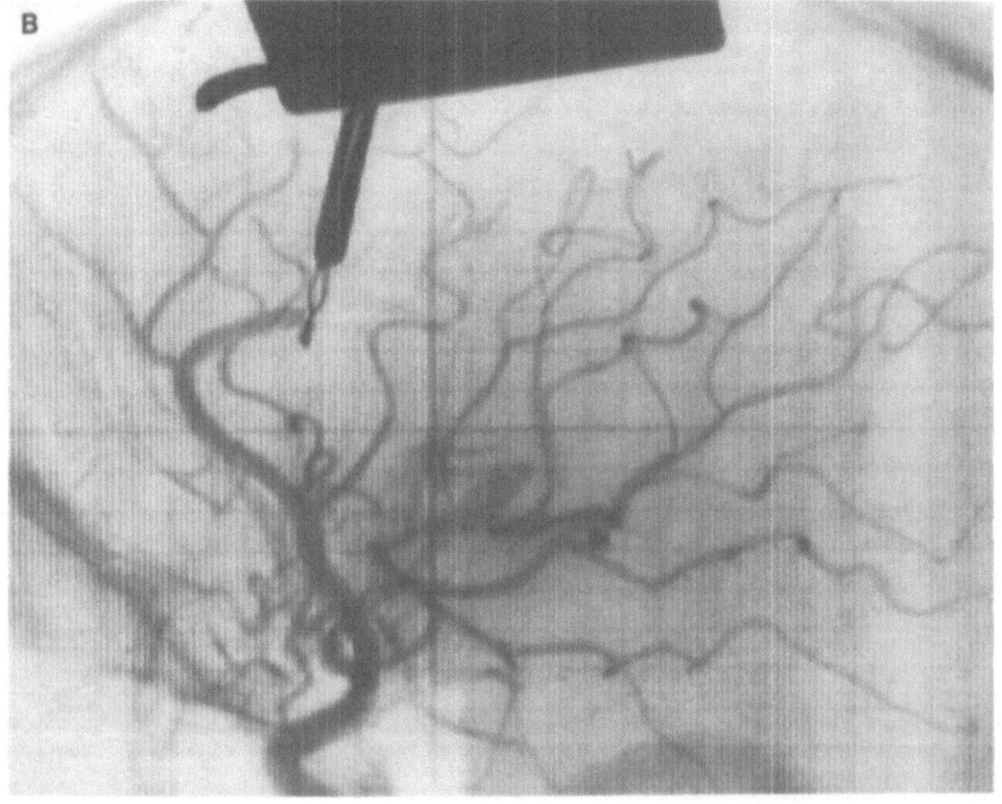
B

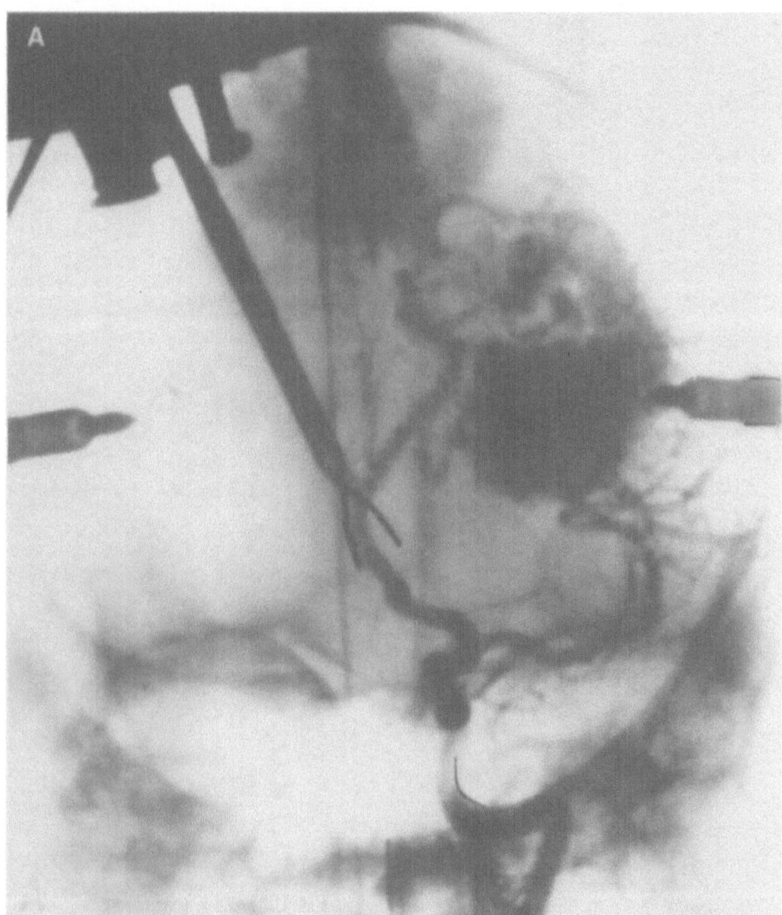

FIGURE 244. Large AVM in the central region of the left hemisphere supplied by hypertrophied left pericallosal and middle cerebral arteries. To avoid damage to large drainage veins, the stereotactic clipping of the left anterior cerebral artery was performed from the right side. Intraoperative AP angiograms (A) before and (B) after clipping. The volume of the AVM is significantly reduced.

He was admitted to our clinic 3 weeks after the last hemorrhage. Angiography disclosed a large AVM involving the CC, the deep medial cerebral structures, and the posterior part of V_3. The aneurysm was supplied mainly by the hypertrophied right pericallosal artery and, to a lesser extent, by branches of the right middle and both posterior cerebral arteries (Fig. 243A). The deep and bilateral location of the aneurysm made it inoperable.

It was decided to clip the main feeder of the aneurysm—the pericallosal artery. Successful stereotactic clipping of the artery was performed. Control angiography showed that the AVM no longer filled through the artery (Fig. 243B). In comparison with the preoperative studies, the CBF markedly decreased. There were no postoperative complications. During 8 years of follow-up, there were no epileptic fits or intracranial hemorrhages. The patient remained in good condition, graduated from secondary school, and is now working in a factory.

In particular cases of AVMs of the anterior cerebral artery, the dangers and technical difficulties are related to the fact that the trajectory of the cannula with the clip is so close to the AVM nidus or great drainage veins that it creates a potential danger of their being damaged. In such a case, we approach the anterior cerebral artery through the contralateral ("healthy") hemisphere, eliminating the possibility of disturbing the AVM vessels. Below we describe a case of such contralateral stereotactic clipping of the anterior cerebral artery.

A 33-year-old woman was admitted to our clinic a

FIGURE 243. A large AVM in deep central part of the right hemisphere supplied mainly by the hypertrophied pericallosal artery. This artery was clipped at a distance from the aneurysm. Intraoperative lateral angiograms (A) before and (B) immediately after clipping. The AVM is nearly completely excluded from the circulation.

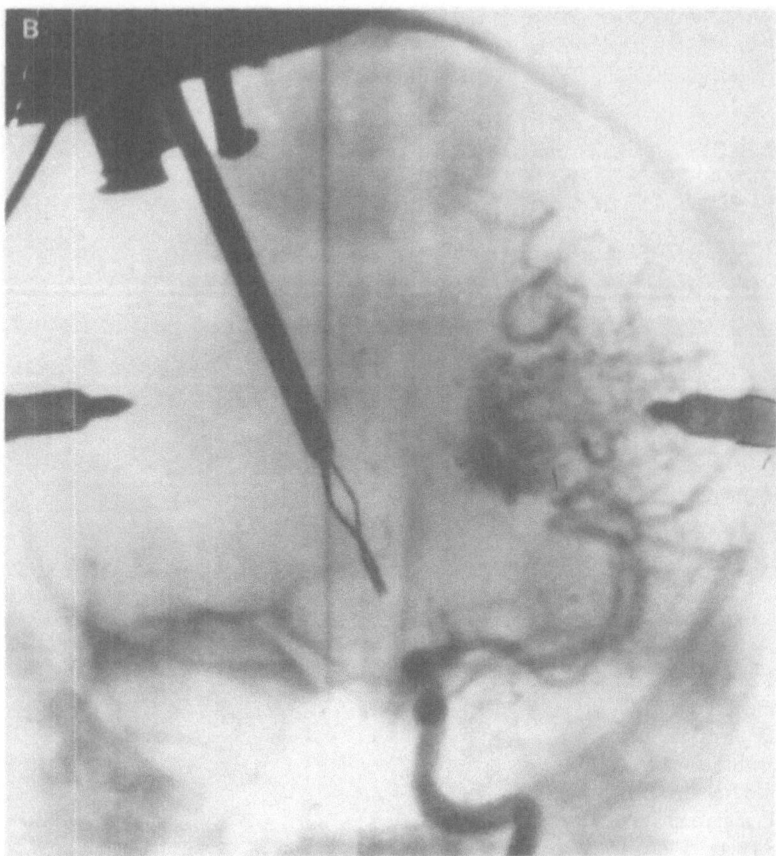

FIGURE 244. *(Continued)*

few months after the sudden onset of a severe headache, loss of consciousness, and coma for about 24 hr. After regaining consciousness, she had paralysis of the right arm and severe paresis of the right leg.

Four-vessel angiography disclosed a large AVM (about 6 cm on a lateral picture) fed by hypertrophied anterior and middle cerebral arteries (Fig. 244A). The malformation was located in the left frontal lobe and was encircled by many large drainage veins. A burr hole was made on the contralateral side in the right parasagittal region near the coronal suture. The left pericallosal artery above the CC, which fed the AVM, was clipped stereotactically. Control angiography showed the filling of only a part of the AVM supplied by small branches of the middle cerebral artery (Fig. 244B). Two and one half years after the operation, the patient's condition is quite good except for a mild right-sided hemiparesis. There were no recurrent hemorrhages.

The treatment of giant inoperable AVMs supplied from both hemispheres remains unsolved. In such cases stereotactic clipping of two or more feeding vessels is only palliative surgical management, but in selected cases, the procedure may produce a satisfactory clinical effect. In the next case involving a giant AVM, bilateral clipping of afferent arteries was performed.

For about 30 years this 45-year-old man had suffered from pulsating headaches, which had become continuous and severe several years before admission. He had had epileptic seizures for about 10 years. At first they were focal but later became generalized and more frequent. A slight left-side hemiparesis developed 2 years prior to admission.

Four-vessel angiography showed a giant AVM (193 cm^3) occupying the right parietal, temporal, and occipital lobes. It was supplied by greatly enlarged arteries from both sides, by the right middle and left anterior cerebral arteries and, to a lesser extent, by branches of the right anterior and posterior cerebral arteries (Fig. 245A). The angioma was evaluated as absolutely inoperable.

Through a burr hole near the coronal suture, ster-

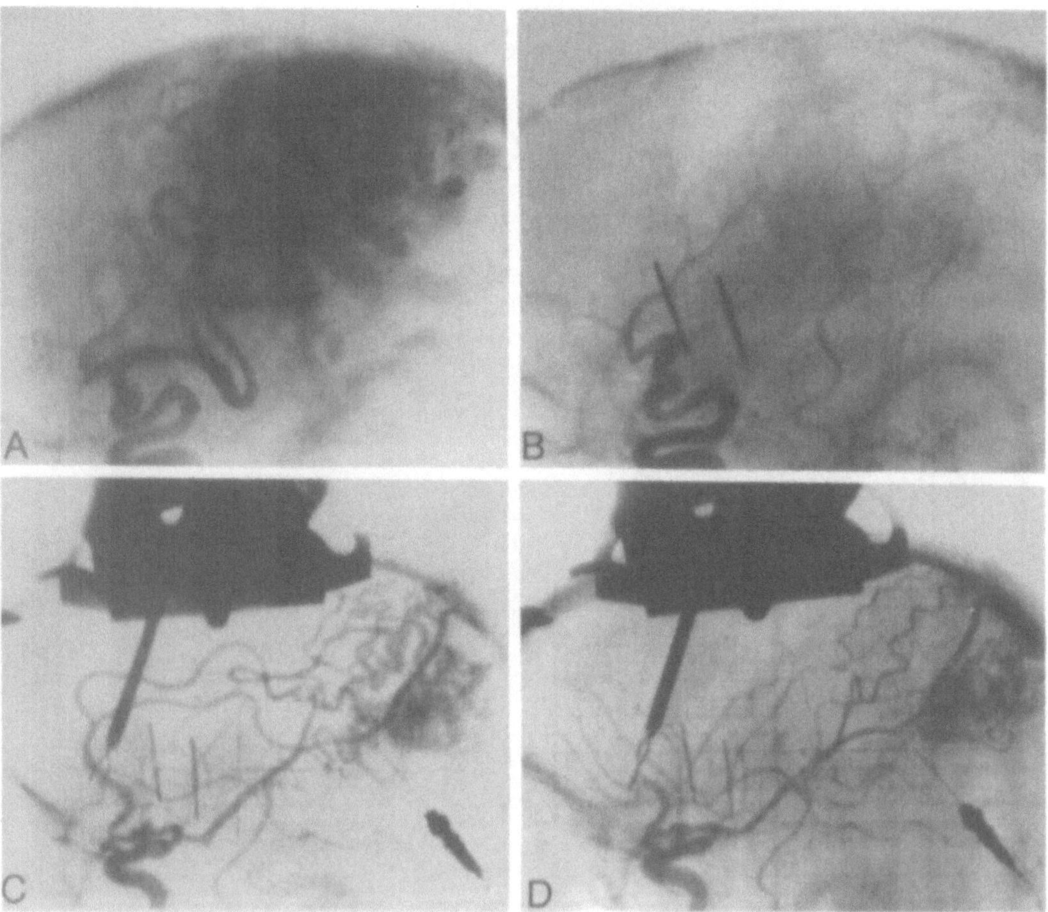

FIGURE 245. (A) Inoperable giant AVM occupying about half of the right hemisphere and supplied by greatly enlarged arteries from both sides. (B) After stereotactic clipping of the two large branches of the right middle cerebral artery, the clipping of the left anterior cerebral artery was performed (C, D, before and after clipping). The volume of the AVM decreased substantially.

eotactic clipping of the enlarged (6-mm) branches of the right middle cerebral artery was carried out (Fig. 245B). Two clips were applied. There were no complications following surgery. Control angiography showed a marked decrease in the size of the aneurysm and of the total and hemispheric CBF. The headaches became less severe.

Six months later, stereotactic clipping of the hypertrophied branch of the anterior cerebral artery (Fig. 245C) was performed on the other side without complications. The clipping was confirmed angiographically, and a consequent substantial decrease in volume was achieved (Fig. 245D). After the second operation, there was a marked clinical improvement: severe headaches and seizures disappeared almost completely in the 6-year follow-up period.

Besides clipping the anterior and middle cerebral arteries, stereotactic clipping of the deep-seated arterial branches supplying AVMs located in the lateral ventricles and paraventricular regions is also technically possible. The following is an example of successful clipping of an anterior choroidal artery feeding a small AVM in the lateral ventricle.

Three years before admission, a 29-year-old man suffered a head injury with subarachnoid hemorrhages from which he appeared to recover. Three years later, a severe subarachnoid–parenchymatous hemorrhage ruptured into the left hemisphere suddenly. A severe right-sided hemiparesis and meningeal irritation were prominent at admission. Lumbar puncture showed bloody CSF.

Angiography disclosed a small AVM in the anterior part of the right lateral ventricle. The AVM was supplied by the anterior choroidal artery, which was

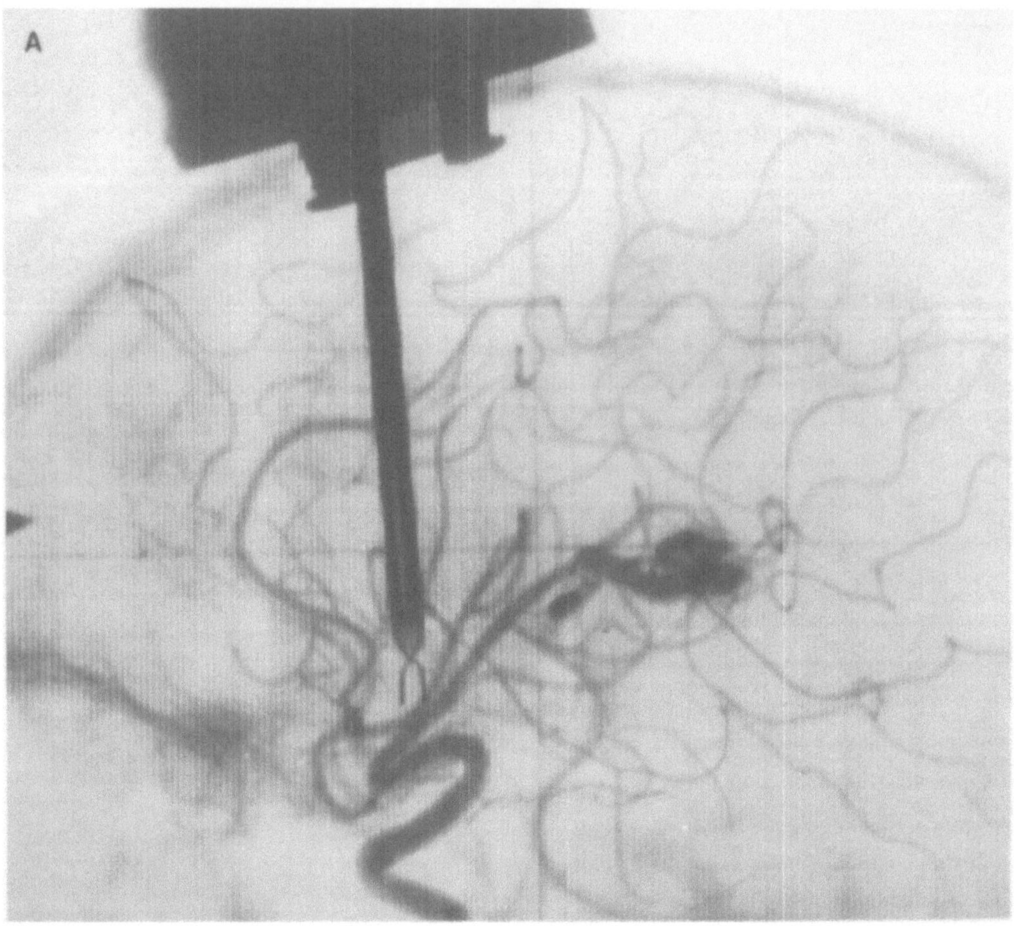

FIGURE 246. A small AVM supplied by a branch of the middle cerebral artery (A) before and (B) after stereotactic clipping of this branch at its origin from the main trunk of the artery. Note the other branch of the same artery, which does not supply the AVM. This branch, located close to the clipped artery, was left undisturbed after clipping.

clipped stereotactically. Intraoperative angiography confirmed the nonfilling of the AVM. Only slight hemiparesis remains 2 years following the operation.

The next case is important for the evaluation of the method. In this patient, clipping was used to manage an AVM that undoubtedly could have been completely removed by classical open surgery. Although AVMs supplied by one feeder are relatively rare, their total extirpation usually may be achieved without great technical difficulties. Nevertheless, to evaluate the possibilities of stereotactic clipping, we used it in a few cases in which the AVM was operable with a radical procedure. It should be emphasized that stereotactic clipping totally eliminated the angiomas from the circulation. An illustrative case follows.

A 26-year-old man was admitted to our clinic several hours after a single subarachnoid–parenchymatous hemorrhage. A severe hemiparesis gradually disappeared. Angiography showed a small AVM (23 cm^3) in the right parietotemporal region with only one small feeding branch arising from the middle cerebral artery (Fig. 246A), which was clipped stereotactically.

There were no postoperative complications. Control angiography 3 weeks after the operation showed the complete exclusion of the malformation from the circulation (Fig. 246B). It is important to note that a second ''normal'' branch of the middle cerebral artery, which was located very close to the first branch supplying the AVM, remained intact. Postoperative CBF studies demonstrated a return to normal figures.

About 8 years after surgery the patient is in good condition without a neurological deficit. He is working full-time. There has been no recurrence of intracranial hemorrhage.

The new technique of stereotactic treatment of cerebral arterial aneurysms and AVMs is undoubtedly

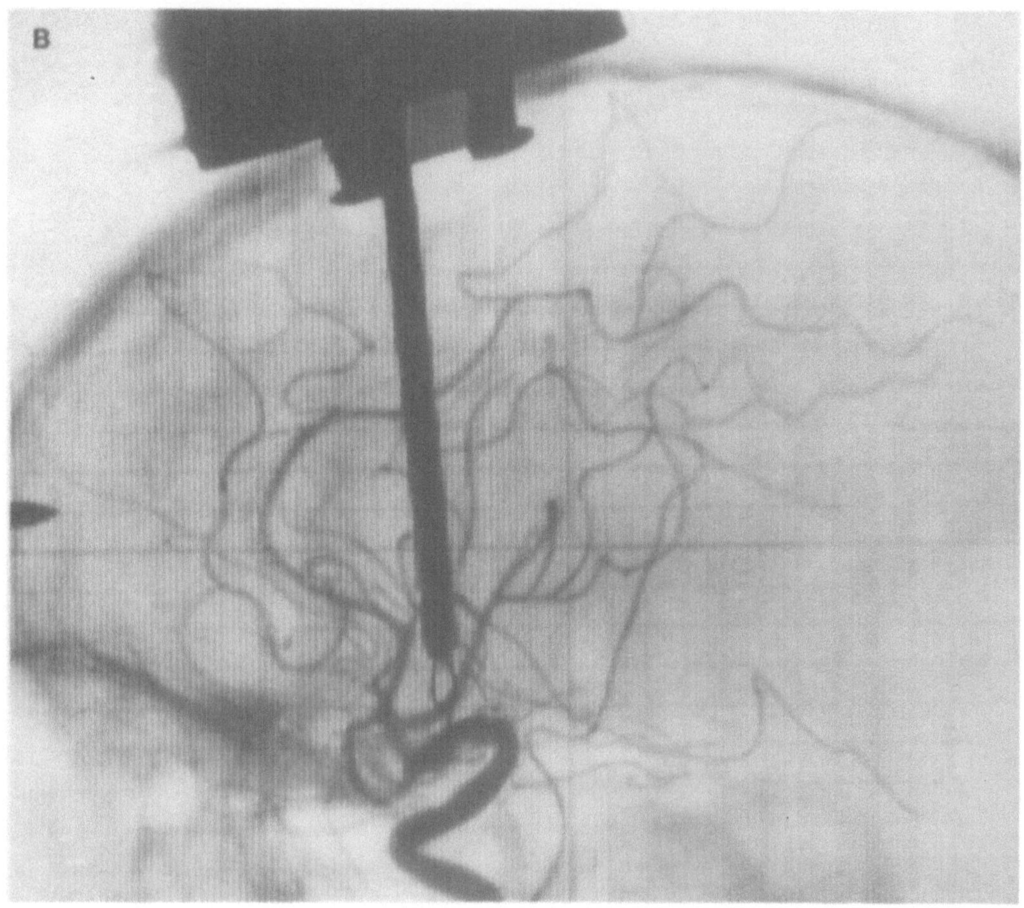

FIGURE 246. *(Continued)*

not universally accepted. As with all new surgical procedures, it is necessary to make prolonged and careful investigations of the indications and possible improvements of the technique. But at this stage of its development, certain important advantages may be considered to be proved: in carefully selected cases the method is less traumatic, decidedly safer than direct surgery by an open approach, has fewer complications, and may be used in the presence of arterial spasm after subarachnoid hemorrhage.

17

Spontaneous Intracerebral Hematomas

1. Introduction

It is generally ackowledged that during the last two decades the incidence of hemorrhagic stroke has substantially decreased because of the successful treatment of hypertensive disease. However, some investigators believe that the incidence of spontaneous intracerebral hematomas has remained stable during the last 20 years, comprising approximately 10% of all strokes (Higgins and Nashold, 1980b).

The surgical treatment of spontaneous (nontraumatic, hypertensive) intracerebral hemorrhages caused by hypertension and cerebral arteriosclerosis has been evolving during the past three decades, yet many important problems remain to be resolved (Lazorthes, 1959; Arutyunov and Pedachenko, 1960; McKissock *et al.*, 1961; Guatico *et al.*, 1965; Pia, 1972; Laine, 1976; Kaneko *et al.*, 1977; Bolander *et al.*, 1983; Kanno *et al.*, 1984). Different and sometimes opposite points of view exist regarding the indications and contraindications for surgery, timing of the operations, surgical techniques, etc. Because of these controversies, the postoperative mortality varies greatly (from 10% to 80%) in different reports. In deeply comatose patients and those with medial (thalamic) hemorrhage, the postoperative mortality is extremely high (up to 80%). At the same time, there are grounds for suggesting that early evacuation of the hematoma may not only increase the chances for saving human life but also improve functional recovery.

The classical method of removing intracerebral hematomas *ad oculus* by craniotomy and encephalotomy is not technically difficult, but as clinical experience has shown, in the most severe cases, the operation is connected with high risk. Many patients do not tolerate open removal of the hematoma. Many neurosurgeons believe that surgery is contraindicated if the patient is in a comatose state.

For many years numerous attempts have been made to remove hematomas radically by a less traumatic technique that would not worsen the patient's grave condition. The simplest way to achieve this is to aspirate the hematoma through a burr hole. But it is known that a few hours after the onset of symptoms, a hematoma consists of liquid blood (about 20% of its volume) and dense clots (about 80%). Many attempts to remove a hematoma through a cannula, even one with a large diameter, were unsuccessful in the past because the evacuation of dense clots was practically impossible. The attempt to aspirate a hematoma through a cannula introduced stereotactically was equally ineffective and not accepted in practice (Benes *et al.*, 1965).

This situation stimulated the search for less traumatic, better tolerated, and safer methods of removing intracerebral hematomas. The development of CT scans in neurosurgery opens up new possibilities for using stereotactic techniques in the treatment of spontaneous intracerebral hemorrhages. A new principle for the stereotactic evacuation of hematomas was proposed by Backlund and von Holst (1978) and used successfully in one case.

The authors used a relatively simple instrument consisting of a cannula 4 mm in diameter surrounding a mandrel not unlike Archimedes' screw, which the surgeon must rotate by hand. The propellerlike action of the screw dislodges and breaks up the thick blood clots. A conventional surgical aspirator is connected with the outer end of the cannula. The small pieces of clots go into the trap bottle that is in the aspirator line.

This original device of Backlund and von Holst had some substantional drawbacks. The manual rotation of the screw is not convenient, and the procedure is practically impossible to perform for any length of time.

Broseta *et al.* (1982a) used the above-mentioned device in 16 patients with hypertensive hematomas. Eleven were in a comatose state, and in nine, the blood had ruptured into the ventricles. The postoperative mortality was extremely high (81%); two patients died from recurrent hemorrhage.

Higgins and Nashold (1980a,b) and Higgins *et al.* (1982) have modified the device by adding a thin tube along the cannula. The tube can maintain suction when the cannula is plugged by clot pieces. It is also possible to inject a saline solution through the tube into the hematoma cavity. This device was used by the authors for operating on three patients, two of them twice. It was shown that after stereotactic evacuation of the intracerebral hematomas the cerebrospinal fluid increased significantly in the subcortical structures of the affected as well as in the "healthy" hemisphere (Tanizaki *et al.*, 1985). The successful stereotactic evacuation of intracerebral hematomas in three cases has been described by Hondo *et al.* (1984).

2. Improved Technique

For stereotactic evacuation of intracranial hematomas, the author and his co-worker Peresedov proposed a new and improved technique (Kandel and Peresedov, 1980, 1985, 1987a).

The main component of our device is a stainless steel cannula 17 cm long and 4 mm in outer diameter, opening at the distal end (Fig. 247). The cannula is introduced into the hematoma cavity with the aid of the stereotactic apparatus (Fig. 248). There are two interchangeable pivots for the cannula. The first is a metallic pin with a blunt tip. The second is a thin stainless

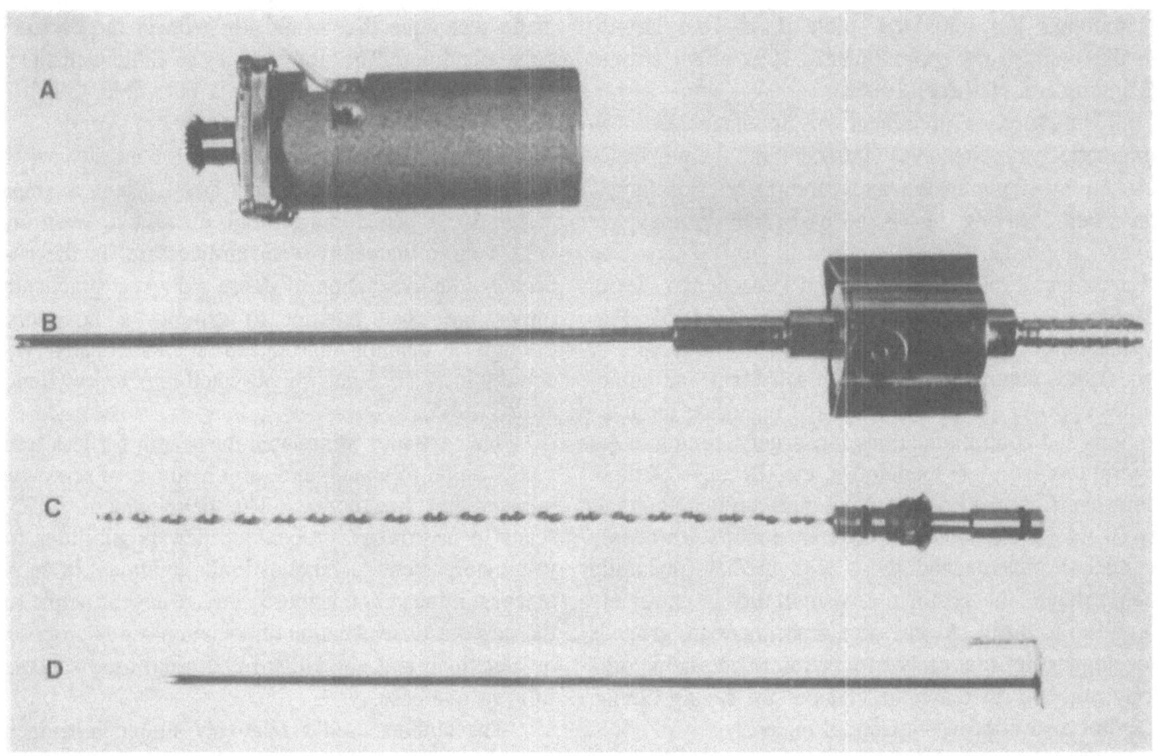

FIGURE 247. Components of our device for stereotactic evacuation of intracerebral hematomas. (A) Small engine for revolving the spiral; (B) metallic cannula with transmission block and connector for a plastic tube going to the aspirator; (C) rotated spiral in the form of an Archimedes' screw for destruction of dense blood clots; and (D) guide stock for hematoma puncture.

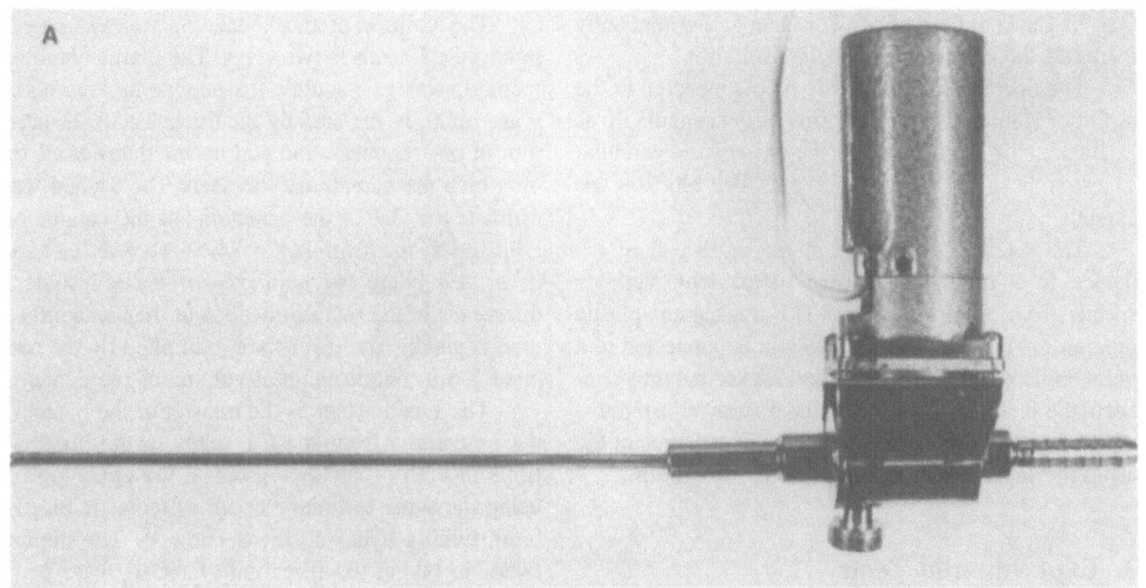

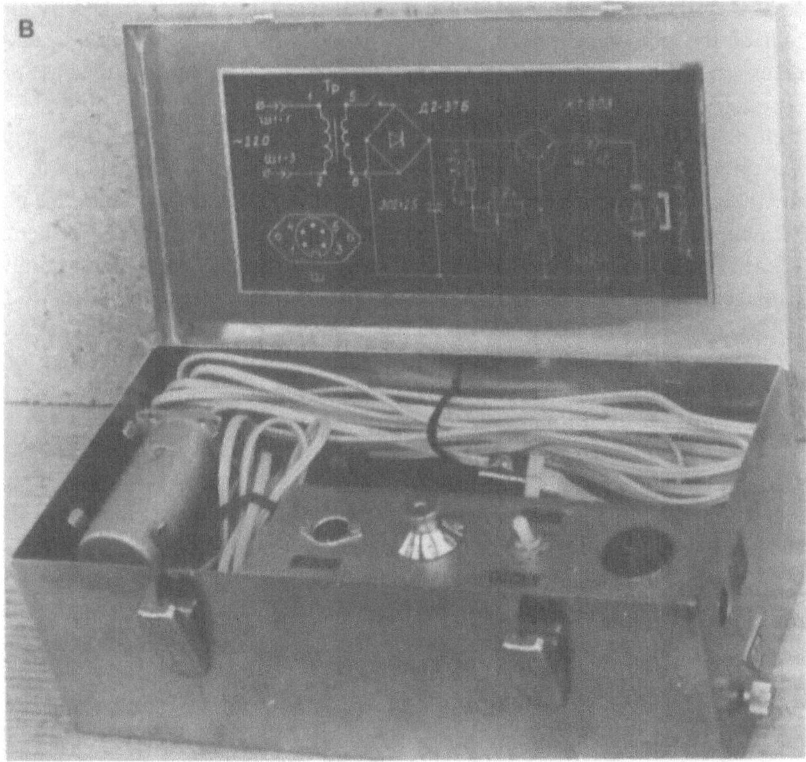

FIGURE 248. (A) Assembled instrument for stereotactic evacuation of intracerebral hematomas and (B) carrying case with regulating knobs for current parameters.

steel Archimedes' screw to break up densely coagulated clots. The tip of the screw is 1.5 mm shorter than the open end of the cannula (see below) and should fill only 7% of the space inside the cannula, much less than in Backlund's device. This thin screw markedly increases the effectiveness of the aspiration.

The opening for the plastic tube connected to the aspirator is located not on the side of the cannula tip as in Backlund's device but on the axis of the cannula. The straight suction of clots prevents their blocking the cannula.

The screw is connected at the outer end of the device to a miniature electric motor with variable speeds from 50 to 200 rpm. The transparent plastic tube on the outer end of the cannula is connected to a glass bottle graduated in milliliters. A second tube connects the bottle with a conventional surgical aspirator, which removes the fragmented clots. The degree of the aspirator vacuum may be changed up to −2 atm.

3. Experimental Tests

Before its clinical use, our device was tested in several series of animal and technical experiments. The aim of the first experiments on rabbit brain was to determine the optimal rotational speed of the screw and optimal vacuum of the aspirator. Small trephinations were made under general anesthesia, the dura mater was incised, and the cannula tip was introduced into the rabbit brain for 5–6 mm. With less than −0.5 atm vacuum on the aspirator, there is no aspiration of brain tissue. Screw speed from 60 to 200 rpm does not damage tissue, but the optimal speed is about 100 rpm.

The device was also tested on human blood stored without preservatives for 2–7 days to investigate the optimal parameters for the suction removal of clots. The optimal speed of rotation for denser clots is up to 120 rpm, for less dense clots, it is about 80 rpm. A vacuum of approximately −0.2 atm is optimal.

In a second series of experiments, it was shown that the optimal result is obtained when the screw tip is 1.5–2 mm shorter than the cannula tip. With that arrangement, there is no damage to the brain tissue during aspiration, and the suction is most effective.

4. Preoperative Calculations

Before the operation we determined on CT scans the size and location of the hematoma, its x-ray density, the degree of brain edema, and the penetration of the blood into ventricles. To exclude the presence of an arterial aneurysm or AVM, angiography was also performed.

The volume of the hematoma was estimated on Polaroid CT scans in two ways. The first, a planimetric method, was to calculate the hematoma area on each scan, multiply the area by the thickness of the scan (8 mm in our scanner), and add up the data for all scans on which the hematoma was seen. The second was to estimate the size of the hematoma as the volume of an ellipsoid by the formula $V = (4/5\ \pi)\ a \times b \times c$, where a, b, and c are the hemiaxes of the ellipsoid. The difference in the two estimations of the hematoma volume is small. The results are multiplied by the coefficient 3.5 to obtain the real volume of the hematoma.

The second stage is the transfer of the contours of the hematoma from the CT scans to the plain x-ray films in both projections taken in the operating room using the same technique as for stereotactic biopsy of brain tumors (Chapter 15, Section 3). The important factor in taking the films is to have the base perpendicular to the scanning planes of both films. To achieve this objective, the base plane of the CT investigation was checked by aligning metallic markers attached to the patient's head. Using TV control, it is not difficult to adjust the base plane perpendicular to both stereotactic plain films.

The next step is to select the target point (or points) within the hematoma on the scans for aiming the cannula. We chose a point 1–2 cm posterior to the center of the hematoma, presuming that during the stereotactic aspiration with the patient in supine position, the clots would sink to the cannula. In the presence of a very large hematoma, we sometimes aspirated from two or three parts of the hematoma, in which case we selected several target points on the plain films.

Selection of the burr hole location depended on the site and volume of the hematoma. As a rule, we made the burr hole near the coronal suture. In deep-lying hematomas, the analysis of the brain structures on the cannula trajectory was made according to the Schaltenbrand and Bailey stereotactic atlas as described in Chapter 16.

5. Operative Technique

As a rule, the operation is performed under endotracheal anesthesia, but sometimes under neuroleptanalgesia. In deeply comatose patients, it is possible to carry out the operation under local anesthesia.

Before the operation, the sagittal line and the point where it crosses the coronal suture are marked on the patient's head. The burr hole site is measured from both marks. The patient's head, in a supine position, is put on the special headrest and fixed in the stereotactic frame. After a skin incision 4 cm long, the burr hole is made using a trephine and burr 25 mm in diameter. After incision of the dura mater, the stereotactic apparatus described in Chapter 3 (Section 5.9) with the aspiration device attached is inserted in the burr hole. Through a second burr hole, made on the other side near the coronal suture, a catheter is introduced into V for continuous monitoring of intraventricular pressure during and after the operation. Plain films in both projections are taken according to the routine stereotactic technique.

Stereotactic calculations, described in Chapter 4 (Section 5) are carried out on both films, and the necessary corrections in degrees are made on the arches of the stereotactic apparatus. The aspirating cannula with its stylet is advanced to the target point in the preselected part of the hematoma. After reaching the target, the stylet is removed from the cannula and replaced by a small-bore catheter. If a test aspiration with an ordinary syringe draws dark liquid blood, the catheter is removed, and the previously described Archimedes' screw is introduced into the cannula.

The next step is to decrease the usually high intracranial pressure by withdrawing CSF from the ventricle on the other side while controlling intraventricular pressure to a level of 15–20 mm Hg. (Fig. 249).

The electric motor and aspirator are switched on using a preselected speed and vacuum level. The above-mentioned experiments have shown that the optimal speed of the motor is about 100 rpm and aspirator vacuum about −0.2 atm. Dark fragmented clots pass through the transparent tube into the graduated vessel at a speed of about 2 ml/min. The effectiveness of hematoma evacuation may be judged by the quantity of liquid blood and clots in the vessel (Fig. 250).

Depending on the speed of clot removal, it is possible to vary the speed of the motor and the negative pressure. When the quantity of clots removed is a few milliliters less than the estimated volume of the hematoma, the aspiration may be stopped. After the hematoma removal has been completed, one should wait about 10–15 min to confirm that there is no fresh bleeding, a complication that we did not encounter.

Unlike Backlund, we usually tried to evacuate the hematoma totally, leaving only several millimeters of blood clot in the cavity. If for some reason it was not possible to remove the hematoma completely, or if recurrent hemorrhage occurred in the postoperative period, the above-mentioned procedure may be repeated.

We see no need to inject air into the cavity of the partially removed hematoma or into the lateral ventricle to control the degree of evacuation, as proposed by Higgins and Nashold (1980a,b). As one can see in the following figures (251, 252, 256), air spontaneously enters the cavity of the hematoma as its content is

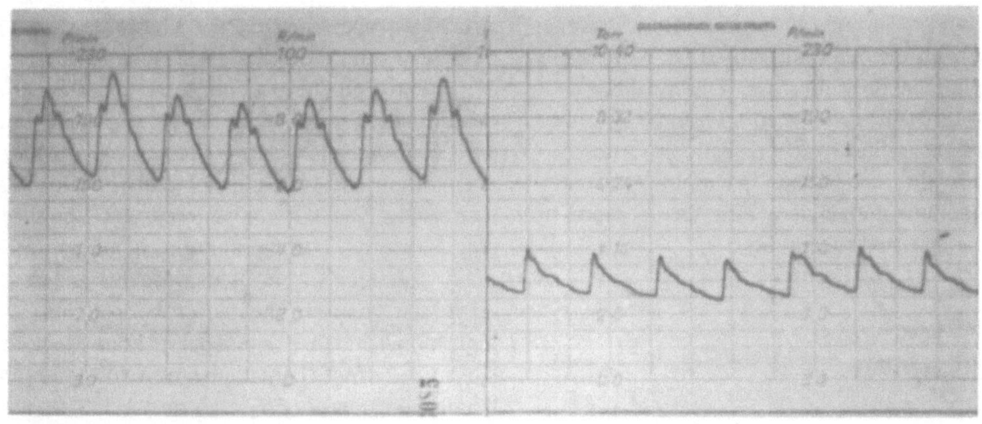

FIGURE 249. Pressure monitoring in the contralateral ventricle in a patient with a large intracerebral hematoma (volume 65 ml). Left: High intraventricular pressure at the beginning of the operation (80 mm Hg). Right: The pressure 2 hr after stereotactic aspiration of the hematoma (25 mm Hg).

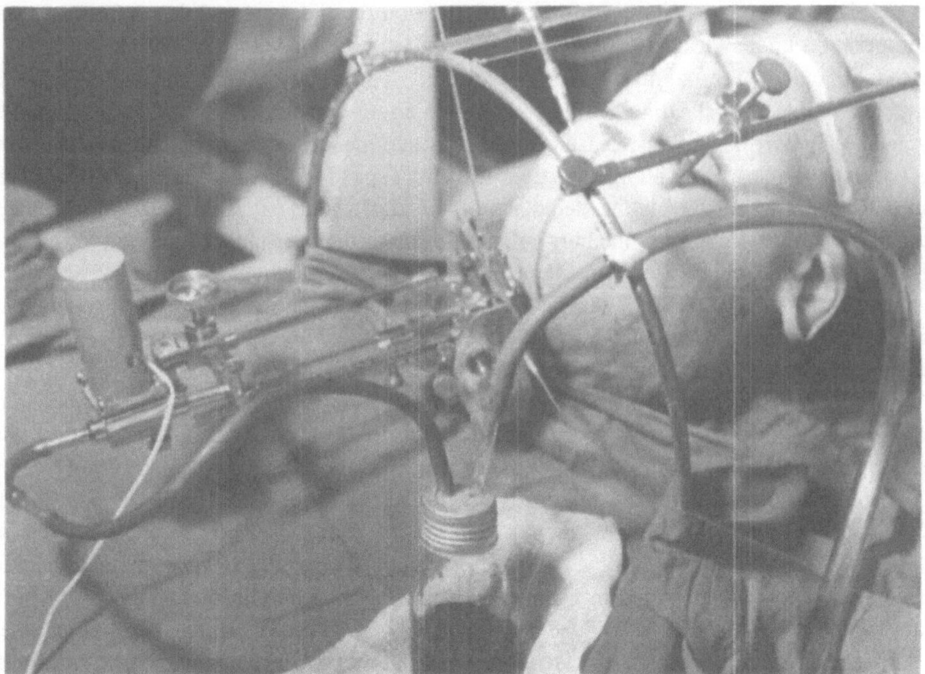

FIGURE 250. General view of the operative setup during stereotactic evacuation of intracerebral hematoma. The device is attached to the stereotactic apparatus. Blood clots from the hematoma are aspirated into the glass bottle. The volume of evacuated clots can be determined by graduations on the bottle and compared with the volume of the brain hematoma calculated from the preoperative CT scans.

evacuated and is clearly visible both on CT scans and ordinary x rays.

The screw is removed from the cannula and replaced by a balloon catheter, which is inflated and left in the hematoma cavity for manometric readings. The cannula is withdrawn from the brain, and the stereotactic apparatus is removed. The balloon catheter is fixed to the skin, and the wound is sutured. The catheter is left in place with the outer end closed for 3–4 days to control the intracavitary pressure. During this period the pressure is gradually reduced by removal of a few milliliters of the saline from the balloon. After 3–4 days of observation, the balloon is withdrawn. Control CT scans are made immediately after the operation on the following and subsequent days.

6. Clinical Results

During the last 4 years we used the method described here in 56 patients (30 men and 26 women) aged 27 to 64. Spontaneous intracerebral hemorrhages caused by arterial hypertension and cerebral arteriosclerosis were observed in 54 cases, and aneurysmal intracerebral hematomas in two cases (one case of arterial aneurysm and one case of AVM). In four cases, stereotactic aspiration was performed twice (altogether, 60 operations). Stereotactic aspiration was carried out at different times after the onset of the stroke: 43 operations in the first 3 days, 14 in 4–9 days, and three after 3 weeks.

Nearly all the patients were admitted in serious condition: 24 were stuporous, and 28 were comatose. Only four patients were in relatively satisfactory general condition. All the patients rapidly developed a grave neurological deficit: complete or near complete hemiplegia, hemianesthesia, aphasia, hemianopsia, etc. In many patients, severe cardiovascular and respiratory disorders occurred. The majority of patients (92%) had high blood pressure and signs of brain stem compression. In 32 cases, hematomas were located in the right hemisphere, and in 24 cases, in the left.

Preoperative CT scans disclosed hematomas of mixed localization (basal ganglia, thalamus, and internal capsule) in 23 cases, lateral (globus pallidus and putamen) in 28 cases, and medial (thalamic) in five cases. Massive extravasation of blood into the ventricular system was noted in 38 cases. Brain edema was present in varying degrees.

Based on the above-mentioned methods of measurement, the hematoma volume varied from 24 to 125 ml (less than 30 ml in 7 cases, 30–50 ml in 16 cases, and more than 50 ml in 33 cases).

In all but seven cases, the hematomas were removed totally as confirmed by CT control immediately after surgery. It is important to note that in the majority of cases it was possible to remove practically all the clots from the ventricles. The patients' condition in the postoperative period was in general serious, as before the operation. All patients required long intensive care, and some required supportive respiratory therapy.

In the few days following surgery, intracerebral hemorrhages recurred in seven patients, in whom the original total removal of the hematoma was confirmed by CT. Four patients were reoperated on by the same method, and two of them survived.

There were 12 fatal outcomes (21%) after the operations. All but one patient had been comatose before the operation. The causes of death were recurrent hemorrhage in five cases, pulmonary thromboembolism in three, renal insufficiency in two, and myocardial infarction in two. Eight patients of the 12 who died were operated on early, within the first 3 days of stroke onset. It is interesting to note that fatal outcome from recurrent hemorrhages occurred in patients operated on within 3 days of the stroke, but from pulmonary embolism after operations performed from 3 to 7 days after the ictus.

A follow-up of 44 patients after 2 months to 5 years has shown that in 39 cases the neurological deficit disappeared or was ameliorated to different degrees. In 7 patients practically complete recovery was observed: patients were employed full- or part-time or worked at home. Thirty-two patients with hemiparesis can walk with the help of a cane and are relatively independent. Five patients are bedridden. There was not a single case of recurrent hemorrhage during the follow-up.

The following two cases illustrate the results of total stereotactic evacuation of intracranial hematomas.

A 51-year-old woman, suffering from arterial hypertension for many years, was admitted to our clinic about 20 hr after the development of deep stupor with right-sided hemiplegia, aphasia, and brainstem signs. The CSF was quite bloody; CT scans disclosed a large hematoma (volume 65 ml) in the left hemisphere lateral to CI (Fig. 251A). Complete stereotactic aspiration of the hematoma was carried out (Fig. 251B). Five days after surgery, active movements in the right extremities appeared, and the patient began to speak a few words. Relatively rapid neurological recovery took place over the next 2 months.

One year after the operation only a slight hemiparesis and mild sensory aphasia remained. The patient could walk with the help of a cane, and perform housework. On control CT scans, a small cystic strip at the site of the hematoma was noted (Fig. 251C).

A 58-year-old woman was admitted in comatose condition with left-sided hemiplegia and signs of secondary brainstem involvement. She had suffered for many years from arterial hypertension and cerebral arteriosclerosis. A CT scan revealed a medium-sized (35-ml) hematoma in the right hemisphere lateral to CI (Fig. 252A). The day after the stroke the hematoma was removed by stereotactic aspiration (Fig. 252B). Six hours following surgery, the patient regained consciousness and began to move the left arm and leg. Three months after surgery, the strength and mobility of the left extremities were restored completely. The patient walked without assistance and is now fully independent.

It is known that medial (thalamic) hemorrhages have an extremely high mortality (up to 80–90%). The following case illustrates a good recovery after stereotactic removal of a thalamic hematoma.

A woman aged 53 had suffered from arterial hypertension for many years. She suddenly developed right-sided hemiplegia and aphasia and was admitted to our clinic in a deeply stuporous state.

Our CT investigations disclosed a relatively small hematoma (25 ml) in the left Th with involvement of the CI (Fig. 253A) as well as extensive bleeding into the lateral and third ventricles. Stereotactic evacuation of the hematoma and partial removal of the blood in the lateral ventricle were carried out on the third day following the stroke (Fig. 253B). Two weeks after surgery, movements of the right extremities and spontaneous speech began to return.

Two and a half years after surgery, only a mild right-sided hemiparesis and slight motor aphasia remained. The patient was able to walk with the aid of a cane.

The following case illustrates the successful stereotactic aspiration of a recurrent hemorrhage shortly after the first operation on a patient in critical condition.

A woman aged 42 with a long history of renal

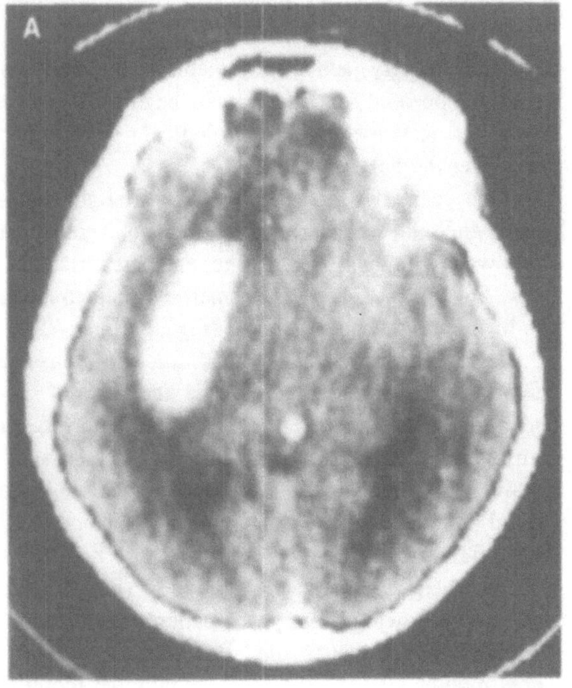

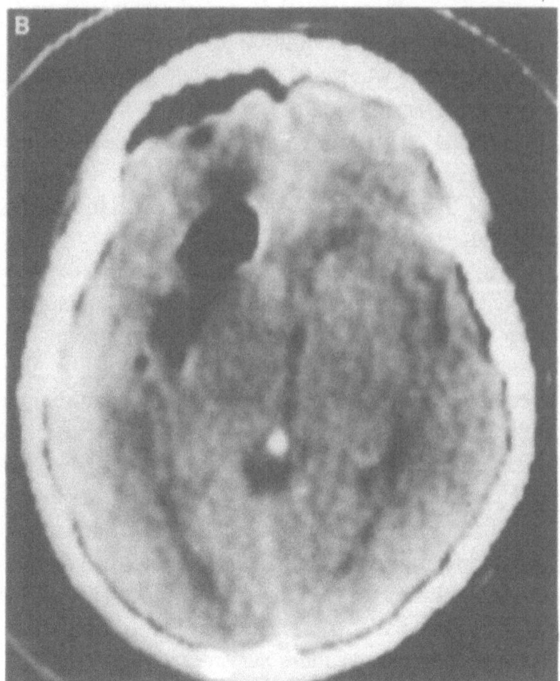

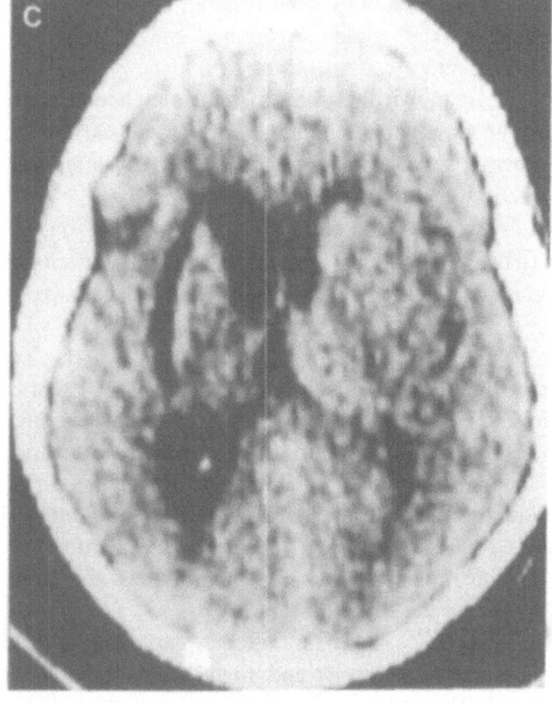

FIGURE 251. Computed tomographic scans in a 51-year-old woman with a large (65-ml) intracerebral lateral hematoma in the left hemisphere (A) before stereotactic aspiration—the ventricular system is not visualized—and (B) immediately after surgery—the cavity of the removed hematoma is filled with air. (C) After one year a small cystic strip at the site of the hematoma is noted.

arterial hypertension with recurrent crises was admitted to our clinic in severe coma with fixed pupils and bilateral extensor plantar reflexes. A very large hematoma (62 ml in volume) was visible on CT scans in the region of CI and adjacent structures in the right hemisphere. The hematoma was removed completely using stereotactic technique 18 hr after the onset of stroke.

The patient's condition improved rapidly; she regained consciousness and responded to simple commands. Three days later, after a brief rise of blood

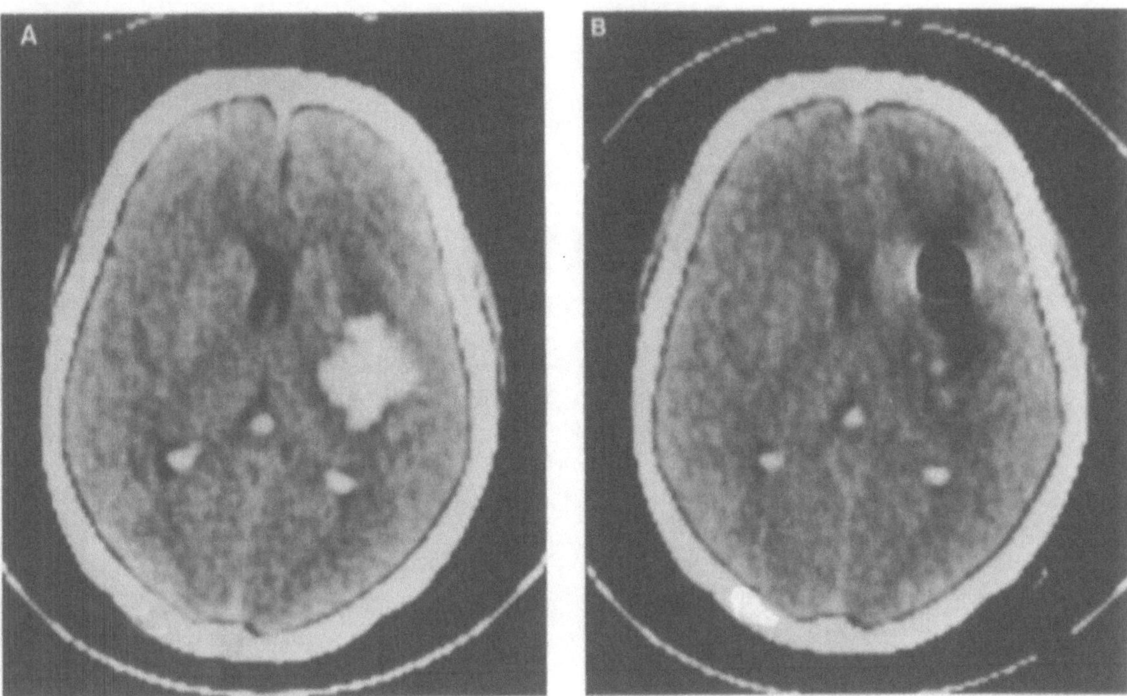

FIGURE 252. Computed tomographic scans in a 58-year-old woman with a medium-sized (35-ml) intracerebral hematoma in the right hemisphere (A) before surgery and (B) immediately after complete stereotactic evacuation of the hematoma.

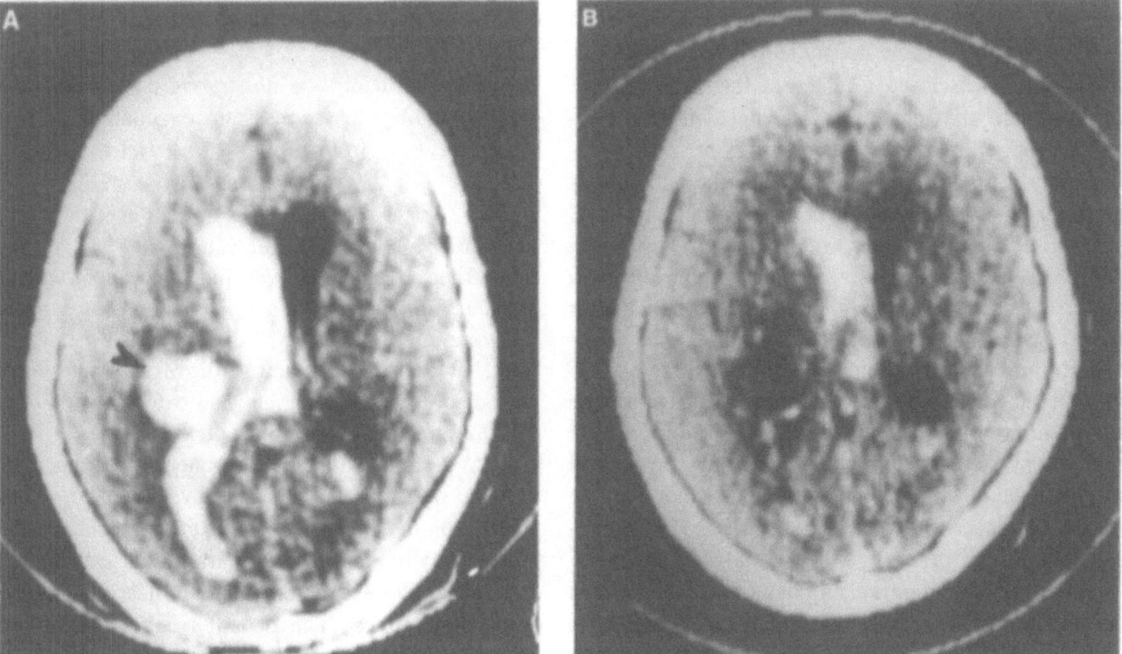

FIGURE 253. Computed tomographic scans in a 53-year-old woman. A relatively small medial thalamic hematoma (arrowhead) ruptured into the ipsilateral ventricle, which is completely filled with blood. (A) Before surgery and (B) shortly after complete evacuation of the hematoma and the blood from the posterior half of the lateral ventricle.

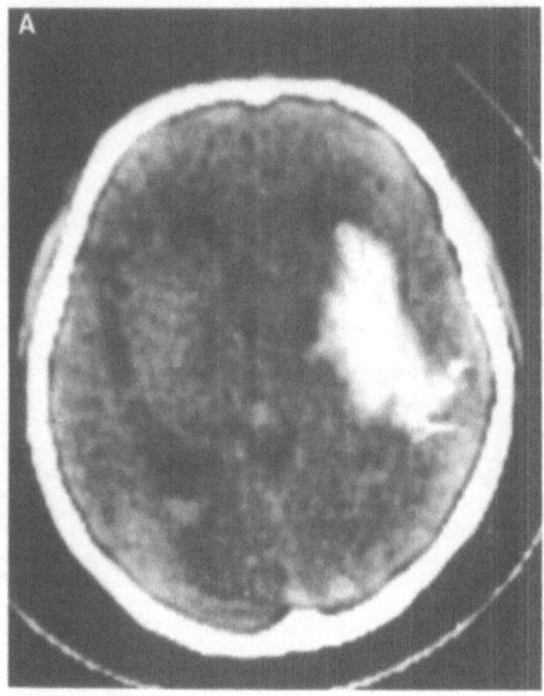

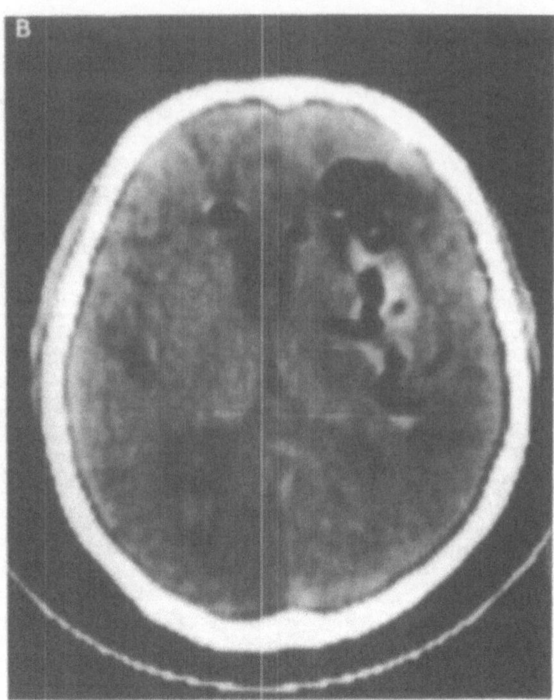

FIGURE 254. Computed tomographic scans in a 42-year-old patient with a large recurrent intracerebral hematoma. (A) The hematoma at the same site 3 days after its initial stereotactic evacuation. (B) Several hours after practically complete stereotactic reevacuation of the hematoma.

pressure to 210/130 mm Hg, her condition worsened rapidly, and she became deeply stuporous. Repeat CT investigations disclosed a recurrent hematoma of the same volume and location. A second complete stereotactic aspiration of the hematoma was performed immediately (Fig. 254). On the next day the patient was conscious but required artificial respiration for about 3 weeks. Her condition improved gradually. Two and a half years after surgery the patient has quite a severe hemiparesis but can walk with assistance and look after herself.

Higgins and Nashold (1980a,b) expressed the opinion that intracerebral hematomas caused by the rupture of arterial aneurysms and AVMs should not be removed by stereotactic methods. We have performed only two such operations but believe that they may be carried out successfully in selected cases. The operation is indicated primarily when a one-stage open removal of the hematoma and clipping of the aneursym is too dangerous or impossible because of the serious condition of the patient, or if pronounced brain edema is present. Undoubtedly, stereotactic aspiration of the hematoma must be performed extremely carefully so as not to trigger repeated bleeding. We believe that in such cases the hematoma should be removed only partially.

The following two illustrative cases demonstrate the possibilities of stereotactic evacuation of hematomas resulting from rupture of arterial aneurysms or AVMs.

A 27-year-old man was admitted in critical condition after a sudden intracerebral and subarachnoid hemorrhage. The patient was in a deep stupor with left-sided hemiplegia and a marked brainstem syndrome ("swimming movements," divergence of the eyes, bilateral Babinski signs, and cardiovascular and respiratory disorders). The patient had a tracheostomy, and artificial respiration was introduced.

Angiography disclosed an aneurysm of the middle cerebral artery, and CT scans revealed a large hematoma (volume 65 ml) deep in the right hemisphere. After a few hours, 35 ml of dark blood and clots were stereotactically evacuated. The clot near the aneurysm was not removed. The patient's condition improved gradually. One month later he began to walk. Angiography showed that the aneurysm did not fill. This was an extremely rare case of spontaneous thrombosis of an arterial aneurysm.

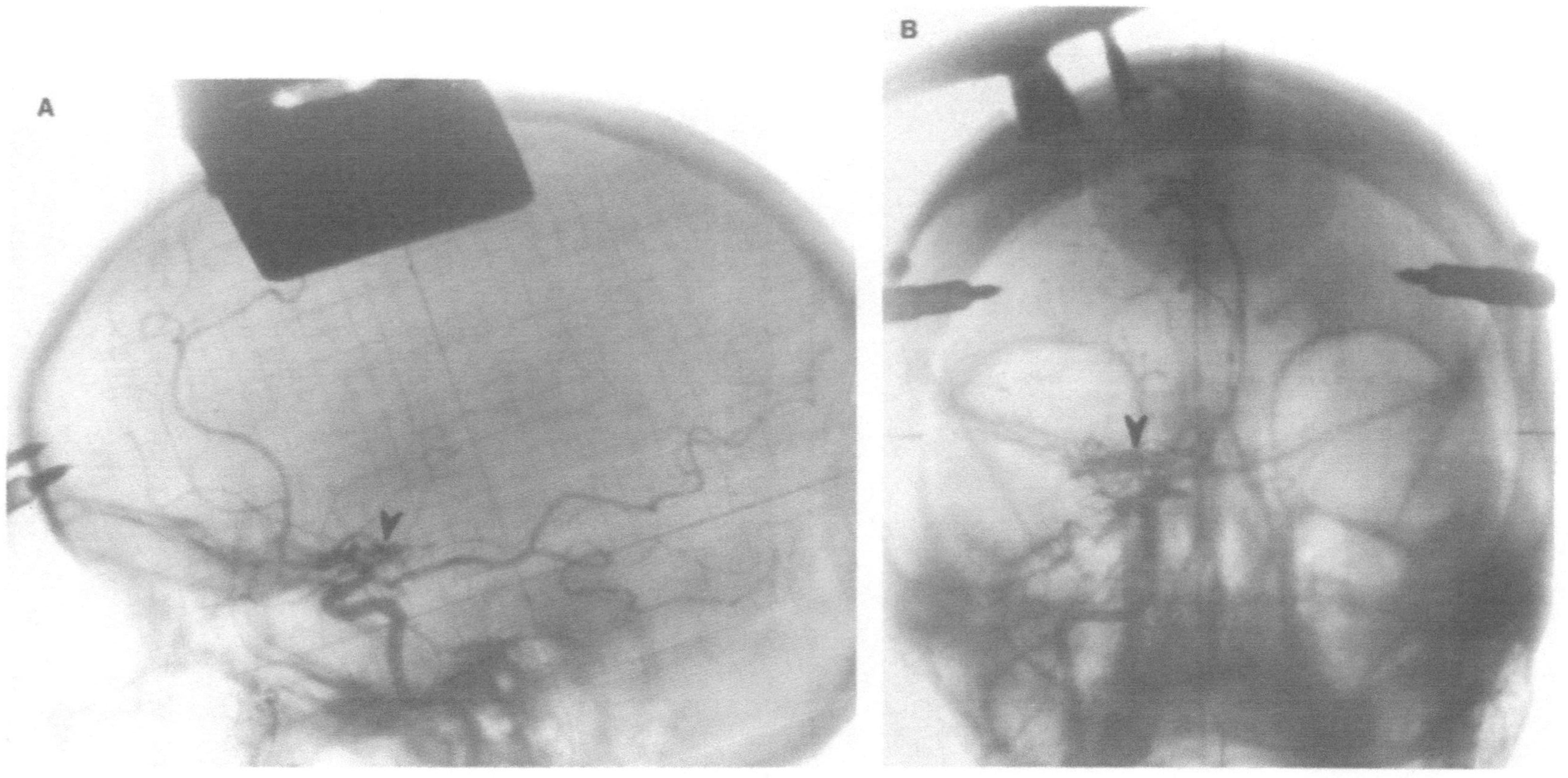

FIGURE 255. Intracerebral hemorrhage from small AVM. (A) Small AVM of the right middle cerebral artery (arrowhead) with severe spasm of the artery (lateral angiogram). (B) The same AVM on AP angiogram. A pronounced shift of the anterior cerebral artery to the other side is caused by the intracerebral hematoma in right hemisphere. (C) Giant hematoma (100 ml) in the right hemisphere on the CT scan. (D) Partial stereotactic evacuation of the hematoma.

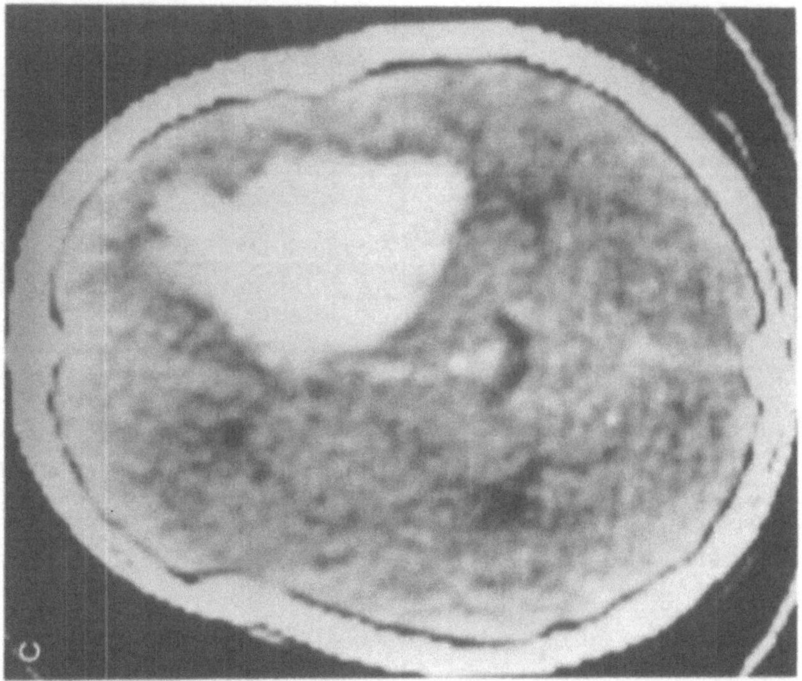

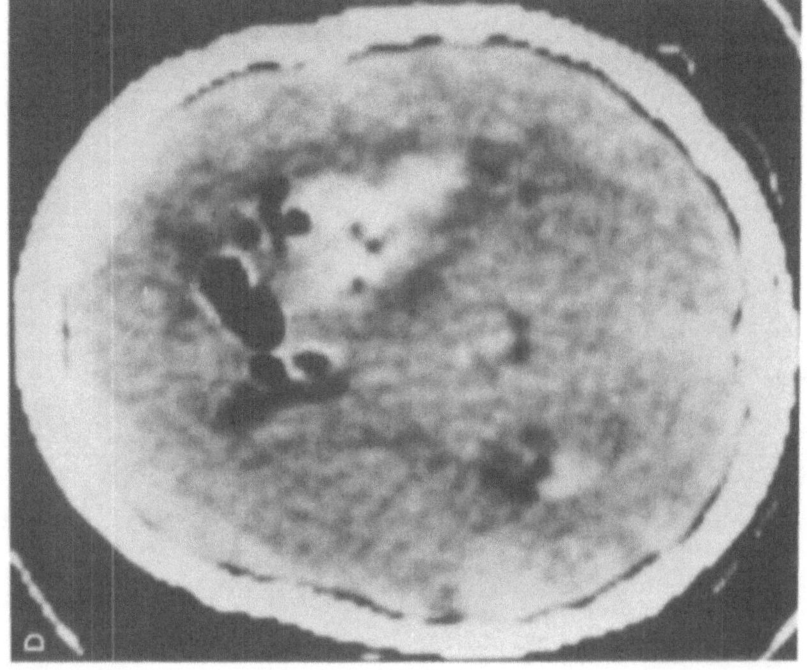

FIGURE 255. *(Continued)*

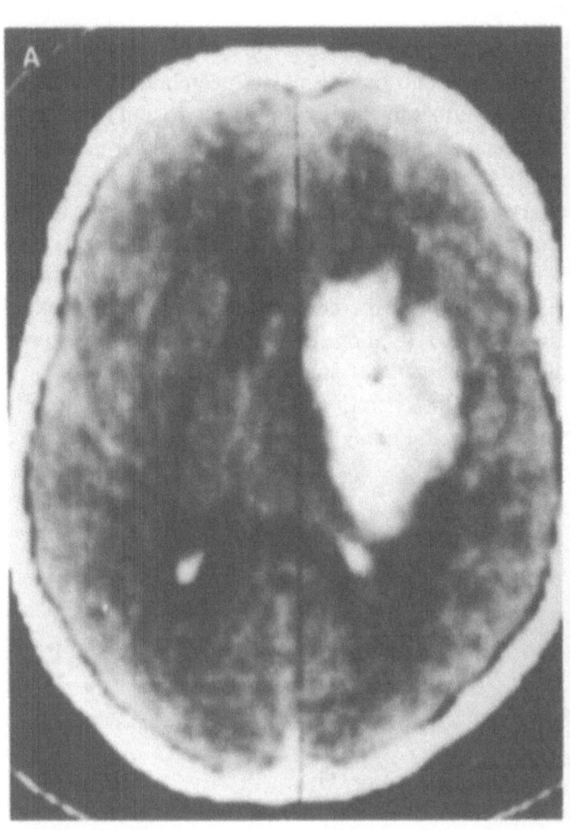

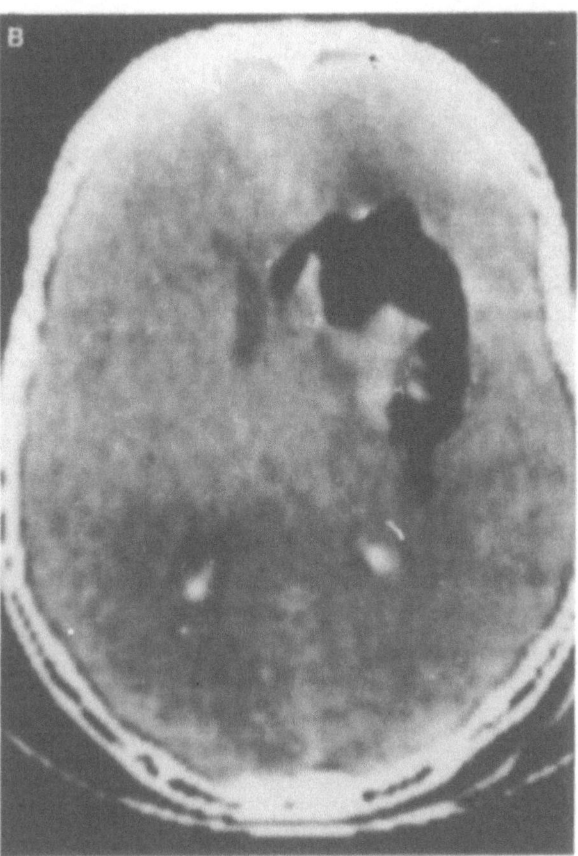

FIGURE 256. (A) Computed tomographic scans of a large intracerebral hematoma before stereotactic evacuation. (B) The cavity of a stereotactically removed large intracerebral hematoma is filled with air.

About 1 year after surgery the patient has only a mild hemiparesis and can walk without a cane.

A 43-year-old normotensive woman suddenly developed a left-sided hemiplegia and became stuporous. On the next day after the stroke, her CSF was bloody.

A small AVM was seen angiographically near the main trunk of the right middle cerebral artery (Fig. 255A,B), which was in marked spasm. The right anterior cerebral artery was markedly shifted to the left. The CT showed a very large hematoma (100 ml) in the frontal and temporal lobes with penetration of blood into the third ventricle (Fig. 255C). Two days after the onset of symptoms, the main portion of the hematoma (about 70 ml) was removed stereotactically. One-third of the hematoma was left near the AVM (Fig. 255D). During the operation, the AG control was carried out.

On the next day, the patient was alert, and active movements in the left extremities appeared. She began to walk 2 weeks after surgery. One month later, movements in the left arm and leg had completely returned. The patient refused a second operation for the removal of the AVM. There was no recurrent bleeding during the 2 years of follow-up.

Recurrent bleeding after hematoma aspiration by any existing method remains an important unsolved problem. Recurrence of hemorrhage took place in 13% of our stereotactic aspirations. As CT control has shown, the hematoma cavity does not collapse immediately after aspiration (Fig. 256). It may be that negative pressure in the cavity is the main predisposing factor to recurrent bleeding.

To prevent this very serious and relatively frequent complication, we developed a new technique with inflation (by saline solution) of a silastic balloon in the cavity after hematoma evacuation. The balloon was deflated gradually while monitoring and controlling ICP and was removed 3 to 4 days after surgery. When the method was used, there was no recurrent hematoma, but our experience is too limited to allow definite conclusions to be drawn.

We cannot agree with the opinion of Backlund and von Holst (1978) that the total evacuation of spontaneous hematomas is not advisable. As our experience has shown, it is possible to remove practically the entire hematoma, leaving in its cavity only a few milliliters of blood. There is no evidence that complete removal of the hematoma increases the risk of recurrent bleeding. But certainly one must be very careful aspirating hematomas after aneurysmal rupture. In such cases, the clot near the aneurysm must not be disturbed, as we have shown in the two cases described above.

7. Summary

Our experience with stereotactic evacuation of spontaneous and aneurysmal intracerebral hemorrhages has shown that this method is effective and offers substantional advantages over conventional surgical treatment. It affords the possibility of total or practically total evacuation of all intracerebral hematomas of different volumes and locations, including aneurysmal hematomas. It also offers the possibility of removing blood clots from the cerebral ventricles. The method allows the aspiration of clots of any consistency through a cannula of small diameter.

Since the operation is nontraumatic and relatively safe, the method may be used in patients who are in a serious, comatose condition.

18

Mental Disorders

I expected to be able to transform the psychic reactions and to relieve the patients thereby.
Egas Moniz

1. Introduction

The history of surgical intervention in the human brain to treat psychic disorders goes back approximately a century. The first procedure was performed by Burckhardt, who, in 1890, extirpated parts of the cerebral cortex for schizophrenia.

In 1910 a Russian neurosurgeon, L. Puusepp, was one of the first to operate for mental disorders by sectioning the brain at the border of the frontal and parietal lobes. However, three such operations that Puusepp (1930) reported much later were not effective.

A milestone in the development of psychosurgery was the prefrontal lobotomy (leukotomy) proposed in 1935 by Egas Moniz, who subsequently, for that advance, was awarded the Nobel Prize. At Moniz's suggestion, the first contemporary psychosurgical operation was performed by Almeida Lima. It consisted of injecting absolute alcohol in the white matter of the frontal lobe. Later, the white matter of the frontal lobe on one or both sides was sectioned or broken up with a special leukotome in order to interrupt the frontothalamic paths. Subsequently, Moniz reported that of 20 mental patients who underwent surgery, seven were cured and another seven displayed an improvement.

As the basis for the new operation, Moniz considered that the emotional component of intellectual activity lay in Th and that the thalamocortical pathways integrated the intellectual and affective aspects of man's psyche. Pathological activity coming from Th to the cerebral cortex was considered to underlie the development of the psychopathological syndromes. That is the reason leukotomy, the objective of which is to sever the frontothalamic pathways, restores normal mental functions.

In the 1950s, Freeman and Watts (1950) performed leukotomy in more than 1000 psychiatric patients. Leukotomy became quite popular in the period after their monograph summing up this experience was published. In a large number of cases this operation produced an indisputable therapeutic effect at a time when modern psychopharmacology did not yet exist. Nevertheless, the destructive nature of that operation and the unwarranted expansion of its indications soon revealed very serious undesirable consequences. Mundinger (1974b) emphasized that approximately 150,000 mm^3 of brain tissue are destroyed in prefrontal leukotomy. Quite often, this led to the development of an irreversible psychoorganic syndrome with reduced intellect, degradation of personality, and so forth. Classical lobotomy sometimes resulted in the development of a typical frontal syndrome with apathy, disorientation, confusion, lapses in memory, and loss of control over bladder and bowel functions.

The negative aspects of leukotomy were shown in striking artistic form in M. Forman's well-known film *One Flew Over the Cuckoo's Nest*. In the initial period, the postoperative mortality from leukotomy (mainly the result of intracerebral hemorrhages) was 3–4%.

Numerous modifications of prefrontal leukotomy were subsequently proposed (see below) for the purpose of decreasing the undesirable after effects—transorbital "undercutting" of the basal frontal cortex, section of the mediobasal quadrant of the frontal lobe, and other procedures. After such operations, the intellectual defects were considerably less pronounced than after leukotomy.

The 1960s witnessed a sharp reduction in the use of leukotomy in all countries of the world. This was the result not only of the serious complications of this operation but also of the remarkable achievements of modern psychopharmacology in producing many highly effective drugs for treating mental diseases. However, one must admit that psychopharmacology has not solved the problem of treating mental diseases, nor has it led to their disappearance. This fact prompted a kind of "renaissance" of psychosurgery* beginning some 10 to 12 years ago. According to the data of the Research Committee, Royal College of Psychiatrists, at least 100,000 psychosurgical operations have been performed in the world over the past 35 years. Some 200 to 300 such operations are performed annually in Great Britain according to the same data. In all likelihood, the Japanese neurosurgeon Hirose (1975) has acquired the greatest experience in the world in the field of neurosurgery; he has performed 523 such operations in 25 years.

It is noteworthy that modern psychosurgery has very little in common with leukotomy, which has been relegated to the history of neurosurgery. The main distinguishing feature of modern psychosurgery is the extensive use of stereotactic surgery for treating mental disorders. Leukotomy, with crude destruction of large areas of brain tissue, has been replaced by strictly local, discrete foci of destruction in various subcortical structures, the selection of which is determined by anatomophysiological criteria and based on contemporary, although limited, knowledge about the anatomic pathology of mental functions.

Stimulation techniques for treating mental disease have been evolved in recent years. Some attempts to stimulate various subcortical structures continuously with implanted electrodes have been successful. Nevertheless, stimulation for treating mental diseases is used quite rarely and with extreme caution (Dieckmann, 1979). The first psychosurgical stereotactic operations were performed by the pioneers of that method, Spiegel and Wycis, who first destroyed the dorsomedial nuclei of Th in 1947.

One cannot assert that modern neurosurgery is based on strictly scientific and firm principles. Nevertheless, there is every reason to believe that the pathogenesis of many mental diseases and syndromes is largely the result of pathologically increased activity of certain neuronal systems in the brain. For example, it has been established that aggressiveness stems from damage to Am and its links with Th and Hypoth.

Since the medial portions of the limbic system (Am, medial frontal cortex) are closely connected with the "activation mechanisms," surgical destruction of these structures leads to a reduction in psychomotor excitation, anxiety, fear, and compulsive behavior (Livingston, 1975). It is also considered that the normal biological and psychological defense mechanisms in case of mental diseases become more active, giving rise to tension, anguish, emotional disturbances, and other symptoms.

A psychosurgical operation breaks this vicious cycle and stabilizes the above-mentioned mechanisms (Hirose, 1975). The establishment of a link between sexual deviation and a damaged Hypoth offers still one more example. In view of this, every stereotactic operation for mental disease has as its objective the selective elimination of individual cerebral structures or their pathways; this determines its curative effect. Fernandes Barahana and da Fonseca Simoes (1982) consider that this is the basis of a "fundamental structure of sociocultural personality."

It should be especially emphasized that psychosurgical operations are indicated only in those cases in which all conservative methods of treatment over many years have proved totally ineffective. The majority of authors believe that psychosurgery is not indicated earlier than 3 to 4 years after the onset of a mental disorder.

So far, there is still no rigid framework for the nomenclature of mental disorders that are indications for surgical treatment. At the present time, symptoms and syndromes rather than the disease process itself can be treated by surgical methods. These include certain forms of schizophrenia, and mental disorders in epilepsy (particularly aggressiveness in temporal lobe epilepsy), obsessive–compulsive neuroses, narcotic dependence, phobias, sexual deviations, depression, depersonalization syndrome, severe senescent pathologies, de la Tourette's syndrome, erethic oligophrenia, anorexia nervosa, certain types of organic pathology of the brain characterized by bursts of fury, aggressiveness, asocial behavior, and so forth. Affective psychoses are also considered to be important indications for psychosurgery, since they are characterized by prolonged depressions with agitation, anguish, suicidal thoughts, anorexia, insomnia, and other behavioral aberrations. Among the indications for such an operation, the literature also mentions such syndromes as psychotic and reactive depression,

*The term "psychosurgery" cannot be considered ideal. However, a more adequate (and as brief) term has not yet been suggested. One must also reckon with the fact that this term has won itself a firm position in international neurosurgical literature.

pseudoneurotic schizophrenia, and agitated states in old age.

As is usually the case with every new and rapidly developing scientific trend, there is much disagreement, vagueness, and controversy about present-day psychosurgery. No clearly defined indications for operation have been reached. The cerebral structures (targets) the destruction of which gives the best effect in various diseases have not been determined, nor have contradictory results of various operations been sufficiently studied. And unfortunately, the complicated moral–ethical problems related to psychosurgery have not as yet been resolved.

Not infrequently in psychosurgery, when various stereotactic and functional operations are used for the same mental disease process, the different operations yield approximately the same result. The attempts to compare objectively the results of different operations on various nuclear structures run into great difficulties, especially when the operations are performed at various neurosurgical clinics.

Scoville and Bettis (1975) quite justifiably emphasize that surgical intervention in the brain may impair its functions; however, a progressing mental disease impairs these functions to an even greater degree. Hence, an exacerbation after a psychosurgical operation, especially in schizophrenia, is often caused not by the operation itself but by further progression of the disease process (Hirose, 1975).

One must also take into account that a psychosurgical operation causes perhaps insignificant but nevertheless irreversible organic damage to the brain, the extent of which cannot be accurately prognosticated. It is precisely for this reason that many specialists, mainly psychiatrists, are actively opposed to psychosurgery, denying in principle that there are medical and ethical grounds for performing surgical operations on the brain in mental diseases. In the 1930s, many psychiatrists spoke out against the new operation proposed by Moniz. Even at the present time, certain Japanese psychiatrists have opposed the psychosurgical operations performed by their compatriot, the prominent neurosurgeon Sano. Their comments were especially critical at the World Congress of Neurological Surgery in Tokyo in 1972.

The viewpoints and arguments of both supporters and opponents of this procedure are presented in a recently published book edited by Valenstein (1980), in which scores of works on the scientific, legal, and ethical aspects of psychosurgery are discussed.

It is only natural that the representatives of different medical specialities should have different opinions regarding the scientific basis and therapeutic value of a new procedure. At the same time, as numerous examples from the history of medicine have demonstrated, the scientific foundation of a new method of treatment may be demonstrated considerably later than its effective therapeutic application. One of the greatest discoveries in medicine—vaccination against smallpox—was made by Jenner many decades before the viral nature of that disease was established. Contemporary psychiatry cannot claim to know the physiological basis of the therapeutic action of insulin shock, which has long been employed with success to treat mental cases.

In these circumstances, the main criterion for determining the expediency and advisability of psychosurgery is the result of its practical application. "Practice is the criterion of truth," and if one considers that psychosurgery really improves the condition of patients who cannot be helped by anything else, then there should be no doubt as to whether the use of this method is justifiable. As noted by Hassler (1982), ". . . the only consideration should be that the patients have the most effective treatment."

It should be considered an inflexible rule that the decision for any psychosurgical operation should be made not by the neurosurgeon himself but in collaboration with a group of expert representatives of various medical specialities (psychiatrists, psychologists, rehabilitation specialists, etc.). Such a committee must thoroughly weigh all the "pros" and "cons" of a psychosurgical operation and evaluate its prospects as well as possible side effects. An important role in the final decision is played by a "Committee on Ethical Problems" that has lately been created at a number of medical institutions and scientific societies.

In view of this, it must be strongly emphasized that any attempts to employ psychosurgery as an instrument to influence people with a normal psyche must be unconditionally condemned as immoral and criminal.

Attempts have been undertaken recently to sum up the results of the development of psychosurgery for the past 10 to 15 years. Heath and Cox (1976), based on the study of numerous published works, came to the conclusion that destruction of various cerebral structures is most effective in states of depression and fear, in particular, in those cases in which these states are the predominant feature of schizophrenia.

To sum up the results of modern psychosurgery, Smith and Kiloh (1975) report that in 75% of patients it is possible to diminish significantly depressions, restlessness, and obsessive states by cingulotomy. A considerable therapeutic effect was observed in more than half of the patients with aggressiveness syndrome and fits of fury after amygdalotomy or hypothalamotomy (see below).

2. Stereotactic Functional Operations

Having discussed these general concerns, we shall now present data on the main psychosurgical operations and the results of treating certain mental diseases and psychopathological syndromes by stereotactic operations on various cerebral structures.

2.1. Cingulotomy

One of the major psychosurgical operations is cingulotomy, i.e., destruction of GC (field 24 Brodmann), which represents an important link in the limbic system. The GC is quite long in the rostral–caudal direction, and that is why destruction is conducted in segments—the medial part, the frontal part, and the part under the knee in the rostral GC (subrostral cingulotomy). Since this operation is also performed to relieve pain, the technique is described in detail in Chapter 13 (Section 5.2.4).

The first stereotactic cingulotomy was performed by Ballantine *et al.* in 1967. Indications for this operation include several mental diseases—schizophrenia, aggressiveness, grave neuroses, narcotics addiction, etc. Laitinen *et al.* (1973) reported the results of anterior cingulotomy in 46 patients with various mental diseases, the majority being schizophrenia, grave neuroses, or narcotism. All the patients had spent many years in psychiatric clinics.

As a rule, the operations were performed under local anesthesia. Lesions in GC were made stereotactically on both sides. The centers of the foci of destruction were approximately 6 mm from the midline, and the size of the lesion varied from 6 × 6 to 12 × 8 mm. The greatest therapeutic effect of the operation was in the diminished agitation in these patients, whereas the influence on tension and depression was negligible. The authors thought that the course of schizophrenia was milder after cingulotomy. The long-term results of cingulotomy (average period of observation 1.5 years) were very promising. Half of the patients were totally or almost totally cured, being able to return to their jobs and functioning without drugs. A definite improvement was noted in 26% of the patients, although all required psychiatric assistance. In the remaining 24% of the cases, the operation did not have a noticeable effect. Other authors have reported that the good effects of cingulotomy gradually diminish (Sano, 1962).

No serious complications, including epileptic fits, were observed after cingulotomy. Psychological tests conducted before and after cingulotomy did not reveal any impairment in intelligence, memory, or attention resulting from the operation. The fact that there are no changes in personality after cingulotomy makes this operation more attractive than the classical leukotomy.

Bailey *et al.* (1977) have had considerable experience with cingulotomy. They operated on 200 patients with severe depressive states and followed them for long periods (up to 13 years). The operation was performed by the open method, making burr holes on both sides rostral to the coronal suture. After puncturing the anterior horn of the lateral ventricle, a small (1.5-cm) incision was made in the cortex, and spatulas were introduced to spread apart the brain and open the anterior horn. Then the inferior part of GC was identified at the point where it bends around the apex of the anterior horn. A graduated cannula was used to measure the distance between the lower margin of the white matter and the orbital surface of the frontal lobe, which is usually 1.5–2 cm. Then GC was transsected along the frontal plane with an incision 12 to 14 cm in length. A similar operation was performed on the other side.

Based on the scale for evaluating the results that was proposed earlier by the authors, the great majority of patients displayed a significant and lasting improvement—93% for two or more grades on that scale, and 80% for three or more grades. The postoperative mortality was only 0.5%. Cingulotomy was most effective in manic–depressive states and phobias. The therapeutic result was less in obsessive–compulsive syndromes and schizoaffective states. The various psychometric tests performed before and after the operation revealed an improvement of intellectual functions in 86% of the patients.

The results of anterior cingulotomy were analyzed by Brown and Lighthill (1968), who operated on 110 patients in three groups: "state of anxiety and tension," obsessive–compulsive psychoneuroses, and manic–depressive psychoses. A significant and stable improvement was noted in more than 90% of the operated patients.

The largest experience with cingulotomy has been accumulated by Ballantine and Giriunas (1979), who operated on 237 patients, mainly with serious depression. In their analysis of the long-term results of bilateral operations in 154 patients, the authors rated 75% of the patients in the category of "significant improvement." There were no serious complications or cases of postoperative mortality. The authors did not clearly classify the patients according to nosology, limiting their descriptions to an enumeration of the most frequently encountered symptoms prior to the operation: depression, 97% of the patients; fear, 77%; insomnia, 77%; suicidal thoughts, 61%; anxiety, 60%; anorexia, 47%; obsessive states, 39%. The most favorable results after the operation were observed in patients suffering from depression, fear, anxiety, and insomnia.

After cingulotomy in 44 patients with various mental disorders, Le Beau (1954) noted that the most positive effects were obtained in the relief of aggressiveness and "increased excitability." Other authors consider that the main result of cingulotomy is the elimination or amelioration of fear, tension, and obsessive states; however, such symptoms as aggressive outbursts diminished only temporarily (Mingrino and Schergna, 1972).

Bilateral cingulotomy gave good results in patients suffering from schizophrenia, depression, and alcoholism (Meyer *et al.*, 1972; Vilkki, 1975; Choi and Kang, 1981). Martin *et al.* (1977) noted a significant improvement in approximately half of 68 patients with depression and schizophrenia over a period of 4 years. At the same time, a good result was obtained in all alcoholic patients.

Consequently, in summary, it may be said that cingulotomy is an effective and practically safe operation. The operation gave good and lasting results in several mental disorders. There were no changes in personality or intelligence after unilateral or even bilateral cingulotomy (or cingulectomy).

2.2. Basal Frontal Lobotomy

This section considers the combined data on several psychosurgical operations that have been described in literature under various names—"limbic leukotomy," "undercutting of the orbital cortex," "subcaudatal tractotomy." It is possible to group these operations since their objective is to destroy or isolate limited zones of the cortex and white matter of the basal frontal lobe on one or both sides. Just as in prefrontal leukotomy, these operations sever the connections of the frontobasal (orbital and medial) cortex with the limbic system (Papez ring), mediobasal nuclei of Th, and Hypoth (Scoville, 1954; Post *et al.*, 1968; Richardson, 1973a,b). Such operations on the basal portions of the frontal lobe are performed by both open and stereotactic methods.

Limbic leukotomy is considered to be a "broader version" of cingulotomy. In this operation, several foci of stereotactic destruction are made in the basal–medial quadrants of both frontal lobes above the gyrus rectus. This lesion is often combined with the destruction of the anterior portion of GC. The results of this operation have been published by several authors (Post *et al.*, 1968; Newcombe, 1975; Goktere *et al.*, 1975). Mitchell-Heggs *et al.* (1977) operated on 66 patients who had spent many years in mental hospitals with various disorders (schizophrenia, depression, obsessive neuroses) considered incurable. The clinical evaluation of the follow-up was supplemented by a large number of psychometric tests, so that there can be no doubt as to the significant positive effects produced by this operation. An improvement of varying degrees was observed after a year and a half in 76% of the patients; 89% of the patients with obsessive–compulsive neuroses were better.

Hassler (1982) believes that this operation, which he has named subcaudatal tractotomy is very effective in depressive states and phobias but rarely of help in obsessive–compulsive neuroses.

In order to limit the undesirable consequences of leukotomy by reducing the amount of damage in the frontal lobe, Scoville (1949) proposed an operation quite similar to the previous one, which he called orbital undercutting. Subsequently, this operation was developed by Busch (1955) and Hirose (1965). It involves an open, limited section of the white matter in the ventral portion of the frontal lobe and the limbic pathways proceeding from the orbital cortex, in particular, those from GC. Orbital undercutting, as a rule, avoids the confusion and amnesia that quite frequently complicate the standard leukotomy.

The operation of undercutting the orbital or lateral cortex of the frontal lobe is usually performed by an open operation simultaneously on the two sides. The burr holes are made to the right and to the left approximately midway between the coronal suture and the superior margin of the orbit at a distance of 3–4 cm from the midline. After opening of the dura mater and coagulation of the cortex for approximately 3 cm, a leukotome is introduced to undercut the white substance of the frontal lobe to a depth of 3–4 cm and laterally for 3–4 cm. The dura mater is then sutured, and the skull plugs are returned to their places.

Over a period of 20 years, Scoville and Bettis

(1975) operated on 107 patients by bilateral orbital undercutting. After a long-term follow-up, the authors very highly recommend this operation, considering that in 75% of cases the results were excellent or good. These operations were most successful in protracted depressions, obsessive neuroses, and "schizoaffective disorders." However, four patients died after the operation, and 12 patients had infrequent epileptic fits. The authors report that many patients returned to normal life and social activity. Psychometric tests did not reveal any changes in personality. It was noted that after an operation involving orbitomedial undercutting of the frontal cortex, clinical improvement was gradual over a long period of time (up to 2 years) (Hirose, 1975).

Although the literature offers some results on undercutting two different portions of the frontal lobe—the orbitomedial and basolateral cortex—the results are not sufficiently different to determine which of these methods is more effective and for what psychopathological syndromes.

In order to diminish the undesirable consequences of Scoville's orbital undercutting by reducing the amount of the white substance damaged, Knight in 1964 proposed subcaudatal tractotomy. At first the author performed this operation via an open approach; later he used a stereotactic method to introduce granules of radioactive yttrium. This operation involves destruction of a 2-cm-wide volume of the mediobasal portion of the frontal lobe at a distance of about 1 cm from the orbit roof. This lesion is very close to the locus of subrostral cingulotomy. Both intracerebral and bony reference points are used for pinpointing the targets in this operation. On a lateral ventriculogram a line is drawn from the tuberculum sella along the bone of the base of the anterior cranial fossa. On this line 5 mm is measured off, and a perpendicular is erected; another 11 mm is measured off to pinpoint the target. On anterior–posterior views, the target is marked at a distance of 10 mm from the midline.

Ström-Olsen and Carlisle (1971) analyzed the long-term results of the operations performed by Knight in 150 of 210 mental patients operated on. Forty-nine percent of the patients were found to be completely or almost completely cured. There was no postoperative morality; the incidence of epilepsy was considerably lower than after open orbital undercutting. It has been established that this operation produces the best results in chronic depression, both endogenous and exogenous. A good effect was also obtained in obsessive–compulsive neuroses.

Another stereotactic target in psychosurgical operations is the substantia innominata (SI) in the subcaudal region near the orbital and medial cortex. An extensive experience in the destruction of this structure (innominotomy) has been presented by Knight (1972). A substantial improvement in aggressiveness was noted after a long follow-up in 90% of 210 patients. Moreover, phobias were found to diminish in 77% of the patients, and grave depression was completely eliminated in 44%. According to Vaernet and Madsen (1972), innominotomy is effective in chronic psychoses with behavioral disorders. In spite of such promising results, the indications for destruction of SI must be more precisely defined, and the evaluation of results requires a longer period of observation.

2.3. Dorsomedial Thalamotomy

The first stereotactic destruction of the dorsomedial thalamic nuclei in mental patients was performed in 1947 by Spiegel and Wycis, who believed that the operation severed the pathways from the prefrontal cortex to the nuclei of Th. In the majority of cases, the effect of this operation was temporary. Subsequently, Hassler (1947, 1949) described in detail the projections of various portions of the thalamic medial nuclei to the different zones of the cortex (prefrontal, orbital, gyrus rectus, etc.). On the basis of these investigations, Hassler and Riechert (1957) selectively destroyed one particular portion of the medial nuclei, depending on the type of the mental disease (caudal portion for aggressiveness, frontal portion for depressive states). However, this method did not give significant results.

Since stereotactic destruction of the medial thalamic nuclei produced only a partial effect on the various psychopathological syndromes, the same authors suggested the additional destruction of the intralaminary thalamic nuclei. In most cases, the operation was unilateral; however, for erethic aggressiveness, it was performed on both sides. Subsequent surgical experience demonstrated that such a combined operation was effective for a wide range of mental disorders (obsessive–compulsive neuroses, catatonic schizophrenia, de la Tourette's syndrome, erethic oligophrenia, phobias, etc.) (Roeder *et al.*, 1971; Hassler and Dieckmann, 1972; Hassler, 1982; Spuler, 1982). However, the combined destruction of the thalamic medial and intralaminar nuclei, especially if bilateral, at times produces undesirable side effects (Korsakoff's and akinetic syndromes, depressive reactions). In view of this, Spuler (1982) recommends that the first operation be done on the nondominant hemisphere and the second side 6 months later.

It has been stated that stereotactic destruction of the anterior and dorsomedial thalamic nuclei for aggressive behavior produces a more pronounced therapeutic effect than posterior hypothalamotomy (Nadvornik *et al.*, 1972c).

2.4. Destruction of CM

Andy (1970) reported good results on behavioral disturbances from stereotactic destruction of another target, CM. After this operation, five patients were improved with respect to "excitation and aggressiveness." In a subsequent work, Andy and Stephan, (1982) reported on the results of this operation in 34 patients with aggressiveness, pathological affectation, and hyperkinesia. The authors pointed out that the destruction of the medial portion of CM considerably decreased aggressiveness, whereas destruction of the lateral portion of that structure diminished pathological affectation and hyperkinesia. Bilateral operations were more effective than unilateral.

A combined destruction of CM and Subth in aggressive epileptics reduced the frequency of seizures and normalized behavior (Mizuno *et al.*, 1972).

2.5. Amygdalotomy

The stereotactic destruction of the deep-seated nuclei of the temporal lobe, Am and Hipp, is described in detail in Chapter 14 (Section 6.5.3c). It should be noted that the aggressiveness typical of temporal lobe epilepsy and the epileptic seizures are both greatly reduced by destruction of Am and Hipp, just as classical temporal lobectomy usually results not only in reducing the frequency or eliminating epileptic fits but also in diminishing aggressiveness and other emotional personality changes.

Siegfried (1977) emphasizes that amygdalotomy to relieve the aggressive syndrome must be bilateral. In six epileptic patients with sudden paroxysms of unprovoked fury and aggressiveness, Smith and Kiloh (1975) implanted fine gold electrodes in both Am. The parameters for stimulating this structure were: monopolar; 40 Hz; duration of impulse, 0.5 msec; voltage, from 2 to 8 V. The authors reported that depending on the localization of the electrodes within Am, stimulation may either produce fits of fury or tranquilize the patient. In the first case, the medial portion of Am was stimulated; in the second, the lateral portion. Subsequent bilateral stereotactic destruction of Am produced good results, and the fits of fury and aggressiveness ceased.

2.6. Destruction of Stria Terminalis

Not many attempts have been made to destroy stereotactically another subcortical structure, the fundus stria terminalis on the base of Am (Hassler and Dieckmann, 1972; Bursaco, 1973; Spuler, 1982). This operation is performed both unilaterally and bilaterally. Judging from limited experience, it seems to be effective in relieving aggressiveness, agitation, and behavioral disturbances, mainly in patients with psychomotor seizures. Spuler (1982), however, believes that the destruction of the stria terminalis is less effective in aggressiveness than amygdalotomy. After destruction of this structure, some patients have had a transitory hyperphagia with a considerable gain in body weight (Bursaco, 1973).

2.7. Posterior Hypothalamotomy

Unilateral or bilateral posterior hypothalamotomy, i.e., stereotactic destruction of the posterior nucleus of Hypoth and the posterior longitudinal fascicle of Schütz is used in contemporary psychosurgery for treating pathological aggressiveness and certain other mental disorders. Since this operation is performed also to treat various pain syndromes, the technique is described in Chapter 13 (Section 5.2.3).

The theoretical basis for the destruction of the posterior Hypoth lies in the assumption that this operation normalizes the "ergotropic" and "trophotropic" balance by diminishing sympathicotonia, which always accompanies fits of fury and aggressiveness (Sano, 1974, 1982). Moreover, the destruction of these nuclei of Hypoth removes the activating effect on the limbic system controlling emotional reactions. This has a sedative effect on aggressive and agitated epileptics.

The very small size of the posterior Hypoth makes it imperative to choose a correspondingly small focus for destruction. It should not be more than 3–4 mm in diameter (30–35 mm^3). The operation is performed consecutively on both sides with an interval of 7–10 days.

Posteromedial hypothalamotomy was introduced and thoroughly studied by Sano (1974, 1982), who performed it on more than 50 mental patients, mainly with aggressive syndromes. Practically all of his patients showed an improvement that was observed over a period of several years. This consisted of a disappearance of aggressiveness, restlessness, or antisocial behavior. Many patients who had been confined to mental institutions for years no longer needed constant care and observation.

Destruction of the posteromedial Hypoth is quite effective in the treatment of infantile erethitic oligophrenia with extreme aggressiveness. In 11 children from 3 to 13 years of age (Arjona, 1974), a considerable improvement was noted in all after the operation. Specifically, constant aggressiveness and the desire for self-mutilation disappeared. Six of the 11 children began to attend special schools and mixed with other children; prior to the operation this had been absolutely out of the question. The author believes that these operations offer the only possible way to treat these children. After surgery, there were no new neurological symptoms or mental and intellectual disorders (Black *et al.*, 1975). There was no protracted endocrine insufficiency, although in the first few days, polyuria was observed. In spite of the close proximity of Hypoth to the chiasma and visual tracts, by accurate stereotactic calculations it was possible to avoid damaging these structures.

Another operation on the hypothalamic nuclei, ventromedial hypothalamotomy, and its results are described in connection with sexual deviations (Section 3.7) and narcotism (Section 3.8).

2.8. Anterior Capsulotomy

Another operation in the arsenal of psychosurgery is anterior capsulotomy, which is a relatively rarely performed stereotactic operation. The objective is to destroy the pathways from Th to the prefrontal and orbital cortex that pass through the anterior limb of CI. It is presumed that the section of these pathways will have therapeutic effect on certain mental disorders.

Stereotactic destruction of the anterior limb of CI is performed simultaneously on both sides. This operation is relatively safe since there is little chance of damaging functionally important structures. For that reason, the size of the focus of destruction may be increased to 16–18 mm in diameter. The coordinates for capsulotomy are: in the lateral projection, 17 mm anterior to CA along the direct continuation of LI; in the A–P projection, 20 mm from the midplane. Other coordinates for capsulotomy were proposed by Kullberg (1977): 5 mm posterior to the apex of the anterior horn and 5 mm above its base; in the A–P projection, 5 mm lateral to the ventricular wall.

Stereotactic destruction of the anterior limb of CI is one of the most effective psychosurgical operations. In 1961, Herner published the long-term results of this operation performed by Leksell in 116 patients. In approximately 80% of the patients, there was a significant improvement of depression, anxiety, and phobias. In 20% of the patients there was a complete disappearance of obsessive–compulsive neurosis, and in many others, the symptoms diminished markedly. Unfortunately, however, transitory states of confusion were quite often observed after this operation.

Even more favorable results after anterior capsulotomy were obtained by Bingley *et al.* (1973), who reported that grave obsessive neurosis was eliminated in nine of 17 patients. Also in this disease, a substantial and lasting improvement after capsulotomy was observed in 85% of patients operated on, half of whom were rated as "completely cured" by Lopez-Ibor and Lopez-Ibor, (1977). Newcombe (1982) considers anterior capsulotomy to be indicated in cases of fear and obsessive–compulsive neuroses but not to be very effective in depression, for which he believes that it is better to perform subcaudatal tractotomy.

One of the few attempts to compare results of two psychosurgical operations was undertaken by Kullberg (1977), who compared the data from 13 cingulotomies and 13 capsulotomies in patients with obsessive–compulsive and other neuroses over a minimal follow-up period of 1 year (some to 9 years). In all cases the operations were bilateral.

A comparison of both groups demonstrated that the number of excellent and good results was twice as high after capsulotomy as compared to cingulotomy. But mildly pronounced changes in personality were somewhat more apparent after capsulotomy.

2.9. Callosotomy

Another target for stereotactic destruction in mental diseases is the knee of the corpus callosum (GCC). This operation, proposed by Laitinen and Vilkki (1973) and Laitinen (1974), was named mesoloviotomy. The author emphasizes that high-frequency stimulation of GCC during the operation is very important for predicting its results. A good effect can be obtained only if the stimulation induces a feeling of relaxation and a "pleasant feeling" in the patient. This operation is most effective in respect to schizophrenia, since it relieves anxiety, tension, and catatonia. On the basis of 55 such operations, Laitinen (1974) states that good results last for a long period of time. It is noteworthy that this operation causes no complications and it is considered safe. At the same time, it should be noted that the operation is not effective in depressions, phobias, and obsessive neuroses.

2.10. Lesions of Internal Medullary Lamina

In order to attain a therapeutic effect for the "syndrome of aggression," a few attempts have been made to destroy another thalamic structure, the internal medullary lamina (IML), a thin strip of white substance in the central part of Th. It has been found that IML has close anatomofunctional ties with the limbic system, the orbital frontal cortex, and the extrapyramidal system, i.e., structures that are directly related to emotional integration. In view of this, it is thought that IML has some "alleviating effect" on the development of mental disorders associated with aggression, and its destruction should have a therapeutic effect on this syndrome (Hassler, 1974).

The stereotactic coordinates for IML relative to LI are: 7 to 8 mm posterior to CA, 2 to 3 mm above LI, and 9 to 10 mm lateral to the midline.

Poblete *et al.* (1970) reported the results of stereotactic lesions of IML in 25 patients with aggressive behavior. The conventional conservative treatment for many years had not been successful. Of 25 operated patients who were observed for an average of a half year, good results were obtained in seven patients, and noticeable improvement in six. The patients, relieved of their psychotic episodes and aggressiveness, were able to return home, and, in a number of cases, returned to supervised jobs. The patients had no complications, and, over a long period of observation, no worsening was observed. Nevertheless, further experience is necessary to evaluate the benefits of IML destruction.

3. Surgical Treatment of Mental Disorders

In the preceding section we described the main stereotactic psychosurgical operations performed at the present time. As can be seen from this discussion, practically all of these operations are used to treat several mental diseases or syndromes. This, quite naturally, makes it difficult to select one of these operations for a specific disease. In order at least partially to overcome these difficulties as well as to follow the nosological principle on which this work is based, we have made an attempt to summarize the literature as well as our own data to determine the most effective procedures for treating specific mental disease or syndromes.

3.1. Aggressive Syndrome

This syndrome is usually not considered a distinct nosological entity, yet aggressiveness may dominate the clinical picture of certain mental diseases. The aggressive syndrome is often combined with various forms of epilepsy, particularly temporal lobe epilepsy (TLE). Fortunately, operations for treating TLE are at the same time the most effective means of treating aggressiveness. Among these operations, one should emphasize stereotactic amygdalotomy and hippocampotomy, the techniques and results of which are presented in the section on TLE (Chapter 14, Section 6.5). These data indicate that surgical treatment of the aggressive syndrome is most effective if the cause is some form of epilepsy.

Other functional and stereotactic operations are also employed to relieve aggressiveness: posterior hypothalamotomy (Section 2.7) (Sano, 1974, 1982; Black *et al.*, 1975), dorsomedial thalamotomy (Section 2.3) (Nadvornik *et al.*, 1973), destruction of striae terminalis (Section 2.6) (Hassler and Dieckmann, 1972; Bursaco, 1973), and destruction of lamina medullaris interna (Section 2.10) (Poblete *et al.*, 1970; Hassler, 1974). The effectiveness of these operations in the aggressive syndrome may be considered proven, but it is difficult to decide which of the above-mentioned operations is the most effective. Cingulotomy performed for this syndrome seems to produce a less satisfactory and less lasting effect (Mingrino and Shergna, 1972).

3.2. Erethic Oligophrenia

In current literature, the terms erethic oligophrenia or erethism are used to refer to severe aggressive syndromes associated with mental deficiency or dementia. As a rule, this disease affects children at the age of 6–8 years. Most frequently this disease is the result of pronounced prenatal, natal, or postnatal injuries to the child's brain. Erethism is often combined with TLE or other epileptic attacks.

The main manifestations of the disease are impaired intellect, hyperactivity, and pronounced aggresiveness. The child continuously performs various movements such as running, jumping, rolling on the floor, flailing the arms, and such motor acts. It is practically impossible to communicate with the child since he attacks people nearby, injures himself (hetero- and autoaggressiveness), breaks or damages objects, simultaneously yelling sounds or words. Quite often it is necessary to bind the hands of such children to prevent them from inflicting grave injuries on themselves. With age, the aggressiveness of such erethic children intensifies. All types of medicinal and other forms of treatment are ineffective, so that surgery is advisable. Spuler (1982) believes that the operation must not be put off indefinitely, since mental deficiency progresses with age.

The data in the literature indicate that psychosurgical operations for erethic oligophrenia are not only effective but at the present time are the only effective method of therapy. A significant improvement may result from one of at least three operations—destruction of the medial and intralaminary thalamic nuclei (Section 2.3) (Hassler and Dieckmann, 1972; Spuler, 1982), destruction of the posteromedial Hypoth (Section 2.7) (Arjona, 1974; Sano, 1974), and finally, amygdalotomy combined with the destruction of the adjacent striae terminalis (Section 2.6) (Hassler and Dieckmann, 1972; Bursaco, 1973; Spuler, 1982). Amygdalotomy is most effective in cases of erethic debility with epilepsy.

3.3. Depression

Depression may be either a distinct entity or a symptom of another mental disease. Since depression is one of the most widespread of mental disorders, its diagnosis, clinical picture, and course have been studied and described in detail in hundreds of articles and journals.

Just as in other mental diseases, mild and moderate depression may respond, to a greater or lesser extent, to conservative treatment, particularly, modern antidepressant drugs. Yet, the treatment of serious forms of the disease remains an acute and unresolved problem. It is known that approximately 15% of patients with affective disorders such as depression commit suicide. The main manifestations of the illness are extremely evident in the severe forms: fear, depression, anxiety, a feeling of hopelessness and anguish, suicidal thoughts, anorexia, and so forth.

The severity of the clinical manifestations of depression, the progression of the disease, and the real possibilities of suicide in all are sufficient grounds for considering psychosurgical operation. Certain authors believe that with the above-mentioned indications, the operation should be performed not later than 2–3 years after the onset of the depression, first, to avert suicide, and second, because a late operation is less effective (Goktere *et al.*, 1975; Newcombe, 1975, 1982).

There are several psychosurgical operations for depression. The most effective, in all probability, is destruction of the medial orbital frontal lobe, particularly subcaudatal tractotomy (Section 2.2) (Knight, 1972; Goktere *et al.*, 1975; Newcombe, 1975, 1982; Mitchell-Heggs *et al.*, 1977; Hirose, 1975). Undoubtedly, good and lasting results are also obtained by cingulotomy, especially by destruction of the anterior portions of GC, the knee and rostrum (Section 2.1) (Meyer *et al.*, 1973; Laitinen *et al.*, 1973; Ballantine and Giriunas, 1979; Teuber *et al.*, 1977). Anterior capsulotomy (Section 2.8) and callosotomy are not indicated in depression, since these operations do not result in lasting improvement.

3.4. Schizophrenia

As is known, schizophrenia is the gravest and most widespread mental disease and, at the same time, the main unresolved problem in contemporary psychiatry. There is no reason to assume that this exceptionally complicated condition can be resolved by resorting to surgery. Since more and more evidence is appearing to indicate that the pathogenesis of schizophrenia is closely related to cerebral metabolic disorders, in particular, abnormalities of the dopaminergic system, there is hope that pharmacological treatment will be effective for this grave disease. However, since such treatment is not yet available, in the most severe, practically incurable forms of schizophrenia, only psychosurgical operations offer some hope for partial alleviation of the condition.

The first such operation was prefrontal lobotomy proposed by Moniz in the 1930s. Over the course of almost a half century, several neurosurgeons have repeatedly attempted to treat this disease surgically with varying degrees of success. It should be noted that the objective of these operations was not to cure schizophrenia but to eliminate or relieve those syndromes that often predominate in the clinical picture: depression (Section 3.3), aggressiveness (Section 3.1), as well as more specific symptoms such as catatonia, schizoaffective conditions, and hallucinations.

It is noteworthy that in schizophrenia neurosurgeons display special caution in the selection of patients for operation. This is explained by the fact that few such operations are performed so that, quite naturally, it is difficult to evaluate long-term results.

One of the most effective psychosurgical operations in treating schizophrenia is stereotactic cingulotomy (Section 2.1) and various modifications of it (Meyer *et al.*, 1973; Laitinen *et al.*, 1973; Choi and Kang, 1981). It has been established that bilateral operations are more effective than unilateral procedures. In certain forms of schizophrenia, in particular, those manifested by fear and phobias, a significant improvement may be obtained by stereotactic or open destruction of the orbitomedial frontal lobe (Mitchell-Heggs *et al.*, 1977; Scoville and Bettis, 1975).

Bilateral stereotactic destruction of the medial thalamic nuclei (Section 2.3) diminishes hallucinations as well as the aggressive behavior of schizophrenics (Roeder *et al.*, 1971). A reduction in the aggressive behavior of such patients was noted by Vaernet and Madsen (1972) after bilateral stereotactic amygdalotomy and by Nadvornik *et al.* (1973) after posterior hypothalamotomy using the Sano method.

Stereotactic destruction of the rostral portion of CC was performed by Vilkki (1975) in 20 chronic schizophrenics, 12 of whom were markedly improved for a prolonged period and three of whom were practically cured. Destruction of the medial portion of CC produced a noticeable improvement in 45% of schizophrenics (Martin *et al.*, 1977).

Several attempts have been made to improve the surgical results in schizophrenia by simultaneously destroying a number of cerebral structures. In 47 chronic patients, Cox and Brown (1977) performed bilateral destruction of Am, GC, as well as bilateral subcaudatal tractotomy. A substantial improvement was noted in the majority of cases. However, it is quite natural to question the expediency and wisdom of destroying six foci in one brain.

Psychosurgical treatment of schizophrenia is indicated in the most severe cases of the disease after careful selection of patients for such operations.

3.5. Obsessive–Compulsive Neuroses

Incessant ritualistic movements and gesticulations, when performed continually, constitute one of the main indications for contemporary psychosurgery. The severe forms of these neuroses are practically incurable, and sufferers must spend their entire lives in psychiatric clinics, since various forms of psycho- and pharmacotherapy (neuroleptics) may be useful only in mild forms of the disorder.

In this disease, which psychiatrists relate to anancastic syndromes, there is, for example, a pathological desire for cleanliness, and such patients continue to wash their hands for hours. It is indicative that such patients are unable to control their inappropriate behavior, although they do realize that it is senseless. The etiology and pathogenesis of these neuroses have not been established.

On the whole, the treatment of obsessive–compulsive neuroses by stereotactic and functional operations is, without a doubt, very effective. Many reports of patients who have been almost completely cured may be found in the literature. Yet, the question of selecting the most appropriate cerebral target remains open; accordingly, it is difficult to compare the results of the different psychosurgical operations.

As already noted, anterior cingulotomy (Section 2.1) is an effective treatment for this disease. It is a safe operation causing no disturbance of memory, intellect, or gnostic functions (Brown and Lighthill, 1968; Paillas *et al.*, 1971; Laitinen, 1974; Ballantine *et al.*, 1979).

In many cases, good results have been reported after open or stereotactic destruction of the basal frontal cortex (Section 2.2) (Tan *et al.*, 1971; Mitchell-Heggs *et al.*, 1977). At the same time, Hassler (1982) believes that this operation does not give consistent results.

Operations involving the combined destruction of the medioorbital portion of the frontal lobe and GC have also been performed for obsessive–compulsive neuroses. The long-term results seem to indicate that these operations are more effective than the destruction of each structure alone (Andy and Stephan, 1982).

The combined destruction of the medial and intralaminary nuclei of Th (Section 2.3), especially on both sides, results in a substantial therapeutic effect (Hassler and Dieckmann, 1972).

There can be no doubt as to the effectiveness of the stereotactic destruction of the anterior limb of CI (Section 2.8), which, according to some data, cures up to 60% of such patients (Bingley *et al.*, 1973; Kullberg, 1977; Lopez-Ibor and Lopez-Ibor, 1977).

Consequently, in severe forms of obsessive–compulsive neuroses, there are several effective operations on different cerebral structures. The fact that the most "effective" structure has not yet been identified is evident from the work of Haaijman *et al.* (1977), who implanted a large number (38) of electrodes in each cerebral hemisphere in five patients for a long period of time (up to 7 months). The bundles of these electrodes were located in various cerebral structures: in the white substance around GC, in the radiation and rostral portion of CC, and in the orbitofrontal white substance. Numerous foci of destruction, each 7 × 4 × 4 mm, were consecutively made in these structures. To varying degrees there was a pronounced improvement in the mental states of all patients; however, the destruction of no specific structure that was effective in any particular disease could be pinpointed.

In conclusion, it is necessary to mention that since psychosurgical operations are not specific or effective, they are not indicated in mild cases (the counting of windows, steps, etc.) that do not handicap the life of the patient, in obsessive neuroses, in manifestations of schizophrenia, in manic–depressive psychosis, or in these neuroses combined with hysteria (Hassler, 1982).

3.6. Gilles de la Tourette's Syndrome

In 1885, Gilles de la Tourette described in nine patients a rare and unusual disease characterized by three peculiarities: onset in childhood; generalized hyperkinesias resembling a generalized tic; sudden, involuntary, often coprolalic loud shouts. de la Tourette gave the name ''maladie des tics convulsifs'' to this syndrome. It was thought to be extremely rare: during the 85 years following the first publication, only 70 cases of this disease (Fernando, 1967) were described in world literature. In recent years, however, there have been quite a number of reports about this disease based on a relatively large number of cases (Shapiro *et al.*, 1973; Melnitchuk *et al.*, 1977; Badaljan *et al.*, 1979). The incidence of the syndrome in the U.S. population is 0.05% (Lawden, 1986).

Two recently published monographs present details of the clinical picture, pathogenesis, and treatment of this syndrome (Shapiro *et al.*, 1973, 1978; Shanko, 1979). A multiauthored volume covers all aspects of Tourette's syndrome (Friedhoff and Chase, 1983). The preliminary data has shown that Tourette's syndrome is connected with the inhibition of dopamine and serotonin metabolism (Lawden, 1986).

The disease usually begins with insidious progressing ticlike involuntary movements in children of 5–10 years. The onset of the illness before the age of 10 years occurred in 90% of cases and in almost all patients by 13 years of age (Shapiro *et al.*, 1973). Uncontrollable shouts come somewhat later. The etiology of this syndrome remains unknown. A number of factors seem to point to the hereditary nature of the disease. For example, Wassman *et al.* (1978) report that 18 of the 21 patients they examined had a similar disease in the family. Hassler (1982) considers that more than 30% of the cases of Tourette's syndrome are familial. Yet other authors do not confirm such a high incidence of the hereditary factor; they indicate that in only 10% of such afflicted children were there analogous or similar phenomena found in parents or close relatives. We have noted a similar disease in the family of only one of our seven patients. Our observations do not confirm the opinion in the literature that Tourette's syndrome is a hereditary disease of the autosomatic dominant type and incomplete gene penetrancy (Lawden, 1986).

The pathological anatomy of this syndrome remains unknown. The only morphological study of this disease in the literature reports changes in Str (Clauss and Balthasar, 1954). Recently, the supposition was presented that the syndrome is caused by the damage of PAG and midbrain tegmentum, but without real morphological evidence (Devinsky, 1983).

Another mystery is that Tourette's syndrome appears in boys three to four times more frequently than in girls. Numerous attempts to discover some abnormality in the premorbid development of these children have proved futile. Their intellectual development does not deviate from standard, and, in fact, quite often it is above average. Emotional disturbances are not observed more frequently than in children ill with other (not mental) diseases. Neurological investigations have also failed to reveal any deviations from normal. Psychopathological symptoms are completely absent.

The hyperkineses in this syndrome are quite different from those of any other extrapyramidal diseases (with the exception of ''tics''). For example, one of our patients, during a conversation with a doctor, quite unexpectedly both to himself and to his physician, flung his arms forward, jerked his whole body, and, at the same time, issued several loud sounds resembling barking. In 30 sec to 1 min, all this was repeated. Another patient, while walking, would suddenly jump up, perform a complicated pirouette in the air, and then continue walking. Still another patient, a modest young woman, along with strange and turbulent body motions resembling ''shuddering,'' loudly shouted obscene words, simultaneously feeling very embarrassed and blushing. The coprolalia is observed in about 50% of the cases (Lawden, 1986). It has been suggested that such patients may, by force of will power, inhibit these hyperkineses and shouting, but only for a brief period of time. When the inner desire for such action gradually increases, the actions become irresistible and are performed, after which a brief feeling of relief is experienced.

An analysis of data in literature indicates that the opinions of authors differ as to whether Tourette's syndrome is a pyschoneurosis (obsessive–compulsive neurosis) or the result of organic damage to the brain. Many authors, mainly psychiatrists, maintain the "psychological" genesis of this syndrome, advancing as etiologic factors "the special position of a child in the family," "inadequate behavior of the parents," "suppressed hatred of one's parents," and so forth. The supposition that Tourette's syndrome is an obsessive neurosis is encountering justified objections based on the presence of a family amamnesis in certain cases or EEG deviations. Hassler (1982) regards Tourette's syndrome as a psychomotor disinhibition manifested by expressive movements and automatisms. Long observations have shown that the syndrome is a lifelong illness. We did not note the presence of remissions in these patients.

All types of therapy for this syndrome are usually futile. Hypnotherapy, electric shock, and insulin shock do not produce any effect. Psychotherapy helps rarely. A few attempts at leukotomy in the 1960s proved unsuccessful. Haloperidol, meprobamate, and other drugs help in only one-quarter of the patients (Hassler, 1982).

In recent years several attempts have been undertaken to use stereotactic destruction of the thalamic nuclei to treating this syndrome (Gajdosova *et al.*, 1972). In particular, the stereotactic destruction of intralaminary and medial nuclei of Th was tested in several cases of this syndrome. A good effect was observed in three of seven patients. The tics and coprolalia almost totally disappeared. However, compulsive shouts were diminished in only two patients (Hassler and Dieckmann, 1972).

In 1982 Hassler reported on 15 patients with this syndrome who underwent psychosurgical operations. For the majority, the operations were successful. The fact that so few of these operations have been reported does not allow one to judge their efficacy. In one case there was an improvement after bilateral dentatotomy (Nadvornik, and Šramka, 1973).

We operated on 12 patients with Tourette's syndrome, seven men and five women between the ages of 14 and 34 years (mean age–22 years). The operations involved stereotactic destruction of VL or both VL and the dorsomedial nuclei of Th. The following case illustrates a good result of surgical treatment of this rare disease with long follow-up.

Patient N, 34 years old, was admitted to our neurosurgical clinic in 1969. Mild hyperkineses had appeared in early childhood. At the age of 19 they became sharply intensified, permanent, and then were accompanied by involuntary shouts, which at first occurred only when he was excited and later also at rest. The patient was treated repeatedly in psychiatric clinics with many drugs but without effect. The disease continued to progress.

Objectively, the patient had normal intellect and no psychic disorders, although he was somewhat euphoric. He had no motor weakness, disturbances of muscle tone, sensation, station, or gait. However, he exhibited almost continuous rapid, disorderly, violent movements, intensifying during conversation or emotional states (Fig. 257A).

These severe irregular hyperkineses involved the torso, extremities, head, neck, and tongue. They occurred frequently, especially during purposeful movements. Quite often there was marked squinting of the eyes and facial grimacing with protrusion and biting of the tongue. One had the impression that many hyperkineses resembled normal motor acts, for example, opening and closing of the fingers to make a fist or a rapid thrust of one or both arms forward. At the same time, the purposelessness and senselessness of these movements were quite obvious (for instance, the patient might suddenly poke the little fingers of both hands into his mouth). There were also sudden purposeless flexing and extending of the legs at the knee and hip joints, retraction of the head, and bending of the torso. The hyperkineses were accompanied by loud, sharp, sudden cries, surprisingly like the barking of a dog. More often than not, these cries included separate sounds and syllables but also at times intelligible words. It was practically impossible for the patient to control these cries and hyperkineses by force of will. Working capacity was completely lost.

The patient underwent stereotactic cryodestruction of VL on the left side. Immediately after the operation, there was an almost complete cessation of hyperkineses, and the involuntary cries disappeared. However, in approximately 6 months the symptoms reappeared, although to a somewhat lesser degree than prior to the operation.

Over the next 2 years the patient underwent, at brief intervals, two more operations—another cryodestruction of VL and dorsomedial nuclei of Th on the

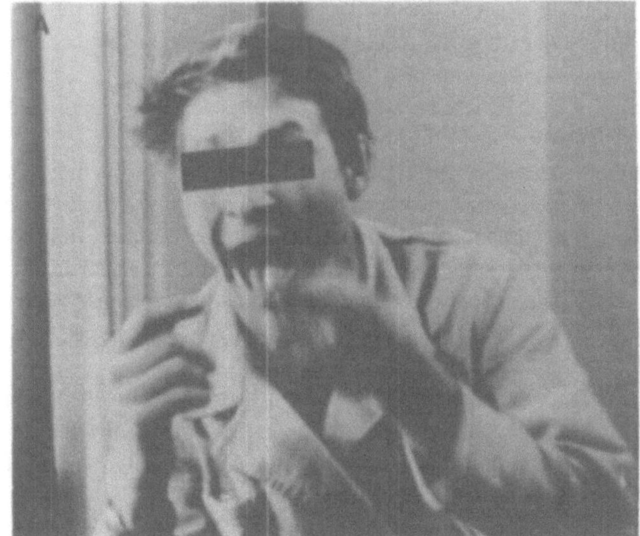

FIGURE 257. Gilles de la Tourette's syndrome. (A) A 34-year-old patient before operation; constant severe hyperkinesias accompanied by involuntary yells. (B) Same patient 12 years after consecutive bilateral stereotactic destruction of VL and dorsomedial nuclei of Th. Hyperkinesias and yells are absent.

left side and then destruction of the same nuclei on the right side. After the third operation, there was a mild hemiparesis, which gradually regressed.

In the 16 years since the three operations, the quite satisfactory result has persisted with practically total disappearance of the motor manifestations (Fig. 257B). No psychological disorders that could be attributed to the operations were observed. The patient is now working as a technician.

Subsequently, we operated on another 11 patients with Tourette's syndrome; in five of them the bilateral stereotactic operations were performed. In these cases, the results differed: good and stable improvement was noted in four patients, moderate improvement was noted in four patients, but the other three patients relapsed at various periods following surgery.

Although a stereotactic operation on the thalamic nuclei or ND so far offers the only real chance of helping the patient, a cure of this mysterious disease still remains an unsolved problem.

3.7. Sexual Deviations

The etiology of the numerous and varied sexual disorders is still unknown. The following have been suggested as etiologic factors alone or in combination: pathological hereditary factors, psychic disturbances, hormonal influence, and organic damage to the brain (Orthner, 1982). The important role of the nuclei of Hypoth in regulating sexual functions has long been known. Numerous experimental and clinical investigations have established that there are two hypothalamic sexual "centers," one which regulates secretion of gonadotropic hormones, and a second that is part of the "extragonadotropic" system of sexual behavior. The first of these mechanisms is connected with the ventromedial nucleus of Hypoth and the tuberomammillary complex, the destruction of which experimentally results in atrophy of the sexual glands. The second mechanism regulating sexual functions and sexual behavior, and which controls the first mechanism, is localized in the oral parts of Hypoth (Dieckmann and Hassler, 1975) and is also linked with the limbic system, especially with the amygdala complex. In experimental animals, bilateral damage of Am and the adjacent piriform cortex causes the Klüver–Bucy syndrome. It has also been established that this syndrome (hypersexuality, sexual perversion), developing after bilateral removal of Am, is abolished as a result of the destruction of ventromedial nuclei of Hypoth (Schreiner and Kling, 1953). Sexual functional disorders (in combination with other symptoms) are observed comparatively frequently in temporal lobe epilepsy (Chapter 14, Section 6.5).

The data presented above served as the rationale for attempts to modify sexual perversions by stereotactic destruction of Hypoth. The first stereotactic operation on this structure to relieve sexual deviations (pedophilia and others) was performed by Roeder (1965a,b), who stereotactically destroyed the tuber cinereum (nucleus ventralis medialis hypothalami) in one patient. A lasting therapeutic effect was obtained after the operation. It was noted that the operation did not cause any changes in metabolism, temperature, or carbohydrate metabolism.

Subsequent stereotactic operations on Hypoth for sexual deviations demonstrated that a "mating center" distinct from neuroendocrine centers is located in that area; for that reason, its destruction usually does not result in any pronounced endocrinometabolic disorders (Orthner, 1982). In the opinion of the author, unilateral hypothalamotomy is quite sufficient to eliminate severe sexual disorders.

Subsequently, Roeder *et al.* (1971, 1972) reported their results of unilateral stereotactic destruction of the ventromedial nucleus of Hypoth, the medial preoptic zone, and the tuberomammillary area in 16 patients with various sexual deviations (pedophilia, homosexualism, and others). In the majority of cases, the operation resulted in a good and lasting therapeutic effect, especially for homosexuality and hypersexuality). There were no relapses after the operation. It is noteworthy that after the operation no changes were observed in respect to intellect, sexual potency was preserved, and there was no obesity or diabetes insipidus.

Dieckmann and Hassler (1975) operated on six patients with sexual deviations (pedophilic homosexuality, hypersexuality). All these patients had been given long prison sentences for sexual crimes and voluntarily consented to an operation. Based on the experimental data above, stereotactic destruction of the basal portion of the ventromedial nucleus of Hypoth and tuberomammillary complex was performed. In patients with hypersexuality, the preoptic area was destroyed, and in one of the patients, the rostral intralaminary nuclei of Th. The destruction in all cases was performed on the right side. In five of six operated cases, the results were evaluated as good, and in one, satisfactory (period of observation from several months to 3.5 years). The pathological sexual behavior that had been observed prior to the operation disappeared almost completely. There were no endocrine or metabolic disturbances following the operation.

Subsequently, the same authors (Dieckmann and Hassler, 1977) reported on stereotactic destruction of Hypoth for hypersexuality that led to sex crimes. They operated on four men who had been sentenced to prison for repeated rape. The operations were performed at the request of the prisoners and with the permission of the court. In all these cases, unilateral stereotactic destruction of Hypoth and the tuberomammillary complex was performed, as a result of which the hormonal and behavioral "mating centers" were "switched off." In all cases a positive effect was obtained, since hypersexuality disappeared.

The experience discussed above indicates that the ventromedial nucleus of Hypoth may be assumed to be situated just above the zone between the infundibular recess and the mammillary bodies, and the tuberomammillary complex lies slightly in front of the ventromedial nucleus. The preoptic zone is localized somewhat rostral to the anterior wall of V_3. In the A–P projection, the ventromedial nucleus and the preoptic zone are 2.5 mm from the lateral wall of V_3, and the tuberomammillary complex lies at a distance of 4.5 mm.

It may be considered established that destruction of the ventromedial Hypoth results in the elimination of hypersexuality and other sexual deviations.

3.8. Narcotic Dependence

In Chapter 13 on the surgical treatment of pain syndromes, we have already mentioned addiction to narcotics that frequently develops in chronic pain. In this section we briefly outline the surgical treatment of "pure" narcotism.

The first, and in all likelihood, preliminary results were obtained in the treatment of the most severe forms of alcoholism by performing bilateral stereotactic destruction of the ventromedial nucleus of Hypoth. Eleven alcoholics who had been treated by all sorts of therapy without result were observed after the above operation by Šramka *et al.* (1977) over a period of from 1 to 5 years. The results of surgery were evaluated as quite satisfactory. Stereotactic cingulotomy for chronic alcoholism yielded good results in all six alcoholics (Martin *et al.*, 1977). However, Meyer *et al.* (1973) reported improvement in only half of 12 alcoholics from such a procedure.

Kanaka and Balasubramaniam (1978) performed stereotactic anterior cingulotomy (Section 2.1) in 73 patients suffering from various addictions (alcohol, morphine, opium). In all cases, psychiatric treatment over a period of many years had been futile. The authors reported that the results of treatment were good: 80% of the morphine addicts and 68% of the alcoholics completely abandoned their addiction. Cingulotomy also produced excellent results in respect to narcotism (morphine, merperidine hydrochloride): six of seven patients were cured with a period of observation from 3 to 5 years. Total withdrawal of narcotics immediately after the operation did not cause any grave abstinence symptoms (Ramamurthi *et al.*, 1982).

3.9. Anorexia Nervosa

In severe forms of anorexia nervosa, which usually ends in death from emaciation or suicide, good results may be obtained by performing prefrontal leukotomy (Crisp and Kalucy, 1973; Post and Schurr, 1977).

Psychosurgery is one of the most complicated, contradictory, but nevertheless rapidly advancing and promising trends in modern functional neurosurgery. There can be no doubt today that several operations, mainly stereotactic, on deep-lying brain structures may provide relief for many years to a great number of patients with various mental diseases and syndromes who could not be helped by any other contemporary method of treatment. One may assume that the further developments of this trend will make it possible within the next 10 to 15 years to demonstrate more evidence of its efficacy and therapeutic value.

19

Spinal Cord Disorders

1. Introduction

The clinical picture, diagnosis, and surgical treatment of disorders of the spinal cord represent a major and important aspect of neurosurgery, the presentation of which is not the objective of this book. That is why we discuss only certain functional operations on the spinal cord, proposed for the treatment of three common and grave syndromes. These syndromes, pyramidal spinal spasticity (often associated with painful spasms), urinary disorders, and respiratory paralysis are of such great clinical importance that their management represents a difficult problem for the neurosurgeon.

2. Spasticity

Muscular spasticity caused by interruption of descending corticospinal and reticulospinal pathways is caused by hyperactivity of the α and γ reflex arcs of the spinal cord (Chapter 2, Section 4). Spasticity occurs as a result of pronounced changes in the functioning of segmental neural circuits released from the inhibitory control of suprasegmental centers. Because of activation of the γ neurons, there is a decrease in the threshold of the α-neuron excitation threshold, which results in a sharp elevation of the stretch reflexes, especially the flexor reflexes. The function of the Renshaw inhibiting interneurons is reduced considerably (Granit *et al.*, 1957). Usually, spinal spasticity involves the leg flexors, which results in flexion–adduction postures and contractures or (more seldom) in severe hypertension. These pathological changes lead to tendinous and muscular contractures, ankyloses of the joints, and bony deformities.

Spasticity, one of the gravest consequences of traumatic and other lesions of the spinal cord and brainstem, has long attracted the attention of neurosurgeons, but it still remains a difficult and unresolved problem. Recently a number of drugs (dantrolene, baclofen) with a relaxing effect on the skeletal muscles, primarily achieved by suppressing discharge of Ca^{2+} in muscle cells, have been proposed. However, the efficacy of these drugs is limited. When drug therapy is not effective, functional neurosurgery is the only alternative treatment.

The surgical treatment of spasticity using ablative techniques or chronic stimulation of different cerebral structures in CP is presented in Chapter 9. This section addresses the use of different surgical methods for the treatment of spasticity resulting from spinal cord lesions of various origins.

The indications for surgical management of spinal spasticity are to be decided strictly on an individual basis. As Sindou (1985) stressed, only relatively regional spastic disorders limited basically to the limbs are proper indications for segmental surgical operations. It is necessary to distinguish spasticity from secondary changes in muscles, tendons, and joints. If there is no functionally meaningful residual motor activity, this kind of surgery is not indicated.

Surgical operations for the relief of spasticity secondary to spinal damage may be performed on the peripheral nerves, the spinal roots, or the spinal cord itself. Operations on the peripheral nerves for the relief of spasticity have been performed for many years. For example, resection of the obturator nerve for the relief of spasticity of the legs was proposed by Lorenz back in 1897. Many other attempts were made to relieve spasticity by section of different peripheral nerves totally or partially (neurotomy by Stoffel, proposed in 1912). These procedures have proved to be ineffective and were abandoned until Gros (1979) recently presented good results of microsurgical selective fascicular neurotomy of the tibial nerve. Recently Sindou (1985) reported that partial resection of the soleus and gastrocnemius nerves under the microscope in equinus deformity and ankle clonus and also the partial resection of the posterior tibial branch or the corresponding fascicles inside the distal tibial trunk gave an excellent and long-lasting clinical improvement in standing and walking in 28 of 31 patients.

Sindou *et al.* (1985b) also reported favorable results of selective neurotomy of the obturator nerve for spasticity of the adductors and flexors of the hip and of sciatic branches for hypertonic flexors of the knee.

Surgical interventions on the spinal cord for relief of spasticity have been performed for many decades. From the point of view of functional surgery, there is no need to discuss destructive operations on the spinal cord for the treatment of spasticity; these include total resection of the cauda equina or removal of the lower parts of the spinal cord (MacCarty and Kiefer, 1949). These operations, which were proposed in the 1950s, produced increased pareses of the legs and severe muscular atrophy and were abandoned.

Of other destructive methods for the treatment of spasticity in the lower limbs and neurogenic bladder, one should mention electrocoagulation of the spinal cord and roots (Gol and Dossman, 1965). This operation is permissible only in cases of total irreversible paraplegia.

A lumbar puncture needle is inserted into the vertebral canal at the level of the lumbar enlargement until the tip of the needle strikes the posterior surface of the vertebral body. The needle is then withdrawn until its tip is in the center of the spinal cord. With an electrode passed through the needle, the spinal cord is destroyed by high-frequency coagulation. Considerable improvement in spasticity and bladder function has been reported after such an operation.

The introduction of alcohol or phenol (Dogliotti, 1931) into the spinal subarachnoid space for the management of spasticity and pain has proved to be ineffective. Since an adhesive arachnoiditis frequently develops after injection of lytic solutions in the subarachnoid space, this procedure has been practically abandoned.

A relatively effective method for relief of spasticity after spinal cord injury, proposed by Babichenko (1966), is hydrothermal destruction of the lumbar enlargement by an intraspinal injection of a small volume of hot distilled water.

2.1. Myelotomy

Among the contemporary techniques for surgical treatment of spinal spasticity, the most commonly used is longitudinal myelotomy, proposed by Bischof in 1951. This operation involves severing the spinal reflex arcs between the anterior and posterior horns of the lumbosacral spinal cord, thus inactivating the segmental neurons involved in the pathogenesis of spasticity as well as the connections of the corticospinal tracts to the neurons of the anterior horns.

The main advantage of the Bischof operation is that the pyramidal tracts and other long pathways of the spinal cord remain intact. The spinal cord is incised between the corticospinal and spinothalamic tracts. In the majority of the patients operated on who have a complete paraplegia of traumatic or other origin, there is no need to preserve the pyramidal tract. However, in patients with some movement, the motor functions must be preserved, although in the presence of severe spasticity it is difficult to assess the amount of voluntary movement remaining. Sometimes after the spasticity has been abolished, the movements in the legs are noticeably better than was presumed before the surgery. As already pointed out, in myelotomy the pyramidal pathways are not damaged appreciably, but after the operation, there may be a slight transitory or sometimes permanent increase in paraparesis. For this reason, great caution must be exercised in advising myelotomy for patients who are able to walk.

Severe spasticity in the legs is often combined with pain, both of which may be relieved by myelotomy.

In the more than 30 years since the pioneer work by Bischof, several modifications of myelotomy have been proposed. The author himself in 1967 suggested a modification of the technique, which is often called Bischof-2 myelotomy. The changes concern the technique of dividing the spinal cord. The laminectomy and exposure of the cord are done in the usual manner.

The patient is placed on the operating table in the prone position, and a catheter is inserted to record pressure in the bladder. A laminectomy of T_{10}–T_{12} or T_{10}–S_1 (checked by x ray) is performed under general anesthesia. A longitudinal dural incision is made to expose the lumbar enlargement of the spinal cord. It is necessary to identify precisely the roots and segments of the spinal cord in order to ascertain the superior and inferior extent of the myelotomy. Laitinen and Singounas (1971) recommend using air myelography before the operation to identify conus medullaris. Root S_1

may be determined visually since it is the thickest root of the cauda equina. For stimulation of the spinal cord, it is rotated somewhat, and electrodes are placed on the anterior roots. The stimulation of the T_{12} root with a weak current (1.5–2V and 60 Hz) causes contraction of the abdominal muscles. Stimulation of S_1 produces foot movements, and S_3 contraction of the bladder detrusor muscle and an increase of bladder pressure. Evoked potentials may also be monitored to control the effectiveness of myelotomy (Cusick *et al.*, 1976).

2.1.1. Lateral Longitudinal Myelotomy

This operation, proposed by Bischof in 1951, is still used (Moyes, 1969; Livshitz *et al.*, 1976; Virozub and Chipko, 1981). The operation involves a deep incision in the lumbar enlargement of the cord in the frontal plane from one or both sides so as to section the anterior commissure. In bilateral myelotomies the incisions are made consecutively so as to divide the spinal cord completely into two halves, ventral and dorsal. In order to determine the depth of the incision, it is necessary to make a preliminary measurement of the width and AP diameter of the lumbar enlargement.

Myelotomy is performed with a special instrument consisting of a metal handle to which is affixed a short, sharp blade at an angle. The length of the blade must be somewhat greater than half of the diameter of the lumbar enlargement. For this blade, a keratome or piece of a safety razor can be used. Three dentate ligaments at the desired segment are cut and held with small forceps so as to rotate the spinal cord around its longitudinal axis by approximately 15–20°.

The blade of the myelotome is introduced into the lateral surface of the spinal cord at the site of attachment of the dentate ligament between the anterior and posterior roots. The incision (usually from top to bottom) is made along the frontal plane, taking into account the slight rotation of the spinal cord. The sharp end of the myelotome should pass through the central canal to the other side of the spinal cord by 1–2 mm. Then, the same myelotomy is performed on the other side, so that the two incisions connect. The length of the incision varies from 60 to 70 mm.

In spastic paraplegia, Bischof recommends bilateral myelotomy from L_1 to S_1 segments and for relief of bladder spasticity continuation of the incision caudally to include S_2–S_3 on one side. After such an operation, spasticity is sometimes still present in the femoral and ilioinguinal muscles. Bischof considers this to be the result of anomalous interrelationships of the vertebrae and spinal segments or anatomic variations of the lumbar enlargement. In order to deafferent further the muscles in the lower extremities, Virozub and Chipko (1981) proposed that the lateral myelotomy be lengthened by two spinal segments, one upward and one downward, so that the section of the spinal cord would be from T_{12} to S_2 segments.

The cuts of the spinal cord from the right and from the left must meet. This is easily checked by passing a thin and narrow spatula, the tip of which is bent at an angle of 90°, through the incision until the tip is seen extruding from the incision of the spinal cord on the other side.

There are usually no hemorrhagic complications after a technically correct myelotomy. If the cord has been erroneously incised posterior or anterior to midfrontal plane, an unlikely occurrence, there may be injuries to the artery of the lumbar enlargement passing along the ventral surface of the spinal cord slightly to the left of the midplane.

In 1962 Tönnis and Bischof published the long-term results (average 4–5 years) of this operation in 20 patients with different lesions of the spinal cord. A satisfactory relief of spasticity was obtained in 17 of 20 cases. Of 18 bedridden patients, only three remained in this state after surgery. Four patients were able to walk, and preoperative leg pain was relieved in all but three cases.

Moyes (1969) summed up the experience of several Canadian neurosurgeons who operated on 21 patients. In the majority, flexor spasticity was the result of spinal cord injury or multiple sclerosis. Immediately after surgery in all the cases spasticity and leg pain were relieved, but some spasticity recurred in four cases. After myelotomy, reflexes mediated by arcs in the spinal cord segments below the level of section may return (Livshitz *et al.*, 1976).

Myelotomy by Bischof's lateral approach was reported by Virozub and Chipko (1975) to improve spasticity in the legs and diminish the tone of a spastic bladder in paraplegics. After this operation 23 of 57 patients could be trained to walk with a split-frame apparatus.

After an extensive experience with Bischof's longitudinal myelotomy, a number of problems have come to light. In patients with some leg movements, there was often an increase in the paraparesis after surgery. In view of the fact that this operation interrupts the connections between the corticospinal pathways and the anterior horns of the spinal cord, there is no increase in muscular strength or enhancement of movements, which, if the fiber tracts were intact, would be expected after relief of spasticity in the legs. If voluntary control of bladder functions was preserved, after myelotomy it might be impaired. Complications arose as the result of rotation of the spinal cord to make the section in the frontal plane. Furthermore, the immediate improvement of spasticity was lost with the passage of time (Laitinen and Singounas, 1971). These factors stimulated the search for ways of improving this operation.

2.1.2. Posterior Longitudinal Myelotomy

As a further development of the principle of longitudinal myelotomy, Bischof (1967) (Bischof-2 operation) suggested a posterior approach through the spinal longitudinal fissure. After a laminectomy and coagulation of the vein lying in the posterior longitudinal fissure, a T-shaped incision is made in the spinal cord (Fig. 258). The first cut is made in the sagittal plane through the posterior longitudinal fissure from the posterior surface of the cord to the central canal at a depth of about 3 mm.

The second incision is made perpendicular to the first in the frontal plane through the white substance on both sides of the midplane incision. In this case, each frontal incision severs approximately three quarters of each half of the spinal cord. The longitudinal incision is made along the entire lumbar enlargement of the spinal cord, extending superiorly from 2.5–3 cm above the conus medullaris. The length of this midline incision is somewhat less than in lateral myelotomy—5.5 cm (Laitinen and Singounas, 1971). The lateral incisions are made with an angular scalpel inserted into the center of the dorsal incision and then passed laterally on both sides. The operation is performed under magnification, usually with a surgical microscope.

The advantage of posterior myelotomy over the lateral operation is the preservation of some connection between the lateral pyramidal tracts and the motorneurons of the anterior horns as well as the posterior horns (Kolliker's anastomoses). These anastomoses (or collaterals) enter the spinal cord via the posterior roots and then proceed upward and downward to the neurons of the anterior horns. With the preservation of these ties, the residual motor function is not disturbed and sometimes even significantly improved.

There are several variants of this operation, which differ little from each other. Cusick *et al.* (1976) suggest the name T-myelotomy for this operation. Since the effect of myelotomy is dependent on the destruction of the spinal pathways in the gray substance, Ivan (1982) proposed a modification, naming it griseotomy. In this operation, an L-shaped instrument with a guard on the sharp tip is introduced in the posterior longitudinal fissure and used to make numerous circular foci in the white substance along the equatorial planes of the spinal cord. A similar technique was used by Yamada *et al.* (1972, 1976), who devised a special instrument for this purpose (Fig. 259). All the modifications mechanically destroy spinal cord tissue.

A technique for the destruction of the white substance by high-frequency coagulation (Ivan, 1982) has been used in five patients. A special electrode needle

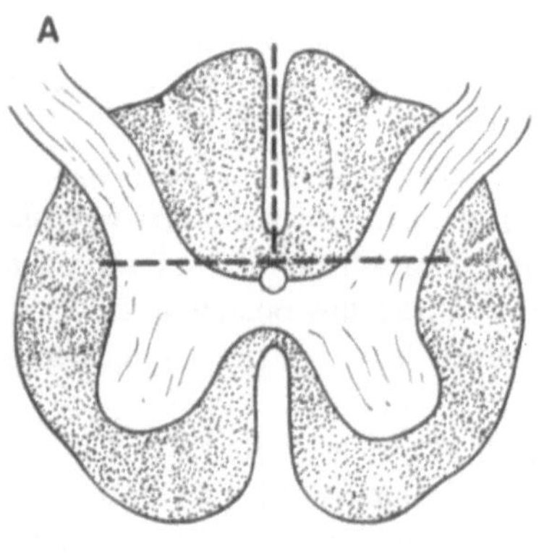

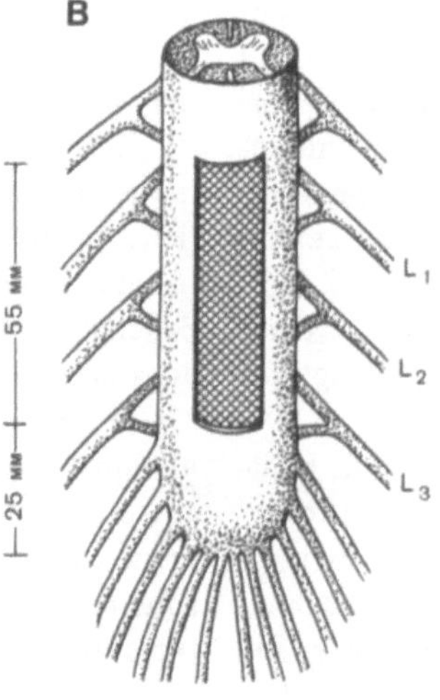

FIGURE 258. Scheme of myelotomy after Bischof. (A) Site of introduction of the myelotome and bilateral section of the spinal cord from the posterior longitudinal fissure are marked with dashed lines. (B) Schematic illustration of the length and size of the myelotomy at the lumbar enlargement of the spinal cord.

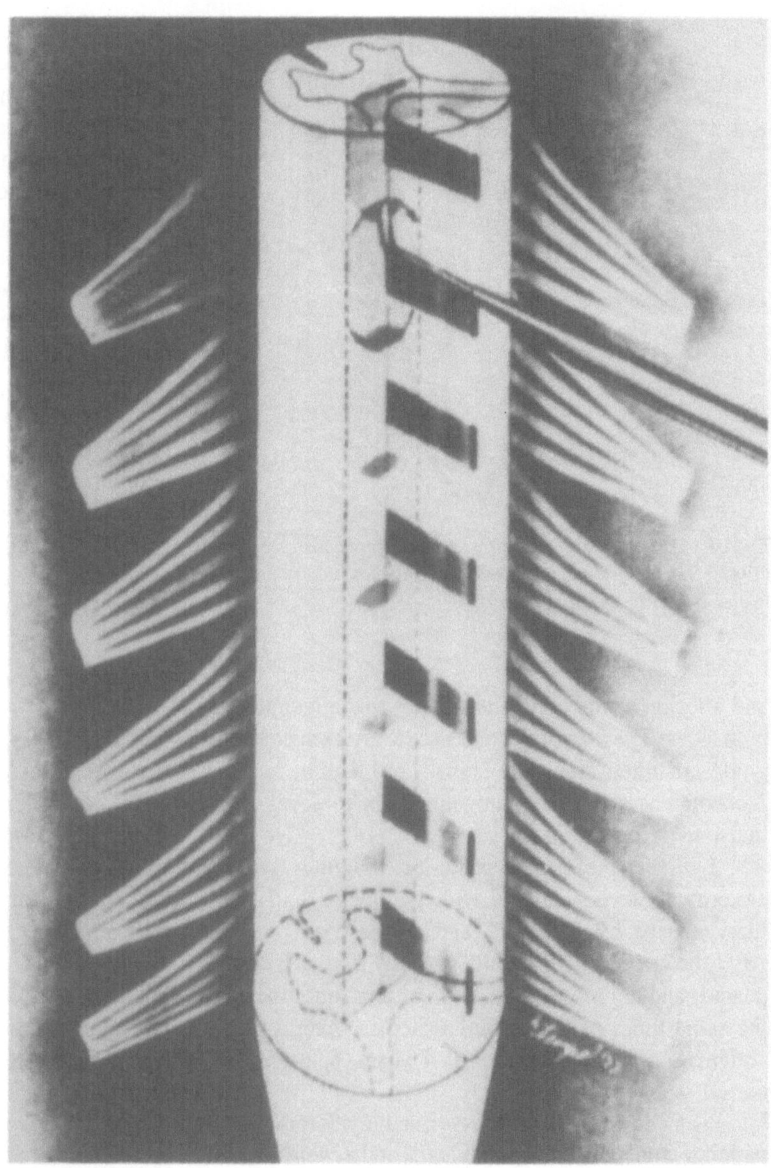

FIGURE 259. Modification of Bischof's myelotomy proposed by Yamada *et al.* (1976).

with a bare sharp tip 0.5 mm long is introduced into the posterior surface of the spinal cord at a distance of 2 mm from the posterior longitudinal fissure. Depending on the AP diameter of the spinal cord, the electrode is inserted to a depth of 3 or 4 mm, and the white substance is subjected to high-frequency coagulation using a Radionics or other lesion generator with a current of 100 mA for 10 sec. The destructive foci are made in a chain at a distance of 4 mm from each other (Fig. 260) from T_{12} to S_1 segments. Altogether, 14 to 18 focal lesions are made.

This operation was further improved by Laitinen *et al.* (1974), who devised a special myelotome. This instrument made it possible to avoid damage to the dorsal vein of the spinal cord as well as to limit the damage to the commissure, thereby preserving the pathways that cross there. The instrument has a relatively simple metallic frame with a guide and needle through which it is possible to introduce a sharp stylet that protrudes laterally. After laminectomy T_{10}–T_{12}, the instrument is fixed securely above and below the laminectomy to the spinous processes of T_9 and L_1. The needle punctures the posterior surface of the spinal cord to a depth of 2.5 mm; then the stylet is inserted,

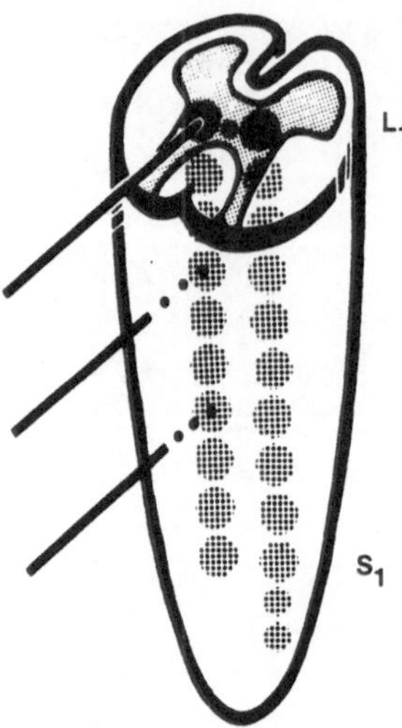

FIGURE 260. Myelotomy (griseotomy) proposed by Ivan (1982). Small radiofrequency lesions are made within spinal gray matter by the posterior approach.

and a lesion of the cord substance is made laterally on both sides. Then the stylet and needle are extracted, the guide is shifted along the spinal cord 8 mm, and the procedure is repeated. Altogether, six to seven such incisions are made in the exposed cord.

Laitinen performed posterior longitudinal myelotomy in seven patients with paraplegia and spasticity caused by MS, CP, and spinal cord injuries. After the operation, spasticity partially or totally disappeared, and leg movements improved in four cases. At the same time, pain was diminished in several of the patients. In follow-up, after 1–4 years, spasticity recurred in a few cases.

Yamada *et al.* (1976) performed myelotomy in 14 patients with para- and tetraplegia and severe painful spasms in leg muscles. After the operation, spasticity in the legs sharply diminished in all cases, painful muscular spasms disappeared, contractures diminished, and bed sores healed. Some of these patients were able to use a wheelchair and even to walk with assistance.

In three of five paraplegics, Ivan (1982) reported that high-frequency coagulation of the spinal gray substance relieved pain and spasticity. Satisfactory results many years after posterior myelotomy for severe spasticity and irreversible paraplegia were noted by Schirmer and Wenken (1981). After the operation, these bedridden patients were able to move about in a wheelchair, which had not been possible before surgery.

Without a doubt, posterior longitudinal myelotomy is an effective method to relieve spasticity. Relapses are rare after the operation. At the same time, it has certain shortcomings. Muscular strength in the legs, greatly reduced in some of these patients (the others are paraplegic), usually is not decreased; however, pain and temperature sensation may be disturbed to a greater degree than before the operation. Nevertheless, posterior myelotomy is preferable to lateral, since decreased sensation is less important for the patient than a greater motor deficit. Bladder functions after the operation also become temporarily worse, but for a relatively short period (Laitinen *et al.*, 1974).

There is every reason to conclude that in severe lesions (especially total interruption) of the spinal cord without hope of restoration of functions, and in the presence of the marked spasticity, contractures, severe pain in the legs, as well as urinary disorders, myelotomy is a worthwhile operation, providing significant relief to such incapacitated patients.

2.2. Posterior Rhizotomy

In 1908 Foerster proposed and carried out for the first time transsection of the posterior roots L_2–L_3 and L_5–S_1 in order to interrupt the afferent path of the spinal reflex arc involved in spasticity of the legs after spinal cord lesions. The rationale of the operation in spastic syndromes is to suppress hyperactive stretch reflexes by decreasing or eliminating the afferent input of the spinal segmental reflex. In a later report, Foerster (1913) described his clinical experience with this operation in 159 patients. Subsequently, the operation named after him was practically abandoned in its original form, since, after total interruption of the lumbar posterior roots, there were disorders of sensation and sphincter functions, trophic alterations, in particular bed sores, as well as relapses of spasticity. At present the operation is only used in the management of pain (Chapter 13, Section 5.3.6).

Forty years after the Foerster operation for spastic paralysis, Munro suggested cutting not the posterior, but the anterior roots from T_{11} to S_1 (Munro, 1945). Soon this operation was also abandoned because of the development of muscular atrophy in the lower extremities, as a result of which bony protuberances formed, leading to the development of bed sores. Urinary disorders developed to an even greater degree. At the same time, there was only mild improvement in the spasticity of the legs. Even section of all roots (anterior and posterior) (Meirowsky *et al.*, 1950) proved no more effective and was abandoned.

Approximately 18 years ago the Foerster operation had a revival. Gros *et al.* (1967) proposed that the posterior roots not be divided completely but only in part—approximately three-quarters of the rootlets of each posterior root. This technique greatly reduced the sensory disorders and trophic disturbances occurring after total rhizotomy. As a rule, each posterior root from L_1 to S_1 is cut with preservation of the radicular vessels supplying the conus medullaris. After the operation, Gros *et al.* (1967, 1981) reported good and lasting results in 62 cases with lower limb spasticity and painful spastic contractures.

At present, total section of the posterior roots (rhizotomy) is used rarely, mainly at the cervical level for spasticity of CP (Chapter 9, Section 5.2). Much more frequently selective posterior rhizotomy is performed, either partial section of the posterior roots or selective section of the rootlets where they enter the spinal cord. Some authors have named the second variant radiculotomy.

Selective rhizotomy offers a number of important advantages over total rhizotomy, but the technically simpler total rhizotomy is still sometimes performed for severe spasticity caused by total transverse interruption of the spinal cord. However, in cases in which some residual spinal function is preserved, total rhizotomy, although relieving spasticity, may significantly impair the residual function, especially superficial and deep sensation. On the other hand, the effect of selective rhizotomy on spasticity is much the same as in total rhizotomy; however, sensation after this operation is not significantly impaired. Another important advantage of selective section of the posterior roots is that there is no deafferentation pain (Sindou *et al.*, 1983).

It has been noted that after the section of the posterior roots or their rootlets at thoracic or lumbar levels, in some cases there is a paradoxical decrease in spasticity in both arms and trunk. The explanation is obscure; however, one may assume that the effect of rhizotomy on spasticity depends not only on segmental deafferentiation but also on changes in the functional state of supraspinal levels regulating muscle tone (Fasano *et al.*, 1977).

Selective rhizotomy at the thoracolumbar level is performed under general anesthesia with the patient in a prone position. Sindou *et al.* (1974b, 1981) stressed that because the identification of the posterior roots is based on the intervertebral foramina, the determination of the level of laminectomy is of paramount importance. It may be made by x ray before surgery with the use of cutaneous markers or by injection of methylene blue into an interspinous ligament (White and Sweet, 1969). After a laminectomy T_{12}–L_1 or T_{12}–L_2 and opening of the dura, the spinal cord and roots of cauda equina are inspected. Approximately 6 cm of the spinal cord to the beginning of filum terminale should be identified (Fraioli and Guidetti, 1977). Then the arachnoid covering the roots is removed, carefully preserving the radicular arteries. At this stage, it is necessary to use a magnifying glass or surgical microscope. Just as in myelotomy, a most important task for the surgeon is to identify the anterior and posterior roots and then each posterior root. This task is much more difficult at the lumbosacral level than at the cervical, thoracic, and upper lumbar levels, where it is easy to follow the root from the intervertebral foramen. For this purpose some reference points are desirable, as well as electrical stimulation of the roots as described for myelotomy (Section 2.1) and for section of posterior rootlets in severe pain (Chapter 13, Section 5.3.6).

Each of the posterior roots to be incised is then separated into individual rootlets where it enters the spinal cord. Each root consists of four to ten rootlets with a diameter of 0.5–1.6 mm that are clearly visible under the microscope. Then, all the rootlets except one in each posterior root from T_{12} or L_1 to S_1 inclusive on both sides are divided (Laitinen and Fugl-Meyer, 1982). It is recommended that the dissection of the rootlets deepened by 1 mm at their entrance zone to the spinal cord (Lissauer tract); moreover, the plane of the section should make an angle of 45° with the surface of the posterior column (Sindou *et al.*, 1974b, 1982a). Fraioli and Guidetti (1977) use another variant of this technique; under a microscope at a distance of 1 mm from the spinal cord, they cut all the rootlets of each posterior root on both sides, but only halfway or two-thirds through. The authors give a simple test for identifying the root of S_1; after section of this root, ankle clonus on the same side disappears. Better relief of spasticity was obtained in patients in whom voluntary movements did not increase postural spasticity. The results in patients with paraplegia were significantly better than those in tetraplegia.

The next stage in the technique of radiculotomy began after Sindou *et al.* (1974b, 1982a), Fasano *et al.* (1976), Privat *et al.* (1976), and Gros *et al.* (1981) introduced electrostimulation as a means of deciding which posterior rootlets to cut. Fasano named this procedure "functional posterior rhizotomy." Certain posterior rootlets on stimulation produced strong tonic contraction of segmental muscles, determined visually or by the surface EMG, whereas stimulation of other rootlets caused weaker or no contractions. It is necessary to section the "active" rootlets and to leave the "inactive" ones intact. Using this criterion in 62 patients, Fasano *et al.* (1976) sectioned from 25 to 90% of the rootlets and in the majority of cases obtained a significant improvement in spasticity of the legs without any sensory deficit.

A similar technique was employed by Laitinen *et al.* (1983) in nine patients with spastic paraplegia, the majority of whom had MS. Using a bipolar stimulus at 60 Hz, 1–6 V, 0.2 msec, and duration 1–3 sec, the authors found that tonic contractions of segmental muscles, and sometimes beyond the segments, occurred on stimulation of 60–80% of the rootlets. These rootlets were sectioned; the rest were left intact. In four of nine patients, spasticity in the extermities disappeared completely, and in the rest there was a significant improvement.

A teleological classification of spasticity has been introduced by Gros *et al.* (1967) and Sindou *et al.* (1982a). They consider that spasticity may be "useful" or "handicapping" (harmful). The goal of the surgeon is to enhance the former and eliminate the latter. The authors showed that it is possible to identify groups of muscles that are affected by spasticity of either the first or second type. With the help of electrostimulation of the rootlets and simultaneous EMG recording, the rootlets of each type may be identified at operation and marked by colored threads. The rootlets responsible for "handicapping" spasticity are selectively sectioned, and the rootlets reponsible for "useful" spasticity are left intact. For example, if bipolar stimulation causes flexion or adduction of the legs ("harmful reaction"), then section of the corresponding rootlets is performed; however, if stimulation leads to extension ("useful reaction"), the rootlets are left intact. Using this technique, Gros *et al.* (1967) obtained good results in the majority of 100 patients with severe spasticity in the legs, and Sindou *et al.* (1982a) employed this technique in 15 cases with spastic and painful paraplegia caused by advanced MS. The authors stressed the importance of sectioning a fairly extensive number of posterior roots. It is necessary, under the operating microscope, to section partially the ventrolateral part of each posterior rootlet. At the junction of these rootlets with the spinal cord, a small 1-mm-deep incision is made to interrupt the small nociceptive fibers that are grouped ventrolaterally and the large myotatic fibers that are located in the central part of the rootlet. The technique leaves intact the majority of the proprioceptive fibers passing to the posterior column, which are situated dorsomedially.

A percutaneous radiofrequency rhizotomy in 25 cases of severe spasticity after head or spinal traumatic lesions was performed by Kasdon and Lathi (1984). The clinical improvement was achieved in a great majority of cases with prolonged relief of spasticity.

Selective posterior rhizotomy at the cervical level for treatment of spasticity in the upper limb has also given satisfactory results in hemiplegia of traumatic or vascular origin. Good and lasting relief of spasticity, defense reactions, and pain was noted by Sindou in 12 of 15 patients in a 1- to 8-year follow-up.

The anterior rhizotomy developed by Monro and later abandoned has been reevaluated recently by Gros *et al.* (1981). The authors recommended cutting a small number of anterior rootlets combined with posterior rhizotomy. Only the anterior rootlets that reproduce the spastic phenomena on intraoperative electrostimulation should be sectioned.

The positive results of DREZ coagulation for the control of central pain in paraplegics are noted in Chapter 13, Section 5.3.7.

2.3. Stimulation Methods

Among the stimulation methods advocated for the relief of spasticity, chronic electrostimulation of the cerebellar cortex or deep Cez nuclei is most frequently used, especially to improve spasticity in CP. It is described in detail in Chapter 9 (Section 5.3).

In recent years, PCS has been successfully used to improve handicapping spasticity and motor disorders. The preliminary testing and technique of PCS as well as the necessary equipment are described in Chapter 13 (Section 5.4.2).

Experimental investigations on decerebrated cats, which was the classical model of γ rigidity, have demonstrated that PCS causes inhibition of the hyperexcitable spinal circuits and stretch reflexes (Chapman and Wiesendanger, 1981). Reduction in the activity of stretch reflexes in patients who underwent PCS (Siegfried, 1980a) has also been reported. The inhibition may be caused by direct activation of the descending inhibitory reticulospinal pathways or by long-loop mechanisms (Barolat-Romana *et al.*, 1985). The exact physiological mechanism involved in reducing spasticity by spinal cord stimulation still remains obscure. In selecting patients for this operation, only test stimulation is a reliably accurate indicator of the therapeutic effect.

Posterior column stimulation was first employed by Cook and Weinstein (1973) in four patients with severe spasticity in the legs from MS. The authors noted a considerable improvement in the spasticity of the extensors as well as a definite improvement in motor functions.

Siegfried *et al.* (1978) and Siegfried (1979, 1980a) implanted endodural (between two leaves of the dura mater) monopolar electrodes at cervical and high thoracic levels in 11 patients with spasticity of the legs from various causes (most frequently MS). After a limited laminectomy using a curved dental instrument, the author separated the dura mater on both sides from the lateral incision of the external dural leaf. An electrode was inserted into the pocket, and the overlying dura mater was sutured. The author assumed that this technique prevents leakage of CSF, inflammatory complications, arachnoiditis, and possible compression of the spinal cord. Percutaneous introduction of so-called floating electrodes into the subarachnoid space was employed by the author temporarily to test the effect of stimulation.

In two patients Siegfried (1980a) inserted electrodes percutaneously in the epidural space. The receiver was implanted subcutaneously in a pocket formed under the clavicle. Shortly after initiation of stimulation, the patients felt paresthesias and reduced muscular tension in the ipsilateral extremities, which in several hours was confirmed by objective examinations. In the majority of patients with PCS, there was a significant improvement in spasticity of the extremities. The results were better in MS patients, much less effective in spinal paraplegics, and absent in cases of CP.

Barolat-Romana *et al.* (1985) used chronic epidural spinal cord stimulation in six cases of intractable spastic spasms resulting from complete and incomplete spinal cord injury. The electrode tip was usually placed at T_{1-2} level in cervical cord lesions, where stimulation produced paresthesias in all extremities. Spasticity was relieved in all six patients. The authors stressed that stimulation produced immediate inhibition of the spasms in three patients. As did other investigators, they noted that the effect of PCS was not limited to.the level of stimulation but spread in both directions along the spinal cord. In one case, there was a marked reduction of extensor spasticity in the arms with the electrode tip at T_3.

By 1981 more than 1000 patients in different countries had been treated by chronic spinal cord stimulation for motor disturbances and spasticity (Lazorthes *et al.*, 1981) (Fig. 261). Summing up the data of many sources, one may conclude that epidural spinal electrostimulation reduces spasticity and improves motor functions in approximately 20–50% of the cases (Illis *et al.*, 1976; Thoden *et al.*, 1977; Dooley *et al.*, 1978; Siegfried *et al.*, 1978; Richardson *et al.*, 1979; Davis *et al.*, 1981; Klingler and Keplinger, 1982; Waltz, 1982; Dimitrijevic *et al.*, 1983; Sharkey *et al.*, 1985). Other authors have noted that

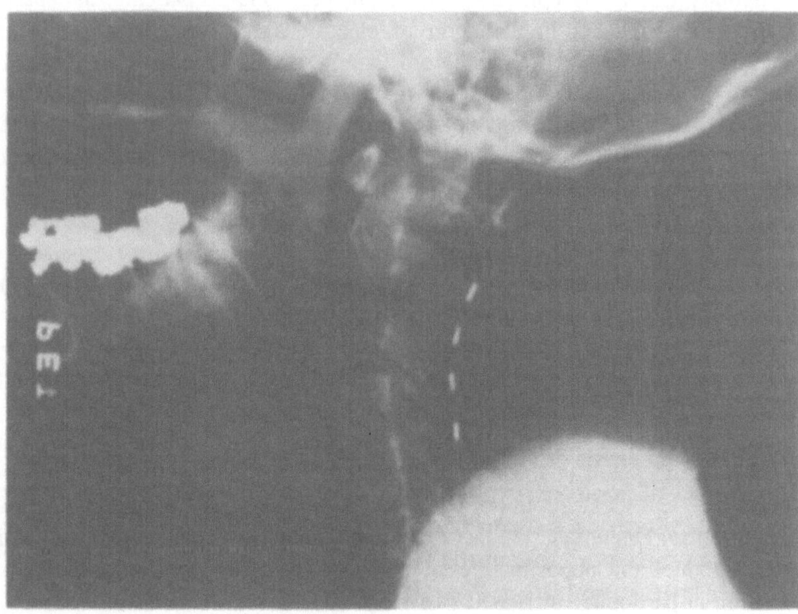

FIGURE 261. Multilevel four-platinum-electrode catheter 1.2 mm in diameter is inserted in the cervical spinal cord for chronic stimulation (Waltz, 1982).

PCS reduces spasticity in 40–70% of the cases (Reynolds and Oakley, 1982).

It has been established that PCS is very effective in so-called cramps or muscular spasms of the legs, which often occur in MS and other spinal lesions. The largest clinical experience and most thorough analysis of long-term results from PCS with a follow-up of several years is a joint report of three neurosurgical clinics in Zurich, Milan, and Toulouse (Siegfried *et al.*, 1981). Of 35 MS patients, 21 were found to have good or very good improvement in spasticity. Prior to the operation only one patient was able to care for himself, but after PCS, 11 were independent. Of 19 patients who prior to the operation were able to move about only in wheelchairs, only 11 needed such assistance after surgery. Of six patients who were bedridden before the operation, only one remained bedridden after surgery. Bladder dysfunctions, which were present in 30 patients before the operation, markedly decreased in 18 and slightly decreased in another nine. In several patients the cerebellar dysfunction was lessened. However, Dooley *et al.* (1978) reported that there was improved motor function of the spinal cord in only 15% of MS patients after PCS.

It should be noted that the criteria for selection of patients with spasticity and disturbed motor functions for PCS are becoming increasingly stringent. Indications for PCS in recent years have been more rigid. According to the data of the previously mentioned cooperative study in three European neurosurgical clinics (Siegfried *et al.*, 1981) of 164 patients who underwent temporary testing by epidural stimulation, only one-third had permanent implantation of electrodes for PCS. Of 15 patients with spinal cord trauma, permanent implantation was done in only three cases.

The first successful attempts at prolonged electrostimulation of the cauda equina for spasticity have been reported. In a patient with spastic paraplegia after midthoracic interruption of the spinal cord, electrodes were implanted in the cauda equina (at the level of L_1–L_3), and a miniature receiver connected to the electrodes was implanted subcutaneously (Richardson and McLone, 1978). For stimulation, a current with the following parameters was used: 1–2 V, 33 Hz, 150–200 msec. After the operation and stimulation, not only was there a sharp reduction in spasticity (by 90%), but the pain in the legs disappeared, erection occurred, and defecation normalized. At the same time, the authors emphasize that excessive stimulation may produce the opposite effects, namely, negative results, in particular, increased spasticity.

Posterior column stimulation was tested in ten patients with familial spastic paraplegia (Strümpel–Lorrain disease). In seven patients, electrodes were permanently implanted, and in five, a significant improvement was noted (Lazorthes, 1981). A long-term follow-up (from 2 to 6 years) showed that spasticity and rigidity were decreased considerably, and the patients began to walk better; in particular, they no longer required assistance in walking.

Only one report has appeared regarding the use of PCS to reduce spasticity in amytrophic lateral sclerosis. In one of two patients so treated, spasticity became somewhat less after stimulation, but the motor function of both patients was without change (Lazorthes *et al.*, 1981).

The data presented in this section indicate that stimulation techniques represent a new and promising trend in the management of spinal spasticity, especially in MS.

2.4. Embolization of Spinal Cord Arteries

A new method was recently proposed for treating severe spasticity in patients with tetra- and paraplegia after total severance of the spinal cord (Shibasaki *et al.*, 1982). The method involves the percutaneous embolization of the artery of Adamkiewicz, which is the sole blood supply to the anterior two-thirds of the lower thoracic segments and lumbar enlargement of the spinal cord. Embolization was performed under x-ray control after selective spinal angiography. In two patients, the spasticity of the flexors and adductors of the legs disappeared almost completely, and in five others, there was noticeable improvement. There were no complications, in particular, no interruption of the automatic function of the bladder. Further surgical experience will be required to evaluate this interesting method.

3. Bladder Dysfunctions

Spastic, hypertonic, hyperactive, neuropathic, and neurogenic bladder are all terms used in the literature to designate specific urinary disorders caused by various diseases and lesions of the spinal cord and its roots. Clinically, a hyperactive bladder is manifested by increased urinary frequency, urgent urination, and uncontrollable voiding. Because of spasticity in the smooth muscles of the bladder, its volume decreases, its walls lose their elasticity, and "detrusor–

sphincter'' reciprocity is disturbed, as the result of which there is more residual urine, and a very dangerous ureteral reflux occurs. Bladder dysfunctions practically always are combined with spasticity in the legs and other symptoms of spinal lesion.

Various methods are employed for the clinical assessment of the bladder functions, in particular, urocystography, cystogasometry, and EMG of the anal sphincter. With these methods, it is possible to obtain information about the pressure in the bladder, force of urination, strength of the bladder contraction and, correspondingly, the state of the detrusor, the bladder–ureteral reflux, and residual urine.

Bladder–ureteral reflux is an important clinical consideration in hypertonic bladder, since it is the cause of many serious complications including renal insufficiencies, pyelonephritis, and urinary sepsis. The relief of this reflux is one of the main goals of the surgical treatment of bladder dysfunctions in spinal cord lesions.

3.1. Myelotomy

Many years ago it was established that denervation of the detrusor muscle improves disturbed function. As already mentioned in Section 2.1, longitudinal myelotomy (Bischof's operation) reduces muscular spasticity not only in the legs but also of the detrusor muscle. It is precisely for this purpose that Bischof proposed extending the myelotomy caudally to the sacral segments on one side of the spinal cord. In a number of cases, this improved the bladder functions as a result of reduced muscular tonicity (Tönnis and Bischof, 1962; Moyes, 1969). However, Laitinen and Singounas (1971) noted such improvement in only two of nine patients.

3.2. Sacral Rhizotomy

For the treatment of a neurogenic bladder, sacral rhizotomy was proposed (Meirowsky *et al.*, 1950). The objective of this operation was to sever all sacral roots on both sides, which frequently led to pronounced dysfunction of the bladder detrusor and sphincter. This led to a number of reports concerning the relative efficacy of severing various sacral roots (anterior, posterior, or both) in order to eliminate hypertonicity of the bladder musculature. Nonselective rhizotomy is permissible only in patients with irreversible paraplegia and is contraindicated in patients with a comparatively minor neurological deficit, since in these conditions the operation may result in increased paresis of the legs and dysfunction of the anal sphincter as well as sexual dysfunction.

In subsequent years sacral rhizotomy has been greatly improved by partial selective section of the rootlets of the anterior sacral roots innervating only the bladder. Such an operation preserves the detrusor reflex and the sphincter function (Nagib *et al.*, 1966; Rockswold *et al.*, 1973, 1978; Toczek *et al.*, 1975). This operation is indicated if all types of conservative therapy have been ineffective, bladder elasticity is preserved, no obstructive lesions are present in the bladder or the ureter, and urination is improved after procaine block of the sacral roots.

During the period when this operation was first employed, selective rhizotomy was performed at the sacral level; however, quite frequently there were relapses of bladder spasticity. The procedure is more effective when the roots are sectioned at the conus medullaris (Toczek *et al.*, 1978). At that level it is easier to identify the anterior and posterior roots.

The operation is performed under mild general anesthesia, less frequently under local anesthesia, with the patient in a prone position. After laminectomy of L_1–L_3 the cauda equina is exposed. A surgical microscope is used to locate the filum terminale and identify the sacral roots that often are very close to the filum. Then the roots S_3–S_4 innervating the bladder are identified, and their rootlets separated. With a weak current (0.5–3 V, 25 Hz), each of these rootlets is stimulated to determine which ones produce detrusor contraction. The effect of stimulation is monitored by cystometry, EMG of the bladder wall, recordings of the anal sphincter (recording of pressure in a balloon introduced rectally), leg movements, as well as the subjective sensations of the patient (distended bladder, frequent urinary urgency) in operations performed under local anesthesia.

At this stage of the operation, the surgeon must carefully separate the rootlets innervating the bladder from those innervating the leg muscles and the anal sphincter. Only then is it possible to section bilaterally and selectively those rootlets that innervate the detrusor muscle. Usually two or three fasciculi on both sides are divided. Investigations have revealed that the main source of innervation of the detrusor is the root of S_3 and to a much lesser degree S_4 (Toczek *et al.*, 1978). It is noteworthy that selective section of higher roots, including S_1, does not improve urinary dysfunction in patients with spastic paraplegia (Laitinen *et al.*, 1983).

Selective rhizotomy for improving urinary function has been done for various lesions of the spinal cord (in MS, traumatic paraplegia, myelopathy, etc.). The clinical experience with this operation is relatively limited; however, in the majority of patients there is improved urinary function—disappearance of urgency, nocturia, and uncontrollable urination. The bladder capacity after the operation increases four- to fivefold, urinary frequency decreases approximately tenfold, and the bladder–ureteral reflux disappears. The operation prevents the development of urinary tract infection. Sexual function and defecation are unchanged.

Selective rhizotomy of S_3 has been successfully used in cases of "primary" enuresis, that is, uncontrollable urination in the absence of any CNS disorder. Selective section of the S_3 root (most frequently on both sides) through a small opening in the sacrum produces a good and lasting effect in such cases (Torrens and Griffith, 1976).

Recently, a new and promising method for treating neurogenic spasticity of the detrusor has been proposed—percutaneous sacral rhizotomy (Lagarrigue *et al.*, 1979; Young and Mulcany, 1980). In order to determine the effectiveness of this operation and thereby its indications, a preliminary bilateral block of the S_2–S_4 roots through the posterior sacral foramina is made. Surgery is considered indicated if, after the block, the bladder capacity is increased by at least 200 ml.

Rhizotomy is performed with a stylet needle introduced into one of the sacral foramina under fluoroscopic control while the bladder pressure is being recorded. Destruction of one or two posterior roots on both sides is performed by high-frequency electrocoagulation at a temperature of 70–80°C. Lagarrigue *et al.* (1979) prefer bilateral coagulation of the S_3 roots. The majority of patients with spinal cord transections operated on by this technique have shown very encouraging results, in particular, a considerable increase in bladder capacity.

3.3. Stimulation Methods

Epidural PCS (Chapter 13, Section 5.4.3) has been intensively developed recently for the management of bladder dysfunction. The first papers on use of this method for neurogenic bladder appeared about 10 years ago (Cook, 1976; Illis *et al.*, 1976). Subsequent publications with precise methods for studying bladder function confirmed the effectiveness of this operation and introduced a number of important modifications (Cook *et al.*, 1979; Meglio *et al.*, 1980).

The mechanism by which urination is improved by PCS is still obscure. It is hard to believe that electric impulses act directly on the "urination center" in the lumbar–sacral spinal cord, since the stimulating electrodes are quite a distance from this level. Meglio *et al.* (1980) presume that stimulation of the posterior columns represents a kind of "substitute for supraspinal influences on the spinal centers of urination." However, it is also possible that the ascending pathways leading to the supraspinal centers of urination are being stimulated.

Usually, electrodes for PCS are introduced epidurally above the spinal level, damage of which causes urinary dysfunction. This method had mainly been used to treat neurogenic bladder of MS, spinal arachnoiditis, spinal cord injury, and spina bifida. In the majority of these patients, urinary dysfunction was combined with spasticity and pain in the legs.

Improved urinary function from PCS may occur in patients with any lesion of the spinal cord, irrespective of the etiology of the neurogenic bladder. With this technique, it is possible to observe a number of significant changes in urination. The frequency over the course of a day is significantly decreased, and nocturnal urination is practically abolished. Urgency is noticeably reduced. Involuntary urination and dysuria are practically absent. The bladder capacity is markedly increased. In a majority of cases, the pressure in the bladder is reduced because of reduced detrusor tone. Accordingly, there is a higher threshold for the detrusor reflex. The residual urine greatly decreases and in some cases disappears completely.

The clinical improvement stabilizes and becomes more evident a month and a half after implantation of the electrodes and the beginning of stimulation (Meglio *et al.*, 1980). In patients with hypotonic bladder, PCS at the thoracic level reduces the urinary frequency and incontinence by approximately 50%. In the hypertonic bladder, PCS was less effective; in only five of 31 patients was there a noticeable improvement several weeks after beginning stimulation (Dooley and Sharkey, 1981). It is interesting to note that PCS has a positive influence only in cases of pathological urination; in patients with normal bladder function, this method does not produce any noticeable changes (Meglio *et al.*, 1980).

To evacuate the bladder in 14 patients with traumatic paraplegia, prolonged electrostimulation of the conus medullaris and at S_1–S_2 level with implanted bipolar electrodes made of platinum–iridium alloy proved to be effective (Nashold *et al.*, 1982). After laminectomy and opening of the dura mater, electrodes were introduced into the gray substance of the spinal cord bilaterally, i.e., from each side of the conus. The electrodes were connected to a receiver implanted subcutaneously, and stimulation was induced by an external high-frequency source.

During stimulation the quantity of residual urine decreased to 30 ml. Bladder capacity increased so that permanent catheterization was not needed. In a number of males, stimulation also caused erection and ejaculation. Improved bladder control was obtained in more than 70% of the patients for 3 to 10 years. In a subsequent paper (Nashold *et al.*, 1982), the authors reported on data from neurosurgeons in the United States, France, and Sweden in 27 patients with paraplegia. A good result was obtained in 55% of the cases; moreover, in many, the spasticity in the legs also decreased. In one patient with traumatic paraplegia, stimulation of the conus over the course of 5 years restored normal urination (Augustinsson *et al.*, 1982). In another patient, such relief continued for 10 years (Nashold *et al.*, 1982). Severe changes in the bladder walls and urinary tract are contraindications for the use of this method.

A comparison of two groups of patients with traumatic injuries of the cervical spine was made by Kiwerski (1986). The first group was treated by surgical decompression with simultaneous implantation of stimulating electrode. The second group was operated upon but without implanted stimulator. Neurological improvement and partial normalization of bladder function were better in the first group with daily stimulation within 3–4 weeks after surgery.

Encouraging results from sacral anterior root stimulation for bladder control in paraplegics are also described (Brindley *et al.*, 1982, 1986).The results were evaluated as good in 39 of 49 patients who are using their implants. Only four patients are dissatisfied while the majority have become continent.

Another way to improve bladder functions in spinal cord injury is stimulation of the bladder wall or the bladder sphincter through the rectum (Livshitz *et al.*, 1976). In recent years, many successful attempts have been made to create an artificial urinary sphincter (Scott *et al.*, 1974; Light, 1985).

3.4. Pudendotomy

Another effective operation in urinary dysfunction after spinal cord trauma is resection of the pudendal nerves (pudendotomy). The operation eliminates the spastic state of the sphincters and improves the disturbed urodynamics. Before the operation a procaine block of the pudendal nerves is made on both sides. If there is relief of urinary function temporarily, one may presume that the operation will be effective. Under local anesthesia, the nerves are divided (on one or both sides) as they lie on the medial surface of the ischial tuberosity. This comparatively simple operation, without a doubt, yields good results in some cases. For example, a noticeable improvement in urodynamics was observed after the operation in 12 of 16 patients (Virozub and Chernovsky, 1976).

4. Phrenic Nerve Stimulation

Phrenic nerve stimulation (PhNS) for artificial respiration is a well-known method that has been used for many years in cases of respiratory paralysis due to cervical spinal cord or brainstem injury, sleep apnea, or Ondine's syndrome (Glenn, 1978; Nashold and Montagno, 1982). Spinal cord stimulation has to be made above the spinal phrenic motor center.

Fodstad *et al.* (1985) have accumulated impressive experience using this method in 16 cases of high cervical or brainstem lesions or the central hypoventilation syndrome. The patients were tetraplegic or tetraparetic and on artificial respiration from 2 months to 8 years. The special pacemaker (Avery Lab, USA) was implanted on the phrenic nerve bilaterally or unilaterally (Fodstad, 1987).

After surgery, the majority of patients may completely omit artificial respiration. Five patients required reoperations because of stimulator failure. Six patients with cervical spinal cord lesion developed some spontaneous respiration after diaphragm pacing. Eleven patients went home after a lengthy stay at the hospital.

Undoubtedly, chronic PhNS is an effective and promising method for treating respiratory insufficiency in cervical spinal cord injury.

To summarize the data presented in this chapter, one must emphasize that contemporary functional neurosurgery offers effective means of managing most of the severe consequences of spinal cord lesions, namely, spasticity and urinary disorders and respiratory paralysis. These new methods require further improvement and evaluation to determine their therapeutic potential.

20

Additional Possibilities of Stereotactic Surgery

1. Introduction

This brief chapter discusses the rare and sometimes unexpected opportunities afforded by stereotactic surgery to remedy exceptional conditions that are seen so infrequently that a surgical form of therapy is generally not accepted.

2. Ventriculoscopy

Iizuka (1975) used a simple stereotactic apparatus for precise introduction of a thin fiberoptic endoscope to check the position of a catheter after shunting operations for various forms of hydrocephalus and for diagnosing certain deep intraventricular tumors (ependymoma, pinealoma, etc.). The technique, termed stereoencephaloscopy, permits a thorough examination of specific parts of the ventricular system, including the chorioid plexus. Stereoencephaloscopy is helpful in obtaining biopsies and in surgical resection of intraventricular tumors.

A modern, improved version of the endoscope has been produced by K. Storz (FRG). The endoscope, which is introduced stereotactically, enables visualization for biopsy, cyst wall puncture and aspiration, as well as that of the ventricles, including FM (Apuzzo *et al.*, 1987).

3. Removal of Foreign Bodies

There are several reports in the literature on the use of stereotactic techniques for removing foreign bodies from deep parts of the brain. Sugita *et al.* (1969) used a stereotactic apparatus of their own design for successful removal of a bullet fired by an air rifle from the brain of a year-old baby through a small burr hole. The bullet was located deep in the left hemisphere near the optic tract and the rostral brainstem.

Mundinger (1968) removed a foreign body from the brain stereotactically through a conventional craniotomy. Riechert (1955) and Fusek and Vladyka (1968) have also reported on the successful use of the stereotactic technique for removal of deep-seated foreign bodies from the brain.

4. Obstructive Hydrocephalus

Stereotactic ventriculocisternostomy, i.e., perforation of the lamina terminalis of V_3 by the stereotactic technique, has been successfully used to relieve obstructive hydrocephalus of different etiologies (tumors of the posterior cranial fossa, stenosis of the Sylvian aqueduct, etc.). Poblete and Zamboni (1975) chose the suprachiasmal recessus as the target point. A communication between V_3 and the interpeduncular cistern was made under control of an electronic amplifier with TV monitoring. In ten patients with obstructive hydrocephalus, the opening made in the lamina terminalis functioned well and gave a good therapeutic result. There was no postoperative mortality. The authors emphasize that this method has a number of advantages over a conventional open ventriculocisternostomy.

5. Reconstruction of the Sylvian Aqueduct

Stereotactic reconstruction of an occluded Sylvian aqueduct producing obstructive hydrocephalus is a new operation (Backlund *et al.*, 1981). In the past, attempts were made to restore the patency of the Sylvian aqueduct by inserting a metal spiral through an open approach via the posterior fossa, but they appeared to be ineffective.

In 1972 Stiner and Leksell were the first to introduce a thin tube stereotactically into the Sylvian aqueduct through a membranous obstruction, which ensured a reliable communication between V_3 and V_4 (unpublished data cited by Backlund *et al.*, 1981).

To cannulate the aqueduct, a thin tube 1.5 mm in diameter and 15–20 mm long of x-ray contrast Teflon or polyethylene is inserted. Short incisions in the rostral end of the tube allow four small "wings" to be folded back and heated to preserve their form. Thus flanged, this end of the tube is fixed in the cavity of V_3. To introduce the tube into the Sylvian aqueduct, a special trocar was designed with a slightly bent end (angle of inclination 10°).

In brief, the technique is as follows. Under local anesthesia the patient, in a sitting position, undergoes pneumoencephalography in order to fill V_4 and the caudal part of the aqueduct with air. The x rays obtained are compared with preoperative ventriculograms to determine the state of V_3 and the oral part of the aqueduct. With the patient in the prone position, a burr hole is made, V_3 is punctured with a small-bore needle, and 0.3 ml of metrizamide is injected. The stereotactic calculations of the target point at the oral end of the Sylvian aqueduct are made on the films in two dimensions. The needle is replaced by a conductor, through which the tube is introduced into the aqueduct. The appearance of the contrast medium in the air-filled V_4 indicates the penetration of the membrane and the restoration of a patent aqueduct. The tube position is checked on the films in two projections, after which the conductor is removed from the brain. A reduction or disappearance of the hydrocephalus on CT scans indicates that the tube is functioning well.

This operation was carried out by Backlund *et al.* (1981) on seven patients with segmental obstruction of the Sylvian aqueduct (Fig. 262). In only three of the cases was the long-term reconstruction of the aqueduct successful. The other four patients underwent a second operation for reposition or restoration of patency of a partially blocked tube. In three patients hydrocephalus was not reduced, so that a ventricular shunt was inserted later, although in two, the patency of the aqueduct was proven. There was no postoperative mortality or serious complications. The authors emphasize that a condition *sine qua non* for successful reconstruction of the aqueduct is an open or intact subarachnoid space and a functioning mechanism for the absorption of cerebrospinal fluid.

6. Relief of Nystagmus

Komai (1967) described a stereotactic operation that he called superior colliculotomy for the relief of nystagmus and vestibular ataxia. He operated on 30 patients for various forms of nystagmus. The operation includes the destruction of the superior part of lamina quadrigemina. According to the author, this operation is effective in horizontal nystagmus of vestibular origin. No serious complications were observed.

7. Treatment of Obesity

On the basis of experimental data, a lesion of the lateral Hypoth causes a loss of appetite and consequently of weight. Proceeding from this, Quaade *et al.* (1974) used electrostimulation with subsequent stereotactic destruction of this subcortical structure to treat severe obesity. They operated on five patients with abnormally voracious appetites who weighed from 118 to 180 kg. An electrode was introduced through the nondominante hemisphere into the target point in the lateral hypothalamic nucleus. The stereotactic coordinates for this nucleus were as follows: 2 mm posterior to CA and 10 mm inferior to LI.

Electrostimulation of that site caused marked sympathetic reactions such as sensation of warmth or cold and changes in the pulse or respiratory rate. Often it was accompanied by a feeling of fear. In two patients, because stimulation failed to produce a feeling of hunger, electrocoagulation was not done but in the other patients in whom electrostimulation induced vegetative feelings, electrocoagulation was followed by disappearance of the feeling of constant hunger. In the postoperative period there was some loss of weight, but in general the effect was insignificant.

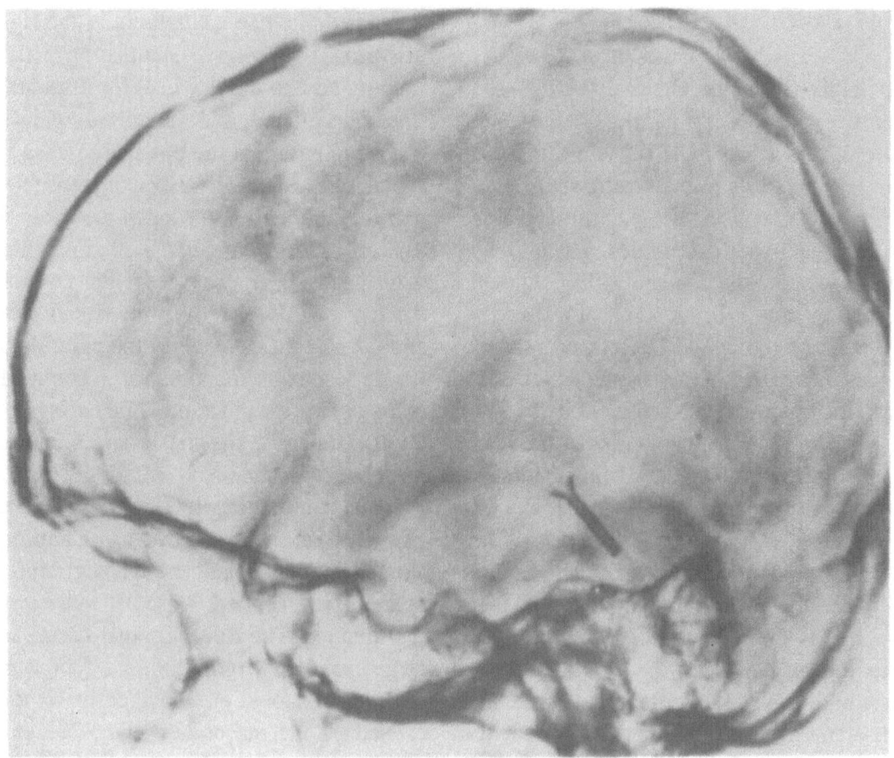

FIGURE 262. Stereotactic reconstruction of Sylvian aqueduct for segmental stenosis. A thin, small tube is introduced into the aqueduct (Backlund *et al.*, 1981).

8. Relief of Writer's Cramp

This rare disorder is characterized by a tonic spasm of the fingers or the whole hand when an attempt is made to write. In such cases, writing or drawing with a pen or pencil is practically impossible.

The etiology and pathogenesis of writer's cramp are unknown. The development of the disorder may be rapid, even sudden, or it may be gradual, over the course of several years. The muscular spasm is often accompanied by tremor of the same hand. However, as a rule, this disorder does not have any signs or symptoms typical of parkinsonism or any other extrapyramidal lesion. In the earlier literature, writer's cramp was considered to be a "psychoneurosis" or, in other words, a functional disease of the CNS. Lately there has been the suggestion that it has an organic etiology. Conservative treatment is generally ineffective.

So far, there have been only a few attempts to treat writer's cramp surgically. One case of stereotactic treatment of writer's cramp with tremor (destruction of VL) was reported by Siegfried *et al.* (1969). After the operation, writer's cramp disappeared completely. In view of this experience, the authors believe that writer's cramp is an organic disease, a peculiar form of dyskinesia. At the time of this operation, microelectrode studies of thalamic neurons showed that some could be activated or inhibited by closing or opening the hand. This suggests that a lesion in thalamic structures may play a specific role in the pathogenesis of writer's cramp.

9. Treatment of Raynaud's Disease

Narabayashi (1962) described one case of bilateral (two stages with a year interval) stereotactic pallidotomy in a grave form of Raynaud's disease. Immediately after the operation, the result was good, and later, the author reported that 10 years after two stereotactic operations there were no symptoms of Raynaud's disease (Narabayashi, 1962).

10. Relief of Erythromelalgia

Erythromelalgia (from the Greek *erythros*—red, *melos*—extremity, and *algos*—pain), or the Weir Mitchell syndrome described in 1872, is a disease of the vegetative nervous system characterized by pain, edema, and trophic lesions in the extremities, more often in the lower than upper extremities. The etiology and pathogenesis of the disease are still unknown. Many authors believe that it is the result of a lesion in the vegetative cells of the spinal lateral horn, which produces changes in the peripheral vasomotor reflexes. In some cases (in spasm of the vessels), it leads to Raynaud's disease, and in other cases (dilatation of the vessels) to erythromelalgia. Typical of this disease is intermittent and then persistent pain in the lower extremities. At the onset of the disease, the pain is not very severe, but it gradually intensifies. The pain is burning, as if hot water were being poured on the feet, and is intensified by heat, compression of the extremity, or letting it hang down. Simultaneously, edema develops in the legs, more often in the shins and feet. The skin of the legs becomes intensely red, the veins dilate, and arteries pulsate. A rise in temperature in the edematous area by 2–4° later spreads to the entire extremity. Frequently there are various trophic lesions (desquamation of the skin, bullous eruption, vesicles, pustules, changes in the nails).

Starting in one leg, the disease, as a rule, spreads to affect the other leg and in some cases also the upper extremities, nose, and ears. Erythromelalgic attacks last for months and, sometimes, for years. Conservative treatment of the disease, for example, with vasotonic drugs as well as operations on the sympathetic nervous system are practically valueless.

As far as we know, there have been no reports in literature on the treatment of erythromelalgia with stereotactic or functional operations. However, we have had two personal observations.

An 11-year-old boy was admitted to our clinic in May, 1981. His mother stated that at 3 years of age, the child began to complain of pain in the legs while walking, particularly in the right leg. The pain gradually intensified and became burning. At 7–8 years of age, the pain was accompanied by reddening of the feet. Periodically, he noted hyperemia of his ears, the tip of the nose, and palms, which became marble-colored. The pain lessened only when the right leg was immersed in cold water.

There was then a temporary remission when the boy complained of having pain only in the evening. In May, 1980 he had a small cut in the right foot, which was closed with three sutures. In 5 days the wound suppurated and did not heal for 2 months. His general state deteriorated, and the burning pain in the right leg was so intense that the boy had to keep it constantly in a pail of cold water. He experienced more frequent burning although less pronounced pain in the left foot as well as in the palms. Sometimes the boy suffered from headaches. Because of the constant immersion, the right ankle and foot developed severe ulceration and edema. At the peak of the painful attacks, the boy would lose consciousness for 3–5 min. In an 8-month course of various therapies, including prolonged extradural tremicaine anesthesia and numerous analgesics, no relief was obtained.

On admittance, the boy was pallid and exhausted. His pulse rate was 120 beats/min and rhythmical. The skin of the hands and the left leg (shin and foot) was cyanotic and spotted. Vascular pulsations were of normal quality. The right inguinal lymph nodes were enlarged, and the right shin and foot were very hyperemic, edematous, and covered by an ulcerous surface caused by the maceration (Fig. 263A). The right leg was kept constantly in a pail of cold water that warmed very rapidly and had to be changed every 20–30 min. When the leg was taken out of the water, the burning pain became so intense that the boy cried and even lost consciousness. No significant neurological findings could be detected. A diagnosis of erythromelalgia was made.

In April, 1981 the boy was subjected to a laminectomy of T_{10}–T_{12} and partial rhizotomy (radiculotomy) L_4–S_1 on the right. After the operation, in spite of the anesthesia in the denervated area, the pain was diminished insignificantly. The boy still had to keep his right leg immersed in cold water. After a month a second operation, stereotactic cryodestruction of the left VLP and CM, was performed. Following that operation, the effect was marked and lasting: the burning pain in the right leg abated, and there was no need to keep it in cold water. The skin on the right leg significantly improved, and the large ulcerated area began to heal.

A year after the two operations, the boy was much improved. The former ulcerous area on the right leg had completely healed. From time to time he suffered short periods of endurable pain in the leg. Three years after surgery the boy had resumed his classes, walked normally, and could run, ski, and skate (Fig. 263B).

The second case is similar to the first, except that in this instance the erythromelalgia was more severe.

A 15-year-old girl was admitted to our clinic with

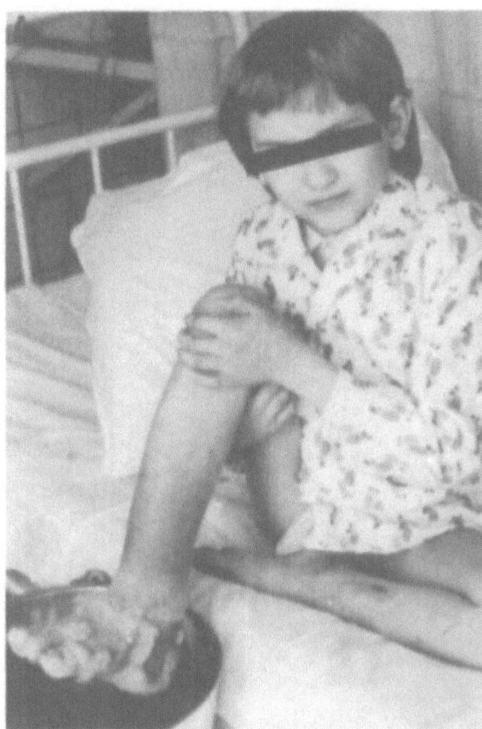

FIGURE 263. A 9-year-old boy suffering from erythromelalgia and intractable burning pain in both legs, extremely severe in the right foot. (A) Before operation, the boy kept the leg in a bucket with cold water continuously for 8 months. (B) The boy 4 years after surgery.

complaints of a persistent, disabling, burning pain in both ankles and feet. The disease began when the girl was 2 years old. At that time, intermittent pain, a sense of burning, and hyperemia in both lower legs appeared after tonsillitis. At the age of 4, the pain became permanent but was not as severe during the summer. The girl's state gradually worsened; the pain in both legs increased, and edema appeared in the distal parts. The child felt relief only after immersing both legs into cold water. Erythromelalgia was first diagnosed in 1983.

The girl's condition worsened progressively, leading to the need to keep both legs in ice-cold water around the clock (Fig. 264A). As a result, edema of the legs increased, and large, deep trophic ulcers appeared with decubiti, skin maceration, and nail lesions. The legs and feet were very hot, red, and cyanotic. A course of x-ray therapy resulted in only a very short improvement.

Left stereotactic cryothalamotomy with destruction of CM and VPL was performed under general anesthesia in June, 1984. There were no complications following surgery. A pronounced improvement was noted immediately: pain and burning disappeared in both legs, and hyperemia and edema diminished. The large ulcerous area and decutibi quickly healed. The girl no longer immersed her legs in cold water. However, about 6 weeks later, all the signs recurred, and trophic ulcers developed again.

The patient was admitted to our clinic a second time in September, 1984. Her general condition was grave. Her legs were kept constantly immersed in a pail of cold water, which warmed rapidly and had to be changed every half hour. The child wept constantly because of intractable pain, which was relieved slightly after regular injections of morphine eight to ten times per day. Edema affected not only the legs but the entire body. The legs were hot and hyperemic, with many ulcers in the distal parts (Fig. 264B) and contractures of knee joints.

Stereotactic cryodestruction of the CM and VPL on the other side (right) was repeated in September, 1984. The girl's condition improved 2–3 days after surgery; the burning pain in the legs diminished, and the edema and hyperemia became less pronounced. However, she vomited blood and collapsed 1 week after the operation. Because of impaired breathing, a

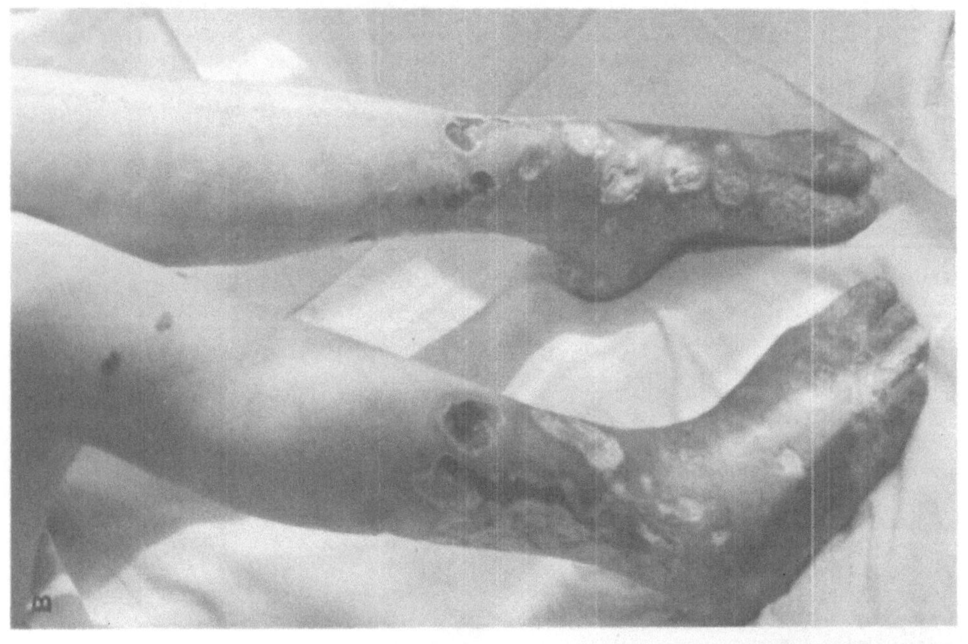

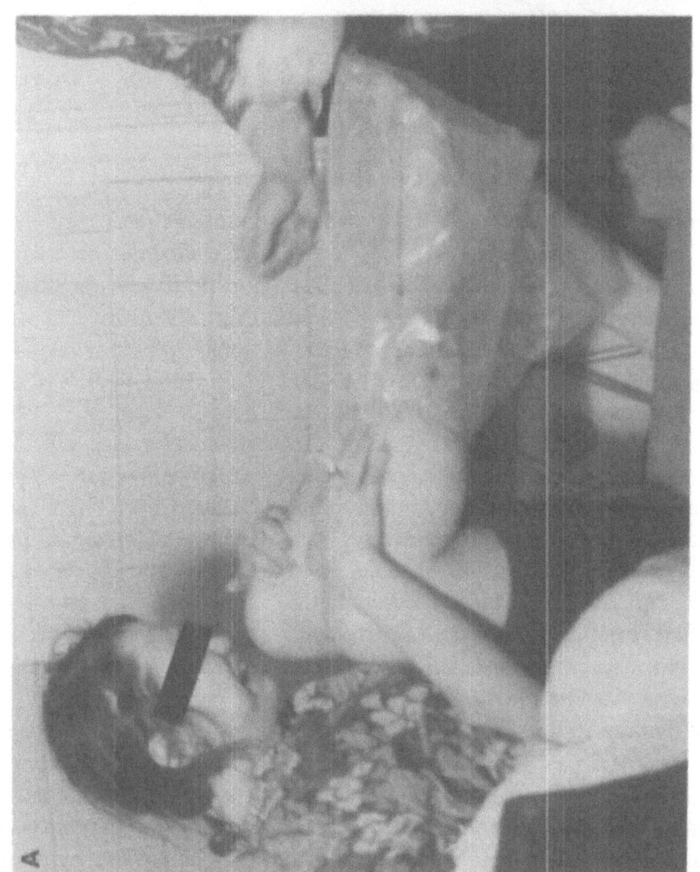

FIGURE 264. (A) An 11-year-old girl with erythromelalgia of 6 years' duration and intractable burning pain in both legs. (B) Severe damage to the soft tissues from keeping both legs constantly in cold water for 2 years. (C,D) The girl 3 years after surgery.

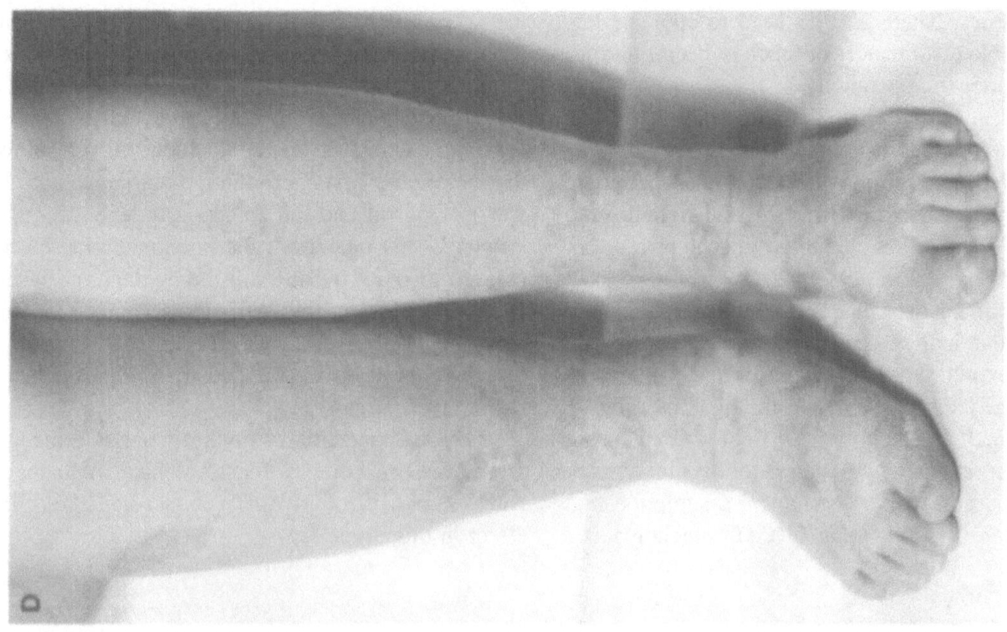

FIGURE 264. *(Continued)*

tracheostomy was performed, and the patient was put on a respirator for about 3 weeks. Following this complication, her condition improved, and normal respiration was restored. Edema and hyperemia of the legs disappeared gradually. Pain diminished to a degree that made it possible to discontinue intake of narcotics and even analgesics. There was no need to cool the legs with water. No disturbances of touch or deep sensation developed after the operation.

Edema, hyperemia, and hyperthermia of the legs disappeared completely 3 months later. The ulcerous lesions healed, and the movements of the knee joints substantially increased. Now, two and a half years after surgery, the girl is able to go to school, run, and dance. (Fig. 264C,D).

The mechanism of the relief of the peripheral vasodilatation that is the basis of erythromelalgia remains unclear. It is not surprising that after each operation, the pain nearly disappeared, but the bilateral effect is not common. It is much more difficult to explain the changes of peripheral circulation in the legs that caused intractible burning pain. The logical assumption may be that severe vasodilatation in erythromelalgia is a consequence of disturbances in segmental vasomotor regulation at the spinal level, which is relieved by the destruction of VPL and CM. It may also be suggested that peripheral vasodilatation is dependent on hyperproduction of a hypothetical vasoactive substance, the production of which is inhibited by thalamotomy. In the second case described above, the catecholamine (epinephrine and norepinephrine) blood levels were determined before and after the operation by liquid chromatography. One might have expected that the concentration would be markedly reduced. It was found that before the first operation the norepinephrine level was 931 pg/ml and the epinephrine level 133 pg/ml. After the first operation, the norepinephrine level decreased to 642.2 pg/ml and the epinephrine level increased to 226.8 pg/ml. After the second operation the norepinephrine level increased again to 971 pg/ml, and the epinephrine level dropped to 172 pg/ml but was still above normal.

The pathogenesis of erythromelalgia remains obscure, but we believe that stereotactic thalamotomy and posterior rhizotomy may produce a marked clinical effect in this disease.

Epilogue

My book *Functional and Stereotactic Neurosurgery* in the Russian language appeared from the Moscow publishing house Meditsina at the end of 1981. The present English edition of this book differs substantially from the previous publication. First, the English edition is more than three times as large as the Russian edition. All chapters have been significantly revised and enlarged. In this edition, I have attempted to incorporate and analyze the vast amount of new information that has appeared in the literature of recent years. Now the book includes a bibliography of more than 2500 references.

It is my hope that this English edition will be read and understood by neurosurgeons and representatives of other medical specialties in many countries. For me, this publication is a great honor and, at the same time, a great responsibility.

Possibly, this book may be read by those neurosurgeons whose efforts brought functional and stereotactic neurosurgery into being and have steadily advanced its therapeutic techniques. I trust that even the leading proponents of this field may find in the volume some information that previously had been unknown to them. In any case, I hope that they will take note of the generalization and systematization of the vast amount of factual material as well as certain scientific and technical advances that I have introduced.

Hopefully, this book will be useful, first of all, to young neurosurgeons who wish to dedicate themselves to functional and stereotactic neurosurgery.
eotactic neurosurgery.

As far as possible, I have endeavored to describe for the reader the present state of the art. How it will develop in the future in our turbulent age of scientific and technological progress is not only difficult but practically impossible to predict. A vivid illustration is the fact that only 4–5 years ago no one would have predicted the emergence of such technical breakthroughs as computed topography and nuclear magnetic resonance.

It is my opinion that the principal achievements in functional and stereotactic neurosurgery of the future cannot be prophesized, yet an extrapolation into the future of certain new trends is possible. There is every reason to believe that the technique of electrostimulation of various cerebral and spinal cord structures as well as of peripheral nerves will develop and improve at a much faster pace than in the past. This augmentative technique offers a means to eliminate serious manifestations of various CNS disorders without destroying nervous tissue. It may be assumed that the number of disorders for which augmentative functional operations will be performed in the not-too-distant future will increase substantially.

At the same time, despite the advances in the application of electrical stimulation to improve the CNS function or to treat its many diseases, a certain skepticism should be keep in mind. Our limited knowledge of the structural and functional organization of the countless brain circuits does not permit us to activate one system and to inhibit the other. Poletti (1987) formulated this thought clearly, pointing out that putting a wire into the brain to deliver electrical pulses offers no greater probability of enhancing CNS function than does putting the same wire into a TV microcircuit to improve picture quality.

On the other hand, we may suppose that the new advances in our knowledge of brain circuits will be the source of information for many new functional brain operations. I also believe that the value of surgical methods for the destruction of nuclear structures and neural pathways (neuroablative procedures) will not lessen. The important advantage of these methods is that they require a single, brief surgical procedure, whereas the implantation of electrodes for stimulation and their repeated stimulation must continue for years. However, as we learn more about the pathogenesis of CNS disease, as well as about the functional organization and connections of cerebral and spinal structures, the destructive foci may be reduced in size and possibly be limited only to groups of neurons. There can be no doubt that methods of destroying nervous tissue will also be substantially improved. The use of laser and ultrasonic energy in combination with fiber optics selectively targeted at nervous tissue also appears to be on the horizon.

Implantable systems providing exact dosages of various substances, in particular, neurotransmitters, for deficient subcortical structures will, in all likelihood, be developed further. The first step in this direction has already been taken in the development of techniques for intrathecal administration of opiates (Chapter 13).

The development of new technologies for teledestruction of nervous tissue through the intact skull may transform the stereotactic operation into a simple procedure that will no longer be called surgery. The focusing of ultrasound as well as laser rays and proton beams seems to be among the most promising of these methods.

Already, computer technology, especially computed tomography, has greatly changed stereotactic and functional operations. Even more amazing results may be expected from the new imaging techniques using nuclear magnetic resonance and positron emission topography. A new generation of such devices with enhanced resolving capabilities appears every 3–4 years, so it is difficult to predict the advances of such equipment by the beginning of the next century, a mere 11 years away.

We shall, without a doubt, witness further technical advances in stereotactic biopsy of brain tumors, particularly malignant and deep-seated tumors, based on imaging methods to reconstruct the volume and shape of tumors in three dimensions. The improvement and miniaturization of electronic sensors will allow the surgeons during a biopsy procedure to obtain information on the histostructure and vascularization of the various parts of the tumor, so that the computer-programmed laser and ultrasonic devices will effect a more radical removal of inoperable gliomas. The means of stereotactically introducing into these tumors photosensitive substances that release free oxygen for destroying tumor cells (photoradiation technique) will be perfected.

One of the most interesting and newer trends is brain grafting into subcortical structures by stereotactically introducing embryonal tissue, adrenal medulla, and possibly embryonic tissue transplants. It is presumed that the transplanted tissues, without causing a rejection reaction, will produce dopamine and possibly other transmitters to compensate for disturbed biochemical ''chains'' that play an extremely important role in the pathogenesis of many CNS diseases. There are bold suppositions that neural tissue transplanted into the brain may be able to form new neuronal ties between cerebral structures, which may establish the functions of these structures. Although the first attempts at such operations in parkinsonism have so far yielded no definite results, the further development of this original idea seems to be quite promising.

The development of functional neural prosthetics, primarily visual and acoustic, is a matter for the more distant future.

There is every reason to hope that functional and stereotactic neurosurgery will advance rapidly and fruitfully in the next few decades. Quite possibly this progress will surpass even the most optimistic forecasts today.

References

Abay, E. O., Laws, E. R., Grado, G. L., Bruckman, J. E., Forbes, G. S., Gomez, M. R., and Scott, M., 1981, Pineal tumors in children and adolescents. Treatment by CSF shunting and radiotherapy, *J. Neurosurg.* **55:**889–895.

Abbe, R., 1889, A contribution to the surgery of the spine, *Med. Rec.* **35:**149–152.

Abrakov, L. V., 1952, *Surgical Treatment of Spasmodic Torticollis,* Thesis, Leningrad (Rus.).

Abrakov, L. V., 1975, *Grounds of Stereotactic Neurosurgery,* Meditsina, Leningrad (Rus.).

Adachi, R., Yoshii, N., Rudo, T., and Shimizu, S., 1971, Stereotaxic thalamic pulvinotomy for pain relief, *Neurol. Med. Chir. (Tokyo)* **2:**339–40.

Adams, J. E., and Rutkin, B. B., 1969, Treatment of temporal-lobe epilepsy by stereotactic surgery, *Confin. Neurol.* **31:**80–85.

Adams, J. E., and Seymour, R. J., 1968, Transsphenoidal cryohypophysectomy in acromegaly, *J. Neurosurg.* **28:**100–104.

Adams, J. E., Hosobuchi, Y., and Fields, H. L., 1974, Stimulation of internal capsule for relief of chronic pain, *J. Neurosurg.* **41:**740–744.

Adams, J. E., Hosobuchi, Y., and Rutkin, B., 1975, Central stimulation in the treatment of pain, *Confin. Neurol.* **37:**2–279.

Adams, J. E., Lippert, R., and Hosobuchi, Y., 1977, Commissural myelotomy, in *Current Techniques in Operative Neurosurgery* (H. Schmidek and W. Sweet, eds.), Grune & Stratton, New York, pp. 427–434.

Adams, J. H., Graham, D. I., and Doyle, D., 1981, *Brain Biopsy: The Smear Technique for Neurosurgical Biopsies,* Lippincott, Philadelphia.

Adson, A. W., 1924, The surgical treatment of glossopharyngeal neuralgia, *Arch. Neurol. Psychiatry* **12:**487–506.

Afshar, F., 1987, Three-dimensional color-coded graphic display of anatomical data for the guidance of stereotactic surgery, in *Stereotactic Surgery* (R. R. Tasker, ed.), Hanley Belfus, Philadelphia, pp. 149–164.

Afshar, F., Watkins, E. S., and Yap, J. C., 1978, *Stereotaxic Atlas of the Human Brainstem and Cerebellar Nuclei,* Raven Press, New York.

Aizerman, M. A., Andreeva, B. L., Kandel, E. I., and Tanenbaum, L. A., 1974, *The Mechanisms of Control of the Muscle Activity,* Nauka, Moscow (Rus.).

Aizerman, M. A., Andreeva, E. A., Kandel, E. I., and Khutorskaya, O. E., 1976, The spectrum of the envelope EMG and its advantages for the analysis of the muscle control system in health and disease, in *Proceedings of the Second All-Union Symposium on the Clinical EMG,* Tbilisi, pp. 102–108 (Rus.).

Akil, H., and Mayer, D. J., 1972, Antagonism of stimulation-produced analgesia by *p*-CPA, a serotonin synthesis inhibitor, *Brain Res.* **44:**692–697.

Akil, H., Mayer, D. J., and Liebeskind, J. C., 1976, Antagonism of stimulation-produced analgesia by naloxone, a narcotic antagonist, *Science* **191:**961–962.

Akil, H., Richardson, D. E., Hughes, J., and Barchas, J. D., 1978, Enkephalin-like material elevated in ventricular cerebrospinal fluid of pain patients after analgetic focal stimulation, *Science* **201:**463–465.

Albé-Fessard, D., Arfel, G., Guiot, G., Hardy, J., Vourch, G., Hertzog, E., Aleonard, P., and Derome, P., 1962, Derivation of spontaneous and evoked activities in deep cerebral structures of man, *Rev. Neurol.* **106**(2):89–105.

Albé-Fessard, D., Arfel., G., Guiot, B., Derome, P., Hertzog, E., Vourch, G., and Aleonard, P., 1966, Electrophysiological studies of some deep cerebral structures in man, *J. Neurol. Sci.* **3:**37–51.

Albé-Fessard, D., Arfel, G., Guiot, G., Derome, P., and Guilbaud, G., 1967a, Thalamic unit activity in man, *Electroencephalogr. Clin. Neurophysiol.* **25:**132–142.

Albé-Fessard, D., Dondey, M., Benita, M., and Besson, J. M., 1967b, Remarks about physiological effects obtained by localized thermal modifications of the central nervous system, *Confin. Neurol.* **29:**208–212.

Albé-Fessard, D., Levante, A., and Lamour, L., 1974, Origin of spinothalamic tract in monkeys, *Brain Res.* **65:**503–509.

Albé-Fessard, D., Tasker, R. R., Yamashiro, K., Chodakiewitz, J., and Dostrovsky, J., 1985, Simultaneous recordings in the human thalamus and cortex of an early evoked component, *Appl. Neurophysiol.* **48:**77–78.

Albert, P., 1982, Personal experience in the treatment of 178 cases of arteriovenous malformations of the brain, *Acta Neurochii.* **61:**207–226.

Alberts, W. W., 1969, Parkinsonian tremor and cerebral potentials, in *Symposium on Parkinson's Disease*, Edinburgh, May 1968, Livingstone, Edinburgh, pp. 146–149.

Alberts, W. W., Feinstein, B., Levin, G., Wright, E. W., Darlaud, M. G., and Scott, E. L., 1965a, Stereotaxic surgery for parkinsonism. Clinical results and stimulation thresholds, *J. Neurosurg.* **23:**174–179.

Alberts, W. W., Libet, E. W., Wright, E. W., and Feinstein, B., 1965b, Physiological mechanisms of tremor and rigidity in parkinsonism, *Confin. Neurol.* **26:**318–327.

Alexander, L., 1942, The vascular supply of the striopallidum in the basal ganglia, *Assoc. Res. Nerv. Ment. Dis.* **21:**77–132.

Alexandrova, L. I., and Stolbun, D. E., 1935, On the role of mental factors in appearance and elimination of pain, *Zh. Nevropatol. Psikhiatr.* **4:**9–10, 29–33 (Rus.).

Alksne, J. F., 1970, Stereotactic thrombosis of intracranial aneurysms, *J. Neurol.* **29:**376–380.

Alksne, J. F., 1972, Progress on the magnetically controlled stereotactic thrombosis of intracranial aneurisms, *Confin. Neurol.* **34:**368–373.

Alksne, J. F., 1980, Microsurgical vascular decompression of cranial nerves, *Am. J. Surg.* **140:**156–157.

Alksne, J. F., and Smith, R. W., 1977, Iron–acrylic compound for stereotaxic aneurysm thrombosis, *J. Neurosurg.* **47:**137–141.

Alksne, J. F., and Smith, R. W., 1980, Stereotaxic occlusion of 22 consecutive anterior communicating artery aneurysms, *J. Neurosurg.* **52:**790–793.

Almay, B. C. L., Johansson, F., Knorring, L., Terenins, L., and Wahlstrom, A., 1978, Endorphins in chronic pain. Difference in CSF endorphin level between organic and psychogenic syndromes, *Pain* **5:**153–162.

Alonso, E., Otero, E., D'Regules, R., and Fugneroa, H. H., 1986, Parkinson's disease: A genetic study, *Can. J. Neurol. Sci.* **13:**248–251.

Alpers, B. J., and Dryer, C. S., 1937, Organic background of some cases of spasmodic torticollis. Report on case with autopsy, *Am. J. Med. Sci.* **193:**378–384.

Altukhov, N. V., 1891, *Encephalometric Investigations of the Brain Relative to the Sex, Age and Skull Indexes*, Moscow.

Amador, L. V., Blundell, J. E., and Wahren, W., 1959, Description of coordinates of the deep structures, in *Introduction to Stereotaxis with an Atlas of the human brain*, Vol. 1 (G. Schaltenbrand and P. Bailey, eds.), G. Thieme Verlag, Stuttgart, pp. 16–28.

Amano, K., Kitamura, R., Sano, K., and Sekino, H., 1976, Relief of intractable pain from neurosurgical point of view with reference to present limits and clinical indications. A review of 100 consecutive cases, *Neurol. Med. Chir. (Tokyo)* **16:**141–153.

Amano, K., Iseki, H., Kawabatake, H., and Notani, M., 1977, Role of computerized transversive axial tomography on stereotactic surgery of the diencephalon, *Appl. Neurophysiol.* **39:**202–211.

Amano, K., Tanikawa, T., Iseki, H., Kawabatake, H., Notani, M., Kawamura, H., and Kitamura, K., 1978a, Single neuron analysis of the human midbrain tegmentum. Rostral mesencephalic reticulotomy for pain relief, *Appl. Neurophysiol.* **41:**66–79.

Amano, K., Tanikawa, T., Iseki, H., Kawabatake, H., Notani, M., Kawamura, H., and Kitamura, K., 1978b, Single neuron analysis of the human midbrain tegmentum, *Appl. Neurophysiol.* **41:**66–78.

Amano, K., Iseki, H., and Notani, M., 1979, Rostral mesencephalic reticulotomy for pain relief with reference to electrode trajectory and clinical results, *Appl. Neurophysiol.* **42:**316–317.

Amano, K., Kitamura, R., and Kawamura, H., 1980, Alterations of immunoreactive beta-endorphin in the third ventricular fluid in response to electrical stimulation of the human periaqueductal gray matter, *Appl. Neurophysiol.* **43:**150–158.

Amano, K., Tanikawa, T., Kawamura, H., Iseki, H., Notani, M., Kawabatake, H., Shiwaku, T., Suda, T., Demura, H., and Kitamura, K., 1982, Endorphins and pain relief. Further observations on electrical stimulation of the lateral part of the periaqueductal gray matter during rostral mesencephalic reticulotomy for pain relief, *Appl. Neurophysiol.* **45:**123–135.

Amano, K., Iseki, H., Tanikawa, T., Kawamura, H., Kawabatake, H., Notani, M., Shiwaku, T., Nagao, T., Kakinoki, G., and Kitamura, K., 1983, A newly devised apparatus for CT guided stereotactic surgery, in *Meeting of the European Society for Stereotactic and Functional Neurosurgery*, Rome, p. 17.

Amano, K., Kavabatake, H., Tanicawa, T., and Kitamura, K., 1985, Long-term follow-up study of stereotactic rostral mesencephalic reticulotomy (RMR) in patients with intractable pain, in *IX Meeting of the World Society for Stereotactic and Fuctional Neurosurgery*, Toronto, p. 26.

Anasuma, C., Thach, W. T., and Jones, E. G., 1983a, Cytoarchitectonic delineation of the ventral lateral thalamic region in monkey, *Brain Res. Rev.* **5:**219–235.

Anasuma, C., Thach, W. T., and Jones, E. G., 1983b, Brain stem and spinal projections of the deep cerebellar nuclei in the monkey, with observations on the brain stem projections of the dorsal column nuclei, *Brain Res. Rev.* **5:**299–322.

Anden, N. W., Carlsson, A., Kerstell, J., Magnusson, T., Olsson, R., Roos, B. E., Steen, B., Steg, G., Svanborg, A., Thieme, G., and Werdinius, B., 1970, Oral L-dopa treatment of parkinsonism, *Acta Med. Scand.* **187:**247–255.

André, N. A., 1876, Observations pratiques sur les maladies del urèthre, et sur plusieurs faits convulsifs, et la guérison de plusieurs maladies chirurgicales, etc., Chez Delaguette, Paris, pp. 311–390.

Andrew, J., and Watkins, E. S., 1969, *A Stereotaxic Atlas of the Human Thalamus and Adjacent Structures*, Williams & Wilkins, Baltimore.

Andrew, J., 1974, The placement of stereotaxic lesions for involuntary movement other than in Parkinson's disease, *Acta Neurochir. (Supp.)* **21:**39–47.

Andrew, J., Fowler, C. J., and Harrison, J. G., 1983, Stereotaxic thalamotomy in 55 cases of dystonia, *Brain* **106:**981–1000.

Andrews, C. J., Burke, D., and Lance, J. W., 1972, The response to muscle stretch and shortening reaction in Parkinsonian rigidity, *Brain* **95**:795–812.

Andreyeva, E. A., and Khutorskaya, O. E., 1987, *The Spectrum Analysis of the Electrical Activity of the Muscles* (M. A. Aizerman and E. I. Kandel, eds.), Nauka, Moscow.

Andreyeva, E. A., Ivanova-Smolenskaya, I. A., Kandel, E. I., and Khutorskaya, O. E., 1985, Envelope EMG spectral analysis in the studies of physiological and pathological tremor, *Electroencephalogr. Clin. Neurophysiol.* **25**:273–293.

Andreyeva, E. A., Kandel, E. I., Ivanova-Smolenskaya, I. A., Smirnova, S. N., and Khutorskaya, O. E., 1986, The spectral analysis of the envelope EMG and its significance in physiological tremor investigation, *J. Neuropatol. Psychiatr.* No. 7:966–969 (Rus.).

Andy, O. J., 1962, Diencephalic coagulation in the treatment of hemiballismus, *Confin. Neurol.* **22**:346–350.

Andy, O. J., 1966, Sensory–motor responses from the diencephalon. Electrical stimulation in man, *J. Neurosurg.* **24**:612–620.

Andy, O. J., 1970, Thalamotomy in hyperactive and aggressive behaviour. *Confin. Neurol.* **32**:322–325.

Andy, O. J., 1973, Successful treatment of long-standing hysterical pain and visceral disturbances by unilateral anterior thalamotomy, *J. Neurosurg.* **39**:252–259.

Andy, O. J., 1975, Development of pain appreciation after thalamotomy, *Confin. Neurol.* **37**:107–112.

Andy, O. J., 1983, Thalamic stimulation for chronic pain, *Appl. Neurophysiol.* **46**:116–123.

Andy, O. J., and Browne, J. S., 1960, Diencephalic coagulation in the treatment of hemiballismus, *Surg. Forum* **10**:795–799.

Andy, O. J., and Stephan, H., 1982, Stereotaxic surgery of the limbic system, in *Stereotaxy of the Human Brain* (G. Schaltenbrand and A. E. Walker, eds.), Georg Thieme Verlag, Stuttgart, New York, pp. 629–644.

Andy, O. J., Jurko, M. F., and Fred, R., 1963, Subthalamotomy in treatment of parkinsonian tremor, *J. Neurosurg.* **20**(10):860–869.

Andy, O. J., Jurko, M. F., and Hughes, J. R., 1975, Amygdalotomy for bilateral temporal lobe seizures, *South. Med. J.* **68**:743–748.

Angel, R. W., Aguilar, J. A., and Hofmann, W. W., 1969, Action tremor and thalamotomy, *Electroencephalogr. Clin. Neurophysiol.* **26**:80–85.

Anichkov, A. D., 1977, Stereotactic apparatus for introducing chronic intracerebral electrodes, *Hum. Physiol.* **3**:370–375 (Rus.).

Anosov, N. N., 1960, The shaking palsy, *Handb. Neurol.* **7**:256–280 (Rus.).

Ansari, K. A., Webster, D. D., and Manning, N., 1972, Spasmodic torticollis and L-dopa, *Neurology (Minneap.)* **32**:670–674.

Antonov, I. P., and Shanko, G. G., 1975, *Hyperkinesias in Children,* Nauka i Tekhnika, Minsk (Rus.).

Apfelbaum, R. I., 1977, A comparison of percutaneous radiofrequency trigeminal neurolysis and micro-vascular decompression of the trigeminal nerve for the treatment of tic douloureux, *Neurosurgery* **1**:16–21.

Apuzzo, M. L. J., 1985, Comments, *Neurosurgery* **17**:18.

Apuzzo, M. L. J., and Sabshin, J. K., 1983, Computed tomographic guidance stereotaxis in the management of intracranial mass lesions, *Neurosurgery* **12**:277–285.

Apuzzo, M. L. J., Petrovich, Z., Luxton, G., Jepson, J. H., Cohen, D., and Breeze, R. E., 1987, Interstitial radiobrachytherapy of malignant cerebral neoplasms: Rationale, methodology, prospects, *Neurol. Res.* **9**:91–100.

Arjona, V. E., 1974, Stereotactic hypothalamotomy in erectic children, in *Advances in Stereotactic and Functional Neurosurgery,* Springer-Verlag, New York, pp. 185–194.

Arkhipova, N. A., and Illinsky, I. A., 1972, Electroencephalographic changes in patients with parkinsonism following anode electrolysis of the ventro-lateral thalamic nucleus, *Vopr. Neirokhir.* **4**:46–51.

Armour, D., 1927, Surgery of the spinal cord and its membranes, *Lancet* **1**:691–697.

Aronson, N., 1966, Stereotactic lesions: A further discussion, *J. Neurosurg.* **24**:459–462.

Arseni, C., and Sandor, G., 1960, Klinisch–Statistische Analyse von 50 operierten Fällen von Torticollis spasticus, *Zentralbl. Neurochir.* **20**(2):91–97.

Artaryan, A. A., and Gurfinkel, V. S., 1971, Tremorography in the diagnosis of cerebellar tumours, *Vopr. Neirokhir.* **4**:50–53 (Rus.).

Artemjeva, E. N., Kudinova, M. P., Zalkind, M. S., Kandel, E. I., Roslyakova, O. E., and Kozlovskaya, N. B., 1977, The study of efferent influence mechanism on the segmental motor levels in the human, *Hum. Physiol.* **3**:913–923 (Rus.).

Arutyunov, A. I., and Pedachenko, G. A., 1960, Intracerebral haemorrhage and its surgical treatment, *Vrach. Delo* **5**:485–490 (Rus.).

Arutyunov, A. I., Kandel, E. I., Konovalov, A. N., and Vasin, N. Ya., 1969, Some results of a ten-year work in the field of stereotaxic neurosurgery, *Vopr. Neirokhir.* **1**:8–11 (Rus.).

Arutyunov, A. I., Blagoveschenskaya, N. S., Lyass, F. M., Soskin, L. S., and Tslaf, Z. Z., 1970a, Treatment of Stsenko–Cushing disease by implantation of yttrium-90 in the hypophysis, *Vopr. Neirokhir.* No. 3:25–29 (Rus.).

Arutyunov, A. I., Kadin, A. L., Konovalov, A. N., Raeva, S. N., Merschikova, T. A., and Seliverstov, V. V., 1970b, Experience of applying the micro-electrode method in the neurosurgical clinical practice during single-stage stereotaxic operations for parkinsonism, *Vopr. Neirokhir.* **6**:12–18 (Rus.).

Asai, K., Hufschmidt, H. J., and Schaltenbrand, G., 1960, The passive muscle relaxation in parkinsonism, *J. Nerv. Ment. Dis.* **130**(6):449–455.

Asenjo, A., Imbernón, S. A., Rocamora, R., Chiorino, R., and Aranda, L., 1964, Tecnica estereotaxica con el aparato Asenjo–Imbernon, *Neurocirurgia (Santiago)* **22**:86–91.

Asenjo, A., Imbernón, A., and Aranda, L., 1966, Tecnica estereotaxica con el aparato Asenjo–Imbernon aplicada a diferentes nucleos talamicos e hipotalamicos, *Neurocirurgia (Santiago)* **24:**99–128.

Augustinsson, L. E., and von Essen, C., 1977, Thalamotomy in patients with multiple sclerosis, in *Third Meeting of the European Society for Stereotactic and Functional Neurosurgery,* Freiburg, p. 24.

Augustinsson, L. E., Carlsson, C. A., and Fall, M., 1982, Autonomic effects of electrostimulation, *Appl. Neurophysiol.* **45:**185–189.

Austin, G., and Tsai, C., 1962, A physiological basis and development of a model for parkinsonian tremor, *Confin. Neurol.* **22:**248–258.

Austin, I., Witmer, E., Sakai, M., and Yokoi, S., 1968, Studies in myoclonus epilepsy (Lafora body form). 1. Isolation and preliminary characterization of Lafora bodies in two cases, *Arch. Neurol.* **19:**15–33.

Avery, R., Comte, P., Haut., H., and Siegfried, J., 1980, Experimental study of a dacron stabilized electrode for epidural spinal cord stimulation, *Proc. Eur. Soc. Artif. Org.* **7:**62–66.

Babchin, I. S., 1934, Surgical treatment of spasmodic torticollis, *Sov. Chir.* No. 6:3–4 (Rus.).

Babchin, I. S., Abrakov, L. V., Garmashov, Yu. A., and Novikov, A. V., 1967, Techniques of stereotactic transnasal diathermocoagulation of the hypophysis, *Vopr. Neirokhir.* No. 6:26–28 (Rus.).

Babichenko, E. I., 1966, Bloodless method of eliminating spasticity of the leg muscles in patients with anatomical interruption of the spinal cord, *Vopr. Neirokhir.* **5:**33–35 (Rus.).

Backlund, E.-O., 1971, A new instrument for stereotaxic brain tumor biopsy. A technical note, *Acta Chir. Scand.* **137:**825–827.

Backlund, E.-O., and Holst, H. von, 1978, Controlled subtotal evacuation of intracerebral hematomas by stereotactic technique, *Surg. Neurol.* **9:**99–101.

Backlund, E.-O., Grepe, A., and Lunsford, D., 1981, Stereotaxic reconstruction of the aqueduct of Sylvius, *J. Neurosurg.* **55**(5)**:**800–810.

Backlund, E.-O., Granberg, P.-O., Hamberger, B., Knutsson, E., Martensson, A., Sedvall, G., Seiger, A., and Olson, L., 1985, Transplantation of adrenal medullary tissue to striatum in parkinsonism, *J. Neurosurg.* **62:**169–173.

Badaljan, L. O., 1975, *Children's Neurology,* Meditsina, Moscow (Rus.).

Badaljan, L. O., Skvortsov, I. A., Kamennykh, L. N., Dadali, E. L., and Temin, P. A., 1979, Paroxysmal generalized tic (Tourette syndrome), *Klin. Med.* **57:**28–34 (Rus.).

Bailey, H. E., Dowling, J. L., and Davies, E., 1977, Cingulotractotomy and related procedures for severe depressive illness (studies in depression IV), in *Neurosurgical Treatment in Psychiatry, Pain, and Epilepsy* (W. H. Sweet, S. Obrador, and J. G. Martin-Rodriguez, eds.), University Park Press, Baltimore, pp. 229–251.

Bailey, P., and Stein, S. U., 1951, *A Stereotaxic Instrument for Use on the Human Brain,* Thomas, Springfield, IL.

Balasubramaniam, V., and Kanaka, T. S., 1975, Amygdalotomy and hypothalamotomy. A comparative study, *Confin. Neurol.* **37:**195–201.

Balasubramaniam, V., Kanaka, T. S., and Ramanujam, P. B., 1974, Stereotaxic surgery for cerebral palsy, *J. Neurosurg.* **40:**577–582.

Ballantine, H. T., and Giriunas, I. E., 1979, Advances in psychiatric surgery, in *Functional Neurosurgery* (T. Rasmussen and R. Marino, eds.), Raven Press, New York. pp. 155–164.

Ballantine, H. T., Cassidy, W. L., Flanagan, N. W., and Marino, R., 1967, Stereotaxic anterior cingulotomy for neuropsychiatric illness and intractable pain, *J. Neurosurg.* **26:**488–495.

Balthasar. K., 1957, Gezielte Kälteschaden in der Grosshirnrinde der Katze, *Dtsch. Z. Nervenheilkd.* **176:**173–199.

Bancaud, J., Talairach, J., Bonis, A., Schaub, C., Szikla, G., Morel, P., and Bordas-Ferrer, M., 1967, *La Stéréo-électroencéphalographie dans l'Epilepsie,* Masson, Paris.

Bancaud, J., Bonis, A., Bordas-Ferrer, M., and Buser, P., 1970, Syndrome de Kojewnikow et accés somato-moteurs, *Encephale* **59:**391–438.

Bantli, H., Bloedel, J., and Tolbert, D., 1976, Activation of neurons in the cerebellar nuclei and ascending reticular formation by stimulation of the cerebellar surface, *J. Neurosurg.* **45:**539–554.

Barbe, D., and Alksne, J. F., 1984, Success of microvascular decompression with and without prior surgical therapy for trigeminal neuralgia, *J. Neurosurg.* **60:**104–107.

Barbeau, A., 1972, Role of dopamine in the nervous system, *Monogr. Hum. Genet.* **6:**114–136.

Barbeau, A., 1975, Résultats à long terme de la L-dopa dans la maladie de Parkinson, *Union Med. Can.* **104:**32–38.

Barbeau, A., 1976, Six years of high-level levodopa therapy in severely akinetic parkinsonian patients, *Arch. Neurol.* **33:**333–338.

Barbeau, A., Jasmin, G., and Duchastel, J., 1963, Biochemistry of Parkinson's disease, *Neurology (Minneap.)* **13:**56–58.

Barbera, J., Barcia-Salorio, J. L., and Broseta, J., 1979, Stereotaxic pontine spinothalamic tractotomy, *Surg. Neurol.* **11:**111–114.

Barchatova, V. P., 1967, Disturbance of amino acid metabolism in some hereditary extrapyramidal diseases, in *Proceedings of the Second All-Union Congress of Neuropathology and Psychiatry,* Meditsina, Moscow, pp. 139–141 (Rus.).

Barchatova, V. P., and Kandel, E. I., 1984, Influence of stereotactic thalamotomy on the cerebral dopamine metabolism in patients with extrapyramidal hyperkinesias, in *XVII Danube Symposium for Neurological Sciences,* Vol. 2, Moscow, p. 19 (Rus.).

Barchatova, V. P., Kandel, E. I., and Demina, E. G., 1981, Approaches to the study of dopamine cerebral metabolism in certain extrapyramidal diseases, *Zh. Nevropatol. Psikhiatr.* **81:**1021–1029 (Rus.).

Barcia-Salorio, J. L., and Broseta, J.. 1976, Stereotactic fornicotomy in temporal epilepsy. Indications and long-term results, *Acta Neurochir. (Wien)* **23:**167–175.

Barcia-Salorio, J. L., Broseta, J., and Hernandez, G., 1982, A new approach for direct CT localization in stereotaxis, *Appl. Neurophysiol.* **45:**383–386.

Barlow, J. S., 1965, Cross-correlation of accelerometer recordings of movement disorders in man, *Q. Prog. Rep.* **76**:293–311.

Barolat-Romana, G., Myklebust, J. B., Hemmy, D. C., Myklebust, B., and Wenninger, W., 1985, Immediate effects of spinal cord stimulation in spinal spasticity, *J. Neurosurg.* **65**:558–562.

Barrash, J. M., and Leavens, M. E., 1973, Dorsal rhizotomy for the relief of intractable pain of malignant tumor origin, *J. Neurosurg.* **38**:755–757.

Barrett, R. E., Yahr, M. D., and Duvoisin, R. C., 1970, Torsion dystonia and spasmodic torticollis results of treatment with L-dopa, *Neurology (Minneap.)* **20**:107–113.

Barük, H., Trotot, M., and Wolfin, M., 1953, Psychochirurgie et neurochirurgie: A propos d'un cas de Parkinson postencephalitique avec hallucinations ayant subi la lobotomie, *Rev. Neurol.* **88**(3):191–196.

Basbaum, A. L., and Fields, H. L., 1978, Endogenous pain control mechanisms: Review and hypothesis, *Ann. Neurol.* **4**:451–462.

Baskin, D. S., Boggan, J. E., and Wilson, C. B., 1982, Transsphenoidal microsurgical removal of growth hormone-secreting pituitary adenomas. A review of 137 cases, *J. Neurosurg.* **56**:634–641.

Bates, J. A. V., and Nathan, P. W., 1980, Transcutaneous electrical nerve stimulation for chronic pain, *Anaesthesia* **35**:807–822.

Bathien, N., and Rondot, P., 1977, Reciprocal continuous inhibition in rigidity of parkinsonism, *J. Neurol. Neurosurg. Psychiatry* **40**:20–24.

Bauer, H. J., 1972, Spasticity—its causes and clinical significance, in *Spasticity—a Topical Survey* (W. Birkmayer, ed.), Hans Huber Publishers, Bern, Stuttgart, Vienna, pp. 31–41.

Bechterev, V. M., 1904, *The Mind and the Life*, St. Petersburg (Rus.).

Bechterev, V. M., 1906, *Grounds of Studies on the Functions of Brain*, St. Petersburg (Rus.).

Bechterev, V. M., 1912, The conjoint-motor reflex method in the investigation of malingering, *Russ. Vrach.* **11**:461–464 (Rus.).

Bechtereva, N. P., 1980, *Healthy and Ill Brain*, Meditsina, Leningrad (Rus.).

Bechtereva, N. P., Bondarchuk, A. N., Smirnov, W. M.. and Trohachev, A. I., 1967, *Physiology and Pathophysiology of the Human Cerebral Deep Structures*, Meditsina, Leningrad (Rus.).

Bechtereva, N. P., Bondarchuk, A. N., and Smirnov, V. M., 1969, Physiological substantiation for operations performed on subcortical structures in hyperkinesis, *Vopr. Neirokhir.* No. 1:1–7 (Rus.).

Bechtereva, N. P., Bondarchuk, A. N., Gretchin, V. B., Iljushina, V. A., Kambarova, D. K., Matveev, J. K., Petushkov, E. P., Pozdeev, V. K., Smirnov, V. M., and Shandurina, A. N., 1972a, Structural–functional organization of the human brain and the pathophysiology of the parkinsonian type hyperkineses, *Confin. Neurol.* **34**:1–4, 14–17 (Rus.).

Bechtereva, N. P., Bondarchuk, A. N., Smirnov, V. M., and Melyucheva, L. A., 1972b, Curative electric stimulation of deep-lying brain structures, *Vopr. Neirokhir.* **1**:7–12 (Rus.).

Bechtereva, N. P., Kambarova, D. K., Smirnov, V. M., and Shandurina, A. N., 1977, Using the brain's latent abilities for therapy: Chronic intracerebral electrical stimulation, in *Neurosurgical Treatment in Psychiatry, Pain, and Epilepsy* (W. H. Sweet, S. Obrador, and J. G. Martin-Rodriguez, eds.), University Park Press, Baltimore, pp. 581–613.

Beck, D. W., and Lobosky, J. M., 1985, Treatment of trigeminal neuralgia and atypical facial pain by percutaneous injection of glycerol into the trigeminal cistern, in *VIII International Congress of Neurological Surgery*, Toronto, p. 143.

Beck, D. W., Olson, J. J., and Urig, E. J., 1986, Percutaneous retrogasserian glycerol rhizotomy for treatment of trigeminal neuralgia, *J. Neurosurg.* **65**:28–31.

Becker, H., Schheider, E., Hachker, H., and Fischer, P.-A., 1979, Cerebral atrophy in Parkinson's disease represented in CT, *Arch. Psychiatr. Nervenkr.* **227**:81–88.

Bedrensky, L. A., and Vasin, N. Ya., 1981, The anterior cerebral commissure as a stereotaxic guiding point (morphological study), *Vopr. Neirokhir.* **4**:30–34 (Rus.).

Bedwell, S. F., 1960, Some observations on hemiballismus, *Neurology* **10**:619–622.

Bein, B. N., and Boreiko, V. B., 1972, Surgical treatment of patients with temporal epilepsy and marked psychic disorders, *Vopr. Neirokhir.* **2**:26–31 (Rus.).

Belmusto, L., Brown, E., and Owens, G., 1963, Clinical observations on respiratory and vasomotor disturbance as related to cervical chordotomies, *J. Neurosurg.* **20**:225–230.

Belyaev, J. I., 1970, *The Clinics, Diagnostics and Surgical Treatment of Temporal Lobe Epilepsy*, Thesis, Sverdlovsk (Rus.).

Belyaev, J. I., and Shevchenko, P. Yu., 1972, Diagnosis of cerebral hemiatrophy and hemispherectomy, *Vopr. Neirokhir.* **5**:44–48 (Rus.).

Belyaev, V. V., Ivannikov, Yu.G., and Usov, V. V., 1965, A method for calculating stereotaxic coordinates in an arbitrary system by converting coordinates on computer, *Vopr. Neirokhir.* **4**:58–61 (Rus.).

Benabid, A. L., Persat, J. C., and Chiroussel, J., 1979, Correlative study between computerized transverse scanning and stereoimpedoencephalography in space-occupying lesions of the brain, *Acta Neurochir. (Wien)* **46**:219–232.

Benabid, A. L., Blond, S., Chasal, J., Cohadon, F., Daumas-Duport, D., Delisle, M., Farnarier, Ph., Frank, F., Gherardi, M., and Gontier, M., 1985, Les biopsies stéréotaxiques des néoformations intra-crâniennes. Reflexions à propos de 3052 cas, *Neurochirurgie* **31**:295–301.

Benes, V., Vladyka, V., and Suerina, E., 1965, Stereotaxic evacuation of typical brain haemorrhage, *Acta Neurochir. (Wien)* **13**:419–426.

Bennett, W. H., 1889, A case in which antispasmodic pain in the left lower extremity was completely relieved by subdural division of the posterior roots of certain spinal nerves, all other treatment having proved useless. Death from sudden collapse and cerebral hemorrhage on the twelfth day after the operation at the commencement of apparent convalescence, *Med. Chir.* **72**:329–348.

Berardelli, A., Sabra, A. F., and Halett, M., 1983, Physiological mechanisms of rigidity in Parkinson's disease, *J. Neurol. Neurosurg. Psychiatry* **46**:45–53.

Berger, M. S., 1986, Ultrasound-guided stereotactic biopsy using a new apparatus, *J. Neurosurg.* **65**:550–554.

Bergland, R. M., Ran, B. S., and Torack, R. M., 1968, Anatomical variations in the pituitary gland and adjacent structures in 225 human autopsy cases, *J. Neurosurg.* **28**(2):93–99.

Bergström, M., and Greitz, T., 1976, Stereotaxic computed tomography, *Am. J. Roentgenol.* **127**:167–170.

Bernheimer, H., and Hornykiewicz, O., 1965, Herabgesatzte Konzentration der Homovanillinsaure im Gehirn von Parkinson-kranken Menschen als Ausdruck der Storung des zentralen Dopaminstoffwechsels, *Klin. Wochenschr.* **43**:711–715.

Bernstein, N. A., 1947, *On Construction of Movements*, Meditsina, Moscow (Rus.).

Bernstein, N. A., 1966, *Essays on Physiology of Movements and Physiology of Activity*, Meditsina, Moscow (Rus.).

Bernstein, S., 1912, Ein Fäll von Torsionskrampf, *Wien. Klin. Wochenschr.* **25**:1567–1571.

Bertrand, C., 1958, A pneumotaxic technique for producing localized cerebral lesions and its use in the treatment of Parkinson's disease, *J. Neurosurg.* **15**:251–264.

Bertrand, C., 1976, The treatment of spasmodic torticollis with particular reference to thalamotomy, in *Current Controversies in Neurosurgery* (E. Morley, ed.), W. B. Saunders, Philadelphia, pp. 455–459.

Bertrand, C., and Martinez, N., 1962, Experimental and clinical surgery in dyskinetic disease, *Confin. Neurol.* **22**:375–383.

Bertrand, C., Hardy, J., Molino-Nergo, P., and Martinez, S. N., 1969, Optimum physiological target for the arrest of tremor, in *Third Symposium on Parkinson's Disease*, S. Livingstone, Edinburgh, London, pp. 251–259.

Bertrand, C., Martinez, S. N., and Hardy, J., 1973, Stereotactic surgery for parkinsonism, *Prog. Neurol. Surg.* **5**:79–112.

Bertrand, C., Oliver, A., and Thompson, C. J., 1974, Computer display of stereotactic brain maps and probe tracts, in *Advances in Stereotactic and Functional Neurosurgery*, Springer-Verlag, New York, pp. 235–244.

Bertrand, C., Molina-Negro, P., Martinez, S. N., and Hardy, J., 1976, Localisation and function of cortico-bulbar and anterior cortico-spinal fibers in the internal capsule, in *Congress of Neurological Surgery*, San Francisco,

Bertrand, C., Molina-Negro, P., and Martinez, S. N., 1978, Combined stereotactic and peripheral surgical approach for spasmodic torticollis, *Appl. Neurophysiol.* **41**:122–133.

Bertrand, C., Molina-Negro, P., and Martinez, S. N., 1979, Stereotaxic targets for dystonias and dyskinesias: Relationship to corticobulbar fibers and other adjoining structures, *Adv. Neurol.* **24**:395–399.

Bertrand, C., Molina-Negro, P., and Martinez, S. N., 1982, Technical aspects of selective peripheral denervation for spasmodic torticollis, *Appl. Neurophysiol.* **45**:326–330.

Bertrand, G., 1979, Computers in functional neurosurgery, in *Functional Neurosurgery* (T. Rasmussen and R. Marino, eds.), Raven Press, New York, pp. 75–87.

Bertrand, G., 1982, Computers in stereotactic surgery, in *Stereotaxy of the Human Brain* (G. Schaltenbrand and A. E. Walker, eds.), Georg Thieme Verlag, Stuttgart, New York, pp. 364–371.

Bertrand, G., and Jasper, H., 1965, Microelectrode recording of unit activity in the human thalamus, *Confin. Neurol.* **26**:205–208.

Bertrand, G., Blundell, J., and Musella, R., 1965, Electrical exploration of the internal capsule and neighbouring structures during stereotaxic procedures, *J. Neurosurg.* **22**:333–343.

Bertrand, G., Jasper, H., and Wong, A.. 1967. Microelectrode study of the human thalamus: Functional organisation in the ventrobasal complex, *Confin. Neurol.* **29**:81–86.

Bertrand, G., Oliver, A., and Thompson, C. J., 1974, The computer brain atlas. Its use in stereotaxic surgery, *Confin. Neurol.* **36**:312–313.

Bés, A., Güell, A., Fabre, N., Dupui, P., Victor, G., and Géraud, G., 1983, Cerebral blood flow studied by xenon-133 inhalation technique in parkinsonism: Loss of hyperfrontal pattern, *J. Cereb. Blood Flow Metab.* **3**:33–37.

Bethlem, J., and Jager, W. A. H., 1960, The incidence and characteristics of Lewy bodies in idiopathic paralysis agitans (Parkinson's disease), *J. Neurol. Neurosurg. Psychiatry* **23**:74–80.

Bettag, W., 1966, Langzeitbeobachtungen nach Schmerzthalamotomien und ihre Bedeutung für die Indikationsstellung sowie die Wahl des Destruktionsortes, *Confin. Neurol.* **27**:234–237.

Bettag, W., 1975, Results of treatment of pain by interruption of the medial pain tract of the brain stem, in *Proceedings of the Third International Congress of Neurological Surgery*, Excerpta Medica, Amsterdam, Oxford, pp. 771–775.

Bettag, W., and Yoshida, T., 1960, Über stereotaktische Schmerzoperationen, *Acta Neurochir. (Wien)* **8**:299–317.

Biljik, V. D., 1971, *Clinic, Pathophysiology and Treatment of Extrapyramidal Hyperkinesias*, Thesis, Vinnitsa (Rus.).

Bingley, T., Leksell, L., Meyerson, B. A., and Rylander, G., 1973, Stereotactic anterior capsulotomy in anxiety and obsessive–compulsive states, in *Surgical Approaches in Psychiatry, Proceedings of the Third International Congress of Psychosurgery* (L. V. Laitinen and K. E. Livingston, eds.), University Park Press, Baltimore, pp. 159–164.

Bird, E. D., and Iversen, L. L., 1974, Huntington's chorea: Postmortem measurement of glutamic acid decarboxylase, choline acetyltransferase and dopamine in basal ganglia, *Brain* **97**:457–472.

Birdsong, J. H., and McKinney, A. S., 1974, Long-range motor performance changes in levodopa-treated patients with Parkinson's disease, *Neurology (Minneap.)* **24**:107–115.

Birg, W., and Mundinger, F., 1973, Computer calculations of target parameters for a stereotaxic apparatus, *Acta Neurochir. (Wien)* **29**:123–129.

Birg, W., and Mundinger, F., 1975, Calculation of the position of a side-protruding electrode tip in stereotactic brain operations using a stereotactic apparatus with polar coordinates, *Acta Neurochir.* **32**:83–87.

Birg, W., and Mundinger, F., 1982, Direct target point determination for stereotactic brain operations from CT data and the calculation of setting parameters for polar-coordinate stereotactic devices, *Appl. Neurophysiol.* **45**(4–5):387–395.

Birg, W., Mundinger, F., and Klar, M., 1977a, A computer programme system for stereotaxic neurosurgery, *Acta Neurochir. (Wien)* **24**:99–108.

Birg, W., Mundinger, F., and Klar, M., 1977b, Computer assistance for stereotactic brain operations, *Adv. Neurosurg.* **4**:287–291.

Birg, W., Mundinger, F., Mohadjer, M., Wiegel, K., and Fuermaier, R., 1985, X-ray and magnetic resonance stereotaxy for functional and nonfunctional neurosurgery, *Appl. Neurophysiol.* **48**:22–29.

Birk, P., Struppler, A., Riescher, H., and Keidel, M., 1985, Somatosensory evoked potentials in the ventrolateral thalamus, *Appl. Neurophysiol.* **48**:151–152.

Birkenfeld, R., and Fischer, R. G., 1963, Successful treatment of causalgia of upper extremity with medullary spinothalamic tractotomy, *J. Neurosurg.* **20**:303–311.

Birkmayer, W. (ed.), 1972, *Spasticity—a Topical Survey,* Hans Huber, Bern, Stuttgart, Vienna.

Birkmayer, W., and Hornykiewicz, O., 1961, Der L-3,4-Dihydroxyphenylalanin (Dopa) Effect bei der Parkinson Akinesie, *Wien Klin. Wochenschr.* **73**:787–788.

Birkmayer, W., and Riederer, P., 1985, *Die Parkinson-Krankheit,* Springer, New York.

Birkmayer, W., Linauer, W., Mentasti, M., and Riederer, P., 1974, Zweijährige Erfahrungen mit einer Kombinationsbehandlung des Parkinson Syndroms mit L-Dopa und dem Dekarboxylasehemmer Benserazid (Ro 4-4602), *Wien. Med. Wochenschr.* **124**(22):340–344.

Bischof, W., 1951, Die longitudinale Myelotomie, *Zentralbl. Neurochir.* **11**:79–88.

Bischof, W., 1967, Zur dorsalen longitudinalen Myelotomie, *Zentralbl. Neurochir.* **28**:123–126.

Bishop, G. H., Clare, M. H., and Price, J., 1948, Patterns of tremor in normal and pathological conditions, *J. Appl. Physiol.* **1**:123–147.

Black, P., Uematsu, S., and Walker, A. E., 1975, Stereotaxic hypothalamotomy for control of violent, aggressive behavior, *Confin. Neurol.* **37**:187–188.

Blaezel, K., and Lazarus, L., 1965, Cryogenic hypophysectomy, *Med. J. Aust.* **54**:148–150.

Blagoveshchenskaya, N. S., Kandel, E. I., and Lyass, F. M., 1968, Application of the radioactive yttrium for treatment of inoperable tumors of the skull base growing into nose and sinuses, *Vestn. Otorinolaringol.* **5**:61–67 (Rus.).

Blinderman, E. E., Weidner, W., and Markham, C. H., 1964, The pneumoencephalogram in Huntington's chorea, *Neurology (Minneap.)* **14**:601–607.

Blocq, P., and Marinesco, G., 1893, Sur un cas de tremblement parkinsonien hemiplégique symptomatique d'une tumeur du pédoncle cérébral, *C. R. Soc. Biol.* **5**:105–111.

Bloedel, J. R., and McGreery, D. B., 1975, Organization of peripheral and central pain pathways, *Surg. Neurol.* **4**:65–81.

Blume, H., Kakelewski, J., Richardson, R., and Rojas, C., 1982a, Radiofrequency denaturation in occipital pain: Results in 450 cases, *Appl. Neurophysiol.* **45**:543–548.

Blume, H., Richardson, R., and Rojas, C., 1982b, Epidural nerve stimulation of the lower spinal cord and cauda equina for the relief of intractable pain in failed low back surgery, *Appl. Neurophysiol.* **45**:456–460.

Boctor, S. M., 1962, A new simple apparatus for stereotaxic surgery in Parkinson's disease, *Alexandria Med. J.* **8**(5):528–532.

Boëthius, J., Lindblom, V., Meyerson, B. A., and Widen, L., 1976, Effects of multifocal stimulation on pain and somatosensory functions, in *Sensory Functions of the Skin in Primates, with Special Reference to Man,* Pergamon Press, Oxford, pp. 531–548.

Bogen, J. E., and Vogel, P. J., 1965, Cerebral commissurotomy in man, *Bull. Los Angeles Neurol. Soc.* **27**:169–172.

Bogen, J. E., Sperry, R. W., and Vogel, P. J., 1969, Commissural section and propagation of seizures, in *Basic Mechanisms of the Epilepsies* (H. H. Jasper, A. A. Ward, and A. Pope, eds.), Little, Brown, Boston, pp. 439–440.

Boggs, J. S., 1976, Selective percutaneous radiofrequency coagulation of Gasserian ganglion for trigeminal neuralgia, *J. Florida Med. Assoc.* **63**:898–899.

Bogolepov, N. K., Arbatskaya, Yu. D., and Gaponova, Yu. G., 1970, Clinical picture, differential diagnostics and work capacity of patients with posttraumatic and postencephalitic parkinsonism, *Zh. Nevropatol. Psychiatr.* **9**:1325–1333 (Rus.).

Bohm, E., 1978, Transcutaneous electrical nerve stimulation in chronic pain after peripheral nerve injuries, *Acta Neurochir. (Wien)* **40**:277–283.

Bohm, E., and Strand, R. R., 1962, Glossopharyngeal neuralgia, *Brain* **85**:371–388.

Bojik, V. P., 1980, Stereotactic lesion of campi Forel for epilepsy treatment with generalized convulsive fits, in *Surgical Treatment of Epilepsy,* Tbilisi, pp. 42–44 (Rus.).

Bolander, H. G., Kourtopoulus, H., Liliequist, B., and Wittbolt, S., 1983, Treatment of spontaneous intracerebral haemorrhage. A retrospective analysis of 74 consecutive cases with special reference to computertomographic data, *Acta Neurochir. (Wien)* **67**:19–28.

Bondarchuk, A. N., 1971, Medical, diagnostical, clinical and physiological possibilities of multiple implanted electrodes technique, in *I All-Union Congress of Neurosurgery,* Vol. 3, Moscow, pp. 173–175 (Rus.).

Bondarchuk, A. N., Borodkin, Y. S., Vakharlovsky, V. G., Neifach, S. A., and Smirnov, V. M., 1975, Complex medical and surgical treatment of hepatolenticular degeneration, *Vopr. Neirokhir.* **1**:40–44 (Rus.).

Bonica, J. J., 1954, *The Management of Pain,* Lea & Febiger, Philadelphia.

Borager, B., 1955, Cerebral pedunculotomy in four cases of hyperkinesia (three cases of choreoathetosis and one of rhythmical tremor), *Acta Psychiatr. Belg.* **30**:107–114.

Borromei, A., Di Nino, G. F., Giancola, L. C., and Kolletzek, M., 1982, La neuroelettrostimolazione transcutanea (T.E.N.S.) nella terapia delle nevralgia trigeminali a diversa etiologia, *Minerva Anestisiol.* **48**:787–789.

Bosch, D. A., 1980, Indications for stereotactic biopsy in brain tumours, *Acta Neurochir. (Wien)* **54**:167–179.

Bosch, D. A., 1983, Stereotactic mesencephalotomy in the treatment of pain syndromes, in *International Symposium of Functional and Stereotactic Neurosurgery*, Bratislava, p. 11.

Bosch, D. A., Rähu, T., and Backlund, E. O., 1978, Treatment of colloid cysts of the third ventricle by stereotactic aspiration, *Surg. Neurol.* **9:**15–18.

Bouchard, G., 1971, Long term results of stereotaxic fornicotomy and fornico-amygdalotomy in patients with temporal lobe epilepsy showing behavioral disturbances, in *Special Topics in Stereotaxis* (W. Umbach, ed.), Hippocrates Verlag, Stuttgart, pp. 53–64.

Bouchard, G., and Kim, Y. K., 1974, Methods in drug-resistant epilepsies, in *Recent Progress in Neurological Surgery,* (K. Sano and S. Ishii, eds.), Excerpta Medica, Amsterdam, pp. 261–264.

Bouchard, G., and Umbach, W., 1972, Indication for the open and stereotactic brain surgery in epilepsy, in *Present Limits of Neurosurgery* (I. Fusek and Z. Zunc, eds.), Avicenum, Czechoslovak Medical Press, Prague, pp. 403–406.

Bouchard, G., Kim, Y. K., and Umbach, W., 1975, Stereotaxic methods in different forms of epilepsy, *Confin. Neurol.* **37:**232–238.

Bousser, M.-G., 1974, Syndrome thalamique, *Rev. Med.* **15:**507–512.

Bouvier, G., Mercier, C., St. Hilaire, Y. M., Giard, N., Labrecque, R., and Beigue, R. A., 1983a, Anterior callosotomy and chronic depth electrode recording in the surgical management of some intractable seizures, *Appl. Neurophysiol.* **46:**52–56.

Bouvier, G., Conilbard, P., Leger, S. L., Lesage, J., Rotent, F., and Beique, R. A., 1983b, Stereotactic biopsy of cerebral space-occupying lesions, *Appl. Neurophysiol.* **43:**227–230.

Bradford, F. K., 1962, A simple instrument for use in stereotaxic surgery, *J. Neurosurg.* **19:**266–267.

Brain, W. R., and Walton, J. N., 1969, *Diseases of the Nervous System,* Oxford University Press, London.

Brandt, F., and Wittkamp, P., 1983, Spätresultate der Thermokoagulation in Ganglion Gasseri beim Tic Douloureux, *Neurochirurgia (Stuttg.)* **26:**133–135.

Bravo, G., Mata, P., and Seiquer, G., 1967, Surgery of bilateral Parkinson's disease, *Confin. Neurol.* **29:**133–138.

Brett, D. C., Ferguson, G. G., Ebers, G. C., and Paty, D. W., 1982, Percutaneous trigeminal rhizotomy. Treatment of trigeminal neuralgia secondary to multiple sclerosis, *Arch. Neurol.* **39:**219–221.

Bricolo, A., 1964, Il problema del dolore e la chirurgia stereotassica del talamo, *Fracastoro* **57:**289–335.

Brierly, J. B., and Beck, E., 1959, The significance in human stereotactic brain surgery of individual variation in the diencephalon and globus pallidus, *J. Neurol. Neurosurg. Psychiatry* **22:**287–298.

Brindley, G. S., Polkey, C. E., and Rushton, D. N., 1982, Sacral anterior root stimulators for bladder control in paraplegia, *Paraplegia* **20:**365–381.

Brindley, G. S., Polkey, C. E., Rushton, D. N., and Cardozo, L., 1986, Sacral anterior root stimulators for bladder control in paraplegia: The first 50 cases, *J. Neurol. Neurosurg. Psychiatry* **49:**1104–1114.

Brion, S., Guiot, G., Derome, P., and Comay, C., 1965, Hemiballismes post operatoires au cours de la chirurgie stereotaxique: À propos de 12 observations dont 2 anatomocliniques dans une serie de 850 interventions, *Rev. Neurol.* **112**(5):410–443.

Brissaud, E., 1894, Tics et spasmes cloniques de la face, *J. Med. Chir. Pract.* **65:**49–64.

Broggi, G., and Franzini, A., 1981, Value of serial stereotactic biopsies and impedance monitoring in the treatment of deep brain tumours, *J. Neurol. Neurosurg. Psychiatry* **44:**397–401.

Broggi, G., and Siegfried, J., 1979, Percutaneous differential radiofrequency rhizotomy of glossopharyngeal nerve in facial pain due to cancer, *Adv. Pain Res. Ther.* **2:**469–473.

Broggi, G., Angelini, L., and Bono, R., 1982, Stereotactic surgery of abnormal movements: Clinical results of 33 cerebral palsy patients, *Appl. Neurophysiol.* **45:**306–311.

Brooks, V. B., Cooke, J. D., and Thomas, J. S., 1972, The continuity of movements, in *Control of Posture and Locomotion* (R. B. Stein, K. G. Pearson, R. S. Smith, and J. B. Redford, eds.), Plenum Press, New York.

Broseta, J., Gonzalez-Garder, J., and Barcia-Salorio, J. Z., 1982a, Stereotactic evacuation of intracerebral hematomas, *Appl. Neurophysiol.* **45**(4–5)**:**443–448.

Broseta, J., Roldan, P., and Gonzales-Darder, J., 1982b, Chronic epidural dorsal column stimulation in the treatment of causalgic pain, *Appl. Neurophysiol.* **45:**190–194.

Broseta, J., Barbera, J., de Vera, J. A., Barsia-Saloria, J. L., Garcia-March, G., Gonzales-Darder, J., Roviana, F., and Joanes, V., 1986, Spinal cord stimulation in peripheral arterial disease. A cooperative study, *J. Neurosurg.* **64:**71–80.

Browder, Y., 1948, Section of the fibers to the anterior limb of the internal capsule in parkinsonism, *Am. J. Surg.* **75:**264–268.

Browder, Y., Kaplan, H. A., and Romber, A. M., 1953, Capsular operation for parkinsonism. Attendant functional changes, *Ann. Surg.* **138:**502–510.

Brown, F. D., Rachlin, J. R., and Rubin, J. M., 1984, Ultrasound guided periventricular stereotaxis, *Neurosurgery* **15:**162–164.

Brown, M. H., 1977, Limbic target surgery in the treatment of intractable pain with drug addiction, in *Neurosurgical treatment in Psychiatry, Pain, and Epilepsy* (W. H. Sweet, S., Obrador, and J. G. Martin-Rodriguez, eds.), University Park Press, Baltimore, pp. 699–706.

Brown, M. H., and Lighthill, J. A., 1968, Selective anterior cingulotomy: A psychosurgical evaluation, *J. Neurosurg.* **29:**513–519.

Brown, R. A., 1979, A computerized tomography—computer graphics approach to stereotaxic localization, *J. Neurosurg.* **50:**715–720.

Brown, R. A., 1981, Stereotactic procedures in computed tomography. II. Computer graphic approach to stereotactic procedures, in *Radiology of the Skull Brain* (T. H. Newton and D. G. Potts, eds.), C. V. Mosby, St. Louis, pp. 4296–4300.

Brünning, F., 1923, Die Chirurgie des vegetativen Nervensystems, *Med. Klin.* **19:**671–675.

Bubnov, A. N., 1975, Neurosurgical anatomy of zona incerta in relation to subthalamotomy, *Vopr. Neirokhir.* **1**:36–40 (Rus.).

Bucy, P. C., 1951, The surgical treatment of extrapyramidal diseases, *J. Neurol. Neurosurg. Psychiatry* **14**:108–117.

Bucy, P. C., and Buchanan, D. N., 1932, Athetosis, *Brain* **55**:179–192.

Bucy, P. C., and Case, T. J., 1939, Tremor: Physiologic mechanism and abolition by surgical means, *Arch. Neurol. Psychiatry* **41**:721–746.

Bullard, D., and Makachinal, T., 1987, Measurement of tissue impedance in conjunction with computed tomography-guided stereotactic biopsies, *J. Neurol. Neurosurg. Psychiatry* **50**:43–51.

Buller, A. I., Dornhorst, A. C., Edwards, R., Kerr, D., and Whelan, R., 1959, Fast and slow muscles in mammals, *Nature* **183**:1516–1519.

Burchiel, K. J., Ojemann, G. A., and Bolender, N., 1980, Localization of stereotaxic centers by computerized tomographic scanning, *J. Neurosurg.* **53**:861–863.

Burckhardt, G., 1890, Über die Rindenexcision als Beitrag zur Operationstherapie der Psychosen, *Allg. Z. Psychiatrie* **47**:488–498.

Burdenko, N. N., and Klosovski, B. N., 1937, The bulbotomy. Report 1. Ceasing of hyperkinetic phenomenas by cutting the extrapyramidal tract in the medulla oblongata, *Vopr. Neirokhir.* **1**:5–16 (Rus.).

Burdenko, N. N., and Klosovski, B. N., 1938, The bulbotomy, in *II Session of the Neurosurgical Council,* Meditsina, Moscow, Leningrad, pp. 246–254 (Rus.).

Bursaco, I. A., 1973, Fundus striae terminalis, an optional target in sedative stereotactic surgery, in *Surgical Approaches in Psychiatry* (L. V. Laitinen, and K. E. Livingston, eds.), Medical and Technical Publishing, Lancaster, pp. 135–137.

Burton, C., 1975, Dorsal column stimulation: Optimization of application, *Surg. Neurol.* **4**:171–179.

Busch, E., 1955, Orbitomedial frontal undercutting in mental disease; follow-up examination of 154 patients, *Dan. Med. Bull.* **2**:10–30.

Buyalsky, I. V., 1850, *Anatomical Surgical Tables,* St. Petersburg.

Cabrini, G. P., Marossero, F., Ettorre, G., and Infuso, L., 1972, Physiopathological observations on the mechanism of stimulus sensitive myoclonus. A stereotactic study, *Confin. Neurol.* **34**:1–4, 64–69.

Cahan, D. L., and Rand, R. W., 1973, Stereotactic coagulation of a paraventricular arteriovenous malformation. Case report, *J. Neurosurg.* **39**:770–775.

Cahan, D. L., and Trombke, B. T., 1975, Computer graphics. Three-dimensional reconstruction of thalamic anatomy from serial sections, *Comput. Prog. Biomed.* **5**:91–98.

Cahan, W., 1964, Cryosurgery of the uterus: Description of technique and potential application, *Am. J. Obstet. Gynecol.* **88**:410–418.

Cail, W. S., and Morris, J. L., 1979, Localization of intracranial lesions from CT scans, *Surg. Neurol.* **11**:35–37.

Cajal, S. R., 1911, *Histologie du Système Nerveux de l'Homme et de Vertébrés,* Paris.

Calne, D. B., 1970, *Parkinsonism: Physiology, Pharmacology and Treatment,* Edward Arnold, London.

Calne, D. B., 1977, Developments in the pharmacology and therapeutics of parkinsonism, *Ann. Neurol.* **1**:111–119.

Calne, D. B., 1978, Bromocriptine and the nigrostriatal system: Parkinsonism, *Triangle* **17**:49–53.

Calne, D. B., 1983, Current views on Parkinson's disease, *Can. J. Neurol. Sci.* **10**:11–15.

Campbell, J. B., Rossi, H. H., Biavati, M. H., and Biavati, B. J., 1962, Production of subcortical lesions by implantation of radioactive substances, *Confin. Neurol.* **22**:178–182.

Campbell, J. W., 1981, Examination of possible mechanisms by which stimulation of the spinal cord in man relieves pain, *Appl. Neurophysiol.* **44**(4):181–186.

Cantor, F. K., 1973, Evoked response studies of the thalamic pain syndrome, *Acta Neurol. Scand.* **49**:280–286.

Caracalos, A., 1972, Results of 103 cryosurgical procedures in involuntary movement disorders, *Confin. Neurol.* **34**:74–81.

Caraceni, T. A., Celano, J., Parati, E., and Girotti, F., 1977, Bromcriptine alone or associated with L-dopa plus benserazide in Parkinson's disease, *J. Neurosurg. Psychiatry* **40**:1142–1146.

Carew, T. J., 1985, Posture and locomotion, in *Principles of Neural Science* (E. I. Kandel and J. H. Schwartz, eds.), Elsevier, New York, pp. 478–486.

Carlos, R., Fukui, M., Hasuo, K., Uchino, A., Matsushima, T., Tamura, S., Kudo, S., Kitamura, K., and Matsuura, K., 1986, Radiological analysis of hemifacial spasm with special reference to angiographic manifestations, *Neuroradiology* **28**:288–295.

Carlsson, A., Lindgvist, M., Magnusson, T., and Waldbeck, B., 1958, On the presence of 3-hydroxytyramine in brain, *Science* **127**:471–478.

Carol, M., 1985, A true "advanced imaging assisted" skull-mounted stereotactic system, *Appl. Neurophysiol.* **48**:69–72.

Carpenter, M. B., 1956, A study of the red nucleus in the rhesus monkey. Anatomic degenerations and physiologic effects resulting from localized lesions of the red nucleus, *J. Comp. Neurol.* **105**:195–249.

Carpenter, M. B., 1961, Brain stem and infratentorial neuraxis in experimental dyskinesia, *Arch. Neurol.* **5**(5):504–524.

Carpenter, M. B., and Whittier, J. R., 1952, Study of methods for producing experimental lesions of the central nervous system with special reference to stereotaxic technique, *J. Compar. Neurol.* **97**:73–132.

Carpenter, M. B., and Stevens, G. H., 1957, Structural and functional relationships between the deep cerebellar nuclei and the brachium conjunctivum in the rhesus monkey, *J. Comp. Neurol.* **107**:109–163.

Carpenter, M. B., Nakano, K., and Kim, R., 1976, Nigrothalamic projections in the monkey demonstrated by autoradiographic technics, *J. Comp. Neurol.* **165**:401–416.

Carrea, R. M. E., and Mettler, F. A., 1955, Function of the primate brachium conjunctivum and related structures, *J. Comp. Neurol.* **102**:151–392.

Carreras, M., Mancia, D., and Pagni, C. A., 1967, Unit discharges recorded from the human thalamus with microelectrodes, *Confin. Neurol.* **29**:87–89.

Carter, R. E., Donovan, W. H., Halstead, L., and Wilkerson, U.

A., 1987, Comparative study of electrophrenic nerve stimulation and mechanical ventilatory support in traumatic spinal cord injury, *Paraplegia* **25**:86–91.

Casey, K. L., 1973a, The neurophysiologic basis of pain, *Postgrad. Med.* **53**(6):58–63.

Casey, K. L., 1973b, Pain. A current view of neural mechanisms, *Am. Sci.* **61**:194–200.

Cauthen, J. C., and Renner, E. J., 1975, Transcutaneous and peripheral nerve stimulation for chronic pain states, *Surg. Neurol.* **4**:102–104.

Cendrowsky, W., 1962, Nevrobol Nervu Trojdzielnego wstwardnieniu Rodzsianym, *Z. Neurol. Neurochir. Psychiatr. Pol.* **12**:773–776.

Chandler, W. F., Knake, J. E., and McGillicuddy, J. E., 1982, Intraoperative use of real-time ultrasonography in neurosurgery, *J. Neurosurg.* **57**:157–163.

Chapman, E., and Wiesendanger, M., 1981, Effect of dorsal column stimulation in experimental spasticity, in *8th Meeting of the World Society for Stereotactic and Functional Neurosurgery*, Zurich, p. 15.

Charcot, J. M., 1876a, Du tremblement sénile, *Prog. Med.* **6**:815–826.

Charcot, J. M., 1876b, Shaking palsy, in *Diseases of Nervous System* (Russian ed.), St. Petersburg, pp. 135–164.

Charcot, J. M., 1887, *Leçons sur les Maladies du Système Nerveux*, Paris, pp. 298–320.

Charcot, J. M., 1888, Encore la chorée chronique. Chorée chronique hémilatérale avec démence chez une femme de 49 ans, quelques remarques à ce propos sur le tremblement héréditaire et le tremblement sénile, in *Leçons du Mardi à la Salpêtriére Policliniques*, Dehbaye, Paris, pp. 563–566.

Chase, T. N., 1970, Biochemical and pharmacologic studies of dystonia, *Neurology (Minneap.)* **20**(II):122–130.

Chayen, M. S., Rudick, V., and Borvine, A., 1980, Pain control with epidural injection of morphine, *Anesthesiology* **53**:338–339.

Chehrazi, B., and Collins, W. F., 1981, Comparison of bipolar and monopolar electrocoagulation, *J. Neurosurg.* **54**:197–203.

Chkhenkeli, S. A., 1978a, Cryostereoencephalotomy in temporal lobe epilepsy, *Rep. Acad. Sci. Georgian SSR* **90**:721–724 (Rus.).

Chkhenkeli, S. A., 1978b, Arresting influence of dentate nucleus electrostimulation on epileptic activity of human amygdala and hippocampus in temporal lobe epilepsy, *Rep. Acad. Sci. Georgian SSR* **4**:406–411 (Rus.).

Chkhenkeli, S. A., 1981, Analysis of some causes of the low efficacy of surgical treatment of epilepsy, *Vopr. Neirockir.* **3**:47–55 (Rus.).

Chkhenkeli, S. A., 1982, Neurophysiological grounds and results of stereotactic combined surgical treatment for complex forms of epileptic fits, *J. Nevropatol. Psikhiatr.* **6**:91–101 (Rus.).

Chkhenkeli, S. A., and Bregvadze, E. S., 1979, Characteristics of deep brain structures local cerebral blood flow in epilepsy, *Hum. Physiol.* **5**:809–817 (Rus.).

Chkhenkeli, S. A., and Bregvadze, E. S., 1980, Polarographic investigation of deep brain structures in patients with bitemporal epilepsy as an additional method of diagnostics for the dominant epileptogenic focus, in *Surgical Treatment of Epilepsy*, Metsniereba, Tbilisi, pp. 198–201 (Rus.).

Chkhenkeli, S. A., Geladze, T. S., and Okudjava, V. M., 1977, Some electrophysiological versions of temporal lobe epilepsy, *J. Nevropatol. Psikhiatr.* **3**:378–384 (Rus.).

Chkhenkeli, S. A., Nadvornik, P., and Šramka, M., 1979, Analysis system of coordinates of intracerebral structures and optimization of stereotaxic calculations in posterior longitudinal approach to hippocampus, *Vopr. Neirokhir.* **5**:30–34 (Rus.).

Choi, C. R., and Kang, S. K., 1981, Bilateral cingulotomy for treatment in psychiatric patients, in *7th International Congress of Neurological Surgery*, Munich, p. 68.

Choi, C. R., and Umbach, W., 1977, Combined stereotaxic surgery for relief of intractable pain, *Neurochirurgia (Stuttg.)* **20**:84–87.

Choteau, P., Gray, F., and Warot, P., 1981, Syndrome de Ramsay-Hunt. Etude anatomique d'un cas, *Rev. Neurol.* **136**(12):837–852.

Clark, R., and Ervin, K., 1958, The use of epidural block as an indication for cordotomy, *Am. Surg.* **24**:467–468.

Clarke, R. H., and Horsley, V., 1906, On a method of investigating the deep ganglia and tracts of the central nervous system (cerebellum), *Br. Med. J.* **2**:1799–1800.

Clauss, J. L., and Balthasar, K., 1954, Zur Kenntnis der striären Tic-Krankheit (maladie des tics, Gilles de la Tourette'sche Krankheit), *Arch. Psychiatr. Nervenkr.* **191**:398–418.

Cloward, R. B., 1958, Anterior approach for removal of ruptured cervical discs, *J. Neurosurg.* **15**:602–606.

Cobb, C. A., and Fung, D., 1983, Quantitative analysis of lesion parameters in radiofrequency trigeminal rhizotomy, *J. Neurosurg.* **58**:388–391.

Cobb, C. A., French, B. N., and Smith, K. A., 1984, Intrathecal morphine for pelvic and sacral pain caused by cancer, *Surg. Neurol.* **22**:63–68.

Coe, J., and Ommaya, A., 1964, Evaluation of the focal lesions of the central nervous system produced by extreme cold, *J. Neurosurg.* **21**:433–444.

Coffey, R. J., and Lunsford, L. D., 1985, Stereotactic surgery for mass lesions of the midbrain and pons, *Neurosurgery* **17**(1):12–18.

Coggeshall, R. S., Appelbaum, M. L., and Fazen, M., 1975, Unmyelinated axons in human ventral roots, a possible explanation for the failure of dorsal rhizotomy to relieve pain, *Brain* **98**:157–166.

Cohadon, F., Rougier, A., Da Silva Nunes Neto, D., Pigneux, J., Cailli, J. M., and Constant, P., 1977, La tomodensitometrie en conditiones stereotaxiques, *Neurochirurgie* **23**:433–452.

Colbassani, H. J., and Wood, J. H., 1986, Management of spasmodic torticollis, *Surg. Neurol.* **25**:153–158.

Coleman, U., 1970, Preliminary remarks on the L-dopa therapy of dystonia, *Neurology (Minneap.)* Suppl. **20**:114–121.

Collier, M., 1890, Spasmodic torticollis treated by nerve ligature, *Lancet* **1**:1354–1355.

Colloff, E., Gleason, C. A., Alberts, W. W., and Wright, E.

W., Jr., 1973, Computer-aided localization techniques for stereotaxic surgery, *Confin. Neurol.* **35**:65–80.

Colombo, F., Dettori, P., Pinna, V., and Benedetti, A., 1981, Stereotactic thalamic lesions studied by CT scanner, *Acta Neurochir.* **57**:205–212.

Colombo, F., Casentini., L., Visona, A., and Benedetti, A., 1982, Stereotactic biopsy-diagnostic problems. Value of cytological and histological examinations, *Zbl. Neurochir.* **43**:309–313.

Colombo, F., Benedetti, A., Pozza, F., Zanardo, A., Avanzo, R. S., Chierego, G., and Marchetti, C., 1985, Stereotactic radiosurgery utilizing a linear accelerator, *Appl. Neurophysiol.* **48**:133–145.

Comte, P., Siegfried, J., and Wieser, H. G., 1983, Multipolar hollow-core electrode for brain recordings, *Appl. Neurophysiol.* **46**:41–46.

Conway, L. W., 1970, Transsphenoidal cryohypophysectomy, technique, *Int. Surg.* **54**:192–203.

Conway, L. W., 1973, Stereotaxic diagnosis and treatment of intracranial tumors including an initial experience with cryosurgery for pinealomas, *J. Neurosurg.* **38**:453–460.

Conway, L. W., 1977, Stereotactic biopsy of deep intracranial tumors, in *Current Techniques in Operative Neurosurgery* (H. Schmidek and W. Sweet, eds.), Grune & Stratton, New York, pp. 187–198.

Conway, L. W., and Collins, W. F., 1969, Transsphenoidal cryohypophysectomy, *N. Engl. J. Med.* **281**:1–7.

Conway, L. W., and Garcia, J. H., 1970, Cryohypophysectomy. Post mortem findings in 16 cases, *J. Neurosurg.* **32**:435–442.

Conway, L. W., O'Foghludha, F. F., and Collins, W. F., 1969, Stereotactic treatment of acromegaly, *J. Neurol. Neurosurg. Psychiatry* **32**(1):48–59.

Cook, A. W., 1976, Electrical stimulation in multiple sclerosis, *Hosp. Pract.* **11**:51–58.

Cook, A. W., and Kawakami, Y., 1977, Commissural myelotomy, *J. Neurosurg.* **47**:1–6.

Cook, A. W., and Weinstein, S., 1973, Chronic dorsal column stimulation in multiple sclerosis. Preliminary report, *N.Y. State J. Med.* **73**:2868–2872.

Cook, A. W., Oygar, A., Baggenstos, P., Pacheco, S., and Kliniga, E., 1976, Vascular disease of extremities, electrical stimulation of spinal cord and posterior roots, *N.Y. State J. Med.* **76**:366–368.

Cook, A. W., Taylor, J. K., and Nidzgorski, F., 1979, Functional stimulation of the spinal cord in multiple sclerosis, *J. Med. Eng. Technol.* **3**:18–23.

Cook, A. W., Taylor, J. K., and Nidzgorski, F., 1981, Results of spinal cord stimulation in multiple sclerosis, *Appl. Neurophysiol.* **44**:55–61.

Cook, B. R., and Jannetta, P. J., 1984, Tic convulsif: Results in 11 cases treated with microvascular decompression of the fifth and seventh cranial nerves, *J. Neurosurg.* **61**:949–951.

Cooper, I. S., 1953a, Anterior choroidal artery ligation for involuntary movements, *Science* **118**:193–196.

Cooper, I. S., 1953b, Ligation of anterior choroidal artery for involuntary movements—parkinsonism, *Psychiatr. Q.* **27**:317–319.

Cooper, I. S., 1954, Intracerebral injection of procaine into the globus pallidus in hyperkinetic disorders, *Science* **119**:417–418.

Cooper, I. S., 1955, Chemopallidectomy: An investigative technique in geriatric parkinsonians, *Science* **121**:217–218.

Cooper, I. S., 1956, Clinical results and follow-up studies in a personal series of 300 operations for parkinsonism, *J. Am. Geriatr. Soc.* **4**:1171–1181.

Cooper, I. S., 1959, Dystonia musculorum deformans alleviated by chemopallidectomy and chemopallidothalamectomy, *Arch. Neurol. Psychiatr.* **81**:8–9.

Cooper, I. S., 1960a, Results of 1000 consecutive basal ganglia operations for parkinsonism, *Ann. Intern. Med.* **52**:483–499.

Cooper, I. S., 1960b, Neurosurgical alleviation of intention tremor of multiple sclerosis and cerebellar disease, *N. Engl. J. Med.* **263**:441–444.

Cooper, I. S., 1961, *Parkinsonism: Its Medical and Surgical Therapy,* Charles C. Thomas, Springfield, IL.

Cooper, I. S., 1962a, Dystonia reversal by operation on basal ganglia, *Arch. Neurol.* **7**(2):132–145.

Cooper, I. S., 1962b, Cryogenic cooling and freezing of the basal ganglia, *Confin. Neurol.* **22**:336–341.

Cooper, I. S., 1963, Cryogenic surgery: New method of destruction or extirpation of benign or malignant tissues, *N. Engl. J. Med.* **268**:743–749.

Cooper, I. S., 1964, Effect of thalamic lesions upon torticollis, *N. Engl. J. Med.* **270**:967–972.

Cooper, I. S., 1965a, Cryogenic surgery for cancer, *Fed. Proc.* **24**:237–240.

Cooper, I. S., 1965b, Clinical and physiologic implications of thalamic surgery for disorders of sensory communication. I. Thalamic surgery for intractable pain, *J. Neurol. Sci.* **2**(6):493–519.

Cooper, I. S., 1965c, Clinical and physiologic implications of thalamic surgery for disorders of sensory communication. II. Intention tremor, dystonia, Wilson's disease and torticollis, *J. Neurol. Sci.* **2**:520–553.

Cooper, I. S., 1965d, Clinical and physiologic implications of thalamic surgery for dystonia and torticollis, *Bull. N.Y. Acad. Med.* **41**(8):870–897.

Cooper, I. S., 1966a, Muscle spindles and motor units, in *Control and Innervation of Skeletal Muscles* (B. L. Andrew, ed.), Livingstone, Edinburgh, pp. 9–16.

Cooper, I. S., 1966b, A cerebellar mechanism in resting tremor, *Neurology (Minneap.)* **16**(10):1003–1015.

Cooper, I. S., 1966c, The relationship of cerebellar intention tremor to the resting tremor of parkinsonism, *J. Am. Geriatr. Soc.* **14**(3):264–271.

Cooper, I. S., 1967, Relief of intention tremor of multiple sclerosis by thalamic surgery, *J.A.M.A.* **199**(10):689–694.

Cooper, I. S., 1969a, *Involuntary Movement Disorders,* Harper & Row, New York.

Cooper, I. S., 1969b, Intention tremor, in *Involuntary Movement Disorders* (I. S. Cooper, ed.), Harper & Row, New York, pp. 95–130.

Cooper, I. S., 1972, Levodopa-induced dystonia, *Lancet* 1317–1318.

Cooper, I. S., 1973, Effect of chronic stimulation of the anterior cerebellum on neurological diseases, *Lancet* **1**:206–209.

Cooper, I. S., 1976, 20-year follow up study of the neurosurgical treatment of dystonia musculorum deformans, in *Dystonia: Advances in Neurology,* Vol. 14 (R. Eldrige and S. Fahn, eds.), Raven Press, New York, pp. 170–177.

Cooper, I. S., 1977, Neurosurgical treatment of the dyskinesias, *Clin. Neurosurg.* **24**:367–390.

Cooper, I. S., 1978, *Cerebellar Stimulation in Man,* Raven Press, New York.

Cooper, I. S., 1982, A general theory of causation and reversibility of involuntary movement disorders, *Appl. Neurophysiol.* **45**:317–323.

Cooper, I. S., and Bravo, G. J., 1957, Alleviation of dystonia musculorum deformans and other involuntary movement disorders of childhood by chemopallidothalamectomy, *Clin. Neurosurg.* **5**:127–149.

Cooper, I. S., and Poloukhine, N., 1960, Neurosurgical relief in intention tremor due to cerebellar disease and multiple sclerosis, *Arch. Phys. Med.* **41**:1–4.

Cooper, I. S., Bergmann, L. L., and Caracalos, A., 1963, Anatomic verification of the lesion which abolishes parkinsonian tremor and rigidity, *Neurology (Minneap.)* **13**:779–787.

Cooper, I. S., Gioino, G., and Terry, R., 1965a, The cryogenic lesion, *Confin. Neurol.* **26**:161–177.

Cooper, R., Winter, A. L., Crow, H. J., and Walter, W. G., 1965b, Comparison of subcortical, cortical and scalp activity using chronically indwelling electrodes in man, *Electroencephalogr. Clin. Neurophysiol.* **18**:217–228.

Cooper, I. S., Gioino, G., and Terry, R., 1966, The cryogenic lesions, in *Advances in Stereoencephalotomy* (E. Spiegel and H. Wycis, eds.), Basel, 161–177.

Cooper, I. S., Amin, I., Chandra, R., and Waltz, J. M., 1973, A surgical investigation of the clinical physiology of the LP pulvinar complex in man, *J. Neurol. Sci.* **18**:89–110.

Cooper, I. S., Riklan, M., and Rakic, P., 1974, *The Pulvinar–LP Complex,* Thomas, Springfield, IL.

Cooper, I. S., Amin, I., Riklan, M., Waltz, J. U., and Poon, P., 1976a, Chronic cerebellar stimulation in epilepsy, *Arch. Neurol.* **33**:559–570.

Cooper, I. S., Cullinan, T., and Riklan, M., 1976b, The natural history of dystonia, in *Dystonia: Advances in Neurology,* Vol. 14 (R. Eldrige and S. Fahn, eds.), Raven Press, New York, pp. 157–169.

Cooper, I. S., Riklan, M., and Amin, I., 1976c, Chronic cerebellar stimulation in cerebral palsy, *Neurology (Minneap.)* **26**:744–753.

Cooper, I. S., Upton, A. R. M., and Amin, I., 1982, Chronic cerebellar stimulation (CCS) and deep brain stimulation (DBS) in involuntary movement disorders, *Appl. Neurophysiol.* **45**(3):209–217.

Corkill, G., and Rosenblatt, R., 1981, Transcutaneous nerve stimulation: Population characteristics of long term users, in *7th International Congress of Neurological Surgery,* p. 65.

Cosman, E. R., Nashold, B. S., and Bedenbaugh, P., 1983, Stereotaxic radiofrequency lesion making, *Appl. Neurophysiol.* **46**:160–166.

Cosman, E. R., Nashold, B. S., and Ovelman-Levitt, J., 1984, Theoretical aspects of radiofrequencey lesions in the dorsal root entry zone, *Neurosurgery* **15**:945–950.

Cotzias, G. C., 1968, L-Dopa for parkinsonism, *N. Engl. J. Med.* **278**:690.

Cowell, T. K., Marsden, C. D., and Owen, D. A. L., 1965, Objective measurement of parkinsonian tremor, *Lancet* **2**:1278–1279.

Cox, A. W., and Brown, M. H., 1977, Results of multi-target limbic surgery in the treatment of schizophrenia and aggressive states, in *Neurosurgical Treatment in Psychiatry, Pain and Epilepsy* (W. H. Sweet, S. Obrador, and J. G. Martin-Rodriges, eds.), University Park Press, Baltimore, pp. 469–482.

Crevier, P. H., 1974, Post-stereotactic intracranial haematomas, in *Advances in Stereotactic and Functional Neurosurgery* (F. Y. Gillingham, E. R. Hitchcock, and Y. W. Turner, eds.), Springer Verlag, New York, pp. 71–76.

Crisp, A. A., and Kalucy, R. S., 1973, The effect of leucotomy in intractable adolescent weight phobia (anorexia nervosa), *Postgrad. Med. J.* **49**:883–893.

Critchley, M., 1929, Arteriosclerotic parkinsonism, *Brain* **52**:23–83.

Critchley, M., 1962, Dyssynergia cerebellaris progressiva, *Trans. Am. Neurol. Assoc.* **87**:81–85.

Cromwell, L. I., and Harris, A. B., 1980, Treatment of cerebral arteriovenous malformations. A combined neurosurgical and neuroradiological approach, *J. Neurosurg.* **52**:705–708.

Crue, B. L., and Todd, E. M., 1964, A simplified technique of sacral rhizotomy for pelvic pain, *J. Neurosurg.* **21**(10):835–837.

Crue, B. L., Todd, E. M., and Carregal, E. J., 1968, Posterior approach for high cervical percutaneous radiofrequency cordotomy, *Confin. Neurol.* **30**:41–52.

Crue, B. L., Todd, E. M., and Carregal, E. J. A., 1970, Percutaneous radiofrequency stereotactic tractotomy, in *Pain and Suffering* (B. L. Crue, ed.), Charles C. Thomas, Springfield, IL, pp. 69–79.

Cunha, L., Goncalves, A. F., Oliveria, C., and Dinis, M., 1983, Homovanillic acid in the cerebrospinal fluid of parkinsonian patients, *Can. J. Neurol. Sci.* **10**:43–46.

Cushing, H., 1920, The major trigeminal neuralgias and their surgical treatment based on experiences with 332 Gasserian operations, *Am. J. Med. Sci.* **160**:157–184.

Cusick, J. F., Larson, S. J., and Sances, A., 1976, The effect of T-myelotomy on spasticity, *Surg. Neurol.* **6**:289–292.

Dandy, W. E., 1927, Glossopharyngeal neuralgia (tic douloureux). Its diagnosis and treatment, *Arch. Surg.* **15**:193–214.

Dandy, W. E., 1928, Removal of right cerebral hemisphere for certain tumors with hemiplegia. Preliminary report, *J.A.M.A.* **90**:823–825.

Dandy, W. E., 1929, Operation for the cure of tic douloureux. Partial section of the sensory root at the pons, *Arch. Surg.* **18**:687–734.

Dandy, W. E., 1930, An operation for the treatment of spasmodic torticollis, *Arch. Surg.* **20**:1021–1032.

Dandy, W. E., 1934, Concerning the cause of trigeminal neuralgia, *Am. J. Surg.* **24**:447–455.

Danner, R., Leino, E., Partanew, J., Sorvari, T., and Riekkinen,

P., 1982, Electromyographical and morphological findings in progressive myoclonus epilepsy, *Acta Neurol. Scand.* **66**:673–680.

Daumas-Duport, C., Monsaingeon, V., Szenthe, L., and Szikla, G., 1982, Serial stereotactic biopsies: A double histological code of gliomas according to malignancy and 3-D configuration, as an aid to therapeutic decision and assessment of results, *Appl. Neurophysiol.* **45**:431–437.

Daumas-Duport, C., Monsaingeon, V., N'Guyen, J. P., Missir, O., and Szikla, G., 1983, Correlations between histological and CT aspects of cerebral gliomas contributing to the choice of significant biopsy tracks, in *VI Meeting of the European Society for Stereotactic and Functional Neurosurgery,* Rome, p. T1.

Davidenkov, S. N., 1919, Approach to the torsion spasm, *Vrach. Delo* **11**:345–353 (Rus.).

Davidenkov, S. N., 1922, Unitarism and dualism in approach to epidemic encephalitis and shaking palsy, *Vrach. Delo* **12**:24–26 (Rus.).

Davidenkov, S. N., 1932, *Hereditary Diseases of the Nervous System,* Meditsina, Moscow (Rus.).

Davidenkov, S. N., 1956, *Clinical Lectures on Neurological Diseases,* Vol. 1, Meditsina, Leningrad, pp. 71–90 (Rus.).

Davidenkov, S. N., 1957, The postencephalitic syndrome of local torsion dystonia. Spasm of antagonists, in *Clinical Lectures on Neurological Diseases,* Vol. III, Meditsina, Leningrad, pp. 154–168 (Rus.).

Davidenkov, S. N., 1960, Spasmodic torticollis, *Handb. Neurol.* **7**:317–322.

Davidenkov, S. N., and Zolotova, N. A., 1921, The torsion spasm family, *Proc. Baku Med. Inst.* **1**:7–13 (Rus.).

Davidson, C., and Goodhart, S. P., 1933, Dystonia musculorum deformans: Clinicopathologic study, *Arch. Neurol. Psychiatry* **29**:1108–1124.

Davidson, C., and Goodhart, S. P., 1938, Dystonia musculorum deformans: Clinicopathologic study, *Arch. Neurol. Psychiatry* **39**:939–972.

Davis, R., and Gray, E., 1980, Technical problems and advances in the cerebellar stimulating systems used for reduction of spasticity and seizures, *Appl. Neurophysiol.* **43**:230–243.

Davis, R., Cullen, R. F., Duenas, D., and Engel, H., 1976, Cerebellar stimulation for cerebral palsy, *J. Florida Med. Assoc.* **63**:910–912.

Davis, R., Cullen, R. F., Flitter, M., Duenas, D., Engle, H., and Ennis, B., 1977, Control of spasticity and involuntary movements, *Neurosurgery* **1**:205–207.

Davis, R., Barolat-Romana, G., and Engle, H., 1980, Chronic cerebellar stimulation for cerebral palsy: Five year study, *Acta Neurochir. (Wien)* **30**:317–332.

Davis, R., Gray, E., and Kurdzma, J., 1981, Beneficial augmentation following dorsal column stimulation in some neurological diseases, *Appl. Neurophysiol.* **44**:37–49.

Davis, R., Engle, H., Kurdzma, J., Gray, E., Ryan, T., and Dusnak, A., 1982, Update of chronic cerebellar stimulation for spasticity and epilepsy, *Appl. Neurophysiol.* **45**:44–50.

Davison, C., 1942, The role of the globus pallidus and substantia nigra in the production of rigidity and tremor, *A. Res. Nerv. Ment. Dis. Proc.* **21**:267–333.

Debrun, G., Vinuela, F., Fox, A., and Drake, C. G., 1982, Embolization of cerebral arteriovenous malformations with bucrylate, *J. Neurosurg.* **56**:615–627.

Dechterev, V. V., 1927, *Shaking Palsy,* Moscow (Rus.).

De Divitius, E., Spaziante, R., Cappablanca, P., Caputí, F., Pettinato, G., and Del Basso De Caro, M., 1983, Reliability of stereotactic biopsy. A model to test the value of diagnoses obtained from small fragments of nervous system tumors, *Appl. Neurophysiol.* **46**:295–303.

Déjerine, J., 1902, *Sémiologie du Système Nerveux,* Paris.

Déjerine, J., and Roussy, G., 1906, Le syndrome thalamique, *Rev. Neurol.* **14**:521–532.

De Jong, D., 1966, Parkinson's diesase: Statistics, *J. Neurosurg.* **24**:149–158.

De Jong, H., 1926, Action tremor, *J. Nerv. Ment. Dis.* **64**:1–17.

De Lange, C., 1945, Dystonia musculorum progressiva (torsiondystonie), *Ann. Paediatr. (Basel)* **164**:169–181.

Delgado, J. M. R., 1959, Prolonged stimulation of brain in awake monkeys, *J. Neurophysiol.* **22**:458–475.

Delgado, J. M. R., 1969, *Physical Control of the Mind,* Harper and Row, New York.

Delgado, J. M. R., and Hamlin, H., 1956, Surface and depth electrography of the frontal lobes in conscious patients, *Electroencephalogr. Clin. Neurophysiol.* **8**:371–384.

Delgado, J. M. R., and Hamlin, H., 1962, Depth electrography, *Confin. Neurol.* **22**:228–235.

Delmas, A., and Pertuiset, B., 1959, *Topométrie Cranio-Encéphalique chez l'Homme,* Masson, Paris.

Delmas-Marsalet, P., and Van Bogaert, L., 1935, Sur un cas de myoclonies rhythmique continues par une intervention chirurgicale sur le tronc cérébrale, *Rev. Neurol.* **64**:728–733.

Delwaide, P. J., and Desseilles, M., 1977, Spontaneous buccolinguofacial dyskinesia in the elderly, *Acta Neurol. Scand.* **24**:256–262.

De Myer, W., 1959, Number of axons and myelin sheaths in adult human medullary pyramids. Study with silver impregnation and iron hematoxylin staining methods, *Neurology (Minneap.)* **9**:42–49.

Deniau, J. M., Hannmond, C., and Riszk, A., 1978, Electrophysiological properties of identified output neurons of the rat substantia nigra (pars compacta and pars reticulata): Evidences for the existence of branched neurons, *Exp. Brain Res.* **32**:409–422.

Dennis, S. G., and Melzack, R., 1977, Pain-signalling systems in the dorsal and ventral spinal cord, *Pain* **4**:97–132.

Denny-Brown, D., 1946, *Diseases of the Basal Ganglia and Subthalamic Nuclei,* Oxford University Press, Oxford.

Denny-Brown, D., 1962, *The Basal Ganglia, and Their Relation to Disorders of Movement,* Oxford University Press, Oxford.

Denny-Brown, D., Adams, R. D., Brenner, C., and Doherty, M. M., 1945, The pathology of injury to nerve induced by cold, *J. Neuropathol. Exp. Neurol.* **4**:305–324.

Dervin, E., Heywood, O. B., and Crossley, T. R., 1974, The use of a small digital computer for stereotactic surgery, *Acta Neurochir. (Wien)* [Suppl.] **21**:245–252.

Devinsky, O., 1983, Neuroanatomy of Gilles de la Tourette's syndrome. Possible midbrain involvement, *Arch. Neurol.* **40**:508–514.

Dieckmann, G., 1976, Traitement stéréotaxique du torticollis extrapyramidal, *Neurochirurgie* **22**:568–571.

Dieckmann, G., 1979, Chronic mediothalamic stimulation for control of phobias, in *Modern Concepts in Psychiatric Surgery* (E. R. Hitchcock, H. T. Ballantine, and B. A. Meyerson, eds.), Elsevier/North-Holland, Amsterdam, pp. 85–93.

Dieckmann, G., and Hassler, R., 1972a, Relief from compulsions and obsessions by combined intralaminar–medial thalamotomy, in *Present Limits of Neurosurgery* (I. Fusek and Z. Kunc, eds.), Avicenum, Czechoslovak Medical Press, Prague, pp. 483–486.

Dieckmann, G., and Hassler, R., 1972b, Stereotaxic treatment of extrapyramidal myoclonus, *Confin. Neurol.* **34:**57–63.

Dieckmann, G., and Hassler, R., 1975, Unilateral hypothalamotomy in sexual delinquents, *Confin. Neurol.* **37:**177–186.

Dieckmann, G., and Hassler, R., 1977, Treatment of sexual violence by stereotactic hypothalamotomy, in *Neurosurgical Treatment in Psychiatry, Pain, and Epilepsy* (W. H. Sweet, S. Obrador, and J. Martin-Rodriguez, eds.), University Park Press, Baltimore, pp. 451–462.

Dieckmann, G., and Krainick, J. U., 1979, Pain relief by chronic mediothalamic stimulation in man, *Adv. Neurosurg.* **7:**172–180.

Dieckmann, G., and Veras, G., 1985, Bipolar spinal cord stimulation for spasmodic torticollis, *Appl. Neurophysiol.* **48:**339–346.

Dieckmann, G., and Witzmann, A., 1982, Initial and long-term results of deep brain stimulation for chronic intractable pain, *Appl. Neurophysiol.* **45:**167–172.

Dieckmann, G., Gabriel, E., and Hassler, R., 1965, Size, form and structural peculiarities of experimental brain lesions obtained by thermocontrolled radiofrequency, *Confin. Neurol.* **26:**134–142.

Dieckmann, G., Veras, G., and Sogabe, K., 1985, Retrogasserian glycerol injection or percutaneous stimulation in the treatment of typical and atypical trigeminal pain, in *VIII International Congress of Neurological Surgery*, Toronto, p. 142.

Diemath, H. E., Heppner, F., and Walker, A. E., 1961, Anterolateral chordotomy for relief of pain, *Postgrad. Med.* **29:**485–495.

Diemath, H. E., 1966, Die stereotaktische vordere Cingulotomie bei Therapie-resistenter generalisierter Epilepsie, *Confin. Neurol.* **27:**124–128.

Dierssen, G., and Marg, E., 1965, The value of impedance measurements to aid in the localization in stereotactic surgery, *Confin. Neurol.* **26:**407–410.

Dierssen, G., and Obrador, S., 1967, The problem of bilateral lesions in stereotaxic surgery, *Confin. Neurol.* **29:**181–185.

Dietz, H., Winkelmüller, W., and Seidel, B. U., 1979, Chronic cerebellar stimulation in cerebral palsy, in *Proceedings, Annual Meeting, Scandinavian Neurosurgical Society*, p. 131.

Dildin, A. S., Kalakutski, L. I., and Nesterov, S. L., 1984, Position of transcutaneous electrostimulation in treatment of patients with chronic pain syndromes, *J. Nevropathol. Psikhiatr.* **4:**539–542 (Rus.).

Dimitri, V., 1935, Segunda observacion anatomoclinica de distonia muscular deformante, *Semin Med. (B. Aires)* **42:**77–95.

Dimitrijevic, M. R., Faganel, J., and Sherwood, A. M., 1983, Spinal cord stimulation as a tool for physiological research, *Appl. Neurophysiol.* **46:**245–253.

Dittmar, C., 1873, Veber die Lage des sogenannten Gefässzentrums in der Medulla oblongata. Berichte Sächsischen Gesellschaft der Wissenschaften, *Leipzig Math. Phys. Klasse* 449–469.

Ditullio, M. V., and Rand, R. W., 1977, Efficacy of cryohypophysectomy in the treatment of acromegaly, *J. Neurosurg.* **46:**1–11.

Djindjian, R., 1972, *L' Embolisation en Neuroradiologie Vascularie, Presse Med.* **1:**2155–2158.

Doerr, M., Krainick, J.-U., and Thoden, U., 1978, Pain perception in man after long term spinal cord stimulation, *J. Neurol.* **217:**261–270.

Dogliotti, A. M., 1931, Traitement des syndromes douloureux de la peripherie par alcoholisation subarachoïdienne des racines postérieures à leur émergence de la moella épiniere, *Presse Med.* **39:**1249–1252.

Dogliotti, M., 1938, First surgical sections, in man, of the lemniscus lateralis at the brain stem, for the treatment of diffused rebellious pain, *Anesth. Analg. (Cleve.)* **17:**143–145.

Donauer, E., Moringlane, J. R., and Ostertag, C. B., 1986, Colloid cysts of the third ventricle. Open operative approach or stereotactic aspiration? *Acta Neurochir.* **83:**24–30.

Dondey, M., Albe-Fessard, D., and Le Beau, J., 1962, Premieres applications neurophysiologiques d'une methode permettant le blocade electif et reversible de structures centrales par refrigeration localisée, *EEG Clin. Neurophysiol.* **14:**758–763.

Dooley, D. M., and Sharkey, J., 1981, Electrical stimulation of the spinal cord in patients with demyelinating and degenerative diseases of the central nervous system, *Appl. Neurophysiol.* **44:**218–224.

Dooley, D. M., Sharkey, J., Keller, W., and Kaspark, M., 1978, Treatment of demyelinating and degenerative diseases by electrostimulation of the spinal cord, *Med. Prog. Technol.* **6:**1–13.

Doshay, L. J., 1960, Parkinson's disease. Its meaning and management, *J.A.M.A.* **174:**1962–1965.

Drake, C. G., 1968, Further experience with surgical treatment of aneurysms of the basilar artery, *J. Neurosurg.* **29:**372–392.

Drake, C. G., 1976, Surgical treatment of vertebrobasilar aneurysms, in *Clinical Microneurosurgery* (G. Ross, ed.), Thieme Verlag, Stuttgart, pp. 213–219.

Drake, C. G., 1979, Cerebral arteriovenous malformations: Consideration for and experience with surgical treatment in 166 cases, *Clin. Neurosurg.* **26:**145–208.

Drake, C. G., and Allcock, J. M., 1973, Postoperative angiography and the "slipped" clip, *J. Neurosurg.* **39:**683–689.

Drake, C. G., and McKenzie, R., 1953, Mesencephalic tractotomy for pain, *J. Neurosurg.* **10:**457–462.

Driollet, R., Schvarcz, J. R., and Orlando, J., 1973a, Optimum target for tremor arrest, in *The Sixth Symposium of the International Society for Research in Stereoencephalotomy*, Vol. II, Tokyo, pp. 3–5.

Driollet, R., Schvarcz, J. R., and Orlando, J., 1973b, Surgical management of torticollis, in *The Sixth Symposium of the*

International Society for Research in Stereoencephalotomy, Vol. II, Tokyo, pp. 10–12.

Driollet, R., Schvarcz, J. R., and Orlando, J., 1975, Surgical management of torticollis, *Confin. Neurol.* **37:**9–19.

Dubenko, V. G., and Nebotov, V. A., 1976, On the periodic structure of physiologic tremor, *J. Nevropatol. Psikhiatr.* **7:**966–972 (Rus.).

Dubois, P. J., Nashold, B. S., and Perry, J., 1982, CT-guided stereotaxis using a modified conventional stereotaxic frame, *A.J.N.R.* **3**(3):345–351.

Dupont, E., 1980, Parkinson's disease and essential tremor: Differential diagnostic and epidemiological aspects, in *Parkinson's Disease. Current Progress, Problems and Management* (U. K. Rinne, M. S. Klinger, and G. Stanm, eds.), Elsevier/North-Holland Biomedical Press, Amsterdam, pp. 165–185.

Durity, F., and Logue, V., 1971, The effect of proximal anterior cerebral occlusion on anterior communicating artery aneurysms. Postoperative radiological survey of 43 cases, *J. Neurosurg.* **35:**16–19.

Dyakonova, E. N., 1968, *Study of the Pathogenesis of the Extrapyramidal Hyperkinesias in Animal Experiment,* Thesis, Moscow.

Dyck, P., Bouzaglon, A., and Gruskin, P., 1987, Stereotactic biopsy and brachytherapy of brain tumors, *Neurol. Res.* **9:**69–90.

Dzerjinski, V. E., 1916, Dystonia musculorum deformans (distonia lordofica progressiva), *Psikhiatr. Gaz.* **16:**319–330 (Rus.).

Dzugaeva, S. B., 1949, *Macroscopic Investigation of Associative, Commissural and Thalamo-cortical Pathways of the Human Brain,* Thesis, Moscow (Rus.).

Eadie, M. J., and Sutherland, J. U., 1964, Arteriosclerosis in parkinsonism, *J. Neurol. Neurosurg. Psychiatry* **27:**237–240.

Ebin, J., 1949, Combined lateral and ventral pyramidotomy in treatment of paralysis agitans, *Arch. Neurol. Psychiatry* **62:**27–35.

Eccles, J. C., 1980, Physiology of motor control in man, *Appl. Neurophysiol.* **44:**5–15.

Echols, D. N., 1969, Sensory rhizotomy following operation for ruptured intervertebral disc: A review of 62 cases, *J. Neurosurg.* **31:**331–338.

Economo, C., von, 1917, Encephalitis lethargica, *Wien. Clin. Wochenschr.* **30:**581–585.

Edner, G., 1975, Stereotaxic brain tumor biopsy. Five years experiences, *Acta Neurochir.* **31:**261–265.

Edner, G., 1981, Stereotaxic biopsy of intracranial space occupying lesions, *Acta Neurochir. (Wien)* **57:**213–234.

Egorov, B. G., 1959, Anatomic and physiologic grounds of surgical approach to brain tumors through the parietal lobe, in *Problems of Contemporary Neurosurgery,* Vol. III, Meditsina, Moscow, pp. 5–28 (Rus.).

Egorov, B. G., Kandel, E. I., and Lyass, F. M., 1961, Technique and method of implanting of the metallic yttrium-90 into brain tumors, in *Proceedings, Central Institute of Medical Radiology,* Leningrad, pp. 43–44 (Rus.).

Ehni, G., and Brenner, B., 1984, Occipital neuralgia and the C_{1-2} arthrosis syndrome, *J. Neurosurg.* **61:**961–965.

Ehringer, H., and Hornykiewicz, O., 1960, Verteilung von Noradrenalin und Dopamin (3-Hydroxytyramin) im Gehirn des Menschen und ihr Verhalten bei Erkrankungen des extrapyramidalen Systems, *Wien. Klin. Wochenschr.* **38:**1236–1239.

Eldridge, R., 1970, The torsion dystonias: Literature review and genetic and clinical studies, *Neurology (Minneap.)* **20**(Part II):1–78.

Eldridge, R., 1976, Variable course of the hereditary dystonias, *Adv. Neurol.* **14:**171–176.

Eldridge, R., and Gottlieb, R., 1976, The primary hereditary dystonias: Genetic classification of 768 families and revised estimate of gene frequency, autosomal recessive form, with selected bibliography, in *Advances in Neurology,* Vol. 14: *Dystonia* (R. Eldridge and S. Fahn, eds.), Raven Press, New York, pp. 457–474.

Eletski, A. G., 1923, On surgical treatment of spastic hyperkinesia with athetosis, *Nov. Chir. Arch.* **3**(2):16–24 (Rus.).

Emmers, R., 1981, *Pain: A Spike-Interval Coded Message in the Brain,* Raven Press, New York.

Emmers, R., and Tasker, R. R., 1975, *The Human Somesthetic Thalamus,* Raven Press, New York.

Erickson, T. C., 1940, Spread of the epileptic discharge: An experimental study of after discharge induced by electrical stimulation of cerebral cortex, *Arch. Neurol. Psychiatry* **43:**429–452.

Eriksson, M. B. E., Sjölund, B. H., Sundbärg, G., 1984, Pain relief from peripheral conditioning stimulation in patients with chronic facial pain, *J. Neurosurg.* **61:**149–155.

Erlander, R. N., Netsky, M. G., and Adelman, L. S., 1975, Location of human pyramidal tract in the internal capsule: Anatomic evidence, *Neurology* **25:**823–826.

Ervin, F. R., Brown, C. E., and Mark, V. H., 1966, Striatal influence of facial pain, *Confin. Neurol.* **27:**75–86.

Esposito, S., Delitala, A., Bruni, P., Hernandez, R., and Callovini, G. M., 1985, Therapeutic protocol in the treatment of trigeminal neuralgia, *Appl. Neurophysiol.* **48:**271–273.

Evans, H. M., Briggs, J. H., and Dixon, J. S., 1966, The physiology and chemistry of growth hormone, in *The Pituitary Gland,* Vol. 1 (G. W. Harris and B. T. Donovan, eds.), University of California Press, Berkeley, Los Angeles, pp. 439–491.

Evarts, E. V., 1971, Activity of thalamic and cortical neurons in relation to learned movement in the monkey, *Int. J. Neurol.* **8:**321–329.

Evarts, E. V., 1979, Brain mechanism of movement, *Sci. Am.* **241**(3):164–179.

Fabinyi, G. C. A., and Adams, C. B. T., 1978, Hemifacial spasm: Treatment by posterior fossa surgery, *J. Neurol. Neurosurg. Psychiatry* **41:**829–833.

Fager, C. A., 1968, Evaluation of thalamic and subthalamic surgical lesions in the alleviation of Parkinson's disease, *J. Neurosurg.* **28:**145–149.

Fahn, S., 1977, Secondary parkinsonism, in *Scientific Approaches to Clinical Neurology* (E. S. Goldensohn and S. Appel, eds.), Lea & Febiger, Philadelphia, pp. 1159–1189.

Fahn, S., and Bressman, S. B., 1984, Should levodopa therapy for parkinsonism be started early or late? Evidence against early treatment, *Can. J. Neurol. Sci.* **11**(Suppl.):200–206.

Fahn, S., and Eldridge, R., 1976, Definition of dystonia and classification of the dystonic states, in *Advances in Neurology,* Vol. 14 (R. Eldridge and S. Fahn, eds.), Raven Press, New York, pp. 1–5.

Fahn, S., Côté, L. J., Snider, S. R., Barrett, R. S., and Isgreen, W., 1979, The role of bromocriptine in the treatment of parkinsonism, *Neurology (Minneap.)* **29:**1077–1083.

Fahn, S., Côté, L. J., Barrett, R. S., and Snider, S. R., 1980, Further experiences with low doses of bromocriptine in Parkinson's disease, *Adv. Biochem. Psychopharmacol.* **23:**255–260.

Fairholm, D., Wu, J.-M., and Lïu, R.-N., 1983, Hemifacial spasm: Results of microvascular relocation, *Can. J. Neurol. Sci.* **10:**187–191.

Fairman, D., 1959, Roentgenologic principles of a new stereotaxic apparatus, *Am. J. Roentgenol. Rad. Ther. Nucl. Med.* **81:**1001–1005.

Fairman, D., 1966, Evaluation of results in stereotactic thalamotomy for the treatment of intractable pain, *Confin. Neurol.* **27:**67–70.

Fairman, D., 1972, Hypothalamotomy as a new perspective for alleviation of intractable pain and regression of metastatic malignant tumors, in *Present Limits of Neurosurgery* (I. Fusek and Z. Kunc, eds.), Avicenum Czechoslovakian Medical Press, Prague, pp. 525–528.

Fairman, D., 1981, Monoaminergic mechanism in stereotactic hypothalamic stimulation for the relief of cancer pain, in *VIII Meeting of the World Society for Stereotactic and Functional Neurosurgery,* Zurich, p. 22.

Fairman, D., 1983, Correlation between monoaminergic mechanism and stereotactic hypothalamic stimulation for neoplastic pain relief, in *International Symposium of Functional and Stereotactic Neurosurgery: Brain Stimulation and Psychosurgery,* Bratislava, p. 3.

Fairman, D., and Perlmutter, I., 1965, Physiological observations during stereotaxic surgery of the basal ganglia, *Confin. Neurol.* **26:**299–305.

Fairman, D., and Perlmutter, J., 1967, Stereotactic technique in the treatment of hyperkinesia and spasticity in cerebral palsy, *Confin. Neurol.* **29:**247–251.

Falconer, M. A., 1953, Surgical treatment of intractable phantom-limb pain, *Gr. Br. Med.* **1:**299–304.

Falconer, M. A., 1965, The surgical treatment of temporal lobe epilepsy, *Neurochirurgia (Stuttg.)* **8**(5):161–172.

Falconer, M. A., 1966, Relief of phantom pain by cordotomy, in *Pain,* Little, Brown, Boston, pp. 273–277.

Falconer, M. A., 1972a, The place of surgery in epileptic children and adolescents with mesial temporal (Ammon's horn) sclerosis, *Israel J. Med. Sci.* **141:**147–161.

Falconer, M. A., 1972b, Problems of surgical treatment of temporal lobe epilepsy in children, *Vopr. Neirokhir.* **36:**21–26 (Rus.).

Falconer, M. A., 1975, The place of surgery in temporal lobe epilepsy in childhood and adolescence, *Confin. Neurol.* **37:**243–253.

Falconer, M. A., and Taylor, D. C., 1968, Surgical treatment of drug-resistant epilepsy due to medial temporal sclerosis. Etiology and significance, *Arch. Neurol.* **19**(4):353–361.

Faria, M. A., and Tindall, G. T., 1982, Transsphenoidal microsurgery for prolactin-secreting pituitary adenomas. Results in 100 women with the amenorrhea–galactorrhea syndrome, *J. Neurosurg.* **56:**33–43.

Fasano, V. A., Broggi, G., and de Nunno, T., 1964, Cryotherapie et neurochirurgie, *Neuro-chirurgie* **10:**172–179.

Fasano, V. A., Barolat-Romana, G., Ivaldi, A., and Sguazzi, A., 1976, La radicotomie postérieure fonctionnelle dans le traitment de la spasticité cerebrale, *Neurochirurgie* **22**(1):23–24.

Fasano, V. A., Urciuoli, R., Broggi, G., 1977, New aspects in the surgical treatment of cerebral palsy, *Acta Neurochir. (Wien) [Suppl.]* **24:**53–57.

Fasano, V. A., Broggi, G., Zeme, S., Lo Russo, G., and Sguazzi, A., 1980, Long-term results of posterior functional rhizotomy, *Acta Neurochir. (Wien) [Suppl.]* **30:**435–439.

Fasano, V. A., Broggi, G., Zeme, S., and Lo Russo, G., 1982, Long term results of cerebellar chronic electrostimulation in involuntary movement disorders, *Appl. Neurophysiol.* **45:**218–220.

Fay, T., and Henney, G. C., 1938, Correlation of body segmental temperature and its relation to the location of carcinomatous metastasis, *Surg. Gynecol. Obstet.* **66:**512–524.

Felsööry, A., and Crue, B. L., 1976, Results of 19 years experience with sacral rhizotomy for perineal and perianal cancer pain, *Pain* **2**(4):431–433.

Fenelon, F., 1950, Essais de traitement neurochirurgical du syndrome parkinsonien par intervention directe sur les voies extrapyramidales immediatement sousstriopallidales (aux lenticulaire), *Rev. Neurol.* **83:**437–440.

Fenelon, F., 1955, Neurosurgery of ansa lenticularis in dyskinesias and Parkinson's disease: Review of principles and techniques of a personal operation, *Sem. Hop. Paris* **31:**1835.

Ferguson, G. G., Brett, D. C., and Peerless, S. J., 1981, Trigeminal neuralgia: A comparison of the results of percutaneous rhizotomy and microvascular decompression, *Can. J. Neurol. Sci.* **8:**207–214.

Fernandes Barahana, H. J., and da Fonseca Simoes, J. L., 1982, General considerations on psychiatric stereotaxic surgery for behavioral disturbances, in *Stereotaxy of the Human Brain* (G. Schaltenbrand and A. E. Walker, eds.), Georg Thieme Verlag, Stuttgart, pp. 565–569.

Fernando, S., 1967, Gilles de la Tourette's syndrome: A report on four cases and a review of published case reports, *Br. J. Psychiatry* **113**(499):607–616.

Fields, H. L., Basbaum, A. I., Clanton, C. H., and Anderson, S. D., 1977, Nucleus raphe magnus inhibition of spinal cord dorsal horn neurons, *Brain Res.* **126:**441–453.

Filatov, J. M., 1972, *Arteriovenous Malformations of Brain Hemispheres (Clinic, Diagnostics and Surgical Treatment),* Thesis, Moscow.

Findley, L. J., Gresty, M. A., and Halmagyi, G. M., 1981, Tremor, the cogwheel phenomenon and clonus in Parkinson's disease, *J. Neurol. Neurosurg.* **44:**534–546.

Fink, R. A., 1984, Neurosurgical treatment of nonmalignant intractable rectal pain: Microsurgical commissural myelotomy with the carbon dioxide laser, *Neurosurgery* **14:**64–65.

Flamm, E. S., and Van Buren, J. M., 1966, The reliability of recontructed ventricular landmarks for localization of depth electrodes in man, *J. Neurosurg.* **25**:67–72.

Flatau, E., and Sterling, W., 1911, Progressiver Torsionspasmus bei Kindern, *Z. Ges. Neurol. Psychiatrie* **7**:586–612.

Flaubert, M., 1872, Memoire sur plusieurs cas de luxation, *Rep. Gen. Anat. Physiol. Pathol. Clin. Chir.* **3**:55–69.

Fodstad, H., 1987, The Swedish experience in phrenic nerve stimulation, *Pace* **10**:246–251.

Fodstad, H., Andersson, G., Blom, S., and Linderholm, H., 1985, Phrenic nerve stimulation (diaphragm pacing) in respiratory paralysis, *Appl. Neurophysiol.* **48**:351–357.

Foerster, O., 1908, Über eine neue operative Methode der Behandlung Lähmungen mittels Resektion hinterer Rückenmarkswurzeln, *Z. Orthop. Chir.* **22**:203–223.

Foerster, O., 1909, Der atonisch-astatische Typus der infantilen Cerebrallähmung, *Dtsch. Arch. Klin. Med.* **98**:216–244.

Foerster, O., 1911, Resection of the posterior nerve roots of spinal cord, *Lancet* **2**:75–79.

Foerster, O., 1913, On the indications and results of the excision of posterior spinal nerve roots in men, *Surg. Gynecol. Obstet.* **16**:463–474.

Foerster, O., 1921, Zur Analyse und Pathophysiologie der striären Bewegungsstörungen, *Z. Ges. Neurol. Psychiatrie* **73**:1–169.

Foerster, O., 1925, Zur Pathogenese und chirurgischen Behandlung der Epilepsie, *Zentralbl. Chir.* **52**:531–549.

Foerster, O., 1926, Operative Behandlung des Torticollis spasticus, *Zentralbl. Chir.* **44**:2804–2805.

Foerster, O., 1933, Mobile spasm of the neck muscles and its pathologic basis, *J. Comp. Neurol.* **58**:725–735.

Foerster, O., and Altenburger, C., 1935, Electrobiologische Vorgänge in der Menschlichen, *Dtsch. Z. Nervenheilkd.* **135**:277.

Foerster, O., and Penfield, W., 1930, The structural basis of traumatic epilepsy and results of radical operation, *Brain* **53**:99–119.

Foix, C., and Thévenard, A., 1922, Les réflexes de posture, *Presse Med.* **30**:765–767.

Foltz, E. L., Knopp, L. M., and Ward, A. A., 1959, Experimental spasmodic torticollis, *J. Neurosurg.* **16**:55–72.

Foltz, E. L., and White, L. E., 1962, Pain "relief" by frontal cingulotomy, *J. Neurosurg.* **19**:899–908.

Forno, L., S., 1966, Pathology of parkinsonism: A preliminary report of 24 cases, *J. Neurosurg.* **2**:266–271.

Fox, A. J., Vinulla, F., Pelz, D. U., Peerless, S. J., Ferguson, G. G., Drake, C. G., and Debrun, G., 1987, Use of detachable balloons for proximal artery occlusion in the treatment of unclippable cerebral aneurysms, *J. Neurosurg.* **66**:40–46.

Fox, J. L., 1968, Percutaneous stereotaxic cordotomy. I. Problems encountered, *Acta Neurochir. (Wien)* **18**:309–317.

Fox, J. L., 1970, Experimental relationship of radiofrequency electrical current and lesion size for application to percutaneous cordotomy, *J. Neurosurg.* **33**(4):415–421.

Fox, J. L., 1973, Percutaneous trigeminal tractotomy for facial pain, *Acta Neurochir. (Wien)* **29**:83–88.

Fox, J. L., 1976a, Percutaneous trigeminal tractotomy for relief of intractable facial pain, in *Clinical Microneurosurgery* (W. T. Koos, F. W. Böck, and R. F. Spetzler, eds.), Georg Thieme, Stuttgart, pp. 161–165.

Fox, J. L., 1976b, Vascular clips for the microsurgical treatment of the stroke, *Stroke* **7**:489–499.

Fox, J. L., and Al-Mefty, D., 1977, Embolization of an arteriovenous malformation of the brain stem, *Surg. Neurol.* **8**:7–9.

Fox, J. L., and Green, R. C., 1968a, Stereotaxic brain surgery. I. Geometric consideration of polar coordinates and polar range finding, *Acta Neurochir. (Wien)* **18**:57–67.

Fox, J. L., and Green, R. C., 1968b, Stereotaxic brain surgery. II. Description of a method using biplane television guidance, *Acta Neurochir. (Wien)* **18**(3):171–185.

Fox, J. L., and Green, R. C., 1968c, Percutaneous stereotaxic cordotomy. II. A guidance technique for the anterior approach, *Acta Neurochir. (Wien)* **18**:318–326.

Fox, J. L., and Green, R. C., 1969a, Electrical resistance in percutaneous cordotomy. Technical note, *Acta Neurochir. (Wien)* **20**:53–58.

Fox, J. L., and Green, R. C., 1969b, New method of stereotaxis, *Radiology* **92**:259–264.

Fox, J. L., and Green, R. C., 1969c, Stereotaxic surgery using a television guidance system. I. Thalamotomy, *Acta Neurochirg. (Wien)* **20**:331–342.

Fox, J. L., and Green, R. C., 1969d, Stereotactic surgery using a television guidance system. II. Percutaneous cordotomy, *Acta Neurochir. (Wien)* **21**:31–42.

Fox, J. L., and Green, R. C., 1969e, Polar coordinates and television monitoring in stereotaxic brain surgery, *Confin. Neurol.* **31**:123–128.

Fraenkel, J., 1912, Dysbasia lordotica progressiva, dystonia musculorum deformans–tortipelvis, *J. Nerv. Ment. Dis.* **39**:361–374.

Fraioli, B., and Guidetti, B., 1975a, Effects of stereotactic lesions of pulvinar and lateralis posterior nucleus on intractable pain and dyskinetic syndromes of man, *Appl. Neurophysiol.* **32**(1):23–30.

Fraioli, B., and Guidetti, B., 1975b, Effects of stereotactic lesions of the dentate nucleus of the cerebellum in man, *Appl. Neurophysiol.* **38**:81–90.

Fraioli, B., and Guidetti, B., 1977, Posterior partial rootlet section in the treatment of spasticity, *J. Neurosurg.* **46**(5):618–626.

Frank, F., Tognetti, F., Gaist, G., Frank, G., Galassi, E., and Sturiale, C., 1982, Stereotaxic rostral mesencephalotomy in treatment of malignant faciothoracobrachial pain syndromes. A survey of 14 treated patients, *J. Neurosurg.* **56**(6):807–811.

Frank, F., Gaist, G., Fabrizi, A. P., and Sturiale, C., 1985a, Deep brain stimulation in the treatment of chronic pain, in *IX Meeting of the World Society for Stereotactic and Functional Neurosurgery*, Toronto, p. 36.

Frank, F., Gaist, G., Piazza, G., Ricci, R. F., Sturiale, C., and Galassi, E., 1985b, Stereotaxic biopsy and radioactive implantation for interstitial therapy of tumors of the pineal region, *Surg. Neurol.* **23**:275–280.

Fraser, J. D., 1975, History and development of cryosurgery, in *Practical Cryosurgery* (H. B. Holden, ed.), Pitman Medical, Oxford, pp. 1–9.

Frazier, C. H., 1930, Spasmodic torticollis, *Ann. Surg.* **91:**848–854.

Freckman, N., Hagenah, R., and Herrmann, H. D., 1981, Treatment of neurogenic torticollis by microvascular lysis of the accessory nerve roots—Indication, technique and first results, *Acta Neurochir.* **59:**167–175.

Freckmann, N., Hagenah, R., Herrmann, H. D., and Müller, D., 1983, Spasmodic torticollis—3 years experience with microsurgical lysis of the spinal accessory nerve root, in *7th European Congress of Neurosurgery*, Brussels, p. 258.

Fredriksen, T. A., Bergmann, S., Hesselberg, J. P., Stolt-Nielsen, A., Ringkjib, R., and Sjaastad, O., 1986, Electrical stimulation in multiple sclerosis. Comparison of transcutaneous electrical stimulation and epidural spinal cord stimulation, *Appl. Neurophysiol.* **49:**4–24.

Freed, W. J., Ro, G. N., and Niehoff, D. L., 1983, Normalization of spiroperidol binding in the denervated rat striatum by homologous grafts of substantia nigra, *Science* **222:**937–939.

Freeman, T. B., Campbell, J. N., and Long, D. M., 1983, Naloxone does not affect pain relief induced by electrical stimulation in man, *Pain* **17:**189–195.

Freeman, W., and Watts, J. W., 1950, *Psychosurgery*, Charles C. Thomas, Springfield, IL.

French, L. A., 1974, High cervical cordotomy: Technique and results, *Clin. Neurosurg.* **21:**239–245.

French, L. A., Story, J. L., Gallich, J. H., and Schultz, E. A., 1962, Some aspects of stimulation and recording from the basal ganglia in patients with abnormal movements, *Confin. Neurol.* **22:**265–273.

Freund, H. J., and Dietz, V., 1978, The relationship between physiological and pathological tremor, in *Physiological Tremor, Pathological Tremor and Clonus* (J. E. Desmedt, ed.), S. Karger, Basel, pp. 66–89.

Friedhoff, A. J., and Chase, T. N. (eds.), 1983, *Gilles de la Tourette Syndrome, Advances in Neurology*, Vol. 35, Raven Press, New York.

Friedlander, W. J., 1956, Characteristics of postural tremor in normal and in various abnormal states, *Neurology (Minneap.)* **6:**716–723.

Friedman, A. H., and Nashold, B. S., 1986, DREZ lesions for relief of pain related to spinal cord injury, *J. Neurosurg.* **65:**465–469.

Friedman, A. H., Nashold, B. S., and Ovelman-Levitt, J., 1984, Dorsal root entry zone lesions for the treatment of post-herpetic neuralgia, *J. Neurosurg.* **60:**1258–1262.

Fröder, M., Seitzer, D., Büren, G., and Dieckmann, G., 1983, Digital radiography for target point evaluation in stereotactic neurosurgery, *Appl. Neurophysiol.* **46:**206–210.

Fry, W. J., and Fry, F. J., 1960, Fundamental neurological research and human neurosurgery using intense ultrasound, *IEE Trans. Med. Electron.* **7:**166–181.

Fry, W. J., and Meyers, R., 1962, Ultrasonic method of modifying brain structures, *Confin. Neurol.* **22:**315–328.

Fry, W. J., Mosberg, W. H., and Barnard, J. W., 1954, Production of focal destructive lesions in the central nervous system with ultrasound, *J. Neurosurg.* **11:**471–478.

Fry, W. J., Goss, I. A., and Patrick, J. T., 1981, Transskull focal lesions in cat brain produced by ultrasound, *J. Neurosurg.* **54**(5)**:**659–663.

Fujita, S., and Cooper, I. S., 1976, Impedance and spontaneous electrical activity as a localizing method in percutaneous spinal surgery, *Acta Neurol. Scand.* **53:**201–208.

Fujita, S., and Hosoda, K., 1984, CT-controlled stereotaxis—application to functional neurosurgical procedures, *Appl. Neurophysiol.* **47:**89–90.

Fujita, Y., Mori, K., and Handa, H., 1971, Spasmodic torticollis-like posture and changes in monoamines related to destruction of the mesencephalic tegmentum in cats, *Neurol. Med. Chir.* **11:**335–338.

Fukamachi, A., Ohye, C., and Narabayashi, H., 1973. Delineation of the thalamic nuclei with a microelectrode in stereotaxic surgery for parkinsonism and cerebral palsy, *J. Neurosurg.* **39:**214–225.

Fukamachi, A., Ohye, C., and Saito, J., 1977, Estimation of the neural noise within the human thalamus, *Acta Neurochir. (Wien) [Suppl.]* **24:**121–136.

Fukushima, T., 1982, Posterior fossa neurovascular decompression for trigeminal neuralgia, *No To Shinkei Geka* **10:**1257–1261.

Furni, T., Mutsuga, N., Kageyama, N., and Takeda, A., 1980, Alteration of methionine-enkephalin-like substance level in ventricular fluid on intracerebral stimulation, *Appl. Neurophysiol.* **43:**340–346.

Fusek, I., and Vladyka, V., 1968, Extrakce cizich tells z mozku stereotaktickoy metodom, *Vojen. Zdrav. Listy* **37:**46–51.

Gabibov, G. A., Anzimirov, V. L., Kuznetsov, A. S., Labutin, V. V., Lyass, F. M., Snegirev, V. S., and Yartsev, V. V., 1972, Clinical evaluation of radiometric and rheometric methods in the surgery of the glial brain tumors, *Vopr. Neirokhir.* **1:**15–19 (Rus.).

Gage, A. A., Koepf, S., and Wehrie, G., 1965, Cryotherapy for cancer of the lip and oral cavity, *Cancer* **18:**1646–1654.

Gage, A., Fazekas, G., and Rilet, E., 1967, Freezing injury to large blood vessels in dogs, *Surgery* **61:**748–754.

Gahbauer, H., Sturm, V., and Schlegel, W., 1983, Combined use of stereotaxic CT and angiography for brain biopsies and stereotaxic irradiation, *A.J.N.R.* **4:**715–726.

Gahm, N., Russman, B., Cerciello, R., Fiorentino, M., and McGrath, D., 1981, Chronic cerebellar stimulation for cerebral palsy: A double-blind study, *Neurology (N.Y.)* **31:**87–90.

Gajdosova, D., Sramka, M., and Nadvornik, P., 1972, Pozorovanie u choroby Gilles de la Tourette, *Cesk. Neurol. Neurochir.* **35:**294–297.

Galanda, M., and Grác, J., 1983, Choice of the optimal place and stimulating program for deep cerebellar stimulation at central movement disorders, in *International Symposium of Functional and Stereotactic Neurosurgery: Brain Stimulation and Psychosurgery*, Bratislava, p. 10.

Galanda, M., Nadvornik, P., and Šramka, M., 1977, Combined transtentorial dentatotomy with pulvinarotomy in cerebral palsy, *Acta Neurochir. (Wien)* **24:**21–26.

Ganev, G., 1976, *Parkinsonism*, Meditsina i Fiskultura, Sofia (Bul.).

Ganglberger, J. A., 1961a, The EEG in parkinsonism and its alteration by stereotaxically produced lesions in pallidum and thalamus, *Electroencephalogr. Clin. Neurophysiol.* **13:**828.

Ganglberger, J. A., 1961b, Vorübergehende Herdveranderungen im EEG nach stereotaktischen Operationen an den Basalganglien, *Arch. Psychiatr.* **201:**528–548.

Ganglberger, J. A., 1976, New possibilities of stereotaxic treatment of temporal lobe epilepsy, *Acta Neurochir. (Wien) [Supp.]* **23:**211–214.

Ganglberger, J. A., Gestring, G. F., and Groll-Knapp, E., 1970a, Comparison of cortical responses to stimulation of thalamus and fornix, *Electroencephalogr. Clin. Neurophysiol.* **28:**324–327.

Ganglberger, J. A., Groll-Knapp, E., and Haider, M., 1970b, Thalamocortical components of reaction time, *Electroencephalogr. Clin. Neurophysiol.* **28:**328–334.

Ganz, E., and Mullan, S., 1977, Percutaneous cordotomy, in *Persistent Pain: Modern Methods of Treatment,* Vol. 1 (S. Lipton, ed.), Academic Press, London, pp. 21–34.

Gardner, W. J., 1949, Surgical aspects of Parkinson's syndrome, *Postgrad. Med.* **5:**107–111.

Gardner, W. J., 1962, Concerning the mechanism of trigeminal neuralgia and hemifacial spasm, *J. Neurosurg.* **19:**947–958.

Gardner, W. J., and Miklos, M. V., 1959, Response of trigeminal neuralgia to "decompression" to sensory root: Discussion of cause of trigeminal neuralgia, *J.A.M.A.* **170:**1773–1776.

Gardner, W. J., and Sava, G. A., 1962, Hemifacial spasm: A reversible pathophysiologic state, **19:**240–247.

Gardner, W. J., Stowell, A., and Dutlinger, R., 1947, Resection of the greater superficial petrosal nerve in the treatment of unilateral headache, *J. Neurosurg.* **4:**105–114.

Garsia Sola, R., Miravet, J., and Parera, C., 1982, Curative or palliative possibilities in the surgical treatment of epileptic patients, *Appl. Neurophysiol.* **45:**471–477.

Gastaut, H., 1963, Epilepsies, in *Encyclopédie Médico-Chirurgicale,* Paris, pp. 44–48.

Gastaut, H., 1968, Sémiologie des myoclonies et nosologie analytique des syndromes myocloniques, *Rev. Neurol.* **119:**1–30.

Gauthier, S., 1986, Idiopathic spasmodic torticollis: Pathophysiology and treatment, *Can. J. Neurol. Sci.* **13:**88–90.

Gelfand, N. M., Gurfinkel, V. S., Kotz, J. M., Krinski, V. N., Zetlin, M. L., and Shick, M. L., 1964, The study of postural activity, *Biofizika* **9:**710–717 (Rus.).

Gerlach, J., 1959, Variations of the human skull, in *Introduction in Stereotaxis with an Atlas of the Human Brain,* Vol. 1 (G. Schaltenbrand and P. Bailey, eds.), Georg Thieme, Stuttgart, pp. 29–41.

Gerlach, J., 1976, Effect of CB 154 (2-bromo-alpha-ergocryptine) on paralysis agitans compared with Madopar in a double-blind, cross-over trial, *Acta Neurol. Scand.* **53:**189–200.

German, W. J., 1942, Surgical treatment of spasmodic facial tic, *Surgery* **11:**912–914.

Giaquinto, S., de Divitüs, E., and Signorelli, C. D., 1971, Peripheral influences on VPM–VPL thalamic nuclei in the human, *Confin. Neurol.* **33**(3)**:**174–185.

Gilbert, G. J., 1972, The medical treatment of spasmodic torticollis, *Arch. Neurol.* **22:**503–506.

Gilbert, G. J., 1977, Familial spasmodic torticollis, *Neurology* **27:**11–13.

Gildenberg, P. L., 1969, Impedance measuring device for detection of penetration of the spinal cord in anterior percutaneous cervical cordotomy. Technical note, *J. Neurosurg.* **30:**87–92.

Gildenberg, P. L., 1971, Angle-meter to indicate the proper angle of insertion in anterior percutaneous cervical cordotomy, *J. Neurosurg.* **34:**244–247.

Gildenberg, P. L., 1972, Stereotaxic lower cervical cordotomy for the treatment of intractable pain, *Confin. Neurol.* **34:**275–278.

Gildenberg, P. L., 1974, Percutaneous cervical cordotomy, *Clin. Neurosurg.* **21:**246–255.

Gildenberg, P. L., 1976, Percutaneous cervical cordotomy, *Appl. Neurophysiol.* **39:**97–113.

Gildenberg, P. L., 1977, Treatment of spasmodic torticollis with dorsal column stimulation, *Acta Neurochir. (Wien) [Suppl.]* **24:**65–66.

Gildenberg, P. L., 1978, Treatment of spasmodic torticollis by dorsal column stimulation, *Appl. Neurophysiol.* **41:**113–121.

Gildenberg, P. L., 1979, The use of pacemakers (electrical stimulation) in functional neurological disorders, in *Functional Neurosurgery* (T. Rasmussen and R. Marino, eds.), Raven Press, New York, pp. 59–74.

Gildenberg, P. L., 1982, Spinal stereotaxic procedures, in *Stereotaxy of the Human Brain* (G. Schaltenbrand and A. E. Walker, eds.), Georg Thieme Verlag, Stuttgart, pp. 469–474.

Gildenberg, P. L., 1983, Stereotactic neurosurgery and computerized tomographic scanning, *Appl. Neurophysiol.* **46:**170–179.

Gildenberg, P. L., 1985, Selective brain surgery procedures for motor disorders, in *Recent Achievements in Restorative Neurology. Upper Motor Neuron Functions and Dysfunctions* (J. Eccles and M. R. Dimitrijevic, eds.), S. Karger, Basel, pp. 71–78.

Gildenberg, P. L., and Hirschberg, R., 1981a, Treatment of cancer pain with limited myelotomy, *Med. J. St. Joseph Hosp. (Houston)* **16:**199–204.

Gildenberg, P. L., and Hirschberg, R. M., 1981b, Limited myelotomy for the treatment of intractable pain, *Neurochirurgie [Suppl.]* **66:**18–26.

Gildenberg, P. L., Lin, P. M., and Polakoff, P. P., 1967, A stereotaxic approach to the spinal cord, *Confin. Neurol.* **29:**252–255.

Gildenberg, P. L., Lin, P. M., Polakoff, P., and Flitter, M. G., 1968, Anterior percutaneous cervical cordotomy: Determination of target point and calculation of angle of insertion, *J. Neurosurg.* **28:**173–176.

Gildenberg, P. L., Kaufman, H. L., and Murthy, K. S. K., 1982, Calculation of stereotactic coordinates from the computed tomographic scan, *Neurosurgery* **10:**580–586.

Gill, W., and Frazer, J., 1968, A look at cryosurgery, *Scot. Med. J.* **13:**268–273.

Gill, W., Frazer, J., and Carter, D., 1968, Repeated freeze–thaw cycles in cryosurgery, *Nature* **219:**410–413.

Gillingham, F. J., 1960, Surgical management of the dyskinesias, *J. Neurol. Neurosurg.* **23:**347–348.

Gillingham, F. J., 1962, Small localised surgical lesions of the internal capsule in the treatment of the motor disorders, *Confin. Neurol.* **22:**385–393.

Gillingham, F. J., 1966a, Depth recording and stimulation, *J. Neurosurg.* **24:**382–387.

Gillingham, F. J., 1966b, Bilateral stereotaxic lesions in the management of parkinsonism, *J. Neurosurg.* **24:**449–453.

Gillingham, F. J., Watson, W. S., Donaldson, A. A., and Naughton, J. A., 1960, The surgical treatment of parkinsonism, *Br. Med. J.* **2:**1395–1402.

Gioino, G. G., Dierssen, G., and Cooper, I. S., 1966, The effect of subcortical lesions on production and alleviation of hemiballic or hemichoreic movements, *J. Neurol. Sci.* **3**(1)**:**10–36.

Giorgi, C., and Broggi, G., 1984, Surgical treatment of glossopharyngeal neuralgia and pain from cancer of the nasopharynx. A 20-year experience, *J. Neurosurg.* **61:**952–955.

Giorgi, C., Garibotto, G., Cerchiari, U., Broggi, G., and Franzini, A., 1983, Neuroanatomical digital image processing in CT-guided stereotactic operations, *Appl. Neurophysiol.* **46:**236–239.

Girmunskaya, E. A., Kandel, E. I., and Pokrovskaya, Z. A., 1971, Analysis of electrosubcorticographic data in stereotaxic operations on basal ganglia of brain, *Zh. Nevropatol. Psychiatr.* **71:**1771–1775 (Rus.).

Gleason, C. A., Wise, B. L., and Feinstein, B., 1978, Stereotactic localization (with computerized tomography scanning), biopsy, and radiofrequency treatment of deep brain lesions, *Neurosurgery* **2:**217–222.

Glenn, W. L., 1978, Diaphragm pacing: Present status, *Pace* **1:**357–370.

Glynn, C. J., 1979, Spinal narcotics and respiratory depression, *Lancet* **2:**356–357.

Goander, M., Soanes, W., and Smith, V., 1964, Experimental prostate cryosurgery, *Invest. Urol.* **1:**610–619.

Goddard, G. V., 1967, Development of epileptic seizures through brain stimulation at low intensity, *Nature* **214:**1020–1024.

Goedhart, Z. D., Francaviglia, N., Feirabend, H. K. P., and Voogd, J., 1984, Technical pitfalls in median commissural myelotomy for malignant sacral pain, *Appl. Neurophysiol.* **47:**216–222.

Goetz, C. G., Tanner, C. M., Glantz, R., and Klawans, H. L., 1983, Pergolide in Parkinson's disease, *Arch. Neurol.* **40:**785–787.

Goktere, E. O., Young, L. B., and Bridges, P. K., 1975, A further review of the results of stereotactic subcandate tractotomy, *Br. J. Psychiatry* **126:**270–280.

Gol, A., 1967, Relief of pain by electrical stimulation of the septal area, *J. Neurol. Sci.* **5:**115–120.

Gol, A., and Dossman, W. F., 1965, Electrocoagulation of cord and nerve roots in the treatment of spasms of bladder and lower limbs, *J. Neurosurg.* **22:**352–353.

Goldring, S., and Gregorie, E. M., 1984, Surgical management of epilepsy using epidural recordings to localize the seizure focus. Review of 100 cases, *J. Neurosurg.* **60:**457–466.

Goldstein, A., Lowney, L. I., and Pal, B. K., 1971, Stereospecific and nonspecific interactions of morphine in subcellular fractions of mouse brain, *Proc. Natl. Acad. Sci. U.S.A.* **69:**1742–1747.

Golubev, V. L., 1974, The structure of personality and its dynamics in parkinsonism, in *Parkinsonism (Questions of Clinic, Pathogenesis and Treatment),* Meditsina, Moscow, pp. 48–56 (Rus.).

Golubeva, L. I., 1962, Approaches to clinic and treatment of essential tremor, in *Essays on Clinical Neurology,* Vol. 1, Meditsina, Moscow, pp. 173–183 (Rus.).

Gönshirt, H., Reuther, R., and Suliridoff, F., 1978, Transitorischer hemiballismus als symptom der Vertebrobasilären Insuffizienz, *Nervenarzt* **49:**730–734.

Gornall, P., Hitchcook, E., and Kirland, I. S., 1975, Stereotaxic neurosurgery in the management of cerebral palsy, *Dev. Med. Child. Neurol.* **17:**279–286.

Gortvai, P., and Teruchkin, S., 1974, The position and extent of the human dentate nucleus, *Acta Neurochir.* **31:**101–110.

Goto, A., Kosaka, K., Kubota, K., Nakamura, R., and Narabayashi, H., 1968, Thalamic potentials from muscle afferents in the human, *Arch. Neurol.* **19:**302–309.

Gouda, K. I., Freidberg, S. R., Baker, R. A., Larsen, C. R., and Silverman, M. L., 1986, Gouda frame redesigned specifically for computed tomographic compatibility, *Appl. Neurophysiol.* **49:**192–200.

Gouda, K. I., Freidberg, S. R., Fager, C. A., Tarlov, E. C., Baker, R. A., Larsen, C. R., and Kott, S. H., 1986, Stereotactic computed tomographic-guided functional neurosurgery using the redesigned Gouda frame, *Appl. Neurophysiol.* **49:**201–212.

Gowers, W. R., 1888, *A Manual of Diseases of the Nervous System,* Vol. II, J. & A. Churchill, London, pp. 228–237.

Gowers, W. R., 1893, *A Manual of Diseases of the Nervous System,* Blakiston, Philadelphia.

Goya, T., Kinoshita, K., Yamakawa, Y., Morita, Y., Veda, T., Mihara, K., and Fukui, M., 1983, Hemifacial spasm. Analysis of 40 cases of neurovascular decompression, *Neurol. Med. Chir. (Tokyo)* **23:**651–658.

Grachev, K. V., and Stepanova, T. S., 1971, Some data contributive to the evaluation of a minimal size of the epileptogenic focus in the thalamus, *Vopr. Neirokhir.* **3:**18–23 (Rus.).

Graf, C. J., 1958, On cervical cordotomy—a new technique, *J. Neurosurg.* **15:**576–580.

Graf, C. J., 1959, Glossopharyngeal neuralgia and ossification of the stylohyoid ligament, *J. Neurosurg.* **16:**448–453.

Granit, R., 1970, *The Basis of Motor Control,* Academic Press, New York.

Granit, R., Pascoe, J. E., and Steg, G., 1957, The behavior of the tonic alpha- and gamma-motoneurones during stimulation of recurrent collaterals, *J. Physiol. (Lond.)* **138:**381–400.

Grant, E., 1923, Localization of brain tumors by determination of the electrical resistance of the growth, *J.A.M.A.* **81:**2169–2171.

Grechin, V. B., 1973, Application of thermoresistors in stereotactic neurosurgery, *Vopr. Neirokhir.* **1:**57–60 (Rus.).

Grechin, V. B., and Borovikova, V. N., 1975, On the possibilities of impedansometry in the implanted intracerebral electrodes, *Vopr. Neirokhir* **2:**39–44 (Rus.).

Greenberg, H. S., Taren, J., Ensminger, W. D., and Doan, K., 1982, Benefit from and tolerance to continuous intrathecal infusion of morphine for intractable cancer pain, *J. Neurosurg.* **57:**360–364.

Greenblatt, S. H., Rayport, M., Savolaine, E. R., Harris, J. H., and Hitchins, M. W., 1982, Computed tomography—guided intracranial biopsy and cyst aspiration, *Neurosurgery* **11:**589–597.

Greenfield, J. C., 1955, The pathology of Parkinson's disease, in *James Parkinson,* Macmillan, London, pp. 219–243.

Greenfield, J. C., and Bosanguet, F. D., 1953, The brain-stem lesions in parkinsonism, *J. Neurol. Neurosurg. Psychiatry* **16:**213–223.

Greenwood, J., Jr., 1955, Two-point or interpolar coagulation. Review after a twelve-year period with notes on addition of a sucker tip, *J. Neurosurg.* **12:**196–197.

Greitz, T., Bergström, M., Boethius, J., Kingsley, D., and Ribbe, T., 1980, Head fixation system for integration of radiodiagnostic and therapeutic procedures, *Neuroradiology* **19:**1–6.

Grigorovich, K. A., 1981, *Surgery of the Peripheral Nerves,* Meditsina, Moscow.

Grindel, O. M., and Bragina, N. N., 1974, Diffuse theta-activity in man's EEG by affection of brain medio-basal structures, in *Main Problems of Brain Electrophysiology,* Meditsina, Moscow, pp. 261–274 (Rus.).

Grindel, O. M., Kandel, E. I., and Raeva, S. N., 1962, Change of electrical activity of the brain in parkinsonian patients in connection with the operations on the basal ganglia, *Vopr. Neirokhir.* **6:**23–28 (Rus.).

Grinker, R. R., and Walker, A. E., 1933, The pathology of spasmodic torticollis with a note on respiratory failure from anaesthesia in chronic encephalitis, *J. Nerv. Ment. Dis.* **78:**630–637.

Grinstein, A. M., 1946, *Pathways and Centres of the Nervous System,* Meditsina, Moscow (Rus.).

Gros, C., 1979, Spasticity: Clinical classification and surgical treatment, in *Advances and Technical Standards in Neurosurgery,* Vol. 6 (H. Krayenbühl, ed.), Springer, Vienna, pp. 55–97.

Gros, C., Vlahovitch, B., and Enjalbert, J. M., 1955, Action des interventions préfrontales (leucotomie, topectomie) sur les manifestations psychiques et kinétiques des syndromes parkinsonies, *Sem. Hop. Paris* **31:**1831–1832.

Gros, C., Vlahovitch, B., and Nghia, N. C., 1964, La zone "cible" dans la chirurgie de la maladie de Parkinson. Etude d'une série de 407 opérations stéréotaxiques, *Neurochirurgie* **10:**413–426.

Gros, C., Vlahovitch, B., and Adib-Yardi, I. S., 1966, Les indications opératoires dans les "cas limites" de maladie de Parkinson, *Ann. Chir.* **20:**223–228.

Gros, C., Ouaknine, G., Vlahovitch, B., and Frerebeau, P., 1967, La radiocotomie sélective postérieure dans le traitement neurochirurgical l'hypertonie pyramidale, *Neurochirurgie* **13:**505–518.

Gros, C., Frerebeau, P., Perez-Dominguez, E., Barin, M., and Privat, J. M., 1976a, Long-term results of stereotactic surgery for infantile dystonia and dyskinesia, *Neurochirurgia (Stuttg.)* **19:**171–178.

Gros, C., Frerebeau, P., and Privat, J. M., 1976b, Chirurgie stéréotaxique dans les dystonies et dyskinésies non parkinsoniennes, *Neurochirurgie* **22:**539–552.

Gros, C., Privat, C., Privat, J. M., Frerebeau, P., and Benezech, J., 1981, Les radicotomies antérieures: Leur place dans le traitement de la spasticité, in *Actualités en Rééducation Fonctionnelle et Réadaptation* (A. Simon, ed.), Masson, Paris, pp. 63–70.

Grundy, B., and Jannetta, P., 1981, BAEP monitoring during cerebellar pontine angle surgery, *Electroencephalogr. Clin. Neurophysiol.* **51:**39–47.

Grunert, V., Witzmann, A., and Grunert, P., 1986, Partielle verticale Nucleotomie als Modification der Tractotomie des Nervus trigeminus, *Neurochirurgie* **29:**109–110.

Guatico, W., Abid, S., and Gaston, P., 1965, Spontaneous intracerebral hematomas, *J. Neurosurg.* **22:**569–575.

Gudmundsson, K., Rhoton, A. L., and Rushton, J. G., 1971, Detailed anatomy of the intracranial portion of the trigeminal nerve, *J. Neurosurg.* **35:**592–600.

Guidetti, B., 1973, Results of 98 intracranial aneurysm operations performed with the aid of an operating microscope, *Acta Neurochir.* **29:**65–71.

Guidetti, B., and Delitala, A., 1980, Intracranial arteriovenous malformations. Conservative and surgical treatment, *J. Neurosurg.* **53:**149–152.

Guidetti, B., and Fraioli, B., 1977, Neurosurgical treatment of spasticity and dyskinesias, *Acta Neurochir. [Suppl.]* **24:**27–39.

Guidetti, B., Ricci, G. B., and Gagliardi, F. M., 1972, Epilepsy with bilateral and independent foci: Pre-operative criteria and results of surgical treatment, in *Present Limits of Neurosurgery* (I. Fusek and Z. Kunc, eds.), Avicenum, Czechoslovak Medical Press, Prague, pp. 411–416.

Guiot, G., 1958a, *Adénomes Hypophysaires,* Masson, Paris.

Guiot, G., 1958b, Le traitement des syndromes parkinsoniens par la destruction du pallidum intérne, *Neurochirurgie* **1:**94–98.

Guiot, G., and Brion, S., 1952, Traitement neuro-chirurgical des syndromes choreo-athétosiques et parkinsonien, *Sem. Hop. Paris* **49:**2095–2099.

Guiot, G., and Brion, S., 1955, La chirurgie pallidale dans les dyskinesis, *Sem. Hop.* **31:**1838–1845.

Guiot, G., and Pecker, J., 1949, Tractotomie mesencephalique anterieure pour tremblement parkinsonien, *Rev. Neurol.* **81:**387–391.

Guiot, G., and Trujillo, J., 1969, Radiology and electrophysiology in stereotactic thalamotomy, in *Third Symposium of Parkinson's Disease,* Livingstone, Edinburgh, London, pp. 221–223.

Guiot, G., Brion, S., Rougerie, J., Hertzog, E., and Sachs, M., 1958, La destruction stéréotaxique du pallidum interne dans les syndromes parkinsoniens. Etude technique, *Ann. Chir.* **12**:1003–1032.

Guiot, G., Sachs, M., Hertzog, E., Brion, S., Rougerie, J., Dalloz, J. C., and Napoléone, F., 1959, Stimulation électrique et lésions chirurgicales de la capsule intérne. Déductions anatomiques et physiologiques, *Neurochirurgie* **5**:17–42.

Guiot, G., Rougerie, J., Sachs, M., and Hertzog, E., 1960, Repérage stéréotaxique des malformations vasculaires profondes intracérébrales, *Neurochirurgie* **6**:266–268.

Guiot, G., Brion, S., and Akerman, M., 1961, Anatomie stereotaxique du pallidum interne, du thalamus et de la capsule interne, *Ann. Chir.* **15**:557–586.

Guiot, G., Hardy, J., and Albé-Fessard, D., 1962, Délimination précise des structures sous-corticales et identification de noyaux thalamiques chez l'homme par l'électrophysiologie stéréotaxique, *Neurochirurgia* **5**(1):1–18.

Guiot, G., Albé-Fessard, D., Arfel, G., and Derome, P., 1964, Derivations d'activité unitaires en cours d'interventions stéréotaxiques, *Neurochirurgie* **10**:427–435.

Guiot, G., Arfel, G., and Derome, P., 1968a, La chirurgie stereotaxique des tremblement de repos et d'attitude, *G.M. France* **75**:4029–4056.

Guiot, G., Arfel, G., Derome, P., and Kahn, A., 1968b, Procédés de controle neurophysiologique pour la thalamotomie stéréotaxique, *Neurochirurgie* **14**:553–566.

Guldberg, H. C., Turner, J. W., Hanieh, A., Asheroft, G. W., Crawford, T. B. B., Perry, W. L. M., and Gillingham, F. J., 1967, On the occurence of homovanillic acid and 5-hydroxyindol-3-ylacetic acid in the ventricular C.S.F. of patients suffering from parkinsonism, *Confin. Neurol.* **29**:73–77.

Gurevich, M. W., and Kartseva, A. G., 1973, Influence of electrical stimulation of the human amygdaloid complex on the hemodymics, *Fiziol. Zh. USSR* **19**:637–641 (Rus.).

Gurevich, M. W., and Kartseva, A. G., 1974, The influence of partial destruction of hippocampus on the hemodynamics in epileptic patients, in *Epileptic Focus and Surgical Treatment of Epilepsy,* Meditsina, Moscow, pp. 261–274 (Rus.).

Gurfinkel, V. S., and Safronov, V. A., 1971, Cerebellar hypotonicity in human, in *Structural and Functional Organization of Cerebellum,* Meditsina, Leningrad, pp. 131–136 (Rus.).

Gurfinkel, V. S., Kandel, E. I., Kotz, L. N., and Shick, L. M., 1963, Application of tremorograms for prognosis of effectiveness of parkinsonism surgical treatment, *Vopr. Neirokhir.* **1**:1–6 (Rus.).

Gurfinkel, V. S., Kandel, E. I., Kotz, J. M., and Shick, L. M., 1965, On the mechanism of generation of parkinsonian tremor, *J. Nevropatol. Psychiatr.* **5**:645–651 (Rus.).

Gutin, P. H., and Dormandy, R. H., 1982, A coaxial catheter system for after-loading radioactive sources for the interstitial irradiation of brain tumors, *J. Neurosurg.* **56**:734–735.

Gutin, P. H., Phillips, T. L., Wara, W. M., Leibel, S. A., Hosobuchi, Y., Levin, V. A., Weaver, K. A., and Lamb, S., 1984, Brachytherapy of recurrent malignant brain tumors with removable high-activity iodine-125 sources, *J. Neurosurg.* **60**:61–68.

Guttman, M., Seeman, P., and Reynolds, G. P., 1986, Dopamine D_2 receptor density remains constant in treated Parkinson's disease, *Ann. Neurol.* **19**:487–492.

Gybels, J. M., 1963, *The Neural Mechanism of Parkinsonian Tremor* (S. A. Arstia, ed.), Bruxelles.

Haaijman, V. P., van Leeuwen, W. S., and van Veelen, C. W. M., 1977, Assessment of behavior modification in patients treated by psychosurgery: Five patients with severe obsessive–compulsive neurosis, in *Neurosurgical Treatment in Psychiatry, Pain and Epilepsy* (W. H. Sweet, S. Obrador, and J. G. Martin-Rodriguez, eds.), University Park Press, Baltimore, pp. 267–286.

Hadley, M. H., Shetter, A. G., and Rob Amos, M., 1985, Use of the Brown–Roberts–Wells stereotactic frame for functional neurosurgery, *Appl. Neurophysiol.* **48**:61–68.

Hahn, I. F., Levy, W. I., and Weinstein, M. I., 1979, Needle biopsy of intracranial lesions guided by computerized tomography, *Neurosurgery* **5**:11–15.

Haines, D. E., 1981, Zones in the cerebellar cortex. Their organization and potential relevance to cerebellar stimulation, *J. Neurosurg.* **55**(2):254–264.

Haines, S. J., Jannetta, P. J., and Zorub, D. S., 1980, Microvascular relations of the trigeminal nerve. An anatomical study with clinical correlation, *J. Neurosurg.* **52**(3):381–386.

Håkanson, S., 1981, Treatment of trigeminal neuralgia by injection of glycerol into the trigeminal cistern, *Neurosurgery* **9**:638–646.

Håkanson, S., 1982, Trigeminal neuralgia treated by retrogasserian injection of glycerol, thesis, Stockholm.

Hall, H. C., 1921, *La Dégénérescence Hépato-Lenticularie,* Masson, Paris.

Halliday, A. M., 1967, The electrophysiological study of myoclonus in man, *Brain* **90**:241–250.

Halliday, A. M., and Redfearn, J. W., 1956, An analysis of the frequencies of finger tremor in health subjects, *J. Physiol. (Lond.)* **184**:600–608.

Hamby, W. B., 1953, Surgical treatment of dystonia musculorum deformans, *J. Neurosurg.* **10**:490–495.

Hamby, W., and Schieffer, S., 1969, Spasmodic torticollis. Results after cervical rhizotomy in 50 cases, *J. Neurosurg.* **31**:323–326.

Hamer, J., 1986, Glossopharyngeusneuralgie und neurovaskuläre Decompression, *Nervenarzt* **57**:302–305.

Hamien, A., and Maloney, A. F. J., 1969, Localization of stereotactic lesions in the treatment of parkinsonism: A clinicopathological comparison, *J. Neurosurg.* **31**:393–399.

Handa, H., Araki, C., Mori, K., Mizawa, I., and Ito, M., 1962, Spasmodic torticollis treated by chemothalamotomy and chemopallidotomy, *Confin. Neurol.* **22**:393–396.

Handa, H., Mori, K., Ito, M., Shimabukuro, H., Yoneda, S., and Mizawa, I., 1971, Spasmodic torticollis: Experimental study and indication to the stereotaxtic surgery, *Neurol. Med. Chir.* **11**:319–320.

Hankinson, J., 1969, Physiological confirmation of stereotactic localization, in *Third Symposium on Parkinson's Disease* (F. J. Gillingham and I. M. L. Donaldson, eds.), Livingstone, Edinburgh, London, pp. 218–221.

Hara, M., Takeuchi, K., Okada, J., Takizawa, T., and Matsumoto, M., 1980, Evaluation of brain tumor laser surgery, *Acta Neurochir. (Wien)* **53**:141–149.

Harbaugh, R. E., Coombs, D. W., Saunders, R. L., Gaylor, M., and Pageau, M., 1982, Implanted continuous epidural morphine infusion system. A preliminary report, *J. Neurosurg.* **56**:803–806.

Harbaugh, R. E., Wilson, D. H., Reeves, A. G., and Gazzaniga, M. S., 1983, Forebrain commissurotomy for epilepsy. Review of 20 consecutive cases, *Acta Neurochir.* **68**:263–275.

Hardy, J., 1966, Electrophysiological localization and identification, *J. Neurosurg.* **24**:410–414.

Hardy, J., 1969, Transsphenoidal microsurgery of the normal and pathological pituitary, *Clin. Neurosurg.* **16**(10):185–217.

Hardy, J., 1975a, Le traitment du cancer du sein métastatique par l'hypophysectomie transphénoidale, *Union Med. Can.* **104**:1557–1562.

Hardy, J., 1975b, Transsphenoidal microsurgical removal of pituitary microadenomata, in *Progress in Neurological Surgery* (H. Krayenbühl, P. E. Maspes, W. H. Sweet, eds.), Karger, Basel, pp. 200–216.

Hardy, J., 1983, Transsphenoidal microsurgery of prolactinomas: Report on 355 cases, in *Prolactin and Prolactinomas* (G. Tolis, ed.), Raven Press, New York, pp. 431–440.

Hardy, J., and Bertrand, C., 1965, Electrophysiological exploration of subcortical structures with microelectrode during stereotaxic surgery, *Confin. Neurol.* **26**:201–203.

Hardy, J., Le Clercq, T., and Mercky, F., 1974, Microsurgical cordotomy by the anterior approach, *J. Neurosurg.* **41**:640–643.

Hardy, T. L., and Koch, J., 1982, Computer-assisted stereotactic surgery, *Appl. Neurophysiol.* **45**:396–398.

Hardy, T. L., Bertrand, G., and Thompson, C. J., 1979a, The position and organization of motor fibers in the internal capsule found during stereotactic surgery, *Appl. Neurophysiol.* **42**(3):160–170.

Hardy, T. L., Bertrand, G., and Thompson, C. J., 1979b, Organization and topography of sensory responses in the internal capsule and nucleus ventralis caudalis found during stereotactic surgery, *Appl. Neurophysiol.* **42**:335–351.

Hardy, T. L., Bertand, G., and Thompson, C. J., 1980a, Position and organization of thalamic cellular activity during diencephalic recording, *Appl. Neurophysiol.* **43**:18–36.

Hardy, T. L., Bertrand, G., and Thompson, C. J., 1980b, The topography of "bilateral-movement-evoked" thalamic cellular activity found during diencephalic recording, *Appl. Neurophysiol.* **43**:67–74.

Hardy, T. L., Koch, J., and Lassiter, A., 1983, Computer graphics with computerized tomography for functional neurosurgery, *Appl. Neurophysiol.* **46**:217–226.

Harik, S. J., and Post, M. J. D., 1981, Computed tomography in Wilson's disease, *Neurology (N.Y.)* **31**:107–110.

Harik, S., La Manna, J. C., Snyder, S., Wetherbee, J. R., and Rosenthal, M., 1982, Abnormalities of cerebral oxidative metabolism in animal models of Parkinson's disease, *Neurology* **32**:382–389.

Harris, W., 1912, Alcohol injection of the Gasserian ganglian for trigeminal neuralgia, *Lancet* **1**:218–221.

Harris, W., 1921, Persistent pain in lesions of the peripheral and central nervous system, *Brain* **44**:557–571.

Harris, W., 1926, *Neuritis and Neuralgia,* Oxford University Press, London, pp. 327–329.

Harrison, M. T., Jennet, W. B., and Cross, J. N., 1970, The use of cryogenic surgery in the treatment of pituitary tumors, *Proc. R. Soc. Med.* **63**:224–225.

Härtel, F., 1914, Die Behandlung der Trigeminusneuralgie mit intracranielen Alkoholeinspritzungen, *Dtsch. Z. Chir.* **126**:429–552.

Hartley, F., 1892, Intracranial neurectomy of the second and third divisions of the fifth nerve, *N.Y. J. Med.* **55**:317–319.

Hartmann von Monakow, K. H., 1960, *Das Parkinson-Syndrom. Klinik und Therapie,* Karger, Basel, New York.

Hartmann von Monakow, K. H., 1962, Histological and clinical correlations in 29 Parkinson patients with stereotaxic surgery, *Confin. Neurol.* **34**:210–217.

Hartog, M., Doyle, F., and Fraser, R., 1965, Partial pituitary ablation with implants of gold-198 and yttrium-90 for acromegaly, *Br. Med. J.* **2**:396–398.

Hartwig, H. G., and Wahren, W., 1982, Anatomy of hypothalamus, in *Stereotaxy of the Human Brain* (G. Schaltenbrand and A. E. Walker, eds.), G. Thieme Verlag, New York, pp. 87–106.

Hasby, J., and Korsgaard, A. G., 1975, Late results of thalamotomy in parkinsonism with and without the influence of levodopa, *Acta Neurochir.* **31**:260–264.

Hass, G. M., and Taylor, C. B., 1948, A quantitative hypothermal method for the production of local injury of tissue, *Arch. Pathol.* **45**:563–580.

Hassler, R., 1947, *Die pathophysiologische Bedeutung des Thalamus für einige psychische Erscheinungen,* Lecture, Tubingen.

Hassler, R., 1949, Über die afferenten Bahnen und Thalamuskerne des motorischen Systems des Grosshirns. I and II, *Arch. Psychiatr. Nervenkrank.* **82**:759–785, 786–818.

Hassler, R., 1950a, Über Kleinhirnprojektion zum Mittelhirn und Thalamus beim Menschen, *Dtsch. Z. Nervenheilkd.* **163**:629–671.

Hassler, R., 1950b, Die Anatomie des Thalamus, *Arch. Psychiatrie* **184**:249–256.

Hassler, R., 1955, The pathological and pathophysiological basis of tremor and parkinsonism, in *Proceedings II International Congress of Neuropathology,* London, pp. I, 29–40, IV, 637–642.

Hassler, R., 1956, Die extrapyramidalen Rindensysteme und die zentrale Regelung der Motorik, *Dtsch Z. Nervenheilkd.* **175**:233–240.

Hassler, R., 1959a, Gezielte Operationen gegen extrapyramidale Bewegungsstörungen, in *Einführung in die stereotaktischen Operationen mit einem Atlas des menschlichen Gehirns,* Vol. 1 (G. Schaltenbrand and P. Bailey, eds.), Thieme, Stuttgart, pp. 472–488.

Hassler, R., 1959b, Anatomy of the thalamus, in *Introduction to Stereotaxis with an Atlas of the Human Brain,* Vol. 1 (G.

Schaltenbrand and P. Bailey, eds.), Thieme, Stuttgart, pp. 230–290.

Hassler, R., 1960, Die zentralen Systeme des Schmerzes, *Acta Neurochir. (Wien)* **8**:353–423.

Hassler, R., 1964, Spezifische und unspezifische systeme des menschichen zwischenhirns, in *Lectures on the Diencephalon* (W. Bergmann and J. P. Schade, eds.), Elsevier, Amsterdam, pp. 1–32.

Hassler, R., 1966a, Thalamic regulation of muscle tone and the speed of movements, in *The Thalamus* (D. Purpura and M. D. Yahr, eds.), Columbia University Press, New York, pp. 419–438.

Hassler, R., 1966b, Discussion of pain symposium, *Confin. Neurol.* **27**:89–92.

Hassler, R., 1968, Extrapyramidal myoclonus treated by stereotaxic coagulation of the dentatothalamic pathway and their physiopathological mechanism, *Rev. Neurol.* **119**(5):409–418.

Hassler, R., 1969, Mioclonies extrapyramidales traitées par coagulation stéreotaxique de la voie dentato-thalamique et leur méchanisme physiopathologique, *Ann. Med. Physiol.* **127**(4):698–701.

Hassler, R., 1972a, Über die Zweiteilung der Schmerzleitung in die Systeme der Schmerzempfindung und des Schmerzgefühes, in *Schmerz. Grundlagen—Pharmakologie—Therapie* (R. Janzen, ed.), Thieme, Stuttgart, pp. 105–124.

Hassler, R., 1972b, Physiopathology of rigidity, in *Parkinson's Disease*, Vol. 2 (J. Siegfried, ed.), Hans Huber, Bern, pp. 19–46.

Hassler, R., 1972c, Sagittal thalamotomy for relief of motor disorders in cases of double athetosis and cerebral palsy, *Confin. Neurol.* **34**:15–18.

Hassler, R., 1974, Fiber connections within the extrapyramidal system, *Confin. Neurol.* **36**:237–255.

Hassler, R., 1975, Central interactions of the system of the rapidly and slowly conducted pain, *Adv. Neurosurg.* **3**:143–149.

Hassler, R., 1976, Wechselwirkungen zwischen dem System der schnellen Schmerzempfindung und dem des langsamen, nachhaltigen Schmerzgefühls, *Langebecks Arch. Chir.* **342**:47–61.

Hassler, R., 1982, Stereotaxic surgery for psychiatric disturbances, in *Stereotaxy of the Human Brain* (G. Schaltenbrand and A. E. Walker, eds.), Georg Thieme Verlag, Stuttgart, New York, pp. 570–590.

Hassler, R., 1985, Pathophysiology of hypertonic and hypotonic motor disorders and of their stereotactic therapy, in *IX Meeting of the World Society for Stereotactic and Functional Neurosurgery*, Toronto,

Hassler, R., and Dieckmann, G., 1967a, Stereotaxic treatment of compulsive and obsessive symptoms, *Confin. Neurol.* **29**:153–158.

Hassler, R., and Dieckmann, G., 1967b, Arrest reaction, delayed inhibition and unusual gaze behaviour resulting from stimulation of the putamen in awake, unrestrained cats, *Brain Res.* **5**:504–508.

Hassler, R., and Dieckmann, G., 1970a, Stereotactic treatment of different kinds of spasmodic torticollis, *Confin. Neurol.* **32**:135–143.

Hassler, R., and Dieckmann, G., 1970b, Traitement stéréotaxique des tics et cris inarticules ou coprolaliques considérés comme phénomène d'obsession matrice an cours de la maladie de Gilles de la Tourette, *Rev. Neurol.* **123**:89–100.

Hassler, R., and Dieckmann, G., 1970c, Die stereotaktische Behandlung des Torticollis aufgrund tierexperimenteller Erfahrungen über die richtungsbestimmen Bewegungen, *Nervenarzt* **41**:473–487.

Hassler, R., and Dieckmann, G., 1972, Violence against oneself and against others as a target for stereotaxic psychosurgery (erethismic imbecility and temporal lobe epilepsy), in *Present Limits of Neurosurgery* (I. Fusek and Z. Kunc, eds.), Avicenum, Czechoslovak Medical Press, Prague, pp. 477–482.

Hassler, R., and Dieckmann, G., 1982, Stereotactic treatment for spasmodic torticollis, in *Stereotaxy of the Human Brain* (G. Schaltenbrand and A. E. Walker, eds.), Georg Thieme Verlag, Stuttgart, New York, pp. 522–531.

Hassler, R., and Hess, W. R., 1954, Experimentelle und anatomische Befunde über die Drehbewegungen und ihre nervösen Apparate, *Arch. Psychiatr. Nervenkr.* **192**:488–526.

Hassler, R., and Riechert, T., 1954, Indikationen und Lokalisationsmethode der gezielten Hirnoperationen, *Nervenarzt* **25**:441–447.

Hassler, R., and Riechert, T., 1957, Über einen Fall von doppelseitiger Fornicotomie bei sonenannter temporaler Epilepsie, *Acta Neurochir.* **5**:330–340.

Hassler, R., and Riechert, T., 1959, Klinische und anatomische Befunde bei stereotacktischen Schmerzoperationen in Thalamus, *Arch. Psychiatr. Nervenkr.* **200**:93–122.

Hassler, R., Riechert, T., and Mundinger, F., 1960, Physiological observations in stereotaxic operations in extrapyramidal motor disturbances, *Brain* **83**:337–351.

Hassler, R., Mundinger, F., and Riechert, T., 1970, Pathophysiology of tremor at rest derived from the correlation of anatomical and clinical data, *Confin. Neurol.* **32**:79–87.

Hassler, R., Bronisch, F., Mundinger, F., and Riechert, F., 1975, Intention myoclonus of multiple sclerosis, its pathoanatomical basis and its stereotactic relief, *Neurochirurgia (Stuttg.)* **18**:90–106.

Hassler, R., Mundinger, R., and Riechert, T., 1979, *Stereotaxis in Parkinson Syndrome*, Springer-Verlag, Berlin.

Hayward, R., 1986, Observations on the innervation of the sternomastoid muscle, *J. Neurol. Neurosurg. Psychiatry* **49**:951–953.

Heath, R. G., and Cox, A. W., 1976, Surgical treatment of psychological disease. Historical review 1950 to 1974, in *Current Controversies in Neurosurgery* (T. P. Morley, ed.), W. B. Saunders, Philadelphia, pp. 709–721.

Heath, R. G., and Mickle, W. A., 1960, Evaluation of seven year's experience with depth electrode studies in human patients, in *Electrical Studies on the Unanesthetized Brain* (A. Ramy and C. O'Doherty, eds.), Hoeber, New York, pp. 214–247.

Hecaen, H., Talairach, J., David, M., and Dell, M. B., 1949,

Coagulations limities du thalamus dans les algies du syndrome thalamique. Resultats therapeutiques et physiologiques, *Rev. Neurol.* **81**:917–931.

Heikkinen, E. R., 1986, Stereotactic neurosurgery: New aspects of an old method, *Ann. Clin. Res.* **18**:73–83.

Heilbrun, M. P., Roberts, T. S., Apuzzo, M. L. J., Wells, T. H., and Sabshin, J. K., 1983, Preliminary experience with the Brown–Roberts–Wells (BRW) computerized tomography stereotaxic guidance system, *J. Neurosurg.* **59**:217–222.

Heilbrun, M. P., Brown, R. A., and McDonald, P. R., 1985, Real-time three-dimensional graphic reconstructions using Brown–Roberts–Wells frame coordinates in a microcomputer environment, *Appl. Neurophysiol.* **48**:7–10.

Heimburger, R. F., 1967, Dentatectomy in the treatment of dyskinetic disorders, *Confin. Neurol.* **29**:101–106.

Heimburger, R. F., 1970a, The cerebellum and spasticity, *Int. J. Neurol.* **7**:232–243.

Heimburger, R. F., 1970b, The role of the cerebellar nuclei in spasticity, *Confin. Neurol.* **32**:105–113.

Heimburger, R. F., 1975a, Putamenotomy as an aid to upper extremity control, *Confin. Neurol.* **37**(1–3):16–23.

Heimburger, R. F., 1975b, Multiple sequential stereotaxic surgery for cerebral palsy, *Confin. Neurol.* **37**:270–278.

Heimburger, R. F., and Whitlock, C. C., 1965, Stereotaxic destruction of the human dentate nucleus, *Confin. Neurol.* **26**:346–358.

Heimburger, R. F., Whitlock, C., and Kalsbeck, J., 1966, Stereotaxic amygdalotomy for epilepsy with aggressive behavior, *J.A.M.A.* **198**:165–169.

Heimburger, R. F., Slominski, O. T., and Griswold, P., 1973, Cervical posterior rhizotomy for reducing spasticity in cerebral palsy, *J. Neurosurg.* **39**:30–34.

Heimburger, R. F., Small, J. F., Small, J. G., Milstein, V., and Moore, D., 1978, Stereotactic amygdalotomy for convulsive and behavioral disorders, *Appl. Neurophysiol.* **41**(1–4):43–51.

Helfand, M. H., Leksell, L., and Strang, R. R., 1965, Experiences with intractable pain treated by stereotaxic mesencephalotomy, *Acta Chir. Scand.* **129**:573–580.

Henderson, W. R., 1965, The anatomy of the Gasserian ganglion and the distribution of pain in relation to injections and operations for trigeminal neuralgia, *Ann. R. Coll. Surg.* **37**:346–373.

Heppner, F., 1978, The laser scalpel on the nervous system, in *Laser Surgery II* (I. Kaplan, ed.), Jerusalem Academic Press, Jerusalem, pp. 79–80.

Herman, R., Freedman, W., and Mayer, N., 1974, Neurophysiologic mechanisms of hemiplegic and paraplegic spasticity: Implications for therapy, *Arch. Phys. Med. Rehabil.* **55**:150–153.

Herner, T., 1961, Treatment of mental disorders with frontal stereotaxis thermo-lesions. A follow-up study of 116 cases, *Acta Psychiatr. Neurol. Scand. [Suppl.]* **158**:28–36.

Herringham, W. P., 1890, On muscular tremor, *J. Physiol. (Lond.)* **11**:478–485.

Herz, E., 1944a, Dystonia: Historical review, analysis of dystonic symptoms and physiologic mechanisms involved, *Arch. Neurol. Psychiatry* **51**:305–318.

Herz, E., 1944b, Pathology and conclusions, *Arch. Neurol. Psychiatry* **59**:20–24.

Herz, E., and Glaser, G. H., 1949, Spasmodic torticollis. II. Clinical evaluation, *Arch. Neurol. Psychiatry* **61**:227–239.

Herz, E., and Hoefer, P. F. A., 1949, Spasmodic torticollis. I. Physiologic analysis of involuntary motor activity, *Arch. Neurol. Psychiatry* **61**:129–136.

Herz, E., and Meyers, R., 1962, The extrapyramidal diseases, in *Clinical Neurology* (A. B. Baker, ed.), Hoeber–Harper, New York, pp. 1285–1337.

Hess, W. R., 1928, Hirnreizungsversuche über den Mechanismus des Schlafes, *Arch. Psychiatr. Nervenkr.* **86**:287–292.

Hess, W. R., 1932, *Die Methodik der localisierten Reizung und Ausschaltung subkortikaler Hirnabschnitte*, Leipzig.

Hesse, E. R., 1929, Eleven cases of sympathectomy in parkinsonism after epidemic encephalitis, *Vestn. Khir.* **53**:214–220 (Rus.).

Hiedl, P., Struppler, A., and Gessler, M., 1979, Local analgesia by percutaneous electrical stimulation of sensory nerves, *Pain* **7**:129–134.

Higgins, A. C., and Nashold, B. S., Jr., 1980a, Modification of instrument for stereotactic evacuation of intracerebral hematoma: Technical note, *Neurosurgery* **7**:604–605.

Higgins, A. C., and Nashold, B. S., Jr., 1980b, Stereotactic evacuation of large intracerebral hematomas, *Appl. Neurophysiol.* **43**:96–103.

Higgins, A. C., Nashold, B. S., and Cosman, E., 1982, Stereotactic evacuation of primary intracerebral hematomas: New instrumentation, *Appl. Neurophysiol.* **45**:438–442.

Hillman, P., and Wall, P. D., 1969, Inhibitory and excitatory factors in influencing the receptive fields of lamina 5 spinal cord cells, *Exp. Brain Res.* **9**:284–306.

Hirai, T., Nagaseki, J., Kawashima, Y., Wada, H., Tsukahara, Y., Imai, S., and Ohye, C., 1982, Large neurons in the thalamic ventrolateral mass in humans and monkeys, *Appl. Neurophysiol.* **45**:245–250.

Hirato, M., Kawashima, Y., Wada, H., Nagaseki, J., Hirai, T., Shibazaki, T., Denda, J., and Ohye, C., 1984, Quantitative analysis of neural activity: Of the thalamic Vim nucleus in patients with thalamic pain, *Appl. Neurophysiol.* **47**:85–86.

Hirose, S., 1965, Orbito-ventromedial undercutting 1957–1963. Follow-up study of 77 cases, *Am. J. Psychiatry* **121**:1194–1202.

Hirose, S., 1975, Psychiatric evaluation of psychosurgery, in *Neurosurgical Treatment in Psychiatry, Pain, and Epilepsy* (W. H. Sweet, S. Obrador, and J. G. Martin-Rodriguez, eds.), University Park Press, Baltimore, pp. 203–209.

Hitchcock, E. R., 1969a, Stereotaxic spinal surgery. A preliminary report, *J. Neurosurg.* **31**:386–392.

Hitchcock, E. R., 1969b, An apparatus for stereotactic spinal surgery, *Lancet* 705–706.

Hitchcock, E. R., 1970, Stereotactic cervical myelotomy, *J. Neurol. Psychiatry* **33**:224–230.

Hitchcock, E. R., 1972, Stereotaxis of the spinal cord, *Confin. Neurol.* **34**:299–310.

Hitchcock, E. R., 1973, Stereotaxic pontine spinothalamic tractotomy, *J. Neurosurg.* **39**:746–752.

Hitchcock, E. R., 1974, Stereotactic myelotomy, *J. R. Soc. Med.* **67**:771–772.

Hitchcock, E. R., 1975, Stereotaxic neurosurgery for cerebral palsy, *Br. Med. J.* **4**(5991):285–289.

Hitchcock, E. R., and Leece, B., 1967, Somatotopic representation of the respiratory pathways in the cervical cord of man, *J. Neurosurg.* **27**:320–329.

Hitchcock, E. R., and Schvarcz, Y. R., 1972, Stereotaxic trigeminal tractotomy for post-herpetic facial pain, *J. Neurosurg.* **37**:412–417.

Hitchcock, E. R., and Teixeira, M. J., 1981, A comparison of results from center–median and basal thalamotomies for pain, *Surg. Neurol.* **15**(5):341–351.

Hitchcock, E. R., Ashcroft, G. W., Cairns, V. M., and Murray, L. G., 1973, Observations on the development of an assessment scheme for amygdalotomy, in *Surgical Approaches in Psychiatry* (L. Laitinen and E. E. Livingston, eds.), Medical and Technical Publishing, Lancaster, pp. 143–155.

Hitchcock, E. R., Kim, M. C., and Sotelo, M. G., 1985, Further experience in stereotactic pontine tractotomy, *Appl. Neurophysiol.* **48**:242–246.

Hitchcock, E. R., Morris, C. S., Sotelo, M. G., and Salmon, M., 1986, Comparison of smear and imprint techniques for rapid diagnosis in neuro-oncology, *Surg. Neurol.* **26**:176–182.

Hockley, A. D., 1975, Proximal occlusion of the anterior cerebral artery for anterior communicating aneurysm, *J. Neurosurg.* **43**:426–431.

Hodge, C. J., Apkarian, A. V., and Stevens, R. T., 1986, Inhibition of dorsal-horn cells responses by stimulation of the Kölliker–Fuse nucleus, *J. Neurosurg.* **65**:825–833.

Hoefer, P. F. A., and Putnam, T. S., 1940, Action potentials of muscles in athetosis and Sydenham's chorea, *Arch. Neurol. Psychiatry* **44**:517–531.

Hoefnagel, D., Allen, F. H., and Falk, G., 1970, Hereditary dystonia musculorum deformans, *Clin. Genet.* **5/6**:258–262.

Hoehn, M. M., and Yahr, M. D., 1967, Parkinsonism: Onset, progression and mortality, *Neurology (Minneap.)* **17**:427–442.

Hoehn, M. M., Crowley, T. I., and Rutledge, C. O., 1976, Dopamine correlates of neurological and psychological status in untreated parkinsonism, *J. Neurol. Neurosurg. Psychiatry* **39**:941–951.

Hondo, H., Soga, T., Tsuda, T., Kageyama, T., and Matsumoto, K., 1984, CT-guided stereotactic evacuation of intracerebellar hematomas—a report of three cases, *Appl. Neurophysiol.* **47**:88–92.

Hood, T. W., and Gebarski, S. S., 1985, Evaluation of thalamotomy by positron emission tomography and magnetic resonance imaging, *Appl. Neurophysiol.* **48**:315–319.

Hood, T. W., Gebarski, S. S., McKeever, P. E., and Venes, J. L., 1986, Stereotactic biopsy of intrinsic lesions of the brain stem, *J. Neurosurg.* **65**:172–176.

Hopf, A., Woringer, E., and Hamon, I., 1968, Postoperative hemiballismus, *Neurochirurgie* **2**:1–18.

Hoppenstein, R., 1975, Percutaneous implantation of chronic spinal cord electrodes for control of intractable pain, *Surg. Neurol.* **4**:195–198.

Hori, T., Fukushima, T., Terao, H., Takakura, K., and Sano, K., 1981, Percutaneous radiofrequency facial nerve coagulation in the management of facial spasm, *J. Neurosurg.* **54**(5):655–658.

Hori, Y., Terada, C., Kanazawa, K., and Miyamoto, S., 1968, The effect of stereotaxic putamectomy for epileptic seizures, *Neurol. Med. Chir. (Tokyo)* **10**:321–326.

Horner, N. B., and Potts, D. G., 1984, A comparison of CT-stereotaxic brain biopsy techniques, *Invest. Radiol.* **19**:367–373.

Hornykiewicz, O., 1966a, Metabolism of brain dopamine in human parkinsonism: Neurochemical and clinical aspects, in *Biochemistry and Pharmacology of the Basal Ganglia* (E. Costa, ed.), Raven Press, New York, pp. 171–185.

Hornykiewicz, O., 1966b, New aspects of the biochemical pharmacology of Parkinson's syndrome, *Wien Z. Nervenheilkd.* **23**(1/3):103–109.

Hornykiewicz, O., 1971, Neurochemical pathology and pharmacology of brain dopamine and acetylcholine: Rational basis for current drug treatment of parkinsonism, *Contemp. Neurol.* **8**:34–38.

Hornykiewicz, O., 1976, Neurohumoral interactions and basal ganglia function and dysfunction, in *The Basal Ganglia* (M. D. Yahr, ed.), Raven Press, New York, pp. 269–278.

Horsley, V., 1886, Epilepsy, *Lancet* **2**:1211–1213.

Horsley, V., 1890, Remarks on the surgery of the central nervous system, *Br. Med. J.* **2**:1286–1292.

Horsley, V., 1909, The Linacre lecture on the function of the so-called "motor" area of the brain, *Br. Med. J.* **21**:125–132.

Horsley, V., and Clarke, R. H., 1908, The structure and functions of the cerebellum examined by a new method, *Brain* **31**:45–123.

Horsley, V., Taylor, J., and Coleman, W. S., 1891, Remarks on the various surgical procedures devised for relief or cure of trigeminal neuralgia (tic douloureux), *Br. Med. J.* 1139–1143.

Hosobuchi, J., 1979, Current status of brain stimulation, in *IV Meeting of the European Society for Stereotactic and Functional Neurosurgery,* Paris.

Hosobuchi, J., 1980, The majority of unmyelinated afferent axons in human ventral roots probably conduct pain, *Pain* **8**:167–180.

Hosobuchi, J., 1982, Analgesia induced by brain stimulation with chronically implanted electrodes, in *Operative Neurosurgical Techniques: Indications and Methods* (H. Schmidek and W. Sweet, eds.), Grune & Stratton, New York, pp. 981–991.

Hosobuchi, J., 1983a, Combined electrical stimulation of periaqueductal gray (PAG) and sensory thalamus (STH), in *VI Meeting of the European Society for Stereotactic and Functional Neurosurgery,* Rome, p. P8.

Hosobuchi, J., 1983b, Combined electrical stimulation of the periaqueductal gray matter and sensory thalamus, *Appl. Neurophysiol.* **46**:112–115.

Hosobuchi, J., 1986, Subcortical electric stimulation for control of intractable pain in humans. Report of 122 cases (1970–1984), *J. Neurosurg.* **64:**543–553.

Hosobuchi, J., Adams, J. E., and Rutkin, B., 1973, Chronic thalamic stimulation for the control of facial anaesthesia dolorosa, *Arch. Neurol.* **29:**158–161.

Hosobuchi, Y., Adams, J. E., and Rutkins, B., 1975, Chronic thalamic and internal capsule stimulation for the control of central pain, *Surg. Neurol.* **4:**91–92.

Hosobuchi, Y., Rossier, J., Bloom, J. E., and Guillemin, R., 1979, Stimulation of human periaqueductal gray for pain relief. Increased immunoreactive beta-endorphin in ventricular fluid, *Science* **203:**279–281.

Houdart, R., le Besnerais, 1963, *Les Aneurysmes Arterio-Veineux des Hémispheres Cérébraux,* Masson, Paris.

Hougaard, R., Oikawa, T., and Sveindottir, E., 1976, Regional cerebral blood flow in focal cortical epilepsy, *Arch. Neurol.* **33:**527–535.

Housepian, E. M., and Carpenter, M. B., 1957, Spatial relationships between the globus pallidus and the anterior commissure, *J. Neurosurg.* **14**(4):363–373.

Howe, J. F., 1983, Phantom limb pain—a reafferentation syndrome, *Pain* **15:**101–107.

Huck, F. R., Radvany, J., and Avilla, J. O., 1980, Anterior callosotomy in epileptics with multiform seizures and bilateral synchronous spike and wave EEG pattern, *Acta Neurochir. (Wien) [Suppl]* 127–132.

Hugenholtz, H., and Morley, T. P., 1972, The results of proximal anterior cerebral artery occlusion for anterior communicating aneurysms, *J. Neurosurg.* **37:**65–70.

Hughes, B., 1965, Involuntary movements following stereotactic operations for parkinsonism with special reference to hemichorea (ballismus), *J. Neurol. Neurosurg. Psychiatry* **28**(4):291–303.

Hughes, B., 1969, Evaluation of the subthalamic lesion in parkinsonism, in *Symposium on Parkinson's Disease* (F. J. Gillingham and I. M. L. Donaldson, eds.), Livingstone, Edinburgh, pp. 259–260.

Huk, W., and Baer, V., 1980, A new targeting device for stereotaxic procedures within the CT scanner, *Neuroradiology* **19:**13–17.

Hunt, J. R., 1914, Dyssynergia cerebellaris progressiva—a chronic progressive form of cerebellar tremor, *Brain* **37:**247–268.

Hunt, J. R., 1921, Dyssynergia cerrebellaris myoclonica—primary atrophy of the dentate system: A contribution to the pathology and symptomatology of the cerebellum, *Brain* **44:**490–538.

Hurt, R. W., and Ballantine, H. T., 1974, Stereotactic anterior cingulate lesions for persistent pain. A report on 68 cases, *Clin. Neurosurg.* **21:**334–351.

Ielasic, F., 1959, Über die Behandlung des Trigeminusneuzalgie mittes mechanischer Kompression des Ganglion Gasseri durch das Foramen ovale, *Acta Neurochir.* **7:**440–445.

Ignelzi, R. J., and Nyquist, J. K., 1979, Excitability changes in peripheral nerve fibers after repetitive electrical stimulation. Implications in pain modulation, *J. Neurosurg.* **51:**826–833.

Iizuka, I., 1975, Development of a stereotaxic endoscopy of the ventricular system, *Confin. Neurol.* **37:**141–149.

Illinsky, I., 1970, Emotional and affective reactions produced by electrostimulation of the ventrolateral nucleus of the thalamus, *Vopr. Neirokhir.* No. 4:42–45.

Illis, L. S., Oygar, A. E., Sedgwick, E. M., and Awadalla, M. A. S., 1976, Dorsal column stimulation in the rehabilitation of patients with multiple sclerosis, *Lancet* 1383–1386.

Ioku, M., Ogawa, M., and Jinnai, D., 1971, Studies on facilitation of the H-reflex in spasticity and rigidity, *Electromyography* **11:**11–23.

Isamat, F., Ferran, E., and Acebes, J., 1981, Selective percutaneous thermocoagulation rhizotomy in essential glossopharyngeal neuralgia, *J. Neurosurg.* **55:**575–580.

Ischia, S., Luzzani, A., Ischia, A., and Maffezzoli, G., 1984a, Bilateral percutaneous cervical cordotomy: Immediate and long-term results in 36 patients with neoplastic disease, *J. Neurol. Neurosurg. Psychiatry* **47:**141–147.

Ischia, S., Luzzani, A., Ischia, A., and Pacini, L., 1984b, Role of unilateral percutaneous cervical cordotomy in the treatment of neoplastic vertebral pain, *Pain* **19:**123–131.

Iseki, H., Amano, K., and Kawamura, H., 1982, Somatotopic arrangement of lateral spinothalamic tract in percutaneous cervical cordotomy, *Appl. Neurophysiol.* **45:**484–491.

Iseki, H., Tanikawa, T., Kawamura, H., and Nagao, T., 1985, A new apparatus for CT guided stereotactic surgery, in *IX Meeting of the World Society for Stereotactic and Functional Neurosurgery,* Toronto, pp. 50–60.

Isgreen, W. P., Fahn, S., Barrett, R. E., Snider, S. R., and Chutorian, A. M., 1976, Carbamazepine in torsion dystonia, in *Advances in Neurology,* Vol. 14, *Dystonia* (R. Eldridge and S. Fahn, eds.), Raven Press, New York, pp. 411–416.

Ishijima, B., Yoshimasu, N., Fukushima, T., Hori, T., Sekino, H., and Sano, K., 1973, Nociceptive neurons in the human thalamus, in *The Sixth Symposium of the International Society for Research in Stereoencephalotomy,* Vol. II, Tokyo, p. 24.

Ivan, L. P., 1982, Longitudinal (Bischof's) myelotomy, in *Operative Neurosurgical Techniques. Indications, Methods and Results,* Vol. II (H. H. Schmidek and W. H. Sweet, eds.), Grune & Stratton, New York, pp. 1163–1176.

Ivan, L. P., and Ventureyra, E. C. G., 1982, Chronic cerebellar stimulation in cerebral palsy, *Appl. Neurophysiol.* **45:**51–54.

Ivan, L. P., Ventureyra, E. C. G., Wiley, J., Pressman, E., Knights, R., Guzman, C., and Uttley, D., 1981, Chronic cerebellar stimulation in cerebral palsy, *Surg. Neurol.* **15:**81–84.

Ivannikov, J. G., 1969, *Computer Application for Stereotactic Brain Operations,* Meditsina, Leningrad (Rus.).

Ivanova-Smolenskaya, I. A., 1978, Approaches to differential diagnostics for parkinsonism and essential tremor, in *Pathogenesis, Clinic and Treatment of Parkinsonism,* Meditsina, Moscow, pp. 131–133 (Rus.).

Ivanova-Smolenskaja, I. A., Kandel, E. I., and Andreeva, E. A., 1986, The spectral EMG analysis of the essential tremor, *J. Neuropathol. Psychiatry* No. 7:975–980 (Rus).

Iwakuma, T., Matsumoto, A., and Nakamura, N., 1982, Hemifacial spasm. Comparison of three different operative procedures in 110 patients, *J. Neurosurg.* **57**:753–756.

Jackson, J. H., 1873, On the anatomical physiological, and pathological investigation of epilepsies, reprinted 1958 in *Selected Writings of John Hughlings Jackson*, Vol. 1 (J. Taylor, ed.), Basic Books, New York, p. 93.

Jacob, A., 1932, Zur Frage der nosologischen und lokalisatorischen Auffassung der torsiondystonischen Krankheitserscheinungen, *Dtsch. Z. Nervenheilkd.* **124**:148–153.

Jacob, C., 1928, Sindrome de hemiballisms, coreiforme cruzado por hemorrhagia en el nucleco hipothalamico, *Arch. Argent. Neurol.* **2**:1–15.

Jaeger, R., 1959, The results of injecting hot water into the Gasserian ganglion: The relief of tic douloureux, *J. Neurosurg.* **16**:656–663.

Jane, J. A., Jashon, D., Becker, D. P., Beatty, R., and Sugar, O., 1968, The effect of destruction of the corticospinal tract in the human cerebral peduncle upon motor function and involuntary movements: Report of 11 cases, *J. Neurosurg.* **29**:581–585.

Janko, M., and Trontely, J. V., 1980, Transcutaneous electrical nerve stimulation: A microneurographic and perceptual study, *Pain* **9**(2):219–230.

Jankovic, J., and Frost, J. D., 1981, Quantitative assessment of parkinsonian and essential tremor: Clinical application of triaxial accelerometry, *Neurology (Minneap.)* **31**:1235–1240.

Jannetta, P. J., 1967, Arterial compression of the trigeminal nerve at the pons in patients with trigeminal neuralgia, *J. Neurosurg.* **26**:159–162.

Jannetta, P. J., 1970, Microsurgical exploration and decompression of the facial nerve in the hemifacial spasm, *Curr. Top. Res.* **2**:217–220.

Jannetta, P. J., 1976, Microsurgical approach to the trigeminal nerve for tic douloureux, *Prog. Neurol. Surg.* **7**:180–200.

Jannetta, P. J., 1977a, Treatment of trigeminal neuralgia by suboccipital and transtentorial cranial operations, *Clin. Neurosurg.* **24**:538–549.

Jannetta, P. J., 1977b, Observations on the etiology of trigeminal neuralgia, hemifacial spasm, acoustic nerve dysfunction and glossopharyngeal neuralgia. Definitive microsurgical treatment and results in 117 patients, *Neurochirurgia (Stuttg.)* **20**(5):145–154.

Jannetta, P. J., 1979a, Microsurgery of cranial nerve cross-compression, *Clin. Neurosurg.* **26**:607–615.

Jannetta, P. J., 1979b, Treatment of trigeminal neuralgia, *Neurosurgery* **4**:93–94.

Jannetta, P. J., 1981a, Hemifacial spasm, in *The Cranial Nerves* (M. Samii and P. J. Jannetta, eds.), Springer-Verlag, Berlin, pp. 484–493.

Jannetta, P. J., 1981b, Vascular decompression in trigeminal neuralgia, in *The Cranial Nerves* (M. Samii and P. J. Jannetta, eds.), Springer-Verlag, Berlin, pp. 331–340.

Jannetta, P. J., and Rand, R. W., 1966, Transtentorial retrogasserian rhizotomy in trigeminal neuralgia by microneurosurgical technique, *Bull. Los Angeles Neurol. Soc.* **31**:93–99.

Jannetta, P. J., Abbasy, M., Maroon, J. C., Ramos, F. M., and Albin, M. S., 1977, Etiology and definitive microsurgical treatment of hemifacial spasm. (Operative techniques and results in 47 patients.), *J. Neurosurg.* **47**:321–328.

Jansen, E. N. H., 1978, Bromcriptine in levodopa response-losing parkinsonism. A double blind study, *Eur. Neurol.* **17**:92–99.

Jasper, H. H., 1962, Changing concepts of focal epilepsy, in *Round Table Conference on the Surgical Treatment of the Epilepsies and its Neurophysiological Aspects* (V. Cernácek and B. Cigánek, eds.), Bratislava, pp. 27–35.

Jasper, H. H., 1966, Recording from microelectrodes in stereotactic surgery for Parkinson's disease, *J. Neurosurg.* **24**:219–221.

Jasper, H. H., and Bertrand, G., 1966, Thalamic units involved in somatic sensation and voluntary and involuntary movements in man, in *The Thalamus* (D. P. Purpura and M. D. Yahr, eds.), Columbia University Press, New York, pp. 438.

Jasper, H. H., Capdeville, G. A., and Rasmussen, T., 1961, Evaluation of EEG and cortical electrographic studies for prognosis of seizures following surgical excision of epileptogenic lesions, *Epilepsia* **2**:130–137.

Jefferson, A., 1963, Trigeminal root and ganglion injections using phenol in glycerine for the relief of trigeminal neuralgia, *J. Neurol. Neurosurg. Psychiatry* **26**:345–352.

Jellinger, K., 1975, The morphology of centrally situated angiomas, in *Cerebral Angiomas: Advance in Diagnosis and Therapy* (H. W. Pia, T. R. W. Gleave, E. Crote, and J. Zierski, eds.), Springer-Verlag, Berlin, pp. 9–20.

Jellinger, K., 1982, Adjuvant treatment of Parkinson's disease with dopamine agonists: Open trial with bromocriptine and CV 32-085, *J. Neurol.* **227**:75–88.

Jellinger, K., and Bliesath, H., 1987, Adjuvant treatment of Parkinson's disease with budipine: A double-blind trial placebo, *J. Neurol.* **234**:280–282.

Jelsma, R. K., Bertrand, C. M., Martinez, S. N., and Molino-Negro, P., 1973, Stereotaxic treatment of frontal-lobe and centrencephalic epilepsy, *J. Neurosurg.* **39**:42–51.

Jensen, I., 1975, Temporal lobe surgery around the world. Results, complications, and mortality, *Acta Neurol. Scand.* **52**:354–373.

Jensen, I., 1976, Temporal lobe epilepsy, *Acta Neurol. Scand.* **53**(5):335–337.

Jinnai, D., 1966, Clinical results and significance of Forel-H-tomy in the treatment of epilepsy, *Confin. Neurol.* **27**:129–136.

Jinnai, D., and Mukawa, J., 1970, Forel H-tomy for the treatment of epilepsy, *Conf. Neurol.* **32**:307–315.

Jinnai, D., Mukawa, J., and Kobajashi, K., 1976, Forel-H-tomy for the treatment of intractable epilepsy, *Acta Neurochirurg. (Wien) [Suppl.]* **23**:159–165.

Johansson, F., Almay, B., Knorring, L., and Terenius, L., 1980, Predictors for the outcome of treatment with high frequency transcutaneous electrical nerve stimulation in patients with chronic pain, *Pain* **9**(1):55–61.

Johansson, G., and Laitinen, L., 1966, Electrical stimulation of the thalamus and subthalamic area in Parkinson's disease, *Confin. Neurol.* **26**:445–450.

Johnson, P. C., West, K. M., and Rutledge, B. J., 1958, Destruction of the hypophysis with radioactive colloidal chromic phosphate in cancer of the prostate, *J. Neurosurg.* **15:**519–527.

Johnson, W., Schwartz, G., and Barbeau, A., 1962, Studies on dystonia musculorum deformans, *Arch. Neurol.* **7**(4)**:**301–303.

Judin, S. S., 1943, *Essays on Military Surgery,* Meditsina, Moscow.

Jung, R., 1941, Physiologische Untersuchungen über den Parkinsontremor und andere Zitterformen beim Menschen, *Z. Ges. Neurol. Psychol.* **173:**263–269.

Jung, R., and Hassler, R., 1960, *Handbook of Physiology.* Section I. *Neurophysiology.* Vol. 2, *The Extrapyramidal Motor System,* Williams & Wilkins, Baltimore.

Jurko, M. F., Andy, O. J., and Foshee, D. P., 1963, Diencephalic influence on tremor mechanisms. A study of parkinson tremor during stereotaxic surgery, *Arch. Neurol.* **9:**358–362.

Kadikov, A. S., 1973, Experience of the treatment of parkinsonian patients with midantan, in *Experimental and Clinical Pharmacotherapy,* Suppl. 5, Sinatne Riga, pp. 21–26 (Rus.).

Kahn, E. A., 1973, The open technique of anterolateral corolotomy, in *Neurological Surgery,* Vol. III (J. R. Yonmans, ed.), W. B. Saunders, Philadelphia, pp. 1754–1757.

Kalinina, L. V., 1966, Torsion dystonia in unioval twins, in *Problems of Pathology of the Nervous System,* Kishinev, pp. 94–99.

Kall, B. A., Kelly, P. J., and Goerss, S., 1985, Geometric methodology and clinical applications for anatomically labelled computed tomographic sections, in *IX Meeting of the World Society for Stereotactic and Functional Neurosurgery,* Toronto, p. 89.

Kalyanaraman, S., 1975, Some observations during stimulation of the human hypothalamus, *Confin. Neurol.* **37:**189–192.

Kalyanaraman, S., and Gillingham, F. J., 1964, Stereotaxic biopsy, *J. Neurosurg.* **21:**854–858.

Kalyanaraman, S., and Ramamurthi, B., 1965, Simultaneous bilateral stereotaxic lesions in the diencephalon, *Confin. Neurol.* **26:**310–314.

Kalyanaraman, S., and Ramamurthi, B., 1970, Stereotaxic surgery for generalised epilepsy, *Neurol. India* **18:**34–41.

Kambarova, D. K., 1981, The possibilities of neurophysiology in study and treatment the mental disorders in epilepsy, *Hum. Physiol.* **7:**483–511 (Rus.).

Kamm, R. F., and Austin, G., 1965, The use of bony landmarks of the skull for localization of the anterior–posterior commissural line, *J. Neurosurg.* **22:**576–580.

Kanaka, T. S., 1972, Stereotaxic dentatotomy, *Ann. Indian Acad. Med. Sci.* **8:**246–254.

Kanaka, T. S., and Balasubramaniam, V., 1975, Dentato-thalamotomy in infantile hemiplegia, *Conf. Neurol.* **37**(1–4)**:**271–276.

Kanaka, T. S., and Balasubramaniam, V., 1978, Stereotactic cingulumotomy for drug addiction, *Appl. Neurophysiol.* **41:**86–92.

Kandel, E. I., 1960, Experience of surgical treatment of parkinsonism and other diseases of the extrapyramidal system, in *The All-Union Conference of Neurosurgery,* Moscow, pp. 62–64 (Rus.).

Kandel, E. I., 1961, Surgical treatment of parkinsonism and other diseases of extrapyramidal system. Report 1. Technique and method of the surgical operation, *Vopr. Neirokhir.* **2:**2–9 (Rus.).

Kandel, E. I., 1963, Stereotactic method, *Med. Encycloped.* **31:**405–417 (Rus.).

Kandel, E. I., 1964, *Parkinsonism and Its Surgical Treatment,* Meditsina, Moscow, p. 266 (Rus.).

Kandel, E. I., 1966, On the role of basal ganglia in pathogenesis of parkinsonian syndrome, in *Deep Structures of the Human Brain,* Nauka, Leningrad, pp. 91–94 (Rus.).

Kandel, E. I., 1967, Development of stereotactic neurosurgery in the Soviet Union, *Vopr. Neirokhir.* **6:**15–19 (Rus.).

Kandel, E. I., 1971a, Contemporary problems of drug and surgical treatment of parkinsonism, in *Atherosclerosis of Brain Vessels and Age,* Zdorovje, Kiev, pp. 312–319 (Rus.).

Kandel, E. I., 1971b, Disappearance of the phantom pain syndrome after stereotaxic destruction of the ventrolateral posterior nucleus of the thalamus, *Vopr. Neirokhir.* **6:**14–16 (Rus.).

Kandel, E. I., 1972, Stereotactic neurosurgery. Clinical problems, in *The First All-Union Congress of Neurosurgeons,* Vol. 5, Moscow, pp. 195–201 (Rus.).

Kandel, E. I., 1974a, Cryosurgical method in stereotactic neurosurgery, in *Cryosurgery* (E. I. Kandel, ed.), Meditsina, Moscow, pp. 96–130 (Rus.).

Kandel, E. I., 1974b, Stereotactic surgery for motor dysfunction: Method and results in tremor, in *Recent Progress in Neurological Surgery* (K. Sano, ed.), Excerpta Medica, Amsterdam, pp. 246–250.

Kandel, E. I. (ed.), 1974c, *Cryosurgery,* Meditsina, Moscow (Rus.).

Kandel, E. I., 1975a, New stereotactic apparatus and cryogenic device for stereotactic surgery, *Confin. Neurol.* **37:**128–132.

Kandel, E. I., 1975b, James Parkinson and modern problems of parkinsonism, *Zh. Nevropatol. Psikhiatr.* No. 11:1721–1725 (Rus.).

Kandel, E. I., 1976a, Catecholamine metabolism in parkinsonism by drug therapy and surgery, in *Epilepsy,* Vol. 7, Tbilisi, pp. 124–135 (Rus.).

Kandel, E. I., 1976b, Immediate and long term results of stereotactic operations for hereditary diseases of central nervous system, in *Medical Genetics and Hereditary Diseases,* Meditsina, Moscow, pp. 77–79 (Rus.).

Kandel, E. I., 1978a, Combined Conray–Myodil ventriculography, *Acta Neurochir. (Wien)* **40:**151–156.

Kandel, E. I., 1978b, Surgical treatment of parkinsonism (twenty years experience), in *Pathogenesis, Clinic and Treatment of Parkinsonism,* Meditsina, Moscow, pp. 212–215 (Rus.).

Kandel, E. I., 1980, Influence of the cerebellar dentate nuclei destruction on characteristics of the voluntary movements control system, *Hum. Physiol.* No. 6(3):461–473 (Rus.).

Kandel, E. I., 1981, *Functional and Stereotactic Neurosurgery,* Meditsina, Moscow (Rus.).

Kandel, E. I., 1982, Treatment of hemihyperkinesias by stereotactic operations on basal ganglia, *Appl. Neurophysiol.* **45**:225–229.

Kandel, E. I., 1983, Main problems of functional and stereotactic neurosurgery, in *Proceedings of III All-Union Congress of Neurosurgeons,* Moscow, pp. 95–100 (Rus.).

Kandel, E. I., 1984, Stereotactic method for treatment of extrapyramidal spasmodic torticollis, *Vopr. Neirokhir.* No. 6:28–34 (Rus.).

Kandel, E. I., 1986a, History of stereotactic neurosurgery, in *The First Arctic Stereotactic Workshop,* Umea, Sweden, pp. 20–21.

Kandel, E. I., 1986b, Treatment of pain by chronic electrostimulation of spinal posterior columns, *Vopr. Neirokhir.* No. 2:41–47 (Rus.).

Kandel, E. I., and Bijesin, O. A., 1970, Cryosurgery of brain tumors, *Vopr. Neirokhir.* **1**:3–9 (Rus.).

Kandel, E. I., and Chebotaryova, N. M., 1972, Conray ventriculography in stereotaxic surgery. Experience with 320 operations, *Confin. Neurol.* **34**:34–40.

Kandel, E. I., and Guluyev, K. A., 1977, Treatment of hemihyperkinesia by stereotaxic operations on the basal cerebral ganglia, *Vopr. Neirokhir.* No. 6:29–33 (Rus.).

Kandel, E. I., and Hondkarian, O. A., 1977, Surgical treatment of hyperkinetic form of multiple sclerosis, in *The New Data on Pathogenesis, Clinic and Treatment of Nervous and Mental Diseases,* Kishinjev, pp. 303–305 (Rus.).

Kandel, E. I., and Hondkarian, O. A., 1985, Surgical treatment of the hyperkinetic form of multiple sclerosis, *Acta Neurol.* **3–4**:345–347.

Kandel, E. I., and Kukin, A. V., 1972, New stereotaxic apparatus, *Vopr. Neirokhir.* No. 2:56–57 (Rus.).

Kandel, E. I., and Kuparadze, G. R., 1964, Cryothalamectomy for diseases of extrapyramidal system, *Vopr. Neirokhir.* No. 4:41–48 (Rus.).

Kandel, E. I., and Nikolaenko, E. M., 1969, Investigation of cerebral blood flow and metabolism during artificial embolization of arterio-venous aneurysms, in *Follow-up of Surgical Treatment of Meningiomas, Acoustic Neurinomas and New Data in Neurosurgery,* Meditsina, Leningrad, pp. 306–309 (Rus.).

Kandel, E. I., and Nikolaenko, E. M., 1971, Effects of an increased cerebro–spinal fluid pressure on cerebral blood flow and energy metabolism of the brain, *Vopr. Neirokhir.* No. **3**:62–63 (Rus.).

Kandel, E. I., and Omorov, T. M., 1984, Vascular factors in etiology of hemihyperkinesias, *Vopr. Neirokhir.* No. 2:19–24 (Rus.).

Kandel, E. I., and Peresedov, V. V., 1975, Stereotaxic clipping of an arterial aneurysm in the brain, *Vopr. Neirokhir.* No. **1**:13–14 (Rus.).

Kandel, E. I., and Peresedov, V. V., 1976, Stereotactic clipping of arterial and arterio-venous aneurysms of the brain, in *Proceedings of II All-Union Congress of Neurosurgeons,* Moscow, pp. 416–418 (Rus.).

Kandel, E. I., and Peresedov, V. V., 1977a, Stereotaxic clipping of arterial aneurysms and arteriovenous malformations, *J. Neurosurg.* **46**:12–23.

Kandel, E. I., and Peresedov, V. V., 1977b, Device for stereotaxic clipping of arterial and arteriovenous aneurysms in the brain, *Vopr. Neirokhir.* No. 2:53–55 (Rus.).

Kandel, E. I., and Peresedov, V. V., 1980, Stereotaxic clipping of cerebral arteriovenous aneurysms, *Vopr. Neirokhir.* No. 3:3–7 (Rus.).

Kandel, E. I., and Peresedov, V. V., 1985, Stereotaxic evacuation of spontaneous intracerebral hematomas, *J. Neurosurg.* **62**:206–213.

Kandel, E. I., and Peresedov, V. V., 1987a, Stereotactic evacuation of the intracerebral hematomas, *Vopr. Neirokhir.* No. 3:16–21 (Rus.).

Kandel, E. I., and Peresedov, V. V., 1987b, Stereotactic clipping of arteriovenous malformation of the brain, *Neurol. Res.* **9**:129–136.

Kandel, E. I., and Plevako, N. S., 1966, Ventriculography with the use of Conray, *Vopr. Neirokhir.* No. 5:42–45 (Rus.).

Kandel, E. I., and Podgornaya, A. Y., 1966, Morphological study of the brain after stereotactic operations for parkinsonism, *Vopr. Neirokhir.* No. 1:28–32 (Rus.).

Kandel, E. I., and Pokrovskaya, Z. A., 1972, Dynamics of the bioelectric activity of the human brain as related to stereotaxic operations on the basal ganglia, *Vopr. Neirokhir.* No. 4:42–46 (Rus.).

Kandel, E. I., and Pukanov, V. S., 1973, On the technique of percutaneous cordotomy, in *Proceedings of Scientific Conference of Neurosurgeons,* Alma-Ata, pp. 365–367 (Rus.).

Kandel, E. I., and Schavinsky, J. V., 1972, Stereotaxic apparatus and operations in Russia in the 19th century, *J. Neurosurg.* **37**:407–411.

Kandel, E. I., and Tsibulnikov, N. D., 1971, Length of the line "posterior edge of foramen Monro–posterior commissure" according to sex, age and degree of hydrocephalus in patients with extrapyramidal hyperkinesias, in *Proceedings of First All-Union Congress of Neurosurgeons,* Vol. 3, Moscow, pp. 209–211 (Rus.).

Kandel, E. I., and Vojtina, S. V., 1968, Erfahrungen mit der chirurgischen Behandlung der hyperkinetischen Form der hepatolentikulären Degeneration, *Česk. Neurol.* **31**:119–125 (Rus.).

Kandel, E. I., and Vojtina, S. V., 1971, *Deforming Muscular (Torsion) Dystonia. Clinical and Surgical Treatment,* Meditsina, Moscow (Rus.).

Kandel, E. I., and Yadgarov, I. S., 1984, Combined surgical and medical treatment of parkinsonism, *J. Neuropathol. Psychiatry* No. 8:1157–1161 (Rus.).

Kandel, E. I., Vichert, T. M., and Lyass, F. M., 1960, Hystopathologic alterations in central nervous system by direct introduction of radioactive gold into the brain, *Arch. Patol.* No. 3:48–54 (Rus.).

Kandel, E. I., Kukin, A. V., Shalnikov, A. I., and Shick, M. L., 1962, Improvement of the technique of local freezing of the subcortical structures in stereotactic brain operations, *Vopr. Neirokhir.* No. 4:51–54 (Rus.).

Kandel, E. I., Gurfinkel, V. S., Kotz, J. M., and Shick, M. L., 1963, Application of tremorography for prognosis of effectiveness of surgical treatment of parkinsonism, *Vopr. Neirokhir.* No. 4:1–6 (Rus.).

Kandel, E. I., Sklyanik, A. J., Atlas, D. V., and Zelman, V. L.,

1970, Stereotactic transnasal cryodestruction of hormone-active hypophysis adenomas, in *Proceedings of I All-Union Conference for Children's Endocrinology,* Ivanovo, pp. 81–84 (Rus.).

Kandel, E. I., Atlas, D. V., and Sklyanik, A. Ya., 1973a, Stereotactic cryodestruction of hypophyseal tumours, *Vopr. Neirokhir.* No. 5:8–13 (Rus.).

Kandel, E. I., Yadgarov, I. S., and Matlina, E. S., 1973b, Influence of L-dopa on urine excretion of catecholamines in parkinsonian patients, in *Clinical Importance of L-Dopa,* Moscow, Basel, pp. 149–158 (Rus.).

Kandel, E. I., Levina, G. J., Markova, E. D., and Vojtina, S. V., 1973c, Morphologic brain alterations in torsion dystonia, in *Approaches in Clinical Neurogenetics,* Moscow, pp. 96–106 (Rus.).

Kandel, E. I., Stolarova, L. G., Yadgarov, I. S., and Kistenev, B. A., 1973d, Results of L-dopa treatment of the parkinsonian patients after stereotactic surgery, in *Clinical Importance of L-Dopa,* Moscow, Basel, pp. 129–134 (Rus.).

Kandel, E.I., Aizerman, M. A., Andreeva, E. A., 1974, Interrelations of resting, postural and action tremor in parkinsonian patients before and after surgery, *Confin. Neurol.* **36:**356–359.

Kandel, E. I., Kozlovskaya, M. B., and Kudinova, M. P., 1980, Influence of cerebellar dentate nucleus destruction on characteristics of voluntary movement control system, *J. Human Physiol.* **3:**464–473 (Rus.).

Kandel, E. I., Vavilov, S. B., Peresedov, V. V., and Saribekyan, A. S., 1981, Stereotaxic biopsy of brain tumours according to the results of computer tomography, *Vopr. Neirokhir.* **4:**3–8 (Rus.).

Kandel, E. I., Andreeva, E. A., Ivanova-Smolenskaja, I. A., and Khutorskaja, O. Y., 1986, The study of tremor pathogenesis by the computerized spectral analysis of envelope EMG, *J. Neuropathol. Psychiatry* No. 7:970–975 (Rus.).

Kandel, E. I., Ogleznev, K. Y., and Dreval, O. S., 1987, Surgical treatment of severe pain after the avulsion of the brachial plexus, *Vopr. Neirokhir.* No. 6:9–15 (Rus.).

Kaneko, M., Koba, T., and Yokoyama, T., 1977, Early surgical treatment for hypertensive intracerebral hemorrage, *J. Neurosurg.* **46:**579–583.

Kanno, T., Sano, H., Shinomiya, J., Katada, K., Nagata, J., Hishino, M., and Mitsuyama, F., 1984, Role of surgery in hypertensive intracerebral hematoma. A comparative study of 305 nonsurgical and 157 surgical cases, *J. Neurosurg.* **61:**1091–1099.

Kaplan, S. D., 1974, Age distribution of patients with Parkinson's disease in 1960 and 1970 in 110 hospitals, *Neurology (Minneap.)* **24:**972–975.

Karlov, V. A., 1978, Generalized epileptic fits by affection of medio-basal region of frontal lobe, *Zh. Nevropatol. Psikhiatr.* **12:**1809–1814 (Rus.).

Kasdon, D. L., and Lathi, E. S., 1984, A prospective study of radio-frequency rhizotomy in the treatment of posttraumatic spasticity, *Neurosurgery* **15:**526–529.

Katayama, Y., Tsubokawa, T., Maejima, S., and Yamamoto, T., 1986, Responses of raphe–spinal neurons to stimulation of the pontine parabrachial region producing behavioral nociceptive suppression in the cat, *Appl. Neurophysiol.* **49:**112–120.

Katz, J., and Levin, A. B., 1977, Treatment of diffuse metastatic cancer pain by instillation of alcohol into the sella turcica, *Anesthesiology* **46:**115–121.

Kawamura, H., Asakura, T., Mihara, T., and Kitamura, K., 1972, A radioanatomical study of the mobility of cervical spinal cord in the canal. An appraisal of percutaneous cervical cordotomy, *Neurol. Med. Chir. (Tokyo)* **12:**354–355.

Keen, W. W., 1891, A new operation for spasmodic wry neck, namely, division or excision of the nerves supplying the posterior rotator muscles of the head, *Ann. Surg.* **13:**44–47.

Kelly, D. L., Goldring, R., and O'Leary, I. L., 1965, Averaged evoked somatosensory responses from cortex of man, *Arch. Neurol.* **13:**1–9.

Kelly, P. J., 1983, Future possibilities in stereotactic neurosurgery, *Surg. Neurol.* **19:**4–9.

Kelly, P. J., 1985, Principles of the functional and anatomical organization of the nervous system, in *Principles of Neural Science* (E. R. Kandel and J. H. Schwartz, eds.), Elsevier, New York, pp. 209–221.

Kelly, P. J., 1987, Computerized guidance for stereotactic treatment of brain tumors: From CT-guided biopsy to computerized resection, in *Stereotactic Surgery* (R. R. Tasker, ed.), Hanley and Belfus, Philadelphia, pp. 165–192.

Kelly, P. J., and Alker, G. J., 1981, A stereotactic approach to deep-seated central nervous system neoplasms using the carbon dioxide laser, *Surg. Neurol.* **15**(5)**:**331–334.

Kelly, P. J., and Gillingham, F. J., 1980, The long-term results of stereotaxic surgery and L-dopa therapy in Parkinson's disease, *J. Neurosurg.* **53:**332–337.

Kelly, P. J., Olson, M. H., and Wright, A. E., 1978, Stereotactic implantation of iridium 192 into CNS neoplasms, *Surg. Neurol.* **10:**349–354.

Kelly, P. J., Alker, G. J., and Zoll, J. G., 1982, A microstereotactic approach to deep-seated arteriovenous malformations, *Surg. Neurol.* **17**(4)**:**260–262.

Kelly, P. J., Kall, B. A., Goerss, S., and Earnest, F., 1985a, Computer assisted stereotactic laser resection of intra-axial tumors: Methodology and clinical experience, in *IX Meeting of the World Society for Stereotactic and Functional Neurosurgery,* Toronto.

Kelly, P. J., Kall, B. A., Goerss, S., and Earnest, F., 1985b, Present and future developments of stereotactic technology, *Appl. Neurophysiol.* **48:**1–6.

Kelly, P. J., Sharbrough, F. W., Kall, B. A., and Goerss, S. J., 1986, MRI-based computer-assisted stereotactic resection of hippocampus and amygdala in patients with temporal lobe epilepsy, in *VII Meeting of European Society for Stereotactic and Functional Neurosurgery,* Birmingham.

Kelly, P. J., Sharbrough, F. W., Kall, B. A., and Goerss, S. J., 1987, Magnetic resonance imaging-based computer-assisted stereotactic resection of the hippocampus and amygdala in patients with temporal lobe epilepsy, *Mayo Clin. Proc.* **62:**103–108.

Kendall, B. E., Pollock, S. S., Bass, N. M., and Valentine, A. R., 1981, Wilson's disease. Clinical correlation with cranial computed tomography, *Neuroradiology* **22:**1–5.

Kepplinger, B., 1983, Neurophysiological assessment and percutaneous replacement of epidural leads in chronic spinal cord stimulation, in *International Symposium of Functional and Stereotactic Neurosurgery: Brain Stimulation and Psychosurgery*, Bratislava, p. 12.

Kerber, C. W., 1980, Flow-controlled therapeutic embolization: A physiological and safe technique, *AJR* **134:**557–561.

Kerr, F. W. L., 1975, Neuroanatomical substrates of nociception in the spinal cord, *Pain* **1:**325–356.

Kessler, I. I., 1978, Parkinson's disease in epidemiologic perspective, in *Neurological Epidemiology* (B. S. Schoenberg, ed.), Raven Press, New York, pp. 355–383.

Ketelaer, P., Swartenbroekx, G., Deltenre, P., Carton, H., Gybels, J., 1979, Percutaneous epidural dorsal cord stimulation in multiple sclerosis, *Acta Neurochir.* **49:**95–101.

Keynes, G., 1952, *The Apologie and Treatise of Ambroise Pare*, University of Chicago Press, Chicago.

Keyserling, H. Van, 1956, Zum familiären Vorkommen der idiopathischen Torsionsdystonie, *Nervenarzt* **27**(1):34–35.

Khilko, V. A., 1966, Artificial thrombosis of arterial aneurysms, in *Actual Problems of Practical Neurosurgery* (B. A. Samotokin, ed.), Meditsina, Leningrad, pp. 97–99 (Rus.).

Khilko, V. A., 1969, Introduction of polystyrol balls to produce embolism in arteriovenous aneurysms, *Vopr. Neirokhir.* **1:**22–26 (Rus.).

Khilko, V. A., 1970, *Intra- and Extracranial Aneurysms and Angiomas*, Thesis, Leningrad (Rus.).

Khilko, V. A., 1974, Immediate and long-term results of brain arteriovenous malformations and fistulas embolization, in *Diagnostics and Surgical Treatment of Cerebral Vascular Diseases*, Meditsina, Leningrad, pp. 155–157 (Rus.).

Khilko, V. A., 1976, Artificial embolism of arteriovenous aneurysms of brain through vertebrobasilar system, *Vopr. Neirokhir.* **3:**15–20 (Rus.).

Kikut, R. P., 1976, Treatment of aneurysms of cerebral arteries with the aid of stereotactically oriented external constant magnetic fields, *Vopr. Neirokhir.* **1:**3–7 (Rus.).

Kikut, R. P., and Kadish, S. A., 1973, Approach to cerebral aneurysm thrombosis with application of magnetic effects, in *Actual Problems of Traumatology and Orthopedics. Proceedings of the Congress Riga*, pp. 167–169 (Rus.).

Kikut, R. P., and Kadish, S. A., 1976, Fixation of thrombi in arterial aneurysms of the brain by means of magnetobiological effects, in *Cerebral Blood Flow and Characteristics of Large Arteries in Normal State and Disease*, Riga, pp. 14–20 (Rus.).

Kim, J. K., and Umbach, W., 1972a, Combined stereotactic lesions for treatment of behaviour disorders and severe pain, in *Surgical Approaches in Psychiatry* (L. W. Laitinen and K. E. Livingston, eds.), Cambridge, pp. 182–188.

Kim, J. K., and Umbach, W., 1972b, Comparative evaluation of different psychosurgical methods, in *Present Limits of Neurosurgery* (I. Fusek and Z. Kunc, eds.), Avicenum Czechoslovak Medical Press, Prague, pp. 465–469.

Kim, J. K., Umbach, W., and Zeytountchian, C., 1970, Gezielte Ventrikeldarstellung bei stereotaktischer Operation, *Dtsch. Med. Wochenschr.* **95:**2211–2214.

King, R. B., 1977, Anterior commissurotomy for intractable pain, *J. Neurosurg.* **47:**7–12.

King, R. B., 1979, Principles of pain management, *J. Neurosurg.* **50:**554–559.

King, R. B., 1980, Pain and tryptophan, *J. Neurosurg.* **53:**44–52.

Kirschner, M., 1932, Electrocoagulation des Ganglion Gasseri, *Zentralbl. Chir.* **47:**2841–2843.

Kirschner, M., 1933, Die Punktionstechnik und die Electrocoagulation des Ganglion Gasseri. Über "gezielte" Operationen, *Arch. Klin. Chir.* **176:**581–620.

Kiwerski, J., 1986, Stimulation of the spinal cord in the treatment of traumatic injuries of cervical spine, *Appl. Neurophysiol.* **49:**166–171.

Kjellberg, R. N., Kochler, A. M., Preston, W. M., and Sweet, W. H., 1962, Stereotaxic instrument for use with the Bragg peak of a proton beam, *Confin. Neurol.* **22:**183–190.

Kjellberg, R. N., Nguyen, N. C., and Kliman, B., 1972, Le Bragg peak protonique en neurochirurgie stéréotaxique, *Neurochirurgie* **18:**225–264.

Klar, E., and Mletzko, J., 1960, Erfahrungen bei 33 Chordotomiens, *Chirurgie* **9:**403–405.

Klemme, R. M., 1940, Surgical treatment of dystonia, paralysis agitans and athetosis, *Arch. Neurol. Psychiatry* **44:**926.

Klemme, R. M., 1942, Surgical treatment of dystonia, with report of 100 cases, *Assoc. Res. Nerv. Ment. Dis.* **21:**596–601.

Klingler, D., and Kepplinger, B., 1982, Quantification of the effect of epidural spinal electrostimulation (ESES) in central motor disorders, *Appl. Neurophysiol.* **45:**221–224.

Klosovski, B. N., and Volzhina, N. S., 1960, Surgical method of simultaneous bilateral complete ablation of thalami optici in dogs, *Sechenov Physiol. J. USSR* **46**(1):117–120 (Rus.).

Klosovski, B. N., Volzhina, N. S., and Vasiljev, G. A., 1959, Approach to physiology of the thalamus, *Vopr. Neirokhir.* **6:**1–6 (Rus.).

Klun, B., 1981, Neuro-vascular relationships in trigeminal neuralgia, *Zentralbl. Neirokhir.* **2/3:**123–128.

Klüver, H., and Bucy, P., 1937, Psychic blindness and other symptoms following bilateral temporal lobectomy in rhesus monkeys, *Am. J. Physiol.* **119:**352–353.

Knight, G. C., 1965, Stereotaxic tractotomy in the surgical treatment of mental illness, *J. Neurol. Neurosurg. Psychiatry* **28:**304–309.

Knight, G., 1972, Neurosurgical aspects of psychosurgery, *Proc. Bog. Soc. Med.* **65:**1099–2004.

Koch, B., Braillier, D., Eng, G., and Binder, H., 1980, Computerized tomography in cerebral-palsied children, *Dev. Med. Child. Neurol.* **22**(5):595–607.

Koella, W. P., 1955, Motor effects from electrical stimulation of basal cerebellum in unrestrained cat, *J. Neurophysiol.* **18:**559–573.

Koeze, T. H., Simpson, B. A., and Watkins, E. S., 1984, Diagnosis and repair of malfunction of implanted central nervous system stimulators, *Appl. Neurophysiol.* **47:**111–116.

Kojevnikoff, A. I., 1895, Eine besondere Form von corticaler Epilepsie, *Zentralbl. Neurol.* **14:**47–48.

Koljubiakin, S. L., 1923, Treatment of cortical epilepsy by injection of alcohol in motor centres, *Arch. Klin. Chir. (Berl.)* **124:**114–119.

Komai, N., 1967, Stereotaktische Operationen zur Behandlung des Nystagmus, der Vestilulären Ataxie und der spastischen Streckmuskeltonuss-teigerung, *Neurochirurgie* **10:**19–34.

Komai, N., Kuriyama, T., Matsumoto, K., Nishina, H., and Kido, T., 1968, The motor responses of the electrical stimulation of the thalamus (oral ventral nucleus), red nucleus, substantia nigra and subthalamic nucleus during stereotaxic operations, *Neurol. Med. Chir.* **10:**313–314.

Konovalov, A. N., 1973, *Surgical Treatment of Arterial Aneurysms of the Brain,* Meditsina, Moscow (Rus.).

Konovalov, A. N., and Kornienko, W. N., 1985, *Computed Tomography in Neurosurgical Clinic,* Meditsina, Moscow (Rus.).

Konovalov, N. V., 1948, *Hepato-Lenticular Degeneration (Pseudosclerosis, Wilson's Disease). Liver and Brain,* Meditsina, Moscow (Rus.).

Konovalov, N. V., 1960, *Hepato-Cerebral Dystrophy,* Meditsina, Moscow (Rus.).

Korein, J., Brudny, J., Grynbaum, B., Sachs-Frankel, B., Weisinger, M., and Levidov, I., 1976, Sensory feedback therapy of spasmodic torticollis and dystonia. Results in treatment of 55 patients, *Adv. Neurol.* **14:**376–402.

Kosary, I. Z., Shacked, I., and Farine, I., 1977, Use of surgical laser in the removal of an osteoma of the skull, *Surg. Neurol.* **8:**151–154.

Koshino, K., Nakano, M., Miki, M., and Matsumure, H., 1975, Stereotaxic operations for the control of infantile epilepsy and associated behavioral disorder, *Confin. Neurol.* **37**(1–3):223–231.

Koslow, M., Abele, M. G., and Griffith, R. C., 1981, Stereotactic surgical system controlled by computer tomography, *Neurosurgery* **8:**72–82.

Koslowa, G. P., 1982, Neurosurgical aspects of the topography of the cerebellar nuclei, Thesis, Leningrad.

Kottke, F. Y., 1970, Modification of athetosis by denervation of the tonic neck reflexes, *Dev. Med. Child. Neurol.* **14:**236–237.

Koulousakis, A., and Nittner, K., 1982, Bilateral C_{1-2} cordotomies. (Can complications be avoided?), *Appl. Neurophysiol.* **45:**500–503.

Krainick, J. U., Thoden, V., and Riechert, T., 1975, Spinal cord stimulation in postamputation pain, *Surg. Neurol.* **4:**167–170.

Krainick, J. U., Lazorthes, J., Probst, C., Siegfried, J., Steude, U., Thoden, U., and Winkelmüller, W., 1979, Long-term follow-up of dorsal cord stimulation for pain in some European clinics, *Dolore* **1:**91–95.

Krainick, J. U., Thoden, V., and Riechert, T., 1980, Pain reduction in amputees by long-term spinal cord stimulation. Long-term follow-up study over 5 years, *J. Neurosurg.* **52**(3):346–350.

Krause, F., 1892, Resection des Trigeminus innerhalb der Schädelhöhle, *Arch. Klin. Chir.* **44:**821–832.

Krayenbühl, H., and Siegfried, J., 1969, Stereotaxic surgery of the dentate nucleus in the treatment of hyperkinetic and spastic conditions, *Dev. Med. Child. Neurol.* **11:**684–685.

Krayenbühl, H., and Siegfried, J., 1972, Dentatotomies or thalamotomies in the treatment of hyperkinesia, *Confin. Neurol.* **34:**29–33.

Krayenbühl, H., Akert, K., Hartmann, K., and Jasargil, M. G., 1964, Etude de la corrélation anatomoclinique chez des malades opérés de parkinsonism, *Neurochirurgie* **10:**397–412.

Kritcheff, I., Madayag, M., and Braunstein, P., 1972, Transfemoral catheter embolization of cerebral and posterior fossa arterivenous malformations, *Radiology* **103:**107–111.

Krynauw, R., 1950, Infantile hemiplegia treated by removing one cerebral hemisphere, *Neurol. Neurosurg. Psychiatry* **13:**243–247.

Kryzhanovsky, G. N., 1976a, Pathogenesis of pain and itch central mechanisms (theory of generator mechanisms), *Zh. Nevropatol. Psikhiatr.* **76:**1090–1098 (Rus.).

Kryzhanovsky, G. N., 1976b, The experimental central pain syndromes: Modeling and general theory, in *Advances in Pain Research and Therapy,* Vol. 1 (J. J. Bonica and D. Albe-Fessard, eds.), Raven Press, New York, pp. 225–240.

Kryzhanovsky, G. N., 1979, Analgesia induced by generator of excitation in midbrain gray matter, in *Advances in Pain Research and Therapy,* Vol. 3 (J. J. Bonica, ed.), Raven Press, New York, pp. 473–482.

Kryzhanovsky, G. N., 1980, *Determinant Structures in Pathophysiology of the Nervous System,* Meditsina, Moscow (Rus.).

Kryzhanovsky, G. N., and Aliev, M. N., 1978, Experimental parkinsonian syndrome, in *Pathogenesis, Clinic and Treatment of Parkinsonism,* Meditsina, Moscow, pp. 26–29 (Rus.).

Kryzhanovsky, G. N., Igonkina, S. I., Grafova, V. N., and Danilova, E. I., 1974a, Experimental trigeminal neuralgia (for conception of the pain's generating mechanism), *Bull. Exp. Biol. Med.* **11:**16–23 (Rus.).

Kryzhanovsky, G. N., Grafova, V. N., Danilova, E. I., and Igonkina, S. I., 1974b, Investigation of pain of the spinal origin (for conception of the pain's generating mechanism), *Bull. Exp. Biol. Med.* **7:**15–23 (Rus.).

Kudinova, M. P., Artemjeva, E. I., and Roslyakova, O. E., 1976, Electromyographic analysis of stereotactic dentatotomy for cerebral palsy, in *Proceedings of II All-Union Symposium on Clinical Electromyography,* Tbilisi, pp. 85–86 (Rus.).

Kuhler, A., 1976, La cordotomie cervicale percutanée, *Neurochirurgie* **22:**261–270.

Kühner, A., 1976, La valeur des interventions sur les racines sacrées dans le traitement des syndromes douloureux du bassin, *Neurochirurgie* **22:**429–443.

Kühner, A., 1979, Möglichkeiten der neurochirurgischen Schmerzbehandlung, *Med. Welt* **30**(20):789–794.

Kukujev, L. A., 1947, Development of striopallidum in onto- and phylogenesis, *Zh. Nevropatol. Psikhiatr.* **16:**38–44 (Rus.).

Kullberg, G., 1975, A clinical study of acute confusion occuring in connection with VL thalamotomy, *Confin. Neurol.* **37**:167–171.

Kullberg, G., 1977, Differences in effect of capsulotomy and cingulotomy, in *Neurosurgical Treatment in Psychiatry, Pain and Epilepsy* (W. H. Sweet, S. Obrador, and J. G. Martin-Rodriguez, eds.), University Park Press, Baltimore, pp. 301–308.

Kunc, Z., 1965, Treatment of essential neuralgia of the 9th nerve by selective tractotomy, *J. Neurosurg.* **23**(5):494–500.

Kunkel, R. S., and Dohn, D. F., 1974, Surgical treatment of chronic migrainous neuralgia, *Cleveland Clin. Q.* **41**(4):189–192.

Kunze, S., and Steiner, H. H., 1987, Trigeminusneuralgie. Ergebnisse der mikrochirurgischen parapontinen Decompression, *Nervenarzt* **58**:33–39.

Kurepina, M. A., 1944, *Interrelation of Structural Formations of the Thalamus Opticus in Onto- and Phylogenesis,* Thesis, Moscow (Rus.).

Kurland, L. T., 1958, Epidemiology: Incidence, geographic distribution and genetic consideration, in *Pathogenesis and Treatment of Parkinsonism,* Charles C. Thomas, Springfield, IL, pp. 5–44.

Kurland, L. T., Hauser, W. A., Okazaki, H., and Nobrega, F. T., 1969, Epidemiologic studies of parkinsonism with special reference to the cohort hypothesis, in *Third Symposium on Parkinson's Disease,* E. S. Livingstone, Edinburgh, pp. 12–16.

Kuroda, R., Takimoto, Y., Taneda, M., Kanai, N., Ioku, T., Mogami, H., Kawai, R., Kuru, Y., and Shigematsu, Y., 1972, Experience of stereotaxic radio-Gasserian-gangliotomy by linac irradiation and its basic experiment, *Neurol. Med. Chir.* **12**:374–375.

Kuroda, R., Nakatani, J., Ioku, M., and Koshino, K., 1985, Clinicoanatomical study of the thalamic stimulation for pain relief, *Appl. Neurophysiol.* **48**:181–190.

Kusske, J. A., and Kelly, W. A., 1972, Embolization and reduction of the "steal" syndrome in cerebral arteriovenous malformations, *J. Neurosurg.* **40**:313–320.

Ladurner, G., Sager, W. D., Dusik, B., and Lechner, H., 1979, Die Bedeutung der Computertomographie in der Diagnose der Epilepsien, *Festschr. Neurol. Psychiatrie* **47**:264–268.

La Fia, D. J., 1969, Hemiballismus as a complication of thalamotomy: Report of two cases, *Confin. Neurol.* **31**:42–47.

Lagarrigue, J., Lazorthes, Y., Verdie, J. C., Alwan, A., Sarramon, J. P., and Rossignol, G., 1979, Thermocoagulation percutanée des racines sacrées dans la traitement des neurovessies spastiques, *Neurochirurgie* **25**:91–95.

Laha, R., and Jannetta, P. J., 1977, Glossopharyngeal neuralgia, *J. Neurosurg.* **46**:316–320.

Laine, E., 1976, Indications et limites du traitment neurochirurgical des hémorragies cérébrales, *J. Sci. Med. Lille* **94**:85–102.

Laitinen, L., 1963, Stereotactic treatment of spasmodic torticollis, *Acta Neurol. Scand.* **39**:231–236.

Laitinen, L., 1966, Thalamic targets in the stereotoxic treatment of Parkinson's disease, *J. Neurosurg.* **24**:82–85.

Laitinen, L. V., 1967, Thalamotomy in progressive myoclonus epilepsy, *Acta Neurol. Scand. [Suppl.]* **31**:170–171.

Laitinen, L., 1971, A new stereoencephalotome, *Zbl. Neurochir.* **32**:67–73.

Laitinen, L., 1972, Surgical treatment, past and present in Parkinson's disease, in *Proceedings of the XII Congress of Scandinavian Neurologists,* Oslo, pp. 43–58.

Laitinen, L., 1974, Differential effects of various psychosurgical approaches, in *Recent Progress in Neurological Surgery,* Excerpta Medica, Amsterdam, pp. 256–260.

Laitinen, L., 1976, Placement des electrodes dans la stimulation transcutanée de la douleur chronique, *Neurochirurgie* **22**:517–526.

Laitinen, L. V., 1977, Anterior pulvinotomy in the treatment of intractable pain, in *Neurosurgical Treatment in Psychiatry, Pain and Epilepsy* (W. H. Sweet, S. Obrador, and J. G. Martin-Rodriguez, eds.), University Park Press, Baltimore, pp. 669–672.

Laitinen, L. V., 1984, Trigeminus stereoguide: An instrument for stereotactic approach through the foramen ovale and foramen jugulare, *Surg. Neurol.* **22**:519–523.

Laitinen, L., 1985a, Brain targets in surgery of Parkinson's disease, *J. Neurosurg.* **62**:349–351.

Laitinen, L. V., 1985b, CT-guided ablative stereotaxis without ventriculography, *Appl. Neurophysiol.* **48**:1–6, 18–21.

Laitinen, L. V., and Fugl-Meyer, A. R., 1982, Assessment of functional effect of epidural electrostimulation and selective posterior rhizotomy in spasticity, *Appl. Neurophysiol.* **45**:331–334.

Laitinen, L. V., and Ohno, J., 1970, Effect of thalamic stimulation and thalamotomy on the H-reflex, *Electroencephalogr. Clin. Neurophysiol.* **28**:586–591.

Laitinen, L., and Singounas, E., 1971, Longitudinal myelotomy in the treatment of spasticity of the legs, *J. Neurosurg.* **35**:536–540.

Laitinen, L., Toivakka, E., 1972, Locating brain tumors through depth EEG probes, *Confin. Neurol.* **34**:101–105.

Laitinen, L., and Toivakka, E., 1979, Depth EEG and electrical stimulation of human amygdala, in *IV Meeting of the European Society for Stereotactic and Functional Neurosurgery,* Paris.

Laitinen, L. V., and Vilkki, J., 1973, Observations on the transcallosal emotional connections, in *Surgical Approaches in Psychiatry* (L. V. Laitinen and K. E. Livingston, eds.), University Park Press, Baltimore, pp. 74–80.

Laitinen, L., Johansson, G. G., and Sipponen, P., 1966, Impedance and phase angle as a locating method in human stereotaxic surgery, *J. Neurosurg.* **25**:628–633.

Laitinen, L., Toivakka, E., Vilkki, U., 1973, Rostral cingulotomy for mental disturbances, *Vopr. Neirokhir.* **1**:23–30 (Rus.).

Laitinen, L., Arsalo, A., and Hänninen, A., 1974, Combination of thalamotomy and longitudinal myelotomy in the treatment of multiple sclerosis, in *Advances in Stereotactic and Functional Neurosurgery* (F. J. Gillingham, E. R. Hitchcock, and J. W. Turner, eds.), Springer, Vienna, New York, pp. 89–92.

Laitinen, L. V., Nilsson, S.. Fugl-Meyer, A. R., 1983, Selective posterior rhizotomy for treatment of spasticity, *J. Neurosurg.* **58**:895–899.

Laitinen, L. V., Liliequist, B., Fagerlund, M., and Eriksson, A. T., 1985, CT guided stereotaxis without ventriculography, *Appl. Neurophysiol.* **48:**18–21.

Lakke, W. F., 1977, Parkinson's disease: Concepts, in *Parkinson's Disease Concepts and Prospects. Symposium, the University of Groningen* (W. F. Lakke, J. Korf, and H. Wesseling, eds.), Excerpta Medica, Amsterdam, pp. 1–7.

Lakke, W. F., de Jong, P. I., Koppejan, T., and van Weerden, T. W., 1980, Observations on postural behavior: Axial rotation in recumbent position in parkinsonian patients after L-dopa treatment, in *Parkinson's Disease. Current Progress, Problems and Management* (U. K. Rinne, M. Klinger, and G. Stamm, eds.), Elsevier/North-Holland Biomedical Press, Amsterdam, pp. 187–197.

Lamotte, C., Pert, C. B., and Snyder, S. H., 1976, Opiate receptor binding in primate spinal cord: Distribution and changes after dorsal root section, *Brain Res.* **112:**407–412.

Lance, J. W., Schwab, R. S., and Peterson, E. A., 1963, Action tremor and the cogwheel phenomenon in Parkinson's disease, *Brain* **86**(Part I):95–110.

Landau, W. M., and Clare, M. H., 1964, Fusimotor function. Part IV. H-reflex, tendon jerk and reinforcement in hemiplegia, *Arch. Neurol.* **10:**128–133.

Landolt, A. M., 1978, Progress in pituitary adenoma biology, in *Advances and Technical Standards in Neurosurgery,* Vol. 4, Springer, New York, pp. 1–50.

Landolt, A. M., and Siegfried, J., 1970, Zur Behandlung maligner, metastasierender Tumoren mit der stereotaktischen transsphenoidalen Elektrokoagulation der Hypophyse, *Schweiz. Med. Wochenschr.* **100:**1297–1306.

Laponogov, O. A., 1965, Dynamics of parkinsonian syndrome after stereotactic operations, *Vrach. Delo* **11:**8–14 (Rus.).

Laponogov, O. A., 1969, Immediate results of stereotaxic operations in the treatment of extrapyramidal hyperkinesias, *Vopr. Neirokhir.* **1:**12–16 (Rus.).

Laponogov, O. A., Tsimbaljuk, V. I., and Matjuk, N. G., 1978, Long-term results of surgical treatment of parkinsonism, in *Pathogenesis, Clinic and Treatment of Parkinsonism,* Meditsina, Moscow, pp. 220–222 (Rus.).

Larson, S. J., and Sances, A., 1966, Evoked potentials in man. Neurosurgical applications, *Am. J. Surg.* **3:**857–861.

Larson, S. J., and Sances, A., 1968, Averaged evoked potentials in stereotaxic surgery, *J. Neurosurg.* **28:**227–233.

Larson, S. J., Sances, A., Cusick, J. F., Meyer, G. A., and Swiontek, T. A., 1975, Comparison between anterior and posterior spinal implant systems, *Surg. Neurol.* **4:**180–186.

Larsson, T., and Sjögren, T., 1966, Dystonia musculorum deformans. A genetic and clinical population study of 121 cases, *Acta Neurol. Scand.* Suppl. 17.

Lasjaunias, P., Manelfe, C., Terbrugge, K., and Ibor, L. L., 1986, Endovascular treatment of cerebral arteriovenous malformations, *Neurosurg. Rev.* **9:**265–275.

Latchaw, J. P., Jr., Hardy, R. W., Jr., and Forsythe, S. B., 1983, Trigeminal neuralgia treated by radiofrequency coagulation, *J. Neurosurg.* **59:**479–484.

Laursen, A. M., 1963, Physiology of the corpus striatum, *Acta Neurol. Scand. [Suppl.]* **39:**61–83.

Law, J. D., and Cocak, R. E., 1982, Progress in developing a CT-compatible adapter for the Leksell stereotach, *Appl. Neurophysiol* **45:**381–382.

Law, J. D., and Miller, L. V., 1982, Importance and documentation of an epidural stimulating position, *Appl. Neurophysiol.* **45:**461–464.

Law, J. D., Sweet, J., and Kirsch, W. M., 1980, Retrospective analysis of 22 patients with chronic pain treated by peripheral nerve stimulation, *J. Neurosurg.* **52:**482–485.

Lawden, M., 1986, Gilles de la Tourette syndrome: A review, *J. R. Soc. Med.* **79:**282–288.

Laws, E. R., Jr., Niedermeyer, E., and Walker, A. E., 1970, Diagnostic significance of scalp and depth EEG findings in patients with temporal and frontal lobe epilepsy, *Johns Hopkins Med. J.* **126:**146–153.

Lazorthes, G., 1959, Surgery of cerebral hemorrhage. Report on the results of 52 surgically treated cases, *J. Neurosurg.* **16:**355–364.

Lazorthes, G., 1981, Electrical spinal cord stimulation for spastic motor disorders, in *7th International Congress of Neurological Surgery,* Toronto, p. 137.

Lazorthes, G., and Salamon, G., 1971, The arteries of the thalamus: An anatomical and radiological study, *J. Neurosurg.* **34:**23–26.

Lazorthes, Y., and Verdie, J.-C., 1979, Radiofrequency coagulation of the petrous ganglion in glosso-pharyngeal neuralgia, *Neurosurgery* **4:**512–516.

Lazorthes, Y., Verdie, J.-C., and Bouyssou, M., 1976a, Interêt d'un cadre "stéréotaxique" dans la thermocoagulation sélective du ganglion de Gasser, *Neurochirurgie* **22:**77–83.

Lazorthes, Y., Verdie, J.-C., and Lagerrigue, J., 1976b, Thermocoagulation percutanée nerfs rachidiens à visée anasique, *Neurochirurgie* **25:**445–453.

Lazorthes, Y., Verdie, Y., and Arbus, L., 1978, Stimulation analgesique médullaire antérière et postérieure par technique d'implantation percutanée, *Acta Neurochir.* **40:**253–276.

Lazorthes, Y., Gonarderes, C., Verdie, J. C., Montsarrat, B., Bastide, R., Campa, L., and Gros, J., 1980, Analgesie par injection intrathecale de morphine. Etude pharmacocinetique et application aux douleurs irreductibles, *Neurochirurgie* **26:**159–164.

Lazorthes, Y., Siegfried, J., and Broggi, G., 1981, Electrical spinal cord stimulation for spastic motor disorders in demyelinating diseases—a cooperative study, in *Indications for Spinal Cord Stimulation* (Y. Hosobuchi and T. Corbin, eds.), Excerpta Medica, Amsterdam, pp. 48–57.

Lazorthes, Y., Bastide, R., Verdie, J., and Lavados, A., 1985, Spinal versus ventricular intrathecal opiate administration for cancer pain, *Appl. Neurophysiol.* **48:**234–241.

Leavens, M. E., Hui, C. S., Cech, D. H., Weyland, J. B., and Weston, J. S., 1982, Intrathecal and intraventricular morphine for pain in cancer patients: Initial study, *J. Neurosurg.* **56:**241–245.

Le Beau, J., 1954, Anterior cingulectomy in man, *J. Neurosurg.* **11:**268–276.

Le Beau, J., and Dondey, M., 1964, Premières observations humaines de reparage de structures cerébrales profondes par refroidissement localise et réversible au cours des interventions stéréotaxiques, *Neurochirurgie* **7:**24–33.

Lee, K. H., Chung, J. M., and Willis, W. D., 1985, Inhibition of primate spinothalamic tract cells by TENS, *J. Neurosurg.* **62**:276–287.

Lee, S. H., Vilafana, T., and Lapayowker, M. S., 1978, CT intracranial localization with a new marker system, *Neuroradiology* **16**:570–571.

Leenders, K. L., Palmer, A. J., Quinn, N., Clark, J. C., Firnau, G., Garnett, E. S., Nahmias, C., Jones, T., and Marsden, C. D., 1986, Brain dopamine metabolism in patients with Parkinson's disease measured with positron emission tomography, *J. Neurol. Neurosurg. Psychiatry* **49**:853–860.

Lees, A. J., Haddad, S., Shaw, K. M., Kohout, L. J., and Stern, G. M., 1978, Bromocriptine in parkinsonism. A long-term study, *Arch. Neurol.* **35**:503–505.

Lehman, R. M., 1972, Related subthalamic structures, *Confin. Neurol.* **34**:200–209.

Leksell, D. G., 1987, Stereotactic radiosurgery, *Neurol. Res.* **9**:60–68.

Leksell, L., 1949, A stereotaxic apparatus for intracerebral surgery, *Acta Chir. Scand.* **99**:229–233.

Leksell, L., 1971a, *Stereotaxis and Radiosurgery: An Operative System*, Charles C. Thomas, Springfield, IL.

Leksell, L., 1971b, Stereotaxic radiosurgery in trigeminal neuralgia, *Acta Chir. Scand.* **137**:311–314.

Leksell, L., and Jernberg, B., 1980, Stereotaxis and tomography. A technical note, *Acta Neurochir.* **52**:1–7.

Leksell, L., Meyerson, B. A., and Forster, D. M. C., 1972, Radiosurgical thalamotomy for intractable pain, *Confin. Neurol.* **34**:275–278.

Leksell, L., Leksell, D., and Schwebel, J., 1985, Stereotaxis and nuclear magnetic resonance, *J. Neurol. Neurosurg. Psychiatry* **48**:14–18.

Lema, J. A., and Hitchcock, E., 1986, Respiratory changes after stereotactic high cervical cord lesions for pain, *Appl. Neurophysiol.* **49**:62–68.

Lenz, F. A., Tasker, R. R., Kwan, H. C., and Murphy, J. T., 1985, Cross-correlation analysis of thalamic neurons and EMG activity in parkinsonian and cerebellar tremor, *Appl. Neurophysiol.* **48**:301–308.

Lenz, F. A., Schneider, S., Tasker, R. R., Kwong, R., Kwan, H. C., Dostrovsky, J. O., and Murphy, J. T., 1986, Application of closed loop system identification techniques to detect feedback in the activity of thalamic "tremor cells" recorded in parkinsonian tremor, in *VII Meeting of European Society for Stereotactic and Functional Neurosurgery*, Birmingham.

Lenzi, A., Galli, G., Gandolfini, M., and Marini, G., 1985, Intraventricular morphine in paraneoplastic painful syndrome of the cervicofacial region: Experience in thirty-eight cases, *Neurosurgery* **17**:6–11.

Leriche, R., 1912, Über chirurgischen Eingriff bei der Parkinsonschen Krankheit, *Neurol. Zentralbl.* **13**:1093–1096.

Levin, A. B., 1985, Experience in the first 100 patients undergoing computerized tomography-guided stereotactic procedures utilizing the Brown–Roberts–Wells guidance system, *Appl. Neurophysiol.* **48**:45–49.

Levin, A. B., and Cosman, E. R., 1980, Thermocouple-monitored cordotomy electrode. Technical note, *J. Neurosurg.* **53**:266–268.

Levin, A. B., Katz, J., and Benson, R. C., 1980, Treatment of pain of diffuse metastatic cancer by stereotactic chemical hypophysectomy: Long-term results and observations on mechanism of action, *Neurosurgery* **6**:258–262.

Levy, A., 1967, Stereotactic brain operations in Parkinson's syndrome and related motor disturbances. Comparison of lesions in the pallidum and thalamus with those in the internal capsule, *Confin. Neurol.* Suppl. 29.

Levy, F. M., 1913, Zur pathologische Anatomie des Paralysis agitans, *Dtsch. Z. Nervenheilkd.* **50**:50–55.

Levy, G., 1922, *Contribution a l'Étude des Manifestations Tardives de l'Encéphalite Épidemique*, Paris,

Levy, R. M., Lambs, S., and Adams, J. E., 1984, Deep brain stimulation for chronic pain: Long-term follow-up in 145 patients from 1972–1984, *Pain* (Suppl.) **2**:S115.

Levy, W. J., Nutkiewcz, A., Ditmore, M., and Watts, C., 1983, Laser-induced dorsal root entry zone lesions for pain control. Report of three cases, *J. Neurosurg.* **59**:884–886.

Lewis, P. D., 1971, Parkinsonism—neuropathology, *Br. Med. J.* **3**:690–693.

Le Witt, P. A., Ward, C. D., Larsen, T. A., Raphaelson, M. I., Newman, R. P., Foster, N., Dambrosia, J. M., and Calne, D. B., 1983, Comparison of pergolide and bromocriptine therapy in parkinsonism, *Neurology (N.Y.)* **33**:1009–1014.

Liberson, W. T., Voris, H. C., and Uematsu, S., 1970, Recording of somatosensory evoked potentials during mesencephalotomy for intractable pain, *Confin. Neurol.* **32**:185–194.

Liddell, E. G. T., and Sherrington, C. S., 1924, Reflexes in response to stretch (myotatic reflexes), *Proc. R. Soc. Med.* **96**:212–242.

Lieberman, A. N., 1974, Parkinson's disease: A clinical review, *Am. J. Med. Sci.* **267**:66–80.

Lieberman, A., Miyamoto, T., Battista, A. F., and Goldstein, M., 1975, Studies on the antiparkinsonian efficacy of lergotrile, *Neurology (Minneap.)* **25**:459–462.

Lieberman, A., 1976a, Treatment of Parkinson's disease with bromocriptine, *N. Engl. J. Med.* **295**:1402–1404.

Lieberman, A., Kupersmith, M., Estey, E., and Goldstein, M., 1976b, Modification on the on–off effect with bromocriptine and lergotrile, *N. Engl. J. Med.* **295**:1400–1401.

Lieberman, A., Lolfghari, M., Boal, D., Hassouri, H., Vogel, B., Battista, A., Fuxe, K., and Goldstein, M., 1976c, the Antiparkinsonism efficacy of bromocriptine, *Neurology (Minneap.)* **26**:405–409.

Lieberman, A., Estey, E., Gopinathan, G., Ohashi, T., Sauter, A., and Goldstein, M., 1978, Comparative effectiveness of two extracerebral dopa decarboxylase inhibitors in Parkinson's disease, *Neurology (Minneap.)* **28**(1):964–968.

Lieberman, A. N., Kupersmith, M., and Gopinathan, G., 1979, Bromocriptine in Parkinson disease, *Neurology (Minneap.)* **29**:363–369.

Light, J. R., 1985, Bladder dysfunction: Restorative procedures, in *Recent Achievements in Restorative Neurology* (J. Eccles and M. R. Dimitrijevic, eds.), Karger, Basel, pp. 128–135.

Lin, P. M., Gildenberg, P. L., and Polakoff, P. P., 1966, An anterior approach to percutaneous lower cervical cordotomy, *J. Neurosurg.* **25:**553–560.

Lindblom, V., and Meyerson, B. A., 1975, Influence on touch, vibration and cutaneous pain of dorsal column stimulation in man, *Pain* **1:**257–270.

Lindblom, V., and Tegner, R., 1979, Are the endorphins active in clinical pain states? Narcotic antagonism in chronic pain patients, *Pain* **7:**65–68.

Lippert, R. G., Hosobuchi, Y., and Nielsen, S. L., 1974, Spinal commissurotomy, *Surg. Neurol.* **2:**373–378.

Lipton, S., Dervin, E., and Heywood, O. B., 1974, A stereotactic approach to the anterior percutaneous electrical cordotomy, in *Advances in Stereotactic and Functional Neurosurgery* (F. J. Gillingham, E. R. Hitchcock, and J. W. Turner, eds.), Springer-Verlag, Vienna, pp. 125–136.

Lipton, S., Miles, J., Williams, N., and Bark-Jones, N., 1978, Pituitary injection of alcohol for widespread cancer pain, *Pain* **5,** 73–82.

Little, W. J., 1961, On the influence of abnormal parturition, difficult labour, premature birth, and asphyxia neonatorum, on the mental and physical condition of the child, especially in relation to deformities, *Trans. Obstet. Soc. Lond.* **3:**293–299.

Livingston, R. E., 1953, Cingulate cortex isolation for the treatment of psychosis and psychoneurosis, *Assoc. Res. Nerv. Ment. Dis.* **74:**374–378.

Livingston, R. E., 1975, Surgical contributions to psychiatric treatment, in *American Handbook of Psychiatry,* Vol. 5, Basic Books, New York, pp. 548–563.

Livshitz, A. V., Volkov, G. M., and Helfand, V. B., 1976, Clinico–electrophysiologic study of the spastic syndrome and its neurosurgical treatment in patients with spinal cord lesions, *Vopr. Neirokhir.* **5:**36–43 (Rus.).

Livshitz, L. J., 1973, A method of electrophysiologic control of the transcutaneous approach to the trigeminal root and device for the nerve stimulation during operations for trigeminal neuralgia, *Vopr. Neirokhir.* **2:**23–25 (Rus.).

Llewellyn, R. C., and Heath, R. G., 1962, A surgical technique for chronic electrode implantation in humans, *Confin. Neurol.* **22:**223–227.

Lobato, R. D., Rivas, J. J., Cabello, A., and Roger, R., 1982, Stereotactic biopsy of brain lesions visualized with computed tomography, *Appl. Neurophysiol.* **45:**426–430.

Lobo-Antunes, J., Yahr, M. D., and Hilal, S. K., 1974, Extrapyramidal dysfunction with cerebral arteriovenous malformations, *J. Neurol. Neurosurg. Psychiatry* **37:**259–265.

Locke, S., 1960, The projection of the medial pulvinar of the macaque, *J. Comp. Neurol.* **115:**155–170.

Loeser, J. D., 1972, Dorsal rhizotomy for the relief of chronic pain, *J. Neurosurg.* **36:**745–750.

Loeser, J. D., Ward, A. A., Jr., and White, L. E., Jr., 1968, Chronic deafferentation of human spinal cord neurons, *J. Neurosurg.* **29:**48–50.

Loeser, J. D., Black, R. G., and Christman, A., 1975, Relief of pain by transcutaneous stimulation, *J. Neurosurg.* **42:**308–314.

Logue, V., 1956, Surgery in spontaneous subarachnoid hemorrhage. Operative treatment of aneurysms on the anterior cerebral and anterior communicating artery, *Br. Med. J.* **1:**473–479.

Logue, V., Durward, M., and Pratt, R. T., 1968, The quality of survival after rupture of an anterior cerebral aneurysm, *Br. J. Psychiatry* **114:**137–160.

Long, D. M., 1973, Electrical stimulation for relief of pain from chronic nerve injury, *J. Neurosurg.* **39:**718–722.

Long, D. M., 1974, Cutaneous afferent stimulation for relief of chronic pain, *Clin. Neurosurg.* **21:**257–268.

Long, D. M., 1980, Surgical therapy of chronic pain, *Neurosurgery* **6:**317–328.

Long, D. M., and Erickson, D. E., 1975, Stimulation of the posterior columns of the spinal cord for relief of intractable pain, *Surg. Neurol.* **4:**134–141.

Long, D. M., Erickson, D., Campbell, J., and North, R., 1981, Electrical stimulation of the spinal cord and peripheral nerves for pain control, *Appl. Neurophysiol.* **44:**207–217.

Lopez-Ibor, J. J., and Lopez-Ibor, A., 1977, Selection criteria for patients who should undergo psychiatric surgery, in *Neurosurgical Treatment in Psychiatry, Pain and Epilepsy* (W. H. Sweet, S. Obrador, and J. G. Martin-Rodriguez, eds.), University Park Press, Baltimore, pp. 151–162.

Lorente de No, R., 1934, Studies on the structure of the cerebral cortex. The area entorhinalis, *J. Psychol. Neurol.* **45:**381–438.

Lorenz, A., 1897, Über die chirurgische Behandlung der angeborenen spastischen Gliederstarre, *Wien Klin. Rundschau* **11:**345,364,378,401,421,454.

Lorenz, R., 1976, Methods of percutaneous spino-thalamic trace section, *Adv. Tech. Stand. Neurosurg.* **3:**123–145.

Losina-Losinsky, L. K., 1972, *Essays on Cryobiology,* Meditsina, Leningrad (Rus.).

Lowenthal, M., and Horsley, V., 1897, On the relation between the cerebellar and other centers (namely cerebral and spinal) with special reference to the action of antagonistic muscles, *Proc. R. Soc.* **61:**20–25.

Luczywek, E., and Mempel, E., 1980, Memory and learning in epileptic patients treated by amygdalotomy and anterior hippocampotomy, *Acta Neurochir. [Suppl.]* **30:**169–175.

Luessenhop, A. J., and Mujica, P. H., 1981, Embolization of segments of the circle of Willis and adjacent branches for management of certain inoperable cerebral arterio–venous malformations, *J. Neurosurg.* **54:**573–582.

Luessenhop, A., and Prosper, J., 1975, Surgical embolization of cerebral arteriovenous malformations through internal carotid and vertebral arteries, *J. Neurosurg.* **42:**443–451.

Luessenhop, A., and Spence, W., 1960, Artificial embolization of the cerebral arteries for the treatment of arteriovenous malformations, *J.A.M.A.* **172:**1153–1155.

Lunsford, L. D., 1982, A dedicated CT system for the stereotactic operating room, *Appl. Neurophysiol.* **45:**374–378.

Lunsford, L. D., 1983, Advanced intraoperative imaging for stereotaxis—the surgical CT scanner, in *VI Meeting of the European Society for Stereotactic and Functional Neurosurgery,* Rome, p. T.24.

Lunsford, L. D., and Bennett, M. H., 1984, Percutaneous retrogasserian glycerol rhizotomy for tic douloureux. Part I. Technique and results in 112 patients, *Neurosurgery* **14**:424–430.

Lunsford, L. D., and Martinez, A. J., 1984, Stereotactic exploration of the brain in the era of computed tomography, *Surg. Neurol.* **22**:222–230.

Lunsford, L. D., Leksell, L., and Jernberg, B., 1983, Probe holder for stereotactic surgery in the CT scanner. A technical note, *Acta Neurochir. (Wien)* **69**:297–304.

Lunsford, L. D., Deutsch, M., and Yoder, V., 1985a, Stereotactic interstitial brachytherapy. Current concepts and concerns in twenty patients, *Appl. Neurophysiol.* **48**:1–6, 117–120.

Lunsford, L. D., Gumerman, L., and Levine, G., 1985b, Stereotactic intracavitary irradiation of cystic neoplasm of the brain, *Appl. Neurophysiol.* **48**:146–150.

Lunsford, L. D., Martinez, A. J., and Latchaw, R. E., 1986, Stereotaxic surgery with a magnetic resonance and computerized tomography-compatible system, *J. Neurosurg.* **64**:872–878.

Lunsford, L. D., Listerud, J. A., Rowberg, A. H., and Latchaw, R. E., 1987, Stereotactic software for the GE 8800 CT scanner, *Neurol. Res.* **9**:118–122.

Luyet, B., 1960, On various phase transitions occurring in aqueous solutions at low temperatures, *Ann. N.Y. Acad. Sci.* **85**:549–569.

Lwoff, M. M., Cornil, L., and Targowla, R., 1922, Spasme de torsion (dystonie lenticulaire) d'origine infectieuse, *Rev. Neurol.* **38**:1429–1434.

Lyass, F. M., 1971, *Intratumoral β-Therapy by Means of Yttrium-90 in Neurosurgical Clinic,* Thesis, Moscow (Rus.).

Lyass, F. M., Kandel, E. I., and Kadin, A. L., 1963, Application of stereotactic technique of yttrium-90 introduction for treatment of basal brain tumors, in *Proceedings of All-Union Conference of Neurosurgeons,* Meditsina, Moscow, pp. 431–434 (Rus.).

Lysachev, A. T., Serbinenko, F. A., and Lyass, F. M., 1982, Reformation of hemodynamics after occlusion of arteriovenous malformations of afferent vessels, *Vopr. Neirokhir.* **2**:3–9 (Rus.).

MacCarty, C. S., and Kiefer, E. J., 1949, Thoracic, lumbar and sacral spinal cordectomy: Preliminary report, *Mayo Clin. Proc.* **24**:108–115.

MacKay, A. R., Gutin, P. H., Hosobuchi, Y., and Norman, D., 1982, Computed tomography-directed stereotaxy for biopsy and interstitial irradiation of brain tumors, *Neurosurgery* **11**:38–42.

Madden, J., 4th, Akil, H., and Patrick, R. L., 1977, Stress-induced parallel changes in central opioid levels and pain responsiveness in the rat, *Nature* **265**:358–360.

Magoun, H. W., and Rhines, R., 1962, *Spasticity: The Stretch Reflex and Extrapyramidal System,* Charles C. Thomas, Springfield, IL, pp. 26–55.

Mahl, G. F., Rothenberg, A., Delgado, J. M. R., and Hamlin, H., 1964, Psychological response in human to intracerebral stimulation, *Psychosom. Med.* **26**:337–368.

Majorchik, V. E., Arkhipova, N. A., Vasin, N. J., and Grochovski, N. P., 1980, Peculiarity of thalamo-cortical relations in phantom pain syndrome (electrophysiological study during stereotactic operations), *Vopr. Neirokhir.* **6**:21–29 (Rus.).

Mandell, S., 1970, The treatment of dystonia with L-dopa and haloperidol, *Neurology* **20**:103–106.

Mankovsky, N. B., Wainshtok, N. B., and Oleinick, L. I., 1974, Experience of prolonged application of Midantan (amantadine hydrochloride) in parkinsonism, *Zh. Nevropathol. Psychiatr.* **74**(3):355–360 (Rus.).

Manrique, M., Vaquero, I., Oya, S., Lozano, A. P., and Bravo, G., 1980, Side effects and long-term results of chronic cerebellar stimulation in man, *Acta Neurochir. (Wien) [Suppl.]* **30**:333–338.

Mansuy, L., Sindou, M., Fischer, G., and Brunon, J., 1976, La cordotomie spinothalamique dans le douleurs cancereuses. Resultats d'une série de 124 malades opérés par abord direct postérieur, *Neurochirurgie* **22**:437–444.

Manzano, G. M., Ragazzo, P. C., Tavares, S. M., and Marino, R., 1986, Anterior zygomatic electrodes: A special electrode for the study of temporal lobe epilepsy, *Appl. Neurophysiol.* **49**:213–217.

Margolin, D. I., and Marsden, C. D., 1982, Episodic dyskinesias and transient cerebral ischemia, *Neurology (N.Y.)* **32**:1379–1380.

Margorin, E. M., 1970, Individual anatomic differences in some subcortical brain formations with special reference to stereoencephalotomy, *Vopr. Neirokhir.* **3**:3–6 (Rus.).

Margulis, M. S., 1940, Epidemic encephalitis (Economo type), *Handb. Neurol.* **5**:664–794 (Rus.).

Marinesco, G., and Nicolesco, M., 1929, Quelques donnée cliniques sur les troubles du tonus dans les dystonies d'attitude, *Rev. Neurol.* **1**:502–507.

Marino, R., 1985, Surgery for epilepsy. Selective partial microsurgical callosotomy for intractable multiform seizures: Criteria for clinical selection and results, *Appl. Neurophysiol.* **48**:404–407.

Marino, R., Camargo, C. H. P., Avila, J. O., Huck, F. R., Radvany, J., Riva, D., Ragazzo, P. C., and Rossi, C., 1981, Neuropsychological changes after anterior callosotomy in patients with intractable multiform seizures, in *VIII Meeting of the World Society for Stereotactic and Functional Neurosurgery,* Zurich, p. 48.

Mark, V. H., and Ervin, F. R., 1965, Role of thalamotomy in treatment of chronic severe pain, *Postgrad. Med.* **37**:563–571.

Mark, V. H., Ervin, F. R., and Hackett, T. P., 1960, Clinical aspects of stereotaxic thalamotomy in the human. Part I. The treatment of chronic pain, *Arch. Neurol.* **3**:351–367.

Mark, V. H., Chato, J. C., and Eastman, F. C., 1961, Localized cooling in the brain, *Science* **134**:1520–1521.

Mark, V. H., Ervin, F. R., and Yakovlev, P. I., 1962, The treatment of pain by stereotaxic methods, *Confin. Neurol.* **22**:238–246.

Mark, V. H., Ervin, F. R., and Yakovlev, P. I., 1963, Stereotactic thalamotomy. III. The verification of anatomical lesion sites in the human thalamus, *Arch. Neurol.* **8**:5.

Markham, C. H., Brown, W. J., and Rand, R. W., 1966, Stereotactic lesions in Parkinson's disease: Clinico–pathological correlations, *Arch. Neurol.* **15**(5):480–497.

Markova, E. D., Gotovtzeva, E. V., and Barchatova, V. P., 1966, Clinic–genetic analysis of the torsion dystonia, in *Questions of Pathology of the Nervous System*, Kishinjev, pp. 362–365 (Rus.).

Markova, E. D., Ivanova-Smolenskaya, I. A., Insarova, N. G., and Barchatova, V. P., 1978, Clinic and treatment of certain parkinsonism-like syndromes of hereditary etiology, *Klin. Med.* **56**:104–111 (Rus.).

Maroon, J. C., Lunsford, L. D., and Deeb, Z. L., 1978, Hemifacial spasm due to aneurysmal compression of the facial nerve, *Arch. Neurol.* **35**:545–546.

Marossero, F., Cabrini, G. P., Ettorre, G., and Infuso, L., 1972, Electromyographic study of motor responses following electrical stimulation of the corticospinal tract in man during stereotaxy, *Confin. Neurol.* **34**:230–236.

Marsden, C. D., 1976, Dystonia. The spectrum of the disease, in *The Basal Ganglia* (M. D. Yahr, ed.), Raven Press, New York.

Marsden, C. D., 1977, The need for alternative therapy in Parkinson's disease, in *Parkinson's Disease, Concepts and Prospects*, Excerpta Medica, Amsterdam, pp. 117–120.

Marsden, C. D., 1978, The mechanisms of physiological tremor and their significance for pathological tremors, in *Physiological Tremor, Pathological Tremor and Clonus* (J. E. Desmedt, ed.), S. Karger, Basel, pp. 1–16.

Marsden, C. D., 1980, "On-off" phenomena in Parkinson's disease, in *Parkinson's Disease* (U. K. Rinne, M. Klinger, and G. Stamm, eds.), Elsevier/North-Holland Biomedical Press, Amsterdam, New York, pp. 241–254.

Marsden, C. D., and Harrison, M. J. G., 1974, Idiopathic torsion dystonia (dystonia musculorum deformans). A review of forty-two patients, *Brain* **97**:793–810.

Marshall, J., 1962, Observations on essential tremor, *J. Neurol. Neurosurg. Psychiatry* **25**(2):122–126.

Marshall, J., and Walsh, E. G., 1956, Physiological tremor, *J. Neurol. Neurosurg. Psychiatry* **19**(4):260–267.

Martin, G. F., Humberston, A. O., and Laxson, C., 1979, Dorsolateral pontospinal system. Possible routes for catecholamine modulation of nociception, *Brain Res.* **163**:333–338.

Martin, J. P., 1927, Hemichorea resulting from a local lesion of the brain, *Brain* **50**:637–651.

Martin, J. P., 1957, Hemichorea (hemiballismus) without lesions in the corpus Huysii, *Brain* **80**:1–10.

Martin, J. P., and McCaul, I. R., 1959, Acute hemiballismus treated by ventrolateral thalamolysis, *Brain* **82**:104–108.

Martin, M. L., McElhaney, M. L., and Meyer, G. A., 1977, Stereotactic cingulotomy, in *Neurosurgical Treatment in Psychiatry, Pain and Epilepsy*, University Park Press, Baltimore, pp. 381–386.

Martinez, S. N., Bertrand, C., Molina-Negro, P., and Perez-Calvo, J. M., 1975, Alteration of pain perception by stereotactic lesions of frontothalamic pathways, *Confin. Neurol.* **37**:113–118.

Martin-Rodriguez, J. G., and Obrador, S., 1975, Evaluation of stereotaxic pulvinar lesions, *Confin. Neurol.* **37**:56–62.

Martins, L. F., and Umbach, W., 1975, Size and position of stereotaxic lesions in comparison with clinical pain relief, *Confin. Neurol.* **37**:80–85.

Marttila, R. J., 1980, Etiology of Parkinson's disease, in *Parkinson's Disease* (U. K. Rinne, M. Klinger, G. Stamm, eds.), Elsevier/North-Holland Biomedical Press, Amsterdam, New York, pp. 3–16.

Marttila, R. J., and Rinne, U. K., 1976, Arteriosclerosis heredity, and some previous infections in the etiology of Parkinson's disease. A case-control study, *Clin. Neurol. Neurosurg.* **79**:45–56.

Marttila, R. J., Arstila, P., Nikoskelainen, J., Halonen, P. E., and Rinne, U. K., 1977, Viral antibodies in the sera from patients with Parkinson disease, *Eur. Neurol.* **15**:25–33.

Mashanski, F. I., 1935, Experience of surgical treatment of involuntary movements in legs in parkinsonism, *Grekows Vestn. Chir.* **39**:110–111.

Maspes, P. E., and Pagni, C. A., 1964, Surgical treatment of dystonia and choreo-athetosis in infantile cerebral palsy by pedunculotomy. Pathophysiological observations and therapeutic results, *J. Neurosurg.* **21**(12):1076–1086.

Masuzawa, T., Shinoda, S., Faruse, M., Nakahara, N., Abe, F., and Sato, F., 1983, Cerebral angiographic changes on serial examination of a patient with migraine, *Neuroradiology* **24**:277–281.

Matsumoto, K., 1986, Reappraisal of ventrolateral thalamotomy for Parkinson's disease, in *Neurosurgery* (R. R. Tasker, ed.), Hanley and Belfus, Philadelphia, pp. 209–234.

Matsumoto, K., Tanigawa, M., and Nishimoto, A., 1971, Combined stereotaxic operation for treatment of deep-seated A–V malformation, *Neurol. Med. Chir. (Tokyo)* **11**:337–338.

Matsumoto, K., Miyamoto, T., Mimura, Y., and Beck, H., 1973, Electrophysiological study on thalamo-cortical relation in man, in *The Sixth Symposium of the International Society for Research in Stereoencephalotomy*, Vol. II, Tokyo, p. 37.

Matsumoto, K., Shichijo, F., Masuda, T., and Miyake, H., 1985, CT-controlled stereotactic surgery, *Appl. Neurophysiol.* **48**:39–44.

Matthews, B., and Cadden, S. W., 1970, Conduction block in peripheral A and C fibers following electrical stimulation, *Neurosci. Abstr.* **5**:306.

Matthews, P. B. C., 1962, The differentation of two types of fusimotor fibre by their effects on the dynamic response of muscle spindle primary endings, *Q. J. Exp. Physiol.* **47**:324–327.

Max, M., Inturris, C. E., and Gradinski, P., 1981, Epidural opiates: Plasma and cerebro-spinal fluid (CSF) pharmacokinetics of morphine, methadone and beta-endorphin, *Pain [Suppl.]* **1**:122–129.

Maxwell, R. E., 1982, Surgical control of chronic migrainous neuralgia by trigeminal gangliorhizolysis, *J. Neurosurg.* **57**:459–466.

Mayanagi, Y., Sano, K., Suzuki, I., Kanazawa, I., Aoyagi, I., and Miyachi, Y., 1981, Stimulation and coagulation of the posteromedial hypothalamus for intractable pain, with reference to beta-endorphin study, in *VIII Meeting of the World Society for Stereotactic and Functional Neurosurgery*, Zurich, p. 21.

Mayanagi, Y., Sano, K., and Suzuki, I., 1982, Stimulation and coagulation of the posteromedial hypothalamus for intractable pain, with reference to beta-endorphins, *Appl. Neurophysiol.* **45**(1–2):136–142.

Mayer, D. J., and Hayes, R. L., 1975, Stimulation-produced analgesia. Development of tolerance and cross-tolerance to morphine, *Science* **188**:941–943.

Mayer, D. J., and Liebeskind, J. C., 1974, Pain reduction by focal electrical stimulation of the brain: An anatomical and behavioral analysis, *Brain Res.* **68**:73–93.

Mayer, D. J., Price, D. D., Becker, D. P., and Young, H. F., 1975, Threshold for pain from anterolateral quadrant stimulation as a predictor of success of percutaneous cordotomy for relief of pain, *J. Neurosurg.* **43**:445–447.

Mazars, G., 1975, Intermittent stimulation of nucleus ventralis posterolateralis for intractable pain, *Surg. Neurol.* **4**:93–95.

Mazars, G., 1976a, *La Chirurgie de la Douleur,* Masson, Paris.

Mazars, G. J., 1976b, Etat actual de la chirurgie de la douleur, *Neurochirurgie* **22**(Suppl. 1):95–98.

Mazars, Y., Mazars, G., and Chiarelli, J., 1960, Refrigeration et congelation corticales; leur application à l'interruption des états de mal corticaux, *Rev. Neurol.* **102**:287–291.

Mazars, G., Chodkiewicz, J. P., Mezienne, L., and Gotusso, C., 1966, Hypophysectomies par congelation l'azote liquide, *Neurochirurgie* **12**(6):683–687.

Mazars, G., Mérienne, L., and Ciolocca, C., 1973, Stimulations thalamiques intermittentes analgiques, *Rev. Neurol.* **128**:273–279.

Mazars, G., Merienne, L., and Ciolocca, C., 1974, Traitment de certains types de douleurs par des stimulateurs thalamiques implantables, *Neurochirurgie* **20**:117–124.

Mazur, P., 1968, Physical–chemical factors underlying cell injury in cryosurgical freezing, in *Cryosurgery* (R. W. Rand, ed.), Thomas, Springfield, pp. 32–51.

Mazziotta, J. C., and Engel, J., 1984, The use and impact of positron-computed tomography scanning in epilepsy, *Epilepsia* [Suppl.] **25**:86–104.

McKenzie, K. G., 1924, Intrameningeal division of the spinal accessory and roots of the upper cervical nerves for the treatment of spasmodic torticollis, *Surg. Gynecol. Obstet.* **39**:5–10.

McKinley, J. C., and Berkwitz, N. J., 1928, Quantitative studies on human muscle tonus, *Arch. Neurol. Psychiatry* **19**:1036–1055.

McKissock, W., Richardson, A., and Taylor, J., 1961, Primary intracerebral hemorrhage: A controlled trial of surgical and conservative treatment in 180 unselected cases, *Lancet* **2**:221–226.

McLellan, D. L., 1973, Dynamic spindle reflexes and the rigidity of parkinsonism, *J. Neurol. Neurosurg. Psychiatry* **36**:342–349.

McLellan, D. L., Chalmers, R. I., and Jonson, R. H., 1975, Clinical and pharmacological evaluation of the effects of piribedil in patients with parkinsonism, *Acta Neurol. Scand.* **51**:74–82.

McLellan, D. L., Wright, G. D. S., and Renouf, F., 1981, Calibration of clinical cerebellar and deep brain stimulation systems, *J. Neurol. Neurosurg. Psychiatry* **44**:392–396.

McNaughton, F. L., and Rasmussen, T., 1975, Criteria for selection of patients for neurosurgical treatment, in *Advances in Neurology,* Vol. 8 (D. D. Purpura, J. K. Penry, and R. D. Walter, eds.), Raven Press, New York, pp. 37–48.

McSherry, I. W., 1982, Intraoperative electroencephalographic and evoked potential studies in operative neurosurgery, in *Operative Neurosurgical Techniques. Indications, Methods, and Results,* Vol. 2 (H. H. Schmidek and W. H. Sweet, eds.), Grune & Stratton, New York, pp. 957–962.

Meerson, J. A., and Tetz, I. S., 1975, Comparative peculiarities of visual identification in patients with temporal lobe epilepsy before and after resection of left or the right temporal lobe, in *Surgical Treatment of Epilepsy with Mental Disorders,* Meditsina, Leningrad, pp. 58–70 (Rus.).

Meglio, M., 1983, Percutaneously implantable chronic electrode for radiofrequency stimulation of the gasserian ganglion. A new perspective in the management of trigeminal pain, in *VI Meeting of the European Society for Stereotactic and Functional Neurosurgery,* Rome, p. 29.

Meglio, M., 1984, A new system for chronic electrical stimulation of the trigeminal ganglion, *Pain [Suppl.]* **2**:227.

Meglio, M., and Cioni, B., 1982, Personal experience with spinal cord stimulation in chronic pain management, *Appl. Neurophysiol.* **45**(1–2):195–200.

Meglio, M., Cioni, B., and D'Amico, E., 1980, Epidural spinal cord stimulation for the treatment of neurogenic bladder, *Acta Neurochir. (Wien)* **54**:191–199.

Meglio, M., Cioni, B., and D'Annunzio, V., 1986a, Percutaneous microcompression of the Gasserian ganglion. Personal experience, in *VII Meeting of the European Society for Stereotactic and Functional Neurosurgery,* Birmingham, p. 31.

Meglio, M., Cioni, B., Rossi, G. F., Sandric, S., and Santarelli, P., 1986b, Spinal cord stimulation affects the central mechanisms of regulation of heart rate, *Appl. Neurophysiol.* **49**:139–146.

Meirowsky, A. M., Scheibert, C. D., and Hinchey, T. R., 1950, Studies on the sacral reflex arc in paraplegia. I. Response of the bladder to surgical elimination of sacral nerve impulses by rhizotomy, *J. Neurosurg.* **7**:33–38.

Melnitchuk, P. V., and Sosnovskaya, L. S., 1973, Treatment of deforming muscular dystonia in children by L-dopa, *Zh. Nevropathol. Psychiatr.* **10**:1495–1498 (Rus.).

Melnitchuk, P. V., Sosnovskaya, L. S., and Chailova, I. M., 1977, De la Tourette syndrome, *Zh. Nevropatol. Psychiatr.* **7**:1148–1152 (Rus.).

Melzack, R., 1971a, Phantom limb pain, implications for treatment of pathologic pain, *Anesthesiology* **35**:409–419.

Melzack, R., 1971b, Phantom limb pain: Concept of a central biasing mechanism, *Clin. Neurosurg.* **18**:188–207.

Melzack, R., 1973, *The Puzzle of Pain,* Penguin Books, New York.

Melzack, R., 1975, Prolonged relief of pain by brief, intense transcutaneous somatic stimulation, *Pain* **1**:357–373.

Melzack, R., and Loeser, J. D., 1978, Phantom body pain in paraplegies: Evidence for a central "pattern generating mechanism" for pain, *Pain* **4**:195–210.

Melzack, R., and Wall, P. D., 1965, Pain mechanisms: A new theory, *Science* **150**:971–979.

Mempel, E., 1971a, The effect of partial amygdalectomy on emotional disturbances and epileptic seizures, *Polish Med. J.* **10:**969–974.

Mempel, E., 1971b, Einfluß partieller Amygdalotomien auf emotionelle Störungen und Epilepsie, *Neurol. Neurochir. Pol.* **21:**81–86.

Mempel, E., 1972, The influence of partial (dorso-medial) amygdalectomy on emotional disturbances and epileptic fits in humans, in *Present Limits of Neurosurgery* (I. Fusek and Z. Kunc, eds.), Avicenum, Czechoslovak Medical Press, Prague, pp. 497–500.

Mempel, E., Sierpinski, S., and Pilipowska, T., 1971, Odlegle wyniki leczenia zespolow parkinsonowskich metoda stereotaktyezna, *Neurol. Neurochir. Psychiatr. Pol.* **5:**11–16.

Mempel, E., Witkiewicz, D., and Standicki, R., 1979, The effect of amygadolotomy and anterior hippocampotomy on behavior and seizures in epileptic patients, in *IV Meeting of the European Society for Stereotactic and Functional Neurosurgery,* Paris, pp. 115–119.

Mendel, K., 1911, *Die Paralysis Agitans,* Berlin.

Mendel, K., 1919, Torsiondystonie (dystonia musculorum deformans, Torsionsspasmus), *Z. Psychiatr. Neuropathol.* **46:**309–361.

Menzel, J., Piotrowski, W., and Penzholz, H., 1975, Long-term results of Gasserian ganglion electrocoagulation, *J. Neurosurg.* **42:**140–143.

Mercuri, S., Contratti, F., Russo, A., and Savino, S., 1980, Trattamento con stimolazione elettrica transcutanea nella terapia del dolore chronico esperienza a distanza su 100 casi, *Rev. Neurol.* **50:**133–141.

Merskey, H., and Boyd, D., 1978, Emotional adjustment and chronic pain, *Pain* **5:**173–178.

Meryman, H. T., 1956, Mechanics of freezing in living cells and tissues, *Science* **124:**515–521.

Meryman, H. T., 1966, *Cryobiology,* Academic Press, London.

Meshcherski, R. M., 1961, *Stereotactic Method,* Meditsina, Moscow.

Messing, R. B., and Lytle, L. D., 1977, Serotonin-containing neurons: Their possible role in pain and analgesia, *Pain* **4:**1–21.

Meyer, F. B., Marsh, W. R., Laws, E. R., and Sharbrough, F. W., 1986, Temporal lobectomy in children with epilepsy, *J. Neurosurg.* **64:**371–376

Meyer, G. A., Martin, W. L., McGraw, C. P., McElhaney, M. L., and Barratt, E. S., 1972, Stereotactic cingulotomy, in *Third World Congress Psychosurgery: Abstracts.*

Meyer, G. A., McElhaney, W., Martin, W., and McGraw, C. P., 1973, Stereotactic cingulotomy with results of acute stimulation and serial psychological testing, in *Surgical Approaches in Psychiatry. Proceedings of the Third International Congress of Psychosurgery* (L. V. Laitinen and K. E. Livingston, eds.), University Park Press, Baltimore, pp. 40–58.

Meyer-Lohmann, J., Conrad, B., Matsunami, K., and Brooks, V. B., 1975, Effects of dentate cooling on precentral unit activity following torque pulse injections into elbow movements, *Brain Res.* **94:**237–251.

Meyers, H. R., 1940, Surgical procedure for postencephalitic tremor with notes on the physiology of the premotor fibers, *Arch. Neurol. Psychiatry* **44:**455–461.

Meyers, R., 1942, The modification of alternating tremors, rigidity and festination by surgery of the basal ganglia, *Res. Publ. Assoc. Nerv. Ment. Dis.* **21:**602–665.

Meyers, R., 1951, Surgical experiments in the therapy of certain extrapyramidal diseases: A current evaluation, *Acta Psychiatr. Neurol. [Suppl.],* **67:**1–42.

Meyers, R., 1955, in *Clinical Neurology,* Vol. 2 (A. B. Bacer, ed.), Harper, New York, p. 1107.

Meyers, R., 1956a, Results of bilateral intermediate midbrain crusotomy in seven cases of severe athetotic and dystonic quadriparesis, *Am. J. Physiol.* **35:**84–105.

Meyers, R., 1956b, Physiological and therapeutic effects of bilateral intermediate midbrain crusotomy for atheto-dystonia (17 cases), *Surg. Forum* **6:**486–488.

Meyers, R., Fry, W. J., Fry, F. J., Dreyer, L. L., Schultz, D. F., and Noyes, R. F., 1959, Early experiences with ultrasonic irradiation of the pallidofugal and nigral-complexes in hyperkinetic and hypertonic disorders, *J. Neurosurg.* **16:**32–54.

Meyers, R., Fry, F. J., and Fry, W. J., 1960, Determination of topologic human brain representations and modifications of signs and symptoms of some neurological disorders by the use of high level ultrasound, *Neurology (Minneap.)* **10:**271–277.

Meyerson, B. A., 1985, Aspects on the present state of intracerebral stimulation for pain, in *Brain Stimulation and Neuronal Plasticity* (T. Tsubokawa, ed.), Tokyo, pp. 33–54.

Meyerson, B. A., and Hakanson, S., 1978, Alleviation of facial pain by stimulation of the Gasserian ganglion via an implanted electrode, in *Annual Meeting of the Scandinavian Neurosurgery Society,* pp. 23–27.

Meyerson, B. A., and Hakanson, S., 1986, Suppression of pain in trigeminal neuropathy by electric stimulation of the Gasserain ganglion, *Neurosurgery* **18:**59–66.

Meyerson, B. A., Boëthius, I., and Carlsson, A. M., 1978, Percutaneous central gray stimulation for cancer pain, *Appl. Neurophysiol.* **41:**57–65.

Meyerson, B. A., Linderoth, B., and Brodin, E., 1985, Possible neurohumoral mechanisms in CNS stimulation for pain suppression, *Appl. Neurophysiol.* **48:**175–180.

Miles, J., 1979, Chemical hypophysectomy, in *Advances in Pain Research and Therapy,* Vol. 2 (J. J. Bonica and V. Ventafridda, eds.), Raven Press, New York, pp. 373–380.

Miles, I., and Lipton, S., 1978, Phantom limb pain treated by electrical stimulation, *Pain* **5:**373–382.

Mingrino, S., 1978, Supratentorial arteriovenous malformations of the brain, in *Advances and Technical Standards in Neurosurgery,* Vol. 5 (H. Krayenbühl, ed.), Springer Verlag, Vienna, pp. 93–123.

Mingrino, S., and Schergna, E., 1972, Stereotaxic anterior cingulotomy in the treatment of severe behavior disorders, in *Psychosurgery* (E. Hitchcock, L. Laitinen, and K. Vaernet, eds.), Charles C. Thomas, Springfield, IL, pp. 258–263.

Minnius, I., 1641, Cited by Finney, J. M. T., and Hughson, W., 1925, Spasmodic torticollis, *Ann. Surg.* **81:**255–269.

Mish, V. M., 1923, The possibility of successful surgical intervention for athetosis, *Nov. Chir. Arch.* **3**(11):419–438 (Rus.).

Mitchell, S. W., 1872, Clinical lecture on certain painful affections of the feet, *Phila. Med. Times* **3**(81):i13–116.

Mitchell, S. W., Morehouse, G. R., and Keen, W. W., 1864, *Gunshot Wounds and Other Injuries of Nerves*, Lippincott, Philadelphia.

Mitchell-Heggs, N., Relly, D., and Richardson, A. E., 1977, Stereotactic limbic leucotomy, in *Neurosurgical Treatment in Psychiatry, Pain and Epilepsy*, University Park Press, Baltimore, pp. 367–379.

Miyazaki, Y., Ervin, F. R., and Siegfried, J., 1963, Localized cooling in the central system, *Arch. Neurol.* **9**:392–399.

Mizokami, T., Yoshii, N., Ushikubo, Y., Kuramitsy, T., Fukuda, S., and Yamada, F., 1981, Comparative study between invaded area and operative effects after pulvinarotomy, in *VIII Meeting of the World Society for Stereotactic and Functional Neurosurgery*, Zurich, p. 57.

Mizuno, M., Koshino, K., Ikeda, K., Matsumura, H., and Sakamoto, Y., 1972, Stereotaxic surgery for the treatment of behavioral disorder, *Neurol. Med. Chir.* **12**:370–374.

Modesti, L. M., and Perl, T., 1982, Observation and radiofrequency trigeminal gangliolysis by stereotactic method, *Appl. Neurophysiol.* **45**:518–519.

Molina-Negro, P., and Hardy, J., 1971, Etude sémiologique des tremblements, *Union Med. Can.* **100**:879–895.

Molina-Negro, P., Bertrand, C., Martinez, S. N., and Hardy, J., 1974, Controlled lesions of VOI and adjoining structures in stereotactic surgery of spasmodic torticollis, in *American Association of Neurological Surgeons Congress*, St. Louis.

Møller, M. B., and Møller, A. R., 1985, Loss of auditory function in microvascular decompression for hemifacial spasm. Results in 143 consecutive cases, *J. Neurosurg.* **63**:17–20.

Moniz, E., 1936a, *Tentatives Opératoires dans le Traitement de Certaines Psychoses*, Masson, Paris.

Moniz, E., 1936b, Essai d'un traitement chirurgical de certaines psychoses, *Bull. Acad. Med.* **115**:385–392.

Monnier, M., 1938, Le torticollis spasmodique, ses variations sous l'influence de diverses sensitives, psychiques et végétatives. Etude clinique–expérimentale sur la pathogénie des dystonies, *Schweiz. Arch. Neurol. Psychiatr.* **40**:345–361.

Monnier, A., 1952, Appareil stéréotaxique et technique de repérage pour la coagulation du relais thalamique de la douleur chez l'homme, *Schweiz. Med. Wochenschr.* **82**:1031–1037.

Monnier, M., and Fischer, R., 1951, Stimulation électrique et coagulation thérapeutic du thalamus chez l'homme (névralgies faciales), *Confin. Neurol.* **11**:282–286.

Moody, R. H., and Poppen, J. L., 1970, Arteriovenous malformations, *J. Neurosurg.* **32**:503–511.

Moran, C. J., Naidlich, T. P., Gado, M. H., and Marchosky, J. A., 1979, Central nervous system lesions biopsied or treated by CT-guided needle placement, *Radiology* **131**:681–684.

Morbus Parkinson, 1983, Sandoz, Basel.

Morell, F., Hoeppner, T., and Whisler, W. W., 1981, The use of intraoperative somatosensory evoked potentials to delineate the postcentral gyrus in man, *Electroencephalogr. Clin. Neurophysiol.* **51**:41–46.

Morello, G., and Borghi, G. P., 1973, Cerebral angiomas, *Acta Neurochir. (Wien)* **28**:135–155.

Morgan, C., de, 1867, A case in which severe spasmodic contraction of cervical muscles is produced by movement, *Lancet* **2**:128–129.

Mori, K., Fujita, Y., and Shimabukuro, H., 1975, Some considerations for treatment of spasmodic torticollis, *Confin. Neurol.* **37**:265–269.

Mori, K., Iwayama, K., Baba, H., Kaminogo, M., Nishimura, S., and Ono, K., 1982, Stereotactic VA thalamotomy for the control of focal seizures: Experimental and Clinical Studies, *Appl. Neurophysiol.* **45**:478–483.

Mori, K., Shimabukuro, H., Yamashiro, K., Miyake, H., and Kawano, T., 1985, Spasmodic head movements produced by destruction of unilateral ventromedial tegmentum in cats, *Appl. Neurophysiol.* **48**:347–350.

Moricca, G., 1977, Pituitary neuroadenolysis in the treatment of intractable pain from cancer, in *Persistent Pain: Modern Methods of Treatment*, Vol. 1 (S. Lipton, ed.), Academic Press, London, Grune & Stratton, New York, pp. 149–173.

Mosinger, M., 1950, Anatomie de l'hypothalamus et du sous-thalamus élargi, *Schweiz. Arch. Neurol. Neurochir. Psychiatrie* **65**:135–180.

Moyes, P. D., 1969, Longitudinal myelotomy for spasticity, *J. Neurosurg.* **31**:615–619.

Mukawa, J., Kimura, T., Nagao, I., Kobayashi, K., Iwata, Y., Koshino, K., Ikeda, T., Kamikawa, K., Mogami, H., and Jinnai, D., 1975, Forel-H-tomy for the treatment of intractable epilepsy, *Confin. Neurol.* **37**(1–3):302–307.

Müke, R., 1974, Chirurgische Therapiemöglichkeiten beim Phantomschmerz, *Therapiewoche* **24**:348–353.

Mukherjee, A., Varma, S. K., Kucheria, K., and Bole, S. V., 1973, Family cerebral palsy, *J. Indian Med. Assoc.* **60**:300–301.

Mullan, S., 1966, Percutaneous cordotomy for pain, *Surg. Clin. North Am.* **46**:3–12.

Mullan, S., 1971, Percutaneous cordotomy, *J. Neurosurg.* **35**:360–365.

Mullan, S., 1974, Experiences with surgical thrombosis of intracranial berry aneurysms and carotid cavernous fistulas, *J. Neurosurg.* **41**:657–671.

Mullan, S., and Hosobuchi, Y., 1968, Respiratory hazards of high cervical percutaneous cordotomy, *J. Neurosurg.* **28**:291–297.

Mullan, S., and Lichter, T., 1985, Treatment of trigeminal neuralgia by percutaneous microcompression: A six year follow-up, in *VIII International Congress of Neurological Surgery*, Toronto, p. 141.

Mullan, S., Harper, P. V., Hekmatpanah, J., Torres, H., and Dabben, B., 1963, Percutaneous interruption of spinal-pain tracts by means of a strontium^{90} needle, *J. Neurosurg.* **20**:931–939.

Mullan, S., Beckman, F., Vailat, G., Karasick, S. A., and Dollen, G., 1964, An experimental approach to the problem of cerebral aneurysm, *J. Neurosurg.* **21**(10):838–846.

Mullan, S., Hekmatpanah, J., Dobben, G., and Beckman, F., 1965a, Percutaneous intramedullary cordotomy utilizing the unipolar anodal electrolytic lesion, *J. Neurosurg.* **22**(6):548–553.

Mullan, S., Mailis, M., Karasick, G., Vailati, G., and Beckman, F., 1965b, A reappraisal of the unipolar anodal electrolytic lesion, *J. Neurosurg.* **22**:531–537.

Mullan, S., Raimondi, A. J., Dobben, G., Vailati, G., and Hekmatpanah, J., 1965c, Electrically induced thrombosis in intracranial aneurysms, *J. Neurosurg.* **22**:539–544.

Mullan, S., Vailati, G., Karasick, J., and Mailis, M., 1967, Thalamic lesions for the control of epilepsy, *Arch. Neurol.* **16**:277–285.

Mullan, S., Kawanaga, H., and Patronas, N. J., 1979, Micro vascular embolization of cerebral arteriovenous malforma tions, *J. Neurosurg.* **51**(5):621–627.

Müller, J., 1840, *Handbuch der Physiologie des Menschen* Hölscher, Koblenz.

Mundinger, F., 1965a, Die Subthalamotomie zur Behandlung Extrapyramidaler Bewegungsstörungen, *Dtsch. Med. Wochenschr.* **90**:2002–2007.

Mundinger, F., 1965b, Ergebnisse der primär- und kombiniert operativstereotaktischen Radioisotopbestrahlung von Hypophysenadenomen, in *Radio-Isotope in der Endokrinologie* (G. Hoffman, ed.), Schattauer, Stuttgart, pp. 397–416.

Mundinger, F., 1965c, Stereotaxic interventions on the zona incerta area for treatment of extrapyramidal motor disturbances and their results, *Confin. Neurol.* **26**:222–230.

Mundinger, F., 1968, *Chirurgie des Gehirns und Rückenmarks im Kindes- und Jugendalter,* Hippokrates-Verlag, Stuttgart, pp. 1011–1067.

Mundinger, F., 1969, Results of 500 subthalamotomies in the region of the zona incerta, in *Third Symposium on Parkinson's Disease* (F. J. Gillingham and I. M. L. Donaldson, eds.), Livingstone, Edinburgh, pp. 261–266.

Mundinger, F., 1973, *Stereotaktische Operationen am Gehirn,* Hippokrates, Stuttgart.

Mundinger, F., 1974a, Stereotaktische Operationen gegen anderweitig unbehandelbare schwere Schmerzzustände, *Z. Allg. Med.* **50**:860–864.

Mundinger, F., 1974b, Psychiatrische Chirurgie, *Z. Allg. Med.* **50**:856–859.

Mundinger, F., 1974c, Stereotactic curie-therapy of pituitary adenomas. A long-term follow-up study, in *Advances in Stereotactic and Functional Neurosurgery,* Springer-Verlag, Vienna, pp. 169–176.

Mundinger, F., 1977, Die Behandlung chronischer Schmerzen mit Hirnstimulation, *Dtsch. Med. Wochenschr.* **102**:1724–1729.

Mundinger, F., 1985, Postoperative and long-term results of 1561 stereotactic operations in parkinsonism, *Appl. Neurophysiol.* **48**:293.

Mundinger, F., and Becker, P., 1975, Long term results of central stereotactic interventions for pain, *Adv. Neurosurg.* **3**:237–241.

Mundinger, F., and Birg, W., 1984, CT-stereotaxy in the clinical routine, *Neurosurg. Rev.* **7**:219–224.

Mundinger, F., and Hoefer, T., 1974, Protracted long-term irradiation of inoperable midbrain tumours by stereotactic Curie-therapy using iridium^{192}, *Acta Neurochir. [Suppl.]* **21**:93 100.

Mundinger, F., and Kuhn, J., 1982, Postoperative and long-term results after stereotactic operations for action myoclonia in cases of encephalomyelitis disseminata, *Appl. Neurophysiol.* **45**:299–305.

Mundinger, F., and Metzel, E., 1970, Interstitial radioisotope therapy of intractable diencephalic tumours by the stereotaxic permanent implantation of iridium-192, including bioptic control, *Confin. Neurol.* **32**:195–202.

Mundinger, F., and Neumüller, H., 1982, Programmed stimulation for control of chronic pain and motor disease, *Appl. Neurophysiol.* **45**:102–111.

Mundinger, F., and Potthof, P., 1961, Messungen im Pneumencephalogramm zur intracerebralen und craniocerebralen Korrelationstopographie bei stereotaktischen Hirnoperationen, unter besonderer Berücksichtung der stereotaktischen Pallidotomie, *Acta Neurochir. (Wien)* **9**(2):196–214.

Mundinger, F., and Riechert, T., 1960, Indications, technique and results of the stereotactic operations upon the hypophysis using radioisotopes, *J. Nerv. Ment. Dis.* **131**:1–9.

Mundinger, F., and Riechert, T., 1961, Ergebnisse der stereotaktischen Hirnoperationen bei extrapyramidalen Bewegungstörungen auf Grund postoperativer und Langzeituntersuchgen, *Dtsch. Z. Nervenheilkd.* **182**:542–576.

Mundinger, F., and Riechert, T., 1962, Stereotaxic irradiation—procedure of brain tumors and pituitary adenomas by means of radio-isotopes and its results, *Confin. Neurol.* **22**:190–204.

Mundinger, F., and Riechert, T., 1963, Die stereotaktischen Hirnoperationen zur Behandlung extrapyramidaler Bewegungsstörungen (Parkinsonismus und Hyperkinesen) und ihre Resultate, *Fortschr. Neurol.* **31**:1–65.

Mundinger, F., and Riechert, T., 1967, *Hypophysentumoren, Hypophysectomie,* Thieme, Stuttgart.

Mundinger, F., and Salomao, J. F., 1980, Deep brain stimulation in mesencephalic lemniscus medialis for chronic pain, *Acta Neurochir.* **30**:245–258.

Mundinger, F., and Uhl, H., 1967, Investigations into possible roentgenographic errors in stereotaxis, *Confin. Neurol.* **29**(2–5):202–207.

Mundinger, F., and Weigel, K., 1983, Indication and results of stereotactic curie-therapy with iridium-192 and iodine-125 for non-resectable tumors of the hypothalamic region, in *VI Meeting of the European Society for Stereotactic and Functional Neurosurgery,* Rome, p. T10.

Mundinger, F., and Zissner, O., 1966, Clinico-experimental studies of stereotaxic thalamotomy of the oral ventral nuclei in extrapyramidal motor system disorders: Somatotopics, optimal lesion sites and angle position of the electrode, *Neurochirurgia (Stuttg.)* **9**(2):41–61.

Mundinger, F., Noetzel, H., Riechert, T., 1959, Erhahrungen mit der lokalisierten Bestrahlung von malignen Hirngeschwülsten mit Radioisotopen, *Acta Neurochir. [Suppl.]* **4**:171–182.

Mundinger, F., Riechert, T., and Gabriel, E., 1960, Untersuchungen zu den physikalischen und technischen Voraussetzungen einer dosierten Hochfrequenzkoagulation bei stereotaktischen Hirnoperationen, *Zentralbl. Chir.* **85**:1051–1063.

Mundinger, F., Riechert, T., and Disselhof, J., 1972, Long-term results of stereotactic treatment of spasmodic torticollis, *Confin. Neurol.* **34**:41–46.

Mundinger, F., Becker, P., Grœbner, E., and Bachschmid, G., 1975, Long term results of stereotactic amygdalotomy and fornicotomy in temporal epilepsy, in *Symposium on Stereotactic Treatment of Epilepsy,* Bratislava, p. 59.

Mundinger, F., Becker, P., and Grœbner, E., 1976, Late results of stereotactic surgery of epilepsy predominantly temporal lobe type, *Acta Neurochir. [Suppl.]* **23:**177–182.

Mundinger, F., Milios, E., and Sistig, W., 1977, Brain stimulation in pyramidal and extrapyramidal motor disturbances, in *III Meeting of the European Society for Stereotactic and Functional Neurosurgery,* Freiburg, p. 15.

Mundinger, F., Birg, W., and Klar, M., 1978a, Computer-assisted stereotactic brain operations including computerized axial tomography, in *Advances in Stereoencephalotomy,* S. Karger, Basel, pp. 169–182.

Mundinger, F., Birg, W., and Osterag, C., 1978b, Treatment of small cerebral gliomas with CT-aided stereotaxic curie-therapy, *Neuroradiology* **16:**564–567.

Mundinger, F., Salomao, F., and Grœbner, E., 1981, Indications and long-term results of stereotactic operations in therapy-resistant epilepsy, *Arch. Psychiatr. Nervenkr.* **231**(1)**:**1–11.

Mundinger, F., Monadjer, M., Gorke, H., and Milios, E., 1986, Long-term results of essential tremor after stereotactic operation, in *VII Meeting of the European Society for Stereotactic and Functional Neurosurgery,* Birmingham.

Munro, D., 1945, The rehabilitation of patients totally paralyzed below the waist: With special reference to making them ambulatory and capable of earning their living. I. Anterior rhizotomy for spastic paraplegia. *N. Engl. J. Med.* **233:**453–461.

Murayama, Y., Tsuda, T., and Sogabe, K., 1979, CT appearances of thalamic lesions in stereotactic surgery, *Appl. Neurophysiol.* **42**(5)**:**307.

Murayama, Y., Sogabe, K., and Matsumoto, K., 1982, CT appearance of thalamic lesions in stereotactic surgery, *Appl. Neurophysiol.* **45**(4)**:**399–403.

Musolino, A., Munari, C., Blond, S., Betti, O., Lajat, Y., Schaub, C., Askienazy, S., and Chadkiewicz, J. P., 1985, Traitment stéréotaxique des cystes expansifs de cranio-pharyngiomes par irradiation endocavitaire beta (Re 186; Au 198; Y 20), *Neurochirurgie* **31:**169–178.

Nadvornik, P., and Šramka, M., 1973, Stereotactic dentatotomy, *Vopr. Neirokhir.* **4:**41–43 (Rus.).

Nadvornik, P., Petr, R., Nemecek, S., Schindlery, C., and Beran, J., 1965, Stereotactic charts of the cerebellar nuclei, *J. Hirnforsch.* **8**(1)**:**67–91.

Nadvornik, P., Fröhlich, J., Jezek, V., and Šramka, M., 1972a, New apparatus for spinal cord stereotaxis and its use in the microsurgery of lumbar enlargement, *Confin. Neurol.* **34:**311–314.

Nadvornik, P., Šramka, M., Lizy, L., and Svicka, J., 1972b, Experiences with dentatotomy, *Confin. Neurol.* **34:**320–324.

Nadvornik, P., Šramka, M., and Pogady, J., 1972c, The results of stereotactic treatment of the aggressive syndrome, in *Psychosurgery* (E. Hitchcock, L. Laitinen, and K. Vaernet, eds.), Charles C. Thomas, Springfield, IL, pp. 126–128.

Nadvornik, P., Pogady, J., and Šramka, M., 1973, Experience with stereotactic interventions for aggressiveness, *Vopr. Neirokhir.* **4:**41–43 (Rus.).

Nadvornik, P., Fröglich, I., Šramka, M., 1974, First experience with the spinal cord stereotactic surgery for treatment of pain, *Vopr. Neirokhir.* **2:**46–49 (Rus.).

Nadvornik, P., Šramka, M., Gajdosova, D., and Kokavec, M., 1975, Longituoinal hippocampectomy, *Confin. Neurol.* **37:**244–248.

Nadvornik, P., Šramka, M., and Ramamurthi, B., 1977, Combined stereotactic operations, *Vopr. Neirokhir.* **2:**23–29 (Rus.).

Nagaseki, Y., Shibazaki, T., Hirai, T., Kawachima, Y., Hirato, M., Wada, H., Miyazaki, M., and Ohye, C., 1986, Long-term follow-up study of selective Vim thalamotomy, *Appl. Neurophysiol.* **65:**296–302.

Nagib, A., Leal, J., and Voris, H. C., 1966, Successful control of selective anterior sacral rhizotomy for treatment of spastic bladder and ureteric reflux in paraplegics, *Med. Serv. J. Can.* **22**(7)**:**576–581.

Nakajima, H., Fukamachi, A., Isobe, I., Miyazaki, M., Shibazaki, T., and Ohye, C., 1978, Estimation of neural noise. Functional anatomy of the human thalamus, *Appl. Neurophysiol.* **41**(1–4)**:**193–202.

Namba, S., Nakao, J., Matsumoto, Y., Ohmoto, T., and Nishimoto, A., 1984, Electrical stimulation of the posterior limb of the internal capsule for treatment of thalamic pain, *Appl. Neurophysiol.* **47:**137–148.

Namba, S., Wani, T., Shimizu, Y., Fujiwara, N., Namba, Y., Nakama, S., and Nishimoto, A., 1985, Sensory and motor responses to deep brain stimulation. Correlation with anatomical structures, *J. Neurosurg.* **63:**224–234.

Naquet, R., Silva-Barrat, C., and Ménini, C., 1986, Role du thalamus dans la physiopathologie des épilepsies, *Rev. Neurol.* **142:**384–390.

Narabayashi, H., 1962, Neurophysiological ideas on pallidotomy and ventrolateral thalamotomy in hyperkinesis, *Confin. Neurol.* **22:**291–303.

Narabayashi, H., 1967, Further results of thalamic surgery (in the sub-ventrolateral area and the CM) in athetosis and spasticity, *Confin. Neurol.* **29:**256–260.

Narabayashi, H., 1968, Functional differentiation in and around the ventrolateral nucleus of the thalamus based on experience in human stereoencephalotomy, *Johns Hopkins Med. J.* **122:**295–300.

Narabayashi, H., 1969, Muscle tone conducting system and tremor concerned structures, in *The Symposium on Parkinson's Disease* (F. J. Gillingham and I. M. L. Donaldson, eds.), Livingstone, Edinburgh, London, pp. 246–251.

Narabayashi, H., 1977, Experiences of stereotactic surgery on cerebral palsy patients, *Acta Neurosurg. [Suppl.]* **24:**3–10.

Narabayashi, H., 1979, Long-range results of medial amygdalotomy on epileptic traits in adult patients, in *Functional Neurosurgery* (T. Rasmussen and R. Marino, eds.), Raven Press, New York, pp. 243–252.

Narabayashi, H., 1982a, Tremor mechanisms, in *Stereotaxy of the Human Brain* (Georg Schaltenbrand and A. E. Walker, eds.), G. Thieme Verlag, Stuttgart, New York, pp. 510–514.

Narabayashi, H., 1982b, Behavioral disturbances in epilepsy, in *Stereotaxy of the Human Brain* (G. Schaltenbrand and A. E. Walker, eds.), Georg Thieme Verlag, Stuttgart, New York, pp. 669–673.

Narabayashi, H., 1982c, Choreoathetosis and spasticity, in *Stereotaxy of the Human Brain* (G. Schaltenbrand and A. E. Walker, eds.), Georg Thieme Verlag, Stuttgart, New York, pp. 532–543.

Narabayashi, H., 1983a, New aspects of pharmacological treatment in parkinson's disease and the role of stereotactic surgery, *Appl. Neurophysiol.* **46:**319.

Narabayashi, H., 1983b, Recent status of stereotactic surgery, *Surg. Neurol.* **19:**493–496.

Narabayashi, H., and Mizutani, T., 1970, Epileptic seizures and the stereotaxic amygdalotomy, *Confin. Neurol.* **32**(25)**:**289–297.

Narabayashi, H., and Nagao, T., 1972, Ten year follow-up of a case of Raynaud's disease treated by pallidotomy, *Confin. Neurol.* **34**(1–4)**:**152–155.

Narabayashi, H., and Shima, F., 1973, Which is the better amygdala target, the medial or lateral nuclei? in *Surgical Approaches in Psychiatry* (L. V. Laitinen and K. E. Livingston, eds.), University Park Press, Baltimore, pp. 129–134.

Narabayashi, H., and Uno, M., 1966, Long range results of stereotaxic amygdalotomy for behavior disorders, *Confin. Neurol.* **27:**168–171.

Narabayashi, H., Yokochi, F., and Nakajima, Y., 1985, Idiopathic oromandibular dyskinesia treated by Vo complex microstereotactic thalamotomy, *Appl. Neurophysiol.* **48:**309–314.

Nash, C. L., Lorig, R. A., and Schatringer, L. A., 1977, Spinal cord monitoring during operative treatment of the spine, *Clin. Orthop.* **126:**100–105.

Nash, T. P., 1986, Percutaneous radiofrequency lesioning of dorsal root ganglia for intractable pain, *Pain* **24:**67–74.

Nashold, B. S., 1972, Extensive cephalic and oral pain relieved by midbrain tractotomy, *Confin. Neurol.* **34:**382–388.

Nashold, B. S., 1975, Dorsal column stimulation for control of pain. A three year follow-up, *Surg. Neurol.* **4:**146–147.

Nashold, B. S., Jr., 1981, Modification of DREZ lesion technique [letter], *J. Neurosurg.* **55:**1012.

Nashold, B. S., Jr., 1982, Brainstem stereotaxic procedures, in *Stereotaxy of the Human Brain* (G. Schaltenbrand and A. E. Walker, eds.), Georg Thieme, Stuttgart, New York, pp. 475–483.

Nashold, B. S., and Bullitt, E., 1981, Dorsal root entry zone lesions to control central pain in paraplegics, *J. Neurosurg.* **55:**414–419.

Nashold, B. S., and Friedman, H., 1972a, Dorsal column stimulation for control of pain, *J. Neurosurg.* **36:**590–597.

Nashold, B. S., and Friedman, H., 1972b, Electromicturition in paraplegia. Implantation of a spinal neuroprothesis, *Arch. Surg.* **104:**195–204.

Nashold, B. S., and Goldner, J. L., 1975, Electrical stimulation of peripheral nerves for relief of intractable chronic pain, *Med. Instrum.* **9**(5)**:**224–225.

Nashold, B. S., and Montagno, E., 1982, Phrenic nerve pacing in man, *Appl. Neurophysiol.* **45:**38–39.

Nashold, B. S., and Ostdahl, R. H., 1979, Dorsal root entry zone lesions for pain relief, *J. Neurosurg.* **51**(1)**:**59–69.

Nashold, B. S., and Ostdahl, R. H., 1980, Pain relief after dorsal root entry zone lesions, *Acta Neurochir.* Suppl. 383–389.

Nashold, B. S., and Slaughter, D. G., 1969, Effects of stimulating or destroying the deep cerebellar regions in man, *J. Neurosurg.* **31:**172–186.

Nashold, B. S., Jr., and Wilson, W. P., 1966, Central pain. Observations in man with chronic implanted electrodes in the midbrain tegmentum, *Confin. Neurol.* **27:**30–44.

Nashold, B. S., Jr., Wilson, W. P., and Slaughter, D. G., 1969a, Sensations evoked by stimulation in the midbrain of man, *J. Neurosurg.* **30:**14–24.

Nashold, B. S., Wilson, W. P., and Slaughter, D. G., 1969b, Stereotaxic midbrain lesion for central dysesthesia and phantom pain, *J. Neurosurg.* **30**(2)**:**116–126.

Nashold, B. S., Stewart, B., and Wilson, W. P., 1972, Depth electrode studies in centrencephalic epilepsy, *Confin. Neurol.* **34:**252–263.

Nashold, B. S., Flanigin, H., Wilson, W. P., and Stewart, B., 1973, Stereotactic evaluation of bitemporal epilepsy with electrodes and lesions, *Confin. Neurol.* **35:**94–100.

Nashold, B. S., Wilson, W. P., and Flanigan, F., 1975, Effects of stimulation in the mesial temporal lobe in persons with bitemporal epilepsy, *Confin. Neurol.* **37**(1–3)**:**241–242.

Nashold, B. S., Urban, B., and Zorub, D. S., 1976, Phantom relief by focal destruction of substantia gelatinosa of Rolando, in *Advances in Pain Research and Therapy,* Vol. 1 (J. J. Bonica and D. Albe-Fessard, eds.), Raven Press, New York, pp. 959–963.

Nashold, B. S., Mullen, J. B., and Avery, R., 1979, Peripheral nerve stimulation for pain relief using a multicontact electrode system, *J. Neurosurg.* **51**(6)**:**872–873.

Nashold, B. S., Goldner, J. L., Mullen, J. B., and Bright, D. S., 1982, Long-term pain control by direct peripheral nerve stimulation, *J. Bone Joint Surg.* **64A:**1–10.

Nashold, B. S., Ostdahl, R. H., Bullitt, E., Friedman, A., and Brophy, B., 1983, Dorsal root entry zone lesions: A new neurosurgical therapy for deafferentation pain, in *Advances in Pain Research and Therapy* (J. Bonica, ed.), Raven Press, New York, pp. 84–96.

Nathan, P. W., 1963, The descending respiratory pathway in man, *J. Neurol. Neurosurg. Psychiatry* **26:**487–499.

Nathan, P. W., and Wall, P. D., 1974, Treatment of post-herpetic neuralgia by prolonged electrical stimulation, *Br. Med. J.* **3:**645–647.

Nauta, H. J. W., Guinto, F. C., and Pisharodi, M., 1984, Arterial bolus contrast medium enhancement for computed tomographically guided stereotactic biopsy, *Surg. Neurol.* **22:**559–564.

Nazarov, N. N., 1927, Über Alkoholinjektionen in die kortikale Hirnsubstanz bei Athetose, *Zentralbl. Chir.* **54:**1478–1481.

Nesterov, L. N., 1967, *Clinical Picture, Pathophysiology and Surgical Treatment of Kojevnikov Epilepsy and Some Diseases of Extrapyramidal System,* Thesis, Sverdlovsk (Rus.).

Nesterov, L. N., Kravtzov, J. I., and Skupchenko, V. V., 1976,

Stereotactic operations for hyperkinetic form of cerebral palsy, *Vopr. Neirokhir.* **5:**14–17 (Rus.).

Netter, F. H., 1975, *Nervous System,* Vol. 1, CIBA Publications, West Caldwell, N.J.

Newcombe, R., 1975, The lesion in stereotactic subcaudate tractotomy, *Br. J. Psychiatry* **126:**478–481.

Newcombe, R., 1982, Depression and schizophrenic syndromes, in *Stereotaxy of the Human Brain* (G. Schaltenbrand and A. E. Walker, eds.), Georg Thieme Verlag, Stuttgart, New York, pp. 591–599.

Nielsen, V. K., 1984, Pathophysiology of hemifacial spasm. I. Ephaptic transmission and ectopic excitation, *Neurology (N.Y.)* **34:**418–426.

Nielson, R. D., Adams, J. E., and Hosobuchi, Y., 1975, Phantom limb pain. Treatment with dorsal column stimulation, *J. Neurosurg.* **42**(3)**:**301–307.

Nielson, R. D., Watts, G., and Clark, W. K., 1976, Peripheral nerve injury from implantation of chronic stimulating electrodes for pain control, *Surg. Neurol.* **5:**51–53.

Niemeyer, P., 1958, The transventricular amygdala–hippocampectomy in temporal lobe epilepsy, in *Temporal Lobe Epilepsy* (M. Baldwin and P. Bailey, eds.), Charles C. Thomas, Springfield, IL, pp. 461–482.

Nieuwenhuys, R., Voogd, J., and Van Huijzen, C., 1978, *The Human Central Nervous System,* Springer-Verlag, Berlin, Heidelberg, New York.

Niizuma, H., Kwak, R., Ikeda, S., Ohyama, H., Suzuki, J., and Saso, S., 1982, Follow-up results of centromedian thalamotomy for central pain, *Appl. Neurophysiol.* **45:**324–325.

Nishimoto, H., Tsubokawa, T., Yamamoto, T., Kitamura, M., Katayama, Y., and Moriyasu, N., 1980, Inhibitory effect of dorsal column stimulation upon thalamic noxious neurons, *Appl. Neurophysiol.* **43:**336–337.

Nishimoto, A., Namba, S., Nakao, Y., Matsumoto, Y., and Ohmoto, T., 1984, Inhibition of nociceptive neurons by internal capsule stimulation, *Appl. Neurophysiol.* **47:**117–127.

Nitter, R., and Scholl, H., 1978, A new apparatus to avoid certain complications in stereotactic operations at the upper cervical spinal cord, in *Advances in Stereoencephalotomy,* S. Karger, Basel, pp. 143–145.

Nittner, K., 1982, Localization of electrodes in cases of phantom limb pain in the lower limbs, *Appl. Neurophysiol.* **45:**205–208.

Norell, H. A., Winternitz, W. W., Wilson, C. B., and Maddy, J. A., 1970, Stereotaxic cryosurgery in the management of acromegaly, *Int. Surg.* **53:**5–10.

Noth, J., 1986, Long loop reflexes: Concepts and consequences, *Appl. Neurophysiol.* **49:**262–268.

Nugent, G. R., 1982a, Technique and results of 800 percutaneous radiofrequency thermocoagulations for trigeminal neuralgia, *Appl. Neurophysiol.* **45:**504–507.

Nugent, G. R., 1982b, Technique and results of 800 percutaneous radiofrequency thermocoagulations for trigeminal neuralgia, in *Modern Neurosurgery* (M. Brock, ed.), Springer-Verlag, Berlin, pp. 469–475.

Nugent, G. R., and Berry, B., 1974, Trigeminal neuralgia treated by differential percutaneous radiofrequency coagulation of the Gasserian ganglion, *J. Neurosurg.* **40:**517–523.

Nyberg, P., Adolfsson, R., Andén, N. E., and Winblad, B., 1982, Concentrations of dopamine and noradrenaline in some limbic and related regions of the human brain, *Acta Neurol. Scand.* **65:**267–273.

Oberbauer, R. W., Heppner, F., and Schröttner, O., 1982, Die mikrochirurgische Dekompression des Nervus trigeminus im Brückenwinkel, *Nervenarzt* **53:**110–113.

Obrador, S., and Dierssen, G., 1959, Results and complications following one hundred subcortical lesions performed in Parkinson's disease and other hyperkinesias, *Acta Neurochir. (Wien)* **7**(2)**:**206–216.

Obrador, S., and Dierssen, G., 1965, Observations on the treatment of intentional and postural tremor by subcortical stereotaxic lesions, *Confin. Neurol.* **26:**250–253.

Obrador. S., and Dierssen, G., 1966, Sensory responses to subcortical stimulation and management of pain disorders by stereotaxic methods, *Confin. Neurol.* **27:**45–51.

Obrador, S., Carrascosa, R., and Carbonell, J., 1961, Study of some motor syndromes (rigidity, tremor, spasticity and hemidecortication) by the carotid amytal test, *J. Neurosurg.* **18**(4)**:**507–511.

Obrador, S., Delgado, J. M. R., and Martin-Rodriguez, J. G., 1974, The future of functional neurosurgery, in *Recent Progress in Neurological Surgery* (K. Sano and S. Ishii, eds.), Excerpta Medica, Amsterdam, pp. 265–269.

O'Brien, W. J., Rutan, F. M., Sanborn, C., and Omer, G. E., 1984, Effect of transcutaneous electrical nerve stimulation on human blood β-endorphin levels, *Phys. Ther.* **64:**1367–1374.

Ogle, W., French, L., and Peyton, W., 1956, Experiences with high cervical cordotomy, *J. Neurosurg.* **13**(1)**:**81–88.

Oh, S., Tok, S., Allemann, J., Prevost, A., and Schmid, U. D., Exhärese bei okzipitalneuralgie, *Neurochirurgia (Stuttg.)* **26:**47–50.

Ohye, C., 1982, Depth microelectrode studies, in *Stereotaxy of the Human Brain* (G. Schaltenbrand and A. E. Walker, eds.), Georg Thieme Verlag, Stuttgart, New York, pp. 372–389.

Ohye, C., 1987, Stereotactic surgery in movement disorders: Choice of patient, localization of lesion with microelectrodes and long-term results, in *Stereotactic Surgery* (R. R. Tasker, ed.), Hanley and Belfus, Philadelphia, pp. 193–208.

Ohye, C., and Narabayashi, H., 1979, Physiological study of presumed ventralis intermedius nucleus in the human thalamus, *J. Neurosurg.* **50**(3)**:**290–297.

Ohye, C., Nakamura, R., Fukamachi, A., and Narabayashi, H., 1975, Recording and stimulation of the ventralis intermedialis nucleus of the human thalamus, *Confin. Neurol.* **37:**258–264.

Ohye, C., Hirai, T., Miyazaki, M., Shibazaki, T., and Nakajima, H., 1982a, Vim thalamotomy for the treatment of various kinds of tremor, *Appl. Neurophysiol.* **45:**275–280.

Ohye, C., Miyazaki, M., Hirai, T., Shibazaki, T., Nakajima, H., and Nagaseki, Y., 1982b, Primary writing tremor treated by stereotactic selective thalamotomy, *J. Neurol. Neurosurg. Psychiatry* **45:**988–997.

Ohye, C., Kawashima, Y., Hirato, M., Wada, H., and Nakajima, H., 1984a, Stereotactic CT scan applied to stereotactic thalamotomy and biopsy, *Acta Neurochir.* **71**(1–2)**:**55–68.

Ojemann, G., 1985, Enhancement of memory with human ventrolateral thalamic stimulation: Effect evident on a dichotic listening task, *Appl. Neurophysiol.* **48:**212–215.

Ojemann, G., and Fedio, P., 1968, Effect of stimulation of the human thalamus and parietal and temporal white matter on short-term memory, *J. Neurosurg.* **29:**51–59.

Ojemann, G. A., and Ward, A. A., 1971, Speech representation in ventrolateral thalamus, *Brain* **94:**669–680.

Ojemann, G. A., and Ward, A. A., 1973, Abnormal movement disorders, in *Neurological Surgery,* Vol. 3 (J. R. Youmans, ed.), Saunders, Philadelphia, pp. 1829–1867.

Ojemann, G. A., and Ward, A. A., 1975, Stereotactic and other procedures for epilepsy, in *Advances in Neurology,* Vol. 8 (D. D. Purpura, J. K. Perry, and R. D. Waller, eds.), Raven Press, New York, pp. 241–263.

Okudjava, V. M., 1969, *Main Neurophysiological Mechanisms of Epileptic Activity,* Nauka, *Tbilisi* (Rus.).

Olivecrona, H., 1931, Der spastische Shiefhals und seine chirurgische Behandlung, *Arch. Klin. Chir.* **167:**293–301.

Olivecrona, H., and Ladenheim, J., 1957, *Congenital Arteriovenous Aneurysms of the Carotid and Vertebral Artery System,* Springer, Berlin.

Oliver, L. C., 1950, Surgery in Parkinson's disease: Complete section of the lateral column of the spinal cord for tremor, *Lancet* **1:**847–848.

Olivier, A., 1986, Double-headed stereotaxic carrier apparatus for insertion of depth electrodes, *J. Neurosurg.* **65:**258–259.

Olivier, A., and Bertrand, G., 1982, Stereotaxic device for percutaneous twist-drill insertion of depth electrodes and for brain biopsy. Technical note, *J. Neurosurg.* **56:**307–308.

Olivier, A., and de Lotbiniere, A., 1987, Stereotactic technique in epilepsy, in *Stereotactic Surgery* (R. R. Tasker, ed.), Hanley and Belfus, Philadelphia, pp. 257–286.

Olivier, A., Peters, T., Bertrand, G., and Mawko, G., 1985a, Stereotactic system for use with CAT scan, digital angiography and NMR, in *IX Meeting of the World Society for Stereotactic and Functional Neurosurgery,* Toronto, pp. 94–96.

Olivier, A., Peters, T. M., Clark, J. A., Mawko, G., and Winfield, J. A., 1985b, The use of digital subtraction angiography in stereotactic surgery, in *IX Meeting of the World Society for Stereotactic and Functional Neurosurgery,* Toronto.

Ommaya, A. K., and Baldwin, M., 1963, Extravascular brain cooling of the brain in man, *J. Neurosurg.* **20:**8–20.

Onofrio, B. M., 1975, Radiofrequency percutaneous Gasserian ganglion lesions: Results in 140 patients with trigeminal pain, *J. Neurosurg.* **42:**132–139.

Onofrio, B. M., and Campa, H. K., 1972, Evaluation of rhizotomy. Review of 12 years' experience, *J. Neurosurg.* **36:**751–755.

Onofrio, B. M., Yaksh, T. L., and Arnold, P. G., 1981, Continuous low-dose intrathecal morphine administration in the treatment of chronic pain of malignant origin, *Mayo Clin. Proc.* **56:**516–520.

Openchowski, P., 1883, Sur l'action localisée du froid applique à la surface de la region cortical du cérveau, *C. R. Soc. Biol. (Paris)* **7:**38–43.

Oppel, F., and Mulch, G., 1979, Selective trigeminal root section via an endoscopic transpyramidal retrolabyrinthine approach, *Acta Neurochir. (Wien)* **28:**565–571.

Oppel, F., Mulch, G., and Brock, M., 1981, Endoscopic section of the sensory trigeminal root, the glossopharyngeal nerve and the cranial part of the vagus for intractable facial pain caused by upper jaw carcinoma, *Surg. Neurol.* **16**(2)**:**92–95.

Oppenheim, H., 1911, Über eine eigenartige Krampfkrankheit des kindlichen und jugendlichen Alters, *Zentralbl. Neurol.* **30:**1090–1107.

Oppenheim, H., 1923, *Lehrbuch der Nervenkrankheiten,* Berlin.

Ordenstein, L., 1867, *Sur la Paralysie Agitante,* Vol. 1, Martinet, Paris.

O'Reilly, S., Loncin, M., and Cooksey, B., 1965, Dopamine and basal ganglia disorders, *Neurology (Minneap.)* **15:**980–984.

Organ, L. W., Tasker, R. R., and Moody, N. F., 1967, The impedance profile of the human brain as a localization technique in stereoencephalotomy, *Confin. Neurol.* **29:**192–196.

Orthner, H., 1982, Sexual disorders, in *Stereotaxy of the Human Brain* (G. Schaltenbrand and A. E. Walker, eds.), Georg Thieme Verlag, Stuttgart, New York, pp. 600–616.

Orthner, H., and Roeder, F., 1966, Further clinical and anatomical experiences with stereotactic operations for relief of pain, *Confin. Neurol.* **27:**418–430.

Ostertag, C. B., Mennel, H. D., and Kiessling, M., 1980, Stereotactic biopsy of brain tumors, *Surg. Neurol.* **14:**275–283.

Pagni, C. A., 1966, Discussion at the Second International Symposium on Stereoencephalotomy, in *Advances in Stereoencephalotomy* (E. A. Spiegel and H. T. Wycis, eds.), S. Karger, Basel, pp. 88–89.

Pagni, C. A., 1976, Central pain and painful anesthesia: Pathophysiology and treatment of sensory deprivation syndromes due to central and peripheral nervous system lesions, in *Progress in Neurological Surgery,* Vol. 8 (H. Krayenbühl, ed.), S. Karger, Basel, pp. 132–257.

Pagni, C. A., and Marossero, F., 1965, Some observations of the human rhinencephalon—a stereoelectroencephalographic study, *Electroencephalogr. Clin. Neurophysiol.* **18:**260–271.

Pagura, J. R., 1983, Percutaneous radiofrequency spinal rhizotomy, *Appl. Neurophysiol.* **46:**138–146.

Pagura, J. R., Schnapp, M., and Passarelli, P., 1983, Percutaneous radiofrequency glossopharyngeal rhizotomy for cancer pain, *Appl. Neurophysiol.* **46:**154–159.

Paillas, J. E., Chabert, V., Alliez, B., and Masquin, L., 1971, Réflexions sur la psychochirurgie dans le traitement des névroses obsessionelles, *Sem. Hop. Paris* **47**(15)**:**944–949.

Pakkenberg, H., 1963, Globus pallidus in parkinsonism, *Acta Neurol. Scand.* **39**[Suppl. 4]**:**139–144.

Pallis, C. A., 1971, Parkinsonism: Natural history and clinical features, *Br. Med. J.* **5776:**683–690.

Pampiglione, G., and Falconer, M., 1960, Electrical stimulation of the hippocampus in man, in *Handbook of Physiology*, Vol. 2, American Physiological Society, Washington, pp. 1391–1395.

Papez, J. W., 1937, A proposed mechanism of emotions, *Arch. Neurol. Psychiatr.* **38:**725–743.

Papez, J. W., 1942, A summary of fiber connections of the basal ganglia with each other and with other portions of the brain, *Res. Publ. Assoc. Res. Nerv. Ment. Dis.* **21:**21–68.

Parkes, J. D., Calver, D. M., Zilkha, R. J., and Knill-Jones, R., 1970, Controlled trial of amantadine hydrochloride in Parkinson's disease, *Lancet* **1:**259–262.

Parkes, J. D., Debono, A. G., and Marsden, C. D., 1976, Bromocriptine in parkinsonism: Long-term treatment, dose response, and comparison with levodopa, *J. Neurol. Neurosurg. Psychiatry* **37:**1101–1108.

Parkinson, D., and Shields, C. B., 1974, Persistent trigeminal artery: Its relationship to the normal branches of the cavernous carotid artery, *J. Neurosurg.* **40**(2)**:**244–248.

Parkinson, J., 1817, *An Essay on the Shaking Palsy*, Whittingham & Rowland, London.

Patil, A. A., 1983, Computer tomography (CT) oriented rotary stereotactic system.. A technical note, *Acta Neurochir. (Wien)* **68**(1–2)**:**19–26.

Patil, A. A., Yamanashi, W. S., Ross-Duggan, J. W., and Lester, P. D., 1986, Magnetic resonance imaging (MRI) stereotaxis using Patil system. A technical note, *Acta Neurochir. (Wien)* **82:**141–143.

Patterson, R., and Little, S., 1943, Spasmodic torticollis, *J. Nerv. Ment. Dis.* **98:**6.

Patti, F., Marano, P., Nicoletti, F., Giammona, G., and Nicoletti, P., 1985, Generalized and focal dystonic syndromes, *Eur. Neurol.* **24:**386–391.

Payne, E. E., 1969, *Atlas of Pathology of the Brain*, Springer, New York.

Pecker, J., Simon, J., Faivre, J., and Scarabin, J. M., 1975, Neuroradiology and biopsy of pineal tumors by stereotaxy, in *Fifth Congress of Neurosurgery*, Oxford, p. 110.

Pecker, J., Scarabin, J. M., Brücher, J. M., and Vallee, B., 1979a, *Démarche Stéréotaxique en Neurochirurgie Tumorale*, Pierre Fabre, Paris.

Pecker, J., Scarabin, J. M., Vallee, B., and Brücher, J. M., 1979b, Treatment in tumours of the pineal region: Value of stereotaxic biopsy, *Surg. Neurol.* **12:**341–345.

Pedachenko, G. A., and Orlov, Yu. A., 1977, Cryosurgery of arteriovenous aneurysms of the brain, *Vopr. Neirokhir.* **6:**3–6 (Rus.).

Pedersen, H., Taudorf, K.,and Melchior, J. C., 1982, Computed tomography in spastic cerebral palsy, *Neuroradiology* **23:**275–278.

Pederzoli, M., Girrotti, F., Scigliano, G., Aiello, G., Carella, F., and Caraceni, T., 1983, L-Dopa long-term treatment in Parkinson's disease: Age-related side effects, *Neurology (N.Y.)* **33:**1518–1522.

Pellettieri, L., 1980, *Surgical versus Conservative Treatment of Intracranial Arteriovenous Malformations*, Springer Verlag, New York.

Peluso, F., and Gybels, J., 1972, Computer calculation of the position of side-protruding electrode tip during penetration in human brain, *Confin. Neurol.* **34**(1–4)**:**94–100.

Penfield, W., and Baldwin, M., 1952, Temporal lobe seizures and the technique of subtotal temporal lobectomy, *Ann. Surg.* **136:**625–634.

Penfield, W., and Baldwin, M., 1953, Temporal lobe seizures and the technique of subtotal temporal lobectomy, *Trans. Am. Surg. Assoc.* **70:**288–297.

Penfield, W., and Erickson, T., 1941, *Epilepsy and Cerebral Localization*, Charles C. Thomas, Springfield, IL.

Penfield, W., and Jasper, H., 1954, *Epilepsy and the Functional Anatomy of the Human Brain*, Little, Brown, Boston.

Penfield, W., and Jasper, H., 1959, *Epilepsy and Functional Anatomy of Human Brain*, Little, Brown, Boston.

Penfield, W., and Steelman, H., 1947, The treatment of focal epilepsy by cortical excision, *Ann. Surg.* **126:**740–762.

Penn, R. D., and Etzel, M. Z., 1977, Chronic cerebellar stimulation and developmental reflexes, *J. Neurosurg.* **46:**506–511.

Penn, R. D., Gottlieb, G. I., and Agarwal, G. C., 1978a, Cerebellar stimulation in man. Quantitative changes in spasticity, *J. Neurosurg.* **48:**779–786.

Penn, R. D., Whisler, W. W., Smith, C. A., and Jasnoff, W. A., 1978b, Stereotactic surgery with image processing of computerized tomographic scans, *Neurosurgery* **3**(2)**:**157–163.

Penn, R. D., Myklebust, B. M., Gottlieb, G. L., Agarwal, G. G., and Etzel, M. E., 1980, Chronic cerebellar stimulation for cerebral palsy, *J. Neurosurg.* **53:**160–165.

Penn, R. D., Paice, J. A., Gottschalk, W., and Ivankovich, A. D., 1984, Cancer pain relief using chronic morphine infusion. Early experience with a programmable implanted drug pump, *J. Neurosurg.* **61:**302–306.

Peresedov, V. V., and Kandel, E. I., 1983, Device for stereotactic aspiration of intracerebral haematomas, *Vopr. Neirokhir.* No. 6:53–55 (Rus.).

Perlow, M. J., Freed, W. J., and Hoffer, B. J., 1979, Brain grafts reduce motor abnormalities produced by destruction of nigrostrial dopamine system, *Science* **204:**643–647.

Pernett, G., and Nishioka, H., 1966, Arteriovenous malformations. An analysis of 545 cases of craniocerebral arteriovenous malformations and fistulae reported to the cooperative study, *J. Neurosurg.* **25:**467–490.

Perria, C., Francaviglia, N., Borzone, M., Chinnici, A., Piano, E., and Pacini, P., 1983, The value and limitations of the CO_2 laser in neurosurgery, *Neurochirurgia (Stuttg.)* **26:**6–11.

Pertuiset, B., 1979, Middle cerebral artery aneurysms: Treatment, in *Cerebral Aneurysms. Advances in Diagnosis and Therapy* (H. W. Pia, C. Langmaid, and J. Zierski, eds.), Springer-Verlag, Berlin, pp. 286–289.

Pertuiset, B., Aneri, D., and Goutorbe, J., 1976, Variations du volume sanguin cérébral local en fonction de la pression artérielle moyenne chez l'homme, *Rev. Neurol.* **132**(3)**:**213–218.

Petelin, L. S., 1965, On the classification and pathogenesis of hyperkinesias, *Zh. Nevropathol. Psikhiatr.* **2:**179–186 (Rus.).

Petelin, L. S., 1970, *Extrapyramidal Hyperkinesias*, Meditsina, Moscow (Rus.).

Petelin, L. S., and Koteneva, V. M., 1975, *Pathochemical*

Grounds for Treatment of Parkinsonism, Meditsina, Moscow (Rus.).

Petelin, L. S., and Pigarov, V. A., 1973, Side-effects of L-dopa, in *Side-Effects of Drugs,* Moscow, pp. 60–62 (Rus.).

Petelin, L. S., Vartanjan, K. Z., and Romensaya, L. C., 1977, Prolonged L-dopa treatment of parkinsonism and its pathochemical aspect, *Zh. Nevropatol. Psikhiatr.* **12:**1810–1813 (Rus.).

Peters, T. M., Olivier. A., Clark, J. A., Mawko, G., Ethier, G., and Dieumegarde, M., 1985, Integration of CT, MRI and digital angiographic techniques in stereotaxic surgical planning, in *IX Meeting of the World Society for Stereotactic and Functional Neurosurgery,* Toronto, p. 42.

Pevsner, P. H., and Doppman, J. L., 1980, Therapeutic embolization with a microballoon catheter system, *Am. J. Roentgenol.* **134:**949–958.

Phillips, C. G., Matthews, P. B. C., and Rushworth, G., 1959, Experimental observations on hypertonus. I. Chairman's general introduction. II. Hypertonus and the gamma motoneurones. III. The nature of the functional disorder in the hypertonic states, *Oxford Univ. Bull.* **7:**1–7.

Pia, H. W., 1972, The surgical treatment of intracerebral and intraventricular haematomas, *Acta Neurochir. (Wien)* **27:**149–164.

Pia, H. W., 1979, Anterior cerebral artery aneurysms: Operative treatment, in *Cerebral Aneurysms. Advances in Diagnosis and Therapy* (H. W. Pia, C. Langmaid, and J. Zierski, eds.), Springer-Verlag, Berlin, pp. 274–281.

Picard, L., Moret, J., and Lepoire, J., 1984, Traitement endovasculaire des angiomes artério-veineux intracérébraux. Technique, indications, résultats, *J. Neuroradiol.* **11:**9–28.

Picaza, J. A., Cannon, B. W., Hunter, S. E., Boyd, A. S., Guma, J., and Maurer, D., 1975, Pain suppression by peripheral nerve stimulation, *Surg. Neurol.* **1:**115–126.

Picaza, J. A., Hunter, S. E., and Cannon, B. W., 1977/78, Pain suppression by peripheral nerve stimulation: Chronic effects of implanted devices, *Appl. Neurophysiol.* **40:**223–234.

Pierrot-Deseilligny, E., 1983, Pathophysiology of spasticity, *Triangle* **22:**165–174.

Pierry, A., and Cameron, M., 1979, Chronic hemifacial spasm from posterior fossa arteriovenous malformation, *J. Neurol. Neurosurg. Psychiatry* **42:**670–672.

Pilon, R. N., and Baker, A. R., 1976, Chronic pain control by means of an epidural catheter, *Cancer* **37:**903–905.

Pineda, A., 1975, Dorsal column stimulation and its prospects, *Surg. Neurol.* **4:**157–163.

Pineda, A., 1977, Complications of dorsal column stimulation, *J. Neurosurg.* **48**(1):64–68.

Pines, L. J., Zurabashvili, A. D., and Kunakov, K. A., 1939, Connections of thalamus opticus with brain cortex in human, *Sov. Psychiatr.* **2:**308–315 (Rus.).

Pirogov, N. I., 1961, *The Grounds of Military Surgery. Collected Works,* Vol. 5, Meditsina, Moscow, p. 92.

Piskun, W. S., Stevens, E. A., La Morgese, J. R., Paullus, W. S., and Meyers, P. W., 1979, A simplified method of CT assisted localization and biopsy of intracranial lesions, *Surg. Neurol.* **11:**413–417.

Plotkin, R., 1982, Results in 60 cases of deep brain stimulation for chronic intractable pain, *Appl. Neurophysiol.* **45:**173–178.

Poblete, M., and Zamboni, R., 1974, Stereotactic pituitary implantation of radioisotopes by transfrontal route, *Acta Neurochir. (Wien) [Suppl.]* **21:**159–163.

Poblete, M., and Zamboni, R., 1975, Stereotaxic third ventriculocisternostomy, *Confin. Neurol.* **37:**150–155.

Poblete, M., Palestini, M., Figueroa, E., Gallardo, R., Rojas, J., Covarrubias, M., and Doyharcabal, Y., 1970, Stereotaxic thalamotomy (lamella medialis) in aggressive psychiatric patients, *Confin. Neurol.* **32**(2–5):326–331.

Podivinsky, F., 1960, Electromyographical analysis of torticollis spastica in surgical practice, *Bratisl. Lek. Listy* **40**(II):15–26.

Podivinsky, F., 1964, *Analýza Mimovalných Pohubov Hlavy (Torticollis Spastica),* Bratislava.

Podivinsky, F., 1968, Torticollis, in *Handbook of Clinical Neurology. Diseases of the Basal Ganglia,* Vol. 6 (P. J. Vinken and G. W. Bruyn, eds.), North Holland, Amsterdam, pp. 567–603.

Poirier, L., 1972, Physiopathology of akinesia, in *Parkinson's Disease,* Vol. 1 (J. Siegfried, ed.), Huber, Bern, pp. 115–126.

Polenov, A. L., 1928, New way to surgical treatment of some forms of hyperkinesias, *Zh. Sov. Chir.* **3**(6)**:**954–974 (Rus.).

Polenov, A. L., 1937, New developments in surgery of central nervous system, *Vestn. Khir.* **49:**223–227 (Rus.).

Poletti, C. E., 1982, Open cordotomy—new techniques, in *Operative Neurosurgical Techniques. Indications, methods and results,* Vol. 2 (H. H. Schmidek and W. H. Sweet, eds.), Grune & Stratton, New York, pp. 1119–1136.

Poletti, C. E., 1987, CNS functional restoration, in *Neurosurgery* (R. R. Tasker, ed.), Hanley and Belfus, Philadelphia.

Poletti, C. E., and Schmidek, H. H., 1982, Pain control with implantable systems for the long-term infusion of intraspinal opioids in man, in *Operative Neurosurgical Techniques. Indications, Methods, and Results,* Vol. 2 (H. H. Schmidek and W. H. Sweet, eds.), Grune & Stratton, New York, pp. 1199–1210.

Poletti, C. E., Cohen, A. M., Todd, D. P., Ojemann, R. G., Sweet, W. H., and Zervas, N. T., 1981, Cancer pain relieved by long-term epidural morphine with permanent indwelling systems for self-administration, *J. Neurosurg.* **55:**581–584.

Pollettieri, L., 1979, Surgical versus conservative treatment of intracranial arteriovenous malformations, *Acta Neurochir. (Wien)* Suppl. 29.

Pollock, L. J., and Davis, L., 1930, Muscle tone in parkinsonian states, *Arch. Neurol. Psychiatry* **23:**319.

Pollock, M., and Hornabrook, R. W., 1966, The prevalence, natural history and dementia of Parkinson's disease, *Brain* **89**(3)**:**429–448.

Pomme, B., Sanny, P., Planche, R., and Gibert, J., 1958, Attitude dystonia with torsion spasm. Marked improvement with peduncular pyramidotomy, *Rev. Neurol.* **99**(6)**:**648–652.

Pool, J. L., 1954, Psychosurgery in older people, *J. Am. Geriatr. Soc.* **2:**456–465.

Pool, J. L., and Potts, D. G., 1965, *Aneurysms and Arteriovenous Anomalies of the Brain. Diagnosis and Treatment,* Harper & Row, New York.

Poskanzer, D. C., and Schwab, R. S., 1963, Cohort analysis of Parkinson's syndrome. Evidence for a single etiology related to subclinical infection about 1920, *J. Chron. Dis.* **16**:961–973.

Poskanzer, D. C., Schwab, R. S., and Fraser, D. W., 1969, The cohort phenomenon in Parkinson's syndrome, in *Third Symposium of Parkinson's Disease* (F. J. Gillingham and I. U. L. Donaldson, eds.), Livingstone, Edinburgh, pp. 12–16.

Post, F., and Schurr, P. H., 1977, Changes in the pattern of diagnosis of patients subjected to psychosurgical procedures, with comments on their use in the treatment of self-mutilation and anorexia nervosa, in *Neurosurgical Treatment in Psychiatry, Pain, and Epilepsy* (W. H. Sweet, S. Obrador, and J. G. Martin-Rodriguez, eds.), University Park Press, Baltimore, pp. 261–266.

Post, F., Rees, L., and Schurr, P. H., 1968, An evaluation of bimedial leucotomy, *Br. J. Psychiatry* **114**:223–244.

Potthoff, P. C., Tetteh, J., and Riechert, T., 1972, Postencephalographic psychoorganic syndrome in parkinsonism, *Confin. Neurol.* **34**:285–294.

Pourple, M. H., 1960, Traitement neuro-chirurgical des contractures chez les paraplégiques posttraumatiques, *Neurochirurgie* **6**:229–236.

Powers, S. R., Adams, J. E., Edwards, M. S. B., Boggan, J. E., and Hosobuchi, Y., 1984a, Pain relief from dorsal root entry zone lesions made with argon and carbon dioxide microsurgical lasers, *J. Neurosurg.* **61**:841–847.

Powers, S. R., Edwards, M. S. B., Boggan, J. E., Pitts, L. H., Gutin, P. H., Hosobushi, Y., Adams, J. E., and Wilson, C. B., 1984b, Use of the argon surgical laser in neurosurgery, *J. Neurosurg.* **60**:523–530.

Poza, M., Perez-Espejo, M. A., Lage, J. M., and Sola, J., 1985, Intracranial tumors biopsy. CT-guided stereotactic surgery, *Appl. Neurophysiol.* **48**:482–487.

Pribram, K. H., 1960, Theory of physiological psychology, *Annu. Rev. Psychol.* **11**:1–40.

Privat, J. M., Benezech, J., Frerebeau, P., and Gros, C., 1976, Sectorial posterior rhizotomy, a new technique of surgical treatment for spasticity, *Acta Neurochir. (Wien)* **35**:181–195.

Privat, J. M., Allieu, Y., Bonnel, F., and De Godebont, J., 1985, Les douleurs après lésions traumatiques du plexus brachial, *Neurochirurgie* **31**:435–441.

Pruvot, P., Bancaud, J., Delandsheer, J. M., Bordas-Ferri, M., and Talairach, J., 1972, Crises èpileptiques généralisées et lésion corticale focale. (A propos d'une epilepsie frontale post-traumatique.), *Rev. Electroencephalogr.* **2**:165–170.

Przuntek, H., and Muhr, H., 1983, Essential familial myoclonus, *J. Neurol.* **230**:153–162.

Pudenz, R. H., Bullara, L. A., and Talalla, A., 1975, Electrical stimulation of the brain. I. Electrodes and electrode arrays, *Surg. Neurol.* **4**:37–42.

Putnam, T. J., 1933, Treatment of athetosis and dystonia by section of the extrapyramidal motor tracts, *Arch. Neurol. Psychiatry* **29**:504–521.

Putnam, T. J., 1934, Myelotomy of the commissure. A new method of treatment for pain in the upper extremities, *Arch. Neurol. Psychiatry* **32**:1189–1193.

Putnam, T. J., 1938, Results of treatment of athetosis by section of extrapyramidal tracts in the spinal cord, *Arch. Neurol. Psychiatry* **39**:258–275.

Putnam, T. J., 1939, Diagnosis and treatment of athetosis and dystonia, *J. Bone Joint Surg.* **21**:948–957.

Putnam, T. J., 1940, Treatment of unilateral paralysis agitans by section of the lateral pyramidal tract, *Arch. Neurol. Psychiatry* **44**:950–962.

Putnam, T. J., 1942, The operative treatment of disease characterized by involuntary movement, *Assoc. Res. Nerv. Ment. Dis.* **21**:666–692.

Puusepp, L. M., 1914, *Operative Treatment of Spastic Palsies,* Prakticheskaya Meditsina, St. Peterburg (Rus.).

Puusepp, L. M., 1930, Cordotomia posterior lateralis (fasc. Burdachi) on account of trembling and hypertonia of the muscles in hand, *Folia Nevropatol.* **10**:62–66.

Puzillo, M. V., 1958, *Pathways Between Thalamus Opticus and Frontal Lobe,* Thesis, Moscow (Rus.).

Quaade, F., 1974, Letter: Stereotaxy for obesity, *Lancet* **1**:267.

Quaade, F., Vaernet, K., and Larsson, S., 1974, Stereotaxic stimulation and electrocoagulation of the lateral hypothalamus in obese humans, *Acta Neurochir. (Wien)* **30**:111–117.

Rack, P. M. H., 1978, Mechanical and reflex factors in human tremor, in *Physiological Tremor, Pathological Tremor and Clonus* (J. E. Desmedt, ed.), S. Karger, Basel, pp. 17–27.

Radermecker, M., Franck, G., and Bostem, F., 1963, Hemichorée et occlusion de l'artère carotide interne, *Acta Neurol. Belg.* **63**:950–954.

Ralston, B. L., 1962, Hemispherectomy and hemithalamectomy in man, *J. Neurosurg.* **19**:909–912.

Ramamurthi, B., and Davidson, A., 1975, Central median lesions and lysis of 89 cases, *Confin. Neurol.* **37**:63–72.

Ramamurthi, B., and Kalyanaraman, S., 1982, Stereotaxic targets for epilepsy, in *Stereotaxy of the Human Brain* (G. Schaltenbrand and A. E. Walker, eds.), Georg Thieme Verlag, Stuttgart, New York, pp. 653–660.

Ramamurthi, B., Ramamurthi, R., and Narayanan, R., 1982, Long-term follow-up of functional neurosurgery in psychiatric disorders—experience of 30 cases, *Appl. Neurophysiol.* **45**:538–539.

Ramani, S. V., Yap, J. C., and Gumnit, R. J., 1980, Stereotactic fields of Forel interruption for intractable epilepsy, *Appl. Neurophysiol.* **43**:104–108.

Ramirez, L. F., and Levin, A. B., 1984, Pain relief after hypophysectomy, *Neurosurgery* **14**:499–504.

Rand, R. W., 1960, Dystonia musculorum deformans alleviated by chemopallidothalamectomy and substantia nigralysis, *J. Neurosurg.* **17**:1093–1099.

Rand, R. W., 1969, *Microneurosurgery,* C. V. Mosby, St. Louis.

Rand, R. W., and Gardner, W. J., 1981, Neurovascular decompression of the trigeminal and facial nerves for tic douloureuse and hemifacial spasm, *Surg. Neurol.* **16**:329–332.

Rand, R. W., and Mosso, J. A., 1972, Treatment of cerebral

aneurysms by stereotaxic ferromagnetic silicone thrombosis, *Bull. Los Angeles Neurol. Soc.* **38:**21–23.

Rand, R. W., Brown, W. J., and Stern, W. E., 1956, Surgical occlusion of anterior choroidal arteries in parkinsonism. Clinical and neuropathologic findings, *Neurology (Minneap.)* **6**(6):390–401.

Rand, R. W., Rinfret, A. P., and von Leden, H. (eds.), 1968, *Cryosurgery*, Thomas, Springfield, IL.

Rand, R. W., Heuser, E., and Dache, A., 1969, Stereotaxic transsphenoidal biopsy and cryosurgery of pituitary tumors, *Am. J. Roentgenol.* **105:**273–286.

Ranson, S. W., and Magoun, H. W., 1933, Respiratory and pupillary reactions induced by electrical stimulation of hypothalamus, *Arch. Neurol. Psychiatry* **29:**1179–1194.

Rap, Z. M., and Mempel, E., 1971, Badania histopatologiczne uszkodzen yader podstrawy mozgu po zastosowaniu metody kryochirurgiczney, *Neurol. Neurochir. Psychiatr. Pol.* **5:**49–54.

Rasmussen, T., 1964, Surgical therapy of frontal lobe epilepsy, *Epilepsia* **4:**181–198.

Rasmussen, T., 1975, Surgery of frontal lobe epilepsy, *Adv. Neurol.* **8:**197–205.

Rasmussen, T., 1979a, Cortical resection for medically refractory focal epilepsy: Results, lessons, and questions, in *Functional Neurosurgery* (T. Rasmussen and R. Marino, eds.), Raven Press, New York, pp. 253–269.

Rasmussen, T., 1976b, Surgical aspects of temporal lobe epilepsy, in *IV Meeting of the European Society for Stereotactic and Functional Neurosurgery*, Paris, pp. 188–199.

Rasmussen, T., 1980, Surgical aspects of temporal lobe epilepsy. Results and problems, *Acta Neurochir. (Wien)* **30:**13–24.

Rasmussen, T., 1983, Surgical treatment of complex partial seizures: Results, lessons and problems, *Epilepsia* [Suppl. 1] **24:**65–76.

Rautakorpi, I., Takala, J., Marttila, R. J., Sievers, K., and Rinne, V. K., 1982, Essential tremor in a Finnish population, *Acta Neurol. Scand.* **66:**58–67.

Ray, C. D., 1975, Control of pain by electrical stimulation. A clinical following up review, in *Advances in Neurosurgery* (M. Brock and J. Hamer, eds.), Springer-Verlag, Berlin, Heidelberg, New York, pp. 64–72.

Ray, C. D., and Burton, C. V., 1980, Deep brain stimulation for severe, chronic pain, *Acta Neurochir. (Wien) [Suppl.]* **30:**289–293.

Rayport, M., Fergusson, S. M., and Corrie, W. S., 1983, Outcomes and indications of corpus callosum section for intractable seizure control, *Appl. Neurophysiol.* **46:**47–51.

Razumovski, V. I., 1913, Approach to surgical treatment of cortical traumatic and nontraumatic epilepsy, *Nevrol. Vestn.* **20**(3):401–416 (Rus.).

Redfearn, J. W. T., 1957, Frequency analysis of physiological and neurotic tremors, *J. Neurol. Neurosurg. Psychiatry* **20:**302.

Reeves, A. G. (ed.), 1985, *Epilepsy and the Corpus Callosum*, Plenum Press, New York.

Regensburg, I., 1930, Zur Klinik des hereditären torsionsdystonischen Symptomkomplexes, *Monatsschr. Psychiatr. Neurol.* **75:**323–345.

Reshetnjak, W. K., and Meizerov, E. E., 1981, *Theory and Practice of Reflexotherapy*, Shtiinza, Kishinew (Rus.).

Rey, L., 1959, Applications of cold to general biology, to medicine and to surgery, *Sci. Med. Lille* **77:**427–436.

Rey, L., 1962, *Preserving of Life by Cold* (Rus. ed.), Meditsina, Moscow.

Reyes-Vazquez, C., and Dafny, N., 1983, Microiontophoretically applied THIP effects upon nociceptive responses of neurons in medial thalamus, *Appl. Neurophysiol.* **46:**254–260.

Reynolds, A. F., and Oakley, J. C., 1982, High frequency cervical epidural stimulation for spasticity, *Appl. Neurophysiol.* **45**(1–2):93–97.

Reynolds, D. V., 1969, Surgery in the rat during electrical analgesia induced by focal brain stimulation, *Science* **164:**444–445.

Rhodes, M. L., Glenn, W. V., Azzawi, Y. M., and Slater, R., 1982, Stereotactic neurosurgery using 3-D image data from computed tomography, *J. Med. Syst.* **6:**105–119.

Rhoton, A. L., 1978, Microsurgical neurovascular decompression for trigeminal neuralgia and hemifacial spasm, *J. Florida Med. Assoc.* **65:**425–428.

Rhoton, A. L., 1981, Microsurgical neurovascular decompression for hemifacial spasm, in *7th International Congress of Neurological Surgery*, Munich, p. 61.

Ribera, A. B., and Cooper, I. S., 1960, The natural history of dystonia musculorum deformans: A clinical study, *Arch. Pediatr.* **77**(2):55–71.

Richards, P., Shawden, H., and Illingworth, R., 1983, Operative findings in microsurgical exploration of the cerebello-pontine angle in trigeminal neuralgia, *J. Neurol. Neurosurg. Psychiatry* **46:**1098–1101.

Richardson, A., 1973a, Stereotactic limbic leucotomy. Surgical techniques, *Postgrad. Med. J.* **49:**860–864.

Richardson, A., 1973b, Stereotactic limbic leucotomy. A follow-up at 16 months, *J. Psychiatry* **128:**226–240.

Richardson, D. E., 1974, Thalamotomy for control of chronic pain, *Acta Neurochir. [Suppl.]* **21:**77–88.

Richardson, D. E., 1981, Long-term follow-up of deep brain stimulation for relief of chronic pain in the human, in *Abstracts of the 7th International Congress of Neurological Surgery*, Georg Thieme Verlag, Stuttgart, p. 64.

Richardson, D. E., 1982a, Analgesia produced by stimulation of various sites in the human beta-endorphin system, *Appl. Neurophysiol.* **45:**116–122.

Richardson, D. E., 1982b, Long-term follow-up of deep brain stimulation for relief of chronic pain in the human, *Mod. Neurosurg.* **1:**449–453.

Richardson, D. E., and Akil, H., 1974, Chronic self-administration of brain stimulation for pain relief in human patients, *Proc. Am. Assoc. Neurol. Surg.*

Richardson, D. E., and Akil, H., 1977, Pain reduction by electrical brain stimulation in man. Chronic self-administration in the periventricular gray matter, *J. Neurosurg.* **47:**178–183.

Richardson, D. E., and Zorub, D. S., 1970, Sensory function of the pulvinar, *Confin. Neurol.* **32:**165–173.

Richardson, R. R., and McLone, D. G., 1978, Percutaneous

epidural neurostimulation for paraplegic spasticity, *Surg. Neurol.* **9**:153–156.

Richardson, R. R., Sequeira, E. B., and Cerullo, L. J., 1979, Spinal epidural neurostimulation for treatment of acute and chronic intractable pain. Initial and long term results, *Neurosurgery* **5**:344–348.

Richter, H., 1923, Beitrage zur Klinik und pathologischen Anatomie der extrapyramidalen Bewegungsstörungen, *Arch. Psychiatr. Nervenkr.* **67**:226–231.

Riechert, T., 1953, Neurochirurgische Therapie, in *Handbuch der Inneren Medizin,* ed. 4, Vol. 5, Part 1, Springer, Berlin, pp. 1472–1543.

Riechert, T., 1955, Entfernung von tiefsitzenden Hirnstechtsplittern mit Hilfe des stereotaktischen Operationsverfahrens, *Z. Neurochir.* **15**:159–164.

Riechert, T., 1959, Über die Technik und einige Indikationen der gezielten Hirnoperationen, *Nervenarzt* **9**:385–391.

Riechert, T., 1960, Die chirurgische Behandlung der zentralen Schmerzustände einschliesclich der stereotaktischen Operationen in Thalamus und Mesencephalon, *Acta Neurochir (Wien)* **8**:136–152.

Riechert, T., 1962, Long term follow-up of results of stereotaxic treatment of extrapyramidal disorders, *Confin. Neurol.* **22**:356–363.

Riechert, T., 1966a, Stereotaxic operations for extrapyramidal motor disturbances with particular regards to age groups, *Confin. Neurol.* **26**:213–217.

Riechert, T., 1966b, Relief of certain types of intractable pain, in *Pain,* Vol. 39 (Knighton and Dumke, eds.), Little, Brown, Boston, pp. 519–529.

Riechert, T., 1972, The stereotactic technique and its application in extrapyramidal hyperkinesia, *Confin. Neurol.* **34**:325–330.

Riechert, T., 1973, Operative therapy of chronic pain by electric stimulation of the dorsal column, *Dtsch. Med. Wochenschr.* **98**:1130–1137.

Riechert, T., 1980, *Stereotactic Brain Operations. Methods, Clinical Aspects, Indications,* Hans Huber, Bern, Stuttgart, Vienna.

Riechert, T., and Mundinger, F., 1955, Beschreibung und Amvendung eines Zielgerätes für stereotaktische Hirnoperationen (II. Modell), *Acta Neurochir. (Wien) [Suppl.]* **3**:308–337.

Riechert, T., and Mundinger, F., 1959, Stereotaktische Geräte, in *Einführung in die Stereotaktischen Operationen mit einem Atlas des menschlichen Gehirns* (G. Schaltenbrand and P. Bailey, eds.), Georg Thieme, Stuttgart.

Riechert, T., and Mundinger, F., 1960, Indications, technique and results of the stereotactic operations upon the hypophysis using radio-isotopes, *J. Nerv. Ment. Dis.* **131**(1):1–9.

Riechert, T., and Mundinger, F., 1964. Combined stereotaxic operation for treatment of deep-seated angiomas and aneurysms, *J. Neurosurg.* **21**:358–363.

Riechert, T., and Richter, D., 1972, Surgical treatment of tremor in multiple sclerosis and essential tremor, *Munch. Med. Wochenschr.* **114**:2025–2028.

Riechert, T., and Spuler, H., 1982, Instrumentation of stereotaxy, in *Stereotaxy of the Human Brain* (G. Schaltenbrand and A. E. Walker, eds.), Georg Thieme, Stuttgart, New York, pp. 350–363.

Riechert, T., and Wolff, M., 1951, Über ein neues Zielgerät zur intrakraniellen elektrische Ableitung und Ausschaltung, *Arch. Psychiatr. Nervenkr.* **186**:225–230.

Riechert, T., Mölbert, E., Gisinger, M. A., 1967, Biopsien während stereotaktischer Operationen beim Parkinsonsyndrom, *Neurochirurgia (Stuttg.)* **10**(3):106–118.

Riechert, T., Hassler, R., and Mundinger, E., 1975, Pathologic–anatomical findings and cerebral localization in stereotactic treatment of extrapyramidal motor disturbances in multiple sclerosis, *Confin. Neurol.* **37**:24–40.

Ries, L., and Tytus, J. S., 1960, Rapid freezing: A surgical technique, *Bull. Mason Clin.* **14**:20–26.

Riklan, M., Cullinan, T., and Cooper, I. S., 1977, A psychometric study of chronic cerebellar stimulation in man, *J. Nerv. Ment. Dis.* **164**:176–181.

Rinfret, A. P., 1968, Cryobiology: Some fundamentals in surgical context, in *Cryosurgery* (R. W. Rand, ed.), Thomas, Springfield, IL, pp. 19–31.

Rinne, U. K., 1980, Recent advances in the treatment of Parkinson's disease, *Acta Neurol. Scand. [Suppl.]* **78**:103–121.

Rinne, U. K., 1981, Treatment of Parkinson's disease: Problems with a progressing disease, *J. Neural. Transm.* **51**(1–2):161–174.

Rinne, U. K., 1982, Parkinson's disease as a model for changes in dopamine receptor dynamics with aging, *Gerontology* **28**(1):35–52.

Rinne, U. K., Sonninen, V., and Hyyppa, M., 1971, Effect of L-dopa on brain monoamines and their metabolites in Parkinson's disease, *Life Sci.* **10**(10):549–557.

Rivas, J. J., and Lobato, R. D., 1985, CT-assisted stereotaxic aspiration of colloid cysts of the third ventricle, *J. Neurosurg.* **62**:238–242.

Robb, P., 1975, Focal epilepsy: The problem, prevalence, and contributing factors, *Adv. Neurol.* **8**:11–22.

Roberts, D. W., Strohbehn, J. W., Hatch, J. F., Murray, W., and Kettenberger, H., 1986, A frameless stereotactic integration of computerized tomographic imaging and the operating microscope, *J. Neurosurg.* **65**:545–549.

Robson, J. T., and Bonica, J., 1950, The vagus nerve in surgical consideration of glossopharyngeal neuralgia, *J. Neurosurg.* **7**:482–484.

Rockswold, G. L., Bradley, W. E., and Chou, S. N., 1973, Differential sacral rhizotomy in the treatment of neurogenic bladder dysfunction. Preliminary report of six cases, *J. Neurosurg.* **38**:748–754.

Rockswold, G. L., Chou, S. N., and Bradley, W. E., 1978, Reevaluation of differential sacral rhizotomy for neurological bladder disease, *J. Neurosurg.* **48**:773–778.

Roeder, F. D., 1965a, Stereotaxic lesions of the tuber cinereum in sexual deviation, *Confin. Neurol.* **27**:162–163.

Roeder, F. D., 1965b, Indicationen und technische Durchführung der stereotaktischen Ausschaltung des Tuber Cinereum bei Krimineller sexueller Triebhaftigkeit, *Excerpta Med.* **94**:246–247.

Roeder, F., Müller, D., and Orthner, H., 1971, Stereotaxic treatment of psychoses and neuroses, in *Special Topics in Stereotaxis* (W. Umbach, ed.), Hippokrates, Stuttgart, pp. 82–105.

Roeder, F., Orthner, H., and Müller, D., 1972, The stereotaxic treatment of pedophilic homosexuality and other sexual deviations, in *Psychosurgery* (E. Hitchcock, L. Laitinen, and K. Vaernet, eds.), Charles C. Thomas, Springfield, IL, pp. 87–111.

Rogulov, V. A., 1968, *Surgical Anatomy of Thalamus Opticus and Its Ventrolateral Region Related to Thalamotomy*, Thesis, Leningrad (Rus.).

Roland, P. E., Larsen, B., Lassen, N. A., and Skinhoj, E., 1980, Supplementary motor area and other cortical areas in organization of voluntary movements in man, *J. Neurophysiol.* **43:**118–136.

Roldan, P., Broseta, J., and Barcia-Salorio, J. L., 1982, Chronic VPM stimulation for anesthesia dolorosa following trigeminal surgery, *Appl. Neurophysiol.* **45**(1–2)**:**112–113.

Romenskaya, L. H., 1976, *Epidemiology, Clinical and Pharmacotherapy of Parkinsonism*, Thesis, Moscow.

Romodanov, A. P. (ed.), 1970, Temporal lobe epilepsy and its surgical treatment, Zdorovja, Kiev (Rus.).

Romodanov, A. P., 1972, Combined surgical interventions of the brain for epilepsy, in *Present Limits of Neurosurgery* (I. Fusek and Z. Kunc, eds.), Avicenum, Prague, pp. 407–408.

Romodanov, A. P., 1980, Modern methods of operations for epilepsy and its results, in *Surgical Treatment for Epilepsy*, Nauka, Tbilisi, pp. 9–14 (Rus.).

Romodanov, A. P., and Laponogov, O. A., 1971, Our experience of 420 stereotactic operations, *Zh. Nevropatol. Psychiatr.* **5:**641–645 (Rus.).

Romodanov, A. P., and Michailovski, V. S., 1965, Thalamocortical tractotomy for treatment of intractable pain, in *Pain Syndromes*, Zdorovja, Kiev, pp. 5–12 (Rus.).

Romodanov, A. P., Ryabokon, N. S., and Bojik, V. P., 1971a, Methods of surgical operations for epilepsy, in *Proceedings of the First All-Union Congress of Neurosurgeons*, Vol. 3, Moscow, pp. 118–120 (Rus.).

Romodanov, A. P., Zozulya, Yu. A., Laponogov, O. A., and Sklyar, A. A., 1971b, Surgical treatment of malignant tumors in deep-seated portions of the brain practised in conjunction with chemotherapy, *Vopr. Neirokhir.* **6:**24–29 (Rus.).

Romodanov, A. P., Rasin, S. D., Ryabokon, N. S., and Lishchilin, M. G., 1974, Clinical and pathophysiologic characteristics of epileptogenic focus, in *Epileptogenic Focus and Surgical Treatment of Epilepsy*, Zdorovja, Kiev, pp. 5–9 (Rus.).

Romodanov, A. P., Stepien, L., and Mempel, E., 1975, Comparative analysis of results of classic and stereotaxic operations for epilepsy, in *Symptoms of Stereotactic Treatment of Epilepsy*, Bratislava, pp. 99–105.

Romodanov, A. P., Zozulja, Y. A., and Shcheglov, V. I., 1979, Balloon catheter occlusion of the feeding vessels of arteriovenous malformations of the brain, *Zbl. Neurochir.* **40:**21–28.

Romodanov, A. P., Laponogov, O. A., Kopjev, O. V., and Tsimbaljuk, V. I., 1983, Ultrastructural alterations of ventro-oral nuclei of the human thalamus in extrapyramidal diseases, in *Modern Problems of Neurosurgery*, Meditsina, Kaunas, pp. 46–52 (Rus.).

Rondot, P., and Bathien, N., 1978, Pathophysiology of parkinsonian tremor. A study of the pattern of motor unit discharges, in *Physiological Tremor, Pathological Tremor and Clonus* (J. E. Desmedt, ed.), S. Karger, Basel, pp. 138–149.

Roquefeuil, B., Benezech, J., Blanchet, P., Batier, C., Frerebeau, P., and Gros, C., 1984, Intraventricular administration of morphine in patients with neoplastic intractable pain, *Surg. Neurol.* **21:**155–158.

Rosanov, V. N., and Chugunov, S. A., 1927, Approach to the surgical treatment of postencephalitic parkinsonism, *Russ. Clin.* **7:**36–42 (Rus.).

Rose, M., 1937, Die morphologische Grundlage der Torsions dystonie, *Arch. Biol. Tow. Nauk.* **6**(Fasc.2)**:**1–19.

Rosen, J. A., and Barsoum, A. N., 1979, Failure of chronic dorsal column stimulation in multiple sclerosis, *Ann. Neurol.* **6:**66–67.

Rosenberg, M., Curtis, L., and Bourke, D. L., 1978, Transcutaneous electrical nerve stimulation for the relief of postoperative pain, *Pain* **5:**129–133.

Rosenthal, C., 1922, Die dysbatisch–dysstatische Form der Torsions dystonie, *Arch. Psychiatr.* **66:**445–472.

Rosomoff, H. L., 1959, Experimental brain injury during hypothermia, *J. Neurosurg.* **16:**177–187.

Rosomoff, H. L., 1969, Bilateral percutaneous cervical radiofrequency cordotomy, *J. Neurosurg.* **31:**41–46.

Rosomoff, H. L., Carroll, F., and Brown, J., 1965, Percutaneous radiofrequency cervical cordotomy; technique, *J. Neurosurg.* **23:**639–644.

Rossi, G. F., 1980, Why, when, and how surgery of epilepsy? *Acta Neurochir. (Wien) [Suppl.]* **30:**7–11.

Rossi, G. F., Walter, R. D., and Crandall, P. H., 1968, Generalized spike and wave discharges and nonspecific thalamic nuclei, *Arch. Neurol.* **19**(2)**:**174–183.

Rossi, G. F., Gentilomo, A., and Colicchio, G., 1974, Le problème de la recherche de la topographie d' origine de l'épilepsie, *Schweiz. Arch. Neurol. Neurochir. Psichiatrie* **115:**229–270.

Rossi, G. F., Colicchio, G., Gentilomo, A., and Scerrati, M., 1978, Discussion on the causes of failure of surgical treatment of partial epilepsies, *Appl. Neurophysiol.* **41:**29–37.

Rossolimo, G. I., 1906–1907, The brain topograph (device for projection of brain parts on the skull surface), *Ezheg. Ekaterin Bolnizi* **1:**63–65 (Rus.).

Rostotskaya, V. I., Motsnaya, M. Ya., and Tyapina, R. S., 1968, Experience gained in employing ventriculography with maiodyl in neurosurgical treatment of children, *Vopr. Neirokhir.* **6:**39–42 (Rus.).

Roth, D. A., and Mark, V. H., 1973, Thalamotomy for relief of pain, in *Neurological Surgery*, Vol. III (J. R. Youmans, ed.), W. B. Saunders, Philadelphia, pp. 1783–1789.

Rothwell, J. C., Ubeso, J. A., Traub, M. M., and Marsden, C. D., 1983, The behavior of the long-latency stretch reflex in patients with Parkinson's disease, *J. Neurol. Neurosurg. Psychiatry* **46:**35–44.

Rozkanski, N. A., and Lagutina, G., 1957, *Essays on Physiology of Nervous System,* Medgiz, Leningrad (Rus.).

Rümler, B., Schaltenbrand, G., Spuler, H., and Wahren, W., 1972, Somatotopic array of the ventro-oral nucleus of the thalamus based on electrical stimulation during stereotactic procedures, *Confin. Neurol.* **34:**197–199.

Runge, W., 1936, Die Erkrankungen des extrapyramidalen motorischen Systems, *Fortschr. Neurol. Psychiatr.* **8**(109):133–159.

Rushton, J. G., Stevens, J. C., and Miller, R. H., 1981, Glossopharyngeal (vagoglossopharyngeal) neuralgia, *Arch. Neurol.* **38:**201–206.

Rushworth, G., 1960, Spasticity and rigidity: An experimental study and review, *J. Neurol. Neurosurg. Psychiatry* **23:**99–118.

Rushworth, G., 1969, The gamma system in parkinsonism, *J. Neurol.* **2:**34–50.

Rushworth, R. G., 1980, Stereotactic guided biopsy in the computerized tomographic scanner, *J. Surg. Neurol.* **14:**451–454.

Rusinko, J., Walker, C. F., and Richardson, D. E., 1985, Computer model of focal deep brain stimulation by multiple scalp electrodes, in *IX Meeting of the World Society for Stereotactic and Functional Neurosurgery,* Toronto, p. 52.

Russell, A., and Brown, M. D., 1979, A computerized tomography–computer graphics approach to stereotaxis localization, *J. Neurosurg.* **50:**715–720.

Ryabokon, N. S., 1980, Long-term results of surgical treatment of epilepsy after ablation of epileptic zone, in *Surgical Treatment of Epilepsy,* Nauka, Tbilisi, pp. 21–26 (Rus.).

Ryan, D. C., 1950, Dystonia musculorum deformans with report of a case in a child, *Med. J. Aust.* **11**(10):360–362.

Sachs, E., 1935, The subpial resection of the cortex in the treatment of Jacksonian epilepsy (Horsley operation), *Brain* **58:**492–503.

Safronov, V. A., 1970, Muscle tone in parkinsonism, *Vopr. Neirokhir.* **3:**11–17 (Rus.).

Safronov, V. A., 1974, Spectromechanomyographic investigation of the muscle tone in akinetic form of parkinsonism, *Vopr. Neirokhir.* **1:**29–34 (Rus.).

Safronov, V. A., 1979, γ-Rigidity in parkinsonism, *Zh. Nevropatol. Psychiatr.* **79**(2):164–168 (Rus.).

Safronov, V. A., and Kandel, E. I., 1975, The shortening reflex (Westphal phenomenon) in dystonia musculorum deformans, *Zh. Nevropatol. Psychiatr.* **8:**1495–1500 (Rus.).

Safronov, V. A., Vasin, N. Ya., and Lesov, N. S., 1978, The effect of dentatotomy on the muscular tonus in infantile cerebral paralysis, *Vopr. Neirokhir.* **1:**24–30 (Rus.).

Sager, O., 1962, *Diencephalon,* Meditsina, Moscow (Rus. ed.).

Saglam, S., Kragt, C. L., Wilson, C. B., and Kaplan, S. L., 1972, Graded cryohypophysectomy in the rhesus monkey. Histopathology and endocrine function, *J. Neurosurg.* **36:**169–177.

Sahni, R. S., Ghatak, N. R., and Young, H. F., 1985, CT-guided stereotactic biopsy of lower brain stem lesions, *Appl. Neurophysiol.* **48**(1–6):488–489.

Saito, Y., and Ohye, C. H., 1974, Automatically-controlled recording and processing of thalamic unit discharges in human stereotactic operations, *Confin. Neurol.* **36:**314–325.

Sakare, K. M., Garmashov, J. A., and Beshlyaga, P. V., 1982, Anatomo–topographic basis of simultaneous stereotactic longitudinal hippocampotomy for treatment of temporal lobe epilepsy, in *Complex Treatment of Epilepsy,* Meditsina, Leningrad, pp. 92–99 (Rus.).

Salar, G., Mingrino, S., Trabucchi, M., Bosio, A., and Semenza, C., 1981, Evaluation of endorphin content in the CSF of patients with trigeminal neuralgia before and after Gasserian ganglion thermocoagulation, *J. Neurosurg.* **55**(6):935–937.

Salar, G., Job, I., Trabucchi, M., Bosio, A., and Mingrino, S., 1982, CSF endorphins level in patients with cancer pain treated by percutaneous cordotomy, in *Modern Neurosurgery* (M. Brock, ed.), Springer-Verlag, Berlin, pp. 464–468.

Salar, G., Baratto, C., Ori, C., Iob, I., and Mingrino, S., 1983a, Percutaneous thermolesion of the glossopharyngeal nerve: Results and anatomophysiological considerations, in *6th Meeting of the European Society for Stereotactic and Functional Neurosurgery,* Rome, p. 28.

Salar, G., Ori, C., Baratto, C., Iob, I., and Mingrino, S., 1983b, Selective percutaneous thermolesions of the ninth cranial nerve by lateral approach: Report of eight cases, *Surg. Neurol.* **20:**276–279.

Samii, M., and Moringlane, R., 1983, Thermocoagulation of the substantia gelatinosa for the treatment of pain, in *7th European Congress of Neurosurgery,* Brussels, p. 144.

Samotokin, B. A., and Khilko, V. A., 1968, Results and nearest perspectives of the surgery of cerebral aneurysms, in *Proceedings of Conference of Neurosurgeons,* Meditsina, Leningrad, pp. 97–99 (Rus.).

Samotokin, B. A., and Khilko, V. A., 1973, *Aneurysms and Arteriovenous Malformations of the Brain,* Meditsina, Leningrad (Rus.).

Sanchez, G., Imparato, A., and Ransohoff, J., 1969, Internal jugular oxygen saturation and arteriovenous oxygen difference during artificial embolizations of arteriovenous malformations, *J. Neurosurg.* **30:**227–232.

Sanders, K. H., Klein, C. E., Mayor, T. E., Heym, C., and Handwerker, H. O., 1980, Differential effects of noxious and non-noxious input on neurones according to location in ventral periaqueductal grey and dorsal raphe nucleus, *Brain Res.* **186**(1):83–87.

Sano, K., 1954, Cingulectomy in the treatment of agitated mental defectives, *Neurol. Med. Chir. (Tokyo)* **6:**146–156.

Sano, K., 1962, Sedative neurosurgery with reference to posteromedial hypothalamotomy, *Neurol. Med. Chir. (Tokyo)* **4:**112–142.

Sano, K., 1974, Surgery of the hypothalamus—in commemoration of O. Foerster, in *Recent Progress in Neurological Surgery,* Excerpta Medica, Amsterdam, pp. 210–218.

Sano, K., 1977, Intralaminar thalamotomy (thalaminotomy) and posteromedial hypothalamotomy in the treatment of intractable pain, in *Progress in Neurological Surgery,* Vol. 8 (H. Krayenbühl, P. E. Maspes, and W. H. Sweet, eds.), S. Karger, Basel, pp. 50–103.

Sano, K., 1982, Aggressiveness, in *Stereotaxy of the Human Brain* (G. Schaltenbrand and A. E. Walker, eds.), Georg Thieme Verlag, Stuttgart, New York, pp. 617–621.

Sano, K., and Yoshioka, M., 1967, Autonomic, somatomotor and electroencephalographic responses upon stimulation of the hypothalamus and rostral brain stem in man, *Confin. Neurol.* **29:**257–261.

Sano, K., Yoshioka, M., Ogashiwa, M., Ishijima, B., and Ohye, C., 1966, Thalamolaminotomy. A new operation for relief of intractable pain, *Confin. Neurol.* **27:**63–66.

Sano, K., Yoshioka, M., Ogashiwa, M., Ishijima, B., Ohye, C., Sekino, H., and Mayanagi, Y., 1967, Central mechanisms of neck movements in the human brain stem, *Confin. Neurol.* **29:**107–110.

Sano, K., Yoshioka, M., Mayanagi, Y., Sekino, H., Yoshimasu, N., and Tsukamoto, Y., 1970a, Stimulation and destruction of and around the interstitial nucleus of Cajal in man, *Confin. Neurol.* **32:**118–125.

Sano, K., Mayanagi, Y., Sekino, H., Ogashiwa, M., and Ishijima, B., 1970b, Results of stimulation and destruction of the posterior hypothalamus in man, *J. Neurosurg.* **33:**689–707.

Sano, K., Sekino, H., Tsukamoto, Y., Yoshimasu, N., and Ishijima, B., 1972, Stimulation and destruction of the region of the interstitial nucleus in cases of torticollis and see–saw nystagmus, *Confin. Neurol.* **34**(5)**:**331–338.

Sano, K., Kimbo, M., Saito, I., and Basugi, N., 1975a, Artificial embolization of inoperable angioma with polymerizing substance, in *Cerebral Angiomas—Advances in Diagnosis and Therapy* (H. W. Pia, J. R. W. Gleave, E. Grote, and J. Zierski, eds.), Springer-Verlag, Berlin, pp. 222–229.

Sano, K., Sekino, H., and Hashimoto, I., 1975b, Posteromedial hypothalamotomy in the treatment of intractable pain, *Confin. Neurol.* **37:**285–290.

Sano, H., Kanno, T., and Katada, K., 1980, Treatment of the dural AVM-embolization using aron alpha, *Neurol. Med. Chir. (Tokyo)* **20:**845–851.

Saradjishvili, P. M., 1980, Neurosurgical aspect of the problem of so-called primary-generalized seizures, in *Surgical Treatment of Epilepsy,* Nauka, Tbilisi, pp. 14–19 (Rus.).

Saradjishvili, P. M., and Geladze, T. S., 1977, *Epilepsy,* Meditsina, Moscow (Rus.).

Saris, S. C., Iacono, R. P., and Nashold, B. S., 1985, Dorsal root entry zone lesions for post-amputation pain, *J. Neurosurg.* **62:**72–76.

Saris, S. C., Silver, J. M., Vieira, J. F. S., and Nashold, B. S., 1986, Sacrococcygeal rhizotomy for perineal pain, *Neurosurgery* **19:**789–793.

Sato, H., 1982, Functional characteristics of human skeletal muscle revealed by spectral analysis of the surface electromyogram, *Electroencephalogr. Clin. Neurophysiol.* **22:**459–516.

Saunders, M. L., Young, H. F., Becker, D. P., Greenberg, R. P., Newlon, P. G., Corales, R. L., Ham, W. T., and Povlishock, J. T., 1980, The use of the laser in neurological surgery, *Surg. Neurol.* **14:**1–10.

Saunders, R. L., and Coombs, D. W., 1982, Comments, in *Operative Neurosurgical Techniques, Indications, Methods and Results* (H. H. Schmidek and W. H. Sweet, eds.), Grune & Stratton, New York, pp. 1211–1212.

Sax, D. S., O'Donnell, B., Butters, N., Menzer, L., Montgomery, K., and Kayne, H., 1983, Computed tomographic, neurologic, and neuropsychological correlates of Huntington's disease, *Int. J. Neurosci.* **18:**21–36.

Scanarini, M., and Mingrino, S., 1980, Functional classification of pituitary adenomas, *Acta Neurochir. (Wien)* **52:**195–202.

Scarabin, J. M., Pecker, J., Brucher, J. M., Vallée, B., Guegan, Y., Faivre, J., and Simon, J., 1978, Stereotactic exploration in 200 supratentorial brain tumors. Its value in addition to computerized tomography, *Neuroradiology* **16:**591–593.

Scerrati, M., Fiorentino, A., Fiorentino, M., and Pola, P., 1984, Stereotaxic device for polar approaches in orthogonal systems. Technical note, *J. Neurosurg.* **61:**1146–1147.

Scerrati, M., Pola, P., Florentino, A., and Florentino, M., 1985, Two devices to improve the flexibility of Talairach's stereotactic apparatus, in *IX Meeting of the World Society for Stereotactic and Functional Neurosurgery,* Toronto, p. 19.

Schaltenbrand, G., 1953, Orthoroentgenography, *Am. J. Roentgenol.* **170:**95–105.

Schaltenbrand, G., and Bailey, P. (eds.), 1959, *Introduction to Stereotaxis with an Atlas of the Human Brain,* Georg Thieme Verlag, Stuttgart, New York.

Schaltenbrand, G., and Bailey, P., 1977, *Atlas for Stereotaxy of the Human Brain,* Georg Thieme, Stuttgart.

Schaltenbrand, G., and Wahren, W. (eds.), *1977, Atlas for Stereotaxy of the Human Brain,* ed. 2, Georg Thieme Verlag, Stuttgart.

Schaltenbrand, G., and Wahren, W., 1982, Electroanatomical observations, in *Stereotaxy of the Human Brain* (G. Schaltenbrand and A. E. Walker, eds.), Georg Thieme Verlag, Stuttgart, New York, pp. 390–409.

Schaltenbrand, G., and Walker, A. E., 1982, *Stereotaxy of the Human Brain,* Georg Thieme Verlag, Stuttgart, New York.

Schaltenbrand, G., Spuler, H., Nadjmi, M., Hoff, H. C., and Wahren, W., 1966, Die stereotaktische Behandlung der Epilepsien, *Confin. Neurol.* **27:**111–113.

Schinz, H. R., Baensch, W. E., Friedel, E., and Vehlinger, E., 1952, *Lehrbuch der Röntgendiagnostik,* Georg Thieme Verlag, Stuttgart.

Schirmer, M., and Wenken, H., 1981, Long-term experiences with Bischof's longitudinal lumbar myelotomy in the treatment of high grade spasticity, in *7th International Congress of Neurological Surgery,* Munich, p. 224.

Schlegel, W., Scharfenberg, H., and Sturm, V., 1981, Direct visualization of intracranial tumours in stereotactic and angiographic films by computer calculation of longitudinal CT-sections: A new method for stereotactic localization of tumour outlines, *Acta Neurochir. (Wien)* **58:**27–35.

Schlösser, C., 1907, *Kongress für Innere Medizin XXIV,* pp. 49–55.

Schmitt, W., and Scholz, W., 1932, Klinischer und pathologisch-anatomischer Beitrag zur Torsiondystonie, *Dtsch. Z. Nervenheilkd.* **126:**53–79.

Schneider, E., and Fischer, P. A., 1982, Long-term experience with bromcriptine in advanced parkinsonism, *J. Neurol.* **228**:249–258.

Schneider, E., Fischer, P. A., and Becker, H., 1977, Relationship between arteriosclerosis and cerebral atrophy in Parkinson's disease, *J. Neurol.* **217**:11–16.

Schneider, E., Fischer, P. A., Jacobi, P., Becker, H., and Beyer, M., 1979, Cerebral atrophy and long-term response to levodopa in Parkinson's disease, *J. Neurol.* **222**:37–43.

Schneider, R. C., Crosby, E. C., and Kahn, E. A., 1963, Certain afferent cortical connections of the rhinencephalon, in *The Rhinencephalon and Related Structures* (Bergmann and Schadé, eds.), Elsevier, Amsterdam, pp. 191–217.

Schott, B., and Lapras, C., 1961, Les mouvements anormaux (Sémiologie, thérapeutique et Physiopathologie générales), *Gas. Med. Fr.* **68**:853–878.

Schott, B., Laurent, B., and Mauguière, F., 1986, Role du thalamus dans la physiopathologie des épilepsies, *Rev. Neurol.* **142**:384–390.

Schreiner, L., and Kling, A., 1953, Behavioral changes following paleocortical injury in cat, *J. Neurophysiol.* **16**:643–659.

Schürmann, K., Butz, M., and Brock, M., 1972, Temporal retrogasserian resection of trigeminal root versus controlled elective percutaneous electrocoagulation of the ganglion of Gasser in the treatment of trigeminal neuralgia, *Acta Neurochir. (Wien)* **26**:33–53.

Schuster, G. D., and Infante, M. C., 1980, Pain relief after low back surgery: The efficacy of transcutaneous electrical nerve stimulation, *Pain* **8**:299–302.

Schvarcz, J. R., 1974, Spinal cord stereotactic surgery, in *Recent Progress in Neurological Surgery* (K. Sano and S. Ishii, eds.), Excerpta Medica, Amsterdam, pp. 234–241.

Schvarcz, J. R., 1975, Stereotactic trigeminal tractotomy, *Confin. Neurol.* **37**:73–77.

Schvarcz, J. R., 1977a, Postherpetic craniofacial dysaesthesiae: Their management by stereotaxic trigeminal nucleotomy, *Acta Neurochir. (Wien)* **38**:65–72.

Schvarcz, J. R., 1977b, Functional exploration of the spinomedullary junction, *Acta Neurochir. (Wien)* **24**:179–185.

Schvarcz, J. R., 1978, Spinal cord stereotactic techniques: Trigeminal nucleotomy and extralemniscal myelotomy, *Appl. Neurophysiol.* **41**:99–112.

Schvarcz, J. R., 1985, Chronic stimulation of area septalis for the relief of intractable pain, *Appl. Neurophysiol.* **48**:191–194.

Schvarcz, J. R., Sica, E. R., and Morita, E., 1981, Electrophysiological changes induced by chronic stimulation of the dentate nucleus, in *VIII Meeting of the World Society for Stereotactic and Functional Neurosurgery,* Zurich, p. 15.

Schvarcz, J. R., Sica, R. S., Morita, E., Bronstein, A., and Sanz, O., 1982, Electrophysiological changes induced by chronic stimulation of the dentate nucleus for cerebral palsy, *Appl. Neurophysiol.* **45**:55–61.

Schwab, R. S., 1964, Problems in the clinical estimation of rigidity (hypertonia), *Clin. Pharmacol. Ther.* **5**:942–946.

Schwab, R. S., 1966, Neurological assessment of the deficits in Parkinson's disease before and after stereotactic surgery, in *Advances in Stereoencephalotomy,* Part 2 (E. A. Spiegel and H. T. Wycis, eds.), S. Karger, Basel, New York, pp. 218–223.

Schwab, R. S., and Cobb, S., 1939, Simultaneous electromyograms and electroencephalograms in paralysis agitans, *J. Neurophysiol.* **2**:36–41.

Schwab, R. S., and England, A. C., 1968, Parkinson syndromes due to various specific causes, in *Handbook of Clinical Neurology,* Vol. 6 (P. J. Vinken and G. W. Bruyn, eds.), North-Holland, Amsterdam, pp. 227–247.

Schwab, R. S., and England, A. C., 1969, Amantadine HCl (Symmetrel) and its relation to *levo*-dopa in the treatment of Parkinson's disease, *Trans. Am. Neurol. Assoc.* **94**:85–90.

Schwab, R. S., England, A. C., and Peterson, E., 1959, Akinesia in Parkinson's disease, *Neurology (Minneap.)* **9**:65–72.

Schwalbe, W., 1908, Eine eigenthümliche tonische Krampform mit hysterischen Symptomen, Berlin.

Schwartz, H. G., 1967, High cervical cordotomy, *J. Neurosurg.* **26**(4):452–455.

Schwartz, J., and Elizan, T. S., 1979, Research of virus fractions and virus-specific products in brain tissue in idiopathic parkinsonism, *Ann. Neurol.* **6**:261–263.

Schwarz, G. A., and Fahn, S., 1970, Newer medical treatment in parkinsonism, *Med. Clin. North Am.* **54**:773–785.

Scott, R. M., Brody, J. A., and Cooper, I. S., 1970, The effect of thalamotomy on the progress of unilateral Parkinson's disease, *J. Neurosurg.* **32**:286–288.

Scott, F., Bradley, W., and Timm, G., 1974, Treatment of urinary incontinence by an implantable prosthetic urinary sphincter, *J. Urol.* **112**:75–80.

Scott, M., 1973, Ligation of an anterior cerebral artery for aneurysms of the anterior communicating artery complex, *J. Neurosurg.* **38**:481–487.

Scoville, W. B., 1949, Selective cortical undercutting as means of modifying and studying frontal lobe function in man, *J. Neurosurg.* **6**:65–73.

Scoville, W. B., 1954, Orbital undercutting in the treatment of psychoneuroses, depressions and senile emotional states, *Dis. Nerv. Syst.* **15**:11–16.

Scoville, W. B., 1966, Extradural spinal surgery. Rhizotomy, *J. Neurosurg.* **25**:94–95.

Scoville, W. B., 1969a, Hearing loss following exploration of cerebellopontine angle in treatment of hemifacial spasm, *J. Neurosurg.* **31**:47–49.

Scoville, W. B., 1969b, Partial extracranial section of seventh nerve for hemi-facial spasm, *J. Neurosurg.* **31**:106–108.

Scoville, W. B., and Bettis, D. B., 1975, Results of orbital undercutting today: A personal series, in *Neurosurgical Treatment in Psychiatry, Pain and Epilepsy* (W. H. Sweet, S. Obrador, and J. G. Martin-Rodriguez, eds.), University Park Press, Baltimore, pp. 189–202.

Scoville, W. B., and Bettis, D. B., 1979, Motor tics of the head and neck, *Acta Neurochir. (Wien)* **48**:47–66.

Seaber, J. H., and Nashold, B. S., 1980, Comparison of ocular motor effects of unilateral stereotactic midbrain lesions in man, *Neuroophthalmology* **1**:95–99.

Sedan, R., and Harter, M., 1966, L'irradiation interstitielle hypophysaire dans l'exophtalmie oedémateuse maligne, *Neurochirurgie* **12**:226–240.

Sedan, R., and Lazorthes, Y., 1978, La neurostimulation éléctrique therapeutique, *Neurochirurgie* **24**(Suppl. 1):54–59.

Sedan, R., Regis, S., Lyagnoli, S., Saxer, J., Lavielle, J., and Vacherai, S., 1969, Ventral posterieur du thalamus. Correlations anatomo-neuro-physiologiques, *Neurochirurgie* **15**(2):131–136.

Sedan, R., Peràgut, J., and Farnarier, P., 1981, Place de la biopsie en condition stereotaque dans les gliomas, *Neurochirurgie* **27**(5):285–286.

Seeger, W., 1975, The artificial embolisation of inoperable angiomas, in *Cerebral Angiomas,* Springer-Verlag, Berlin, Heidelberg, New York, pp. 213–222.

Segal, M., and Sandberg, D., 1977, Analgesia produced by electrical stimulation of catecholamine nuclei in the rat brain, *Brain Res.* **123**:369–372.

Seitzer, D., Fröder, M., Schäfer, H., and Dieckman, G., 1980, Ein kompaktes Tischerechnersystem zur Zielpunktsberechnung in der stereotaktischen Neurochirurgie, *Neurochirurgia (Stuttg.)* **23**:62–66.

Sekhar, L. N., Heros, R. C., and Kerber, C. W., 1979, Carotid–cavernous fistula following percutaneous retrogasserian procedures, *J. Neurosurg.* **51**:700–706.

Selby, G., 1967, Stereotactic surgery for the relief of Parkinson's disease. A critical review, *J. Neurol. Sci.* **5**:315–342.

Selby, G., 1968, Parkinson's disease, in *Handbook of Clinical Neurology. Diseases of the Basal Ganglia,* Vol. 6. (P. J. Vinken and G. W. Bruyn, eds.), North-Holland, Amsterdam.

Sem-Jacobsen, C. W., 1966, Depth electrographic observation related to Parkinson's disease, *J. Neurosurg.* **24**:388–402.

Sem-Jacobsen, C. W., Petersen, M. C., Lazarte, J. A., Dodge, H. W., and Holman, C. B., 1955, Electroencephalographic rhythms from the depths of the frontal lobe in 60 psychiatric patients, *Clin. Neurophysiol.* **7**:193–210.

Sem-Jacobsen, C. W., Petersen, M. C., Dodge, H. W., Jr., Lazarte, J. A., and Holman, C. B., 1956, Electroencephalographic rhythms from the depths of the parietal, occipital, and temporal lobes in man, *Electroencephalogr. Clin. Neurophysiol.* **8**:263–278.

Serbinenko, F. A., 1971a, Catheterization and occlusion of main cerebral vessels, in *The First All-Union Congress of Neurosurgeons,* Vol. 1, Moscow, pp. 114–119 (Rus.).

Serbinenko, F. A., 1971b, Surgical treatment of fistulas between meningeal arteries and cavernous sinus, in *The First All-Union Congress of Neurosurgeons,* Vol. 1, Moscow, pp. 119–123 (Rus.).

Serbinenko, F. A., 1971c, Balloon occlusion of a cavernous portion of the carotid artery as a method of treating carotid–cavernous fistulas, *Vopr. Neirokhir.* **6**:3–9 (Rus.).

Serbinenko, F. A., 1971d, Catheterization and occlusion of major cerebral vessels and prospects for the development of vascular neurosurgery, *Vopr. Neirokhir.* **5**:17–27 (Rus.).

Serbinenko, F. A., 1974a, Balloon catheterization and occlusion of major cerebral vessels, *J. Neurosurg.* **41**:125–126 (Rus.).

Serbinenko, F. A., 1974b, Occlusion and ballooning of arterial aneurysms of the cerebral arteries, *Vopr. Neirockir.* No. 4:8–15.

Serbinenko, F. A., 1976, Five years experience with endovascular neurosurgery, in *The II All-Union Congress of Neurosurgeons,* Vol. 2, Moscow, pp. 516–519 (Rus.).

Serbinenko, F. A., Filatov, Yu. M., Smirnov, N. A., Lysachev, A. G.. and Sazanova, O. V., 1978, The remote results of balloon occlusion of afferent vessels of arteriovenous aneurysms, *Vopr. Neirokhir.* **3**:3–8 (Rus.).

Seymour, R. J., Levin, S., Tyrell, B., and Forsham, P. H., 1978. Long-term results of cryohypophysectomy for the treatment of acromegaly, in *Treatment of Pituitary Adenomas,* Georg Thieme, Stuttgart, pp. 253–260.

Shah, Y., and Dostrovsky, J. O., 1980. Electrophysiological evidence for a projection of the periaqueductal gray matter to nucleus raphe magnus in cat and rat, *Brain Res.* **193**:534–538.

Shalnikov, A. I., Kandel, E. I., and Kukin, A. V., 1970, Subsequent improvement of the device for local freezing of subcortical structures in stereotactic operations, *Vopr. Neirokhir.* **3**:51–52 (Rus.).

Shanko, G. G., 1979, *Generalized Tic (De La Tourette Disease) in Children,* Nauka i Technika, Minsk (Rus.).

Shapiro, A. K., Shapiro, E., and Wayne, H., 1973, Organic factors in Gilles de la Tourette's syndrome, *Br. J. Psychiatry* **122**:659–664.

Shapiro, A. K., Shapiro, E. S., Broun, R. D., and Sweet, R. D., 1978, *Gilles de la Tourette Syndrome,* Raven Press, New York.

Sharkey, P. C., Dimitrijevic, M. R., Nakajima, K., and Shwerwood, A. M., 1985, Spinal cord stimulation for the treatment of muscle hypertonia in patients with chronic spinal cord injury, *Appl. Neurophysiol.* **48**:105–106.

Shaw, K. M., Hunter, K. R., and Stern, G. M., 1972, Medical treatment of spasmodic torticollis, *Lancet* **1**:1399.

Shaw, K. M., Lees, A. J., and Stern, G. M., 1980, The impact of treatment with levodopa on Parkinson's disease, *Q. J. Med.* **49**(195):283–293.

Shcheglov, V. I., 1979, Modern possibilities of endovascular operations with the aid of detachable balloon-catheter for treatment of some vascular diseases of the brain, in *Clinic and Surgical Treatment of the Brain Vascular Affections in Diseases of the Nervous System,* Meditsina, Leningrad, pp. 19–21 (Rus.).

Shealy, C. N., 1975, Dorsal column stimulation: Optimization of application, *Surg. Neurol.* **4**:142–145.

Shealy, C. N., and Maura, D., 1974, Transcutaneous nerve stimulation for control of pain, *Surg. Neurol.* **5**:45–49.

Shealy, C. N., Mortimer, J. T., and Reswick, J. B., 1967, Electrical inhibition of pain by stimulation of the dorsal columns. Preliminary clinical report, *Anesth. Analg.* **46**:489–491.

Shealy, C. N., Mortimer, J. T., and Hagfors, N. R., 1970, Dorsal column electroanalgesia, *J. Neurosurg.* **32**:560–564.

Shefer, D. G., and Nesterov, L. N., 1964, Pallidectomy and thalamectomy in the treatment of some affections of the extrapyramidal system, *Vopr. Neurochir.* **3**:35–39 (Rus.).

Shefer, D. G., and Nesterov, A. N., 1965, Thalamotomy for treatment of Kojevnikov epilepsy, *Vopr. Neirokhir.* **1**:8–11 (Rus.).

Shefer, D. G., and Nesterov, L. N., 1966, Thalamotomy in causalgia, *Vopr. Neurochir.* **5**:52–54 (Rus.).

Shefer, D. G., Belyaev, J. I., Ivanov, E. V., and Bein, B. N., 1968, Our experience with surgical treatment of epilepsy, in *Clinic, Diagnostics and Surgical Treatment of Epilepsy,* Meditsina, Leningrad, pp. 9–12 (Rus.).

Shefer, D., Belyaev, Yu. J., Bein, V. N., and Boreiko, V. B., 1970a, Remote results subsequent to surgical treatment of temporal epilepsy through partial resection of the temporal lobe, *Vopr. Neirokhir.* **3**:17–24 (Rus.).

Shefer, D. G., Ivanov, E. V., and Gurevich, V. L., 1970b, Polarographic determination of the oxygen tension in deeply-seated structures of the temporal lobe during stereotaxic operations in epilepsy, *Vopr. Neurochir.* **4**:20–25 (Rus.).

Shefer, D. G., Starikov, A. S., and Kokoreva, N. S., 1971, Some biochemical indices in hepatocerebral dystrophy before and after stereotaxic surgery, *Vopr. Neirokhir.* **6**:16–20 (Rus.).

Shefer, D. G., Bein, B. N., Boreiko, V. B., Skryabin, V. V., and Ivanova, A. S., 1972a, Indications for surgical treatment of resistant forms of temporal epilepsy, *Vopr. Neurochir.* **6**:16–20 (Rus.).

Shefer, D. G., Bein, B. N., Boreiko, V. B., Skryabin, V. V., and Ivanova, A. S., 1972b, Indications for surgical treatment of resistant forms of temporal epilepsy, *Vopr. Neurochir.* **6**:48–53 (Rus.).

Shefer, D. G., Bein, B. N., Boreiko, V. D., Obraztsova, R. G., Kokoreva, N. S., and Skryabin, V. V., 1974, Evolution of notions on possibilities of surgical treatment for epilepsy, in *Epileptogenic Focus and Surgical Treatment of Epilepsy,* Zdorovje, Kiev, pp. 27–31 (Rus.).

Shelden, C. H., Paul, F., Jacques, D. B., and Pudenz, R. H., 1975, Electrical stimulation of the nervous system, *Surg. Neurol.* **4**:127–132.

Shelden, C. H., McCann, G., and Jacques, S., 1980, Development of a computerized microstereotaxic method for localization and removal of minute CNS lesions under direct 3-D vision, *J. Neurosurg.* **52**:21–27.

Shelden, C. H., Jacques, S., and McCann, G., 1982, The Shelden CT-based microneurosurgical stereotactic system: Its application to CNS pathology, *Appl. Neurophysiol.* **45**:341–346.

Sheljakin, Y. A., 1971, Neurosurgical anatomy of the amygdaloid complex, in *Questions of Brain Neurosurgical Anatomy,* Meditsina, Leningrad, pp. 41–47 (Rus.).

Sheljakin, Y. A., 1973, Stereotactic topography of the amygdaloid complex on the frontal and sagittal planes, in *Actual Problems of Modern Medicine,* Meditsina, Leningrad, pp. 115–116 (Rus.).

Sherrington, C. S., 1898, Decerebrate rigidity and reflex coordination of movements, *J. Physiol. (Lond.)* **22**:319–332.

Shetter, A. G., Bertuccini, T. V., and Pittman, H. W., 1977, Closed needle biopsy in the diagnosis of intracranial mass lesions, *Surg. Neurol.* **8**:341–345.

Shibasaki, K., Nakai, S., and Higuchi, M., 1982, Percutaneous embolisation of major spinal cord artery as a treatment for intractable spasticity, *Paraplegia* **20**:158–168.

Shichijo, F., Fukami, T., Sakamoto, M., Tsuda, T., Murayama, Y., Sogabe, K., and Matsumoto, K., 1983, A 10-year follow-up study of bilateral VL thalamotomy for Parkinson's disease, *Appl. Neurophysiol.* **46**:317–321.

Shieff, C., and Nashold, B. S., 1987, Stereotactic mesencephalic tractotomy for thalamic pain, *Neurol. Res.* **9**:101–104.

Shimabukuro, H., and Mori, K., 1969, The role of the mesencephalic tegmentum on the spasmodic torticollis-like posture and ocular symptoms in cats, *Arch. Jpn. Chir.* **38**:626–632.

Shimazu, H., Hongo, T., Kubota, K., and Narabayashi, H., 1962, Rigidity and spasticity in man, *Arch. Neurol.* **6**:10–17.

Shubin, N. V., 1960, *Ixodic Encephalitis in the West Siberia,* Thesis, Irkutzk (Rus.).

Sicard, R., and Robineau, V., 1920, Algie vélo-pharyngée essentielle. Traitement chirurgical, *Rev. Neurol.* **36**:256–257.

Sicuteri, F., Anselmi, B., and Del Bianco, P. L., 1978, Systemic nonorganic central pain: A new syndrome with decentralization supersensitivity, *Headache* **18**(3):133–136.

Siegfried, J., 1968, *Die Parkinsonische Krankheit und ihre Behandlung,* Vol. 1, Springer, New York.

Siegfried, J., 1971, Stereotaxic cerebellar surgery, *Confin. Neurol.* **33**:350–356.

Siegfried, J., 1972a, Die functionelle Neurochirurgie, *Schweiz. Arch. Neurol. Neurochir. Psychiatrie* **2**:435–442.

Siegfried, J., 1972b, Stereotactic treatment of hypertoxicity, in *Present Limits of Neurosurgery,* Vol. 1 (I. Fusek and Z. Kunic, eds.), Avicenum, Prague.

Siegfried, J., 1974, Methods and results in hyperkinesia and hypertonicity, in *Recent Progress in Neurological Surgery* (K. Sano and S. Ishii, eds.), Excerpta Medica, Amsterdam, pp. 251–255.

Siegfried, J., 1976, The treatment of parkinsonian tremor, in *Advances in Parkinsonism* (W. Birkmayer and O. Hornykiewicz, eds.), Editiones Roche, Basle, pp. 319–327.

Siegfried, J., 1977, 500 percutaneous thermocoagulations on the Gasserian ganglion for trigeminal pain, *Surg. Neurol.* **8**:126–131.

Siegfried, J., 1978, The neurosurgical treatment of Parkinson's disease, *Br. J. Hosp. Med.* **20**:666–670.

Siegfried, J., 1979, Neurosurgical treatment of spasticity, in *Functional Neurosurgery* (T. Rasmussen and R. Marino, eds.), Raven Press, New York, pp. 123–129.

Siegfried, J., 1980a, Treatment of spasticity by dorsal cord stimulation, *Int. Rehabil. Med.* **2**(1):31–34.

Siegfried, J., 1980b, Neurosurgical treatment for intractable pain of terminal cancer patients, in *The Continuing Care of Terminal Cancer Patients* (R. G. Twycross and V. Ventafridda, eds.), Pergamon, Oxford, New York, pp. 369–376.

Siegfried, J., 1980c, Neurosurgical treatment of Parkinson's disease. Present indications and value, in *Parkinson's Disease: Current Progress, Problems and Management* (U. K. Rinne, M. Klinger, and G. Staum, eds.), Elsevier/North-Holland Biomedical Press, Amsterdam, pp. 369–376.

Siegfried, J., 1981, Percutaneous controlled thermocoagulation of Gasserian ganglion in trigeminal neuralgia. Experiences with 1000 cases, in *The Cranial Nerves* (M. Samü and P. J. Jannetta, eds.), Springer, Berlin, pp. 322–330.

Siegfried, J., 1982, Monopolar electrical stimulation of nucleus ventroposteromedialis thalami for postherpetic facial pain, *Appl. Neurophysiol.* **45**:179–184.

Siegfried, J., 1983, Long-term results of electrical stimulation in the treatment of pain by means of implanted electrodes (epidural spinal cord and deep brain stimulation), in *Pain Therapy* (R. Rizzi and M. Visentin, eds.), Elsevier Biomedical Press, pp. 463–475.

Siegfried, J., and Ben-Shmuel, A., 1973, Long-term assessment of stereotactic amygdalotomy for aggressive behavior, in *Surgical Approaches in Psychiatry* (L. V. Laitinen and K. E. Livingstone, eds.), Medical and Technical Publishing, Lancaster, pp. 138–141.

Siegfried, J., and Bernoulli, C., 1976, Stereo-electroencephalographic exploration and epilepsia partialis continua, *Acta Neurochir. (Wien) [Suppl.]* **23**:183–191.

Siegfried, J., and Braendli-Graber, S., 1980, Repérage radiologique simple du trou de Monro sur les radiographies craniennes à vide, *Neurochirurgie* **26**:387–389.

Siegfried, J., and Cetinalp, E., 1981, Neurosurgical treatment of phantom limb pain, in *Phantom and Stump Pain* (J. Siegfried and M. Zimmerman, eds.), Springer-Verlag, Berlin, Heidelberg, New York, pp. 148–155.

Siegfried, J., and Hood, T., 1983, Current status of functional neurosurgery, in *Advances and Technical Standards in Neurosurgery*, Vol. 10 (H. Krayenbühl, ed.), Springer-Verlag, New York, pp. 20–79.

Siegfried, J., and Lazorthes, Y., 1982, Long-term follow-up of dorsal cord stimulation for chronic pain syndrome after multiple lumbar operations, *Appl. Neurophysiol.* **45**:201–204.

Siegfried, J., and Lindenberger, J., 1979, Trigeminal neuralgia in multiple sclerosis. Therapeutical considerations, *Dolore* **1**(3):183–186.

Siegfried, J., and Verdie, Y. C., 1977, Long-term assessment of stereotactic dentatotomy for spasticity and other disorders, *Acta Neurochir. (Wien) [Suppl.]* **24**:41–48.

Siegfried, J., and Wieser, H. G., 1978, Effets de la stimulation de la substance grise périaqueductale chez l'homme sur l'activité spontanée et évoquée, *Neurochirurgie* **24**:407–414.

Siegfried, J., and Zumstein, H., 1976, Thalamotomies stéréotaxiques pour troubles fonctionelles chez les personnes agées, *Neurochirurgie* **22**:536–539.

Siegfried, J., Ervin, F. R., Miyazaki, J., and Mark, V. H., 1962, Localised cooling of the central nervous system. I. Neurophysiological studies in experimental animals, *J. Neurosurg.* **19**(10):840–852.

Siegfried, J., Crowell, R., and Perrett, E., 1969, Cure of tremulous writer's cramp by stereotaxic thalamotomy, *J. Neurosurg.* **30**:182–185.

Siegfried, J., Esslen, E., Gretener, U., Ketz, E., and Perret, E., 1970, Functional anatomy of the dentate nucleus in the light of stereotaxic operations, *Confin. Neurol.* **32**:1–10.

Siegfried, J., Krainick, J. U., Haas, H. L., Adorjani, C., Meier, M., and Thoden, U., 1978, Electrical spinal cord stimulation on spastic movement disorders, *Appl. Neurophysiol.* **41**:134–141.

Siegfried, J., Lazorthes, Y., and Sedan, R., 1980, Indications and ethical considerations of deep brain stimulation, *Acta Neurochir. (Wien) [Suppl.]* **30**:269–274.

Siegfried, J., Lazorthes, Y., and Broggi, G., 1981, Electrical spinal cord stimulation for spastic movement disorders, *Appl. Neurophysiol.* **44**(1–3):77–92.

Sigua, O. A., and Chkhenkeli, S. A., 1976, Stereotactic amygdalo- and amygdalohippocampotomy in the treatment of patients with temporal lobe epilepsy with evident behavioral disorders, *Vopr. Neirokhir.* **1**:17–22 (Rus.).

Sigwald, J., Bouttier, D., and Solignac, J., 1959, Fixed or slowly evolving forms of the parkinsonian syndrome (with respect to 90 cases), *Rev. Neurol.* **101**(5):663–664.

Simonov, L. N., 1866, Experimental evidences of "reflexes arresting centers" existence in mammals, *Mil. Med. J.* **97**:1–31 (Rus.).

Sindou, M., 1985, Microsurgical selective procedures in the peripheral nerves and the posterior root–spinal cord junction for the treatment of localized and severe spastic disorders, in *Abstracts of the IX Meeting of the World Society for Stereotactic and Functional Neurosurgery*, Toronto.

Sindou, M., and Keravel, Y., 1979, Thermocoagulation percutanée du trijumeau dans le traitment de la neuralgie faciale essentielle. Résultats en fonction du siége de la thermolésion, *Neurochirurgie* **25**(3):116–172.

Sindou, M., and Lapras, C., 1982, Neurosurgical treatment of pain in the Pancoast–Tobias syndrome: Selective posterior rhizotomy and open anterolateral C_2-cordotomy, in *Advances in Pain Research and Therapy*, Vol. 4 (J. Bonica, ed.), Raven Press, New York, pp. 199–206.

Sindou, M., Fischer, G., Goutelle, A., and Mansuy, L., 1974a, La radicellotomie postérieure sélective. Premiers résultats dans la chirurgie de la douleur, *Neurochirurgie* **20**:399–408.

Sindou, M., Fischer, G., Goutelle, A., Schott, B., and Mansuy, L., 1974b, La radicellotomie postérieure sélective dans le traitement des spasticités, *Rev. Neurol.* **130**:201–216.

Sindou, M., Fischer, G., Goutelle, A., and Allegre, G. E., 1981, Long-term results of selective posterior rhizotomy in the treatment of severe spastic disorders, in *Abstracts of 7th International Congress of Neurological Surgeons*, Munich, p. 223.

Sindou, M., Millet, M. F., Mortamais, J., and Byssette, M., 1982, Results of selective posterior rhizotomy in the treatment of painful and spastic paraplegia secondary to multiple sclerosis, **45**:335–340.

Sindou, M., Monib, H., Perrin, G., and Goutelle, A., 1983, Results of selective posterior rhizotomy in pain syndromes (89 cases), in *7th European Congress of Neurosurgery*, Brussels, p. 146.

Sindou, M., Abdennebi, B., and Sharkey, P., 1985a, Microsurgical selective procedures in the peripheral nerves and the posterior root–spinal cord junction for the treatment of localized and severe spastic disorders, in *IX Meeting of the World Society for Stereoactic and Functional Neurosurgery*, Toronto, pp. 96–104.

Sindou, M., Pregelj, R., Boisson, D., Eyssette, M., and Goutelle, A., 1985b, Surgical selective lesions of nerve fibers and myelotomies for modifying muscle hypertonia, in *Recent Achievements in Restorative Neurology* (J. Eccles and M. R. Dimitrijevic, eds.), S. Karger, Basel, pp. 10–26.

Sindou, M., Keravel, Y., Abdennebi, B., and Szapiro, J., 1987, Traitement neuro-chirurgical de la névralgie trigéminale. Abord direct ou méthode percutanée, *Neurochirurgie* **33**:89–111.

Sjölander, V., Lindgren, P. G., and Hugosson, R., 1983, Ultrasound sector scanning for the localization and biopsy of intracerebral lesions, *J. Neurosurg.* **58**:7–10.

Sjöqvist, O., 1937, Eine neue Operationsmethode bei Trigeminusneuralgie: Durchschneidung des Tractus spinalis trigemini, *Zentralbl. Neurochir.* **2**:274–281.

Sjöqvist, O., 1938, Studies on pain conduction in the trigeminal nerve. A contribution to the surgical treatment of facial pain, *Acta Psychiatr. Scand.* **17**:1–138.

Skryabin, V. V., Bein, B. N., and Ivanov, E. V., 1976, Stereotaxic surgery of temporal epilepsy, *Vopr. Neirokhir.* **4**:9–16 (Rus.).

Slaughter, D. G., and Nashold, B. S., 1968, Stereotactic coordinates for the human dentate nucleus, *Confin. Neurol.* **30**:375.

Small, I. F., Heimburger, R. F., Small, L. G., Milgteyn, V., and Moore, D. F., 1977, Follow-up of stereotaxic amygdalotomy for seizure and behaviour disorders, *Biol. Psychiatry* **12**(3):401–411.

Smirnov, V. M., 1976, *Stereotactic Neurology,* Meditsina, Leningrad (Rus.).

Smirnov, V. M., and Iovlev, B. V., 1974, Mathematical–statistical analysis of changes in parkinsonian tremor during diagnostic electric stimulations of deep-seated brain structures, *Vopr. Neirokhir.* **5**:40–46 (Rus.).

Smith, H. P., McWhorter, J. M., and Challa, V. R., 1981, Radiofrequency neurolysis in a clinical model. Neuropathological correlation, *J. Neurosurg.* **55**(2):246–253.

Smith, J. S., and Kiloh, L. G., 1975, Sydney Symposium on Psychosurgery and Society, *Bull. Int. Soc. Psychol. Surg.* **6**:8–9.

Smith, M. C., 1967, Stereotactic operations for Parkinson's disease: Anatomical observations, *Mod. Trends Neurol.* **4**:21–52.

Smith, R., 1973, Outlining the cervical spinal cord with tantalum powder: Application to percutaneous cordotomy. Technical note, *J. Neurosurg.* **38**:257–259.

Sogabe, K., Dieckmann, G., and Veras, G., 1985, Complications in cases of epidural stimulation electrode implants, *Appl. Neurophysiol.* **48**:377–379.

Sólyom, A., Tóth, S., Holezinger, I., Vajada, J., Tóth, Z., and Káluanchey, B., 1985, The spread of somatosensory-evoked potentials within the nervous system, *Appl. Neurophysiol.* **48**:222–225.

Sonoda, H., Takahashi, S., Kodama, K., Hara, N., Umeki, Y., Nakagawa, M., and Yoshitake, J., 1987, Percutaneous thermocoagulation of trigeminal ganglion, *Neurol. Med. Chir. (Tokyo)* **27**:18–23.

Sorensen, B. F., and Hamby, W. B., 1966, Spasmodic torticollis: Results in 71 surgically treated patients, *Neurology (Minneap.)* **16**:868–878.

Šourek, K., 1969, Commissural myelotomy, *J. Neurosurg.* **31**:524–527.

Spaziante, R., Gerone, A., and Cappabianka, P., 1986, Simplified method to implant chronic stimulatory electrodes in the Gasserian ganglion, *Appl. Neurophysiol.* **49**:1–3.

Speransky, A. D., 1937, *Elements of the Theory of Medicine,* Meditsina, Moscow.

Spiegel, E. A., 1965, Methodological problems in stereoencephalotomy, *Confin. Neurol.* **26**:125–132.

Spiegel, E. A., 1969, Indications for stereoencephalotomies. A critical assessment, *Confin. Neurol.* **31**:1–2,5–10.

Spiegel, E. A., 1982a, Observations and perspectives related to subcortical procedures, *Appl. Neurophysiol.* **45**:1–7.

Spiegel, E. A., 1982b, *Guided Brain Operations,* S. Karger, Basel.

Spiegel, E. A., and Wycis, H. T., 1948, Mesencephalotomy for relief of pain, in *Anniversary Volume for O. Poetzl,* Springer, Vienna, p. 438.

Spiegel, E. A., and Wycis, H. T., 1950a, Pallido-thalamotomy in chorea, *Arch. Neurol. Psychiatry.* **64**:495–496.

Spiegel, E. A., and Wycis, H. T., 1950b, Thalamic recordings in man with special reference to seizure discharges, *Electroencephalogr. Clin. Neurophysiol.* **2**:23–27.

Spiegel, E. A., Wycis, H. T., 1952a, Stereoencephalotomy, *J.A.M.A.* **148**:446–461.

Spiegel, E. A., and Wycis, H. T., 1952b, *Stereoencephalotomy. Part I. Methods and Stereotaxic Atlas of Human Brain,* Grune & Stratton, New York.

Spiegel, E. A., and Wycis, H. T., 1953, Mesencephalotomy in treatment of intractable pain, *Arch. Neurol. Psychiatry* **69**:1–13.

Spiegel, E. A., and Wycis, H. T., 1954, Ansotomy in paralysis agitans, *Arch. Neurol. Psychiatry* **71**:598–601.

Spiegel, E. A., and Wycis, H. T., 1958, Pallido–ansotomy: Anatomic–physiologic foundation and histopathologic control, in *Pathogenesis and Treatment of Parkinsonism* (S. Field, ed.), Charles C. Thomas, Springfield, IL, pp. 86–105.

Spiegel, E. A., and Wycis, H. T., 1960, The basal ganglia in parkinsonism, *J. Nerv. Ment. Dis.* **131**(4):310–317.

Spiegel, E. A., and Wycis, H. T., 1962a, Long-term results in the treatment of intractable pain by stereotaxic midbrain surgery, *J. Neurosurg.* **19**(2):101–107.

Spiegel, E. A., and Wycis, H. T., 1962b, *Stereoencephalotomy, Thalamotomy and Related Procedures. Part II. Clinical and Physiological Applications,* Grune & Stratton, New York, p. 504.

Spiegel, E. A., and Wycis, H., 1963, Campotomy in various extrapyramidal disorders, *J. Neurosurg.* **20**:871–884.

Spiegel, E. A., and Wycis, H. T., 1966, Present status of stereoencephalotomies for pain relief, in *Advances in Stereoencephalotomy,* Part II (E. A. Spiegel and H. T. Wycis, eds.), S. Karger, Basel, pp. 7–17.

Spiegel, E. A., and Wycis, H. T., 1967a, Multiple representation of various functions in the human subcortex. Conduction and perception of pain, *Confin. Neurol.* **29**:163–167.

Spiegel, E. A., and Wycis, H. T., 1967b, The possible role of hypothalamus in parkinsonian brady- and akinesia, *Contin. Neurol.* **29**:262–264.

Spiegel, E. A., Wycis, H. T., Marks, M., and Lee, A., 1947, Stereotaxic apparatus for operations on the human brain, *Science* **106:**349–350.

Spiegel, E. A., Wycis, H. T., and Reyes, V., 1951a, Diencephalic mechanisms in petit mal epilepsy, *Electroencephalogr. Clin. Neurophysiol.* **3:**473–475.

Spiegel, E. A., Wycis, H. T., and Thur, C., 1951b, The stereoencephalotome, *J. Neurosurg.* **8:**452–453.

Spiegel, E. A., Wycis, H. T., and Good, R., 1956, Studies in stereoencephalotomy V. A universal stereoencephalotome (model V), *J. Neurosurg.* **3:**305–309.

Spiegel, E. A., Wycis, H., and Baird, H., 1958b, Long range effects of electropallidoansotomy in extrapyramidal and convulsive disorders, *Neurology (Minneap.)* **8:**738–743.

Spiegel, E. A., Wycis, H. T., Szekeley, E. G., Gildenberg, P. L., and Zanes, C., 1964a, Combined dorsomedial intralaminar and basic thalamotomy for relief of so-called intractable pain, *J. Int. Coll. Surg.* **42:**160–168.

Spiegel, E. A., Wycis, H. T., Szekely, E. G., Soloff, L., Adams, J., Gildenberg, P., and Zenes, C., 1964b, Stimulation of Forel's field during stereotaxic operations in the human brain, *Electroencephalogr. Clin. Neurophysiol.* **16:**537–548.

Spiegel, E. A., Wycis, H. T., Szekely, E. G., and Gildenberg, P. L., 1966, Medial and basal thalamotomy in so-called intractable pain, in *Pain* (R. S. Knighton and P. R. Dumke, eds.), Boston, Little, Brown, pp. 503–517.

Spiller, W. G., and Martin, E., 1912, The treatment of persistent pain of organic origin in the lower part of the body by division of the anterolateral column of the spinal cord, *J.A.M.A.* **58:**1489–1490.

Spuler, H., 1982, Erethic oligophrenia, in *Stereotaxy of the Human Brain* (G. Schaltenbrand and A. E. Walker, eds.), Georg Thieme Verlag, Stuttgart, New York, pp. 622–628.

Šramka, M., 1985, *Stereotaktické Liecenie Temporálnej Epilepsie,* Veda, Bratislava.

Šramka, M., and Nadvornik, P., 1975a, Stereotactic longitudinal hippocampotomy and its prospects for treatment of epilepsy, *Vopr. Neirokhir.* **4:**37–41 (Rus.).

Šramka, M., and Nadvornik, P., 1975b, Surgical complication of posterior hypothalamotomy, *Confin. Neurol.* **37:**193–194.

Šramka, M., Nadvornik, P., and Lisy, L., 1972, Myoclonic disease of Ramsey-Hunt treated by transtentorial bilateral dentatotomy from the right side, *Confin. Neurol.* **34:**315–319.

Šramka, M., Fritz, G., Galanda, M., and Nadvornik, P., 1976, Some observations on stimulation treatment of epilepsy, *Acta Neurochir. [Suppl.]* **23:**257–262.

Šramka, M., Nadvornik, P., Pogády, J., and Huba, J., 1977, Stereotactic treatment of alcoholism by anterior hypothalamotomy, *Excerpta Med. Int. Cong. Ser.* **418:**158–159.

Šramka, M., Fritz, G., and Nadvornic, P., 1979, Results of therapeutic stimulation for epilepsy, in *IV Meeting of the European Society for Stereotactic and Functional Neurosurgery,* Paris.

Šramka, M., Fritz, G., and Gajdosova, D., 1980, Central stimulation treatment of epilepsy, *Acta Neurochir. (Wien)* **30:**183–187.

Šramka, M., Chckenkeli, S. A., and Nadvornik, P., 1982, Physiological aspects of central stimulation, *Appl. Neurophysiol.* **45**(1–2)**:**98–101.

Šramka, M., Drlickova, V., and Nadvornik, P., 1983, The stimulation of caudate nucleus in the treatment of epilepsy, in *VI Meeting of the European Society for Stereotactic and Functional Neurosurgery,* Rome, p. E13.

Starikov, A. S., 1978, *Clinic and Surgical Treatment of Extrapyramidal System Diseases,* Thesis, Kiev (Rus.).

Starikov, A. S., 1979, Electrophysiological indices in athetosis torsion dystonia before and after stereotaxic operations, *Vopr. Neirokhir.* **4:**36–43 (Rus.).

Starikov, A. S., and Staduchin, V. O., 1970, Results of surgical treatment of hemiathetosis, in *Proceedings of Conference of Neurosurgeons,* Sverdlovsk, pp. 177–178 (Rus.).

Steg, G., 1964b, Alpha-rigidity in reserpinized rats, *Experientia* **20:**79–80.

Stein, B., and Wolpert, S., 1977, Surgical and embolic treatment of cerebral arteriovenous malformation, *Surg. Neurol.* **7:**359–369.

Steiner, L., Leksell, L., and Forster, D., 1974, Stereotactic radiosurgery in intracranial arteriovenous malformations, in *Advances in Stereotactic and Functional Neurosurgery* (F. J. Gillingham, E. R. Hitchcock, and J. W. Turner, eds.), Springer-Verlag, New York, pp. 195–210.

Stellar, S., and Cooper, I. S., 1968, Mortality and morbidity in cryothalamectomy for Parkinson's disease. A stereotactical study of 2868 consecutive operations, *J. Neurosurg.* **28:**459–467.

Stepanova, T. S., and Grachev, K. V., 1976, Stereoelectrosubcorticography in epilepsy, the focus and epileptogenic system, *Acta Neurochir. [Suppl.]* **23:**27–31.

Stephan, H., and Andy, O. J., 1982, Anatomy of the limbic system, in *Stereotaxy of the Human Brain* (G. Schaltenbrand and A. E. Walker, eds.), Georg Thieme Verlag, Stuttgart, New York, pp. 269–293.

Stepien, L., Bacia, T., and Bidzinski, J., 1972, Age, intellectual and EEG changes as limiting factors in surgical treatment of temporal lobe epilepsy, in *Present Limits of Neurosurgery* (I. Fusek and Z. Kunc, eds.), Avicenum, Czechoslovak Medical Press, Prague, pp. 423–424.

Stepien, L., Bidzinski, E., and Bache, T., 1980, Long-term results of epileptic foci removing, in *Surgical Treatment of Epilepsy,* Nauka, Tbilisi, pp. 33–35 (Rus.).

Stern, J., and Ward, A. A., 1960, Inhibition of muscle spindle discharges by ventrolateral thalamic stimulation, *Arch. Neurol.* **3:**193–204.

Stern, J., and Ward, A. A., 1962, Supraspinal and drug modulation of the alpha motor system, *Arch. Neurol.* **6:**404–413.

Stern, J., and Ward, A. A., 1963, The relationship of the alpha and gamma motor systems to the efficacy of the surgical therapy of parkinsonism, *J. Neurosurg.* **20:**185–187.

Steude, U., 1978, Percutaneous electrostimulation of the trigeminal nerve in patients with atypical trigeminal neuralgia, *Neurochirurgia (Stuttg.)* **21:**66–69.

Steven, J. L., 1966, Postoperative angiography in treatment of intracranial aneurysms, *Acta Radiol. Diagn.* **5**(1–6):536–548.

Stoffel, A., 1912, The treatment of spastic contracture, *Am. J. Orthop. Surg.* **10**:611–622.

Stolyarova, L. G., Kadikov, A. S., Kistenjev, B. A., and Pivovarova, V. M., 1979, *Rehabilitation of Parkinsonian Patients,* Meditsina, Moscow (Rus.).

Story, J. L., French, L. A., Chou, S. N., and Meier, M. J., 1966, Experiences with subthalamic lesions in patients with movement disorders, *Confin. Neurol.* **26**(3–5):218–225.

Stowell, A., 1963, The physiological mechanisms and treatment of headache, *Trans. Am. Neurol. Assoc.* **88**:297–301.

Strain, G. M., Babb, T. L., Soper, H. V., Perryman, K. M., Lieb, J. P., and Crandall, P. H., 1979, Effects of chronic cerebellar stimulation on chronic limbic seizures in monkeys, *Epilepsia* **20**:651–664.

Strait, T. A., and Hunter, S. E., 1981, Intraspinal extradural sensory rhizotomy in patients with failure of lumbar disc surgery, *J. Neurosurg.* **54**(2):193–197.

Ström-Olsen, R., and Carlisle, S., 1971, Bifrontal stereotactic tractotomy. A follow-up study of its effects on 210 patients, *Br. J. Psychiatrv* **118**:141–154.

Strümpel, A., 1898, Über die Westphalische Pseudosklerose und über diffuse Hirnsklerose, insbesondere bei Kindern, *Dtsch. Z. Nervenheilkd.* **12**:115–126.

Struppler, A., 1973, Summary of main lectures on rigidity published in Volume 1, in *Parkinson's Disease,* Vol. 2 (J. Siegfried, ed.), Hans Huber, Bern, Stuttgart, Vienna, pp. 19–25.

Struppler, A., Lucking, C. H., and Erbel, F., 1972, Neurophysiological findings during stereotactic operations in thalamus and subthalamus, *Confin. Neurol.* **34**(1–4):70–73.

Struppler, A., Velko-Groneberg, P., and Claussen, M., 1976, Clinical and pathophysiology of tremor, in *Advances in Parkinsonism* (W. Birkmayer and O. Honykiewicz, eds.), Editiones Roche, Basel, pp. 287–302.

Sturm, V. M., Pastyr, O., Schlegel, W., Scharfenberg, H., Zabel, H. J., Netzeband, G., Schabbert, S., and Barberich, W., 1983, Stereotactic computer tomography with a modified Riechert-Mundinger device as the basis for integrated stereotactic investigations, *Acta Neurochir.* **68**:11–17.

Sugar, O., 1978, Changing attitudes toward psychosurgery, *Surg. Neurol.* **9**:331–335.

Sugita, K., and Doi, T., 1967, The effects of electrical stimulation on the motor and sensory system during stereotaxic operation, *Confin. Neurol.* **29**:224–229.

Sugita, K., and Kobayashi, S., 1982, General considerations of pain, in *Stereotaxy of the Human Brain* (G. Schaltenbrand and A. E. Walker, eds.), Georg Thieme Verlag, Stuttgart, New York, pp. 462–468.

Sugita, K., and Murata, K., 1966, A stereotaxic apparatus for brain surgery and high frequency coagulation with automatic thermocontrol, *Nagoya J. Med. Sci.* **28**:126–141.

Sugita, K., Doi, T., Sato, O., Takaoka, J., Mutsuga, N., and Tsugane, R., 1969, Successful removal of intracranial air-gun bullet with stereotaxic apparatus, *J. Neurosurg.* **30**:177–181.

Sugita, K., Doi, T., Mutsuga, N., and Takaoka, J., 1971, Clinical study of fornicotomy for psychomotor epilepsy and behavior disorder, in *Special Topics in Stereotaxis* (W. Umbach, ed.), Hippokrates, Stuttgart, pp. 42–52.

Sugita, K., Mutsuga, N., Takaoka, Y., and Doi, T., 1972, Results of stereotaxic thalamotomy for pain, *Confin. Neurol.* **34**:265–274.

Sugita, K., Mutsuga, N., Takaoka, Y., Hirota, T., Shibuya, U., and Doi, T., 1975, Stereotaxic exploration of parathird ventricle tumors, *Confin. Neurol.* **37**:156–162.

Sukoff, M. H., and Ragatz, R. E., 1980, Cerebellar stimulation for chronic extensor–flexor rigidity and opisthotonus secondary to hypoxia. Report of two cases, *J. Neurosurg.* **53**:391–396.

Sunderland, S., 1976, Pain mechanisms in causalgia, *J. Neurol. Neurosurg. Psychiatry* **39**:471–480.

Sunder-Plassmann, M., and Grunert, V., 1976, Commissural myelotomy for drug-resistant pain, in *Clinical Microneurosurgery* (W. T. Koos, F. W. Böck, and R. F. Spetzler, eds.), Georg Thieme, Stuttgart, pp. 165–170.

Sunder-Plassmann, M., Grunnert, V., and Böck, F., 1971, Zur Frage des vasculären Kompression des Tractus spinalis nervi trigemini als mögliche Ursache der Trigeminus-neuralgie, *Nervenartz* **42**:323–325.

Sutton, G. G., and Sykes, K., 1967, The variation of hand tremor with force in healthy subjects, *J. Physiol. (Lond.)* **191**:699–711.

Swanson, S. E., and Farhat, S. M., 1982, Neurovascular decompression with selective partial rhizotomy of the trigeminal nerve for tic douloureux, *Surg. Neurol.* **18**(1):3–6.

Sweet, W. H., 1972, Controlled thermocoagulation of trigeminal ganglion and rootlets for differential destruction of pain fibers, in *Present Limits of Neurosurgery* (I. Fusek and Z. Kunc, eds.), Avicenum, Czechoslovak Medical Press, Prague, p. 529.

Sweet, W. H., 1976, Treatment of facial pain by percutaneous differential thermal trigeminal rhizotomy, *Prog. Neurol. Surg.* **7**:153–179.

Sweet, J. E., and Law, J. D., 1983, Analgesia with peripheral nerve stimulation: Absence of a peripheral mechanism, *Pain* **15**:55–70.

Sweet, W. H., and Mark, V. H., 1953, Unipolar anodal electrolytic lesions in the brain of man and cat, *Arch. Neurol. Psychiatry* **70**:224–234.

Sweet, W. H., and Poletti, C. E., 1982, Retrogasserian glycerol injection as treatment for trigeminal neuralgia, in *Operative Neurosurgical Technique. Indications, Methods and Results,* Vol. 2 (H. H. Schmidek and W. H. Sweet, eds.), Grune & Stratton, New York, pp. 1107–1117.

Sweet, W. H., and Poletti, C. E., 1985, Problems with retrogasserian glycerol in the treatment of trigeminal neuralgia, *Appl. Neurophysiol.* **48**:252–257.

Sweet, W. H., and Wepsic, J. G., 1968, Treatment of chronic pain by stimulation of fibers of primary afferent neuron, *Trans. Am. Neurol. Assoc.* **93**:103–107.

Sweet, W. H., and Wepsic, J. G., 1974, Controlled thermocoagulation of trigeminal ganglion and rootlets for differential destruction of pain fibers. Part I. Trigeminal neuralgia, *J. Neurosurg.* **40**:143–156.

Sylvius de la Boe, cited by Molina-Negro, P., and Hardy, J., 1971, Etude sémiologique des tremblements, *Union Med. Can.* **100**:879–895.

Szapiro, J., Jr., Sindou, M., and Szapiro, J., 1985, Prognostic factors in microvascular decompression for trigeminal neuralgia, *Neurosurgery* **17**:920–929.

Szekely, E. G., Spuler, H., and Spiegel, E. A., 1962, Stimulation of ventrolateral area of the thalamus after degeneration of fibers originating in frontal lobe, sigmoid gyri and pallidum, *Confin. Neurol.* **22**:304–309.

Szikla, G., 1979, Stereotactic neuroradiology and functional neurosurgery: Localization of cortical structures by three-dimensional angiography, in *Functional Neurosurgery* (T. Rasmussen and R. Marino, eds.), Raven Press, New York, pp. 197–217.

Szikla, G., Bouvier, G., Hori, T., and Petrov, V., 1977a, *Angiography of the Human Brain Cortex. Atlas,* Springer-Verlag, Berlin.

Szikla, G., Bouvier, G., Hori, T., and Petrov, V., 1977b, Radiologic exploration of the brain cortex in epileptic patients: Stereotactic angiography as a tool for cortical localization, in *The 7th Symposium of the World Society for Stereotactic and Functional Neurosurgery,* São Paulo, p. 35.

Taarnhøj, P., 1954, Decompression of the trigeminal root, *J. Neurosurg.* **11**:299–305.

Tabaddor, K., Wolfson, L. I., and Sharpless, N. S., 1978, Diminished ventricular fluid dopamine metabolites in adult-onset dystonia, *Neurology (Minneap.)* **28**:1254–1258.

Taiwo, Y. O., Fabian, A., and Pazoles, C. J., 1985, Potentiation of morphine antinociception by monoamine reuptake inhibitors in the rat spinal cord, *Pain* **21**:329–337.

Takeda, F., 1984, Neurosurgical treatment of chronic pain, *Postgrad. Med. J.* **60**:905–913.

Takeda, F., Fujii, T., Uki, J., Fuse, Y., Tozawa, R., Kitani, Y., and Fujita, T., 1983a, Cancer pain relief and tumor regression by means of pituitary neuroadenolysis and surgical hypophysectomy, *Neurol. Med. Chir.* **23**:41–49.

Takeda, F., Fujii, T., Uki, J., Tozawa, R., Fuse, Y., Kitani, Y., and Fujita, T., 1983b, Alterations of hypothalamopituitary interaction and pain threshold following pituitary neuroadenolysis, *Neurol. Med. Chir.* **23**:551–560.

Takeda, F., Uki, J., Fujii, T., Kitani, Y., and Fujita, T., 1983c, Pituitary neuroadenolysis to relieve cancer pain: Observation of spread of ethanol instilled into the sella turcica and subsequent changes of the hypothalamopituitary axis at autopsy, *Neurol. Med. Chir.* **23**:50–54.

Takeda, F., Uki, J., Fuse, Y., Kitani, Y., and Fujita, T., 1985, The pituitary as a target of antalgic treatment of chronic cancer pain. A possible mechanism of pain relief through pituitary ablation, in *VIII International Congress of Neurological Surgery,* Toronto, p. 140.

Takeuchi, J., Handa, H., Taki, W., and Yamagani, T., 1982, ThNd:YAG laser in neurological surgery, *Surg. Neurol.* **18**:140–142.

Takusagawa, Y., Wakiya, K., and Fakushima, T., 1985, Treatment of trigeminal neuralgia by percutaneous gasserian ganglion glycerol injection, in *VIII International Congress of Neurological Surgery,* Toronto, p. 143.

Talairach, J., 1952, Destruction du noyau ventral anterier thalamique dans le traitement des maladies mentales, *Rev. Neurol.* **87**:352.

Talairach, J., 1955, Chirurgie stéréotaxique du thalamus, in *VI Congress Latin-American Neurochirurgia,* Montevideo, 21–24, 3, 865–925.

Talairach, J., and David, M., 1952, Etudes stéréotaxiques des structures encéphaliques profondes chez l'homme, *Presse Med.* **60**:605.

Talairach, J., and Szikla, G., 1965, Destruction partielle amygdalohippocampique par l'yttrium 90 dans le traitement de certaines epilepsies a expression rhinencephalique, *Neurochirurgie* **11**:233–240.

Talairach, J., and Tournoux, P., 1955, Appareil de stéréotaxie hypophysaire pour voie d'abord nasale, *Neurochirurgie* **1**:127–131.

Talairach, J., Hecaen, H., David, M., Monnies, M., and Ajuriaguerra, J., 1949, Recherches sur la coagulation therapeutique des structures sous corticales chez l'homme, *Rev. Neurol.* **81**:4–24.

Talairach, J., Paillas, J., and David, M., 1950, Dyskinesie de type hemiballique traitée par cortectomie frontale limitée, puis par coagulation de l'anse lenticulaire et de la portion interne du globus pallidus, *Rev. Neurol.* **83**:440–451.

Talairach, J., Aboulker, J., Tournoux, P., and David, M., 1956, Technique stéréota .ique de la chirurgie hypophysaire par voie nasale, *Neurochirurgia (Stuttg.)* **2**:3–20.

Talairach, J., David, M., Tournoux, P., Corredor, H., and Kvasina, T., 1957, *Atlas d'Anatomie Stéréotaxique,* Masson, Paris.

Talairach, J., David, M., and Tournoux, P., 1958, *L'Exploration Chirurgicale Stéréotaxique du Lobe Temporal dans l'Epilepsie Temporale,* Masson, Paris.

Talairach, J., Tournoux, P., Bancaud, J., and Djahanchahi, D., 1959, Exploration stéréotaxique du lobe pariétal dans trois cas de syndrome douloureux consécutif a des plaies des nerfs. Destruction therapeutique de fibres thalamo-pariétales au niveau du pied de la couronne rayonnante, *Neurochirurgie* **5**(1):130–133.

Talairach, J., Tournoux, P., and Bancaud, J., 1960, Chirurgie pariétale de la douleur, *Acta Neurochir. (Wien)* **8**(2–3):153–250.

Talairach, J., Szikla, G., Tournoux, P., Bonis, P., and Bancaud, J., 1962, La chirurgie stéréotaxique hypophysaire, *Confin. Neurol.* **22**:204–217.

Talairach, J., Szikla, G., Tournoux, P., Prossolentis, A., and Bordas-Ferrer, M., 1967, *Atlas d'Anatomie Stéréotactique du Télencéphale,* Masson, Paris.

Talairach, J., Bancaud, J., Bonis, B., Szikla, G., and Tournoux, P., 1969, Functional stereotaxis of epilepsy, *Confin. Neurol.* **22**:328–331.

Talairach, J., Bonis, G., Szikla, G., Schaub, G., Bancaud, J., Covello, L., and Bordas-Ferrer, F., 1970, Stereotaxic implantation of radioactive isotopes in functional pituitary surgery. Techniques and results, in *Radionuclide Applications in Neurology and Neurosurgery* (Yen Wang and Paoletti, Eds.), Charles C. Thomas, Springfield, IL, pp. 267–327.

Talairach, J., Bancaud, J., Bonis, A.. and Szikla, G., 1974a,

Approche Nouvelle de la Neurochirurgie de l'Epilepsie, Masson, Paris.

Talairach, J., Bancaud, J., Szikla, G., Bonis, A., Geier, S., and Vedrenne, C., 1974b, Approche nouvelle de la neurochirurgie de l'épilepsie. Méthodologie stéréotaxique et résultats thérapeutiques, *Neurochirurgie* **20**[Suppl. 1]:1–240.

Tallis, R. C., Illis, L. S., Sedgwick, E. M., Hardwidge, C., and Garfield, J. S., 1983, Spinal cord stimulation in peripheral vascular disease, *J. Neurol. Neurosurg. Psychiatry* **46**:478–484.

Tan, E., Marks, J. M., and Marset, P., 1971, Bimedial leucotomy in obsessive–compulsive neurosis. A controlled serial enquiry, *Br. J. Psychiatry* **118**:155–164.

Tanizaki, Y., Sugita, K., Toriyama, T., and Hokama, M., 1985, New CT-guided stereotactic apparatus and clinical experience with intracerebral hematomas, *Appl. Neurophysiol.* **48**:11–17.

Taptas, N., 1911, Les injections d'alcool dans le ganglion de Gasser à travers le trou ovale, *Presse Med.* **19**:798–802.

Taren, J. A., and Kahn, E. A., 1962, Anatomic pathways related to pain in face and neck, *J. Neurosurg.* **19**:116–119.

Taren, J., Guiot, G., Derome, P., and Trigo, J. C., 1968, Hazards of stereotaxic thalamectomy. Added safety factor in corroborating x-ray target localization with neurophysiological methods, *J. Neurosurg.* **29**:173–182.

Taren, J. A., Davis, R., and Crosby, E. C., 1969a, Target physiologic corroboration in stereotaxic cervical cordotomy, *J. Neurosurg.* **30**:569–584.

Taren, J., Guiot, G., and Derome, P., 1969b, Delimitation of the ventral posterior nucleus of the thalamus. Comparison of radiologic and electrophysiologic techniques, *Acta Radiol. Diag.* **9**:209–218.

Taren, J. A., Kahn, E. A., and Humphrey, T., 1969c, The surgery of pain, in *Correlative Neurosurgery* (E. A. Kahn, E. C. Crosby, and R. C. Schneider, eds.), Charles C. Thomas, Springfield, IL.

Tarlov, E., 1969, The postural effect of lesions of the vestibular nuclei. A note on species differences among primates, *J. Neurosurg.* **31**:187–195.

Tarlov, E., 1970, On the problem of spasmodic torticollis in man, *J. Neurol. Neurosurg. Psychiatry* **33**:457–463.

Tasker, R. R., 1982a, Percutaneous cordotomy—the lateral high cervical technique, in *Operative Neurosurgical Techniques. Indications, Methods, and Results,* Vol. 2 (H. H. Schmidek and W. H. Sweet, eds.), Grume & Stratton, New York, pp. 1137–1153.

Tasker, R. R., 1982b, Thalamic stereotaxic procedures, in *Stereotaxy of the Human Brain* (G. Schaltenbrand and A. E. Walker, eds.), Georg Thieme, Stuttgart, New York, pp. 484–497.

Tasker, R. R., 1985, Physiological confirmation of stereotactic biopsy, *Appl. Neurophysiol.* **48**:444–447.

Tasker, R. R., and Emmers, R., 1969, A double somatotopic representation in the human thalamus. Its application in localization during thalamotomy for Parkinson's disease, in *Proceedings 3rd Symposium on Parkinson's Disease* (F. J. Gillingham and I. M. L. Donaldson, eds.), Livingstone, Edinburgh, pp. 94–100.

Tasker, R. R., and Organ, L. W., 1973, Percutaneous cordotomy, *Confin. Neurol.* **35**:110–117.

Tasker, R. R., Richardson, P., Rewcastle, B., and Emmers, R., 1972, Anatomical correlations of detailed thalamic sensory mapping in man, *Confin. Neurol.* **34**:184–196.

Tasker, R. R., Organ, L. W., and Smith, K. S., 1974, Physiological guidelines for the localization of lesion by percutaneous cordotomy, *Acta Neurochir. (Wien) [Suppl.]* **21**:111–117.

Tasker, R. R., Rowe, I. H., Hawrylyshyn, P., and Organ, L. W., 1976, Computer mapping of brain stem sensory centers in man, *J. Neurosurg.* **44**:458–464.

Tasker, R. R., Howrylyshyn, P., Rowe, I. H., and Organ, L. W., 1977, Computerized graphic display of results of subcortical stimulation during stereotactic surgery, *Acta Neurochir. (Wien) [Suppl.]* **24**:85–98.

Tasker, R. R., Organ, L. W., and Hawrylyshin, P., 1982a, Investigation of the surgical target for alleviation of involuntary movement disorders, *Appl. Neurophysiol.* **45**(3):261–274.

Tasker, R. R., Organ, L. W., and Hawrylyshin, P. A., 1982b, *The Thalamus and Midbrain of Man. A Physiological Atlas Using Electrical Stimulation,* Charles C. Thomas, Springfield, IL.

Tasker, R. R., Siqueira, J., Hawrylyshin, P., and Organ, L. W., 1983, What happened to VIM thalamotomy for Parkinson's disease? *Appl. Neurophysiol.* **46**(1–4):68–83.

Tasker, R. R., Lenz, F., Yamashiro, K., Corecki, J., Hirajama, T., and Dostrovsky, J. O., 1987, Microelectrode techniques in localization of stereotactic targets, *Neurol. Res.* **9**:105–112.

Tator, C. H., and Rowed, D. W., 1976, Fluoroscopy of foramen ovale as an aid to thermocoagulation of the Gasserian ganglion, *J. Neurosurg.* **44**(2):254–257.

Taub, A., and Collins, W. F., 1973, Physiological anatomy of pain, in *Neurological Surgery,* Vol. III (J. R. Youmans, ed.), W. B. Saunders, Philadelphia, pp. 1587–1614.

Taylor, A. S., 1915, Operations on peripheral nerves, in *Operative Therapeusis* (A. B. Johnson, ed.), D. Appleton, New York, London.

Taylor, P. E., 1962, Traumatic intradural avulsion of the nerve roots of the brachial plexus, *Brain* **85**:579–602.

Teddy, P. J., Adams, C. B. T., Briggs, M., Jamous, M. A., and Kerr, J. H., 1981, Extradural diamorphine in the control of pain following lumbar laminectomy, *J. Neurol. Neurosurg.* **44**:1074–1078.

Teräväinen, H., Fogelholm, R., and Larsen, A., 1976, Effect of propranolol on essential tremor, *Neurology (Minneap.)* **26**(1):27–30.

Tessitore, A., Battaglia, A., and Celentano, I., 1983, L'utilita della TNS nelle sindromi dolorose del capo (stimolazione nervosa transcutanea), *Rass. Int. Clin. Ter.* **63**:543–547.

Teuber, J. L., Corkin, S., and Twitchell, T. F., 1977. Study of cingulotomy in man. A summary, in *Neurosurgical Treatment in Psychiatry, Pain and Epilepsy* (W. H. Sweet, S. Obrador, and J. G. Martin-Rodriguez, eds.), University Park Press, Baltimore, pp. 355–362.

Tew, J. M., and Keller, J. T., 1977, The treatment of trigeminal neuralgia by percutaneous radiofrequency technique, *Clin. Neurosurg.* **24**:557–578.

Tew, J. M., and Tobler, W. D., 1982, Percutaneous rhizotomy in the treatment of intractable facial pain (trigeminal, glossopharyngeal and vagal nerves), in *Operative Neurosurgical Techniques. Indications, Methods and Results,* Vol. 2 (H. H. Schmidek and W. H. Sweet, eds.), Grune & Stratton, New York, pp. 1083–1100.

Tew, J. M., Keller, J. T., and Williams, D. S., 1978, Application of stereotactic principles to the treatment of trigeminal neuralgia, *Appl. Neurophysiol.* **41**(1–4):146–156.

Thoden, V., Krainick, J.-U., Strasburg, H. M., and Zimmerman, H., 1977, Influence of dorsal column stimulation on spastic movement disorders, *Acta Neurochir. (Wien)* **39**(3–4):233–240.

Thoden, V., Doerr, M., Dieckmann, G., and Krainick, J.-U., 1979, Medial thalamic permanent electrodes for pain control in man—an electrophysiological and clinical study, *Electroencephalogr. Clin. Neurophysiol.* **47**:582–591.

Thomalske, G., 1975, Operative Behandlung der Epilepsie, *Dtsch. Arztebl.* **49**:3359–3368.

Thomas, A., and Long-Landry, M., 1928, Deux cas de tremblement d'attitude du membre supérieur, *Rev. Neurol.* **1**:585–587.

Thomas, D. G. T., and Sheehy, J. P. R., 1983, Dorsal root entry zone lesions (Nashold's procedure) for pain relief following brachial plexus avulsion, *J. Neurol. Neurosurg. Psychiatry* **46**:924–928.

Thomas, J. E., Reagen, T. J., and Class, D. W., 1977, Epilepsia partialis continua. A review of 32 cases, *Arch. Neurol.* **34**:266–275.

Thulin, C. A., Essen, C., and Zeuchner, E., 1972, Displacements of brain due to positional changes during stereotactic operations, *Confin. Neurol.* **34**:348–354.

Tindall, G. T., Nixon, D. W., Christy, J. H., and Neill, J., 1977, Pain relief in metastatic cancer other than breast and prostate gland following transsphenoidal hypophysectomy. A preliminary report, *J. Neurosurg.* **47**:659–662.

Tkachev, R. A., Markova, E. D., and Barchatova, V. P., 1973, L-Dopa treatment of some extrapyramidal hereditary diseases, in *Clinical Importance of L-Dopa,* Moscow, Basel, pp. 141–148 (Rus.).

Tobler, W. D., Tew, J. M., Cosman, E., Keller, J. T., and Quallen, B., 1983, Improved outcome in the treatment of trigeminal neuralgia by percutaneous stereotactic rhizotomy with a new, curved tip electrode, *Neurosurgery* **12**:313–317.

Toczek, S. K., McCullough, D. C., Gargour, C. W., Kachman, R., Baker, R., and Luessenhop, A. J., 1975, Selective sacral rootlet rhizotomy for hypertonic neurogenic bladder, *J. Neurosurg.* **42**(5):567–574.

Toczek, S. K., McCullough, D. C., and Boggs, J. S., 1978, Sacral rootlet rhizotomy at the conus medullaris for hypertonic neurogenic bladder, *J. Neurosurg.* **48**(2):193–196.

Todd, E. M., 1972, *Stereotaxy. Procedural Aspects,* Wells, Pasadena, CA.

Todd, E. M., 1983, *The Neuroanatomy of Leonardo da Vinci,* Capra Press, Santa Barbara.

Todd, E. M., Crue, B. L., and Carregal, E. J., 1969, Posterior percutaneous tractotomy and cordotomy, *Confin. Neurol.* **31**:106–110.

Todd, J., 1853, *Clinical Lectures on Paralysis,* London.

Tönnis, W., and Bischof, W., 1962, Ergebnisse der lumbalen Myelotomie nach Bischof, *Zentralbl. Neurochir.* **23**(1):29–36.

Tönnis, W., and Walter, W., 1967, Differentialdiagnose, Klinik und Behandlung der cerebral Gefasmisbildungen, in *Biologie und Klinik des Zentralnervensystems,* Sandoz, Basel.

Törmá, T., and Troupp, H., 1958, Spasmodic torticollis, *Ann. Chir.* **47**:401–410.

Török, P., 1960, Parietale Topektomie bei Phantomschmerz-Therapeutischer Erfolg und psychopathologische Bedeutung, *Acta Neurochir.* **8**:293–298.

Torrens, M. J., and Griffith, H. B., 1976, Management of the uninhibited bladder by selective sacral neurectomy, *J. Neurosurg.* **44**:176–185.

Torvik, A., 1959, Sensory, motor and reflex changes in two cases of intractable pain after stereotactic mesencephalic tractotomy, *J. Neurol. Neurosurg. Psychiatry* **22**:299–305.

Toth, S., 1961, The effect of the removal of the nucleus dentatus on the parkinsonian syndrome, *J. Neurol. Neurosurg. Psychiatry* **24**(2):143–147.

Toth, S., 1972, Surgical results of hyperkinetic dyskinesias. Some electrophysiological aspects of the different target points, *Confin. Neurol.* **34**:51–54.

Toth, S., and Tomka, I., 1968, Responses of the human thalamus and pallidum to high frequency stimulations, *Confin. Neurol.* **30**:17–40.

Toth, S., and Vajda, J., 1980, Multitarget technique in parkinson surgery, *Appl. Neurophysiol.* **43**:109–113.

Toth, S., Zarand, P., and Lazar, L., 1974, The role of the cortex and subcortical ganglia in the evoked rhythmic motor activity, *Acta Neurochir. (Wien) [Suppl.]* **21**:25–33.

Tourette, G. de la, 1885, Etude sur une affection nerveuse caractérisée par de l'incoordination matrice accompagneé d'écholalie et de coprolalie, *Arch. Neurol.* **9**:158–200.

Tower, D. B., 1960, *Neurochemistry of Epilepsy. Seizure Mechanisms and Their Management,* Charles C. Thomas, Springfield, IL.

Tretjakoff, K., 1919, *Contribution à l'Étude de l'Anatomie Pathologique du Locus Niger avec Quelques Déduction Rélatives a la Pathologenie de la Maladie de Parkinson,* Thesis, Paris.

Troupp, H., Martilla, I., and Halonen, V., 1970, Arteriovenous malformations of the brain. Prognosis without operation, *Acta Neurosurg. (Wien)* **22**:125–128.

Tsubokawa, T., and Moriyasu, N., 1975a, Lateral pallidotomy for relief of ballistic movements, *Confin. Neurol.* **37**:10–15.

Tsubokawa, T., and Moriyasu, N., 1975b, Follow-up results of centre median thalamotomy for relief of intractable pain, *Confin. Neurol.* **37**:280–284.

Tsubokawa, T., Kotani, A., and Nishimoto, H., 1976, Thalamic relay nucleus stimulation as a preventive method against recurrent tendency following thalamotomy for intractable pain, *Neurol. Med. Chir. (Tokyo)* **16**:247–254.

Tsubokawa, T., Yamamoto, T., and Katayama, Y., 1982a, Deep

brain stimulation for relief of intractable pain. Clinical results of thalamic relay stimulation, *Neurol. Med. Chir. (Tokyo)* **22**(3):211–218.

Tsubokawa, T., Yamamoto, T., Katayama, Y., and Moriyasu, N., 1982b, Clinical results and physiological basis of thalamic relay nucleus stimulation for relief of intractable pain with morphine resistance, *Appl. Neurophysiol.* **45**:143–155.

Tsubokawa, T., Yamamoto, T., Katayama, Y., Hirayama, T., and Sibuya, H., 1984, Thalamic relay nucleus stimulation for relief of intractable pain. Clinical results and β-endorphin immunoreactivity in the cerebrospinal fluid, *Pain* **18**:115–126.

Tsubokawa, T., Yamamoto, T., Katayama, Y., and Hirayama, T., 1985, Deafferentation pain and stimulation of the thalamic relay nucleus. Experimental and clinical study, in *IX Meeting of the World Society for Stereotactic and Functional Neurosurgery*, Toronto, pp. 166–171.

Tsutsumi, Y., Andoh, Y., and Inoue, N., 1982, Ultrasound-guided biopsy for deep-seated brain tumors, *J. Neurosurg.* **57**:164–167.

Turnbull, I. M., Shulman, R., and Woodhurst, W. B., 1980, Thalamic stimulation for neuropathic pain, *J. Neurosurg.* **52**:486–493.

Turner, J. W., Nath, F. P., Wyper, D., Hadley, D., and Grossart, K., 1986, The use of CT and magnetic resonance imaging in stereotactic surgery, in *VII Meeting of the European Society for Stereotactic and Functional Neurosurgery*, Birmingham.

Tytus, J. S., 1968, Cryosurgery: Its history and development, in *Cryosurgery* (R. W. Rand, A. P. Rinfret, and H. von Leden, eds.), Charles Thomas, Springfield, IL, pp. 3–18.

Tzonos, T., Bergleiter, R., and Pampus, F., 1975, Experiences in the use of artificial embolization as a method of treating cerebral angiomas, in *Cerebral Angiomas* (H. W. Pia, C. Langmaid, and J. Zierski, eds.), Springer-Verlag, Berlin, Heidelberg, New York, pp. 206–213.

Uematsu, S., 1977, Percutaneous electrothermocoagulation of spinal nerve trunk, ganglion, and rootlets, in *Current Techniques in Operative Neurosurgery* (H. Schmidek and W. H. Sweet, eds.), Grune & Stratton, New York, pp. 469–490.

Uematsu, S., 1982, Percutaneous electrothermocoagulation of spinal nerve trunk, ganglion and rootlets, in *Operative Neurosurgical Techniques. Indications, Methods and Results*, Vol. 2 (H. H. Schmidek and W. H. Sweet, eds.), Grune & Stratton, New York, pp. 1177–1198.

Uematsu, S., Udvarhelyi, G. B., Benson, D. W., and Siebens, A. A., 1974, Percutaneous radiofrequency rhizotomy, *Surg. Neurol.* **5**:319–325.

Ugrjumov, V. M., and Zotov, J. V., 1970, Surgical treatment of epilepsy in adults, in *Epilepsy*, Vol. 2, Nauka, Tbilisi, pp. 30–61 (Rus.).

Ugrjumov, V. M., Vvedenskaya, I. V., Grachev, K. V., Stepanova, T. S., and Yatsuk, S. L., 1967, Principles underlying diagnosis and treatment of patients with extrapyramidal disorders and epilepsy by using the method of deeply implanted indwelling electrodes, *Vopr. Neirokhir.* **6**:20–25 (Rus.).

Ugrjumov, V. M., Zotov, Y. V., and Schedrenok, V. V., 1969, Changes of some gas exchange and cerebral circulation indices in patients with cranium and brain injuries, *Vopr. Neirokhir.* **6**:43–49 (Rus.).

Ugrjumov, V. M., Stepanova, T. S., and Grachev, K. V., 1974, Stereotactic treatment for epilepsy in adults and children, in *Epileptogenic Focus and Surgical Treatment of Epilepsy*, Zdorovje, Kiev, pp. 39–43 (Rus.).

Umbach, W., 1960, *Differenzialdiagnose und Therapie der Gesichtsneuralgien*, Georg Thieme, Stuttgart.

Umbach, W., 1966, Long-term results of fornicotomy for temporal epilepsy, *Confin. Neurol.* **27**:121–123.

Umbach, W., and Ehrhardt, K. J., 1965, Ableitungen mit Mikroelektroden in den Stammganglien des Menschen, *Arch. Psychiatr. Nervenkr.* **207**:106–113.

Umbach, W., and Ehrhardt, K., 1966, Micro-electrode registration in the basal ganglia during stereotaxic operations, in *Advances in Stereoencephalotomy* (E. A. Spiegel and H. T. Wycis, eds.), S. Karger, Basel, pp. 315–318.

Umbach, W., and Riechert, T., 1964, Elektrophysiologische und Klinische Ergebnisse stereotaktischer Eingriffe im limbischen System bei temporale Epilepsie, *Nervenarzt* **35**:482–488.

Umbach, W., Bouchard, G., and Kim, N., 1973, The effectiveness of surgical treatment of focal epilepsy, *Vopr. Neirokhir.* **5**:43–50 (Rus.).

Upton, A. R. M., Cooper, I. S., and Amin, I., 1982, Quantitative assessment of stereotactic and functional neurosurgery, *Appl. Neurophysiol.* **45**:281–290.

Urban, B. J., and Nashold, B. S., 1978, Percutaneous epidural stimulation of the spinal cord for relief of pain, *J. Neurosurg.* **48**:323–328.

Urban, B. J., and Nashold, B. S., 1982, Combined epidural and peripheral nerve stimulation for relief of pain. Description of technique and preliminary results, *J. Neurosurg.* **57**:365–369.

Vaernet, K., 1972, Stereotaxic amygdalotomy in temporal lobe epilepsy, *Confin. Neurol.* **34**(1–4):176–183.

Vaernet, K., and Madsen, A., 1972, Lesions in the amygdala and the substantia innominata in aggressive psychotic patients, in *Psychosurgery* (E. Hitchcock, L. Laitinen, and K. Vaernet, eds.), Charles C. Thomas, Springfield, IL, pp. 187–194.

Vainshtock, A. B., 1974, *The Characteristics, Clinic and Physiology of the Idiopathic, Postencephalic and Vascular Parkinsonism*, Thesis, Leningrad.

Valenstein, E. S. (ed.), 1980, *The Psychosurgery Debate. Scientific, Legal and Ethical Perspectives*, W. H. Freeman, San Francisco.

Van Buren, J. M., 1972, *Variations of the Human Diencephalon*, Vol. 2, Springer, New York.

Van Buren, J. M., 1974, The neuroanatomical basis for functional neurosurgery, in *Recent Progress in Neurological Surgery*, Excerpta Medica, Amsterdam, pp. 219–233.

Van Buren, J. M., and Borke, K. C., 1972a, *The Nuclei and Cerebral Connections of the Human Thalamus*, Springer, Berlin.

Van Buren, J. M., and Borke, R. C., 1972b, *Variations and Connections of the Human Thalamus*, Springer, New York.

Van Buren, J. M., and McCubbin, D. A., 1962a, A standard method of plotting loci in human depth stimulation and electrography with an estimation of errors, *Confin. Neurol.* **22:**259–264.

Van Buren, J. M., and McCubbin, D. A., 1962b, An outline atlas of the human basal ganglia with estimation of anatomical variants, *J. Neurosurg.* **19:**811–839.

Van Buren, J. M., and Ratcheson, R. A., 1973, Principles of stereotaxic surgery, in *Neurological Surgery*, Vol. 3 (J. R. Yomans, ed.), W. B. Saunders, Philadelphia, pp. 1793–1828.

Van Buren, J. M., Li, C. L., and Shapiro, D. J., 1973, A qualitative and quantitative evaluation of parkinsonians three to six years following thalamotomy, *Confin. Neurol.* **35:**202–235.

Van Buren, J. M., Ajmone-Marsan, C., Matsuba, N., and Sagowsky, D., 1975, Surgery of temporal lobe epilepsy, *Adv. Neurol.* **8:**155–196.

Vanderark, G. D., and McGrath, K. A., 1975, Transcutaneous electrical stimulation in treatment of postoperative pain, *Am. J. Surg.* **130:**338–340.

Van Hoof, J. J. M., Horstink, M. W. I., Berger, H. J. C., Van Spaendonck, K. P. M., and Cools, A. P., 1987, Spasmodic torticollis: The problem of pathophysiology and assessment, *J. Neurol.* **224:**322–327.

Van Loveren, H., Tew, J. M., Keller, J. T., and Nurre, M. A., 1982, A 10-year experience in the treatment of trigeminal neuralgia. Comparison of percutaneous stereotaxic rhizotomy and posterior fossa exploration, *J. Neurosurg.* **57**(6):757–764.

Van Manen, J., 1967, *Stereotactic Methods and Their Applications in Disorders of the Motor System*, Charles C. Thomas, Springfield, IL.

Van Manen, J., 1969, Postural instability after ventrolateral thalamic lesions, in *Third Symposium on Parkinson's disease* (F. J. Gillingham and I. M. L. Donaldson, eds.), E. & S. Livingstone, Edinburgh, London, pp. 237–241.

Van Manen, J., 1974, Stereotactic operations in cases of hereditary and intention tremor, in *Advances in Stereotactic and Functional Neurosurgery* (F. J. Gillingham, E. R. Hitchcock, and J. W. Turner, eds.), Springer-Verlag, Vienna, New York, pp. 49–56.

Van Manen, J., and Van Hoytema, G. J., 1962, A Dutch stereotaxis apparatus, *Psychiatry Neurol. Neurochir.* **65:**81–91.

Van Wagenen, W. P., and Herren, R. Y., 1940, Surgical division of commissural pathways in the corpus callosum. Relation to spread of an epileptic attack, *Arch. Neurol. Psychiatry* **44:**740–759.

Vaquero, J., Oya, S., Manrique, M., Bujan, J., Lozano, A. P., and Bravo, G., 1979, Medial frontal epilepsy: Histological, histochemical and ultrastructural study. Preliminary observations, *Acta Neurochir. (Wien)* **46:**233–241.

Vasin, N. Ya., 1973, Methods and results of phenol blocks of the gasserian ganglion in grave forms of facial algesia, *Vopr. Neirokhir.* **2:**16–23 (Rus.).

Vasin, N. Ya., and Ratsa, V. I., 1971a, Some main problems of surgical treatment of pain by operations on the thalamus, in *All-Union Congress of Neurosurgery*, Vol. 3, Moscow, pp. 184–186 (Rus.).

Vasin, N. Ya., and Ratsa, V. I., 1971b, Stereotaxic destruction of the sensitive thalamic nucleus as a method for surgical treatment of intractable pain, *Vopr. Neirokhir.* **3:**3–9 (Rus.).

Vasin, N. Ya., Shakhnovich, A. R., Iljinsky, J. A., and Milovanova, L. S., 1970, H-reflex and capsular effect in stereotaxic operations on the ventrolateral thalamic nucleus, *Vopr. Neirokhir.* **3:**6–10 (Rus.).

Vasin, N. Ya., Iljinsky, I. A., and Safronov, V. A., 1971, Determination of the optimal area of destruction in the ventrooral group of thalamic nucleus, based on the effects of electrostimulation in parkinsonism, *Vopr. Neirokhir.* **6:**9–13 (Rus.).

Vasin, N. Ya., Lesov, N. S., Maiorchik, V. E., Arkhipova, N. A., and Safronov, V. A., 1977, Clinico-physiological effect of electrostimulation and destruction of the ventrolateral parts of the cerebellar dentate nuclei in different forms of motor pathology, *Vopr. Neirokhir.* **4:**12–21 (Rus.).

Vasin, N. Ya., Nadvornik, P., Lesov, N., Kadin, A. L., Shramka, M., Golonda, N., and Nadvornik, V., 1979, Stereotaxic combined dentato-thalamotomy in the treatment of spastico-hyperkinetic forms of subcortical dyskinesia, *Vopr. Neirokhir.* **6:**22–27 (Rus.).

Vein, A. M,. Golubev, V. L., and Jachno, N. N., 1975, Role of the brain "nonspecific system" disturbances in pathogenesis of parkinsonism, *Zh. Nevropatol. Psikhiatr.* No. 7:970–972.

Vein, A. M., Golubev, V. L., and Berzinsh, J. E., 1981, *Parkinsonism: Clinic, Etiology, Pathogenesis, Treatment*, Zinatne, Riga (Rus.).

Veki, A., Uno, M., Anderson, M., and Yoshida, M., 1977, Monosynaptic inhibition of thalamic neurons produced by stimulation of the substantic nigra, *Experientia* **33:**1480–1482.

Velasco, F., and Molina-Negro, P., 1973, Electrophysiological topography of the human diencephalon, *J. Neurosurg.* **38:**204–214.

Velasco, F. G., Molina-Negro, P., Bertrand, C., and Hardy, J., 1972, Further definition of the subthalamic target for arrest of tremor, *J. Neurosurg.* **36:**184–191.

Venna, N., Sabin, T. D., Ordia, J. I., and Mark, V. K., 1984, Treatment of severe Parkinson's disease by intraventricular injection of dopamine, *Appl. Neurophysiol.* **47:**62–64.

Verdie, J. C., and Lazorthes, Y., 1982, Thermocoagulation percutanée analgésique des racines rachidiennes, *Nouv. Press. Med.* **11:**2131–2134.

Vihlein, A., Weerasooriya, L. A., and Holman, C. B., 1969, Percutaneous electric cervical cordotomy for the relief of intractable pain, *Mayo Clin. Proc.* **44:**176–183.

Vikhert, T. M., Kandel, E. J., Kukin, A. V., and Kuparadze, G. R., 1968, A study of an intracerebral lesions following local congelation with liquid nitrogen, *Vopr. Neirokhir.* **3:**45–48 (Rus.).

Vilkki, J., 1975, Late psychological and clinical effects of subrostral cingulotomy and anterior mesoloviotomy in psychiatric illness, in *Neurosurgical Treatment in Psychiatry, Pain, and Epilepsy* (W. H. Sweet, S. Obrador, and J. G. Martin-Rodriguez, eds.), University Park Press, Baltimore, pp. 253–259.

Virozub, J. D., and Chernovsky, V. J., 1976, Treatment of neurogenic disorders in micturition by way of dividing the pudendal nerves in patients with vertebral and spinal cord injuries, *Vopr. Neirokhir.* **4**:6–9 (Rus.).

Virozub, J. D., and Chipko, S. S., 1975, Our experience with 60 frontal myelotomies in the treatment of patients with spastic manifestations due to the damage of the vertebral column and spinal cord, *Vopr. Neirokhir.* **2**:2–26 (Rus.).

Virozub, J. D., and Chipko, S. S., 1981, *Spastic Phenomena in Spinal Cord Lesions,* Zdorovja, Kiev (Rus.).

Visca, A., and Bernasconi, V., 1972, A radiographic technique for trigeminal neurolytic injections, *Acta Neirokhir.* **26**:55–60.

Vlahovitch, B., Fuentes, J. M., Choucair, Y., Moreau, P., and Pascal, M., 1976, Gestes complémentaires à l'opération de Sjoqvist dans les algies cancéreuses cervico-faciales, *Neurochirurgie* **22**:503–515.

Vogt, C., 1909, La myéloarchitecture du thalamus du cercopithéque, *J. Psychol. Neurol.* **12**:285–324.

Vogt, C., 1911, Quelques considérations générales a propos du syndrome du corps strié, *J. Psychol. Neurol.* **18**:479–488.

Vogt, C., and Oppenheim, H., 1911, Wesen und Lokalisation der kongenitalen und infantilen Pseudobulbaerparalyse, *J. Psychol. Neurol.* **18**:1–4.

Vogt, C., and Vogt, O., 1919–1920, Zur Lehre der Erkrankungen der striären Systems, *J. Psychol. Neurol.* **25**:627–846.

Vogt, C., and Vogt, O., 1937, Sitz und Wesen der Krankheiten im Liechte der topistischen Himforschung und des Varierens der Tiere, *J. Psychol. Neurol.* **47**:237–245.

Vogt, C., and Vogt, O., 1942, Morphologische gestaltungen unter normalen und pathogen Bedingungen. Ein hirnanatomischer Beitrag zu ihrer Kenntnis, *J. Psychol. Neurol.* **50**:161–524.

Vogt, O., and Vogt, C., 1948, Thalamusstudien, I–III, *J. Psychol. Neurol.* **50**:33–154.

Volkov, A. A., and Yatsuk, S. L., 1963, The pallidotomy technique by stereotactic device, *Exp. Chir. Anesthesiol.* **1**:12–18 (Rus.).

Voris, H. G., 1951, Ipsilateral sensory loss following cordotomy. Report of a case, *Arch. Neurol. Psychiatry* **65**:95–96.

Voris, H. G., and Whisler, W. W., 1975, Results of stereotaxic surgery for intractable pain, *Confin. Neurol.* **37**:86–96.

Voronin, G. V., and Kandel, E. I., 1971, Is there supraspinal "rhythm pacemaker" of tremor in parkinsonism? *Zh. Nevropatol. Psikhiatr.* **7**:967–172 (Rus.).

Wada, T., 1953, A modified stereotaxic apparatus for brain surgery of deep structures, *Tohoku J. Exp. Med.* **58**:299–301.

Wada, J. A., and Sato, M., 1974, Generalized convulsive seizures induced by daily electrical stimulation of the amygdala in cats, correlative electrographic and behavioral features, *Neurology (Minneap.)* **24**:565–574.

Wada, J. A., Sato, M., and Corcoran, M. E., 1974, Persistent seizure susceptibility and recurrent spontaneous seizures in kindled cats, *Epilepsia* **15**:465–478.

Waga, S., Morikawa, A., and Kojima, T., 1979, Trigeminal neuralgia: Compression of the trigeminal nerve by an elongated and dilated basilar artery, *Surg. Neurol.* **11**:13–16.

Wagner, H. N., Burns, L. D., Dannals, R. F., Wong, D. F., Langstron, B., Duelfer, T., Frost, J. J., Ravert, H. T., Links, J. M., Rosenbloom, S. B., Lukas, S. E., Kramer, A. V., and Kuhar, M. J., 1983, Imaging dopamine receptors in the human brain by positron tomography, *Science* **221**:1264–1266.

Walder, H., 1968, Experimental cryosurgery, in *Cryosurgery* (R. W. Rand, A. P. Rinfret, and H. von Leden, eds.), Charles Thomas, Springfield, pp. 133–186.

Walder, H. A., 1975, Freezing arteriovenous anomalies in the brain, in *Cerebral Angioma* (H. V. Pia, ed.), Springer Verlag, Berlin, Heidelberg, New York, pp. 194–198.

Walder, H., Jasper, H. H. J., and Meijer, E., 1970, Application of cryotherapy in cerebrovascular anomalies, *Psychiatr. Neurol. Neurochir.* **73**:471–486.

Waldman, A. V., and Ignatov, Yu. D., 1976, *Central Mechanisms of Pain,* Nauka, Moscow (Rus.).

Walker, A. E., 1938, *The Primate Thalamus,* University of Chicago Press, Chicago.

Walker, A. E., 1940, The spinothalamic tract in man, *Arch. Neurol. Psychiatry* **43**:284–298.

Walker, A. E., 1942a, Relief of pain by mesencephalic tractotomy, *Arch. Neurol. Psychiatry* **48**:865–883.

Walker, A. E., 1942b, Somatotopic localization of spinothalamic and secondary trigeminal tracts in mesencephalon, *Arch. Neurol. Psychiatry* **48**:884–889.

Walker, A. E., 1943, Central representation of pain, *Res. Publ. Assoc. Res. Nerv. Ment. Dis.* **23**:63–85.

Walker, A. E., 1949a, Cerebral pedunculotomy for the relief of involuntary movements. Hemiballismus, *Acta Psychiatr.* **24**:723–729.

Walker, A. E., 1949b, *Posttraumatic Epilepsy,* Charles C. Thomas, Springfield, IL.

Walker, A. E., 1952, Cerebral pedunculotomy for the relief of voluntary movements. II. Parkinsonian tremor, *J. Nerv. Ment. Dis.* **116**:766–775.

Walker, A. E., 1957, Physiological principles and results of neurosurgical interventions in extrapyramidal diseases, in *Premier Congress International Society of Neurology,* Pergamon Press, Brussels, pp. 118–137.

Walker, A. E., 1959, Normal and pathological physiology of the thalamus, in *Introduction to Stereotaxis,* Vol. 1 (G. Schaltenbrand and P. Bailey, eds.), Georg Thieme, Stuttgart, pp. 291–316.

Walker, A. E., 1962, Stereotaxic methods for the study of subcortical activity in epilepsy, *Confin. Neurol.* **22**:217–223.

Walker, A. E., 1966, Internal structure and afferent–efferent relations of the thalamus, in *The Thalamus* (D. Purpura and M. Yahr, eds.), Raven Press, New York, pp. 1–24.

Walker, A. E., 1967a, Temporal lobectomy, *J. Neurosurg.* **26**:642–649.

Walker, A. E., 1967b, Surgery of the pituitary gland, in *Anales del XII Congreso Latinoamericano de Neurocirurgia,* Lima, pp. 1033–1038.

Walker, A. E., 1969, The current concepts of pathogenesis of tremor in Parkinson's disease, in *Third Symposium on Parkinson's Disease* (F. J. Gillingham and I. M. L. Donaldson, eds.), Livingstone, Edinburgh, London, pp. 100–105.

Walker, A. E., 1972, The technique of temporal lobectomy for psychomotor epilepsy and postoperative sexual changes, *Vopr. Neirokhir.* **3**:9–16 (Rus.).

Walker, A. E., 1974, Surgery for epilepsy, in *Handbook of Clinical Neurology,* Vol. 15. *The Epilepsies* (O. Magnus and A. M. Lorentz de Haas, eds.), North-Holland, Amsterdam, pp. 739–757.

Walker, A. E., 1982, General principles of stereotaxic surgery for epilepsy, in *Stereotaxy of the Human Brain* (G. Schaltenbrand and A. E. Walker, eds.), Georg Thieme Verlag, Stuttgart, New York, pp. 645–652.

Walker, A. E., and Burton, C. V., 1966, Radiofrequency telethermocoagulation, *J.A.M.A.* **197**(9):700–704.

Walker, A., and Ribstein, M., 1957, Enregistrements et stimulations des formations rhinencephalique avec electrodes profondes a demeure chez l'homme, *Rev. Neurol.* **96**:453–459.

Walker, A. E., Uematsu, S., Niedermeyer, E., and MacDonald, L., 1982, Depth recording, in *Stereotaxy of the Human Brain* (G. Schaltenbrand and A. E. Walker, eds.), Georg Thieme Verlag, Stuttgart, New York, pp. 661–668.

Wall, P. D., 1969, Organization of cord cells which transmit sensory information, in *The Skin Senses* (D. R. Kenshalo, eds.), Charles C. Thomas, Springfield, IL, pp. 512–533.

Wall, P. D., 1978, The gate control theory of pain mechanisms: A reexamination and a restatement, *Brain* **101**:1–16.

Wall, P. D., 1980, The role of substantia gelatinosa as a gate control, in *Pain* (J. Bonica, ed.), Raven Press, New York, pp. 205–231.

Walsh, L. S., 1976, Spasmodic torticollis, in *Institute of Neurology, Madras, Proceedings,* Vol. 6, pp. 36–41.

Waltregny, A., 1976, Apport des méthodes stéréotaxique dans le diagnostic et le traitement des tumeurs cérébrales, *Ann. Radiol.* **19**:241–252.

Waltregny, A. J. M., 1982, A stereotactic frame for trigeminal ganglionectomy, *Appl. Neurophysiol.* **45**:516–517.

Waltz, J. M., 1981, Computerized percutaneous multi-level spinal cord stimulation in motor disorders, in *VIII Meeting of the World Society for Stereotactic and Functional Neurosurgery,* Zurich, p. 17.

Waltz, J. M., 1982, Computerized percutaneous multi-level spinal cord stimulation in motor disorders, *Appl. Neurophysiol.* **45**:73–92.

Waltz, J. M., Stellar, S., and Cooper, I. S., 1966, Cryothalamectomy for Parkinson's disease. A statistical analysis, *Neurology (Minneap.)* **16**:994–1002.

Waltz, J. M., Reynolds, L. O., and Riklan, M., 1981, Multi-lead spinal cord stimulation for control of motor disorders, *Appl. Neurophysiol.* **44**(4):244–257.

Waltz, J. M., Scozzari, C. A., and Hunt, D. P., 1985, Spinal cord stimulation in the treatment of spasmodic torticollis, *Appl. Neurophysiol.* **48**:334–338.

Wang, J. K., 1985, Intrathecal morphine for intractable pain secondary to cancer of pelvic organs, *Pain* **21**:99–102.

Ward, A. A., 1948, The anterior cingulate gyrus and personality, *Res. Publ. Assoc. Nerv. Ment. Dis.* **27**:438–445.

Ward, A. A., 1961, Physiological mechanism in parkinsonism, *Rev. Can. Biol.* **20**:345–380.

Ward, A. A., 1975, Theoretical basis for surgical therapy of epilepsy, *Adv. Neurol.* **8**:23–35.

Wassman, E. R., Eldridge, R., Abuzzaheb, F. S., and Nee, L., 1978, Gilles de la Tourette syndrome: Clinical and genetic studies in a midwestern city, *Neurology (Minneap.)* **28**:304–307.

Watanabe, E., Mayanagi, Y., and Sano, K., 1980, Influence of the periaqueductal gray and posterior hypothalamic area on neuronal activities of the spinal dorsal horn, *Appl. Neurophysiol.* **43**:338–339.

Watkins, E. S., 1965, Heat gains in brain during electrocoagulative lesions, *J. Neurosurg.* **23**:319–323.

Watkins, E. S., 1975, Stereotaxic thalamotomy for intractable pain, *Confin. Neurol.* **37**:278–288.

Watkins, L. R., and Mayer, D. Y., 1982, Organization of endogenous opiate and nonopiate pain control systems, *Science* 1185–1192.

Watson, C. P., Morley, T. P., Richardson, J. C., Schutz, H., and Tasker, R. R., 1983, The surgical treatment of chronic cluster headache, *Headache* **23**:289–295.

Webster, D. D., 1972, Clinical aspects of rigidity, in *Parkinson's Disease: Rigidity, Akinesia, Behavior,* Vol. 1 (J. Siegfried, ed.), Hans Huber, Bern, Stuttgart, Vienna, pp. 65–92.

Wechsler, I. S., and Brock, S., 1922, Dystonia musculorum deformans with especial reference to a myostatic form and the occurrence of decerebrate rigidity phenomena, *Arch. Neurol. Psychiatry* **8**:538–552.

Weigel, K., and Mundinger, F., 1986, Computerized tomography-guided stereotactic dentatotomy, *Appl. Neurophysiol.* **49**:301–306.

Weil, S. M., Van Loveren, H. R., Tomsick, T. A., Quallen, B. L., and Tew, J. M., 1987, Management of inoperable cerebral aneurysms by the navigational balloon technique, *Neurosurgery* **21**:296–302.

Weingarten, S. M., Cherlow, D. G., and Halgren, E., 1977, Relationship of hallucinations to the depth structures of the temporal lobe, in *Neurosurgical Treatment in Psychiatry, Pain, and Epilepsy* (W. H. Sweet, S. Obrador, and J. G. Martin-Rodriguez, eds.), University Park Press, Baltimore, pp. 553–568.

Weisenburg, T. H., 1910, Cerebello-pontine tumor diagnosed for six years as tic douloureux. The symptoms of irritation of ninth and tenth cranial nerves, *J.A.M.A.* **54**:1600–1604.

Wenderowich, E., and Nikitin, M., 1913, Ueber die Verbreitung der Faserdegenerationen bei amyotrophischer Lateralsklerose mit besonderer Berücksichtigung der Veranderungen im Grosshirn, *Arch. Psychiatr. Nervenkr.* **111**:300–334.

Wepsic, J. G., 1976, Complications of percutaneous surgery for pain, *Clin. Neurosurg.* **23**:454–464.

Wertheimer, P., and Bonniot, A., 1926, *Chirurgie du Sympathique Chirurgie du Tonus Musculaire. La Section des Rameaux Communicants,* Mason, Paris.

Wertheimer, P., and Lecuire, J., 1953, La myélotomie commissurale postérieure. A propos de 107 observations, *Acta Chir. (Belg.)* **52**:568–574.

Wertheimer, P., and Mansuy, L., 1950, La pyramidotomie pédunculaire dans le traitement des dyskinésies. A propos de deux observations, *Rev. Neurol.* **1953, 83**:433–437.

Westphal, C., 1883, Über eine dem Bilde der cerebrospinalen

grauen Degeneration ähnliche Erkrankung des centralen Nervensystems, ohne Anatomischen Befund, nebst einigen Bemerkungen über paradoxe Kontraktion, *Arch. Psychiatr. Nervenkr.* **14**:87–134.

Whisler, W. W., and Voris, H. C., 1978, Mesencephalotomy for intractable pain due to malignant disease, *Appl. Neurophysiol.* **41**:52–56.

White, J. C., 1963, Anterolateral cordotomy—its effectiveness in relieving pain of non-malignant disease, *Neurochirurgia (Stuttg.)* **6**(3):83–102.

White, J. C., 1964, La rhizotomie postérieure: Évaluation de son efficacité et de ses limites dans le contrôle de la douleur due à des affections non malignes, *Neurochirurgie* **10**(6):648–649.

White, J. C., 1973, Neuralgias of peripheral nerves, in *Neurological Surgery,* Vol. III (J. R. Youmans, ed.), W. B. Saunders, Philadelphia, pp. 1702–1716.

White, J. C., and Sweet, W. H., 1955, *Pain: Its Mechanisms and Neurosurgical Control,* Vol. 1, Charles C. Thomas, Springfield, IL.

White, J., and Sweet, W., 1969, *Pain and the Neurosurgeon, a 40-Year Experience,* Charles C. Thomas, Springfield, IL.

Whittaker, C. K., 1980, Cerebellar stimulation for cerebral palsy, *J. Neurosurg.* **52**:648–653.

Whitty, C. W. M., Duffield, J. E., Tow, P. M., and Cairns, H., 1952, Anterior cingulectomy in the treatment of mental disease, *Lancet* **1**:475–481.

Wiesendanger, M., 1986, Motor functions of the basal ganglia, *Appl. Neurophysiol.* **49**:269–277.

Wieser, H., and Yasargil, M., 1982, Selective amygdalohippocampectomy as a surgical treatment of mesiobasal limbic epilepsy, *Surg. Neurol.* **17**:445–457.

Wieser, H. G., Graf, H. P., Bernoulli, C., and Siegfried, J., 1978, Quantitative analysis of intracerebral recordings in epilepsia partialis continua, *Electroencephalogr. Clin. Neurophysiol.* **44**:14–22.

Wieser, H. G., Meles, H. P., and Bernoulli, C., 1980, Clinical and chronotopographic psychomotor seizure patterns (SEEG study with reference to postoperative results), *Acta Neurochir. (Wien)* **30**:103–112.

Wilder, B., 1971, Electroencephalogram activation in medically intractable epileptic patients, activation technique including surgical follow-up, *Arch. Neurol.* **25**:415–422.

Williams, D., 1965, The thalamus and epilepsy, *Brain* **88**:539–548.

Williams, D., and Parsons-Smith, G., 1949, The spontaneous electrical activity of the human thalamus, *Brain* **72**:450–482.

Willis, W. D., and Grossman, R. G., 1973, *Medical Neurobiology. Neuroanatomical and Neurophysiological Principles Basic to Clinical Neuroscience,* C. V. Mosby, St. Louis.

Wilson, C. B., and Dempsey, L. C., 1978, Transsphenoidal microsurgical removal of 250 pituitary adenomas, *J. Neurosurg.* **48**:13–22.

Wilson, D. H., and Chang, A. E., 1974, Bilateral anterior cingulectomy for the relief of intractable pain. Report of 23 patients, *Confin. Neurol.* **36**:61–68.

Wilson, D. H., Culver, C., and Waddington, M., 1975, Disconnection of the cerebral hemispheres. An alternative to hemispherectomy for the control of intractable seizures, *Neurology (Minneap.)* **25**:1149–1153.

Wilson, D. H., Reeves, A., and Gazzanica, M., 1978, Division of the corpus callosum for uncontrollable epilepsy, *Neurology (Minneap.)* **28**:649–653.

Wilson, S. A. K., 1912, Progressive lenticular degeneration: A familial nervous disease associated with cirrhosis of the liver, *Brain* **34**:295–509.

Wilson, S. A. K., 1914, An experimental research into the anatomy and physiology of the corpus striatum, *Brain* **36**:427–492.

Wilson, S. A. K., 1954, Friedrich's disease, *Neurology (Minneap.)* **2**:1078–1091.

Wilson, V. J., Talbot, W. H., and Kato, M., 1964, Inhibitory convergence upon Renshaw cells, *J. Neurophysiol.* **27**(6):1063–1079.

Wimmer, A., 1929, Le spasm de torsion, *Rev. Neurol.* **36**:904–915.

Winkelmüller, W., 1981, Experience with the control of low back pain by the dorsal column stimulation (DCS) and by the peridural electrode system (PISCES), in *Spinal Cord Stimulation* (Y. Hosobuchi and J. Cobrin, eds.), Excerpta Medica, Amsterdam, pp. 34–40.

Winkelmüller, W., Seidel, B. U., and Dietz, H., 1977, Über die Beenflussung von Bevegungsstörungen bei der infantilen Gerebralparese durch chronische Kleinhirnstimulation, *Neurochirurgia (Stuttg.)* **20**:179–185.

Winther, K., 1973, Central pains and treatment by stereotaxic thalamotomies, *Eur. Neurol.* **10**:65–74.

Wise, B. L., and Gleason, C., 1979, CT directed stereotactic surgery for diagnosis and treatment of deep cerebral lesions, in *Annual Meeting of A.A.N.S.,* Los Angeles.

Witzmann, A., and Dieckmann, G., 1982, Intracranial facial nerve decompression in the management of hemifacial spasm, *Appl. Neurophysiol.* **45**:291–294.

Wolfenden, R. N., and Williams, E., 1888, A note on the rhythm and character of certain tremors, *Br. Med. J.* **1**:1049–1050.

Wong, P., Hoffman, H., Froese, A., Platt, L., Blacha, S., and Gear, J., 1979, Cerebellar stimulation in the management of cerebral palsy: Clinical and physiological studies, *Neurosurgery* **5**:217–224.

Wood, J. H., Raymond, C. L., Ziegler, M. G., and Van Buren, J. M., 1977, Neurophysiological and neurochemical alterations during electrical stimulation of human caudate nucleus, *J. Neurosurg.* **46**(3):361–368.

Wooten, G. F., Eldridge, R., and Axelrod, J., 1973, Elevated plasma dopamine-β-hydroxylase activity in autosomal dominant torsion dystonia. *N. Engl. J. Med.* **288**:284–289.

Woroschiloff, C., 1874, Der Verlauf der motorisierten und sensiblen Bahnen durch das Lendenmark des Kaninchens, *Ber. Sachs. Ges. Wiss. Leipzig [Math.-Physik.]* **26**:248–304.

Wright, B. M., 1971, Nitrous-oxide-cooled neurosurgical cryoprobe, *Lancet* 951.

Wycis, H., and Gildenberg, P. L., 1979, Long-range evaluation of the surgical treatment of spasmodic torticollis, *Excerpta Med.* **193**:97–101.

Wycis, H. T., and Spiegel, E. A., 1962, Long-range results in the treatment of intractable pain by stereotaxic midbrain surgery, *J. Neurosurg.* **19:**101–106.

Wycis, H. T., and Spiegel, E. A., 1969, Campotomy in myoclonia, *J. Neurosurg.* **30**(6):708–713.

Wycis, H. T., Lee, A. J., and Spiegel, E. A., 1949, Simultaneous records of thalamic and cortical potentials in schizophrenics and epileptics, *Confin. Neurol.* **9:**264–272.

Wycis, H. T., Robbins, R., Spiegel-Adolf, M., Meszoros, J., and Spiegel, E. A., 1954, Studies in stereoencephalotomy. III. Treatment of a cystic craniopharyngioma by injection of radioactive P^{32}, *Confin. Neurol.* **14:**193–202.

Wycis, H. T., Szekely, E. G., and Spiegel, E. A., 1957, Tremor on stimulation of midbrain tegmentum after degeneration of the brachium conjunctivum, *J. Neuropathol. Exp. Neurol.* **16:**79–84.

Wycis, H., Baird, H., and Spiegel, E., 1966, Long range results following pallidotomy and pallidoamygdalotomy in certain types of convulsive disorders, *Confin. Neurol.* **27:**114–120.

Yahr, M. D., 1976, Evaluation of long-term therapy in Parkinson's disease: Mortality and therapeutic efficacy, in *Advances in Parkinsonism* (W. Birkmayer and O. Hornykiewicz, eds.), Editiones Roche, Basel, pp. 435–443.

Yahr, M. D., 1979, Overview of present-day treatment of Parkinson's disease, *J. Neurol.* **227:**75–88.

Yahr, M. D., Duvoisin, R. C., Schear, M. J., Barrett, R. E., and Hoehn, M. M., 1969, Treatment of parkinsonism with levodopa, *Arch. Neurol.* (Chicago) **21:**343–354.

Yaksh, T. L., and Rudy, T. A., 1976, Analgesia mediated by a direct spinal action of narcotics, *Science* **192:**1357–1358.

Yamada, S., Mitchell, O. C., and Hargest, T. S., 1972, Evaluation of Bischof's myelotomy for control of mass spasms, in *Present Limits of Neurosurgery* (I. Fusek and Z. Kunc, eds.), Avicenum, Czechoslovak Medical Press, Prague, pp. 543–544.

Yamada, S., Perot, P. L., Ducker, T. B., and Lockard, I., 1976, Myelotomy for control of mass spasms in paraplegia, *J. Neurosurg.* **45**(6):683–691.

Yamashita, J., Handa, H., Ishikawa, M., Tsukahara, T., and Mori, K., 1981, Treatment of causalgia by PISCES (percutaneously inserted spinal cord electrical stimulation), *Neurol. Med. Chir. (Tokyo)* **21:**413–418.

Yanagisawa, N., and Goto, A., 1971, Dystonia musculorum deformans. Analysis with electromyography, *J. Neurol. Sci.* **13:**39–65.

Yasargil, M. G., 1962, Ergebnisse der stereotaktischen Operationen bei Hyperkinesien, *Schweiz. Med. Wochenschr.* **92**(48):1550–1555.

Yasargil, M. G., 1969, Reconstructive and constructive surgery of the cerebral arteries in man, in *Microsurgery* (M. G. Yasargil, ed.), Georg Thieme, Stuttgart, pp. 95–117.

Yasargil, M. G., 1984, *Microneurosurgery*, Georg Thieme Verlag, Stuttgart, New York.

Yasargil, M. G., and Fox, J. L., 1975, The microsurgical approach to intracranial aneurysms, *Surg. Neurol.* **3:**7–14.

Yeates, A., Enzmann, D. R., Britt, R. H., and Silverberg, G., 1982, Simplified and accurate CT-guided needle biopsy of central nervous system lesions, *J. Neurosurg.* **57:**390–393.

Yeoman, P. M., 1968, Cervical myelography in traction injuries of the brachial plexus, *J. Bone Joint Surg.* **50B:**253–260.

Yoshida, M., and Kuramota, S., 1987, Neural noise measurement as a real time guide for tumor site verification during stereotaxic biopsy of brain tumors, in *VIII European Congress of Neurosurgery*, Barcelona, p. 356.

Yoshida, M., Yanagisawa, N., Shimazu, H., Givre, A., and Narabayashi, H., 1964, Physiological identification of the thalamic nucleus, *Arch. Neurol.* **11**(4):435–443.

Yoshii, N., and Fukuda, S., 1979, Effects of unilateral and bilateral invasion of thalamic pulvinar for pain relief, *Tohoku J. Exp. Med.* **127:**81–84.

Yoshii, N., Kudo, T., and Shimizu, S., 1975, Clinical and experimental studies of thalamic pulvinotomy, *Confin. Neurol.* **37:**97–98.

Yoshii, N., Mizokami, T., Ushikubo, T., Kuramitsu, T., and Fukuda, S., 1980, Long-term follow-up study after pulvinotomy for intractable pain, *Appl. Neurophysiol.* **43:**128–132.

Young, B., and Mulcany, J., 1980, Percutaneous sacral rhizotomy for neurogenic detrusor hyperreflexia, *J. Neurosurg.* **53:**85–87.

Young, R. F., 1978, Evaluation of dorsal column stimulation in the treatment of chronic pain, *Neurosurgery* **3:**373–378.

Young, R. F., and Chambi, V. I., 1987, Pain relief by electrical stimulation of the periaqueductal and periventricular gray matter. Evidence for a non-opioid mechanism, *J. Neurosurg.* **66:**364–371.

Young, R. F., Kroening, R., Fulton, W., Feldman, R. A., and Chambi, I., 1985, Electrical stimulation of the brain in treatment of chronic pain. Experience over 5 years, *J. Neurosurg.* **62:**389–396.

Yukharev, S. P., 1979, Neurosurgical anatomy of Forel's H-field as applied to campotomy, *Vopr. Neirokhir.* **2:**48–52 (Rus.).

Zapletal, B., 1965, Nigrotomy in treatment of extrapyramidal diseases, *Acta Neurochir. (Wien)* **13**(3–4):388–392.

Zapletal, B., 1969, Open mesencephalotomy and thalamotomy for intractable pain, *Acta Neurochir. (Wien) [Suppl.]* **18:**118–126.

Zelondjev, L. V., 1968, *Comparative Characteristics of the Different Methods of Destruction of Subcortical Structures in Stereotactic Operations*, Thesis, Kharkov (Rus.).

Zeman, W., 1976, Dystonia: An overview, in *Advances in Neurology*, Vol. 14 (R. Eldridge and S. Fahn, eds.), Raven Press, New York, pp. 91–103.

Zeman, W., and Dyken, P., 1967, Dystonia musculorum deformans (clinical, genetic and pathoanatomical studies), *Psychiatr. Neurol. Neurochir.* **70**(2):77–121.

Zeman, W., and Dyken, P., 1968, Dystonia musculorum deformans, in *Handbook of Clinical Neurology*, Vol. 6 (P. J. Vinken and G. W. Bryan, eds.), North-Holland, Amsterdam, pp. 517–543.

Zeman, W., Kaelbling, R., and Pasamanick, B., 1960, Idiopathic dystonia musculorum deformans. Part 2 (the formes frustes), *Neurology (Minneap.)* **10:**1068–1075.

Zemskaya, A. G., 1961, The surgical treatment of epilepsy in children, *Vopr. Neirokhir.* **6:**3–8 (Rus.).

Zemskaya, A. G., 1971, *Focal Epilepsy in Children*, Meditsina, Leningrad (Rus.).

Zemskaya, A. G., 1972, Focal epilepsy (uni- and multifocal) in children and some aspects of its surgical treatment, in *Present Limits of Neurosurgery* (I. Fusek and Z. Kunc, eds.), Avicenum, Czechoslovak Medical Press, Prague, pp. 409–410.

Zemskaya, A. G., Garmashov, Yu, A., and Ryabukha, N. P., 1975, Application of the stereotaxic method in combination with classic craniotomy in the treatment of focal epilepsy, *Vopr. Neirokhir.* **2**:33–38 (Rus.).

Zenz, M., Schappler-Scheele, B., and Neuhaus, R., 1981, Long term peridural morphine analgesia in cancer pain (letter), *Lancet* **1**:91.

Zernov, D. N., 1889, Encephalometer. Device for estimation of parts of brain in human, in *Proceedings of the Society of Physicomedicine*, Vol. 2, Moscow University, Moscow, pp. 70–80 (Rus.).

Zervas, N. T., 1965, Technique of radio-frequency hypophysectomy, *Confin. Neurol.* **26**(3–5):157–160.

Zervas, N. T., 1970, Paramedial cerebellar nuclear lesions, *Confin. Neurol.* **32**:114–117.

Zervas, N. T., 1977, Long-term review of dentatectomy in dystonia musculorum deformans and cerebral palsy, *Acta Neurochir. (Wien) [Suppl]* **24**:49–51.

Zervas, N. T., and Hamlin, H., 1974, Stereotaxic thermal pituitary ablation, *Acta Neurochir. (Wien) [Suppl.]* **21**:165–168.

Zervas, N. T., and Hamlin, H., 1978, Stereotactic radiofrequency hypophysectomy, *Appl. Neurophysiol.* **41**:219–222.

Zervas, N. T., Horner, F. A., and Pickren, K. S., 1967, The treatment of dyskinesia by stereotaxic dentatectomy, *Confin. Neurol.* **29**:93–100.

Ziegler, M. G., Lake, C. R., Eldridge, R., and Kopin, I. J., 1976, Plasma norepinephrine and dopamine-β-hydroxylase in dystonia, in *Advances in Neurology*, Vol. 14 (R. Eldridge and S. Fahn, eds.), Raven Press, New York, pp. 307–318.

Ziehen, T. H., 1911, Ein Fall von tonischer Torsions-neurose, *Neurol. Zentralbl.* **30**:109–110.

Zimbaljuk, V. I., 1984, Surgical treatment of spasticity in patients with extrapyramidal lesion, *Neurosurgery* (Kiev) **18**:8–12 (Rus.).

Zimmerman, M., 1979, Peripheral and central nervous mechanisms of nociception, pain and pain therapy, facts and hypothesis, in *Advances in Pain Research and Therapy*, Vol. 3 (J. Bonica, eds.), Raven Press, New York, pp. 3–32.

Zimmerman, M., 1981, Physiological mechanisms of pain and pain therapy, *Triangle* **20**:7–18.

Zlatos, J., and Cierny, G., 1975, Method of determining topometric values of some structures of the human spinal cord, *Appl. Neurophysiol.* **38**:104–109.

Zlotnik, E. I., and Sekach, S. F., 1976, Occlusion of carotid–cavernous fistula with reconstruction of internal carotid artery after E. A. Serdinenko, *Vopr. Neirokhir.* **4**:60–64 (Rus.).

Zlotnik, E. I., Tsemachov, V. G., and Tjappo, S. A., 1971, Surgical treatment of aneurysms of anterior cerebral–anterior communicating arteries, in *First All-Union Congress of Neurosurgeons*, Vol. 1, Moscow, pp. 52–53 (Rus.).

Zlotnik, E. I., Sekatch, S. F., and Oleshkevitch, F. A., 1976, Microsurgical technique in neurosurgery, in *Proceedings of II All-Union Congress of Neurosurgeons*, Moscow, pp. 402–403 (Rus.).

Zorub, D. S., Nashold, B. S., and Cook, W. A., 1974, Avulsion of the brachial plexus. I. A review with implications on the therapy of intractable pain, *Surg. Neurol.* **3**:347–353.

Zotov, J. V., 1971, Method of surgical treatment of focal epilepsy, in *The First All-Union Congress of Neurosurgeons*, Vol. 3, Moscow, pp. 60–64 (Rus.).

Zotov, J. V., 1976, Frontal or generalized epilepsy, *Zh. Nevropatol. Psikhiatr.* **76**:1794–1998 (Rus.).

Zotov, J. V., 1977, Results of surgical treatment of focal epilepsy in adults, *Vopr. Neirokhir.* **6**:17–23 (Rus.).

Zozulya, Yu. A., and Shcheglov, V. J., 1976, Experience in employment of intravascular interventions with a balloon catheter in some types of cerebral pathology, *Vopr. Neirokhir.* **1**:7–11 (Rus.).

Zubkov, J. N., 1973a, Catheterization of brain vessels, *Proc. Leningrad Polenov Neurosurg. Res. Inst.* **5**:218–223 (Rus.).

Zubkov, J. N., 1973b, Intravascular surgical operation for carotid–cavernous fistulas, in *Proc. Leningrad Polenov Neurosurg. Res. Inst.* **5**:228–233 (Rus.).

Zubkov, J. N., 1974, *Intravascular Operations for Carotid–Cavernous Fistulas and Aneurysms of Brain Vessels*, Thesis, Leningrad (Rus.).

Zubkov, J. N., and Matzko, D. E., 1982, The clinic–anatomical analysis of the complications of endovascular interventions in the vascular neurosurgery, *Vopr. Neirokhir.* **2**:12–17 (Rus.).

Zucker, M. B., 1960, Torsion dystonia, *Handb. Neurol.* **7**:249–255 (Rus.).

Zucker, M. B., 1963, *Infectious Diseases of Nervous System in Children*, Meditsina, Moscow (Rus.).

Zukic, A., and Kelly, P. J., 1983, Neuropathic analgesia for stereotactic surgery, *Appl. Neurophysiol.* **46**(1–4):167–169.

Zumpano, B. J., and Saunders, R. L., 1976, Percutaneous epidural dorsal column stimulation, *J. Neurosurg.* **45**:459–460.

Zumstein, H., and Siegfried, J., 1976, Mortality among Parkinson patients treated with L-dopa combined with a decarboxylase inhibitor, *Eur. Neurol.* **14**:321–327.

Zurabashvili, A. D., 1957, Centro-cephalic theory and certain data on the development of nuclei of the optic thalami in man, *Zh. Nevropatol. Psikhiatr.* **56**:701–705 (Rus.).

Zvorikin, V. P., 1982, Variability of size of the neostriatum in connection with age and sex, *J. Neurol. Psychiatry* No. 8:1174–1178 (Rus.).

Index